T0363904

SEEDS

SEEDS

Ecology, Biogeography, and Evolution of Dormancy and Germination

Carol C. Baskin
Jerry M. Baskin

School of Biological Sciences
University of Kentucky
Lexington, Kentucky 40506-0225

ACADEMIC PRESS

An Imprint of Elsevier

San Diego   San Francisco   New York   Boston   London   Sydney   Tokyo

This book is printed on acid-free paper.

Copyright © 2001 by ACADEMIC PRESS
An Imprint of Elsevier
All Rights Reserved.
No part of this publication may be reproduced or transmitted in any form or by any
means, electronic or mechanical, including photocopy, recording, or any information
storage and retrieval system, without permission in writing from the publisher.

Permissions may be sought directly from Elsevier's Science and Technology Rights
Department in Oxford, UK. Phone: (44) 1865 843830, Fax: (44) 1865 853333, e-mail:
permissions@elsevier.co.uk. You may also complete your request on-line via the
Elsevier homepage: http://www.elsevier.com by selecting "Customer Support" and
then "Obtaining Permissions".

Academic Press
An Imprint of Elsevier
525 B Street, Suite 1900, San Diego, California 92101-4495, USA
http://www.academicpress.com

Academic Press
Harcourt Place, 32 Jamestown Road, London NW1 7BY, UK
http://www.academicpress.com

Library of Congress Catalog Card Number: 00-97-80574

ISBN-13: 978-0-12-080263-0
ISBN-10: 0-12-080263-5

Printed and bound in the United Kingdom

Transferred to Digital Printing, 2010

To Drs. Lela V. Barton and Marianna G. Nikolaeva,
two ladies who distinguished themselves by their
contributions to our knowledge of seed
germination biology

Contents

10. A Geographical Perspective on Germination Ecology: Temperate and Arctic Zones

11. Germination Ecology of Plants with Specialized Life Cycles and/or Habitats

12. Biogeographical and Evolutionary Aspects of Seed Dormancy

Preface

Our involvement in seed germination began in spring 1966, when, as graduate students at Vanderbilt University in Nashville, Tennessee (USA), we took a course in plant autecology under Professor Elsie Quarterman. She came to class one day with a list of research projects and went around the room assigning one of them to every two students. We happened to be sitting next to each other and thus were given the same project: germination studies on caryopses of the two summer annual grasses *Sporobolus vaginiflorus* and *Aristida longespica*.

The plant autecology course gave us an awareness of the general ecological question: Why do plants grow where they do? Further, it was clear that a knowledge of seed dormancy and germination requirements could be an important part of the answer to this question. We were intrigued by studies on seeds of desert plants by F. W. Went in the southwestern United States and by M. Evenari, D. Koller, and others in Israel. Equally interesting were the results from studies by B. Quinlivan on legume seeds in Australia and those of D. Ratcliffe and of E. Newman on winter annuals in England. In all these studies, seed and environmental factors interacted to control the timing of germination in nature.

Our excitement about seeds was stimulated further by Dr. Lela V. Barton, who came to Vanderbilt University for a visit. She patiently answered questions and told us fascinating things about seeds, such as the different dormancy-breaking/germination requirements of ray and disk achenes in some Asteraceae and double dormancy in seeds of *Trillium*. A tremendous stimulus to our interest in the different kinds of seed dormancy and how these are broken in nature was provided by the publication of the English translation of Dr. Marianna G. Nikolaeva's (1967) book "Physiology of Deep Dormancy in Seeds" in 1969.

Over the years, it has been our pleasure to study the seed germination ecology of numerous species. We have concentrated on herbaceous species but occasionally have investigated shrubs. These studies have included nonnative (weeds) and native geographically widespread, as well as narrowly restricted (endemic), species. Most seeds have come from the eastern United States, and they have been collected from plants growing in arable fields, lawns, pastures, roadsides, forests, prairies, rock outcrops, and mudflats and other wetlands. However, some seeds have come from species of deserts (New Mexico and Arizona), mountains (Colorado and Utah), and semiarid habitats (Texas and Utah) in the western United States. To expand our knowledge of the field, we have made great efforts over the years to become familiar with, and collect copies of, the past and current literature on seed germination ecology.

In autumn, 1989, we started writing a book on seed dormancy and germination from an ecological, biogeographical, and evolutionary perspective. To our surprise, we discovered that many articles that had been published on these aspects of seed biology were not in our files; thus, much time has been devoted to obtaining copies of them. As we wrote this book, new papers were appearing in the literature almost continuously, and at times it has felt like we were on a fast-paced treadmill (*i.e.*, new information was being published faster than it could be incorporated into the manuscript). However, the quantity and quality of new material being published on seed dormancy and germination is very exciting, and it testifies to the interest in this subject of scientists all over the world.

Work has been cited only if we have examined a copy of it. It should be made clear, however, that not every publication on seed germination ecology has been (or could be) included in this book. Some omissions are due to our ignorance, and for that we apologize in advance to the authors and hope they will send us copies of (or references to) their work. Other studies may not have been mentioned because (1) they were very similar to representative ones already cited or (2) the methods were not clearly stated and thus the meaning of the results was unclear. We thank the librarians at the University of Kentucky for all their help in finding and/or obtaining copies of references. Special thanks are due to Bradley O. Grissom, Biology Librarian, and to various people in the Interlibrary Loan Department.

Thanks are extended to Gar W. Rothwell for reviewing Chapter 12; Jeffrey L. Walck for his very valuable assistance in proofing this book; and Charles R. Crumly, Melanie E. Gross, and Dana L. Polachowski at Academic Press for help in preparing the book for publication. Last, but not least, we gratefully acknowledge support from Austin Peay State University, the Swedish University of Agricultural Sciences, and the University of Kentucky.

We hope that this book will (1) provide people who are just beginning to learn about seed germination a comprehensive overview of seed germination ecology, biogeography, and evolution and (2) give active researchers in the field a sense of what we know and do not know about these aspects of seed biology and thus what research needs to be done on them in the future.

Introduction

I. PURPOSE

Each chapter in this book begins with a brief summary of its purpose(s); thus, in keeping with this format, one objective of Chapter 1 is to explain the philosophy for the whole book. The original concept for this book was to synthesize available information related to seed germination ecology, including evolutionary aspects. However, as the mass of accumulated material on seed germination grew, it became clear that the stage was now set to combine the many pieces of data to (1) determine if plants with specialized life cycle and/or habitats were unique with regard to seed dormancy and germination characteristics, (2) obtain an understanding of the world biogeography of seed dormancy, and (3) formulate hypotheses concerning the origin and relationships of the various types of dormancy in seeds. Thus, this book attempts to cover the present state of knowledge of the ecology, biogeography, and evolution of seed dormancy and germination.

Other objectives of this introductory chapter are to explain what is meant by seed germination ecology, provide an overview of the kinds of information needed to understand the seed germination ecology of a species, and comment on the values of such studies. Also, attention will be given to reasons why this book on seed germination is needed and how it differs from previous ones written on the subject.

II. SEED GERMINATION ECOLOGY

Much information is available in the literature on the seed germination phenology of individual species, and these data show that each species has a characteristic germination season (or seasons). For many species, the time of year when germination is possible is quite limited, e.g., only in autumn, spring, or the wet season. In contrast, the germination season for other species is long, e.g., throughout the growing season. Thus, the broad objective of a seed germination ecologist is to explain how the timing of germination is controlled in nature, and the ecological and evolutionary origins and consequences of this timing.

From a careful reading of Evenari's (1980–1981) paper on the history of germination research, one comes to the conclusion that Theophrastus (c. 372–287 B.C.) should be called the father of seed germination ecology. This remarkable early Greek philosopher and scientist knew that (1) food reserves are stored in seeds, (2) the environmental conditions under which seeds mature affect their germination characteristics, and (3) germination can be influenced by climatic factors, inhibitors, seed age, and seed coats.

Determining what controls the timing of seed germination in the field requires information on the seed, environmental conditions in the habitat, and how the two interact from time of seed maturation to germination. The best way to study the seed germination ecology of a species is to break the problem into a series of questions. (1) When do seeds mature? (2) When are they dispersed? (3) What is the dormancy state of seeds at the time of maturation and at dispersal? (4) What are the environmental conditions in the habitat between time of maturation and germination? (5) What environmental conditions are required to break dormancy and to induce it? (6) What conditions are required to promote germination of nondormant seeds?

In the study of seed germination ecology, other kinds of questions come to mind. For example, how do genetics and/or the climate under which seeds develop influence the requirements for dormancy break and germination? How much variation is there within a species? Does the species form long-lived seed banks? Do buried seeds undergo changes in their dormancy state? What role does timing of germination play in the fitness of a plant?

Since the days of Theophrastus, a vast amount of information related to seed germination has accumulated; however, as shown in subsequent chapters, much remains to be learned about some aspects of the field. There are two reasons why people have been (are) interested in learning how the timing of seed germination is controlled under natural conditions: economics and academics. These two reasons for doing seed research are not independent, and discoveries by people with one point of view frequently benefit those with the other way of approaching the subject.

Information on seed germination potentially has great monetary value. A knowledge of what controls the timing of germination enhances (1) planning for the effective control of weeds; (2) propagation of native economically important trees, shrubs, vines, forbs, and grasses; and (3) reclamation of damaged ecosystems. Thus, much has been learned about seed germination from the research efforts of agronomists, foresters, rangeland specialists, and reclamation ecologists.

Information on seed germination also is important because it contributes to a better understanding of certain biological concepts such as plant reproductive strategies, life history traits, adaptation to habitats, and physiological processes. The physiological processes that have received considerable research attention include seed development and maturation, storage and mobilization of food reserves, aging, cellular and biochemical events in germination, and responses of nondormant seeds to various environmental factors, such as temperature, light, soil moisture, nutrients, and salinity. Thus, much has been learned about seed germination from the research of plant anatomists, biochemists, ecologists, geneticists, molecular biologists, and physiologists.

III. A NEW BOOK ON SEED GERMINATION

A. Need

Because information on seed germination is derived from the research efforts of people in a wide variety of disciplines, it should not be surprising that data are published in many different kinds of journals, in countries all over the world, and in a variety of languages. Thus, the general subject matter of journals that publish papers dealing with some aspect of seed germination ranges from agronomy and crop science to biochemistry and molecular biology, botany, forestry, genetics, horti-

culture, morphology, physiology, range management, and weed science. However, a few journals, e.g., *Journal of Seed Technology, Seed Science and Technology*, and *Seed Science Research*, are devoted exclusively to seed research. Information on seed germination also is found in agriculture research station bulletins, dissertations and theses, and proceedings of symposia and conferences, as well as in reviews (e.g., *Annual Review of Ecology and Systematics, Annual Review of Plant Physiology*, and *Encyclopedia of Plant Physiology*) and in books.

Some books on seeds emphasize (1) chemical composition, physiology, and biochemistry (Khan, 1977, 1982; Bewley and Black, 1982, 1983, 1994; Murray, 1984), (2) types of dormancy and how they are broken (Nikolaeva, 1969; Bradbeer, 1988), (3) structure (Corner, 1976), and (4) morphology (Davis, 1993). Others cover various aspects of seed physiology and/or dormancy and how environmental factors affect germination (Crocker and Barton, 1957; Heydecker, 1972; Kozlowski, 1972; Fenner, 1985; Mayer and Poljakoff-Mayber, 1989; Kigel and Galili, 1995). Books have been written (or compiled) on special topics such as seed aging (Priestly, 1986), storage and longevity (Roberts, 1972), soil seed banks (Leck *et al.*, 1989), germination of desert plants (Gutterman, 1993) and grasses (Simpson, 1990), and the role of the seed in the regeneration of plant communities (Fenner, 1992). It is not uncommon for books to contain one or more chapters on the germination of special groups of plants such as orchids (Arditti, 1977, 1982; Rasmussen, 1995), halophytes (Ungar, 1991), parasites (Parker and Riches, 1993), or weeds (Benech-Arnold and Sanchez, 1995). A number of handbooks have been published that provide detailed instructions on how to germinate seeds of weeds (Andersen, 1968), woody plants (Young and Young, 1992), native plants of a country (e.g., Langkamp, 1987), and plants of interest to gardeners (Deno, 1994).

Although much information obviously is available on seed germination ecology, it is scattered throughout the scientific literature. Thus, no single comprehensive source of data is available on the subject, and a synthesis is needed. Such a synthesis would accomplish several things: (1) facilitate and help focus research efforts, (2) identify questions that need to be addressed by additional research, (3) make it possible to analyze the presence of the various kinds of dormancy in different life forms of plants (i.e., trees, shrubs, lianas, and herbaceous species) in the major vegetation types of the world, and (4) provide a foundation to which other kinds of information (such as fossil history and phylogenetic positions of families) can be added to formulate hypotheses on the evolution of the various kinds of seed dormancy.

B. Contents

Our book differs from others on seed germination in six important ways. (1) It contains a chapter on procedures for doing germination studies so that laboratory and greenhouse results can be extrapolated to the field. (2) Information on many topics (such as dormancy-breaking requirements of seeds with each type of dormancy and genetics of seed dormancy and germination) is placed into an ecological context. (3) A critical evaluation of methodology used in soil seed bank studies of plant communities is presented which shows that many researchers have sampled mixtures of persistent and transient seed banks rather than persistent seed banks only. (4) Data were compiled on species with different life forms growing in the major vegetation types on earth to gain a world perspective on geographical/ecological relationships of the dormancy-breaking and germination requirements of seeds. (5) Data on environmental conditions required to break seed dormancy and stimulate germination have been synthesized for species with specialized life cycles and/or habitats, including parasites, saprophytes, orchids, carnivorous plants, aquatics, halophytes, and psammophytes; this information has been put into an ecological context. (6) For the first time, an attempt has been made to unravel the evolutionary/phylogenetic origins and relationships of the various kinds of seed dormancy and to explain the conditions under which each may have evolved.

References

Andersen, R. N. (1968). "Germination and Establishment of Weeds for Experimental Purposes." W. F. Humphrey Press, Inc., Geneva, NY.

Arditti, J. (ed.) (1977). "Orchid Biology: Reviews and Perspectives," I. Cornell Univ. Press, Ithaca, NY.

Arditti, J. (ed.) (1982). "Orchid biology: Reviews and Perspective," II. Cornell Univ. Press, Ithaca, NY.

Benech-Arnold, R. L., and Sanchez, R. A. (1995). Modeling weed seed germination. *In,* "Seed Development and Germination" (J. Kigel and G. Galili, eds.), pp. 545–566. Dekker, New York.

Bewley, J. D., and Black, M. (1982). "Physiology and Biochemistry of Seeds in Relation to Germination," Vol. 2. Springer-Verlag, Berlin.

Bewley, J. D., and Black, M. (1983). "Physiology and Biochemistry of Seeds in Relation to Germination," Vol. 1. Springer-Verlag, Berlin.

Bewley, J. D., and Black, M. (1994). "Seeds: Physiology of Development and Germination," 2nd Ed. Plenum Press, New York.

Bradbeer, J. W. (1988). "Seed Dormancy and Germination." Blackie, Glasgow.

Corner, E. J. H. (1976). "The Seeds of Dicotyledons," Vols. 1 and 2. Cambridge Univ. Press, London.

Crocker, W., and Barton, L. V. (1957). "Physiology of Seeds." Chronica Botanica Co., Waltham, MA.

Davis, L. W. (1993). "Weed Seeds of the Great Plains." Univ. of Kansas Press, Lawrence.

Deno, N. C. (1994). "Seed Germination Theory and Practice." Published by the author. State College, PA.

Evenari, M. (1980–1981). The history of germination research and the lesson it contains for today. *Israel J. Bot.* **29,** 4–21.

Fenner, M. (1985). "Seed Ecology." Chapman and Hall, London.

Fenner, M. (ed.) (1992). "Seeds: The Ecology of Regeneration in Plant Communities." CAB International, Wallingford, UK.

Gutterman, Y. (1993). "Seed Germination in Desert Plants." Springer-Verlag, Berlin.

Heydecker, W. (ed.) (1972). "Seed Ecology." Pennsylvania State Univ. Press, University Park, PA.

Khan, A. A. (ed.) (1977). "The Physiology and Biochemistry of Seed Dormancy and Germination." North-Holland, Amsterdam.

Khan, A. A. (ed.) (1982). "The Physiology and Biochemistry of Seed Development, Dormancy and Germination." Elsevier Biomedical Press, Amsterdam.

Kigel, J., and Galili, G. (eds.) (1995). "Seed Development and Germination." Dekker, New York.

Kozlowski, T. T. (ed.) (1972). "Seed Biology." Academic Press, New York.

Langkamp, P. (ed.) (1987). "Germination of Australian Native Plant Seeds." Inkata Press, Melbourne.

Leck, M. A., Parker, V. T., and Simpson, R. L. (eds.) (1989). "Ecology of Soil Seed Banks." Academic Press, San Diego.

Mayer, A. M., and Poljakoff-Mayber, A. (1989). "The Germination of Seeds," 4th Ed. Pergamon Press, Oxford.

Murray, D. R. (ed.) (1984). "Seed Physiology," Vols. 1 and 2. Academic Press, Sydney.

Nikolaeva, M. G. (1969). "Physiology of Deep Dormancy in Seeds." Izdatel'stvo "Nauka." Leningrad. [Translation from Russian by Z. Shapiro, National Science Foundation, Washington, DC.]

Parker, C., and Riches, C. R. (1993). "Parasitic Weeds of the World: Biology and Control." CAB International, Wallingford, UK.

Priestley, D. A. (1986). "Seed Ageing: Implications for Seed Storage and Persistence in the Soil. Cornell University Press, Ithaca, NY.

Rasmussen, H. N. (1995). "Terrestrial Orchids from Seed to Mycotrophic Plant." Cambridge Univ. Press, Cambridge, UK.

Roberts, E. H. (ed.) (1972). "Viability of Seeds." Syracuse University Press, Syracuse, NY.

Simpson, G. M. (1990). "Seed Dormancy in Grasses." Cambridge Univ. Press, Cambridge, UK.

Ungar, I. A. (1991). "Ecophysiology of Vascular Halophytes." CRC Press, Boca Raton, FL.

Young, J. A., and Young, C. G. (1992). "Seeds of Woody Plants in North America." Dioscorides Press, Portland.

2

Ecologically Meaningful Germination Studies

I. PURPOSE

Numerous germination studies have been done with the goal of achieving a better understanding of how germination is controlled in nature. Frequently, however, the research was performed in such a way that the results cannot be extrapolated to the field situation, or extrapolation must be done with great caution. The purpose of this chapter is to show how seed germination studies can be done so that data can be used to help explain the timing and control of seed germination of species in nature.

An understanding of seed germination ecology is enhanced by knowledge of the (1) physiological (germination responses), morphological (development of the embryo), and physical (permeability of coats) states of seeds at the time they are matured; (2) changes in physiological, morphological, and physical states of seeds that must precede germination; (3) environmental conditions required for these changes to take place; and (4) environmental conditions occurring in the habitat between the time of maturation and germination. To obtain this information, observations need to be made on the species' life cycle, especially the seed dispersal and germination phases, in relation to seasonal changes in environmental conditions such as temperature and precipitation. Further, germination experiments need to be conducted during the course of the natural dormancy-breaking period. The questions are what kinds of experiments are needed and how should they be done?

By pooling data from many authors and using our own experience, some guidelines for conducting studies on seed germination ecology have been developed. First, we will consider things that need to be kept in mind when laboratory experiments are conducted, and then we will discuss how laboratory studies supplement those done in the field, slathouse, nonheated greenhouse, or transplant garden.

II. GUIDELINES FOR LABORATORY STUDIES ON GERMINATION ECOLOGY

A. Collect Seeds at Maturity

Seeds should not be collected until they are fully ripened. Immature seeds of many species will not germinate, and if they are placed on a moist substrate they quickly become covered with fungi. However, seeds of some species are capable of germinating a few days following anthesis (Gill, 1938; McAlister, 1943; Hume, 1984). It should be noted that these immature seeds will die if they are allowed to dry out (Harrington, 1972). Germination requirements and percentages of immature seeds may be different from those of mature seeds of the same species. Immature seeds of *Avena fatua* germinated to 62%, but mature ones gave no germination (were dormant); development of dormancy in this species is correlated with a decline in embryo water content (McIntyre and Hsiao, 1985). In other species, immature seeds may germinate faster than mature ones (see Lang, 1965). Mature seeds of some species do not germinate because seed coats are impermeable to water, and Hyde (1954) and Helgeson (1932) have shown that this seed coat impermeability develops as seeds dry. Thus, if seeds of hard-seeded species are collected before they have a chance to dry on the mother plant, germination is likely (Sidhu and Cavers, 1977; Helgeson, 1932).

Seeds of many species turn some color other than green when they are mature, and they are no longer "milky" when pinched or cut into halves. The vast majority of seeds can be dried to 2–5% (Roberts, 1973) and sometimes even down to 0.5% moisture content without a loss of viability (Joseph, 1929; Osborne, 1981); these are called orthodox seeds (Chin *et al.*, 1989). Thus, a low moisture content is a good clue in many species that seeds are mature and that it is time to collect them for germination studies.

After seeds on the mother plant stop increasing in dry weight (= mass maturity), an abscission layer forms in orthodox seeds, cutting off water supplies. In the absence of water, maturation drying occurs (Hay and Probert, 1995). Drying is a prerequisite for germination in some species, and higher germination percentages are obtained if seeds dry while attached to the mother plant than if they are dried after collection (Brown, 1965). Seeds of some species collected at mass maturity and dried to a low moisture content exhibit high viability (Harrington, 1972). However, seeds of other species do not attain the potential for maximum longevity until after the mass maturity date. For example, 50% (or less) of the *Digitalis purpurea* seeds collected 0, 4, 8, and 12 days after the mass maturity date were alive after about 1, 3, 7, and 9 days, respectively, of ageing at 50% relative humidity (RH) at 50°C (Hay and Probert, 1995).

In some species, mature seeds do not dry out on the mother plant, and their moisture content is near 100% when they are dispersed. Further, if seed moisture content drops below 30–65%, depending on the species, viability is lost (Chin *et al.*, 1989); these are recalcitrant seeds (Table 2.1). Size, color, and relative ease of detachment from the mother plant must be used, instead of degree of dryness, to determine when to collect recalcitrant seeds.

A good general rule is to collect seeds when natural dispersal begins. Seeds of some species are not dispersed until several months after maturity, during which time the dry seeds remain attached to the mother plant. In these species, nondispersed seeds should be collected at maturity and at regular intervals until dispersal occurs. Because nondispersed seeds may undergo changes in their germination responses, they may germinate over a broader or narrower range of environmental conditions than they did initially. For example, seeds of the winter annual *Sedum pulchellum* mature in early summer, when plant senescence occurs, but they are not dispersed until autumn. At the time of maturation, only a few *S. pulchellum* seeds germinate at low temperatures (Table 2.2). During summer, there is an increase in the maximum temperatures at which seeds can germinate, as well as an increase in germination percentages at all temperatures. In contrast, dry seeds of *Geum canadense* (Table 2.3), eight species of *Salix* (Densmore and Zasada, 1984), and *Frasera caroliniensis* (Baskin and

Baskin, 1986a) remaining on the mother plant for several months following maturation show a decrease in the temperature range for germination. In some species, however, dry nondispersed seeds exhibit no changes in their dormancy-breaking or germination requirements (e.g., Baskin and Baskin, 1984c).

B. Use Seeds Immediately after Harvesting

Germination studies need to be started shortly after the seeds are collected, preferably within 7–10 days. One reason for using seeds immediately is that they may undergo changes in their germination responses during dry storage at room temperatures. Fresh seeds of *Eucalyptus pauciflora* germinated to about 45% at 20°C, but after 1 year of dry storage none of them germinated; they were still viable (Beardsell and Mullett, 1984). Similarly, fresh seeds of *Uniola paniculata* germinated to about 10% in water and 50% in gibberellic acid, but after 4 months of dry storage maximum germination was 2% (Westra and Loomis, 1966). Further, dry storage of *Corylus avellana* nuts for 8 weeks decreased germination from 64 to 10% (Bradbeer, 1968), but germination of dry-stored seeds increased when they subsequently were given a dormancy-breaking treatment (Bradbeer and Colman, 1967). Fresh seeds of *Viola rafinesquii* did not germinate at any temperature, but after 4 months of dry storage they germinated to a maximum of 94% (Table 2.4). Fresh seeds of *Arthropodium cirratum* did not germinate, but after 6 months of dry storage they germinated to about 95%. However, after 9 months of storage only about 55% of the seeds germinated; 95% of them were viable (Conner and Conner, 1988).

Results from germination studies initiated after seeds have been stored dry for several months (e.g., Varshney, 1968; Henson, 1970; Young and Evans, 1980; Hester and Mendelssohn, 1987) or, even worse, for unspecified periods of time (e.g., Kadman-Zahavi, 1955; Kumar and Irvine, 1971; Willemsen and Rice, 1972) are of little ecological value. That is, there is no way to know what the germination responses of the fresh seeds were or how they may have changed through time.

It is possible to slow the rate of physiological changes in many seeds by storing them dry at low temperatures. Thus, seeds sometimes are stored in freezers or refrigerators until people are ready to do experiments. If this is done, it is advisable to test seeds before and after the storage period to make sure they have not undergone changes in their germination requirements. Dry storage at low temperatures has resulted in differences in the germination responses of some seeds. Germination percentages increased for seeds of *Lepidium virginicum* during storage at −18°C (Toole *et al.*, 1957a) and for those of *Dactylis glomerata* stored at −75°C (Probert *et al.*,

TABLE 2.1 Selected Examples of Temperate and Tropical Species with Recalcitrant Seeds

Species	Family	Reference
Temperate		
Acer saccharinum	Aceraceae	Jones (1920)
Aesculus hippocastanum	Hippocastanaceae	Tompsett and Prichard (1993)
Carya spp.	Juglandaceae	Schopmeyer (1974)
Castanea spp.	Fagaceae	Schopmeyer (1974)
Corylus spp.	Corylaceae	Schopmeyer (1974)
Fagus spp.	Fagaceae	Schopmeyer (1974)
Juglans spp.	Juglandaceae	Schopmeyer (1974)
Potamogeton spp.	Potamogetonaceae	Muenscher (1936)
Quercus spp.	Fagaceae	Holmes and Busezewicz (1958)
Populus spp.	Salicaceae	Schopmeyer (1974)
Sagittaria latifolia	Alismataceae	Muenscher (1936)
Salix spp.	Salicaceae	Schopmeyer (1974)
Spartina anglica	Poaceae	Probert and Longley (1989)
Zizania aquatica	Poaceae	Muenscher (1936)
Tropical		
Araucaria hunsteinii	Araucariaceae	Tompsett (1987a)
Artocarpus heterophyllus	Moraceae	Chin *et al.* (1984)
Avicennia marina	Avicenniaceae	Farrant *et al.* (1986)
Camellia oleifera	Theaceae	Koopman (1963)
Chrysalidocarpus leutescens	Arecaceae	Becwar *et al.* (1982)
Coffea canephora	Rubiaceae	Huxley (1964)
Dipterocarpus spp.	Dipterocarpaceae	Yap (1981)
Hevea brasiliensis	Euphorbiaceae	Kidd (1914)
Hopea odorata	Dipterocarpaceae	Corbineau and Come (1988)
Mangifera indica	Anacardiaceae	Koopman (1963)
Montezuma speciossima	Bombacaceae	Barton (1945)
Podocarpus henkelii	Podocarpaceae	Dodd and van Sladen (1981)
Telfairia occidentalis	Curcurbitaceae	Akoroda (1986)
Theobroma cacao	Sterculiaceae	Barton (1965)
Trichilia dregeana	Meliaceae	Choinski (1990)

1985). *Veronica anthelmintica* seeds stored dry at 30–35% RH at 0.6°C for 24 weeks did not come out of dormancy, but 9–51% of them came out of dormancy after 96 weeks, depending on the genetic line (White and Bass, 1971).

Recalcitrant seeds need to be used immediately after collection or they may die. Much effort has been expended to find ways to store recalcitrant seeds without a loss of viability. Seeds of aquatic genera such as *Butomus, Eleocharis, Najas, Nymphaea, Orontium, Potamo-*

TABLE 2.2 Germination Percentages (Mean ± SE) of Seeds of the Winter Annual *Sedum pulchellum*[a,b]

Month of collection	Seed age (month)	Test temperature regimes (°C)			
		15/6	20/10	30/15	35/20
July	0	7 ± 4	5 ± 1	1 ± 1	0
August	1	5 ± 2	1 ± 1	0	0
September	2	24 ± 7	19 ± 3	31 ± 2	5 ± 1
October	3	11 ± 2	24 ± 6	83 ± 3	43 ± 2

[a]Seeds were collected at various times from dead, upright mother plants in the field and immediately incubated over a range of alternating (12/12 hr) temperature regimes at a 14-hr daily photoperiod for 15 days.
[b]Modified from Baskin and Baskin (1977a).

TABLE 2.3   Germination Percentages (Mean ± SE) of Seeds of
Geum canadense[a,b]

Month of collection	Seed age (month)	Test temperature regimes (°C)				
		15/6	20/10	25/15	30/15	35/20
September	0	12 ± 2	100	94 ± 2	90 ± 2	8 ± 2
November	2	0	87 ± 1	58 ± 8	10 ± 6	8 ± 2
December	3	0	83 ± 2	19 ± 6	5 ± 1	1 ± 1

[a]Seeds were collected at various times from dead shoots of mother plants in the field and immediately incubated over a range of alternating (12/12 hr) temperature regimes at a 14-hr daily photoperiod for 15 days.
[b]Modified from Baskin and Baskin (1985a).

geton, *Sagittaria, Scirpus,* and *Vallisneria* (Muenscher, 1936) have been stored successfully in water at 3°C, and those of temperate woody species, including *Juglans* spp. (Schopmeyer, 1974) and *Corylus* spp. (Koopman, 1963), have been stored at low (5°C) temperatures and high relative humidities (85%) without a loss of viability. Seeds of a few tropical species, e.g., *Theobroma cacao* (King and Roberts, 1982), *Coffea arabica* (Valio, 1976), *Dipterocarpus obtusifolia,* and *D. turbinatus* (Tompsett, 1987b), have been stored successfully at room temperatures at high moisture levels. Basically, it appears that the environmental conditions required for the survival of recalcitrant seeds are similar to those found in the natural habitat during the time between dispersal and germination.

C. Check for Imbibition of Water

Before germination studies are initiated, it is important to know if seeds will imbibe water. The terms "hard" and "impermeable" refer to water-impermeable seeds. However, just because a seed (or fruit) feels hard to the touch or has a tough seed (or fruit) coat does

not necessarily mean that it is impermeable to water. The way to determine if seeds (or fruits) are impermeable to water is to place them on moist filter paper at room temperatures. Then, at time 0 and at hourly (or shorter) intervals for 8–12 hr, remove the seeds from the wet paper, blot them dry, and weigh them. An increase in seed weight indicates that seeds (or fruits) have permeable coats (Fig. 2.1), whereas no increase in weight indicates that they have impermeable coats (Bansal *et al.,* 1980).

D. Use Intact Natural Dispersal Units

As noted by Koller (1955), germination ecology of a species cannot be understood unless the natural dispersal unit is studied. In many plant families, the natural dispersal unit is a seed, which should be kept intact during germination studies. Strangely, researchers sometimes scarify (i.e., cut holes or soak in acid) seeds that naturally are permeable to water, and scarification may cause changes in germination responses. Some researchers say that the seed (fruit) coat is hard (impermeable), but this conclusion is not accurate.

TABLE 2.4   Germination Percentages (Mean ± SE) of
Viola rafinesquii Seeds[a,b]

Month tested	Seed age (month)	Test temperature regime (°C)			
		15/6	20/10	30/15	35/20
June	0	0	0	0	0
July	1	54 ± 10	35 ± 2	0	0
August	2	55 ± 5	65 ± 9	39 ± 9	0
September	3	67 ± 1	90 ± 3	56 ± 1	0
October	4	78 ± 5	94 ± 2	61 ± 4	0

[a]Seeds were stored dry in a closed glass bottle at room temperatures. At monthly intervals, some seeds were removed from storage and incubated over a range of alternating (12/12 hr) temperature regimes at a 14-hr daily photoperiod for 15 days.
[b]Modified from Baskin and Baskin (1972a).

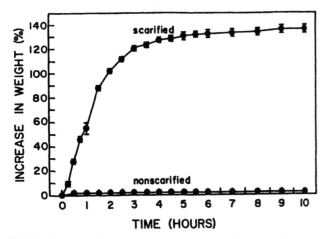

FIGURE 2.1 Imbibition curves for scarified and nonscarified seeds of *Dalea foliosa* (Fabaceae). From Baskin and Baskin (1998).

Scarification overcame the light requirement for the germination of *Ocimum americanum* (Varshney, 1968), *Portulaca oleracea* (Vengris *et al.,* 1972), and *Amaranthus deflexus* (Felippe and Polo, 1983) seeds and removed the dark requirement for the germination of *Cucumis anguria* seeds (Cardoso and Felippe, 1988). Seeds of *Phytolacca americana* soaked in 95% sulfuric acid for 5 min germinated to higher percentages than the controls (Farmer and Hall, 1970), and holes cut in *Elaeis guineensis* seeds reduced the time to germination from 60 or more days to 1 day (Nwankwo, 1981). Scarification increased germination rates and decreased sensitivity to continuous white light and hypoxia in *Amaranthus caudatus* seeds (Gutterman *et al.,* 1992). A small hole cut in *Citrullus lanatus* seeds at the radicle end increased germination in darkness at 25°C and reduced sensitivity to far-red light (Thanos and Mitrakos, 1992). Removal of the caruncle promoted the germination of *Ricinus communis* seeds (Lagoa and Pereira, 1987).

In the case of impermeable seeds, there is a great temptation to scarify them because this usually results in immediate germination. However, results from studies using scarified seeds do not help explain how seeds become permeable in nature or how the timing of their germination is controlled. The germination ecology of seeds (or fruits) with impermeable coats is discussed in detail in Chapter 6.

In many families, e.g., Aceraceae, Asteraceae, Cyperaceae, Poaceae, and Polygonaceae, the dispersal unit is a seed covered by adhering fruit structures. If the dispersal unit has holes cut in it, is soaked in acid or has layers removed from it, germination responses may be changed. For example, germination percentages are increased in numerous grasses, including *Aristida contorta* (Mott, 1974), *Dactylis glomerata* (Junttila, 1977), *Digitaria milanjiana* (Baskin *et al.,* 1969), *Echinochloa turn-*

erana (Conover and Geiger, 1984), *Oryzopsis hymenoides* (Jones and Nielson, 1992), *Paspalum notatum* (Andersen, 1953), *Setaria lutescens* (Rost, 1975), *Stipa trichotoma* (Joubert and Small, 1982), and *Zizania aquatica* (Oelke and Albrecht, 1978), by removal of the palea and lemma. Removal of the glumes promoted seed germination in the grass *Themeda triandra* (Baxter and van Staden, 1993).

Removing pericarp hairs from achenes of *Anemone coronaria* reduced the time for germination by 1 or 2 days (Bullowa *et al.,* 1975). Cutting a hole in achenes of *Cyperus inflexus* (Baskin and Baskin, 1971a) and removing the testa from samaras of *Acer pseudoplatanus* (Webb and Wareing, 1972) substituted for cold stratification, i.e., fruits did not have to be given a cold, moist treatment to break dormancy. Removal of the pericarp from achenes of *Polygonum* spp. reduced the length of the cold stratification period required for germination (Justice, 1941), and soaking achenes of *Rumex crispus* in sulfuric acid caused them to become more sensitive to light, cold stratification, temperature shifts, and gibberellic acid (Hemmat *et al.,* 1985).

E. Replications

All too frequently, germination studies are done using one sample of seeds at each test condition. Obviously, statistical analyses cannot be done on "experiments" of this kind. Germination tests must be replicated, and it is better to have several small replications than one large sample of seeds at each test condition.

In 1966, we had the privilege of meeting and asking the late Dr. Lela V. Barton some questions about seed germination. One thing we wanted to know about was replication of germination experiments. She explained that she once had asked the statisticians at the Boyce Thompson Institute for Plant Research (then at Yonkers, New York, now at Ithaca, New York) the same question. They suggested to Dr. Barton that she use three replications of 50 seeds each for each test condition, and we have faithfully followed this advice.

F. Statistical Analyses

In many germination studies, data are arcsin transformed, and the significance (if any) of differences between treatments is determined by analysis of variance. However, various other ways to analyze germination data, including curve fitting (Goodchild and Walker, 1971; Janssen, 1973; Richter and Switzer, 1982; Berry *et al.,* 1988; Torres and Frutos, 1989), fitting of the logistic function to cumulative germination curves (Schimpf *et al.,* 1977; Brown and Mayer, 1988b; Torres and Frutos,

1990), and probit analysis (Campbell and Sorensen, 1979; Berrie and Taylor, 1981), have been developed. A discussion of the various methods of statistical analyses is beyond the scope of this book, but see Scott *et al.* (1984) for a concise summary of various methods. These authors also explain how and when different types of techniques can (and cannot) be used to analyze germination data.

G. Petri Dishes

A variety of small containers can be used for germination studies, but petri dishes have the most advantages. The petri dish was devised in 1887 by Richard Petri, who was an assistant to the microbiologist Robert Koch (Bullock, 1938). Petri dishes have covers that retard loss of water and are transparent, thus light reaches the seeds. Also, petri dishes, including plastic ones, can be washed and reused many times. If plant hormones or other chemicals are used in germination studies, glass dishes are recommended. If plant hormones are placed in plastic dishes, the dishes probably need to be discarded; small amount of chemicals might adhere to the plastic, even after washing.

H. Substrate

A variety of substrates, including soil, white builders sand, peat moss, filter paper, and blotters, can be used for germination experiments. Each of these substrates has advantages and disadvantages. Filter paper is especially nice for extremely small seeds because seeds can

be easily found and checked for germination. However, filter paper can be expensive. If sand or soil is used for tiny seeds, there is a danger of burying them when water is added. Sand and soil have advantages over filter paper because with a 1- to 2-cm layer of sand or soil in the petri dish there is a larger reservoir of water for the seeds than with filter paper. Consequently, seeds on sand or soil do not have to be watered as often as those on filter paper. Also, if materials such as sugars, amino acids, or germination inhibitors leach from seeds, they will be more diluted in sand- or soil-filled dishes (because of the greater volume of water) than in those with moist filter paper. A dilution of leachates helps reduce the chances of seeds being attacked by fungi or germination being reduced by inhibitors. Sand is a little easier to handle than soil, and its light color facilitates finding and checking seeds for germination.

A good substrate for germination tests is soil collected from the habitat; however, there is a possibility that seeds of the species under investigation are in the soil. Thus, control dishes of soil (to which no seeds of the species are added) should be used at all test conditions.

Although seeds of most species germinate equally well on a variety of substrates, those of some species do not. Both freshly matured and cold-stratified seeds of *Campanula americana* germinated to higher percentages and over a wider range of thermoperiods on soil than on sand (Table 2.5). Seeds of *Alliaria petiolata* (Baskin and Baskin, 1992) and *Conium maculatum* (Baskin and Baskin, 1990) also germinated to higher percentages on soil than on sand. Seeds of *Trientalis borealis* germinated to higher percentages on sand than

TABLE 2.5   Effect of Substrate on the Germination of *Campanula americana* Seeds[a,b]

Treatment	Test temperature regime (°C)				
	15/6	20/10	25/15	30/15	35/20
Fresh seeds					
Soil (light)	88 ± 2	90 ± 3	89 ± 3	55 ± 2	21 ± 2
Sand (light)	6 ± 1	31 ± 5	81 ± 3	69 ± 2	28 ± 3
Soil (dark)	15 ± 1	54 ± 1	69 ± 3	31 ± 1	2 ± 1
Sand (dark)	0	0	0	0	0
Cold stratified (12 weeks)					
Soil (LL)[c]	90 ± 2	85 ± 3	85 ± 4	68 ± 3	65 ± 2
Sand (LL)[c]	5 ± 1	13 ± 2	38 ± 1	40 ± 2	9 ± 1
Soil (DD)[d]	51 ± 2	49 ± 1	60 ± 1	26 ± 2	2 ± 1
Sand (DD)[d]	2 ± 1	1 ± 1	0	2 ± 1	0

[a]Germination percentages (mean ± SE) of freshly matured and cold-stratified (12 weeks) seeds were incubated over a range of alternating (12/12 hr) temperature regimes at a 14-hr daily photoperiod and in continuous darkness for 30 days.
[b]Modified from Baskin and Baskin (1984a).
[c]Seeds cold stratified in light and tested in light.
[d]Seeds cold stratified in darkness and tested in darkness.

on filter paper (Anderson and Loucks, 1973). Seeds of *Gilia capitata, Oenothera micrantha, Mimulus bolanderi* (Sweeney, 1956), and dormoats, which is a hybrid between *Avena sativa* and *A. fatua* (Andrews and Burrows, 1974), germinated better on soil than on filter paper. Seeds of *Treculia africana* germinated to higher percentages on sand and humus than on soil (Mabo *et al.,* 1988).

Seeds of *Digitalis purpurea* incubated at 20°C in darkness or in 5% leaf-filtered daylight germinated to higher percentages on filter paper than on soil (van Baalen, 1982). *Lythrum salicaria* and *Epilobium hirsutum* seeds germinated to higher percentages on filter paper than on sand or soil and to higher percentages on sand than on soil (Shamsi and Whitehead, 1974). *Eragrostis lehmanniana* seeds germinated to a higher percentage on Whatman No. 2 or Eaton and Dikman No. 617 filter papers than on Whatman Nos. 4 and 5 or Schleicher and Schuell No. 597 filter papers (Wright, 1973). Different lots of Whatman No. 1 filter paper influenced the germination of *Arabidopsis thaliana* seeds (Rehwaldt, 1968).

Regardless of the germination substrate, seeds need to be distributed evenly in the petri dish. Clumping of seeds may increase (Linhart, 1976; Waite and Hutchings, 1978; Linhart and Pickett, 1973), decrease (Bergelson and Perry, 1989), or have no effect on germination (Ismail, 1985).

Orientation of seeds with regard to the substrate can cause differences in germination percentages (Bosy and Aarssen, 1995). Thus, all seeds need to have the same degree of contact with the substrate. Orientation is not a problem with small, round seeds. However, if the dispersal unit is longer than broad or if it has appendages (e.g., pappus or wings), just dropping them on sand in a petri dish could result in a diversity of orientations and thus a variation in germination percentages. If a portion of the unit is covered by the substrate, germination percentages also may be decreased (Wajid and Shaukat, 1993).

I. Water

One question concerning water, especially in a closed container such as a petri dish, is how much to use? Studies with crop species indicate that seeds must attain a certain minimum species-specific moisture content, e.g., soybeans, 50%; sugar beets, 31.0%; corn, 30.5%; and rice, 26.5%, before they will germinate (Hunter and Erickson, 1952). Thus, germination may be inhibited if the amount of water is too low (Springfield, 1968; Williams and Shaykewich, 1971; Edgar, 1977; Botha *et al.,* 1984), but it also can be inhibited if too much water is present (Negbi *et al.,* 1966; Heydecker and Orphanos, 1968; Coumans *et al.,* 1979; Phaneendranath, 1980;

Mabo *et al.,* 1988; Blank and Young, 1992). However, seeds of some aquatic plants such as *Typha latifolia* (Morinaga, 1926) germinate better under water than on moist filter paper, and soaking seeds (caryopses) of the grass *Scolochloa festucacea* in water prior to planting in soil stimulated them to germinate (Smith, 1972). Seeds of herbaceous species collected from wetland habitats in England showed great reductions in germination at soil water potentials below −0.05 MPa, whereas those of some (but not all) species from dry habitats germinated at −1.0 or −1.5 MPa (Evans and Etherington, 1990).

The amount of water can interact with temperature, light, and substrate texture to control germination. Seeds of *Lotus corniculatus* germinated to 50% or more at 10, 15, 20, and 25°C over a range of osmotic stresses from 0 to −0.8 MPa, but at 30 and 35°C they germinated to 50% or more only at 0 to −0.05 MPa (Woods and MacDonald, 1971). Seeds of *Salvia reflexa* germinated to 50% or more at 0, −0.3, −0.6, and −0.9 MPa at 27/18 and 33/24°C but only at 0 and −0.3 MPa at 21/12 or 15/6°C (Weerakoon and Lovett, 1986). Under moisture stress, white light inhibited germination in *Avena fatua* (Hsiao and Simpson, 1971a), *Lactuca sativa* cv. Grand Rapids (Hsiao and Simpson, 1971b), and *Sinapis alba* seeds (MacDonald and Hart, 1981), but it promoted the germination of *Eucalyptus occidentalis* seeds (Zohar *et al.,* 1975). The lowest water potential at which seeds of *Trifolium subterraneum* germinated was −0.6, −0.8, and −1.2 MPa on sand, loam, and clay substrates, respectively (Young *et al.,* 1970).

According to Chilton and Isely (1961), 10 ml of water was required to fill all the intergranular spaces of 40 g of flint-shot sand, resulting in 100% substrate hydration. These authors found that final germination percentages of *Lycopersicon esculentum, Bromus inermis, Melilotus officinalis,* and *Lactuca sativa* seeds tested on sand moistened to 50–100% saturation did not vary more than about 5%. At 25% saturation, however, there was a significant reduction in germination percentages.

Care should be taken not to add so much water to the germination substrate that a film of water forms around the seeds (Association of Official Seed Analysts, 1960). If sand or soil is the germination substrate, water is added until the last grains of soil or sand no longer look light colored. In wetting filter paper, add water until the paper is saturated and then hold the dish on its side to allow excess water to drain from the paper (Steinbauer and Grigsby, 1957).

After seeds have been placed on a moist substrate, petri dishes can be wrapped with clear plastic film to retard evaporation. Dishes incubated in light need to be checked at about 5-day intervals and water added if needed. Dishes in darkness should not be opened until the end of the germination test (see Section II,M).

J. pH

Effects of pH on germination have been studied in a number of species (see review by Justice and Reece, 1954). Seeds of many species germinate to high percentages over a wide range of pH's (Studdendieck, 1974; Singh *et al.*, 1975; Mabo *et al.*, 1988; Rivard and Woodard, 1989; Arts and van der Heijden, 1990), but those of others germinate to high percentages only at specific pH's. The optimum pH for the germination of *Calluna vulgaris* is 4.0 (Poel, 1949), but it is 7.0 for seeds of *Bidens biternata* (Ahlawat and Dagar, 1980) and 8.0 for those of *Tridax procumbens* (Ramakrishnan and Jain, 1965) and *Euphorbia thymifolia* (Ramakrishnan, 1965).

The International Seed Testing Association (1985) recommends that a water–sand system for seed germination should have a pH of 6.0 to 7.5. Even calcifuge (acid-loving) species, including *Alopecurus pratensis, Deschampsia flexuosa, Festuca pratensis,* and *Lolium perenne,* germinate well at a pH of 6.5 (Hackett, 1964). Seeds of *Paulownia tomentosa* did not germinate at pH's of 1.5 to 3.5, but they germinated to 79 and 98% at pH 4.0 and 7.0, respectively (Turner *et al.*, 1988). Seeds of *P. tomentosa* transferred from pH's 3.0 and 3.5 to pH 6.5 germinated to 59 and 74%, respectively; however, those transferred from pH's 1.5 and 2.5 did not germinate at 6.5, indicating that low pH probably killed them (Turner *et al.*, 1988). Gibberellic acid (GA$_3$) and abscisic acid are more effective in promoting the germination of *Lactuca sativa* seeds at a pH of 3.2 to 4.2 than at 7.0 (Zagorski and Lewak, 1984). A pH of 4.0 significantly reduced seed germination of the calcicolous species *Crithmum maritimum, Spergularia rupicola, Daucus carota* ssp. *gummifer,* and *Lavatera arborea* (Okusanya, 1978).

One area of germination ecology that needs further investigation is the pH requirements for the germination of species that grow in fens and other habitats with high pH's. Rudolfs (1922, 1925) has shown that seeds of a number of species allowed to imbibe in salt solutions of various kinds have the ability to increase the hydrogen ion concentration in their immediate surroundings and thus lower the pH. Do seeds of species growing in alkaline habitats require (or tolerate) high pH's for germination, or can they decrease the pH of their germination environment?

Browning of the radicle tip is a clue that a given pH is not optimal for germination (Justice and Reece, 1954). A pH problem can be solved by using soil from the natural habitat of the species or by adjusting the pH of the substrate with buffers, according to the methods of Dawson and Elliott (1959). Seeds of *Tephrosia candida* germinated to higher percentages at pH's maintained with potassium than with sodium acetate buffers (Gupta and Basu, 1988).

K. Disinfectants

Generally, we do not recommend the use of disinfectants in studies on the ecology of seed germination. After all, the goal is to learn what controls the timing of germination in the field, and seeds in or on the soil may be exposed to various kinds of pathogenic and nonpathogenic organisms. Crocker and Barton (1957) noted that seeds of wild plants germinated on a mixture of sand, peat moss, and steam-sterilized sod compost rarely were attacked by fungi. Although there are exceptions, it also has been our experience that seeds of most species are not attacked by fungi. Fungi problems usually can be avoided by using fully developed, healthy seeds; inferior ones quickly become covered with fungi. In fact, fungi do a very thorough job of testing for inferior or dead seeds!

L. Constant versus Alternating Temperature Regimes

Although Conner and Conner (1988) found that seeds of *Arthropodium cirratum* germinated to higher percentages at constant than at alternating temperatures, alternating temperatures usually are more favorable for germination than constant ones (Harrington, 1923; Morinaga, 1926; Steinbauer and Grigsby, 1957; Matumura *et al.*, 1960; Thompson and Grime, 1983). In fact, some species will germinate only at alternating temperature regimes (e.g., Steinbauer and Grigsby, 1957; Thompson, 1969; Goedert and Roberts, 1986; Pons and Schroder, 1986; Pegtel, 1988). In the natural habitat, seeds obviously are exposed to alternating, not constant, temperatures.

Morinaga (1926) found that the daily period of exposure to either high or low temperatures must be 4.5 to 8 hr to obtain high germination percentages for seeds with an alternating temperature requirement. Morinaga (1926) and Pons and Schroder (1986) reported that for a high percentage of the seeds to germinate, the difference between high and low temperatures must be 10°C or more. In some species, however, differences of only 1°C may stimulate germination (Thompson and Grime, 1983). The optimum high temperature depends on the species and varies from 22–25°C in *Apium graveolens* (Thompson, 1974) to 45°C in *Sorghum halepense* (Harrington, 1923). There are no critical maximum or minimum daily temperatures for the germination of *Lycopus europaeus* seeds, if the difference between the two temperatures exceeds 7°C (Thompson, 1969). In their model of seed germination responses to alternating temperatures, Murdoch *et al.* (1989) identified seven primary variables: number of cycles, time per cycle above mid-temperature, time per cycle below mid-temperature, maximum temperature, minimum temperature, rate of warming, and rate of cooling. Secondary variables were temperature duration, periodic time (time taken to complete one cycle of temperatures), mean temperature, temperature, amplitude, and rate of temperature change.

Alternating temperatures interact with other environmental factors, including light (Toole *et al.*, 1955; Probert *et al.*, 1986; Voesenek *et al.*, 1992), nitrates (Vincent and Roberts, 1977; Williams, 1983), thiourea (Probert *et al.*, 1987), substrate moisture level (Hegarty, 1975), oxygen concentrations (Morinaga, 1926), and with seed age (Akamine, 1947; Junttila, 1977) and level of the active (Pfr) form of phytochrome (Probert and Smith, 1986; Probert *et al.*, 1987) to control germination.

Because the objective of germination ecology is to understand germination in the field, seeds need to be incubated at alternating temperatures simulating those in the natural habitat. Daily alternating temperature regimes can be achieved (1) by moving seeds from one temperature to another each morning and evening or (2) by using thermogradient bars, incubators, or plant growth chambers. Thermogradient bars provide a wonderful array of alternating temperature regimes, and several designs for constructing them have been published (see Larson, 1971; Grime and Thompson, 1976; Thompson and Whatley, 1984). However, unless several bars are available, doing research on more than one species at a time may be difficult due to space limitations. Incubators provide a lot of space; however, a different incubator is required for each alternating temperature regime, i.e., unless seeds are moved twice each day.

Regardless of how alternating temperature cycles are maintained, the choice of regimes to use needs to be based on temperatures occurring in the species' habitat. Ideally, one would like to have enough alternating temperature regimes to simulate each month of the growing season. If long-term soil or air temperature data are not available from the population site(s) of the study species, information from the nearest government-operated weather station will provide a good approximation of monthly means of daily maximum and minimum air temperatures, which can be used in choosing incubation temperature regimes.

M. Light and Darkness

Nondormant seeds of many species germinate equally well in light and darkness (Baskin and Baskin, 1988), those of others germinate to higher percentages in light than in darkness (Grime *et al.*, 1981; Baskin and Baskin, 1988), and those of a relatively few germinate to higher percentages in darkness than in light (Table 2.6).

Seeds of many light-requiring species germinate in the spring, after they have been exposed to low winter temperatures (Baskin and Baskin, 1977b, 1988), whereas those of others germinate in autumn, after they have been exposed to high summer temperatures (Baskin and Baskin, 1982, 1988). An important question is when can the light requirement be fulfilled? If seeds are exposed to light during the dormancy-breaking period, will they germinate in darkness during the subsequent germination season? Seeds of *Cyperus inflexus* exposed to light during winter only did not germinate in darkness in spring (Baskin and Baskin, 1971c), but those of *Aster pilosus* exposed to light only in winter germinated in darkness in spring (Baskin and Baskin, 1985c). Seeds of *Diamorpha cymosa* (Baskin and Baskin, 1972b) and *Draba verna* (Baskin and Baskin, 1972c) exposed to light during the summer dormancy-

TABLE 2.6 Examples of Species Whose Seeds Germinate to Higher Percentages in Darkness than in Light

Species	Reference
Anemone coronaria	Bullowa *et al.* (1975)
Amaranthus blitoides	Kadman-Zahavi (1955)
Asphodelus microcarpus	Hammouda and Bakr (1969)
Astragalus sieberi	Hammouda and Bakr (1969)
Bromus sterilis	Hilton (1982)
Calligonum comosum	Koller (1956)
Centaurea alexandrina	Hammouda and Bakr (1969)
Citrullus colocynthia	Koller *et al.* (1963)
Cucumus anguria	Noronha *et al.* (1978)
Cynoglossum officinale	van Breeman (1984)
Echinops spinosissimus	Hammouda and Bakr (1969)
Echium sericeum	Hammouda and Bakr (1969)
Enarthrocarpus strangulatus	Hammouda and Bakr (1969)
Eremobium microcarpa	Hammouda and Bakr (1969)
Eryngium creticum	Hammouda and Bakr (1969)
Eschscholzia californica	Goldthwaite *et al.* (1971)
Galium spurium	Malik and Vanden Born (1987)
Glaucium flavum	Thanos *et al.* (1989)
Haloxylon salicornicum	Hammouda and Bakr (1969)
Launaea glomerata	Datta (1965)
Nemophilia insignis	Chen (1968)
Nigella damascens	Isikawa (1957)
Phacelia tanacetifolia	Chen and Thimann (1966)
Spinifex sericeus	Maze and Whalley (1992)

breaking period germinated in darkness in autumn. Seeds of *Solidago shortii* exposed to white light for 4 weeks in autumn at 15/6°C and subsequently given 12 weeks cold stratification in darkness at 5°C germinated to 70% in darkness in spring at 20/10°C; dark controls germinated to only 45% (Walck *et al.*, 1997). Seeds of *Lactuca sativa* retained a light stimulus during 1 year of dry storage, presumably at room temperatures (Vidaver and Hsiao, 1972). Retention of the light stimulus in *Cirsium palustre* seeds depended on the temperatures to which seeds were exposed after they are placed in darkness. At 4°C, *C. palustre* seeds retained the light stimulus, but at 22°C they lost it (Pons, 1983).

In studying the light requirement for germination, it is important to test seeds at a daily photoperiod and in continuous darkness at each of the daily alternating temperature regimes. Seeds need to be tested in light and darkness when they are freshly matured and at regular intervals during the dormancy-breaking period because their light requirement may change as they come out of dormancy. Low winter temperatures (cold stratification) removed the light requirement for the germination of *Picea mariana* (Farmer *et al.*, 1984), *Pinus palustrus* (McLemore and Hansbrough, 1970), and *Spergularia media* seeds (Ungar and Binet, 1975). However, high (28, 40°C) temperatures removed the light requirement for the germination of *Hygrophila auriculata* seeds (Amritphale *et al.*, 1989). Also, seeds subjected to seasonal temperature changes in the temperate zone may exhibit annual cycles in their light requirement for germination (Fig. 2.2). Seeds held in dry storage at room temperatures for several months may lose some or all of their light requirement for germination (Amritphale *et al.*, 1984; Viana and Felippe, 1986, 1989; Lima and Felippe, 1988).

The light requirement for germination may vary with temperature. Seeds of *Lactuca sativa* cv. Grand Rapids germinated to >80% in light at temperatures of 10 to 30°C, whereas germination in darkness exceeded 45% only at temperatures of 10 to about 22°C; it was near 0% at 30°C (Evenari, 1952). Seeds of *Bidens pilosa* (Felippe, 1978), *Cynodon dactylon*, *Deschampsia caespitosa* (Thompson *et al.*, 1977), *Nicotiana tabacum* (Toole *et al.*, 1957b), and *Typha latifolia* (Morinaga, 1926) required light to germinate at constant temperatures, but they germinated in light and in darkness at alternating temperatures.

Cool white fluorescent tubes are a better light source for germination studies than incandescent light bulbs. Cool white fluorescent tubes emit considerable red but very little far-red light (Toole, 1963), whereas incandescent lights emit lots of far-red light and heat (Steinbauer and Grigsby, 1957). Absorption of red light converts the inactive Pr form of phytochrome (which inhibits

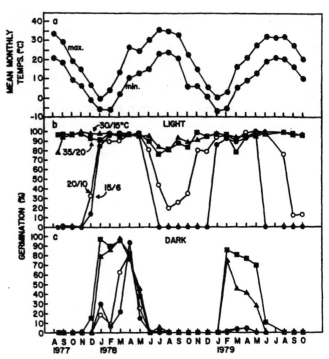

FIGURE 2.2 (a) Mean daily maximum and minimum monthly air temperatures in a nonheated greenhouse in Lexington, Kentucky. (b) Germination of *Verbascum blattaria* seeds exhumed after 0 to 26 months of burial at near-natural temperature conditions in the nonheated greenhouse and incubated on moist sand at a 14-hr daily photoperiod for 15 days. (c) Germination of exhumed seeds incubated in continuous darkness for 15 days. From Baskin and Baskin (1981b).

germination) to the active Pfr form (which promotes germination); the absorption of far-red light converts Pfr to Pr (Kendrick, 1976). However, red light (and thus Pfr) inhibits seed germination in a few species, including *Bromus sterilis* (Hilton, 1982, 1984) and *Aristida murina* (Ginzo, 1978). In *B. sterilis,* moisture stress and low temperatures (≤15°C) increased the inhibitory effect of Pfr and how long it persisted (Hilton, 1984). Toledo *et al.* (1990) designed a light gradient bar (something like a thermogradient bar) with varying red:far-red ratios to investigate responses of seeds to light quality.

The International Seed Testing Association (1985) recommends an illuminance (frequently and erroneously called light intensity) of approximately 750–1250 lux from cool white fluorescent lamps for germination studies. An illuminance of 750–1250 lux is roughly equal to a fluence rate of 20 μmol m^{-2} sec^{-1}, 400–700 nm of cool white fluorescent light. According to Ellis *et al.* (1986a), 750–1250 lux would be about 0.29–0.48 mol m^{-2} day^{-1} for an 8-hr day; however, these authors recommend daily doses of light of only about 0.14–0.21 mol m^{-2} day^{-1}. This means that one or two 20-W cool white fluorescent tubes 15–20 cm above each incubator shelf will provide enough light for germination, if the lights are on for several hours each day.

The high irradiance reaction (HIR) [also called the high energy reaction (HER)] has received considerable attention in seed germination studies. In the HIR, light of ". . . almost any quality inhibits germination if applied at high irradiances for sustained periods" (Roberts *et al.*, 1987). The maximum photon dose (= incident photon flux area density × duration of exposure) of light that can be given to seeds without inhibition of germination varies with the species (Ellis *et al.*, 1986a) and among seeds of the same species (Ellis *et al.*, 1986b). The minimum photon dose that inhibited the germination of *Barbarea verna*, *Brassica chinensis*, *B. juncea*, *B. oleracea*, and *Camelina sativa* seeds was 0.5 mol m^{-2} day^{-1} (Ellis *et al.*, 1989). However, Thompson (1989) found that photon doses of 19.2 mol m^{-2} day^{-1} were not inhibitory to any of the 19 species he studied except *Bromus sterilis* and *B. erectus*. Photon doses above 10$^{-0.5}$ mol m^{-2} day^{-1} inhibited germination of four cultivars of *Echinochloa colonum*, and those above 10$^{-0.5}$ mol m^{-2} day^{-1} (0.316 mol m^{-2} day^{-1}) inhibited germination of two cultivars of this species (Ellis *et al.*, 1990).

Photoperiod, potassium nitrate, and temperature may also influence the maximum photon dose for promotion or inhibition of germination. At daily photoperiods of 1 min, 1 hr, or 8 hr, germination of *Echinochloa turnerana* seeds increased with increases in the photon dose up to 5 × 10^{-1} mol m^{-2} day^{-1}, but in continuous light a photon dose above 10^{-1} mol m^{-2} day^{-1} reduced germination (Ellis *et al.*, 1986a). Potassium nitrate reduced the optimum photon dose for the germination of *Eragrostis tef* seeds (Ellis *et al.*, 1986a). Photon doses above 3 × 10^{-1} mol m^{-2} day^{-1} inhibited the germination of *Bromus sterilis* seeds at 15°C, whereas doses of 3 × 10^{-2} mol m^{-2} day^{-1} were inhibitory at 25°C (Ellis *et al.*, 1986b).

Because the HIR overrides the reversible phytochrome reactions (Borthwick *et al.*, 1969; Bartley and Frankland, 1982), exposure to continuous light during germination tests may inhibit germination, i.e., continuous light may result in a photon dose that is high enough to be inhibitory. Even if the low-energy reaction of phytochrome does control germination, continuous white light can inhibit it, e.g., *Citrullus lanatus* (Thanos and Mitrakos, 1992). Thus, as little as 1 hr of darkness per day can promote germination (Borthwick *et al.*, 1964). In studies of germination ecology, seeds need to be incubated in light for a maximum of only 12–14 hr each day, and the daily light period should coincide with the daily high temperature period. In our studies, we allow the lights to come on in the incubators 1 hr before the beginning of the 12-hr high temperature and remain on for 1 hr after the beginning of the 12-hr low temperature period.

Some light-stimulated seeds exhibit photoperiodic responses (Black and Wareing, 1954; Isikawa, 1954; Felippe *et al.*, 1971); for a list of species with short- or long-day seeds see Black (1967). Short-day seeds have a certain maximum photoperiod favorable for germination, and germination is inhibited by continuous light. However, the germination of long-day seeds increases with increases in the length of the daily light period, and continuous light is not inhibitory (Evenari, 1965).

The photoperiodic response of *Betula pubescens* (Black and Wareing, 1955; Vaartaja, 1956), *B. verrucosa* (Vaartaja, 1956), *Nemophila insignis* (Black and Wareing, 1960), *Chenopodium botrys* (Cumming, 1963), and *Catalpa speciosa* seeds (Fosket and Briggs, 1970) was dependent on temperature. For example, freshly matured seeds of *B. pubescens* given an 8-hr photoperiod germinated to 34 and 93% at 15 and 20°C, respectively, whereas those given a 20-hr photoperiod germinated to 90 and 93%, respectively (Black and Wareing, 1955). Seeds of some species, including *Betula pubescens* (Black and Wareing, 1955; Vaartaja, 1956), *B. verrucosa* (Vaartaja, 1956), *Tsuga canadensis* (Stearns and Olson, 1958), and *Eragrostis ferruginea* (Fujii and Isikawa, 1962), do not exhibit photoperiodic responses after they are cold stratified. A photoperiodic response was present after seeds of *Cyperus inflexus* had been cold stratified and it was temperature independent (Baskin and Baskin, 1976). Thus, studies on photoperiodic responses of seeds need to be preceded by preliminary experiments to determine if dormancy-breaking treatments and test temperatures cause a variation in responses.

Illuminance can modify photoperiodic responses. For example, nondormant seeds of *Cyperus inflexus* exposed to high illuminance (13,770 lux) germinated to higher percentages at 1- or 4-hr than at 20-hr daily photoperiods, whereas seeds at low illuminance (1372 lux) germinated to higher percentages at 20-hr than at 1- or 4-hr daily photoperiods (Baskin and Baskin, 1976).

An easy way to incubate seeds in darkness is to wrap the seed-containing dishes with two layers of aluminum foil. If dishes are wrapped first with plastic film, the aluminum foil will not corrode. Various types of light-tight containers also can be used for the dark incubation of seeds. After the substrate has been moistened, seeds to be incubated in darkness need to be placed there immediately. After only 10 min of hydration, seeds of *Lactuca sativa* showed a partial response to red light (McArthur, 1978). Seeds tested in darkness should not be exposed to any light until the end of the germination experiment. To monitor germination rates in darkness, use the destructive sampling method, i.e., once dishes containing dark-incubated seeds are exposed to light and germination percentages are determined, they are discarded.

Seeds of some species require only a very small amount of light for germination. Seeds of *Epilobium cephalostigma* required 864×10^5 m-candle seconds (= lux × seconds of exposure) (Isikawa and Shimogawara, 1954), whereas those of *Nicotiana* sp. required only 100 m-candle seconds (Isikawa, 1952). Because some seeds require such small amounts of light to stimulate germination, it is impossible to check them for germination (even in dim light) without fulfilling their light requirement. For example, two 1-min exposures to 753 lux of cool white fluorescent light fulfilled the light requirement in 48% of *Helenium amarum* seeds (Baskin and Baskin, 1975). Thus, any report in the literature of seeds germinating in darkness probably should be discounted if the investigator exposed them to any light at all during the incubation period (e.g., Ratcliffe, 1961; Caplenor, 1967; Vickery, 1967).

It is possible to check seed germination in darkness by using a green safe light. However, green light stimulates the seeds of some species to germinate, including those of *Bidens pilosa* (Valio *et al.,* 1972), *Plantago major* (Blom, 1978), and *Stellaria media* (Baskin and Baskin, 1979a). As seeds are exposed to dormancy-breaking treatments, it seems reasonable that they might gain the ability to germinate in response to green light. For example, seeds of *Solidago altissima* receiving 0 and 8 weeks of cold stratification germinated to 3 and 97%, respectively, following a 5-min exposure to green light (ca. 3.2 μmol m^{-2} sec^{-1}, 500–600 nm) (Raynolds *et al.,* unpublished results). Thus, if a green light is used to monitor germination in darkness, periodic checks need to be made to make sure seeds have not gained the ability to respond to it.

N. Store Seeds under Natural or Simulated Habitat Conditions

At the time freshly matured seeds are tested for germination, the remaining seeds should be placed under natural or simulated habitat conditions. Therefore, if seeds are dormant they will become nondormant. The progress of dormancy loss can be followed by regularly removing some of the seeds from the natural or simulated habitat conditions and testing them for germination in light and darkness at the same thermoperiods used to test freshly matured seeds.

One way to store seeds is to place them in fine-mesh bags and return them to the habitat where they were collected. Mesh bags also allow seeds to be exposed to natural temperature and soil moisture conditions. Sometimes logistical problems make it difficult, if not impossible, to return seeds to the native habitat. Also,

it is desirable to keep temperature records, which may be difficult if the habitat is in a remote location. An alternative to placing seeds in the native habitat is to place them outside in situations that are as similar to the habitat as possible (e.g., garden, field plots, wood lots, slathouse, nonheated greenhouse). Temperature and precipitation records need to be kept so that the conditions to which the seeds have been exposed are known. Seeds can be subjected to simulated field conditions by moving them to a different incubator or growth chamber each month (or by changing the temperature in a chamber) so that the alternating temperature regime corresponds to mean monthly maximum and minimum soil temperatures in the habitat, or to those in the general geographical area, for that month.

In trying to simulate nature in the laboratory, attention should be paid to the time when the seeds are dispersed. If seeds are dispersed in autumn, they need to be given cold stratification, whereas if they are dispersed in spring, they need to be given warm stratification treatments. What are cold and warm stratification treatments? This means that seeds are placed on a moist substrate at representative winter (cold stratification) or summer (warm stratification) temperatures.

The effective temperatures for cold stratification are from about 0 to 10°C, with about 5°C being optimal for many species (Stokes, 1965; Nikolaeva, 1969). However, the optimum stratification temperature was 0°C for seeds of *Dioscorea japonica* and *D. septemloba* (Okagami and Kawai, 1982), *Eriodictyon* spp. (Roof, 1988), and *Pyrus malus* (Nguyen and Come, 1984); 2°C for those of *Primula sieboldii* (Washitani and Kabaya, 1988); and 0–3°C for those of *Ferula* spp. (Nikolaeva, 1969). An alternating temperature regime of 1 and 10°C sometimes was more effective in promoting the germination of *Rhodotypos kerrioides* seeds than a constant temperature of 5°C (Flemion, 1933). Seeds of *Geum urbanum* cold stratified outside in England germinated to higher percentages over a range of temperatures than those cold stratified at 5°C (Pons, 1983).

In a first attempt to promote germination with a cold stratification treatment, place seeds on a moist substrate and keep them at 5°C for about 12 weeks. If 5°C does not work, try temperatures slightly above and below 5°C and/or a daily alternating temperature regime of 5/1 or 6/1°C. Sometimes dry seeds are placed in a freezer to give them a "cold stratification treatment" (e.g., Choate, 1940). However, this treatment usually does not promote loss of dormancy because (1) seeds are not imbibed and (2) temperatures in a freezer are below the effective range for cold stratification. During a cold stratification treatment, the substrate needs to be checked from time to time to make sure it remains moist. If a period of drying occurs either during or after cold stratification treatment, seeds of some species will not germinate (Flemion, 1931; Stokes, 1965) or they may germinate only in light (Campbell, 1982) or at high temperatures (Allen, 1962). In other species, a period of drying given after the cold stratification treatment does not reduce germination when seeds subsequently are imbibed and placed at appropriate germination temperatures (Danielson and Tanaka, 1978; Griesbach and Voth, 1957; Bazzaz, 1970). Germination was not inhibited when cold-stratified seeds of *Panicum anceps* were dried for 1 month, but it was reduced from 58 to 22% by a 2-month drying period prior to sowing (Garmon and Barton, 1946).

The range of effective temperatures for warm stratification is 20–35°C (Baskin and Baskin, 1986b), with 20–25°C being optimal (Nikolaeva, 1969). Thus, for warm stratification treatments place seeds on a moist substrate at simulated summer temperatures (e.g., 25/15, 30/15, or 35/20°C) in growth chambers or incubators for about 12 weeks. In nature, however, seeds may be dry, or usually alternately wet and dry, during summer. Many seeds that require exposure to high summer temperatures before they can germinate in autumn also will germinate following several months of dry storage at room temperatures (Crocker and Barton, 1957; Baskin and Baskin, 1971b, 1972b, 1982).

The relative humidity of dry-stored seeds may influence loss of dormancy. Following several months exposure to various levels of RH, germination percentages of *Draba verna* seeds incubated at 10, 15, 15/6, and 20/10°C were: 0–30% RH, 0–13% germination; 40% RH, 25–39%; 50–60% RH, 49–65%; and 60–100% RH, 0% (Baskin and Baskin, 1979b). Most of the seeds died at relative humidities of 70–100% (Baskin and Baskin, 1979b). To prevent growth of fungi, seeds to be stored dry at room temperatures need to be dried for several days before they are placed in a closed container. Dry storage of seeds at room temperatures does not simulate the natural world, and even in deserts seeds experience daily fluctuations of RH as well as temperatures (Kappen *et al.*, 1975). However, germination responses of seeds following dry storage at room temperatures and those given alternate wet–dry treatments at summer temperatures can be very similar (Baskin and Baskin, 1983).

O. Test Seeds Frequently

Seeds stored at natural or simulated habitat conditions need to be tested for germination in light and darkness at regular intervals during summer or winter (depending on the species) over the same range of thermoperiods used in the initial germination test. Consequently, the more often seeds are tested, the finer the

resolution of the study will be, but tests at 2- to 4-week intervals usually are sufficient.

The purpose of testing seeds at short intervals of time is to monitor the rate of dormancy loss and the pattern of changes (if any) in germination responses of seeds as they come out of dormancy. For example, seeds of *Bidens polylepis* buried in moist soil and exposed to cold stratification during winter gained the ability to germinate at high (30/15, 35/20°C) temperatures by midwinter (Fig. 2.3). By late winter, the minimum temperature for germination had decreased, and by spring seeds germinated to high percentages at thermoperiods of 15/6, 20/10, 25/15, 30/15, and 35/20°C. If germination tests of *B. polylepis* seeds had been done only in spring instead of at monthly intervals from late autumn through spring, there would be no way to know that temperature requirements for germination changed during the course of the winter cold stratification period. These changes in temperature requirements for germination are important in controlling the timing of germination (Chapter 4).

P. Rates of Germination

As reviewed by Northam and Callihan (1994), opinions differ on how to determine whether a seed has

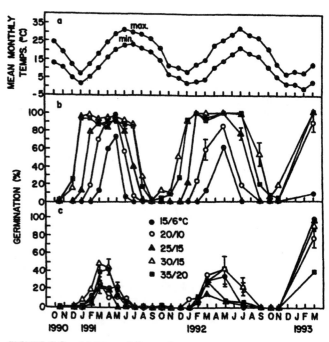

FIGURE 2.3 (a) Mean daily maximum and minimum monthly air temperatures in a nonheated greenhouse in Lexington, Kentucky. (b) Germination of *Bidens polylepis* seeds exhumed after 0 to 28 months of burial in the nonheated greenhouse and incubated for 15 days on moist sand at a 14-hr daily photoperiod. (c) Germination of exhumed seeds incubated in continuous darkness for 15 days. From Baskin *et al.* (1995).

germinated. Usually, emergence of the radicle is the criterion for germination. However, the shoot may appear first (Baskin and Baskin, 1971a) or the root and shoot may emerge simultaneously (Northam and Callihan, 1994). Radicle emergence preceded that of the shoot when *Taeniatherum caput-medusae* seeds were incubated at 8°C, but radicle and shoot emergence occurred at the same time at 18°C (Northam and Callihan, 1994).

Regardless of the criterion chosen for germination, determining germination percentages at frequent intervals may help explain the timing of seed germination in the field. For example, seeds of the winter annual *Phacelia dubia* var. *interior* germinated at simulated summer temperatures (33/20°C) in August, but 24 days were required for more than 5% of them to germinate (Fig. 2.4). Thus, germination does not occur in the field during summer because soils dry within a week following a rain. By autumn, however, temperatures have decreased, germination rates, as well as the periods of time that soils remain moist, have increased, and germination occurs (Baskin and Baskin, 1971d).

Germination rates provide valuable information about the degree of dormancy loss and the favorability of germination conditions. Therefore, both rates and percentages sometimes are used to calculate a germination index. Various germination indices are available (e.g., Timson, 1965; Czabator, 1962; Brown and Mayer, 1986, 1988a), but Brown and Mayer (1988a) have questioned the value of using a single number to quantify cumulative germination. Nonetheless, germination indices, especially Timson's index, are used frequently in germination studies. Timson's index is

$$\Sigma n,$$

where n is the cumulative daily germination percentage for each day of the study.

To calculate Timson's index, germination percentages are determined daily and, at the end of the germination test, are summed. With a 10-day germination test period, the possible values for Timson's index range from 0 (if no seeds germinate) to 1000 (if all seeds germinate the first day). The Timson's index test can be used for test periods of any length, but Brown and

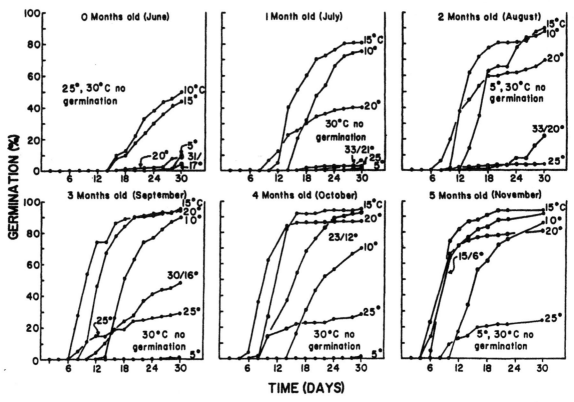

FIGURE 2.4 Cumulative germination percentages of *Phacelia dubia* var. *interior* seeds stored dry at room temperatures during the summer. At monthly intervals, some seeds were removed from storage and incubated at a 14-hr daily photoperiod at constant temperatures of 5, 10, 15, 20, 25, and 30°C and at an alternating (12/12 hr) temperature regime simulating air temperatures in the field for that particular month: June, 31/17; July, 33/21; August, 33/20; September, 30/16; October, 23/12; and November, 15/6°C. From Baskin and Baskin (1971d).

Mayber (1988a) caution that it is important not to extend the test period beyond a time when germination is mostly completed. After the Timson's index has been calculated for each treatment, the results can be analyzed by statistical tests (Timson, 1965).

Q. Length of Germination Test Period

A germination test has to be long enough to allow seeds sufficient time for germination; however, it should not be so long that seeds can receive enough warm (or cold) stratification to promote germination. Thus, in view of the fact that considerable germination can occur following only a month of dry storage at room temperatures (Table 2.4), germination tests should be terminated after about 2 weeks. Two-week germination tests have worked well in the majority of our studies, with most of the seeds germinating within 10 days or less. If germination percentages are still increasing at the end of 2 weeks, consideration should be given to extending the germination period to a maximum of 3 or 4 weeks.

R. Test Viability of Ungerminated Seeds

At the conclusion of a germination test, ungerminated seeds need to be checked for viability. Seeds that are covered with fungi and collapse when pinched gently with forceps are dead. Seeds that do not collapse when pinched and have a firm, white embryo probably are viable. Seeds with gray, yellow, or brownish embryos usually are dead. A quick and easy method of checking the viability of seeds is to use the tetrazolium test (Cottrell, 1947). Embryos are dissected from imbibed seeds or seeds are cut so that the embryo is bisected, and then embryos or seeds are placed in a 0.1% solution of 2,3,5-triphenyl-2H-tetrazolium chloride (TTC). If seeds are not cut through the embryo, a 1.0% solution of TTC, is recommended (Grabe, 1970). Viable embryos release hydrogen ions during respiration, which combine with TTC, causing it to turn red or pink; therefore, viable embryos turn red or pink. This method of determining viability works for dormant as well as nondormant embryos (Flemion and Poole, 1948). Other methods used to determine seed viability include tests for catalase activity and reducing substances in leachates and conductivity of electrolytes in leachates of imbibed seeds (Freeland, 1976).

Sometimes seeds that are normal in size, have endosperm, are firm when pinched, and are not infected with fungi do not have embryos! Embryoless seeds are quite common in Apiaceae (Flemion and Waterbury, 1941; Flemion and Henrickson, 1949), but they also occur in *Ginkgo,* corn, wheat, rye, barley, rice, and castor bean (see Flemion and Waterbury, 1941).

III. GERMINATING SEEDS YOU KNOW NOTHING ABOUT

The first thing to consider is phenology of the seed phase of the life cycle. In particular, when do the seeds mature, when are they dispersed, and when do they germinate? The next question is what are the environmental conditions in the habitat during each of these stages? It is especially important to have information about temperatures during the time between maturation and germination. Are seeds exposed to warm stratification, cold stratification, or both before they germinate? Are these treatments necessary for germination?

Much information about the germination ecology of a species can be obtained from germination phenology studies. In these studies, seeds are collected at the time of natural dispersal and are sown on soil where they receive natural temperature and soil moisture conditions either outside or in a nonheated greenhouse. Regardless of where the studies are done, seeds need to be checked for germination at short, regular intervals. Temperature records greatly enhance germination phenology studies. These data tell us what temperatures seeds are exposed to before they germinate and temperatures during the time of germination. If temperatures in the habitat or in the general area are known, germination phenology studies can be done in incubators or growth chambers by subjecting seeds to a sequence of monthly temperature regimes that simulate those in nature. If seeds are not placed under natural conditions, care should be taken to approximate soil moisture conditions in the habitats during various seasons of the year.

Information on temperatures occurring in the habitat for the time between maturation and germination permit one to formulate testable questions about the dormancy-breaking requirements of seeds. For example, is warm stratification a prerequisite for the germination of seeds that mature in spring and germinate in autumn? Seeds of the winter annuals *Thlaspi perfoliata, Draba verna,* and *Holosteum umbellatum* are dispersed in spring and germinate in autumn and they require exposure to high summer temperatures before they can germinate at autumn temperatures in autumn (Baskin and Baskin, 1986b). Seeds of *Isopyrum biternatum* also mature in spring and germinate in autumn, but they do not require exposure to high summer temperatures before they can germinate at autumn temperatures in autumn. Freshly matured seeds of this species will germinate at simulated autumn temperatures; thus, the only effect high summer temperatures have on its seeds is to prevent germination until autumn (Baskin and Baskin, 1986c).

In temperate regions, seeds of summer annuals and many perennials are dispersed in autumn and germinate the following spring and/or summer. Is cold stratification a prerequisite for germination? Seeds of many species will not germinate at spring temperatures unless they receive several weeks of cold stratification during winter (Baskin and Baskin, 1977b, 1987, 1988). However, at the time of maturity in autumn, seeds of *Helenium amarum* (Baskin and Baskin, 1973) and *Galinsoga* spp. (Baskin and Baskin, 1981a) can germinate at simulated spring temperatures. Thus, the only effect low winter temperatures have on seeds of these two species is to delay germination until spring.

Seeds of some species are dispersed in spring, but they do not germinate until the following spring. Thus, seeds are exposed to warm (summer) followed by cold (winter) stratification before they germinate. Are both treatments necessary for germination? To answer this question, seeds are given a warm followed by a cold stratification treatment and only a cold stratification treatment (control). Freshly matured seeds of *Stylophorum diphyllum* that are cold stratified will germinate (Baskin and Baskin, 1984b). Thus, high summer temperatures do not play a role in breaking seed dormancy in this species. However, freshly matured seeds of *Erythronium albidum* that are cold stratified will not germinate unless they are first exposed to a period of warm stratification (Baskin and Baskin, 1985b). Thus, both summer and winter conditions in the habitat play a role in the germination of *E. albidum*.

References

Ahlawat, A. S., and Dagar, J. C. (1980). Effect of different pH, light qualities and some growth regulators on seed germination of *Bidens biternata* (Lour.) Merr and Sherff. *Indian For.* **106**, 617–620.

Akamine, E. K. (1947). Germination of *Asystasia gangetica* L. seed with special reference to the effect of age on the temperature requirement for germination. *Plant Physiol.* **22**, 603–607.

Akoroda, M. O. (1986). Seed desiccation and recalcitrance in *Telfairia occidentalis*. *Seed Sci. Technol.* **14**, 327–332.

Allen, G. S. (1962). Factors affecting the viability and germination behavior of coniferous seed. VI. Stratification and subsequent treatment, *Pseudotsuga menziesii* (Mirb.) Franco. *For. Chron.* **38**, 485–496.

Amritphale, D., Iyengar, S., and Sharma R. K. (1989). Effect of light and storage temperature on seed germination in *Hygrophila auriculata* (Schumach.) Haines. *J. Seed Technol.* **13**, 39–43.

Amritphale, D., Mukhiya, Y. K., Gupta, J. C., and Iyengar, S. (1984). Effect of storage, photoperiod and mechanical scarification on seed germination in *Ocimum americanum*. *Physiol. Plant.* **61**, 649–652.

Andersen, A. M. (1953). The effect of the glumes of *Paspalum notatum* Flugge on germination. *Proc. Assoc. Offic. Seed Anal.* **43**, 93–100.

Anderson, R. C., and Loucks, O. L. (1973). Aspects of the biology of *Trientalis borealis* Raf. *Ecology* **54**, 798–808.

Andrews, C. J., and Burrows, V. D. (1974). Increasing winter survival of dormoat seeds by a treatment inducing secondary dormancy. *Can. J. Plant Sci.* **54**, 565–571.

Arts, G. H. P., and van der Heijden, R. A. J. M. (1990). Germination ecology of *Littorella uniflora* (L.) Aschers. *Aquat. Bot.* **37**, 139–151.

Association of Official Seed Analysts. (1960). Rules for testing seeds. *Proc. Assoc. Offic. Seed Anal.* **49**, 1–71.

Bansal, R. P., Bhati, P. R., and Sen, D. N. (1980). Differential specificity in water imbibition of Indian arid zone seeds. *Biol. Plant.* **22**, 327–331.

Bartley, M. R., and Frankland, B. (1982). Analysis of the dual role of phytochrome in the photoinhibition of seed germination. *Nature* **300**, 750–752.

Barton, L. V. (1945). A note on the viability of seeds of maga, *Montezuma speciosissima*. *Contrib. Boyce Thomp. Inst.* **13**, 423–426.

Barton, L. V. (1965). Viability of seeds of *Theobroma cacao* L. *Contrib. Boyce Thomp. Inst.* **23**, 109–122.

Baskin, C. C., and Baskin, J. M. (1988). Germination ecophysiology of herbaceous plant species in a temperate region. *Am. J. Bot.* **75**, 286–305.

Baskin, C. C., Baskin, J. M., and Chester, E. W. (1995). Role of temperature in the germination ecology of the summer annual *Bidens polylepis* Blake (Asteraceae). *Bull. Torrey Bot. Club* **122**, 275–281.

Baskin, J. M., and Baskin, C. C. (1971a). Germination of *Cyperus inflexus* Muhl. *Bot. Gaz.* **132**, 3–9.

Baskin, J. M., and Baskin, C. C. (1971b). Germination ecology and adaptation to habitat in *Leavenworthia* spp. (Cruciferae). *Am. Midl. Nat.* **85**, 22–35.

Baskin, J. M., and Baskin, C. C. (1971c). The possible ecological significance of the light requirement for germination in *Cyperus inflexus*. *Bull. Torrey Bot. Club* **98**, 25–33.

Baskin, J. M., and Baskin, C. C. (1971d). Germination ecology of *Phacelia dubia* var. *dubia* [actually var. interior] in Tennessee glades. *Am. J. Bot.* **58**, 98–104.

Baskin, J. M., and Baskin, C. C. (1972a). Physiological ecology of germination of *Viola rafinesquii*. *Am. J. Bot.* **59**, 981–988.

Baskin, J. M., and Baskin, C. C. (1972b). Germination characteristics of *Diamorpha cymosa* seeds and an ecological interpretation. *Oecologia* **10**, 17–28.

Baskin, J. M., and Baskin, C. C. (1972c). The light factor in the germination ecology of *Draba verna*. *Am. J. Bot.* **59**, 756–759.

Baskin, J. M., and Baskin, C. C. (1973). Ecological life cycle of *Helenium amarum* in central Tennessee. *Bull. Torrey Bot. Club* **100**, 117–124.

Baskin, J. M., and Baskin, C. C. (1975). Do seeds of *Helenium amarum* have a light requirement for germination? *Bull. Torrey Bot. Club* **102**, 73–75.

Baskin, J. M., and Baskin, C. C. (1976). Effect of photoperiod on germination of *Cyperus inflexus* seeds. *Bot. Gaz.* **137**, 269–273.

Baskin, J. M., and Baskin, C. C. (1977a). Germination ecology of *Sedum pulchellum* Michx. (Crassulaceae). *Am. J. Bot.* **64**, 1242–1247.

Baskin, J. M., and Baskin, C. C. (1977b). Germination of common milkweed (*Asclepias syriaca* L.) seeds. *Bull. Torrey Bot. Club* **104**, 167–170.

Baskin, J. M., and Baskin, C. C. (1979a). Promotion of germination of *Stellaria media* seeds by light from a green safe lamp. *New Phytol.* **82**, 381–383.

Baskin, J. M., and Baskin, C. C. (1979b). Effect of relative humidity on afterripening and viability in seeds of the winter annual *Draba verna*. *Bot. Gaz.* **140**, 284–287.

Baskin, J. M., and Baskin, C. C. (1981a). Temperature relations of seed germination and ecological implications in *Galinsoga parviflora* and *G. quadriradiata*. *Bartonia* **48**, 12–18.

Baskin, J. M., and Baskin, C. C. (1981b). Seasonal changes in germination responses of buried seeds of *Verbascum thapsus* and *V. blattaria* and ecological implications. *Can. J. Bot.* **59**, 1769–1775.

Baskin, J. M., and Baskin, C. C. (1982). Germination ecophysiology of *Arenaria glabra*, a winter annual of sandstone and granite outcrops of southeastern United States. *Am. J. Bot.* **69**, 973–978.

Baskin, J. M., and Baskin, C. C. (1983). Germination ecology of *Veronica arvensis. J. Ecol.* **71**, 57–68.

Baskin, J. M., and Baskin, C. C. (1984a). The ecological life cycle of *Campanula americana* in northcentral Kentucky. *Bull. Torrey Bot. Club* **111**, 329–337.

Baskin, J. M., and Baskin, C. C. (1984b). Germination ecophysiology of an eastern deciduous forest herb *Stylophorum diphyllum. Am. Midl. Nat.* **111**, 390–399.

Baskin, J. M., and Baskin, C. C. (1984c). Germination ecophysiology of the woodland herb *Osmorhiza longistylis* (Umbelliferae). *Am. J. Bot.* **71**, 687–692.

Baskin, J. M., and Baskin, C. C. (1985a). Role of dispersal date and changes in physiological responses in controlling timing of germination in *Geum canadense* Jacq. *Can. J. Bot.* **63**, 1654–1658.

Baskin, J. M., and Baskin, C. C. (1985b). Seed germination ecophysiology of the woodland spring geophyte *Erythronium albidum. Bot. Gaz.* **146**, 130–136.

Baskin, J. M., and Baskin, C. C. (1985c). The light requirement for germination of *Aster pilosus* seeds: Temporal aspects and ecological consequences. *J. Ecol.* **73**, 765–773.

Baskin, J. M., and Baskin, C. C. (1986a). Change in dormancy status of *Frasera caroliniensis* seeds during overwintering on parent plant. *Am. J. Bot.* **73**, 5–10.

Baskin, J. M., and Baskin, C. C. (1986b). Temperature requirements for after-ripening in seeds of nine winter annuals. *Weed Res.* **26**, 375–380.

Baskin, J. M., and Baskin, C. C. (1986c). Germination ecophysiology of the mesic deciduous forest herb *Isopyrum biternatum. Bot. Gaz.* **147**, 152–155.

Baskin, J. M., and Baskin, C. C. (1987). Temperature requirements for after-ripening in buried seeds of four summer annual weeds. *Weed Res.* **27**, 385–389.

Baskin, J. M., and Baskin, C. C. (1990). Seeds germination ecology of poison hemlock, *Conium maculatum. Can J. Bot.* **68**, 2018–2024.

Baskin, J. M., and Baskin, C. C. (1992). Seed germination biology of the weedy biennial *Alliaria petiolata. Nat. Areas J.* **12**, 191–197.

Baskin, J. M., and Baskin, C. C. (1998). Greenhouse and laboratory studies on the ecological life cycle of *Dalea foliosa* (Fabaceae), a federal-endangered species. *Nat. Areas J.* (in press).

Baskin, J. M., Schank, S. C., and West, S. H. (1969). Seed dormancy in two species of *Digitaria* from Africa. *Crop Sci.* **9**, 584–586.

Baxter, B. J. M., and van Staden, J. (1993). Coat imposed and embryo dormancy in *Themeda triandra* Fors[s]k. *In* "4th International Workshop on Seeds" (C. Come and F. Corbineau, eds.), Vol. 2, pp. 677–682. Universite Pierre et Marie Curie, Paris, France.

Bazzaz, F. A. (1970). Secondary dormancy in seeds of common ragweed *Ambrosia artemisiifolia. Bull. Torrey Bot. Club* **97**, 302–305.

Beardsell, D., and Mullett, J. (1984). Seed germination of *Eucalyptus pauciflora* Sieb. ex Spreng. from low and high altitude populations in Victoria. *Aust. J. Bot.* **32**, 475–480.

Becwar, M. R., Stanwood, P. C., and Roos, E. R. (1982). Dehydration effects on imbibitional leakage from desiccation-sensitive seeds. *Plant Physiol.* **69**, 1132–1135.

Bergelson, J., and Perry, R. (1989). Interspecific competition between seeds: Relative planting date and density affect seedling emergence. *Ecology* **70**, 1639–1644.

Berrie, A. M. M., and Taylor, G. C. D. (1981). The use of population parameters in the analysis of germination of lettuce seed. *Physiol. Plant.* **51**, 229–233.

Berry, G. J., Cawood, R. J., and Flood, R. G. (1988). Curve fitting of germination data using the Richards function. *Plant Cell Environ.* **11**, 183–188.

Black, M. (1967). Photoperiodic control of germination. *In* "Physiology, Ecology, and Biochemistry of Germination" (H. Borris, ed.), pp. 147–154. Proc. Int. Symp. held at the Bot. Inst., Ernst-Moritz-Arndt-University of Greifswald.

Black, M., and Wareing, P. F. (1954). Photoperiodic control of germination in seed of birch (*Betula pubescens* Ehrh.). *Nature* **174**, 705–706.

Black, M., and Wareing, P. F. (1955). Growth studies in woody species VII. Photoperiodic control of germination in *Betula pubescens* Ehrh. *Physiol. Plant.* **8**, 300–316.

Black, M., and Wareing, P. F. (1960). Photoperiodism in the light-inhibited seeds of *Nemophilia insignis. J. Exp. Bot.* **11**, 28–39.

Blank, R. R., and Young, J. A. (1992). Influence of matric potential and substrate characteristics on germination of Nezpar Indian ricegrass. *J. Range Manage.* **45**, 205–209.

Blom, C. W. P. M. (1978). Germination, seedling emergence and establishment of some *Plantago* species under laboratory and field conditions. *Acta Bot. Neerl.* **27**, 257–271.

Borthwick, H. A., Hendricks, S. B., Schneider, M. J., Taylorson, R. B., and Toole, V. K. (1969). The high-energy light action controlling plant responses and development. *Proc. Natl. Acad. Sci., USA* **64**, 479–486.

Borthwick, H. A., Toole, E. H., and Toole, V. K. (1964). Phytochrome control of *Paulownia* seed germination. *Israel J. Bot.* **13**, 122–133.

Bosy, J., and Aarssen, L. W. (1995). The effect of seed orientation on germination in a uniform environment: Differential success without genetic or environmental variation. *J. Ecol.* **83**, 769–773.

Botha, F. C., Grobbelaar, N., and Small, J. G. C. (1984). The effect of water stress on the germination of *Citrullus lanatus* seeds. *S. Afr. J. Bot.* **3**, 111–114.

Bradbeer, J. W. (1968). Studies in seed dormancy. IV. The role of endogenous inhibitors and gibberellin in the dormancy and germination of *Corylus avellana* L. seeds. *Planta* **78**, 266–276.

Bradbeer, J. W., and Colman, C. (1967). Studies in seed dormancy. I. The metabolism of [2-^{14}C] acetate by chilled seeds of *Corylus avellana* L. *New Phytol.* **66**, 5–15.

Brown, R. (1965). Physiology of seed germination. *In* "Encyclopedia of Plant Physiology" (W. Ruhland, ed.), Vol. 15/2, pp. 894–908. Springer-Verlag, Berlin/Heidelberg/New York.

Brown, R. F., and Mayer, D. G. (1986). A critical analysis of Maguire's germination rate index. *J. Seed Technol.* **10**, 101–110.

Brown, R. F., and Mayer, D. G. (1988a). Representing cumulative germination. 1. A critical analysis of single-value germination indices. *Ann. Bot.* **61**, 117–125.

Brown, R. F., and Mayer, D. G. (1988b). Representing cumulative germination. 2. The use of the Weibull function and other empirically derived curves. *Ann. Bot.* **61**, 127–138.

Bulloch, W. (1938). "The History of Bacteriology." Oxford University Press, London.

Bullowa, S., Negbi, M., and Ozeri, Y. (1975). Role of temperature, light and growth regulators in germination in *Anemone coronaria* L. *Aust. J. Plant Physiol.* **2**, 91–100.

Campbell, R. K., and Sorensen, F. C. (1979). A new basis for characterizing germination. *J. Seed Technol.* **2**, 24–34.

Campbell, T. E. (1982). Imbibition, desiccation, and reimbibition effects on light requirements for germinating southern pine seeds. *For. Sci.* **28**, 539–543.

Caplenor, D. (1967). Temperature control of germination of *Helenium amarum. Ecology* **48**, 661–664.

Cardoso, V. J. M., and Felippe, G. M. (1988). The action of the testa upon the germination of seeds of *Cucumis anguria* L. *Biol. Plant.* **30**, 48–52.

Chen, S. S. C. (1968). Germination of light-inhibited seed of *Nemophila insignis. Am. J. Bot.* **55**, 1177–1183.

Chen, S. S. C., and Thimann, K. V. (1966). Nature of seed dormancy in *Phacelia tanacetifolia*. *Science* **153**, 1537–1539.

Chilton, M. W., and Isely, D. (1961). Moisture control in seed laboratory germination. *Proc. Assoc. Offic. Seed Anal.* **51**, 155–164.

Chin, H. F., Hor, Y. L., and Lassim, M. B. M. (1984). Identification of recalcitrant seeds. *Seed Sci. Technol.* **12**, 429–436.

Chin, H. F., Krishnapillay, B., and Stanwood, P. C. (1989). Seed moisture: Recalcitrant vs. orthodox seeds. *Crop Sci. Soc. Amer. Special Publ. No.* **14**, 15–22.

Choate, H. A. (1940). Dormancy and germination in seeds of *Echinocystis lobata*. *Am. J. Bot.* **27**, 156–160.

Choinski, J. S., Jr. (1990). Aspects of viability and post-germinative growth in seeds of the tropical tree, *Trichilia dregeana* Sonder. *Ann. Bot.* **66**, 437–442.

Conner, A. J., and Conner, L. N. (1988). Germination and dormancy of *Arthropodium cirratum* seeds. *New Zeal. Nat. Sci.* **15**, 3–10.

Conover, D. G., and Geiger, D. R. (1984). Germination of Australian channel millet [*Echinochloa turnerana* (Domin.) J. M. Black] seeds. I. Dormancy in relation to light and water. *Aust. J. Plant Physiol.* **11**, 395–408.

Corbineau, F., and Come, D. (1988). Storage of recalcitrant seeds of four tropical species. *Seed Sci. Technol.* **16**, 97–103.

Cottrell, H. J. (1947). Tetrazolium salt as a seed germination indicator. *Nature* **159**, 748.

Coumans, M., Ceulemans, E., and Gaspar, T. (1979). Stabilized dormancy in sugarbeet fruits. III. Water sensitivity. *Bot. Gaz.* **140**, 389–392.

Crocker, W., and Barton, L. V. (1957). "Physiology of Seeds." Chronica Botanica Co., Waltham, MA.

Czabator, F. J. (1962). Germination value: An index combining speed and completeness of pine seed germination. *For. Sci.* **8**, 386–396.

Cumming, B. G. (1963). The dependence of germination on photoperiod, light quality, and temperature, in *Chenopodium* spp. *Can. J. Bot.* **41**, 1211–1233.

Danielson, H. R., and Tanaka, Y. (1978). Drying and storing stratified Ponderosa pine and Douglas-fir seeds. *For. Sci.* **24**, 11–16.

Datta, S. C. (1965). Germination of seeds of two arid zone species. *Bull. Bot. Soc. Bengal* **19**, 51–53.

Dawson, R. M. C., and Elliott, W. H. (1959). Buffers and physiological media. *In* "Data for Biochemical Research" (R. M. C. Dawson, D. C. Elliott, and K. M. Jones, eds.), pp. 192–209. Oxford Univ. Press, London.

Densmore, R., and Zasada, J. (1984). Seed dispersal and dormancy patterns in northern willows: Ecological and evolutionary significance. *Can. J. Bot.* **61**, 3207–3216.

Dodd, M. C., and van Staden, J. (1981). Germination and viability studies on the seeds of *Podocarpus henkelii* Stapf. *S. Afr. J. Sci.* **77**, 171–174.

Edgar, J. G. (1977). Effects of moisture stress on germination of *Eucalyptus camaldulensis* Dehnh. and *E. regans* F. Muell. *Aust. For. Res.* **7**, 241–245.

Ellis, R. H., de Barros, M. A., Hong, T. D., and Roberts, E. H. (1990). Germination of seeds of five cultivars of *Echinochloa colonum* (L.) Link in response to potassium nitrate and white light of varying photon flux density and photoperiod. *Seed Sci. Technol.* **18**, 119–130.

Ellis, R. H., Hong, T. D., and Roberts, E. H. (1986a). Quantal response of seed germination in *Brachiaria humidicola, Echinochloa turnerana, Eragrostis tef* and *Panicum maximum* to photon dose for the low energy reaction and the high irradiance reaction. *J. Exp. Bot.* **37**, 742–753.

Ellis, R. H., Hong, T. D., and Roberts, E. H. (1986b). The response of seeds of *Bromus sterilis* L. and *Bromus mollis* L. to white light of varying photon flux density and photoperiod. *New Phytol.* **104**, 485–496.

Ellis, R. H., Hong T. D., and Roberts, E. H. (1989). Quantal response of seed germination in seven genera of Cruciferae to white light and varying photon flux density and photoperiod. *Ann. Bot.* **63**, 145–158.

Evans, C. E. and Etherington, J. R. (1990). The effect of soil water potential on seed germination of some British plants. *New Phytol.* **115**, 539–548.

Evenari, M. (1952). The germination of lettuce seeds. I. Light, temperature and coumarin as germination factors. *Palestine J. Bot.* **5**, 138–160.

Evenari, M. (1965). Light and seed dormancy. *In* "Encyclopedia of Plant Physiology" (W. Ruhland, ed.), Vol 15/2, pp. 804–847. Springer-Verlag, Berlin/Heidelberg/New York.

Farmer, R. E., Jr., and Hall, G. C. (1970). Pokeweed seed germination: Effects of environment, stratification, and chemical growth regulators. *Ecology* **51**, 894–898.

Farmer, R. E., Charrette, P., Searle, I. E., and Tarjan, D. P. (1984). Interaction of light, temperature, and chilling in the germination of black spruce. *Can. J. For. Res.* **14**, 131–133.

Farrant, J. M., Pammenter, N. W., and Berjak, P. (1986). The increasing desiccation sensitivity of recalcitrant *Avicennia marina* seeds with storage time. *Physiol. Plant.* **67**, 291–298.

Felippe, G. M. (1978). Estudos de germinacao, drescimento e floracao de *Bidens pilosa* L. *Rev. Mus. Paulista* **25**, 183–217.

Felippe, G. M., Giulietti, A. M., and Lucas, N. M. C. (1971). Estudos de germinacao em *Porophyllum lanceolatum* DC. I. Efeito de luz, temperatura e fotoperiodo. *Hoehnea* **1**, 1–9.

Felippe, G. M., and Polo, M. (1983). Germinacao de ervas invasoras: Efeito de luz e escarificacao. *Rev. Brasil. Bot.* **6**, 55–60.

Flemion, F. (1931). After-ripening, germination, and vitality of seeds of *Sorbus aucuparia* L. *Contrib. Boyce Thomp. Inst.* **3**, 413–439.

Flemion, F. (1933). Physiological and chemical studies of after-ripening of *Rhodotypos kerrioides* seeds. *Contrib. Boyce Thomp. Inst.* **5**, 143–159.

Flemion, F., and Poole, H. (1948). Seed viability tests with 2,3,5-triphenyl tetrazolium chloride. *Contrib. Boyce Thomp. Inst.* **15**, 243–258.

Flemion, F., and Henrickson, E. T. (1949). Further studies on the occurrence of embryoless seeds and immature embryos in the Umbelliferae. *Contrib. Boyce Thomp. Inst.* **15**, 291–297.

Flemion, F., and Waterbury, E. (1941). Embryoless dill seeds. *Contrib. Boyce Thomp. Inst.* **12**, 157–161.

Fosket, E. B., and Briggs, W. R. (1970). Photosensitive seed germination in *Catalpa speciosa*. *Bot. Gaz.* **131**, 167–172.

Freeland, P. W. (1976). Test for the viability of seeds. *J. Biol. Edu.* **10**, 57–64.

Fujii, T., and Isikawa, S. (1962). Effects of after-ripening on photoperiodic control of seed germination in *Eragrostis ferruginea* Beauv. *Bot. Mag. Tokyo* **75**, 296–301.

Garmon, H. R., and Barton, L. V. (1946). Germination of seeds of *Panicum anceps* Michx. *Contrib. Boyce Thomp. Inst.* **14**, 117–122.

Gill, N. T. (1938). The viability of weed seeds at various stages of maturity. *Ann. Appl. Biol.* **25**, 447–456.

Ginzo, H. D. (1978). Red and far red inhibition of germination in *Aristida murina* Cav. *Z. Pflanzenphysiol.* **90**, 303–307.

Goedert, C. O., and Roberts, E. H. (1986). Characterization of alternating-temperature regimes that remove seed dormancy in seeds of *Brachiaria humidicola* (Rendle) Schweickerdt. *Plant Cell Environ.* **9**, 521–525.

Goldthwaite, J. J., Bristol, J. C., Gentile, A. C., and Klein, R. M. (1971). Light-suppressed germination of California poppy seed. *Can. J. Bot.* **49**, 1655–1659.

Goodchild, N. A., and Walker, M. G. (1971). A method of measuring seed germination in physiological studies. *Ann. Bot.* **35**, 615–621.

Grabe, D. F. (ed.) (1970). Tetrazolium testing handbook for agricultural seeds. Contribution No. 29 to the Handbook on Seed Testing. *Assoc. Offic. Seed Anal.* (no location given).

Griesbach, R. A., and Voth, P. D. (1957). On dormancy and seed germination in *Hemerocallis. Bot. Gaz.* **118**, 223–237.

Grime, J. P. (1981). The role of seed dormancy in vegetation dynamics. *Ann. Appl. Biol.* **98**, 555–558.

Grime, J. P., Mason, G., Curtis, A. V., Rodman, J., Band, S. R., Mowforth, M. A. G., Neal A. M., and Shaw, S. (1981). A comparative study of germination characteristics in a local flora. *J. Ecol.* **69**, 1017–1059.

Grime, J. P., and Thompson, K. (1976). An apparatus for measurement of the effect of amplitude of temperature fluctuation upon the germination of seeds. *Ann. Bot.* **40**, 795–799.

Gupta, I., and Basu, P. K. (1988). Role of pH on natural regeneration of *Tephrosia candida*, an endangered species of North Bengal. *Environ. Ecol.* **6**, 537–541.

Gutterman, Y., Corbineau, F., and Come, D. (1992). Interrelated effects of temperature, light and oxygen on *Amaranthus caudatus* L. seed germination. *Weed Res.* **32**, 111–117.

Hackett, C. (1964). Ecological aspects of the nutrition of *Deschampsia flexuosa* (L.) Trin. I. The effect of aluminium, manganese and pH on germination. *J. Ecol.* **52**, 159–167.

Hammouda, M. A., and Bakr, Z. Y. (1969). Some aspects of germination of desert seeds. *Phyton (Austria)* **13**, 183–201.

Harrington, G. T. (1923). Use of alternating temperatures in the germination of seeds. *J. Agric. Res.* **23**, 295–332.

Harrington, J. F. (1972). Seed storage and longevity. *In* "Seed Biology" (T. T. Kozlowski, ed.), Vol. III, pp. 145–245. Academic Press, New York.

Hay, F. R., and Probert, R. J. (1995). Seed maturity and the effects of different drying conditions on desiccation tolerance and seed longevity in foxglove (*Digitalis purpurea* L.). *Ann. Bot.* **76**, 639–647.

Hegarty, T. W. (1975). Effects of fluctuating temperature on germination and emergence of seeds in different moisture environments. *J. Exp. Bot.* **26**, 203–211.

Helgeson, E. A. (1932). Impermeability in mature and immature sweet clover seeds as affected by conditions of storage. *Trans. Wis. Acad. Sci. Arts Letts.* **27**, 193–206.

Hemmat, M., Zeng, G.-W., and Khan, A. A. (1985). Responses of intact and scarified curly dock (*Rumex crispus*) seeds to physical and chemical stimuli. *Weed Sci.* **33**, 658–664.

Henson, I. E. (1970). The effects of light, potassium nitrate and temperature on the germination of *Chenopodium album* L. *Weed Res.* **10**, 27–39.

Hester, M., and Mendelssohn, I. (1987). Seed production and germination response of four Louisiana populations of *Uniola paniculata* (Gramineae). *Am. J. Bot.* **74**, 1093–1101.

Heydecker, W., and Orphanos, P. I. (1968). The effect of excess moisture on the germination of *Spinacia oleracea* L. *Planta* **83**, 237–247.

Hilton, J. R. (1982). An unusual effect of the far-red absorbing form of phytochrome: Photoinhibition of seed germination in *Bromus sterilis* L. *Planta* **155**, 524–528.

Hilton, J. R. (1984). The influence of temperature and moisture status on the photoinhibition of seed germination in *Bromus sterilis* L. by the far-red absorbing form of phytochrome. *New Phytol.* **97**, 369–374.

Holmes, G. D., and Buszewicz, G. (1958). The storage of seed of temperate forest tree species. *For. Abst.* 19 (Nos. 3–4), 313–322, 455–476.

Hsiao, A. I.-H., and Simpson, G. M. (1971a). Dormancy studies in seed of *Avena fatua*. 7. The effects of light and variation in water regime on germination. *Can. J. Bot.* **49**, 1347–1357.

Hsiao, A. I.-H., and Simpson, G. M. (1971b). An influence of water regime on the light response in germination of lettuce (*Lactuca sativa* var. Grand Rapids) seed. *Can. J. Bot.* **49**, 1359–1362.

Hume, L. (1984). The effect of seed maturity, storage on the soil surface, and burial on seeds of *Thlaspi arvense* L. *Can. J. Plant Sci.* **64**, 961–969.

Hunter, J. R., and Erickson, A. E. (1952). Relation of seed germination to soil moisture tension. *Agron. J.* **44**, 107–109.

Huxley, P. A. (1964). Investigations on the maintenance of viability of robusta coffee seed in storage. *Proc. Intl. Seed Test. Assoc.* **29**, 423–444.

Hyde, E. O. C. (1954). The function of the hilum in some Papilionaceae in relation to the ripening of the seed and the permeability of the testa. *Ann. Bot.* **18**, 241–256.

Ikuma, H., and Thimann, K. V. (1963). The role of the seed-coats in germination of photosensitive lettuce seeds. *Plant Cell Physiol.* **4**, 169–185.

International Seed Testing Association. (1985). International rules for seed testing. Annexes 1985. *Seed Sci. Technol.* **13**, 356–513.

Isikawa, S. (1952). On the light-sensitivity of tobacco seeds. 1. Change of the light-sensitivity with the time of imbibition. *Bot. Mag. Tokyo* **65**, 771–772.

Isikawa, S. (1954). Light-sensitivity against the germination. I. "Photoperiodism" of seeds. *Bot. Mag. Tokyo* **67**, 51–56.

Isikawa, S. (1957). Interaction of temperature and light in the germination of *Nigella* seeds I. *Bot. Mag. Tokyo* **70**, 264–275.

Isikawa, S., and Shimogawara, G. (1954). Effects of light upon the germination of forest tree seeds. I. Light-sensitivity and its degree. *J. Jap. For. Soc.* **36**, 318–323.

Ismail, A. M. A. (1985). A comparative study of the effects of clumping, substrate, light, depth of burial and soil moisture upon the germination of seeds of weeds and crops grown in the Sudan Gezira. *Qatar Univ. Sci. Bull.* **5**, 153–167.

Janssen, J. G. M. (1973). A method of recording germination curves. *Ann. Bot.* **37**, 705–708.

Jones, H. A. (1920). Physiological study of maple seeds. *Bot. Gaz.* **69**, 127–152.

Jones, T. A., and Nielson, D. C. (1992). Germination of prechilled mechanically scarified and unscarified Indian ricegrass seed. *J. Range Manage.* **45**, 175–179.

Joseph, H. C. (1929). Germination and keeping quality of parsnip seeds under various conditions. *Bot. Gaz.* **87**, 195–210.

Joubert, D. C., and Small, J. G. C. (1982). Seed germination and dormancy of *Stipa trichotoma* (Nassella tussock). I. Effect of dehulling, constant temperatures, light, oxygen, activated charcoal and storage. *S. Afr. J. Bot.* **1**, 142–146.

Junttila, O. (1977). Dormancy in dispersal units of various *Dactylis glomerata* populations. *Seed Sci. Technol.* **5**, 463–471.

Justice, O. L. (1941). A study of dormancy in seeds of *Polygonum. Cornell Univ. Agric. Exp. Stn. Mem.* 235. 43 pp.

Justice, O. L., and Reece, M. H. (1954). A review of literature and investigation on the effects of hydrogen-ion concentration on the germination of seeds. *Proc. Assoc. Offic. Seed Anal.* **44**, 144–149.

Kadman-Zahavi, A. (1955). The effect of light and temperature on the germination of *Amaranthus blitoides* seeds. *Bull. Res. Counc. Israel (D)* **4**, 370–374.

Kappen, L., Oertli, J. J., Lange, O. L., Schulze, E.-D., Evenari, M., and Buschbom, U. (1975). Seasonal and diurnal courses of water relations of the arido-active plant *Hammada scoparia* in the Negev Desert. *Oecologia* **21**, 175–192.

Kendrick, R. E. (1976). Photocontrol of seed germination. *Sci. Prog. Oxford* **63**, 347–367.

Kidd, F. (1914). The controlling influence of carbon dioxide in the maturation, dormancy and germination of seeds. II. *Proc. Royal Soc. Lond.* **87**, 609–625.

King, M. W., and Roberts, E. H. (1982). The imbibed storage of cocoa (*Theobroma cacao*) seeds. *Seed Sci. Technol.* **10**, 535–540.

Koller, D. (1955). Germination regulating mechanisms in some desert seeds. I. *Bull. Res. Counc. Israel (D)* **4**, 379–381.

Koller, D. (1956). Germination regulating mechanisms in some desert seeds. III. *Calligonum comosum* L'Her. *Ecology* **37**, 430–433.

Koller, D., Poljakoff-Mayber, A., Berg, A., and Diskin, T. (1963). Germination-regulating mechanisms in *Citrullus colocynthis*. *Am. J. Bot.* **50**, 597–603.

Koopman, M. J. F. (1963). Results of a number of storage experiments conducted under controlled conditions (other than agriculture seeds). *Proc. Int. Seed Test. Assoc.* **28**, 853–860.

Kumar, V., and Irvine, D. E. G. (1971). Germination of seeds of *Cirsium arvense* (L.) Scop. *Weed Res.* **11**, 200–203.

Lagoa, A. M. M. A., and de Fatima A. Pereira, M. (1987). The role of the caruncle in the germination of seeds of *Ricinus communis*. *Plant Physiol. Biochem.* **25**, 125–128.

Lang, A. (1965). *In* "Encyclopedia of Plant Physiology" (W. Ruhland, ed.), Vol. 15/2, pp. 848–891. Springer-Verlag, Berlin/Heidelberg/New York.

Larson, A. L. (1971). Two-way thermogradient plate for seed germination research: Construction plans and procedures. *USDA ARS* 51–41.

Lima, R. F., and Felippe, G. M. (1988). Efeitos de luz e temperatura na germinacao de *Talinum patens*. *Anais V Congr. SBSP.* 15–21.

Linhart, Y. B. (1976). Density-dependent seed germination strategies in colonizing versus non-colonizing plant species. *J. Ecol.* **64**, 375–380.

Linhart, Y. B., and Pickett, R. A., II. (1973). Physiological factors associated with density-dependent seed germination in *Boisduvalia glabella* (Onagraceae). *Z. Pflanzenphysiol.* **70**, 367–370.

Mabo, O. O., Lakanmi, O. O., and Okusanya, O. T. (1988). Germination ecology of *Treculia africana* (Decne). *Nigerian J. Bot.* **1**, 66–72.

MacDonald, I. R., and Hart, J. W. (1981). An inhibitory effect of light on the germination of mustard seed. *Ann. Bot.* **47**, 275–277.

Malik, N., and Vanden Born, W. H. (1987). Germination response of *Galium spurium* L. to light. *Weed Res.* **27**, 251–258.

Matumura, M., Takase, N., and Hirayoshi, I. (1960). Physiological and ecological studies on germination of *Digitaria* seeds. 1. Difference in response to germinating conditions and dormancy among individual plants. *Res. Bull. Fac. Agric. Gifu Univ. Japan* **12**, 89–96.

Maze, K. M., and Whalley, R. D. B. (1992). Germination, seedling occurrence and seedling survival of *Spinifex sericeus* R.Br. (Poaceae). *Aust. J. Ecol.* **17**, 189–194.

McAlister, D. F. (1943). The effect of maturity on the viability and longevity of the seeds of western range and pasture grasses. *Agron. J.* **35**, 442–453.

McArthur, J. A. (1978). Light effects upon dry lettuce seeds. *Planta* **144**, 1–5.

McIntyre, G. I., and Hsiao, A. I. (1985). Seed dormancy in *Avena fatua*. II. Evidence of embryo water content as a limiting factor. *Bot. Gaz.* **146**, 347–352.

McLemore, B. F., and Hansbrough, T. (1970). Influence of light on germination of *Pinus palustris* seeds. *Physiol. Plant.* **23**, 1–10.

Morinaga, T. (1926). Effect of alternating temperatures upon the germination of seeds. *Am. J. Bot.* **13**, 148–158.

Mott, J. J. (1974). Mechanisms controlling dormancy in the arid zone grass *Aristida contorta*. I. Physiology and mechanisms of dormancy. *Aust. J. Bot.* **22**, 635–645.

Muenscher, W. C. (1936). Storage and germination of seeds of aquatic plants. *Cornell Univ. Agric. Exp. Stn. Bull.* 642.

Murdoch, A. J., Roberts, E. H., and Goedert, C. O. (1989). A model for germination responses to alternating temperatures. *Ann. Bot.* **63**, 97–111.

Negbi, M., Rushkin, E., and Koller, D. (1966). Dynamic aspects of water-relations in germination of *Hirschfeldia incana* seeds. *Plant Cell Physiol.* **7**, 363–376.

Nguyen, X. V., and Come, D. (1984). Opposite effects of temperature on breaking of dormancy and on induction of frost resistance in apple embryos. *Physiol. Plant.* **62**, 79–82.

Nikolaeva, M. G. (1969). "Physiology of Deep Dormancy in Seeds." Izdatel'stvo Nauka, Leningrad. [Translated from Russian by Z. Shapiro, National Science Foundation, Washington, DC]

Noronha, A., Vicente, M., and Felippe, G. M. (1978). Photocontrol of germination of *Cucumis anguria* L. *Biol. Plant.* **20**, 281–286.

Northam, F. E., and Callihan, R. H. (1994). Interpreting germination results based on differing embryonic emergence criteria. *Weed Sci.* **42**, 474–481.

Nwankwo, B. A. (1981). Facilitated germination of *Elaeis guineensis* var. *pisifera* seeds. *Ann. Bot.* **48**, 251–254.

Oelke, E. A., and Albrecht, K. A. (1978). Mechanical scarification of dormant wild rice seed. *Agron. J.* **70**, 691–694.

Okagami, N., and Kawai, M. (1982). Dormancy in *Dioscorea*: Differences of temperature responses in seed germination among six Japanese species. *Bot. Mag. Tokyo* **95**, 155–166.

Okusanya, O. T. (1978). The effect of acid soil on the germination and early growth of some maritime cliff species. *Oikos* **30**, 549–554.

Osborne, D. J. (1981). Dormancy as a survival stratagem. *Ann. Appl. Biol.* **98**, 525–531.

Pegtel, D. M. (1988). Germination in declining and common herbaceous plant populations co-occurring in an acid peaty heathland. *Acta Bot. Neerl.* **37**, 215–223.

Phaneendranath, B. R. (1980). Influence of amount of water in the paper towel on standard germination tests. *J. Seed Technol.* **5**, 82–87.

Poel, L. W. (1949). Germination and development of heather and the hydrogen ion concentration of the medium. *Nature* **163**, 647–648.

Pons, T. L. (1983). Significance of inhibition of seed germination under the leaf canopy in ash coppice. *Plant Cell Environ.* **6**, 385–392.

Pons, T. L., and Schroder, H. F. J. M. (1986). Significance of temperature fluctuation and oxygen concentration for germination of the rice field weeds *Fimbristylis littoralis* and *Scirpus juncoides*. *Oecologia* **68**, 315–319.

Probert, R. J., Gajjar, K. H. and Halam, I. K. (1987). The interactive effects of phytochrome, nitrate and thiourea on the germination response to alternating temperatures in seeds of *Ranunculus sceleratus* L.: A quantal approach. *J. Exp. Bot.* **38**, 1012–1025.

Probert, R. J., and Longley, P. L. (1989). Recalcitrant seed storage physiology in three aquatic grasses (*Zizania palustris*, *Spartina angelica* and *Porteresia coarctata*). *Ann. Bot.* **63**, 53–63.

Probert, R. J., and Smith, R. D. (1986). The joint action of phytochrome and alternating temperatures in the control of seed germination in *Dactylis glomerata*. *Physiol. Plant.* **67**, 299–304.

Probert, R. J., Smith, R. D., and Birch, P. (1985). Germination responses to light and alternating temperatures in European populations of *Dactylis glomerata* L. IV. The effects of storage. *New Phytol.* **101**, 521–529.

Probert, R. J., Smith, R. D., and Birch, P. (1986). Germination responses to light and alternating temperatures in European populations of *Dactylis glomerata* L. V. The principle components of the alternating temperature requirement. *New Phytol.* **102**, 133–142.

Ramakrishnan, P. S. (1965). Studies on edaphic ecotypes in *Euphorbia thymifolia* L. *J. Ecol.* **53**, 157–162.

Ramakrishnan, P. S., and Jain, R. S. (1965). Germinability of the seeds of the edaphic ecotypes in *Tridax procumbens* L. *Trop. Ecol.* **6**, 47–55.

Ratcliffe, D. (1961). Adaptation to habitat in a group of annual plants. *J. Ecol.* **49**, 187–203.

Rehwaldt, C. A. (1968). Filter paper effect on seed germination of *Arabidopsis thaliana*. *Plant Cell Physiol.* **9**, 609–611.

Richter, D. D., and Switzer, G. L. (1982). A technique for determining quantitative expressions of dormancy in seeds. *Ann. Bot.* **50**, 459–463.

Rivard, P. G., and Woodard, P. M. (1989). Light, ash, and pH effects on the germination and seedling growth of *Typha latifolia* (cattail). *Can. J. Bot.* **67**, 2783–2787.

Roberts, E. H. (1973). Predicting the storage life of seeds. *Seed Sci. Technol.* **1**, 499–514.

Roberts, E. H., Murdoch, A. J., and Ellis, R. H. (1987). The interaction of environmental factors on seed dormancy. *Brit. Crop Protect. Conf. Weeds* **7C-1**, 687–694.

Roof, J. (1988). Germination studies of some *Eriodictyon* species. *The Four Seasons* **8**, 19–34.

Rost, T. L. (1975). The morphology of germination in *Setaria lutescens* (Gramineae): The effects of covering structures and chemical inhibitors on dormant and non-dormant florets. *Ann. Bot.* **39**, 21–30.

Rudolfs, W. (1922). Effect of seeds upon hydrogen-ion concentration of solutions. *Bot. Gaz.* **74**, 215–220.

Rudolfs, W. (1925). Effect of seeds upon hydrogen-ion concentration equilibrium in solution. *J. Agric. Res.* **30**, 1021–1026.

Schimpf, D. J., Flint, S. D., and Palmblad, I. G. (1977). Representation of germination curves with the logistic function. *Ann. Bot.* **41**, 1357–1360.

Schopmeyer, C. S. (technical coordinator) (1974). "Seeds of Woody Plants in the United States." United States Department of Agriculture, Forest Service, Washington, DC. Agriculture Handbook No. 450.

Scott, S. J., Jones, R. A., and Williams, W. A. (1984). Review of data analysis methods for seed germination. *Crop Sci.* **24**, 1192–1199.

Shamsi, S. R. A., and Whitehead, F. H. (1974). Comparative ecophysiology of *Epilobium hirsutum* L. and *Lythrum salicaria* L. I. General biology, distribution and germination. *J. Ecol.* **62**, 279–290.

Sidhu, S. S., and Cavers, P. B. (1977). Maturity-dormancy relationships in attached and detached seeds of *Medicago lupulina* L. (black medick). *Bot. Gaz.* **138**, 174–182.

Singh, V. P., Mall, S. L., and Billore, S. K. (1975). Effect of pH on germination of four common grass species of Ujjain (India). *J. Range Manage.* **28**, 497–498.

Smith, A. L. (1972). Factors influencing germination of *Scolochloa festucacea* caryopses. *Can. J. Bot.* **50**, 2085–2092.

Springfield, H. W. (1968). Germination of winterfat seeds under different moisture stresses and temperatures. *J. Range Manage.* **21**, 314–316.

Stearns, F., and Olson, J. (1958). Interactions of photoperiod and temperature affecting seed germination in *Tsuga canadensis*. *Am. J. Bot.* **45**, 53–58.

Steinbauer, G. P., and Grigsby, B. (1957). Interaction of temperature, light, and moistening agent in the germination of weed seeds. *Weeds* **5**, 175–182.

Stokes, P. (1965). Temperature and seed dormancy. *In* "Encyclopedia of Plant Physiology" (W. Ruhland, ed.), Vol. 15/2, pp. 746–803. Springer-Verlag, Berlin/Heidelberg/New York.

Stubbendieck, J. (1974). Effect of pH on germination of three grass species. *J. Range Manage.* **27**, 78–79.

Sweeney, J. R. (1956). Responses of vegetation to fire: A study of the herbaceous vegetation following chaparral fires. *Univ. Calif. Publ. Bot.* **28**, 143–216 + 27 plates.

Thanos, C. A., Georghious, K., and Skarou, F. (1989). *Glaucium flavum* seed germination: An ecophysiological approach. *Ann. Bot.* **63**, 121–130.

Thanos, C. A., and Mitrakos, K. (1992). Watermelon seed germination. 1. Effects of light, temperature and osmotica. *Seed Sci. Res.* **2**, 155–162.

Thompson, K. (1989). A comparative study of germination responses to high irradiance light. *Ann. Bot.* **63**, 159–162.

Thompson, K., and Grime, J. P. (1983). A comparative study of germination responses to diurnally-fluctuating temperatures. *J. Appl. Ecol.* **20**, 141–156.

Thompson, K., Grime, J. P., and Mason, G. (1977). Seed germination in response to diurnal fluctuations of temperature. *Nature* **267**, 147–149.

Thompson, K., and Whatley, J. C. (1984). Thermogradient bar apparatus for the study of the germination requirements of buried seeds *in situ*. *New Phytol.* **96**, 459–471.

Thompson, P. A. (1969). Germination of *Lycopus europaeus* L. in response to fluctuating temperatures and light. *J. Exp. Bot.* **20**, 1–11.

Thompson, P. A. (1974). Germination of celery (*Apium graveolens* L.) in response to fluctuating temperatures. *J. Exp. Bot.* **25**, 156–163.

Timson, J. (1965). New method of recording germination data. *Nature* **207**, 216–217.

Toledo, J. R., Rincon, E., and Vazquez-Yanes, C. (1990). A light quality gradient for the study of red : far red ratios on seed germination. *Seed Sci. Technol.* **18**, 277–282.

Tompsett, P. B. (1987a). The effect of desiccation on the longevity of seeds of *Araucaria hunsteinii* and *A. cunninghamii*. *Ann. Bot.* **50**, 693–704.

Tompsett, P. B. (1987b). Desiccation and storage studies on *Dipterocarpus* seeds. *Ann. Appl. Biol.* **110**, 371–379.

Tompsett, P. B., and Prichard, H. W. (1993). Water status changes during development in relation to the germination and desiccation tolerance of *Aesculus hippocastanum* L. seeds. *Ann. Bot.* **71**, 107–116.

Toole, E. H., Toole, V. K., Borthwick, H. A., and Hendricks, S. B. (1955). Interaction of temperature and light in germination of seeds. *Plant Physiol.* **30**, 473–478.

Toole, E. H., Toole, V. K., Borthwick H. A., and Hendricks, S. B. (1957a). Changing sensitivity of seeds to light. *Plant Physiol.* **32**, xi. [Abstract]

Toole, E. H., Toole, V. K., Hendricks, S. B., and Borthwick, H. A. (1957b). Effect of temperature on germination of light-sensitive seeds. *Proc. Int. Seed Test. Assoc.* **22**, 196–204.

Toole, V. K. (1963). Light control of seed germination. *Assoc. Offic. Seed Anal.* **53**, 124–143.

Torres, M., and Frutos, G. (1989). Analysis of germination curves of aged fennel seeds by mathematical models. *Environ. Exp. Bot.* **29**, 409–415.

Torres, M., and Frutos, G. (1990). Logistic function analysis of germination behaviour of aged fennel seeds. *Environ. Exp. Bot.* **30**, 383–390.

Turner, G. D., Lau, R. R., and Young, D. R. (1988). Effect of acidity on germination and seedling growth of *Paulownia tomentosa*. *J. Appl. Ecol.* **25**, 561–567.

Ungar, I. A., and Binet, P. (1975). Factors influencing seed dormancy in *Spergularia media* (L.) C. Presl. *Aquat. Bot.* **1**, 45–55.

Vaartaja, O. (1956). Photoperiodic response in germination of seed of certain trees. *Can. J. Bot.* **34**, 377–388.

Valio, I. F. M. (1976). Germination of coffee seeds (*Coffea arabica* L. cv. mundo novo). *J. Exp. Bot.* **27**, 983–991.

Valio, I. F. M., Kirszenzaft, S. L., and Rochia, R. F. (1972). Germination of achenes of *Bidens pilosa* L. I. Effect of light of different wavelengths. *New Phytol.* **71**, 677–682.

van Baalen, J. (1982). Germination ecology and seed population dynamics of *Digitalis purpurea. Oecologia* **53**, 61–67.

van Breeman, A. M. M. (1984). Comparative germination ecology of three short-lived monocarpic Boraginaceae. *Acta Bot. Neerl.* **33**, 283–305.

Varshney, C. K. (1968). Germination of the light-sensitive seeds of *Ocimum americanum* Linn. *New Phytol.* **67**, 125–129.

Vengris, J., Dunn, S., and Stacewicz-Sapuncakis, M. (1972). Life history studies as related to weed control in the Northeast. Common purslane. *Univ. Mass. Agric. Exp. Stn. Res. Bull.* 598.

Viana, A. M., and Felippe, G. M. (1986). Efeitos da luz e da temperatura na germinacao de sementes de *Dioscorea composita. Rev. Brasil. Bot.* **9**, 109–115.

Viana, A. M., and Felippe, G. M. (1989). Reproducao de *Dioscorea composita. Pesq. Agropec. Bras. Brasilia.* **24**, 573–578.

Vickery, R. K., Jr. (1967). Ranges of temperature tolerance for germination of *Mimulus* seeds from diverse populations. *Ecology* **48**, 647–651.

Vidaver, W., and Hsiao, A. I.-H. (1972). Persistence of phytochrome-mediated germination control in lettuce seeds for 1 year following a single monochromatic light flash. *Can. J. Bot.* **50**, 687–689.

Vincent, E. M., and Roberts, E. H. (1977). The interaction of light, nitrate and alternating temperature in promoting the germination of dormant seeds of common weed species. *Seed Sci. Technol.* **5**, 659–670.

Voesenek, L. A. C. J., de Graaf, M. C. C., and Blom, C. W. P. M. (1992). Germination and emergence of *Rumex* in river flood-plains. II. The role of perianth, temperature, light and hypoxia. *Acta Bot. Neerl.* **41**, 331–343.

Waite, S., and Hutchings, M. J. (1978). The effects of sowing density, salinity and substrate upon the germination of seeds of *Plantago coronopus* L. *New Phytol.* **81**, 341–348.

Wajid, A., and Shaukat, S. S. (1993). Effects of seed dispersal unit and the position of seed relative to substrate on germination of three composites. *Pakistan J. Bot.* **25**, 118–126.

Walck, J. L., Baskin, J. M., and Baskin, C. C. (1997). A comparative study of the germination biology of a narrow endemic and two geographically-widespread species of *Solidago* (Asteraceae). 3. Photoecology of germination. *Seed Sci. Res.* **7**, 293–301.

Washitani, I., and Kabaya, H. (1988). Germination responses to temperature responsible for the seedling emergence seasonality of *Primula sieboldii* E. Morren in its natural habitat. *Ecol. Res.* **3**, 9–20.

Webb, D. P., and Wareing, P. F. (1972). Seed dormancy in *Acer pseudoplatanus* L.: The role of the covering structures. *J. Exp. Bot.* **23**, 813–829.

Weerakoon, W. L., and Lovett, J. V. (1986). Studies of *Salvia reflexa* Hornem. III. Factors controlling germination. *Weed Res.* **26**, 269–276.

Westra, R. N., and Loomis, N. E. (1966). Seed dormancy in *Uniola paniculata. Am. J. Bot.* **53**, 407–411.

White, G. A., and Bass, L. N. (1971). *Vernonia anthelmintica*: A potential seed oil source of epoxy acid. III. Effects of line, harvest date, and seed storage on germination. *Agron. J.* **63**, 439–441.

Willemsen, R. W., and Rice, E. L. (1972). Mechanism of seed dormancy in *Ambrosia artemisiifolia. Am. J. Bot.* **59**, 248–257.

Williams, E. D. (1983). Effects of temperature, light, nitrate and prechilling on seed germination of grassland plants. *Ann. Appl. Biol.* **103**, 161–172.

Williams, J., and Shaykewich, C. F. (1971). Influence of soil water matric potential and hydraulic conductivity on the germination of rape (*Brassica napus* L.). *J. Exp. Bot.* **22**, 586–597.

Woods, L. E., and MacDonald, H. A. (1971). The effects of temperature and osmotic moisture stress on the germination of *Lotus corniculatus. J. Exp. Bot.* **22**, 575–585.

Wright, L. N. (1973). Seed dormancy, germination environment, and seed structure of Lehmann lovegrass, *Eragrostis lehmanniana* Nees. *Crop Sci.* **13**, 432–435.

Yap, S. K. (1981). Collection, germination and storage of dipterocarp seeds. *Malayan For.* **44**, 281–300.

Young, J. A., and Evans, R. A. (1980). Germination of desert needlegrass. *J. Seed Technol.* **5**, 40–46.

Young, J. A., Evans, R. A., and Kay, B. L. (1970). Germination of cultivars of *Trifolium subterraneum* L. in relation to matric potential. *Agron. J.* **62**, 743–745.

Zagorski, S., and Lewak, S. (1984). Are effects of gibberellic and abscisic acids on lettuce seed germination pH-dependent? *Acta Physiol. Plant.* **6**, 27–32.

Zohar, Y., Waisel, Y., and Karschon, R. (1975). Effects of light, temperature and osmotic stress on seed germination of *Eucalyptus occidentalis* Endl. *Aust. J. Bot.* **23**, 391–397.

3

Types of Seed Dormancy

I. PURPOSE

Nikolaeva was the first seed biologist to develop a comprehensive classification scheme for the various types of seed dormancy, and it was published in Russian in 1967 and in English 2 years later (Nikolaeva, 1969). In 1977, Nikolaeva expanded and revised the system; this version is still the best general classification scheme for seed dormancy types available today. In this chapter, seed dormancy will be defined and an overview of each of Nikolaeva's six broad categories of dormancy will be presented. Also, types of dormancy will be considered in relation to types of seeds.

II. DEFINITION OF DORMANCY

To many people, seed dormancy simply means that a seed has not germinated, but we will soon see that this definition is inadequate. Unfavorable environmental conditions are one reason for lack of seed germination. That is, seeds could be in a paper bag on the laboratory shelf (i.e., lack of water), buried in mud at the bottom of a lake (i.e., insufficient oxygen and/or light), or exposed to temperatures that are above or below those suitable for plant growth. These obviously unfavorable conditions for germination are examples of how the environment rather than some factor associated with the seed per se prevents germination.

A second reason why seeds may not germinate is that some property of the seed (or dispersal unit) prevents it. Thus, the lack of germination is a seed rather than an environmental problem. Dormancy that results from some characteristic of the seed is called organic dormancy, and this type of dormancy usually is of most interest to seed biologists and ecologists. In fact,

throughout the remainder of this book, we will be concerned with organic seed dormancy.

According to Nikolaeva (1969, 1977) there are two general types of organic seed dormancy: endogenous and exogenous (Table 3.1). In endogenous dormancy, some characteristic of the embryo prevents germination, whereas in exogenous dormancy, some characteristic of structures, including endosperm (sometimes perisperm), seed coats, or fruit walls, covering the embryo prevents germination. For example, seeds may not germinate because the seed (or fruit) coats are impermeable to water. Before seeds with either endogenous or exogenous dormancy can germinate, changes must occur in seeds that remove the block(s) to germination. The challenge of germination ecologists is to define the environmental conditions required to bring about changes in seeds that result in a release from dormancy and to correlate these with factors that promote dormancy break in nature.

III. TYPES OF SEEDS

Before discussing the types of dormancy, we will take a look at the various types of seeds because frequently there is a relationship between the two. Martin (1946) distinguished 12 types of seeds based on embryo morphology, relative amount of endosperm, and position of the embryo in relation to the endosperm (Table 3.2).

Linear and spatulate are the only types of seeds found in extant gymnosperms (Table 3.2). The Gnetaceae is the only gymnosperm family whose seeds have spatulate embryos, although some of the more advanced members of the Pinaceae have slightly expanded cotyledons (Martin, 1946).

TABLE 3.1   Simplified Version of Nikolaeva's (1977) Classification Scheme of Organic Seed Dormancy Types

Type	Cause	Broken by
Endogenous dormancy		
Physiological	Physiological inhibiting mechanism (PIM) of germination	Warm and/or cold stratification
Morphological	Underdeveloped embryo	Appropriate conditions for embryo growth/germination
Morphophysiological	PIM of germination and underdeveloped embryo	Warm and/or cold stratification
Exogenous dormancy		
Physical	Seed (fruit) coats impermeable to water	Opening of specialized structure
Chemical	Germination inhibitors	Leaching
Mechanical	Woody structures restrict growth	Warm and/or cold stratification

Six types of seeds are found among the monocots (Table 3.2). With respect to seeds with linear embryos, some interesting differences are found between monocots and dicots. Linear embryos in most dicots are surrounded completely by endosperm (Fig. 3.1a), but in some species the radicular end of the embryo touches the base of the seed (Fig. 3.1b). The radicular end of the embryo also touches the base of the seed in a large number of monocots (Fig. 3.1c). The linear embryo extends to both ends of the seed in the monocot family Pontederiaceae (Fig. 3.1d), and in the Cannaceae (Fig. 3.1e), Zingiberaceae, Sparganiaceae, Pontederiaceae, and in some Arecaceae and Commelinaceae the radicular end of the embryo extends into a depression at the base of the seed (Martin, 1946). No endosperm is present in seeds of various monocot families, including the Cymodoceaceae, Hydrocharitaceae, Najadaceae, Posidoniaceae, Potamogetonaceae, Zannichelliaceae, Zosteraceae, and some Araceae, and food reserves are stored in the greatly enlarged hypocotyl (Goebel, 1905; Dahlgren and Clifford, 1982); these are called macropodous embryos (Fig. 3.1f). Because macropodous em-

TABLE 3.2   Characteristics of the 12 Seed Types[a]

Type of seed	Embryo position, size/shape	Representative family	Relative size		Phylogenetic occurrence[b]		
			Seed	Embryo			
Broad		Juncaceae	Large	Small		M	D
Capitate		Cyperaceae	Large	Small		M	
Lateral		Poaceae	Large	Small		M	
Peripheral		Caryophyllaceae	Large	Large			D
Rudimentary		Ranunculaceae	Large	Small		M	D
Dwarf		Ericaceae	Small	Small			D
Micro		Orchidaceae	Minute	Minute		M	D
Linear		Liliaceae	Large	Small	G	M	D
Spatulate		Lamiaceae	Large	Large	G		D
Investing		Rhamnaceae	Large	Large			D
Bent		Fabaceae	Large	Large			D
Folded		Malvaceae	Large	Large			D

[a] From Martin (1946). Black, embryo; white, endosperm.
[b] G, gymnosperm; M, monocot; and D, dicot.

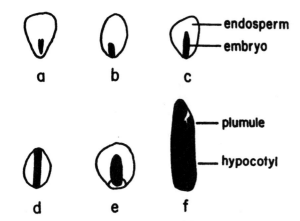

FIGURE 3.1   Linear embryos in seeds. (a) Dicot with linear embryo surrounded by endosperm (Apiaceae). (b) Dicot with linear embryo touching base of seed (Lardizabalaceae). (c) Monocot with linear embryo touching base of seed (Amaryllidaceae). (d) Monocot with linear embryo touching both poles of seed (Pontederiaceae). (e) Monocot with embryo extending into a depression at base of seed (Cannaceae). (f) Monocot with macropodous (linear) embryo (Hydrocharitaceae). a—e modified from Martin (1946), f modified from Dahlgren and Clifford (1982).

bryos are longer than broad, Martin (1946) placed seeds with this type into the linear category.

Ten types of seeds occur among dicots, but five of them are not found in monocots (Table 3.2). However, two types, capitate and lateral, occur in monocots but not in dicots. In the dicot families Cactaceae, Clusiaceae, and Lecythidaceae, food reserves may be stored in the hypocotyl rather than in the cotyledons or endosperm (Goebel, 1905). A hypocotyl with stored food is enlarged, and the cotyledons are minute, resulting in something that looks like macropodous embryos of monocots. Martin (1946) placed Cactaceae seeds into the peripheral category. Although the hypocotyl of Cactaceae embryos is enlarged (and the cotyledons tiny), a small amount of endosperm is present in seeds of some genera, and the embryo is external to it.

IV. OVERVIEW OF TYPES OF SEED DORMANCY

A. Physiological Dormancy

Physiological dormancy (PD) occurs in representatives of all 12 types of seeds, so the type of seed is not the cause of PD. Most seeds with PD are permeable to water, but there are a few exceptions (see Section IV,E). PD is caused by a physiological inhibiting mechanism of the embryo that prevents radicle emergence. However, structures that cover the embryo, including endosperm, seed coats, and indehiscent fruit walls, may play a role in

preventing germination. Nikolaeva (1977) distinguished three levels of PD: nondeep, intermediate, and deep.

1. Nondeep Physiological Dormancy: Description

Nondeep PD is common in seeds of most weeds, vegetables, many garden flowers, and some woody plants. Freshly matured seeds with nondeep PD either cannot germinate at any temperature or they germinate only over a very narrow range of temperatures. Embryos excised from seeds usually grow, and the resulting seedlings are normal. Nondeep PD in some species is broken by relatively short periods of cold stratification ranging from 5 days in *Triticum* sp. (Nikolaeva, 1969) to 60–90 days in *Impatiens biflora* (Crocker, 1948). Seeds stored dry at room temperatures come out of dormancy (afterripen). However, the time required for seeds to afterripen usually is much longer than that required for dormancy loss during cold stratification. Seeds of *Digitaria ischaemum* required 8 weeks of cold stratification at 3°C or 1 year of dry storage at room temperature to become nondormant (Toole and Toole, 1941). Seeds of *Ambrosia trifida* came out of dormancy during 3 months of cold stratification at 5°C, whereas some seeds stored dry in the laboratory remained dormant after 1 or more years (Davis, 1930).

In other species, nondeep PD is broken by exposure to high (≥15°C) temperatures, and the dormancy break is incomplete, or does not occur at all, if seeds are cold stratified (Baskin and Baskin, 1986a). The time required for dormancy break at high temperatures ranges from several weeks to many months, depending on the species. At 35/20°C (simulated summer field temperatures), imbibed seeds of *Lamium purpureum* (Baskin and Baskin, 1984a) became nondormant after 8 weeks, whereas those of *L. amplexicaule* (Baskin and Baskin, 1984b) required 12 weeks to come out of dormancy. Seeds of *Echinochloa turnerana* required 28 weeks of dry storage at room temperatures (28°C) for loss of dormancy (Conover and Geiger, 1984). Dormancy loss may occur at high temperatures regardless of whether seeds are imbibed. Rates of dormancy loss in seeds of the winter annuals *Draba verna* and *Holosteum umbellatum* stored dry at room temperatures were about the same as those of seeds imbibed at 35/20°C (Baskin and Baskin, unpublished results).

Nondeep PD can be broken by chemicals, including potassium nitrate (Toole, 1941), thiourea (Garman and Barton, 1946; Baskin and Baskin, 1971a), kinetin (Reynolds and Thompson, 1973), ethylene (Egley, 1982), and gibberellins (Khan et al., 1957; Dunwell, 1981; Watkins and Cantliffe, 1983a). The amount of exogenous gibberellic acid (GA) required for the germination of *Avena fatua* seeds decreased with an increase in the afterripening period (Hsiao and Quick, 1985). Embryos excised from dormant seeds of *A. fatua* line M73 did not grow unless treated with GA and/or fructose, whereas those excised from fully afterripened (nondormant) seeds grew without any treatments. Although embryos excised from dormant *A. fatua* seeds germinated when treated with GA and/or fructose, rates of germination and seedling growth were less than those for embryos from nondormant seeds (Myers *et al.*, 1997).

A light requirement for germination is another manifestation of nondeep PD (Nikolaeva, 1977). Seeds of some species lose their light requirement for germination as they come out of dormancy in response to cold stratification (Baskin and Baskin, 1994) or high summer temperatures (Baskin and Baskin, 1982), but those of other species have a light requirement for germination after they have received dormancy-breaking treatments (Baskin and Baskin, 1976).

2. Nondeep Physiological Dormancy: Causes

Many questions remain to be answered about nondeep PD. If removal or disruption of structures that cover the embryo, such as the endosperm, seed coats, fruits coats, and bracts, including palea and lemma (or "hulls") of grasses, results in germination, is dormancy caused by covering structures? If intact seeds germinate after they have been given dormancy-breaking treatments, such as warm or cold stratification, is dormancy caused by the embryo or by an interaction between the embryo and its covering structures?

a. Covering Structures: Oxygen Concentration

Nikolaeva (1969) attributed nondeep PD to the low permeability of embryo covers to oxygen. In fact, good evidence suggests that these structures can restrict the movement of oxygen to the embryo (Come and Tissaoui, 1972; Brown and van Staden, 1973; Dungey and Pinfield, 1980). Also, in some seeds, phenolic compounds in embryo covers fix oxygen by oxidation (Come and Tissaoui, 1972; Coumans *et al.*, 1976), thus making it unavailable for the embryo. Further support for the inhibition of germination due to low rates of oxygen diffusion to the embryo comes from experiments in which a portion of the embryo cover is removed or the covers are pricked with a pin. In these cases, germination percentages have been increased (Atwood, 1914; Black and Wareing, 1959; Brown and van Staden, 1973; Probert *et al.*, 1985; Hatterman-Valenti *et al.*, 1996), supposedly because oxygen supply to the embryo was increased. Also, increasing the oxygen concentration in the atmosphere surrounding seeds may increase germination percentages (Atwood, 1914; Edwards, 1968a; Gay *et al.*, 1991).

However, several pieces of evidence suggest that lack of germination in some species with nondeep PD is not due to low rates of oxygen diffusion into seeds. (1) Seed coat permeability to oxygen did not increase as seeds of *Xanthium pensylvanicum* came out of dormancy (Porter and Wareing, 1974). (2) Oxygen diffuses to both the upper and the lower seed in the dispersal unit (bur) of *X. pensylvanicum* at a greater rate than it is used by the embryo. Thus, oxygen could not be the limiting factor for germination (Porter and Wareing, 1974). (3) Rates of oxygen consumption were the same in dormant and nondormant seeds of *Phacelia tanacetifolia* (Chen, 1970). (4) Although GA stimulated high percentages of dormant *Sinapis arvensis* seeds to germinate, embryo covers remained highly resistant to the diffusion of oxygen (Edwards, 1968a). (5) If the hole cut in seed coats of imbibed caryopsis of *Avena fatua* was covered with wet paper (to minimize oxygen uptake), germination rates increased (Hsiao *et al.*, 1983). (6) The removal of husks (palea and lemma) from freshly matured seeds of Japonica cultivars of *Oryza sativa* decreased germination. Further, germination of freshly matured dehusked seeds of these cultivars decreased with an increase in oxygen from 0 to 50%. However, after seeds had lost dormancy in dry storage, dehusking did not inhibit germination (Takahashi, 1985).

b. Covering Structures: Inhibitors

Embryo covers may (1) prevent leaching of inhibitors from embryos; (2) retard the entrance of oxygen, which can oxidize (inactivate) (Wareing and Foda, 1957; Black and Wareing, 1959) or prevent production of inhibitors (Edwards, 1968b, 1969); and (3) contain growth inhibitors (Witcombe *et al.*, 1969). Unfortunately, these hypotheses are difficult to address experimentally because the tolerance of embryos to inhibitors may change as physiological dormancy is broken. For example, seed coats from both dormant and nondormant seeds of *Sinapis arvensis* soaked in water yielded an inhibitor that prevented the growth of embryos excised from dormant seeds of this species, but embryos from dormant seeds grew less than those from nondormant seeds in test solutions derived from either dormant or nondormant seeds (Witcombe *et al.*, 1969). That is, after seed dormancy is broken, the inhibitor is present in the seed coats, but the embryo is insensitive to it.

c. Physical Restriction and Embryo Growth Potential

Another possible explanation for inhibitory effects of embryo covers on germination of seeds is that they mechanically restrict embryo growth. The force required to break seed coats ranges from 9.9 MPa in *Pancratium maritimum* to 133.2 MPa in *Iris lorteti* (Blumenthal *et al.*, 1986). Thus, germination may be inhibited because embryos lack a sufficient growth potential to break open seed coats or other structures (Khan and Saminy, 1982). Germination of embryos excised from *Setaria faberi* seeds depended on the stage of embryo development and tissues left in association with them (Dekker *et al.*, 1996). For example, the presence of caryopsis tissue inhibited germination (relative to excised embryos) in 85% of seeds 8 days after anthesis, but it inhibited germination in only 35% of seeds 12 days after anthesis (Dekker *et al.*, 1996).

If seeds are placed under appropriate dormancy-breaking and/or germination conditions, the growth potential of the embryo increases and germination occurs. Esashi and Leopold (1968) measured the physical thrust generated by dormant and nondormant seeds of *Xanthium pensylvanicum* and found that nondormant ones developed twice the thrust of dormant ones. Depending on the species, cold stratification (Carpita *et al.*, 1983), GA (Baskin and Baskin, 1971b), incubation temperatures (Junttila, 1973), light (Scheibe and Lang, 1965), or darkness (Chen, 1968) may increase the growth potential of the embryo enough for the radicle to push through the seed coat and thus for seeds to germinate.

One way to study the growth potential of embryos is to place excised embryos in osmotica. If the osmotic potential of the solution is greater than the growth potential of the embryo, germination does not occur. However, if the osmotic potential is equal to or less than the growth potential of the embryo, germination occurs. In negatively photoblastic seeds of *Phacelia tanacetifolia* (Chen and Thimann, 1966) and *Nemophila insignis* (Chen, 1968), the inhibitory effect of light on the germination of nondormant seeds was overcome if the endosperm was removed from the tip of the radicle. However, if seeds were placed in a solution with a high osmotic potential after a piece of the endosperm was removed, the dark requirement for germination was reinstated. Scarification of positively photoblastic seeds of cucumber and sunflower resulted in about a 50% increase in germination in light at 22°C in 0.45 M (1.13 MPa) and 0.3 M (0.75 MPa) mannitol solutions, respectively, indicating that removal of the mechanical constraint allowed seeds to germinate at an increased osmotic potential (McDonough, 1967).

Exposure of half seeds, i.e., embryonic axis and part of the cotyledons (Scheibe and Lang, 1965), and excised embryos (Carpita *et al.*, 1979) of *Lactuca sativa* to red light increased growth potential, whereas far-red light decreased it in both half seeds and embryos. Embryos in *Lycopersicon esculentum* seeds did not show a decrease in growth potential after exposure to far-red light if endosperm over the tip of the radicle was removed (Nomaguchi *et al.*, 1995). To obtain 50% germination in seeds of *L. sativa* incubated in NaCl solutions, the duration of exposure to red irradiation had to be increased with each increase in NaCl concentration (Scorer *et al.*, 1985). Thus, additional red light had to be given to increase the growth potential of the embryo enough to overcome the osmotic potential of the NaCl solution.

Nabors and Lang (1971) used gravimetric techniques to determine the water potential of embryos germinating in osmotica. *Lactuca sativa* embryos exposed to red light developed lower water potentials and thus imbibed more water (equals an increase in growth potential) than those kept in darkness. The difference between the water potential of embryos treated with red light and that of embryos kept in darkness was equal to the osmotic potential of a 0.3 M (0.75 MPa) mannitol solution. Further, the force (in terms of osmotic potential) required for a radicle to break through the seed coats was equal to that of a 0.16 M (0.4 MPa) to 0.38 M (0.95 MPa) mannitol solution. Thus, seeds germinated in red light but not in darkness. The reason for this is that the growth potential of embryos in seeds exposed to red light increased enough to overcome the inhibiting force of the seed coats, whereas those in darkness did not (Nabors and Lang, 1971). In another study, the restraining force of seed coats in lettuce seeds was estimated to be equal to that of a 0.4 M (1.0 MPa) mannitol solution (Takeba and Matsubara, 1979).

d. Changes in Embryo-Covering Structures

As seeds come out of dormancy and/or when environmental conditions become favorable for germination, do embryo covers become less resistent to penetration by the radicle? Not much is known about the effects of dormancy-breaking treatments on changes in the resistance of endosperm, seed coats, or other structures to penetration by the radicle.

The force required to punch a hole in the megagametophyte/nucellus isolated from dormant and nondormant seeds of *Picea glauca* was not significantly different. However, when dormant and nondormant seeds were given 21 days of cold stratification, the force required to puncture the megagametophyte/nucellus decreased significantly in both types of seeds (Downie and Bewley, 1996). Resistance of the endosperm remained the same when *Syringa reflexa* seeds were exposed to low (dormancy breaking) or high (dormancy inducing) temperatures. However, if nondormant seeds were placed at favorable germination temperatures, resistance of the endosperm declined just before the radicle emerged (Junttila, 1973).

Endosperm rather than seed coats is the main force restricting embryo growth and thus germination in some species (e.g., Brown and Bridglall, 1987). In *Olea europaea,* removal of the woody endocarp of the fruit did not promote germination, and embryos did not grow until they were excised from the endosperm (Mitrakos and Diamantoglou, 1984). Studies on several species discussed later indicate that endosperm (and sometimes seed coat) resistance declines just before germination occurs. However, seeds of these species did not require either warm or cold stratification or afterripening before they would germinate; thus, we must conclude that these studies on changes in resistance were done on nondormant seeds. That is, germination occurred as soon as seeds were placed under appropriate environmental conditions, depending on the species.

Resistance has been shown to vary depending on the temperature and light: dark conditions under which seeds were imbibed. Endosperm resistance to radicle emergence decreased faster if *Capsicum annuum* seeds were imbibed at 25°C than if they were imbibed at 15°C (Watkins and Cantliffe, 1983b). Also, the force needed to fracture the seed coat of *Onopordum nervosum* decreased as the imbibition period increased, and resistance decreased faster at 25 than at 15°C; germination percentages were higher at 25 than at 15°C (Perez-Garcia and Pita, 1989). Light and GA caused decreases in the puncture resistance of endosperm in *Lactuca sativa* seeds (Tao and Khan, 1979). Mobilization of storage materials and vacuolation of the cytoplasm occurred in cells of *L. sativa* endosperm adjacent to the radicular end of the embryo under appropriate light–dark, temperature, and osmotic conditions for germination. These changes did not occur in endosperm cells when conditions were unsuitable for germination (Georghiou *et al.*, 1983).

Much research has been done in attempting to understand how the endosperm becomes less restrictive to radicle growth during germination (Black, 1996). Endosperm breakdown did not occur in *Capsicum annuum* seeds until 1 day before germination occurred. This activity was concentrated in the area of the endosperm covering the radicle and was promoted by GA (Watkins *et al.*, 1985). Breakdown of endosperm at the tip of the radicle also preceded germination in *Lycopersicon esculentum* seeds, and GA was required. Further, seeds of a GA-deficient dwarf-mutant line *ga1* of *L. esculentum* did not germinate unless GA was supplied or if endosperm at the radicle tip was removed (Groot and Karssen, 1987). When endosperm from seeds of the GA-deficient *gib1* mutant of *L. esculentum* was treated with GA, the production of endo-β-mannanase was induced and activities of mannohydrolase and α-galactosidase increased (Groot *et al.*, 1988). Galactomannan-hydrolyzing enzymes were active in endosperm at the radicle tip of *L. esculentum* seeds 1 day prior to germination and in the remainder of the endosperm after germination had occurred (Nomaguchi *et al.*, 1995). The abil-

ity of seeds of this species to germinate at 12°C was correlated with the presence of endomannanase activity in endosperm at the radicle tip (Leviatov *et al.*, 1995). Subsequent studies showed that endo-β-mannanase is present only in endosperm cells adjacent to the radicular end of the embryo (Nonogaki and Morohashi, 1996). In developing seeds, however, various isoforms of this enzyme are present in the embryo and also in the endosperm adjacent to the cotyledons (Voigt and Bewley, 1996).

Exposure to red light promoted the germination of *Datura ferox* seeds by increasing the growth potential of the embryo and softening the endosperm. However, increases in growth potential of the embryo were not enough to promote germination unless endosperm softening also occurred (de Miguel and Sanchez, 1992). Cellulase activity increased in *D. ferox* seeds prior to germination, and this was correlated with softening of the endosperm over the radicle. The peak of cellulase activity occurred 36 hr after the red-light treatment, and far-red light given 15 and 24 hr after the red light decreased cellulase activity to near that of dark controls (Sanchez *et al.*, 1986). In addition, walls of endosperm cells in the region over the radicle were softened in *D. ferox* and *D. stramonium* seeds due to a decrease in 4-linked mannose, which is one of the primary cell wall constituents; the enzyme was not specified (Sanchez *et al.*, 1990).

Rupture of the seed coat in *Nicotiana tabacum* seeds began after 30 hr of imbibition in light, which was followed by an increase in the activity of β-1,3-glucanase, especially in the micropylar region where the radicle emerges. An increase in β-1,3-glucanase activity was correlated with emergence of the radicle from the endosperm, suggesting that this enzyme was involved in softening the endosperm (Leubner-Metzger *et al.*, 1995).

One of the conclusions from studies on endosperm softening is that this process is very sensitive to environmental factors, i.e., seeds must be imbibed and subjected to appropriate temperature and light: dark conditions for germination before enzymes in the portion of the endosperm over the radicle are synthesized and/or activated. Thus, studies on endosperm softening have contributed greatly to understanding the process of germination per se in nondormant seeds.

e. Interaction between Embryo and Covering Structures

The next challenge is to understand why freshly matured seeds of many species will not germinate under a certain set of environmental conditions, but germinate to high percentages at the same set of test conditions after they have received a dormancy-breaking treatment. What has changed? Do dormancy-breaking treatments cause changes in the endosperm? In seeds that lack endosperm, do dormancy-breaking treatments cause changes in the seed coats and/or other structures? Do dormancy-breaking treatments cause changes in the embryo? What effect does the embryo in dormant vs nondormant seeds have on its covering structures?

Is it possible that one of the results of dormancy-breaking treatments in some seeds, especially those with endosperm, is the production of some kind of signal (chemical?) by the embryo? Jacobsen *et al.* (1976) suggested that GA was released by the growing embryo in *Apium graveolens* seeds and that it stimulated the production of hydrolases in the endosperm. Subsequently, hydrolases broke down much of the endosperm before germination occurred. Application of GA to endosperm from *A. graveolens* seeds caused cell separation (Jacobsen *et al.*, 1976). However, if the softening of covering layers is ultimately controlled by the embryo, it is doubtful that GA is the only signal. In fact, endo-β-mannanase activity in endosperm at the radicle tip in *Lycopersicon esculentum* seeds is not induced by GA (Toorop *et al.*, 1996).

Other possible roles of the embryo might be to (1) remove or absorb products of enzyme hydrolysis that inhibit enzyme activity if present in high concentrations (Spyropoulos and Reid, 1985; Zambou and Spyropoulos, 1990) and/or (2) regulate inhibitors in the endosperm, e.g., saponin-like substances that affect the production of α-galactosidase (Zambou *et al.*, 1993). Also, after endosperm softening has occurred in *Lycopersicon esculentum* seeds, germination does not occur unless the embryo generates enough growth potential to overcome resistance of the seed coats (Hilhorst and Downie, 1995).

Some hydrolytic reactions occur in endosperm tissue after it is separated from the embryo, e.g., endo-β-mannanase was active in isolated imbibed endosperm from the legumes *Trigonella foenumgraecum* (Reid and Davies, 1977; Kontos *et al.*, 1996) and *Ceratonia siliqua* (Kontos *et al.*, 1996). This observation suggests another reason why the embryo may be involved in controlling germination. If hydrolytic enzymes in the endosperm were not controlled by the embryo, what would prevent all seeds with endosperm from germinating as soon as they are imbibed under temperature and light : dark conditions suitable for germination? It generally is recognized that dormancy-breaking treatments are needed to promote the germination of many seeds with endosperm.

It seems reasonable that although embryos excised from seeds with nondeep PD grow normally, they probably are involved in controlling germination of intact seeds. Some possible roles of the embryo have been mentioned earlier. The assumption that embryos help

control germination is based on the facts that (1) many seeds with nondeep PD require dormancy-breaking treatments before they will germinate (Chapter 4) and (2) dormancy-breaking treatments per se do not seem to have much effect on covering structures. That is, the force required to break covering structures is about the same in dormant and nondormant seeds, especially if nondormant seeds are under unfavorable conditions for germination (Junttila, 1973). The resistance of endosperm changes when nondormant seeds are under favorable germination conditions (Watkins and Cantliffe, 1983; Georghiou *et al.*, 1983). The exact cause of nondeep PD probably varies, depending on the species and the type of structures covering the embryo. In seeds of many species, however, an interaction between the embryo and its covering structures may be the best explanation for the cause of nondeep PD. Future studies on the role of the endosperm in controlling germination will need to take into account the fact that Cranston *et al.* (1996) concluded that wound-induced ethylene synthesis may promote the growth of embryos excised from dormancy caryopses of *Avena fatua*.

3. Intermediate Physiological Dormancy

Examples of species whose seeds have intermediate PD include *Acer negundo* (Nikolaeva, 1969), *A. pseudoplatanus* (Pinfield and Stobart, 1972), *A. saccharum* (Webb and Dumbroff, 1969), *Corylus avellana* (Frankland and Wareing, 1966), *Fagus sylvatica* (Frankland and Wareing, 1966), *Ferula karatavica* (Nikolaeva, 1969), *Fraxinus americana, F. pennsylvanica* (Steinbauer, 1937), *Melampyrum lineare* (Curtis and Cantlon, 1963), and *Polygonum* spp. (Ransom, 1935).

Embryos isolated from seeds with intermediate PD will grow and the resulting seedlings are normal (Nikolaeva, 1977). The dormancy of intact seeds or dispersal units is broken by cold stratification (Nikolaeva, 1969), but up to 6 months of this dormancy-breaking treatment may be required (Choate, 1940), depending on the species. Dry storage at room temperatures reduces the length of the cold stratification treatment needed to break dormancy in various species, including *Amelanchier canadensis* (Crocker and Barton, 1931), *Echinocystis lobata* (Choate, 1940), *Polygonum* spp. (Ransom, 1935), *Talinum calcaricum* (Ware and Quarterman, 1969), and *Vitis bicolor* (Flemion, 1937).

GA substituted for the cold stratification requirement for dormancy break in seeds of various species, including *Polygonum convolvulus* (Timson, 1966) and *Stachys alpina* (Pinfield *et al.*, 1972). Although GA did not promote the germination of intact nuts of *Corylus avellana* or *Fagus sylvatica*, it caused dormant seeds of these species to germinate after the pericarp was removed. Thiourea and kinetin also stimulated excised seeds of *C. avellana* and *F. sylvatica* to germinate, but they inhibited root growth (Frankland and Wareing, 1966). GA had no effect on the germination of intact samaras of *Acer pseudoplatanus,* but kinetin induced 45% of the seeds to germinate (Pinfield and Stobart, 1972). Intact samaras of *A. saccharum* did not respond to GA or kinetin. However, both GA and kinetin decreased the cold stratification requirement for germination when samara walls were removed and the seed coat was pricked with a pin (Webb and Dumbroff, 1969). In *Melampyrum lineare,* GA substituted for the period of dry storage at room temperatures that must precede cold stratification (Curtis and Cantlon, 1963).

4. Deep Physiological Dormancy

Examples of species whose seeds have deep PD include *Impatiens parviflora* (Nikolaeva, 1969), *Malus domestica* (Harrington and Hite, 1923), *Sorbus aucuparia* (Flemion, 1931), *Rhodotypos kerrioides* (Flemion, 1933a,b), *Prunus persica, Crataegus* sp. (Flemion, 1934), *Acer tartaricum, Euonymous europaea,* and *Acer platanoides* (Pinfield *et al.,* 1974).

Embryos isolated from seeds with deep PD either do not grow or they produce abnormal seedlings (Nikolaeva, 1977). The only treatment that overcomes the dormancy of intact seeds (or dispersal units) is a relatively long period of cold stratification. The length of the cold stratification period required to break dormancy varies from 7 weeks in *Acer platanoides* (Pinfield *et al.,* 1974) to 14 weeks in *Prunus persica* (Crocker and Barton, 1931) to 18 weeks in *Impatiens parviflora* (Nikolaeva, 1969). Dry storage at room temperature prior to cold stratification increased germination percentages of *Euonymous europaea* seeds, but it did not decrease the length of the cold stratification period required to break dormancy (Nikolaeva, 1969).

Although GA stimulates the germination of seeds with nondeep and intermediate PD, it does not break deep PD in intact dispersal units (Nikolaeva *et al.,* 1973; Nikolaeva, 1977). In some species, GA stimulated the growth of embryos excised from seeds with deep PD, e.g., *Malus arnoldiana* (Barton, 1956), *Prunus persica* (Gray, 1958), and *Euonymous europaea* (Singh, 1985), but in others such as *Acer platanoides* (Pinfield *et al.,* 1974) it did not. Although GA stimulated the growth of excised embryos of *Euonymous europaea,* it did not cause intact seeds to germinate. Germination of *E. europaea* occurs in two phases. The first phase requires relatively high (9–10°C) temperatures, and the embryo enlarges and splits the seed coat. The second phase requires cold stratification at 0–3°C for several weeks, and then the radicle emerges at 0–3°C. GA substitutes

for the first but not the second phase of germination (Nikolaeva *et al.,* 1973).

Kinetin also has been used in attempts to break deep PD, but usually it is ineffective unless used with GA or cold stratification. Neither kinetin nor GA broke the dormancy of *Acer tartaricum* seeds at 20°C; however, kinetin or GA plus kinetin enhanced the dormancy break during cold stratification. In *Euonymous europaea,* neither kinetin nor GA stimulated germination, but when used together 50% of the seeds at 9–10°C germinated after a 2-month period (Nikolaeva *et al.,* 1973).

In a number of species, especially certain members of the Rosaceae, the removal of embryos from seeds that have not been cold stratified results in abnormal, slow-growing plants (Flemion, 1933a, 1934; Nikolaeva, 1969). These plants are dwarfs with very short internodes and small, malformed leaves, and sometimes they have a whorl or rosette of leaves at the tip of the shoot (called a "perched rosette") (Tukey and Carlson, 1945). Production of dwarf plants from seeds that have not been cold stratified is called nanism. The dwarf condition of these plants is long persisting; a dwarfed peach seedling kept in a warm greenhouse for 10 years (and maybe longer) never grew normally (Flemion, 1959).

Under high temperatures and continuous light, lateral buds of dwarf plants may initiate growth and produce normal branches (Lammerts, 1943); however, growth produced by the terminal bud is not normal (Flemion, 1959). One way to stimulate dwarf plants to grow normally is to give them a cold treatment at about 5°C for 4–8 weeks (Tukey and Carlson, 1945; Flemion, 1959). GA was effective in overcoming the dwarf condition in *Malus arnoldiana* (Barton, 1956), *Prunus persica,* apricot (*P. armeniaca*), plum (*P. domestica* ?) (Blommaert and Hurter, 1959), and *Rhodotypos kerrioides* (Flemion, 1959), but it had no stimulatory effects on dwarfs of *Euonymous europaea* (Nikolaeva, 1969). However, plants of *Rhodotypos kerrioides* started producing short internodes again when GA treatments were discontinued (Flemion, 1959).

With grafting studies, Flemion and Waterbury (1945) showed that the shoot and not the root of a dwarf plant is dormant. Because cold stratification is required for the germination of seeds with deep PD, this raises a question that has not been answered: Is it necessary to cold stratify the whole seed or only the plumule (shoot) to break deep PD?

B. Morphological Dormancy

At the time of dispersal, seeds of some species have embryos in which a radicle and cotyledon(s) can be distinguished, i.e., the embryo is differentiated, but it is not fully grown (underdeveloped). Thus, embryo growth is required before germination occurs. In seeds of other species, the embryo is just a mass of cells at the time of dispersal, i.e., the embryo is not differentiated, and germination does not take place until both differentiation and growth occur. In both types of seeds, germination is prevented at the time of maturity due to morphological characteristics of the embryo, hence the term morphological dormancy.

1. Differentiated Embryos

Morphological dormancy occurs in seeds with rudimentary and linear embryos (see Table 3.2). Most of the interior of these relatively large seeds is occupied

by endosperm, and the embryo may be only 1.0% of the size (volume) of the seed, or less (Nikolaeva, 1969). Although rudimentary and linear embryos are differentiated, they frequently are referred to collectively as underdeveloped embryos (Grushvitzky, 1967). These embryos are underdeveloped in the sense that they are small and consequently have to grow before the seeds can germinate. Underdeveloped embryos occur in a number of plant families (Table 3.3), and Martin (1946) placed seeds with rudimentary embryos at the base of a family tree of seed phylogeny; those with linear embryos are higher on the tree than those with rudimentary embryos.

Growth of underdeveloped embryos takes place after seeds have been dispersed from the mother plant. Requirements for growth are a moist substrate and suitable

TABLE 3.3 Plant Families in Which One to Many Species Has (Have) Seeds with Rudimentary or Linear Embryo(s)[a]

Family	Primary region of geographical distribution[b]	Family	Primary region of geographical distribution[b]
Amaryllidaceae	Tropical or subtropical	Hydrophyllaceae	Cosmopolitan, except Australia
Amborellaceae	Tropical	Illiciaceae	Tropical
Annonaceae	Tropical (especially Old World)	Iridaceae	Tropical and temperate
Apiaceae	Northern temperate	Lactoridaceae	Tropical
Aquifoliaceae	Tropical and temperate	Lardizabalaceae	Temperate
Araceae	Tropical and temperate	Loranthaceae	Tropical and temperate
Araliaceae	Tropical	Liliaceae	Warm temperate and tropical
Arecaceae	Tropical and subtropical	Magnoliaceae	Temperate and tropical
Aristolochiaceae	Tropical and warm temperate	Melanthiaceae	Warm temperate and tropical
Berberidaceae	Northern temperature and tropical mountains	Menyanthaceae	Temperate and boreal
Buxaceae	Tropical and temperate	Monimiaceae	Southern tropical
Canellaceae	Tropical	Myristicaceae	Tropical
Cannaceae	Tropical	Nandinaceae	Northern temperate
Caprifoliaceae	Northern temperate and tropical mountains	Oleaceae	Temperate and tropical
Chloranthaceae	Tropical and subtropical	Paeoniaceae	Northern temperate
Convallariaceae	Temperate	Papaveraceae	Northern temperate
Cycadaceae	Tropical	Piperaceae	Tropical
Daphniphyllaceae	Tropical	Pittosporaceae	Tropical
Degeneriaceae	Tropical	Podocarpaceae	Southern temperate
Dilleniaceae	Tropical and subtropical	Ranunculaceae	Northern temperate
Escalloniaceae	Southern temperate	Santalaceae	Tropical and temperate
Eupomatiaceae	Tropical	Sarraceniaceae	Temperate and tropical
Fumariaceae	Northern temperate	Schisandraceae	Temperate and tropical
Garrayaceae	Warm temperate	Smilacaceae	Tropical and temperate
Ginkgoaceae	Temperate	Taxaceae	Temperate and tropical
Grossulariaceae	Temperate	Trochodendraceae	Northern temperate
Haemodoraceae	Temperate and tropical	Winteraceae	Tropical

[a] From Martin (1946), Grushvitzky (1961, 1967), Corner (1976), and Zomlefer (1994).
[b] Information from Willis (1966).

temperatures, and some species have specific light : dark requirements. For many species with morphological dormancy, optimum temperatures for embryo growth and germination are from 15 to 30°C (Grushvitzky, 1967; Lush *et al.*, 1984; Lohotska and Moravcova, 1989; Baskin and Baskin, 1986b), but those of the oil palm, *Elaeis guineensis,* germinate best at temperatures of 35–40°C, depending on the variety (Hussey, 1958). Seeds of *Apium graveolens* require light for germination (Pressman *et al.*, 1977; Jacobsen and Pressman, 1979), those of *Conium maculatum* germinate to higher percentages in light than in darkness (Baskin and Baskin, 1990a), those of the cultivated de Caen type of *Anemone coronaria* germinate faster and to higher percentages in darkness than in light (Bullowa *et al.*, 1975), and those of *Pulsatilla slavica* germinate equally well in light and darkness (Lhotska and Moravcova, 1989). When morphologically dormant seeds are placed under favorable conditions, the time required for 50% of them to germinate varies from 6 days in *Apium graveolens* (Jacobsen and Pressman, 1979) to 3.5 to 5.5 months in *E. guineensis,* depending on the variety (Hussey, 1958).

GA increased germination percentages of *Anemone coronaria* (de Caen type) seeds at supraoptimum (25°C) but not at optimum (10–20°C) temperatures (Bullowa *et al.*, 1975). Also, GA increased the rate but not the final percentage of germination of *Clematis microphylla* seeds (Lush *et al.*, 1984).

Underdeveloped embryos occur in seeds of a number of families that primarily are tropical in distribution, including the Annonaceae, Arecaceae, Degeneriaceae, Lactoridaceae, Monimiaceae, Myristicaceae, and Winteraceae (Grushvitzky, 1967). Under tropical conditions, underdeveloped embryos grow slowly after seeds are dispersed, and the seeds eventually germinate (Grushvitzky, 1967). If any physiological dormancy is associated with the morphological dormancy in these seeds, it has not been reported. However, physiological dormancy in association with morphological dormancy may help explain the germination of *Elaeis guineensis.* Seeds of the oil palm varieties *dura* and *tenera* held at 39.5°C for 100 days germinated to a maximum of 42 and 53%, respectively, whereas those given 50 or more days at 39.5°C and then moved to room temperatures (28°C) germinated to 64–90 and 60–88%, respectively. A room temperature control was not used, but *dura* and *tenera* seeds receiving 10 days at 39.5°C prior to being placed at 28°C germinated to only 1 and 0%, respectively (Rees, 1959). An increase in germination percentages of seeds of both varieties at 28°C following warm stratification at 39.5°C suggests that some seeds have physiological dormancy (which was broken at 39.5°C) in addition to morphological dormancy (which was broken at 28°C).

Seeds of some tropical species with morphological dormancy placed under appropriate conditions for embryo growth and germination require 1–3 months to germinate, e.g., *Annona squamosa* (Hayat, 1963) and *A. crassiflora* (Rizzini, 1973). More studies are needed to determine if these seeds require warm stratification to break physiological dormancy before morphological dormancy is broken (i.e., embryo growth). However, physiological dormancy may be lacking, in which case embryo growth is very slow in seeds of these species. Studies to monitor the rates of embryo growth during the 1- to 3-month period required for germination would be very informative.

Underdeveloped embryos occur in many plant families in temperate regions; however, in most cases morphological dormancy is associated with physiological dormancy. Morphological dormancy per se has been documented in only two temperate families: Apiaceae and Ranunculaceae. In the Apiaceae, studies on morphological dormancy have been done on seeds of *Apium graveolens* (Jacobsen and Pressman, 1979), *Conium maculatum* (Baskin and Baskin, 1990a), and *Pastinaca sativa* (Baskin and Baskin, 1979) and in the Ranunculaceae on seeds of the de Caen type of *Anemone coronaria* (Bullowa *et al.*, 1975), *Isopyrum biternatum* (Baskin and Baskin, 1986b), and *Pulsatilla slavica* (Lhotska and Moravcova, 1989). It should be noted that seeds of the cultivated de Caen type of *Anemone coronaria* have morphological dormancy, but those of the wild type have both morphological and physiological dormancy (Horovitz *et al.*, 1975).

Some seeds that have morphological dormancy when they are freshly matured can develop physiological dormancy if they are exposed to changes in environmental conditions. For example, morphologically dormant seeds of *Conium maculatum* develop physiological dormancy during cold stratification (Baskin and Baskin, 1990a).

2. Undifferentiated Embryos

One or more (sometimes all) genera in a number of plant families have undifferentiated embryos (Table 3.4); however, this type of embryo is found only in micro and dwarf seeds. Micro seeds are usually less than 0.2 mm in length, and dwarf seeds generally are from 0.3 to 2.0 mm in length (Martin, 1946). Embryos of micro seeds can have from 25 to 100 cells (Martin, 1946), but they may have as few as 2 (Olson, 1980). In the germination of seeds with undifferentiated embryos, a cotyledon(s) and radicle per se are not formed. Instead, either the radicular or the plumular pole of the embryo elongates and eventually emerges from the seed. The tissue that emerges from the seed enlarges to form a

TABLE 3.4 Examples of Plant Families with Genera Whose Seeds Have Undifferentiated Embryos[a]

Family	Source of nutrition	Type of seed	Embryo size (No. of cells)
Balanophoraceae	Parasitic	Micro?	4–12
Burmanniaceae	Mycoheterotrophs	Micro	4–10
Ericaceae	Mycoheterotrophs	Dwarf	2–3
Gentianaceae	Mycoheterotrophs	Dwarf	5–24
Hydnoraceae	Parasitic	Dwarf?	18 or more
Lennoaceae	Parasitic	Dwarf?	12 or more
Monotropaceae	Mycoheterotrophs	Micro	No data
Orchidaceae	Autotrophs, Mycoheterotrophs	Micro	2 or more
Orobanchaceae	Parasitic	Dwarf	9
Pyrolaceae	Mycoheterotrophs	Micro	9
Rafflesiaceae	Parasitic	Dwarf?	4–10

[a] From Martin (1946), Rangaswamy (1967), Kuijt (1969), Kumar (1977), Olson (1980), and Natesh and Rau (1984).

haustorium or protocorm, depending on the species (see Chapter 11, Sections II,A and IV,A).

C. Morphophysiological Dormancy

Morphophysiological dormancy (MPD) occurs in seeds with rudimentary or linear embryos, and as the name indicates it is a combination of morphological and physiological dormancy, i.e., the underdeveloped embryos have physiological dormancy. MPD is found in a number of plant families, including the Apiaceae, Aquifoliaceae, Araceae, Araliaceae, Aristolochiaceae, Berberidaceae, Fumariaceae, Illiciaceae, Lardizabalanceae, Liliaceae, Magnoliaceae, Papaveraceae, Ranunculaceae, and Schisandraceae (Grushvitzky, 1967). However, there are many families whose seeds have underdeveloped embryos (Table 3.3), but their germination has not been studied. Thus, it is not known if the seeds have morphological or morphophysiological dormancy.

Two general kinds of things must happen before seeds with MPD can germinate: (1) the embryo must grow to a species-specific critical size and (2) physiological dormancy of the embryo must be broken. The secret to germinating seeds with MPD is to figure out what environmental conditions promote each event. In some species, embryo growth and dormancy break are promoted by the same environmental conditions, whereas in others they require different conditions. Depending on the species, embryo growth and dormancy break may require (1) warm (≥15°C) stratification only (Baskin and Baskin, 1990b), (2) cold (0–10°C) stratification only (Baskin and Baskin, 1984d), (3) warm followed by cold stratification (Baskin and Baskin, 1984c),

or (4) cold followed by warm followed by cold stratification (Nikolaeva, 1977). In some species, embryo dormancy is broken and then growth occurs (Baskin and Baskin, 1984c, 1985), whereas in others dormancy break and embryo growth occur at the same time (Baskin and Baskin, 1984d). The various types of morphophysiological dormancy and how they are broken in nature will be discussed in Chapter 5.

D. Physical Dormancy

In this type of dormancy, the primary reason for the lack of germination is the impermeability of seed (or fruit) coats to water. Physical dormancy is present in at least 15 families of angiosperms (Table 3.5); however, it would not be surprising if some members of the families Bixaceae, Melastomataceae, and Winteraceae had physical dormancy. These families have a palisade layer of lignified cells in the seed coat (Vazquez-Yanes and Perez-Garcia, 1976; Corner, 1976), a characteristic frequently found in seeds with physical dormancy. It should be noted, however, that just because a family has seeds with physical dormancy does not mean that all of its members have this type. For example, although many genera of Fabaceae have physical dormancy, seeds in some members of several others, including *Andira, Arachis, Bauhinia, Brownea, Castanospermum, Dialium, Inocarpus, Mora, Pentaclethra, Pithecellobium,* and *Saraca,* have permeable seed coats (Corner, 1951).

Based on embryo morphology, seven types of seeds are found among the 15 families known to have members with physical dormancy (Table 3.5). With the exception of the Cannaceae, Musaceae, and Nelumbonaceae, seeds of all families that contain taxa with physical dormancy have large embryos (bent, folded, investing, or spatulate), and most of the food reserves are stored in the embryo rather than in endosperm. Seeds of the Cannaceae, Musaceae, and Nelumbonaceae have linear, capitate, and broad embryos, respectively, and much endosperm (Martin, 1946).

Seed coat impermeability usually is associated with the presence of one or more layers of impermeable palisade cells. For example, impermeable seed coats in the Fabaceae have one palisade layer (Rolston, 1978), whereas those of *Cuscuta pedicellata* and *C. campestris* (Convolvulaceae) have two palisade layers; only the inner one is impermeable (Lyshede, 1992). Impermeable palisade layers are composed of sclereid cells that have thick lignified secondary walls. The most common type of sclereid cell in palisade layers of seeds is the macrosclereid or Malpighian cell (Esau, 1964).

In the Anacardiaceae, the seed coat is not well differentiated, and the embryo is protected by the fruit wall (Corner, 1976). Fruits of some members of this family

TABLE 3.5   Plant Families in Which at Least Some Members Have Seeds with Physical Dormancy[a]

Family	Seed type[b]	Specialized structure	Reference
Anacardiaceae	Bent	Operculum; endocarp slit	von Teichman and Robbertse (1986a,b)
Bombacaceae	Spatulate	Micropyle?[c]	
Cannaceae	Linear	Lid on raphe	Grootjen and Bouman (1988)
Cistaceae	Bent	Chalazal plug	Thanos and Georghiou (1988), Corral *et al.* (1989)
Convolvulaceae	Folded	Plug near micropyle	Koller and Cohn (1959)
Curcurbitaceae	Spatulate	Micropyle?[c]	
Fabaceae			
Caesalpinioideae	Investing	Hilum	Jones and Geneve (1995)
Mimosoideae	Investing	Strophiole	Cavanagh (1980), Dell (1980), Hanna (1984), Serrato-Valenti *et al.* (1995)
Papilionoideae	Bent	Strophiole	Hagon and Ballard (1970), Rolston (1978), Manning and van Staden (1987)
Geraniaceae	Folded	Hilum?[c]	
Malvaceae	Folded	Chalazal plug	Christiansen and Moore (1959), Winter (1960), Egley *et al.* (1986)
Musaceae	Capitate	Micropylar lid-like structure	Boesewinkel and Bouman (1984), Graven *et al.* (1996)
Nelumbonaceae	Broad	Protuberance	Ohga (1926)
Rhamnaceae	Investing	—[d]	
Sapindaceae	Folded	—[d]	
Sterculiaceae	Spatulate	—[d]	
Tiliaceae	Folded	—[d]	

[a] Seed type and specialized structure on seed (or fruit) coat where water enters is given for each family, if known.
[b] Seed type information from Martin (1946).
[c] The opening is our speculation based on examination of drawings or photographs of longitudinal sections of seeds.
[d] No data available.

have physical dormancy because the endocarp layer of the pericarp is impermeable to water. The endocarp of *Rhus lancea* fruits consists of four layers: macrosclereids, osteosclereids, brachysclereids, and crystal cells (von Teichman and Robbertse, 1986a; von Teichman, 1989). The pericarp wall of *Nelumbo nucifera* (Nelumbonaceae) has a middle sclerenchymatous layer composed of macrosclereids (Ohga, 1926).

When macrosclereids are viewed under a microscope, a line appears to be running across them; this is called the light line (Rolston, 1978; Kumar and Singh, 1991). According to Rolston (1978), a line is seen at the same place in each macrosclereid, giving the impression of a continuous line. The light line is due to differences in refraction of light by the top and bottom portions of each cell, which differ somewhat in chemical composition.

Macrosclereids are impermeable to water because they are impregnated with water-repellent substances, including cutin, lignin, quinones, pectic-insoluble materials, suberin, and wax (Rolston, 1978; Werker, 1980–1981). High concentrations of a water-repellent substance called callose in the light line also may contribute to seed coat impermeability (Serrato-Valenti *et al.*, 1993, 1994). For example, seed coat impermeability in the legume *Stylosanthes scabra* was attributed to the pres-

ence of phenolics, callose, and hydrophobic lipid substances in the palisade layer and to a high concentration of callose in the light-line region (Serrato-Valenti *et al.*, 1993).

Seed impermeability in *Cercis siliquastrum* is attributed to total imperviousness of the hilum, a cuticle layer composed of lipids and pectins, a lipid layer between the integument and the endosperm, and water-repellent palisade and hypodermal cells (Riggio-Bevilacqua *et al.*, 1985). The water barrier in seeds of *Prosopis tamarugo* is hydrophobic materials (lipids) located in the superficial portion of palisade cells (Valenti *et al.*, 1986).

Impermeable seed coats develop in *Pisum elatius* during seed dehydration. As the dehydration process begins, catechol oxidase increases eightfold in the seed coats, but it decreases as seeds reach their final dry weight (Marbach and Mayer, 1975). These authors suggest that catechol oxidase plays a role in melanin formation and tanning reactions, which are correlated with the development of seed coat impermeability.

In addition to the development of impermeable layers in the seed (or fruit) coats, all the natural seed openings, including the micropyle, hilum, and chalazal area, also become impermeable to water. For example, the hilum region in legume seeds has an additional layer of macrosclereids, called the counter palisade (Sanchez-Yelamo *et al.,* 1992). Before seeds can germinate, however, an opening or passage through the palisade or other impermeable layer(s) must be formed, whereby water enters the seed. The type of opening through the impermeable layers varies with the plant family. In the Cistaceae and Malvaceae families, water enters after the chalazal plug is disrupted (Fig. 3.2a), and in the Convolvulaceae family there is a plug-like structure near the micropyle (Table 3.5). In seeds of the Papilionoideae and Mimosoideae, the site of water entry is the strophiole (Fig. 3.2b), which is a swelling on the raphe of the seed coat between the hilum and chalaza (Table 3.5). Seeds of the Cannaceae have an "imbibition lid" on the raphe, and the micropyle of Musaceae seeds develops into a lid-like structure called the operculum (Fig. 3.2c) that "pops out" prior to germination. Ohga (1926) suggested that the most likely site of water entry into fruits of Nelumbonaceae was the protuberance on the stylar end. The specialized structures listed in Table 3.5 for Bombacaceae, Curcubitaceae, and Geraniaceae are only speculations, and thus much work is needed to

explain how physical dormancy is broken in these families. For four families, insufficient information is available to be able to speculate on the type of opening that permits water entry. Chapter 6 will consider the environmental factors that cause removal or displacement of the plugs or coverings of these natural openings.

E. Physical plus Physiological Dormancy

In the majority of species with impermeable seed coats, the embryo is nondormant, but there are some species whose seeds have impermeable coats and dormant embryos (see Table 6.10). The presence of physical and physiological dormancy in a single seed sometimes is referred to as combined dormancy (Nikolaeva, 1969), and germination does not occur until both types have been broken. In some species, physical dormancy is broken before physiological dormancy (e.g., Barton, 1934), but in others physiological dormancy is broken before physical dormancy (e.g., McKeon and Mott, 1984). For a discussion of germination of seeds with physical and physiological dormancy, see Chapter 6, Section VII.

F. Chemical Dormancy

1. Description

According to Nikolaeva (1969, 1977), chemically dormant seeds do not germinate due to the presence of inhibitors in the pericarp. Further, chemical dormancy is broken by removal of the pericarp or leaching of the fruits. This definition of chemical dormancy has been broadened to include compounds that are either produced in or translocated to the seed, where they block embryo growth. Numerous studies have demonstrated that germination in petri dishes is inhibited by a variety of compounds found in many plant families (Evenari, 1949; Ketring, 1973). However, it is quite a different task to show that these compounds are translocated to seeds prior to dispersal and that they prevent embryo growth and germination.

Much attention has been devoted to trying to find germination inhibitors in mature seeds. In the early days of studies on chemical dormancy, extracts of whole seeds frequently were made. These extracts were chromatographed, and the resulting fractions were tested on seeds of *Lactuca sativa, Triticum, Lepidium,* and other species to determine if they inhibited germination (see review by Wareing, 1965). From these studies there was no way to know if the inhibitor (1) came from the embryo and/ or other seed parts or (2) would prevent embryo growth of nondormant seeds of the species from which it was extracted.

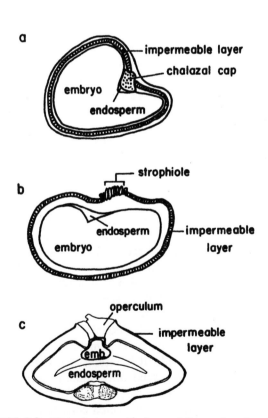

FIGURE 3.2 Water-impermeable layer of the seed coat, embryo, endosperm, and specialized structure on seed coat (chalazal cap, strophiole, or operculum) that moves or becomes dislodged, allowing entry of water: (a) Malvaceae, chalazal cap; (b) Fabaceae (Mimosoideae and Papilionoideae), strophiole; and (c) Musaceae, operculum. Musaceae drawing is with permission of Graven *et al.* (1996).

Germination inhibitors have been found in the embryo, endosperm, and seed coats of seeds and in structures that sometimes are dispersed along with the seeds of some species. Lists of inhibitory compounds and the seed parts in which they occur are given by Bewley and Black (1982, 1994) and Bradbeer (1988). Unfortunately, in many of the studies in which the location of the inhibitor was determined, the investigator used seeds of *Lactuca, Triticum,* or other cultivated species in the bioassay rather than nondormant seeds of the species from which the compound was extracted. However, inhibitors isolated from embryos of dormant *Acer pseudoplatanus* seeds inhibited the germination of nondormant seeds of this species (Webb and Wareing, 1972), and extracts of dormant *Fraxinus excelsior* embryos inhibited the growth of nondormant embryos (Nikolaeva and Vorob'eva, 1979). Leachate from seeds of *Lactuca sativa* induced into dormancy by high temperatures reduced the germination of nondormant seeds of this species (Small and Gutterman, 1991).

Using Bewley and Black's (1982, 1994) and Bradbeer's (1988) lists of species whose seeds have germination inhibitors, it appears that inhibitors are found in species with all types of seeds, except broad, capitate, dwarf, and micro. However, the absence of reports of inhibitors in these types of seeds could be due to lack of studies rather than to lack of inhibitors per se.

A complication in the study of chemical dormancy is that most seeds from which germination inhibitors have been isolated also exhibit physiological dormancy (e.g., Steinbauer, 1937; Webb and Dumbroff, 1969). Further, when physiological dormancy is broken, there is no evidence that chemicals in the seed prevent germination. For example, leachates from dormant achenes of *Rosa rugosa* inhibited the germination of embryos excised from dormant seeds of this species, but they did not inhibit the germination of embryos from cold-stratified (nondormant) seeds (Jackson and Blundell, 1963). Seeds of *Hyoscyamus muticus* contain at least one water-soluble compound that inhibits germination; however, afterripening resulted in an increase in germination percentages (El Hajzein *et al.,* 1995). What happened to the inhibitors in *R. rugosa* and *H. muticus* seeds? There are three possibilities: (1) Inhibitors were removed by leaching or inactivation. With the exception of abscisic acid (ABA), however, effects of dormancy-breaking treatments on the concentration of inhibitors in most seeds have not been studied intensively. (2) Embryos became less sensitive to inhibitors as they came out of physiological dormancy. For example, the concentration of ABA required to inhibit the germination of *Pyrus malus* seeds increased with an increase in cold stratification (Rudnecki, 1969). Also, the inhibitor content of the embryo in *Fraxinus excelsior* seeds

changed little during 6 months of cold stratification, although the seeds became nondormant (Villiers and Wareing (1960). (3) Germination-promoting chemicals (e.g., GA) were produced and counteracted the effects of the inhibitor. Although endogenous GA has been found in seeds of only a few species, Karssen *et al.* (1989) have shown, using GA-deficient mutants of *Arabidopsis thaliana* and *Lycopersicon* sp., that GA absolutely is required for germination. However, there was little direct evidence that GA blocked germination inhibitors. In *Corylus avellana,* ABA declined during cold stratification at 5°C, but most of the GA was not produced until after seeds were transferred from 5 to 20°C (Ross and Bradbeer, 1971). Also, when ABA and GA were applied exogenously to *C. avellana* seeds, ABA antagonized GA (Bradbeer, 1968), i.e., GA did not override ABA! GA overcame the inhibitory effects of ABA in intact caryopses of *Hordeum vulgare* but not in isolated embryos (Dunwell, 1981). Obviously, much research remains to be done on the relationship between GA and ABA (and other germination inhibitors) and their role in the induction, maintenance, and breaking of physiological dormancy.

In conclusion, it is hard to be sure if there are any cases of true chemical dormancy because the effects of inhibitors in many studies have been tested on seeds after PD was broken. Thus, it is unclear whether inhibitors would have prevented germination of nondormant seeds (e.g., Went, 1955). We suggest that chemical dormancy be used to describe only those species whose seeds lack physiological dormancy, and thus the factor preventing germination is a chemical that can be leached out of the seed or somehow deactivated.

2. Abscisic Acid

Abscisic acid is found in various plant parts, including the seeds of numerous species, and it inhibits the germination of nondormant seeds of many species when applied exogenously (Milborrow, 1974; Zigas and Coombe, 1977). Levels of ABA in seeds generally increase during the first half of seed development, when dry weight is increasing, and they decrease in the second half, when water content is declining (Hilhorst, 1995). Because the embryo is dormant at maturity in *Phaseolus vulgaris* (legume) seeds, it is interesting to follow the levels of ABA during seed development. Peak ABA levels in *P. vulgaris* embryos occurred from the 18th to 29th day following anthesis, which corresponds to the period of maximum increase in embryo weight but little decrease in water content. From days 29 to 42, there was very little increase in embryo weight, but water content decreased greatly. The delay (lag time) in the germination of embryos excised from developing seeds increased with an increase in their ABA content (Prevost and Le Page-Degivry, 1985). Thus, ABA may play a role in preventing germination until seed drying occurs (Finkelstein et al., 1985).

The direct role, if any, of ABA in dormancy induction is not clear. In freshly matured seeds, ABA levels may be higher in dormant than in nondormant seeds (e.g., Sondheimer et al., 1968; Walker-Simmons, 1987), but some nondormant seeds have high levels of ABA (Braun and Khan, 1975; Nikolaeva et al., 1978). In some cases, e.g., *Malus domestica*, embryo dormancy is induced before ABA levels rise (Balboa-Zavala and Dennis, 1977). Further, the level of ABA in some seeds may decline before development is completed (Walton, 1980). ABA prevented germination in achenes of *Lactuca sativa* incubated at 25°C and thus facilitated development of secondary dormancy. However, ABA was not required to maintain secondary dormancy, and when ABA levels decreased, dormancy was not broken (Karssen, 1982). Thus, a high concentration of ABA in seeds does not necessarily mean that ABA induces dormancy. Further, the effects of ABA may be influenced by incubation temperatures (Walker-Simmons, 1988) and/or a variation in sensitivity of embryos to ABA at different stages of development (Welbaum et al., 1990).

In a number of species, dormancy develops as seeds dry on the mother plant (Chapter 2). This phenomenon suggests that the role of ABA in dormancy induction may not be induction of dormancy per se but prevention of germination until dormancy is induced. That is, ABA is important in preventing precocious germination of the developing embryo (Zeevaart and Creelman, 1988; Benech Arnold et al., 1991). In some species, ABA applied to excised embryos prevents precocious germination (Kermode et al., 1989). Embryos from mutant *Zea mays* seeds that germinate prior to dispersal from the mother plant (vivipary) are less sensitive to ABA than those from seeds of nonviviparous strains (Robichaud and Sussex, 1986).

One way in which ABA may prevent germination is to inhibit radicle growth, as, for example, in seeds of *Chenopodium album* (Karssen, 1976) and *Sinapis alba* (Schopfer et al., 1979). Radicle growth of *S. alba* stops because ABA inhibits the uptake of water (Schopfer et al., 1979). Further, ABA lowers the ability of the embryo to take up water when it is subjected to osmotic stress (Schopfer and Placky, 1984). Water uptake is controlled by cell wall loosening rather than by changes in osmotic potential or water conductance (Schopfer and Placky, 1985). Thus, a peak in ABA may signal the initiation of embryo/seed drying. Sussex (1975) suggested that RNA and protein synthesis and embryo growth cease after ABA accumulates.

If the primary role of ABA in dormancy induction is to prevent embryo growth until dormancy is induced during drying, then it is easy to understand why (1) nondormant seeds could have high levels of ABA in them and (2) some dormant seeds have low levels of it. In the former case, the seeds dried before ABA was broken down, whereas in the latter case it was broken down by the time seed drying was completed. It would be interesting to determine the amount of ABA in recalcitrant seeds of both temperate and tropical species during the time of seed maturation and dispersal.

Galau et al. (1991) concluded that an understanding of the role of ABA in the regulation of embryogenesis and germination may not be possible until more is known about seed development at the molecular level. Certainly, research on *Arabidopsis thaliana* is moving in that direction (see McCarty, 1995). Crossing experiments between an ABA-deficient mutant line of *A. thaliana* and a wild type of this species have shown that both maternal tissues and the embryo/endosperm produce ABA, but only an increase in ABA of the embryo/endosperm was correlated with the development of seed dormancy. After dormancy developed in seeds, however, ABA was not required for its maintenance (Karssen et al., 1983). Seeds of the *A. thaliana* ABA-deficient and ABA-insensitive double mutant were viable, and they germinated if placed on a wet substrate and died if allowed to dry rapidly (Ooms et al., 1993). However, desiccation tolerance was induced by (1) slow drying, (2) osmotic stress, or (3) incubation in 100 μM ABA. Desiccation tolerance occurred because either the ABA-responsive- or the dehydration-responsive genes were activated (Ooms et al., 1994).

Because cold stratification breaks dormancy in seeds of many species, a number of studies have been done attempting to correlate a decrease in ABA with loss of dormancy during chilling. ABA could not be found in seeds of *Pyrus malus* that had been cold stratified at 2–4°C for 21 days; however, seeds required 84 days of cold stratification for a complete loss of dormancy (Rudnicki, 1969). ABA levels decreased in *Fraxinus americana* seeds during cold stratification, but they did not decline in nonchilled ones (Sondheimer et al., 1968). ABA decreased 98 and 37% for seeds of *Acer saccharum* incubated at 5 and 20°C, respectively, but seeds became nondormant only at 5°C (Webb et al., 1973). However, ABA levels in *Corylus avellana* seeds decreased 61% at both 5 and 20°C (Williams et al., 1973), but only seeds at 5°C became nondormant. Likewise, ABA levels in seeds of *Malus domestica* decreased greatly at both 5 and 20°C, but only those at 5°C became nondormant (Balboa-Zavala and Dennis, 1977). The story is the same in achenes of *Rosa rugosa* (Tillberg, 1983) and in intact seeds of *Acer platanoides* (Pinfield et al., 1989) and *A.*

velutinum (Pinfield and Stutchbury, 1990). Thus, dormancy-breaking conditions are not required for a decrease in the concentration of ABA in some seeds. However, a decrease in ABA does not mean that seeds will become nondormant.

G. Mechanical Dormancy

According to Nikolaeva (1969), mechanical dormancy is due to the presence of a hard, woody fruit wall. The woody structure usually is endocarp, but sometimes the mesocarp also is woody (Hill, 1933). Stony endocarps are found in (some or all members of) many families, including Anacardiaceae, Apocynaceae, Arecaceae, Burseraceae, Cornaceae, Elaeagnaceae, Elaeocarpaceae, Juglandaceae, Lecythidaceae, Meliaceae, Nyssaceae, Oleaceae, Pandaceae, Rhamnaceae, Rosaceae, Sapotaceae, Tiliaceae, and Zygophyllaceae (Nikolaeva, 1969; Hill, 1933, 1937). In the Anacardiaceae, Rhamnaceae, and Tiliaceae, the endocarp can be impermeable to water, resulting in the fruits (seeds) having physical dormancy (Table 3.5). In other families, the endocarp is permeable to water, but germination does not occur until fruits receive a dormancy-breaking treatment. Some fruits with stony endocarps [e.g., some Rosaceae in the subfamily Prunoideae (*sensu* Mabberley, 1987)] have embryos with deep physiological dormancy; consequently, long periods of cold stratification are required for loss of dormancy (Nikolaeva, 1969). In tropical families such as the Burseraceae, Lecythidaceae, and Sapotaceae, it is conceivable that fruits would require warm stratification before they could germinate, but this has not been documented.

Nikolaeva (1969, 1977) worked only with stony fruits that had a cold stratification requirement for germination. She noted that there was no evidence that the endocarp ". . . acts as a mechanical obstacle to a germinating embryo." In other words, once dormancy of the embryo is broken, it has enough growth potential to push through the endocarp. Thus, it appears that mechanical dormancy in these species should be viewed as an aspect or manifestation of physiological dormancy.

Hill (1933, 1937) investigated the manner in which endocarps open at the time of germination. In Rosaceae, Oleaceae, and Juglandaceae, the endocarp splits into two halves, but in many families only a portion of the endocarp is pushed aside by the emerging radicle, something like the opening of a shutter or lid. In some cases, the stony seed-bearing structure represents the fusion of two or more fruits, each with a single seed. Thus, the stony structure may have more than one opening after all the seeds have germinated: two in *Cornus* (Cornaceae), three in *Canarium* (Burseraceae), four in *Tectona* (Verbenaceae), and five in *Davidia* (Davidiaceae).

In *Parinari* (Rosaceae), *Dracontomelon* (Anacardiaceae), and *Bertholletia* (Lecythidaceae), the opening is more like a cork or stopper than a lid (Hill, 1937).

References

Atwood, W. M. (1914). A physiological study of the germination of *Avena fatua*. *Bot. Gaz.* **57,** 386–414.

Balboa-Zavala, O., and Dennis, F. G., Jr. (1977). Abscisic acid and apple seed dormancy. *J. Am. Soc. Hort. Sci.* **102,** 633–637.

Barton, L. V. (1934). Dormancy in *Tilia* seeds. *Contrib. Boyce Thomp. Inst.* **6,** 69–89.

Barton, L. V. (1956). Growth response of physiologic dwarfs of *Malus arnoldiana* Sarg. to gibberellic acid. *Contrib. Boyce Thomp. Inst.* **18,** 311–317.

Baskin, C. C., and Baskin, J. M. (1994). Germination requirements of *Oenothera biennis* seeds during burial under natural seasonal temperature cycles. *Can. J. Bot.* **72,** 779–782.

Baskin, J. M., and Baskin, C. C. (1971a). Germination of *Cyperus inflexus* Muhl. *Bot. Gaz.* **132,** 3–9.

Baskin, J. M., and Baskin, C. C. (1971b). Effect of chilling and gibberellic acid on growth potential of excised embryos of *Ruellia humilis*. *Planta* **100,** 365–369.

Baskin, J. M., and Baskin, C. C. (1976). Effect of photoperiod on germination of *Cyperus inflexus* seeds. *Bot. Gaz.* **137,** 269–273.

Baskin, J. M., and Baskin, C. C. (1979). Studies on the autecology and population biology of the weedy monocarpic perennial, *Pastinaca sativa*. *J. Ecol.* **67,** 601–610.

Baskin, J. M., and Baskin, C. C. (1982). Ecological life cycle and temperature relations of seed germination and bud growth of *Scutellaria parvula*. *Bull. Torrey Bot. Club* **109,** 1–6.

Baskin, J. M., and Baskin, C C. (1984a). Role of temperature in regulating timing of germination in soil seed reserves of *Lamium purpureum* L. *Weed Res.* **24,** 341–349.

Baskin, J. M., and Baskin, C. C. (1984b). Effect of temperature during burial on dormant and non-dormant seeds of *Lamium amplexicaule* L. and ecological implications. *Weed Res.* **24,** 333–339.

Baskin, J. M., and Baskin, C C. (1984c). Germination ecophysiology of the woodland herb *Osmorhiza longistylis* (Umbelliferae). *Am. J. Bot.* **71,** 687–692.

Baskin, J. M., and Baskin, C. C. (1984d). Germination ecophysiology of an eastern deciduous forest herb *Stylophorum diphyllum*. *Am. Midl. Nat.* **111,** 390–399.

Baskin, J. M., and Baskin, C. C. (1985). Seed germination ecophysiology of the woodland spring geophyte *Erythronium albidum*. *Bot. Gaz.* **146,** 130–136.

Baskin, J. M., and Baskin, C. C. (1986a). Temperature requirements for after-ripening in seeds of nine winter annuals. *Weed Res.* **26,** 375–380.

Baskin, J. M., and Baskin, C. C. (1986b). Germination ecophysiology of the mesic deciduous forest herb *Isopyrum biternatum*. *Bot. Gaz.* **147,** 152–155.

Baskin, J. M., and Baskin, C. C. (1990a). Seed germination ecology of poison hemlock, *Conium maculatum*. *Can. J. Bot.* **68,** 2018–2024.

Baskin, J. M., and Baskin, C. C. (1990b). Germination ecophysiology of seeds of the winter annual *Chaerophyllum tainturieri*: A new type of morphophysiological dormancy. *J. Ecol.* **78,** 993–1004.

Benech Arnold, R. L., Fenner, M., and Edwards, P. J. (1991). Changes in germinability, ABA content and ABA embryonic sensitivity in developing seeds of *Sorghum bicolor* (L.) Moench. induced by water stress during grain filling. *New Phytol.* **118,** 339–347.

Bewley, J. D., and Black, M. (1982). "Physiology and Biochemistry of Seeds in Relation to Germination," Vol. 2. Springer-Verlag, Berlin.

Bewley, J. D., and Black, M. (1994). "Seeds: Physiology of Development and Germination," 2nd Ed. Plenum Press, New York/London.

Black, M. (1996). Liberating the radicle: A case for softening-up. *Seed Sci. Res.* **6**, 39–42.

Black, M., and Wareing, P. F. (1959). The role of germination inhibitors and oxygen in the dormancy of the light-sensitive seed of *Betula* spp. *J. Exp. Bot.* **10**, 134–145.

Blommaert, K. L. J., and Hurter, N. (1959). Growth response of physiologic dwarf seedlings of peach, apricot and plum to gibberellic acid. *S. Afr. J. Agric. Sci.* **2**, 409–411.

Blumenthal, A., Lerner, H. R., Werker, E., and Poljakoff-Mayber, A. (1986). Germination preventing mechanisms in iris seeds. *Ann. Bot.* **58**, 551–561.

Boesewinkel, F. D., and Bouman, F. (1984). The seed: structure. *In* "Embryology of Angiosperms" (B. M. Johri, ed.), pp. 567–610. Springer-Verlag, Berlin/Heidelberg/New York/Tokyo.

Bradbeer, J. W. (1968). Studies in seed dormancy. IV. The role of endogenous inhibitors and gibberellin in the dormancy and germination of *Corylus avellana* L. seeds. *Planta* **78**, 266–276.

Bradbeer, J. W. (1988). "Seed Dormancy and Germination." Chapman and Hall, New York.

Braun, J. W., and Khan, A. A. (1975). Endogenous abscisic acid levels in germinating and nongerminating lettuce seed. *Plant Physiol.* **56**, 731–733.

Brown, N. A. C., and Bridglall, S. S. (1987). Preliminary studies of seed dormancy in *Datura stramonium. S. Afr. J. Bot.* **53**, 107–109.

Brown, N. A. C., and van Staden, J. (1973). Studies on the regulation of seed germination in the South African Proteaceae. *Agroplantae* **5**, 111–116.

Bullowa, S., Negbi, M., and Ozeri, Y. (1975). Role of temperature, light and growth regulators in germination in *Anemone coronaria* L. *Aust. J. Plant Physiol.* **2**, 91–100.

Carpita, N. C., Nabors, M. W., Ross, C. W., and Petretic, N. L. (1979). The growth physics and water relations of red-light-induced germination in lettuce seeds. *Planta* **144**, 217–224.

Carpita, N. C., Sharia, A., Barnett, J. P., and Dunlap, J. R. (1983). Cold stratification and growth of radicles of loblolly pine (*Pinus taeda*) embryos. *Physiol. Plant.* **59**, 601–606.

Cavanagh, A. K. (1980). A review of some aspects of the germination of Acacias. *Proc. Roy. Soc. Vict.* **91**, 161–180.

Chen, S. S. C. (1968). Germination of light-inhibited seed of *Nemophilia insignis. Am. J. Bot.* **55**, 1177–1183.

Chen, S. S. C. (1970). Influence of factors affecting germination on respiration of *Phacelia tanacetifolia* seeds. *Planta* **95**, 330–335.

Chen, S. S. C., and Thimann, K. V. (1966). Nature of seed dormancy in *Phacelia tanacetifolia. Science* **153**, 1537–1539.

Christiansen, M. N., and Moore, R. P. (1959). Seed coat structural differences that influence water uptake and seed quality in hard seed cotton. *Agron. J.* **51**, 582–584.

Choate, H. A. (1940). Dormancy and germination in seeds of *Echinocystis lobata. Am. J. Bot.* **27**, 156–160.

Come, D., and Tissaoui, T. (1972). Interrelated effects of imbibition, temperature and oxygen on seed germination. *In* "Seed Ecology" (W. Heydecker, ed.), pp. 157–167. Proceedings of the 19th Easter School in Agriculture, Univ. of Nottingham. Pennsylvania State Univ. Press, University Park, PA.

Conover, D. G., and Geiger, D. R. (1984). Germination of Australian channel millet [*Echinochloa turnerana* (Domin) J. M. Black] seeds. I. Dormancy in relation to light and water. *Aust. J. Plant Physiol.* **11**, 395–408.

Corner, E. J. H. (1951). The leguminous seed. *Phytomorphology* **1**, 117–150.

Corner, E. J. H. (1976). "The Seeds of Dicotyledons," Vols. I and II. Cambridge Univ. Press, Cambridge/London/New York/Melbourne.

Corral, R., Perez-Garcia, F., and Pita, J. M. (1989). Seed morphology and histology in four species of *Cistus* L. (Cistaceae). *Phytomorphology* **39**, 75–80.

Coumans, M., Come, D., and T. Gaspar, T. (1976). Stabilized dormancy in sugar beet fruits. I. Seed coats as a physicochemical barrier to oxygen. *Bot. Gaz.* **137**, 274–278.

Cranston, H. J., Kern, A. J., Gerhardt, S. A., and Dyer, W. E. (1996). Wound-induced ethylene and germination of embryos excised from dormant *Avena fatua* L. caryopses. *Int. J. Plant Sci.* **157**, 153–158.

Crocker, Wm. (1948). "Growth of Plants," Chap. 3. Reinhold, New York.

Crocker, Wm., and Barton, L. V. (1931). After-ripening, germination, and storage of certain Rosaceous seeds. *Contrib. Boyce Thomp. Inst.* **3**, 385–404.

Curtis, E. J. C., and Cantlon, J. E. (1963). Germination of *Melampyrum lineare*: Interrelated effects of afterripening and gibberellic acid. *Science* **140**, 406–408.

Dahlgren, R. M. T., and Clifford, H. T. (1982). "The Monocotyledons: A Comparative Study." Academic Press, New York.

Davis, W. E. (1930). Primary dormancy, after-ripening and the development of secondary dormancy in embryos of *Ambrosia trifida. Am. J. Bot.* **17**, 58–76.

Dekker, J., Dekker, B., Hilhorst, H., and Karssen, C. (1996). Weedy adaptation in *Setaria* spp. IV. Changes in the germinative capacity of *S. faberii* (Poaceae) embryos with development from anthesis to after abscission. *Am. J. Bot.* **83**, 979–991.

Dell, B. (1980). Structure and function of the strophiolar plug in seeds of *Albizia lophantha. Am. J. Bot.* **67**, 556–563.

de Miguel, L., and Sanchez, R. A. (1992). Phytochrome-induced germination, endosperm softening and embryo growth potential in *Datura ferox* seeds: Sensitivity to low water potential and time to escape to FR reversal. *J. Exp. Bot.* **43**, 969–974.

Downie, B., and Bewley, J. D. (1996). Dormancy in white spruce (*Picea glauca* [Moench.] Voss.) seeds is imposed by tissues surrounding the embryo. *Seed Sci. Res.* **6**, 9–15.

Dungey, N. O., and Pinfield, N. J. (1980). The effect of temperature on the supply of oxygen to embryos of intact *Acer pseudoplatanus* seeds. *J. Exp. Bot.* **31**, 983–992.

Dunwell, J. M. (1981). Dormancy and germination in embryos of *Hordeum vulgare* L.: Effect of dissection, incubation temperature and hormone application. *Ann. Bot.* **48**, 203–213.

Edwards, M. M. (1968a). Dormancy in seeds of charlock. II. The influence of the seed coat. *J. Exp. Bot.* **19**, 583–600.

Edwards, M. M. (1968b). Dormancy in seeds of charlock. III. Occurrence and mode of action of an inhibitor associated with dormancy. *J. Exp. Bot.* **19**, 601–610.

Edwards, M. M. (1969). Dormancy in seeds of charlock. IV. Interrelationships of growth, oxygen supply and concentration of inhibitor. *J. Exp. Bot.* **20**, 876–894.

Egley, G. H. (1982). Ethylene stimulation of weed seed germination. *Agric. For. Bull., Univ. Alberta* **5**(2), 13–18.

Egley, G. H., Paul, R. N., Jr., and Lax, A. R. (1986). Seed coat imposed dormancy: Histochemistry of the region controlling onset of water entry into *Sida spinosa* seeds. *Physiol. Plant.* **67**, 320–327.

El Hajzein, B., Geslot, A., and Mairone, Y. (1995). La germination des graines de *Hyoscyamus muticus* (L.) Schrad. I. Influence des enveloppes seminales et dormance embryonnaire relative. *Rev. Cytol. Biol. Veget. Bot.* **18**, 21–37.

Esashi, Y., and Leopold, A. C. (1968). Physical forces in dormancy and germination of *Xanthium* seeds. *Plant Physiol.* **43**, 871–876.

Esau, K. (1964). "Anatomy of Seed Plants." Wiley, New York.

Evenari, M. (1949). Germination inhibitors. *Bot. Rev.* **15**, 153–194.

Finkelstein, R. R., Tenbarge, K. M., Shumway, J. E., and Crouch, M. L. (1985). Role of ABA in maturation of rapeseed embryos. *Plant Physiol.* **78**, 630–636.

Flemion, F. (1931). After-ripening, germination, and vitality of seeds of *Sorbus aucuparia* L. *Contrib. Boyce Thomp. Inst.* **3**, 413–439.

Flemion, F. (1933a). Physiological and chemical studies of after-ripening of *Rhodotypos kerrioides* seeds. *Contrib. Boyce Thomp. Inst.* **5**, 143–159.

Flemion, F. (1933b). Dwarf seedlings from non-after-ripened embryos of *Rhodotypos kerrioides*. *Contrib. Boyce Thomp. Inst.* **5**, 161–165.

Flemion, F. (1934). Dwarf seedlings from non-after-ripened embryos of peach, apple, and hawthorne. *Contrib. Boyce Thomp. Inst.* **6**, 205–209.

Flemion, F. (1937). After-ripening at 5°C favors germination of grape seeds. *Contrib. Boyce Thomp. Inst.* **9**, 7–15.

Flemion, F. (1959). Effect of temperature, light and gibberellic acid on stem elongation and leaf development in physiologically dwarfed seedlings of peach and *Rhodotypos*. *Contrib. Boyce Thomp. Inst.* **20**, 57–70.

Flemion, F., and Waterbury, E. (1945). Further studies with dwarf seedlings of non-after-ripened peach seeds. *Contrib. Boyce Thomp. Inst.* **13**, 415–422.

Frankland, B., and Wareing, P. F. (1966). Hormonal regulation of seed dormancy in hazel (*Corylus avellana* L.) and beech (*Fagus sylvatica* L.). *J. Exp. Bot.* **17**, 596–611.

Galau, G. A., Jakobsen, K. S., and Hughes, D. W. (1991). The controls of late dicot embryogenesis and early germination. *Physiol. Plant.* **81**, 280–288.

Garman, H. R., and Barton, L. V. (1946). Response of lettuce seeds to thiourea treatments as affected by variety and age. *Contrib. Boyce Thomp. Inst.* **14**, 229–241.

Gay, C., Corbineau, F., and Come, D. (1991). Effects of temperature and oxygen on seed germination and seedling growth in sunflower (*Helianthus annuus* L.). *Environ. Exp. Bot.* **31**, 193–200.

Georghiou, K., Psaras, G., and Mitrakos, K. (1983). Lettuce endosperm structural changes during germination under different light, temperature, and hydration conditions. *Bot. Gaz.* **144**, 207–211.

Goebel, K. (1905). "Organography of plants," Part II. Clarendon Press, Oxford.

Graven, P., de Koster, C. G., Boon, J. J., and Bouman, F. (1996). Structure and macromolecular composition of the seed coat of the Musaceae. *Ann. Bot.* **77**, 105–122.

Gray, R. A. (1958). Breaking the dormancy of peach seeds and crab grass seeds with gibberellins. *Plant Physiol.* **33**, Suppl. xl–xli. [Abstract]

Groot, S. P. C., and Karssen, C. M. (1987). Gibberellins regulate seed germination in tomato by endosperm weakening: A study with gibberellin-deficient mutants. *Planta* **171**, 525–531.

Groot, S. P. C., Kieliszewska-Rokicka, B., Vermeer, E., and Karssen, C. M. (1988). Gibberellin-induced hydrolysis of endosperm cell walls in gibberellin-deficient tomato seeds prior to radicle protrusion. *Planta* **174**, 500–504.

Grootjen, C. J., and Bouman, F. (1988). Seed structure in Cannaceae: Taxonomic and ecological implications. *Ann. Bot.* **61**, 363–371.

Grushvitzky, I. V. (1961). Role of embryo underdevelopment in the evolution of flowering plants. *Komarovskie Chteniya* **4**, 1–46.

Grushvitzky, I. V. (1967). After-ripening of seeds of primitive tribes of Angiosperms, conditions and peculiarities. *In* "Physiologie, Okologie und Biochemie der Keimung" (H. Borriss, ed.), pp. 329–336. Ernst-Moritz-Arndt-Universitat, Greifswald.

Hagon, M. W., and Ballard, L. A. T. (1970). Reversibility of strophiolar permeability to water in seeds of subterranean clover (*Trifolium subterraneum* L.). *Aust. J. Biol. Sci.* **23**, 519–528.

Hanna, P. J. (1984). Anatomical features of the seed coat of *Acacia kempeana* (Mueller) which relate to increased germination rate induced by heat treatment. *New Phytol.* **96**, 23–29.

Harrington, G. T., and Hite, B. C. (1923). After-ripening and germination of apple seeds. *J. Agric. Res.* **23**, 153–161.

Hatterman-Valenti, H., Bello, I. A., and Owen, M. D. K. (1996). Physiological basis of seed dormancy in woolly cupgrass (*Eriochloa villosa* [Thunb.] Kunth.). *Weed Sci.* **44**, 87–90.

Hayat, M. A. (1963). Morphology of seed germination and seedling in *Annona squamosa*. *Bot. Gaz.* **124**, 360–362.

Hill, A. W. (1933). The method of germination of seeds enclosed in a stony endocarp. *Ann. Bot.* **47**, 873–887.

Hill, A. W. (1937). The method of germination of seeds enclosed in a stony endocarp. II. *Ann. Bot.* **51**, 239–256.

Hilhorst, H. W. M. (1995). A critical update on seed dormancy. I. Primary dormancy. *Seed Sci. Res.* **5**, 61–73.

Hilhorst, H. W. M., and Downie, B. (1995). Primary dormancy in tomato (*Lycopersicon esculentum* c.v. moneymaker): Studies with the *sitiens* mutant. *J. Exp. Bot.* **47**, 89–97.

Horovitz, A., Bullowa, S., and Negbi, M. (1975). Germination characters in wild and cultivated *Anemone coronaria* L. *Euphytica* **24**, 213–220.

Hsiao, A. I., McIntyre, G. I., and Hanes, J. A. (1983). Seed dormancy in *Avena fatua*. I. Induction of germination by mechanical injury. *Bot. Gaz.* **144**, 217–222.

Hsiao, A. I., and Quick, W. A. (1985). Wild oats (*Avena fatua* L.) seed dormancy as influenced by sodium hypochlorite, moist storage and gibberellin A$_3$. *Weed Res.* **25**, 281–288.

Hussey, G. (1958). An analysis of the factors controlling the germination of the seed of the oil palm, *Elaeis guineensis* (Jacq.). *Ann. Bot.* **22**, 259–284 + 2 plates.

Jackson, G. A. D., and Blundell, J. B. (1963). Germination in *Rosa*. *J. Hort. Sci.* **38**, 310–320.

Jacobsen, J. V., and Pressman, E. (1979). A structural study of germination in celery (*Apium graveolens* L.) seed with emphasis on endosperm breakdown. *Planta* **144**, 241–248.

Jacobsen, J. V., Pressman, E., and Pyliotis, N. A. (1976). Gibberellin-induced separation of cells in isolated endosperm of celery seed. *Planta* **129**, 113–122.

Jones, R. O., and Geneve, R. L. (1995). Seedcoat structure related to germination in eastern redbud (*Cercis canadensis* L.). *J. Am. Soc. Hort. Sci.* **120**, 123–127.

Junttila, O. (1973). The mechanism of low temperature dormancy in mature seeds of *Syringa* species. *Physiol. Plant.* **29**, 256–263.

Karssen, C. M. (1976). Two sites of hormonal action during germination of *Chenopodium album* seeds. *Physiol. Plant.* **36**, 264–270.

Karssen, C. M. (1982). Indirect effect of abscisic acid on the induction of secondary dormancy in lettuce seeds. *Physiol. Plant.* **54**, 258–266.

Karssen, C. M., Brinkhorst-van der Swan, D. L. C., Breekland, A. E., and Koornneef, M. (1983). Induction of dormancy during seed development by endogenous abscisic acid: Studies on abscisic acid deficient genotypes of *Arabidopsis thaliana* (L.) Heynh. *Planta* **157**, 158–165.

Karssen, C. M., Zagorski, S., Kepcynski, J., and Groot, S. P. C. (1989). Key role for endogenous gibberellins in the control of seed germination. *Ann. Bot.* **63**, 71–80.

Kermode, A. R., Oishi, M. Y., and Bewley, J. D. (1989). Regulatory roles for desiccation and abscisic acid in seed development: A comparison of the evidence from whole seeds and isolated embryos. *Crop Sci. Soc. Am. Spec. Publ. No.* **14**, 23–50.

Ketring, D. L. (1973). Germination inhibitors. *Seed Sci. Technol.* **1**, 305–324.

Khan, A., Goss, J. A., and Smith, D. E. (1957). Effect of gibberellin on germination of lettuce seed. *Science* **125**, 645–646.

Khan, A. A., and Samimy, C. (1982). Hormones in relation to primary and secondary seed dormancy. *In* "The Physiology and Biochemistry of Seed Development, Dormancy and Germination" (A. A. Khan, ed.), pp. 203–241. Elsevier Biomedical Press, Amsterdam/New York/Oxford.

Koller, D., and Cohen, D. (1959). Germination-regulating mechanisms in some desert seeds. VI. *Convolvulus lanatus* Vahl, *Convolvulus negevensis* Zoh. and *Convolvulus secundus* Desr. *Bull. Res. Counc. Israel* **7D**, 175–180.

Kontos, F., Spyropoulos, C. G., Griffen, A., and Bewley, J. D. (1996). Factors affecting endo-β-mannanase activity in the endosperms of fenugreek and carob seeds. *Seed Sci. Res.* **6**, 23–29.

Kuijt, J. (1969). "The Biology of Parasitic Flowering Plants." Univ. of California Press, Berkeley, CA.

Kumar, U. (1977). Morphogenetic regulation of seed germination in *Orobanche aegyptiaca* Pers. *Can. J. Bot.* **55**, 2613–2621.

Kumar, P., and Singh, D. (1991). Development and structure of seed coat in *Malva* L. *Phytomorphology* **41**, 147–153.

Lammerts, W. E. (1943). Effect of photoperiod and temperature on growth of embryo-cultured peach seedlings. *Am. J. Bot.* **30**, 707–711.

Leubner-Metzger, G., Frundt, C., Vogeli-Lange, R., and Meins, F., Jr. (1995). Class 1 β-1,3-glucanases in the endosperm of tobacco during germination. *Plant Physiol.* **109**, 751–759.

Leviatov, S., Shoseyov, O., and Wolf, S. (1995). Involvement of endo-mannanase in the control of tomato seed germination under low temperature conditions. *Ann. Bot.* **76**, 1–6.

Lhotska, M., and Moravcova, L. (1989). The ecology of germination and reproduction of less frequent and vanishing species of the Czechoslovak flora. II. *Pulsatilla slavica* Reuss. *Folia Geobot. Phytotax.* **24**, 211–214.

Lush, W. M., Kaye, P. E., and Groves, R. H. (1984). Germination of *Clematis microphylla* seeds following weathering and other treatments. *Aust. J. Bot.* **32**, 121–129.

Lyshede, O. B. (1992). Studies on mature seeds of *Cuscuta pedicellata* and *C. campestris* by electron microscopy. *Ann. Bot.* **69**, 365–371.

Mabberley, D. J. (1987). "The Plant-Book: A Portable Dictionary of the Higher Plants." Cambridge Univ. Press, Cambridge, UK.

Manning, J. C., and van Staden, J. (1987). The functional differentiation of the testa in seed of *Indigofera parviflora* (Leguminosae: Papilionoideae). *Bot. Gaz.* **148**, 23–34.

Marbach, I., and Meyer, A. M. (1975). Changes in catechol oxidase and permeability to water in seed coats of *Pisum elatius* during seed development and maturation. *Plant Physiol.* **56**, 93–96.

Martin, A. C. (1946). The comparative internal morphology of seeds. *Am. Midl. Nat.* **36**, 513–660.

McCarty, D. R. (1995). Genetic control and integration of maturation and germination pathways in seed development. *Annu. Rev. Plant Physiol. Plant Mol. Biol.* **46**, 71–93.

McDonough, W. T. (1967). Dormant and non-dormant seeds: Similar germination responses when osmotically inhibited. *Nature* **214**, 1147–1148.

McKeon, G. M., and Mott, J. J. (1984). Seed biology of *Stylosanthes*. *In* "The Biology and Agronomy of *Stylosanthes*" (H. M. Stace and L. A. Edye, eds.), pp. 311–332. Academic Press, Sydney.

Milborrow, B. V. (1974). The chemistry and physiology of abscisic acid. *Annu. Rev. Plant Physiol.* **25**, 259–307.

Mitrakos, K., and Diamantoglou, S. (1984). Endosperm dormancy breakage in olive seeds. *Physiol. Plant.* **62**, 8–10.

Myers, S. P., Foley, M. E., and Nichols, M. B. (1997). Developmental

differences between germinating after-ripened and dormant excised *Avena fatua* L. embryos. *Ann. Bot.* **79**, 19–23.

Nabors, M. W., and Lang, A. (1971). The growth physics and water relations of red-light-induced germination in lettuce seeds. I. Embryos germinating in osmoticum. *Planta* **101**, 1–25.

Natesh, S., and Rau, M. A. (1984). The embryo. *In* "Embryology of Angiosperms" (B. M. Johri, ed.), pp. 377–443. Springer-Verlag, Berlin.

Nikolaeva, M. G. (1969). Physiology of deep dormancy in seeds. Izdatel'stvo "Nauka," Leningrad [Translated from Russian by Z. Shapiro, National Science Foundation, Washington, DC]

Nikolaeva, M. G. (1977). Factors controlling the seed dormancy pattern. *In* "The Physiology and Biochemistry of Seed Dormancy and Germination" (A. A. Khan, ed.), pp. 51–74. North-Holland, Amsterdam/New York.

Nikolaeva, M. G., Daletskaya, T. V., Razumova, M. V., and Kofanova, N. N. (1973). Effects of gibberellin and kinetin on embryo growth and seed germination in spindle tree and tatar maple. *Sov. Plant Physiol.* **20**, 600–605.

Nikolaeva, M. G., Polyakova, E. N., Razumova M. V., and Askochenskaya, N. A. (1978). Mechanism governing inhibition of germination in seeds of Siberian pea tree. *Sov. Plant Physiol.* **25**, 991–999.

Nikolaeva, M. G., and Vorob'eva, N. S. (1979). The role of abscisic acid and indolic compounds in dormancy of the seeds of ash species. *Sov. Plant Physiol.* **26**, 105–113.

Nomaguchi, M., Nonogaki, H., and Morohashi, Y. (1995). Development of galactomannan-hydrolyzing activity in the micropylar endosperm tip of tomato seed prior to germination. *Physiol. Plant.* **94**, 105–109.

Nonogaki, H., and Morohashi, Y. (1996). An endo-β-mannanase develops exclusively in the micropylar endosperm of tomato seeds prior to radicle emergence. *Plant Physiol.* **110**, 555–559.

Ohga, I. (1926). On the structure of some ancient, but still viable fruits of Indian lotus, with special reference to their prolonged dormancy. *Jpn. J. Bot.* **3**, 1–19.

Olson, A. R. (1980). Seed morphology of *Monotropa uniflora* L. (Ericaceae). *Am. J. Bot.* **67**, 968–974.

Ooms, J. J. J., Leon-Kloosterziel, K. M., Bartels, D., Koorneei M., and Karssen, C. M. (1993). Acquisition of desiccation tolerance and longevity in seeds of *Arabidopsis thaliana*. *Plant Physiol.* **102**, 1185–1191.

Ooms, J. J. J., van der Veen, R., and Karssen, C. M. (1994). Abscisic acid and osmotic stress or slow drying independently induce desiccation tolerance in mutant seeds of *Arabidopsis thaliana*. *Physiol. Plant.* **92**, 506–510.

Perez-Garcia, F. and Pita, J. M. (1989). Mechanical resistance of the seed coat during germination of *Onopordum nervosum* Boiss. *Seed Sci. Technol.* **17**, 277–282.

Pinfield, N. J., Davies, H. V., and Stobart, A. K. (1974). Embryo dormancy in seeds of *Acer platanoides*. *Physiol. Plant.* **32**, 268–272.

Pinfield, N. J., Martin, M. H., and Stobart, A. K. (1972). The control of germination in *Stachys alpina* (L.). *New Phytol.* **71**, 99–104.

Pinfield, N. J., and Stobart, A. K. (1972). Hormonal regulation of germination and early seedling development in *Acer pseudoplatanus* (L.). *Planta* **104**, 134–145.

Pinfield, N. J., and Stutchbury, P. A. (1990). Seed dormancy in *Acer*: The role of testa-imposed and embryo dormancy in *Acer velutinum*. *Ann. Bot.* **66**, 133–137.

Pinfield, N. J., Stutchbury, P. A., Bazaid, S. A., and Gwarazimba, V. E. E. (1989). Seed dormancy in *Acer*: The relationship between seed dormancy, embryo dormancy and abscisic acid in *Acer platanoides* L. *J. Plant Physiol.* **135**, 313–318.

Porter, N. G., and Wareing, P. F. (1974). The role of oxygen permeability of the seed coat in the dormancy of seed of *Xanthium pensylvanicum* Wallr. *J. Exp. Bot.* **25**, 583–594.

Pressman, E., Negbi, M., Sachs, M., and Jacobsen, J. V. (1977). Varietal differences in light requirements for germination of celery (*Apium graveolens* L.) seeds and the effects of thermal and solute stress. *Aust. J. Plant Physiol.* **4**, 821–831.

Prevost, I., and Le Page-Degivry, M. Th. (1985). Inverse correlation between ABA content and germinability through the maturation and the *in vitro* culture of the embryo of *Phaseolus vulgaris*. *J. Exp. Bot.* **36**, 1457–1464.

Probert, R. J., Smith, R. D., and Birch, P. (1985). Germination responses to light and alternating temperatures in European populations of *Dactylis glomerata* L. *New Phytol.* **100**, 447–455.

Rangaswamy, N. S. (1967). Morphogenesis of seed germination in Angiosperms. *Phytomorphology* **17**, 477–487.

Ransom, E. R. (1935). The inter-relations of catalase, respiration, after-ripening, and germination in some dormant seeds of the Polygonaceae. *Am. J. Bot.* **22**, 815–825.

Rees, A. R. (1959). The germination of oil palm seed: The cooling effect. *J. W. Afr. Inst. Oil Palm Res.* **3**, 76–82.

Reid, J. S. G., and Davies, C. (1977). Endo-β-mannanase, the leguminous aleurone layer and the storage galactomannan in germinating seeds of *Trigonella foenum-graecum* L. *Planta* **133**, 219–222.

Reynolds, T., and Thompson, P. A. (1973). Effects of kinetin, gibberellins and (±) abscisic acid on the germination of lettuce (*Lactuca sativa*). *Physiol. Plant.* **28**, 516–522.

Riggio-Bevilacqua, L., Roti-Michelozzi, G., and Serrato-Valenti, G. (1985). Barriers to water penetration in *Cercis siliquastrum*. *Seed Sci. Technol.* **13**, 175–182.

Rizzini, C. T. (1973). Dormancy in seeds of *Annona crassiflora* Mart. *J. Exp. Bot.* **24**, 117–123.

Robichaud, C., and Sussex, I. M. (1986). The response of viviparous-1 and wild type embryos of *Zea mays* to culture in the presence of abscisic acid. *J. Plant Physiol.* **126**, 235–242.

Rolston, M. P. (1978). Water impermeable seed dormancy. *Bot. Rev.* **44**, 365–396.

Ross, J. D., and Bradbeer, J. W. (1971). Studies in seed dormancy. V. The content of endogenous gibberellins in seeds of *Corylus avellana* L. *Planta* **100**, 288–302.

Rudnicki, R. (1969). Studies on abscisic acid in apple seeds. *Planta* **86**, 63–68.

Sanchez, R. A., de Miguel, L., and Mercuri, O. (1986). Phytochrome control of cellulase activity in *Datura ferox* L. seeds and its relationship with germination. *J. Exp. Bot.* **37**, 1574–1580.

Sanchez, R. A., Sunell, L., Labavitch, J. M., and Bonner, B. A. (1990). Changes in the endosperm cell walls of two *Datura* species before radicle protrusion. *Plant Physiol.* **93**, 89–97.

Sanchez-Yelamo, M. D., Tortosa, M. E., Perez-Garcia, F., and Cuquerella, A. (1992). Variability among seed coats in some species of the genus *Onobrychis* Miller (Leguminosae-Fabaceae). *Phytomorphology* **42**, 257–265.

Scheibe, J., and Lang, A. (1965). Lettuce seed germination: Evidence for a reversible light-induced increase in growth potential and for phytochrome mediation of the low temperature effect. *Plant Physiol.* **40**, 485–492.

Schopfer, P., and Plachy, C. (1984). Control of seed germination by abscisic acid. II. Effect on embryo water uptake in *Brassica napus* L. *Plant Physiol.* **76**, 155–160.

Schopfer, P., and Plachy, C. (1985). Control of seed germination by abscisic acid. III. Effect on embryo growth potential (minimum turgor pressure) and growth coefficient (cell wall extensibility) in *Brassica napus* L. *Plant Physiol.* **77**, 676–686.

Schopfer, P., Bajracharya, D., and Plachy, C. (1979). Control of seed

germination by abscisic acid. I. Time course of action in *Sinapis alba* L. *Plant Physiol.* **64**, 822–827.

Scorer, K. N., Epel, B. L., and Waisel, Y. (1985). Interactions between mild NaCl stress and red light during lettuce (*Lactuca sativa* L. cv Grand Rapids) seed germination. *Plant Physiol.* **79**, 149–152.

Serrato-Valenti, G., Cornara, L., Ferrando, M., and Modenesi, P. (1993). Structural and histochemical features of *Stylosanthes scabra* (Leguminosae; Papilionoideae) seed coat as related to water entry. *Can. J. Bot.* **71**, 834–840.

Serrato-Valenti, G., Cornara, L., Ghisellini, P., and Ferrando, M. (1994). Testa structure and histochemistry related to water uptake in *Leucaena leucocephala* Lam. (De Wit). *Ann. Bot.* **73**, 531–537.

Serrato-Valenti, G., de Vries, M., and Cornara, L. (1995). The hilar region in *Leucaena leucocephala* Lam. (De Wit) seed: Structure, histochemistry and the role of the lens in germination. *Ann. Bot.* **75**, 569–574.

Singh, C. P. (1985). A comparison between low temperature and gibberellic acid removed dormancy of *Euonymous europaeus* L. embryos with respect to seedling growth and development. *Phyton (Buenos Aires)* **45**, 143–148.

Small, J. G. C., and Gutterman, Y. (1991). Evidence for inhibitor involvement in thermodormancy of grand rapids lettuce seeds. *Seed Sci. Res.* **1**, 263–267.

Sondheimer, E., Tzou, D. S., and Galson, E. C. (1968). Abscisic acid levels and seed dormancy. *Plant Physiol.* **43**, 1443–1447.

Spyropoulos, C. G., and Reid, J. S. G. (1985). Regulation of α-galactosidase activity and the hydrolysis of galactomannan in the endosperm of the fenugreek (*Trigonella foenum-graecum* L.) seed. *Planta* **166**, 271–275.

Steinbauer, G. P. (1937). Dormancy and germination of *Fraxinus* seeds. *Plant Physiol.* **12**, 813–824.

Sussex, I. (1975). Growth and metabolism of the embryo and attached seedling of the viviparous mangrove, *Rhizophora mangle*. *Am. J. Bot.* **62**, 948–953.

Takahashi, N. (1985). Inhibitory effects of oxygen on the germination of *Oryza sativa* L. seeds. *Ann. Bot.* **55**, 597–600.

Takeba, G., and Matsubara, S. (1979). Measurement of growth potential of the embryo in New York lettuce seed under various combinations of temperature, red light and hormones. *Plant Cell Physiol.* **20**, 51–61.

Tao, K.-L., and Khan, A. A. (1979). Changes in the strength of lettuce endosperm during germination. *Plant Physiol.* **63**, 126–128.

Thanos, C. A., and Georghiou, K. (1988). Ecophysiology of fire-stimulated seed germination in *Cistus incanus* ssp. *creticus* (L.) Heywood and *C. salvifolius* L. *Plant Cell Environ.* **11**, 841–849.

Tillberg, E. (1983). Levels of endogenous abscisic acid in achenes of *Rosa rugosa* during dormancy release and germination. *Physiol. Plant.* **58**, 243–248.

Timson, J. (1966). The germination of *Polygonum convolvulus* L. *New Phytol.* **65**, 423–428.

Toole, E. H., and Toole, V. K. (1941). Progress of germination of seed of *Digitaria* as influenced by germination temperature and other factors. *J. Agric. Res.* **63**, 65–90.

Toole, V. K. (1941). Factors affecting the germination of various dropseed (*Sporobolus* spp.). *J. Agric. Res.* **62**, 691–715.

Toorop, P. E., Bewley, J. D., and Hilhorst, H. W. M. (1996). Endo-β-mannanase isoforms are present in the endosperm and embryo of tomato seeds, but are not essentially linked to the completion of germination. *Planta* **200**, 153–158.

Tukey, H. B., and Carlson, R. F. (1945). Morphological changes in peach seedlings following after-ripening treatments of the seeds. *Bot. Gaz.* **106**, 431–440.

Valenti, G. S., Modenesi, P., Roti-Michelozzi, G., and Bevilacqua, L. (1986). Structural and histochemical characters of the *Prosopis tamarugo* Phil. seed coat, in relation to its hardness. *Acta Bot. Neerl.* **35**, 475–487.

Vazquez-Yanes, C., and Perez-Garcia, B. (1976). Notas sobre la morfologia y la anatomia de la testa de las semillas de *Ochroma lagopus* Sw. *Turrialba* **26**, 310–311.

Villiers, T. A., and Wareing, P. F. (1960). Interaction of growth inhibitor and a natural germination stimulator in the dormancy of *Fraxinus excelsior* L. *Nature* **185**, 112–114.

Voigt, B., and Bewley, J. D. (1996). Developing tomato seeds when removed from the fruit produce multiple forms of germinative and post-germinative endo-B-mannanase: Responses to desiccation, abscisic acid and osmoticum. *Planta* **200**, 71–77.

von Teichman, I. (1989). Reinterpretation of the pericarp of *Rhus lancea* (Anacardiaceae). *S. Afr. J. Bot.* **55**, 383–384.

von Teichman, I., and Robbertse, P. J. (1986a). Development and structure of the pericarp and seed of *Rhus lancea* L. fil. (Anacardiaceae), with taxonomic notes. *Bot. J. Linnean Soc.* **93**, 291–306.

von Teichman, I., and Robbertse, P. J. (1986b). Development and structure of the drupe in *Sclerocarya birrea* (Richard) Hochst. subsp *caffra* Kokwaro (Anacardiaceae), with special reference to the pericarp and the operculum. *Bot. J. Linn. Soc.* **92**, 303–322.

Walker-Simmons, M. (1987). ABA levels and sensitivity in developing wheat embryos of sprouting resistant and susceptible cultivars. *Plant Physiol.* **84**, 61–66.

Walker-Simmons, M. (1988). Enhancement of ABA responsiveness in wheat embryos by high temperature. *Plant Cell Environ.* **11**, 769–775.

Walton, D. C. (1980). Biochemistry and physiology of abscisic acid. *Annu. Rev. Plant Physiol.* **31**, 453–489.

Ware, S. A., and Quarterman, E. (1969). Seed germination in cedar glade *Talinum. Ecology* **50**, 137–140.

Wareing, P. F. (1965). Endogenous inhibitors in seed germination and dormancy. *In* "Encyclopedia of Plant Physiology" (W. Ruhland, ed.), vol. 15/2, pp. 909–924. Springer-Verlag, Berlin/Heidelberg/New York.

Wareing, P. F., and Foda, H. A. (1957). Growth inhibitors and dormancy in *Xanthium* seed. *Physiol. Plant.* **10**, 266–280.

Watkins, J. T., and Cantliffe, D. J. (1983a). Hormonal control of pepper seed germination. *HortScience* **18**, 342–343.

Watkins, J. T., and Cantliffe, D. J. (1983b). Mechanical resistance of the seed coat and endosperm during germination of *Capsicum annuum* at low temperature. *Plant Physiol.* **72**, 146–150.

Watkins, J. T., Cantliffe, D. J., Huber, D. J., and Nell, T. A. (1985). Gibberellic acid stimulated degradation of endosperm in pepper. *J. Am. Soc. Hort. Sci.* **110**, 61–65.

Webb, D. P., and Dumbroff, E. B. (1969). Factors influencing the stratification process in seeds of *Acer saccharum. Can. J. Bot.* **47**, 1555–1563.

Webb, D. O., and Wareing, P. F. (1972). Seed dormancy in *Acer:* Endogenous germination inhibitors and dormancy in *Acer pseudoplatanus* L. *Planta* **104**, 115–125.

Webb, D. O., van Staden, J., and Wareing, P. F. (1973). Seed dormancy in *Acer:* Changes in endogenous cytokinins, gibberellins and germination inhibitors during the breaking of dormancy in *Acer saccharum* Marsh. *J. Exp. Bot.* **24**, 105–116.

Welbaum, G. E., Tissaoui, T., and Bradford, K. J. (1990). Water relations of seed development and germination in muskmelon (*Cucumis melo* L.). *Plant Physiol.* **92**, 1029–1037.

Went, F. W. (1955). The ecology of desert plants. *Sci. Am.* **192**(4), 68–75.

Werker, E. (1980–1981). Seed dormancy as explained by the anatomy of embryo envelopes. *Israel J. Bot.* **29**, 22–44.

Williams, P. M., Ross, J. D., and Bradbeer, J. W. (1973). Studies in seed dormancy. VII. The abscisic acid content of the seeds and fruits of *Corylus avellana* L. *Planta* **110**, 303–310.

Willis, J. C. (revised by H. K. A. Shaw) (1966). "A Dictionary of the Flowering Plants and Ferns," 7th Ed. Cambridge Univ. Press, Cambridge, UK.

Winter, D. M. (1960). The development of the seed of *Abutilon theophrasti.* II. Seed coat. *Am. J. Bot.* **47**, 157–162.

Witcombe, J. R., Hillman, J. R., and Whittington, W. J. (1969). Growth inhibitor in the seed coat of charlock. *Nature* **222**, 1200–1201.

Zambou, K., and Spyropoulos, C. G. (1990). D-galactose uptake by fenugreek cotyledons. *Plant Physiol.* **93**, 1417–1421.

Zambou, K., Spyropoulos, C. G., Chinou, I., and Kontos, F. (1993). Saponin-like substances inhibit α-galactosidase production in the endosperm of fenugreek seeds. *Planta* **189**, 207–212.

Zeevaart, J. A. D., and Creelman, R. A. (1988). Metabolism and physiology of abscisic acid. *Annu. Rev. Plant Physiol.* **39**, 439–473.

Zigas, R. P., and Coombe, B. G. (1977). Seedling development in peach, *Prunus persica* (L.) Batsch. II. Effects of plant growth regulators and their possible role. *Aust. J. Plant Physiol.* **4**, 359–369.

Zomlefer, W. B. (1994). "Guide to Flowering Plant Families." Univ. of North Carolina Press, Chapel Hill.

4

Germination Ecology of Seeds with Nondeep Physiological Dormancy

I. PURPOSE

It is well recognized that dormant seeds must become nondormant before they can germinate. However, some species whose seeds have nondeep physiological dormancy (PD) at maturity can cycle back and forth between dormancy and nondormancy many times before they eventually germinate. The change from dormancy to nondormancy or from nondormancy to dormancy is gradual and it occurs in response to changes in environmental factors. Thus, during the course of a year seeds may go from dormancy to nondormancy and then back to dormancy, thereby exhibiting an annual cycle. As far as we know, seeds with nondeep PD are the only ones that can reenter dormancy and thus exhibit annual cycles in their dormancy states.

This chapter will discuss the changes in dormancy states of seeds with nondeep PD and how these changes in dormancy states are affected by various environmental factors. Emphasis will be placed on how changes in germination responses of the seeds interact with the environment to control the timing of seed germination in the field.

II. DISCOVERY OF DORMANCY CYCLES

Because most weeds of arable lands have nondeep PD, this type of dormancy probably has received more research attention than any other type. In fact, much of what is known about physiological dormancy (and nondeep PD in particular) comes from attempts to better understand the timing of weed seed germination in arable lands as a part of control strategies.

It has been known for decades that buried seeds of many species need to be brought to the soil surface before they will germinate (e.g., Odum, 1965; Wesson and Wareing, 1969a). Thus, it is understandable why it would have been thought that buried seeds were in a constant state of readiness to germinate and that they would do so if brought to the soil surface sometime during the growing season and exposed to light (Wesson and Wareing, 1967, 1969a). However, data from studies in which soil was plowed or disturbed at regular intervals throughout the year showed that seeds of many species germinated only at specific times of the year (Brenchley and Warington, 1930; Chepil, 1946; Roberts, 1964; Karssen, 1982). Germination periodicity of exhumed seeds (Fig. 4.1) indicated that buried seeds could not germinate during certain parts of the year, even if soil disturbance resulted in them being brought to the surface. In other studies, seeds of various weedy species were buried at different depths in the field, but the soil was not disturbed. Here too, germination, and thus seedling emergence, showed strong seasonal periodicity (Stoller and Wax, 1973). The timing of germination of buried seeds obviously was not due solely to the seed being brought to the soil surface; therefore, there were seasonal differences in how seeds responded to the environment. Studies on seasonal emergence of weeds have continued into the 1990s (e.g., Egley and Williams, 1991; Popay et al., 1995; Sans and Masalles, 1995).

Courtney (Fig. 4.2), Schafer and Chilcote (1970), and Taylorson (1970) obtained the first direct evidence that buried seeds undergo changes in their dormancy state. These authors buried seeds in soil outside and exhumed samples at regular intervals and tested them for germination. At some times of the year, exhumed seeds germinated to high percentages, but at other times little or no germination was obtained. Since these early studies, seeds of many species have been buried in soil under natural (or near-natural) temperature conditions, and

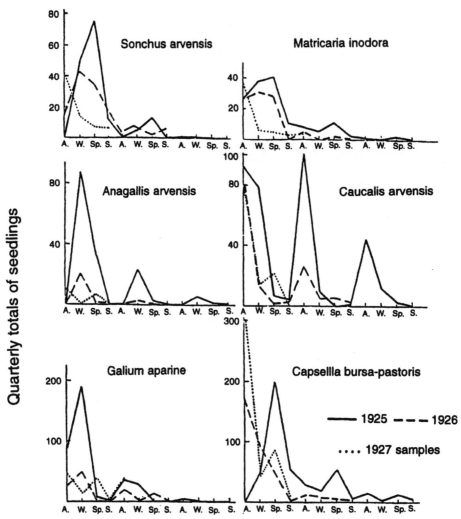

FIGURE 4.1   Periodicity of seed germination in soil samples removed from the field and kept in a nonheated greenhouse in England. From Brenchley and Warington, 1930.

III. CHANGES IN THE STATE OF DORMANCY

at regular intervals samples of them have been exhumed and tested for germination, usually over a range of temperatures in light and darkness. Thus, much is known about changes in dormancy states of buried seeds, especially those of annual weeds of the temperate region.

If freshly matured seeds fail to germinate when incubated over a range of test conditions, they are in primary innate dormancy. However, "primary innate dormancy" usually is shortened to "dormancy" in discussions of germination (Table 4.1). As seeds come out of dormancy, they first enter conditional dormancy, during which they germinate only over a narrow range of conditions. During the progression of dormancy loss, how-

ever, this range widens until seeds finally germinate over the full range of conditions possible for the population or taxon, at which point they are nondormant (Fig. 4.3). The transitional state (actually a series of states) between dormancy and nondormancy is called conditional dormancy, and it means the same thing as the relative dormancy of Vegis (1964, 1973).

Nondormant seeds of some species may reenter dormancy if environmental conditions are unfavorable for germination. As seeds reenter dormancy, they first become conditionally dormant, during which they germinate over almost the full range of conditions possible for the taxon. During progression through conditional dormancy into dormancy, however, this range narrows until seeds finally do not germinate at any condition, at which point they are dormant. This dormancy is referred to as secondary dormancy. Dormancy induction in seeds

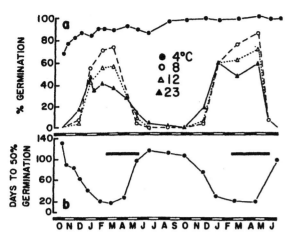

FIGURE 4.2 Germination of *Polygonum aviculare* seeds at various constant temperatures (light : dark condition for tests not given) after 0–20 months of burial in soil in the field in Warwick, England. Seeds were incubated at 8, 12, and 23°C for 4 weeks. Seeds remained at 4°C until monthly increments of germination percentage were no more than 2%. (b) Number of days at 4°C for seeds to reach 50% germination at 4°C. Horizontal bars represent the period of seedling emergence observed in the field. From Courtney (1968).

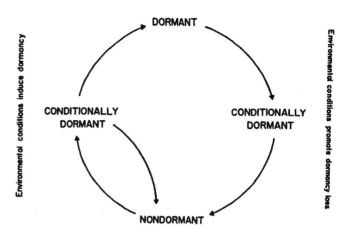

FIGURE 4.3 Diagram of annual changes in dormancy states of seeds. Modified from Baskin and Baskin (1989d).

of many species stops at conditional dormancy, and thus they lose their ability to germinate at some but not all test conditions. In still other species, however, nondormant seeds do not enter either conditional dormancy or dormancy; they remain nondormant throughout the year.

Seeds of some species annually cycle between dormancy and nondormancy and those of others between conditional dormancy and nondormancy (Table 4.2). Seeds may go through many cycles before environmental conditions become favorable for germination, but the maximum number of times they can do this is unknown. Seeds of *Ambrosia artemisiifolia* become nondormant during winter and, if conditions are unfavorable for germination in spring, they reenter dormancy. In a 28-month study of this species, buried seeds exposed to natural seasonal temperature cycles in the temperate

zone reentered dormancy twice (Baskin and Baskin, 1980). A few seeds of *A. artemisiifolia* germinated when exhumed after 40 years of burial in the Beal buried seed experiment (Darlington, 1922), indicating that they may have gone through 40 annual dormancy/nondormancy cycles.

In species exhibiting annual changes in their dormancy state, the time of year when seeds are in each state varies with the life cycle (Fig. 4.4). Seeds of winter annuals such as *Arabidopsis thaliana* are dormant in late winter and spring, come out of dormancy during summer, and are nondormant in autumn (Baskin and Baskin, 1983a). Seeds of summer annuals such as *A. artemisiifolia* are dormant in summer and autumn, come out of dormancy during winter, and are nondormant in late winter and early spring (Baskin and Baskin, 1980). Nondormant seeds of some winter and summer annuals enter conditional dormancy and then cycle between conditional dormancy and nondormancy until they eventually either die or germinate (Baskin and Baskin, 1987a). Seeds of winter annuals with conditional dormancy/nondormancy cycles are conditionally dormant during spring (Baskin and Baskin, 1983b), whereas those of summer annuals with conditional dormancy/nondormancy cycles are in conditional dormancy during autumn (Baskin and Baskin, 1978; Baskin *et al.*, 1994).

Not all seeds that are dormant or conditionally dormant when freshly matured exhibit annual changes in their dormancy states. After seeds come out of dormancy or conditional dormancy, they do not reenter either conditional dormancy or dormancy (Table 4.2, Fig. 4.4d). For example, seeds of *Potentilla recta* (Baskin and Baskin, 1990a) and *Hottonia inflata* (Baskin *et al.*,

TABLE 4.1 Various States of Physiological Dormancy and a Shortened Version of Terminology Frequently Used in Discussions of Annual Changes in Dormancy States

Dormancy state	Term(s) usually used
Nondormant	Nondormant
Primary innate dormancy	Dormancy or primary dormancy
Primary conditional dormancy	Conditional dormancy
Secondary innate dormancy	Dormancy or secondary dormancy
Secondary conditional dormancy	Conditional dormancy

TABLE 4.2 Type of Seed Dormancy Cycles in Species with Various Kinds of Life Cycles

Species	Life cycle[a]	Type of dormancy cycle[b]	Reference
Alopecurus myosuroides	WA	—[c]	Froud-Williams *et al.* (1984)
Amaranthus retroflexus	SA	D/ND	Egley (1989)
Ambrosia artemisiifolia	SA	D/ND	Baskin and Baskin (1980)
A. trifida	SA	—[d]	Stoller and Wax (1974)
Aphanes arvensis	WA	D/ND[e]	Roberts and Neilson (1982b)
Arabidopsis thaliana	WA	D/ND	Baskin and Baskin (1983a)
Arenaria serpyllifolia	WA	—[d]	Pons (1991b)
Barbarea vulgaris	WA, SA	None[f]	Baskin and Baskin (1989c)
Bidens polylepis	SA	D/ND	Baskin *et al.* (1995)
Capsella bursa-pastoris	WA	CD/ND	Baskin and Baskin (1989b)
Carex canescens	PP	—[d]	Schuetz (unpublished results)
C. elongata	PP	—[d]	Schuetz (unpublished results)
C. paniculata	PP	—[d]	Schuetz (unpublished results)
C. remota	PP	—[d]	Schuetz (unpublished results)
Carlina vulgaris	MP	None[g]	Pons (1991b)
Chenopodium album	SA	CD/ND	Bouwmeester and Karssen (1993a), Baskin and Baskin (unpublished results)
C. bonus-henricus	SA	CD/ND[c]	Khan and Karssen (1980)
Commelina communis	SA	—[d]	Watanabe and Hirokawa (1975)
Crupina vulgaris	WA	—[d]	Thill *et al.* (1985)
Cyperus erythrorhizos	SA	None	Baskin *et al.* (1993b)
C. flavicomus	SA	CD/ND	Baskin *et al.* (1993b)
C. inflexus	SA	CD/ND	Baskin and Baskin (1978)
C. odoratus	PP	None	Baskin *et al.* (1989)
Datura ferox	SA	—[d]	Reisman-Berman and Kigel (1991)
D. stramonium	SA	—[d]	Stoller and Wax (1974)
Daucus carota	MP	—[d]	Pons (1991b)
Digitaria ciliaris	A	—[d]	Marks and Nwachuku (1986)
Echinochloa crus-galli	SA	—[d]	Taylorson (1970), Watanabe and Hirokawa (1975)
Emex australis	WA	—[d]	Panetta and Randall (1993)
Eriastrum diffusum	WA	D/ND	Baskin *et al.* (1993)
Eriogonum abertianum	WA	CD/ND	Baskin *et al.* (1993)
Eupatorium odoratum	PP	None[g]	Marks and Nwachuku (1986)
Fimbristylis autumnalis	SA	CD/ND	Baskin *et al.* (1993b)
F. vahlii	SA	CD/ND	Baskin *et al.* (1993b)
Gratiola viscidula	PP	CD/ND	Baskin *et al.* (1989)

(continues)

1996) are dormant at maturity in late spring or early summer and become nondormant during summer and remain nondormant regardless of seasonal changes.

Although seeds of many species are dormant at maturity, those of others are conditionally dormant (Fig. 4.4). Seeds that are conditionally dormant at maturity may become nondormant and then (1) cycle between dormancy and nondormancy (Baskin and Baskin, unpublished results), (2) cycle between nondormancy and conditional dormancy (Baskin and Baskin, 1989b), or (3) remain nondormant (Baskin *et al.*, 1989).

The dormancy seeds possess when they are freshly matured is called primary dormancy, and seeds have either primary innate dormancy or primary conditional dormancy. These terms are usually shortened to dormancy or conditional dormancy, respectively (Table

TABLE 4.2—*Continues*

Species	Life cycle[a]	Type of dormancy cycle[b]	Reference
Hottonia inflata	WA	None	Baskin *et al.* (1996)
Krigia oppositifolia	WA	D/ND	Baskin *et al.* (1991)
Lamium amplexicaule	WA, SA	CD/ND	Baskin and Baskin (1981a)
L. purpureum	WA	D/ND	Baskin and Baskin (1984a)
Leptochloa panicoides	SA	CD/ND	Baskin *et al.* (1993c)
Lesquerella lescurii	WA	D/ND	Baskin *et al.* (1992)
L. stonensis	WA	D/ND	Baskin and Baskin (1990c)
Leucospora multifida	SA	CD/ND	Baskin *et al.* (1994)
Linum cartharticum	MP	—[d]	Pons (1991b), Milberg (1994c)
Lobelia inflata	SA, MP	CD/ND	Baskin and Baskin (1992)
Lychnis flos-cuculi	PP	—[d]	Milberg (1994a)
Mariscus alternifolius	PP	None[g]	Marks and Nwachuku (1986)
M. flabelliformis	PP	None[g]	Marks and Nwachuku (1986)
Matricaria perforata	SA, WA, PP	None	Bowes *et al.* (1995)
Nemophila aphylla	WA	D/ND	Baskin *et al.* (1993a)
Oenothera biennis	MP	CD/ND	Baskin and Baskin (1994)
Origanum vulgare	PP	None[g]	Pons (1991b)
Panicum capillare	SA	D/ND	Baskin and Baskin (1986b)
P. dichotomiflorum	SA	D/ND[e]	Baskin and Baskin (1983c)
Penstemon palmeri	PP	CD/ND	Meyer and Kitchen (1992)
Penthorum sedoides	PP	None	Baskin *et al.* (1989)
Phacelia ranunculacea	WA	D/ND	Baskin *et al.* (1993a)
Phytolacca americana	PP	CD/ND	Baskin and Baskin (unpublished results)
Plantago virginica	WA	CD/ND	Baskin *et al.* (unpublished results)
Poa trivialis	WA	—[d]	Froud-Williams *et al.* (1986)
Polygonum aviculare	SA	D/ND	Courtney (1968)
		D/ND[e]	Baskin and Baskin (1990b)
P. lapathifolium	SA	—[d]	Watanabe and Hirokawa (1975)
P. pensylvanicum	SA	—[d]	Taylorson (1972)
P. persicaria	SA	D/ND	Bouwmeester and Karssen (1992)
Portulaca smallii	SA	None	Baskin *et al.* (1987)
Potentilla recta	PP	None	Baskin and Baskin (1990a)
Primula veris	PP	—[d]	Milberg (1994b)
Rumex crispus	PP	None	Baskin and Baskin (1985b)
R. obtusifolius	PP	D/ND, CD/ND	van Assche and Verleberghe (1989)
Sabatia angularis	MP	CD/ND	Baskin and Baskin (unpublished results)

(*continues*)

4.1). Seeds that are either dormant or conditionally dormant at maturity and then become nondormant can be induced into dormancy or conditional dormancy under certain environmental conditions (Karssen, 1980–1981b; Baskin and Baskin, 1985a); this is called secondary innate or secondary conditional dormancy. These terms frequently are shortened to dormancy and conditional dormancy, respectively (Table 4.1).

The terms "induced dormancy" and "enforced dormancy" sometimes are found in the literature, and we need to know what they mean and how they relate to dormancy terminology already used in this chapter. According to Harper (1957), various environmental factors cause some seeds that are nondormant at maturity to enter dormancy, which he called induced dormancy. However, we are unaware of a species whose seeds are nondormant at maturity that have been induced into dormancy. Thus, until someone induces dormancy in

TABLE 4.2—*Continues*

Species	Life cycle[a]	Type of dormancy cycle[b]	Reference
Scabiosa columbaria	PP	None[g]	Pons (1991b)
Scirpus lineatus	PP	CD/ND	Baskin *et al.* (1989)
Senecio vulgaris	SA	—[d]	Karssen (1980–1981b)
Setaria faberi	SA	—[d]	Taylorson (1972)
S. glauca	SA	CD/ND	Baskin *et al.* (1996)
Sisymbrium officinale	SA	D/ND	Bouwmeester and Karssen (1993c)
Solanum mauritianum	PP	—[d]	Campbell *et al.* (1992)
S. nigrum	SA	CD/ND	Roberts and Lockett (1978b)
S. sarrachoides	SA	D/ND[e]	Roberts and Boddrell (1983)
Solenostemon monostachyus	PP	None[g]	Marks and Nwachuku (1986)
Solidago altissima	PP	CD/ND	Walck *et al.* (1997)
S. nemoralis	PP	CD/ND	Walck *et al.* (1997)
S. shortii	PP	CD/ND	Walck *et al.* (1997)
Specularia perfoliata	WA	D/ND	Baskin and Baskin (unpublished results)
Spergula arvensis	SA, WA	D/ND[e]	Bouwmeester and Karssen (1993b)
Talinum parviflorum	PP	CD/ND	Baskin and Baskin (unpublished results)
Thlaspi arvense	WA	D/ND	Baskin and Baskin (1989a)
Tridax procumbens	A	—[d]	Marks and Nwachuku (1986)
Verbascum blattaria	MP	CD/ND	Baskin and Baskin (1981c)
V. thapsus	MP	CD/ND	Baskin and Baskin (1981c)
Veronica arvensis	WA	CD/ND	Baskin and Baskin (1983b)
V. hederifolia	WA	D/ND[e]	Roberts and Lockett (1978a)
V. peregrina	WA	CD/ND	Baskin and Baskin (1983d)
V. persica	WA	—[d]	Froud-Williams *et al.* (1984)
Viola arvensis	WA	D/ND, CD/ND	Baskin and Baskin (1995)

[a] WA, winter annual; SA, summer annual; PP, polycarpic perennial; MP, biennial or monocarpic perennial; A, annual.

[b] D, dormancy; ND, nondormancy; CD, conditional dormancy.

[c] Only one test temperature was used, but seeds appeared to have a CD/ND cycle.

[d] Only one test temperature was used; it is not possible to determine if seeds had a CD/ND or D/ND cycle.

[e] Most seeds had a D/ND cycle, but some had a CD/ND cycle.

[f] Most seeds lacked an annual cycle, but some had a CD/ND cycle.

[g] Only one test temperature was used. Buried seeds appeared not to have a cycle, but some could have an annual CD/ND cycle.

seeds that are completely nondormant at maturity, we have no reason to use the term "induced dormancy." Further, the term induced dormancy is inappropriate for seeds that are dormant or conditionally dormant at maturity and subsequently become nondormant and then reenter secondary dormancy. In this situation, the meaning of the term induced dormancy overlaps with that of secondary dormancy. Harper (1957) defined enforced dormancy as lack of germination due to unfavorable environmental conditions. Care should be used when talking about enforced dormancy to make sure that lack of germination is due entirely to an unfavorable environment, i.e., enforced dormancy applies to nondormant seeds. Enforced dormancy is equivalent to quiescence, which is the arrested growth of nondormant seeds

or organs due to unfavorable environmental conditions (Vegis, 1963). According to Karssen (1980–1981a), enforced dormancy should be referred to as the environmental inhibition of germination.

IV. THE DORMANCY CONTINUUM

In the early 1900s, scientists realized that dormancy does not begin or end abruptly (Vegis, 1963); this concept is still accepted (see Gordon, 1973; Baskin and Baskin, 1985a; Vleeshouwers *et al.,* 1995). Atterberg (1899) showed that there was a gradual widening of the temperature range for germination as seeds (caryopses) of cereals came out of dormancy, and Fuchs (1941)

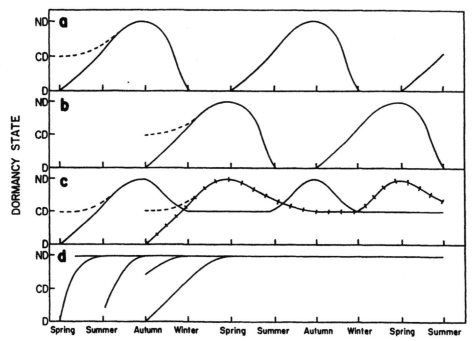

FIGURE 4.4 Diagrams of seasonal changes in states of physiological dormancy. (a) Winter annual or perennial seeds mature in spring and have an annual dormancy/nondormancy cycle. (b) Summer annual or perennial seeds mature in autumn and have an annual dormancy/nondormancy cycle. (c) A spring-maturing (—) and a summer-maturing (+–+–) annual or perennial whose seeds have an annual nondormancy/conditional dormancy cycle. (d) Species with dormant seeds, or with seeds in various states of conditional dormancy, that become nondormant and remain nondormant, or a species with nondormant seeds that remain nondormant. Modified from Baskin and Baskin (1989d).

found a decrease in maximum temperature for germination as dormancy was induced in caryopses maturing on the mother plant.

Vegis (1963, 1964, 1973) greatly expanded our knowledge of changes in temperature requirements for germination, as well as bud growth, as primary dormancy is broken. He did not study induction or breaking of secondary dormancy. During the period of dormancy loss in some species (see Fig. 4.5), the maximum temperature for germination increases (type 1), whereas in others the minimum temperature for germination decreases (type 2). In a few species, seeds first germinate at an intermediate temperature and then maximum and minimum temperatures increase and decrease, respectively (type 3). Reversal of these responses to temperature occurs as nondormant seeds are induced into dormancy (Vegis, 1963, 1964, 1973).

Since Vegis described types 1, 2, and 3 of nondeep PD, two other types have been identified. In type 4, seeds germinate only at relatively high temperatures when they become nondormant, e.g., *Aristida ramosa, Bothriochloa macra* (Lodge and Whalley, 1981), and *Callicarpa americana* (Baskin and Baskin, unpublished results). In type 5, seeds germinate only at relatively low temperatures when they become nondormant, e.g.,

Eriastrum diffusum (Baskin *et al.*, 1993) and *Androcymbium europaeum* (Schuetz, unpublished results). Dormancy loss occurs at high temperatures in seeds with type 4 or 5 identified thus far.

Conditional dormancy is a sequence or continuum of states between dormancy and nondormancy and between nondormancy and dormancy (Baskin and

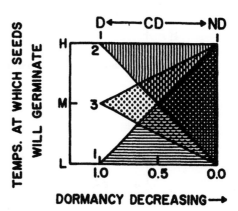

FIGURE 4.5 Three patterns (types 1, 2, and 3) of changes in temperature requirements for germination as seeds come out of nondeep physiological dormancy. This diagram is based on the conceptual models of Vegis (1963, 1964, 1973). From Baskin *et al.* (1994).

Baskin, 1985a). To illustrate this point, let us consider the germination responses of seeds of the winter annual *Veronica arvensis* as they come out of dormancy (Fig. 4.6). When *V. arvensis* seeds mature in June (early summer) in northcentral Kentucky, most of them are dormant, but a low percentage is conditionally dormant and consequently will germinate at low temperatures. By October, the seeds are nondormant and they germinate to high percentages over a range of temperatures. As seeds come out of dormancy from June to October, they pass from one state of conditional dormancy to the next and the maximum temperature at which germination is possible, as well as the percentage and rate of germination, increases. Although each seed becomes less dormant with time, an individual seed can only germinate once. Thus, the curves in Fig. 4.6 represent what is happening to the population of seeds through time.

In addition to increases in the temperature range for germination and in germination percentages and rates, other physiological responses are correlated with the continuum of dormancy states (i.e., conditional dormancy) occurring between dormancy and nondormancy (Table 4.3). As seeds come out of dormancy, they become increasingly sensitive to various factors such as light and hormones; therefore, very small amounts of light or hormones will promote germination. Also, seeds become decreasingly sensitive to substrate moisture; thus, they will germinate at increased water stresses or perhaps when a substrate is saturated or oversaturated with water. Seeds exhibit a continuum of responses as

they go from nondormancy to dormancy. As seeds reenter dormancy, they become less sensitive to light and hormones; therefore, increased amounts of light or hormones are required to stimulate germination. Also, seeds become more sensitive to soil moisture; thus, they will not germinate under water stress or on a water-saturated substrate. At any point in time, however, a population of seeds will exhibit a normal distribution in their sensitivity to an environmental factor (Bradford, 1996).

V. ENVIRONMENTAL FACTORS CAUSING CHANGES IN DORMANCY STATES

Seeds of a few species, e.g. *Mesembryanthemum nodiflorum* (Gutterman 1980–1981) and *Poa trivialis* (Froud-Williams *et al.*, 1986), stored dry at ambient laboratory temperatures have an annual rhythm of dormancy and nondormancy that is maintained for many years. For the most part, however, changes in dormancy states occur in response to changes in environmental factors. The major environmental factor causing changes in dormancy states is temperature, but other factors, including darkness, light, gases, chemicals, and water, may be important (Baskin and Baskin, 1987a). These factors can cause changes in dormancy states or prevent germination until dormancy is induced by changes in temperature. In this discussion, effects of each environmental factor on changes in dormancy states will be

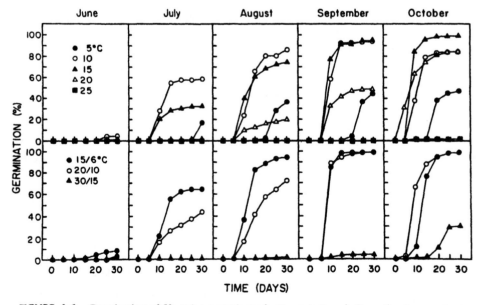

FIGURE 4.6 Germination of *Veronica arvensis* seeds at constant and alternating temperatures at a 14-hr daily photoperiod after 0–4 months of storage at 50% relative humidity at 25 °C. Modified from Baskin and Baskin (1983b).

TABLE 4.3   Physiological Correlates of the Dormancy Continuum

Physiological correlate	Decrease in dormancy (D → CD → ND)	Increase in dormancy (ND → CD → D)	Reference
Germination %	I[a]	D[b]	Baskin and Baskin (1983b)
Germination rate	I	D	Baskin and Baskin (1983b)
Temperature range of germination	I	D	Vegis (1963)
Germination at constant temperatures	I	D	Ghersa et al. (1992)
Sensitivity to light			
Light-promoted seeds	I	D	Taylorson (1970)
Light-inhibited seeds	(D)[c]	(I)[c]	?[c]
Sensitivity to			
Substrate moisture	D	I	Pemadasa and Lovell (1975)
Hormones	I	D	Vidaver and Hsiao (1974)
Light and hormones	I	D	Speer et al. (1974)
Anesthetics	I	D	Adkins et al. (1984)
Growth potential	I	D	Powell et al. (1984)

[a] Increase.
[b] Decrease.
[c] Data not available.

evaluated in terms of how timing of germination is controlled.

A. Temperature

1. Dormancy Loss at High Temperatures

Seeds of obligate (or strict) winter annuals (Table 4.4) germinate in autumn, and plants overwinter as rosettes or semirosettes and flower and set seeds the following spring and/or summer. However, most seeds of facultative winter annuals germinate in autumn, but some also germinate in spring. Plants from seeds that germinate in autumn flower, set seeds, and die the following growing season, like those of obligate winter annuals. Plants from seeds that germinate in spring flower, set seeds, and die in late spring and/or summer, thus behaving as short-lived summer annuals or ephemerals. Seeds of most obligate and facultative winter annuals mature in spring and/or early summer and are either dormant or conditionally dormant. If seeds are conditionally dormant, they germinate only at temperatures (e.g., 5, 10, 15, 15/6, 20/10°C) that do not occur in the habitat in late spring or summer (Baskin and Baskin, 1983b; Baskin et al., 1993).

Dormant or conditionally dormant seeds of obligate or facultative winter annuals become nondormant at high (e.g., 25/15, 30/15, 35/20°C), but not at low (e.g., 5, 15/6°C) temperatures (Baskin and Baskin, 1986a). Thus, seeds of winter annuals must be exposed to high summer temperatures for several months to germinate at autumn temperatures in autumn (Baskin and Baskin, 1984a,b; Roberts and Neilson, 1982a; Standifer and Wilson,

1988). Further, dormancy loss in seeds of the winter annual *Avena fatua* is optimal at seed moisture contents of 7–22% (Foley, 1994). Thus, both the high temperature and the low soil moisture conditions in habitats of many winter annuals during summer (e.g., Evans et al., 1975) probably promote loss of dormancy. Because seeds of winter annuals receive dormancy-breaking

TABLE 4.4   Examples of Obligate and Facultative Winter Annuals

Type of life cycle	Species	Reference
Obligate		
	Arabidopsis thaliana	Baskin and Baskin (1983a)
	Krigia oppositifolia	Baskin et al. (1981)
	Lamium purpureum	Baskin and Baskin (1984a)
	Leavenworthia exigua	Baskin et al. (unpublished results)
	Lesquerella lescurii	Baskin et al. (1992)
	L. stonensis	Baskin and Baskin (1990c)
	Phacelia dubia	Baskin and Baskin (1973b)
	Specularia perfoliata	Baskin and Baskin (unpublished results)
Facultative		
	Aphanes arvensis	Roberts and Neilson (1982b)
	Capsella bursa-pastoris	Baskin and Baskin (1989b)
	Lepidium virginicum	Baskin and Baskin (unpublished results)
	Lamium amplexicaule	Baskin and Baskin (1981a)
	Thlaspi arvense	Baskin and Baskin (1989a)
	Veronica hederifolia	Roberts and Lockett (1978a)
	V. arvensis	Baskin and Baskin (1983b)

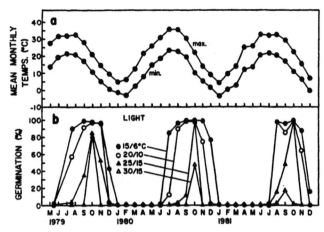

FIGURE 4.7 Annual dormancy/nondormancy cycle in buried seeds of the winter annual *Arabidopsis thaliana*. (a) Mean daily maximum and minimum monthly temperatures that seeds were exposed to during burial. (b) Germination of seeds incubated at a 14-hr daily photoperiod for 15 days following various periods of burial. From Baskin and Baskin (1985a). Seeds incubated in darkness did not germinate.

treatments in the field during summer, they are nondormant by autumn and can germinate if light and soil moisture are nonlimiting.

As dormancy loss occurs in seeds of winter annuals, they first gain the ability to germinate at low (e.g., 15/6, 20/10°C) temperatures, and then with additional dormancy loss also at high (e.g., 25/15, 30/15°C) temperatures, i.e., they have a type 1 response pattern (Tables 2.2 and 2.4, Fig. 4.7). As dormancy loss progresses, the rate of germination increases (Allen *et al.*, 1995) and seeds of some species lose their light requirement for germination (Corbineau *et al.*, 1992). Germination is prevented during summer because the maximum temperatures at which seeds can germinate are below those occurring in the habitat (Baskin and Baskin 1982a). Seeds germinate in autumn because the maximum tem-

perature for germination has increased and habitat temperatures have declined until there is an overlap between the two (Fig. 4.8). For example, the autumn germination season begins for seeds of *Avena* spp. when maximum and minimum air temperatures decrease to below 20 and 9°C, respectively (Aibar *et al.*, 1991). After temperatures decrease in autumn, germination begins as soon as soil moisture becomes nonlimiting (Popay, 1981).

If environmental conditions (e.g., burial in soil) prevent seeds of obligate winter annuals from germinating in autumn, low temperatures during winter induce them into dormancy (Baskin and Baskin, 1973b, 1984a; Haferkamp *et al.*, 1994). Consequently, viable seeds that fail to germinate in autumn cannot germinate in spring because they are dormant. As seeds reenter dormancy, they first lose the ability to germinate at high and then at low temperatures (Fig. 4.7), just the opposite of what happens as they come out of dormancy.

The seasonal temperature cycle is responsible for the annual dormancy cycle in seeds of *Arabidopsis thaliana*, and germination increased after exposure to high summer temperatures and decreased after exposure to low winter temperatures (Derkx and Karssen, 1994). Also, seeds became increasingly sensitive to light as they came out of dormancy and decreasingly sensitive to it as they reentered dormancy. Gibberellic acid (GA) biosynthesis does not play a role in the annual dormancy cycles of this species; seasonal dormancy cycles were the same in seeds of wild-type and GA-deficient mutants (Derkx and Karssen, 1994).

As seeds of facultative winter annuals come out of dormancy, they first gain the ability to germinate at low temperatures, and with an additional loss of dormancy they also germinate at high temperatures. However, if seeds fail to germinate in autumn, they lose the ability

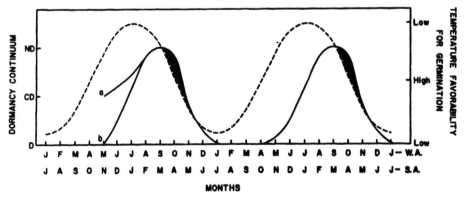

FIGURE 4.8 The annual dormancy cycle in buried seeds of obligate winter annuals (W.A.) and of spring-germinating summer annuals (S.A.). Hatched area shows when germination is possible. D, dormant; CD, conditionally dormant; ND, nondormant; solid line, dormancy continuum; dotted line, temperature favorability for germination. From Baskin and Baskin (1985a).

to germinate at high, but not at low, temperatures during winter and/or early spring (Fig. 4.9). Seeds of many facultative winter annuals enter (secondary) conditional dormancy when exposed to low (e.g., 5°C) winter temperatures (Baskin and Baskin, 1984b). However, seeds of the facultative winter annual *Capsella bursa-pastoris* entered conditional dormancy as temperatures increased in spring (Baskin and Baskin, 1989b). The ability to germinate at low temperatures means seeds can germinate at prevailing habitat temperatures during spring, if light and soil moisture are nonlimiting (Baskin *et al.*, 1986). Montegut (1975) lists a number of temperate and Mediterranean annuals whose seeds germinate in the field only in autumn and others that germinate mostly in autumn but also in spring. However, studies have not been done to determine what kind of dormancy cycle these seeds have.

In some facultative winter annuals, high percentages of the seeds enter conditional dormancy during winter, and the remainder enter dormancy. In other species, only a low percentage of the seeds enters conditional dormancy during winter, with the remainder entering dormancy (Baskin and Baskin, 1983b). However, some *Viola arvensis* seed lots have an annual dormancy/non-

dormancy cycle, whereas others have an annual conditional dormancy/nondormancy cycle (Baskin and Baskin, 1995). Further, the type of cycle found in seeds of this species during the first year of burial persisted during subsequent years.

Seeds of the facultative winter annual *Lamium amplexicaule* produced in spring require exposure to high summer temperatures to germinate to 80–100% at autumn temperatures in autumn. However, seeds given cold stratification for 5 months germinated to 65% at 15/6°C (Baskin and Baskin, 1984b). Plants from *L. amplexicaule* seeds that germinate in early autumn can grow large enough to produce some seeds before the onset of winter. About 80% of these autumn-produced seeds came out of dormancy during cold stratification in winter and germinated (at low temperatures) in early spring (Baskin and Baskin, 1981a). Seeds from low dormancy subpopulations of *Ranunculus sceleratus* were much more likely to come out of dormancy during cold stratification than those from high dormancy subpopulations. Cold stratification apparently induced dormancy in seeds from high dormancy subpopulations (Probert *et al.*, 1989).

Spergula arvensis has characteristics of both winter and summer annuals, with dormancy break occurring at relatively low temperatures and dormancy induction at relatively high temperatures (Bouwmeester and Karssen, 1993b). These authors found that dormancy loss took place in spring at temperatures of 10–15°C, which are lower than the optimum dormancy-breaking temperatures for most winter annuals. Further, dormancy induction in *S. arvensis* seeds occurred in autumn, at temperatures (10–15°C) that broke dormancy in spring. These dormancy-inducing temperatures are lower than those that usually induce dormancy in seeds of summer annuals (Bouwmeester and Karssen, 1993b).

Seeds of some facultative winter annuals, e.g., *Helenium amarum* (Baskin and Baskin, 1973a), mature in late summer and are nondormant. If these nondormant seeds are dispersed in early autumn, they germinate immediately, and plants behave as winter annuals. However, if seed dispersal occurs in late autumn, germination is delayed until spring, and plants behave as summer annuals. Seeds overwintering in the field are not induced into dormancy by the low temperatures.

Less information is available on dormancy breaking and germination requirements of seeds of perennials that become nondormant during summer than for winter annuals. Only 6 of 73 (8%) species of herbaceous polycarpic perennials (*Delphinium virescens*, *Lychnis alba*, *Plantago lanceolata*, *Potentilla recta*, *Satureja glabella*, and *Scutellaria parvula*) and 1 of 18 (6%) species of monocarpic perennials (*Dianthus armeria*) with non-deep PD studied in temperate eastern North America

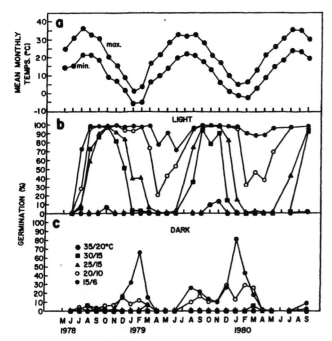

FIGURE 4.9 Annual conditional dormancy/nondormancy cycle in buried seeds of the facultative winter annual *Lamium amplexicaule*. (a) Mean daily maximum and minimum monthly temperatures that seeds were exposed to during burial. (b) Germination of seeds incubated at a 14-hr daily photoperiod for 15 days following various periods of burial. (c) Germination of seeds incubated in darkness for 15 days following various periods of burial. Modified from Baskin and Baskin (1981a).

(Baskin and Baskin, 1988) had seeds that became non-dormant during summer. Seeds of *Narcissus bulbocodium* from the Iberian Peninsula, southwestern France, and northwestern Africa required high temperatures for loss of dormancy and germinated in the autumn at the start of the moist season (Thompson, 1977). Seeds of *Cenchrus ciliaris* required high temperatures for loss of dormancy (60°C for 12 weeks promoted about 80% germination), and germination occurred in the field in Australia during autumn (Hacker, 1989). Seeds of *Chrysopogon fallax, C. latifolius, Sorghum plumosum,* and *Themeda australis* came out of dormancy during dry storage at 60/20°C or in the field during the dry season in tropical Australia (Mott, 1978). Achenes of six perennial Asteraceae from southcentral Texas came out of dormancy when subjected to simulated summer habitat temperatures (27/15 to 35/20°C) and germinated to 50–100% in autumn (Baskin *et al.,* 1994).

Little is known about what happens to seeds of polycarpic and monocarpic perennials if they come out of dormancy in summer but fail to germinate in autumn. Buried seeds of the polycarpic perennial *Potentilla recta* became nondormant during summer and remained nondormant during 27 months of burial (Baskin and Baskin, 1990a).

2. Dormancy Loss at Low Temperatures

Seeds of summer annuals germinate in spring or summer, and the resulting plants complete their life cycle before, or at the time of, frost in autumn. At maturity in autumn, seeds are dormant or conditionally dormant, depending on the species. If seeds of summer annuals are conditionally dormant at maturity in autumn, they will germinate at high (35/20, 30/15°C) but not at low (15/6, 20/10°C) temperatures. Thus, conditionally dormant seeds do not germinate in autumn because habitat temperatures are below those required for germination.

During winter, dormant seeds of summer annuals become conditionally dormant and then eventually nondormant. When seeds first become conditionally dormant, they will germinate only at high temperatures. With additional dormancy loss, however, the minimum temperature at which seeds can germinate declines (Baskin and Baskin, 1983d); this is a type 2 response pattern (Fig. 4.5). When seeds become nondormant, they can germinate over a range of temperatures, including those in the habitat in spring. Seeds that are conditionally dormant in autumn also come out of dormancy during winter, and as they do so the minimum temperature for germination declines. Thus, by spring these seeds germinate at temperatures (e.g., 15/6°C) that were too low for germination the preceding autumn (Baskin and Baskin, 1981b). In temperate regions, seeds of some

species of summer annuals germinate only in spring, whereas those of other species germinate in spring and summer (Table 4.5). Seeds of both groups of summer annuals are either dormant or conditionally dormant in autumn, depending on the species.

In temperate regions, seeds of summer annuals that germinate in spring as well as those that germinate in spring and summer require exposure to low winter temperatures to come out of dormancy (Fig. 4.10). Seeds of the spring- and summer-germinating summer annual *Chenopodium album* stratified at a low temperature (5°C) germinated over a range of temperatures, whereas those stratified at a high temperature (35/20°C) germinated only at high temperatures (Fig. 4.10). In contrast, seeds of *C. album* from Japan stratified at low temperatures germinated only at low temperatures, whereas those stratified at high temperatures germinated only at high temperatures (Matsuo and Kubota, 1988).

If seeds of spring-germinating species of summer annuals fail to germinate in spring, they reenter dormancy in spring or early summer and must be cold stratified again the following winter to come out of dormancy (Courtney, 1968; Watanabe and Hirokawa, 1975; Baskin and Baskin, 1980). Reinduction of dormancy began in March (early spring) and was completed by early June in buried seeds of *Ambrosia artemisiifolia* exposed to natural seasonal temperature changes (Fig. 4.11). In the laboratory, nondormant seeds of *A. artemisiifolia* held at 15/6°C (March temperatures) for 1 month lost the

TABLE 4.5 Examples of Spring-Germinating and Spring- and Summer-Germinating Summer Annuals[a]

Germination season	Species
Spring	*Ambrosia artemisiifolia*
	Aristida longespica
	Croton capitatus
	Diodia teres
	Helianthus annuus
	Heliotropium tenellum
	Isanthus brachiatus
	Polygonum pensylvanicum
	Salvia reflexa
	Sporobolus vaginiflorus
Spring and summer	*Amaranthus* spp.
	Artemisia annua
	Chenopodium album
	Cyperus inflexus
	Euphorbia maculata
	Galinsoga spp.
	Leucospora multifida
	Mollugo verticillata
	Panicum capillare
	Portulaca oleracea

[a] From Baskin and Baskin (1988).

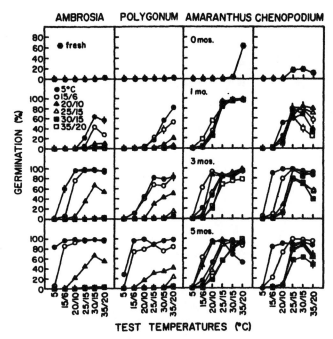

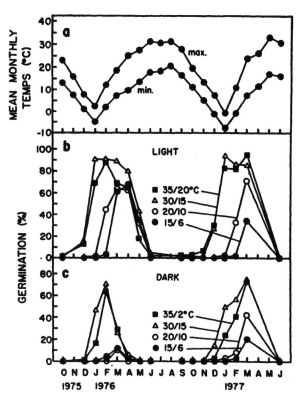

FIGURE 4.10 Germination of seeds of the spring-germinating summer annuals *Ambrosia artemisiifolia* and *Polygonum pensylvanicum* and the spring- and summer-germinating summer annuals *Amaranthus hybridus* and *Chenopodium album*. Seeds were buried in soil at 5, 15/6, 20/10, 25/15, 30/15, and 35/20°C for 0, 1, 3, and 5 months and then exhumed and tested at a 14-hr photoperiod at the six temperature regimes. From Baskin and Baskin (1987b).

FIGURE 4.11 The annual dormancy/nondormancy cycle in buried seeds of the spring-germinating summer annual *Ambrosia artemisiifolia*. (a) Mean daily maximum and minimum monthly temperatures seeds were exposed to during burial. (b) Germination of seeds incubated at a 14-hr daily photoperiod for 15 days following various periods of burial. (c) Germination of seeds incubated in darkness for 15 days following various periods of burial. Redrawn and modified from Baskin and Baskin (1985a).

ability to germinate in darkness. Then, when seeds sequentially were subjected to 20/10°C (April temperatures) for 1 month and 25/15°C (May temperatures) for 1 month, they lost the ability to germinate in light (Baskin and Baskin, 1980). Thus, in the field, seeds are dormant by late spring–early summer and cannot germinate even if exposed to light and adequate soil moisture. Field studies in Indiana support this model. Soil disturbance in autumn or winter resulted in good populations of *A. artemisiifolia* becoming established in spring, but when soil disturbance was delayed until early summer, no seeds germinated (Squires, 1989).

Seeds of spring- and summer-germinating species of summer annuals do not reenter dormancy in spring; this is why they can germinate during summer. However, seeds of *Amaranthus hybridus* (not *A. retroflexus,* as reported) (Baskin and Baskin, 1977), *Bidens polylepis* (Baskin *et al.,* 1995), *Chenopodium album* (Baskin and Baskin, 1977), *Cyperus inflexus* (Baskin and Baskin, 1978), *Euphorbia supina* (now *E. maculata*) (Baskin and Baskin, 1979a), *Solanum nigrum* (Roberts and Lockett, 1978b), *S. sarrachoides* (Roberts and Boddrell, 1983), *Panicum dichotomiflorum* (Baskin and Baskin, 1983d), *P. capillare* (Baskin and Baskin, 1986b), and *Portulaca smallii* (Baskin *et al.,* 1987) lose the ability to germinate

at some, or all, temperatures by autumn (Fig. 4.12). The consequence of seeds entering dormancy or conditional dormancy by autumn is that germination is prevented at the end of the growing season.

Induction of dormancy or conditional dormancy in seeds of spring- and summer-germinating summer annuals is controlled by temperature. Maximum soil temperature was the factor inducing dormancy in summer annuals that stopped germinating in late July or in late August (summer) in Japan, and seeds entered dormancy earlier at 0–5 cm than at 10- to 15-cm soil depths (Watanabe and Hirokawa, 1975). Nondormant, upper seeds from *Xanthium pensylvanicum* burs (each bur has an upper and a lower seed) entered dormancy at 27–28°C, and 3 months of cold stratification at 5°C were required for loss of dormancy (Thornton, 1935). Seeds (achenes) of *Helianthus annuus* were induced into dormancy after 6 days at 45°C (Corbineau *et al.,* 1988). Dormancy induction occurred in *Polygonum persicaria* seeds buried in soil in The Netherlands when temperatures exceeded 10–15°C, and dormancy break took place at tempera-

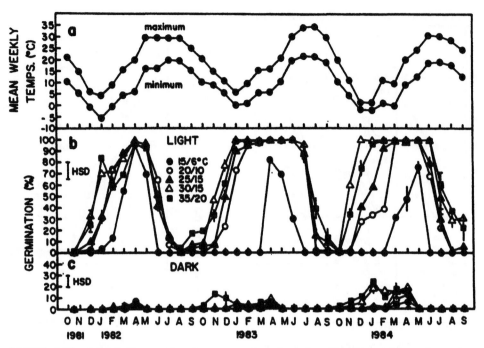

FIGURE 4.12 Annual dormancy/nondormancy cycle in buried seeds of the spring- and summer-germinating summer annual *Panicum capillare*. (a) Mean daily maximum and minimum monthly temperatures seeds were exposed to during burial. (b) Germination of seeds incubated at a 14-hr daily photoperiod for 15 dyas following various periods of burial. (c) Germination of seeds incubated in darkness for 15 days following various periods of burial. From Baskin and Baskin (1986b).

tures lower than these. Further, changes in dormancy states of *P. persicaria* (Bouwmeester and Karssen, 1992) and *Sisymbrium officinale* (Bouwmeester and Karssen, 1993c) seeds were not influenced by nitrate or soil moisture content. Low temperatures broke and high temperatures induced dormancy in *S. officinale* seeds (Bouwmeester and Karssen, 1993c). Dormancy induction in *Bidens polylepis* seeds occurred during exposure to natural temperatures for most of the summer or during exposure to 25/15 or 30/15°C for 24 and 12 weeks, respectively (Baskin *et al.*, 1995).

Although seasonal temperature cycles are the primary reason for changes in dormancy states of buried seeds, environmental factors such as desiccation and/or nitrate enrichment may influence the dormancy pattern of seeds by effectively extending the germination season (Bouwmeester and Karssen, 1989). In the case of *Spergula arvensis*, it is predicted that seeds will germinate in moist, nitrate-poor sites in autumn, whereas in alternately wet and dry, nitrate-rich sites they will germinate from early spring to late autumn (Karssen *et al.*, 1988).

Many herbaceous polycarpic and monocarpic perennials in temperate regions have seeds with nondeep PD, which become nondormant during winter, exhibiting a decrease in the minimum temperature for germination, i.e., type 2 response (Baskin and Baskin, 1988). Thus,

seeds would be nondormant in spring. However, if seeds of perennials fail to germinate the first spring following dispersal, what happens to them? Do they reenter dormancy? A few studies have been done to determine whether buried seeds of perennials that come out of dormancy in winter undergo seasonal changes in their germination responses. Buried seeds of the polycarpic perennials *Rumex crispus* (Baskin and Baskin, 1985b), *Cyperus odoratus,* and *Penthorum sedoides* (Baskin *et al.,* 1989) came out of dormancy and remained nondormant, whereas those of *Gratiola viscidula, Scirpus lineatus* (Baskin *et al.,* 1989), and *Rumex obtusifolius* (van Assche and Vanlerberghe, 1989) had an annual nondormancy/conditional dormancy cycle. Buried seeds of the monocarpic perennials *Verbascum thapsus, V. blattaria* (Baskin and Baskin, 1981c; Vanlerberghe and van Assche, 1986), and *Oenothera biennis* (Baskin and Baskin, 1994) had an annual conditional dormancy/nondormancy cycle in light but an annual dormancy/nondormancy cycle in darkness (Fig. 4.13). Exhumed seeds of *Lychnis flos-cuculi* incubated in light and darkness at 20/8°C germinated to about 10% throughout the year but exhibited a cyclic pattern of germination in darkness, with the peak occurring in spring (Milberg, 1994a).

Seeds of *R. crispus* entered dormancy when held at constant temperatures from 10 to 37°C (Le Deunff and

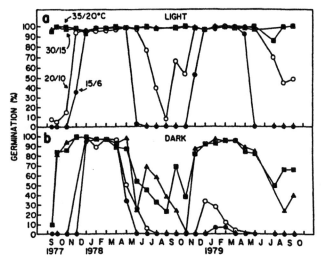

FIGURE 4.13 Annual conditional dormancy/nondormancy cycles in buried seeds of the monocarpic perennial *Verbascum thapsus*. (a) Germination of seeds incubated at a 14-hr daily photoperiod for 15 days following various periods of burial. (b) Germination of seeds incubated in darkness for 15 days following various periods of burial. From Baskin and Baskin (1981c).

Chaussant, 1968; Taylorson and Hendricks, 1973). However, in nature, seeds are unlikely to be exposed to constant temperatures. *Verbascum thapsus* seeds cold stratified at 4°C became nondormant, whereas those exposed to 10 and 20°C entered conditional dormancy; the higher the temperature, the deeper the degree of conditional dormancy. Conditional dormancy subsequently was reversed in seeds cold stratified at 4°C for 1 month (Vanlerberghe and van Assche, 1986). Imbibed seeds of *V. blattaria* kept in darkness at 27°C for 24 days lost the ability to germinate at 25°C in light; however, after 72 hr of cold stratification at 5°C they germinated to nearly 100% at 25°C in light (Kivilaan, 1975). Seeds of *Rosa setigera* cv. Beltsville and *Rosa reversa* germinated to 80% or more after 120 days of cold stratification at 4.4°C; however, seeds given five cycles of 30 days at 4.4°C and then 15 days at 15.6–18.3°C in a greenhouse germinated to only about 25 and 45%, respectively (Semeniuk and Stewart, 1962). Little or no additional germination occurred in seeds receiving cold–warm cycles after they had received three cycles. Thus, the authors interpreted lack of any increase in germination after the fourth and fifth cycle to mean that seeds had entered secondary dormancy.

Seeds of many trees and shrubs in temperate regions have physiological dormancy (probably nondeep PD) and come out of dormancy in winter (see Tables 10.8–10.11). Temperature requirements for germination decrease in seeds of some species following cold stratification (Junttila, 1976). No studies have been done on buried seeds of trees or shrubs to determine if they have annual cycles in their dormancy states.

Temperatures unfavorable for germination may delay germination until subsequent environmental conditions induce dormancy. For example, the amplitude of the daily temperature regime plays an important role in determining whether nondormant seeds will germinate. In seeds of *Rumex obtusifolius*, a 15-min exposure to 45°C or a daily alternating temperature regime with an amplitude of 10°C or more (25°C was daily high temperature) resulted in high germination percentages (Felippe, 1978). Van Assche and Vanlerberghe (1989) viewed the alternating temperature requirement for germination in seeds of this species to be a mechanism to detect depth in soil, presence of gaps in vegetation, and change in seasons. In fact, sensitivity to temperature fluctuations in darkness may be a mechanism for depth detection in buried seeds of many species (Thompson and Grime, 1983). A single high temperature (32–40°C) treatment of *Mallotus japonicus* seeds resulted in germination, thus high fluctuating temperatures may be a means of gap detection in seeds of this species (Washitani and Takenaka, 1987).

Amplitude of the daily temperature fluctuation, as well as magnitude of maximum daily temperature, is important in the germination of *Sorghum halepense* seeds, and they regulate germination at increasing depths (Ghersa *et al.*, 1992) and under leaf canopies (Benech Arnold *et al.*, 1988). However, effects of daily temperature cycles are additive in seeds of *S. halepense*; after exposure to many cycles, seeds can germinate at constant temperatures (Benech Arnold *et al.*, 1990a). Soil temperatures have been used to construct a predictive model of germination of buried *S. halepense* seeds. This model uses the number of alternating temperature cycles rather than the amplitude of daily temperature fluctuations (Benech Arnold *et al.*, 1990b). As seeds of *S. halepense* age, however, they become less sensitive to temperature fluctuations and germinate while buried in soil at depths of 0.5, 5.0, and 15.0 cm (Ghersa *et al.*, 1992).

3. Models Related to Changes in Dormancy States and Temperature

Hilhorst's (1993) model attempts to explain the annual dormancy cycle in light-requiring seeds. During the dormancy-breaking treatment (winter or summer environmental conditions, depending on the species), phytochrome receptors are formed in cell membranes. When temperatures become favorable for germination, membrane consistency changes, allowing receptors to move to the surface, where they are activated by nitrate. An activated receptor binds with the pigment phytochrome, which subsequently becomes activated during irradiation of the seed. GA is produced after phytochrome is activated and binds with a GA receptor. The GA–receptor complex produces a (chemical?) signal that stimulates germination, thus seeds are nondormant. When temperatures become too low or high for germination for extended periods of time, the synthesis of phytochrome receptors stops and seeds reenter dormancy.

A model based on Hilhorst's physiological model was used to simulate the annual dormancy cycle for buried seeds of the summer annual *Polygonum persicaria*. Data from germination studies on *P. persicaria* seeds exhumed at regular intervals for 3 years were very similar to the simulated results (Vleeshouwers and Bouwmeester, 1993). In this model, temperature was the most important factor causing changes in the dormancy state.

Other models have considered changes that occur in germination responses of winter annual seeds during the high temperature dormancy-breaking period. Favier (1995) has examined increases in seed germination rates in winter annuals, especially *Hordeum vulgare*, during the dormancy-breaking period as a function of temperature. Mean germination time during a germination test is

$$\bar{t}_g(m) = t_m - \sum_{i=1}^{m-1} G_i/G_m,$$

where $\bar{t}_g$ is the mean germination time, t_m is the cumulative test time, G_m is the cumulative germination after the mth (final) interval, and G_i is the cumulative germination after the ith interval.

An equation for the temperature function looked at distribution of the dormancy period and the rate of change of germination time within the seed population:

$$\log_{10}\beta = K_{tg} + C_d T,$$

where K_{tg} is $(\log_{10}\alpha_1) + K_b$; β is $1/\alpha_1\sigma_d$; α_1 is a constant depending on genetic and/or phenological factors; C_d is a constant, which is a function of genotype; K_b is a constant, which is a function of phenotypic factors; σ_d is the standard deviation of distribution of the dormancy period among seeds in the population; and T is seed temperature (°C).

Favier (1995) concluded that temperature controls the removal of the inhibitive factor(s) for germination. After this inhibition is removed (in response to dormancy-breaking temperatures), seeds can germinate under favorable environmental conditions, and rate and percentage of germination are affected by the test temperature. Further, the influence of temperature in removing the inhibitive factor(s) for germination is normally distributed within the seed population.

Probit regression models were used to consider ef-

fects of temperature and water potential on dormancy loss and control of timing of germination in seeds of the winter annual *Bromus tectorum* (Christensen *et al.*, 1996). Probit (g) was regressed on $\Psi - \Theta_{HT}/[(T - T_b)t_g]$, which was equivalent to $\Psi_b(g)$. $\Psi_b(50)$ was allowed to vary with incubation temperature and stage of dormancy loss, and the authors regressed probit (g/g_m) for $0.05 < g/g_m < 0.95$ on $[\Psi - \Theta_{HT}/((T - T_b)t_g)] - \Psi_{adj}$, where g_m is the fraction of viable seeds in the population; Θ_{HT} is the amount of hydrothermal time (MPa-degree-day) that a seed must accumulate to germinate; Ψ is the water potential of incubation; Ψ_{adj} is the adjustment term, which varies with each storage duration/incubation temperature combination; $\Psi_b(g)$ is the threshold water potential at or below which germination of the g fraction of the population will not occur; $\Psi_b(50)$ is the mean (median) base water potential of the population, which varies as a function of duration of dry storage and the test temperature; T is the temperature (°C) of incubation; T_b is the minimum threshold temperature at which germination does not occur; and T_g is the actual time required for germination of the g fraction.

Germination time–course curves from experimental data for *Bromus tectorum* generally fit those generated by the models. Thus, changes in germination time–course curves during the dormancy-breaking period were explained to a large degree by decreases in the mean base water potential (Christensen *et al.*, 1996).

B. Darkness

One of the most obvious factors associated with the burial environment is darkness. Light penetrates only the top few millimeters of soil, and depth of penetration depends on soil particle size, moisture content, and color. Transmission of light through soil is much greater (1) with an increase in the size of soil particles (Mandoli *et al.*, 1982; Bliss and Smith, 1985; Benvenuti, 1995), (2) in light- than in dark-colored soils (Kasperbauer and Hunt, 1988; Benvenuti, 1995), and (3) in dry than in wet silty loam soils (Woolley and Stoller, 1978). Long light waves penetrate deeper into soil than short ones. Thus, little red light reaches the zone immediately above permanent darkness, and here the red : far-red ratio is reduced (Bliss and Smith, 1985; Tester and Morris, 1987). The depth to which enough light penetrates the soil to stimulate germination also is dependent on the species. Seeds of *Lactuca sativa* germinated under 2 but not 6 mm of soil (Woolley and Stoller, 1978); *Senecio jacobea* under 4 but not 8 mm of sand (Van der Meijen and van der Waals-Kooi, 1979); *Artemisia monosperma* under 2 but not 3 mm of natural desert sand (Koller *et al.*, 1964); *Chenopodium album* under 2 but not 4 mm of sand (Bliss and Smith, 1985); and *Rumex obtusifolius* under 4 but not 6 mm of sand (Bliss and Smith, 1985). However, seeds of *Digitalis purpurea* germinated to about 70% under 10 mm of sand at light levels of only 0.026 μmol m^{-2} sec^{-1}; dark controls germinated to 4% (Bliss and Smith, 1985).

Seeds that have the potential to germinate in darkness at some temperature(s) are induced into secondary dormancy if they are kept in darkness for prolonged periods of time at an inappropriate temperature for germination (Table 4.6). This type of secondary dormancy is called skotodormancy because seeds do not respond to red light; they also do not respond to GA (Powell *et al.*, 1983). The effective temperature for the induction of skotodormancy varies with the species and ranges from 15°C in *Lamium amplexicaule* (Taylorson and Hendricks, 1976) to 37°C in *Apium graveolens* (Biddington and Thomas, 1979). Also, the rate of induction into skotodormancy can be decreased by repeated exposures to red light (Kristie *et al.*, 1981).

Seeds of some species cannot be induced into skotodormancy, whereas those of others can be induced into it only at certain times of the year. Nondormant seeds of some species do not germinate in darkness at any temperature (Grime *et al.*, 1981), i.e., they have an absolute light requirement for germination. Seeds with an absolute light requirement for germination are not induced into dormancy by darkness, and annual changes in their dormancy state mostly are due to the seasonal temperature cycle. Seeds of some species germinate in darkness over a narrow range of temperatures (Evenari, 1952) and can be induced into dormancy if incubated in darkness at an unfavorable temperature for germination. Seeds of other species exhibit annual cycles in their ability to germinate in darkness (Baskin and Baskin, 1980, 1981a,c; Karssen, 1980, 1981b; Pons, 1991b), and it is only during the time of year when seeds can germinate in darkness that incubation in darkness at an unfavorable temperature for germination will induce them into skotodormancy.

Skotodormancy was induced in seeds of *Hygrophila auriculata* held in darkness at 28°C for 7 days, and it was characterized by a reduction in the growth potential of the embryo (Sharma and Amritphale, 1989). These authors suggested that the growth potential of the embryo decreased due to a loss of K$^+$ and other ions. Further, treatments with KNO$_3$ and KH$_2$PO$_4$ increased germination in light but not in darkness (Sharma and Amritphale, 1989). Seeds of *H. auriculata* incubated in darkness at 28°C for 0, 5, 10, and 20 days germinated to 70, 31, 14, and 17%, respectively, following a 10-min exposure to red light. These results indicate that skotodormancy increased with an increase in the length of the period of dark imbibition (Amritphale *et al.*, 1993). Seeds of *Lactuca sativa* induced into skotodormancy by darkness and supraoptimal temperatures first lost the ability to respond to GA and then to red light (Bewley, 1980). Because embryos dissected from these dormant seeds germinated (Bewley, 1980), skotodormancy was due to a loss of growth potential of the embryo. Also, lettuce seeds in skotodormancy germinated following a treatment with 1 N HCl, which weakened the embryo covers, thus permitting the embryo to grow (Hsiao *et al.*, 1984).

TABLE 4.6 Species Whose Seeds Enter Secondary Dormancy in Darkness at Temperatures Unfavorable for Germination in Darkness

Species	Temperature (°C)	Reference
Ambrosia artemisiifolia	25/15	Baskin and Baskin (1980)
Apium graveolens	37	Biddington and Thomas, 1979
Chenopodium album	25	Karssen (1967)
C. bonus-henricus	29	Khan and Karssen (1980)
Eragrostis curvula	24/18	Voigt (1973)
Hygrophila auriculata	28	Sharma and Amritphale (1989)
Kalanchoe blossfeldiana	20	Rethy *et al.* (1983)
Lactuca sativa	20	Vidaver and Hsiao (1974), Powell *et al.* (1983)
Lamium amplexicaule	15	Taylorson and Hendricks (1976)
Limnanthes alba	25	Nyunt and Grabe (1987)
Ocium americanum	25	Varshney (1968)
Oldenlandia corymbosa	35	Corbineau and Come (1985)
Plantago maritima	18	Arnold (1973)
Portulaca oleracea	35	Duke *et al.* (1977)
Rumex crispus	30	Taylorson and Hendricks (1973)

In addition to its role in the induction of skotodormancy, darkness could be a factor preventing germination until dormancy is induced by temperature and/or other factors. In some species, the inhibitory effects of darkness are enhanced by decreased oxygen and increased carbon dioxide levels (Popay and Roberts, 1970a). Although buried seeds of *Sinapis arvensis* were induced into a deeper state of dormancy than those on the soil surface, buried seeds retained the ability to respond to light. However, seeds of this species induced into dormancy by high carbon dioxide concentrations lost the ability to respond to light (Frankland, 1975).

Wesson and Wareing (1969b) reported that burial induced a light requirement for germination in seeds of 11 weedy species. One way in which a light requirement for germination could be induced in buried seeds is the conversion of Pfr to Pr (Pons, 1991a). However, the rate of induction of a light requirement for germination varies with the species and depends on preexistent Pfr in the seeds, whether they are exposed to light just prior to burial, the threshold level for Pfr action, and temperature (conversion of Pfr to Pr is faster at high than at low temperatures) (Pons, 1991a). One determinant of preexisting levels of Pfr in seeds is whether they were covered by green maternal tissues (fruits, bracts) (thus exposure to low red : far-red, low Pfr) as they matured or were exposed to white light (thus high Pfr) immediately prior to drying (Cresswell and Grime, 1981).

Burial increased light sensitivity in seeds of eight annual weeds, but there were seasonal variations in responses of exhumed seeds to light quality and quantity (Froud-Williams *et al.*, 1984). Light sensitivity of *Datura ferox* seeds increased 10,000-fold after 2 months of burial, consequently millisecond exposures to light stimulated 50% germination. Increased light sensitivity means that brief light exposures such as those associated with plowing could stimulate germination (Scopel *et al.*, 1991). In fact, cultivating soil at night (vs in the day) decreased the percentage of buried weed seeds that germinated (Scopel *et al.*, 1994) and the kinds and numbers of weeds growing in fields (Hartmann and Nezadal, 1990).

C. Light

Light can prevent the germination of negatively photoblastic (light inhibited) seeds (see Table 2.6). For example, the germination of *Citrullus lanatus* seeds was inhibited by white and far-red light but was stimulated by red light (Botha *et al.*, 1982a). Ethylene overcame the germination inhibition of *C. lanatus* seeds in white light (Botha *et al.*, 1982b) and promoted the germination of seeds in darkness (Botha *et al.*, 1983). Germination

TABLE 4.7 Some of the Many Species in Which Seed Germination Is Inhibited by Light Filtered through Green Leaves

Species	Reference
Agropyron repens	Gorski *et al.* (1977)
Amaranthus caudatus	Fenner (1980a)
A. patulus	Washitani and Saeki (1984)
Arenaria serpyllifolia	King (1975)
Bidens pilosa	Fenner (1980a)
Cecropia glaziovi	Valio and Joly (1979)
Cerastium holosteoides	King (1975)
Cirsium palustre	Pons (1983)
Erigeron alpinus	Gorski *et al.* (1977)
Lactuca sativa	Gorski (1975)
Nicotiana tobacum	Taylorson and Borthwick (1969)
Oenothera biennis	Gross (1985)
Pinus densiflora	Washitani and Saeki (1986)
Piper auritum	Vazquez-Yanes (1980)
Prunella vulgaris	Silvertown (1980)
Ruellia tuberosa	van der Veen (1970)
Rumex crispus	Gorski *et al.* (1977)
Taraxacum officinale	Gorski (1975)
Tussilago farfara	Gorski *et al.* (1977)
Urera caracasana	Orozco-Segovia *et al.* (1987)
Veronica arvensis	King (1975)

of *Cucumis anguria* seeds was inhibited by white, blue, and far-red light but was promoted by darkness and red light (Noronha *et al.*, 1978). In the field, exposure to light could prevent the germination of negatively photoblastic seeds until after factors such as high or low temperatures induced them into secondary dormancy.

Direct solar radiation may inhibit the germination of some seeds. For example, natural sunlight at high irradiances inhibited the germination of *Lactuca sativa* seeds, but low irradiances promoted it. Seeds at high irradiances germinated after transfer to darkness or very low irradiances (Gorski and Gorska, 1979). *Oldenlandia corymbosa* seeds germinated to high percentages at high irradiances of white light at 32, 35, and 40°C but not at 17, 20, 24, and 26°C. However, seeds incubated at high irradiances at 20 and 25°C germinated at these temperatures after they were transferred to darkness (Corbineau and Come, 1982). Some nonimbibed seeds of *Bromus sterilis* exposed to near-full sunlight were induced into dormancy; however, it was broken in seeds subsequently incubated on a wet substrate in darkness (Pollard, 1982).

Sunlight filtered through green leaves inhibits the seed germination of many species (Table 4.7) because a much higher proportion of red than far-red light is absorbed thus reducing the red : far-red ratio (Fig. 4.14). Cold stratification decreased the inhibitory effects of leaf-filtered sunlight on the germination of *Plantago major* seeds. Also, at low red : far-red ratios, germination increased with an increase in amplitude of daily temperature fluctuations and with a decrease in water stress (Pons, 1986). *Poa trivialis* seeds collected from grasslands were not as strongly inhibited by leaf-filtered light as were those collected from open arable habitats (Hilton *et al.*, 1984).

Seeds of some species are not inhibited by leaf-filtered light (Gorski *et al.*, 1977). This is not surprising, however, in view of the fact that far-red light has been shown to promote (e.g., Downs, 1964; Dixit and Amritphale, 1996) and red light to inhibit seed germination in a few species (Hilton, 1984a). Low red : far-red ratios decreased the germination of freshly matured seeds of *Scabiosa columbaria* and *Carlina vulgaris* more at 15 than at 20°C, but had little or no effect on the germination of those of *Picris hieracioides* and *Daucus carota*. Cold stratification did not change the effect of low red : far-red ratios on *D. carota* and *S. columbaria* seeds, eliminated the effect of low ratios on *C. vulgaris* seeds, and increased the sensitivity of *P. hieracioides* seeds to low ratios (Senden *et al.*, 1986).

Only 1–2 hr of leaf-filtered sunlight were required to induce a light requirement for germination in *Bidens pilosa* seeds (Fenner, 1980b), and a germination inhibitor diffused from *Amaranthus patulus* seeds exposed to filtered sunlight but not to dim white light (Washitani, 1985). *Avena fatua* (Hou and Simpson, 1990) and *Lactuca sativa* (Blaauw-Jansen, 1981) seeds enter dormancy

during long exposures to far-red light, and leaf-filtered sunlight inhibits the germination of *L. sativa* seeds (Gorski, 1975). Thus, seeds of a few species appear to be induced into secondary dormancy by filtered sunlight. However, the usual effect of filtered sunlight on seeds is the prevention of germination rather than the induction of dormancy. It seems possible that during the inhibition of germination by filtered sunlight other environmental factors could induce seeds into dormancy.

Leaf-filtered sunlight may make seeds more sensitive to other factors such as soil moisture. An increase in the osmotic concentration of the germination medium of *Chenopodium album* seeds resulted in an increase in the amount of Pfr required for germination (Karssen, 1970), and a reduction in the water content of *Rumex crispus* seeds decreased the effectiveness of Pfr in promoting germination (Duke, 1978). Red light increasingly stimulated the production of Pfr in seeds of *Lactuca sativa* as moisture content rose from 8 to 15%, whereas far-red light stimulated the production of Pr at seed moisture contents of 4–32% (Vertucci *et al.*, 1987). Seeds of *Origanum vulgare* and *Plantago major* became less responsive to treatment with red light with an increase in imbibition time in a polyethylene glycol (PEG) solution at −1.2 MPa (Pons, 1991a). Thus, in leaf-filtered light there may not be formation of enough Pfr under low soil moisture conditions for germination to occur.

Treatment with simulated leaf-filtered light had a persistent inhibitory effect on the germination of *Agrostis capillaris* seeds, and the inhibition increased with a decrease in the amplitude of daily temperature fluctuations (Williams, 1983). That is, simulated gaps (high amplitude of daily temperature fluctuation) promoted higher germination percentages than simulated canopy conditions. Thus, far-red light coupled with a low amplitude of daily temperature fluctuations may prevent the germination of some seeds under forest canopies.

D. Gases

Oxygen, carbon dioxide, and ethylene occur in the gaseous environment of soils, and each may influence seed dormancy and/or germination. Oxygen is required for the breaking (Brennan *et al.*, 1978) and induction of dormancy (Davis, 1930a,b; Le Deunff, 1973; Vidaver and Hsiao, 1975) in seeds of many species. However, temporary anaerobiosis may promote germination in dormant seeds of some species (Come *et al.*, 1985; Lonchamp and Gora, 1979) or substitute for cold stratification (Come *et al.*, 1991; Corbineau and Come, 1995). The upper seed from *Xanthium pensylvanicum* burs soaked in water at 23°C for 22 days and then kept under

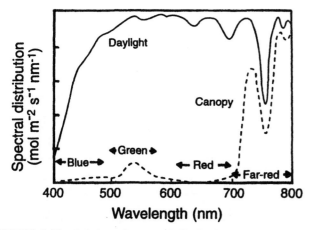

FIGURE 4.14 Relative solar spectral distribution of photons above and within a vegetation canopy. Arrows indicate the main spectral regions. Note the almost complete lack of red and blue light within the canopy and the relative enrichment of far-red. Modified from Smith (1993).

anaerobic conditions for 4 days at this temperature germinated to near 100% at 23°C following 1 day of cold stratification. However, seeds not given 1 day of cold stratification germinated to only about 25% (Esashi *et al.*, 1978). The response of nondormant seeds to low oxygen levels is variable. For example, whereas seeds of *Alopecurus myosuroides, Rumex obtusifolius, Matricaria chamomilla, Poa annua, Apera spica-venti*, and *Veronica persica* germinated to about 75% at 4.5–8% oxygen, those of *Galium aparine, Sonchus aspera, Chenopodium album, Stellaria media*, and *Avena fatua* required 12–16% oxygen for comparable germination (Mullverstedt, 1963).

Low oxygen levels may cause seeds to enter secondary dormancy. Exposure of seeds of *Limnanthes alba* to 2% oxygen for 9 days caused a 20% reduction in germination, indicating that some seeds may have entered secondary dormancy (Nyunt and Grabe, 1987). Oxygen deficiency induced seeds of *Lobelia dortmanna* into secondary dormancy, and cold stratification was required to break it (Farmer and Spence, 1987). Seeds of *Echinochloa crus-galli* were induced into secondary dormancy by oxygen deficiency treatments (submerged in water) at 25°C but not at 7°C. In fact, seeds submerged in water at 7°C came out of dormancy (Honek and Martinkova, 1992). It should be noted that flooding reduces oxygen levels to almost 0% (Burges, 1967).

Seeds of *Oldenlandia corymbosa* induced into secondary dormancy by temperatures ≥30°C lose their responsiveness to light, and this happened faster at 0% than at 7% oxygen (Corbineau and Come, 1985). Seeds of *Xanthium pensylvanicum* entered secondary dormancy in atmospheres of pure nitrogen, hydrogen, carbon dioxide, or in mixtures of carbon dioxide and nitrogen, but dormancy induction was more effective if as little as 1% oxygen was present (Thornton, 1935). A higher percentage of *Viola* sp. and *Veronica hederaefolia* seeds entered dormancy at 4 and 8% than at 0 and 2% oxygen, but the dormancy induction of *V. persica* seeds increased with a decrease in oxygen from 21 to 0% (Lonchamp and Gora, 1979). All seeds of *Tragopogon dubius* and *T. pratensis* were induced into secondary dormancy by soaking in deoxygenated water for 2 days in darkness at 25°C (Qi and Upadhyaya, 1993).

An aspirated nitrogen atmosphere at 20°C for 9 hr caused seeds (caryopses) of *Avena fatua* to enter secondary dormancy, showing that they were very sensitive to anoxia. Seeds of this species remained in secondary dormancy following 2 weeks of dry storage at 26°C and 25% relative humidity (Symons *et al.*, 1986). The level of germination inhibitors in caryopses of *A. fatua* increased in seeds induced into secondary dormancy by anaerobic conditions (Black, 1959). Lack of oxygen (nitrogen gas atmosphere) prevented the induction of sko-

todormancy in *Lactuca serriola* seeds imbibed in darkness at 25°C for 12 days, but it did not prevent the induction of dormancy in seeds imbibed in darkness at 38°C for 8 days (Small and Gutterman, 1992). Seeds of *Datura stramonium* were induced into secondary dormancy at 25°C in jars flushed with nitrogen gas (Benvenuti and Macchia, 1995).

Seeds of *Sisymbrium officinale* incubated in darkness at 24, 15, 10, or 6°C showed an increase and then a decline in oxygen uptake. However, the decline in oxygen uptake took place before the induction of secondary dormancy. Further, oxygen uptake remained constant as seeds came out of secondary dormancy at 2°C. Thus, the annual dormancy cycle was not correlated with changes in oxygen uptake (Derkx *et al.*, 1993).

A higher percentage of the seeds of some aquatic (Simpson, 1966; McIntyre *et al.*, 1989), mudflat (Baskin and Baskin, unpublished results), and vernal pool (Griggs, 1976) species come out of dormancy under flooded than under nonflooded conditions. Further, seeds of *Peltandra virginica* (Edwards, 1933), *Typha latifolia* (Bonnewell *et al.*, 1983), *Fimbristylis littoralis, Scirpus juncoides* (Pons and Schroder, 1986), and *Echinochloa crus-galli* (Kennedy *et al.*, 1980) will germinate under water (see Chapter 11,VI).

The level of carbon dioxide in soils depends on depth, temperature, moisture levels, porosity, amount of biotic activity, and rates of gas exchange with the air (Egley, 1986). Seeds in the top 10 cm of soil are exposed to carbon dioxide levels of about 0.1% (Egley, 1986; Richter and Markewitz, 1995); however, the concentration of carbon dioxide increases with depth (Fig. 4.15) and may reach 8% at 50 cm. Carbon dioxide levels at a 20-cm depth may reach 5–7% when biological activity peaks in the spring and summer, but they decrease to 0.1–2% by

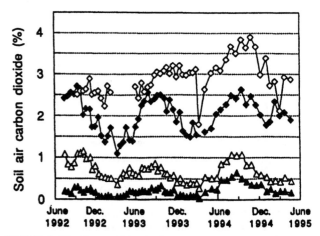

FIGURE 4.15 Seasonality of soil carbon dioxide at depths of 0.15 (▲), 0.6 (△), 1.75 (◆), and 4.0 (◇) m at the Calhoun Experimental Forest in South Carolina. From Richter and Markewitz (1995).

autumn (Egley, 1986). Maximum carbon dioxide levels in desert soils occur in the wet season, the period of the year with maximum biological activity (Buyanovsky *et al.*, 1982). Seeds located adjacent to decaying organic material could be subjected to high levels of carbon dioxide. Kidd (1914a) found carbon dioxide levels of 20% about 7.5 cm away from decaying green grass that he had buried in the soil. High levels of carbon dioxide could also result from the root respiration of actively growing plants (Wagner and Buyanovsky, 1983).

In general, carbon dioxide levels of 2–5% stimulate germination (Toole *et al.*, 1964; Hart and Berrie, 1966; Jones and Hall, 1979; Taylorson, 1980; Schonbeck and Egley, 1981a). Thus, carbon dioxide levels greater than those usually found in soils could promote germination. The stimulatory effect of carbon dioxide on the germination of *Helianthus annuus* seeds may be due to the promotion of ethylene synthesis by the seeds (Corbineau *et al.*, 1990). Carbon dioxide levels in excess of 5% inhibit germination (Kidd, 1914a,b; Bibbey, 1948; Roberts, 1972) in a number of species. However, Thornton (1936) found that 40–80% carbon dioxide stimulated the germination of *Lactuca sativa* seeds in both darkness and light at 35°C and prevented the induction of dormancy in seeds held in darkness at 35°C. Also, carbon dioxide concentrations of 1–16% enhanced the loss of dormancy in seeds of *Polygonum scandens* (Justice, 1941). Seeds of *Avena fatua* germinated to 90% at 15.0–15.5% oxygen and 24–26% carbon dioxide (Bibbey, 1948). Kidd's (1914a) work suggests that seeds buried in soil with high biotic activity possibly may be induced into dormancy. He found that seeds of *Brassica alba* were dormant after being exposed to carbon dioxide (10–22%) arising from decaying grass. To break dormancy and stimulate germination, seeds of *B. alba* had to be dried and then rewet (Kidd, 1917).

The stimulatory effects of 2–5% carbon dioxide indicate that soil carbon dioxide levels probably are not very important in causing changes in dormancy states. Seeds of *Amaranthus viridis, Cyperus difformis, C. iria, Echinochloa crus-galli, Eleusine indica, Fimbristylis miliacea, Monochoria vaginalis, Paspalum conjugatum,* and *Portulaca oleracea* were buried at depths of 0, 2.5, 7.5, 15, and 25 cm in soil in Taiwan for 10 to 365 days; exhumed after 10, 20, 30, 60, 90, 120, 180, 240, 300, and 365 days; and tested for germination in light at room temperatures. In general, seeds of each species from the soil surface germinated to lower percentages than those that had been buried. However, seeds of *C. difformis* from both the surface and various soil depths germinated to 86–100% after 10–120 days of burial. Further, each time seeds of a species were exhumed, there were no significant differences in germination percentages between seeds buried at 2.5 cm and the other depths

(Horng and Leu, 1978). Thus, soil depth (and indirectly carbon dioxide levels) did not affect the rate of dormancy loss.

Van Assche and Vanlerberghe (1989) also concluded that depth of burial had no effect on the dormancy cycle in *Rumex obtusifolius* seeds. However, seeds of *Datura ferox* buried at a soil depth of 5 cm had annual dormancy cycles, whereas those buried at 20 cm came out of dormancy and remained nondormant, regardless of seasonal temperature changes (Reisman-Berman *et al.*, 1991). Thus, seeds buried to depths of 20 cm or more may not exhibit dormancy cycles due to increased carbon dioxide levels.

Germination rates of *Setaria faberi* and *Xanthium pensylvanicum* seeds in darkness at 23°C were promoted by 30 mmol mol^{-1} carbon dioxide. This concentration of carbon dioxide increased germination rates under hypoxia, abscisic acid (ABA), or water stress. However, seeds that failed to germinate at 30 mmol mol^{-1} carbon dioxide under any of these three stress conditions entered secondary dormancy. Further, although carbon dioxide stimulated germination in field-buried seeds of *S. faberi* exhumed during the winter dormancy-breaking period, it accelerated induction into the secondary dormancy of seeds exhumed during summer (Yoshioka *et al.*, 1995). Thus, the effects of carbon dioxide on germination ecophysiology are influenced by other environmental factors and by the dormancy state of seeds.

E. Water

The level of seed hydration plays a role in the breaking and induction of dormancy. Seeds requiring cold stratification to become nondormant must be imbibed while they are exposed to low temperatures, otherwise dormancy is not broken (Stokes, 1965). Further, seeds of *Ambrosia trifida* (Davies, 1930a), *Avena ludoviciana, A. fatua* (Quail and Carter, 1969), and *Draba verna* (Baskin and Baskin, 1979b) come out of dormancy during dry storage and require a minimum relative humidity of 30–40% for dormancy loss to occur.

Seeds of *Ambrosia trifida* (Davis, 1930a), *Malus* (apple), *Prunus* (cherries) (Haut, 1932), *Polygonum* spp. (Justice, 1941, 1944; Staniforth and Cavers, 1979), and *Pyrus* spp. (Westwood and Bjornstad, 1968) become nondormant during cold stratification, but they reenter dormancy if air dried at room temperatures. Dormancy can be broken in these seeds by cold stratification. However, seeds of *Ambrosia artemisiifolia* (Bazzaz, 1979), *Impatiens glandulifera* (Mumford, 1988), and a number of other species (see Nikolaeva, 1969) that come out of dormancy during cold stratification can be air dried without reentering dormancy. Seeds of *Lactuca sativa* soaked in distilled water entered dormancy when they were allowed to dry in darkness at 20–35 °C, and those dried slowly became more dormant than those dried rapidly (Nutile, 1945).

Although air-dried seeds of some species enter dormancy, this response in the laboratory may indicate what happens in the field. Thus, PEG solutions have been used to create water stresses similar to those seeds receive in the field. In general, the water potential of soils at the permanent wilting percentage is about −1.5 MPa. Dormancy was induced in seeds of *Chenopodium bonus-henricus* with PEG solutions at −0.86 MPa in darkness at 15°C (Khan and Karssen, 1980) and in those of *Rumex crispus* at −1.57 MPa in light at 15°C (Samimy and Khan, 1983). Thus, dormancy can be induced in seeds of some species at water potentials that could occur in the field. However, seeds may have to be at a certain level of hydration before they can be induced into secondary dormancy. A 0.6 M (−1.49 MPa) mannitol solution inhibited the germination of *Citrullus lanatus* seeds, but secondary dormancy was not induced (Thanos and Mitrakos, 1992). At suboptimal levels of hydration for germination, seeds of *Lactuca sativa* were not induced into secondary dormancy by high temperatures and darkness, and those of *Phacelia tanacetifolia* were not induced into secondary dormancy by continuous light (McDonough, 1968).

Seeds of *Sisymbrium altissimum* (and probably those of *S. officinale*) have mucilaginous seed coats (Young and Evans, 1973), and the mucilage expands in the presence of water, forming an oxygen barrier (Witztum *et al.,* 1969). Thus, presumably oxygen levels were low in seeds of *S. officinale* at high moisture levels, which may help explain why seeds entered secondary dormancy faster at high than at low soil moisture levels (Karssen, 1980–1981a). Soaking seeds of *Datura ferox* and *D. stramonium* in water inhibits germination because water becomes trapped in tissues between the embryo and seed coat, creating an oxygen barrier (Reisman-Berman *et al.,* 1989). Although soaking in water apparently did not induce secondary dormancy in seeds of *Datura* spp., excess water may delay germination until some other environmental factor(s) induces dormancy.

Alternate wetting and drying (wet–dry cycles) appear to be required for loss of dormancy in the mucilaginous seeds of the winter annual *Lesquerella filiformis*. Seeds kept continuously moist during summer germinated to 3% at 20/10°C in autumn, whereas those given wet–dry cycles germinated to 51% (Baskin and Baskin, unpublished results). A desiccation treatment of exhumed seeds of *Chenopodium album, Sisymbrium officinale, Spergula arvensis,* and *Polygonum lapathifolium* promoted germination, with the stimulatory effect being highest at the beginning of induction of secondary dormancy (Bouwmeester, 1990). Seeds of *Rumex acetosella*

came out of dormancy while on moist sand at room temperatures, but most of them did not germinate until after they had been dried and reimbibed (Hintikka, 1990). Wet–dry cycles overcame dormancy in *Rumex crispus* seeds at alternating temperatures, but they had no effect at constant temperatures (Vincent and Cavers, 1978). Also, one or two drying treatments reduced the level of skotodormancy in *Lactuca sativa* seeds and prevented nondormant seeds from reentering skotodormancy (Hsiao, 1992).

After seeds become nondormant in the field, they may be alternately imbibed and dried a number of times before conditions become suitable for germination. Depending on the species, wet–dry cycles may result in no change, decreases, or increases in germination rates and percentages (Griswold, 1936; Maynard and Gates, 1963; Cocks and Donald, 1973; Lush and Groves, 1981; Lush *et al.,* 1981; Baskin and Baskin, 1982b; Taylorson, 1986); however, wet–dry cycles did not induce seeds into secondary dormancy in these studies.

F. Inorganic Chemicals

Although the effects of many kinds of inorganic ions on seed germination have been investigated (see review by Egley and Duke, 1985), only nitrate and nitrite significantly influence dormancy states. Hydrogen ions influence germination rates and percentages, but they are not known to change the dormancy state of seeds. However, a pH of 3 increased the effectiveness of nitrite in promoting the germination of *Oryza sativa* seeds (Cohn *et al,* 1983). Seeds of *Cynoglossum officinale* germinated over a range of pH's in the presence of nitrate (Freijsen *et al.,* 1980).

Nitrate and nitrite are effective in overcoming dormancy and/or promoting germination in many light-sensitive seeds (Toole *et al.,* 1956); however, their effects are influenced by various factors. Potassium, sodium, and ammonium nitrates (10^{-1} and 10^{-2} M) and nitrites (10^{-2} and 10^{-3} M) promoted the germination of *Capsella bursa-pastoris* seeds at alternating but not at constant temperatures (Popay and Roberts, 1970a), and potassium nitrate (0.2%) stimulated the germination of *Polypogon monspeliensis* seeds at alternating but not at constant temperatures (Toole, 1938). Cold stratification breaks the dormancy of *Sisymbrium officinale* seeds at 165 mg NO_3^-/kg soil but not at 6 mg NO_3^-/kg soil (Karssen, 1980–1981b). Further, nitrate inhibited induction of a secondary light requirement for germination in seeds of this species (Karssen, 1980–1981a). Potassium nitrate (0.2%) promoted the germination of *Digitaria ischaemum* seeds at alternating temperature regimes of 25/15 and 30/20°C but not at 35/20 and 40/20°C (Toole and Toole, 1941), and 1000 and 500 ppm ammonium nitrate stimulated 63–87% of the seeds of *Dactyloctenium sindicum* and *Eragrostis tremula*, respectively, to germinate at 30 and 32°C but not at 24 and 27°C (Kumari *et al.,* 1987).

When seeds of *Sisymbrium officinale* containing different levels of nitrate were buried in the field in early winter in The Netherlands, nitrate rapidly leached from them (Bouwmeester *et al.,* 1994). Seeds of *S. officinale* exhumed at various times during winter showed an increased sensitivity to nitrate treatments, and those cold stratified at 2°C became more sensitive to nitrate as dormancy was broken. Seeds of *Avena fatua* exhumed from soil throughout the year in England were sensitive to nitrate from late autumn to late winter or early spring (Murdoch and Carmona, 1993). Thus, sensitivity to nitrate increased with a decrease in dormancy. Nitrate promoted the germination of dormant *Avena fatua* seeds in darkness, but the stimulatory effect decreased as temperatures increased (Saini *et al.,* 1985a).

The promotive effects of nitrate can be increased by various environmental factors. Germination percentages of seeds of *Chenopodium album, Capsella bursa-pastoris* (Roberts and Benjamin, 1979), *Avena fatua* (Hilton, 1984b, 1985), *Sinapis arvensis* (Goudey *et al.,* 1987), and *Hygrophila auriculata* (Sharma and Amritphale, 1989) incubated in light were increased by nitrate. Red light and nitrate stimulated the germination of *Sisymbrium officinale* seeds (Hilhorst *et al.,* 1986; Hilhorst and Karssen, 1989), and germination in a number of species is improved by light, nitrate, and alternating temperatures (Taylorson and McWhorter, 1969; Vincent and Roberts, 1977).

Ethylene and nitrate act synergistically in promoting the dark germination of nitrate-deficient seeds of *Chenopodium album* (Saini *et al.,* 1985b; Carmona and Murdock, 1995) and in breaking dormancy of *Avena fatua* (Saini *et al.,* 1985a; Carmona and Murdoch, 1995) and *Matricaria maritima* (Mekki and Leroux, 1991) seeds in petri dishes. However, seedling emergence was variable when KNO_3 was applied to soil containing seeds of *A. fatua,* and no synergistic effect was detected (Adkins and Adkins, 1994). Red light, ethylene, and nitrate promoted the germination of *Spergula arvensis* seeds (Olatoye and Hall, 1973), whereas nitrate and ethylene or nitrite and ethylene stimulated *Portulaca oleracea* seeds to germinate in darkness at 35°C (Egley, 1984).

Reasons for variation in nitrate levels in soils include moisture, temperature, soil type, microbial activity, agricultural practices (Young and Aldag, 1982), and presence or absence of vegetation (Pons, 1989). Popay and Roberts (1970b) observed that peaks in germination of *Capsella bursa-pastoris* seeds were correlated with increases in nitrate levels in the soil. In chalk grasslands of The Netherlands, *Plantago lanceolata* regenerates much better from seeds in gaps than in areas where vegetation cover is intact; nitrate levels in gaps (0.2–1.1 m*M*) were higher than those in closed (0.1 m*M*) vegetation. Because nitrates are absorbed by plants in vegetated sites, the high nitrate requirement for the germination of *P. lanceolata* may serve as a gap-detecting mechanism (Pons, 1989). An increase in nitrification often occurs in forests following soil disturbance; therefore, nitrate levels could indicate reduced competition from other plants (Hintikka, 1987).

Levels of nitrogen (N)-containing compounds in soils also are increased by the addition of N-containing fertilizers to farmland, and the effects of these compounds on the germination of weed seeds have received considerable attention. Breaking of seed dormancy by nitrate or other N-containing compounds would be of great benefit to farmers in their efforts to control weeds in crops. Applications of nitrate-containing fertilizers to the soil stimulated the germination of *Setaria glauca* (Schimpf and Palmblad, 1980), *Avena fatua* (Sexsmith and Pittman, 1963), and *Prunus pensylvanica* (Auchmoody, 1979) seeds; further, *Chenopodium album* seeds produced in fertilized plots were less dormant than those produced in nonfertilized plots (Fawcett and Slife, 1978). In general, however, nitrate fertilizers do not increase the germination of weed seeds (Fawcett and Slife, 1978; Schimpf and Palmblad, 1980; Hilton, 1984b; Saini *et al.,* 1985a; Hurtt and Taylorson, 1986). One reason why nitrate fertilizers may not stimulate large numbers of seeds to germinate is that many seeds require both nitrates (or nitrites) and light to germinate. Thus, nitrate treatments would be ineffective on seeds buried in soil. Perhaps nitrogen fertilizers would be of more benefit in promoting germination if they were applied after tillage of the soil had exposed weed seeds to light than if applied prior to tilling (Hilton, 1984b; Goudy *et al.,* 1987).

Another aspect of trying to use nitrogen fertilizers to stimulate the germination of weed seeds is that high concentrations of nitrate in seeds can inhibit germination (Peterson and Bazzaz, 1978; Goudey *et al.,* 1988). Subsequent decreases in nitrate levels, however, permit germination. Thus, the germination of weed seeds potentially could be inhibited in early spring when fertilizers are applied at the time crop seeds are sown, but it might be promoted after the crop is established. Any change in the timing of germination of weed seeds could have management implications.

Liquid ammonia is used as a nitrogen fertilizer in various parts of the world and has been shown to break dormancy in seeds of *Avena fatua* and several other grass weed species but not of dicotyledonous species (Cairns and de Villiers, 1986). Consequently, farmers who use this nitrogen source to fertilize crops should be alert for bursts of germination of weedy grasses after it is applied to the soil.

The stimulatory effects of boron on the germination of *Themeda triandra* seeds (Cresswell and Nelsen, 1972) show how little is known about the influence of various elements on germination, especially in nature. Seeds of *T. triandra,* however, were not stimulated by cobalt, copper, iron, manganese, molybdenum, or zinc.

G. Organic Chemicals

Numerous organic compounds are present in soils, and their sources include seeds, soil organisms, plants rooted in the soil, and remains of plants that have died. Regardless of origin, these compounds have the potential to stimulate or inhibit germination and thus play a role in its timing.

In a variety of habitats, including old fields (Wilson and Rice, 1968), the Great Basin desert (Schlattere and Tisdale, 1969), chaparral (Chou and Muller, 1972), annual grasslands (Bell and Muller, 1973), Ponderosa pine forests (Rietveld, 1975), and cedar glades (Turner and Quarterman, 1975), leachates from various plant parts and/or root exudates from some plant species are capable of inhibiting seed germination of other members of the community. Further, leachates and/or exudates from weedy species can inhibit the germination of crop species (Rice, 1979; Shaukat *et al.,* 1985; Noor and Khan, 1994) and vice versa (Lockerman and Putnam, 1979). These compounds collectively are called allelochemicals, and much effort has gone into trying to isolate and identify them (Chou and Patrick, 1976; Karssen and Hilhorst, 1992). Although many organic compounds released into the soil inhibit germination in petri dishes (e.g., Chou and Muller, 1972), none of them have been shown to actually change the dormancy state of seeds. However, these compounds could play a role in inhibiting germination until other environmental factors induce dormancy.

One consequence of seeds being buried in soil is that rates of gaseous exchange between seeds and atmosphere are reduced; therefore, low levels of oxygen may develop in the immediate vicinity of seeds, causing them to undergo fermentation metabolism. Further, low rates of gaseous exchange can cause toxic volatile fermentation products to accumulate around buried seeds and inhibit germination (Benvenuti and Macchia, 1995). Thus, decreases in the germination of seeds buried in soil may be due to the presence of toxic compounds instead of lack of oxygen per se. Burial in soil inhibited the germination of *Datura stramonium* seeds, but germination occurred when the soil was flushed with nitrogen gas (Benvenuti and Macchio, 1995).

Wesson and Wareing (1969b) concluded that a gaseous inhibitor in the soil, other than carbon dioxide, prevents the germination of *Spergula arvensis* seeds. Further, they maintained that the inhibitor(s) came from the seeds themselves. Using gas chromatography, Hall *et al.* (1987) identified acetone, ethanol, and acetaldehyde from the atmosphere surrounding buried seeds of *S. arvensis;* when these compounds (at concentrations of $10^{-8}\ M$) were applied to seeds, germination was inhibited. Holm (1972) found that acetone, ethanol, and acetaldehyde were produced by buried seeds of *Abutilon theophrasti, Ipomoea purpurea,* and *Brassica kaber* and that they reduced germination percentages of *I. purpurea* seeds. Acetaldehyde was more inhibitory to germination than acetone and ethanol.

Soils contain a number of volatile hydrocarbons, including ethane, ethylene, propane, propylene, isobutane, *n*-butane, and butene (Smith and Dowdell, 1973), and some of them such as ethylene and propylene (an analog of ethylene) can stimulate germination (Taylorson, 1979). Ethylene is of special interest to seed ecologists because its concentration in soil air may reach 18 ppmv (Campbell and Moreau, 1979), which is high enough to inhibit root growth (Smith and Russell, 1969) and stimulate seed germination (Taylorson, 1979). The most favorable conditions for the production of ethylene in soils are (1) a low oxygen concentration, as would occur when soils become waterlogged (Smith and Russell, 1969); (2) high temperatures (Smith and Restall, 1971); (3) the presence of organic matter (Smith and Russell, 1969); (4) a soil moisture potential of -0.5 MPa or higher (Smith and Cook, 1974); and (5) a low pH (Goodlass and Smith, 1978). Ethylene production is suppressed by high nitrate concentrations (Smith and Restall, 1971; Hunt *et al.,* 1980), but it is little affected by sulfate and phosphate (Smith and Restall, 1971). Evidence indicates that more ethylene is produced by soil fungi than by bacteria (Lynch, 1975).

Ethylene also can be produced by seeds themselves (Katoh and Esashi, 1975b; Kepczynski *et al.,* 1977; Rudnicki *et al.,* 1978; Cardoso and Felippe, 1983, 1987; Corbineau *et al.,* 1990; Esashi, 1991). Nondormant seeds of *Xanthium pensylvanicum* produced more ethylene than dormant ones (Katoh and Esashi, 1975a), and ethylene production increased in excised embryos of *Malus domestica* as seed dormancy was broken (Kepczynski *et al.,* 1977). High-vigor seeds of *Pisum sativum* and *X. pensylvanicum* produce more ethylene than low-vigor ones (Gorecki *et al.,* 1991). No ethylene production could be detected by gas chromatography in *Rumex obtusifolius* seeds, and germination in this species was not promoted by ethylene (Takaki *et al.,* 1982).

Ethylene treatments at normal atmospheric concentrations of carbon dioxide (0.0355% by volume) stimulate seeds of a number of species to germinate (Egley and Dale, 1970; Olatoye and Hall, 1973; Egley, 1978, 1980; Taylorson 1979; Jones and Hall, 1979; Keegan *et al.*, 1989), but the effects of ethylene frequently are enhanced when the carbon dioxide concentration is raised above 0.0355% (Negm *et al.*, 1973; Katoh and Esashi, 1975b; Esashi and Katoh, 1975; Esashi *et al.*, 1976). However, ethylene in combination with 10% carbon dioxide inhibited the germination of *Striga lutea* seeds (Egley and Duke, 1985). The dormancy-breaking action of ethylene on *Chenopodium album* seeds increased with an increase in the nitrate level in the seeds (Saini *et al.*, 1986; Saini and Spencer, 1987).

Ethylene causes changes in dormancy states primarily by breaking dormancy. However, not all seeds respond to ethylene, and some are inhibited by it (Olatoye and Hall, 1973; Taylorson, 1979; Suzuki and Taylorson, 1981). Seeds that respond to ethylene also germinate to higher percentages in light than in darkness, but ethylene does not promote germination in all light-stimulated seeds (Taylorson, 1979). Ethylene can break dormancy (Schonbeck and Egley, 1980a) or substitute for a light requirement for germination (Schonbeck and Egley, 1980a,b). Also, ethylene can break secondary dormancy induced by darkness and/or high temperatures (Schonbeck and Egley, 1981b; Corbineau *et al.*, 1988). In some species, ethylene-treated seeds germinated at higher temperatures (Schonbeck and Egley, 1980a) and under greater water stresses (Schonbeck and Egley, 1980b, 1981a; Esashi *et al.*, 1989) than nontreated ones. Ethylene increased the osmotic potential in cells of *Xanthium pensylvanicum* embryos (Esashi and Ishizawa, 1989), thereby increasing their growth potential. In seed-priming studies on this species, promotive effects of ethylene increased with water stress and with increased oxygen and decreased carbon dioxide levels (Esashi *et al.*, 1990).

The role of ethylene in the annual dormancy/nondormancy cycles of buried seeds has not been investigated. However, studies on *X. pensylvanicum* suggest that an increase in ethylene production is one of the things that happens during loss of dormancy (Katoh and Esashi, 1975a). Schonbeck and Egley (1981c) reported that seeds buried in soil over winter or during summer became increasingly sensitive to ethylene. A question that needs to be investigated is whether high concentrations of ethylene in soil only at a particular time of the year can stimulate seeds to germinate at some future date. For example, suppose soils become waterlogged, and seeds are exposed to relatively high concentrations of ethylene. However, at the time of exposure to stimulatory concentrations of ethylene, temperatures were unfavorable for germination. When temperatures or other factors became favorable for germination, would the previous ethylene treatment stimulate germination? If

it does, ethylene may have a greater role in regulating the timing of germination than previously realized.

Olatoye and Hall (1973) found that *Spergula arvense* seeds dried up to 144 hr after treatment with ethylene retained 65% of the stimulatory effect when they were reimbibed. Further, seeds of *S. arvensis* had to be exposed to ethylene for a minimum of only 6 hr to obtain a stimulatory effect. Ethylene-treated seeds of *Ricinodendron rautaneii* dried at room temperatures for 8 days germinated to higher percentages than nontreated seeds (Keegan *et al.*, 1989). Thus, short-term increases in soil levels of ethylene potentially can stimulate germination even if seeds are dried before conditions become suitable for germination. Carbon dioxide promoted the germination of secondarily dormant seeds of *X. pensylvanicum* given an ethylene treatment prior to being induced into secondary dormancy (Esashi *et al.*, 1988).

Some organic compounds, including pesticides, synthetic germination stimulators, and herbicides, are deliberately applied to arable soils, and they may influence seed germination of weedy species. The pesticide DDT had little effect on the germination of lima bean, cucumber, rye, or squash seeds (Hopkins and Toole, 1950), but the pesticides chlorphyrifos, dimethoate, and iprodione inhibited the germination of some weed seeds. Chlorpyrifos reduced germination in an annual grass and an annual forb; dimethoate, in an annual grass and six annual forbs; and iprodione, in a perennial forb (Gange *et al.*, 1992).

Efforts have been made to identify germination stimulators that could be applied to arable soils and promote germination of weed seeds; in which case, seedlings could be destroyed before crops were sown (Taylorson, 1987). Gibberellins are too expensive for widespread use in agriculture; consequently, there is interest in finding inexpensive compounds with stimulatory properties similar to those of GA. A substituted phthalimide, AC-94377, is such a compound and has 10% the activity of GA in plant growth bioassays (Suttle and Schreiner, 1982). AC-94377 was as effective as GA_3 in promoting the germination of dormant seeds of *Avena fatua*, *Brassica kaber*, *Rumex crispus*, *Thlaspi arvense*, *Phacelia tanacetifolia* (Metzger, 1983), and *Solanum nigrum* (Bond and Burch, 1990). This compound substitutes for light and alternating temperature regimes in promoting the germination of some species (Bond and Burch, 1990). AC-94377 and AC-99524 applied to the soil surface in greenhouse studies stimulated the germination of *Sinapis arvensis* (Donald and Hoerauf, 1985) and AC-94377 that of *Solanum nigrum* seeds (Bond and Burch, 1990). AC-94377 shows some promise in promoting seeds to germinate in the field (Bond and Burch, 1990; Donald and Tanaka, 1993).

Because large amounts of herbicides are used in agri-

TABLE 4.8　Effect of Herbicides on Seed Germination

Species	Herbicide	Effect[a]	Reference
Abutilon theophrasti	Butylate	+	Fawcett and Slife (1975)
	Chlorpropham	+	
	CDEC[b]	+	
	Diallate	+	
	EPTC[c]	+	
	Vernolate	+	
Amaranthus retroflexus	2,4-D[d]	−, +	Rojas-Garciduenas and Kommendahl (1960)
Asystasis gangetica	Glufosinate	−	Sahid and Kalithasan (1994)
	Terbuthylazine	−	
Barbarea stricta	Chlormequat	−	Hintikka (1988)
	Daminozide	−	
B. vulgaris	Chlormequat	−	Hintikka (1988)
	Daminozide	−	
Bouteloua curtipendula	Chlopyralid	0, −	Huffman and Jacoby (1984)
	Triclopyr	−	
	2,4-D	−	
	Picloram	−	
B. gracilis	Chlopyralid	0, −	Huffman and Jacoby (1984)
	Triclopyr	−	
	2,4-D	−	
	Picloram	−	
Brassica japonica	2,4-D	−	Mitchell and Marth (1946)
B. napus	2,4-D	−	Hamner *et al.* (1946)
B. oleracea	2,4-D	−	Hamner *et al.* (1946)
Bromus arvense	2,4-D	−	Hamner *et al.* (1946)
Buchloe dactyloides	Chlopyralid	0, −	Huffman and Jacoby (1984)
	Triclopyr	−	
	2,4-D	−	
	Picloram	−	
Capsella bursa-pastoris	Chlormequat	−	Hintikka (1988)
	Daminozide	−	
Carthamus tinctorius	Picloram	−	Chang and Foy (1971)
Celosia argentea	2,4-D	+	Pandya and Baghela (1973)
Chenopodium album	Butylate	+	Fawcett and Slife (1975)
	CDEC	+	
	EPTC	+	
Festuca elatior	2,4-D	0	Hamner *et al.* (1946)
F. rubra	Picloram	0	Flater *et al.* (1974)
Glycine max	Picloram	−	Chang and Foy (1971)
Helianthus annuus	Basalin	−	Saxena and Srivastava (1994)
Hordeum vulgare	Picloram	0	Chang and Foy (1971)
Hordeum sp.	2,4-D	−	Mitchell and Marth (1946)
Lactuca sativa	SAN 9789[e]	+	Widell *et al.* (1985)
Linum usitatissimum	Basalin	−	Saxena and Srivastava (1994)
Lolium rigidum	Sulfometuron	−	Burnet *et al.* (1994)

(continues)

TABLE 4.8—*Continued*

Species	Herbicide	Effect[a]	Reference
Lycopersicon esculentum	Picloram	0	Flater *et al.* (1974)
	2,4-D	0	Hamner *et al.* (1946)
Melilotus alba	2,4-D	−	Hamner *et al.* (1946)
Oryza sativa	Bulachlor-2,4-D	−	Mabbayad and Moody (1992)
Paspalum conjugatum	Glufosinate	−	Sahid and Kalithasan (1994)
Phaseolus vulgaris	2,4-D	−	Hamner *et al.* (1946)
Phleum pratense	2,4-D	0	Hamner *et al.* (1946)
Pinus resinosa	CDAA[f]	−, +	Kozlowski and Torrie (1965)
	CDEC	0, −, +	Sasahi *et al.* (1968)
	DCPA[g]	0, +	Sasaki and Kozlowski (1968)
	EPTC	0, −, +	
	Ipazine	0	
	Monuron	0	
	Prometryne	0	
	Propazine	0	
	Simazine	0, +	
	Triazine	0	
	2,4-D	−	
Pisum sativum	2,4-D	−	Hamner *et al.* (1946)
Raphanus sativus	Picloram	0	Chang and Foy (1971)
Secale cereale	2,4-D	0	Hamner *et al.* (1946)
Setaria faberii	Butylate	+	Fawcett and Slife (1975)
Sorghum vulgare	2,4-D	−	Hamner *et al.* (1946)
Thlaspi arvense	Chlormequat	−	Hintakka (1988)
	Daminozide	−	
Trifolium hybridum	Picloram	−	Flater *et al.* (1974)
	2,4-D	−	Hamner *et al.* (1946)
Triticum aestivum	Picloram	0	Flater *et al.* (1974)
T. vulgare	2,4-D	0	Hamner *et al.* (1946)
Vicia villosa	2,4-D	−	Hamner *et al.* (1946)
Vigna mugo	2,4-D	−	Shaukat *et al.* (1980)
	2,4,5-T, MCPB[h]	−	
	s-Triazine	−	
	Terbutryne	−	
V. radiata	2,4-D	−	Shaukat *et al.* (1980)
	2,4,5-T, MCPB	−	Shaukat *et al.* (1980)
	s-Trizaine	−	
	Terbutryne	−	

[a] +, herbicide stimulated germination; −, herbicide reduced germination; 0, herbicide had no effect on germination.
[b] 2-Chloroallyl diethylthiocarbamate.
[c] Ethyl-*N-N*-di-*n*-propylthiocarbamate.
[d] 2,4-Dichlorophenoxyacetic acid.
[e] 4-Chloro-5-methylamino-2-(α,α,α-trifluoro-*m*-tolyl)-3(2*H*)-pyridazinone.
[f] 2-Chloro-*N,N*-diallylacetamide.
[g] Dimethyl-2,3,5,6-tetrachloroterephthalate.
[h] Methyl-chlorophenoxybutyric acid.

culture, it is reasonable to ask if these compounds have any effect on seed germination. Depending on the type of herbicide and plant species, germination may be increased, decreased, or not affected (Table 4.8). However, the method of herbicide application (Bond and Burch, 1990), concentration (Chang and Foy, 1971; Pandya and Baghela, 1973; Fawcett and Slife, 1975; Shaukat *et al.,* 1980; Morash and Freedman, 1989), length of exposure (Rojas-Garciduenas and Kommedahl, 1960), other chemicals applied with the herbicide (Widell *et al.,* 1985; Shaukat, 1974), and temperatures at the time of seed exposure (Widell *et al.,* 1985) may influence germination percentages. SAN 9789 promoted the germination of *Lactuca sativa* seeds in darkness (Widell *et al.,* 1985), and chlomequat and daminozide induced seeds of *Barbarea stricta* and *B. vulgaris* into secondary dormancy (Hintikka, 1988).

The use of herbicides may cause seed banks in arable land to have a high concentration of seeds of species that are relatively herbicide tolerant (Ball and Miller, 1990). Further, the use of herbicides in summer crops may select for winter annual forms of species (Putwain *et al.,* 1982). Thus, although the control of weeds may be enhanced by using herbicides, in many cases it cannot be achieved by this method exclusively. Information on (1) the biology of weeds (Chancellor, 1981), (2) annual dormancy cycles of buried seeds, and (3) responses of seeds to soil nitrogen levels, light exposures during cultivation, and various organic compounds (Dyer, 1995) needs to be used along with herbicides to develop effective control programs at the seed level for most weeds.

References

Adkins, S. W., and Adkins, A. L. (1994). Effect of potassium nitrate and ethephon on fate of wild oat (*Avena fatua*) seed in soil. *Weed Sci.* **42,** 353–357.

Adkins, S. W., Naylor, J. M., and Simpson, G. M. (1984). The physiological basis of seed dormancy in *Avena fatua*. V. Action of ethanol and other organic compounds. *Physiol. Plant.* **62,** 18–24.

Aibar, J., Ochoa, M. J., and Zaragoza, C. (1991). Field emergence of *Avena fatua* L. and *A. sterilis* ssp. *ludoviciana* (Dur.) Nym. in Aragon, Spain. *Weed Res.* **31,** 29–32.

Allen, P. S., Meyer, S. E., and Beckstead, J. (1995). Patterns of seed after-ripening in *Bromus tectorum. J. Exp. Bot.* **46,** 1737–1744.

Amritphale, D., Gutch, A., and Hsiao, A. I. (1993). Acidification, growth promoter and red light effects on germination of skotodormant seeds of *Hygrophila auriculata. Envir. Exp. Bot.* **33,** 471–477.

Arnold, S. M. (1973). Interactions of light and temperature on the germination of *Plantago maritima* L. *New Phytol.* **72,** 583–593.

Atterberg, A. (1899). Om sadesvarornas eftermognad. *Kgl. Landbruk-sakad. Handl. och Tidskr.* **38,** 227–250.

Auchmoody, L. R. (1979). Nitrogen fertilization stimulates germination of dormant pin cherry seed. *Can J. For. Res.* **9,** 514–516.

Ball, D. A., and Miller, S. D. (1990). Weed seed population response to tillage and herbicide use in three irrigated cropping sequences. *Weed Sci.* **38,** 511–517.

Baskin, C. C., and Baskin, J. M. (1988). Germination ecophysiology

of herbaceous plant species in a temperate region. *Am. J. Bot.* **75,** 286–305.

Baskin, C. C., and Baskin, J. M. (1994). Germination requirements of *Oenothera biennis* seeds during burial under natural seasonal temperature cycles. *Can. J. Bot.* **72,** 779–782.

Baskin, C. C., Baskin, J. M., and Chester, E. W. (1991). Temperature response pattern during afterripening of achenes of the winter annual *Krigia oppositifolia* (Asteraceae). *Plant Species Biol.* **6,** 111–115.

Baskin, C. C., Baskin, J. M., and Chester, E. W. (1993a). Seed germination ecology of two mesic woodland winter annuals, *Nemophila aphylla* and *Phacelia ranunculacea* (Hydrophyllaceae). *Bull. Torrey Bot. Club* **120,** 29–37.

Baskin, C. C., Baskin, J. M., and Chester, E. W. (1993b). Seed germination ecophysiology of four summer annual mudflat species of Cyperaceae. *Aquat. Bot.* **45,** 41–52.

Baskin, C. C., Baskin, J. M., and Chester, E. W. (1993c). Germination ecology of *Leptochloa panicoides,* a summer annual grass of seasonally dewatered mudflats. *Acta Oecol.* **14,** 693–704.

Baskin, C. C., Baskin, J. M., and Chester, E. W. (1994). Annual dormancy cycle and influence of flooding in buried seeds of mudflat populations of the summer annual *Leucospora multifida. Eco-Science* **1,** 47–53.

Baskin, C. C., Baskin, J. M., and Chester, E. W. (1995). Role of temperature in the germination ecology of the summer annual *Bidens polylepis* Blake (Asteraceae). *Bull. Torrey Bot. Club* **122,** 275–281.

Baskin, C. C., Baskin, J. M., and Chester, E. W. (1996). Seed germination ecology of the aquatic winter annual *Hottonia inflata. Aquat. Bot.* **54,** 51–57.

Baskin, C. C., Baskin, J. M., and El-Moursey, S. A. (1996). Seasonal changes in germination responses of buried seeds of the weedy summer annual grass *Setaria glauca. Weed Res.* **36,** 319–324.

Baskin, C. C., Baskin, J. M., and Van Auken, O. W. (1994). Germination response patterns during dormancy loss in achenes of six perennial Asteraceae from Texas, USA. *Plant Species Biol.* **9,** 113–117.

Baskin, C. C., Chesson, P. L., and Baskin, J. M. (1993). Annual seed dormancy cycles in two desert winter annuals. *J. Ecol.* **81,** 551–556.

Baskin, J. M., and Baskin, C. C. (1973a). Ecological life cycle of *Helenium amarum* in central Tennessee. *Bull. Torrey Bot. Club* **100,** 117–124.

Baskin, J. M., and Baskin, C. C. (1973b). Delayed germination in seeds of *Phacelia dubia* var. *dubia. Can. J. Bot.* **51,** 2481–2486.

Baskin, J. M., and Baskin, C. C. (1977). Role of temperature in the germination ecology of three summer annual weeds. *Oecologia* **30,** 377–382.

Baskin, J. M., and Baskin, C. C. (1978). Seasonal changes in the germination response of *Cyperus inflexus* seeds to temperature and their ecological significance. *Bot. Gaz.* **139,** 231–235.

Baskin, J. M., and Baskin, C. C. (1979a). Timing of seed germination in the weedy summer annual *Euphorbia supina. Bartonia* **46,** 63–68.

Baskin, J. M., and Baskin, C. C. (1979b). Effect of relative humidity on afterripening and viability in seeds of the winter annual *Draba verna. Bot. Gaz.* **140,** 284–287.

Baskin, J. M., and Baskin, C. C. (1980). Ecophysiology of secondary dormancy in seeds of *Ambrosia artemisiifolia. Ecology* **61,** 475–480.

Baskin, J. M., and Baskin, C. C. (1981a). Seasonal changes in the germination responses of buried *Lamium amplexicaule* seeds. *Weed Res.* **21,** 299–306.

Baskin, J. M., and Baskin, C. C. (1981b). Temperature relations of seed germination and ecological implications in *Galinsoga parviflora* and *G. quadriradiata. Bartonia* **48,** 12–18.

Baskin, J. M., and Baskin, C. C. (1981c). Seasonal changes in germination responses of buried seeds of *Verbascum thapsus* and *V. blattaria* and ecological implications. *Can. J. Bot.* **59**, 1769–1775.

Baskin, J. M., and Baskin, C. C. (1982a). Germination ecophysiology of *Arenaria glabra*, a winter annual of sandstone and granite outcrops of southeastern United States. *Am. J. Bot.* **69**, 973–978.

Baskin, J. M., and Baskin, C. C. (1982b). Effects of wetting and drying cycles on the germination of seeds of *Cyperus inflexus*. *Ecology* **63**, 248–252.

Baskin, J. M., and Baskin, C. C. (1983a). Seasonal changes in the germination responses of buried seeds of *Arabidopsis thaliana* and ecological interpretation. *Bot. Gaz.* **144**, 540–543.

Baskin, J. M., and Baskin, C. C. (1983b). Germination ecology of *Veronica arvensis*. *J. Ecol.* **71**, 57–68.

Baskin, J. M., and Baskin, C. C. (1983c). Seasonal changes in the germination responses of fall panicum to temperature and light. *Can. J. Plant Sci.* **63**, 973–979.

Baskin, J. M., and Baskin, C. C. (1983d). Seasonal changes in the germination responses of seeds of *Veronica peregrina* during burial, and ecological implications. *Can. J. Bot.* **61**, 3332–3336.

Baskin, J. M., and Baskin, C. C. (1984a). Role of temperature in regulating timing of germination in soil seed reserves of *Lamium purpureum* L. *Weed Res.* **24**, 341–349.

Baskin, J. M., and Baskin, C. C. (1984b). Effect of temperature during burial on dormant and non-dormant seeds of *Lamium amplexicaule* L. and ecological implications. *Weed Res.* **24**, 333–339.

Baskin, J. M., and Baskin, C. C. (1985a). The annual dormancy cycle in buried weed seeds: A continuum. *BioScience* **35**, 492–498.

Baskin, J. M., and Baskin, C. C. (1985b). Does seed dormancy play a role in the germination ecology of *Rumex crispus*? *Weed Sci.* **33**, 340–343.

Baskin, J. M., and Baskin, C. C. (1986a). Temperature requirements for after-ripening in seeds of nine winter annuals. *Weed Res.* **26**, 375–380.

Baskin, J. M., and C. C. Baskin. (1986b). Seasonal changes in the germination responses of buried witchgrass (*Panicum capillare*) seeds. *Weed Sci.* **34**, 22–24.

Baskin, J. M., and Baskin, C. C. (1987a). Environmentally induced changes in the dormancy states of buried weed seeds. *Proc. 1987 Brit. Crop Prot. Conf. Weeds* **7C-2**, 695–706.

Baskin, J. M., and Baskin, C. C. (1987b). Temperature requirements for after-ripening in buried seeds of four summer annual weeds. *Weed Res.* **27**, 385–389.

Baskin, J. M., and Baskin, C. C. (1989a). Role of temperature in regulating timing of germination in soil seed reserves of *Thlaspi arvense* L. *Weed Res.* **29**, 317–326.

Baskin, J. M., and Baskin, C. C. (1989b). Germination responses of buried seeds of *Capsella bursa-pastoris* exposed to seasonal temperature changes. *Weed Res.* **29**, 205–212.

Baskin, J. M., and Baskin, C. C. (1989c). Seasonal changes in the germination responses of buried seeds of *Barbarea vulgaris*. *Can. J. Bot.* **67**, 2131–2134.

Baskin, J. M., and Baskin, C. C. (1989d). Physiology of dormancy and germination in relation to seed bank ecology. *In* "Ecology of Soil Seed Banks" (M. A. Leck, V. T. Parker, and R. L. Simpson, eds.), pp. 53–66. Academic Press, San Diego.

Baskin, J. M., and Baskin, C. C. (1990a). Role of temperature and light in the germination ecology of buried seeds of *Potentilla recta*. *Ann. Appl. Biol.* **117**, 611–616.

Baskin, J. M., and Baskin, C. C. (1990b). The role of light and alternating temperatures on germination of *Polygonum aviculare* seeds exhumed on various dates. *Weed Res.* **30**, 397–402.

Baskin, J. M., and Baskin, C. C. (1990c). Seed germination biology of the narrowly endemic species *Lesquerella stonensis* (Brassicaceae). *Plant Species Biol.* **5**, 205–213.

Baskin, J. M., and Baskin, C. C. (1992). Role of temperature and light in the germination ecology of buried seeds of weedy species of disturbed forests. I. *Lobelia inflata*. *Can. J. Bot.* **70**, 589–592.

Baskin, J. M., and Baskin, C. C. (1995). Variation in the annual dormancy cycle in buried seeds of the weedy winter annual *Viola arvensis*. *Weed Res.* **35**, 353–362.

Baskin, J. M., Baskin, C. C., and Chester, E. W. (1992). Seed dormancy pattern and seed reserves as adaptations of the endemic winter annual *Lesquerella lescurii* (Brassicaceae) to its floodplain habitat. *Nat. Areas J.* **12**, 184–190.

Baskin, J. M., Baskin, C. C., and McCormick, J. F. (1987). Seasonal changes in germination responses of buried seeds of *Portulaca smallii*. *Bull. Torrey Bot. Club* **114**, 169–172.

Baskin, J. M., Baskin, C. C., and Parr, J. C. (1986). Field emergence of *Lamium amplexicaule* L. and *L. purpureum* L. in relation to the annual seed dormancy cycle. *Weed Res.* **26**, 185–190.

Baskin, J. M., Baskin, C. C., and Spooner, D. M. (1989). Role of temperature, light and date seeds were exhumed from soil on germination of four wetland perennials. *Aquat. Bot.* **35**, 387–394.

Bazzaz, F. A. (1970). Secondary dormancy in the seeds of the common ragweed *Ambrosia artemisiifolia*. *Bull. Torrey Bot. Club* **97**, 302–305.

Bell, D. T., and Muller, C. H. (1973). Dominance of California annual grasslands by *Brassica nigra*. *Am. Midl. Nat.* **90**, 277–299.

Benech Arnold, R. L., Ghersa, C. M., Sanchez, R. A., and Garcia Fernandez, A. E. (1988). The role of fluctuating temperatures in the germination and establishment of *Sorghum halepense* (L.) Pers. regulation of germination under leaf canopies. *Funct. Ecol.* **2**, 311–318.

Benech Arnold, R. L., Ghersa, C. M., Sanchez, R. A., and Insausti, P. (1990a). Temperature effects on dormancy release and germination rate in *Sorghum halepense* (L.) Pers. seeds: A quantitative analysis. *Weed Res.* **30**, 81–89.

Benech Arnold, R. L., Ghersa, C. M., Sanchez, R. A., and Insausti, P. (1990b). A mathematical model to predict *Sorghum halepense* (L.) Pers. seedling emergence in relation to soil temperature. *Weed Res.* **30**, 91–99.

Benvenuti, S. (1995). Soil light penetration and dormancy of jimsonweed (*Datura stramonium*) seeds. *Weed Sci.* **43**, 389–393.

Benvenuti, S., and Macchia, M. (1995). Effect of hypoxia on buried weed seed germination. *Weed Res.* **35**, 443–451.

Bewley, J. D. (1980). Secondary dormancy (skotodormancy) in seeds of lettuce (*Lactuca sativa* cv. Grand Rapids) and its release by light, gibberellic acid and benzyladenine. *Physiol. Plant.* **49**, 277–280.

Bibbey, R. O. (1948). Physiological studies on weed seed germination. *Plant Physiol.* **23**, 467–484.

Biddington, N. L., and Thomas, T. H. (1979). Residual effects of high temperature pre-treatment on the germination of celery seeds (*Apium graveolens*). *Physiol. Plant.* **47**, 211–214.

Blaauw-Jansen, G. (1981). Differences in the nature of thermodormancy and far-red dormancy in lettuce seeds. *Physiol. Plant.* **53**, 553–557.

Black, M. (1959). Dormancy studies in seed of *Avena fatua*. I. The possible role of germination inhibitors. *Can. J. Bot.* **37**, 393–402.

Bliss, D., and Smith, H. (1985). Penetration of light into soil and its role in the control of seed germination. *Plant Cell Environ.* **8**, 475–483.

Bond, W., and Burch, P. J. (1990). Stimulation of weed seed germination by 1-(3-chlorophthalimido) cyclohexanecarboxamide (AC 94377). *Ann. Appl. Biol.* **116**, 119–130.

Bonnewell, V., Koukkari, W. L., and Pratt, D. C. (1983). Light, oxygen, and temperature requirements for *Typha latifolia* seed germination. *Can. J. Bot.* **61**, 1330–1336.

Botha, F. C., Grobbelaar, N., and Small, J. G. C. (1982a). Seed germination in *Citrullus lanatus*. 1. Effect of white light and growth substances on germination. *S. Afr. J. Bot.* **1**, 10–13.

Botha, F. C., Small, J. G. C., and Grobbelaar, N. (1982b). Seed germination in *Citrullus lanatus*. 2. The involvement of phytochrome and ethylene in controlling light sensitivity. *S. Afr. J. Bot.* **1**, 131–133.

Botha, F. C., Small, J. G. C., Grobbelaar, N., and Eller, B. M. (1983). Seed germination in *Citrullus lanatus*. 4. The inhibition of germination within the fruits by light. *S. Afr. J. Bot.* **2**, 181–183.

Bouwmeester, H. J. (1990). "The Effect of Environmental Conditions on the Seasonal Dormancy Pattern and Germination of Weed Seeds." Ph.D dissertation, Agricultural University, Wageningen, The Netherlands.

Bouwmeester, H. J., and Karssen, C. M. (1989). Environmental factors influencing the expression of dormancy patterns in weed seeds. *Ann. Bot.* **63**, 113–120.

Bouwmeester, H. J., and Karssen, C. M. (1992). The dual role of temperature in the regulation of the seasonal changes in dormancy and germination of seeds of *Polygonum persicaria* L. *Oecologia* **90**, 88–94.

Bouwmeester, H. J., and Karssen, C. M. (1993a). Seasonal periodicity in germination of seeds of *Chenopodium album* L. *Ann. Bot.* **72**, 463–473.

Bouwmeester, H. J., and Karssen, C. M. (1993b). The effect of environmental conditions on the annual dormancy pattern of seeds of *Spergula arvensis*. *Can. J. Bot.* **71**, 64–73.

Bouwmeester, H. J., and Karssen, C. M. (1993c). Annual changes in dormancy and germination in seeds of *Sisymbrium officinale* (L.) Scop. *New Phytol.* **124**, 179–191.

Bouwmeester, H. J., Derks, L., Keizer, J. J., and Karssen, C. M. (1994). Effects of endogenous nitrate content of *Sisymbrium officinale* seeds on germination and dormancy. *Acta Bot. Neerl.* **43**, 39–50.

Bowes, G. G., Thomas, A. G., and Lefkovitch, L. P. (1995). Changes with time in the germination of buried scentless chamomile (*Matricaria perforata* Merat) seeds. *Can. J. Plant Sci.* **75**, 277–281.

Bradford, K. J. (1996). Population-based models describing seed dormancy behaviour: Implications for experimental design and interpretation. *In* "Plant Dormancy: Physiolotgy, Biochemistry and Molecular Biology" (G. A. Lang, ed.), pp. 313–339. CAB International, Wallingford.

Brenchley, W. E., and Warington, K. (1930). The weed seed population of arable soil. I. Numerical estimation of viable seeds and observations on their natural dormancy. *J. Ecol.* **18**, 235–272.

Brennan, T., Willemsen, R., Rudd, T., and Frenkel, C. (1978). The interaction of oxygen and ethylene in the release of ragweed seeds from dormancy. *Bot. Gaz.* **139**, 46–49.

Burges, A. (1967). The soil system. *In* "Soil Biology" (A. Burges and F. Raw, eds.), pp. 1–13. Academic Press, London.

Burnet, M. W. M., Christopher, J. T., Holtum, J. A. M., and Powles, S. B. (1994). Identification of two mechanisms of sulfonylurea resistance within one population of rigid ryegrass (*Lolium rigidum*) using a selective germination medium. *Weed Sci.* **42**, 468–473.

Buyanovsky, G., Dicke, M., and Berwick, P. (1982). Soil environment and activity of soil microflora in the Negev desert. *J. Arid Environ.* **5**, 13–28.

Cairns, A. L. P., and de Villiers, O. T. (1986). Breaking dormancy of *Avena fatua* L. seed by treatment with ammonia. *Weed Res.* **26**, 191–197.

Campbell, P., van Staden, J., Stevens, C., and Whitwell, M. I. (1992). The effects of locality, season and year of seed collection on the germination of bugweed (*Solanum mauritianum* Scop.) seeds. *S. Afr. J. Bot.* **58**, 310–316.

Campbell, R. B., and Moreau, R. A. (1979). Ethylene in a compacted field soil and its effect on growth, tuber quality, and yield of potatoes. *Am. Potato J.* **56**, 199–210.

Cardosa, V. J. M., and Felippe, G. M. (1983). Endogenous hormones and the germination of *Cucumis anguria* L. *Revta Brasil. Bot.* **6**, 29–31.

Cardosa, V. J. M., and Felippe, G. M. (1987). Endogenous ethylene and the germination of *Cucumis anguria* L. Seeds. *Revta Brasil. Bot.* **10**, 29–32.

Carmona, R., and Murdoch, A. J. (1995). Interactions of temperature and dormancy-relieving compounds on the germination of weed seeds. *Seed Sci. Res.* **5**, 227–236.

Chancellor, R. J. (1981). The manipulation of weed behaviour for control purposes. *Phil. Trans. R. Soc. Lond. B* **295**, 103–110.

Chang, I.-K., and Foy, C. L. (1971). Effect of picloram on germination and seedling development of four species. *Weed Sci.* **19**, 58–64.

Chepil, W. S. (1946). Germination of weed seeds. I. Longevity, periodicity of germination, and vitality of seeds in cultivated soil. *Sci. Agric.* **26**, 307–346.

Chou, C.-H., and Muller, C. H. (1972). Allelopathic mechanisms of *Arctostaphylos glandulosa* var. *zacaensis*. *Am. Midl. Nat.* **88**, 324–347.

Chou, C.-H., and Patrick, Z. A. (1976). Identification and phytotoxic activity of compounds produced during decomposition of corn and rye residues in soil. *J. Chem. Ecol.* **2**, 369–387.

Christensen, M., Meyer, S. E., and Allen, P. S. (1996). A hydrothermal time model of seed after-ripening in *Bromus tectorum*. *Seed Sci. Res.* **6**, 155–163.

Cocks, P. S., and Donald, C. M. (1973). The germination and establishment of two annual pasture grasses (*Hordeum leporinum* Link and *Lolium regidum* Gaud.). *Aust. J. Agric. Res.* **24**, 1–10.

Cohn, M. A., Butera, D. L., and Hughes, J. A. (1983). Seed dormancy in red rice. III. Response to nitrite, nitrate, and ammonium ions. *Plant Physiol.* **73**, 381–384.

Come, D., Corbineau, F., and Soudain, P. (1991). Beneficial effects of oxygen deprivation on germination and plant development. *In* "Plant Life under Oxygen Deprivation" (M. B. Jackson, D. D. Davies, and H. Lambers, eds.), pp. 69–83. SPB Academic Publ., The Hague, The Netherlands.

Come, D., Perino, C., and Ralambosoa, J. (1985). Oxygen sensitivity of apple (*Pyrus malus* L.) embryos in relation to dormancy. *Israel J. Bot.* **34**, 17–23.

Corbineau, F., Bagnio, S., and Come, D. (1990). Sunflower (*Helianthus annuus* L.) seed dormancy and its regulation by ethylene. *Israel J. Bot.* **39**, 313–325.

Corbineau, F., Belaid, D., and Come, D. (1992). Dormancy of *Bromus rubens* L. seeds in relation to temperature, light and oxygen effects. *Weed Res.* **32**, 303–310.

Corbineau, F., and Come, D. (1982). Effect of the intensity and duration of light at various temperatures on the germination of *Oldenlandia corymbosa* L. seeds. *Plant Physiol.* **70**, 1518–1520.

Corbineau, F., and Come, D. (1985). Effect of temperature, oxygen, and gibberellic acid on the development of photosensitivity in *Oldenlandia corymbosa* L. seeds during their incubation in darkness. *Plant Physiol.* **79**, 411–414.

Corbineau, F., and Come, D. (1995). Control of seed germination and dormancy by the gaseous environment. *In* "Seed Development and Germination" (J. Kigel and G. Galili, eds.), pp. 397–424. Dekker, New York.

Corbineau, F., Rudnicki, R. M., and Come, D. (1988). Induction of secondary dormancy in sunflower seeds by high temperature. Possible involvement of ethylene biosynthesis. *Physiol. Plant.* **73**, 368–373.

Courtney, A. D. (1968). Seed dormancy and field emergence in *Polygonum aviculare*. *J. Appl. Ecol.* **5**, 675–684.

Cresswell, C. F., and Nelson, H. (1972). The effect of boron on the breaking, and possible control of dormancy of seed of *Themeda triandra* Fors[s]k. *Ann. Bot.* **36**, 771–780.

Cresswell, E. G., and Grime, J. P. (1981). Induction of a light requirement during seed development and its ecological consequences. *Nature* **291**, 583–585.

Darlington, H. T. (1922). Dr. W. J. Beal's seed-viability experiment. *Am. J. Bot.* **9**, 266–269.

Davis, W. E. (1930a). Primary dormancy, after-ripening, and the development of secondary dormancy in embryos of *Ambrosia trifida*. *Am. J. Bot.* **17**, 58–76.

Davis, W. E. (1930b). The development of dormancy in seeds of cocklebur (*Xanthium*). *Am. J. Bot.* **17**, 77–87.

Derkx, M. P. M., and Karssen, C. M. (1994). Are seasonal dormancy patterns in *Arabidopsis thaliana* regulated by changes in seed sensitivity to light, nitrate and gibberellin? *Ann. Bot.* **73**, 129–136.

Derkx, M. P. M., Smidt, W. J., Van der Plas, L. H. W., and Karssen, C. M. (1993). Changes in dormancy of *Sisymbrium officinale* seeds do not depend on changes in respiratory activity. *Physiol. Plant.* **89**, 707–718.

Dixit, S., and Amritphale, D. (1996). Very low fluence and low fluence response in the induction and inhibition of seed germination in *Celosia argentea*. *Seed Sci. Res.* **6**, 43–48.

Donald, W. W., and Hoerauf, R. A. (1985). Enhanced germination and emergence of dormant wild mustard (*Sinapis arvensis*) seed by two substituted phthalimides. *Weed Sci.* **33**, 894–902.

Donald, W. W., and Tanaka, F. S. (1993). The germination stimulant AC94377 reduces seed survival of wild mustard (*Sinapis arvensis*). *Weed Sci.* **41**, 185–193.

Downs, R. J. (1964). Photocontrol of germination of the seeds of Bromeliaceae. *Phyton (Buenos Aires)* **21**, 1–6.

Duke, S. O. (1978). Interactions of seed water content with phytochrome-initiated germination of *Rumex crispus* (L.) seeds. *Plant Cell Physiol.* **19**, 1043–1049.

Duke, S. O., Egley, G. H., and Reger, B. J. (1977). Model for variable light sensitivity in imbibed dark-dormant seeds. *Plant Physiol.* **59**, 244–249.

Dyer, W. E. (1995). Exploiting weed seed dormancy and germination requirements through agronomic practices. *Weed Sci.* **43**, 498–503.

Edwards, T. I. (1933). The germination and growth of *Peltandra virginica* in the absence of oxygen. *Bull. Torrey Bot. Club* **60**, 573–581.

Egley, G. H. (1978). The ethylene-stimulated germination of light-sensitive common purslane seeds. *Plant Physiol. (Suppl.)* **61**, 16. [Abstract]

Egley, G. H. (1980). Stimulation of common cocklebur (*Xanthium pensylvanicum*) and redroot pigweed (*Amaranthus retroflexus*) seed germination by injections of ethylene into soil. *Weed Sci.* **28**, 510–514.

Egley, G. H. (1984). Ethylene, nitrate and nitrite interactions in the promotion of dark germination of common purslane seeds. *Ann. Bot.* **53**, 833–840.

Egley, G. H. (1986). Stimulation of weed seed germination in soil. *Rev. Weed Sci.* **2**, 67–89.

Egley, G. H. (1989). Some effects of nitrate-treated soil upon the sensitivity of buried redroot pigweed (*Amaranthus retroflexus* L.) seeds to ethylene, temperature, light and carbon dioxide. *Plant Cell Environ.* **12**, 581–588.

Egley, G. H., and Dale, J. E. (1970.) Ethylene, 2-chloroethylphosphonic acid, and witchweed germination. *Weed Sci.* **18**, 586–589.

Egley, G. H., and Duke, S. O. (1985). Physiology of weed seed dormancy and germination. *In* "Weed Physiology" (S. O. Duke, ed.), Vol. I, pp. 27–64. CRC Press, Boca Raton, FL.

Egley, G. H., and Williams, R. D. (1991). Emergence periodicity of six summer annual weed species. *Weed Sci.* **39**, 595–600.

Esashi, Y. (1991). Ethylene and seed germination. *In* "The Plant Hormone Ethylene" (A. K. Mattoo and J. C. Suttle, eds.), pp. 134–157. CRC Press, Boca Raton, FL.

Esashi, Y., Abe, Y., Ashino, H., Ishizawa, K., and Saitoh, K. (1989). Germination of cocklebur seed and growth of their axial and cotyledonary tissues in response to C_2H_4, CO_2 and/or O_2 under water stress. *Plant Cell Environ.* **12**, 183–190.

Esashi, Y., and Ishizawa, K. (1989). Oxygen-independent ethylene action in cocklebur seed germination in relation to osmoregulation. *In* "Biochemical and Physiological Aspects of Ethylene Production in Lower and Higher Plants" (H. Clijsters, M. DeProft, R. Marcelle, and M. Van Poucke, eds), pp. 73–80. Kluwer Academic, Dordrecht, The Netherlands.

Esashi, Y., and Katoh, H. (1975). Dormancy and impotency of cocklebur seeds. III. CO_2- and C_2H_4-dependent growth of the embryonic axis and cotyledon segments. *Plant Cell Physiol.* **16**, 707–718.

Esashi, Y., Kawabe, K., Isuzugawa, K., and Ishizawa, K. (1988). Interrelations between carbon dioxide and ethylene on the stimulation of cocklebur seed germination. *Plant Physiol.* **86**, 39–43.

Esashi, Y., Matsuyama, S., Hoshina, M., Ashino, H., and Ishizawa, K. (1990). Mechanism of action of ethylene in promoting the germination of cocklebur seeds. I. Osmoregulation. *Aust. J. Plant Physiol.* **17**, 537–550.

Esashi, Y., Okazaki, M., and Watanabe, K. (1976). The role of C_2H_4 in anaerobic induction of cocklebur seed germination. *Plant Cell Physiol.* **17**, 1151–1158.

Esashi, Y., Tsukada, Y., and Ohhara, Y. (1978). Interrelation between low temperature and anaerobiosis in the induction of germination of cocklebur seed. *Aust. J. Plant Physiol.* **5**, 337–345.

Evans, R. A., Kay, B. L., and Young, J. A. (1975). Microenvironment of a dynamic annual community in relation to range improvement. *Hilgardia* **43**, 79–102.

Evenari, M. (1952). The germination of lettuce seed. I. Light, temperature and coumarin as germination factors. *Palestine J. Bot. Jerusalem Ser.* **5**, 138–160.

Farmer, A. M., and Spence, D. H. N. (1987). Flowering, germination and zonation of the submerged aquatic plant *Lobelia dortmanna* L. *J. Ecol.* **75**, 1065–1076.

Favier, J. F. (1995). A model for germination rate during dormancy loss in *Hordeum vulgare*. *Ann. Bot.* **76**, 631–638.

Fawcett, R. S., and Slife, F. W. (1975). Germination stimulation properties of carbamate herbicides. *Weed Sci.* **23**, 419–424.

Fawcett, R. S., and Slife, F. W. (1978). Effects of field applications of nitrate on weed seed germination and dormancy. *Weed Sci.* **26**, 594–596.

Felippe, G. M. (1978). Effects of temperature on germination of *Rumex obtusifolius*. *Rev. Museu Paul.* **25**, 173–181.

Fenner, M. (1980a). Germination tests on thirty-two east African weed species. *Weed Res.* **20**, 135–138.

Fenner, M. (1980b). The induction of a light requirement in *Bidens pilosa* seeds by leaf canopy shade. *New Phytol.* **84**, 103–106.

Flater, R. L., Yarish, W., and Vaartnou, H. (1974). Effects of picloram on germination and development of six crop species. *Can. J. Plant Sci.* **54**, 219–221.

Foley, M. E. (1994). Temperature and water status of seed affect afterripening in wild oat (*Avena fatua*). *Weed Sci.* **42**, 200–204.

Frankland, B. (1975). Phytochrome control of seed germination in relation to the light environment. *In* "Light and Plant Development" (H. Smith, ed.), pp. 477–491. Butterworths, London/Boston.

Freijsen, A. H. J., Troelstra, S. R., and van Kats, M. J. (1980). The effect of soil nitrate on the germination of *Cynoglossum officinale* L. (Boraginaceae) and its ecological significance. *Acta Oecol.* **1**, 71–79.

Froud-Williams, R. J., Drennan, D. S. H., and Chancellor, R. J. (1984). The influence of burial and dry-storage upon cyclic changes in dormancy, germination and response to light in seeds of various arable weeds. *New Phytol.* **96**, 473–481.

Froud-Williams, R. J., Hilton, J. R., and Dixon, J. (1986). Evidence for an endogenous cycle of dormancy in dry stored seeds of *Poa trivialis* L. *New Phytol.* **102**, 123–131.

Fuchs, W. H. (1941). Keimungsstudien an getreide. I. Keimungstemperatur und reifezustand. *Z. Pflanzenzuchtg.* **24**, 165–185.

Gange, A. C., Brown, V. K., and Farmer, L. M. (1992). Effects of pesticides on the germination of weed seeds: Implications for manipulative experiments. *J. Appl. Ecol.* **29**, 303–310.

Ghersa, C. M., Benech Arnold, R. L., and Martinez-Ghersa, M. A. (1992). The role of fluctuating temperatures in germination and establishment of *Sorghum halepense:* Regulation of germination at increasing depths. *Funct. Ecol.* **6**, 460–468.

Goodlass, G., and Smith, K. A. (1978). Effect of pH, organic matter content and nitrate on the evolution of ethylene from soils. *Soil. Biol. Biochem.* **10**, 193–199.

Gordon, A. G. (1973). The rate of germination. *In* "Seed Ecology" (W. Heydecker, ed.), pp. 391–409. Pennsylvania State Univ. Press, University Park, PA.

Gorecki, R. J., Ashino, H., Satoh, S., and Esashi, Y. (1991). Ethylene production in pea and cocklebur seeds of differing vigour. *J. Exp. Bot.* **42**, 407–414.

Gorski, T. (1975). Germination of seeds in the shadow of plants. *Physiol. Plant.* **34**, 342–346.

Gorski, T., and Gorska, K. (1979). Inhibitory effects of full daylight on the germination of *Lactuca sativa* L. *Planta* **144**, 121–124.

Gorski, T., K. Gorski, K., and Nowicki, J. (1977). Germination of seeds of various herbaceous species under leaf canopy. *Flora* **166**, 249–259.

Goudey, J. S., Saini, H. S., and Spencer, M. S. (1987). Seed germination of wild mustard (*Sinapis arvensis*): Factors required to break primary dormancy. *Can. J. Bot.* **65**, 849–852.

Goudey, J. S., Saini, H. S., and Spencer, M. S. (1988). Role of nitrate in regulating germination of *Sinapis arvensis* L. (wild mustard). *Plant Cell Environ.* **11**, 9–12.

Griggs, F. T. (1976). Life history strategies of the genus *Orcuttia* (Gramineae). *In* "Vernal Pools, Their Ecology and Conservation" (S. Jain, ed.), pp. 57–62. Institute of Ecology Publication No. 9, Univ. of California, Davis.

Grime, J. P., Mason, G., Curtis, A. V., Rodman, J., Band, S. R., Mowforth, M. A. G., Neal, A. M., and Shaw, S. (1981). A comparative study of germination characteristics in a local flora. *J. Ecol.* **69**, 1017–1059.

Griswold, S. M. (1936). Effect of alternate moistening and drying on germination of seeds of western range plants. *Bot. Gaz.* **98**, 243–269.

Gross, K. L. (1985). Effects of irradiance and spectral quality on the germination of *Verbascum thapsus* L. and *Oenothera biennis* L. seeds. *New Phytol.* **101**, 531–541.

Gutterman, Y. (1980–1981). Annual rhythm and position effect in the germinability of *Mesembryanthemum nodiflorum. Israel J. Bot.* **29**, 93–97.

Hacker, J. B. (1989). The potential for buffel grass renewal from seed in 16-year-old buffel grass-siratro pastures in southeast Queensland. *J. Appl. Ecol.* **26**, 213–222.

Haferkamp, M. R., Karl, M. G., and MacNeil, M. D. (1994). Influence of storage, temperature, and light on germination of Japanese brome seed. *J. Range. Manage.* **47**, 140–144.

Hall, M. A., Acaster. M. A., Cantrell, I. C., Smith, A. R., and Yousif, O. A. F. (1987). The manipulation of weed seed dormancy. *Brit. Crop Prot. Conf. Weeds* **7C-4**, 719–724.

Hamner, C. L., Moulton, J. E., and Tukey, H. B. (1946). Effect of

treating soil and seeds with 2,4-dichlorophenoxyacetic acid on germination and development of seedlings. *Bot. Gaz.* **107**, 352–361.

Harper, J. L. (1957). The ecological significance of dormancy and its importance in weed control. *Proc. Int. Cong. Crop Protect.* **4**, 415–420.

Hart, J. W., and Berrie. A. M. M. (1966). The germination of *Avena fatua* under different gaseous environments. *Physiol. Plant.* **19**, 1020–1025.

Hartmann, K. M., and Nezadal, W. (1990). Photocontrol of weeds without herbicides. *Naturwissenschaften* **77**, 158–163.

Haut, I. C. (1932). The influence of drying on the after-ripening and germination of fruit tree seeds. *Proc. Am. Soc. Hort. Sci.* **29**, 371–374.

Hilhorst, H. W. M. (1993). New aspects of seed dormancy. *In* "Proceedings of the 4th International Workshop on Seeds: Basic and Applied Aspects of Seed Biology" (E. Come and F. Corbineau, eds.), pp. 551–579. Universite Pierre et Marie Curie, Paris.

Hilhorst, H. W. M., and Karssen, C. M. (1989). Nitrate reductase independent stimulation of seed germination in *Sisymbrium officinale* L. (hedge mustard) by light and nitrate. *Ann. Bot.* **63**, 131–137.

Hilhorst, H. W. M., Smitt, A. I., and Karssen, C. M. (1986). Gibberellin-biosynthesis and -sensitivity mediated stimulation of seed germination of *Sisymbrium officinale* by red light and nitrate. *Physiol. Plant.* **67**, 285–290.

Hilton, J. R. (1984a). The influence of temperature and moisture status on the photoinhibition of seed germination in *Bromus sterilis* L. by the far-red absorbing form of phytochrome. *New Phytol.* **97**, 369–374.

Hilton, J. R. (1984b). The influence of light and potassium nitrate on the dormancy and germination of *Avena fatua* L. (wild oat) seed and its ecological significance. *New Phytol.* **96**, 31–34.

Hilton, J. R. (1985). The influence of light and potassium nitrate on the dormancy and germination of *Avena fatua* L. (wild oat) seed stored buried under natural conditions. *J. Exp. Bot.* **36**, 974–979.

Hilton, J. R., Froud-Williams, R. J., and Dixon, J. (1984). A relationship between phytochrome photoequilibrium and germination of seeds of *Poa trivialis* L. from contrasting habitats. *New Phytol.* **97**, 375–379.

Hintikka, V. (1987). Germination ecology of *Galeopsis bifida* (Lamiaceae) as a pioneer species in forest succession. *Silva Fennica* **21**, 301–313.

Hintikka, V. (1988). Induction of secondary dormancy in seeds of *Barbarea stricta* and *B. vulgaris* by chlormequat and daminozide, and its termination by gibberellic acid. *Weed Res.* **28**, 7–11.

Hintikka, V. (1990). Germination ecology and survival strategy of *Rumex acetosella* (Polygonaceae) on drought-exposed rock outcrops in South Finland. *Ann. Bot. Fennici* **27**, 205–215.

Holm, R. E. (1972). Volatile metabolites controlling germination in buried weed seeds. *Plant Physiol.* **50**, 293–297.

Honek, A., and Martinkova, Z. (1992). The induction of secondary seed dormancy by oxygen deficiency in a barnyard grass *Echinochloa crus-galli. Experientia* **48**, 904–906.

Hopkins, H. T., and Toole, E. H. (1950). Effect of DDT on germination of certain seeds. *Bot. Gaz.* **112**, 130–132.

Horng, L. C., and Leu, L. S. (1978). The effects of depth and duration of burial on the germination of ten annual weed seeds. *Weed Sci.* **26**, 4–10.

Hou, J. Q., and Simpson, G. M. (1990). Phytochrome action and water status in seed germination of wild oats (*Avena fatua*). *Can. J. Bot.* **68**, 1722–1727.

Hsiao, A. I. (1992). Effects of repetitive drying, acid immersion, and red light treatments on phytochrome- and gibberellin A₃-mediated germination of skotodormant lettuce seeds. *J. Exp. Bot.* **43**, 741–746.

Hsiao, A. I., Vidaver, W., and Quick, W. A. (1984). Acidification, growth promoter, and red light effects on germination of skotodormant lettuce seeds (*Lactuca sativa*). *Can. J. Bot.* **62**, 1108–1115.

Huffman, A. H., and Jacoby, P. W., Jr. (1984). Effects of herbicides on germination and seedling development of three native grasses. *J. Range Manage.* **37**, 40–43.

Hunt, P. G., Campbell, R. B., and Moreau, R. A. (1980). Factors affecting ethylene accumulation in a Norfolk sandy loam soil. *Soil Sci.* **129**, 22–27.

Hurtt, W., and Taylorson, R. B. (1986). Chemical manipulation of weed emergence. *Weed Res.* **26**, 259–267.

Jones, J. F., and Hall, M. A. (1979). Studies on the requirement for carbon dioxide and ethylene for germination of *Spergula arvensis* seeds. *Plant Sci. Lett.* **16**, 87–93.

Junttila, O. (1976). Seed germination and viability in five *Salix* species. *Universitetsforlaget Tromso Norway* **9**, 19–24.

Justice, O. L. (1941). A study of dormancy in seeds of *Polygonum*. *Cornell Univ. Agric. Exp. Sta. Mem.* 235.

Justice, O. L. (1944). Viability and dormancy in seeds of *Polygonum amphibium* L., *P. coccineum* Muhl. and *P. hydropiperoides* Michx. *Am. J. Bot.* **31**, 369–377.

Karssen, C. M. (1967). The light promoted germination of the seeds of *Chenopodium album* L. I. The influence of the incubation time on quantity and rate of the response to red light. *Acta Bot. Neerl.* **16**, 156–160.

Karssen, C. M. (1970). The light promoted germination of the seeds of *Chenopodium album* L. IV. Effects of red, far-red and white light on non-photoblastic seeds incubated in mannitol. *Acta Bot. Neerl.* **19**, 95–108.

Karssen, C. M. (1980–1981a). Environmental conditions and endogenous mechanisms involved in secondary dormancy of seeds. *Israel J. Bot.* **29**, 45–64.

Karssen, C. M. (1980–1981b). Patterns of change in dormancy during burial of seeds in soil. *Israel J. Bot.* **29**, 65–73.

Karssen, C. M. (1982). Seasonal patterns of dormancy in weed seeds. *In* "The Physiology and Biochemistry of Seed Development, Dormancy and Germination" (A. A. Khan, ed.), pp. 243–270. Elsevier Biomedical Press, Amsterdam.

Karssen, C. M., Derkx, M. P. M., and Post, B. J. (1988). Study of seasonal variation in dormancy of *Spergula arvensis* L. seeds in a condensed annual temperature cycle. *Weed Res.* **28**, 449–457.

Karssen, C. M., and Hilhorst, H. W. M. (1992). Effect of chemical environment on seed germination. *In* "Seeds: The Ecology of Regeneration in Plant Communities" (M. Fenner, ed.), pp. 327–348. CAB International, Wallingford, UK.

Kasperbauer, M. J., and Hunt, P. G. (1988). Biological and photometric measurement of light transmission through soils of various colors. *Bot. Gaz.* **149**, 361–364.

Katoh, H. and Esashi, Y. (1975a). Dormancy and impotency of cocklebur seeds. I. CO_2, C_2H_4, O_2 and high temperature. *Plant Cell Physiol.* **16**, 687–696.

Katoh, H., and Esashi, Y. (1975b). Dormancy and impotency of cocklebur seeds. II. Phase sequence in germination process. *Plant Cell Physiol.* **16**, 697–706.

Keegan, A. B., Kelly, K. M., and van Staden, J. (1989). Ethylene involvement in dormancy release of *Ricinodendron ratanenii* seeds. *Ann. Bot.* **63**, 229–234.

Kennedy, R. A., Barrett, S. C. H., Vander Zee, D., and Rumpho, M. E. (1980). Germination and seedling growth under anaerobic conditions in *Echinochloa crus-galli* (barnyard grass). *Plant Cell Environ.* **3**, 243–248.

Kepczynski, J., Rudnicki, R. M., and Khan, A. A. (1977). Ethylene requirement for germination of partly after-ripened apple embryo. *Physiol. Plant.* **40**, 292–295.

Khan, A. A., and Karssen, C. M. (1980). Induction of secondary dormancy in *Chenopodium bonus-henricus* L. seeds by osmotic and high temperature treatments and its prevention by light and growth regulators. *Plant Physiol.* **66**, 175–181.

Kidd, F. (1914a). The controlling influence of carbon dioxide in the maturation, dormancy, and germination of seeds. I. *Proc. Roy. Soc. Lond.* **87**, 408–421.

Kidd, F. (1914b). The controlling influence of carbon dioxide in the maturation, dormancy and germination of seeds. II. *Proc. Roy. Soc. Lond.* **87**, 609–625.

Kidd, F. (1917). The controlling influence of carbon dioxide. IV. On the production of secondary dormancy in seeds of *Brassica alba* following treatment with carbon dioxide, and the relation of this phenomenon to the question of stimuli in growth processes. *Ann. Bot.* **31**, 457–487.

King, T. J. (1975). Inhibition of seed germination under leaf canopies in *Arenaria serpyllifolia*, *Veronica arvensis* and *Cerastium holosteoides*. *New Phytol.* **75**, 87–90.

Kivilaan, A. (1975). Skotodormancy in *Verbascum blattaria* seed. *Flora* **164**, 1–5.

Koller, D., Sachs, M., and Negbi, M. (1964). Germination-regulating mechanisms in some desert seeds. VIII. *Artemisia monosperma*. *Plant Cell Physiol.* **5**, 85–100.

Kozlowski, T. T., and Torrie, J. H. (1965). Effect of soil incorporation of herbicides on seed germination and growth of pine seedlings. *Soil Sci.* **100**, 139–146.

Kristie, D. N., Bassi, P. K., and Spencer, M. S. (1981). Factors affecting the induction of secondary dormancy in lettuce. *Plant Physiol.* **67**, 1224–1229.

Kumari, J., Thomas, T. P., and Sen, D. N. (1987). Breaking seed dormancy in some grasses of Indian arid zone. *Geobios* **14**, 131–133.

Le Deunff, M. Y. (1973). Interactions entre l'oxygene et allumiere dans la germination et l'induction d'une dormance secondaire chez les semences de *Rumex crispus* L. *C. R. Acad. Sci. Paris. Ser. D* **276**, 2381–2384.

Le Deunff, Y., and Chaussat, R. (1968). Etude de la dormance secondaire des semences chez *Rumex crispus* L. *Ann. Physiol. Veg.* **10**, 227–236.

Lockerman, R. H., and Putnam, A. R. (1979). Evaluation of allelopathic cucumbers (*Cucumis sativus*) as an aid to weed control. *Weed Sci.* **27**, 54–57.

Lodge, G. M., and Whalley, R. D. B. (1981). Establishment of warm- and cool-season native perennial grasses on the north-west slopes of New South Wales. I. Dormancy and germination. *Aust. J. Bot.* **29**, 111–119.

Lonchamp, J.-P., and Gora, M. (1979). Influence d'anoxies partielles sur la germination de semences de mauvaises herbes. *Oecol. Plant.* **14**, 121–128.

Lush, W. M., and Groves, R. H. (1981). Germination, emergence and surface establishment of wheat and ryegrass in response to natural and artificial hydration-dehydration cycles. *Aust. J. Agric. Res.* **32**, 731–739.

Lush, W. M., Groves, R. H., and Kaye, P. E. (1981). Presowing hydration-dehydration treatments in relation to seed germination and early seedling growth of wheat and ryegrass. *Aust. J. Plant Physiol.* **8**, 409–425.

Lynch, J. M. (1975). Ethylene in soil. *Nature* **256**, 576–577.

Mabbayad, M. O., and Moody, K. (1992). Herbicide seed treatment for weed control in wet-seeded rice. *Trop. Pest Manage.* **38**, 9–12.

Mandoli, D. F., Waldron, L., Nemson, J. A., and Briggs, W. R. (1982). Soil light transmission: Implications for phytochrome-mediated responses. *Carnegie Inst. Wash. Yrbk.* **81**, 32–34.

Marks, M. K., and Nwachuku, A. C. (1986). Seed-bank characteristics in a group of tropical weeds. *Weed Res.* **26,** 151–157.

Matsuo, K., and Kubota, T. (1988). Effect of temperature and light conditions on the breaking of seed dormancy and the germination of *Chenopodium album* L. *Weed Res. (Japan)* **33,** 293–300.

Maynard, M. L., and Gates, D. H. (1963). Effects of wetting and drying on germination of crested wheatgrass seed. *J. Range Manage.* **16,** 119–121.

McDonough, W. T. (1968). Secondary dormancy in seeds: Effects of substrate hydration. *Bot. Gaz.* **129,** 361–364.

McIntyre, S., Mitchell, D. S., and Ladiges, P. Y. (1989). Seedling mortality and submergence in *Diplachne fusca*: A semi-aquatic weed of rice fields. *J. Appl. Ecol.* **26,** 537–549.

Mekki, M., and Leroux, G. D. (1991). False chamomile seed germination requirements and its enhancement by ethephon and nitrate. *Weed Sci.* **39,** 385–389.

Metzger, J. D. (1983). Promotion of germination of dormant weed seeds by substituted phthalimides and gibberellic acid. *Weed Sci.* **31,** 285–289.

Meyer, S. E., and Kitchen, S. G. (1992). Cyclic seed dormancy in the short-lived perennial *Penstemon palmeri*. *J. Ecol.* **80,** 115–122.

Milberg, P. (1994a). Annual dark dormancy cycle in buried seeds of *Lychnis flos-cuculi*. *Ann. Bot. Fennici* **31,** 163–167.

Milberg, P. (1994b). Germination ecology of the polycarpic grassland perennials *Primula veris* and *Trollius europeaus*. *Ecography* **17,** 3–8.

Milberg, P. (1994c). Germination ecology of the grassland biennial *Linum cartharticum*. *Acta Bot. Neerl.* **43,** 261–269.

Mitchell, J. W., and Marth, P. C. (1946). Germination of seeds in soil containing 2,4-dichlorophenoxyacetic acid. *Bot. Gaz.* **107,** 408–416.

Montegut, J. (1975). Ecologie de la germination des mauvaises herbes. *In* "La Germination des Semences" (R. Chaussat and Y. Le Deunff, eds.), pp. 193–217. Bordas, Paris.

Morash, R., and Freedman, B. (1989). The effects of several herbicides on the germination of seeds in the forest floor. *Can. J. For. Res.* **19,** 347–350.

Mott, J. J. (1978). Dormancy and germination in five native grass species from savannah woodland communities of the Northern Territory. *Aust. J. Bot.* **26,** 621–631.

Mullverstedt, R. (1963). Untersuchungen uber die keimung von unkrautsamen in abhangigkeit vom sauerstoffpartialdruck. *Weed Res.* **3,** 154–163.

Mumford, P. M. (1988). Alleviation and induction of dormancy by temperature in *Impatiens glandulifera* Royle. *New Phytol.* **109,** 107–110.

Murdoch, A. J., and Carmona, R. (1993). The implications of the annual dormancy cycle of buried weed seeds for novel methods of weed control. *Brighton Crop Protect. Conf. Weeds* **4B-10,** 329–334.

Negm, F. B., Smith, O. E., and Kumamoto, J. (1973). The role of phytochrome in an interaction with ethylene and carbon dioxide in overcoming lettuce seed thermodormancy. *Plant Physiol.* **51,** 1089–1094.

Nikolaeva, M. G. (1969). "Physiology of Deep Dormancy in Seeds." Izdatel'stvo "Nauka," Leningrad. [Translated from Russian by Z. Shapiro, National Science Foundation, Washington, DC]

Noor, M., and Khan, M. A. (1994). Allelopathic potential of *Albizia samans* Merr. *Pakistan J. Bot.* **26,** 139–147.

Noronha, A., Vicente, M., and Felippe, G. M. (1978). Photocontrol of germination of *Cucumis anguria* L. *Biol. Plant.* **20,** 281–286.

Nutile, G. E. (1945). Inducing dormancy in lettuce seed with coumarin. *Plant Physiol.* **20,** 433–442.

Nyunt, S., and Grabe, D. F. (1987). Induction of secondary dormancy in seeds of meadowfoam (*Limnanthes alba* Benth.). *J. Seed Technol.* **11,** 103–110.

Odum, S. (1965). Germination of ancient seeds: Floristical observations and experiments with archaeologically dated soil samples. *Dansk Bot. Ark.* **24**(2), 1–70.

Olatoye, S. T., and Hall, M. A. (1973). Interaction of ethylene and light on dormant weed seeds. *In* "Seed Ecology" (W. Heydecker, ed.), pp. 233–249. Pennsylvania State Univ. Press, University Park.

Orozco-Segovia, A., Vazquez-Yanes, C., Coates-Estrada, R., and Perez-Nasser, N. (1987). Ecophysiological characteristics of the seed of the tropical forest pioneer *Urea caracasana* (Urticaceae). *Tree Physiol.* **3,** 375–386.

Pandya, S. M., and Baghela, N. (1973). Ecological studies of *Celosia argentea* Linn., a weed. I. Seed germination, seedling emergence and growth performance in different soil types. *Trop. Ecol.* **14,** 39–51.

Panetta, F. D., and Randall, R. P. (1993). Variation between *Emex australis* populations in seed dormancy/non-dormancy cycles. *Aust. J. Ecol.* **18,** 275–280.

Pemadasa, M. A., and Lovell, P. H. (1975). Factors controlling germination of some dune annuals. *J. Ecol.* **63,** 41–59.

Peterson, D. L., and Bazzaz, F. A. (1978). Life cycle characteristics of *Aster pilosus* in early successional habitats. *Ecology* **59,** 1005–1013.

Pollard, F. (1982). Light induced dormancy in *Bromus sterilis*. *J. Appl. Ecol.* **19,** 563–568.

Pons, T. L. (1983). Significance of inhibition of seed germination under the leaf canopy in ash coppice. *Plant Cell Environ.* **6,** 385–392.

Pons, T. L. (1986). Response of *Plantago major* seeds to the red/far-red ratio as influenced by other environmental factors. *Physiol. Plant.* **68,** 252–258.

Pons, T. L. (1989). Breaking of seed dormancy by nitrate as a gap detection mechanism. *Ann. Bot.* **63,** 139–143.

Pons, T. L. (1991a). Induction of dark dormancy in seeds: Its importance for the seed bank in the soil. *Funct. Ecol.* **5,** 669–675.

Pons, T. L. (1991b). Dormancy, germination and mortality of seeds in a chalk-grassland flora. *J. Ecol.* **79,** 765–780.

Pons, T. L., and Schroder, H. F. J. M. (1986). Significance of temperature fluctuation and oxygen concentration for germination of the rice field weeds *Fimbristylis littoralis* and *Scirpus juncoides*. *Oecologia* **68,** 315–319.

Popay, A. I. (1981). Germination of seeds of five annual species of barley grass. *J. Appl. Ecol.* **18,** 547–558.

Popay, A. I., Cox, T. I., Ingle, A., and Kerr, R. (1995). Seasonal emergence of weeds in cultivated soil in New Zealand. *Weed Res.* **35,** 429–436.

Popay, A. I., and Roberts, E. H. (1970a). Factors involved in the dormancy and germination of *Capsella bursa-pastoris* (L.) Medik. and *Senecio vulgaris* L. *J. Ecol.* **58,** 103–122.

Popay, A. I., and Roberts, E. H. (1970b). Ecology of *Capsella bursa-pastoris* (L.) Medik. and *Senecio vulgaris* L. in relation to germination behaviour. *J. Ecol.* **58,** 123–139.

Powell, A. D., Dulson, J., and Bewley, J. D. (1984). Changes in germination and respiratory potential of embryos of dormant Grand Rapids lettuce seeds during long-term imbibed storage, and related changes in the endosperm. *Planta* **162,** 40–45.

Powell, A. D., Leung, D. W. M., and Bewley, J. D. (1983). Long-term storage of dormant Grand Rapids lettuce seeds in the imbibed state: Physiological and metabolic changes. *Planta* **159,** 182–188.

Probert, R. J., Dickie, J. D., and Hart, M. R. (1989). Analysis of the effect of cold stratification on the germination response to light and alternating temperatures using selected seed populations of *Ranunculus sceleratus* L. *J. Exp. Bot.* **40,** 293–301.

Putwain, P. D., Scott, K. R., and Holliday, R. J.. (1982). The nature of resistance to triazine herbicides: Case histories of phenology and population studies. *In* "Herbicide resistance in Plants"

(H. M. LeBaron and J. Gressel, eds.), pp. 99–115. Wiley, New York.

Qi, M., and Upadhyaya, M. K. (1993). Seed germination ecophysiology of meadow salsify (*Tragopogon pratensis*) and western salsify (*T. dubius*). *Weed Sci.* **41**, 362–368.

Quail, P. H., and Carter, O. G. (1969). Dormancy in seeds of *Avena ludoviciana* and *A. fatua. Aust. J. Agric. Res.* **20**, 1–11.

Reisman-Berman, O., Kigel, J., and Rubin, B. (1989). Short soaking in water inhibits germination of *Datura ferox* L. and *D. stramonium* L. seeds. *Weed Res.* **29**, 357–363.

Reisman-Berman, O., Kigel, J., and Rubin, B. (1991). Dormancy patterns in buried seeds of *Datura ferox* and *D. stramonium. Can. J. Bot.* **69**, 173–179.

Rethy, R., Dedonder, A., Frederocq, H., and Greef, J. De. (1983). Factors affecting the induction and release of secondary dormancy in *Kalanchoe* seeds. *Plant Cell Environ.* **6**, 731–738.

Rice, E. L. (1979). Allelopathy: An update. *Bot. Rev.* **45**, 15–109.

Richter, D. D., and Markewitz, D. (1995). How deep is soil? *BioScience* **45**, 600–609.

Rietveld, W. J. (1975). Phytotoxic grass residues reduce germination and initial root growth of Ponderosa pine. *USDA For. Serv. Res. Pap.* RM-153.

Roberts, E. H. (1972). Dormancy: A factor affecting seed survival in the soil. *In* "Viability of Seeds" (E. H. Roberts, ed.), pp. 321–359. Syracuse Univ. Press, Syracuse, NY.

Roberts, E. H., and Benjamin, S. K. (1979). The interaction of light, nitrate and alternating temperature on the germination of *Chenopodium album, Capsella bursa-pastoris* and *Poa annua* before and after chilling. *Seed Sci. Technol.* **7**, 379–392.

Roberts, H. A. (1964). Emergence and longevity in cultivated soil of seeds of some annual weeds. *Weed Res.* **4**, 296–307.

Roberts, H. A., and Boddrell, J. E. (1983). Field emergence and temperature requirements for germination in *Solanum sarrachoides* Sendt. *Weed Res.* **23**, 247–252.

Roberts, H. A., and Lockett, P. M. (1978a). Seed dormancy and periodicity of seedling emergence in *Veronica hederifolia* L. *Weed Res.* **18**, 41–48.

Roberts, H. A., and Lockett, P. M. (1978b). Seed dormancy and field emergence in *Solanum nigrum* L. *Weed Res.* **18**, 231–241.

Roberts, H. A., and Neilson, J. E. (1982a). Role of temperature in the seasonal dormancy of seeds of *Veronica hederifolia* L. *New Phytol.* **90**, 745–749.

Roberts, H. A., and Neilson, J. E. (1982b). Seasonal changes in the temperature requirements for germination of buried seeds of *Aphanes arvensis* L. *New Phytol.* **92**, 159–166.

Rojas-Garciduenas, M., and Kommedahl, T. (1960). The effect of 2,4-D on germination of pigweed seed. *Weeds* **8**, 1–5.

Rudnicki, R. M., Braun, J. W., and Khan, A. A. (1978). Low pressure and ethylene in lettuce seed germination. *Physiol. Plant.* **43**, 189–194.

Russell, E. W. (1961). "Soil Conditions and Plant Growth," 9th Ed. Longmans, London.

Sahid, I., and K. Kalithasan. (1994). Effects of glufosinate-ammonium and terbuthylazine on germination and growth of two weed species. *Plant Protect. Quart.* **9**, 15–19.

Saini, H. S., Bassi, P. K., and Spencer, M. S. (1985a). Interactions among ethephon, nitrate, and after-ripening in the release of dormancy of wild oat (*Avena fatua*) seed. *Weed Sci.* **34**, 43–47.

Saini, H.S., Bassi, P. K., and Spencer, M. S. (1985b). Seed germination in *Chenopodium album* L.: Further evidence for the dependence of the effects of growth regulators on nitrate availability. *Plant Cell Environ.* **8**, 707–711.

Saini, H. S., Bassi, P. K., and Spencer, M. S. (1986). Use of ethylene

and nitrate to break seed dormancy of common lambsquarters (*Chenopodium album*). *Weed Sci.* **34**, 502–506.

Saini, H. S., and Spencer, M. S. (1987). Manipulation of seed nitrate content modulates the dormancy-breaking effect of ethylene on *Chenopodium album* seed. *Can. J. Bot.* **65**, 876–878.

Samimy, C., and Khan, A. A. (1983). Secondary dormancy, growth-regulator effects, and embryo growth potential in curly dock (*Rumex crispus*) seeds. *Weed Sci.* **31**, 153–158.

Sans, F. X., and Masalles, R. M. (1995). Phenological patterns in an arable land weed community related to disturbance. *Weed Res.* **35**, 321–332.

Sasaki, A., and Kozlowski, T. T. (1968). Effects of herbicides on seed germination and early seedling development of *Pinus resinosa. Bot. Gaz.* **129**, 238–246.

Sasaki, A., Kozlowski, T. T., and Torrie, J. H. (1968). Effect of pretreatment of pine seeds with herbicides on seed germination and growth of young seedlings. *Can. J. Bot.* **46**, 255–262.

Saxena, S., and Srivastava, S. (1994). Cytotoxicity of the herbicide basalin (fluchloralin) in *Helianthus* and *Linum. J. Environ. Sci. Health. Part B* **29**, 1137–1152.

Schafer, D. E., and Chilcote, D. O. (1970). Factors influencing persistence and depletion in buried seed populations. II. The effects of soil temperature and moisture. *Crop Sci.* **10**, 342–345.

Schimpf, D. J., and Palmblad, I. G. (1980). Germination response of weed seeds to soil nitrate and ammonium with and without simulated overwintering. *Weed Sci.* **28**, 190–193.

Schlatterer, E. F., and Tisdale, E. W. (1969). Effects of litter of *Artemisia, Chrysothamnus*, and *Tortula* on germination and growth of three perennial grasses. *Ecology* **50**, 869–873.

Schonbeck, M. W., and Egley, G. H. (1980a). Redroot pigweed (*Amaranthus retroflexus*) seed germination responses to afterripening, temperature, ethylene, and some other environmental factors. *Weed Sci.* **28**, 543–548.

Schonbeck, M. W., and Egley, G. H. (1980b). Effects of temperature, water potential, and light on germination responses of redroot pigweed seeds to ethylene. *Plant Physiol.* **65**, 1149–1154.

Schonbeck, M. W., and Egley, G. H. (1981a). Phase-sequence of redroot pigweed seed germination responses to ethylene and other stimuli. *Plant Physiol.* **68**, 175–179.

Schonbeck, M. W., and Egley, G. H. (1981b). Changes in sensitivity of *Amaranthus retroflexus* L. seeds to ethylene during preincubation. I. Constant temperatures. *Plant Cell Environ.* **4**, 229–235.

Schonbeck, M. W., and Egley, G. H. (1981c). Changes in sensitivity of *Amaranthus retroflexus* L. seeds to ethylene during preincubation. II. Effects of alternating temperature and burial in soil. *Plant Cell Environ.* **4**, 237–242.

Scopel, A. L., Ballare, C. L. and Radosevich, S. R. (1994). Photostimulation of seed germination during soil tillage. *New Phytol.* **126**, 145–152.

Scopel, A. L., Ballare, C. L., and Sanchez, R. A. (1991). Induction of extreme light sensitivity in buried weed seeds and its role in the perception of soil cultivations. *Plant Cell Environ.* **14**, 501–508.

Semeniuk, P., and Stewart, R. N. (1962). Temperature reversal of after-ripening of rose seeds. *Proc. Am. Soc. Hort. Sci.* **80**, 615–621.

Senden, J. W., Schenkeveld, A. J., and Verkaar, H. J. (1986). The combined effect of temperature and red : far-red ratio on the germination of some short-lived chalk grassland species. *Acta Oecol.* **7**, 251–259.

Sexsmith, J. J., and Pittman, U. J. (1963). Effect of nitrogen fertilizers on germination and stand of wild oats. *Weeds* **11**, 99–101.

Sharma, R. K., and Amritphale, D. (1989). Effect of light, seed transfer and potassium salts on germination of skotodormant seeds of *Hygrophila auriculata* (Schumach.) Haines. *Indian J. Exp. Biol.* **27**, 822–823.

Shaukat, S. S. (1974). The effects of 2,4-D and coumarin in conjunction with other compounds on germination of *Pinus nigra*. *J. Sci. (Karachi, Pakistan)* **3**, 45–50.

Shaukat, S. S., Khan, N. A., and Ahmed, F. (1980). Herbicide influence on germination and seedling growth of *Vigna mungo* (L.) Hepper and *V. radiata* (L.) Wilczek. *Pakistan J. Bot.* **12**, 97–106.

Shaukat, S. S., Perveen, G., Kahn, D., and Ahmad, M. (1985). Phytotoxic effects of *Citrullus colocynthis* (L.) Schrad on certain crop plants. *Pakistan J. Bot.* **17**, 235–246.

Silvertown, J. (1980). Leaf-canopy-induced seed dormancy in a grassland flora. *New Phytol.* **85**, 109–118.

Simpson, G. M. (1966). A study of germination in the seed of wild rice (*Zizania aquatica*). *Can. J. Bot.* **44**, 1–9.

Small, J. G. C., and Gutterman, Y. (1992). A comparison of thermo- and skotodormancy in seeds of *Lactuca serriola* in terms of induction, alleviation, respiration, ethylene and protein synthesis. *J. Plant Grow. Reg.* **11**, 301–310.

Smith, A. M., and Cook, R. J. (1974). Implications of ethylene production by bacteria for biological balance of soil. *Nature* **252**, 703–705.

Smith, H. (1994). Sensing the light environment: The function of the phytochrome family. *In* "Photomorphogenesis in Plants." (R. E. Kendrick and G. H. M. Kronenberg, eds.), 2nd Ed., pp. 377–416. Kluwer Academic, Dordrecht, The Netherlands.

Smith, K. A., and Dowell, R. J. (1973). Gas chromatographic analysis of the soil atmosphere: Automatic analysis of gas samples for O_2, N_2, Ar, CO_2, N_2O and C_1-C_4 hydrocarbons. *J. Chromatogr. Sci.* **11**, 655–658.

Smith, K. A., and Restall, S. W. F. (1971). The occurrence of ethylene in anaerobic soil. *J. Soil Sci.* **22**, 430–443.

Smith, K. A., and Russell, R. S. (1969). Occurrence of ethylene, and its significance, in anaerobic soil. *Nature* **222**, 769–771.

Speer, H. L., Hsiao, A. I., and Vidaver, W. (1974). Effects of germination-promoting substances given in conjunction with red light on the phytochrome-mediated germination of dormant lettuce seeds (*Lactuca sativa* L.). *Plant Physiol.* **54**, 852–854.

Squires, E. R. (1989). The effects of seasonal timing of disturbance on species composition in a first-year oldfield. *Bull. Torrey Bot. Club* **116**, 356–363.

Standifer, L. C., and Wilson, P. W. (1988). A high temperature requirement for after ripening of imbibed dormant *Poa annua* seeds. *Weed Res.* **28**, 365–371.

Standiforth, R. J., and Cavers, P. B. (1979). Field and laboratory germination responses of achenes of *Polygonum lapathifolium*, *P. pensylvanicum*, and *P. persicaria*. *Can. J. Bot.* **57**, 877–885.

Stokes, P. (1965). Temperature and seed dormancy. *In* "Encyclopedia of Plant Physiology" Vol. 15/2, pp. 746–803. Springer-Verlag, New York/Heidelberg/Berlin..

Stoller, E. W., and Wax, L. M. (1973). Periodicity of germination and emergence of some annual weeds. *Weed Sci.* **21**, 574–580.

Stoller, E. W., and Wax, L. M. (1974). Dormancy changes and fate of some annual weed seeds in the soil. *Weed Sci.* **22**, 151–155.

Stoutjesdijk, Ph. (1972). Spectral transmission curves of some types of leaf canopies with a note on seed germination. *Acta. Bot. Neerl.* **21**, 185–191.

Suttle, J. C., and Schreiner, D. R. (1982). The biological activity of AC 94,377[1-3-chlorophthalimido)-cyclohexanecarboxamide]. *J. Plant Grow. Reg.* **1**, 139–146.

Susuki, S., and Taylorson, R. B. (1981). Ethylene inhibition of phytochrome-induced germination in *Potentilla norvegica* L. seeds. *Plant Physiol.* **68**, 1385–1388.

Symons, S. J., Naylor, J. M., Simpson, G. M., and Adkins, S. W. (1986). Secondary dormancy in *Avena fatua*: Induction and characteristics in genetically pure dormant lines. *Physiol. Plant.* **68**, 27–33.

Takaki, M., Dietrich, S. M. C., Felippe, G. M., and Cardoso, V. J. M. (1982). Cytokinin and ethylene during the germination of *Rumex obtusifolius* L. *Revta Brasil. Bot.* **5**, 29–32.

Taylorson, R. B. (1970). Changes in dormancy and viability of weed seeds in soils. *Weed Sci.* **18**, 265–269.

Taylorson, R. B. (1972). Phytochrome controlled changes in dormancy and germination of buried weed seeds. *Weed Sci.* **20**, 417–422.

Taylorson, R. B. (1979). Response of weed seeds to ethylene and related hydrocarbons. *Weed Sci.* **27**, 7–10.

Taylorson, R. B. (1980). Aspects of seed dormancy in fall panicum (*Panicum dichotomiflorum*). *Weed Sci.* **28**, 64–67.

Taylorson, R. B. (1986). Water stress-induced germination of giant foxtail (*Setaria faberi*) seeds. *Weed Sci.* **34**, 871–875.

Taylorson, R. B. (1987). Environmental and chemical manipulation of weed seed dormancy. *Rev. Weed Sci.* **3**, 135–154.

Taylorson, R. B. and Borthwick, H. A. (1969). Light filtration by foliar canopies: Significance for light-controlled weed seed germination. *Weed Sci.* **17**, 48–51.

Taylorson, R. B., and Hendricks, S. B. (1973). Phytochrome transformation and action in seeds of *Rumex crispus* L. during secondary dormancy. *Plant Physiol.* **52**, 475–479.

Taylorson, R. B., and Hendricks, S. B. (1976). Interactions of phytochrome and exogenous gibberellic acid on germination of *Lamium amplexicaule* L. seeds. *Planta* **132**, 65–70.

Taylorson, R. B., and McWhorter, C. G. (1969). Seed dormancy and germination in ecotypes of Johnsongrass. *Weed Sci.* **17**, 359–361.

Tester, M., and Morris, C. (1987). The penetration of light through soil. *Plant Cell Environ.* **10**, 281–286.

Thanos, C. A., and Mitrakos, K. (1992). Watermelon seed germination. 2. Osmomanipulation of photosensitivity. *Seed Sci. Res.* **2**, 163–168.

Thill, D. C., Zamora, D. L., and Kambitsch, D. L. (1985). Germination and viability of common crupina (*Crupina vulgaris*) achenes buried in the field. *Weed Sci.* **33**, 344–348.

Thompson, K., and Grime, J. P. (1983). A comparative study of germination responses to diurnally-fluctuating temperatures. *J. Appl. Ecol.* **20**, 141–156.

Thompson, P. A. (1977). A note on the germination of *Narcissus bulbocodium* L. *New Phytol.* **79**, 287–290.

Thornton, N. C. (1935). Factors influencing germination and development of dormancy in cocklebur seeds. *Contrib. Boyce Thomp. Inst.* **7**, 477–496.

Thornton, N. C. (1936). Carbon dioxide storage. IX. Germination of lettuce seeds at high temperatures in both light and darkness. *Contrib. Boyce Thomp. Inst.* **8**, 25–40.

Toole, V. K. (1938). Germination requirements of the seed of some introduced and native range grasses. *Proc. Assoc. Offic. Seed Anal.* **30**, 227–243.

Toole, V. K., Bailey, W. K., and Toole, E. H. (1964). Factors influencing dormancy of peanut seeds. *Plant Physiol.* **39**, 822–832.

Toole, E. H., Hendricks, S. B., Borthwick, H. A., and Toole, V. K. (1956). Physiology of seed germination. *Annu. Rev. Plant Physiol.* **7**, 299–324.

Toole, E. H., and Toole, V. K. (1941). Progress of germination of seed of *Digitaria* as influenced by germination temperature and other factors. *J. Agric. Res.* **63**, 65–90.

Turner, B. H., and Quarterman, E. (1975). Allelochemic effects of *Petalostemon gattingeri* on the distribution of *Arenaria patula* in cedar glades. *Ecology* **56**, 924–932.

Valio, I. F. M., and Joly, C. A. (1979). Light sensitivity of the seeds on the distribution of *Cecropia glaziovi* Snethlage (Moraceae). *Z. Pflanzenphysiol.* **91**, 371–376.

Van Assche, J. A., and Vanlerberghe, K. A. (1989). The role of temperature on the dormancy cycle of seeds of *Rumex obtusifolius* L. *Funct. Ecol.* **3**, 107–115.

Van der Meijden, E., and van der Waals-Kooi, R. E. (1979). The population ecology of *Senecio jacobaea* in a sand dune system. I. Reproductive strategy and the biennial habit. *J. Ecol.* **67**, 131–153.

Van der Veen, R. (1970). The importance of the red-far red antagonism in photoblastic seeds. *Acta Bot. Neerl.* **19**, 809–812.

Vanlerberghe, K. A., and van Assche, J. A. (1986). Dormancy phases in seeds of *Verbascum thapsus* L. *Oecologia* **68**, 479–480.

Varshney, C. K. (1968). Germination of the light-sensitive seed of *Ocimum americanum* Linn. *New Phytol.* **67**, 125–129.

Vazquez-Yanes, C. (1980). Light quality and seed germination in *Cecropia obtusifolia* and *Piper auritum* from a tropical rain forest in Mexico. *Phyton (Buenos Aires)* **38**, 33–35.

Vegis, A. (1963). Climatic control of germination, bud break, and dormancy. *In* "Environmental Control of Plant Growth" (L. T. Evans, ed.), pp. 265–287. Academic Press, New York.

Vegis, A. (1964). Dormancy in higher plants. *Annu. Rev. Plant Physiol.* **15**, 185–224.

Vegis, A. (1973). Dependence of the growth processes on temperature. *In* "Temperature and Life" (H. Precht, J. Christophersen, H. Hensel, and W. Larcher, eds.), pp. 145–170. Springer-Verlag, New York/Heidelberg/Berlin.

Vertucci, C. W., Vertucci, F. A., and Leopold, A. C. (1987). Water content and the conversion of phytochrome regulation of lettuce dormancy. *Plant Physiol.* **84**, 887–890.

Vidaver, W., and Hsiao, A. I. (1974). Actions of gibberellic acid and phytochrome on the germination of Grand Rapids lettuce seeds. *Plant Physiol.* **53**, 266–268.

Vidaver, W., and Hsiao, A. I. (1975). Secondary dormancy in light-sensitive lettuce seeds incubated anaerobically or at elevated temperature. *Can. J. Bot.* **53**, 2557–2560.

Vincent, E. M., and Cavers, P. B. (1978). The effects of wetting and drying on the subsequent germination of *Rumex crispus*. *Can. J. Bot.* **56**, 2207–2217.

Vincent, E. M., and Roberts, E. H. (1977). The interaction of light, nitrate and alternating temperature in promoting the germination of dormant seeds of common weed species. *Seed Sci. Technol.* **5**, 659–670.

Vleeshouwers, L. M., and Bouwmeester, H. J. (1993). A simulation model for the dormancy cycle of weed seeds in the seed bank. *In* "8th EWRS Symposium: Quantitative Approaches in Weed and Herbicide Research and Their Practical Application," Vol. 2, pp. 593–600. Braunschweig, Germany.

Vleeshouwers, L. M., Bouwmeester, H. J., and Karssen, C. M. (1995). Redefining seed dormancy: An attempt to integrate physiology and ecology. *J. Ecol.* **83**, 1031–1037.

Voigt, P.W. (1973). Induced seed dormancy in weeping lovegrass, *Eragrostis curvula*. *Crop Sci.* **13**, 76–79.

Wagner, G. H., and Buyanovsky, G. A. (1983). Use of gas sampling tubes for direct measurement of $^{14}CO_2$ in soil air. *Int. J. Appl. Radiat.* **34**, 645–648.

Walck, J. L., Baskin, J. M., and Baskin, C. C. (1997). A comparative study of the seed germination biology of a narrow endemic and two geographically-widespread species of *Solidago* (Asteraceae).

2. Germination responses of buried seeds in relation to seasonal temperature cycles. *Seed Sci. Res.* **7**, 209–220.

Washitani, I. (1985). Field fate of *Amaranthus patulus* seeds subjected to leaf-canopy inhibition of germination. *Oecologia* **66**, 338–342.

Washitani, I., and Saeki, T. (1984). Leaf-canopy inhibition of germination as a mechanism for the disappearance of *Amaranthus patulus* Bertol. in the second year of a secondary succession. *Jap. J. Ecol.* **34**, 55–61.

Washitani, I., and Saeki, T. (1986). Germination responses of *Pinus densiflora* seeds to temperature, light and interrupted imbibition. *J. Exp. Bot.* **37**, 1376–1387.

Washitani, I., and Takenaka, A. (1987). Gap-detecting mechanism in the seed germination of *Mallotus japonicus* (Thunb.) Muell. Arg., a common pioneer tree of secondary succession in temperate Japan. *Ecol. Res.* **2**, 191–201.

Watanabe, Y., and Hirokawa, F. (1975). Ecological studies on the germination and emergence of annual weeds. 4. Seasonal changes in dormancy status of viable seeds in cultivated and uncultivated soil. *Weed Res. (Japan)* **19**, 20–24. [In Japanese with English summary]

Wesson, G., and Wareing, P. F. (1967). Light requirements of buried seeds. *Nature* **213**, 600–601.

Wesson, G., and Wareing, P. F. (1969a). The role of light in the germination of naturally occurring populations of buried weed seeds. *J. Exp. Bot.* **20**, 402–413

Wesson, G., and Wareing, P. F. (1969b). The induction of light sensitivity in weed seeds by burial. *J. Exp. Bot.* **20**, 414–425.

Westwood, M. N., and Bjornstad, H. O. (1968). Chilling requirements of dormant seeds of 14 pear species as related to their climatic adaptation. *Proc. Am. Soc. Hort. Sci.* **92**, 141–149.

Widell, K.-O., Sundqvist, C., and Virgin, H. I. (1985). Characterization of SAN 9789-stimulated lettuce (*Lactuca sativa*) seed germination. *Weed Sci.* **33**, 160–164.

Williams, E. D. (1983). Effects of temperature fluctuation, red and far-red light and nitrate on seed germination of five grasses. *J. Appl. Ecol.* **20**, 923–935.

Wilson, R. E., and Rice, E. L. (1968). Allelopathy as expressed by *Helianthus annuus* and its role in old-field succession. *Bull. Torrey Bot. Club* **95**, 432–448.

Witztum, A., Gutterman, Y., and Evenari, M. (1969). Integumentary mucilage as an oxygen barrier during germination of *Blepharis persica* (Burn.) Kuntze. *Bot. Gaz.* **130**, 238–241.

Woolley, J. T., and Stoller, E. W. (1978). Light penetration and light-induced seed germination in soil. *Plant Physiol.* **61**, 597–600.

Yoshioka, T., Ota, H., Segawa, K., Takeda, Y., and Esashi, Y. (1995). Contrasted effects of CO_2 on the regulation of dormancy and germination in *Xanthium pennsylvanicum* and *Setaria faberi* seeds. *Ann. Bot.* **76**, 625–630.

Young, J. A., and Evans, R. A. (1973). Mucilaginous seed coats. *Weed Sci.* **21**, 52–54.

Young, J. L., and Aldag, R. W. (1982). Inorganic forms of nitrogen in soil. *In* "Nitrogen in Agricultural Soils" (F. J. Stevenson, ed.), pp. 43–66. American Society of Agronomy No. 22, Madison, WI.

5

Germination Ecology of Seeds with Morphophysiological Dormancy

I. PURPOSE

As discussed in Chapter 3, some seeds have both morphological and physiological dormancy, and thus they have morphophysiological dormancy (MPD). This chapter will discuss the dormancy breaking and germination requirements of seeds with MPD. Particular attention will be given to the phenology of embryo growth, germination phenology, and conditions (especially temperature) in the habitat between the time of seed maturation and the time of germination. Eight types of MPD are known (Table 5.1), and each will be considered in as much detail as possible. It will become obvious, however, that more research is needed on some of the types.

II. TYPES OF MORPHOPHYSIOLOGICAL DORMANCY

A. Nondeep Simple Morphophysiological Dormancy

This type of MPD (Table 5.2) was not mentioned by Nikolaeva (1977), and it was described first in seeds of the winter annual member of the Apiaceae *Chaerophyllum tainturieri* (Baskin and Baskin, 1990). Seeds of this species mature in late spring and potentially can germinate the following autumn if both physiological and morphological dormancies are broken. Physiological dormancy (PD) in seeds of these species is broken at high (25/15, 30/15, and 35/20°C) but not at low (5, 15/6, and 20/10°C) temperatures. Thus, PD is broken in the field during summer.

Morphological dormancy (MD) cannot be broken in *Chaerophyllum tainturieri* seeds until after PD is bro-

ken; thus, embryo growth does not occur until autumn. Further, seeds must be on a moist substrate at suitable germination temperatures (e.g., 25/15°C) and exposed to light before embryos in 85–95% of them will grow. If seeds that have lost their PD are placed under appropriate moisture, temperature, and light conditions for germination, embryos grow rapidly, doubling in length in only 6 days. Seeds can be exposed to light in summer, when PD is just beginning to be broken, and embryo growth and seed germination will occur in darkness in autumn.

Most buried seeds of *Chaerophyllum tainturieri* fail to germinate in autumn because the light requirement for embryo growth has not been fulfilled. If seeds do not germinate in autumn, they reenter nondeep PD; effective temperatures for dormancy induction are those characteristic of the habitat in mid- to late autumn (e.g., 20/10, 15/6°C). Embryo growth will not occur after seeds reenter nondeep PD, even if seeds are incubated at optimum temperature and light conditions (Fig. 5.1). Seeds come out of PD during the following summer, and MD can be broken the subsequent autumn if light, water, and temperature are nonlimiting. Seeds can go into and out of nondeep PD many times, but they cannot germinate until embryo growth occurs (Baskin and Baskin, 1990).

Nondeep simple MPD is found in seeds of temperate region annual and perennial species that have underdeveloped embryos and germinate in autumn. Seeds come out of PD during summer, and embryo growth and germination occur in autumn. In tropical and subtropical regions, seeds of some plants such as *Zamia* spp. (Whitte, 1977; Dehgan and Schutzman, 1983) and *Anona crassiflora* (Rizzini, 1973) require many months of warm stratification before they will germinate. Presumably, these seeds have nondeep simple MPD, and a long

TABLE 5.1   Types of Morphophysiological Dormancy (MPD), Temperature Requirements for
Dormancy Break and Embryo Growth, and One Example of a Species with Each Type of MPD

| Type of MPD | Temperature required[a] | | Gibberellic acid overcomes dormancy[b] | Example |
	To break seed dormancy	At time of embryo growth		
Nondeep simple[c]	W	W	+	*Chaerophyllum tainturieri*
Intermediate simple	W + C	W	+	*Aralia mandshurica*
Deep simple	W + C	W	+/−[d]	*Jeffersonia diphylla*
Deep simple epicotyl	W + C	W	+/−[e]	*Hydrophyllum macrophyllum*
Deep simple double	C + W + C	W	?	*Smilacina racemosa*
Nondeep complex	W + C	C	+/−[d]	*Erythronium albidum*
Intermediate complex	C	C	+	*Stylophorum diphyllum*
Deep complex	C	C	−	*Frasera caroliniensis*

[a]W, warm stratification; C, cold stratification.
[b]+, yes; −, no.
[c]Seeds will come out of physiological dormancy during dry storage at room temperatures.
[d]GA substitutes for warm but not for cold stratification.
[e]GA promotes growth of epicotyl after radicle has emerged, but it does not promote radicle growth.

period of warm stratification breaks both PD and MD. It seems likely that seeds of other tropical/subtropical species with underdeveloped embryos have nondeep simple MPD.

Gibberellic acid (GA) promoted loss of dormancy in seeds of *Chaerophyllum tainturieri* incubated at a constant temperature of 20°C at a 14-hr daily photoperiod (Baskin and Baskin, 1990). This response to GA is to be expected, as GA overcomes dormancy in many seeds with nondeep PD (Chapter 3).

TABLE 5.2   Species Whose Seeds Have Nondeep Simple Morphophysiological Dormancy

Species	Life cycle[a]	Geographical distribution[b]	Reference
Anemone coronaria[c]	P	Trop.	Horovitz *et al.* (1975)
Anona crassiflora[d]	P	Trop.	Rizzini (1973)
Chaerophyllum procumbens	WA	Temp.	Baskin and Baskin (unpublished results)
C. tainturieri	WA	Temp.	Baskin and Baskin (1990)
Corydalis flavula	WA	Temp.	Baskin and Baskin (1994)
C. ledebouriana	P	Temp.	Liden and Staaf (1995)
C. solida	P	Temp.	Liden and Staaf (1995)
Hyacinthoides non-scripta[d]	P	Temp.	Thompson and Cox (1978)
Paeonia californica[d]	P	Temp.	Schlising (1976)
Spermolepias echinata	WA	Temp.	Baskin *et al.* (unpublished results)
Stangeria eriopus[d]	P	Trop.	Dyer (1965)
Zamia sp.[d]	P	Trop.	Whitte (1977)
Zamia furfuracea[d]	P	Trop.	Dehgan and Schutzman (1983)

[a]P, perennial; WA, winter annual.
[b]Trop, tropical/subtropical; Temp, temperate.
[c]Seeds of the cultivated de Caen type of *A. coronaria* have only morphological dormancy.
[d]Detailed studies of embryo growth were not done; seeds are presumed to have nondeep simple MPD.

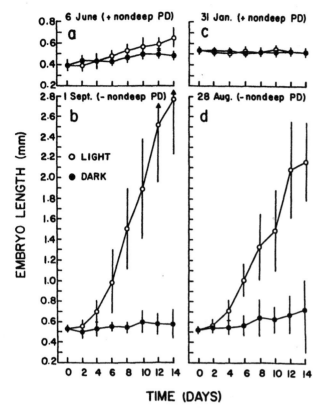

FIGURE 5.1 Embryo growth (mean ± SE) in seeds of *Chaerophyllum tainturieri* at a 12/12-hr daily thermoperiod of 25/15°C in light (O) and in darkness (●). Embryo growth rates were determined in freshly matured seeds and in seeds that had been buried in soil and exposed to natural seasonal temperature changes for various periods of time. Embryo growth rates were for (a) freshly matured seeds in which nondeep physiological dormancy (PD) had not been broken, (b) seeds in which nondeep PD had been broken, (c) seeds that had reentered nondeep PD, and (d) seeds in which nondeep PD had been broken the second time. Modified from Baskin and Baskin (1990).

B. Intermediate Simple Morphophysiological Dormancy

Seeds with intermediate simple MPD (Table 5.3) require warm stratification for loss of PD, which occurs during summer. After PD is broken, embryo growth occurs in autumn at temperatures of 15–20°C. However, seeds with elongated embryos require cold stratification

TABLE 5.3 Species Whose Seeds Have Intermediate Simple Morphophysiological Dormancy

Species	Reference
Aralia mandshurica	Nikolaeva (1977)
Dendropanax japonicum	Grushvitzky (1967)

before they can germinate; thus, germination occurs in the field in spring. Seeds of *Aralia mandshurica* require 4 months at 18–20°C followed by 4 months at 0–3°C for maximum germination. However, the time required for loss of dormancy in this species was reduced from 8 to 4 months by alternately giving imbibed seeds 1 day at 15–20°C and 2 days at 0–3°C (Nikolaeva, 1977).

GA substitutes for cold stratification in seeds with intermediate simple MPD. GA-treated seeds of *Aralia mandshurica* and *Dendropanax japonicum* germinated after 1.5 and 2 months of warm stratification, respectively (Grushvitzky, 1967).

C. Deep Simple Morphophysiological Dormancy

Seeds with deep simple MPD (Table 5.4) require warm followed by cold stratification before they will germinate. However, as pointed out by Nikolaeva (1977), the second part of the warm stratification treatment should be at a lower temperature than the first part. Thus, these various dormancy-breaking requirements are fulfilled in temperate regions by the summer (warm stratification at relatively high temperatures), autumn (warm stratification at relatively low temperatures), and winter (cold stratification) sequence of temperatures, and germination occurs in spring.

The dormancy-breaking requirements for seeds with deep simple MPD are illustrated in *Jeffersonia diphylla*. Seeds of this species ripened and were dispersed in late

TABLE 5.4 Species Whose Seeds Have Deep Simple Morphophysiological Dormancy

Species	Reference
Fraxinus excelsior	Villiers and Wareing (1964)
F. mandshurica var. *japonica*	Asakawa (1956)
F. nigra	Steinbauer (1937), Vanstone and LaCroix (1975)
Ilex opaca[a]	Ives (1923), Barton and Thornton (1947)
Jeffersonia diphylla	Baskin and Baskin (1989)
Panax ginseng	Grushvitzky (1967), Choi and Takahashi (1977)
P. quinquefolius	Stoltz and Snyder (1985)
P. pseudoginseng[a]	Joshi *et al.* (1991)
P. trifolium	Baskin *et al.* (unpublished results)
Taxus baccata[a]	Devillez (1978)
T. mairei	Kuo-Hunag *et al.* (1996)

[a]More work is needed on this species, especially on requirements for embryo growth.

May and early June (early summer), and fresh seeds sown on soil under natural seasonal temperature changes germinated the following March and April (Baskin and Baskin, 1989). Embryo growth began in early September and continued until mid-January, with the peak of growth occurring from mid-September to mid-November, when mean maximum and minimum weekly temperatures were 22.5 and 6.7°C, respectively (Fig. 5.2). Embryos grew slowly, or not at all (and seeds did not germinate), when freshly matured seeds were placed on a moist substrate at simulated autumn (20/10, 15/6°C) or winter (5°C) temperatures (Fig. 5.3). However, embryos grew rapidly at 20/10°C if seeds were given 12 weeks at 30/15°C prior to being transferred to 20/10°C, but they did not grow at 5°C if seeds were given 12 weeks at 30/15°C prior to being transferred to 5°C. Thus, autumn temperatures were required for embryo growth. However, if seeds were transferred from 15/6 or 20/10 to 5°C after embryo growth was initiated, it continued at 5°C. Seeds of *J. diphylla* did not germinate until spring because cold stratification was required after embryo growth was completed (Baskin and Baskin, 1989). Prolonged warm stratification of *Fraxinus excelsior* seeds at 23°C promoted embryo growth and decreased the cold stratification requirement for germination (Krauss and Kohler, 1985).

Two types of PD as well as MD appear to be involved in deep simple MPD. High (e.g., 30°C) summer tempera-

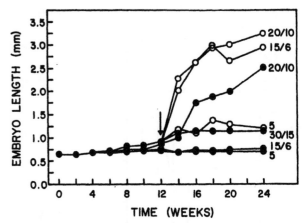

FIGURE 5.3   Embryo growth (mean length) of *Jeffersonia diphylla* seeds placed at 5, 15/6, 20/10, or 30/15°C for 24 weeks (●) and of seeds placed at 30/15°C for 12 weeks and then transferred to 5, 15/6, or 20/10°C for 12 weeks (○). From Baskin and Baskin, (1989).

tures are required to break one kind of PD, whereas low (e.g., 5°C) winter temperatures are required to break a second one. MD cannot be broken until (1) high temperatures overcome the first type of PD and (2) seeds are subjected to autumn temperatures (15, 20°C). After MD is broken, seeds cannot germinate until low temperatures overcome the second type of PD. GA substituted for warm stratification and thus stimulated embryo growth in seeds of *Fraxinus excelsior* (Wcislinska, 1977), *Jeffersonia diphylla* (Baskin and Baskin, 1989), *Panax ginseng* (Choi, 1977), and *P. quinquifolius* (Stoltz and Snyder, 1985). However, GA did not substitute for cold stratification and consequently did not promote germination. Thus, the first type of PD may be nondeep, whereas the second one may be deep. Nondeep PD is broken by high or low temperatures and GA, whereas deep PD is broken only by low temperatures (Chapter 3).

D. Deep Simple Epicotyl Morphophysiological Dormancy

This type of MPD is referred to as "epicotyl dormancy" in much of the seed germination literature, and we follow that tradition. However, as seen later, both the epicotyl and the radicle can be dormant. In general, epicotyl dormancy means emergence of the radicle in autumn and emergence of the shoot the following spring (Fig. 5.4). Thus far, the species with epicotyl dormancy that have been studied (Table 5.5) grow in temperate regions of the world. However, some species that have underdeveloped embryos and a long lag period between radicle and epicotyl emergence could occur in tropical/subtropical regions.

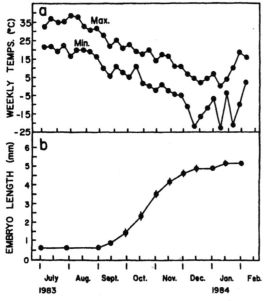

FIGURE 5.2   Phenology of embryo growth (mean length ± SE) of *Jeffersonia diphylla* seeds kept in a fine-mesh bag on moist soil under leaves in a shady location in a garden in Lexington, Kentucky. Mean maximum and minimum weekly temperatures in the garden are shown for the duration of the study. From Baskin and Baskin (1989).

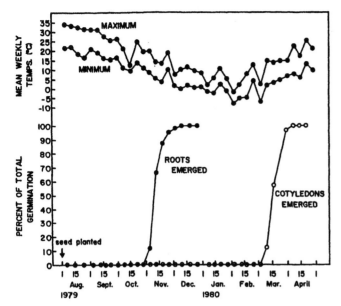

FIGURE 5.4 Germination phenology of *Hydrophyllum macrophyllum* seeds sown on soil in a nonheated greenhouse. From Baskin and Baskin (1983)

The word "epicotyl" refers to the portion of a seedling above the cotyledons (i.e., the stem or shoot), and it may refer to only the apical meristem or bud. Seeds with epicotyl dormancy do not exhibit growth of the upper portion (or shoot) of the plant immediately after radicle emergence. Although Saunders (1918) observed that epicotyl growth was delayed after radicles emerged from seeds of *Paeonia suffruticosa*, Barton (1933) actually proved that the epicotyl in seeds of this species was dormant. Thus, she was the first to discover this type of MPD, and chose the name "epicotyl dormancy."

The time (and thus temperature) when radicle emergence occurs in autumn varies with the species. For example, radicles emerged from seeds of (1) *Trillium flexipes* and *T. sessile* mostly in September, when mean daily maximum and minimum temperatures were 28.2 and 16.8°C, respectively (Baskin and Baskin, unpublished results); (2) *Asarum canadense* (Baskin and Baskin, 1986a) and *Allium tricoccum* (Baskin and Baskin, unpublished results) in October, 21.2 and 11.1°C; and (3) *Cimicifuga racemosa, Hepatica acutiloba* (Baskin and Baskin, 1985a), *Hydrophyllum macrophyllum* (Baskin and Baskin, 1983), and *H. appendiculatum* (Baskin and Baskin, 1985b) in November, 13.7 and 5.6°C. After radicle emergence, a root system develops slowly during late autumn and winter, and by early spring the main root and several laterals may be up to 4–7 cm long, depending on the species.

Emergence of the cotyledon(s) in spring occurs as soon as temperatures are high enough for leaf growth. In our nonheated greenhouse in Kentucky, emergence

of the cotyledon(s) in most species with epicotyl dormancy occurs mainly in March, when mean daily maximum and minimum temperatures are 15.6 and 4.2°C, respectively (temperature data from our nonheated greenhouse, 1969–1987). However, in *T. flexipes* and *T. sessile,* the cotyledon emerged in November and early December (Baskin and Baskin, unpublished results).

Barton (1933) found that epicotyl dormancy could be broken by exposing *Paeonia suffruticosa* seeds, with the roots 4–5 cm long, to temperatures ranging from 1 to 10°C; 5°C was the best. That is, temperatures that are effective for cold stratification of seeds broke dormancy of the epicotyl. After doing several studies on

TABLE 5.5 Species Whose Seeds Have Deep Simple Epicotyl Morphophysiological Dormancy

Species	References
Actaea pachypoda	Baskin and Baskin (1988)
A. spicata	Eriksson (1994)
Allium burdickii	Baskin and Baskin (unpublished results)
A. tricoccum	Baskin and Baskin (unpublished results)
A. ursinum	Ernst (1979)
Asarum canadense	Barton (1944)
A. heterotropoides	Liu *et al.* (1993)
Cimicifuga racemosa	Baskin and Baskin (1985a)
C. rubifolia	Cook (1993)
Disporum lanuginosum	Baskin and Baskin (unpublished results)
Fritillaria ussuriensis	Liu *et al.* (1993)
Hepatica acutiloba	Baskin and Baskin (1985a)
Hydrophyllum appendiculatum	Baskin and Baskin (1985b)
H. macrophyllum	Baskin and Baskin (1983)
H. virginianum	Baskin and Baskin (1988)
Lilium auratum	Barton (1936)
L. canadense	Barton (1936)
L. japonicum	Barton (1936)
Paeonia suffruticosa	Barton (1933)
Polygonatum commutatum[a]	Barton (1944)
Sanguinaria canadensis[b]	Barton (1944)
Trillium flexipes	Baskin and Baskin (1988)
T. sessile	Baskin and Baskin (1988)
Viburnum acerifolium, dentatum, dilatatum, lentago, opulus, prunifolium, rufidulum	Giersbach (1937)

[a]58% of the seeds had double dormancy.
[b]49% of the seeds had double dormancy.

seeds with epicotyl dormancy (Barton, 1933, 1936; Giersbach, 1937), Barton (1939) concluded that seeds with an emergent radicle needed 0.5 to 4 months of cold stratification to break epicotyl dormancy. An example of the effect of cold stratification on the breaking of epicotyl dormancy is seen in *Hydrophyllum macrophyllum* seeds (Fig. 5.5). Cotyledon emergence was 100% when seeds with roots 2–3 cm in length were given cold stratification at 5°C for 6 weeks. If seeds were cold stratified longer than 6 weeks, cotyledons began emerging at 5°C.

Although Barton (1944) discovered that the epicotyl of *Asarum canadense* embryos was dormant, she did not realize that the radicle was dormant. The reason Barton did not discover radicle dormancy in this species is that she incubated seeds at temperatures (e.g., 20/10°C) suitable for breaking radicle dormancy and promoting radicle emergence. This also has happened in other studies of the dormancy-breaking requirements of seeds with epicotyl dormancy (e.g., Barton, 1933, 1936; Giersbach, 1937). In fact, in describing the conditions required for complete germination of a number of species whose seeds have epicotyl dormancy, Barton (1939) said that seeds required from 3 to 17 months at temperatures of 20, 30/15, or 32/20°C for radicle emergence, depending on the species. Thus, various researchers probably have broken radicle dormancy in seeds without realizing radicles were dormant.

Embryos in *Hydrophyllum macrophyllum* seeds have dormant radicles in addition to dormant epicotyls, and radicle dormancy is broken by warm stratification (Fig. 5.6). Seeds of *H. appendiculatum* (Baskin and Baskin, 1985b) and *Asarum canadense* (Baskin and Baskin, 1986a) also require warm stratification to break radicle

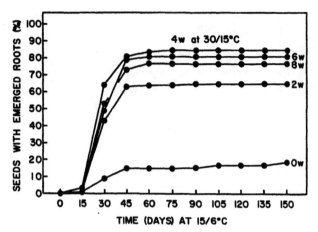

FIGURE 5.6   Root emergence in seeds of *Hydrophyllum macrophyllum* at 15/6°C following 0–8 weeks of warm stratification at 30/15°C. From Baskin and Baskin (1983).

dormancy. Presumably, seeds of other species with epicotyl dormancy have dormant radicles, but dormancy-breaking and emergence requirements of radicles have not been determined.

Because both the radicle and the epicotyl are dormant, "epicotyl dormancy" is not an accurate description of the kind of dormancy found in seeds of *Hydrophyllum macrophyllum, Asarum canadense,* and other species. However, the term is found throughout seed germination literature, and Nikolaeva (1977) used it in her system of classification of dormancy types. Thus, the benefits of a name more accurate than epicotyl dormancy probably would never counterbalance all of the confusion created by a name change.

The epicotyl appears to be sensitive to cold stratification only after the radicle has elongated. Although seeds of some species do not have an emergent radicle until the second, third or later autumn after sowing (Table 5.6), root production always precedes shoot emergence. The response of epicotyls in *Paeonia suffruticosa* seeds to cold stratification depends on the degree of root development. Eighty-five percent of *P. suffruticosa* seedlings with roots 4–5 cm long produced emergent cotyledons after 7 weeks of cold stratification at 5°C, whereas only 40% of those with roots 2–3 cm long produced emergent cotyledons (Barton and Chandler, 1957). Although we have observed root production in *Sanguinaria canadensis* seeds only in autumn, Barton (1944) found that cold stratification at 5°C for 3 months increased root production (percentage not given) over the control (51%) kept at 20°C.

Seeds with epicotyl dormancy have underdeveloped embryos of either the rudimentary or the linear type, but the phenology of embryo growth has not been monitored in any species. However, general observations

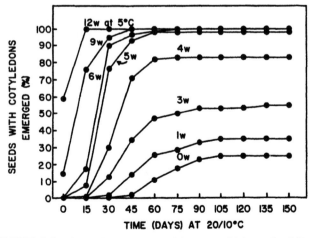

FIGURE 5.5   Cotyledon emergence (shoot growth) in seeds of *Hydrophyllum macrophyllum* at 20/10°C following 0–12 weeks of cold treatment at 5°C. From Baskin and Baskin (1983).

TABLE 5.6 Radicle Emergence in Species Whose Seeds Have Epicotyl Dormancy[a]

Species	Autumn after sowing										
	1	2	3	4	5	6	7	8	9	10	11
Allium burdickii	x	x	x	x							
A. tricoccum[b]		x									
Asarum canadense	x	x	x	x	x						
Cimicifuga racemosa	x	x	x								
Disporum lanuginosum	x										
Hepatica acutiloba	x										
Hydrophyllum appendiculatum	x	x	x	x	x	x	x	x	x	x	x
H. macrophyllum	x	x	x	x	x	x	x	x	x		
H. virginianum	x	x									
Sanguinaria canadensis	x	x	x	x	x	x	x	x			
Trillium flexipes[c]	x	x	x	x	x	x					
T. sessile	x	x	x	x							

[a]Seeds were sown on soil, covered with dead *Quercus* (oak) leaves, and kept in a nontemperature-controlled greenhouse in Lexington, Kentucky (Baskin and Baskin, unpublished results).

[b]Similar results were obtained by Nault and Gagnon (1993) for seeds of this species sown in four different years.

[c]No seed bank was reported for *Trillium grandiflorum* (Hanzawa and Kalisz, 1993).

have been made on embryo growth in *Hydrophyllum appendiculatum* seeds. The embryo in seeds of this species is located at the outer edge of the endosperm, and the radicle is pointed toward the seed coat (Fig. 5.7). Thus, the radicle emerges from the seed, after it grows ≤1 mm in autumn. Cotyledons elongate during cold stratification of seeds with emergent radicles, and they are almost as long as the seed at the time of their emergence in spring; most of the endosperm disappears by the time cotyledons are fully elongated.

The type of PD in radicles is nondeep, as it was broken by high temperatures. Because cold stratification is required to break epicotyl dormancy, the type of PD in epicotyls could be nondeep, intermediate, or deep. However, GA substituted for cold stratification in breaking the PD in epicotyls of *P. suffruticosa* embryos (Barton and Chandler, 1957) and overcame epicotyl dormancy in seeds of *Viburnum trilobum* (= *V. opulus*)

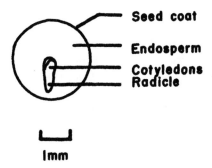

FIGURE 5.7 Longitudinal section through a freshly matured seed of *Hydrophyllum appendiculatum*.

(Fedec and Knowles, 1973). These results indicate that the type of PD in epicotyls is either nondeep or intermediate. Epicotyl dormancy was broken in *V. trilobum* seeds when cotyledons were removed, and shoots grew normally (Knowles and Zalik, 1958). Thus, inhibition of shoot growth appears to be in the cotyledons and not in the epicotyl per se.

E. A Special Type of Simple Epicotyl Morphophysiological Dormancy

Seeds of *Hydrastis canadensis* (Ranunculaceae) have some characteristics of both deep simple and deep simple epicotyl MPD (Fig. 5.8). Fruits of *H. canadensis* mature in mid- to late July (summer) and are dispersed shortly thereafter. Embryos in freshly matured seeds were 0.6±0.02 mm (mean±SE) in length and grew very little in seeds placed under natural temperature regimes until after October 15; by January 1 they were 3.7±0.10 mm in length. In November, the black seed coats began splitting open at the radicle end of seeds, revealing the chartreuse-colored endosperm. Although the embryo protruded slightly beyond the end of the seed, it remained covered by the endosperm, which stretched like a rubber glove. The radicle, followed immediately by the cotyledons, emerged from the endosperm in mid- to late March. *Hydrastis canadensis* seeds are like those with deep simple MPD because embryo growth occurred in autumn, but unlike those with deep simple MPD the seed coats split open in autumn. Seeds of *H. canadensis* are somewhat like those with epicotyl

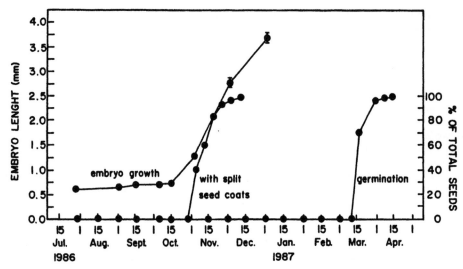

FIGURE 5.8   Phenology of embryo growth (mean length ± SE, if ≥0.06 mm), seed coat splitting, and germination of *Hydrastis canadensis* seeds in a nontemperature-controlled greenhouse, Lexington, Kentucky. Baskin and Baskin, (unpublished results).

dormancy because the radicle grows beyond the limits of the seed coats, but unlike those with epicotyl dormancy, the radicle remains covered by endosperm until spring. Perhaps *H. canadensis* represents a transitional stage between deep simple and deep simple epicotyl MPD.

F. Deep Simple Double Morphophysiological Dormancy

This type of MPD (Table 5.7) usually is referred to as "double dormancy," and the seeds sometimes are called "2-year seeds." These terms refer to the fact that 2 years (or at least two winters and one summer) are required to complete germination. With the exception of *Caulophyllum thalictroides* and *Sanguinaria canadensis,* all species known to have double dormancy are monocots, and all except *Arisaema dracontium, A. triphyllum, C. thalictroides,* and *S. canadensis* belong to the Liliaceae.

Pickett (1913) seems to be the first person who studied seeds with double dormancy. He found that the radicle emerged from seeds of *Arisaema triphyllum* and *A. dracontium* one spring, but epicotyl growth did not occur until the following spring. Thus, cold stratification preceded both radicle and epicotyl emergence. At the time of radicle emergence, the lower end of the single cotyledon elongates and pushes the radicle, hypocotyl, shoot bud (plumule), and part of the cotyledon to the outside of the seed (Fig. 5.9). However, the tip of the cotyledon remains inside the seed. The hypocotyl swells, becoming a "corm," and the radicle grows downward,

forming a root. Energy for corm development and production of additional roots come from food reserves in the endosperm, which are absorbed by the cotyledon. In some *Arisaema* seedlings, the bud at the top of the corm produces a leaf a few weeks after corm formation (Fig. 5.9f); however, most seedlings do not produce a leaf until the following (second) spring. The cotyledon,

TABLE 5.7   Species Whose Seeds Have Deep Simple Double Morphophysiological Dormancy

Species	Reference
Arisaema dracontium	Pickett (1913)
A. triphyllum	Pickett (1913)
Caulophyllum thalictroides	Barton (1944)
Clematis albicoma	Platt (1951)
C. viticaulis	Platt (1951)
Convallaria majalis	Barton and Schroeder (1941)
Polygonatum biflorum	Baskin and Baskin (unpublished results)
P. commutatum[a]	Barton (1944)
Sanguinaria canadensis[b]	Barton (1944)
Smilacina racemosa	Barton and Schroeder (1941)
Trillium erectum	Barton (1944)
T. grandiflorum	Barton (1944)
Uvularia grandiflora	Baskin and Baskin (unpublished results)
U. perfoliata	Whigham (1974)

[a]42% of the seeds had epicotyl dormancy.
[b]51% of the seeds had epicotyl dormancy.

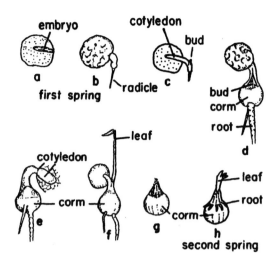

FIGURE 5.9 Germination morphology of seeds of *Arisaema dracontium*. From Pickett (1913), with views c and h added by authors.

roots on the corm, and leaf, if present, die after food reserves are depleted, leaving the corm with its bud covered by scales (Pickett, 1913). In the second spring, absorbing roots are produced at the top of the corm, near the base of the bud, and the shoot bud gives rise to a single trifoliate leaf (Fig. 5.9h).

In contrast to *Arisaema*, seedlings of other species with double dormancy may have a persistent root system. For example, the cotyledon pushes the radicle and small shoot bud to the outside of *Smilacina racemosa* seeds in early summer (May or early June), and a persistent root system develops during summer (Fig. 5.10). The shoot bud of *S. racemosa* is 0.5 cm in length by the second spring, and it elongates rapidly in mid-March, giving rise to a single leaf.

Barton and Schroeder (1941) were the first seed physiologists to study germination requirements of seeds with double dormancy. They found that cold stratification at 5°C for 3 months increased root production in seeds of *Convallaria majalis,* and cold stratification was required for root production in *Smilacina racemosa* seeds. Further, dormancy break of the shoot bud (epicotyl dormancy) required 3–5 months at 5°C for *C.*

majalis and 5 months at 10°C for *S. racemosa.* However, the second cold treatment was ineffective in breaking bud dormancy, unless the young seedlings previously were given a 2- to 3-month period in a warm greenhouse. During the warm period, root systems developed in both species, and the shoot bud grew until it broke through the cotyledonary sheath and then stopped. Bud growth resumed after seedlings were cold stratified and returned to the heated greenhouse. In contrast, epicotyl dormancy was broken if seeds with deep simple epicotyl MPD were cold stratified immediately after the root was produced (Crocker, 1948). That is, in species with deep simple epicotyl dormancy, the bud did not have to grow to a minimum size before its dormancy was overcome by cold stratification, but in species with deep simple double MPD the bud had to reach a minimum size before its dormancy was broken by cold stratification.

Barton's (1944) work on *Trillium grandiflorum* probably is the most famous study on seeds with double dormancy. Using her data on this species, it appears that for maximum germination in the shortest period of time, freshly matured seeds of *T. grandiflorum* should be placed on a moist substrate and cold stratified at 5 or 10°C for 3 months (to simulate winter) to break radicle dormancy. After radicle dormancy is broken, seeds should be exposed to temperatures of 20–30°C in a greenhouse for 3 months (to simulate spring and summer) to permit emergence of the radicle, production of a root system, and formation and growth of the shoot bud. A 4-month period of cold stratification at 5°C will break dormancy of the epicotyl (shoot bud), after which seedlings should be placed in a cool greenhouse at 15–20°C (spring) for growth of the shoot. Barton (1944) found that temperatures of 10 and 15°C were also effective in breaking dormancy of the shoot bud.

No seeds with double dormancy have been tested with GA3; thus, we cannot speculate on the type(s) of PD they have.

G. A Special Type of Simple Double Morphophysiological Dormancy

Freshly matured seeds of *Medeola virginiana* (Liliaceae) sown on soil in a nonheated greenhouse in Lexington, Kentucky, on September 6 (late summer) 1986 did not germinate until April (spring) 1988, when both the radicle and cotyledon emerged (Baskin and Baskin, unpublished results). Further, seeds subjected to a sequence of simulated monthly habitat temperatures in incubators did not germinate until they had received two periods of cold stratification. However, 37 months of continuous cold stratification at 5°C did not break dormancy. In both the greenhouse and incubators, seeds

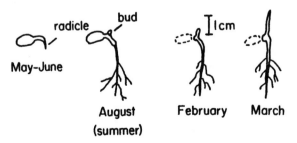

FIGURE 5.10 Germination morphology of seeds of *Smilacina racemosa*.

TABLE 5.8 Species Whose Seeds Have Nondeep
Complex Morphophysiological Dormancy

Species	Reference
Eranthis hiemalis	Frost-Christensen (1974)
Erythronium albidum	Baskin and Baskin (1985c)
E. americanum	Baskin and Baskin (unpublished results)
E. rostratum	Baskin and Baskin (unpublished results)
Osmorhiza claytonii	Baskin and Baskin (1991)
O. longistylis	Baskin and Baskin (1984a)
O. occidentalis	Baskin *et al.* (1995)

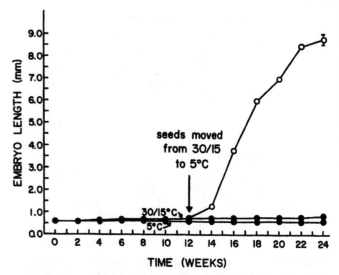

FIGURE 5.11   Embryo lengths (mean ± SE, if ≥0.2 mm) of freshly matured seeds of *Osmorhiza longistylis* incubated at (1) 30/15 (warm stratification) and/or 5°C (cold stratification) for 24 weeks (●) or (2) 30/15°C for 12 weeks and then moved to 5°C for 12 weeks (○). Modified from Baskin and Baskin (1984a).

received warm (autumn 1986) + cold + warm + cold stratification before they germinated. Embryos in freshly matured seeds were 0.32±0.01 mm in length and did not grow in seeds kept on moist soil in the nonheated greenhouse until after May 1987 (first spring following dispersal). Embryos were 0.32±0.01 mm in length on May 1, 1987, but after this date no seeds were available with which to monitor growth. It appears that seeds of *M. virginiana* are 2-year seeds, i.e., two winters and a summer are required to break dormancy. However, unlike species whose seeds have a traditional double dormancy, *M. virginiana* seeds do not produce a root system during the warm interval between the two cold stratification treatments. Does this represent a transition from deep simple double MPD to some type of complex MPD?

H. Nondeep Complex Morphophysiological Dormancy

Nikolaeva (1977) did not include this type of MPD in her classification system of dormancy types, and it first was reported by Frost-Christensen (1974) in seeds of *Eranthis hiemalis*. Since its discovery in seeds of *E. hiemalis*, nondeep MPD has been found in seeds of several species (Table 5.8), but a name was not proposed for it until 1991 (Baskin and Baskin, 1991).

Cold stratification is required for embryo growth in seeds with nondeep complex MPD, but it is ineffective unless seeds first receive a period of warm stratification (Fig. 5.11). The length of the warm stratification pretreatment required for seeds to germinate to 70–80% after they subsequently are cold stratified varies from 2 weeks in seeds of *Erythronium albidum* (Fig. 5.12a) and *Osmorhiza claytonii* (Baskin and Baskin, 1991) to 4 weeks in those of *O. longistylis* (Baskin and Baskin, 1984a). The length of the cold stratification period required for 70–80% germination in seeds that previously were warm stratified ranges from 14 to 6 weeks in *O.*

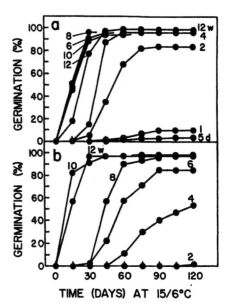

FIGURE 5.12   (a) Germination of *Erythronium albidum* seeds in light at 15/6°C after 0, 1, 3, and 5 days or 2–12 weeks of warm stratification at 30/15°C followed by 12 weeks of cold stratification at 5°C. None of the seeds receiving 0 or 3 days of warm stratification plus 12 weeks of cold stratification germinated. (b) Germination of *E. albidum* seeds in light at 15/6°C after 12 weeks of warm stratification at 30/15°C followed by 0–12 weeks of cold stratification at 5°C. None of the seeds receiving 12 weeks of warm stratification and 0 weeks of cold stratification germinated. From Baskin and Baskin (1985c).

longistylis (Baskin and Baskin, 1984a) and *E. albidum* (Fig. 5.12b), respectively.

The warm plus cold stratification requirement for germination may have effects on germination phenology, especially if dispersal is delayed. For example, seeds of *O. longistylis* mature in July (summer), but dispersal does not begin until September. Seeds dispersed in September receive warm followed by cold stratification; consequently, they germinate the following spring. Although most of the seeds are dispersed in late autumn and winter, some can remain on the dead, upright flowering shoots for up to 18 months. Because seeds on dead flowering shoots are not imbibed, they cannot be either warm or cold stratified. Seeds imbibe water after they fall to the soil surface and thus can be warm and/or cold stratified. Seeds dispersed in October through March do not receive a warm stratification treatment because temperatures in the habitat during these months are below those required for warm stratification. Consequently, these seeds would not germinate until the second spring following maturation, after they had gone through a summer (warm stratification) and a winter (cold stratification) on the soil (Baskin and Baskin, 1984a).

Because nondeep PD is the only type of PD known to be broken by warm stratification, the dormancy broken during warm stratification probably is nondeep PD. Consistent with this conclusion is the fact that GA was effective in substituting for warm stratification in seeds of *Eranthis hiemalis* (Frost-Christensen, 1974) and partially effective in those of *Osmorhiza claytonii* (Baskin and Baskin, 1991). It is difficult to know what kind of PD, if any, is broken during cold stratification. GA did not substitute for cold stratification in seeds of *E. hiemalis* or *O. claytonii*, thus the type of dormancy broken during cold stratification could be intermediate or deep PD. However, in view of the rapid rate of embryo growth when warm-stratified seeds were placed at temperatures effective for cold stratification (Fig 5.11), it is possible that seeds do not have any PD after they have received warm stratification. In other words, cold stratification may not break any type of PD in these seeds, and its promotive effects on germination are related to the breakdown of food reserves in the endosperm, making them available for growing embryos (see Stokes, 1952a). To help resolve this question, it would be interesting to excise embryos from warm-stratified (and nonwarm-stratified) seeds and incubate them on nutrient-enriched agar at 15–20°C.

I. Intermediate Complex Morphophysiological Dormancy

Seeds with intermediate complex MPD (Table 5.9) require only cold stratification for loss of PD and MD

TABLE 5.9 Species Whose Seeds Have Intermediate Complex Morphophysiological Dormancy

Species	Reference
Aralia continentalis	Nikolaeva (1977)
Ginkgo biloba[a]	West *et al.* (1970)
Stylophorum diphyllum	Baskin and Baskin (1984b)
Trollius altaicus	Nikolaeva *et al.* (1986)
T. apertus	Nikolaeva *et al.* (1986)
T. asiaticus	Nikolaeva *et al.* (1986)
T. chinensis	Nikolaeva *et al.* (1986)
T. dschungaricus	Nikolaeva *et al.* (1986)
T. europaeus	Nikolaeva *et al.* (1986)
T. laxus	Nikolaeva *et al.* (1986)
T. ledebouri	Hepher and Roberts (1985)
T. pulcher	Nikolaeva *et al.* (1986)
T. ranunculinus	Nikolaeva *et al.* (1986)
T. riederianus	Nikolaeva *et al.* (1986)
T. yunnanensis	Nikolaeva *et al.* (1986)

[a]Some of the seeds had only morphological dormancy.

of the embryo. Nikolaeva (1977) used the name "intermediate complex MPD" because GA substitutes for cold stratification in these seeds. Embryo growth begins in freshly matured seeds shortly after they are subjected to cold stratification. Regardless of when seeds mature in nature, however, dormancy break and embryo growth do not occur until temperatures decline to the effective range (0–10°C) for cold stratification in late autumn and winter. For example, if seeds of *Stylophorum diphyllum* are placed under natural temperature conditions at the time of seed maturation in June (early summer), embryo growth is not initiated until late autumn, when daily minimum temperatures are within the range of those effective for cold stratification. However, embryo length doubled in about 7 weeks in freshly matured seeds cold stratified at 5°C (Baskin and Baskin, 1984b).

J. Deep Complex Morphophysiological Dormancy

Seeds with deep complex MPD (Table 5.10) require only cold stratification for loss of PD and MD of the embryo. Thus, in the habitat seeds become nondormant in response to cold stratification during winter and germinate in early spring.

Stokes (1952b) demonstrated that low temperatures were necessary to make food reserves in the endosperm of *Heracleum sphondylium* (Apiaceae) seeds available

TABLE 5.10 Species Whose Seeds Have Deep Complex
Morphophysiological Dormancy

Species	Reference
Cryptotaenia canadensis	Baskin and Baskin (1988)
Delphinium tricorne	Baskin and Baskin (1994)
Erythronium grandiflorum	Baskin et al. (1995)
Frasera albacaulis	Florance (1994)
F. caroliniensis[a]	Threadgill et al. (1981)
F. umpquaensis	Florance (1994)
Fritillaria eduardii	Pozdova and Razumova (1994)
F. raddeana	Pozdova and Razumova (1994)
Heracleum sphondylium	Stokes (1952a)
Magnolia acuminata	Afanasiev (1937)
M. grandiflora	Evans (1933)
Ornithogalum arcuatum	Pozdova and Razumova (1994)
Osmorhiza chilensis	Baskin et al. (1995)
O. occidentalis	Baskin et al. (1995)
Thaspium pinnatifidum	Baskin et al. (1992)
Tulipa greigii	Pozdova and Razumova (1994)
T. tarda	Nikolaeva (1977)

[a] Seeds remaining nondispersed over winter can develop nondeep complex MPD (Baskin and Baskin, 1986b).

to the growing embryo. Low temperatures (2°C) stimulated the breakdown of proteins into soluble nitrogenous compounds (Stokes, 1953a) and of formation of the amino acids glycine and arginine, which are beneficial for embryo growth on synthetic media (Stokes, 1953b). At high (15°C) temperatures, soluble nitrogenous compounds were not available, and alanine (which does not stimulate growth of excised embryos) was the most important amino acid present.

Results obtained from Stokes's studies on *Heracleum sphondylium* raise the question: Is embryo growth the only thing that happens during cold stratification of seeds with deep complex MPD? Studies on *Thaspium pinnatifidum* (Apiaceae) seeds showed that cold stratification does something to promote germination, in addition to stimulating embryo growth (Baskin et al., 1992). Embryos in *T. pinnatifidum* seeds were 1.5 mm in length after 4 weeks of cold stratification at 5°C, and seeds moved to 15/6°C germinated to 61% after 4 weeks. However, embryos in seeds of this species kept continuously at 15/6°C were 2.3 mm in length after 4 weeks, but only 2% of the seeds germinated after 8 weeks.

Because GA did not promote the germination of *Tulipa tarda* seeds, Nikolaeva (1977) concluded that they had deep PD, in addition to MD. Subsequent studies have shown that GA does not stimulate the germination of seeds of *Delphinium tricorne* (Baskin and Baskin,

1994), *Frasera caroliniensis* (Baskin and Baskin, unpublished results), *Heracleum lanatum* (=*H. sphondylium*) (McDonough, 1969), or *Thaspium pinnatifidum* (Baskin et al., 1992). Thus, deep complex MPD may be a combination of deep PD and MD.

References

Afanasiev, M. (1937). A physiological study of dormancy in seed of *Magnolia acuminata. Cornell Univ. Agric. Exp. Sta. Mem.* 208.

Asakawa, S. (1956). Studies on the delayed germination of *Fraxinus mandshurica* var. *japonica* seeds. 2. Pregermination of *F. mandshurica* var. *japonica* seeds. Physiological properties of embryos of *Fraxinus* seeds. *Ringyo Shikenjo Kenkyu Hokoku (Bull. Govern. For. Exp. Sta., Tokyo)* **83**, 19–28.

Barton, L. V. (1933). Seedling production of tree peony. *Contrib. Boyce Thomp. Inst.* **5**, 451–460.

Barton, L. V. (1936). Germination and seedling production in *Lilium* sp[p]. *Contrib. Boyce Thomp. Inst.* **8**, 297–309.

Barton, L. V. (1939). Experiments at Boyce Thompson Institute on germination and dormancy in seeds. *Sci. Hort.* **7**, 186–193.

Barton, L. V. (1944). Some seeds showing special dormancy. *Contrib. Boyce Thomp. Inst.* **13**, 259–271.

Barton, L. V., and Chandler, C. (1957). Physiological and morphological effects of gibberellic acid on epicotyl dormancy of tree peony. *Contrib. Boyce Thomp. Inst.* **19**, 201–214.

Barton, L. V., and Schroeder, E. M. (1941). Dormancy in seeds of *Convallaria majalis* L. and *Smilacina racemosa* (L.) Desf. *Contrib. Boyce Thomp. Inst.* **12**, 277–300.

Barton, L. V., and Thornton, N. C. (1947). Germination and sex population studies of *Ilex opaca* Ait. *Contrib. Boyce Thomp. Inst.* **14**, 405–410.

Baskin, C. C., and Baskin, J. M. (1988). Germination ecophysiology of herbaceous plant species in a temperate region. *Am. J. Bot.* **75**, 286–305.

Baskin, C. C., and Baskin, J. M. (1994). Deep complex morphophysiological dormancy in seeds of the mesic woodland herb *Delphinium tricorne* (Ranunculaceae). *Int. J. Plant Sci.* **15**, 738–743.

Baskin, C. C., Chester, E. W., and Baskin, J. M. (1992). Deep complex morphophysiological dormancy in seeds of *Thaspium pinnatifidum* (Apiaceae). *Int. J. Plant Sci.* **153**, 565–571.

Baskin, C. C., Meyer, S. E., and Baskin, J. M. (1995). Two types of morphophysiological dormancy in seeds of two genera (*Osmorhiza* and *Erythronium*) with an Arcto-Tertiary distribution pattern. *Am. J. Bot.* **82**, 293–298.

Baskin, J. M., and Baskin, C. C. (1983). Germination ecophysiology of eastern deciduous forest herbs: *Hydrophyllum macrophyllum. Am. Midl. Nat.* **109**, 63–71.

Baskin, J. M., and Baskin, C. C. (1984a). Germination ecophysiology of the woodland herb *Osmorhiza longistylis* (Umbelliferae). *Am. J. Bot.* **71**, 687–692.

Baskin, J. M., and Baskin, C. C. (1984b). Germination ecophysiology of an eastern deciduous forest herb *Stylophorum diphyllum. Am. Midl. Nat.* **111**, 390–399.

Baskin, J. M., and Baskin, C. C. (1985a). Epicotyl dormancy in seeds of *Cimicifuga racemosa* and *Hepatica acutiloba. Bull. Torrey Bot. Club* **112**, 253–257.

Baskin, J. M., and Baskin, C. C. (1985b). Germination ecophysiology of *Hydrophyllum appendiculatum*, a mesic forest biennial. *Am. J. Bot.* **72**, 185–190.

Baskin, J. M., and Baskin, C. C. (1985c). Seed germination ecophysiology of the woodland spring geophyte *Erythronium albidum. Bot. Gaz.* **146**, 130–136.

Baskin, J. M., and Baskin, C. C. 1986a). Seed germination ecophysiology of the woodland herb *Asarum canadense*. *Am. Midl. Nat.* **116**, 132–139.

Baskin, J. M., and Baskin, C. C. (1986b). Change in dormancy status of *Frasera caroliniensis* seeds during overwintering on parent plant. *Am. J. Bot.* **73**, 5–10.

Baskin, J. M., and Baskin, C. C. (1988). The ecological life cycle of *Cryptotaenia canadensis* (L.) DC. (Umbelliferae), a woodland herb with monocarpic ramets. *Am. Midl. Nat.* **119**, 165–173.

Baskin, J. M., and Baskin, C. C. (1989). Seed germination ecophysiology of *Jeffersonia diphylla*, a perennial herb of mesic deciduous forests. *Am. J. Bot.* **76**, 1073–1080.

Baskin, J. M., and Baskin, C. C. (1990). Germination ecophysiology of seeds of the winter annual *Chaerophyllum tainturieri*: A new type of morphophysiological dormancy. *J. Ecol.* **78**, 993–1004.

Baskin, J. M., and Baskin, C. C. (1991). Nondeep complex morphophysiological dormancy in seeds of *Osmorhiza claytonii* (Apiaceae). *Am. J. Bot.* **78**, 588–593.

Baskin, J. M., and Baskin, C. C. (1994). Nondeep simple morphophysiological dormancy in seeds of the mesic woodland winter annual *Corydalis flavula* (Fumariaceae). *Bull. Torrey Bot. Club* **121**, 40–46.

Choi, K. G. (1977). Studies on seed germination in *Panax ginseng* C. A. Meyer. 2. The effect of growth regulators on the dormancy breaking. *Bull. Inst. Agric. Res. Tohoku Univ.* **28**, 159–170.

Choi, K. G., and Takahashi, N. (1977). Studies of seed germination in *Panax ginseng* C. A. Meyer. 1. The effect of germination inhibitors in fruits on dormancy breaking. *Bull. Inst. Agric. Res. Tohoku Univ.* **28**, 145–157.

Cook, R. A. (1993). "The Population Biology and Demography of *Cimicifuga rubifolia* Kearney and the Genetic Relationships among the North American *Cimicifuga* Species." Ph.D. thesis, University of Tennessee, Knoxville.

Crocker, W. (1948). "Growth of Plants," pp. 67–138. Reinhold, New York.

Dehgan, B., and Schutzman, B. (1983). Effect of H_2SO_4 and GA_3 on seed germination of *Zamia furfuracea*. *HortScience* **18**, 371–372.

Devillez, F. (1978). Influence de la température sur la postmaturation et la germination des graines de l'if (*Taxus baccata* L.). *Bull. Acad. Roy. Elg. Classe des Sciences* **54**, 203–218.

Dyer, R. A. (1965). The cycads of southern Africa. *Bothalia* **8**, 405–515.

Eriksson, O. (1994). Seedling recruitment in the perennial herb *Actaea spicata* L. *Flora* **189**, 187–191.

Ernst, W. H. O. (1979). Population biology of *Allium ursinum* in northern Germany. *J. Ecol.* **67**, 347–362.

Evans, C. R. (1933). Germination behavior of *Magnolia grandiflora*. *Bot. Gaz.* **94**, 729–754.

Fedec, P., and Knowles, R. H. (1973). Afterripening and germination of seeds of American highbush cranberry (*Viburnum trilobum*). *Can. J. Bot.* **51**, 1761–1764.

Florance, E. R. (1994). Structure, germination, and dormancy of seeds from *Frasera albacaulis* and *F. umpquaensis*. *Am. J. Bot.* **81** (Suppl. to No. 6), 70. [Abstract]

Frost-Christensen, H. (1974). Embryo development in ripe seeds of *Eranthis hiemalis* and its relation to gibberellic acid. *Physiol. Plant.* **30**, 200–205.

Giersbach, J. (1937). Germination and seedling production of species of *Viburnum*. *Contrib. Boyce Thomp. Inst.* **9**, 79–90.

Grushvitzky, I. V. (1967). After-ripening of seeds of primitive tribes of angiosperms, conditions and peculiarities. *In* "Physiologie, Okologie and Biochemie der Keimung" (H. Borris, ed.), Vol. 1, pp. 329–345. Ernst-Moritz-Arndt Universitat, Greifswald.

Hanzawa, F. M., and Kalisz, S. (1993). The relationship between age,

size, and reproduction in *Trillium grandiflorum* Liliaceae). *Am. J. Bot.* **80**, 405–410.

Hepher, A., and Roberts, J. A. (1985). The control of seed germination in *Trollius ledebouri*: The breaking of dormancy. *Planta* **166**, 314–320.

Horovitz, A., Bullowa, S., and Negbi, M. (1975). Germination characters in wild and cultivated *Anemone coronaria* L. *Euphytica* **24**, 213–220.

Iven, S. A. (1923). Maturation and germination of seeds of *Ilex opaca*. *Bot. Gaz.* **76**, 60–77.

Joshi, G. C., Tiwari, K. C., Tiwari, R. N., and Uniyal, M. R. (1991). Conservation and large scale cultivation strategy of Indian ginseng: *Panax pseudoginseng* Wall. *Indian For.* **117**, 131–134.

Knowles, R. H., and Zalik, S. (1958). Effects of temperature treatment and of a native inhibitor on seed dormancy and of cotyledon removal on epicotyl growth in *Viburnum trilobum* Marsh. *Can. J. Bot.* **36**, 561–566.

Krauss, N., and Kohle, K.-H. (1985). Ein beitrag zur kenntnis uber die stratifikation und keimung von echensamen (*Fraxinus excelsior* L.). *Flora* **177**, 91–105.

Kuo-Huang, L.-L., Chien, C.-T., and Lin, T.-P. (1996). Ultrastructural study on *Taxus mairei* seed during the germination promotion by combination of warm and cold stratification. *Am. J. Bot.* **83** (Suppl. to No. 6), 45. [Abstract]

Liden, M., and Staaf, R. (1995). Embryo growth in tuberous *Corydalis* species. *Bull. Torrey Bot. Club* **122**, 312–313.

Liu, M., Li, R.-J., and Liu, M.-Y. (1993). Adaptive responses of roots and root systems to seasonal changes. *Environ. Exp. Bot.* **33**, 175–188.

McDonough, W. T. (1969). Effective treatments for the induction of germination in mountain rangeland species. *Northwest Sci.* **43**, 18–22.

Nault, A., and Gagnon, D. (1993). Ramet demography of *Allium tricoccum*, a spring ephemeral, perennial forest herb. *J. Ecol.* **81**, 101–119.

Nikolaeva, M. G. (1977). Factors controlling the seed dormancy pattern. *In* "The Physiology and Biochemistry of Seed Dormancy and Germination" (A. A. Kahn, ed.), pp. 51-74. North-Holland, Amsterdam.

Nikolaeva, M. G., Daletskaya, T. V., Pozdova, L. M., Rasumova, M. V., and Tikhonova, V. L. (1986). Germination conditions of dormant seeds of the genus *Trollius* (Ranunculaceae) species and some other rare species, requiring protection. *Bot. Zhur.* **72**, 238–244. [In Russian]

Pickett, F. L. (1913). The germination of seeds of *Arisaema*. *Indiana Acad. Sci. Proc.* **1913**, 125–128.

Platt, R. B. (1951). An ecological study of the mid-Appalachian shale barrens and of the plants endemic to them. *Ecol. Monogr.* **21**, 269–300.

Pozdova, L. M., and Razumova, M. V. (1994). Dormancy and germination in some rare plants of the family Liliaceae. *Bot. Zhur.* **79**, 69–73. [In Russian with English abstract]

Rizzini, C. T. (1973). Dormancy in seeds of *Anona crassiflora* Mart. *J. Exp. Bot.* **24**, 117–123.

Saunders, A. P. (1918). A method of hastening germination of hard coated seeds. *Bull. Am. Peony Soc.* **6**, 17–19.

Schlising, R. A. (1976). Reproductive proficiency in *Paeonia californica* (Paeoniaceae). *Am. J. Bot.* **63**, 1095–1103.

Steinbauer, G. P. (1937). Dormancy and germination of *Fraxinus* seeds. *Plant Physiol.* **12**, 813–824.

Stokes, P. (1952a). A physiological study of embryo development in *Heracleum sphondylium* L. I. The effect of temperature on embryo development. *Ann. Bot.* **16**, 441–447.

Stokes, P. (1952b). A physiological study of embryo development in

Heracleum sphondylium L. II. The effect of temperature on after-ripening. *Ann. Bot.* **16**, 571–576.

Stokes, P. (1953a). A physiological study of embryo development in *Heracleum sphondylium* L. III. The effect of temperature on metabolism. *Ann. Bot.* **17**, 157–169.

Stokes, P. (1953b). The stimulation of growth by low temperature in embryos of *Heracleum sphondylium* L. *J. Exp. Bot.* **4**, 222–234.

Stoltz, L. P., and Snyder, J. C. (1985). Embryo growth and germination of American ginseng seed in response to stratification temperatures. *HortScience* **20**, 261–262.

Thompson, P. A., and Cox, S. A. (1978). Germination of the bluebell (*Hyacinthoides non-scripta* (L.) Chouard) in relation to its distribution and habitat. *Ann. Bot.* **42**, 51–62.

Threadgill, P. F., Baskin, J. M., and Baskin, C. C. (1981). Dormancy in seeds of *Frasera caroliniensis* (Gentianaceae). *Am. J. Bot.* **68**, 80–86.

Vanstone, D. E., and LaCroix, L. J. (1975). Embryo immaturity and dormancy of black ash. *J. Am. Soc. Hort. Sci.* **100**, 630–632.

Villiers, T. A., and Wareing, P. F. (1964). Dormancy in fruits of *Fraxinus excelsior* L. *J. Exp. Bot.* **15**, 359–367.

Wcislinska, B. (1977). The role of gibberellic acid (GA_3) in the removal of dormancy in *Fraxinus excelsior* L. seeds. *Biol. Plant.* **19**, 370–376.

West, W. C., Frattarelli, F. J., and Russin, K. J. (1970). Effect of stratification and gibberellin on seed germination in *Ginkgo biloba*. *Bull. Torrey Bot. Club* **97**, 380–384.

Whigham, D. (1974). An ecological life history study of *Uvularia perfoliata* L. *Am. Midl. Nat.* **91**, 343–359.

Witte, W. T. (1977). Storage and germination of *Zamia* seed. *Proc. Florida State Hort. Soc.* **90**, 89–91.

6

Germination Ecology of Seeds with Physical Dormancy

I. PURPOSE

Natural openings that permit entry of water into seeds with physical dormancy have been described in many plant families (Table 3.5). The "unplugging" of these openings in the seed or fruit coat is not a random or haphazard event, and the timing of germination of seeds with physical dormancy is environmentally controlled in nature. Depending on species and habitat, various environmental factors cause seeds to become permeable during a certain time(s) of the year; thus, timing of germination is predictable.

This chapter presents, examples of germination phenologies of seeds with physical dormancy and discusses the role of drying in the development and maintenance of physical dormancy. Much information on the germination requirements of seeds (after physical dormancy is broken) comes from studies in which various laboratory techniques have been used to make seeds permeable. Thus, methods for artificially breaking physical dormancy, as well as germination requirements for seeds after they become permeable, will be surveyed. The role of various environmental factors in the breaking of physical dormancy will be examined. Finally, attention will be given to how physiological dormancy is broken in those seeds with both physical and physiological dormancy.

II. GERMINATION PHENOLOGY

When seeds with impermeable seed (or fruit) coats are sown on soil and exposed to natural seasonal temperature changes, germination may be spread over a number of years, depending on the species (Table 6.1).

Thus, many species whose seeds have physical dormancy have the potential to form long-lived seed reserves; however, a discussion of this topic will be delayed until Chapter 7.

Although seeds of a species may germinate over a period of years, the germination season is about the same each year (Fig. 6.1). The germination period in some species is restricted to a few weeks. For example, seeds of *Melilotus alba* germinate in winter and early spring (Fig. 6.1), and it is surprising to find healthy seedlings with radicles 1–2 cm in length lying on soil that is frozen solid. In many species, a few seeds germinate throughout the growing season, but there is a peak of germination at some specific time of the year (Brenchley and Warington, 1930; Roberts and Boddrell, 1985a,b). For example, seeds of *Geranium carolinianum* may germinate to low percentages in spring and summer, but exhibit a strong peak of germination in autumn (Fig. 6.1). In still other species, seeds germinate throughout the growing season. For example, seeds of *Napaea dioica* (Fig. 6.1) germinate from late winter through autumn; however, much of the germination takes place in spring. Germination of this species is sporadic during summer, often occurring after periods of unusually high temperatures.

The examples of germination phenology of seeds with physical dormancy shown in Fig. 6.1 are for species growing in a temperate climate with hot, moist summers and cold, wet winters. Seeds of many species with physical dormancy growing in subtropical/tropical regions with annual wet and dry seasons germinate at the beginning of the wet season (Bhardwaj and Prabhakar, 1990; Nazrul-Islam and Hoque, 1990; Papavassiliou *et al.*, 1994; Wagner and Spira, 1994), but those of a few species germinate at the beginning of the dry season (Lonsdale and Abrecht, 1989).

TABLE 6.1 Germination of Seeds with Physical Dormancy[a]

Species	Year sown	Year of last germination	No. of years that seed germinated
Abutilon theophrasti	1989	1996[b]	7
Anoda cristata	1990	1996[b]	6
Astragalus canadensis	1989	1994	5
A. tennesseensis	1989	1996[b]	7
Baptisia australis	1969	1977	8
Callirhoe bushii	1988	1994	6
Cardiospermum halicacabum	1990	1992	2
Chamaecrista fasciculata	1990	1994	4
Dalea foliosa	1969	1977	8
D. gattingeri	1969	1977	8
Desmanthus illinoensis	1969	1979	9
Evolvulus nuttallianus	1982	1989	7
Geranium carolinianum	1988	1995	7
Hibiscus palustris	1989	1993	4
Iliamna remota	1986	1995	9
Ipomoea purpurea	1989	1996[b]	7
Lespedeza capitata	1986	1996[b]	10
L. leptostachya	1986	1989	3
Medicago lupulina	1989	1995	6
Melilotus alba	1987	1992	5
M. officinalis	1989	1993	4
Napaea dioica	1986	1994	8
Orbexilum onobrychis	1987	1993	6
Rhus aromatica	1989	1995	6
Senna marilandica	1989	1996[b]	7
Sida hermaphrodita	1985	1995	10
S. spinosa	1989	1996[b]	7
Trifolium campestre	1993	1995	2

[a]Seed were sown on soil in a nontemperature-controlled greenhouse in Lexington, Kentucky. Germination was considered to be completed if no additional seeds germinated after 1 year (Baskin and Baskin, unpublished results).

[b]Studies are still in progress, as of spring 1997.

III. INTERNAL MOISTURE CONDITIONS OF SEEDS

If seeds (or fruits) that are capable of developing impermeable coats are collected at the time of embryo maturity but before any drying occurs, they will germinate (Chapter 2). However, the number of days from anthesis until seeds reach physiological maturity and maximum weight (i.e., when they first would germinate if removed from the plant and placed on a moist surface) ranges from 12–16 days in *Sida spinosa* (Egley, 1976)

to 82 days in *Acacia auriculiformis* (Pukittayacamee and Hellum, 1988). The impermeability of coats develops as seeds dry, and the moisture content of seeds at the time they become impermeable ranges from 2.0 to 21% (Table 6.2).

The amount of water lost from seeds depends on the relative humidity (RH) during the time of drying. If rainfall is high during the natural time of seed drying in the field, seeds may germinate prior to dispersal. For example, observations of germinated seeds in pods of *Tephrosia purpurea* (Shra and Sen, 1984) probably mean that seeds were drying at a very slow rate. If seeds with permeable coats are stored at high RH, they do not develop impermeable seed coats (Barrett-Lennard and Gladstones, 1964). Thus, the rate and degree of development of seed coat impermeability are controlled by atmospheric moisture (Quinlivan, 1971). However, other factors, such as stage of seed development when drying starts (Aitken, 1939) and genetics (Lebedeff, 1943) help determine the proportion of seeds in a given seed crop that develops impermeable coats.

The moisture content of seeds with impermeable coats has implications for germination ecology because it is related to the rate of loss of physical dormancy. Jones (1928) found that most seeds of *Vicia villosa* held at 25°C and 75–80% RH became permeable if their moisture content was above 14%. Seeds of *Lupinus digitatus* (Gladstones, 1958), *L. varius* (Quinlivan, 1968a), *L. angustifolius* and *L. luteus* (Quinlivan, 1970) with moisture contents of 6–10% were impermeable to water and did not become permeable when placed on a moist substrate. However, seeds of these species with moisture contents of 11–15% eventually imbibed water when they were placed on a moist substrate. Some seeds of *Trifolium pratense*, *T. hybridum*, and *Melilotus alba* with a moisture content of 16.6, 16.2, and 16.2%, respectively, became permeable at 32 and 81% RH (Nakamura, 1962). Storage of *T. repens* seeds with a moisture content of 22.4% at 32 and 81% RH resulted in a 30 and 20% increase in impermeable seeds, respectively. Seeds of *T. repens* var. *latum* with a moisture content of 20.7% exhibited a 10% decrease in impermeable seeds at 81% RH but a 35% increase in impermeable seeds at 32% RH. Further, when seeds of *Vigna unguiculata* with initial moisture contents varying from 4.6 to 12.6% were stored at 70% RH at 25°C for 3 weeks, germination increased significantly in those with initial moisture contents of 9.2 to 12.6% (Murphy *et al.*, 1986).

Although measurements made in the early 1900s showed that seeds with physical dormancy continued to lose water after they were harvested, no one knew exactly how this happened. Hyde (1954) found that the moisture content of *Trifolium repens*, *T. pratense*, and *Lupinus arboreus* (all are members of the Papilio-

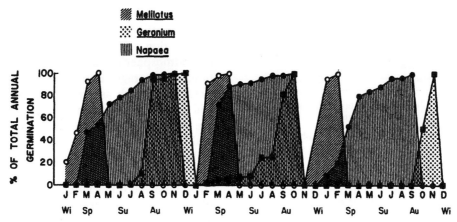

FIGURE 6.1 Germination phenology of seeds of *Geranium carolinianum, Melilotus alba,* and *Napaea dioica* sown on soil in a nontemperature-controlled greenhouse in Lexington, Kentucky. Wi, winter; Sp, spring; Su, summer; and Au, autumn.

noideae, a subfamily of the Fabaceae) seeds at the time of embryo maturity was 150% of the dry weight and declined to about 25% by water loss from the palisade epidermis. When seed moisture content drops below 25%, the hilum (Fig. 6.2) acts as a hygroscopic valve. The hilum in the Papilionoideae has a layer of heavily thickened cells, called the counter palisade, on each side of the hilum fissure. When the RH of the surrounding air becomes lower than any RH previously experienced by the seed, the counter palisade cells dry and shrink, and the hilum fissure opens. When the fissure is open, water vapor diffuses from the seed. As RH of the surrounding air increases, cells in the counter palisade swell, causing the fissure to close, and water vapor no longer diffuses out of the seed. Thus, the moisture content of the seed decreases each time there is a further decrease in RH of the air, but it never increases. Hyde (1954) found that seeds could absorb water vapor from the air if RH was increased so slowly that the hilum did not close. Over a period of 71 days, he raised the RH from 45 to 70%, and seed moisture content increased from about 10 to 13%. Such a slow rise in RH probably seldom, if ever, occurs in nature.

Although seeds of species other than those belonging to the Papilionoideae lose water (Table 6.2) during the development of seed coat impermeability, no studies have been done to determine how this happens. According to Rolston (1978), other subfamilies of legumes (Caesalpinioideae and Mimosoideae) generally do not have a counter palisade in the hilum. Thus, an understanding of seed drying in these subfamilies will require information on the structure and function of the hilum. One reason so much is known about impermeable seed coats in the Papilionoideae is that many of the important pasture legumes (e.g., Lupinus spp., *Trifolium* spp.,

Medicago spp., and *Stylosanthes* spp.) in Australia belong to this subfamily (Quinlivan, 1971).

When seed coats become permeable under natural conditions, the imbibition of water usually occurs through an unplugged natural opening. However, water may be imbibed across the whole seed coat, depending on the moisture content of the seeds. For example, seeds of *Lupinus varius* at moisture contents of 12–14% imbibe water over the surface of the seed, whereas those with a moisture content of 8.5% or less will imbibe water only after the strophiole has been rendered permeable (Quinlivan, 1968a). Moisture content may influence the germination phenology of *L. varius* seeds in the summer-dry, winter-moist pasture ecosystems of Australia. Seeds with a high moisture content can imbibe water following summer showers and thus germinate sporadically over several months, whereas those with low moisture percentages cannot germinate until after high fluctuating summer temperatures make the strophiole permeable. Consequently, these seeds with low moisture contents germinate over a very short period of time following rains in autumn (Quinlivan, 1968a).

After seeds with physical dormancy become permeable, they generally either germinate or rot; however, there may be some exceptions. Hagon and Ballard (1970) artificially made the strophiole on seeds of *Trifolium subterraneum* permeable (by shaking seeds in a glass bottle) and then subjected the seeds to varying levels of RH. The strophiole closed after 12 days at 0 or 5% RH, but it remained open at ≥20% RH. The minimum number of days for the strophiole to close at 0 and 5% RH was not determined. These data imply that under prolonged, very low RH conditions, the strophiole would close; consequently, permeable seeds would become impermeable again. However, drying

TABLE 6.2 Species In Which Seed Coat Impermeability Increases as Seed Moisture Content Decreases, and Moisture Content of Seeds at the Time They Became Impermeable

Species	% Seed moisture	Reference
Acacia auriculiformis	—[a]	Pukittayacamee and Hellum (1988)
A. senegal	—	Kaul and Manohar (1966), Danthu *et al.* (1992)
A. suaveolens	—	Auld (1986a)
Astragalus sinicus	—	Nakamura (1962)
Cassia acutifolia	—	Bhatia *et al.* (1977)
C. angustifolia	—	Bhatia *et al.* (1977)
Convolvulus arvensis	13	Swan (1980)
Crotolaria spectabilis	11	Egley (1979)
Cuscuta campestris		Hutchison and Ashton (1979)
Gossypium hirsutum	12	Patil and Andrews (1985)
Gymnocladus dioica	5–8	Raleigh (1930)
Ipomoea pes-tigridis	2.5–10	Bhati and Sen (1978)
I. turbinata	8.5	Chandler *et al.* (1977)
Lathyrus maritimus	—	Dinnis and Jordan (1939)
Medicago lupulina	—	Sidhu and Cavers (1977)
Lupinus arboreus	11	Hyde (1954)
Melilotus alba	—	Helgeson (1932)
Merremia aegyptia	2.0	Sharma and Sen (1974)
M. dissecta	4.3	Sharma and Sen (1974)
Rhus aromatica	16.1	Li (unpublished results)
R. glabra	9.3	Li (unpublished results)
Sesbania bispinosa	9.6	Graaff and van Staden (1987)
Sida spinosa	20	Egley (1976)
Strophostyles helvola	21	Hutton and Porter (1937)
Tephrosia hamiltonii	—	Bhardwaj and Prabbakar (1990)
Trifolium pratense	14	Hyde (1954)
T. repens	14	Hyde (1954)
Vicia villosa	<14	Jones (1928)

[a]Data not given.

conditions of the length and severity used by Hagon and Ballard (1970) are unlikely to occur in nature.

IV. ARTIFICIAL SOFTENING OF WATER-IMPERMEABLE SEED (OR FRUIT) COATS

Horticulturalists, foresters, farmers, gardeners, and researchers have developed a number of techniques for quickly making physically dormant seeds permeable. These methods include mechanical scarification, acid scarification, enzymes, organic solvents, percussion, high atmospheric pressures, hot water, dry heat, radiation, dry storage, ultrasound, and low temperatures. Treatments such as mechanical and acid scarification will almost always make a seed permeable to water, but the success of other methods varies with the species and treatment intensity and duration. Studies involving artificial methods of breaking physical dormancy have contributed greatly to our understanding of (1) how the strophiole and other natural openings work; (2) effects of various environmental factors such as drying, heating, and freezing on the loss of physical dormancy; and (3) the rate and path of water entry into seeds that have become permeable.

A. Mechanical Scarification

A file, sand paper, knife, razor blade, scalpel, or needle can be used to make a small hole in the seed coat, and water enters through this opening. However, this

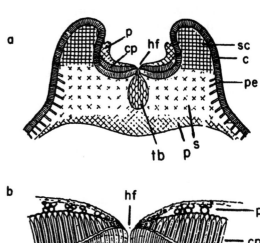

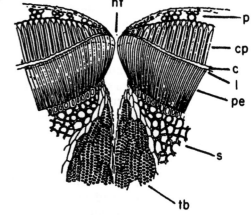

FIGURE 6.2 (a) Transmedian section of the hilum and adjacent tissues in a seed of *Lupinus arboreus*. (b) Transmedian section of the hilum in a seed of *Lupinus arboreus*. p, parenchyma; cp, counterpalisade; hf, hilum fissure; sc, sclerenchyma; c, cuticle; pe, palisade epidermis; tb, tracheid bar; l, light line; s, stellate cells. From Hyde (1954).

can be a very time-consuming method of making seeds permeable to water, especially if large numbers of scarified seeds are required. Thus, machines that roll or blow seeds against an abrasive surface such as glass splinters or sand paper in some kind of a container have been built (Porter, 1949; Townsend and McGinnies, 1972). Cavanagh (1987) points out, however, that while these machines work fine for small, thin-coated seeds like those of *Trifolium subterraneum*, they may not work well for thick-coated seeds like those of *Acacia* spp.

B. Acid Scarification

Seeds (or fruits) usually are soaked in concentrated H_2SO_4 and then washed several times to remove the acid. (Remember to add acid to water; never add water to acid.) Trials have to be run to determine the appropriate period of time for the acid to break through the seed coat but not damage the embryo; this varies with the species (Table 6.3).

Because seeds are immersed in H_2SO_4, both the seed coat and the plugged natural opening are subject to being destroyed. Lumens of macrosclereids in the seed coat of *Coronilla varia* were exposed by acid treatment (Brant *et al.*, 1971). However, Tran and Cavanagh (1984) noted that Brant *et al.* (1971) did not check to see if water could pass through the lower side of the macrosclereids. Acid treatment of *Rhus ovata* fruits for 3 hr caused areas around the micropyle and hilum to be destroyed (Stone and Juhren, 1951).

Acid scarification for 3 hr partially destroyed the counter palisade cells of the hilum in *Lupinus angustifolia* (Burns, 1959) seeds. Consequently, the hilum could not close when *L. angustifolia* seeds were placed on a moist surface, and water entered through it. If acid-treated seeds of this species subsequently were redried and then placed on a wet surface, they took up water through the strophiole. The strophiole was the major region affected when seeds of *Astragalus cicer* were acid scarified for 20 min (Miklas *et al.*, 1987). Acid dissolved the cuticle over the strophiole as well as portions of the subtending Malpighian cells, leaving a small circular cavity with a large groove in the bottom. Consequently, most of the water entered *A. cicer* seeds through the strophiole. A 30- to 90-min period of acid scarification destroyed a plug-like structure in the bottom of the micropylar depression in seeds of *Convolvulus lanatus, C. negevensis,* and *C. secundus,* and this is where water entered (Koller and Cohen, 1959).

C. Enzymes

A few attempts have been made to use enzymes to overcome physical dormancy, but they usually do not work very well. After 24 hr of soaking in hemicellulase and pectinase, 52 and 46%, respectively, of the seeds of *Coronilla varia* were permeable; 32% of the control seeds were permeable (Brant *et al.*, 1971). The site of water entry was not determined.

D. Organic Solvents

Barton (1947) found that seeds of *Cassia, Cercidium, Gleditsia, Gymnocladus, Cercis* (all members of the Caesalpinioideae), and *Acacia greggii* (Mimosoideae) became permeable after they were soaked in absolute ethyl alcohol, but those of Papilionoideae did not. Ether caused "a large percentage" of *Prosopis juliflora* (Mimosoideae) seeds to germinate (Crocker, 1909), a very low percentage of the impermeable fruits of *Nelumbo lutea* to imbibe water (Shaw, 1929), and a 52% (above the control) increase in the germination of *Discaria toumatou* (Rhamnaceae) seeds (Keogh and Bannister, 1994). Small but significant percentages of *Coronilla varia* seeds became permeable after soaking in acetone and petroleum ether (Brant *et al.*, 1971). Soaking in ethyl alcohol and acetone stimulated 10–20% germination in seeds of *Acacia nilotica* and 25–60% in those of *A. torilis;* the point of water entry was the hilum (Brown and Booysen, 1969). Seeds of *Sphaeralcea grossulariaefolia* germinated to 0 and 69% after soaking in diethyl dioxide (dioxan) for 0 and 4 hr, respectively (Page *et al.*, 1966).

Other solvents, including acetone, carbon tetrachloride, chloroform, ethylene chloride, and ethyl alcohol, resulted in 0–14% germination of *P. stephaniana* seeds; the water control germinated to 6% (Khudairi, 1956). In an atmosphere of butylamine and at a constant temperature of 20°C, the seed permeability of *Trifolium subterraneum* was 40% higher than the control (Fairbrother, 1991). Ethyl alcohol and ether removed the plug over the chalaza aperture in *Gossypium hirsutum* seeds and allowed water to enter (Christiansen and Moore, 1959).

E. Percussion

Hamly (1932) observed that the Malpighian cells in the strophiole of an impermeable *Melilotus officinalis* seed appeared to be in a strained condition, i.e., the cells were bent. However, the Malpighian cells in the strophiole of a permeable seed were split apart, and they were only slightly bent. Hamly (1932) surmised that a physical blow on the strophiole would release the tension and cause the cells to pull apart. Consequently, he put some seeds in a glass bottle and shook them at a rate of three times per second for several minutes, for a calculated total of 3600 blows on each seed. He was correct. Impaction (later to be called percussion) caused

TABLE 6.3 Species in Which Physical Dormancy Has Been Broken by
Sulfuric Acid, and Duration of Treatment for Maximum Germination

Species	Time (min) in H_2SO_4	Reference
Abutilon theophrasti	15	Steinbauer and Grigsby (1959)
Acacia alba	30	Teketay (1996)
A. farnesiana	90	Scifres (1974)
A. cyanophylla	90	Jones (1963)
A. senegal	14	Palma *et al.* (1995a)
A. sieberiana	120	Teketay (1996)
Apeiba membranacea	2, 10	Acuna and Garwood (1987)
Aspalathus linearis	120	Kelly and van Staden (1985)
Astragalus cicer	20	Miklas *et al.* (1987)
Caesalpinia spinosa	30	Teketay (1996)
Cassia fistula	10	Bhattacharya and Saha (1990)
Chordospartium stevensonii	10	Conner and Conner (1988)
Clitoria ternatea	30–40	Mullick and Chatterji (1967)
Convolvulus arvensis	45–60	Brown and Porter (1942)
Crotalaria medicaginea	30	Bohra and Sen (1974)
Cuscuta indecora	30	Allred and Tingey (1964)
Citrullus colocynthis	4.5	Sen and Bhandari (1974)
Delonix regia	60	Teketay (1996)
Discaria toumatou	20	Keogh and Bannister (1994)
Entada abyssinica	60	Teketay (1996)
Erodium botrys	1	Young *et al.* (1975)
Erythrina burana	60	Teketay (1994)
E. caffra	120	Small *et al.* (1977)
E. lysistemon	60	Teketay (1996)
Grazuma ulmnifolia	2, 10	Acuna and Garwood (1987)
Gymnocladus dioicus	120	Liu *et al.* (1981)
Hibiscus trionum	20	Everson (1949)
Indigofera linifolia	10	Rao and Reddy (1981)
Ipomoea crassicaulis	720	Misra (1963)
I. hederacea	60	Gomes *et al.* (1978)
Lathyrus martimus	20	Lemmon *et al.* (1943)
Leucaena leucocephala	60	Teketay (1996)
Lupinus angustifolius	180	Burns (1959)
Nelumbo lutea	300	Jones (1928)
Parkia auriculata	15	Coutinho and Struffaldi (1971)
Parkinsonia aculeata	45	Everitt (1983)
Prosopis farcta	25	Dafni and Negbi (1978)
Rhus glabra	120	Farmer *et al.* (1982)
Senna marilandica	60	Nan (1992)
S. obtusifolia	30	Nan (1992)
Sesbania drummondii	240	Eastin (1984)

92% of the seeds to become permeable; only 0.5% of the seeds in the control germinated. Percussion now has been shown to induce permeability in seeds of a number of species (Table 6.4).

With the use of osmic acid, which turns black when it comes into contact with cytoplasm, Hamly (1932) showed that water enters permeable seeds of *Melilotus officinalis* through the strophiole. Ballard (1973) demonstrated that water also enters *Medicago truncatula* seeds at the strophiole (Fig. 6.3). Because the seed coat and strophiole are parts of an integrated system, cutting, filing, piercing, squeezing or striking the seed coat at some point away from the strophiole caused the strophiole on *Medicago scutellata, Stylosanthes humilis, Trifolium hirtum,* and *T. subterraneum* seeds to become permeable to water (Ballard, 1976).

F. High Atmospheric Pressures

Davies (1926, 1928a,b) put *Medicago sativa* and *Melilotus alba* seeds in a container of water inside a sealed chamber and then applied hydrostatic pressures of 50.7 and 202.6 MPa. A pressure of 202.6 MPa was more effective than 50.7 MPa in breaking physical dormancy, and at 202.6 MPa 1 and 10 min were optimal for the subsequent germination of *M. sativa* and *M. alba* seeds, respectively. Further, pressure was more effective when applied at room temperatures (18±2°C) than at 0°C.

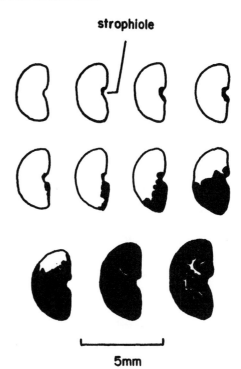

strophiole

5mm

FIGURE 6.3 Stages in the imbibition of a single seed of *Medicago truncatula* made permeable by percussion. Water entered at the strophiole, and its pattern of movement can be seen because the seed was placed in a 0.003 M Fe^{2+} solution which stains imbibed palisade cells black. Redrawn from Ballard (1973).

TABLE 6.4 Species with Physical Dormancy Whose Seeds Become Permeable after Being Shaken in a Glass Bottle (Percussion)

Species	Reference
Acacia greggii	Barton (1947)
Amorpha fruticosa	Hutton and Porter (1937)
Aspalathus linearis	Kelly and van Staden (1987)
Cladrastis lutea	Barton (1947)
C. amurensis	Barton (1947)
Glycine usuriensis	Porter (1949)
Lespedeza capitata	Hutton and Porter (1937)
L. virginica	Hutton and Porter (1937)
Lathyrus odoratus	Porter (1949)
Medicago truncatula	Ballard (1973)
Melilotus alba	Barton (1947)
M. officinalis	Hamly (1932)
Parkinsonia microphylla	Barton (1947)
Pisum sativum	Porter (1949)
Prosopis velutina	Barton (1947)
Robinia pseudo-acacia	Barton (1947)
Trifolium subterraneum	Hagon and Ballard (1970)

Seeds of *Cladrastis lutea* showed an increase in germination (i.e., more seeds became permeable) with an increase in pressure from 6.7 to 206.7 MPa, but ≥310.1 MPa resulted in decreased germination. A pressure of 68.9 MPa applied for 10, 1, and 1 min at 0, 25, and 50°C, respectively, gave 100% germination. The seed coat broke in the region of the hilum, and this is where water entered (Rivera *et al.,* 1937). Seeds of *Gymnocladus dioica* germinated to 90% after being exposed to 6.9 MPa of pressure for 1 min; pressures lower than 6.9 MPa were ineffective and those higher than 6.9 MPa injured the embryos (Rivera *et al.,* 1937).

G. Wet Heat

Immersion in hot water causes impermeable seeds of a number of species to become permeable (Table 6.5). In these studies, seeds are placed in a cloth bag or something like a tea strainer and then dipped into hot water for the required period of time. Seeds usually are allowed to cool to room temperature, but sometimes they are plunged into cold water.

The period of time that seeds can be kept in hot water before they are killed decreases with an increase in treatment temperature. For example, treatment times

TABLE 6.5   Species in Which Physical Dormancy of the Seeds is Broken by
Wet Heat Treatments

Species	Temperature (°C)	Duration	Reference
Abutilon theophrasti	70	1 hr	Horowitz and Taylorson (1984)
Acacia albida	100	5 sec	Teketay (1996)
A. falcata	100	5 sec	Clemens *et al.* (1977)
	80	600 sec	Clemens *et al.* (1977)
A. lebbeck	100	5 sec	Teketay (1996)
A. nilotica	100	10 min	Brown and Booysen (1969)
A. sieberiana	100	45 sec	Teketay (1996)
A. suaveolens	80	200 sec	Clemens *et al.* (1977)
A. terminalis	100	30 sec	Clemens *et al.* (1977)
A. tortilis	100	10 min	Brown and Booysen (1969)
Apeiba membranacea	62–70	10 min	Acuna and Garwood (1987)
Cassia nictitans	80	4 min	Martin and Cushwa (1966)
Coronilla varia	100	30 sec	Brant *et al.* (1971)
Crotolaria sericea	65	10 min	Saha and Takahashi (1981)
Ceanothus spp.	100	5–20 sec	Quick and Quick (1961)
Entada abyssinica	100	5 sec	Teketay (1996)
Erythrina brucei	100	5 sec	Teketay (1994)
Gossypium hirsutum	80	1 min	Christiansen and Moore (1959)
Grazuma ulmnifolia	62–70	2, 10 min	Acuna and Garwood (1987)
Heliocarpus donnell-smithii	60, 80	1 min	Vazquez-Yanes (1981)
Indigofera linifolia	56	360 min	Rao and Reddy (1981)
Lespedeza cyrtobotrya	70	0.5–3 min	Iwata (1966)
Leucaena leucocephala	100	15 sec	Teketay (1996)
Ochroma lagopus	100	15 sec	Vazquez-Yanes (1974)
Parkinsonia aculeata	100	5 sec	Teketay (1996)
Phaseolus mungo	100	7 min	Rao and Mukherjee (1978)
Prosopis juliflora	100	5 sec	Teketay (1996)
Robinia pseudo-acacia	100	1 min	Wilson (1944)
R. hispida	100	1 min	Wilson (1944)
Sesbania punicea	98	30 sec	Graaff and Van Staden (1983)
Stylosanthes macrocephala	100	1 sec	Silva and Felippe (1986)

of up to 600 sec at 80°C resulted in high germination percentages of *Acacia falcata, A. terminalis,* and *A. suaveolens* seeds (Fig. 6.4). However, treatment times of 200, 100, or 20 sec at 100°C for seeds of *A. falcata, A. terminalis,* and *A. suaveolens,* respectively, decreased germination, possibly indicating that embryos had been damaged.

Boiling water caused the palisade layer in seeds of *Acacia* spp. to soften and separate from the underlying mesophyll; consequently, cracks in the seed coat occurred at nonlocalized areas, allowing water to enter at many sites (Brown and Booysen, 1969). Boiling water treatments also caused cracks to develop in seed coats of *Sesbania punicea* (Graaff and Van Staden, 1983) and caused macrosclereids in the palisade layer of *Coronilla varia* seeds to separate from each other (Brant *et al.,* 1971). The strophiole erupted or was uplifted in *Albizia lophantha* seeds (Dell, 1980) and in *Acacia kempeana* (Hanna, 1984) seeds dipped in boiling water for a few seconds. Heat-treated seeds of *A. kempeana* did not imbibe water if the uplifted strophiole was covered with petroleum jelly to block the entry of water (Hanna,

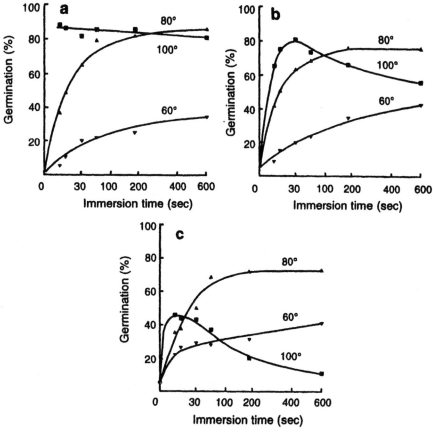

FIGURE 6.4 Final germination percentages of seeds of three species of *Acacia* that were given hot water treatments at various temperatures for 0 to 600 sec. (a) *A. falcata,* (b) *A. terminalis,* and (c) *A. suaveolens.* From Clemens *et al.* (1977).

1984). Hot water removed the chalazal plug in *Gossypium hirsutus* seeds, and water entered through the chalazal region (Christiansen and Moore, 1959). Water at 65°C caused the strophiole of *Crotalaria sericea* seeds to turn golden brown in color (Saha and Takahashi, 1981).

H. Dry Heat

The most common method for giving impermeable seeds a dry-heat treatment is to place them in an oven at the specified temperature. Dry heat is effective in breaking physical dormancy in seeds of a number of species (Table 6.6), but it does not work in all of them. The appropriate dry-heat treatment for a given species is determined by experimentation to find the combination of temperature and duration of treatment that results in high germination percentages but not in death of the embryo. In general, there is a decrease in the length of time required to break physical dormancy with an increase in temperature. Only a few minutes of exposure to temperatures of ≥100°C are required to

break physical dormancy in seeds of most species, but those of *Tephrosia appolina* had to be heated at 120°C for 3 days (Narang and Bhardwaja, 1974).

Another way of subjecting seeds to dry heat is to allow them to slide (Lunden and Kinch, 1957) or bounce (Mott *et al.,* 1982) down a heated inclined plane. Also, Mott (1979) placed seeds of *Stylosanthes* spp. on a tray heated to various temperatures (125, 145, 200, and 260°C) and shook the tray vertically so that seeds bounced 3 cm off the hot surface for 15, 30, or 60 sec. Infrared radiation, radio waves, and microwaves also are methods of giving seeds a dry-heat treatment (see later).

Dry heat caused seed coats of *Tephrosia appolina* (Narang and Bhardwaja, 1974) and *Acacia* spp. (Brown and Booysen, 1969) seeds to develop cracks and the palisade layer of the strophiole on seeds of *Acacia* spp. to split (Brown and Booysen, 1969). Cavanagh (1987) expressed doubts that cracks induced by dry heat were deep enough to allow water to reach the embryo, and he suggested that heat treatments promote the germination of legume seeds because they made the strophiole permeable to water. An alternating temperature regime

TABLE 6.6   Species In Which Dry Heat Overcomes Physical Dormancy of the Seeds

Species	Temperature (°C)	Duration	Reference
Acacia longifolia	140	1 min	Mott *et al.* (1982)
A. nilotica	120	10 min	Brown and Booysen (1969)
A. saligna	100	30 min	Jeffrey *et al.* (1988)
Cassia nictitans	70–80	4 min	Martin *et al.* (1975)
Caesalpinia decapetala	80	15 min	Teketay (1996)
Cistus albidus	90	9 min	Trabaud and Oustric (1989)
C. incanus creticus	100	15 min	Thanos and Georghiou (1988)
C. ladanifer	100	1 min	Valbuena *et al.* (1992)
	100	30 min	Corral *et al.* (1990)
C. laurifolius	100	5 sec	Valbuena *et al.* (1992)
C. monspeliensis	150	1 min	Trabaud and Oustric (1989)
C. salvifolius	110	9 min	Trabaud and Oustric (1989)
Cytisus scoparius	65	2 min	Bossard (1993)
Entada lysistemon	80	30 min	Teketay (1996)
Halimium halimifolium	100	15 min	Thanos *et al.* (1992)
H. pilosum	100	15 min	Thanos *et al.* (1992)
Lespedeza hedysaroides	90	4 min	Martin *et al.* (1975)
Leucaena leucocephala	80	60 min	Teketay (1996)
Medicago sativa	104	4 min	Rincker (1954)
Ochroma lagopus	95	5 min	Vazquez-Yanes (1974)
Podalyria calyptrata	60	5 min	Jeffrey *et al.* (1988)
	100	1 min	Jeffrey *et al.* (1988)
Pueraria phaseoloides	50	4 hr	Wycherley (1960)
Rhus javanica	65–70	30–120 min	Washitani (1988)
Sida grewioides	90	24 hr	Chawan (1971)
S. rhombifolia	90	12 hr	Chawan (1971)
S. spinosa	90	12 hr	Chawan (1971)
S. veronicifolia	70	24 hr	Chawan (1971)
Stylosanthes spp.	85–95	1 hr	Mott and McKeon (1979)
Tephrosia appolina	120	3 days	Narang and Bhardwaja (1974)
Trifolium pratense	104	4 min	Rincker (1954)
Tuberaria lignosa	100	15 min	Thanos *et al.* (1992)

of 40/20°C caused seeds of *Astragalus sinicus* to become permeable, and the strophiole was opaque in permeable seeds and transparent in impermeable ones (Ueki and Suetsugu, 1958).

I. Dry Storage

In some species, seeds with physical dormancy stored dry at room temperatures for several months (or years) become permeable (e.g., Egley, 1976; Silva and Felippe, 1986; Morrison *et al.*, 1992). Cavanagh (1987) concluded that seeds of some legumes, especially those belonging

to the Papilionoideae, became permeable during dry storage because cells in the strophiole broke, allowing water to enter. However, seeds of various Australian legumes belonging to the tribes Bossiaeeae and Phaseoleae become permeable during dry storage, and water entry was not restricted to the strophiole. The mechanism of increase in seed coat permeability in these two tribes during dry storage has not been identified, but it may be related to the fact that seed coats were only 75% as thick as those of species in other tribes (Mirbelieae, Acacieae) that failed to become permeable during dry storage (Morrison *et al.*, 1992).

J. Radiation

Infrared and gas-plasma radiation, radio frequencies, and ultrahigh radio frequencies (microwaves) have been used to break physical dormancy. Much of the work has been done in an effort to find an easy and efficient way to make impermeable seeds of economically important members of the Papilionoideae (e.g., *Medicago* spp., *Trifolium* spp., *Stylosanthes* spp.) permeable to water. All of these treatments are from electrically generated radiation, causing seed temperatures to increase.

Infrared radiation supplied by a 250-W infrared bulb produced a temperature of 104°C on the surface of *Medicago sativa* seeds. After 1.5 min of exposure, germination was increased 47%; after 5 min most of the seeds were dead (Rincker, 1954). A Philips and an Osram infrared lamp with a 90- and 250-W bulb, respectively, produced a temperature of 50 and 43°C, respectively, at the position of test seeds (Wycherley, 1960). Regardless of the type of lamp, germination of *Flemingia congesta* and *Pueraria phaseoloides* seeds increased by about 30 and 50% (over the controls), respectively. Maximum germination of *F. congesta* seeds occurred after a 15-min and 1-hr exposure to the Philips and Osram lamps, respectively, and maximum germination of *P. phaseoloides* seeds was after a 4-hr exposure to both types of lamps. With a 500-W infrared quartz lamp, 110 and 130 V for 1.41 and 1.01 sec, respectively, caused 65–100% of the seeds of seven varieties of *Medicago sativa* and *Trifolium pratense* to become permeable (Works, 1964).

Gas-plasma radiation (glow discharge) caused seeds of Narragansett, Ranger, and Alaskan varieties of *Medicago sativa* to become permeable, with treatments being more effective if the seed moisture content was low (Pettibone, 1965). Infrared radiation, a radio frequency at 39 MHz, and gas-plasma radiation were about equally effective in increasing permeability in seeds of three varieties of *Medicago sativa* (Nelson et al., 1964).

Radio frequencies also increase the temperature of seeds because seeds are poor conductors of electrical charges (i.e, they are dielectric substances). Seeds absorb some of the energy when they are subjected to alternating electromagnetic fields, which causes heating. The amount of temperature rise depends on several factors, especially seed moisture content (Ballard et al., 1976). In an early study, the germination of seeds of *Medicago sativa* placed between two electrodes and exposed to alternating electromagnetic fields at 27 MHz for 25 to 30 sec increased by 10–36%; optimum seed temperature for breaking physical dormancy was 56°C (Eglitis and Johnson, 1957).

Medicago sativa seeds subjected to radio frequencies of 39 MHz in a dielectric oven (i.e., exposed to an alternating electromagnetic field) reached temperatures of 66 to 78°C, and germination increased from 42–80% (controls) to 65–88%, depending on the cultivar (Nelson et al., 1968). Radio frequencies of 5, 10, and 39 MHz in a dielectric oven increased temperatures of *M. sativa* seeds to 71–77°C, and germination ranged from 60 to 95%, regardless of radio wave frequency; controls germinated to 46–53%, depending on the cultivar (Nelson and Wolf, 1964). A hot-air oven, a dielectric oven at a radio frequency of 39 MHz and a microwave oven at 2450 MHz were equally effective in overcoming the physical dormancy of *M. sativa* seeds, if the seed temperature reached 66–88°C (Stetson and Nelson, 1972). Additional studies on *M. sativa* seeds at normal moisture contents showed that radio frequency treatments break physical dormancy without killing the embryo when temperatures rise to 70–80°C (Nelson et al., 1977). Seeds of *Medicago sativa, M. scutellata, M. truncatula, Stylosanthes humilis, Trifolium hirtum,* and *T. subterraneum* subjected to radio frequencies of 39 and 2450 MHz reached temperatures of 60–80°C before they became permeable. At about 80°C, however, seed viability began to decline. In seeds that reached 60–80°C, water entered through the strophiole. As the seed temperature increased to 120–130°C, water also entered through random cracks that developed in the seed coat (Ballard et al., 1976).

The strophiole on *Acacia longifolia* and *A. sophorae* seeds exposed to 2450 MHz in a microwave oven was raised, golden in color, and permeable to water after seed temperatures reached 93–96°C, and 74–90% of the seeds germinated (Tran, 1979). However, species as well as length of microwaving treatment may influence the degree of loss of physical dormancy. After 130 sec of microwaving, the strophiole was raised and golden in color in 85 of 87 (97%) *Acacia longifolia* seeds and 83 of the 85 seeds imbibed water. After 110 sec of microwaving, *A. sophorae* seeds had raised, golden strophioles, but only 56 of the 68 seeds imbibed water (Tran and Cavanagh, 1980).

K. Ultrasound

Seeds of various species have been subjected to sonication (e.g., Weinberger and Burton, 1981), but little is known about the effects of this treatment on breaking of physical dormancy. However, sonication increased the germination of *Medicago sativa* (Kolokol'tseva and Profof'ev, 1974) and *Cassia holosericea* (Faruqi et al., 1974) seeds.

L. Low Temperatures

The physical dormancy of several species has been broken by freezing at very low temperatures (Table 6.7):

TABLE 6.7   Species in Which Physical Dormancy Is Broken
by Freezing of the Seeds

Species	Temperature (°C)	Reference
Coronilla varia	−196	Brant *et al.* (1971)
Lotus corniculata	−185, −196	Eynard (1958)
Medicago sativa	−80	Busse (1930)
	−185, −196	Eynard (1960)
Melilotus alba	−196	Barton (1947)
Trifolium arvense	−196	Pritchard *et al.* (1988)
T. hybridum	−185, −196	Eynard (1960)
T. pratense	−185, −196	Eynard (1960)
T. repens	−185, −196	Eynard (1958)

−195.8 (liquid nitrogen), −185 (liquid oxygen), −190 (liquid air), and −90°C (solid carbon dioxide). Seeds of *Trifolium repens* and *Lotus corniculatus* placed in liquid nitrogen and oxygen for 1, 2, 5, 10, 20, and 60 min germinated as well after 5 min of freezing as they did after the longer treatments (Eynard, 1958). Seeds of *T. hybridum, T. pratense,* and *M. sativa* placed in liquid nitrogen or oxygen for 10 min germinated to about the same percentages as those treated for 60 min (Eynard, 1960). Seeds of *Coronilla varia* frozen at −195.8°C germinated to about the same percentages, regardless of the number of times they were plunged into liquid nitrogen or how long they stayed there (Brant *et al.*, 1971). However, four dips of 30 sec each into liquid nitrogen were more effective in decreasing the impermeability of *Melilotus alba* seeds than one dip of 1 or 5 min (Barton, 1947).

Alternately dipping into boiling water and liquid nitrogen broke the physical dormancy in seeds of *Cytisus scoparius,* but the heating/freezing treatment was more effective if seeds were dipped into the boiling water before they were frozen (Abdallah *et al.,* 1989). Slow cooling and warming before and after being plunged into the liquid nitrogen, respectively, were as effective in breaking the dormancy of *Trifolium arvense* seeds as rapid cooling and warming (Pritchard *et al.,* 1988). These authors found channels through the layer of Malpighian cells in the strophiole of *T. arvense* seeds that had been frozen at −195.8°C; therefore, water entered through the strophiole. Cracks also developed in the seed coats of seeds that had been frozen, but water did not enter through them.

Some seedlings from seeds frozen in liquid nitrogen are abnormal because the petiole is broken, resulting in a detachment of the cotyledons (Pritchard *et al.,* 1988; Wiesner *et al.,* 1994). However, if seeds of *Medicago sativa* were mechanically scarified prior to being frozen

with liquid nitrogen, cotyledon detachment was reduced (Wiesner *et al.,* 1994).

Seeds of *Convolvulus arvensis* stored on dry filter paper in petri dishes in the dark at 5°C for 0, 21, and 42 days germinated to 10, 55, and 85%, respectively, in darkness at 20°C. Scanning electron microscopy showed that the palisade sclerenchyma layer in nonchilled seeds was tightly packed and that the parenchyma cell region below the palisade layer had few, if any, pores. However, seeds chilled at 5°C for 42 days exhibited a breakdown of cells in both the palisade layer and the subtending parenchyma layer (Jordan and Jordan, 1982). It is hard to explain how "cell digestion" occurred in dry seeds at 5°C.

V. GERMINATION REQUIREMENTS OF PERMEABLE SEEDS

Although some seeds with physical dormancy also have physiologically dormant embryos (see Section VII), most seeds with physical dormancy have nondormant embryos. Thus, the purpose of this section is to discuss the germination requirements of the latter type of seeds after physical dormancy has been broken.

Seeds that become permeable by either artificial or natural means must imbibe water before they can germinate. Even if water enters seeds (or fruits) only through some unplugged natural opening, imbibition can be rapid. Imbibition is completed in less than a day at about room temperature (20–25°C). Weight increased to 100% or more in 7 h in mechanically scarified seeds of *Acacia* spp. allowed to imbibe at 23°C (Fig. 6.5), and imbibition curves for all species had plateaued after 14 hr. It should be noted, however, that imbibition rates of permeable seeds increase with an increase in temperature (Brown and Worley, 1912).

Temperature requirements for the germination of permeable seeds in light have been determined for a number of species. Seeds germinate over a wide range of temperatures, but the maximum and minimum temperatures vary with the species. Low temperatures inhibit seed germination in a number of species: *Cassia obtusifolia,* 15°C (Creel *et al.,* 1968); *C. tora,* 15°C (Nazrul-Islam and Hoque, 1990); *Convolvulus* spp., 5–7°C (Koller and Cohen, 1959); *Cuscuta campestris,* 7°C (Allred and Tingey, 1964); *Desmanthus velutinus,* 15/5°C (Haferkamp *et al.,* 1984); *Erythrina burana,* 10°C (Teketay, 1994); *Prosopis farcta,* 10°C (Dafni and Negbi, 1978); *Sesbania drummondii,* 10°C (Eastin, 1984), and *Zornia reticulata,* 5°C (Felippe, 1984). However, permeable seeds of some species can germinate at low temperatures: *Erodium botrys,* 2/2 and 2/5°C (Young *et al.,* 1975); *Ipomoea crassicaulis,* 10°C (Misra, 1963); *Malva*

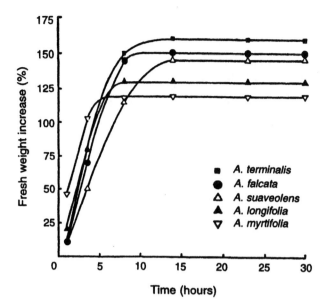

FIGURE 6.5 Imbibition curves for mechanically scarified seeds of five species of *Acacia*. One hundred percent of the scarified seeds imbibed, but only 0–5% of those in the nonscarified controls (not shown on graph) did so. From Clemens *et al.* (1977).

pusilla, 5°C (Blackshaw, 1990); *Rhus glabra* and *R. coppallina*, 5/15°C (Farmer *et al.*, 1982); and *Ulex europaeus*, 4.2°C (Ivens, 1983).

The maximum temperature for the germination of permeable seeds may be relatively high, e.g., *Prosopis farcta*, 40°C (Dafni and Negbi, 1978); *Sesbania drummondii*, 40°C (Eastin, 1984); *Stylosanthes humilis*, 40°C (McKeon, 1985); *Erodium botrys*, 40/30°C (Young *et al.*, 1975); and *Parkia auriculata*, 42°C (Coutinho and Struffaldi, 1971). Some mechanically scarified seeds of *Aeschynomene americana*, *Calopogonium mucunoides*, *Centrosema pubescens*, *Desmodium heterocarpon*, *D. intortum*, *Macroptilium atropurpureum*, *Stylosanthes guianensis*, *S. hamata*, and *S. viscosa* germinated at 46°C, but those of all these species, except *A. americana*, germinated to higher percentages at either 30 or 38°C than at 46°C (Gomes and Kretschmer, 1978). The optimum germination temperature for some species tested only in darkness are: *Anthyllis hermanniae*, 15°C; *A. vulneraria*, 15°C; *Convolvulus elegantissimus*, 10°C; and *Fumana thymifolia*, 15°C (Doussi and Thanos, 1993).

Frequently, seeds are tested only in light, thus there is no way to know how well they would have germinated in darkness. However, permeable seeds of *Acacia aneura* (Preece, 1971), *A. farnesiana* (Scifres, 1974), *Chordospartium stevensonii* (Conner and Conner, 1988), *Ulex europaeus* (Ivens, 1983), *Apeiba membranacea*, *Greaguma ulmnifolia* (Acuna and Garwood, 1987), *Helianthemum vesicarium*, *H. ventosum* (Gutterman and Agami, 1987), and *Merremia aegyptica* (Sharma and Sen, 1975) germinated equally well in light and darkness. Seeds of *Dichrostachys cinerea* (Bell and van Staden, 1993) and *Sicyos deppei* (Brechu-Franco *et al.*, 1992) germinated to higher percentages in light than in darkness, but as seeds afterripened in dry storage germination in darkness increased. Seeds of *Zornia reticulata* germinated (1) only in darkness at 10°C; (2) to a higher percentage in darkness than in light at 15°C; (3) faster in darkness than in light at 20, 25, and 30°C; and (4) equally well in darkness and light at 35°C (Felippe, 1984). Seeds of *Sida spinosa* germinated to 100% in both light and darkness at 25/15, 30/15, 35/20, and 40/25°C, but they germinated to 8 and 42% in light at 15/6 and 20/10°C, respectively, and to 95 and 97% in darkness, respectively (Baskin and Baskin, 1984).

Seeds of *Sida grewioides* germinated to about 70% in continuous darkness and in far-red light but to only about 38% in red light (Chawan and Sen, 1973), whereas those of *S. spinosa* (Chawan and Sen, 1973) and *Merriamia aegyptica* (Sharma and Sen, 1995) germinated to 75–100% in darkness and in far-red and red light. Seeds of *Cistus albidus* and *C. monspeliensis* germinated to higher percentages (and to faster rates) at a red/far-red ratio of 1.1 than at 0.7 (Roy and Sonie, 1992). The germination of *Trifolium repens* seeds under water stress (-0.3 MPa) was reduced by a 3-hr exposure to various kinds of light; seeds exposed to blue, far-red, and red light and no light germinated to about 40, 35, 25, and 59%, respectively (Niedzwiedz-Siegien and Lewak, 1992).

Some germination (10% or more) is possible at increased water stress: *Acacia farnesiana*, -0.18 MPa (Scifres, 1974); *A. senegal*, -1.38 MPa (Palma *et al.*, 1995b); *A. tortilis*, -1.0 MPa (Coughenour and Delting, 1986); *Cassia obtusifolia*, -0.5 MPa (Daiya *et al.*, 1980); *C. occidentalis*, -0.5 MPa (Daiya *et al.*, 1980); *Chordospartium stevensonii*, -1.5 MPa (Conner and Conner, 1988); *Malva pusilla*, -1.53 MPa (Blackshaw, 1990); *Medicago sativa*, -1.0 MPa (Delaney *et al.*, 1986); *Oxytropis riparia*, -2.0 MPa (Delaney *et al.*, 1986); *Sida rhombifolia*, -0.8 MPa (Smith *et al.*, 1992); and *S. spinosa*, -0.6 MPa (Smith *et al.*, 1992).

NaCl solutions of 0.325 (-1.48 MPa) to 0.5 *M* (-2.27 MPa) caused a 50% reduction in the germination of *Prosopis farcta* seeds, depending on geographical location in Israel where seeds were collected (Dafni and Negbi, 1978). Seeds of *Desmodium cephalotus*, *D. gangeticum*, *D. gyrans*, and *D. pulchellum* germinated to 50, 38, 83, and 45%, respectively, at 0.2 M NaCl (-0.91 MPa) (the highest concentration tested) (Datta and Sen, 1987), and those of *Stylosanthes humilis* germinated to about 25–55%, depending on the location in Brazil where seeds were collected, at 0.268 M (-1.22 MPa) NaCl (the highest concentration tested) (Lovata *et al.*,

1994). At 0.17 M (−0.77 MPa) NaCl, seeds of *Acacia schaffneri* and *Parkinsonia aculeata* germinated to about 60 and 80%, respectively (Everitt, 1983).

Permeable seeds of most species germinate over a broad range of pH values. For example, seeds of *Acacia schaffneri* and *Parkinsonia aculeata* germinated from pH 3 to 11 but not at pH 12 (Everitt, 1983). Seeds of *Mimosa bimucronata* did not germinate well at pH 4 or 8 (Ferreira, 1976).

VI. ENVIRONMENTAL CONTROL OF BREAKING PHYSICAL DORMANCY

The purpose of this section is to survey what is known about the breaking of physical dormancy under natural conditions, i.e., what environmental factors are required for the chalaza, hilum, strophiole, and so forth to become permeable to water?

A. High and High Fluctuating Temperatures

Our understanding of the importance of daily temperature fluctuations in the breaking of physical dormancy was enhanced by the results of two studies published in 1982 actually documenting environmental conditions in the habitat when physical dormancy was broken. (1) The percentage of impermeable seeds of *Stylosanthes humilis* and *S. hamata* began to decline in a northern Australian pasture in September (early spring), when mean monthly maximum and minimum temperatures were about 67 and 28°C, respectively (Fig. 6.6). The number of physically dormant seeds decreased until December (early summer), when rains stimulated all the permeable ones to germinate. Physical dormancy was not broken during January to August, when mean daily maximum and minimum temperatures were ≤55 and 25°C, respectively, (McKeon and Mott, 1982). (2) More seeds of *Heliocarpus donnell-smithii* (Tiliaceae) became permeable and germinated when placed in the center, or near the center, of a gap (clearing) in a rain forest in eastern Mexico than when placed at the edge of the gap or in the adjacent forest (Fig. 6.7). The amplitude of daily temperature fluctuations was about 15°C in the center of the gap and less than 5°C at the edge of the gap and in the forest. Maximum germination was obtained in laboratory studies at a 15°C amplitude of daily temperature fluctuations, when the daily high temperature was between 32 and 39°C for 6 hr each day (Vazquez-Yanes and Orozco-Segovia, 1982). The response of seeds of *H. donnell-smithii* to fluctuating temperatures ensures that they germinate in gaps where solar irradiance is high and not in the shade of mature trees, where seedlings are unlikely to survive. Thus, a

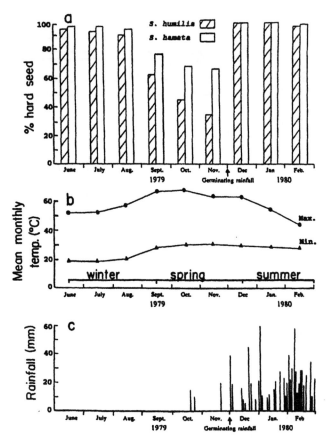

FIGURE 6.6 (a) Change in percentage of impermeable seeds of *Stylosanthes humilis* and *S. hamata*. (b) Mean monthly soil surface temperatures and (c) rainfall in a northern Australian pasture, where seeds were in the soil seed bank. From McKeon and Mott (1982).

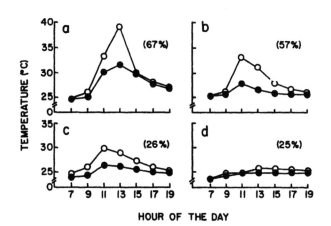

FIGURE 6.7 Soil temperatures at a depth of 2 cm (a) in the center of a gap, (b) near the center of a gap, (c) at the edge of a gap, and (d) in adjacent rainforest in Veracruz, Mexico. The germination percentage of *Heliocarpus donnell-smithii* seeds at each site is given in parentheses. From Vazquez-Yanes and Orozco-Segovia (1982). ●, mean temperatures; ○, maximum temperatures.

high fluctuating temperature requirement to overcome seed dormancy effectively is a means of gap detection by the species.

A gap-detecting mechanism also is implied by the responses of physically dormant seeds of other species. Many seedlings of *Ulex europaeus* (Ivens, 1978) and *Acacia melanoxylon* (Farrell and Ashton, 1978) immediately appeared at sites after mature plants of each species were cut and removed. These observations indicate that seeds of both species germinated in response to an increase in the amplitude of daily temperature fluctuations. Fruits of *Rhus javanica* also germinated after the forest canopy was removed mechanically or after it was destroyed by fire. Laboratory studies showed that physical dormancy was broken in the fruits by brief exposures to temperatures of 48–74°C, which could occur in the field following removal of the canopy (Washitani and Takenaka, 1986).

Dormant seeds of *Erodium botrys* and *E. brachycarpum* were placed in nylon bags and buried 1 cm deep in a bare site (no vegetation or litter), under litter, and on a gopher mound for 3 months in the Central Valley of California (Rice, 1985). Soil temperature data showed that the amplitude of daily temperature fluctuations at the three sites was 36, 21, and 17°C, respectively. Subsequent tests showed that germination was highest for seeds that had been buried in the bare site and lowest for those buried on the gopher mound (Rice, 1985). Thus, seeds of *Erodium* spp. became permeable during exposure to natural summer temperatures. However, did the fluctuating temperatures cause seeds to become permeable or was physical dormancy broken by exposure to the absolute maximum temperature? Using alternating temperature regimes recorded at field sites where seeds of *Erodium* spp. were buried and high constant temperatures, Rice (1985) demonstrated that temperature fluctuations were more important than high constant temperatures in overcoming dormancy.

Impermeable seeds of the typical variety of *Indigofera glandulosa* were placed in cotton bags on the soil surface in a field in Ujjain, India. Only 20% of the seeds exposed to natural temperature fluctuations during the rainy season (June–September) and winter (October–February) became permeable by March; however, 85% of them became permeable during summer (March–May) and germinated in May (Bhat, 1968). Dry storage for 1 month at 60°C or for 4 months at 35 or 60/35°C resulted in 92, 77, and 79% germination, respectively; those stored at 30°C did not become permeable. Therefore, Bhat (1968) concluded that high temperatures in the habitat during summer caused seeds of *I. glandulosa* to become permeable.

Quinlivan (1961) subjected impermeable seeds of *Lupinus digitatus, L. luteus, Medicago tribulus,* and *Tri-folium subterraneum* to alternating temperature regimes of 46/15, 60/15, and 74/15°C (maximum soil surface temperatures during summer in New South Wales, Australia, on cloudy days were 43–49°C and on clear days were 71–77°C) and to constant temperatures of 15 and 60°C for 0 to 5 months. Some seeds of the four species became permeable at all temperatures, but the highest germination percentages were for those exposed to 60/15°C. A daily alternating temperature regime of 60/15°C was effective in breaking physical dormancy in seeds of *Ornithopus compressus,* with 90% of the seeds germinating after 4 months; 28% of the seeds in the controls germinated (Barrett-Lennard and Gladstones, 1964). The number of *Trifolium subterraneum* seeds remaining impermeable at the end of the summer increased, if litter (including standing dead plants of *T. subterraneum*) shaded the soil surface (Quinlivan and Millington, 1962; Quinlivan, 1965). Shading decreased the amplitude of the daily temperature fluctuation by as much as 17°C, which resulted in fewer seeds becoming permeable. Temperature fluctuations also are reduced if seeds become buried. For example, *Medicago truncatula* seeds (still in the fruits) buried 2 cm below the soil surface showed only a 17% reduction in physical dormancy after 27 days, whereas 97% of those on the soil surface became permeable (Kirchner and Andrew, 1971).

Quinlivan (1966) found that dormancy loss in seeds of *Trifolium subterraneum* was determined by the maximum daily temperature, if there was a minimum daily temperature fluctuation of at least 15°C. The maximum daily temperature required for the loss of physical dormancy varies with the species: *T. subterraneum,* 30°C; *Lupinus varius,* 60°C; *T. dubium,* 30°C; *T. hirtum, T. cherleri, T. glomeratum,* and *T. cernuum,* 40°C; *Medicago truncatula, M. littoralis, M. polymorpha,* and *M. scutella,* 50°C (Quinlivan, 1968b); and *Stylosanthes* spp., 50°C (McKeon and Brook, 1983). Loss of physical dormancy in seeds of *Mimosa pigra* required a temperature fluctuation of 20°C, which occurred at alternating temperature regimes ranging from 30/10 to 50/30°C, with 40/20°C being optimal (Dillon and Forcella, 1985). In the Central Valley of California, seeds of *T. subterraneum* and *T. incarnatum* became permeable during summer, when temperatures of 60°C were recorded in dead plant material on the soil surface (Williams and Elliott, 1960).

Loss of physical dormancy in seeds of *Trifolium subterraneum* occurs in two stages (Taylor, 1981). (1) The preliminary or preconditioning phase will occur if seeds are at constant temperatures, and the rate at which this stage is completed increases with an increase in temperature. Seeds prevented from drying (by blocking the hilum) during the first stage are more likely to become permeable in the second stage than those that dehydrate further during stage one. (2) The second stage (when seeds actually become permeable) requires fluctuating temperatures for a maximum loss of dormancy, but constant temperatures of $\geq 50°C$ are somewhat effective. Both stages of seed coat softening take place simultaneously as seeds are exposed to high daily temperature fluctuations during summer. Taylor (1981) suggested that thermal degradation occurs during the first stage, which results in a weakening of the strophiole. In the second stage, physical expansion and contraction associated with daily increases and decreases in temperature cause cells in the strophiole to separate.

Seeds of *Sida spinosa* incubated on moist sand at 15/6, 20/10, 25/15, 30/15, 35/20, and 40/20°C for 30, 90, or 180 days germinated to higher percentages when seeds at each regime were shifted to all temperature regimes higher than each respective regime, (e.g., seeds at 15/6 were shifted to 20/10, 25/15, 30/15, 35/20, and 40/25°C), than those kept continuously at each regime. Seeds germinated to 90% or more when moved from 20/10 and 25/15°C to 35/20 and 40/25°C and from 30/15 to 40/25°C after 180 days. Thus, although seeds were at relatively low thermoperiods, they were conditioned for rapid increases in permeability when shifted to relatively high thermoperiods (Baskin and Baskin, 1984). The response of conditioned seeds to an increase in temperatures is a means of detecting a shift from burial in the soil to the surface.

B. Drying at High Temperatures

Seeds can be exposed to dry-heat treatments in nature due to a lack of precipitation during summer and/or to a fire. In either case, drying at high temperatures indirectly influences the timing of germination of many species. For example, seeds of *Lupinus varius* require a maximum of 60°C in the daily temperature cycle to become permeable (Quinlivan, 1968b). However, high daily temperature fluctuations are ineffective in breaking dormancy if the seed moisture content is above 8.5–9% (Quinlivan, 1968a). Thus, if seeds have a low moisture content or if they are on the soil surface where drying occurs via water loss through the hilum, they will become permeable and germinate when rains come in autumn. If seeds are buried or covered with litter, drying and subsequent loss of physical dormancy may be inhibited. Drying probably also is important in breaking the physical dormancy in seeds of other Papilionoideae, but these studies remain to be done.

At alternating temperatures, more seeds of *Trifolium subterraneum* became permeable at high (74.1–77.6%) than at low (0%) RH, whereas at constant temperatures, RH had no effect on the loss of impermeability (Fairbrother, 1991). Thus, Fairbrother (1991) concluded that rupture of the strophiole eventually occurred at a high RH because cell wall fibers swelled at high temperatures and shrunk at low temperatures. High fluctuating temperatures under wet conditions also promoted the loss of physical dormancy. All seeds of *Leucaena pulverulenta* kept on a wet substrate became permeable after 50 days at a daily 50/30°C temperature regime, whereas only about 10% of those kept dry did so (Owens *et al.,* 1995).

Drying, as well as high summer temperatures, is required for the loss of physical dormancy in seeds of the winter annual *Geranium carolinianum.* Thus, seeds become permeable by late summer and germinate in autumn when the soil moisture becomes nonlimiting (Baskin and Baskin, 1974). Substrate drying and high temperature fluctuations also play a role in the germination of *Neptunia oleracea.* Seeds of this aquatic angiosperm mature in autumn in Bharatpur, India, and are dispersed into the water. In summer (March to mid-June), water levels recede, and seeds are exposed to daily temperature fluctuations of 55/22°C. Seeds become permeable during summer but do not germinate until pools begin to fill with water at the start of the rainy season in mid-June. In laboratory studies, seeds germinated to 100% after being subjected to an alternating temperature regime of 60/20°C for 2 days (Sharma *et al.,* 1984). If the water fails to recede, seeds are not exposed to sufficiently high alternating temperature regimes to become permeable.

It has been suggested that the germination of *Abutilon theophrasti* seeds follows intensive drying, which causes cracks to develop in the seed coat. Although no proof was given that this actually happens in the field, drying over calcium chloride for 5 weeks increased germination from 11 to 63% (LaCroix and Staniforth, 1964). Flushes of *A. theophrasti* germination occurred in a field in Michigan following extended periods when showers of rain did not exceed 12.5 mm (Dekker and Meggitt, 1986). Although the results of this study suggest that drying was important in breaking physical dormancy, the effects of high summer soil surface temperatures were not considered.

The role, if any, of alternate wetting and drying in breaking physical dormancy needs to be investigated. However, wetting (by amounts of precipitation too small to allow germination) followed by drying can influence germination after physical dormancy is broken. For example, alternate wetting (imbibed) and drying of permeable seeds of *Stylosanthes humilis* increased the final germination from 46 to 70% (McKeon, 1984).

C. Fire

Fire is a natural part of many ecosystems, e.g., dry sclerophyll woodlands (Purdie, 1977), matorral

TABLE 6.8 Species with Physical Dormancy Whose Seeds Germinate in the Field after Fire

Species	Family	Reference
Acacia spp.	Fabaceae	Cremer and Mount (1965), Floyd (1966), Purdie and Slatyer (1976), Farrell and Ashton (1978), Shea *et al.* (1979), Monk *et al.* (1981), Griffin and Friedel (1984), Sabiiti and Wein (1988)
Albizia lophantha	Fabaceae	Dell (1980)
Bonamia grandiflora	Convolvulaceae	Hartnett and Richardson (1989)
Bossiaea aquifolium	Fabaceae	Shea *et al.* (1979)
Cassia nemophila	Fabaceae	Moore (1973)
Ceanothus megacarpus	Rhamnaceae	Hadley (1961)
C. sanguineus	Rhamnaceae	Orme and Leege (1976)
Cistus spp.	Cistaceae	Arianoutsou and Margaris (1981), Juhren (1966), Montgomery and Strid (1976)
Commersonia fraseri	Sterculiaceae	Floyd (1976)
Convolvulus occidentalis	Convolvulaceae	Horton and Kraebel (1955)
Daviesia mimosoides	Fabaceae	Purdie and Slatyer (1976)
Dilluynia retorta	Fabaceae	Purdie and Slatyer (1976)
Dodonaea triquetra	Sapindaceae	Floyd (1966)
Geranium bicknellii	Geraniaceae	Abrams and Dickmann (1982)
Helianthemum scoparium	Cistaceae	Borchert (1989)
Indigofera hilaris	Fabaceae	Martin (1966)
I. stricta	Fabaceae	Martin (1966)
Iliamna corei	Malvaceae	Caljouw *et al.* (1994)
I. remota	Malvaceae	Schwegman (1990)
I. rivularis	Malvaceae	Steele and Geier-Hayes (1989)
Kennedia coccinea	Fabaceae	Shea *et al.* (1979)
K. prostrata	Fabaceae	Shea *et al.* (1979)
K. rubicunda	Fabaceae	Floyd (1966)
Lotus hemistrata	Fabaceae	Sweeney (1956)
L. scoparius	Fabaceae	Hanes (1971)
Pomaderris apetala	Rhamnaceae	Cremer and Mount (1965)
Pultenaea procumbens	Fabaceae	Purdie (1977)
Rhus javanica	Anacardiaceae	Kamada *et al.* (1987)
R. typhina	Anacardiaceae	Marks (1979)
Stylosanthes spp.	Fabaceae	Mott (1982)
Tephrosia capensis	Fabaceae	Martin (1966)
Thermopsis macrophylla	Fabaceae	Borchert (1989)
Ulex europaeus	Fabaceae	Ivens (1982)

(Sweeney, 1956), and pine forests (Cushwa *et al.,* 1968), and there is much interest in the regeneration of various plant species following a fire. One way in which species increase in numbers following a fire is via germination of seeds buried in the soil at the site. From lists of species germinating within a few months following a fire, it is easy to find representatives of many families whose seeds have physical dormancy (Table 6.8). Because

these seeds germinated after the habitat was burned, the conclusion frequently reached is that heat from the fire made them permeable. This idea seems logical, especially in view of the previous discussion on the effects of dry heat on the breaking of physical dormancy (see Table 6.6). Will temperatures that occur on/in soils during fires break physical dormancy, without killing the seeds?

Depending on the amount of combustible material, seeds on the soil surface may be exposed to a range of temperatures: >600°C (Lonsdale and Miller, 1993), 600°C (Sweeney, 1956), 386°C (Floyd, 1966), 227–424°C (Mott, 1982), 156°C (Heyward, 1938), and 81–213°C (Beadle, 1940). Temperatures simulating those that occur on the soil surface during fires are lethal to seeds, even after short periods of exposure (Wright, 1931; Stone and Juhren, 1951; Cushwa *et al.*, 1968; Zabkiewicz and Gaskin, 1978; Auld, 1986b). Seeds of *Acacia aneura, Cassia nemophila,* and *Dodonaea viscosa* at 0, 1, or 2 cm in the soil were killed by slow-burning litter fires when temperatures exceeded 80°C (Hodgkinson and Oxley, 1990). Seed death probably can be attributed to the fact that high temperatures were maintained for much longer periods of time under slow- than under fast-burning fires in this study.

Soil temperature profiles have been constructed as various types of vegetation were burned (for methods see Sackett and Haase, 1992). Soil is a good insulator, and with increases in depth, temperatures decline (Fig. 6.8) and the duration of high temperatures increases (e.g., Portlock *et al.*, 1990; Bradstock *et al.*, 1992; Bradstock and Auld, 1995). Further, with an increase in the amount of water in the soil, there is a reduction in the conduction of heat (Beadle, 1940; Portlock *et al.*, 1990). Some temperatures recorded at soil depths of 2–3 cm during fires are 112°C (Beadle, 1940), 111°C (Shea *et al.*, 1979), 110°C (Sweeney, 1956), 85°C (Floyd, 1966), 59–67°C (Beadle, 1940), and 40–85°C (Auld, 1986b). Thus, while seeds may be exposed to lethal temperatures on the soil surface, they are exposed to much lower temperatures a few centimeters beneath the surface.

Temperatures approximating those recorded at soil depths of 2–3 cm during fires have been used to simulate the effects of fire on breaking physical dormancy. Dry heat at temperatures of 60–100°C causes seeds of many species to become permeable (Table 6.9). At low temperatures, the duration of heat is not as critical as the temperature per se in breaking dormancy. However, at high temperatures (100, 120°C), seeds become sensitive to temperature duration (Auld and O'Connell, 1991). By knowing the soil temperature profiles to be expected under various kinds of fuel (litter) loads and temperatures required for dormancy breaking, it is possible to predict the germination of seeds in (and on) the soil following a fire (Auld and O'Connell, 1991). In fact, fire frequently is used to promote the germination of seeds with physical dormancy; consequently, it is an important tool in programs to help control invasive species (Pieterse and Cairns, 1986; Lonsdale and Miller, 1993) and to promote reproduction of economically valuable ones (Portlock *et al.*, 1990).

Exposure to dry heat at 100°C for 5 min caused the first and second layers of the impermeable endocarp in *Rhus ovata* fruits to develop many small cracks, especially near the micropyle. However, when the first and second layers of the endocarp were removed with a dental drill, no cracks were found in the third layer in the region of the micropyle (Stone and Juhren, 1951). Thus, seeds were still impermeable to water. Dry heat caused the hilum fissure to open in seeds of *Ceanothus sanguineus* (Orme and Leege, 1976) and caused the hilum to open in fruits of *Rhus lanceolata* and also the development of cracks leading from the hilum (Rasmussen and Wright, 1988).

Wet-heat treatments sometimes are used to simulate burning under wet conditions. Warcup (1980) subjected soil samples collected in southeastern Australian forests to steam at 60 or 71°C for 30 min and obtained increased germination of species in the Geraniaceae, Fabaceae, Convolvulaceae, and Rhamnaceae. Dormancy was broken by soaking seeds of *Lespedeza cyrtobotrya* in water at 70°C (Iwata, 1966) and those of *Cassia nicticans* at 80°C (Cushwa *et al.*, 1968). Dipping seeds of *Acacia decurrens* (Beadle, 1940), *A. extensa, A. myrtifolia, A. pulchella, A. strigosa,* and *Mirbelia dilatata* (Shea *et al.*, 1979) into boiling water caused them to become permeable (see Table 6.5).

Another aspect of fire at a habitat site is the removal of living plant parts and litter; consequently, solar radiation increases temperatures at the soil surface. Samples

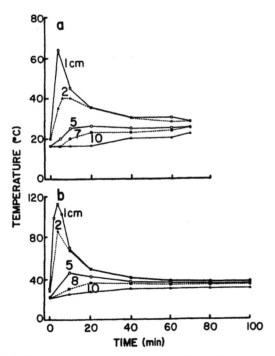

FIGURE 6.8 Soil temperatures at various depths during and after a cool (a) and hot (b) burn. From Auld (1986b).

TABLE 6.9 Species with Physical Dormancy Whose Seeds Have Been Made Permeable with Dry Heat Treatments, Using Temperatures Recorded in Fires

Species	Optimum temperature (°C) for loss of dormancy	Reference
Acacia mangium	60	Hopkins and Graham (1984)
A. suaveolens	60–80	Auld (1986a)
Acacia spp.	70	Floyd (1976)
Bossiaea heterophylla	60	Auld and O'Connell (1991)
Ceanothus megacarpus	100	Hadley (1961)
C. sanguineus	105	Gratkowski (1973)
Cistus albidus	100	Vuillemin and Bulard (1981)
C. monspeliensis	100	Vuillemin and Bulard (1981)
Commersonia fraseri	80	Floyd (1976)
Daviesia alata	80	Auld and O'Connell (1991)
Dillwynia brunioides	80	Auld and O'Connell (1991)
Dodonaea triquetra	90	Floyd (1976)
Geranium bicknellii	90	Abrams and Dickmann (1984)
G. solanderi	62	Warcup (1980)
Glycine clandestina	80	Auld and O'Connell (1991)
Gompholobium glabratum	80	Auld and O'Connell (1991)
Hardenbergia violacea	90	Auld and O'Connell (1991)
Kennedia rubicunda	80	Floyd (1976)
Mirbelia platyloboides	80	Auld and O'Connell (1991)
Platylobium formosum	80	Auld and O'Connell (1991)
Pultenaea dephanoides	80	Auld and O'Connell (1991)
P. linophylla	40	Auld and O'Connell (1991)
Rhamnus californica	71–82	Wright (1931)
Rhus javanica	55	Washitani (1988)
R. lanceolata	82	Rasmussen and Wright (1988)
Seringia arborescens	80	Floyd (1976)
Sphaerolobium vimineum	100	Auld and O'Connell (1991)
Trifolium dubium	62	Warcup (1980)
Ulex europaeus	60–80	Zabkiewicz and Gaskin (1978)

of dry forest soil placed in direct sunlight in southeastern Australia for 6 days reached a maximum of 62°C, and temperatures were above 50°C for several hours each day. These increased temperatures stimulated a twofold increase in the germination of *Geranium solanderi, Dichondra repens,* and *Trifolium dubium* seeds (Warcup, 1980). Soil surface temperatures were 20–40°C higher in burned than in nonburned plots in Idaho the spring following a fire, and this probably is one reason why seeds of *Ceanothus sanguineus* continued to germinate long after the fire occurred (Orme and Leege, 1976). Postfire soil temperatures in New South Wales, Australia, in summer were ≥40°C for several hours each day down to depths of 45 mm and exceeded 60°C down to

4 mm; however, soil temperatures did not reach 40°C in burned sites in winter or in nonburned sites in summer (Auld and Bradstock, 1996). Soil heating at 40°C broke the dormancy in some seeds of eight legumes and at 60°C broke it in those of 15 species (Auld and O'Connell, 1991).

D. Low Winter Temperatures

Two observations, especially on *Melilotus alba* and *Trifolium pratense,* have caused researchers to be quite interested in the role of temperate-zone winter environmental conditions in breaking physical dormancy (Dunn, 1939). (1) Seeds germinate in late winter and/

or early spring before they are exposed to the high temperature fluctuations of summer. (2) Seeds sown in spring do not become permeable until the following late winter and/or spring, after they have been subjected to winter conditions. Various studies have attempted to determine what factor(s) of the winter environment cause(s) seeds to become permeable.

Alternately freezing at about −5 or −15°C and thawing seeds of *Medicago sativa* was only moderately (23% germination) effective in breaking physical dormancy; controls germinated to 8%. Germination increased in *M. sativa* seeds after the first freeze, but subsequent freezing and thawing increased germination very little (Midgley, 1926). Storage of *Melilotus alba* seeds on moist and on dry substrates at 5, −10, and 22°C failed to improve germination. In fact, moist substrates at low temperatures caused the few seeds that became permeable to imbibe, and they subsequently died.

Moist treatments at 5 and −10°C were effective in breaking physical dormancy in seeds of *Vicia villosa* (Dunn, 1939). However, seedlings slowly decayed at 5°C, and partially imbibed seeds were killed by freezing at −10°C (Dunn, 1939). The germination of four selections of *Trifolium repens* increased at 20 and 30°C after 5 months of exposure to low (5–15°C) temperatures and high (60–100%) RH (Burton, 1940). A higher percentage of *Lespedeza cyrtobotrya* seeds was permeable in spring if they overwintered on the ground under snow in Japan than if they overwintered on upright dead plants. However, the rate of seed decay also increased under the snow (Iwata, 1966).

Seeds of *Melilotus* spp. stored dry (1) at room temperatures and (2) at 85% RH at 7°C did not become permeable, whereas those stored dry in an open shed (in Wisconsin or Iowa) became permeable (Helgeson, 1932). In a series of experiments conducted in Iowa (Martin, 1945), *Melilotus* spp. seeds kept wet or dry at −3, 2, 10, and 15–20°C did not lose their physical dormancy. However, seeds became permeable when sown on soil in the field or when kept on a porch and in an open garage in both cotton-stoppered jars and closed jars at a high RH. Few seeds on the soil or in the two types of jars became permeable from November through early March, but permeability reached 72–97% between March 15 and April 10 in a number of years between 1929 and 1942. Martin (1945) thought that fluctuations of temperature in the realm of freezing were responsible for breaking dormancy. However, alternate freezing and thawing did not promote a loss of physical dormancy in seeds of *Dalea foliosa, Pediomelum subacaule* (Fabaceae, Papilionoideae; Baskin and Baskin, 1998b, unpublished results), *Senna marilandica, S. obtusifolia* (Fabaceae, Caesalpinioideae; Nan, 1992), *Desmanthus illinoensis* (Fabaceae, Mimosoideae; Letting, 1961), *Ili-*

amna corei (Malvaceae; Baskin and Baskin, 1997), and *Sida spinosa* (Malvaceae; Baskin and Baskin, 1984).

E. Microbial Action

One reason frequently given for the loss of impermeability of seeds is the action of microbes (e.g., Krefting and Roe, 1949; Mayer and Poljakoff-Mayber, 1989; Janzen, 1981; Fenner, 1985; Lonsdale *et al.*, 1988), but data to support this idea usually are not presented. Little is known about the effects of microbes on seeds with physical dormancy. However, Gogue and Emino (1979) found that impermeable seeds of *Albizia julibrissin* incubated in unsterilized and sterilized natural soil germinated to 30 and 11%, respectively, after 7 days. After 30 days, seeds of *A. julibrissin* germinated to 1, 3, 3, and 2% in the presence of the soil fungi *Fusarium* sp., *Rhizoctonia* sp., or *Phythium* sp. and in the control, respectively; after 60 days seeds germinated to 10, 22, 5, and 3%, and after 90 days to 21, 38, 7, and 9%. The *Rhizoctonia* sp. was the most effective in breaking physical dormancy. Photomicrographs revealed that hyphae had altered the appearance of the seed coat, presumably due to the secretion of enzymes.

Seeds of *Vigna minima* imbibed after 186 hr when placed on blotters soaked with soil suspensions but failed to imbibe after 30 days on blotters moistened with distilled water (Gopinathan and Babu, 1985). These authors hypothesized that microbes softened the seed coat of *V. minima* seeds at the strophiole. No significant differences were found between germination percentages of *Trifolium subterraneum* seeds incubated for 4 and 12 months in the presence of fungi, root nodule bacteria, and seed coat saprophytes and in those incubated in sterile conditions (Aitken, 1939).

Seeds of *Abutilon theophrasti* have a characteristic group of fungi associated with the seed coat, which includes *Cladosporium cladosporioides, Alternaria alternata, Epicoccum purpurascens,* and *Fusarium* sp. (Kremer *et al.,* 1984). These fungi inhibit the establishment of soil microbes on the surface of seeds of this species placed on soil (Kremer, 1986a). Another factor preventing the establishment of soil microbes is the production of antimicrobial compounds, including phenolics, by the seed coat (Kremer, 1986b). The fact that Kremer (1987) found bacteria within the subpalisade cell layer in the seed coat that exhibit antifungal activity makes this story even more intriguing. Thus, fungi associated with the seed coat in some species may prevent the growth of bacteria (see Chapter 7, Section V,G,2).

F. Effects of Animals on Seeds with Physical Dormancy

Observations on intact seeds in feces or in regurgitated material of animals (e.g., Timmons, 1942; Jackson and Gartlan, 1965; Lieberman *et al.*, 1979; Gill, 1985) have caused many biologists to ask: Does being eaten have any subsequent effects on seed germination? There are several possibilities. (1) Seeds are destroyed, i.e., digested (Gardener *et al.* 1993a). (2) Seeds germinate while they are in the animal's digestive tract, but the resulting seedlings die (e.g., Janzen, 1981; Janzen *et al.*, 1985; Gardener *et al.*, 1993a). Germination during passage through a digestive tract may occur if seeds are infected by insect larvae. For example, infestation of *Acacia albida* seeds by bruchid beetles promotes germination. Thus, if infected seeds are eaten by cattle, they germinate during passage through the animal, and subsequently the seedlings die (Hauser, 1994). (3) Dormancy is broken, and defecated or regurgitated seeds germinate to higher percentages than those that have not been ingested: *Acacia constricta* (Cox *et al.*, 1993), *A. cyclops* (Glyphis *et al.*, 1981; Gill, 1985), *A. erioloba* (Hoffman *et al.*, 1989), *A. nilotica* (Miller, 1995), *A. tortilis* (Lamprey, 1967; Miller, 1995), *Biserrula pelecinus* (Malo and Suarez, 1995), *Cistus ladanifer* (Malo and Suarez, 1996), *Medicago sativa* (Swank, 1944), *Rhus glabra* (Krefting and Roe, 1949), *R. hirta* (Swank, 1944), *Trifolium campestre* (Russi *et al.*, 1992), *T. pratense, T. sativa* (Swank, 1944), *T. stellatum, T. tomentosum* (Russi *et al.*, 1992), and *Robinia pseudoacacia* (Swank, 1944).

The way in which a digestive system breaks physical dormancy is unknown (Cavanagh, 1980), but it is assumed to be via acid (Lamprey *et al.*, 1974) and/or mechanical (Cavanagh, 1980) scarification. Germination percentages increased from 20 to 31% when time of retention of *Trifolium campestre* seeds in the digestive system of sheep increased from 24 to 48 hr, but they decreased from 50 to 46% for *T. stellatum* and from 25 to 14% for *T. tomentosum* (Russi *et al.*, 1992). Even prolonged exposure to the acidic contents of a digestive system may not result in a loss of impermeability of 100% of the seeds. For example, *Leucaena leucocephala*, *Stylosanthes scabra*, and *Macroptilium atropurpureum* seeds germinated to about 90, 45, and 15%, respectively, after 240 hr in the rumen of cattle (Gardener *et al.*, 1993b). Malo and Suarez (1995) examined seeds of *Biserrula pelecinus* under a microscope and found no scratches on the seed coats after seeds passed through the digestive system of cattle, indicating that mechanical scarification had not occurred. However, the number of seeds of *Trifolium campestre* recovered from sheep dung was higher than that of *T. stellatum* (Russi *et al.*, 1992). Because seeds of *T. campestre* are smaller than those of *T. stellatum*, Russi *et al.* (1992) reasoned that there was a higher probability of large seeds being destroyed by mastication and rumination than small ones.

In evaluating germination data of seeds with physical dormancy separated from fecal or regurgitated material, attention should be paid to the amount of time between deposition of the material and the retrieval of seeds. Fermentation of fecal material could increase the temperature; consequently, seeds would receive a wet-heat treatment after they are deposited. Also, because waste materials usually are dropped on the soil surface, seeds are exposed to the extremes of daily temperature fluctuations that occur in the habitat. These daily temperature changes may be high enough to break physical dormancy. Gill (1985) suggested that the large fluctuation and high temperatures experienced by seeds of *Acacia clyclops* deposited in open sunny sites contribute to the breaking of physical dormancy.

There is an interaction among the seeds of some *Acacia* spp., bruchid beetles that lay their eggs on the seeds, and mammals and birds that eat the seeds. Gazelles feed on the indehiscent pods of *Acacia gerrardii*, *A. raddiana*, and *A. tortilis* in the Negev desert of Israel and serve as a dispersal agent for the seeds. Further, seeds germinate better after they have passed through the animal's digestive system than if sown on the soil, and seeds infested with larvae of a bruchid beetle germinate to slightly higher percentages than noninfested ones. Thus, both the bruchid beetle larvae and the gazelles enhance germination, but a high percentage of the seeds is lost to each predator (Halevy, 1974). In the Namib Desert of southern Africa, a similar relationship exists among seeds of *A. erioloba*, a bruchid beetle, and mammalian herbivores. Freshly matured seeds eaten by an animal have an increased chance of germinating before the bruchid beetle larva eats enough of the embryo to kill it than seeds not eaten for several weeks or months after maturation (Hoffman *et al.*, 1989). Seeds of *A. tortilis* in central Africa (Tanzania) are eaten by elephants, impalas, dikdiks, and gazelles, and passage through the digestive system improves germination. Therefore, seeds containing a bruchid beetle larvae are more likely to germinate before the parasite destroys the embryo if seeds are eaten by a mammal than if they are not eaten (Lamprey *et al.*, 1974; Pellew and Southgate, 1984). Seeds of the swollen-thorn species of *Acacia* in Central America are eaten and dispersed by birds. Seeds containing a developing larva of a bruchid beetle are broken up by the bird's digestive system, whereas uninfested ones are unlikely to be destroyed during passage through the digestive system. However, seeds removed by birds immediately after maturation may be dispersed before they are predated by the beetles (Janzen, 1969).

Although animals may serve as effective dispersal agents of seeds and passage through their digestive systems may overcome physical dormancy, we should not lose sight of the fact that physical dormancy can be (and probably usually is) broken without the aid of animals. That is, if seeds are not eaten by animals, environmental factors will result in the loss of physical dormancy. Thus, just because seeds are eaten by an animal does not mean that they are dependent on the animal for the breaking of physical dormancy. For example, consider the case of *Enterolobium cylocarpum* (Mimosoideae). This Central American tree has received much attention because its fruits apparently were eaten (and its seeds dispersed) by large herbivores in the early part of the Pleistocene. However, these animals became extinct about 10,000 years ago (Janzen and Martin, 1982), and today introduced cattle and horses eat the fruits and disperse the seeds (Janzen, 1981). The species is not dependent on the herbivores for germination because seeds will germinate in response to dry-heat treatments of 45–50°C (Hunter, 1989), temperatures that frequently occur at the soil surface in the tropics.

Feeding activities of animals may result in holes being cut in the impermeable seed coat; thus, seeds are mechanically scarified. Seed bugs and weevils break the impermeable seed coats of some *Gossypium thurberi* seeds without actually destroying them; however, some seeds are killed by these insects. Seeds scarified by the insects germinated to higher percentages than those that were not attacked (Karban and Lowenberg, 1992). Seeds of *Lotus corniculatus* also were scarified by a weevil seed predator; however, even when as much as one-third of the cotyledon tissue was eaten, about 50% of the seeds germinated and 50–70% of the resulting seedlings survived (Ollerton and Lack, 1996). Seeds of *Acacia sieberiana* with bruchid exit holes in them sown in soil germinated to 17%, whereas none of the control seeds germinated (Mucunguzi, 1995). Seeds of *A. tortilis* and *A. nilotica* chewed by the rodent *Mastomys natalensis* germinated to higher percentages than those in the controls (Miller, 1995).

G. Effects of Animals on Seeds without Physical Dormancy

Some seeds that do not have physical dormancy also are eaten by animals, and they are intact and viable after passing through the digestive system. Germination percentages of seeds separated from fecal material may be equal to (Barnea *et al.*, 1990, 1992; Clergeau, 1992; Clout and Tilley, 1992; Izhaki *et al.*, 1995), lower than (Ocumpaugh *et al.*, 1993; Crossland and Vander Kloet, 1996; Wallander *et al.*, 1995), or higher than (see later) those of the controls. Seeds exhibiting enhanced germination after passing through a digestive system belong to various families, including the Balanitaceae (Lieberman

and Lieberman, 1987; Chapman *et al.*, 1992), Cactaceae (Timmons, 1942), Caprifoliaceae (Fukui, 1995), Ebenaceae (Lieberman and Lieberman, 1987), Euphorbiaceae, Lauraceae (Bustamante *et al.*, 1992, 1993), Meliaceae (Lieberman *et al.*, 1979), Moraceae (Antonio de Figueiredo, 1993; Wrangham *et al.*, 1994), Myristicaceae (Lieberman and Lieberman, 1987), Najadaceae (Agami and Waisel, 1986, 1988), Oleaceae (Fukui, 1995), Phytolaccaceae (Fukui, 1995), Rosaceae (Robinson, 1986), Rubiaceae (Lieberman *et al.*, 1979), Ruppiaceae (Agami and Waisel, 1988), Sapotaceae (Temple *et al.*, 1977, 1979), Solanaceae (Rick and Bowman, 1961; Murray, 1988; Barnea *et al.*, 1990; Clergeau, 1992), and Zygophyllaceae (Nobel, 1975). The beneficial effects on the germination of permeable seeds passing through a digestive system are unknown. Witmer (1991) suggested that the removal of fruit material from around seeds may prevent them from being destroyed by bacteria before germination occurred. Also, removal of fruit pulp by hand significantly improved the germination of *Callicarpa dichtoma, Eurya japonica, Idesia polycarpa, Parthenocissus triscuspidata* (Fukui, 1995), *Hedera helix, Rubus fruticosus,* and *Sambucus nigra* (Clergeau, 1992) seeds, suggesting that germination inhibitors might be present.

The dodo bird and tambalacoque tree (= *Calvaria major,* in the Sapotaceae) were described as an example of an obligate seed–animal mutualism (Temple, 1977). Temple thought that tambalacoque seeds would not germinate until after they had passed through the dodo bird's digestive tract. Therefore, because the bird is now extinct, seeds of the tree cannot germinate in nature and is near extinction. However, it turns out that seeds can and do germinate without passing though a bird's digestive system (Witmer, 1991). Thus, it is unlikely that the germination of tambalacoque seeds was dependent on passage through the dodo bird's digestive tract. Seeds (actually drupes with a woody endocarp) of the Sapotaceae are not known to have physical dormancy (see Table 3.5). Thus, the dodo bird did not make tambalacoque seeds permeable to water because they were already permeable. Also, the dodo bird had grinding stones in its gizzard; consequently, any tambalacoque seeds it ate may have been destroyed (Witmer, 1991). According to Hill (1941), the pressure generated by the swelling embryo in a *C. major* seed causes the upper portion of the endocarp (called the valve) to separate from the remainder of the endocarp. The radicle then emerges, followed by development of a strong hypocotyl and emergence of the cotyledons. Perhaps the benefit, if any, of dodo birds eating the fruits of *C. major* was simply to remove the exocarp and thereby reduce the possibility that bacteria would destroy the embryo before the seed germinated.

VII. SEEDS WITH PHYSICAL PLUS PHYSIOLOGICAL DORMANCY

A number of species occurring in various plant families have seeds with a combination of physical and physiological dormancy (Table 6.10). The breaking of physical dormancy under natural conditions already has been discussed; thus, emphasis now will be placed on how physiological dormancy of the embryo is broken.

Some species with both physical and physiological dormancy, including *Geranium carolinianum, Malva parviflora, Ornithopus compressus, Stylosanthes* spp., and *Trifolium subterraneum,* are winter annuals, and their seeds are produced in spring. Both physical and physiological dormancy are broken during summer, and seeds germinate in autumn, when soil moisture becomes nonlimiting. Embryos have nondeep physiological dormancy, which is broken within 1–4 months after maturation (Barrett-Lennard and Gladstones, 1964; Baskin and Baskin, 1974; Gardener, 1975). Physiological dormancy

was lost in seeds of *Ornithocarpus compressus* at 60/16 but not at 20°C (Barrett-Lennard and Gladstones, 1964) and it was lost in those of *Trifolium subterraneum* at 40, 40/15, 60/15, but not at 15°C (Quinlivan and Nicol, 1971). Physiological dormancy of the embryo usually is broken before seeds become permeable; thus, the only way to monitor loss of physiological dormancy is to scarify seeds at intervals during summer and test for germination. Like seeds of winter annuals that have permeable seed coats, those with physical dormancy germinate to higher percentages at low (5, 10, 15, 15/6°C) than at high (20/25, 20/10, 25/15°C) temperatures (Nakamura, 1962; Baskin and Baskin, 1974) if they are made permeable in the early stages of the loss of physiological dormancy. With a decrease in physiological dormancy, scarified seeds exhibit an increase in rate (Ballard, 1958) and maximum temperature (Baskin and Baskin, 1974) of germination.

Seeds of the winter annual *Trifolium subterraneum* made permeable shortly after they matured germinated

TABLE 6.10 Species Whose Seeds Have Impermeable Seed Coats (Physical Dormancy) and Physiologically Dormant Embryos and the Temperature Treatment Required to Break Physiological Dormancy (PD)

Species	Family	Treatment required to break PD[a]	Reference
Ceanothus sanguineus	Rhamnaceae	C	Gratkowski (1973)
Ceanothus spp.	Rhamnaceae	C	Quick and Quick (1961)
Cercis canadensis	Fabaceae	C	Afanasiev (1944)
C. siliquastrum	Fabaceae	C	Profumo *et al.* (1978)
Dichrostachys cinerea	Fabaceae	W	Bell and van Staden (1993)
Discaria toumatou	Rhamnaceae	C	Keogh and Bannister (1994)
Geranium carolinianum	Geraniaceae	W	Baskin and Baskin (1974)
Koelreuteria paniculata	Sapindaceae	C	Garner (1979)
Malva parviflora	Malvaceae	W	Sumner and Cobb (1967)
Ornithopus compressus	Fabaceae	W	Barrett-Lennard and Gladstones (1964)
Parkia pendula	Fabaceae	W?	Rizzini (1977)
Rhus aromatica	Anacardiaceae	C	Heit (1967)
R. trilobata	Anacardiaceae	C	Heit (1967)
Sicyos deppei	Curcurbitaceae	W	Brechu-Franco *et al.* (1992)
Stylosanthes spp.	Fabaceae	W	McKeon and Mott (1984)
Tilia americana	Tiliaceae	C	Barton (1934)
T. cordata	Tiliaceae	C	Heit (1967)
T. europaea	Tiliaceae	C	Heit (1967)
T. japonica	Tiliaceae	C	Heit (1967)
T. platyphyllos	Tiliaceae	C	Nagy and Szalai (1973)
T. tomentosa	Tiliaceae	C	Heit (1967)
Trifolium subterraneum	Fabaceae	W	Quinlivan and Nicol (1971)

[a] C, cold stratification; W, dry at summer temperatures.

to 8.5% and at a slow rate (Ballard, 1958). However, germination percentages increased greatly when imbibed seeds with dormant embryos were exposed to increased CO_2 concentration (0.3 to about 5%) (Ballard, 1958). The implication of these results is that embryo dormancy could be lost more quickly in permeable seeds that become buried than in those remaining on the soil surface. This is assuming that CO_2 levels are a little higher in the soil than they are in the air above the soil. Low concentrations of O_2 (0.1–1.0%) also break embryo dormancy in seeds of *T. subterraneum* (Ballard and Grant Lipp, 1969), but it is doubtful that O_2 would decline to these levels unless the soil is flooded. Under laboratory conditions, physiological dormancy in permeable seeds of *Stylosanthes humilis* has been overcome by cadmium, copper, and zinc ions (Delatorre and Barros, 1996); ethrel plus benzyladenine; and thiourea (Burin *et al.,* 1987; Vieira and Barros, 1994).

The presence of embryo dormancy possibly could prevent the germination of any seeds that became permeable during early to midsummer. Seedlings from seeds that germinate after a summer rain are likely to die due to drought stress when the soil dries out again (Baskin and Baskin, 1971). However, McKeon and Mott (1984) point out that embryo dormancy in *Stylosanthes* spp. never has been shown to inhibit germination in the field because most seeds are impermeable while the embryo is dormant. Embryo dormancy is lost by the time *Trifolium subterraneum* seeds become permeable (Quinlivan and Nicol, 1971). Gardener (1975) thought embryo dormancy would prevent the germination of seeds in late spring before they dry to about 8–10% moisture content, at which point the seed coats become impermeable to water.

Seeds of the woody shrub *Dichrostachys cinerea,* and possibly those of *Parkia pendula,* also come out of dormancy while they are exposed to high temperatures (Table 6.10). Therefore, it seems likely that seeds of these two species would germinate at the beginning of the wet season in tropical/subtropical regions.

In some species with both physical and physiological dormancy, seeds must be imbibed and given a cold stratification treatment to break physiological dormancy. Length of the cold stratification treatment required to break physiological dormancy ranges from 1 month in *Rhus aromatica* (Heit, 1967) to 3 months in *Tilia* spp. (Barton, 1934; Heit, 1967; Nagy and Szalai, 1973), *Ceanothus* spp. (Quick, 1935; Gratkowski, 1973), and *Koelreuteria paniculata* (Garner, 1979). In *Cercis canadensis,* growth potential of the embryo increased, and penetration resistance of the testa decreased during cold stratification (Geneve, 1991).

Because seeds must be imbibed before embryo dormancy can be broken by low (0–10°C) temperatures

during winter, seeds sown or dispersed in autumn may not germinate until at least the second spring after maturation (Heit, 1967). One reason for such a long delay in germination of some species is that appropriate environmental conditions to overcome physical dormancy do not occur in the habitat until the following summer. If seed coats become permeable in response to fluctuating temperatures during summer, seeds can be cold stratified during winter and germinate the following spring. Obviously, one could shorten the time required for germination by acid scarifying the seeds and then cold stratifying them (Barton, 1934; Afanasiev, 1944; Heit, 1967). Hot water treatments or mechanical scarification prior to cold stratification also are effective (Quick, 1935; Afanasiev, 1944; Garner, 1979).

Although all species of some genera such as *Tilia* appear to have physical and physiological dormancy, only some of the species in other genera have both types of dormancy. Some *Rhus* spp. have only physical dormancy, but others have both physical and physiological dormancy (Heit, 1967). *Ceanothus* spp. from the high montane zone in California have both types of dormancy, whereas maritime species mostly have only physical dormancy (Quick, 1935).

References

Abdallah, M. M. F., Jones, R. A., and El-Beltagy, A. S. (1989). An efficient method to overcome seed dormancy in Scotch broom (*Cytisus scoparius*). *Environ. Exp. Bot.* **29,** 499–505.

Abrams, M. D., and Dickmann, D. I. (1982). Early revegetation of clear-cut and burned jack pine sites in northern lower Michigan. *Can. J. Bot.* **60,** 946–954.

Abrams, M. D., and Dickmann, D. I. (1984). Apparent heat stimulation of buried seeds of *Geranium bicknellii* on jack pine sites in northern lower Michigan. *Michigan Bot.* **23,** 81–88.

Acuna, P. I., and Garwood, N. C. (1987). Efect de la luz y de la escarificación en la germinación de las semillas de cinco especies de arboles tropicales secundarios. *Rev. Biol. Trop.* **35,** 203–207.

Afanasiev, M. (1944). A study of dormancy and germination of seeds of *Cercis canadensis*. *J. Agric. Res.* **69,** 405–419.

Agami, M., and Waisel, Y. (1986). The role of mallard ducks (*Anas platyrhynchos*) in distribution and germination of seeds of the submerged hydrophyte *Najas marina* L. *Oecologia* **68,** 473–475.

Agami, M., and Waisel, Y. (1988). The role of fish in distribution and germination of seeds of the submerged macrophytes *Najas marina* L. and *Ruppia maritima* L. *Oecologia* **76,** 83–88.

Aitken, Y. (1939). The problem of hard seeds in subterranean clover. *Proc. Roy. Soc. Victoria* **51,** 187–213.

Allred, K. R., and Tingey, D. C. (1964). Germination and spring emergence of dodder as influenced by temperature. *Weeds* **12,** 45–48.

Antonio de Figueiredo, R. (1993). Ingestion of *Ficus enormis* seeds by howler monkeys (*Alouatta fusca*) in Brazil: Effects on seed germination. *J. Trop. Ecol.* **9,** 541–543.

Arianoutsou, M., and Margaris, N. S. (1981). Early stages of regeneration after fire in a phryganic ecosystem (East Mediterranean). I. Regeneration by seed germination. *Biol. Ecol. Medit.* **8,** 119–128.

Auld, T. D. (1986a). Dormancy and viability in *Acacia suaveolens* (Sm.) Willd. *Aust. J. Bot.* **34,** 463–472.

Auld, T. D. (1986b). Population dynamics of the shrub *Acacia suaveolens* (Sm.) Willd.: Fire and the transition to seedlings. *Aust. J. Ecol.* **11,** 373–385.

Auld, T. D., and Bradstock, R. A. (1996). Soil temperatures after the passage of a fire: Do they influence the germination of buried seeds? *Aust. J. Ecol.* **21,** 106–109.

Auld, T. D., and O'Connell, M. A. (1991). Predicting patterns of post-fire germination in 35 eastern Australian Fabaceae. *Aust. J. Ecol.* **16,** 53–70.

Ballard, L. A. T. (1958). Studies of dormancy in the seeds of subterranean clover (*Trifolium subterraneum* L.). I. Breaking of dormancy by carbon dioxide and by activated carbon. *Aust. J. Biol. Sci.* **11,** 246–260.

Ballard, L. A. T. (1973). Physical barriers to germination. *Seed Sci. Technol.* **1,** 285–303.

Ballard, L. A. T. (1976). Strophiolar water conduction in seeds of the Trifolieae induced by action on the testa at nonstrophiolar sites. *Aust. J. Plant Physiol.* **3,** 465–469.

Ballard, L. A. T., and Grant Lipp, A. E. (1969). Studies of dormancy in the seeds of subterranean clover (*Trifolium subterraneum* L.). III. Dormancy breaking by low concentrations of oxygen. *Aust. J. Biol. Sci.* **22,** 279–288.

Ballard, L. A.T., Nelson, S. O., Buchwald, T., and Stetson, L. E. (1976). Effects of radiofrequency electric fields on permeability to water of some legume seeds, with special reference to strophiolar conduction. *Seed Sci. Technol.* **4,** 257–274.

Barnea, A., Yom-Tov, Y., and Friedman, J. (1990). Differential germination of two closely related species of *Solanum* in response to bird ingestion. *Oikos* **57,** 222–228.

Barnea, A., Yom-Tov, Y., and Friedman, J. (1992). Effect of frugivorous birds on seed dispersal and germination of multi-seeded fruits. *Acta Oecol.* **13,** 209–219.

Barrett-Lennard, R. A., and Gladstones, J. S. (1964). Dormancy and hard-seededness in western Australian serradella (*Ornithopus compressus* L.). *Aust. J. Agric. Res.* **15,** 895–904.

Barton, L. V. (1934). Dormancy in *Tilia* seeds. *Contrib. Boyce Thomp. Inst.* **6,** 69–89.

Barton, L. V. (1947). Special studies on seed coat impermeability. *Contrib. Boyce Thomp. Inst.* **14,** 355–362.

Baskin, J. M., and Baskin, C C. (1971). Germination of winter annuals in July and survival of the seedlings. *Bull. Torrey Bot. Club* **98,** 272–276.

Baskin, J. M., and Baskin, C. C. (1974). Some eco-physiological aspects of seed dormancy in *Geranium carolinianum* L. from central Tennessee. *Oecologia* **16,** 209–219.

Baskin, J. M., and Baskin, C. C. (1984). Environmental conditions required for germination of prickly sida (*Sida spinosa*). *Weed Sci.* **32,** 786–791.

Baskin, J. M., and Baskin, C. C. (1997). Methods of breaking seed dormancy in the federal endangered species *Iliamna corei* (Malvaceae), with special attention to heating. *Nat. Areas J.* (in press).

Baskin, J. M., and Baskin, C. C. (1998). Greenhouse and laboratory studies on the ecological life cycle of *Dalea foliosa* (Fabaceae), a federal-endangered species. *Nat. Areas J.* (in press).

Beadle, N. C. W. (1940). Soil temperatures during forest fires and their effect on the survival of vegetation. *J. Ecol.* **28,** 180–192.

Bell, W. E., and van Staden, J. (1993). Seed structure and germination of *Dichrostachys cinerea*. *S. Afr. J. Bot.* **59,** 9–13.

Bhardwaj, N., and Prabhakar. (1990). Seed coat regulated germination in *Tephrosia hamiltonii* Drumm and its significance. *In* "International Symposium on Environmental Influences on Seed and Germination Mechanism: Recent Advances in Research and Technol-

ogy" (D. N. Sen, S. Mohammed, P. K. Kasera, and T. P. Thomas, eds.), pp. 55–56. University of Jodhpur, Jodhpur, India.

Bhat, J. L. (1968). Seed coat dormancy in *Indigofera glandulosa* Willd. *Trop. Ecol.* **9,** 42–51.

Bhatia, P., and Sen, D. N. (1978). Temperature responses of seeds in *Ipomoea pes-tigridis* L. *Biol. Plant.* **20,** 221–224.

Bhatia, R. K., Chawan, D. D., and Sen, D. N. (1977). Environment controlled seed dormancy in *Cassia* spp. *Geobios* **4,** 208–210.

Bhattacharya, A., and Saha, P. K. (1990). Ultrastructure of seed coat and water uptake pattern of seeds during germination in *Cassia* sp[p]. *Seed Sci. Technol.* **18,** 97–103.

Blackshaw, R. E. (1990). Influence of soil temperature, soil moisture, and seed burial depth on the emergence of round-leaved mallow (*Malva pusilla*). *Weed Sci.* **38,** 518–521.

Bohra, P. N., and Sen, D. N. (1974). Seed patterns and germination behaviour in *Crotalaria medicaginea* Lamk. growing in Indian arid zone. *Curr. Sci.* **43,** 591–592.

Borchert, M. (1989). Postfire demography of *Thermopsis macrophylla* H.A. var. *agnina* J. T. Howell (Fabaceae), a rare perennial herb in chaparral. *Am. Midl. Nat.* **122,** 120–132.

Bossard, C. C. (1993). Seed germination in the exotic shrub *Cytisus scoparius* (scotch broom) in California. *Madrono* **40,** 47–61.

Bradstock, R. A., and Auld, T. D. (1995). Soil temperatures during experimental bushfires in relation to fire intensity: Consequences for legume germination and fire management in south-eastern Australia. *J. Appl. Ecol.* **32,** 76–84.

Bradstock, R. A., Auld, T. D., Ellis, M. E., and Cohn, J. S. (1992). Soil temperatures during bushfires in semi-arid, mallee shrublands. *Aust. J. Ecol.* **17,** 433–440.

Brant, R. E., McKee, G. W., and Cleveland, R. W. (1971). Effect of chemical and physical treatment on hard seed of penngift crown-vetch. *Crop Sci.* **11,** 1–6.

Brechu-Franco, A., Cruz-Garcia, F., Marquez-Guzman, J., and Laguna-Hernandez, G. (1992). Germination of *Sicyos deppei* (Curcubitaceae) seeds as affected by scarification and light. *Weed Sci.* **40,** 54–56.

Brenchley, W. E., and Warington, K. (1930). The weed seed population of arable soil. I. Numerical estimation of viable seeds and observations on their natural dormancy. *J. Ecol.* **18,** 235–272.

Brown, A. J., and Worley, F. P. (1912). The influence of temperature on the absorption of water by seeds of *Hordeum vulgare* in relation to the temperature coefficient of chemical change. *Proc. Roy. Soc. Lond. B* **85,** 546–553.

Brown, E. O., and Porter, R. H. (1942). The viability and germination of seeds of *Convolvulus arvensis* L. and other perennial weeds. *Iowa Agric. Exp. Sta. Res. Bull.* 294.

Brown, N. A. C., and de V. Booysen, P. (1969). Seed coat impermeability in several *Acacia* species. *Agroplantae* **1,** 51–60.

Burin, M., Barros, R. S., and Rena, A. B. (1987). Chemical regulation of endogenous dormancy in seeds of *Stylosanthes humilis* H.B.K. *Turrialba* **37,** 281–285.

Burns, R. E. (1959). Effect of acid scarification on lupine seed impermeability. *Plant Physiol.* **34,** 107–108.

Burton, G. W. (1940). Factors influencing the germination of seed of *Trifolium repens*. *J. Am. Soc. Agron.* **32,** 731–738.

Busse, W. F. (1930). Effect of low temperatures on germination of impermeable seeds. *Bot. Gaz.* **89,** 169–179.

Bustamante, R. O., Grez, A. A, Simonetti, J. A., Vasquez, R. A., and Walkowiak, A. M. (1993). Antagonistic effects of frugivores on seeds of *Cryptocarya alba* (Mol.) Looser (Lauraceae): Consequences on seedling recruitment. *Acta Oecol.* **14,** 739–745.

Bustamante, R. O., Simonetti, J. A., and Mella, J. E. (1992). Are foxes legitimate and efficient seed dispersers? A field test. *Acta Oecol.* **13,** 203–208.

Caljouw, C. A., Lipscomb, M. V., Adams, S., and St. Clair, M. (1994). Prescribed burn and disturbance history studies at the Narrows: Habitat studies for the endangered Peters Mountain mallow. Natural Heritage Technical Report Number 94–8 submitted to the Virginia Department of Agriculture and Consumer Services, Virginia Department of Conservation and Recreation, Richmond.

Cavanagh, A. K. (1980). A review of some aspects of the germination of Acacias. Proc. Roy Soc. Victoria 91, 161–180.

Cavanagh, T. (1987). Germination of hard-seeded species (order Fabales). In "Germination of Australian Native Plant Seed" (P. Langkamp, ed.). Inkata Press, Melbourne.

Chandler, J. M., Munson, R. L., and Vaughan, C. E. (1977). Purple moonflower: Emergence, growth, reproduction. Weed Sci. 25, 163–167.

Chapman, L. J., Chapman, C. A., and Wrangham, R. W. (1992). Balanites wilsoniana: Elephant dependent dispersal? J. Trop. Ecol. 8, 275–282.

Chawan, D. D. (1971). Role of high temperature pretreatments on seed germination of desert species of Sida (Malvaceae). Oecologia 6, 343–349.

Chawan, D. D., and Sen, D. N. (1973). Action of light in the germination of seeds and seedling growth in two desert species of Sida. Broteria 62, 141–148.

Christiansen, M. N., and Moore, R. P. (1959). Seed coat structural differences that influence water uptake and seed quality in hard seed cotton. Agron. J. 51, 582–584.

Clemens, J., Jones, P. G., and Gilbert, N. H. (1977). Effect of seed treatments on germination in Acacia. Aust. J. Bot. 25, 269–276.

Clergeau, P. (1992). The effect of birds on seed germination of fleshy-fruited plants in temperate farmland. Acta Oecol. 13, 679–686.

Clout, M. N., and Tilley, J. A. V. (1992). Germination of miro (Prumnopitys ferruginea) seeds after consumption by New Zealand pigeons (Hemiphaga novaeseelandiae). New Zeal. J. Bot. 30, 25–28.

Conner, L. N., and Conner, A. J. (1988). Seed biology of Chordospartium stevensonii. New Zeal. J. Bot. 26, 473–475.

Corral, R., Pita, J. M., and Perez-Garcia, F. (1990). Some aspects of seed germination in four species of Cistus L. Seed Sci. Technol. 18, 321–325.

Coughenour, M. B., and Detling, J. K. (1986). Acacia tortilis seed germination responses to water potential and nutrients. Afr. J. Ecol. 24, 203–205.

Coutinho, L. M., and Struffaldi, Y. (1971). Observacoes sobre a germinacao das sementes e o crescimento das plantulas de uma leguminosa da mata amazonica de igapo (Parkia auriculata Spruce Mss.). Phyton (Buenos Aires) 28, 149–159.

Cox, J. R., de Alba-Avila, A., Rice, R. W., and Cox, J. N. (1993). Biological and physical factors influencing Acacia constricta and Prosopis velutina establishment in the Sonoran Desert. J. Range Manage. 46, 43–48.

Creel, J. M., Jr., Hoveland, C. S., and Buchanan, G. A. (1968). Germination, growth, and ecology of sicklepod. Weeds 16, 396–400.

Cremer, K. W., and Mount, A. B. (1965). Early stages of plant succession following the complete felling and burning of Eucalyptus regnans forest in the Florentine Valley, Tasmania. Aust. J. Bot. 13, 303–322.

Crocker, W. (1909). Longevity of seeds. Bot. Gaz. 47, 69–72.

Crossland, D. R., and Vander Kloet, S. P. (1996). Berry consumption by the American robin, Turdus migratorius, and the subsequent effect on seed germination, plant vigour, and dispersal of the lowbush blueberry, Vaccinium angustifolium. Can. Field Nat. 110, 303–309.

Cushwa, C. T., Martin, R. E., and Miller, R. L. (1968). The effects of fire on seed germination. J. Range Manage. 21, 250–254.

Dafni, A., and Negbi, M. (1978). Variability in Prosopis farcta in

Israel: Seed germination as affected by temperature and salinity. Israel J. Bot. 27, 147–159.

Daiya, K. S., Sharma, H. K., Chawan, D. D., and Sen, D. N. (1980). Effect of salt solutions of different osmotic potential on seed germination and seedling growth in some Cassia species. Folia Geobot. Phytotax. 15, 149–153.

Danthu, P., Roussel, J., Dia, M., and Sarr, A. (1992). Effect of different pretreatments on the germination of Acacia senegal seeds. Seed Sci. Technol. 20, 111–117.

Datta, S. C., and Sen, S. (1987). A comparison of the germination characters of Desmodium species. Acta Bot. Hung. 33, 125–131.

Davies, P. A. (1926). Effect of high pressure on germination of seeds (Medicago sativa and Melilotus alba). J. Gen. Physiol. 9, 805–809.

Davies, P. A. (1928a). High pressure and seed germination. Am. J. Bot. 15, 149–156.

Davies, P. A. (1928b). The effect of high pressure on the percentages of soft and hard seeds of Medicago sativa and Melilotus alba. Am. J. Bot. 15, 433–436.

Dekker, J., and Meggitt, W. F. (1986). Field emergence of velvetleaf (Abutilon theophrasti) in relation to time and burial depth. Iowa St. J. Res. 61, 65–80.

Delaney, R. H., Abernethy, R. H., and Johnson, D. W. (1986). Temperature and water stress effects on germination of Ruby Valley pointvetch (Oxytropis riparia Litv.). Crop Sci. 26, 161–165.

Delatorre, C. A., and Barros, R. S. (1996). Germination of dormant seeds of Stylosanthes humilis as related to heavy metal ions. Biol. Plant. 38, 269–274.

Dell, B. (1980). Structure and function of the strophiolar plug in seeds of Albizia lophantha. Am. J. Bot. 67, 556–563.

Dillon, S. P., and Forcella, F. (1985). Fluctuating temperatures break seed dormancy of catclaw mimosa (Mimosa pigra). Weed Sci. 33, 196–198.

Dinnis, E. R., and Jordan, S. (1939). The germination of freshly harvested and of stored seeds of sea pea. South-East. Agric. Coll. Wye, Kent 44, 140–142.

Doussi, M. A., and Thanos, C. A. (1993). The ecophysiology of fire-induced germination in hard-seeded plants. In "Fourth International Workshop on Seeds: Basic and Applied Aspects of Seed Biology" (C. Come and F. Corbineau, eds), pp. 455–460. Angers, France, ASFIS, Paris.

Dunn, L. E. (1939). Influence of low temperature treatments on the germination of seeds of sweet clover and smooth vetch. Agron. J. 31, 687–694.

Eastin, E. F. (1984). Drummond rattlebox (Sesbania drummondii) germination as influenced by scarification, temperature, and seedling depth. Weed Sci. 32, 223–225.

Egley, G. H. (1976). Germination of developing prickly sida seeds. Weed Sci. 24, 239–243.

Egley, G. H. (1979). Seed coat impermeability and germination of showy crotalaria (Crotalaria spectabilis) seeds. Weed Sci. 27, 355–361.

Eglitis, M., and Johnson, F. (1957). Control of hard seed of alfalfa with high-frequency energy. Phytopathology 47, 9. [Abstract]

Everitt, J. H. (1983). Seed germination characteristics of two woody legumes (retama and twisted acacia) from south Texas. J. Range Manage. 36, 411–414.

Everson, L. (1949). Preliminary studies to establish laboratory methods for the germination of weed seed. Proc. Assoc. Offic. Seed Anal. 39, 84–89.

Eynard, I. (1958). Effect of very low temperatures on germination of hard seeds. Herb. Abst. 28, 1027.

Eynard, I. (1960). Effect of liquid N_2 and O_2 on the germinating capacity of hard seeds of Trifolium hybridum, Medicago sativa and Trifolium pratense. Herb. Abst. 30, 635.

Fairbrother, T. E. (1991). Effect of fluctuating temperatures and humidity on the softening rate of hard seed of subterranean clover (*Trifolium subterraneum* L.). *Seed Sci. Technol.* **19**, 93–105.

Farmer, R. E., Lockley, G. C., and Cunningham, M. (1982). Germination patterns of the sumacs, *Rhus glabra* and *Rhus coppalina*: Effect of scarification time, temperatures and genotype. *Seed Sci. Technol.* **10**, 223–231.

Farrell, T. P., and Ashton, D. H. (1978). Population studies on *Acacia melanoxylon* R. Br. I. Variation in seed and vegetative characteristics. *Aust. J. Bot.* **26**, 365–379.

Faruqi, M. A., Kahan, M. K., and Uddin, M. G. (1974). Comparative studies on the effects of ultrasonics, red light and gibberellic acid, on the germination of *Cassia holosericea* Fres seeds. *Biol. Abstr.* **58**, 22360.

Felippe, G. M. (1984). Germinacao de *Zornia reticulata*, uma especie dos cerrados. Anais IV Cong. SBSP, pp. 7–13.

Fenner, M. (1985). "Seed Ecology." Chapman and Hall, London.

Ferreira, A. G. (1976). Germinacao de sementes de *Mimosa bimucronata* (DC.) OK. (Marica). I. Efeito da escarificacao e do pH. *Cien. Cult.* **28**, 1200–1204.

Floyd, A. G. (1966). Effect of fire upon weed seeds in the wet sclerophyll forests of northern New South Wales. *Aust. J. Bot.* **14**, 243–256.

Floyd, A. G. (1976). Effect of burning on regeneration from seeds in wet sclerophyll forest. *Aust. For.* **39**, 210–220.

Fukui, A. W. (1995). The role of the brown-eared bulbul *Hypsypetes amaurotis* as a seed dispersal agent. *Res. Popul. Ecol.* **37**, 211–218.

Gardener, C. J. (1975). Mechanisms regulating germination in seeds of *Stylosanthes*. *Aust. J. Agric. Res.* **26**, 281–294.

Gardener, C. J., McIvor, J. G., and Jansen, A. (1993a). Passage of legume and grass seeds through the digestive tract of cattle and their survival in faeces. *J. Appl. Ecol.* **30**, 63–74.

Gardener, C. J., McIvor, J. G., and Jansen, A. (1993b). Survival of seeds of tropical grassland species subjected to bovine digestion. *J. Appl. Ecol.* **30**, 75–85.

Garner, J. L. (1979). Overcoming double dormancy in golden-rain tree seeds. *Plant Propagator* **25**(2), 6–8.

Geneve, R. L. (1991). Seed dormancy in eastern redbud (*Cercis canadensis*). *J. Am. Soc. Hort. Sci.* **116**, 85–88.

Gill, M. (1985). *Acacia cyclops* G. Don (Leguminosae-Mimosaceae) in Australia: Distribution and dispersal. *J. Roy. Soc. W. Aust.* **67**, 59–65.

Gladstones, J. S. (1958). The influence of temperature and humidity in storage on seed viability and hard-seededness in the west Australian blue lupin, *Lupinus digitatus* Forsk. *Aust. J. Agric. Res.* **9**, 171–181.

Glyphis, J. P., Milton, S. J., and Siegfried, W. R. (1981). Dispersal of *Acacia cyclops* by birds. *Oecologia* **48**, 138–141.

Gogue, G. J., and Emino, E. R. (1979). Seed coat scarification of *Albizia julibrissin* Durazz. by natural mechanisms. *J. Am. Soc. Hort. Sci.* **104**, 421–423.

Gomes, D. T., and Kretschmer, A. E., Jr. (1978). Effect of three temperature regimes on tropical legume seed germination. *Soil Crop Sci. Soc. Florida* **37**, 61–73.

Gomes, L. F., Chandler, J. M., and Vaughan, C. E. (1978). Aspects of germination, emergence, and seed production of three *Ipomoea* taxa. *Weed Sci.* **26**, 245–248.

Gopinathan, M. C., and Babu, C. R. (1985). Structural diversity and its adaptive significance in seeds of *Vigna minima* (Roxb.) Ohwi & Ohashi and its allies (Leguminosae-Papilionoideae). *Ann. Bot.* **56**, 723–732.

Graaff, J. L., and van Staden, J. (1983). The effect of different chemical and physical treatments on seed coat structure and seed germination of *Sesbania*. *Z. Pflanzenphysiol.* **112**, 221–230.

Graaff, J. L., and van Staden, J. (1987). The relationship between storage conditions and seed germination in two species of *Sesbania*. *S. Afr. J. Bot.* **53**, 143–146.

Gratkowski, H. (1973). Pregermination treatments of redstem *Ceanothus* seeds. *USDA For. Serv. Res. Paper* PNW-156.

Griffin, G. F., and Friedel. M. H. (1984). Effects of fire on central Australian rangelands. II. Changes in tree and shrub populations. *Aust. J. Ecol.* **9**, 395–403.

Gutterman, Y., and Agami, M. (1987). A comparative germination study of seeds of *Helianthemum vesicarium* Boiss. and *H. ventosum* Boiss., perennial desert shrub species inhabiting two different neighbouring habitats in the Negev desert highlands, Israel. *J. Arid Environ.* **12**, 215–221.

Hadley, E. B. (1961). Influence of temperature and other factors on *Ceanothus megacarpus* seed germination. *Madrono* **16**, 132–138.

Hagon, M. W., and Ballard, L. A. T. (1970). Reversibility of strophiolar permeability to water in seeds of subterranean clover (*Trifolium subterraneum*). *Aust. J. Biol. Sci.* **23**, 519–528.

Haferkamp, M. R., Kissock, D. C., and Webster, R. D. (1984). Impact of presowing seed treatments, temperature and seed coats on germination of velvet bundleflower. *J. Range Manage.* **37**, 185–188.

Halevy, G. (1974). Effects of gazelles and seed beetles (Bruchidae) on germination and establishment of *Acacia* species. *Israel J. Bot.* **23**, 120–126.

Hamly, D. H. (1932). Softening of the seeds of *Melilotus alba*. *Bot. Gaz.* **93**, 345–375 + 2 plates.

Hanes, T. L. (1971). Succession after fire in the chaparral of southern California. *Ecol. Monogr.* **41**, 27–52.

Hanna, P. J. (1984). Anatomical features of the seed coat of *Acacia kempeana* (Mueller) which relate to increased germination rate induced by heat treatment. *New Phytol.* **96**, 23–29.

Harnett, D. C., and Richardson, D. R. (1989). Population biology of *Bonamia grandiflora* (Convolvulaceae): Effects of fire on plant and seed bank dynamics. *Am. J. Bot.* **76**, 361–369.

Hauser, T. P. (1994). Germination, predation and dispersal of *Acacia albida* seeds. *Oikos* **71**, 421–426.

Heit, C. E. (1967). Propagation from seed. *Am. Nurseryman* **125**(12), 10–11, 37–41, 44–45.

Helgeson, E. A. (1932). Impermeability in mature and immature sweet clover seeds as affected by conditions of storage. *Trans. Wisc. Acad. Sci. Arts Letts.* **27**, 193–206.

Heyward, F. (1938). Soil temperatures during forest fires in the longleaf pine region. *J. For.* **36**, 478–491.

Hill, A. W. (1941). The genus *Calvaria*, with an account of the stony endocarp and germination of the seed and description of a new species. *Ann. Bot.* **5**, 587–606.

Hodgkinson, K. C., and Oxley, R. E. (1990). Influence of fire and edaphic factors on germination of the arid zone shrubs *Acacia aneura*, *Cassia nemophila* and *Dodonaea viscosa*. *Aust. J. Bot.* **38**, 269–279.

Hoffman, M. T., Cowling, R. M., Douie, C., and Pierce, S. M. (1989). Seed predation and germination of *Acacia erioloba* in the Kuiseb River Valley, Namib Desert. *S. Afr. J. Bot.* **55**, 103–106.

Hopkins, M. S., and Graham, A. W. (1984). Viable soil seed banks in disturbed lowland tropical rainforest sites in North Queensland. *Aust. J. Ecol.* **9**, 71–79.

Horowitz, M., and Taylorson, R. B. (1984). Hardseededness and germinability of velvetleaf (*Abutilon theophrasti*) as affected by temperature and moisture. *Weed Sci.* **32**, 111–115.

Horton, J. S., and Kraebel, C. J. (1955). Development of vegetation after fire in the chamise chaparral of southern California. *Ecology* **36**, 244–262.

Hunter, J. R. (1989). Seed dispersal and germination of *Enterolobium cyclocarpum* (Jacq.) Griseb. (Leguminosae: Mimosoideae): Are megafauna necessary? *J. Biogeogr.* **16**, 369–378.

Hutchison, J. M., and Ashton, F. M. (1979). Effect of desiccation and scarification on the permeability and structure of the seed coat of *Cuscuta campestris. Am. J. Bot.* **66**, 40–46.

Hutton, M. E.-J., and Porter, R. H. (1937). Seed impermeability and viability of native and introduced seeds of Leguminosae. *Iowa St. Coll. J. Sci.* **12**, 5–24.

Hyde, E. O. C. (1954). The function of the hilum in some Papilionaceae in relation to the ripening of the seed and the permeability of the testa. *Ann. Bot.* **18**, 241–256.

Ivens, G. W. (1978). Some aspects of seed ecology of gorse. *Proc. New Zeal. Weed Pest Control Conf.* **31**, 53–57.

Ivens, G. W. (1982). Seasonal germination and establishment of gorse. *Proc. New Zeal. Weed Pest Control Conf.* **35**, 152–156

Ivens, G. W. (1983). The influence of temperature on germination of gorse (*Ulex europaeus* L.). *Weed Res.* **23**, 207–216.

Iwata, E. (1966). Germination behaviour of shrubby Lespedeza (*Lespedeza cyrtobotry* Miq.) seeds with special reference to burning. *Ecol. Rev.* **16**, 217–227.

Izhaki, I., Korine, C., and Arad, Z. (1995). The effect of bat (*Rousettus aegyptiacus*) dispersal on seed germination in eastern Mediterranean habitats. *Oecologia* **101**, 335–342.

Jackson, G., and Gartlan, J. S. (1965). The flora and fauna of Lolu Island, Lake Victoria. A study of vegetation, men and monkeys. *J. Ecol.* **53**, 573–597.

Janzen, D. H. (1969). Birds and the ant x *Acacia* interaction in Central America, with notes on birds and other myrmecophytes. *Condor* **71**, 240–256.

Janzen, D. H. (1981). *Enterolobium cyclocarpum* seed passage rate and survival in horses, Costa Rican Pleistocene seed dispersal agents. *Ecology* **62**, 593–601.

Janzen, D. H., Demment, M. W., and Robertson, J. B. (1985). How fast and why do germinating Guanacaste seeds (*Enterolobium cyclocarpum*) die inside cows and horses? *Biotropica* **17**, 322–325.

Janzen, D. H., and Martin, P. S. (1982). Neotropical anachronisms: The fruits the gomphotheres ate. *Science* **215**, 19–27.

Jeffery, D. J., Holmes, P. M., and Rebelo, A. G. (1988). Effects of dry heat on seed germination in selected indigenous and alien legume species in South Africa. *S. Afr. J. Bot.* **54**, 28–34.

Jones, J. A. (1928). Overcoming delayed germination of *Nelumbo lutea. Bot. Gaz.* **85**, 341–343.

Jones, J. P. (1928). A physiological study of dormancy in vetch seed. *Cornell Univ. Agri. Exp. Sta. Mem.* **120**, 1–50.

Jones, R. M. (1963). Preliminary studies of the germination of seed of *Acacia cyclops* and *Acacia cyanophylla. S. Afr. J. Sci.* **59**, 296–298.

Jordan, L. S., and Jordan, J. L. (1982). Effects of pre-chilling on *Convolvulus arvensis* L. seed coat and germination. *Ann. Bot.* **49**, 421–423.

Juhren, M. C. (1966). Ecological observations on *Cistus* in the mediterranean vegetation. *For. Sci.* **12**, 415–426.

Kamada, M., Nakagoshi, N., and Takahashi, F. (1987). Effects of burning on germination of viable seeds in a slash and burn agriculture soil. *Jap. J. Ecol.* **37**, 91–100.

Karban, R., and Lowenberg, G. (1992). Feeding by seed bugs and weevils enhances germination of wild *Gossypium* species. *Oecologia* **92**, 196–200.

Kaul, R. N., and Manohar, S. (1966). Germination studies on arid zone tree seeds. *Indian For.* **92**, 499–503.

Kelly, K. M., and van Staden, J. (1985). Effect of acid scarification on seed coat structure, germination and seedling vigour of *Aspalathus linearis. J. Plant Physiol.* **121**, 37–45.

Kelly, K. M., and van Staden, J. (1987). The lens as the site of perme-

ability in the papilionoid seed, *Aspalathus linearis. J. Plant Physiol.* **128**, 395–404.

Keogh, J. A., and Bannister, P. (1994). Seed structure and germination in *Discaria toumatou* (Rhamnaceae). *Weed Res.* **34**, 481–490.

Khudairi, A. K. (1956). Breaking the dormancy of *Prosopis* seeds. *Physiol. Plant.* **9**, 452–461.

Kirchner, R., and Andrew, W. D. (1971). Effect of various treatments on hardening and softening of seeds in pods of barrel medic (*Medicago truncatula*). *Aust. J. Exp. Agric. Anim. Husb.* **11**, 536–540.

Koller, D., and Cohen, D. (1959). Germination-regulating mechanisms in some desert seeds. VI. *Convolvulus lanatus* Vahl., *Convolvulus negevensis* Zoh. and *Convolvulus secundus* Desr. *Bull. Res. Counc. Israel* **7D**, 175–180.

Kolokol'tseva, L. S., and Prokof'ev, M. K. (1974). Germination of hard alfalfa seeds under the effect of ultrasound. *Biol. Abstr.* **57**, 35043.

Krefting, L. W., and Roe, E. I. (1949). The role of some birds and mammals in seed germination. *Ecol. Monogr.* **19**, 269–286.

Kremer, R. J. (1986a). Microorganisms associated with velvetleaf (*Abutilon theophrasti*) seeds on the soil surface. *Weed Sci.* **34**, 233–236.

Kremer, R. J. (1986b). Antimicrobial activity of velvetleaf (*Abutilon theophrasti*) seeds. *Weed Sci.* **34**, 617–622.

Kremer, R. J. (1987). Identity and properties of bacteria inhabiting seeds of selected broadleaf weed species. *Microb. Ecol.* **14**, 29–37.

Kremer, R. J., Hughes, I. B., Jr., and Aldrich, R. J. (1984). Examination of microorganisms and deterioration resistance mechanisms associated with velvetleaf seed. *Agron. J.* **76**, 745–749.

LaCroix, L. J., and Staniforth, D. W. (1964). Seed dormancy in velvetleaf. *Weeds* **12**, 171–174.

Lamprey, H. F. (1967). Notes on the dispersal and germination of some tree seeds through the agency of mammals and birds. *E. Afr. Wildl. J.* **5**, 179–180.

Lamprey, H. F., Halevy, G., and Makacha, S. (1974). Interactions between *Acacia*, bruchid seed beetles and large herbivores. *E. Afr. Wildl. J.* **12**, 81–85.

Latting, J. (1961). The biology of *Desmanthus illinoensis. Ecology* **42**, 487–493.

Lebedeff, G. A. (1943). Inheritance of hard-seed production in common beans (*Phaseolus vulgaris*). *Genetics* **28**, 80. [Abstract]

Lieberman, D., and Lieberman, M. (1987). Notes on seeds in elephant dung from Bia National Park, Ghana. *Biotropica* **19**, 365–369.

Lieberman, M., Hall, J. B., and Swaine, M. D. (1979). Seed dispersal by baboons in the Shai Hills, Ghana. *Ecology* **60**, 65–75.

Lemmon, P. E., Brown, R. L., and Chapin, W. E. (1943). Sulfuric acid seed treatment of beach pea, *Lathyrus maritimus*, and silvery pea, *L. littoralis*, to increase germination, seedling establishment, and field stands. *J. Am. Soc. Agron.* **35**, 177–191.

Liu, N. Y., Khatamian, H., and Fretz, T. A. (1981). Seed coat structure of three woody legume species after chemical and physical treatments to increase seed germination. *J. Am. Soc. Hort Sci.* **106**, 691–694.

Lonsdale, W. M., and Abrecht, D. G. (1989). Seedling mortality in *Mimosa pigra*, an invasive tropical shrub. *J. Ecol.* **77**, 371–385.

Lonsdale, W. M., Harley, K. L. S., and Gillett, J. D. (1988). Seed bank dynamics in *Mimosa pigra*, an invasive tropical shrub. *J. Appl. Ecol.* **25**, 963–976.

Lonsdale, W. M., and Miller, I. L. (1993). Fire as a management tool for a tropical woody weed: *Mimosa pigra* in Northern Australia. *J. Environ. Manage.* **39**, 77–87.

Lovato, M. B., Martins, P. S., and de Lemos Filho, J. P. (1994). Germination in *Stylosanthes humilis* populations in the presence of NaCl. *Aust. J. Bot.* **42**, 717–723.

Lunden, A. O., and Kinch, R. C. (1957). The effect of high temperature contact treatment on hard seeds of alfalfa. *Agron. J.* **49**, 151–153.

Malo, J. E., and Suarez, F. (1995). Cattle dung and the fate of *Biserrula pelecinus* L. (Leguminosae) in a Mediterranean pasture: Seed dispersal, germination and recruitment. *Bot. J. Linnean Soc.* **118**, 139–148.

Malo, J. E., and Suarez, F. (1996). *Cistus ladanifer* recruitment: Not only fire, but also deer. *Acta Oecol.* **17**, 55–60.

Marks, P. L. (1979). Apparent fire-stimulated germination of *Rhus typhina* seeds. *Bull. Torrey Bot. Club* **106**, 41–42.

Martin, A. R. H. (1966). The plant ecology of the Grahamstown Nature Reserve. II. Some effects of burning. *S. Afr. J. Bot.* **32**, 1–39.

Martin, J. N. (1945). Germination studies of sweet clover seed. *Iowa St. Coll. J. Sci.* **19**, 289–300.

Martin, R. E., and Cushwa, C. T. (1966). Effects of heat and moisture on leguminous seed. *Proc. Tall Timbers Fire Ecol. Conf.* **5**, 159–175.

Martin, R. E., Miller, R. L., and Cushwa, C. T. (1975). Germination response of legume seeds subjected to moist and dry heat. *Ecology* **56**, 1441–1445.

Mayer, A. M., and Poljakoff-Mayber, A. (1989). "The Germination of Seeds," 4th Ed. Pergamon Press, Elmsford, NY.

McKeon, G. M. (1984). Field changes in germination requirements: Effect of natural rainfall on potential germination speed and light requirement of *Stylosanthes humilis*, *Stylosanthes hamata* and *Digitaria ciliaris*. *Aust. J. Agric. Res.* **35**, 807–819.

McKeon, G. M. (1985). Pasture seed dynamics in a dry monsoonal climate. II The effect of water availability, light and temperature on germination speed and seedling survival of *Stylosanthes humilis* and *Digitaria ciliaris*. *Aust. J. Ecol.* **10**, 149–163.

McKeon, G. M., and Brook, K. (1983). Establishment of *Stylosanthes* species: Changes in hardseededness and potential speed of germination at Katherine, N.T. *Aust. J. Agric. Res.* **34**, 491–504.

McKeon, G. M., and Mott, J. J. (1982). The effect of temperature on the field softening of hard seed of *Stylosanthes humilis* and *S. hamata* in a dry monsoonal climate. *Aust. J. Agric. Res.* **33**, 75–85.

McKeon, G. M., and Mott, J. J. (1984). Seed Biology of *Stylosanthes*. In "The Biology and Agronomy of *Stylosanthes*" (H. M. Stace and L. A. Edye, eds.), pp. 311–332. Academic Press, Sydney.

Midgley, A. R. (1926). Effect of alternate freezing and thawing on the impermeability of alfalfa and dodder seeds. *J. Am. Soc. Agron.* **18**, 1087–1098.

Miklas, P. N., Townsend, C. E., and Ladd, S. L. (1987). Seed coat anatomy and the scarification of cicer milkvetch seed. *Crop Sci.* **27**, 766–772.

Miller, M. F. (1995). *Acacia* seed survival, seed germination and seedling growth following pod consumption by large herbivores and seed chewing rodents. *Afr. J. Ecol.* **33**, 194–210.

Misra, B. N. (1963). Germination of seeds of *Ipomoea crassicaulis* (Benth) Robinson. *J. Indian Bot. Soc.* **42**, 358–366.

Monk, D., Pate, J. S., and Loneragan, W. A. (1981). Biology of *Acacia pulchella* R. Br. with special reference to symbiotic nitrogen fixation. *Aust. J. Bot.* **29**, 579–592.

Montgomery, K. R., and Strid, T. W. (1976). Regeneration of introduced species of *Cistus* (Cistaceae) after fire in southern California. *Madrono* **23**, 417–427.

Moore, C. W. E. (1973). Some observations on ecology and control of woody weeds on mulga lands in northwestern New South Wales. *Trop. Grasslands* **7**, 79–88.

Morrison, D. A., Auld, T. D., Rish, S., Porter, C., and McClay, K. (1992). Patterns of testa-imposed seed dormancy in native Australian legumes. *Ann. Bot.* **70**, 157–163.

Mott, J. J. (1979). High temperature contact treatment of hard seed in *Stylosanthes*. *Aust. J. Agric. Res.* **30**, 847–854.

Mott, J. J. (1982). Fire and survival of *Stylosanthes* spp. in the dry savanna woodlands of the Northern Territory. *Aust. J. Agric. Res.* **33**, 203–211.

Mott, J. J., Cook, S. J., and Williams, R. J. (1982). Influence of short duration, high temperature seed treatment on the germination of some tropical and temperate legumes. *Trop. Grasslands* **16**, 50–55.

Mott, J. J., and McKeon, G. M. (1979). Effect of heat treatments in breaking hardseededness in four species of *Stylosanthes*. *Seed Sci. Technol.* **7**, 15–25.

Mucunguzi, P. (1995). Effects of bruchid beetles on germination and establishment of *Acacia* species. *Afr. J. Ecol.* **33**, 64–70.

Mullick, P., and Chatterji, U. N. (1967). Eco-physiological studies on seed germination: Germination experiments with the seeds of *Clitoria ternatea* Linn. *Trop. Ecol.* **8**, 115–125.

Murphy, T. R., Gossett, B. J., and Toler, J. E. (1986). Dormancy and field burial of cowpea (*Vigna unguiculata*) seed. *Weed Sci.* **34**, 260–265.

Murray, K. G. (1988). Avian seed dispersal of three neotropical gap-dependent plants. *Ecol. Monogr.* **58**, 271–298.

Nagy, M., and Szalai, I. (1973). Dormancy in fruits of *Tilia platyphyllos* Scop. I. *Acta Biol. Szeged.* **19**, 71–77.

Nakamura, S. (1962). Germination of legume seeds. *Proc. Int. Seed Test. Assoc.* **27**, 695–709.

Nan, X. (1992). "Comparison of Some Aspects of the Ecological Life History of an Annual and a Perennial Species of *Senna* (Leguminosae: Section *Chamaefistula*), with Particular Reference to Seed Dormancy." M.S. thesis, University of Kentucky, Lexington.

Narang, A. K., and Bhardwaja, N. (1974). Seed coat regulated germination in *Tephrosia appolina* DC. and its significance. *Int. J. Ecol. Env. Sci.* **1**, 47–51.

Nazrul-Islam, A. K. M., and Hoque, A. E. (1990). Germination ecophysiology of *Cassia tora*. *In* "International Symposium on Environmental Influences on Seed and Germination Mechanism: Recent Advances in Research and Technology" (D. N. Sen, S. Mohammed, P. K. Kasera, and T. P. Thomas, eds.), p. 71. Univ. of Jodhpur, Jodhpur, India. [Abstract]

Nelson, S. O., Kehr, W. R., Stetson, L. E., and Wolf, W. W. (1977). Laboratory germination and sand emergence responses of alfalfa seed to radiofrequence electrical treatment. *Crop Sci.* **17**, 534–538.

Nelson, S. O., Stetson, L. E., Stone, R. B., Webb, J. C., Pettibone, C. A, Works, D. W., Kehr, W. R., and VanRiper, G. E. (1964). Comparison of infrared, radiofrequency, and gas-plasma treatments of alfalfa seed for hard-seed reduction. *Trans. Am. Soc. Agric. Eng.* **7**, 276–280.

Nelson, S. O., Stetson, L. E., and Works, D. W. (1968). Hard-seed reduction in alfalfa by infrared and radio-frequency electrical treatments. *Trans. Am. Soc. Agric. Eng.* **11**, 728–730.

Nelson, S. O., and Wolf, W. W. (1964). Reducing hard seed in alfalfa by radio-frequency electrical seed treatment. *Trans. Am. Soc. Agric. Eng.* **7**, 116–119.

Niedzwiedz-Siegien, and Lewak, S. (1992). Involvement of a high irradiance response in photoinhibition of white clover seed germination at low water potential. *Physiol. Plant.* **86**, 293–296.

Nobel, J. C. (1975). The effects of emus (*Dromaius novaehollandiae* Latham) on the distribution of the nitre bush (*Nitraria billardieri* DC.). *J. Ecol.* **63**, 979–984.

Ocumpaugh, W. R., Stuth, J. W., and Archer, S. R. (1993). Recovery and germination of switchgrass seed fed to cattle. *Proc. Int. Grassland Congr.* **17**, 318–319.

Ollerton, J., and Lack, A. (1996). Partial predispersal seed predation in *Lotus corniculatus* L. (Fabaceae). *Seed Sci. Res.* **6**, 65–69.

Orme, M. L., and Leege, T. A. (1976). Emergence and survival of redstem (*Ceanothus sanguineus*) following prescribed burning. *Proc. Tall Timbers Fire Ecol. Confr.* **14**, 391–420.

Owens, M. K., Wallace, R. B. and Archer, S. (1995). Seed dormancy and persistence of *Acacia berlandieri* and *Leucaena pulverulenta* in a semi-arid environment. *J. Arid. Environ.* **29**, 15–23.

Page, R. J., Goodwin, D. L., and West, N. E. (1966). Germination requirements of scarlet globemallow. *J. Range Manage.* **19,** 145–146.

Palma, B., Vogt, G., and Neville, P. (1995a). Environmental factors that influence germination of seeds of *A. senegal,* Willd. *Phyton (Buenos Aires)* **57,** 103–112.

Palma, B., Vogt, G., and Neville, P. (1995b). Endogenous factors that limit seed germination of *Acacia senegal* Willd. *Phyton (Buenos Aires)* **57,** 97–102.

Papavassilious, S., Arianoutsou, M., and Thanos, C. A. (1994). Aspects of the reproductive biology of fire following species of Leguminosae in a *Pinus halepensis* Mill. forest. *Second Int. Conf. Forest Fire Res. Coimbra Portugal. Contribution No. D. 24.*

Patil, V. N., and Andrews, C. H. (1985). Development and release of hardseeded dormancy in cotton (*Gossypium hirsutum*). *Seed Sci. Technol.* **13,** 691–698.

Pellew, R. A., and Southgate, B. J. (1984). The parasitism of *Acacia tortilis* seeds in the Serengeti. *Afr. J. Ecol.* **22,** 73–75.

Pettibone, C. A. (1965). Some effects of gas-plasma radiation of seeds. *Trans. Am. Soc. Agric. Eng.* **8,** 319–323.

Pieterse, P. J., and Cairns, A. L. P. (1986). The effect of fire on an *Acacia longifolia* seed bank in the south-western Cape. *S. Afr. J. Bot.* **52,** 233–236.

Porter. R. H. (1949). Recent developments in seed technology. *Bot. Rev.* **15,** 221–344.

Portlock, C. C., Shea, S. R., Majer, J. D., and Bell, D. T. (1990). Stimulation of germination of *Acacia pulchella*: Laboratory basis for forest management options. *J. Appl. Ecol.* **27,** 319–324.

Preece, P. B. (1971). Contributions to the biology of mulga. II. Germination. *Aust. J. Bot.* **19,** 39–49.

Pritchard, H. W., Manger, K. R., and Prendergast, F. G. (1988). Changes in *Trifolium arvense* seed quality following alternating temperature treatment using liquid nitrogen. *Ann. Bot.* **62,** 1–11.

Profuma, P., Gastaldo, P., and Parisi, V. (1978). Azione della fusicoccina sulla dormienza dei semi di *Cercis siliquastrum* L. *Accad. Nazion. dei Lincei* **63,** 135–140.

Pukittayacamee, P., and Hellum, A. K. (1988). Seed germination in *Acacia auriculiformis*: developmental aspects. *Can. J. Bot.* **66,** 388–393.

Purdie, R. W. (1977). Early stages of regeneration after burning in dry sclerophyll vegetation. II. Regeneration by seed germination. *Aust. J. Bot.* **25,** 35–46.

Purdie, R. W., and Slatyer, R. O. (1976). Vegetation succession after fire in sclerophyll woodland communities in south-eastern Australia. *Aust. J. Ecol.* **1,** 223–236.

Quick, C. R. (1935). Notes on the germination of *Ceanothus* seeds. *Madrono* **3,** 135–140.

Quick, C. R., and Quick, A. S. (1961). Germination of *Ceanothus* seeds. *Madrono* **16,** 23–30.

Quinlivan, B. J. (1961). The effect of constant and fluctuating temperatures on the permeability of the hard seeds of some legume species. *Aust. J. Agric. Res.* **12,** 1009–1022.

Quinlivan, B. J. (1965). The influence of the growing season and the following dry season on the hardseededness of subterranean clover in different environments. *Aust. J. Agric. Res.* **16,** 277–291.

Quinlivan, B. J. (1966). The relationship between temperature fluctuations and the softening of hard seeds of some legume species. *Aust. J. Agric. Res.* **17,** 625–631.

Quinlivan, B. J. (1968a). The softening of hard seeds of sand-plain lupin (*Lupinus varius* L.). *Aust. J. Agric. Res.* **19,** 507–515.

Quinlivan, B. J. (1968b). Seed coat impermeability in the common annual legume pasture species of western Australia. *Aust. J. Exp. Agric. Anim. Husb.* **8,** 695–701.

Quinlivan, B. J. (1970). The interpretation of germination tests on seeds of *Lupinus* species which develop impermeability. *Proc. Int. Seed Test. Assoc.* **35,** 349–359.

Quinlivan, B. J. (1971). Seed coat impermeability in legumes. *J. Aust. Inst. Agri. Sci.* **37,** 283–295.

Quinlivan, B. J., and Millington, A. J. (1962). The effect of a Mediterranean summer environment on the permeability of hard seeds of subterranean clover. *Aust. J. Agric. Res.* **13,** 377–387.

Quinlivan, B. J., and Nicol, H. I. (1971). Embryo dormancy in subterranean clover seeds. I. Environmental control. *Aust. J. Agric. Res.* **22,** 599–606.

Raleigh, G. J. 1930. Chemical conditions in maturation, dormancy, and germination of seeds of *Gymnocladus dioica. Bot. Gaz.* **89,** 273–294.

Rao, S. P., and Mukherjee, R. K. (1978). Dormancy studies in black gram (*Phaseolus mungo* L.). *Biol. Plant.* **20,** 81–85.

Rao, P. N., and Reddy, B. V. N. (1981). Autecological studies in *Indigofera linifolia* (L. f.) Retz. I. Germination behaviour of the seeds. *J. Indian Bot. Soc.* **60,** 51–57.

Rasmussen, G. A., and Wright, H. A. (1988). Germination requirements of flameleaf sumac. *J. Range Manage.* **41,** 48–52.

Rice, K. J. (1985). Responses of *Erodium* to varying microsites: The role of germination cueing. *Ecology* **66,** 1651–1657.

Rick, C. M., and Bowman, R. I. (1961). Galapagos tomatoes and tortoises. *Evolution* **15,** 407–417.

Rincker, C. M. (1954). Effect of heat on impermeable seeds of alfalfa, sweet clover, and red clover. *Agron. J.* **46,** 247–250.

Rivera, R., Popp, H. W., and Dow, R. B. (1937). The effect of high hydrostatic pressures upon seed germination. *Am. J. Bot.* **24,** 508–513.

Rizzini, C. T. (1977). Nota sobre um embriao dormente em leguminosa esclerodermica. *Rodriguesia* **29**(42), 33–39.

Roberts, H. A., and Boddrell, J. E. (1985a). Seed survival and seasonal pattern of seedling emergence in some Leguminosae. *Ann. Appl. Biol.* **106,** 125–132.

Roberts, H. A., and Boddrell, J. E (1985b). Seed survival and seasonal emergence in some species of *Geranium, Ranunculus* and *Rumex. Ann. Appl. Biol.* **107,** 231–238.

Robinson, W. A. (1986). Effect of fruit ingestion on *Amelanchier* seed germination. *Bull. Torrey Bot. Club* **113,** 131–134.

Rolston, M. P. (1978). Water impermeable seed dormancy. *Bot. Rev.* **44,** 365–396.

Roy, J., and Sonie, L. (1992). Germination and population dynamics of *Cistus* species in relation to fire. *J. Appl. Ecol.* **29,** 647–655.

Russi, L., Cocks, P. S., and Roberts, E. H. (1992). The fate of legume seeds eaten by sheep from a Mediterranean grassland. *J. Appl. Ecol.* **29,** 772–778.

Sabiiti, E. N., and Wein, R. W. (1988). Fire behaviour and the invasion of *Acacia sieberiana* into savanna grassland openings. *Afr. J. Ecol.* **26,** 301–313.

Sackett, S. S., and Haase, S. M. (1992). Measuring soil and tree temperatures during prescribed fires with thermocouple probes. *USDA For. Serv. Gen. Tech. Rep.* PSW-131.

Saha, P. K., and Takahashi, N. (1981). Seed dormancy and water uptake in *Crotolaria sericea* Retz. *Ann. Bot.* **47,** 423–425.

Schwegman, J. (1990). Preliminary results of a program to monitor plant species for management purposes. *In* "Ecosystem Management: Rare Species and Significant Habitats." (R. S. Mitchell, C. J. Sheviak, and D. J. Leopold, eds.), pp. 113–116. Proc. 15th Annual Natural Areas Conf., New York State Museum Bull. No. 471.

Scifres, C. J. (1974). Salient aspects of huisache seed germination. *Southw. Nat.* **18,** 383–392.

Sen, D. N., and Bhandari, M. C. (1974). On the ecology of a perennial cucurbit in Indian arid zone: *Citrullus colocynthis* (Linn.) Schrad. *Int. J. Biometeorol.* **18**, 113–120.

Sharma, K. P., Khan, T. I., and Bhardwaj, N. (1984). Temperature-regulated seed germination in *Neptunia oleracea* Lour. and its ecological significance. *Aq. Bot.* **20**, 185–188.

Sharma, S. S., and Sen, D. N. (1974). A new report on secondary dormancy in certain arid zone seeds. *Curr. Sci.* **43**, 386.

Sharma, S. S., and Sen, D. N. (1975). Effect of light on seed germination and seedling growth of *Merremia* species. *Folia Geobot. Phytotax.* **10**, 265–269.

Shaw, M. F. (1929). A microchemical study of the fruit coat of *Nelumbo lutea*. *Am. J. Bot.* **16**, 259–276 + 2 plates.

Shea, S. R., McCormick, J., and Portlock, C. C. (1979). The effect of fires on regeneration of leguminous species in the northern jarrah (*Eucalyptus marginata* Sm) forest of western Australia. *Aust. J. Ecol.* **4**, 195–205.

Shra, R. K. M., and Sen, D. N. (1984). A new report on "vivipary" in *Tephrosia purpurea* Linn. from Indian arid zone. *Geobios New Repts.* **3**, 94–95.

Sidhu, S. S., and Cavers, P. B. (1977). Maturity-dormancy relationships in attached and detached seeds of *Medicago lupulina* L. (Black Medick). *Bot. Gaz.* **138**, 174–182.

Silva, J. C. S., and Felippe, G. M. (1986). Germination of *Stylosanthes macrocephala*. *Revta. Brasil. Bot.* **9**, 263–268.

Small, J. G. C., McNaughton, J. E., and Greeff, J. H. (1977). Physiological studies on the germination of *Erythrina caffra* Thunb. seeds. *S. Afr. J. Bot.* **43**, 213–222.

Smith, C. A., Shaw, D. R., and Newson, L. J. (1992). Arrowleaf sida (*Sida rhombifolia*) and prickly sida (*Sida spinosa*): Germination and emergence. *Weed Res.* **32**, 103–109.

Steele, R., and Geier-Hayes, K. (1989). The Douglas-fir/ninebark habitat type in central Idaho: Succession and management. *USDA For. Serv. Gen. Tech. Rept.* INT-252.

Steinbauer, G. P., and Grigsby, B. (1959). Methods of obtaining field and laboratory germination of seeds of bindweeds, lady's thumb and velvet leaf. *Weeds* **7**, 41–46.

Stetson, L. E., and Nelson, S. O. (1972). Effectiveness of hot-air, 39-MHz dielectric, and 2450-MHz microwave heating for hard-seed reduction in alfalfa. *Trans. Am. Soc. Agric. Eng.* **15**, 530–535.

Stone, E. C., and Juhren, G. (1951). The effect of fire on the germination of the seed of *Rhus ovata* Wats. *Am. J. Bot.* **38**, 368–372.

Sumner, D. C. and Cobb, R. D. (1967). Germination characteristics of cheeseweed (*Malva parviflora* L.) seeds. *Agron. J.* **59**, 207–208.

Swan, D. G. (1980). Field bindweed, *Convolvulus arvensis* L. *Wash. St. Univ. Coll. Agric. Res. Center Bull.* 0888.

Swank, W. G. (1944). Germination of seeds after ingestion by ring-necked pheasants. *J. Wildl. Manage.* **8**, 223–231.

Sweeney, J. R. (1956). Responses of vegetation to fire. A study of the herbaceous vegetation following chaparral fires. *Univ. Calif. Publ. Bot.* **28**, 143–216 + plates 12–27.

Taylor, G. B. (1981). Effect of constant temperature treatments followed by fluctuating temperatures on the softening of hard seeds of *Trifolium subterraneum* L. *Aust. J. Plant Physiol.* **8**, 547–548.

Teketay, D. (1994). Germination ecology of two endemic multipurpose species of *Erythrina* from Ethiopia. *For. Ecol. Manage.* **65**, 81–87.

Teketay, D. (1996). Germination ecology of twelve indigenous and eight exotic multipurpose leguminous species from Ethiopia. *For. Ecol. Manage.* **80**, 209–223.

Temple, S. A. (1977). Plant-animal mutualism: Coevolution with dodo leads to near extinction of plant. *Science* **197**, 885–886.

Temple, S. A. (1979). The dodo and the tambalacoque tree. *Science* **203**, 1363–1364.

Thanos, C. A., Georghiou, K., Kadis, C., and Pantazi, C. (1992). Cistaceae: A plant family with hard seeds. *Israel J. Bot.* **41**, 251–263.

Thanos, C. A., and Georghiou, K. (1988). Ecophysiology of fire-stimulated seed germination in *Cistus incanus* ssp. *creticus* (L.) Heywood and *C. salvifolius*. *Plant Cell Environ.* **11**, 841–849.

Timmons, F. L. (1942). The dissemination of prickley pear seed by jack rabbits. *J. Am. Soc. Agron.* **34**, 513–520.

Townsend, C. E., and McGinnies, W. J. (1972). Mechanical scarification of cicer milkvetch (*Astragalus cicer* L.) seed. *Crop. Sci.* **12**, 392–394.

Trabaud, L., and Oustric, J. (1989). Heat requirements for seed germination of three *Cistus* species in the garrigue of southern France. *Flora* **183**, 321–325.

Tran, V. N. (1979). Effects of microwave energy on the strophiole, seed coat and germination of *Acacia* seeds. *Aust. J. Plant Physiol.* **6**, 277–287.

Tran, V. N., and Cavanagh, A. K. (1980). Taxonomic implications of fracture load and deformation histograms and the effects of treatments on the impermeable seed coat of *Acacia* species. *Aust. J. Bot.* **28**, 39–51.

Tran, V. N., and Cavanagh, A. K. (1984). Structural aspects of dormancy. *In* "Seed Physiology" Vol. 2, (D. R. Murray, ed.), pp. 1–44. Academic Press, Australia.

Ueki, C., and Suetsugu, I. (1958). The identification of hard seeds in some leguminous forage crops. I. Genge (*Astragalus sinicus* L.). *Proc. Int. Seed Test. Assoc.* **23**, 69–72.

Valbuena, L., Tarrega, R., and Luis, E. (1992). Influence of heat on seed germination of *Cistus laurifolius* and *Cistus ladanifer*. *Int. J. Wildland Fire* **2**, 15–20.

Vazquez-Yanes, C. (1974). Studies on the germination of seeds of *Ochroma lagopus* Swartz. *Turrialba* **24**, 176–179.

Vazquez-Yanes, C. (1981). Germinacion de dos especies de Tiliaceas arboreas de la vegetacion secundaria tropical: *Belotia campbellii* y *Heliocarpus donnell-smithii*. *Turrialba* **31**, 81–83.

Vazquez-Yanes, C., and Orozco-Segovia, A. (1982). Seed germination of a tropical rain forest pioneer tree (*Heliocarpus donnell-smithii*) in response to diurnal fluctuation of temperature. *Physiol Plant.* **56**, 295–298.

Vieira, H. D., and Barros, R. S. (1994). Responses of seed of *Stylosanthes humilis* to germination regulators. *Physiol. Plant.* **92**, 17–20.

Vuillemin, J., and Bulard, C. (1981). Ecophysiologie de la germination de *Cistus albidus* L. et *Cistus monspeliensis* L. *Natur. Monspeliensia* **46**, 1–11.

Wagner, L. K., and Spira, T. P. (1994). Germination, recruitment and survival in the weedy annual *Medicago polymorpha* in successive wet and dry years. *Am. Midl. Nat.* **131**, 98–108.

Wallander, R. T., Olson, B. E., and Lacey, J. R. (1995). Spotted knapweed seed viability after passing through sheep and mule deer. *J. Range Manage.* **48**, 145–149.

Warcup, J. H. (1980). Effect of heat treatment of forest soil on germination of buried seed. *Aust. J. Bot.* **28**, 567–571.

Washitani, I. (1988). Effects of high temperatures on the permeability and germinability of the hard seeds of *Rhus javanica* L. *Ann. Bot.* **62**, 13–16.

Washitani, I., and Takenaka, A. (1986). "Safe sites" for the seed germination of *Rhus javanica*: A characterization by responses to temperature and light. *Ecol. Res.* **1**, 71–82.

Weinberger, P., and Burton, C. (1981). The effect of sonication on the growth of some tree seeds. *Can. J. For. Res.* **11**, 840–844.

Wiesner, L. E., Laufmann, J. E., Stanwood, P. C, and Wheeler, L. J. (1994). The effect of liquid nitrogen on alfalfa seed viability, emergence, and broken cotyledons. *J. Seed Tech.* **18**, 1–6.

Williams, W. A., and Elliott, J. R. (1960). Ecological significance of seed coat impermeability to moisture in crimson, subterranean and rose clovers in a Mediterranean-type climate. *Ecology* **41,** 731–742.

Wilson, J. K. (1944). Immersing seeds of species of *Robinia* in boiling water hastens germination. *J. For.* **42,** 453–454.

Witmer, M. C. (1991). The dodo and the tambalacoque tree: An obligate mutualism reconsidered. *Oikos* **61,** 133–137.

Works, D. W. (1964). Infrared irradiation for water-impermeable seeds. *Trans. Am. Soc. Agric. Eng.* **7,** 235–237.

Wrangham, R. W., Chapman, C. A., and Chapman, L. J. (1994). Seed dispersal by forest chimpanzees in Uganda. *J. Trop. Ecol.* **10,** 355–368.

Wright, E. (1931). The effect of high temperatures on seed germination. *J. For.* **29,** 679–687.

Wycherley, P. R. (1960). Seed germination of some tropical legumes. *J. Rubber Res. Inst. Malaya* **16,** 99–117.

Young, J. A., Evans, R. A., and Kay, B. L. (1975). Dispersal and germination dynamics of broadleaf filaree, *Erodium botrys* (Cav.) Bertol. *Agron. J.* **67,** 54–57.

Zabkiewicz, J. A., and Gaskin, R. E. (1978). Effect of fire on gorse seeds. *Proc. New Zeal. Weed Pest Control Conf.* **31,** 47–52.

CHAPTER

7

Germination Ecology of Seeds in the Persistent Seed Bank

I. PURPOSE

A reserve of viable, ungerminated seeds in a habitat is called a seed bank, and according to Thompson and Grime (1979) there are two general types: transient and persistent. These authors defined a transient seed bank as one in which none of the seeds produced in a given year remain viable in the habitat for more than 1 year, whereas seeds in a persistent seed bank are 1 or more years old. However, Walck *et al.* (1996) have suggested that these two types of seed banks should be described in terms of germination seasons rather than age per se. Thus, a transient seed bank is composed of seeds that do not live until the second germination season following maturation. Seeds of some species are in a transient seed bank for several months, during which time dormancy loss occurs and/or environmental conditions in the habitat are unfavorable for germination. Seeds of other species may be nondormant at maturity; consequently, they are in the transient seed bank for only a few days or weeks, depending on the suitability of environmental conditions for germination. A persistent seed bank is composed of seeds that live until the second (or some subsequent) germination season. Biologists usually are more interested in persistent than in transient seed banks; thus, persistent seed banks are the main focus of this chapter.

If a species forms a persistent seed bank, its germination ecology cannot be fully understood until the dynamics of the seeds in the seed bank are elucidated. This chapter covers aerial and buried persistent seed banks, longevity of seeds in the soil, what can happen to seeds while they are in the soil, and effects of seed banks on populations of species and persistence of communities.

II. AERIAL SEED BANKS

According to Lamont (1991), the term "serotinous" means that part, or all, of the seed crop is retained on the mother plant long after seed maturity, whereas "nonserotinous" means that seeds are dispersed at maturity. Further, he suggested that the term "*delayed desiscence*" (not dehiscence) be used to describe serotinous species in which seeds are dispersed in the absence of specific environmental cues. Lamont also proposed terms to describe the opening of seed-holding structures (e.g., cones, capsules) in response to various environmental factors: (1) *necriscense,* death of the structure—usually the stem producing the structure dies cutting off the supply of water (e.g., Ewart, 1907; Lev-Yadun, 1995); (2) *hygriescence,* dead structure is soaked by rain; (3) *soliscence,* dead structure is heated by the sun; (4) *pyriscence,* dead structure is heated by fire; and (5) *pyrohygriscence,* dead structure is heated by fire and then subjected to wet–dry cycles.

A. Development of Serotiny

Lamont *et al.* (1991) formulated various hypotheses to explain why serotiny developed in habitats with a high fire frequency. Serotiny (1) maximizes seed availability when conditions are favorable for germination and seedling establishment, i.e., after a fire; (2) ensures that seeds are available after a fire, even if none was produced the previous year; (3) results in high seedling/juvenile densities, which could reduce the growth of competitors and thus increase chances of survival; (4) indirectly causes seeds to fall on the optimum substrate (seed bed) for germination and establishment; (5) synchronizes seed dispersal, therefore, seed predators are satiated;

(6) promotes seed dispersal at a time when conditions are optimal for wind dispersal; (7) minimizes the time between seed dispersal and when conditions are favorable for germination, thus increasing the chances of seed germination before seeds lose viability; (8) increases the possibility of some postfire dispersed seeds reaching depressions where rain water accumulates; and (9) protects seeds from being killed by fire. Further, as predation of seedlings may be higher in nonburned than in burned areas (Bradstock, 1991), serotiny would help ensure that germination did not occur until after a fire, when predation levels were decreased.

Species with serotiny may grow in regions with low rainfall. For example, in the Namib Desert, serotinous species are more common in areas with 51–100 mm of precipitation per year than in those with more than 100 mm per year. At less than 50 mm of precipitation per year, however, the level of serotiny decreases, compared to areas with 51–100 mm (Gunster, 1992). Aerial seed banks of 1121 and 305 seeds m^{-2} of ground surface and an annual rainfall of 351 and 585 mm, respectively, occurred in two study sites in the kwongan (shrub-heath) vegetation of Western Australia (Bellairs and Bell, 1990), indicating that decreased rainfall was a factor in determining the proportion of species with serotiny. However, fire also is a factor in habitats of woody species with serotiny. In a climatic gradient extending from Perth 500 km north to Northampton, Western Australia, serotiny in three *Banksia* spp. increased with decreased rainfall and increased temperature; fire was very important in the most xeric community, where maximum serotiny occurred (Cowling and Lamont, 1985a).

Serotiny varies between populations of the same species, including *Pinus banksiana* (Gauthier *et al.*, 1996), *P. contorta* (Muir and Lotan, 1985), and *P. rigida* (Givnish, 1981), and it increases in sites frequently disturbed by fire. In *P. contorta*, serotiny is controlled genetically by one locus and two alleles (Perry and Lotan, 1979).

B. How Long Is Seed Dispersal Delayed?

Lamont (1991) divided serotinous species into two categories: (1) "seeds released later" and (2) "seeds retained indefinitely." Seeds retained on the mother plant for relatively short periods are not dispersed in response to specific environmental cues such as fire or rainfall. How long do seeds have to remain on the mother plant before the species is considered to have short-term serotiny? In temperate eastern North America, retention of dry, mature seeds (or fruits) on dead flowering shoots (or plants) for 3 months or more can change the germination phenology (e.g., Baskin and Baskin, 1984a, 1985b) or dormancy state (e.g., Baskin and Baskin, 1975a, 1977b, 1986, 1990). A 1-month delay

in dispersal of *Campanula americana* seeds shifted the germination season from autumn to spring. Further, plants from autumn-germinating seeds behaved as winter annuals, and those from spring-germinating seeds were biennials (Baskin and Baskin, 1984c).

If a 3-month delay in dispersal is used as the minimum period of time to define short-term serotiny, then various species with this type of serotiny can be identified: *Allium artemisietorum,* 6 month (Kamenetsky and Gutterman, 1993); *Celtis occidentalis,* 4.4 (Stapanian, 1982); *Conium maculatum,* 5 (Baskin and Baskin, 1990); *Daucus carota,* 6 (Lacey, 1982); *Frasera caroliniensis,* 7 (Baskin and Baskin, 1986b); *Geum canadense,* 9 (Baskin and Baskin, 1985b); *Juniperus virginiana,* 5.4 (Stapanian, 1982); *Osmorhiza longistylis,* 18 (Baskin and Baskin, 1984a); *Rhus glabra,* 6–12 (Li, unpublished results); *Sedum pulchellum,* 4 (Baskin and Baskin, 1977a); *Symphoricarpus orbiculatus,* 7 (Hidayati, unpublished results); and *Torilis japonica,* 4 (Baskin and Baskin, 1975a).

In those species with long-term serotiny whose seed-holding structures open in response to water, seeds apparently remain undispersed and viable for a number of years (Gutterman, 1972; Gutterman and Ginott, 1994). The structures open each time it rains, and some (but not all) of the seeds are released. Seeds of *Blepharis persica* remained viable on dead mother plants in the Negev Desert in Israel for 10 or more years (Gutterman, 1972). However, for most species with hygriescence the maximum period of seed viability for undispersed seeds is not known. One reason for this lack of information is that there is no easy way to quickly age the dead plants, which are mostly desert annuals.

However, it is relatively easy to age branches of trees and shrubs with unopened seed-holding structures. Viable seeds remain in the canopy for various periods of time: 3 years in *Pinus halepensis* (Daskalakou and Thanos, 1994); 4–6 in *Protea repens, Leucadendron conicum* (Bond, 1985), *Banksia menziesii, B. prionotes* (Cowling *et al.*, 1987), *Pinus clausa* (Cooper, 1951), and *P. radiata* (Warren and Fordham, 1978); 8 in *Widdringtonia cupressoides* (Midgley *et al.*, 1995); 9–12 in *Banksia burdettii* (Lamont and Barker, 1988), *B. hookeriana* (Enright *et al.*, 1996), and *Pinus pungens* (McIntyre, 1929); 9–15 in *Banksia ericifolia* and *B. pulchella* (Bradstock and O'Connell, 1988); 10–20 in *Allocasuarina distyla* (Pannell and Myerscough, 1993), *Banksia leptophylla, B. attenuata* (Cowling *et al.*, 1987), *Callistemon rigida* (Ewart, 1907), *Cupressus sempervirens* (Lev-Yadun, 1995), *Pinus banksiana* (Cayford and MacRae, 1983), and *Hakea* spp. (Bond, 1985); and 10–30 in *Pinus contorta* var. *murrayana* (Blumer, 1910).

C. Seed-Holding Structures

Specialized seed-holding structures have not evolved in species with short-term serotiny described earlier, but they are well developed in species with long-term serotiny. Further, in long-term serotiny the seed-holding structures that open in response to water (rainfall) differ from those that open in response to fire.

Many annuals in hot semideserts and deserts, belonging to families such as the Aizoaceae, Asteraceae, Acanthaceae, Brassicaceae, Fabaceae, Lamiaceae, and Plantaginaceae, have seeds stored in the canopy of the dead plant, and they are dispersed in response to rainfall (Table 9.20). The seed-holding structures include capsules (Gutterman *et al.*, 1967) and inflorescences (heads) surrounded by bracts (Gutterman and Ginott, 1994). Sometimes, the dried skeleton of the mother plant is folded around the seed-holding structures (Ellner and Shmida, 1981; Gunster, 1992). In general, the water-sensitive structures surrounding the seeds are thin and have much surface area.

In contrast, shrubs and trees in various families, including the Casuarinaceae, Cupressaceae, Ericaceae (Lamont *et al.*, 1991), Myrtaceae (Wellington and Nobel, 1985), Pinaceae (Coker, 1909; Warren and Fordham, 1978), and Proteaceae (Bond, 1985; Cowling *et al.*, 1987; Richardson *et al.*, 1987), have seeds stored in the canopy of living plants. The seed-holding structures (cones, capsules) in these families are thick, massive, and woody, and they open in response to fire. In addition, Coutinho (1977) reported that fire promotes fruit opening in *Anemopaegma arvensis* (Bignoniaceae), *Jacaranda decurrens* (Bignoniaceae), and *Nautonia nummularia* (Asclepidaceae) and infrutescence opening in *Gomphrena macrocephala* (Amaranthaceae); all of these species have massive seed-holding structures.

D. Size of Seed Bank

The number of seeds held in the canopy of woody species may vary with plant age (Bradstock, 1990; Enright *et al.*, 1996), disturbance by animals (Linhart, 1978; Lamont and van Leeuwen, 1988; Witkowski *et al.*, 1991; Enright *et al.*, 1996), predispersal seed predation (Coetzee and Giliomee, 1987a,b; Cowling and Lamont, 1987b; Lamont and van Leeuwen, 1988; Lamont and Barker, 1988; Zammitt and Westoby, 1988; Andersen, 1989), and rate of natural opening of seed-holding structures (Wellington, 1989; Enright *et al.*, 1996). For most species, the number of seeds held in the canopy gradually increases as plants reach reproductive maturity, peaks, and then declines as plants reach old age. For example, the aerial seed bank of *Banksia speciosa* and *B. baxteri* peaked when plants were 16 years old (Witkowski *et al.*, 1991). The size of the aerial seed bank has been estimated to be 830 seeds per plant, *B. burdettii* (Lamont and Barker, 1988); 17,105 seeds per plant, *B. cuneata* (Lamont *et al.*, 1991); 1098 seeds m^{-2} projected

canopy cover, *B. ericifolia* (Honig *et al.*, 1992); 700 seeds per plant, *B. hookeriana* (Enright *et al.*, 1996); 84 seeds per plant, *B. triscupis* (Lamont and van Leeuwen, 1988); 40–50 million seeds per plant, *Callistemon rigida* (Ewart, 1907); and 525 seeds m^{-2} projected canopy cover, *Leucadendron laureolum* (Honig *et al.*, 1992).

Species with an aerial seed bank have few, if any, viable seeds in the soil (Richardson *et al.*, 1987; Wellington, 1989; Edwards and Whelan, 1995). Seeds present in the soil may be short-lived (Daskalakou and Thanos, 1994), rapidly losing viability after they are dispersed (Cowling and Lamont, 1987b; Enright and Lamont, 1989b). However, if dispersed seeds of *Banksia* spp. become covered by soil or litter, death rates are reduced (Enright and Lamont, 1989b). Another reason for low numbers of seeds on/in the soil is the high predation levels that occur following dispersal (Wellington and Nobel, 1985; Wellington, 1989). Seeds of *Protea neriifolia* placed in trays in the field and protected from rodents germinated to 83–88% after a 1- to 4-month exposure to spring–summer habitat conditions in South Africa (Le Maitre and Botha, 1991).

E. Opening of Woody Structures

Although fire is important in promoting the opening of woody structures in many species, some of them open spontaneously over a period of years before fire occurs (Bond, 1985; Cowling and Lamont, 1987; Cowling *et al.*, 1987; Lamont and van Leeuwen, 1988; Bradstock, 1990; Daskalakou and Thanos, 1994; Enright *et al.*, 1996). In other species, however, little opening occurs until structures are exposed to the heat of fires (Cooper, 1951; Cayford and McRae, 1983; Richardson *et al.*, 1987; Wellington, 1989). Opening may occur because heat (1) kills the tissues, resulting in desiccation and subsequent pulling apart of the structure (Richardson *et al.*, 1987) or (2) destroys or melts the resinous materials that prevented parts of the structure, especially cones, from separating (Cooper, 1951; Beaufait, 1960; Ahlgren, 1974; Hellum and Pelchat, 1979).

If drying of the fruit was a prerequisite for dispersal, seed release may not occur for several days or weeks following a fire. For example, follicle drying preceded seed release in *Banksia ericifolia*, and about 50, 65, and 70% of the seeds were dispersed 33, 52, and 95 days, respectively, after a fire (Bradstock and Myerscough, 1981). Obviously, if fire promoted the opening of fruits or cones that were already dry, seed release would occur immediately. When serotinous cones of *Pinus* were heated, they opened within seconds (Beaufait, 1960), and seed fall in a *P. clausa* forest began within 9 hr after a fire and was completed within 3 weeks (Cooper, 1951). In a South African fynbos community, 63% of the *Leuca-*

dendron laureoleum and 6% of the *Banksia ericifolium* seeds were released within 1 week after follicles were subjected to fire; those of *B. ericifolium* still were releasing seeds after 12 weeks (Honig *et al.*, 1992).

The intensity of fire also influences the rate of seed dispersal. Most *Banksia cuneata* seeds were released within 24 hr following a hot fire (pyriscence), but after a mild fire seed release did not occur until follicles received three wet–dry cycles (pyrohygriscence) over a 10-day period (Lamont *et al.*, 1991). Wet–dry cycles also promoted seed release from heat-treated follicles of *B. attenuata, B. leptophylla, B. prionotes,* and *B. menziesii.* In *B. leptophylla,* the wet–dry cycles were more effective when conducted at 15°C than at 25 or 30°C (Cowling and Lamont, 1985b).

Information is available on temperature requirements for the opening of woody seed-holding structures of only a few species: *Pinus banksiana,* 49–60°C (Beaufait, 1960); *Banksia tricuspia,* 175°C; *B. prionotes,* 330°C; *B. leptophylla,* 340°C; *B. elegans,* 350°C; *B. menziesii,* 375°C; *B. grossa* and *B. candolleana,* 400°C; and *B. attenuata, B. hookeriana,* and *B. micrantha,* 500°C (Enright and Lamont, 1989b). The temperature requirement for the opening of *Banksia ornata* follicles increased with an increase in moisture content (Gill, 1976).

The heat generated when the dead floral parts remaining around the infructescences ("cones") of *Banksia* spp. are burned was sufficient to stimulate opening of the follicles (Lamont and Cowling, 1984). Temperatures inside infructescences of *B. hookeriana* and *B. leptophylla* reached a maximum of 288 and 187°C, respectively, if they were burned with dead floral parts intact, but there was little increase in temperatures if infructescences were burned after floral parts were removed. Lamont and Cowling (1984) found that most of the nonserotinous *Banksia* spp. did not retain the dead floral remnants around the infructescences.

F. Lethal Temperatures

The beneficial effect of fire on the opening of woody structures has to be viewed from the perspective that fire can, and does, kill seeds of species with serotiny (Purdie, 1977; Lamont and Barker, 1988; Bradstock *et al.*, 1994). With an increase in temperature, the period of time before seed death occurs decreases. Thus, lethal temperatures for seeds depend on the species and duration of exposure: *Eucalyptus obliqua,* 290°C for 1.5 min (Ashton, 1986); *E. regnans,* 350°C for 15 sec (Judd, 1993); *Kunzia ambigua,* 450°C for 15 sec (Judd, 1993); *Leptospermum juniperinum,* 380°C for 4 min (Ashton, 1986); *L. laevigatum,* 600°C for 15 sec (Judd, 1993), *L. myrsinoides,* 500°C for 15 sec (Judd, 1993); and *Pinus*

contorta, 45–50°C for 15 min (Knapp and Anderson, 1980).

Maximum seed temperatures (and their duration) during fires depend on the thickness of fruit walls (Bradstock *et al.*, 1994), capsule size (Bradstock *et al.*, 1994; Judd, 1994), capsule water content (Judd, 1994), and height in the plant canopy (Bradstock and Myerscough, 1981; Bradstock *et al.*, 1994). In a model looking at the relationship between temperature and its duration on death of seeds in woody fruits, Mercer *et al.* (1994) showed that fruit radius was important in predicting lethal temperatures. Further, the position of seeds within the fruit was not critical for their survival, as long as they were in the central core of the fruit.

G. Seed Dormancy and Germination

Seeds in long-term aerial seed banks usually are nondormant and germinate readily when dispersed; this is a major reason why soil seed banks are not well developed. Germination immediately following seed dispersal has been observed in various species, including *Banksia ericifolia* (Honig *et al.*, 1992), *Blepharis persica* (Gutterman, 1972), *Hakea* spp. (Richardson *et al.*, 1987), *Leucadendron coniferum* (Mustart and Cowling, 1991), *L. laureolum* (Honig *et al.*, 1992), *L. meridianum* (Mustart and Cowling, 1991), *Protea obtusifolia,* and *P. susannae* (Mustart and Cowling, 1991). However, most seeds of *Grevillea barklyana* were either dormant or an incubation temperature of 21°C was inhibitory for germination (Edwards and Whelan, 1995). Germination rates (but not percentages) of *Protea neriifolia* seeds decreased with an increase in time they were held on the mother plant (Le Maitre, 1990).

Seeds of *Blepharis persica* from several populations germinated to 100% at temperatures from 10 to 40°C (Gutterman, 1972), whereas those of *Protea obtusifolia, P. susannae, Leucadendron coniferum* and *L. meridianum* germinated to 90% or more at 20/10 and 10/10°C but to only 1–44% at 30/15°C (Mustart and Cowling, 1991). Optimum daily maximum and minimum temperatures for the germination of *Leucospermum cordifolium* seeds were 24 and 9°C, respectively, and for *Serruria florida* seeds were 20 and 7°C, respectively (Brits, 1986). Mean maximum temperatures of the five coolest months at a soil depth of 5 mm in South African fynbos communities that were burned, unburned lightly shaded, and unburned heavily shaded were 24.4, 17.9, and 15.8°C, respectively, and mean minimum temperatures were 6.3, 8.8, and 9.1°C, respectively (Brits, 1986). Thus, temperatures under unburned lightly and heavily shaded soil were not suitable for the germination of *L. cordifolium* and *S. florida* seeds. Temperatures measured at soil depths of 0, 20, and 40 mm during winter were close to those required for the germination of *L. cordifolium* seeds in burned (but not in unburned) areas (Brits, 1987). Seeds of *Banksia* spp. also have a low temperature requirement for germination (Sonia and Heslehurst, 1978; Cowling and Lamont, 1987a,b), which helps explain why they do not germinate until winter, even if they are dispersed in spring or autumn (Enright and Lamont, 1989a).

The timing of rainfall is important in controlling the timing of germination after seeds have been dispersed (Bradstock and Bedward, 1992). The occurrence of rainfall during the low-temperature season of the year greatly enhances favorable soil moisture conditions for germination because rates of evaporation are decreased (Le Maitre, 1990). A long period of favorable moisture conditions is critical for the establishment of some species. For example, seeds of *Protea neriifolia* require 3 weeks of continuously moist soil for germination, but another week of moist soil is required for seedling growth, if seedling establishment is to be successful (Le Maitre, 1990). More seeds of *Protea susannae, P. obtusifolia, Leucadendron coniferum,* and *L. meridianum* germinated on limestone-derived soil than on colluvial sands (Mustart and Cowling, 1993). Imbibed seeds of *Leucadendron* spp. dried faster than those of *Protea* spp., and after 6 hr of drying, seeds of *Protea* spp. were significantly heavier than those of *Leucadendron* spp. Seeds of *P. susannae* and *L. coniferum* germinated to 98% following two short wet (4 days imbibed)/dry (10 days dry) cycles, but they germinated to 96 and 10%, respectively, following two long wet (11 days imbibed)/dry (10 days dry) cycles (Mustart and Cowling, 1993).

Little is known about the light:dark requirements for the germination of seeds in aerial seed banks. Seeds of *Blepharis persica* germinated to 100% in both light and darkness (Gutterman, 1972), but darkness (burial) reduced germination rates and percentages of *Protea neriifolia* seeds (Le Maitre, 1990). Seeds of *Leucospermum cordifolium* germinated after being buried by ants at soil depths of 1–100 mm (Brits, 1987), indicating that they can germinate in darkness.

III. SOIL SEED BANKS

A. Formation

According to Grime (1979) most of the soil seed bank consists of buried seeds; however, some seeds are on the soil surface (Roberts, 1981) or in the litter, duff, or humus (Komarova, 1985). Seeds can fall into cracks in the soil, be covered by sediment during flooding, or have particles blown over them by the wind. Particle size is important in trapping seeds, especially those

moved along the ground surface by wind. In field studies, the number of trapped seeds (or fruits) of all species increased with increased soil particle size up to a species-specific threshold. More small seeds were trapped by small than by large soil particles, and more large seeds were trapped by large than by small particles (Chambers *et al.,* 1991).

1. Animals

Many vertebrates, including badgers, birds, coatis, gophers, groundhogs, porcupines, rabbits, rodents, snakes, and voles, bury seeds either intentionally or accidently when they dig in the soil (Garwood, 1989; Gutterman, 1993). Sometimes, the activities of several kinds of animals may bury seeds. For example, Bonis and Lepart (1994) found that when livestock, wild pigs, and birds were excluded from a wetland in France, seeds of *Callitriche truncata* were concentrated (98%) in the upper 0–1 cm of sediment, whereas in the presence of these animals only 47% of the seeds were in the 0- to 1-cm layer; 34% were in the 1- to 2-cm layer.

Invertebrates, including ants, beetles, and worms, also bury seeds (Garwood, 1989). Seedling excavation studies reveal that ants bury seeds at various depths, e.g., 9 cm (Bond and Stock, 1989), 4–7 cm (Bond and Slingsby, 1983), 4.3 cm (Bond and Slingsby, 1984), and 3–4.5 cm (Brits, 1987). The kinds of seeds collected (and buried) depend on the species of ant and its food preferences and on seed availability (Holldobler and Wilson, 1990). After 6 months in field plots (sandy soil) with digging beetles, seeds of *Juncus bufonius* and *Rorippa palustris* were moved from the surface to depths of 5–10 cm, and those of *J. articulatus* were moved from the surface to depths of 2–5 cm (Bernhardt, 1994). Tubificid worms "churn up" sediments, and thus seeds of aquatic plants such as *Typha* spp. could be buried. After 30 days of tubificid worm activity, 100% of the surface of experimental soil cores was covered by ≥3 mm of sediment (Grace, 1984).

Some earthworms, including the common *Lumbricus terrestris,* come to the soil surface to feed. Here, they forage within a radius of 10–15 cm around the burrow entrance, consuming dead leaves, seeds, dung, and other plant materials and/or pulling food materials into the burrow (Gerard, 1963). Earthworms are selective as to what kinds of seeds they will eat (Willems and Huijsmans, 1994). When seeds of eight species of grasses were sown on the soil surface, *L. terrestris* took more seeds of *Poa pratensis* and *P. trivialis* than of the other six species, whereas the earthworm *Allolobophora longa* took more seeds of *P. pratensis* and *Agrostis tenuis* than of the other six species (Grant, 1983). In addition to ingesting food, earthworms eat large quantities of soil, and eventually waste products and soil are deposited outside the body in masses of material called casts. Depending on the species, casts are deposited on the soil surface or in the burrows (Lee, 1985), and many viable seeds can be found in them (McRill and Sagor, 1973; Hurka and Haase, 1982; Grant, 1983). As earthworms densities may be 74 to about 2000 m^{-2} (Lee, 1985), they are important in the burial of large numbers of seeds. Also, they can deposit seeds well below the soil surface, e.g., earthworms carried seeds of *Capsella bursa-pastoris* from the soil surface to depths of 17–18 cm (Hurka and Haase, 1982).

2. Plowing

In arable lands, plowing is a major way in which seeds become covered with soil. However, plowing does not spread seeds evenly throughout a field. Studies with colored glass beads in arable soils revealed that "seeds" on the soil surface or at a depth of 2 cm had a horizontal movement of 46–71 cm and that they were dispersed throughout the top 5 cm of soil during tractor rototilling. Seeds buried at a depth of 6 cm moved 14–17 cm horizontally, and 89–96% of them were recovered from the 5- to 10-cm layer of soil (Wilson *et al.*, 1989). A moldboard plowing system caused a more uniform distribution of seeds to a depth of 15 cm than did chisel or ridge-till plowing or no plowing (Clements *et al.*, 1996). Weed seeds remain relatively close to the soil surface following chisle (vs moldboard) plowing (Ball, 1992), e.g., *Setaria faberi* seeds were concentrated in the top 0–2.5 cm after soil was chisle plowed (Schreiber, 1992). A no-tillage system in Wisconsin resulted in 60% of the weed seeds being in the top 1 cm of soil, and densities decreased logarithmically to a depth of 19 cm (Yenish *et al.*, 1992). In the same study, chisel plowing resulted in 30% of the seeds being in the top 1 cm of soil, and densities decreased linearly with depth. Moldboard plowing caused an even distribution of seeds throughout the top 19 cm of soil.

Using a matrix model to predict the vertical movement of seeds in arable soils, Cousens and Moss (1990) concluded that patterns of vertical distribution of seeds in soil vary with cultivation methods; however, after a number of years of moldboard or rigid tine cultivation, a stable, relatively even vertical distribution is reached. Stability of vertical distribution was reached sooner with moldboard than with tine cultivation.

B. Methodology of Studies

1. Kinds and Numbers of Seeds

In documenting the presence of a persistent seed bank, samples of soil are removed from the habitat.

Then, the number of seeds of each species in the samples is determined by (1) separating the seeds from the soil and counting them or (2) allowing them to germinate and counting the seedlings. Various methods of washing (Thorsen and Crabtree, 1977; Fay and Olson, 1978), air blowing (Standifer, 1980), flotation (Malone, 1967), or combinations thereof (Roberts, 1981) have been developed to rapidly separate seeds from soil samples. Even a photometric method has been devised (Kondrat'ev *et al.*, 1986). Buhler and Maxwell (1993) substituted K$_2$CO$_3$ in a modification of Malone's (1967) MgSO$_4$ floatation method and then counted the seeds with an automated image capture system.

After seeds are isolated, their identification and viability must be determined. Also, one is faced with the problem of whether the buried seeds would have remained viable until the next germination season. As noted earlier, seeds that germinate or die before their second germination season after dispersal are not a part of the persistent seed bank.

Allowing seeds to germinate in the soil eliminates some of the problems of trying to decide if they are viable and if they would have lived until the time of the natural germination season in the habitat. For example, Forcella (1992) concluded that many weed seeds dispersed in autumn lose viability in the soil during winter. Gross (1990) compared the number of species and the number of seeds of each species by (1) placing soil samples directly in a heated greenhouse, (2) giving samples a cold stratification treatment and then placing them in a greenhouse, and (3) washing samples through sieves to isolate the seeds. More species were identified in samples given cold stratification than in those not cold stratified or washed through sieves. Seed densities were higher if samples were washed through sieves than if seeds were allowed to germinate in the greenhouse (with or without a cold stratification treatment). However, when the number of dead seeds was subtracted there were no differences in the methods of estimating seed densities. Mean numbers of *Ambrosia artemisiifolia* and of *Setaria faberi* seeds separated from soil samples by flotation or determined by the germination method were not significantly different (Rothrock *et al.*, 1993).

Ter Heerdt *et al.* (1996) recommended washing samples through sieves to concentrate the seeds (similar to the technique used by Brenchley and Warington, 1930) and then allowing them to germinate in the greenhouse. These authors obtained 81–100% germination for viable seeds present in the samples. However, it should be noted that their samples were collected in the Netherlands in "early spring" and thus probably contained seeds in the transient seed bank as well as those in the persistent seed bank.

Concentrating the seeds and placing them on the soil surface would make it possible to expose all of them to the same light environment. Dalling *et al.* (1994) found that seed densities were underestimated when soil layers were ≥10 mm; they recommended that soil be ≤5 mm deep. In some studies, soil samples have been stirred periodically to increase the possibility that seeds would be exposed to light (e.g., Brenchley and Warington, 1930; Roberts, 1964).

In addition to automated image capture systems, computers will, no doubt, become increasingly useful in seed bank studies. Benoit *et al.* (1992) demonstrated that computer contour mapping permits one to visualize spatial patterns of change in populations of weed seeds during the growing season. Also, canonical discriminant analysis can be used to consider differences in species abundance, patchiness, and frequency between habitats (Benoit *et al.*, 1992).

In some studies, soil samples were kept in heated greenhouses or growth chambers (Koniak and Everett, 1982; Stieperaere and Timmerman, 1983; McGraw, 1987; Scheiner, 1988; Welling *et al.*, 1988a; Grilz and Romo, 1995) and thus were not subjected to seasonal temperature changes. It is especially important that soil samples from habitats in the temperate zone be given cold stratification treatments, but it also is important that they receive simulated summer conditions. In a heated greenhouse, samples would receive simulated summer, but not winter temperature regimes. Thus, only seeds that are nondormant at the time of sampling and/ or those that come out of dormancy at high temperatures will germinate in the samples. Seeds that require cold stratification to come out of dormancy will not germinate; consequently, the species will go undetected.

Probably the best way to qualify and quantify persistent seed banks in a plant community is to place soil samples in a nonheated greenhouse, slathouse, or well-ventilated shed. Then, water, stir, and check soil samples regularly for newly emerged seedlings over a several-year period, like Brenchley and Warington (1930) did in their classic seed bank studies. Samples kept under natural temperature conditions for less than 1 year (Fyles, 1989) may result in an underestimation of the persistent seed bank.

2. Time of Sample Collection

A review of the buried seed literature revealed that soil samples in many studies probably contained a mixture of transient and persistent or only transient seed banks (Table 7.1). A major problem in some studies is that soil samples were collected after the majority of the species in the community had dispersed their seeds, but before the first germination season was completed.

Thus, the samples may have contained many (mostly?) seeds in the transient seed bank. In communities dominated by perennials and/or summer annuals, samples to determine the persistent seed bank should be taken in summer—after germination is completed but before seed maturation and dispersal begin. However, if one wishes to sample the transient plus persistent seed bank, samples should be collected in winter or early spring—before germination begins (Warr *et al.*, 1994). In communities dominated by winter annuals, samples to determine the persistent seed bank should be collected in winter or early spring—after germination is completed but before seed maturation and dispersal begin.

In communities with both summer and winter annuals, e.g., temperate zone agricultural or ruderal sites, samples collected in summer would contain many freshly dispersed (transient) seeds of winter annuals. Also, summer-collected samples may contain some viable seeds of summer annuals "left over" from the previous spring–early summer germination season(s) e.g., Paatela and Ervio, 1971; Dessaint *et al.*, 1991). Thus, in communities with summer and winter annuals the winter annual component is sampled most accurately by collecting soil in winter or early spring (before seed dispersal), and the summer annual component is sampled most accurately by collecting soil in summer or early autumn (before seed dispersal).

Soil samples collected each month (Zaman and Khan, 1992) or season (Gilfedder and Kirkpatrick, 1993) would permit one to accurately plot the seasonal distribution of seed densities of each species; however, time and space limitations usually make this prohibitive. Sampling for more than 1 year may reveal year-to-year variations in the size of the persistent seed bank (Coffin and Lauenroth, 1989). In some seed bank studies (see Table 7.1), soil samples have been collected at several times, but unfortunately data have been pooled for presentation in publications. Thus, information from the "correct" collection time of individual species cannot be distinguished. No collection date is given for some seed bank studies!

It is difficult to interpret data from some seed bank studies in the tropics because no information is supplied about the phenology of seed dispersal and germination in the habitat (e.g., Cheke *et al.*, 1979; Hodgkinson *et al.*, 1980; Hopkins and Graham, 1983, 1984b; Di Stefano G. and Chaverri, 1992; Terrados, 1993). Thus, one cannot be sure if seeds found in the samples represent only a transient seed bank, a combination of transient and persistent seed banks, or only a persistent seed bank.

Another approach is to study seed banks of individual species after information has been obtained on the time of seed dispersal and germination in the field. If studies were initiated after this kind of information was avail-

TABLE 7.1 Studies in Which There Is a High Probability That the Soil Samples Contained a Mixture of Transient and Persistent or Only Transient Seed Banks

Habitat	Problem[a]	Reference	Habitat	Problem[a]	Reference
Abandoned arable land	1	Kitajima and Tilman (1996)	Montane forests, Mexico	1,4	Williams-Linera (1993)
Abandoned arable land	1	Symonides (1986)			
Agathis forests, New Zealand	3	Sem and Enright (1995)	Oak forest	1	Schiffman and Johnson (1992)
			Oak woodlands	1	Maranon and Bartolome (1989)
Arable land	1	Albrecht and Forster (1996)			
Arable land	1	Forcella *et al.* (1992)	Pasture	1,2	Champness and Morris (1948)
Arable land	1	Mohler and Calloway (1995)	Pasture	2	Dore and Raymond (1942)
Arable land	3	Warwick (1984)	Pasture	1	King (1976)
Arctic	1	Archibold (1984)	Pasture	3	Prince and Hodgdon (1946)
Beech–maple forests	1	Roberts *et al.* (1984)	Pinyon woodlands	1	Koniak and Everett (1982)
Boreal forests	1?, 3	Komarova (1985)	Ponderosa pine	1	Pratt *et al.* (1984)
Boreal forests	1	Komulainen *et al.* (1994)	Prairie	1	Abrams (1988)
Boreal forests	1	Morin and Payette (1988)	Prairie	1	Archibold (1981)
Boreal forests	1?	Vieno *et al.* (1993)	Prairie	1	Johnston and Anderson (1986)
Bunchgrass	1	Major and Pyott (1966)			
Chalk quarries	2	Jefferson and Usher (1987)	Prairie	1	Johnson *et al.* (1969)
Coastal marine sediments	1	Kautsky (1990)	Prairie	1	Lippert and Hopkins (1950)
			Prairie	1	Meiners and Gorchov (1994)
Cold desert	3	Bespalova (1972)	Prairie	1	Rabinowitz (1981)
Coniferous forests	1	Ingersoll and Wilson (1990)	Prairie	1	Rosburg *et al.* (1994)
Coniferous forests	1	McGee and Feller (1993)	Prairie marsh	1	van der Valk and Davis (1979)
Coniferous forests	1	Nakagoshi (1984a)			
Coniferous forests	1	Nakagoshi and Suzuki (1977)	Rice fields	1	McIntyre (1985)
Coniferous forests	1	Strickler and Edgerton (1976)	Ruderal, eastern Mediterranean	1	Schneider and Kehl (1987)
Coniferous forests–bunchgrass	3	Qi *et al.* (1996)	Salt marsh	1	Baldwin *et al.* (1996)
Coniferous/deciduous forests	1	Olmsted and Curtis (1947)	Salt marsh	1	Hopkins and Parker (1984)
			Sandy soil communities	1	Zimmergren (1980)
Coniferous–seral shrubs	3	Morgan and Neuenschwander (1988)	Subalpine	1	Howard and Ashton (1967)
			Subalpine	3	Payette *et al.* (1982)
Coppicewoods	1,2	Brown and Oosterhuis (1981)	Successional series	1,2	Beatty (1991)
Deciduous forests	1	Brown (1992)	Successional series	1,2	Livingston and Allessio (1968)
Deciduous forests	3	Graber and Thompson (1978)			
Deciduous forests	1	Kjellsson (1992)	Successional series	1	Oosting and Humphreys (1940)
Deciduous forests	1	Piroznikow (1983)			
Deciduous and coniferous forests	1	Moore and Wein (1977)	Successional series	1	Partridge (1989)
			Sugar pine forests	1	Quick (1956)
Desert shrub, coastal	3	Gulzar and Khan (1994)	Tropical cloud forests	3	Lawton and Putz (1988)
Dunes	1	Planisek and Pippen (1984)	Tropical deciduous forests	1	Rico-Gray and Garcia-Franco (1992)
Eucalyptus forests	1	Carroll and Ashton (1965)			
Fens and floating forest	1	van der Valk and Verhoeven (1988)	Tropical rainforests	3	Chin (1973)
			Tropical rainforests	2	Graham and Hopkins (1990)
Forest plantations	1	Hill and Stevens (1981)	Tropical rainforests	3	Hopkins *et al.* (1990)
Granite outcrop comm.	2	Houle and Phillips (1988)	Tropical rainforests	1	Hopkins and Graham (1984b)
Grasslands, alvar	1	Bakker *et al.* (1996)	Tropical rainforests	3	Symington (1932)
Heath	1	Stieperaere and Timmerman (1983)	Tropical rainforests	3	Uhl and Clark (1983)
			Tropical savanna	1	Navie *et al.* (1996)
Lake sediments	2	Skoglund and Hytteborn (1990)	Tropical savanna	2	Odgers (1994)
			Tundra	1	Fox (1983)
Lake shore	1	Collins and Wein (1995)	Tundra	1	Gartner *et al.* (1983)
Marsh	1	Leck and Graveline (1979)	Tundra	3	Roach (1983)
Marsh	1	Parker and Leck (1985), Leck and Simpson (1995)	Wetlands	1	Neely and Wiler (1993)
			Wetlands, playa	1	Haukos and Smith (1993, 1994)
Marsh	1	Smith and Kadlec (1985)			
Matorral, Chile	1	Jimenez and Armesto (1992)	Wetlands, subtropical swamp	1	Gerritsen and Greening (1989)
Matorral, USA	1	Zammit and Zedler (1988, 1994)	Wetlands, surface mines	1	Cole (1991)

[a] 1, soil samples were collected after seed dispersal but before the germination season ended; 2, data were pooled so that samples collected at the right time (i.e., after germination season and before dispersal) can not be distinguished from those that contain a transient seed bank; 3, no collection data are given; 4, sampling time probably was correct to show that some species had a persistent seed bank; however, species names were not given.

able, the investigator would know when to collect, i.e., when the germination season was over and before new seeds were dispersed. Further, he/she could know the environmental conditions required for loss of seed dormancy and germination. Seed bank studies have been done on individual species, including *Acacia* spp. (Tybirk *et al.,* 1994), *Avena ludoviciana* (Bhatia and Sandhu, 1990), *Buchloe dactyloides* (Coffin and Lauenroth, 1989), *Bouteloua gracilis* (Coffin and Lauenroth, 1989), *Cecropia obtusifolia* (Alvarez-Buylla and Matinez-Ramos, 1990), *Centaurea maculosa* (Davis *et al.,* 1993), *Cirsium vulgare* (Doucet and Cavers, 1996), *Cymodocea nodosa* (Terrados, 1993), *Erica lusitanica* (Mather and Williams, 1990), *Eupatorium odoratum* (Epp, 1987), *Festuca octoflora* (Coffin and Lauenroth, 1989), *Halophila decipiens* (McMillan, 1988), *H. wrightii* (McMillan, 1981), *Hordeum jubatum* (Badger and Ungar, 1994), *Isanthus brachiatus* (Baskin and Baskin, 1975b), *Koenigia islandica* (Reynolds, 1984), *Leavenworthia stylosa* (Baskin and Baskin, 1978), *Lepidium densiflorum* (Coffin and Lauenroth, 1989), *Lythrum salicaria* (Welling and Becker, 1990), *Panicum miliaceum* (Colosi *et al.,* 1988), *Peltandra virginica* (Whigham *et al.,* 1979), *Polygonum confertiflorum, P. douglasii* (Reynolds, 1984), *Primula vulgaris* (Valverde and Silvertown, 1995), *Rutidosis leptorrhynchoides* (Morgan, 1995), *Sorghum intrans* (Andrew and Mott, 1983; Andrew, 1986), *S. stipoideum* (Andrew and Mott, 1983), *Spergularia marina* (Ungar, 1988), *Sporobolus cryptandrus* (Coffin and Lauenroth, 1989), *Syringodium filiforme* (McMillan, 1981), *Triodia basedowii* (Westoby *et al.,* 1988), *Vallisneria americana* (Kimber *et al.,* 1995), *Zostera marina,* and *Z. noltii* (Hootsman *et al.,* 1989). The individual species approach, however, quickly can result in major space limitations if several species are studied at the same time.

In many species, there is a definite germination season, but seeds of some species germinate throughout the growing season, making it difficult to know when to collect soil samples. One suggestion is to cover the sites where soil samples will be collected with fine-mesh woven wire to prevent seed dispersal into the sites, and then collect soil samples at the end of the growing season (e.g., Leck and Simpson, 1994). Another possibility is to collect the samples in late spring and not count any seedlings that emerge until the beginning of the next growing season.

Some investigators have collected soil samples in the autumn and have attempted to remove the transient seed bank by scraping away the duff, humus, and top layer of mineral soil (Chippendale and Milton, 1934; Olmsted and Curtis, 1947; Planisek and Pippen, 1984; Warr *et al.,* 1993). This method of eliminating the transient seed bank is not recommended for three reasons.

(1) There is no way of knowing how deep the current year's seeds have become buried via action of soil organisms or by falling into cracks. (2) Transient seeds may be pushed into the soil during the scraping activity. (3) Some persistent seeds may be in the duff, humus, or top layer of mineral soil.

3. Sample Sizes

Frequently, data from soil seed bank studies have large standard deviations, and one reason for this is that seeds are not evenly distributed horizontally in a given habitat. Seed banks of most species are aggregated or clustered (Thompson, 1986; Chauvel *et al.,* 1989; Matlack and Good, 1990; Dessaint *et al.,* 1991), and these clusters have an irregular distribution (Bigwood and Inouye, 1988).

A reason given by Thompson (1986) for variation in sampling data is that too few samples are collected. Bigwood and Inouye (1988) found that the precision of seed number estimates was improved greatly by taking many small samples rather than a few large ones. In constructing a species area curve for buried viable seeds, increasing the area of the soil surface sampled beyond 200 cm^2 decreased the rate at which new species were detected (Forcella, 1984). Species richness of the buried seed bank in an Australian pasture did not increase significantly after the combined surface area of replicated samples from randomly distributed plots exceeded 1000 cm^2 (Forcella, 1984). In arable soils in Italy, 200 soil cores (3.4 cm in diameter and 15 cm deep) were required to accurately document the majority of species present in a 169-m^2 plot (Zanin *et al.,* 1989). In a cultivated field in Ontario, Canada, 60 samples were required to quantify the seed bank for *Chenopodium* spp., and sample variance decreased with an increase in sample size; soil augers 1.9, 2.7, and 3.3 cm in diameter were used (Benoit *et al.* 1989). In Michigan, 15–20 sampling locations were required to determine the number of species in the soil seed bank; at each location 20 cores (2.5 cm in diameter $\times$ 15 cm deep) were taken (Gross, 1990). Forcella (1992) suggested that the minimum amount of soil per sample to detect a single species should be 100 g.

C. Types of Persistent Seed Banks

Thompson and Grime (1979) described two transient (types I and II; Fig. 7.1) and two persistent (types III and IV; Fig. 7.1) seed banks. In the type III seed bank, many seeds germinate soon after dispersal, but a small reserve of viable seeds remains ungerminated. In type IV, only a small proportion of the seeds germinates immediately after dispersal, and a large reserve of viable

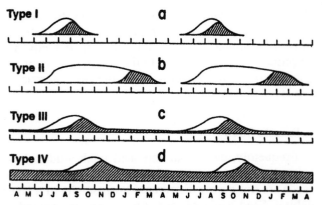

FIGURE 7.1 Four types of seed banks found for herbaceous species in contrasting habitats in England. Unshaded areas, seeds alive but not capable of germinating at 20/15°C; shaded areas, seeds capable of germinating at 20/15°C. Type I, transient, seeds germinate in autumn; type II, transient, seeds germinate in spring; type III, persistent, seeds germinate primarily in autumn and the seed reserve is small; and type IV, persistent, seeds germinate in autumn and the seed reserve is large. From Thompson and Grime (1979), with permission.

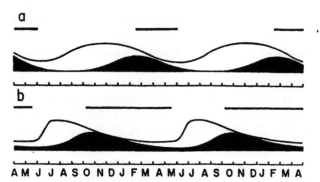

FIGURE 7.2 Persistent seed banks in the spring-germinating (a) summer annual *Polygonum aviculare* and (b) facultative winter annual *Veronica hederifolia*. Black areas, seeds capable of germinating at favorable laboratory conditions; nonshaded areas, seeds not capable of germinating at favorable laboratory conditions. Diagrams begin with freshly matured seeds. From Roberts (1981) with permission.

seeds remains ungerminated. Some germination may occur at any time during the growing season if seeds with types III or IV are brought to the soil surface, but most of it takes place in autumn.

Grime (1981) subdivided the type III seed bank into two categories: type IIIa and type IIIb. Seeds in type IIIa, such as those of *Agrostis tenuis, Digitalis purpurea, Epilobium hirsutum, Holcus lanatus,* and *Poa annua,* germinate in autumn, unless they become buried in soil. Because light is required for germination, burial causes an inhibition of germination, which results in the formation of a small seed bank. Winter annuals, including *Arabidopsis thaliana, Arenaria serpyllifolia,* and *Cardamine hirsuta,* form type IIIb seed banks. In this type, seeds become buried during the summer dormancy-breaking period, and a light requirement for germination prevents germination in autumn. Using data from studies on *Polygonum aviculare* (Courtney, 1968) and *Veronica hederifolia* (Roberts and Lockett, 1978), Roberts (1981) diagramed two types of persistent seed banks (Fig. 7.2). The seed bank for *P. aviculare* (Fig. 7.2a) is not like any of the persistent seed bank types presented by Thompson and Grime (1979) or Grime (1981), whereas the one for *V. hederifolia* (Fig. 7.2b) is like type III (Fig. 7.1) of Thompson and Grime (1979).

Three additional types of persistent seed banks have been identified (Fig. 7.3) based on data obtained from studies on *Lamium purpureum* (Baskin and Baskin, 1984b), *Panicum capillare* (Baskin and Baskin, 1986a), and *Rumex crispus* (Baskin and Baskin, 1985a). Thus, at present, six types of persistent seed banks are known. The *L. purpureum* type is similar to type II of Thompson

and Grime (Fig. 7.1); however, seeds of *L. purpureum* brought to the soil surface can germinate only in autumn. Thus, type III of Thompson and Grime (1979) nicely describes a facultative winter annual, whereas the *L. purpureum* type is for an obligate winter annual. The *P. capillare* type is similar to that of *Polygonum aviculare* [type a of Roberts (Fig. 7.2a)], but *P. capillare* seeds are dispersed in autumn and those of *P. aviculare* throughout the growing season. The *R. crispus* type (Fig. 7.3c) is similar to type IV of Thompson and Grime (Fig. 7.1), but the former is for seeds dispersed in summer and the latter for those dispersed in autumn.

Persistent seed banks have been described as short-term persistent or long-term persistent (Bakker *et al.,* 1996). Seeds in a short-term persistent seed bank live for at least 1 year but less than 5 years, whereas those

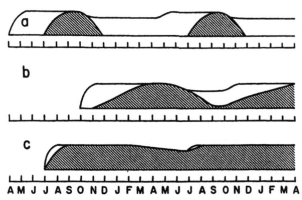

FIGURE 7.3 Persistent seed banks in the (a) obligate winter annual *Lamium purpureum,* (b) spring- and summer-germinating summer annual *Panicum capillare,* and (c) polycarpic perennial *Rumex crispus.* Shaded areas, seeds capable of germinating at favorable conditions in incubators; nonshaded areas, seeds not capable of germinating at otherwise favorable conditions in incubators. Diagrams begin with freshly matured seeds.

in a long-term persistent seed bank live for at least 5 years. However, in our discussion of persistent seed banks below, a persistent seed bank is defined as seeds that live until the second (or some subsequent) germination season.

D. Habitats and Seed Density

In a survey of 171 published papers on seed banks in various kinds of plant communities, the persistent seed bank was only sampled in 78 (46%) of them. Thus, soil samples in 93 (54%) of the studies may have contained a mixture of transient and persistent seed banks or only transient seed banks (Table 7.1). Persistent seed banks have been found in many kinds of habitats (Table 7.2), ranging from the tropics to the arctic tundra and from deserts to marshes (Table 7.2). Also, persistent seed banks occur in disturbed as well as nondisturbed sites.

In 70 of the 78 studies of only persistent seed banks (Table 7.2), seed densities were expressed as number of seeds m^{-2} of surface area sampled (or data could be converted to this unit). Seed densities in the 70 studies varied from 0–9 m^{-2} in subarctic forests to 1935–24,393 m^{-2} in pastures in England (Table 7.2). About half the reported seed densities ranged from 1 to 1000 seeds m^{-2}, and the number exceeded 10,000 seeds m^{-2} in only seven studies (Table 7.2). Thompson (1978) suggested that the densities of buried seeds declined with increasing altitude, latitude, and successional age. He further stated that small numbers of seeds should occur in tropical and in arctic and alpine habitats and that large numbers should be found in disturbed soils.

Arctic and boreal/subalpine plant communities and coniferous and deciduous forests have relatively few persistent seeds, whereas pastures and arable soil have large numbers (Table 7.2). Also, tropical savannas have more persistent seeds than most rainforests, and temperate grasslands or prairies have more than coniferous and deciduous forests. In humid tropical forests in the Western Ghats of Kerala, India, the primary forest species do not form persistent seed banks. Seeds in soil samples collected at 2-month intervals lived for only 2–6 months after they were dispersed (Chandrashekara and Ramakrishnan, 1993).

The exceedingly large numbers of seeds in seed banks sometimes seen in the literature frequently are from studies of arable lands, in which soil samples were collected after seed dispersal in autumn and thus contained transient plus persistent or only transient seed banks (Table 7.3). Of course, it probably is of little concern to a farmer if the weed seeds germinating in his cropland are from the transient or persistent seed bank!

Brenchley and Warington's (1930) work provides some insight into the number of seeds in the transient and persistent seed banks of arable soils. Samples of soil collected from the Rothamsted wheat field in England just before the autumn plowing in 1925 produced 29,383 seedlings m^{-2} in a nonheated greenhouse during 1926. Most of these seeds probably were from the transient seed bank. During 1927 and 1928, 6,563 and 3,046 seedlings m^{-2}, respectively, appeared in the soil samples, and these were from seeds in the persistent seed bank. The number of additional seeds in the persistent seed bank is unknown because the study was not continued beyond 1928.

E. Kinds of Seeds

In general, seeds (or fruits) with a high probability of forming persistent seed banks are small and have smooth seed coats, whereas those with a low probability of forming persistent seed banks are relatively large and have hooks, awns, spines, or other kinds of projections on the seed coat (Thompson and Grime, 1979). However, some relatively large seeds (or fruits), such as those of *Galium palustre*, *G. saxatile*, *Potentilla erecta*, and *Trifolium repens*, form persistent seed banks (Thompson and Grime, 1979). A study of the weights and sizes/shapes (expressed as variance from a sphere) of seeds of 44 species and fruits of 53 species in the United Kingdom showed that compact (low variance) dispersal units weighing <3 mg are persistent, whereas noncompact (high variance) ones weighing >3 mg are short-lived (Thompson *et al.*, 1993).

Information on the plant families, genera, and species found in persistent seed banks has been compiled from various sources (Table 7.4). A total of 155 families was recorded, and 71 of them were represented by two or more genera. The Apiaceae, Asteraceae, Brassicaceae, Cyperaceae, Fabaceae, Lamiaceae, Poaceae, Rosaceae, Rubiaceae, Scrophulariaceae, and Urticaceae each had 15 or more genera, and 20 families had 15 or more species.

Twenty-nine of the families (Amaryllidaceae, Annonaceae, Apiaceae, Aquifoliaceae, Araceae, Araliaceae, Arecaceae, Berberidaceae, Caprifoliaceae, Dilleniaceae, Fumariaceae, Grossulariaceae, Haemodoraceae, Hydrophyllaceae, Illiciaceae, Iridaceae, Liliaceae, Magnoliaceae, Oleaceae, Papaveraceae, Piperaceae, Pittosporaceae, Podocarpaceae, Ranunculaceae, Schisandraceae, Stylidiaceae, Taxaceae, Trochodendraceae, and Winteraceae) found in persistent seed banks are known to have species with underdeveloped embryos (Table 3.3). Thus, members of these 29 families could have morphological and/or morphophysiological dormancy (see Chapter 3, IV). Ten of the families (Anacardiaceae, Cannaceae, Convolvulaceae, Fabaceae, Gera-

TABLE 7.2 Habitats and Densities of Persistent Seed Banks

Habitat	Seeds per m^2	Reference	Habitat	Seeds per m^2	Reference
Abandoned arable land	—[a]	Lavorel et al. (1993)	Montane, Ethiopia	13,700–24,000	Teketay and Granstrom (1995)
Acacia forest	2,931	Howard (1973)			
Annual grasslands	900–1750	Young et al. (1981)	Nothofagus forests	407	Howard (1973)
Arable land, Belize	7,623–9,520	Kellman (1974b, 1980)	Oak woodland	4660	Warr et al. (1994)
Arable land, UK	3,037–6,563	Brenchley and Warington (1930)	Pastures	Volume	Chippindale and Milton (1934)
Arctic coastal plain	144–408	Ebersole (1989)	Pastures	26,752–63,694	Gilfedder and Kirkpatrick (1993)
Arctic heath, fen	56–131	Freedman et al. (1982)			
Birch woodland	4,880	Warr et al. (1994)	Pastures	840–18,780	Hayashi and Numata (1971)
Bog	12,874–377,041	McGraw (1987)			
Boreal forests	239–763	Granstrom (1982)	Pastures	5,226–12,960	Kellman (1974b, 1980)
Boreal forests	426	Archibold (1979)	Pastures	1,935–24,313	Milton (1936)
Coastal desert	5193	Ohga (1992)	Pine forests	130–1,590	Nakagoshi (1984a)
Coastal dunes	213	Altamirano and Guevara (1982)	Prairie	113–927	Iverson and Wali (1982)
			Prairie wetland	Volume	Poiani and Johnson (1988)
Coastal dunes	1,000–1,296	Houle (1996)			
Coniferous forests	206	Kellman (1974a)	Rainforest, Cote d'Ivoire	2051	deRouw and van Oers (1988)
Coniferous forests	1065	Kramer and Johnson (1987)	Rainforest, Malaysia	131	Putz and Appanah (1987)
Coniferous/deciduous forests	431	Marquis (1975)	Rainforest, Mexico	862–3,706	Guevara S. and Gomez-Pompa (1972)
Coniferous/deciduous forests	2–4	Mladenoff (1990)	Rainforest, New Guinea	1,325	Enright (1985)
Deciduous forests	96	Houle (1992)	Rainforest, New Guinea	994	Saulei and Swaine (1988)
Desert grassland	159–486	Dwyer and Aguirre V. (1978)			
Desert, saline	26–936	Aziz and Khan (1995)	Riverine swamp	25–76	Titus (1991)
Desert, saline	1–220	Khan (1993)	Ruderal, mowed sites	540–9,900	Ohtsuka and Ohsawa (1994)
Desert, saline	27–31	Zaman and Khan (1992)			
Douglas fir–hemlock forests	280	Kellman (1970)	Sagebrush semidesert	13–28	Hassan and West (1986)
			Saline pan	Volume	Ungar and Riehl (1980)
Dry forests	Volume	Barbour and Lang (1967)	Salt marsh	0–334	Hartman (1988)
			Salt marsh	13–173	Hutchings and Russell (1989)
Evergreen oak forests	1,091–4,403	Naka and Yoda (1984)			
Flood meadow	4000	Raffaele (1996)	Salt marsh	3,491–4,450	Ungar and Woodell (1993)
Floodplain, by creek	3,800–15,400	Finlayson et al. (1990)			
Freshwater marsh	140–350	Leck and Simpson (1987)	Secondary forests, B.C., Canada	2,612	Kellman (1974)
Freshwater marsh	5,915	Wilson et al. (1993)	Secondary forests, Costa Rica	3,561–6,322	Young et al. (1987)
Freshwater tidal marsh	1,675–5,967	Leck and Simpson (1994)	Southern United States forests	233–4159	Schneider and Sharitz (1986)
Grassland	5–11	Bertiller (1992)	Subalpine forest/ meadow	3,112	Ingersoll and Wilson (1993)
Grassland	2,252–4,117	Kinucan and Smeins (1992)	Subalpine forests	3–53	Whipple (1978)
Grassland	2,276	Howard (1973)	Subalpine forests	0–9	Johnson (1975)
Jarrah forest	377–1,579	Vlahos and Bell (1986)	Successional series	87–4,202	Donelan and Thompson (1980)
Lake sediments	1,413–4,332	Fiore and Putz (1992)			
Lake sediments	Volume	Hayashi et al. (1978)	Successional series	Volume	Numata et al. (1964)
Lake sediments	48	Hytteborn et al. (1991)	Temperate rainforest	565	Enright and Cameron (1988)
Lake sediments	38,583	Nicholson and Keddy (1983)			
			Temperate rainforest	52	Sem and Enright (1996)
Lake shore	25–300	Keddy and Reznicek (1982)	Temperate rainforest	218–4,158	Nakagoshi (1984b)
			Tropical savanna	1,490–3,610	McIvor and Gardener (1991)
Lake shore	8511	Wisheu and Keddy (1991)			
Mojave desert	1,948–3,989	Nelson and Chew (1977)	Tropical wet forest	393–1,812	Young (1985)

[a] Number of seeds was not given.

TABLE 7.3   Maximum Numbers of Seeds in Samples from Arable Soil[a]

Location	Depth (cm)	Seeds m^{-2}	Reference
Argentina	7	55,750	D'Angela et al. (1988)
Czechoslovakia	15	31,672	Kropac (1966)
Australia	10	646,000	McIntyre (1985)
Denmark	20	496,000	Jensen (1969)
England	15	69,490	Brenchley and Warington (1930)
England	15	28,519	Brenchley and Warington (1933)
England	22.5	32,450	Roberts and Dawkins (1967)
England	23	12,752	Roberts and Feast (1973a)
England	10	65,580	Roberts and Ricketts (1979)
England	15	12,000	Williams (1984)
Poland	14.5	44,569	Symonides (1986)
United States	25	137,700	Schweizer and Zimdahl (1984)

[a] Seeds were collected after seed dispersal in autumn and before (or during) the end of the spring germination season, i.e., transient and persistent seed banks are mixed.

niaceae, Malvaceae, Rhamnaceae, Sapindaceae, Sterculiaceae, and Tiliaceae) found in persistent seed banks are known to have species with physical dormancy (Table 3.5). Thus, members of these 10 families could have physical dormancy. Members of the other 116 families represented in persistent seed banks probably have physiological dormancy (see Chapter 3, IV).

IV. LONGEVITY OF SEEDS IN THE SOIL

A. Inferred Age

The longevity of seeds in the soil is of great interest to basic and applied biologists and to agriculturalists, and the subject has been reviewed periodically (e.g., Major and Pyott, 1966; Roberts, 1981; Toole, 1986; Milberg, 1990). Various methods have been devised to estimate the age of seeds in soil.

One way to obtain information on the longevity of seeds in the soil is to collect soil samples at regular (usually yearly) intervals in areas where all plants of a given species were killed and reestablishment prevented. Viable seeds have been found in soil 4–36 years after the last plants of a species produced seeds at the site: *Abutilon theophrasti,* 4 years (Lueschen and Andersen, 1980); *Acacia cyclops,* 12 (Holmes, 1988); *Avena fatua* and *A. ludoviciana,* 5–6 (Thurston, 1966); *Capsella bursa-pastoris,* 16 (Roberts and Neilson, 1981); *Cheno-*

podium album, 5 (Roberts and Dawkins, 1967); *Melilotus* sp., 14 (Stoa, 1933); *Trifolium repens,* 20; *Ranunculus bulbosus,* 14; *Rumex obtusifolius,* 12; *Plantago lanceolata,* 8 (Chancellor, 1985); *Poa annua,* 16; *Polygonum aviculare,* 16; *Senecio vulgaris,* 16; *Solanum nigrum,* 16; *Stellaria media,* 16 (Roberts and Neilson, 1981); and *Ulex europaeus,* 36 (Moss, 1959). After 17 years of continuous alfalfa cover or of chemical fallow, 15 and 25%, respectively, of the original *Abutilon theophrasti* seeds in the soil were viable (Lueschen *et al.,* 1993). In Swedish grasslands, eight species that had not been observed in the extant vegetation for 35 years were found in the soil seed bank (Milberg, 1992), indicating that viable seeds may have been in the soil for at least 35 years.

Another way to determine seed longevity is to sow seeds in field plots (soil is not sterilized), prevent plants from setting seeds, and sample the soil over a period of time to see how many seeds are still alive. For example, some seeds of *Aethusa cynapium, Amaranthus retroflexus, Capsella bursa-pastoris, Chenopodium album, Euphorbia exigua, Fallopia convolvulus, Kickxia spuria, Papaver rhoeas, Polygonum persicaria, Sinapis arvensis,* and *Viola arvensis* were alive after 5 years in the soil (Barralis *et al.,* 1988). Some seeds of each species may have been present in the soil when the study was initiated. Thus, a few seeds found in the fifth year may have been more than 5 years old. Seeds also have been sown in pots of soil sunk in the ground and the emergence of seedlings monitored for 5 years (Roberts, 1979, 1986a; Roberts and Neilson, 1981; Roberts and Boddrell, 1983a).

In Denmark, Odum (1965) collected soil samples from underneath old buildings, walls, and pavements that had been dated by archaeologists. Great care was taken to avoid contamination by wind-blown seeds, but there is no way to be absolutely sure that soil-dwelling animals such as earthworms had not moved recently matured seeds into the areas that were sampled. Seeds of several species germinated in soils that had not been disturbed for 100–600 years, and those of *Chenopodium album* and *Spergula arvensis* were in soils undisturbed for 1700 years. In a later study, Odum (1974) found seeds of *Verbascum thapsiforme* at an 850-year-old site.

Seeds of 11 species germinated in samples taken from topsoils buried by volcanic ash in the Krakatau Islands, Indonesia, in 1930–1933, suggesting that the seeds were about 60 years old (Whittaker *et al.,* 1995). Samples from topsoils buried by 1–3 m of volcanic deposits in northern Japan for 10 years contained viable seeds of 25 species. There was a significant positive correlation between depth of burial and seed viability in *Alopecurus aequalis, Poa annua, Rumex obtusifolius,* and *Viola grypoceras* (Tsuyuzaki, 1991).

Ages of seeds in the soil have been inferred from

TABLE 7.4   Families in Which Persistent Seed Banks Have Been Found[a]

Family	No. of genera	No. of species	Family	No. of genera	No. of species
Aceraceae	1	4	Fagaceae	4	4
Actinidiaceae	1	2	Fumariaceae	1	. 2
Agavaceae	1	1	Gentianaceae	2	2
Aizoaceae	1	1	Geraniaceae	2	6
Alismataceae	1	1	Grossulariaceae	1	4
Altingiaceae	1	1	Haemodoraceae	1	1
Amaranthaceae	7	12	Haloragidaceae	3	3
Amaryllidaceae	1	1	Hydrangeaceae	1	2
Anacardiaceae	1	2	Hydrocharitaceae	3	3
Annonaceae	1	1	Hydrophyllaceae	4	4
Apiaceae	21	24	Illiciaceae	1	1
Apocynaceae	4	4	Iridaceae	2	4
Aquifoliaceae	1	5	Iteaceae	1	1
Araceae	2	2	Juncaceae	2	28
Araliaceae	6	9	Lamiaceae	22	29
Arecaceae	1	1	Lauraceae	3	5
Asclepiadaceae	1	1	Lentibulariaceae	1	1
Asteraceae	73	131	Liliaceae	7	7
Balsaminaceae	1	1	Linaceae	1	1
Begoniaceae	1	1	Loasaceae	1	1
Beberidaceae	1	1	Loganiaceae	2	2
Betulaceae	2	9	Lythraceae	2	2
Bischofiaceae	1	1	Magnoliaceae	2	2
Boraginaceae	7	10	Malvaceae	7	12
Brassicaceae	28	46	Melastomataceae	5	5
Callitrichaceae	1	1	Meliaceae	1	1
Campanulaceae	8	11	Meliosmaceae	1	2
Cannaceae	1	1	Menispermaceae	1	1
Capparidaceae	1	1	Menyanthaceae	1	1
Caprifoliaceae	5	9	Moraceae	3	11
Caricaceae	1	1	Musaceae	1	1
Carpinaceae	1	3	Myricaceae	1	2
Caryophyllaceae	12	25	Myrsinaceae	3	5
Celastraceae	1	1	Myrtaceae	5	5
Centrotepidaceae	1	1	Najadaceae	1	1
Chenopodiaceae	10	21	Naucleaceae	2	2
Cleomaceae	1	2	Nelumbonaceae	1	1
Clethraceae	1	1	Nolanaceae	1	2
Clusiaceae	4	15	Nyctaginaceae	3	. 3
Combretaceae	1	1	Nymphaeaceae	1	2
Convolvulaceae	5	12	Nyssaceae	1	2
Cornaceae	1	4	Oleaceae	2	3
Corylaceae	1	1	Onagraceae	5	19
Costaceae	1	1	Oxalidaceae	1	1
Crassulaceae	1	2	Pandanaceae	1	1
Cupressaceae	2	2	Papaveraceae	3	8
Cymodoceaceae	3	3	Phytolaccaceae	1	4
Cyperaceae	17	83	Peperomiaceae	1	1
Dilleniaceae	1	2	Pinaceae	2	3
Dipsacaceae	3	3	Piperaceae	2	4
Droseraceae	1	2	Pittosporaceae	1	2
Elaeagnaceae	1	2	Plantaginaceae	1	4
Empetraceae	1	2	Poaceae	75	157
Epacridaceae	1	1	Podocarpaceae	1	1
Ericaceae	12	21	Polemoniaceae	1	1
Eriocaulaceae	1	1	Polygalaceae	1	1
Euphorbiaceae	10	21	Polygonaceae	6	32
Fabaceae	32	65	Pontederiaceae	3	3

(continues)

TABLE 7.4—*Continued*

Family	No. of genera	No. of species	Family	No. of genera	No. of species
Portulacaceae	2	3	Sterculiaceae	5	6
Potaliaceae	1	1	Stylidiaceae	2	2
Potamogetonaceae	1	1	Symplocaceae	1	4
Primulaceae	5	6	Taxaceae	1	1
Proteaceae	1	1	Taxodiaceae	2	2
Ranunculaceae	3	8	Tetramelaceae	1	1
Resedaceae	1	2	Theaceae	1	2
Rhamnaceae	2	11	Tiliaceae	6	6
Rosaceae	21	51	Trochodendraceae	1	1
Ruppiaceae	1	1	Typhaceae	1	4
Rubiaceae	17	26	Ulmaceae	3	5
Rutaceae	5	9	Urticaceae	15	19
Salicaceae	2	2	Valerianaceae	1	1
Sapindaceae	1	1	Verbenaceae	6	10
Saxifragaceae	5	9	Violaceae	3	13
Schisandraceae	1	1	Vitidaceae	4	9
Scrophulariaceae	21	35	Winteraceae	1	1
Smilacaceae	1	1	Zannichelliaceae	1	1
Solanaceae	6	24	Zygophyllaceae	2	3
Spigeliaceae	1	1			

[a] See references in Tables 7.2 and 7.5 and Badger and Ungar, 1994; Barralis *et al.*, 1988; Bonis *et al.*, 1995; Cavers, 1983; Cavers *et al.*, 1992; Coffin and Lauenroth, 1989; Colosi *et al.*, 1988; Davis *et al.*, 1993; Doucet and Cavers, 1996; Epp, 1987; Froud-Williams *et al.*, 1983; Grice and Westoby, 1987; Houle, 1991; Lunt, 1995; Mather and Williams, 1990; McMillan, 1981, 1988; Middleton *et al.*, 1991; Milberg, 1992; Mukherjee *et al.*, 1980; Pavlik *et al.*, 1993; Perez-Nasser and Vazquez-Yanes, 1986; Reynolds, 1984; Roberts, 1986a,b; Roberts and Boddrell, 1984a,b; Roberts and Chancellor, 1979; Roberts and Feast, 1972; Shen-Miller *et al.*, 1995; Sivori *et al.*, 1968; Spence, 1990; Suzuki, 1993; Terrados, 1993; Tingley, 1961; Tsuyuzaki, 1994; Tybirk *et al.*, 1994; Ungar, 1988; Valverde and Silvertown, 1995; Welling *et al.*, 1988b; Welling and Becker, 1990; Wendel, 1972; Whigham *et al.*, 1979; Whittaker *et al.*, 1995; Zhang and Maun, 1994.

samples collected from the various stages in the successional series from cropland to mature forests. Soil samples were taken from sites of different ages (since time of abandonment of the cultivated fields). Of special interest in these studies was the age of the site when seeds of weedy species of arable land were no longer found. Also, how long did seeds of species that appeared the first, second, and so on year after abandonment remain in the soil? Some species present in 0-age sites were found in soil seed bank samples taken from sites 15, 33, 55, 85, and 112 years old (Oostings and Humphreys, 1940); 5, 15, 36, 41, 47, and 80 years old (Livingston and Allessio, 1968); 2, 10, 50, 90, and 200 years old (Roberts and Vankat, 1991); and 20 years old (Kiirikki, 1993). The presence of seeds in soils at these sites implies that they have been there for long periods of time, but the exact age is unknown.

Radiocarbon dating of associated materials has been used to help determine the age of *Nelumbo nucifera* fruits recovered from a peat deposit at Kemigawa (Chiba Prefecture) in Japan. Because Professor Ohga germinated all the fruits, none was left for radiocarbon dating. However, wood from an old boat found at the same level in the peat deposit was dated at 3052 and 3277 years (Libby, 1955). Godwin and Willis (1964) obtained an age of 3196 years for the wood, but they rightfully argued that this was not necessarily the age of the fruits.

Seeds of *Lupinus arcticus* and the skull of *Dicrostonyx groenlandicus* (collared lemming) were collected near Miller Creek in the Yukon Territory (Canada) from rodent burrows and nests discovered in frozen silt during a mining operation. The collared lemming is an animal of arctic and alpine tundra and no longer lives in the Miller Creek area. Thus, Porsild *et al.* (1967) reasoned that since the lemming probably disappeared from the area about 10,000 years ago, when postglacial warming began, the seeds were at least 10,000 years old. Seeds were germinated, resulting in normal plants. Unfortunately, none of the seeds was radiocarbon dated.

In a somewhat similar scenario, a necklace made of walnuts (*Juglans australis*) was found at an archeological site in Argentina. Inside each nut was a seed of, what later proved to be, *Canna compacta*, which had been inserted before the nut had matured. Bones from cameloids taken from the site were radiocarbon dated as 530 years old, and thus it was assumed that the seeds were at least this old (Sivori *et al.*, 1968). Later, one of the nutshells was dated as 620 years old, indicating that the *Canna* seeds were older than previously thought (Lerman and Cigliano, 1971).

B. Actual Age

Radiocarbon dating was used to determine the age of *Nelumbo nucifera* fruits found by Ohga (1923) in a Holocene peat bed in the Pulantien Basin in South Manchuria, China. Based on the geology and history of the area, Ohga (1926) reasoned that the fruits probably were 200 or even 400 years old. Libby (1951) radiocarbon dated a sample of the fruits as 1040 years old; however, Godwin and Willis (1964) obtained an age of 100 years for the same samples that Libby used in his determination. However, Priestley and Posthumus (1982) dated two fruits from the Pulantien site at 340 and 430 years, respectively; Shen-Miller *et al.* (1983) radiocarbon dated a freshly germinated fruit from the site at 705 years. Priestley (1986) suggested that one reason for the varying dates for fruits from the Pulantien site was that some were produced and deposited more recently than others.

Until recently, a scientist with ancient seeds in hand had to decide if it was more important to determine seed age than to grow a plant. The radiocarbon-dating procedure required so much plant material that it was impossible to age an ancient seed, if he/she wished to grow a plant from it. The compromise was to age some seeds and grow plants from others; however, one could never be sure that the seeds used for age determination were the same age as those used to grow plants. Recently, however, dating with accelerator mass spectrometry has made it possible to age seeds using samples of <10 mg. Thus, a piece of pericarp from a germinating *Nelumbo* fruit is a sufficient amount of material for aging. Shen-Miller *et al.* (1995) obtained a radiocarbon age of 1350 years (= 1288 calendar years) for material from a germinating *N. nucifera* fruit; a normal plant was grown from the same fruit.

Viable seeds collected from soil buried by a large solifluction lobe (caused by soil "creeping" down hill) in the tundra of Alaska have been germinated. In addition, accelerator mass spectrometry of seed coats from the seedlings showed that the seeds were 1297 years old (McGraw *et al.*, 1991).

A number of studies have been done in which freshly matured viable seeds were buried in soil under natural conditions, and their viability was determined after various periods of time. Of the 32 studies listed in Table 7.5, only 10 of them were continued for 10 or more years; three are still in progress. Studies by Egley and Chandler (1983) in Mississippi and those of Conn and Deck (1995) in Alaska are planned for 50 years, whereas

TABLE 7.5 Studies in Which Viability Has Been Determined for Seeds Buried in Soil under Natural Conditions for Known Periods of Time

Duration of study (years)	No. species with live seeds at beginning/end	Reference
1	6/3	Lunt (1995)
2	50/29	Hopkins and Graham (1987)
2	7/6	Perez-Nasser and Vazquez-Yanes (1986)
2.5	7/7	Zhang and Maun (1994)
3	9/4	Rampton and Ching (1966)
3	1/1	Wendel (1972)
4	4/1	Clark (1962)
5	4/1	Schwerzel (1974)
5	12/6	Goss (1939)
5	14/11	Granstrom (1987)
5	29/22	Kjaer (1940)
5	7/5	Roberts (1979)
5	10/10	Roberts and Neilson (1980)
5	12/11	Roberts and Neilson (1981)
5	8/8	Roberts and Boddrell (1983a)
5	10/10	Roberts and Boddrell (1983b)
5	11/6	Roberts and Boddrell (1985)
6	13/7	Dorph-Petersen (1925)
6	23/10	Juliano (1940)
5.5[a]	20/17	Egley and Chandler (1983)
7	13/4	Rampton and Ching (1970)
8	12/2	Bruch (1961)
9.7[a]	17/11	Conn and Deck (1995)
10	12/11	Burnside *et al.* (1981)
10	1/1	Hintikka (1987)
11	44/25	Salzmann (1954)
15	3/0	Dawson and Burns (1975)
17	41/20	Burnside *et al.* (1996)
20	33/15	Lewis (1973)
20	1/1	Martin (1970)
39	107/36	Toole and Brown (1946)
40	1/1	Quick (1975)
100[a]	23/3	Kivilaan and Bandurski (1981)

[a]The study is still in progress.

the one started by Beal already has been in progress for over 100 years (Kivilaan and Bandurski, 1981).

The length of time that buried seeds remain viable varies with the species. While no seeds of various species live until the second germination season (e.g., Andrew, 1986; Andrew and Mott, 1983; Bhatia and Sandhu, 1990; Hootsman *et al.*, 1987; Milberg, 1994b; Morgan, 1995), some seeds of a large number of species live until the second germination season (or longer) and thus form persistent seed banks (Table 7.5). In the study initiated by W. J. Beal at Michigan Agricultural College (now Michigan State University, in East Lansing, Michigan) in the autumn of 1879, 20 bottles of soil and seeds (50 seeds of each of 21 weed species per jar) were buried. The 14th bottle buried by Beal was exhumed in 1980, after 100 years of burial (Kivilaan and Bandurski, 1981), and some seeds of three (8%) of the species (*Malva rotundifolia, Verbascum blattaria,* and *V. thapsus*) were alive (Table 7.6).

C. Survivorship Curves

In arable lands in England, the total number of seeds for all species, as well as the number for individual species (Fig. 7.4), declined in a negative exponential fashion, i.e., a Deevey type II curve (Roberts and Dawkins, 1967; Roberts and Feast, 1973a,b). This is the type of survivorship curve usually shown for seeds buried in soil (e.g., Sarukhan, 1974; Watanabe and Hirokawa, 1975; Mukerjee *et al.*, 1980; Burnside *et al.*, 1981; Barralis *et al.*, 1988; Moss, 1985; Lonsdale, 1988; Holmes, 1989; Egley and Williams, 1990; Callihan *et al.*, 1993; Lunt, 1995; Qi *et al.*, 1996). These Deevey type II curves mean that the living seeds in the soil have a constant probability of death (Lonsdale, 1988). However, curves for a few species, such as *Chionochloa macra* (Spence, 1990) and *Leptorhynchos squamathus* (Lunt, 1995), are Deevey type I, indicating a slow initial decline in viable seeds followed by a rapid decline.

Using the formula for a rectangular hyperbola and the amount of germination the first year after burial, Burnside *et al.* (1981) predicted the decline in viability of seeds of 12 weed species; they assumed that viability (= germination) would eventually drop to 1%. In short-lived seeds (≤ 5 years) such as *Kochia scoparia, Helianthus annuus, Apocynum cannabinum,* and *Ampelamus albidus,* actual and predicted viabilities were essentially the same. In the other species, actual and predicted viabilities were within 10% of each other after 10 years of burial. However, when Donald (1993) tried this method of modeling seed survival data from two burial studies of *Sinapis arvensis,* he found that it worked for

data in trial 1 (seeds buried in autumn 1983) but not for those from trial 2 (seeds buried in autumn 1984). During the first year of burial, more seeds died in trial 1 than in trial 2. A negative exponential model was the best fit for data in trial 1, and a negative linear model was best for those in trial 2. Donald (1993) was able to describe both data sets with a mixed linear-negative exponential model. The point is that seeds of the same species buried in different years in the same place may have different patterns of survivorship in the soil (Donald, 1993).

V. WHAT CAN HAPPEN TO SEEDS IN THE SOIL?

A. Models of Seed Dynamics in the Soil

A model of the parameters involved in the persistence and depletion of seeds buried in the soil was presented by Schafer and Chilcote (1969):

$$S = P_{ex} + P_{end} + D_g + D_n,$$

where S is the total number of seeds buried in the soil at a point in time; P_{ex} are persistent seeds that are prevented from germinating by exogenous or environmental factors, e.g., darkness or inhibitory CO_2/O_2 ratios; P_{end} are persistent seeds that are prevented from germinating by endogenous factors, are dormant and will not germinate even if they are exhumed and placed under otherwise optimal conditions for germination, and may either have endogenous dormancy when freshly matured (i.e., innate dormancy) or acquire it after dispersal (i.e., induced dormancy); D_g are seeds that germinate *in situ* and thus are no longer a part of the persistent seed bank; and D_n are seeds that were dead at the time of burial, as well as those that died during burial as a result of aging or predation.

Schafer and Chilcote (1970) found that *Lolium multiflorum* seeds, which were nondormant at the time of burial, became dormant during winter when soils were cold and wet. Consequently, these authors presented a diagram (Fig. 7.5) showing that buried seeds may be nondormant, dormant, or nonviable and that dormancy and nondormancy were reversible. Also, either dormant or nondormant seeds may lose viability during burial.

Roberts (1972b) expanded the equation developed by Schafer and Chilcote (1969):

$$S = P_{inn} + P_{ind} + P_{enf} + D_{gd} + D_{ge} + D_{ni} + D_{na} + D_{np}.$$

Roberts divided endogenous dormancy (P_{end}) of Schafer and Chilcote into innate (P_{inn}) and induced (P_{ind}) dormancy, and he called the exogenous dormancy (P_{ex}) of Schafer and Chilcote enforced dormancy (P_{enf}).

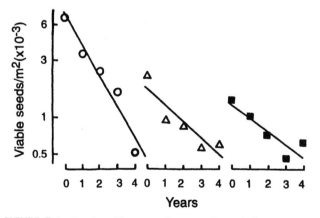

FIGURE 7.4 Survivorship curves for natural populations of seeds of *Poa annua* (O), *Capsella bursa-pastoris* (Δ), and *Chenopodium album* (■) in undisturbed arable soil in England. From Roberts and Feast (1973a).

TABLE 7.6 Results after 5–100 Years in the Beal Buried Seed Experiment[a]

Species tested	Duration	5th year 1884	10th year 1889	15th year 1894	20th year 1899	25th year 1904	30th year 1909	35th year 1914	40th year 1920	50th year 1930	60th year 1940	70th year 1950	80th year 1960	90th year 1970	100th year 1980
Agrostemma githago	1	0	0	0	0	0	0	0	0	0	0	0	0	0	0
Amaranthus retroflexus	1	+[b]	+	+	+	+	+	0	+	0	0	0	0	0	0
Ambrosia artemisifolia	1	0	0	0	+	0	0	0	+	0	0	0	0	0	0
Anthemis cotula	1	+	+	+	0	+	+	+	0	0	0	0	0	0	0
Brassica nigra	1	0	+	0	+	0	+	0	+	+	0	0	0	0	0
Bromus secalinus	1	0	0	0	0	0	0	+	0	0	0	0	0	0	0
Capsella bursa-pastoris	1	+	+	+	+	+	+	0	0	0	0	0	0	0	0
Erechtites hieracifolia	1	0	0	0	0	0	0	0	0	0	0	0	0	0	0
Euphorbia maculata	1	0	0	0	0	0	0	0	0	0	0	0	0	0	0
Lepidium virginicum	1	+	+	+	+	+	+	+	+	0	0	0	0	0	0
Malva rotundifolia	1 or 2	+	0	0	+	0	0	0	0	0	0	0	0	0	1 (2)
Plantago major	perennial	0	0	+	0	0	0	0	+	0	0	0	0	0	0
Polygonum hydropiper	1	+	+	+	+	+	?	0	0	+	0	0	0	0	0
Portulaca oleracea	1	+	+	+	+	+	0	0	+	0	0	0	0	0	0
Setaria glauca	1	+	+	+	0	+	+	0	0	0	0	0	0	0	0
Stellaria media	1	+	+	0	0	0	+	0	0	0	0	0	0	0	0
Trifolium repens	perennial	+	0	0	+	0	0	0	0	0	0	0	0	0	1 (2)
Verbascum thapsus[c]	2	+	?	+	+	+	0	+	0	0	0	0	0	0	0
Oenothera biennis	2	+	+	+	+	+	+	0	19 (38)	19 (38)	12 (24)	7 (14)	5 (10)	0	0
Rumex crispus	2	+	+	+	+	+	+	+	9 (18)	26 (52)	2 (4)	4 (8)	1 (2)	0	0
Verbascum blattaria[c]	2	+	+	+	+	+	+	+		31 (62)	34 (68)	37 (74)	35 (70)	10 (20)	21 (42)

[a] From Kivilaan and Bandurski (1981).

[b] The plus signs following each species, indicate that one or more seeds of that species germinated for the year shown. The number indicates the number of seeds germinating, whereas the number in parentheses indicates the percentage germinating.

[c] There is some question concerning the identification of Verbascum plants in the early period (1884–1920) as V. thapsus rather than V. blattaria (Darlington and Steinbauer, 1961).

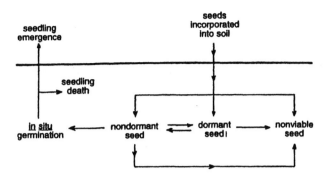

FIGURE 7.5 Diagram illustrating the changes seeds in a population undergo after they become buried in soil. From Schafer and Chilcote, (1970) with permission.

Further, he partitioned *in situ* germination (D_g) into seeds that germinate so deep in the soil that the seedlings fail to emerge (D_{gd}) and those that germinate close enough to the soil surface that seedlings can emerge (D_{ge}). Also, Roberts subdivided death from causes other than *in situ* germination (D_g) into seeds that lose viability before they become a part of the seed bank (D_{ni}), die as a result of physiological aging (D_{na}), and are killed by predators or pathogens (D_{np}).

B. Studies on Seed Bank Dynamics

Since the development of models for the dynamics of buried seed populations, much effort has been expended by various scientists to gather data to use in the equations. In these studies, data have been obtained for various components of the models for persistence and depletion of soil seed banks. Zorner *et al.* (1984a) buried seeds of *Avena fatua* at depths of 1, 3, 5, 10, 15, and 30 cm and determined the number of dead, germinated, and dormant seeds after 1, 2, 4, 6, 9, 12, 18, and 20 months. *In situ* germination decreased with an increase in depth of burial, but seed death (D_n) increased. The persistence of seeds was due to induced dormancy (P_{ind}); however, all seeds at each depth of burial were dead after 24 months. Dormant and nondormant seeds of *Kochia scoparia* buried at depths of 1, 3, 5, 10, 15, and 30 cm were exhumed after 1, 2, 4, 6, 9, 12, 18, 24, 30, and 36 months, and the number of dead, germinated, and dormant seeds was determined (Zorner *et al.,* 1984b). After 30 months of burial, <1 to 3% of the initially dormant and initially nondormant seeds were alive, regardless of depth of burial. Ninety-two percent of the seeds that lost viability died after germination *in situ*.

Many *Rumex crispus* and *R. obtusifolius* seeds sown on the surface of soil-filled concrete cylinders in the field and then "plowed" to a depth of 15 cm were lost from the seed bank by germination or death (D_n)

(Weaver and Cavers, 1979). About 48 and 31% of *R. crispus* and *R. obtusifolius* seeds, respectively, germinated and seedlings emerged; 40 and 65% of the seeds, respectively, died from unknown causes.

Jansen and Ison (1995) found that losses from the seed bank of *Trifolium balansae* and *T. resupinatum* were due to *in situ* germination (36.3 and 31.7%, respectively), ant predation (4.3 and 8.3%), and fungal attack (0.3 and 0.7%). The fate of buried *Medicago lupulina* seeds was monitored by exhuming bags of seeds at various times of the year: September 13, 1979; October 21, 1979; November 16, 1979; March 31, 1980; September 12, 1980; and June 21, 1980. The proportion of (1) germinated seeds was 0.8, 0.8, 0.8, 0.1, 39.5, and 48.3 for each sampling date, respectively; (2) dead seeds was 3.4, 3.4, 3.4, 7.6, 45.2, and 30.1; (3) seeds in enforced dormancy was 0.8, 0.8, 3.5, 63.8, 1.6, and 0; and (4) seeds in innate dormancy was 95.8, 95.8, 92.3, 28.5, 13.7, and 21.5 (Pavone and Reader, 1982).

Two hundred *Tragopogon pratensis* seeds each were placed in soil-filled mesh bags on the soil surface and at depths of 2 and 5 cm in 1989 and in 1990 (Qi *et al.,* 1996). After various periods of time for a total of 12 and 13 months for 1989 and 1990 seeds, respectively, seeds were separated from the soil and tested for germination in darkness at 25°C. The number of seedlings and dead seeds was determined for each bag, and the number of seeds in enforced dormancy and induced dormancy was determined from germination tests. At the end of 12–13 months, exhumed bags contained about 1–13 (out of 200 seeds) seedlings, 184–199 dead seeds, 1–4 seeds in enforced dormancy, and 0–1 seed in induced dormancy; the cause of seed death was not determined.

Seeds of the legumes *Acacia berlandieri* and *Leucaena pulverulenta* were placed in bags, buried in soil at depths of 3–5 cm in Texas, exhumed after 0, 1.5, 6, and 12 months, and tested for germination (Owens *et al.,* 1995). Seeds of *A. berlandieri* were permeable when the study was initiated, and 98% of them were dead after 1.5 months. Seeds of *L. pulverulenta* were impermeable when they were buried, and 71% of them were alive and impermeable after 12 months. Seed death was attributed to loss of physical dormancy followed by germination and/or decomposition.

A "balance sheet" approach has been taken in some seed bank studies. For example, seed input (or production), seed bank size at one time of year (March), and number of seedlings emerging were determined for six nonsprouting dominant shrubs in a fire-prone dune fynbos plant community of South Africa (Pierce and Cowling, 1991a). Seed input ranged from 0 to 16,698 seeds m^{-2}, depending on species, location, and year. High levels of predation of seeds on the soil surface and death of seeds due to *in situ* germination and decay resulted in all species having relatively small soil seed banks, ranging from 2.3 to 629 seeds m^{-2}.

Data on seed input; size of seed bank in October, February, and November; induction of dormancy; and seedling emergence were obtained for *Onopordum illyricum-bracteatum* at three sites in New South Wales, Australia (Cavers *et al.*, 1995). Seed banks in a pasture, oat field, and open *Eucalyptus albens* woodland in October 1987 (just before seed dispersal) were 5024, 712, and 2790 seeds m^{-2}, respectively. Seed production determined in these habitats at time of seed maturation (December 1987) was 172, 0, and 1788 m^{-2}, and seedling emergence in summer/autumn was 534, 137, and 461 m^{-2}. The number of strongly dormant seeds in the seed bank in the three habitats in November 1988 (before seed dispersal) was 891, 248, and 822 seeds m^{-2}.

Annual seed input, size of the seed bank, and removal of seeds from the hills via erosion were determined in a study of the badlands in southeastern Spain (Garcia-Fayos *et al.*, 1995). Seed input was 52 and 37 m^{-2} year^{-1} at the two sites during the first year and 53 and 54 m^{-2} year^{-1} the second year. Soil seed banks were 280 and 266 seeds m^{-2} in the two sites. Seed losses due to erosion following rainfall were 12.5 and 5.6% of the total seed bank in each area. Other seeds apparently died, but unfortunately damaged and empty seeds were disregarded in all aspects of the study.

A comprehensive study in some African savanna grasslands included seed production, viability of fresh seeds, seed bank size, and length of dormancy period (O'Connor and Pickett, 1992). These authors concluded that seeds of the perennial grasses did not live longer than 2 years and that the size of the seed bank depended on annual seed production. Studies in a Mediterranean grassland showed that grasses had smaller persistent seed banks than legumes (Russi *et al.*, 1992). Annual seed production was not determined in the Mediterranean grassland, but soil seed banks were determined seven times each year for 4 years, making it possible to follow annual fluctuations in the size of the seed bank. The seed bank was depleted due to seed consumption by cattle in summer and by germination in autumn–spring.

The pioneer rainforest tree *Cecropia obtusifolia* has high annual seed production but low levels of seed survival in the soil, partly due to predation and attack by pathogenic organisms (Alvarez-Buylla and Garcia-Barrios, 1991). Using models derived from a generalized Lefkovitch model, these authors showed that survival of pioneer species, like *C. obtusifolia,* that lack persistent seed banks is favored more by the creation of relatively small gaps (i.e., slash and burn patches) than by the extensive cutting of large tracts of forests.

Seeds of the winter annual *Collinsia verna* were placed in soil-filled fiberglass mesh bags in spring immediately following maturation, and after 1, 2, and 3 years some of the bags were destructively harvested to determine seed emergence, persistence, or disappearance (Kalisz, 1991). After 1, 2, and 3 years, 16, 12, and 6%, respectively, of the seeds were viable; 36, 6, and 3% of the seeds germinated in the first, second, and third autumns, respectively, after burial. Seeds were lost from the seed bank by *in situ* germination and death from unexplained causes (Kalisz, 1991).

Seed death during 10 months of burial was about 3, 60, and 42% in three families (progeny from three crosses) of the dioecious perennial *Silene latifolia* (Purrington and Schmitt, 1995). Whereas 51, 60, and 53% of the plants from 0-month-old seeds (which were nondormant) in the three families were females, 59, 70, and 59% of the plants from 10-month-old seeds, respectively, were females. Thus, there was a significant increase in females in plants resulting from 10-month-old seeds of the three families. The increase in females in two of the families may be due to more deaths of male than female seeds during the 10-month period of burial. However, in one family, where only about 3% of the seeds died, the change in the female : male ratio must be due to something other than the differential death of seeds.

C. Changes in Dormancy States

Neither the model of Schafer and Chilcote (1969, 1970) nor even the expanded version of this model published by Roberts (1972b) represents the dynamics of soil seed banks in terms of the annual dormancy cycle known to occur in buried seeds of many species (Chapter 4). Seeds with an annual dormancy/nondormancy or conditional dormancy/nondormancy cycle can germinate when they are nondormant if light : dark conditions and soil moisture are nonlimiting (Fig. 7.6). However, buried seeds of a few species remain continuously nondormant (e.g., Baskin and Baskin, 1985a). Responses of buried seeds to seasonal temperature changes (dormancy cycles) keep seeds of many species synchronized with changes in favorability of environmental conditions in the habitat for seedling survival. That is, buried seeds are nondormant when conditions are favorable for seedling establishment and eventual completion of the life cycle or annual growth cycle, and they are dormant at the time of year when conditions are unfavorable for these events.

Is the type of annual dormancy cycle, or lack thereof, related to the longevity of seeds in the soil seed bank? Information on changes in dormancy states is available for 13 of the species in the Duvel and/or Beal buried seed experiments that were alive after 39 or 40 years (Table 7.7). In all but two of the species (*Lepidium*

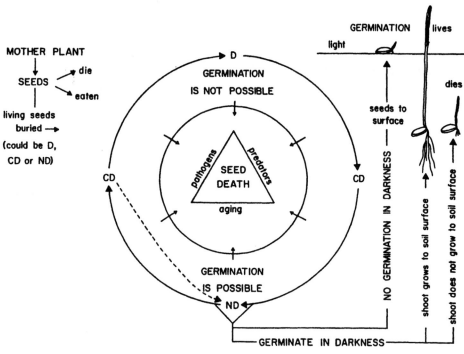

FIGURE 7.6 Conceptual model of the dynamics of the soil seed bank in relation to the annual dormancy cycle. From Baskin and Baskin, (1998), with permission.

TABLE 7.7 Species That Lived for 39 or 40 Years in the Duvel and/or Beal Buried Seed Studies

Species	Changes[a]	Reference
Amaranthus retroflexus	CD/ND/CD →	Baskin and Baskin (1977b)
Ambrosia artemisiifolia	D/CD/ND/CD/D →	Baskin and Baskin (1980a)
Capsella bursa-pastoris	D/CD/ND/CD/ND →	Baskin and Baskin (1989a)
Chenopodium album	CD/ND/CD →	Baskin and Baskin (unpublished results)
Lepidium virginicum	D/CD/ND	Baskin and Baskin (unpublished results)
Oenothera biennis	CD/ND/CD →	Baskin and Baskin (1994)
Phytolacca americana	CD/ND/CD →	Baskin and Baskin (unpublished results)
Rumex crispus	CD/ND	Baskin and Baskin (1985a)
R. obtusifolius	CD/ND →	van Assche and Vanlerberghe (1989)
Setaria glauca	D/CD/ND/CD/ND →	Baskin *et al.* (1996)
Thlaspi arvense	D/CD/ND/CD/D →	Baskin and Baskin (1989b)
Verbascum blattaria	CD/ND/CD →	Baskin and Baskin (1981)
V. thapsus	CD/ND/CD →	Baskin and Baskin (1981)

[a] Data are available on their changes in dormancy states during burial. An arrow (→) indicates that the cycle is repeated each year.

virginicum and *Rumex crispus*), buried seeds undergo annual changes in dormancy states. In *L. virginicum*, freshly matured seeds are dormant, they become conditionally dormant (CD) and then nondormant (ND), and they remain ND regardless of seasonal changes in environmental conditions (Baskin and Baskin, unpublished results). Fresh seeds of *R. crispus* are CD, but they quickly become ND and remain in this state (Baskin and Baskin, 1985a). Of the 11 species that undergo annual changes in dormancy states, 9 of them cycle between CD and ND, and the other two cycle between dormancy (D) and ND (Table 7.7). Seeds of two of the three species that have remained viable for 100 years in the Beal experiment, *Verbascum blattaria* and *V. thapsus,* cycle between CD and ND each year (Baskin and Baskin, 1981). Seeds of the third species, *Malva rotundifolia,* have physical dormancy.

D. *In Situ* Germination

One prerequisite for longevity of seeds in the soil is that they do not germinate while buried. Although germination is possible only when seeds are nondormant (Fig. 7.6), being nondormant does not necessarily mean that they will germinate. Various environmental factors, including darkness, CO_2/O_2 ratios, volatile metabolites, and dampened temperature fluctuations, that can prevent germination of buried seeds have been discussed at length in Chapter 4. In addition, Froud-Williams *et al.* (1984) found that burial of small-seeded arable weeds such as *Agrostis gigantea*, *Plantago major*, *Stellaria media,* and *Veronica arvensis* delayed germination, whereas burial of large-seeded species such as *Avena fatua* and *Galium aparine* enhanced germination. Perhaps the responses of small and large seeds after burial are related to their light requirements for germination, with light-requiring seeds being smaller than those that do not require light (Pons, 1991).

In many species, the length of time seeds remain viable and ungerminated in the soil increases with an increase in depth of burial (Dorph-Peterson, 1925; Toole and Brown, 1946; Tingley, 1961; Rampton and Ching, 1966, 1970; Banting, 1966; Taylorson, 1970; Dawson and Burns, 1975; Barralis and Chadoeuf, 1980; Howe and Chancellor, 1983; Zorner *et al.,* 1984b; Cheam, 1986; Harradine, 1986; Froud-Williams, 1987; Noble and Weiss, 1989; Bridgemohan *et al.,* 1991; Sahoo *et al.,* 1994). In a few cases, however, viability is not affected by depth of burial (Egley and Chandler, 1983) or it decreases with depth (Schwerzel, 1974; Mukherjee *et al.,* 1980; Zorner *et al.,* 1984a; Lonsdale *et al.,* 1988). Shallow burial of *Raphanus raphanistrum* seeds enhanced loss of dormancy and promoted *in situ* germination (Cheam, 1986). Decreases in seedling emergence

with increases in depth of burial reported for some species (e.g., Grundy *et al.,* 1996) are hard to interpret because no information is provided on what happened to the seeds/seedlings. That is, lack of seedling emergence could mean that seeds failed to germinate *in situ* or that they germinated but seedlings died before their shoots reached the soil surface.

Lack of soil disturbance also is important for longevity of buried seeds of many species. In fact, one way to reduce the seed bank in arable soils is to fallow a field for 1 or more years (Brenchley and Warington, 1936, 1945; Chepil, 1946; Chancellor, 1964; Froud-Williams *et al.,* 1984; Warnes and Andersen, 1984; Bridges and Walker, 1985). During fallowing, the land is cultivated or disturbed a number of times to prevent seed production and to stimulate the germination of buried seeds.

Overall, the number of buried, viable seeds for individual species, as well as the total for all species, declines faster in cultivated than in nondisturbed soil (Roberts and Feast, 1973b; Roberts and Dawkins, 1967). However, frequency (Fig. 7.7) and thoroughness of cultivation, rainfall, temperature at time of cultivation, life cycles of the species (Brenchley and Warington, 1945), and depth of disturbance (Chepil, 1946; Dessaint *et al.,* 1997) each play a role in determining what proportion of the seeds in the seed bank germinates. Further, the cropping system [e.g., maize (corn) vs wheat (van Esso *et al.,* 1986)] or a rotation of corn, soybeans, and wheat (Schreiber, 1992), season of year (Sharma *et al.,* 1983), kind of plow, i.e., tine vs. moldboard (Wilson, 1985), and soil type (Cardina *et al.,* 1991) can influence the rate of decline in the number of buried weed seeds associated with increased disturbance. Using transitional matrices to model seed survival, Jordan *et al.* (1995) showed that crop rotation greatly decreased seed banks of *Setaria viridis,* but it had little effect on those of *Abutilon theophrasti.*

It should be noted that an increase in seed germina-

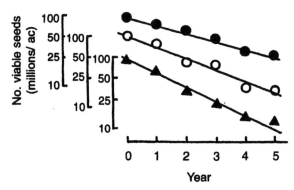

FIGURE 7.7 Number of viable seeds of five species of arable weeds in the top 22.5 cm of soil in plots cultivated 0 (●), 2 (○), and 4 (▲) times each year. From Roberts and Dawkins (1967), with permission.

tion as a result of soil disturbance does not change the periodicity of germination (Froud-Williams *et al.,* 1984). Further, germination of a few species such as *Gnaphalium uliginosum, Sonchus asper, Taraxacum officinale, Juncus bufonius,* and *Sagina procumbens* decreases with an increase in the frequency of cultivation (Chancellor, 1964).

If seeds germinate while they are buried, survival of the seedlings depends on whether their photosynthetically active parts are able to emerge from the soil. The emergence of seedlings varies with depth of burial and size (amount of stored food reserves) of the seed. Germination of buried seeds, followed by death of seedlings before emergence, has been documented in various species, including *Agathosma stenopetala* (Pierce and Cowling, 1991a), *Avena fatua* (Zorner *et al.,* 1984a), *Carlina vulgaris* (Pons, 1991), *Centaurea scabiosa* (Pons, 1991), *Felicia echinata* (Pierce and Cowling, 1991a), *Gentianella germanica* (Pons, 1991), *Leontodon hispidus* (Pons, 1991), *Lolium multiflorum, L. perenne* (Schafer and Chilcote, 1970), *Metalasia muricata* (Pierce and Cowling, 1991a), *Origanum vulgare* (Pons, 1991), *Rhinanthus alectorolophus, R. minor* (Pons, 1991), *Rumex crispus, R. obtusifolius* (Weaver and Cavers, 1979), *Scabiosa columbaria* (Pons, 1991), *Trifolium balansae, T. resupinatum* (Jansen and Ison, 1995), and *Vulpia fasciculata* (Watkinson, 1978).

E. Germination on the Soil Surface

Soil disturbances may enhance *in situ* germination and/or result in seeds being brought to the surface, where they are exposed to light, improved oxygen levels, and so on that promote germination. In some plant communities, e.g., seminatural grasslands in Sweden (Milberg, 1993), few seedlings appear (presumably only a few seeds germinate) unless the vegetation is disturbed. Similarity between seed rain and seed bank may increase with an increase in the disturbance of a community (Pierce and Cowling, 1991b; Chambers, 1993). In addition to plowing, seeds could be exposed to favorable germination conditions by tree falls (Putz, 1983), drainage of wetlands (Wienhold and van der Valk, 1989), or digging by animals (Chambers and MacMahon, 1994).

Earthworms are responsible for bringing many seeds to the soil surface *via* casts, which may accumulate on the soil surface at rates of 2–4 to 15–20 mm per year (Lee, 1985) or 750 g m^{-2} per year (Willems and Huijsmans, 1994). Thus, the potential for seed movement is enormous. Seeds of various numbers of species have been found in casts collected in different countries: 25, Wales (Grant, 1983); 2, The Netherlands (van Tooren and During, 1988); 19, The Netherlands (Willems and Huijsmans, 1994); and 12, England (Thompson *et al.*

1994). Cast production ceases in summer (Grant, 1983; Lee, 1985) and winter (Lee, 1985), which are seasons when one would expect poor, or no, germination of exhumed seeds. After seeds reach the soil surface, they may or may not germinate, depending on (1) the dormancy state (see above) and (2) whether or not temperature and soil moisture requirements are fulfilled for germination.

F. Seed Death Due to Aging

To appreciate natural senescence of seeds in the soil, some information is needed on what happens to them if they die during prolonged dry storage. In dry storage, the moisture content of seeds reaches an equilibrium with the relative humidity (RH) inside the storage container (Villiers, 1980); however, the equilibrium reached at a given RH varies with the species and depends on the chemical composition of the seeds. Proteins absorb large quantities of water, starches, and cellulose; an intermediate amount of water; and no lipids (Harrington, 1973).

Harrington (1973) suggested that a 1% reduction in seed moisture, or a 5°C reduction in temperature, doubles the length of life of seeds in dry storage. However, reductions of moisture content below 2–6% do not increase longevity (Roberts and Ellis, 1989). In fact, seeds can be dried so much [≤0.08 g water (g dry weight)$^{-1}$] that water has little or no mobility, which may increase the rate of membrane deterioration (Vertucci, 1993). Vertucci and Ross (1990) suggest that allowing seeds to equilibrate at relative humidities between 19 and 27% is optimal for long-term seed storage. The actual moisture content of seeds in dry storage frequently is within the range of 5–10% (Roberts, 1988).

If healthy seeds are dried to moisture levels of 5–15% at about room temperatures, cell membranes may become discontinuous, and the plasma membrane separates from the cell wall (Webster and Leopold, 1977). X-ray diffraction shows that when membranes become discontinuous, their phosphoplipids may change from having the hydrocarbon tails (which are hydrophobic) pointed inward to having them pointed outward (Simon, 1974). In dry seeds, the nucleus, mitochondria, plastids, protein bodies, and lipid droplets are recognizable, but the endoplasmic reticulum (ER) is not evident; Golgi bodies are scarce (Villiers, 1980). When dry (15% moisture content) seeds of *Glycine max* were allowed to imbibe water for 20 min, the plasma membrane became relatively intact and continuous, the ER appeared throughout the cytoplasm and around the edges of protein bodies, the distorted mitochondria became oval and filled with cristae, and starch grains appeared in proplastids (Webster and Leopold, 1977). Thus, dry seeds rapidly become structurally functional after they are allowed to imbibe water.

If seeds are kept in dry storage for long periods of time (or subjected to accelerated aging treatments, i.e., seed moisture contents of 12–18% and temperatures of 25–40°C), they lose viability. The amount of time required for seed death varies with the species and dry storage conditions. Some seeds of *Agrostis alba* and *Oryzopsis hymenoides* were alive after 20 years of dry storage in wooden sheds in Idaho and Utah, *Trifolium repens* after 25 years, *Lotus corniculatus* after 28 years, and *Erodium cicutarium* after 37 years (Hull, 1973). Seeds of wheat were alive after 31–33 years of dry storage in a wooden shed in Washington (Haferkamp *et al.*, 1953). Seeds of *Agrostemma githago, Avena sativa, Hordeum vulgare, Lolium temulentum, Sinapis alba, S. arvensis,* and *Vaccaria hispanica* germinated after storage for 100 years in hermetically sealed containers at 10–15°C (Steiner and Ruckenbauer, 1995). *Geranium bohemicum* seeds 129 years old germinated, but those older than 129 years did not (Milberg, 1994a). Tetrazolium tests indicated that seeds of *Hordeum leporinum* (200 years old), *Malva parviflora* (200, 195, 183), *Melilotus indicus* (183), *Chenopodium murale* (183), and *C. album* (143) taken from adobe bricks in California and northern Mexico were viable and one seed of *Medicago polymorpha* (200) germinated (Spira and Wagner, 1983).

A number of things happen as seeds lose viability. Macromolecules such as nucleic acids, membrane components, and enzymes are damaged (Berjak and Villiers, 1972a; Osborne *et al.*, 1980–1981), the structural integrity of membranes is lost (Berjak and Villiers, 1972b), and organelles are damaged and/or fail to differentiate into functional structures (Berjak and Villiers, 1972a,b,c). Usually damage at the subcellular level becomes even more noticeable after the start of imbibition (Berjak and Villiers, 1972c). If damage is too serve, hydrolytic enzymes in the lysosomes are released into the cytoplasm, causing increased damage (Villiers, 1973). Some repairs can be made at the beginning of the imbibition period, which accounts for the delay in germination of some old seeds (Berjak and Villiers, 1972c). Partial hydration or soaking in an osmoticum prior to allowing old seeds to imbibe further improves the amount of repair that may occur (Rao *et al.*, 1987a). As the aging process continues, the leakage of electrolytes increases (Parrish and Leopold, 1978), water uptake and respiratory activities decrease (Hallam *et al.*, 1973), and synthesis of protein, RNA (Sen and Osborne, 1977), DNA (Hallam *et al.*, 1973), and poly(A)polymerase (Grilli *et al.*, 1995) decreases.

Roberts and Abdalla (1968) found that survivorship curves for seeds in dry storage are negative cumulative

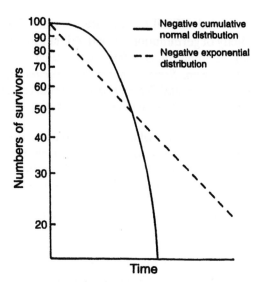

FIGURE 7.8 Comparison of survivorship curves for seeds stored dry in the laboratory (negative cumulative normal distribution) and imbibed in the soil (negative exponential distribution). From Cook (1980), with permission.

normal distributions (Fig. 7.8). A viability equation for seeds in dry storage was devised by Ellis and Roberts (1980):

$$v = K_i - p/10^{K_E - C_W \log m - C_H t - C_Q t^2},$$

where v is the probit of percentage viability, p is time in storage, m is the moisture content of seeds, t is the temperature in °C, K_i is the seed lot constant (seed quality as determined by genotype and prestorage environment), and K_E, C_W, C_H, and C_Q are constants within a species.

The storage conditions causing a loss of seed viability also lead to an accumulation of aberrant cells in the embryo (Abdalla and Roberts, 1968; Roberts, 1972a). These cells are identified on the basis of chromosome aberrations appearing during the first mitosis following seed hydration. The number of chromosome aberrations observed is a measure of how much damage has occurred during the storage period (Abdalla and Roberts, 1968). If enough genetic damage occurs, i.e., if most cells in the embryo are unable to divide properly, the seed does not germinate (Murata *et al.*, 1981) and may die (Roberts, 1988).

Cheah and Osborne (1978) found that DNA was fragmented during dry storage, and they suggested that loss of DNA integrity was the reason for chromosome aberrations. However, many observations on cells in germinating seedlings showed chromatid-type rather than chromosome-type aberrations. As Roberts (1988) explains, lesions occur in the DNA during dry storage, with the number depending on temperature and moisture content of the seeds. At the start of imbibition, some lesions in cells in the G_1 phase of the cell cycle are repaired; however, after cell replication starts, unrepaired lesions develop into chromatid-type aberrations. During the second mitosis, some aberrations will reappear as chromosome-type aberrations. If cells are in the G_2 phase of the cell cycle at the start of imbibition, chromatid-type aberrations will appear at the end of the second mitosis. At a critical moisture level (e.g., 13–18%) for each species, no cells have chromosome aberrations, but there is a high loss of seed viability. Below the critical moisture level and down to about 5.5%, the number of cells with chromosome aberrations observed at a given loss of viability of the seeds increases with each decrease in moisture content (Rao *et al.*, 1987b, 1988).

In contrast to the loss of viability at moisture contents of 10–20%, imbibed seeds at moisture contents of about 50% (Villiers, 1980) remained viable either in the soil (Barton, 1961) or in the laboratory (Villiers, 1971; Vazquez-Yanes and Orozco-Segovia, 1996) for many years. These imbibed seeds did not germinate because they were dormant or environmental conditions were unfavorable for germination. In an attempt to explain why imbibed seeds of *Lactuca sativa* were alive after 105 days and those stored dry in a high humidity chamber at 30°C were dead after 42 days, Toole and Toole (1953) suggested that ". . . dormancy of the imbibed seeds held in check or at least suppressed life processes that lead to seed deterioration." Villiers (1971) showed in short-term experiments that imbibed (but nongerminating) seeds could carry on a variety of activities, including interconversions of stored food materials, membrane synthesis, organelle production, and maintenance of macromolecules. Thus, genetic and structural damages do not accumulate in imbibed seeds, and some molecular damage, such as that sustained from gamma irradiation, can be repaired (Villiers, 1974; Villiers and Edgcumbe, 1975). If seeds are alternately dried and imbibed, damage occurring during the dry period is repaired when seeds are imbibed (Villiers and Edgcumbe, 1975).

DNA is synthesized at low, rather constant levels during long-term imbibition of seeds (Elder and Osborne, 1993; Gerth and Bernhardt, 1995); however, DNA synthesis increases greatly after germination begins (Gerth and Bernhardt, 1995). DNA synthesis during long periods of imbibition would repair damaged DNA and maintain the integrity of the genetic information (Osborne, 1980; Elder and Osborne, 1993; Gerth and Bernhardt, 1995). However, DNA is not replicated, and thus growth does not occur (Elder and Osbourne, 1993). DNA repair, but lack of replication, occurs in imbibed dormant as well as in nondormant seeds (Elder and Osbourne, 1993). In seeds stored dry for long periods of time, DNA repair precedes DNA replication (Osborne *et al.*, 1980–1981).

A number of cytological changes occurred in seeds of *Fraxinus excelsior* kept imbibed at 22–25°C for 6 years: lipid bodies disappeared, lobes formed on the nucleus, mitochondria enlarged, Golgi bodies increased, amyloplasts developed, cytolysome-like structures appeared, and rough ER became associated with protein bodies (Villiers, 1971). After dormancy was broken by cold stratification, germination was delayed 5–7 days in seeds previously kept at 22–25°C for 6 years but only 1–2 days in control seeds. During the 5- to 7-day delay period, cytological damage was repaired in the 6-year-old seeds, and the resulting seedlings were normal (Villiers, 1972). It is assumed that similar cytological damage could occur in seeds buried in soil for many years, but this information is not available. The next time seeds are exhumed in the Beal buried experiment, it would be interesting if someone would study the subcellular structure of the cells and determine if any chromosome aberrations have occurred. Levin (1990) proposed that seed banks might be a good place to find new genetic material (i.e., mutations), but this idea has not been confirmed.

If molecular and membrane repairs occur in imbibed seeds, why would seeds ever die? There are various possibilities (Osborne, 1980; Roos, 1989). (1) Food reserves are exhausted, and thus cells have no energy supply with which to make repairs. (2) Food reserves are altered chemically; therefore, they are no longer usable by the embryo. (3) So many lesions gradually accumulate in the DNA that cell division and subsequent growth become impossible. (4) Membranes become nonfunctional.

G. Seed Death Due to Activities of Soil Organisms

Soil organisms can be divided into three groups, depending on how they interact with buried seeds: (1) animals that predate or eat seeds, (2) fungi and bacteria that attack seeds, and (3) organisms that do not use seeds as a food source. We are concerned only with those organisms that can cause death of seeds.

1. Animals

According to Hole (1981), as many as a million species of animals live in the soil; however, only economically important pests such as the potato beetle and species influencing soil formation and structure have received very much attention by researchers. Animals live in the litter on the soil surface (Isopoda, Glomeridae, Diplopoda, Colembola), burrow in the soil (ants, badgers, bees, beetles, birds, crayfish, crickets, earthworms, lizards, millipedes, mole crickets, moles, potworms, snakes, spiders, and termites), or occupy the pores between soil particles (mites, protozoans, nematodes, and tardigrades) (Hole, 1981). Soil animals also can be subdivided according to size (Lal, 1987): microfauna (<0.2 mm), mesofauna (0.2–10 mm), macrofauna (10–20 mm), and megafauna (>20 mm).

Numerous studies have been done on the effects of pre- and postdispersal predation by various kinds of animals on the number of seeds that potentially enter seed banks (see review by Louda, 1989). However, in most cases, observations have been made on seeds only while they were in the transient seed bank (e.g., Castellanos and Molina, 1990). By the time seeds are old enough to be considered part of the persistent seed bank, their numbers have decreased and they may have become buried, making it difficult to monitor the effects of animal predation. Further, it is assumed that after seeds become buried they are safe from predators that could destroy them on the soil surface (Thompson, 1987). In deserts of the southwestern United States, however, many species of rodents are able to find and thus eat seeds buried in the soil (Clark and Comanor, 1973; Reichman, 1979; Johnson and Jorgensen, 1981; Price and Podolsky, 1989).

Based on size, it seems reasonable that macrofauna (e.g., Enchytraedae, Symphyla, Diptera larvae, Coleoptera, Diplopoda, Gastropoda, Chilopoda, Araneae, and other insects) and megafauna (e.g., earthworms and millipedes) could eat buried seeds. However, a literature search for information on organisms that feed on buried seeds only provided documentation of the ingestion (destruction) of seeds by Mollusca, i.e., slugs (Newell, 1967), Lumbridicidae, i.e., earthworms (Hurka and Haase, 1982; Grant, 1983), and ants (Holldobler and Wilson, 1990). Also, many species of ants in various plant communities throughout the world forage on seeds and store some of them in subterranean chambers. Depending on the species, ants store viable seeds or destroy the radicle, thus preventing the buried seeds from germinating (Holldobler and Wilson, 1990). Some Enchytraidae (oligochaetes) (O'Connor, 1967), Symphyla (arthopods), Diplopoda (millipedes), Isoptera (termites), Coleoptera (beetles), and Isopoda (woodlice) (Raw, 1967) eat plant remains in the soil; thus, it seems possible that they might eat some buried seeds.

2. Microorganisms

The most abundant soil bacteria are coccoid rods (globiforme bacteria); the majority of them belong to the family Corynebacteriaceae, which includes the genus *Cellulomonas*. This genus is of interest because it breaks down cellulose. Other important soil bacteria include the actinomycetes (Clark, 1967), which help decompose plant and animal residues by attacking cellulose, polysaccharides, and hemicelluloses, as well as other compounds (Kuster, 1967). Many kinds of fungi, including Phycomycetes, Ascomycetes, Basidiomycetes and Fungi Imperfecti, occur in soils, and they use a variety of organic materials as food sources (Warcup, 1967).

Although soil bacteria and fungi appear to be prime suspects in causing the death of some buried seeds, few field studies have addressed this problem. As reviewed by Kremer (1993), the death of buried seeds of a number of species has been attributed to fungi, but little or no proof is available. Fungicide treatments of field plots for buried *Mimosa pigra* seeds increased viability by only 10–16%, indicating that death due to soil fungi may not be very important in this species (Lonsdale, 1993). Kremer (1993) concluded that, "Only a small portion of weed seeds in soil succumb to microbial attack." In studies in a shrub-steppe ecosystem in Wyoming, 14–94% of buried seeds of *Artemisia tridentata, Poa canbyi,* and *Bromus tectorum* were decomposed (as opposed to *in situ* germination) during winter, whereas 0–7% of the seeds of *Oryzopsis hymenoides* and *Purshia tridentata* were decomposed (Crist and Friese, 1993).

Most information on the destruction of seeds by microorganisms comes from studies of economically important species such as *Glycine max, Zea mays, Arachis hypogaea, Triticum* spp., *Linum,* and *Sorghum* held in dry storage for long periods of time (Halloin, 1986). Fungi are the main organisms that attack these seeds. Some fungi, such as *Alternaria, Cladosporium, Fusarium,* and *Helminthosporium,* invade seeds while they are still in the field, but others, such as *Aspergillus* and *Penicillium,* attack seeds after they are placed in storage bins (Christensen and Kaufmann, 1969).

Fungi can damage/kill seeds by the production of (1) enzymes, including cellulases, pectinases, amylases, lipases, proteases, and nucleases, that destroy specific compounds in seeds and (2) toxins such as phytotoxins, mycotoxins, and tentoxin, which can cause the inhibition of germination, breakdown of membranes, and increased solute leakage from seeds (see review by Halloin, 1986). Water-soluble exudates, as well as volatile compounds, from seeds can promote the growth of fungi (Harman, 1983). For example, volatile carbonyls produced by *Pisum sativum* seeds aged at 30°C and 92% RH stimulated the germination of spores of various fungi, whereas those from seeds that were not aged did not (Harman *et al.,* 1978). Thus, as seeds age in dry storage, they produce compounds that increase their chances of being destroyed by fungi. It is unknown if the amount of volatiles increases as seeds buried in the soil age. Kremer (1993) proposed that the depletion of seeds in the persistent seed bank could be enhanced by finding ways to promote the growth of seed pathogens; this also means learning more about the seeds.

Microorganisms and/or their metabolic products are known to influence germination. For example, the germination of *Hordeum vulgare* was inhibited by a 4-hr exposure to culture medium containing nitrate in which *Azotobacter chroococcum* was growing; germination was promoted if the medium was nitrogen free (Harper and Lynch, 1980). Exposure of *H. vulgare* seeds to metabolic products from *Gliocladium roseum* and fermented straw also inhibited germination (Harper and Lynch, 1980). Microorganisms naturally associated with the fruit coat promoted the germination of *Cirsium vulgare* achenes if humus was present at the time of germination (van Leeuwen, 1981). The germination of *Fragaria ananassa* achenes was promoted by the presence of the fungi *Ulocladium charatarum*, *Apiospora montagnei*, *Arthrinium* sp., and *Cladosporium* sp., (Guttridge *et al.*, 1984).

Many seed-borne microorganisms reduce germination and/or kill seeds (see TeKrony *et al.*, 1987); however, the presence of one microorganism may prevent the growth of another. The bacterium *Pseudomonas* sp. from caryopses of the grass *Tripsacum dactyloides* inhibited growth of the fungi *Penicillium chrysogenum*, *Rhizopus stolonifer*, and *Trichoderma viride* (Anderson *et al.*, 1980). Inoculation of soybean seeds with *Cercospora kikuchii* reduced infection by *Diaporthe phaseolorum* (both organisms are fungi), thereby increasing germination percentages (Roy and Abney, 1977).

Many of the bacteria isolated from seed coats of *Abutilon theophrasti*, *Datura stramonium*, *Ipomoea hederacea*, and *Glycine max* inhibited the growth of various fungi (Kremer, 1987). Fungi, including *Alternaria alternata*, *Cladosporium cladosporioides*, *Epicoccum purpurascens*, and *Fusarium* spp., associated with seed coats of *A. theophrasti* prevented the growth of soil fungi (Kremer, 1986a). However, bacteria from within *A. theophrasti* seeds were antagonistic to fungi associated with the seed surface (Kremer *et al.*, 1984).

Other examples of antagonistic relationships between fungi that might play a role in seed germination include (1) *Helminthosporium victoriae* and *Chaeetomium* sp. (Tveit and Moore, 1954), (2) *Aspergillus flavus* and *A. niger* (Joffe, 1969), (3) *A. flavus* and *Penicillium funiculosum*, *P. rubrum*, and *Fusarium solani* (Joffe, 1969), and (4) *A. niger*, *P. funiculosum*, *P. rubrum*, and *F. solani* (Joffe, 1969).

If *Abutilon theophrasti* seeds are scarified, the growth of fungi associated with the seed coats may result in death of the embryo (Kremer *et al.*, 1984). In fact, the tiny holes made by insects feeding on developing *A. theophrasti* seeds allow entry of microorganisms and cause many seeds to be killed (Kremer and Spencer, 1989a). Feeding activities of insects on both pre- and postharvest seeds of economically important species are responsible for the introduction of pathogenic microorganisms into many seeds and eventual seed death (see review by Mills, 1983). Seeds of *A. theophrasti* that had been attacked by scentless plant bugs prior to dispersal were more susceptible to fungal attack and death after they were buried than were nonattacked seeds (Kremer and Spencer, 1989b). Although few buried seeds of *Oryzopsis hymenoides* were lost due to decomposition (most were lost by *in situ* germination), the presence of fungi inhibited the ant *Pogonomyrmex occidentalis* from taking the seeds (Crist and Friese, 1993).

Impermeable seed coats (i.e., species with physical dormancy) obviously would be a strong deterrent to fungi, but seed coats that are permeable to water also help resist the invasion of fungi (Halloin, 1983). Permeable seed coats can (1) serve as a mechanical barrier to the growth of hyphae (Halloin, 1983); (2) reduce the diffusion of compounds, including sugars, amino acids, ions, and proteins (Short and Lacy, 1976; Abdel Samad and Pearce, 1978; Duke *et al.*, 1983), that promote the growth of fungi in the vicinity of the seed (Agnihotri and Vaartaja, 1968); and (3) produce chemicals that inhibit the growth of fungi (Kraft, 1977).

Compounds with antifungal properties occur in seed coats of various species, including *Abutilon theophrasti* (Kirkpatrick and Bazzaz, 1979; Kremer, 1986b), *Colutea* spp. (Aguinagalde *et al.*, 1990), *Digitalis purpurea*, *Hypericum pulchrum* (Warr *et al.*, 1992), *Ipomoea hederacea* (Kirkpatrick and Bazzaz, 1979), *Medicago arborea*, and *M. strasseri* (Perez-Garcia *et al.*, 1992), and in the hulls of *Avena sativa* (Picman *et al.*, 1984). Compounds associated with decreased germination and/or growth of fungi include flavonoids (Paszkowski and Kremer, 1988; Aguinagalde *et al.*, 1990; Perez-Garcia *et al.*, 1992), phenolic acids (Picman *et al.*, 1984), and phenolic compounds (Kremer, 1986b). Tannins also may play a role in preventing infection of seed by fungi (Halloin, 1982; Kremer *et al.*, 1984). Germinating seeds of *Digitalis purpurea* produced a compound called cardenolide that inhibited the growth of fungi (Jacobsohn and Jacobsohn, 1985). Seeds of *Arachis hypogaea* and *Cicer arietinum* produced phytoalexins when they were incubated in the presence of fungi naturally associated with their seed coats. The phytoalexins potentially could prevent growth of other microorganisms (Keen, 1975).

Soil fungi separated from the soil and grown in closed containers in the laboratory produce nonvolatile and volatile inhibitors that inhibit the growth of test fungi (see review by Lockwood, 1977). Thus, it seems reasonable that these compounds might inhibit seed germination in the field, especially in poorly aerated soils. However, soil fungi (Hanke and Dollwet, 1976) and bacteria (Smith and Cook, 1974) produce ethylene, which can stimulate seed germination. Much remains to be learned about the chemical environment of buried seeds and how this influences persistence of seeds in the soil.

H. Seed Death: Effects of Fire or Flooding

The size of the seed bank in Amazonian caatinga forests was reduced by fire; however, some seeds of woody successional species survived, resulting in one woody plant m^{-2} only 4 months after the fire (Uhl *et al.*, 1981). Fires reduced the size of soil seed banks of shrubs belonging to the Restionaceae and Epacridaceae in southwestern Australia (Meney *et al.*, 1994) and of the grass *Sorghum intrans* in tropical regions of northern Australia (Watkinson *et al.*, 1989).

Fire had no effect on the number of buried viable grass seeds in eucalyptus forests of southeastern Queensland, Australia (Odgers, 1996), and no effect on soil seed banks in forest or fynbos plant communities in South Africa (Manders, 1990). In *Eragrostis eriopoda* tussock grasslands in Australia, fire significantly reduced the size of the soil seed bank of the invasive shrub *Dodonaea attenuata*. During fires, the *D. attenuata* shrubs were killed, and seed crops on plants were destroyed. Thus, seed banks declined due to a reduction in seed input but not via the death of seeds in the soil (Harrington and Driver, 1995).

Although seeds of many species are tolerant of flooding, e.g., Shull (1914), Nicholson and Keddy (1983), and Bonis *et al.* (1995), seeds of some species, e.g., Digitaria ischaemum (Comes *et al.*, 1978) and *Rumex acetosa* (Voesenek and Blom, 1992), are killed by flooding.

VI. EFFECTS OF SEED BANKS ON POPULATIONS

A. Survival in Risky Environments

Some habitats are risky for the survival of plant species, especially for annuals, because conditions may become so severe in some years that all individuals die before they have a chance to reproduce. For example, lack of soil moisture may result in no seed production by a species at a given site in a particular year (Epling *et al.*, 1960; Baskin and Baskin, 1975b, 1980b). As noted by Cohen (1966), one way for a species to survive in risky environments is to have a persistent seed bank. Thus, if a species fails to produce seeds in one year, the presence of a seed bank ensures that the species can persist at the site without immigration.

Cohen (1968) modeled the conditions needed to optimize the long-term growth rate of a population in a randomly varying environment. That is, the habitat is sometimes unfavorable for seed production, but there is no way of knowing in advance when reproduction will be prevented. This model can be used to find the optimal size of the seed bank:

$$S_{N+1} = S_N[(1 - G)V_N + G(Y_N + V_N)],$$

where S_N is the size of the seed bank at the beginning of interval N, G is the fraction of seeds in the seed bank that germinates, V_N is the survival fraction of seeds in the soil, Y_N is the average number of seeds produced by a germinated seed, and N is the interval.

If there is a high probability of failure to produce seeds, survival of the population requires that (1) plants produce many seeds when they do reproduce, (2) seeds remain viable for many years in the soil, and (3) only a relatively small proportion of the seeds in the soil germinates each year. If there is a high probability for successful seed production, the life span of seeds in the soil is less critical and a relatively large proportion of them can germinate (MacDonald and Watkinson, 1981). According to the model, the fraction of seeds that germinates is about equal to the probability of the species having a high seed yield in a given year (Cohen, 1966). To maximize survival of a species at a habitat site, Cohen (1968) determined that the optimal germination fraction decreases when (1) expectation of a high seed yield decreases, (2) variability of the yield increases, and (3) expectation of seed survival increases.

In another study, Cohen (1967) constructed a model for organisms that optimize their reproduction in a randomly varying environment, but can receive and respond to clues from the habitat:

$$S_{N+1} = S_N[V_N + G_k(Y_N - V_N)].$$

According to the model (Cohen, 1967), population growth rates increase as more information becomes available from the environment. The optimum strategy, however, takes into account the long-term probability of reproductive success and degree of correlation between cues from the environment and the subsequent production of seeds.

The model can be applied to seed banks where S is the size of the seed bank, G_k is the fraction of seeds in a seed bank germinating in response to a certain set of environmental conditions, N is the interval, V is the survival fraction of seeds in the soil each year, and Y is the average number of seeds produced by a germinating seed.

Seeds receive information about their environment in various ways. Changes in the magnitude of daily temperature fluctuations indicate that seeds have been brought to, or near, the soil surface or that a gap has been formed in the plant canopy (Chapter 6), exposure to light means that they are on the soil surface (Chapter 4), changes in the red:far-red ratio indicate that a gap has been formed in the plant canopy (Chapter 4), and the presence of stimulatory chemicals reveals the presence of roots of host plants (Musselman, 1980).

If S_N, V_N, and Y_N are held constant, an increase in G_k results in an increase in the size of the seed bank. That is, as more information becomes available from the environment, more seeds in the seed bank germinate. Plants from these germinating seeds produce seeds, resulting in a net increase in the size of the seed bank.

The interaction between seed dispersal and dormancy strategies can be influenced by the presence of seed banks (Cohen and Levin, 1985). Generally, maximum fitness of a species involves a decrease in seed dormancy with an increase in dispersal and vice versa (see Chapter 12, Section III). However, if seed banks are formed, they ". . . reduce the variability of the densities of germinating seeds in the patches" (Cohen and Levin, 1985). Consequently, the optimal number of seeds that should be dispersed for maximum fitness to occur is decreased in proportion to the size of the seed bank (Cohen and Levin, 1985).

B. Population Stability

Seed banks can have various effects on population dynamics, especially of annual species. These effects can include a speeding up of the oscillatory return to equilibrium or a slowing down of the exponential return to equilibrium (MacDonald and Watkinson, 1981).

In a year when the population growth rate of *Collinsia verna* was increasing, inclusion of a persistent seed bank did not affect the population growth rate estimated by bootstrapped analysis (Kalisz and McPeek, 1992). However, when the population growth rate was decreasing, inclusion of a persistent seed bank in the analysis significantly lowered the estimate. Thus, the seed bank had a greater demographic effect in a poor than in a good year for population growth.

Kalisz and McPeet (1993) used data from long-term studies of the population dynamics of *Collinsia verna* in simulations to examine the effects of good and bad years and the autocorrelation between conditions in consecutive years on the population growth rate, extinction rate, and time to extinction. The frequency of good years varied from 0.10 to 0.90 and autocorrelations varied between −0.90 and +0.90. Population growth rates were increased by the presence of a seed bank unless the frequency of good years was ≥0.7 and autocorrelations were ≤0.7. As expected, the presence of seed banks decreased time to extinction, with maximum delay oc-

curring at a good year frequency of 0.5 and an autocorrelation of near 0.0.

C. Population Variability

Although the actual number of individuals of a species might be relatively small at a population site, genetic variation is increased by the presence of a seed bank (Stewart and Porter, 1995). In fact, genetic variations stored in the seed bank may decrease the genetic change and thus decrease the rate of evolution (Hariston and De Stasio, 1988). In the winter annual *Limnanthes parryae,* which grows in the Mojave Desert of California, the population in any given year may be composed of individuals derived from seeds that matured over the past 10 years (Epling *et al.,* 1960). Because of the seed bank, local genetic deviations in populations of *L. parryae* are quickly returned to normal (Epling *et al.,* 1960).

Gottlieb (1974) found that a seed bank prevented or retarded genetic responses in the summer annual *Stephanomeria exigua* ssp. *coronaria* in eastern Oregon. Regardless of size of the mating group each year, the allelic frequency of the offspring fluctuated within a very narrow amplitude, and rare alleles were not lost from the population of *S. exigua* ssp. *coronaria.* After a drought year in which few seeds of the winter annual *Phacelia dubia* were produced, the allelic composition, heterozygosity, and mixed mating system of the population remained constant, due in part to the presence of a seed bank (Del Castillo, 1994).

Disturbance of a population site may be correlated with an increase in genotypes. For example, more genotypes are found in frequently disturbed habitats of the weedy winter annual *Capsella bursa-pastoris* than in infrequently disturbed ones. Apparently, buried seeds (and thus genotypes) are more likely to be brought to the soil surface in highly disturbed than in infrequently disturbed sites (Bosbach *et al.,* 1982).

Using starch gel electrophoresis to genetically analyze seeds in the soil and seedlings of *Lesquerella fendleri,* Cabin (1996) found differences in allele frequencies and distribution of multilocus heterozygosity (but not in mean heterozygosity) between seedlings and seeds. He suggested that the seedlings represented a nonrandom genetic subset of the *L. fendleri* soil seed bank. In *Plantago lanceolata,* allele frequencies for four of the five polymorphic loci studied differed between seeds in the seed bank and adult plants (Tonsor *et al.,* 1993). The frequency of specific alleles changed (and homozygosity decreased) with a progression of life cycle stages in *P. lanceolata* from seeds in the seed bank to seedlings to adults. Thus, although certain genetic variations may occur in the seed bank, they may be lost via plant death before plants reach reproductive maturity.

Divergence in allele frequencies (F_{ST}) was significantly lower for seeds in the soil seed bank of *Cecropia obtusifolia* than for those on the tree or in the seed rain (Alvarez-Buylla *et al.*, 1996). The authors suggested that the lack of genetic structure among seeds in the soil seed bank of *C. obtusifolia* might be due to (1) inadequate sampling, (2) secondary dispersal of seeds, or (3) differential accumulation of certain genotypes in the soil seed bank.

Plants were grown from recent and old (197 ± 89 years) seeds of *Luzula parviflora* buried by a solifluction lobe in Alaska. Plants from young seeds were larger (Bennington *et al.*, 1991), had more leaves, grew better at high temperatures, and were more sensitive to increased plant densities than those from old seeds (Vavrek *et al.*, 1991). According to Bennington *et al.* (1991) and Vavrek *et al.* (1991), genetic differences in growth characteristics of *L. parviflora* were preserved in the seed bank. However, there is a possibility that differences in growth characteristics were due to genetic deterioration in the old seeds. Plants of *Eriophorum vaginatum* grown from seeds in the seed bank differed from established plants with regard to morphological, growth, and flowering characteristics; the differences were attributed to genetics (McGraw, 1993).

A demonstration of the population genetic effect of a species not having a seed bank is found in the genus *Clarkia*. Lewis (1962) concluded that periodic extinction and recolonization have played important roles in the establishment of genetic deviants and evolution of new species of *Clarkia*, i.e., catastrophic selection was a major factor in speciation. An important condition for catastrophic speciation is that persistent seed banks are not formed; Lewis (1962) found evidence for lack of seed banks in *C. deflexa*. Thus, after the *Clarkia* population at a site was prevented from producing seeds by some catastrophic event, the habitat was available for occupation by unique genotypes because seed banks were not present. If seed banks were present, the "normal" genotypes in the seed bank would, no doubt, swamp out the unique ones and prevent speciation from occurring.

It turns out that some species of *Clarkia* do form seed banks. For example, low- and high-elevation populations of *C. williamsonii* in the Sierra Nevada Mountains, California, had seed banks (Price *et al.*, 1985). Further, translocation heterozygosity was higher in the low- than in the high-elevation populations of *C. williamsonii*, and genetic variation for a number of loci increased with an increase in translocation heterozygosity. A reduction in the number of plants during drought years in the low-elevation populations caused inbreeding, the effects of which were buffered by translocation heterozygosity. Some of the translocation heterozygosity was conserved via the seed bank. High-elevation populations were considered to be "pioneers," thus accounting for their reduced translocation heterozygosity. Also, high-elevation populations were much less likely to undergo catastrophic reduction in plant numbers than low-elevation ones (Price *et al.*, 1985).

A source of new genetic material in seed banks of some species could be transgenes. Genetic engineering in crop species, such as *Brassica rapa*, could result in cultivars that become aggressive seed bank-forming weeds or in new genes being introduced into wild relatives via the formation of hybrid populations and hybrid back-crossing with wild populations (Adler *et al.*, 1993; Linder and Schmitt, 1994; Schmitt and Linder, 1994). Weed biologists are concerned that transgenes will increase the vigor and aggressiveness of the weedy relatives. Needless to say, the "pros" and "cons" of transgenic crop plants have been debated (see Crawley, 1993; Kareiva, 1993). Gene flow from crop plants to wild relatives has occurred in *Helianthus* (Rieseberg *et al.*, 1996), *Fragaria* (Spira *et al.*, 1996), and *Brassica* (Jorgensen, 1996). Thus, gene flow from transgenic plants to wild relatives is possible (Brown *et al.*, 1996). Also, transgenes can be stored in the seed bank (Linder and Schmitt, 1995), which probably would make it almost impossible to eradicate them from the population.

D. Effect of Seed Banks on Selection of Other Traits

Levin and Wilson (1978) and Templeton and Levin (1979) examined the effects of seed banks on the evolution of traits that control the reproductive success of annual species. It was assumed that each trait was controlled by a single locus with two alleles. When loci interact with the environment, seed banks may act as evolutionary filters that increase the selective importance in some years and decrease it in others. Some genotypes are better adapted to good years and others to poor (usually drier) years. Each year, the genotype best adapted to the environment will contribute more seeds to the seed bank than genotypes less well adapted to the environment of that year. However, in years with suboptimal growth conditions, the best-adapted genotypes will produce fewer seeds than they do in years with optimal growing conditions.

The consequence of production of more seeds in good years than in poor years is that the genotypes best adapted to good years can make a larger contribution to the seed bank than those best adapted to poor years. Thus, the seed bank with its preponderance of seeds produced by the best genotypes in good years minimizes the genetic impact of seeds produced by the best genotypes in poor years (Levin and Wilson, 1978; Templeton and Levin, 1979) and thus prevents the population from responding genetically to the environment each year (Templeton and Levin, 1979).

Brown and Venable (1986) used an integrated model that allowed between-year dormancy and nonseed bank traits that affect specialization to conditions found in different kinds of years to evolve simultaneously. In other words, seed dormancy and postgermination adaptations evolve simultaneously. More evolutionary shifts occur in dormancy and postgermination adaptations if the probability of seed longevity in the seed bank changes than if environmental conditions reduce the reproductive success of the plants. As the probability of bad years increases, annual species initially adapt through changes in nonseed bank traits, but eventually a seed bank strategy is favored. If an annual species has a seed bank, little ecotypic differentiation is expected along aridity gradients, and in the same habitat, species with seed banks have more traits adapted to mesic conditions than those without seed banks (Brown and Venable, 1986).

Evans and Cabin (1995) suggested an experimental design to determine if seed dormancy has influenced the evolution of postgermination traits in the winter annual *Lesquerella fendleri.* For this species, they found (1) yearly variations in habitat environmental conditions influence germination and seedling survival, (2) differences in growth and seed production between individuals in populations, (3) persistent seed banks, and (4) differences in allozyme frequencies between seeds in the seed bank and those in seedlings. However, the authors note that the question they need to answer is "What genetic and phenotypic differences exist between individuals that germinate and those that remain dormant?"

E. Density-Escaping Characteristics

Venable and Brown (1988) assumed that seed (bank) dormancy, size, and dispersibility were genetically variable traits. They then showed that various things, including the number of independent environmental patches, probability of favorable conditions, spatial and temporal autocorrelation of environmental conditions (i.e., degree to which conditions in neighboring patches of habitat are the same), and dispersal radius, can alter the selection of seed traits. Seed dormancy (seed bank), size, and dispersal are adaptations that reduce the risk of population extinction and the effects of local crowding or competition with siblings (Venable, 1989). Risk-reducing properties of these three seed traits evolve when there is global temporal variance (Venable and Brown, 1988). However, density-escaping properties of seed dormancy, size, and dispersal may evolve even in environments that lack uncertainty if there is local spatial and/or temporal variation in environmental conditions or spatial structure of the habitat (Venable, 1989). Spatial, temporal, or structural variation in habitat favorability for growth results in opportunities for seedlings to escape effects of sibling competition. Models to show how seed banks allow individuals of a species to escape the effects of density have been developed for a single patch environment (Ellner, 1985a,b) and for a globally risk-free environment (Levin et al., 1984).

F. Coexistence of Species

Although a group of species may not be able to coexist in a constant environment, it can do so in a variable one (Chesson, 1986). Variability in environmental quality in both time and space creates patchiness, which greatly increases the potential for coexistence. Thus, if patchiness is increased by disturbance or by some other means, the diversity of species will increase (Levin et al., 1984).

The presence of a seed bank plays an important role in the coexistence of species at a habitat site (Chesson, 1986; Rees and Long, 1992). Because of interspecific competition, each species may have a number of years when seed production is quite low (Chesson, 1986). However, even if the seed set is low, species are able to persist at the site because of the seed bank. For each species, there will be occasional years when environmental variation causes competition to be reduced, and a large number of seeds will be produced. Thus, the seed bank is replenished (Chesson, 1986).

Coexisting species vary in the proportion of seeds germinating each year (Venable et al., 1993; Pake and Venable, 1996). Habitat disturbance plays a role in species coexistence (Lavorel and Chesson, 1995), with species differing in their germination responses to disturbance (e.g., Armesto and Pickett, 1985). Also, the timing of favorable conditions for germination, e.g., occurrence of rainfall (Hobbs and Mooney, 1991) or drawdown of water levels to expose lake shores (Schneider, 1994), can cause varying proportions of seeds of each species to germinate each year, resulting in yearly differences in community composition.

Ellner (1987) points out that in models where life history parameters are static, coexistence is promoted at either minimum or maximum levels of environmental variation. In models where evolutionary dynamics are considered, however, coexistence is maximized at intermediate levels of environmental variation or intermittent stress. Coexistence is promoted by the evolution of a trade-off between seed survivorship and seed yield (Ellner, 1987).

VII. PERSISTENCE OF COMMUNITIES

Seed banks are important in the long-term survival of individual species, as well as plant communities. However, not all species in a community are represented in the seed bank (e.g., Oosting and Humphreys, 1940; Major and Pyott, 1966; Barbour and Lange, 1967; Donelson and Thompson, 1980; Hopkins and Graham, 1983; Vahlos and Bell, 1986; Partridge, 1989; Bakker *et al.,* 1991; Rico-Gray and Garcia-Franco, 1992; Chambers *et al.,* 1993; Kinucan and Smeins, 1992; Ingersoll and Wilson, 1993; Milberg and Persson, 1994; Milberg, 1995; Milberg and Hansson, 1994; Ungar and Woodell, 1996). However, some species are present in seed banks that do not occur in the extant vegetation (e.g., Cheke *et al.,* 1979; Brown and Oosterhuist, 1981; Rabinowitz, 1981; Jefferson and Usher, 1987; Abrams, 1988; Wienhold and van der Valk, 1989; Matlack and Good, 1990; Middleton *et al.,* 1991; Milberg, 1992; Khan, 1993; Milberg and Hansson, 1994).

If habitat disturbance results in the death of plants, seed banks obviously ensure continuation of a species at the site (e.g., Hayashi and Pancho, 1978; Zimmergren, 1980; Hill and Stevens, 1981; Uhl *et al.,* 1981; Miller and Cummins, 1987; Lawton and Putz, 1988; Ebersole, 1989; Bonis *et al.,* 1995). Also, seed banks are receiving increasing attention as a means of restoration of various kinds of plant communities, including eucalyptus forests (Bell *et al.,* 1990), temperate grasslands (Bakker *et al.,* 1991; McDonald, 1993), tundra grasslands (Gartner *et al.,* 1983), freshwater wetlands (Vivian-Smith and Handel, 1996; van der Valk *et al.,* 1992), dry tropical ecosystems (Skoglund, 1992), and heathlands (Putwain and Gillham, 1990). Seed banks are important in revegetating lands that have been severely disturbed by mining activities (e.g., Iverson and Wali, 1982; Vivian-Smith and Handel, 1996).

It should be emphasized that the seed banks of some plant communities do not contain a good representation of the species in the community and/or total numbers of seeds are low. Therefore, seed banks may not be very helpful in restoration efforts in various communities, including Afromontane forests (Teketay and Granstrom, 1995), alvars of Oland, Sweden (Bakker *et al.,* 1996), *Eucalyptus marginatus* forests in Australia (Bell *et al.,* 1990), humid tropical forests in India (Chandrashekara and Ramakrishnan, 1993), salt marshes (Hutchings and Russell, 1989), sand dunes (Planisek and Pippen, 1984), semiarid grasslands (Coffin and Lauenroth, 1989), and subarctic forests (Johnson, 1975).

References

Abdalla, F. H., and Roberts, E. H. (1968). Effects of temperature, moisture, and oxygen on the induction of chromosome damage in seeds of barley, broad beans, and peas during storage. *Ann. Bot.* **32,** 119–136.

Abdel Samad, I. M., and Pearce, R. S. (1978). Leaching of ions, organic molecules, and enzymes from seeds of peanut (*Arachis hypogea* L.) imbibing without testas or with intact testas. *J. Exp. Bot.* **29,** 1471–1478.

Abrams, M. D. (1988). Effects of burning regime on buried seed banks and canopy coverage in a Kansas tallgrass prairie. *Southw. Nat.* **33,** 65–70.

Adler, L. S., Wikler, K., Wyndham, F. S., Linder, C. R., and Schmittt, J. (1993). Potential for persistence of genes escaped from canola: Germination cues in crop, wild, and crop-wild hybrid *Brassica rapa. Funct. Ecol.* **7,** 736–745.

Agnihotri, V. P., and Vaartaja, O. (1968). Seed exudates from *Pinus resinosa* and their effects on growth and zoospore germination of *Pythium afertile. Can. J. Bot.* **46,** 1135–1141.

Aguinagalde, I., Perez-Garcia, F., and Gonzalez, A. E. (1990). Flavonoids in seed coats of two *Colutea* species: Ecophysiological aspects. *J. Basic Microbiol.* **30,** 547–553.

Ahlgren, C. E. (1974). Effects of fires on temperate forests: North Central United States. In: "Fire and Ecosystems" (T. T. Kozlowski and C. E. Ahlgren, eds.), pp. 195–223. Academic Press, New York.

Albrecht, H., and Forster, E.-M. (1996). The weed seed bank of soils in a landscape segment in southern Bavaria. I. Seed content, species composition and spatial variability. *Vegetatio* **125,** 1–10.

Altamirano, R. M., and Guevara, S. (1982). Ecologia de la vegetacion de dunas costeras: Semillas en el suelo. *Biotica* **7,** 569–575.

Alvarez-Buylla, E. R., Chaos, A., Pinero, D., and Garay, A. A. (1996). Demographic genetics of a pioneer tropical tree species: Patch dynamics, seed dispersal, and seed banks. *Evolution* **50,** 1155–1166.

Alvarez-Buylla, E. R., and Garcia-Barrios, R. (1991). Seed and forest dynamics: A theoretical framework and an example from the neotropics. *Am. Nat.* **137,** 133–154.

Alvarez-Buylla, E. R., and Martinez-Ramos, M. (1990). Seed bank versus seed rain in the regeneration of a tropical pioneer tree. *Oecologia* **84,** 314–325.

Andersen, A. N. (1989). Pre-dispersal seed losses to insects in species of *Leptospermum* (Myrtaceae). *Aust. J. Ecol.* **14,** 13–18.

Anderson, R. C., Liberta, A. E., Packheiser, J., and Neville, M. E. (1980). Inhibition of selected fungi by bacterial isolates from *Tripsacum dactyloides. Plant Soil* **56,** 149–152.

Andrew, M. H. (1986). Population dynamics of the tropical annual grass *Sorghum intrans* in relation to local patchiness in its abundance. *Aust. J. Ecol.* **11,** 209–217.

Andrew, M. H., and Mott, J. J. (1983). Annuals with transient seed banks: The population biology of indigenous *Sorghum* species of tropical north-west Australia. *Aust. J. Ecol.* **8,** 265–276.

Archibold, O. W. (1979). Buried viable propagules as a factor in postfire regeneration in northern Saskatchewan. *Can J. Bot.* **57,** 54–58.

Archibold, O. W. (1981). Buried viable propagules in native prairie and adjacent agricultural sites in central Saskatchewan. *Can. J. Bot.* **59,** 701–706.

Archibold, O. W. (1984). A comparison of seed reserves in arctic, subarctic, and alpine soils. *Can. Field Nat.* **98,** 337–344.

Armesto, J. J., and Pickett, S. T. A. (1985). Experiments on disturbance in old-field plant communities: Impact on species richness and abundance. *Ecology* **66,** 230–240.

Ashton, D. H. (1986). Viability of seeds of *Eucalyptus obliqua* and *Leptospermum juniperinum* from capsules subjected to a crown fire. *Aust. For.* **49,** 28–35.

Aziz, S., and Khan, M. A. (1995). Role of disturbance on the seed bank and demography of *Leucus urticifolia* (Labiatae) populations in a maritime subtropical desert of Pakistan. *Int. J. Plant Sci.* **156**, 834–840.

Badger, K. S., and Ungar, I. A. (1994). Seed bank dynamics in an inland salt marsh, with special emphasis on the halophyte *Hordeum jubatum* L. *Int. J. Plant Sci.* **155**, 66–72.

Bakker, J. P., Bakker, E. S., Rosen, E., Verweij, G. L., and Bekker, R. M. (1996). Soil seed bank composition along a gradient from dry alvar grassland to *Juniperus* shrubland. *J. Veg. Sci.* **7**, 165–176.

Bakker, J. P., Bos, A. F., Hoogveld, J., and Muller, H. J. (1991). The role of the seed bank in restoration management of semi natural grasslands. *In* "Terrestrial and Aquatic Ecosystems: Perturbation and Recovery" (O. Ravera, ed.), pp. 449–455. Ellis Horwood, New York.

Bakker, J. P., Poschlod, P., Strykstra, R. J., Bekker, R. M., and Thompson, K. (1996). Seed banks and seed dispersal: Important topics in restoration ecology. *Acta Bot. Neerl.* **45**, 461–490.

Baldwin, A. H., McKee, K. L. and Mendelssohn, I. A. (1996). The influence of vegetation, salinity, and inundation on seed banks of oligohaline coastal marshes. *Am. J. Bot.* **83**, 470–479.

Ball, D. A. (1992). Weed seedbank response to tillage, herbicides, and crop rotation sequence. *Weed Sci.* **40**, 654–659.

Banting, J. D. (1966). Studies on the persistence of *Avena fatua. Can. J. Plant Sci.* **46**, 129–140.

Barbour, M. G., and Lange, R. T. (1967). Seed populations in some natural Australian topsoils. *Ecology* **48**, 153–155.

Barralis, G., and Chadoeuf, R. (1980). Etude de la dynamique d'une communaute adventice. I. Evolution de la flore adventice au cours du cycle vegetatif d'une culture. *Weed. Res.* **20**, 231–237.

Barralis, G., Chadoeuf, R., and Lonchamp, J. P. (1988). Longevite des semences de mauvaises herbes annuelles dans un sol cultive. *Weed Res.* **28**, 407–418.

Barton, L. V. (1961). "Seed Preservation and Longevity." Interscience, New York.

Baskin, C. C., and Baskin, J. M. (1998). Ecology of seed dormancy and germination in grasses. *In* "Population Ecology of Grasses" (G. P. Cheplick ed.). Cambridge Univ. Press, Cambridge (in press).

Baskin, C. C., Baskin, J. M., and El-Moursey, S. A. (1996). Seasonal changes in germination responses of buried seeds of the weedy summer annual grass *Setaria glauca. Weed Res.* **36**, 319–324.

Baskin, J. M., and Baskin, C. C. (1975a). Ecophysiology of seed dormancy and germination in *Torilis japonica* in relation to its life cycle strategy. *Bull. Torrey Bot. Club* **102**, 67–72.

Baskin, J. M., and Baskin, C. C. (1975b). Seed dormancy in *Isanthus brachiatus* (Labiatae). *Am. J. Bot.* **62**, 623–627.

Baskin, J. M., and Baskin, C. C. (1977a). Germination ecology of *Sedum pulchellum* Michx. (Crassulaceae). *Am. J. Bot.* **64**, 1242–1247.

Baskin, J. M., and Baskin, C. C. (1977b). Role of temperature in the germination ecology of three summer annual weeds. *Oecologia* **30**, 377–382.

Baskin, J. M., and Baskin, C. C. (1978). The seed bank in a population of an endemic plant species and its ecological significance. *Biol. Conserv.* **14**, 125–130.

Baskin, J. M., and Baskin, C. C. (1980a). Ecophysiology of secondary dormancy in seeds of *Ambrosia artemisiifolia. Ecology* **61**, 475–480.

Baskin, J. M., and Baskin, C. C. (1980b). Role of seed reserves in the persistence of a local population of *Sedum pulchellum*: A direct field observation. *Bull. Torrey Bot. Club* **107**, 429–430.

Baskin, J. M., and Baskin, C. C. (1981). Seasonal changes in germination responses of buried seeds of *Verbascum thapsus* and *V. blattaria* and ecological implications. *Can. J. Bot.* **59**, 1769–1775.

Baskin, J. M., and Baskin, C. C. (1984a). Germination ecophysiology of the woodland herb *Osmorhiza longistylis* (Umbelliferae). *Am. J. Bot.* **71**, 687–692.

Baskin, J. M., and Baskin. C. C. (1984b). Role of temperature in regulating timing of germination in soil seed reserves of *Lamium purpureum* L. *Weed Res.* **24**, 341–349.

Baskin, J. M., and Baskin, C. C. (1984c). The ecological life cycle of *Campanula americana* in northcentral Kentucky. *Bull. Torrey Bot. Club* **111**, 329–337.

Baskin, J. M., and Baskin, C. C. (1985a). Does seed dormancy play a role in the germination ecology of *Rumex crispus*? *Weed Sci.* **33**, 340–343.

Baskin, J. M., and Baskin, C. C. (1985b). Role of dispersal date and changes in physiological responses in controlling timing of germination in achenes of *Geum canadense. Can. J. Bot.* **63**, 1654–1658.

Baskin, J. M., and Baskin, C. C. (1986a). Seasonal changes in the germination responses of buried witchgrass (*Panicum capillare*) seeds. *Weed Sci.* **34**, 22–24.

Baskin, J. M., and Baskin, C. C. (1986b). Change in dormancy status of *Frasera caroliniensis* seeds during overwintering on parent plant. *Am. J. Bot.* **73**, 5–10.

Baskin, J. M., and Baskin, C. C. (1989a). Germination responses of buried seeds of *Capsella bursa-pastoris* exposed to seasonal temperature changes. *Weed Res.* **29**, 205–212.

Baskin, J. M., and Baskin, C. C. (1989b). Role of temperature in regulating timing of germination in soil seed reserves of *Thlaspi arvense* L. *Weed Res.* **29**, 317–326.

Baskin J. M., and Baskin, C. C. (1990). Seed germination ecology of poison hemlock, *Conium maculatum. Can. J. Bot.* **68**, 2018–2024.

Baskin, J. M., and Baskin, C. C. (1994). Germination requirements of *Oenothera biennis* seeds during burial under natural seasonal temperature cycles. *Can. J. Bot.* **72**, 779–782.

Beatty, S. W. (1991). Colonization dynamics in a mosaic landscape: The buried seed pool. *J. Biogeogr.* **18**, 553–563.

Beaufait, W. R. (1960). Some effects of high temperatures on the cones and seeds of jack pine. *For. Sci.* **6**, 194–198.

Bell, D. T., Vlahos, S., and Bellairs, S. M. (1990). Seed ecology in relation to reclamation: Lessons from mined lands in Western Australia. *Proc. Ecol. Soc. Aust.* **16**, 531–535.

Bellairs, S. M., and Bell, D. T. (1990). Canopy-borne seed store in three Western Australian plant communities. *Aust. J. Ecol.* **15**, 299–305.

Bennington, C. C., McGraw, J. B., and Vavrek, M. C. (1991). Ecological genetic variation in seed banks. II. Phenotypic and genetic differences between young and old subpopulations of *Luzula parviflora. J. Ecol.* **79**, 627–643.

Benoit, D. L., Derksen, D. A., and Panneton, B. (1992). Innovative approaches to seedbank studies. *Weed Sci.* **40**, 660–669.

Benoit, D. L., Kenkel, N. C., and Cavers, P. B. (1989). Factors influencing the precision of soil seed bank estimates. *Can. J. Bot.* **67**, 2833–2840.

Berjak, P., and Villiers, T. A. (1972a). Ageing in plant embryos. IV. Loss of regulatory control in aged embryos. *New Phytol.* **71**, 1069–1074.

Berjak, P., and Villiers, T. A. (1972b). Ageing in plant embryos. V. Lysis of the cytoplasm in non-viable embryos. *New Phytol.* **71**, 1075–1079.

Berjak, P., and Villiers, T. A. (1972c). Ageing in plant embryos. II. Age-induced damage and its repair during early germination. *New Phytol.* **71**, 135–144.

Bernhardt, K.-G. (1994). Seed burial by soil burrowing beetles. *Nord. J. Bot.* **15**, 257–260.

Bertiller, M. B. (1992). Seasonal variation in the seed bank of a Patagonian grassland in relation to grazing and topography. *J. Veg. Sci.* **3**, 47–54.

Bespalova, Z. G. (1972). Seed distribution in arid zone soils in northern Turan and northern Gobi. *In* "USSR Academy of Sciences, Soviet

National Committee for International Biological Programme. Ecophysiological Foundation of Ecosystems Productivity in Arid Zone," pp. 165–168. International symposium USSR, 7–19 June 1972. Publishing House "Nauka," Leningrad.

Bhatia, R. K., and Sandhu, K. S. (1990). Soil seed reserve and dynamics of wild oat (*Avena ludoviciana* Dur.) in arable land. *In* "International Symposium on Environmental Influences on Seed and Germination Mechanism: Recent Advances in Research and Technology" (D. N. Sen, S. Mohammed, P. K. Kasera, and T. P. Thomas, eds.), pp. 41–42. Univ. of Jodhpur, Jodhpur, India.

Bigwood, D. W., and Inouye, D. W. (1988). Spatial pattern analysis of seed banks: An improved method and optimized sampling. *Ecology* **69**, 497–507.

Blumer, J. C. (1910). The viability of pine seed in serotinous cones. *Torreya* **10**, 108–111.

Bond, W. J. (1985). Canopy-stored seed reserves (serotiny) in Cape Proteaceae. *S. Afr. J. Bot.* **51**, 181–186.

Bond, W. J., and Slingsby, P. (1983). Seed dispersal by ants in shrublands of the Cape Province and its evolutionary implications. *S. Afr. J. Sci.* **79**, 231–233.

Bond, W. J., and Slingsby, P. (1984). Collapse of an ant-plant mutualism: The Argentine ant (*Iridomyrmex humilis*) and myrmecochorous Proteaceae. *Ecology* **65**, 1031–1037.

Bond, W. J., and Stock, W. D. (1989). The costs of leaving home: Ants disperse myrmecochorous seeds to low nutrient sites. *Oecologia* **81**, 412–417.

Bonis, A., and Lepart, J. (1994). Vertical structure of seed banks and the impact of depth of burial on recruitment in two temporary marshes. *Vegetatio* **112**, 127–139.

Bonis, A., Lepart, J., and Grillas, P. (1995). Seed bank dynamics and coexistence of annual macrophytes in a temporary and variable habitat. *Oikos* **74**, 81–92.

Bosbach, K., Hurka, H., and Haase, R. (1982). The soil seed bank of *Capsella bursa-pastoris* (Cruciferae): Its influence on population variability. *Flora* **172**, 47–56.

Bradstock, R. A. (1990). Demography of woody plants in relation to fire: *Banksia serrata* L.f. and *Isopogon anemonifolius* (Salisb.) Knight. *Aust. J. Ecol.* **15**, 117–132.

Bradstock, R. A. (1991). The role of fire in establishment of seedlings of serotinous species from the Sydney region. *Aust. J. Bot.* **39**, 347–356.

Bradstock, R. A., and Bedward, M. (1992). Simulation of the effect of season of fire on post-fire seedling emergence of two *Banksia* species based on long-term rainfall records. *Aust. J. Bot.* **40**, 75–88.

Bradstock, R. A., Gill, A. M., Hastings, S. M., and Moore, P. H. R. (1994). Survival of serotinous seedbanks during bushfires: Comparative studies of *Hakea* species from southeastern Australia. *Aust. J. Ecol.* **19**, 276–282.

Bradstock, R. A., and Myerscough, P. J. (1981). Fire effects on seed release and the emergence and establishment of seedlings in *Banksia ericifolia* L.f. *Aust. J. Bot.* **29**, 521–531.

Bradstock, R. A., and O'Connell, M. A. (1988). Demography of woody plants in relation to fire: *Banksia ericifolia* L.f. and *Petrophile pulchella* (Schrad) R. Br. *Aust. J. Ecol.* **13**, 505–518.

Brenchley, W. E., and Warington, K. (1930). The weed seed population of arable soil. I. Numerical estimation of viable seeds and observations on their natural dormancy. *J. Ecol.* **18**, 235–272.

Brenchley, W. E., and Warington, K. (1933). The weed seed population of arable soil. II. Influence of crop, soil and methods of cultivation upon the relative abundance of viable seeds. *J. Ecol.* **21**, 103–127.

Brenchley, W. E., and Warington, K. (1936). The weed seed population of arable soil. III. The re-establishment of weed species after reduction by fallowing. *J. Ecol.* **24**, 479–501.

Brenchley, W. E., and Warington, K. (1945). The influence of periodic fallowing on the prevalence of viable weed seeds in arable soil. *Ann. Appl. Biol.* **32**, 285–296.

Bridgemohan, P., Brathwaite, R. A. I., and McDavid, C. R. (1991). Seed survival and patterns of seedling emergence studies of *Rottboellia cochinchinensis* (Lour.) W. D. Clayton in cultivated soils. *Weed Res.* **31**, 265–272.

Bridges, D. C., and Walker, R. H. (1985). Influence of weed management and cropping systems on sicklepod (*Cassia obtusifolia*) seed in the soil. *Weed Sci.* **33**, 800–804.

Brits, G. J. (1986). Influence of fluctuating temperatures and H_2O_2 treatment on germination of *Leucospermum cordifolium* and *Serruria florida* (Proteaceae) seeds. *S. Afr. J. Bot.* **52**, 286–290.

Brits, G. J. (1987). Germination depth vs. temperature requirements in naturally dispersed seeds of *Leucospermum cordifolium* and *L. cuneiforme* (Proteaceae). *S. Afr. J. Bot.* **53**, 119–124.

Brown, A. H. F., and Oosterhuis, L. (1981). The role of buried seed in coppicewoods. *Biol. Conserv.* **21**, 19–38.

Brown, D. (1992). Estimating the composition of a forest seed bank: A comparison of the seed extraction and seedling emergence methods. *Can. J. Bot.* **70**, 1603–1612.

Brown, J., Thill, D. C., Brown, A. P., and Brammer, T. A. (1996). Gene transfer between genetically engineered canola (*Brassica napus* L.) and related weeds. *Am. J. Bot.* **83** (Supplement to No. 6), 55 [Abstract]

Brown, J. S., and Venable, D. L. (1986). Evolutionary ecology of seed-bank annuals in temporally varying environments. *Am. Nat.* **127**, 31–47.

Bruch, E. C. (1961). 1932–California Department of Agriculture buried seed project–1960. *Calif. Dept. Agric. Bull.* **50**, 29–30.

Buhler, D. D., and Maxwell, B. D. (1993). Seed separation and enumeration from soil using K_2CO_3-centrifugation and image analysis. *Weed Sci.* **41**, 298–302.

Burnside, O. C., Fenster, C. R., Evetts, L. L., and Mumm, R. F. (1981). Germination of exhumed weed seed in Nebraska. *Weed Sci.* **29**, 577–586.

Burnside, O. C., Wilson, R. G., Weisberg, S., and Hubbard, K. G. (1996). Seed longevity of 41 weed species buried 17 years in eastern and western Nebraska. *Weed Sci.* **44**, 74–86.

Cabin, R. J. (1996). Genetic comparisons of seed bank and seedling populations of a perennial desert mustard, *Lesquerella fendleri*. *Evolution* **50**, 1830–1841.

Callihan, R. H., Prather, T. S., and Northam, F. E. (1993). Longevity of yellow starthistle (*Centaurea solstitialis*) achenes in soil. *Weed Technol.* **7**, 33–35.

Cardina, J., Regnier, E., and Harrison, K. (1991). Long-term tillage effects on seed banks in three Ohio soils. *Weed Sci.* **39**, 186–194.

Carroll, E. J., and Ashton, D. H. (1965). Seed storage in soils of several Victorian plant communities. *Vict. Nat.* **82**, 102–110.

Castellanos, A. E., and Molina, F. E. (1990). Differential survivorship and establishment in *Simmondsia chinensis* (jojoba). *J. Arid Environ.* **19**, 65–76.

Cavers, P. B. (1983). Seed demography. *Can. J. Bot.* **61**, 3578–3590.

Cavers, P. B., Groves, R. H., and Kaye, P. E. (1995). Seed population dynamics of *Onopordum* over 1 year in southern New South Wales. *J. Appl. Ecol.* **32**, 425–433.

Cavers, P. B., Kane, M., and O'Toole, J. J. (1992). Importance of seedbanks for establishment of newly introduced weeds: A case study of proso millet (*Panicum miliaceum*). *Weed Sci.* **40**, 630–635.

Cayford, J. H., and McRae, D. J. (1983). The ecological role of fire in jack pine forests. *In* "The Role of Fire in Northern Circumpolar Ecosystems" (R. W. Wein and D. A. MacLean, eds.), pp. 183–199. Wiley, Chichester.

Chambers, J. C. (1993). Seed and vegetation dynamics in an alpine herb field: Effects of disturbance type. *Can. J. Bot.* **71,** 471–485.

Chambers, J. C., and MacMahon, J. A. (1994). A day in the life of a seed: Movements and fates of seeds and their implications for natural and managed systems. *Annu. Rev. Ecol. Syst.* **25,** 263–292.

Chambers, J. C., MacMahon, J. A., and Haefner, J. H. (1991). Seed entrapment in alpine ecosystems: Effects of soil particle size and diaspore morphology. *Ecology* **72,** 1668–1677.

Champness, S. S., and Morris, K. (1948). The population of buried viable seeds in relation to contrasting pasture and soil types. *J. Ecol.* **36,** 149–173.

Chancellor, R. J. (1964). Emergence of weed seedlings in the field and the effects of different frequencies of cultivation. *Proc. Brit. Weed Cont. Conf.* **7**(2), 599–606.

Chancellor. R. J. (1985). Changes in the weed flora of an arable field cultivated for 20 years. *J. Appl. Ecol.* **22,** 491–501.

Chandrashekara, U. M., and Ramakrishnan, P. S. (1993). Germinable soil seed bank dynamics during the gap phase of a humid tropical forest in the Western Ghats of Kerala, India. *J. Trop. Ecol.* **9,** 455–467.

Chauvel, B., Gasquez, J., and Darmency, H. (1989). Changes of weed seed bank parameters according to species, time and environment. *Weed Res.* **29,** 213–219.

Cheah, K. S. E., and Osborne, D. J. (1978). DNA lesions occur with loss of viability in embryos of ageing rye seed. *Nature* **272,** 593–599.

Cheam, A. H. (1986). Seed production and seed dormancy in wild radish (*Raphanus raphanistrum* L.) and some possibilities for improving control. *Weed Res.* **26,** 405–413.

Cheke, A. S., Nanakorn, W., and Yankoses, C. (1979). Dormancy and dispersal of seeds of secondary forest species under the canopy of a primary tropical forest in northern Thailand. *Biotropica* **11,** 88–95.

Chepil, W. S. (1946). Germination of weed seeds. II. The influence of tillage treatments on germination. *Sci. Agri.* **26,** 347–357.

Chesson, P. L. (1986). Environmental variation and the coexistence of species. *In* "Community Ecology" (J. Diamond and T. J. Case, eds.), pp. 240–256. Harper & Row, New York.

Chin, L. T. (1973). Occurrence of seeds in virgin forest top soil with particular reference to secondary species in Sabah. *Malay. For.* **36,** 185–193.

Chippindale, H. G., and Milton, W. E. J. (1934). On the viable seeds present in the soil beneath pastures. *J. Ecol.* **22,** 508–531.

Christensen, C. M., and Kaufmann, H. H. (1969). "Grain Storage; the Role of Fungi in Quality Loss." Univ. Minnesota Press, Minneapolis.

Clark, F. B. (1962). White ash, hackberry, and yellow-poplar seed remain viable when stored in the forest litter. *Proc. Indiana Acad. Sci.* **72,** 112–114.

Clark, F. E. (1967). Bacteria in soil. *In* "Soil Biology" (A. Burges and R. Raw, eds.), pp. 15–49. Academic Press, London.

Clark, W. H., and Comanor, P. L. (1973). The use of western harvester ant, *Pogonomyrmex occidentalis* (Cresson), seed stores by heteromyid rodents. *Biol. Soc. Nevada Occas. Paps.* **34,** 1–6.

Clements, D. R., Benoit, D. L., Murphy, S. D., and Swanton, C. J. (1996). Tillage effects on weed seed return and seedbank composition. *Weed Sci.* **44,** 314–322.

Coetzee, J. H., and Giliomee, J. H. (1987a). Seed predation and survival in the infructescences of *Protea repens* (Proteaceae). *S. Afr. J. Bot.* **53,** 61–64

Coetzee, J. H., and Giliomee, J. H. (1987b). Borers and other inhabitants of the inflorescences and infructescences of *Protea repens* in the Western Cape. *Phytophylactica* **19,** 1–6.

Coffin, D. P., and Lauenroth, W. K. (1989). Spatial and temporal variation in the seed bank of a semiarid grassland. *Am. J. Bot.* **76,** 53–58.

Cohen, D. (1966). Optimizing reproduction in a randomly varying environment. *J. Theoret. Biol.* **12,** 119–129.

Cohen, D. (1967). Optimizing reproduction in a randomly varying environment when a correlation may exist between the conditions at the time a choice has to be made and the subsequent outcome. *J. Theoret. Biol.* **16,** 1–14.

Cohen, D. (1968). A general model of optimal reproduction in a randomly varying environment. *J. Ecol.* **56,** 219–228.

Cohen, D., and Levin, S. A. (1985). The interaction between dispersal and dormancy strategies in varying and heterogeneous environments. *In* "Lecture Notes in Biomathematics: Mathematical Topics in Population Biology, Morphogenesis and Neurosciences" (E. Teramoto and M. Yamaguti, eds.), pp. 110–122. Springer-Verlag, Berlin.

Coker, W. C. (1909). Vitality of pine seeds and the delayed opening of cones. *Am. Nat.* **43,** 677–681.

Cole, C. A. (1991). The seedbank of a young surface mine wetland. *Wetlands Ecol. Manage.* **1,** 173–184.

Collins, B., and Wein, G. (1995). Seed bank and vegetation of a constructed reservoir. *Wetlands* **15,** 374–385.

Colosi, J. C., Cavers, P. B., and Bough, M. A. (1988). Dormancy and survival in buried seeds of proso millet (*Panicum miliaceum*). *Can. J. Bot.* **66,** 161–168.

Comes, R. D., Burns, V. F., and Kelley, A. D. (1978). Longevity of certain weed and crop seeds in fresh water. *Weed Sci.* **26,** 336–344.

Conn, J. S., and Deck, R. E. (1995). Seed viability and dormancy of 17 weed species after 9.7 years of burial in Alaska. *Weed Sci.* **43,** 583–585.

Cook, R. (1980). The biology of seeds in the soil. *In* "Demography and Evolution in Plant Populations" (O. T. Solbrig, ed.), pp. 107–129 Blackwell, Oxford.

Cooper, R. W. (1951). Release of sand pine seed after a fire. *J. For.* **49,** 331–332.

Courtney, A. D. (1968). Seed dormancy and field emergence in *Polygonum aviculare*. *J. Appl. Ecol.* **5,** 675–684.

Cousens, R., and Moss, S. R. (1990). A model of the effects of cultivation on the vertical distribution of weed seeds within the soil. *Weed Res.* **30,** 61–70.

Coutinho, L. M. (1977). Aspectos ecologicos do fogo no cerrado. II. As queimadas e a dispersao de sementes em algumas especies anemocoricas do estrato herbaceo-subarbustivo. *Bol. Bot. Univ. S. Paulo* **5,** 57–64.

Cowling, R. M., and Lamont, B. B. (1985a). Variation in serotiny of three *Banksia* species along a climatic gradient. *Aust. J. Ecol.* **10,** 345–350.

Cowling, R. M., and Lamont, B. B. (1985b). Seed release in *Banksia*: The role of wet-dry cycles. *Aust. J. Ecol.* **10,** 169–171.

Cowling, R. M., and Lamont, B. B. (1987). Post-fire recruitment of four co-occurring *Banksia* species. *J. Appl. Ecol.* **24,** 645–658.

Cowling, R. M., Lamont, B. B., and Pierce, S. M. (1987). Seed bank dynamics of four co-occurring *Banksia* species. *J. Ecol.* **75,** 289–302.

Crawley, M. J., Hails, R. S., Rees, M., Kohn, D., and Buxton, J. (1993). Ecology of transgenic oilseed rape in natural habitats. *Nature* **363,** 620–623.

Crist, T. O., and Friese, C. F. (1993). The impact of fungi on soil seeds: Implications for plants and granivores in a semiarid shrubsteppe. *Ecology* **74,** 2231–2239.

Dalling, J. W., Swaine, M. D., and Garwood, N. C. (1994). Effect of soil depth on seedling emergence in tropical soil seed-bank investigations. *Funct. Ecol.* **9,** 119–121.

D'Angela, E., Facelli, J. M., and Jacobo, E. (1988). The role of the permanent soil seed bank in early stages of a postagricultural succession in the inland Pampa, Argentina. *Vegetatio* **74**, 39–45.

Daskalakou, E. N., and Thanos, C. A. (1994). Aleppo pine (*Pinus halepensis*) postfire regeneration: The role of canopy and soil seed banks. *Int. Conf. Forest Fire Res.* **2**, 1079–1088.

Davis, E. S., Fay, P. K., Chicoine, T. K., and Lacey, C. A. (1993). Persistence of spotted knapweed (*Centaurea maculosa*) seed in soil. *Weed Sci.* **41**, 57–61.

Dawson, J. H., and Bruns, V. F. (1975). Longevity of barnyardgrass, green foxtail, and yellow foxtail seeds in soil. *Weed Sci.* **23**, 437–440.

Del Castillo, R. F. (1994). Factors influencing the genetic structure of *Phacelia dubia,* a species with a seed bank and large fluctuations in population size. *Heredity* **72**, 446–458.

deRouw, A., and van Oers, C. (1988). Seeds in a rainforest soil and their relation to shifting cultivation in Cote d'Ivoire. *Weed Res.* **28**, 373–381.

Dessaint, F., Chadoeuf, R., and Barralis, G. (1991). Spatial pattern analysis of weed seeds in the cultivated soil seed bank. *J. Appl. Ecol.* **28**, 721–730.

Dessaint, F., Chadoeuf, R., and Barralis, G. (1997). Nine years' soil seed bank and weed vegetation relationships in an arable field without weed control. *J. Appl. Ecol.* **34**, 123–130.

Di Stefano G., J. F., and Chaverri, L. G. (1992). Potencial de germinacion de semillas en un basque secundario premontano en San Pedro de Montes de Oca, Costa Rica. *Rev. Bio. Trop.* **40**, 7–10.

Donald, W. W. (1993). Models and sampling for studying weed seed survival with wild mustard (*Sinapis arvensis*) as a case study. *Can. J. Plant Sci.* **73**, 637–645.

Donelan, M., and Thompson, K. (1980). Distribution of buried viable seeds along a successional series. *Biol. Conserv.* **17**, 297–311.

Dore, W. G., and Raymond, L. C. (1942). Pasture studies. XXIV. Viable seeds in pasture soil and manure. *Sci. Agri.* **23**, 69–79.

Dorph-Petersen, K. (1925). Examinations of the occurrence and vitality of various weed seed species under different conditions, made at the Danish State Seed Testing Station during the years 1896–1923. *Int. Seed Test. Congress (Cambridge) Rept.* **4**, 124–138.

Doucet, C., and Cavers, P. B. (1996). A persistent seed bank of the bull thistle *Cirsium vulgare. Can. J. Bot.* **74**, 1386–1391.

Duke, S. H., Kakefuda, G., and Harvey, T. M. (1983). Differential leakage of intracellular substances from imbibing soybean seeds. *Plant Physiol.* **72**, 919–924.

Dwyer, D. D., and Aguirre V., E. (1978). Plants emerging from soils under three range condition classes of desert grassland. *J. Range Manage.* **31**, 209–212.

Ebersole, J. J. (1989). Role of the seed bank in providing colonizers on a tundra disturbance in Alaska. *Can. J. Bot.* **67**, 466–471.

Edwards, W., and Whelan, R. (1995). The size, distribution and germination requirements of the soil-stored seed-bank of *Grevillea barklyana* (Proteaceae). *Aust. J. Ecol.* **20**, 548–555.

Egley, G. H., and Chandler, J. M. (1983). Longevity of weed seeds after 5.5 years in the Stoneville 50-year buried-seed study. *Weed Sci.* **31**, 264–270.

Egley, G. H., and Williams, R. D. (1990). Decline of weed seeds and seedling emergence over five years as affected by soil disturbances. *Weed Sci.* **38**, 504–510.

Elder, R. H., and Osborne, D. J. (1993). Function of DNA synthesis and DNA repair in the survival of embryos during early germination and in dormancy. *Seed Sci. Res.* **3**, 43–53.

Ellis, R. H., and Roberts, E. H. (1980). Improved equations for the prediction of seed longevity. *Ann. Bot.* **45**, 13–30.

Ellner, S. (1985a). ESS germination strategies in randomly varying environments. I. Logistic-type models. *Theor. Pop. Biol.* **28**, 50–79.

Ellner, S. (1985b). ESS germination strategies in randomly varying environments. II. Reciprocal yield-law models. *Theor. Pop. Biol.* **28**, 80–116.

Ellner, S. (1987). Alternate plant life history strategies and coexistence in randomly varying environments. *Vegetatio* **69**, 199–208.

Ellner, S., and Shmida, A. (1981). Why are adaptations for long-range seed dispersal rare in desert plants? *Oecologia* **51**, 133–144.

Enright, N. (1985). Existence of a soil seed bank under rainforest in New Guinea. *Aust. J. Ecol.* **10**, 67–71.

Enright, N. J., and Cameron, E. K. (1988). The soil seed bank of a kauri (*Agathis australis*) forest remnant near Auckland, New Zealand. *New Zeal. J. Bot.* **26**, 223–236.

Enright, N. J., and Lamont, B. B. (1989a). Seed banks, fire season, safe sites and seedling recruitment in five cooccurring *Banksia* species. *J. Ecol.* **77**, 1111–1122.

Enright, N. J., and Lamont, B. B. (1989b). Fire temperatures and follicle-opening requirements in 10 *Banksia* species. *Aust. J. Ecol.* **14**, 107–113.

Enright, N. J., Lamont, B. B., and Marsula, R. (1996). Canopy seed bank dynamics and optimum fire regime for the highly serotinous shrub, *Banksia hookeriana. J. Ecol.* **84**, 9–17.

Epling, C., Lewis, H., and Ball, F. M. (1960). The breeding group and seed storage: A study in population dynamics. *Evolution* **14**, 238–255.

Epp, G. A. (1987). The seed bank of *Eupatorium odoratum* along a successional gradient in a tropical rain forest in Ghana. *J. Trop. Ecol.* **3**, 139–149.

Evans, A. S., and Cabin, R. J. (1995). Can dormancy affect the evolution of post-germination traits? The case of *Lesquerella fendleri. Ecology* **76**, 344–356.

Ewart, A. J. (1907). The delayed dehiscence of *Callistemon regida,* R. Br. *Ann. Bot.* **21**, 135–137.

Fay, P. K., and Olson, W. A. (1978). Technique for separating weed seed from soil. *Weed Sci.* **26**, 530–533.

Finlayson, C. M., Cowie, I. D., and Bailey, B. J. (1990). Sediment seedbanks in grassland on the Magela Creek floodplain, northern Australia. *Aquat. Bot.* **38**, 163–176.

Fiore, E. B., and Putz, F. E. (1992). Buried dormant seeds in bottom sediments of Newnans Lake, Florida. *Florida Scient.* **55**, 157–159.

Forcella, F. (1984). A species-area curve for buried viable seeds. *Aust. J. Agric. Res.* **35**, 645–652.

Forcella, F. (1992). Prediction of weed seedling densities from buried seed reserves. *Weed Res.* **32**, 29–38.

Forcella, F., Wilson, R. G., Renner, K. A., Dekker, J., Harvey, R. G., Alm, D. A., Buhler, D. D., and Cardina, J. (1992). Weed seedbanks of the U.S. corn belt: Magnitude, variation, emergence, and application. *Weed Sci.* **40**, 636–644.

Fox, J. F. (1983). Germinable seed banks of interior Alaskan tundra. *Arctic Alpine Res.* **15**, 405–411.

Freedman, B., Hill, N., Svoboda, J., and Henry, G. (1982). Seed banks and seedling occurrence in a high Arctic oasis at Alexandra Fjord, Ellesmere Island, Canada. *Can. J. Bot.* **60**, 2112–2118.

Froud-Williams, R. J. (1987). Survival and fate of weed seed populations: Interaction with cultural practice. *Brit. Crop Protect. Conf. Weeds* **7C-3**, 707–718.

Froud-Williams, R. J., Chancellor, R. J., and Drennan, D. S. H. (1983). Influence of cultivation regime upon buried weed seeds in arable cropping systems. *J. Appl. Ecol.* **20**, 199–208.

Froud-Williams, R. J., Chancellor, R. J., and Drennan, D. S. H. (1984). The effects of seed burial and soil disturbance on emergence and survival of arable weeds in relation to minimal cultivation. *J. Appl. Ecol.* **21**, 629–641.

Fyles, J. W. (1989). Seed bank populations in upland coniferous forests in central Alberta. *Can. J. Bot.* **67**, 274–278.

Garcia-Fayos, P., Recatala, T. M., Cerda, A., and Calvo, A. (1995).

Seed population dynamics on badland slopes in southeastern Spain. *J. Veg. Sci.* **6**, 691–696.

Gartner, B. L., Chapin, F. S., III, and Shaver, G. R. (1983). Demographic patterns of seedling establishment and growth of native graminoids in an Alaskan tundra disturbance. *J. Appl. Ecol.* **20**, 965–980.

Garwood, N. C. (1989). Tropical soil seed banks: A review. *In* "Ecology of Soil Seed Banks" (M. A. Leck, V. T. Parker and R. L. Simpson eds.), pp. 149–209. Academic Press, San Diego.

Gauthier, S., Bergeron, Y., and Simon, J.-P. (1996). Effects of fire regime on the serotiny level of jack pine. *J. Ecol.* **84**, 539–548.

Gerard, B. M. (1963). The activities of some species of Lumbricidae in pasture-land. *In* "Soil Organisms" (J. Doeksen and J. van der Drift, eds.), pp. 49–52. North-Holland Amsterdam.

Gerritsen, J., and Greening, H. S. (1989). Marsh seed banks of the Okefenokee swamp: Effects of hydrologic regime and nutrients. *Ecology* **70**, 750–763.

Gerth, U., and Bernhardt, D. (1995). A comparison of the synthesis of DNA, RNA and proteins in the embryos of after-ripened and thermo- or FR-dormant *Agrostemma githago* L. seeds. *Seed Sci. Res.* **5**, 87–97.

Gilfedder, L., and Kirkpatrick, J. B. (1993). Germinable soil seed and competitive relationships between a rare native species and exotics in a semi-natural pasture in the Midlands, Tasmania. *Biol. Conserv.* **64**, 113–119.

Gill, A. M. (1976). Fire and the opening of *Banksia ornata* F. Muell. follicles. *Aust. J. Bot.* **24**, 329–335.

Givnish, T. J. (1981). Serotiny, geography, and fire in the pine barrens of New Jersey. *Evolution* **35**, 101–123.

Godwin, H., and Willis, E. H. (1964). The viability of lotus seeds (*Nelumbium nucifera,* Gaertn.). *New Phytol.* **63**, 410–412.

Goss, W. L. (1939). Germination of buried weed seeds. *Calif. Dept. Agric. Bull.* **28**, 132–135.

Gottlieb, L. D. (1974). Genetic stability in a peripheral isolate of *Stephanomeria exigua* ssp. *coronaria* that fluctuates in population size. *Genetics* **76**, 551–556.

Graber, R. E., and Thompson, D. F. (1978). Seeds in the organic layers and soil of four beech-birch-maple stands. *USDA For. Serv. Res. Paper* NE-401.

Grace, J. B. (1984). Effects of tubificid worms on the germination and establishment of *Typha*. *Ecology* **65**, 1689–1693.

Graham, A. W., and Hopkins, M. S. (1990). Soil seed banks of adjacent unlogged rainforest types in north Queensland. *Aust. J. Bot.* **38**, 261–268.

Granstrom, A. (1982). Seed banks in five boreal forest stands originating between 1810 and 1963. *Can. J. Bot.* **60**, 1815–1821.

Granstrom, A. (1987). Seed viability of fourteen species during five years of storage in a forest soil. *J. Ecol.* **75**, 321–331.

Grant, J. D. (1983). The activities of earthworms and the fates of seeds. *In* "Earthworm Ecology" (J. E. Satchell, ed.), pp. 107–122. Chapman, London.

Grice, A. C., and Westoby, M. (1987). Aspects of the dynamics of the seed-banks and seedling populations of *Acacia victoriae* and *Cassia* spp. in arid western New South Wales. *Aust. J. Ecol.* **12**, 209–215.

Grilli, I., Bacci, E., Lombardi, T., Spano, C., and Floris, C. (1995). Natural ageing: Poly(A) polymerase in germinating embryos of *Triticum durum* wheat. *Ann. Bot.* **76**, 15–21.

Grilz, P. L., and Romo, J. T. (1995). Management considerations for controlling smooth brome in fescue prairie. *Nat. Areas J.* **15**, 148–156.

Grime, J. P. (1979). "Plant Strategies and Vegetation Processes." Wiley, Chichester.

Grime, J. P. (1981). The role of seed dormancy in vegetation dynamics. *Ann. Appl. Biol.* **98**, 555–558.

Gross, K. L. (1990). A comparison of methods for estimating seed numbers in the soil. *J. Ecol.* **78**, 1079–1093.

Grundy, A. C., Mead, A., and Bond, W. (1996). Modelling the effect of weed-seed distribution in the soil profile on seedling emergence. *Weed Res.* **36**, 375–384.

Guevara S., S., and Gomez-Pompa, A. (1972). Seeds from surface soils in a tropical region of Veracruz, Mexico. *J. Arnold Arbor.* **53**, 312–335.

Gulzar, S., and Khan, M. A. (1994). Seed banks of coastal shrub communities. *Ecoprint* **1**, 1–6.

Gunster, A. (1992). Aerial seedbanks in the central Namib: Distribution of serotinous plants in relation to climate and habitat. *J. Biogeogr.* **19**, 563–572.

Gutterman, Y. (1972). Delayed seed dispersal and rapid germination as survival mechanisms of the desert plant *Blepharis persica* (Burm.) Kuntze. *Oecologia* **10**, 145–149.

Gutterman, Y. (1993). Seed Germination in Desert Plants. Springer-Verlag, Berlin.

Gutterman, Y., and Ginott, S. (1994). Long-term protected 'seed bank' in dry inflorescences of *Asteriscus pygmaeus;* achene dispersal mechanism and germination. *J. Arid Environ.* **26**, 149–163.

Gutterman, Y., Witztum, A., and Evenari, M. (1967). Seed dispersal and germination in *Blepharis persica* (Burm.) Kuntze. *Israel J. Bot.* **16**, 213–234.

Guttridge, C. G., Woodley, S. E., and Hunter, T. (1984). Accelerating strawberry seed germination by fungal infection. *Ann. Bot.* **54**, 223–230.

Haferkamp, M. E., Smith, L., and Nilan, R. A. (1953). Studies on aged seeds. I. Relation of age of seed to germination and longevity. *Agron. J.* **45**, 434–437.

Hallam, N. D., Roberts, B. E., and Osborne, D. J. (1973). Embryogenesis and germination in rye (*Secale cereale* L.). III. Fine structure and biochemistry of the non-viable embryo. *Planta* **110**, 279–290.

Halloin, J. M. (1982). Localization and changes in catechin and tannins during development and ripening of cottonseed. *New Phytol.* **90**, 651–657.

Halloin, J. M. (1983). Deterioration resistance mechanisms in seeds. *Phytopathology* **73**, 335–339.

Halloin, J. M. (1986). Microorganisms and seed deterioration. *In* "Physiology of Seed Deterioration," pp. 89–99. Crop Science Society of America Spec. Pub. No. 11.

Hanke, M., and Dollwet, H. H. A. (1976). The production of ethylene by certain soil fungi. *Sci. Biol. J.* **2**, 227–230.

Hairston, N. G., Jr., and De Stasio, B. T., Jr. (1988). Rate of evolution slowed by a dormant propagule pool. *Nature* **336**, 239–242.

Harman, G. E. (1983). Mechanisms of seed infection and pathogenesis. *Phytopathology* **73**, 326–329.

Harman, G. E., Nedrow, B., and Nash, G. (1978). Stimulation of fungal spore germination by volatiles from aged seeds. *Can. J. Bot.* **56**, 2124–2127.

Harper, S. H. T., and Lynch, J. M. (1980). Microbial effects on the germination and seedling growth of barley. *New Phytol.* **84**, 473–481.

Harradine, A. R. (1986). Seed longevity and seedling establishment of *Bromus diandrus* Roth. *Weed Res.* **26**, 173–180.

Harrington, G. N., and Driver, M. A. (1995). The effect of fire and ants on the seed-bank of a shrub in a semi-arid grassland. *Aust. J. Ecol.* **20**, 538–547.

Harrington, J. F. (1973). Problems of seed storage. *In* "Seed Ecology" (H. Heydecker ed.), pp. 251–262. Pennsylvania State Univ. Press, University Park, PA.

Hartman, J. M. (1988). Recolonization of small disturbance patches in a New England salt marsh. *Am. J. Bot.* **75,** 1625–1631.

Hassan, M. A., and West, N. E. (1986). Dynamics of soil seed pools in burned and unburned sagebrush semi-deserts. *Ecology* **67,** 269–272.

Haukos, D. A., and Smith, L. M. (1993). Seed-bank composition and predictive ability of field vegetation in playa lakes. *Wetlands* **13,** 32–40.

Haukos, D. A., and Smith, L. M. (1994). Composition of seed banks along an elevational gradient in playa wetlands. *Wetlands* **14,** 301–307.

Hayashi, I., and Numata, M. (1971). Viable buried-seed population in the *Miscanthus* and *Zoysia* type grasslands in Japan—Ecological studies on the buried-seed population in the soil related in plant succession VI. *Jpn. J. Ecol.* **20,** 243–252.

Hayashi, I., Pancho, J. V., and Sastroutomo, S. S. (1978). Preliminary report on the buried seeds of floating islands and bottom of Lake Rawa Pening, central Java. *Jpn. J. Ecol.* **28,** 325–333.

Hellum, A. K., and Pelchat, M. (1979). Temperature and time affect the release and quality of seed from cones of lodgepole pine from Alberta. *Can. J. For. Res.* **9,** 154–159.

Hill, M. O., and Stevens, P. A. (1981). The density of viable seed in soils of forest plantations in upland Britain. *J. Ecol.* **69,** 693–709.

Hintikka, V. (1987). Germination ecology of *Galeopsis bifida* (Lamiaceae) as a pioneer species in forest succession. *Silva Fenn.* **21,** 301–313.

Hobbs, R. J., and Mooney, H. A. (1991). Effects of rainfall variability and gopher disturbance on serpentine annual grassland dynamics. *Ecology* **72,** 59–68.

Hodgkinson, K. C., Harrington, G. N., and Miles, G. E. (1980). Composition, spatial and temporal variability of the soil seed pool in a *Eucalyptus populnea* shrub woodland in central New South Wales. *Aust. J. Ecol.* **5,** 23–29.

Hole, F. D. (1981). Effects of animals on soil. *Geoderma* **25,** 75–112.

Holldobler, B., and Wilson, E. O. (1990). "The Ants." The Belknap Press of Harvard Univ. Press, Cambridge, MA.

Holmes, P. M. (1988). Implications of alien *Acacia* seed bank viability and germination for clearing. *S. Afr. J. Bot.* **54,** 281–284.

Holmes, P. M. (1989). Decay rates in buried alien *Acacia* seed populations of different density. *S. Afr. J. Bot.* **55,** 299–303.

Honig, M. A., Cowling, R. M., and Richardson, D. M. (1992). The invasive potential of Australian banksias in South African fynbos: A comparison of the reproductive potential of *Banksia ericifolia* and *Leucadendron laureolum. Aust. J. Ecol.* **17,** 305–314.

Hootsmans, M. J. M., Vermaat, J. E., and van Vierssen, W. (1987). Seed-bank development, germination and early seedling survival of two seagrass species from The Netherlands: *Zostera marina* L. and *Zostera noltii* Hornem. *Aquat. Bot.* **28,** 275–285.

Hopkins, M. S., and Graham, A. W. (1983). The species composition of soil seed banks beneath lowland tropical rainforests in North Queensland, Australia. *Biotropic* **15,** 90–99.

Hopkins, M. S., and Graham, A. W. (1984a). Viable soil seed banks in disturbed lowland tropical rainforest sites in North Queensland. *Aust. J. Ecol.* **9,** 71–79.

Hopkins, M. S., and Graham, A. W. (1984b). The role of soil seed banks in regeneration in canopy gaps in Australian tropical lowland rainforest—Preliminary field experiments. *Malay. For.* **47,** 146–158.

Hopkins, M. S., and Graham, A. W. (1987). The viability of seeds of rainforest species after experimental soil burials under tropical wet lowland forest in northeastern Australia. *Aust. J. Ecol.* **12,** 97–108.

Hopkins, D. R., and Parker, V. T. (1984). A study of the seed bank of a salt marsh in northern San Francisco Bay. *Am. J. Bot.* **71,** 348–355.

Hopkins, M. S., Tracey, J. G., and Graham, A. W. (1990). The size and composition of soil seed-banks in remnant patches of three structural rainforests types in north Queensland. *Aust. J. Ecol.* **15,** 43–50.

Houle, G. (1991). Regenerative traits of tree species in a deciduous forest of northeastern North America. *Holarc. Ecol.* **14,** 142–151.

Houle, G. (1992). The reproductive ecology of *Abies balsamea, Acer saccharum* and *Betula alleghaniensis* in the Tantare Ecological Reserve, Quebec. *J. Ecol.* **80,** 611–623.

Houle, G. (1996). Environmental filters and seedling recruitment on a coastal dune in subarctic Quebec (Canada). *Can. J. Bot.* **74,** 1507–1513.

Houle, G., and Phillips, D. L. (1988). The soil seed bank of granite outcrop plant communities. *Oikos* **52,** 87–93.

Howard, T. M. (1973). *Nothofagus cunninghamii* ecotonal stages: Buried viable seed in north west Tasmania. *Proc. Roy. Soc. Victoria* **86,** 137–142.

Howard, T., and Ashton, D. H. (1967). Studies of soil seed in snow gum woodland. *Victorian Nat.* **84,** 331–335.

Howe, C. D., and Chancellor, R. J. (1983). Factors affecting the viable seed content of soils beneath lowland pastures. *J. Appl. Ecol.* **20,** 915–922.

Hull, A. C., Jr. (1973). Germination of range plant seeds after long periods of uncontrolled storage. *J. Range Manage.* **26,** 198–200.

Hurka, H., and Haase, R. (1982). Seed ecology of *Capsella bursa-pastoris* (Cruciferae): Dispersal mechanism and the soil seed bank. *Flora* **172,** 35–46.

Hutchings, M. J., and Russell, P. J. (1989). The seed regeneration dynamics of an emergent salt marsh. *J. Ecol.* **77,** 615–637.

Hytteborn, H., Rydin, H., and Skoglund, J. (1991). Viable seeds in sediments in Lake Hjalmaren. *Aquat. Bot.* **40,** 289–293.

Ingersoll, C. A., and Wilson, M. V. (1990). Buried propagules in an old-growth forest and their response to experimental disturbances. *Can. J. Bot.* **68,** 1156–1162.

Ingersoll, C. A., and Wilson, M. V. (1993). Buried propagule bank of a high subalpine site: Microsite variation and comparisons with aboveground vegetation. *Can. J. Bot.* **71,** 712–717.

Iverson, L. R., and Wali, M. K. (1982). Buried, viable seeds and their relation to revegetation after surface mining. *J. Range Manage.* **35,** 648–652.

Jacobsohn, M. K., and Jacobsohn, G. M. (1985). Production of a fungistat and the role of fungi during germination of *Digitalis purpurea* L. cv. gloxiniaflora seeds. *Ann. Bot.* **56,** 543–552.

Jansen, P. I., and Ison, R. L. (1995). Factors contributing to the loss of seed from the seed-bank of *Trifolium balansae* and *Trifolium resupinatum* over summer. *Aust. J. Ecol.* **20,** 248–256.

Jefferson, R. G., and Usher, M. B. (1987). The seed bank in soils of disused chalk quarries in the Yorkshire Wolds, England: Implications for conservation management. *Biol. Conserv.* **42,** 287–302.

Jensen, H. A. (1969). Content of buried seeds in arable soil in Denmark and its relation to the weed population. *Dansk Bot. Ark.* **27,** 1–56.

Jimenez, H. E., and Armesto, J. J. (1992). Importance of the soil seed bank of disturbed sites in Chilean matorral in early secondary succession. *J. Veg. Sci.* **3,** 579–586.

Joffe, A. Z. (1969). Relationships between *Aspergillus flavus, A. niger* and some fungi in the mycoflora of groundnut kernels. *Plant Soil* **31,** 57–64.

Johnson, E. A. (1975). Buried seed populations in the subarctic forest east of Great Slave Lake, Northwest Territories. *Can. J. Bot.* **53,** 2933–2941.

Johnson, R. G. and Anderson, R. C. (1986). The seed bank of a tallgrass prairie in Illinois. *Am. Midl. Nat.* **115,** 123–130.

Johnson, T. K., and Jorgensen, C. D. (1981). Ability of desert rodents to find buried seeds. *J. Range Manage.* **34,** 312–314.

Johnston, A., Smoliak, S., and Stringer, P. W. (1969). Viable seed populations in Alberta prairie topsoils. *Can. J. Plant Sci.* **49**, 75–82.

Jordan, N., Mortensen, D. A., Prenzlow, D. M., and Cox, K. C. (1995). Simulation analysis of crop rotation effects on weed seedbanks. *Am. J. Bot.* **82**, 390–398.

Jorgensen, R. B, Hauser, T. P., Lando, L., and Mikkelsen, T. R. (1996). Transgenic weed-like plants from spontaneous interspecific crop-weed introgression in *Brassica. Am. J. Bot.* **83** (Suppl. to No. 6), 56. [Abstract]

Judd, T. S. (1993). Seed survival in small myrtaceous capsules subjected to experimental heating. *Oecologia* **93**, 576–581.

Judd, T. S. (1994). Do small myrtaceous seed-capsules display specialized insulating characteristics which protect seed during fire? *Ann. Bot.* **73**, 33–38.

Juliano, J. B. (1940). Viability of some Philippine weed seeds. *Philippine Agric.* **29**, 313–326.

Kalisz, S. (1991). Experimental determination of seed bank age structure in the winter annual *Collinsia verna. Ecology* **72**, 575–585.

Kalisz, S., and McPeek, M. A. (1992). Demography of an age-structured annual: Resampled projection matrices, elasticity analyses, and seed bank effects. *Ecology* **73**, 1082–1093.

Kalisz, S., and McPeek, M. A. (1993). Extinction dynamics, population growth and seed banks: An example using an age-structured annual. *Oecologia* **95**, 314–320.

Kamenetsky, R., and Gutterman, Y. (1993). Synaptospermy as a strategy of seed dispersal in desert geophytes, mainly from the genus *Allium* L. *In* "Fourth International Workshop on Seeds. Basic and Applied Aspects of Seed Biology." (D. Come and F. Corbineau eds.), Vol. 2, pp. 437–442. Universite Pierre et Marie Curie, Paris.

Kareiva, P. (1993). Transgenic plants on trial. *Nature* **363**, 580–581.

Kautsky, L. (1990). Seed and tuber banks of aquatic macrophytes in the Asko area, northern Baltic proper. *Holarc. Ecol.* **13**, 143–148.

Keddy, P. A., and Reznicek, A. A. (1982). The role of seed banks in the persistence of Ontario's coastal plain flora. *Am. J. Bot.* **69**, 13–22.

Keen, N. T. (1975). The isolation of phytoalexins from germinating seeds of *Cicer arietinum, Vigna sinensis, Arachis hypogaea,* and other plants. *Phytopathology* **65**, 91–92.

Kellman, M. C. (1970). The viable seed content of some forest soil in coastal British Columbia. *Can. J. Bot.* **48**, 1383–1385.

Kellman, M. C. (1974a). Preliminary seed budgets for two plant communities in coastal British Columbia. *J. Biogeogr.* **1**, 123–133.

Kellman, M. C. (1974b). The viable weed seed content of some tropical agricultural soils. *J. Appl. Ecol.* **11**, 669–678.

Kellman, M. C. (1980). Geographic patterning in tropical weed communities and early secondary successions. *Biotropica* **12** (Suppl.), 34–39.

Khan, M. A. (1993). Relationship of seed bank to plant distribution in saline arid communities. *Pakistan J. Bot.* **25**, 73–82.

Kiirikki, M. (1993). Seed bank and vegetation succession in abandoned fields in Karkali Nature Reserve, southern Finland. *Ann. Bot. Fennici* **30**, 139–152.

Kimber, A., Korschgen, C. E., and van der Valk, A. G. (1995). The distribution of *Vallisneria americana* seeds and seedling light requirements in the upper Mississippi River. *Can. J. Bot.* **73**, 1966–1973.

King, T. J. (1976). The viable seed contents of ant-hill and pasture soil. *New Phytol.* **77**, 143–147.

Kinucan, R. J., and Smeins, F. E. (1992). Soil seed bank of a semiarid Texas grassland under three long-term (36-years) grazing regimes. *Am. Midl. Nat.* **128**, 11–21.

Kirkpatrick, B. L., and Bazzaz, F. A. (1979). Influence of certain fungi on seed germination and seedling survival of four colonizing annuals. *J. Appl. Ecol.* **16**, 515–527.

Kitajima, K., and Tilman, D. (1996). Seed banks and seedling establishment on an experimental productivity gradient. *Oikos* **76**, 381–391.

Kivilaan, A., and Bandurski, R. S. (1981). The one hundred-year period for Dr. Beal's seed viability experiment. *Am. J. Bot.* **68**, 1290–1291.

Kjaer, A. (1940). Germination of buried and dry stored seeds. I. 1934–1939. *Proc. Int. Seed Test. Assoc.* **12**, 167–188.

Kjellsson, G. (1992). Seed banks in Danish deciduous forests: Species composition, seed influx and distribution pattern in soil. *Ecography* **15**, 86–100.

Knapp, A. K., and Anderson, J. E. (1980). Effect of heat on germination of seeds from serotinous lodgepole pine cones. *Am. Midl. Nat.* **104**, 370–372.

Koller, D. (1972). Environmental control of seed germination. *In* "Seed Biology" (T. T. Kozlowski, ed.), Vol. 2, pp. 1–101. Academic Press, New York.

Komarova, T. A. (1985). Role of forest fires in germination of seeds dormant in the soil. *Soviet J. Ecol.* **16**, 311–315.

Komulainen, M., Vieno, M., Yarmishko, V. T., Kaletskaja, T. D., and Maznaja, E. A. (1994). Seedling establishment from seeds and seed banks in forests under long-term pollution stress: A potential for vegetation recovery. *Can. J. Bot.* **72**, 143–149.

Kondrat'ev, K. Ya., Korzov, V. I., and Rudneva, L. B. (1986). Determination of soil weediness by the photometric method. *Doklady Bot. Sci.* **284**, 99–102.

Koniak, S., and Everett, R. L. (1982). Seed reserves in soils of successional stages of pinyon woodlands. *Am. Midl. Nat.* **108**, 295–303.

Kraft, J. M. (1977). The role of delphinidin and sugars in the resistance of pea seedlings to fusarium root rot. *Phytopathology* **67**, 1057–1061.

Kramer, N. B., and Johnson, F. D. (1987). Mature forest seed banks of three habitat types in central Idaho. *Can. J. Bot.* **65**, 1961–1966.

Kremer, R. J. (1986a). Microorganisms associated with velvetleaf (*Abutilon theophrasti*) seeds on the soil surface. *Weed Sci.* **34**, 233–236.

Kremer, R. J. (1986b). Antimicrobial activity of velvetleaf (*Abutilon theophrasti*) seeds. *Weed Sci.* **34**, 617–622.

Kremer, R. J. (1987). Identity and properties of bacteria inhabiting seeds of selected broadleaf weed species. *Microb. Ecol.* **14**, 29–37.

Kremer, R. J. (1993). Management of weed seed banks with microorganisms. *Ecol. Appl.* **3**, 42–52.

Kremer, R. J., Hughes, I. B., Jr., and Aldrich, R. J. (1984). Examination of microorganisms and deterioration resistance mechanisms associated with velvetleaf seed. *Agron. J.* **76**, 745–749.

Kremer, R. J., and Spencer, N. R. (1989a). Impact of a seed-feeding insect and microorganisms on velvetleaf (*Abutilon theophrasti*) seed viability. *Weed Sci.* **37**, 211–216.

Kremer, R. J., and Spencer, N. R. (1989b). Interaction of insects, fungi, and burial on velvetleaf (*Abutilon theophrasti*) seed viability. *Weed Technol.* **3**, 322–328.

Kropac, Z. (1966). Estimation of weed seeds in arable soil. *Pedobiologia* **6**, 105–128.

Kuster, E. (1967). The actinomycetes. *In* "Soil Biology" (A. Burges and F. Raw, eds.), pp. 111–127. Academic Press, London.

Lacey, E. P. (1982). Timing of seed dispersal in *Daucus carota. Oikos* **39**, 83–91.

Lal, R. (1987). "Tropical Ecology and Physical Edaphology." Wiley, Chichester.

Lamont, B. B. (1991). Canopy seed storage and release: What's in a name? *Oikos* **60**, 266–268.

Lamont, B. B., and Barker, M. J. (1988). Seed bank dynamics of a serotinous, fire-sensitive *Banksia* species. *Aust. J. Bot.* **36**, 193–203.

Lamont, B. B., Connell, S. W., and Bergl, S. M. (1991). Seed bank and population dynamics of *Banksia cuneata*: The role of time, fire, and moisture. *Bot. Gaz.* **152,** 114–122.

Lamont, B. B., and Cowling, R. M. (1984). Flammable infrutescences in *Banksia*: A fruit-opening mechanism. *Aust. J. Ecol.* **9,** 295–296.

Lamont, B. B., Le Maitre, D. C., Cowling, R. M., and Enright, N. J. (1991). Canopy seed storage in woody plants. *Bot. Rev.* **57,** 277–317.

Lamont, B. B., and van Leeuwen, S. J. (1988). Seed production and mortality in a rare *Banksia* species. *J. Appl. Ecol.* **25,** 551–559.

Lavorel, S., and Chesson, P. (1995). How species with different regeneration niches coexist in patchy habitats with local disturbances. *Oikos* **74,** 103–114.

Lavorel, S., Debussche, M., Lebreton, J.-D., and Lepart, J. (1993). Seasonal patterns in the seed bank of Mediterranean old-fields. *Oikos* **67,** 114–128.

Lawton, R. O., and Putz, F. E. (1988). Natural disturbance and gap-phase regeneration in a wind-exposed tropical cloud forest. *Ecology* **69,** 764–777.

Leck, M. A., and Graveline, K. J. (1979). The seed bank of a freshwater tidal marsh. *Am. J. Bot.* **66,** 1006–1015.

Leck, M. A., and Simpson, R. L. (1987). Seed bank of a freshwater tidal wetland: Turnover and relationship to vegetation change. *Am. J. Bot.* **74,** 360–370.

Leck, M. A., and Simpson, R. L. (1994). Tidal freshwater wetland zonation: Seed and seedling dynamics. *Aquat. Bot.* **47,** 61–75.

Leck, M. A., and Simpson, R. L. (1995). Ten-year seed bank and vegetation dynamics of a tidal freshwater marsh. *Am. J. Bot.* **82,** 1547–1557.

Lee, K. E. (1985). "Earthworms, Their Ecology and Relationships with Soils and Land Use." Academic Press, Sydney.

Le Maitre, D. C. (1990). The influence of seed ageing on the plant on seed germination in *Protea neriifolia* (Proteaceae). *S. Afr. J. Bot.* **56,** 49–53.

Le Maitre, D. C., and Botha, S. A. (1991). The effects of exposure to different environments on the viability of *Protea neriifolia* seeds. *S. Afr. J. Bot.* **57,** 226–228.

Lerman, J. C., and Cigliano, E. M. (1971). New carbon-14 evidence for six hundred years old *Canna compacta* seed. *Nature* **232,** 568–570.

Levin, D. A. (1990). The seed bank as a source of genetic novelty in plants. *Am. Nat.* **135,** 563–572.

Levin, D. A., Cohen, D., and Hastings, A. (1984). Dispersal strategies in patchy environments. *Theor. Pop. Biol.* **26,** 165–191.

Levin, D. A., and Wilson, J. B. (1978). The genetic implications of ecological adaptations in plants. *In* "Structure and Functioning of Plant Populations" (A. H. J. Freysen and J. W. Woldendorp eds.), pp. 75–98. North-Holland, Amsterdam.

Lev-Yadun, S. (1995). Living serotinous cones in *Cupressus sempervirens. Int. J. Plant Sci.* **156,** 50–54.

Lewis, H. (1962). Catastrophic selection as a factor in speciation. *Evolution* **16,** 257–271.

Lewis, J. (1973). Longevity of crop and weed seeds: Survival after 20 years in soil. *Weed Res.* **13,** 179–191.

Libby, W. F. (1951). Ratiocarbon dates, II. *Science* **114,** 291–296.

Libby, W. F. (1955). "Radiocarbon Dating," 2nd ed. Univ. of Chicago Press, Chicago.

Linder, C. R., and Schmitt, J. (1994). Assessing the risks of transgene escape through time and crop-wild hybrid persistence. *Mol. Ecol.* 3, 23–30.

Linder, C. R., and Schmitt, J. (1995). Potential persistence of escaped transgenes: Performance of transgenic, oil-modified *Brassica* seeds and seedlings. *Ecol. Appl.* **5,** 1056–1068.

Linhart, Y. B. (1978). Maintenance of variation in cone morphology in California closed-cone pines: The roles of fire, squirrels and seed output. *Southw. Nat.* **23,** 29–40.

Lippert, R. D., and Hopkins, H. H. (1950). Study of viable seeds in various habitats in mixed prairie. *Trans. Kansas Acad. Sci.* **53,** 355–364.

Livingston, R. B., and Allessio, M..L. (1968). Buried viable seed in successional field and forest stands, Harvard Forest, Massachusetts. *Bull. Torrey Bot. Club* **95,** 58–69.

Lockwood, J. L. (1977). Fungistasis in soils. *Biol. Rev.* **52,** 1–43.

Lonsdale, W. M. (1988). Interpreting seed survivorship curves. *Oikos* **52,** 361–364.

Lonsdale, W. M. (1993). Losses from the seed bank of *Mimosa pigra*: Soil micro-organisms vs. temperature fluctuations. *J. Appl. Ecol.* **30,** 654–660.

Lonsdale, W. M., Harley, K. L. S., and Gillett, J. D. (1988). Seed bank dynamics in *Mimosa pigra,* an invasive tropical shrub. *J. Appl. Ecol.* **25,** 963–976.

Louda, S. M. (1989). Predation in the dynamics of seed regeneration. *In* "Ecology of soil seed banks" (M. A. Leck, V. T. Parker, and R. L. Simpson, eds.), pp. 25–51. Academic Press, San Diego.

Lueschen W. E., and Andersen, R. N. (1980). Longevity of velvetleaf (*Abutilon theophrasti*) seeds in soil under agricultural practices. *Weed Sci.* **28,** 341–346.

Lueschen, W. E., Andersen, R. N., Hoverstad, T. R., and Kanne, B. K. (1993). Seventeen years of cropping systems and tillage affect velvetleaf (*Abutilon theophrasti*) seed longevity. *Weed Sci.* **41,** 82–86.

Lunt, I. D. (1995). Seed longevity of six native forbs in a closed *Themeda triandra* grassland. *Aust. J. Bot.* **43,** 439–449.

MacDonald, N., and Watkinson, A. R. (1981). Models of an annual plant population with a seedbank. *J. Theor. Biol.* **93,** 643–653.

Major, J., and Pyott, W. T. (1966). Buried, viable seeds in two California bunchgrass sites and their bearing on the definition of a flora. *Vegetatio* **13,** 253–282.

Malone, C. R. (1967). A rapid method for enumeration of viable seeds in soil. *Weeds* **15,** 381–382.

Manders, P. T. (1990). Soil seed banks and post-fire seed deposition across a forest-fynbos ecotone in the Cape Province. *J. Veg. Sci.* **1,** 491–498.

Maranon, T., and Bartolome, J. W. (1989). Seeds and seedling populations in two contrasted communities: Open grassland and oak (*Quercus agrifolia*) understory in California. *Acta Oecol.* **10,** 147–158.

Marquis, D. A. (1975). Seed storage and germination under northern hardwood forests. *Can. J. For. Res.* **5,** 487–484.

Martin, S. C. (1970). Longevity of velvet mesquite seed in the soil. *J. Range Manage.* **23,** 69–70.

Mather, L. S., and Williams, P. A. (1990). Phenology, seed ecology, and age structure of Spanish heath (*Erica lusitanica*) in Canterbury, New Zealand. *New Zeal. J. Bot.* **28,** 207–215.

Matlack, G. R., and Good, R. E. (1990). Spatial heterogeneity in the soil seed bank of a mature coastal plain forest. *Bull. Torrey Bot. Club* **117,** 143–152.

McDonald, A. W. (1993). The role of seedbank and sown seeds in the restoration of an English flood-meadow. *J. Veg. Sci.* **4,** 395–400.

McGee, A., and Feller, M. C. (1993). Seed banks of forested and disturbed soils in southwestern British Columbia. *Can. J. Bot.* **71,** 1574–1583.

McGraw, J. B. (1987). Seed-bank properties of an Appalachian sphagnum bog and a model of the depth distribution of viable seeds. *Can. J. Bot.* **65,** 2028–2035.

McGraw, J. B. (1993). Ecological genetic variation in seed banks. IV. Differentiation of extant and seed bank-derived populations of *Eriophorum vaginatum. Arctic Alpine Res.* **25,** 45–49.

McGraw, J. B., Vavrek, M. C., and Bennington, C. C. (1991). Ecological genetic variation in seed banks. I. Establishment of a time transect. *J. Ecol.* **79**, 617–625.

McIvor, J. G., and Gardener, C. J. (1991). Soil seed densities and emergence patterns in pastures in the seasonally dry tropics of northeastern Australia. *Aust. J. Ecol.* **16**, 159–169.

McIntyre, A. C. (1929). A cone and seed study of the mountain pine (*Pinus pungens* Lambert). *Am. J. Bot.* **16**, 402–406.

McIntyre, S. (1985). Seed reserves in temperate Australian rice fields following pasture rotation and continuous cropping. *J. Appl. Ecol.* **22**, 875–884.

McMillan, C. (1981). Seed reserves and seed germination for two seagrasses, *Halodule wrightii* and *Syringodium filiforme,* from the western Atlantic. *Aquat. Bot.* **11**, 279–296.

McMillan, C. (1988). The seed reserve of *Halophila decipiens* Ostenfeld (Hydrocharitaceae) in Panama. *Aquat. Bot.* **31**, 177–182.

McRill, M., and Sagar, G. R. (1973). Earthworms and seeds. *Nature* **243**, 482.

Meiners, S. J., and Gorchov, D. L. (1994). The soil seed pool of Huffman Prairie, a degraded Ohio prairie, and its potential in restoration. *Ohio J. Sci.* **94**, 82–86.

Meney, K. A., Nielssen, G. M., and Dixon, K. W. (1994). Seed bank patterns in Restionaceae and Epacridaceae after wildfire in kwongan in southwestern Australia. *J. Veg. Sci.* **5**, 5–12.

Mercer, G. N., Gill, A. M., and Weber, R. O. (1994). A time-dependent model of fire impact on seed survival in woody fruits. *Aust. J. Bot.* **42**, 71–81.

Middleton, B. A., van der Valk, A. G., Mason, D. H., Williams, R. L., and Davis, C. B. (1991). Vegetation dynamics and seed banks of a monsoonal wetland overgrown with *Paspalum distichum* L. in northern India. *Aquat. Bot.* **40**, 239–259.

Midgley, J. J., Bond, W. J., and Geldenhuys, C. J. (1995). The ecology of southern African conifers. *In* "Ecology of the Southern Conifers" (N. J. Enright and R. S. Hill eds.), pp. 64–80. Smithsonian Institution Press, Washington.

Milberg, P. (1990). Hur lange kan ett fro leva? *Svensk Bot. Tidskr.* **84**, 323–352. [In Swedish with English abstract]

Milberg, P. (1992). Seed bank in a 35-year-old experiment with different treatments of a semi-natural grassland. *Acta Oecol.* **13**, 743–752.

Milberg, P. (1993). Seed bank and seedlings emerging after soil disturbance in a wet semi-natural grassland in Sweden. *Ann. Bot. Fennici* **30**, 9–13.

Milberg, P. (1994a). Germination of up to 129-year old, dry-stored seeds of *Geranium bohemicum* (Geraniaceae). *Nord. J. Bot.* **14**, 27–29.

Milberg, P. (1994b). Germination ecology of the endangered grassland biennial *Gentianella campestris*. *Biol. Conserv.* **70**, 287–290.

Milberg, P. (1995). Soil seed bank after eighteen years of succession from grassland to forest. *Oikos* **72**, 3–13.

Milberg, P. and Hansson, M. L. (1994). Soil seed bank and species turnover in a limestone grassland. *J. Veg. Sci.* **5**, 35–42.

Milberg, P., and Persson, T. S. (1994). Soil seed bank and species recruitment in road verge grassland vegetation. *Ann. Bot. Fennici* **31**, 155–162.

Miller, G. R. and Cummins, R. P. (1987). Role of buried viable seeds in the recolonization of disturbed ground by heather (*Calluna vulgaris* [L.] Hull) in the Cairngorm Mountains, Scotland, U.K. *Arctic Alpine Res.* **19**, 396–401.

Mills, J. T. (1983). Insect-fungus associations influencing seed deterioration. *Phytopathology* **73**, 330–334.

Milton, W. E. J. (1936). The buried viable seeds of enclosed and unenclosed hill land. *Bull. Welsh Plant Breed. Expt. Sta Ser. H* **14**, 58–73.

Mladenoff, D. J. (1990). The relationship of the soil seed bank and understory vegetation in old-growth northern hardwood-hemlock treefall gaps. *Can. J. Bot.* **68**, 2714–2721.

Mohler, C. L., and Callaway, M. B. (1995). Effects of tillage and mulch on weed seed production and seed banks in sweet corn. *J. Appl. Ecol.* **32**, 627–639.

Moore, J. M., and Wein, R. W. (1977). Viable seed populations by soil depth and potential site recolonization after disturbance. *Can. J. Bot.* **55**, 2408–2412.

Morgan, J. W. (1995). Ecological studies of the endangered *Rutidosis leptorrhyncoides*. I. Seed production, soil seed bank dynamics, population density and their effects on recruitment. *Aust. J. Bot.* **43**, 1–11.

Morgan, P., and Neuenschwander, L. F. (1988). Seed-bank contributions to regeneration of shrub species after clear-cutting and burning. *Can. J. Bot.* **66**, 169–172.

Morin, H., and Payette, S. (1988). Buried seed populations in the montane, subalpine, and alpine belts of Mont Jacques-Cartier, Quebec. *Can. J. Bot.* **66**, 101–107.

Moss, G. R. (1959). The gorse seed problem. *Proc. New Zeal. Weed Pest Cont. Conf.* **12**, 59–64.

Moss, S. R. (1985). The survival of *Alopecurus myosuroides* Huds. seeds in soil. *Weed Res.* **25**, 201–211.

Muir, P. S. and Lotan, J. E. (1985). Disturbance history and serotiny of *Pinus contorta* in western Montana. *Ecology* **66**, 1658–1668.

Mukherjee, U., Tripathi, R. S., and Yadav, A. S. (1980). Fate of buried seeds of *Alysicarpus monilifer* L. and *Indigofera enneaphylla* under sward situation. *Indian J. Ecol.* **7**, 88–95.

Murata, M., Roos, E. E., and Tsuchiya, T. (1981). Chromosome damage induced by artificial seed aging in barley. I. Germinability and frequency of aberrant anaphases at first mitosis. *Can. J. Genet. Cytol.* **23**, 267–280.

Musselman, L. J. (1980). The biology of *Striga, Orobanche,* and other root-parasitic weeds. *Annu. Rev. Phytopath.* **18**, 463–489.

Mustart, P. J., and Cowling, R. M. (1991). Seed germination of four serotinous Agulhas Plain Proteaceae. *S. Afr. J. Bot.* **57**, 310–313.

Mustart, P. J., and Cowling, R. M. (1993). Effects of soil and seed characteristics on seed germination and their possible roles in determining field emergence patterns of four Agulhas Plain (South Africa) Proteaceae. *Can. J. Bot.* **71**, 1363–1368.

Naka, K., and Yoda, K. (1984). Community dynamics of evergreen broadleaf forests in southwestern Japan. II. Species composition and density of seeds buried in the soil of a climax evergreen oak forest. *Bot. Mag. Tokyo* **97**, 61–79.

Nakagoshi, N. (1984a). Ecological studies on the buried viable seed population in soil of the forest communities in Miyajima Island, southwestern Japan. *Hikobia* **9**, 109–122.

Nakagoshi, N. (1984b). Buried viable seed populations in forest communities on the Hiba Mountains, southwestern Japan. *J. Sci. Hiroshima Univ.* **19**, 1–56.

Nakagoshi, N., and Suzuki, H. (1977). Ecological studies on the buried viable seed population in soil of the forest communities in Miyajima Island, southwestern Japan. *Hikobia* **8**, 180–192.

Navie, S. C., Cowley, R. A., and Rogers, R. W. (1996). The relationship between distance from water and the soil seed bank in a grazed semi-arid subtropical rangeland. *Aust. J. Bot.* **44**, 421–431.

Neely, R. K., and Wiler, J. A. (1993). The effect of sediment loading on germination from the seed bank of three Michigan wetlands. *Michigan Bot.* **32**, 199–208.

Nelson, J. F., and Chew, R. M. (1977). Factors affecting seed reserves in the soil of a Mojave Desert ecosystem, Rock Valley, Nye County, Nevada. *Am. Midl. Nat.* **97**, 300–320.

Newell, P. F. (1967). Mollusca. *In* "Soil Biology" (A. Burges and F. Raw eds.) pp. 413–433. Academic Press, London.

Nicholson, A., and Keddy, P. A. (1983). The depth profile of a shoreline seed bank in Matchedash Lake, Ontario. *Can. J. Bot.* **61**, 3293–3296.

Noble, I. R., and Weiss, P. W. (1989). Movement and modelling of buried seed of the invasive perennial *Chrysanthemoides monilifera* in coastal dunes and biological control. *Aust. J. Ecol.* **14**, 55–64.

Numata, M., Hayashi, I., Komura, T., and Oki, K. (1964). Ecological studies on the buried-seed population in the soil as related to plant succession I. *Jap. J. Ecol.* **14**, 207–217.

O'Connor, F. B. (1967). The Enchytraeidae. *In* "Soil Biology" (A. Burges and F. Raw, eds.), pp. 213–257. Academic Press, London.

O'Connor, T. G., and Pickett, G. A. (1992). The influence of grazing on seed production and seed banks of some African savanna grasslands. *J. Appl. Ecol.* **29**, 247–260.

Odgers, B. M. (1994). Seed banks and vegetation of three contrasting sites in an urban eucalypt forest reserve. *Aust. J. Bot.* **42**, 371–382.

Odgers, B. M. (1996). Fire, buried germinable seed banks and grass species establishment in an urban eucalypt forest reserve. *Aust. J. Bot.* **44**, 413–419.

Odum, S. (1965). Germination of ancient seeds. *Dansk Bot. Ark.* **24**, 7–70.

Odum, S. (1974). Seeds in ruderal soils, their longevity and contribution to the flora of disturbed ground in Denmark. *Proc. Brit. Weed Cont. Conf.* **12**, 1131–1144.

Ohga, I. (1923). On the longevity of seeds of *Nelumbo nucifera*. *Bot. Mag. (Tokyo)* **37**, 87–95.

Ohga, I. (1926). On the structure of some ancient, but still viable fruits of Indian lotus, with special reference to their prolonged dormancy. *Jpn. J. Bot.* **3**, 1–20.

Ohga, N. (1992). Buried seed population in the herbaceous lomas on Loma Ancon in the coastal desert of central Peru. *Ecol. Res.* **7**, 341–353.

Ohtsuka, T., and Ohsawa, M. (1994). Accumulation of buried seeds and establishment of ruderal therophytic communities in disturbed habitat, central Japan. *Vegetatio* **110**, 83–96.

Olmsted, N. W., and Curtis, J. D. (1947). Seeds of the forest floor. *Ecology* **28**, 49–52.

Oosting, H. J., and Humphreys, M. E. (1940). Buried viable seeds in a successional series of old field and forest soil. *Bull. Torrey Bot. Club* **67**, 253–273.

Osborne, D. J. (1980). Senescence in seeds. *In* "Senescence in Plants" (K. V. Thimann, ed.), pp. 14–37. CRC Press, Boca Raton, FL.

Osborne, D. J. (1981). Dormancy as a survival stratagem. *Ann. Appl. Biol.* **98**, 525–562.

Osborne, D. J., Sharon, R., and Ben-Ishai, R. (1980–1981). Studies on DNA integrity and DNA repair in germinating embryos of rye (*Secale cereale*). *Israel J. Bot.* **29**, 259–272.

Owens, M. K., Wallace, R. B., and Archer, S. (1995). Seed dormancy and persistence of *Acacia berlandieri* and *Leucaena pulverulenta* in a semi-arid environment. *J. Arid Environ.* **29**, 15–23.

Paatela, J., and Ervio, L.-R. (1971). Weed seeds in cultivated soils in Finland. *Ann. Agric. Fenniae* **10**, 144–152.

Pake, C. E., and Venable, D. L. (1996). Seed banks in desert annuals: Implications for persistence and coexistence in variable environments. *Ecology* **77**, 1427–1435.

Pannell, J. R., and Myerscough, P. J. (1993). Canopy-stored seed banks of *Allocasuarina distyla* and *A. nana* in relation to time since fire. *Aust. J. Bot.* **41**, 1–9.

Parker, V. T., and Leck, M. A. (1985). Relationships of seed banks to plant distribution patterns in a freshwater tidal wetland. *Am. J. Bot.* **72**, 161–174.

Parrish, D. J., and Leopold, A. C. (1978). On the mechanism of aging in soybean seeds. *Plant Physiol.* **61**, 365–368.

Partridge, T. R. (1989). Soil seed banks of secondary vegetation on the Port Hills and Banks Peninsula, Canterbury, New Zealand, and their role in succession. *New Zeal. J. Bot.* **27**, 421–436.

Paszkowski, W. L., and Kremer, R. J. (1988). Biological activity and tentative identification of flavonoid components in velvetleaf (*Abutilon theophrasti* Medik.). *J. Chem. Ecol.* **14**, 1573–1582.

Pavlik, B. M., Ferguson, N., and Nelson, M. (1993). Assessing limitations on the growth of endangered plant populations, II. Seed production and seed bank dynamics of *Erysimum capitatum* ssp. *angustatum* and *Oenothera deltoides* ssp. *howellii*. *Biol. Conserv.* **65**, 267–278.

Pavone, L. V., and Reader, R. J. (1982). The dynamics of seed bank size and seed state of *Medicago lupulina*. *J. Ecol.* **70**, 537–547.

Payette, S., Deshaye, J., and Gilbert, H. (1982). Tree seed populations at the treeline in Riviere aux Feuilles area, northern Quebec, Canada. *Arctic Alpine Res.* **14**, 215–221.

Perez-Garcia, F., Ceresuela, J. L., Gonzalez, A. E., and Aguinagalde, I. (1992). Flavonoids in seed coats of *Medicago arborea* and *M. strasseri* (Leguminosae): Ecophysiological aspects. *J. Basic Microbiol.* **32**, 241–248.

Perez-Nasser, N., and Vazquez-Yanes, C. (1986). Longevity of buried seeds from some tropical rainforest trees and shrubs of Veracruz, Mexico. *Malay. For.* **49**, 352–356.

Perry, D. A., and Lotan, J. E. (1979). A model of fire selection for serotiny in lodgepole pine. *Evolution* **33**, 958–968.

Picmam, A. K., Giaccone, R., Ivarson, K. C., and Altosaar, I. (1984). Antifungal properties of oat hulls. *Phytoprotection* **65**, 9–15.

Pierce, S. M., and Cowling, R. M. (1991a). Dynamics of soil-stored seed banks of six shrubs in fire-prone dune fynbos. *J. Ecol.* **79**, 731–747.

Pierce, S. M., and Cowling, R. M. (1991b). Disturbance regimes as determinants of seed banks in coastal dune vegetation of the southeastern Cape. *J. Veg. Sci.* **2**, 403–412.

Piroznikow, E. (1983). Seed bank in the soil of stabilized ecosystem of a deciduous forest (Tilio-Carpinetum) in the Bialowieza National Park. *Ekol. Pol.* **31**, 145–172.

Planisek, S. L., and Pippen, R. W. (1984). Do sand dunes have seed banks? *Michigan Bot.* **23**, 169–177.

Poiani, K. A., and Johnson, W. C. (1988). Evaluation of the emergence method in estimating seed bank composition of prairie wetlands. *Aquat. Bot.* **32**, 91–97.

Pons, T. L. (1991). Dormancy, germination and mortality of seeds in a chalk-grassland flora. *J. Ecol.* **79**, 765–780.

Porsild, A. E., Harington, C. R., and Mulligan, G. A. (1967). *Lupinus arcticus* Wats. grown from seeds of Pleistocene age. *Science* **158**, 113–114.

Pratt, D. W., Black, R. A., and Zamora, B. A. (1984). Buried viable seed in a ponderosa pine community. *Can. J. Bot.* **62**, 44–52.

Price, M. V., and Podolsky, R. H. (1989). Mechanisms of seed harvest by heteromyid rodents: Soil texture effects on harvest rate and seed size selection. *Oecologia* **81**, 267–273.

Price, S. C., Sward, W. L., and Wedberg, H. L. (1985). Genetic variation, translocation heterozygosity and seed dormancy in *Clarkia williamsonii*. *Bot. Gaz.* **146**, 150–156.

Priestley, D. P. (1986). "Seed Aging: Implications for Seed Storage and Persistence in the Soil." Cornell Univ. Press, Ithaca, NY.

Priestley, D. A., and M. A. Posthumus. (1982). Extreme longevity of lotus seeds from Pulantien. *Nature* **299**, 148–149.

Prince, F. S., and Hodgdon, A. R. (1946). Viable seeds in old pasture soils. *New Hampshire Agri. Exp. Sta. Bull.* **89**, 3–16.

Purdie, R. W. (1977). Early stages of regeneration after burning in dry sclerophyll vegetation. II. Regeneration by seed germination. *Aust. J. Bot.* **25**, 35–46.

Purrington, C. B., and Schmitt, J. (1995). Sexual dimorphism of dormancy and survivorship in buried seeds of *Silene latifolia*. *J. Ecol.* **83**, 795–800.

Putwain, P. D., and Gillham, D. A. (1990). The significance of the dormant viable seed bank in the restoration of heathlands. *Biol. Conserv.* **52**, 1–16.

Putz, F. E. (1983). Treefall pits and mounds, buried seeds, and the importance of soil disturbance to pioneer trees on Barro Colorado Island, Panama. *Ecology* **64**, 1069–1074.

Putz, F. E., and Appanah, S. (1987). Buried seeds, newly dispersed seeds, and the dynamics of a lowland forest in Malaysia. *Biotropica* **19**, 326–333.

Qi, M., Upadhyaya, M. K., and Turkington, R. (1996). Dynamics of seed bank and survivorship of meadow salsify (*Tragopogon pratensis*) populations. *Weed Sci.* **44**, 100–108.

Quick, C. R. (1956). Viable seeds from the duff and soil of sugar pine forests. *For. Sci.* **2**, 36–42.

Quick, C. R. (1975). Seed longevity of the Sierra gooseberry. *Madrono* **23**, 236.

Rabinowitz, D. (1981). Buried viable seeds in a North American tall-grass prairie: The resemblance of their abundance and composition to dispersing seeds. *Oikos* **36**, 191–195.

Raffaele, E. (1996). Relationship between seed and spore banks and vegetation of a mountain flood meadow (mallin) in Patagonia, Argentina. *Wetlands* **16**, 1–9.

Rampton, H. H., and Ching, T. M. (1966). Longevity and dormancy in seeds of several cool-season grasses and legumes buried in soil. *Agron. J.* **58**, 220–222.

Rampton, H. H., and Ching, T. M. (1970). Persistence of crop seeds in soil. *Agron. J.* **62**, 272–277.

Rao, N. K., Roberts, E. H., and Ellis, R. H. (1987a). The influence of pre and post-storage hydration treatments on chromosomal aberrations, seedling abnormalities, and viability of lettuce seeds. *Ann. Bot.* **60**, 97–108.

Rao, N. K., Roberts, E. H., and Ellis, R. H. (1987b). Loss of viability in lettuce seeds and the accumulation of chromosome damage under different storage conditions. *Ann. Bot.* **60**, 85–96.

Rao, N. K., Roberts, E. H., and Ellis, R. H. (1988). A comparison of the quantitative effects of seed moisture content and temperature on the accumulation of chromosome damage and loss of seed viability in lettuce. *Ann. Bot.* **62**, 245–248.

Raw, F. (1967). Anthopoda (except Acari and Collembola). *In* "Soil biology," (A. Burges and F. Raw, eds.), pp. 323–362. Academic Press, London.

Rees, M., and Long, M. J. (1992). Germination biology and the ecology of annual plants. *Am. Nat.* **139**, 484–508.

Reichman, O. J. (1979). Desert granivore foraging and its impact on seed densities and distributions. *Ecology* **60**, 1085–1092.

Reynolds, D. N. (1984). Population dynamics of three annual species of alpine plants in the Rocky Mountains. *Oecologia* **62**, 250–255.

Richardson, D. M., Van Wilgen, B. W., and Mitchell, D. T. (1987). Aspects of the reproductive ecology of four Australian *Hakea* species (Proteaceae) in South Africa. *Oecologia* **71**, 345–354.

Rico-Gray, V., and Garcia-Franco, J. G. (1992). Vegetation and soil seed bank of successional stages in tropical lowland deciduous forest. *J. Veg. Sci.* **3**, 617–624.

Rieseberg, L. H., Whitton, J., Linder, C. R., and Snow, A. A. (1996). Introgression of crop genes into wild sunflower populations. *Am. J. Bot.* **83** (Suppl. to No. 6), 57. [Abstract]

Roach, D. A. (1983). Buried seed and standing vegetation in two adjacent tundra habitats, northern Alaska. *Oecologia* **60**, 359–364.

Roberts, E. H. (1972a). Cytological, genetical, and metabolic changes associated with loss of viability. *In* "Viability of Seeds" (E. H. Roberts, ed.), pp. 253–306. Syracuse Univ. Press, Syracuse, NY.

Roberts, E. H. (1972b). Dormancy: A factor affecting seed survival in the soil. *In* "Viability of Seeds" (E. H. Roberts, ed.), pp. 321–357. Syracuse Univ. Press, Syracuse, NY.

Roberts, E. H. (1988). Seed aging: The genome and its expression. *In* "Senescence and Aging in Plants" (L. D. Nooden and A. C. Leopold, eds.), pp. 465–498. Academic Press, San Diego.

Roberts, E. H., and Abdalla, F. H. (1968). The influence of temperature, moisture, and oxygen on period of seed viability in barley, broad beans, and peas. *Ann. Bot.* **32**, 97–117.

Roberts, E. H., and Ellis, R. H. (1989). Water and seed survival. *Ann. Bot.* **63**, 39–52.

Roberts, H. A. (1964). Emergence and longevity in cultivated soil of seeds of some annual weeds. *Weed Res.* **4**, 296–307.

Roberts, H. A. (1979). Periodicity of seedling emergence and seed survival in some Umbelliferae. *J. Appl. Ecol.* **16**, 195–201.

Roberts, H. A. (1981). Seed banks in soils. *Adv. Appl. Biol.* **6**, 1–55.

Roberts, H. A. (1986a). Seed persistence in soil and seasonal emergence in plant species from different habitats. *J. Appl. Ecol.* **23**, 639–656.

Roberts, H. A. (1986b). Persistence of seeds of some grass species in cultivated soil. *Grass Forage Sci.* **41**, 273–276.

Roberts, H. A., and Boddrell, J. E. (1983a). Seed survival and periodicity of seedling emergence in eight species of Cruciferae. *Ann. Appl. Biol.* **103**, 301–309.

Roberts, H. A., and Boddrell, J. E. (1983b). Seed survival and periodicity of seedling emergence in ten species of annual weeds. *Ann. Appl. Biol.* **102**, 523–532.

Roberts, H. A., and Boddrell, J. E. (1984a). Seed survival and periodicity of seedling emergence in four weedy species of *Papaver*. *Weed Res.* **24**, 195–200.

Roberts, H. A., and Boddrell, J. E. (1984b). Seed survival and seasonal emergence of seedlings of some ruderal plants. *J. Appl. Ecol.* **21**, 617–628.

Roberts, H. A., and Boddrell, J. E. (1985). Seed survival and seasonal pattern of seedling emergence in some Leguminosae. *Ann Appl. Biol.* **106**, 125–132.

Roberts, H. A., and Chancellor, R. J. (1979). Periodicity of seedling emergence and achene survival in some species of *Carduus, Cirsium* and *Onopordum*. *J. Appl. Ecol.* **16**, 641–647.

Roberts, H. A., and Dawkins, P. A. (1967). Effect of cultivation on the numbers of viable weed seeds in soil. *Weed Res.* **7**, 290–301.

Roberts, H. A., and Feast, P. M. (1972). Fate of seeds of some annual weeds in different depths of cultivated and undisturbed soil. *Weed Res.* **12**, 316–324.

Roberts, H. A., and Feast, P. M. (1973a). Changes in the numbers of viable weed seeds in soil under different regimes. *Weed Res.* **13**, 298–303.

Roberts, H. A., and Feast, P. M. (1973b). Emergence and longevity of seeds of annual weeds in cultivated and undisturbed soil. *J. Appl. Ecol.* **10**, 133–143.

Roberts, H. A., and Lockett, P. M. (1978). Seed dormancy and periodicity of seedling emergence in *Veronica hederifolia* L. *Weed Res.* **18**, 41–48.

Roberts, H. A., and Neilson, J. E. (1980). Seed survival and periodicity of seedling emergence in some species of *Atriplex, Chenopodium, Polygonum* and *Rumex*. *Ann. Appl. Biol.* **94**, 111–120.

Roberts, H. A., and Neilson, J. E. (1981). Seed survival and periodicity of seedling emergence in twelve weedy species of Compositae. *Ann. Appl. Biol.* **97**, 325–334.

Roberts, H. A., and M. E. Ricketts. (1979). Quantitative relationships between the weed flora after cultivation and the seed population in the soil. *Weed Res.* **19**, 269–275.

Roberts, T. L., Carson, W. P., and Vankat, J. L. (1984). The seed bank and the initial revegetation of disturbed sites in Hueston Woods State Nature Preserve. *In* "Hueston Woods State Park and Nature Preserve, Proceedings of a Symposium" (G. E. Willeke ed.), pp. 150–155. Miami Univ. Oxford, OH.

Roberts, T. L., and Vankat, J. L. (1991). Floristics of a chronosequence corresponding to old field-deciduous forest succession in southwestern Ohio. II. Seed banks. *Bull. Torrey Bot. Club* **118**, 377–384.

Roos, E. E. (1989). Long-term seed storage. *Plant Breed. Rev.* **7**, 129–158.

Rosburg, T. R., Jurik, T. W., and Glenn-Lewin, D. C. (1994). Seed banks of communities in the Iowa loess hills: Ecology and potential contribution to restoration of native grassland. *In* "Spirit of the Land, our Prairie Legacy" (R. G. Wickett, P. D. Lewis, A. Woodliffe, and P. Pratt, eds.), pp. 221–237. Proc. 13th North American Prairie Conf., Windsor, Ontario, Canada, Dept. Parks and Recreation, Windsor, Ontario, Canada.

Rothrock, P. E., Squiers, E. R., and Sheeley, S. (1993). Heterogeneity and size of a persistent seedbank of *Ambrosia artemisiifolia* L. and *Setaria faberi* Herrm. *Bull. Torrey Bot. Club* **120**, 417–422.

Roy, K. W., and Abney, T. S. (1977). Antagonism between *Cerospora kikuchii* and other seedborne fungi of soybeans. *Phytopathology* **67**, 1062–1066.

Russi, L., Cocks, P. S., and Roberts, E. H. (1992). Seed bank dynamics in a Mediterranean grassland. *J. Appl. Ecol.* **29**, 763–771.

Sahoo, U. K., Tripathi, R. S., Pandey, H. N., and Misra, J. (1994). Population dynamics of buried weed seeds as influenced by shifting and terrace cultivation in the humid subtropics of India. *Weed Res.* **34**, 157–165.

Salzmann, R. (1954). Untersuchungen uber die lebersdauer von unkrautsamen im boden. *Mitteil. Schweiz. Landwirtschaft.* **2**, 170–176.

Sarukhan, J. (1974). Studies on plant demography: *Ranunculus repens* L., *R. bulbosus* L. and *R. acris*. II. Reproductive strategies and seed population dynamics. *J. Ecol.* **62**, 151–177.

Saulei, S. M., and Swaine, M. D. (1988). Rain forest seed dynamics during succession at Gogol, Papua New Guinea. *J. Ecol.* **76**, 1133–1152.

Schafer, D. E., and Chilcote, D. O. (1969). Factors influencing persistence and depletion in buried seed populations. I. A model for analysis of parameters of buried seed persistence and depletion. *Crop Sci.* **9**, 417–419.

Schafer, D. E., and Chilcote, D. O. (1970). Factors influencing persistence and depletion in buried seed populations. II. The effects of soil temperature and moisture. *Crop Sci.* **10**, 342–345.

Scheiner, S. M. (1988). The seed bank and above-ground vegetation in a upland pine-hardwood succession. *Michigan Bot.* **27**, 99–106.

Schiffman, P. M., and Johnson, W. C. (1992). Sparse buried seed bank in a southern Appalachian oak forest: Implications for succession. *Am. Midl. Nat.* **127**, 258–267.

Schmitt, J., and Linder, C. R. (1994). Will escaped transgenes lead to ecological release? *Mol. Ecol.* **3**, 71–74.

Schneider, R. (1994). The role of hydrologic regime in maintaining rare plant communities of New York's coastal plain pondshores. *Biol. Conserv.* **68**, 253–260.

Schneider, R. L., and Sharitz, R. R. (1986). Seed bank dynamics in a southeastern riverine swamp. *Am. J. Bot.* **73**, 1022–1030.

Schneider, U., and Kehl, H. (1987). Samenbank und vegetationsaufnahmen ostmediterraner therophytenfluren im vergleich. *Flora* **179**, 345–354.

Schreiber, M. M. (1992). Influence of tillage, crop rotation, and weed management on giant foxtail (*Setaria faberi*) population dynamics and corn yield. *Weed Sci.* **40**, 645–653.

Schweizer, E. E., and Zimdahl, R. L. (1984). Weed seed decline in irrigated soil after rotation of crops and herbicides. *Weed Sci.* **32**, 84–89.

Schwerzel, P. J. (1974). The effect of depth of burial in soil on the survival of some common Rhodesian weed seeds. *Rhodesian Agric.* **73**, 97–99.

Sem, G., and Enright, N. J. (1995). The soil seed bank in *Agathis*

australis (D. Don) Lindl. (kauri) forests of northern New Zealand. *New Zeal. J. Bot.* **33**, 221–235.

Sem, G., and Enright, N. J. (1996). The relationship between seed rain and the soil seed bank in a temperate rainforest stand near Auckland, New Zealand. *New Zeal. J. Bot.* **34**, 215–226.

Sen, S., and Osborne, D. J. (1977). Decline in ribonucleic acid and protein synthesis with loss of viability during the early hours of imbibition of rye (*Secale cereale* L.) embryos. *Biochem. J.* **166**, 33–38.

Sharma, M. P., McBeath, D. K., and Vanden Born, W. H. (1983). Effect of fall and spring tillage on wild oat germination. *Can. J. Plant Sci.* **63**, 561–562.

Shen-Miller, J., Mudgett, M. B., Schopf, J. W., Clarke, S., and Berger, R. (1995). Exceptional seed longevity and robust growth: Ancient sacred lotus from China. *Amer. J. Bot.* **82**, 1367–1380.

Shen-Miller, J., Schopf, J. W., and Berger, R. (1983). Germination of a *ca.* 700 year-old lotus seeds from China: Evidence of exceptional longevity of seed viability. *Amer. J. Bot.* **70**(5), Part 2, p. 78. [Abstract]

Short, G. E., and Lacy, M. L. (1976). Carbohydrate exudation from pea seeds: Effect of cultivar, seed age, seed color, and temperature. *Phytopathology* **66**, 182–187.

Shull, F. H. (1914). The longevity of submerged seeds. *Plant World* **17**, 329–337.

Simon, E. W. (1974). Phospholipids and plant membrane permeability. *New Phytol.* **73**, 377–420.

Sivori, E., Nakayama, F., and Cigliano, E. (1968). Germination of achira seed (*Canna* sp.) approximately 550 years old. *Nature* **219**, 1269–1270.

Skoglund, J. (1992). The role of seed banks in vegetation dynamics and restoration of dry tropical ecosystems. *J. Veg. Sci.* **3**, 357–360.

Skoglund, J., and Hytteborn, H. (1990). Viable seeds in deposits of the former lakes Kvismaren and Hornborgasjon, Sweden. *Aquat. Bot.* **37**, 271–290.

Smith, A. M., and Cook, R. J. (1974). Implications of ethylene production by bacteria for biological balance of soil. *Nature* **252**, 703–705.

Smith, L. M., and Kadlec, J. A. (1985). Predictions of vegetation change following fire in a Great Salt Lake marsh. *Aquat. Bot.* **21**, 43–51.

Sonia, L., and Heslehurst, M. R. (1978). Germination characteristics of some *Banksia* species. *Aust. J. Ecol.* **3**, 179–186.

Spence, J. R. (1990). A buried seed experiment using caryopses of *Chionochloa macra* Zotov (Danthonieae: Poaceae), South Islands, New Zealand. *New Zealand J. Bot.* **28**, 471–474.

Spira, T. P., and Wagner, L. K. (1983). Viability of seeds up to 211 years old extracted from adobe brick buildings of California and northern Mexico. *Amer. J. Bot.* **70**, 303–307.

Spira, T. P., Sossey-Alaqui, K., Rajapakse, S., Miller, M. B, Abbott, A. G., and Tonkyn, D. W. (1996). Gene flow from cultivated to wild strawberry (*Fragaria* spp.). *Amer. J. Bot.* **83** (Suppl. to No. 6), 57–58. [Abstract]

Standifer, L. C. (1980). A technique for estimating weed seed populations in cultivated soil. *Weed Sci.* **28**, 134–139.

Stapanian, M. A. (1982). Evolution of fruiting strategies among fleshy-fruited plant species of eastern Kansas. *Ecology* **63**, 1422–1431.

Steiner, A. M., and Ruckenbauer, P. (1995). Germination of 110-year-old cereal and weed seeds, the Vienna sample of 1877. Verification of effective ultra-dry storage at ambient temperature. *Seed Sci. Res.* **5**, 195–199.

Stewart, C. N., Jr., and Porter, D. M. (1995). RAPD profiling in biological conservation: An application to estimating clonal variation in rare and endangered *Iliamna* in Virginia. *Biol. Conserv.* **74**, 135–142.

Stieperaere, H., and Timmerman, C. (1983). Viable seeds in the soils of some parcels of reclaimed and unreclaimed heath in the Flemish district (northern Belgium). *Bull. Soc. Roy. Bot. Belg.* **116**, 62–73.

Stoa, T. E. (1933). Persistence of viability of sweet clover seed in a cultivated soil. *J. Amer. Soc. Agron.* **25**, 177–181.

Strickler, G. S., and Edgerton, P. J. (1976). Emergent seedlings from coniferous litter and soil in eastern Oregon. *Ecology* **57**, 801–807.

Suzuki, W. (1993). Germination of *Rubus palmatus* var. *coptophyllus* and *R. microphyllus* seeds buried in soil for 7.5 years. *Ecol. Res.* **8**, 107–110.

Symington, C. F. (1932). The study of secondary growth on rain forest sites in Malaya. *Malay. For.* **2**, 107–117.

Symonides, E. (1986). Seed bank in old-field successional ecosystems. *Ekol. Pol.* **34**, 3–29.

Taylorson, R. B. (1970). Changes in dormancy and viability of weed seeds in soils. *Weed Sci.* **18**, 265–269.

Teketay, D., and Granstrom, A. (1995). Soil seed banks in dry Afromontane forests of Ethiopia. *J. Veg. Sci.* **6**, 777–786.

TeKrony, D. M., Egli, D. B., and White, G. M. (1987). Seed production and technology. *In* "Soybeans: Improvement, Production, and Uses," 2nd Ed. pp. 295–353. Agron. Monogr. No. 16.

Templeton, A. R., and Levin, D. A. (1979). Evolutionary consequences of seed pools. *Amer. Nat.* **114**, 232–249.

Ter Heerdt, G. N. J., Verweij, G. L., Bekker, R. M., and Bakker, J. P. (1996). An improved method for seed-bank analysis: Seedling emergence after removing the soil by sieving. *Funct. Ecol.* **10**, 144–151.

Terrados, J. (1993). Sexual reproduction and seed banks of *Cymodocea nodosa* (Ucria) Ascherson meadows on the southeast Mediterranean coast of Spain. *Aquat. Bot.* **46**, 293–299.

Thompson, K. (1978). The occurrence of buried viable seeds in relation to environmental gradients. *J. Biogeogr.* **5**, 425–430.

Thompson, K. (1986). Small-scale heterogeneity in the seed bank of an acidic grassland. *J. Ecol.* **74**, 733–738.

Thompson, K. (1987). Seeds and seed banks. *New Phytol.* **106** (Suppl.), 23–34.

Thompson, K., Band. S. R., and Hodgson, J. G. (1993). Seed size and shape predict persistence in soil. *Funct. Ecol.* **7**, 236–241.

Thompson, K., Green, A., and Jewels, A. M. (1994). Seeds in soil and worm casts from a neutral grassland. *Funct. Ecol.* **8**, 29–35.

Thompson, K., and Grime, J. P. (1979). Seasonal variation in the seed banks of herbaceous species in ten contrasting habitats. *J. Ecol.* **67**, 893–921.

Thorsen, J. A., and Crabtree, G. (1977). Washing equipment for separating weed seed from soil. *Weed Sci.* **25**, 41–42.

Thurston, J. M. (1966). Survival of seeds of wild oats (*Avena fatua* L. and *Avena ludoviciana* Dur.) and charlock (*Sinapis arvensis* L.) in soil under leys. *Weed Res.* **6**, 67–80.

Tingley, D. C. (1961). Longevity of seeds of wild oats, winter rye, and wheat in cultivated soil. *Weeds* **9**, 607–611.

Titus, J. H. (1991). Seed bank of a hardwood floodplain swamp in Florida. *Castanea* **56**, 117–127.

Tonsor, S. J., Kalisz, S., Fisher, J., and Holtsford, T. P. (1993). A life-history based study of population genetic structure: Seed bank to adults in *Plantago lanceolata. Evolution* **47**, 833–843.

Toole, E. H. and Brown, E. (1946). Final results of the Duvel buried seed experiment. *J. Agric. Res.* **72**, 201–210.

Toole, V. K. (1986). Ancient seeds; seeds longevity. *J. Seed Technol.* **10**, 1–23.

Toole, V. K., and Toole, E. H. (1953). Seed dormancy in relation to seed longevity. *Proc. Int. Seed Test. Assoc.* **18**, 325–328.

Tsuyuzaki, S. (1991). Survival characteristics of buried seeds 10 years after the eruption of the Usu volcano in northern Japan. *Can. J. Bot.* **69**, 2251–2256.

Tsuyuzaki, S. (1994). Fate of plants from buried seeds on Volcano Usu, Japan, after the 1977–1978 eruptions. *Amer. J. Bot.* **81**, 395–399.

Tveit, M., and Moore, M. B. (1954). Isolates of *Chaetomium* that protect oats from *Helminthosporium victoriae. Phytopathology* **44**, 686–689.

Tybirk, K, Schmidt, L. H., and Hauser, T. (1994). Notes on soil seed banks of African acacias. *Afr. J. Ecol.* **32**, 327–330.

Uhl, C., and Clark, K. (1983). Seed ecology of selected Amazon Basin successional species. *Bot. Gaz.* **144**, 419–425.

Uhl, C., Clark, K., Clark, H., and Murphy, P. (1981). Early plant succession after cutting and burning in the upper Rio Negro region of the Amazon Basin. *J. Ecol.* **69**, 631–649.

Ungar, I. A. (1988). A significant seed bank for *Spergularia marina* (Caryophyllaceae). *Ohio J. Sci.* **88**, 200–202.

Ungar, I. A., and Reihl, T. E. (1980). The effect of seed reserves on species composition in zonal halophyte communities. *Bot. Gaz.* **141**, 447–452.

Ungar, I. A., and Woodell, S. R. J. (1993). The relationship between the seed bank and species composition of plant communities in two British salt marshes. *J. Veg. Sci.* **4**, 531–536.

Ungar, I. A., and Woodell, S. R. J. (1996). Similarity of seed banks to aboveground vegetation in grazed and ungrazed salt marsh communities on the Gower Peninsula, South Wales. *Int. J. Plant Sci.* **157**, 746–749.

Valverde, T., and Silvertown, J. (1995). Spatial variation in the seed ecology of a woodland herb (*Primula vulgaris*) in relation to light environment. *Funct. Ecol.* **9**, 942–950.

van Assche, J. A., and Vanlerberghe, K. A. (1989). The role of temperature on the dormancy cycle of seeds of *Rumex obtusifolius* L. *Funct. Ecol.* **3**, 107–115.

van der Valk, A. G., and Davis, C. B. (1979). A reconstruction of the recent vegetational history of a prairie marsh, Eagle Lake, Iowa, from its seed bank. *Aquat. Bot.* **6**, 29–51.

van der Valk, A. G., Pederson, R. L., and Davis, C. B. (1992). Restoration and creation of freshwater wetlands using seed banks. *Wetlands Ecol. Manage.* **1**, 191–197.

van der Valk, A. G., and Verhoeven, J. T. A. (1988). Potential role of seed banks and understory species in restoring quaking fens from floating forests. *Vegetatio* **76**, 3–13.

van Esso, M. L., Ghersa, C. M., and Soriano, A. (1986). Cultivation effects on the dynamics of a johnson grass seed population in the soil profile. *Soil Tillage Res.* **6**, 325–335.

van Leeuwen, B. H. (1981). Influence of micro-organisms on the germination of the monocarpic *Cirsium vulgare* in relation to disturbance. *Oecologia* **48**, 112–115.

van Tooren, B. F., and During, H. J. (1988). Viable plant diaspores in the guts of earthworms. *Acta Bot. Neerl.* **37**, 181–185.

Vavrek, M. C., McGraw, J. B., and Bennington, C. C. (1991). Ecological genetic variation in seed banks. III. Phenotypic and genetic differences between young and old seed populations of *Carex bigelowii. J. Ecol.* **79**, 645–662.

Vazquez-Yanes, C., and Orozco-Segovia, A. (1996). Comparative longevity of seeds of five tropical rain forest woody species stored under different moisture conditions. *Can. J. Bot.* **74**, 1635–1639.

Venable, D. L. (1989). Modeling the evolutionary ecology of seed banks. *In* "Ecology of Soil Seed Banks" (M. A. Leck, V. T. Parker, and R. L. Simpson eds), pp. 67–87. Academic Press, San Diego

Venable, D. L., and Brown, J. S. (1988). The selective interactions of dispersal, dormancy, and seed size as adaptations for reducing risk in variable environments. *Am. Nat.* **131**, 360–384.

Venable, D. L., Pake, C. E., and Caprio, A. C. (1993). Diversity and coexistence of Sonoran Desert winter annuals. *Plant Species Biol.* **8**, 207–216.

Vertucci, C. W. (1993). Towards a unified hypothesis on seed aging.

In "Basic and Applied Aspects of Seed Biology" (D. Come and F. Corbineau, eds.), pp. 739–746. Fourth International Workshop on Seeds, Universite Pierre et Marie Curie, Paris.

Vertucci, C. W., and Roos, E. E. (1990). Theoretical basis of protocols for seed storage. *Plant Physiol.* **94,** 1019–1023.

Vieno, M., Komulainen, M. and Neuvonen, S. (1993). Seed bank composition in a subarctic pine-birch forest in Finnish Lapland: Natural variation and the effect of simulated acid rain. *Can. J. Bot.* **71,** 379–384.

Villiers, T. A. (1971). Cytological studies in dormancy. I. Embryo maturation during dormancy in *Fraxinus excelsior. New Phytol.* **70,** 751–760.

Villiers, T. A. (1972). Cytological studies in dormancy. II. Pathological ageing changes during prolonged dormancy and recovery upon dormancy release. *New Phytol.* **71,** 145–152.

Villiers, T. A. (1973). Ageing and the longevity of seeds in field conditions. *In* W. Heydecker (ed.). "Seed ecology." Pp. 265–286. Pennsylvania State Univ. Press, University Park, PA.

Villiers, T. A. (1974). Seed aging: Chromosome stability and extended viability of seeds stored fully imbibed. *Plant Physiol.* **53,** 875–878.

Villiers, T. A. (1980). Ultrastructural changes in seed dormancy and senescence. *In* "Senescence in Plants." K. V. Thimann (ed.). pp. 39–66. CRC Press, Boca Raton, FL.

Villiers, T. A., and Edgcumbe, D. J. (1975). On the case of seed deterioration in dry storage. *Seed Sci. Technol.* **3,** 761–774.

Vivian-Smith, G., and Handel, S. N. (1996). Freshwater wetland restoration of an abandoned sand mine: Seed bank recruitment dynamics and plant colonization. *Wetlands* **16,** 185–196.

Vlahos, S., and Bell, D. T. (1986). Soil seed-bank components of the northern jarrah forest of western Australia. *Aust. J. Ecol.* **11,** 171–179.

Voesenek, L. A. C. J., and Blom, C. W. P. M. (1992). Germination and emergence of *Rumex* in river flood-plains. I. Timing of germination and seedbank characteristics. *Acta Bot. Neerl.* **41,** 319–329.

Walck, J. L., Baskin, J. M., and Baskin, C. C. (1996). An ecologically and evolutionarily meaningful definition of a persistent seed bank in *Solidago. Amer. J. Bot.* **83** (Supplement to No. 6), 78–79. [Abstract]

Warcup, J. H. (1967). Fungi in soil. *In* A. Burges and F. Raw (Eds.). "Soil biology," pp. 51–110. Academic Press, London.

Warnes, D. D., and Andersen, R. N. (1984). Decline of wild mustard (*Brassica kaber*) seeds in soil under various cultural and chemical practices. *Weed Sci.* **32,** 214–217.

Warr, S. J., Kent, M., and Thompson, K. (1994). Seed bank composition and variability in five woodlands in south-west England. *J. Biogeogr.* **21,** 151–168.

Warr, S. J., Thompson, K., and Kent, M. (1992). Antifungal activity in seed coat extracts of woodland plants. *Oecologia* **92,** 296–298.

Warr, S. J., Thompson, K., and Kent, M. (1993). Seed banks as a neglected area of biogeographic research: A review of literature and sampling techniques. *Progr. Phys. Geogr.* **17,** 329–347.

Warren, R., and Fordham, A. J. (1978). The fire pines. *Arnoldia* **38**(1), 1–11.

Warwick, M. A. (1984). Buried seeds in arable soils in Scotland. *Weed Res.* **24,** 261–268.

Watanabe, Y., and Hirokawa, F. (1975). Ecological studies on the germination and emergence of annual weeds. 3. Changes in emergence and viable seeds in cultivated and uncultivated soil. *Weed Res., Japan* **19,** 14–19 (in Japanese with English summary).

Watkinson, A. R. (1978). The demography of a sand dune annual: *Vulpia fasciculata.* II. The dynamics of seed populations. *J. Ecol.* **66,** 35–44.

Watkinson, A. R., Lonsdale, W. M. and Andrew, M. H. (1989). Model-

ling the population dynamics of an annual plant *Sorghum intrans* in the wet-dry tropics. *J. Ecol.* **77,** 162–181.

Weaver, S. E., and Cavers, P. B. (1979). Dynamics of seed populations of *Rumex crispus* and *Rumex obtusifolius* (Polygonaceae) in disturbed and undisturbed soil. *J. Appl. Ecol.* **16,** 909–917.

Webster, B. D., and Leopold, A. C. (1977). The ultrastructure of dry and imbibed cotyledons of soybean. *Amer. J. Bot.* **64,** 1286–1293.

Welling, C. H., and Becker, R. L. (1990). Seed bank dynamics of *Lythrum salicaria* L.: Implications for control of this species in North America. *Aquat. Bot.* **38,** 303–309.

Welling, C. H., Pederson, R. L., and van der Valk, A. G. (1988a). Temporal patterns in recruitment from the seed bank during drawdowns in a prairie wetland. *J. Appl. Ecol.* **25,** 999–1007.

Welling, C. H., Pederson, R. L., and van der Valk, A. G. (1988b). Recruitment from the seed bank and the development of zonation of emergent vegetation during a drawdown in a prairie wetland. *J. Ecol.* **76,** 483–496.

Wellington, A. B. (1989). Seedling regeneration and the population dynamics of eucalypts. *In* "Mediterranean Landscapes in Australia" (J. C. Noble and R. A. Bradstock eds.), pp. 155–167. CSIRO Publications, Melbourne.

Wellington, A. B., and Noble, I. R. (1985). Seed dynamics and factors limiting recruitment of the malee *Eucalyptus incrassata* in semiarid, south-eastern Australia. *J. Ecol.* **73,** 657–666.

Wendel, G. W. (1972). Longevity of black cherry seed in the forest floor. *USDA For. Serv. Res. Note* NE-149.

Westoby, M., Rice, B., Griffin, G., and Friedel, M. (1988). The soil seed bank of *Triodia basedowii* in relation to time since fire. *Aust. J. Ecol.* **13,** 161–169.

Whigham, D. F., Simpson, R. L., and Leck, M. A. (1979). The distribution of seeds, seedlings, and established plants of arrow arum (*Peltandra virginica* (L.) Kunth) in a freshwater tidal wetland. *Bull. Torrey Bot. Club* **106,** 193–199.

Whipple, S. A. (1978). The relationship of buried, germinating seeds to vegetation in an old-growth Colorado subalpine forest. *Can. J. Bot.* **56,** 1505–1509.

Whittaker, R. J., Partomihardjo, T., and Riswan, S. (1995). Surface and buried seed banks from Krakatau, Indonesia: Implications for the sterilization hypothesis. *Biotropica* **27,** 346–354.

Wienhold, C. E., and van der Valk, A. G. (1989). The impact of duration of drainage on the seed banks of northern prairie wetlands. *Can. J. Bot.* **67,** 1878–1884.

Willems, J. H., and Huijsmans, K. G. A. (1994). Vertical seed dispersal by earthworms: A quantitative approach. *Ecography* **17,** 124–130.

Williams, E. D. (1984). Changes during 3 years in the size and composition of the seed bank beneath a long-term pasture as influenced by defoliation and fertilizer regime. *J. Appl. Ecol.* **21,** 603–615.

Williams-Linera, G. (1993). Soil seed banks in four lower montane forests of Mexico. *J. Trop. Ecol.* **9,** 321–337.

Wilson, B. J. (1985). Effect of seed age and cultivation on seedling emergence and seed decline of *Avena fatua* L. in winter barley. *Weed Res.* **25,** 213–219.

Wilson, M. V., Ingersoll, C. A. and Roush, M. L. (1989). Measuring seed movement in soil. *Bull. Ecol. Soc. Am.* **70**(suppl. to No. 2), 300–301.

Wilson, S. D., Moore, D. R. J., and Keddy, P. A. (1993). Relationships of marsh seed banks to vegetation patterns along environmental gradients. *Freshwater Biol.* **29,** 361–370.

Wisheu, I. C., and Keddy, P. A. (1991). Seed banks of a rare wetland plant community: Distribution patterns and effects of human-induced disturbance. *J. Veg. Sci.* **2,** 181–188.

Witkowski, E. T. F., Lamont, B. B., and Connell, S. J. (1991). Seed bank dynamics of three co-occurring banksias in south coastal

Western Australia: The role of plant age, cockatoos, senescence and interfire establishment. *Aust. J. Bot.* **39**, 385–397.

Yenish, J. P., Doll, J. D., and Buhler, D. D. (1992). Effects of tillage on vertical distribution and viability of weed seed in soil. *Weed Sci.* **40**, 429–433.

Young, J. A., Evans, R. A., Raguse, C. A., and Larson, J. R. (1981). Germinable seeds and periodicity of germination in annual grasslands. *Hilgardia* **49**(1), 1–37.

Young, K. R. (1985). Deeply buried seeds in a tropical wet forest in Costa Rica. *Biotropica* **17**, 336–338.

Young, K. R., Ewel, J. J., and Brown, B. J. (1987). Seed dynamics during forest succession in Costa Rica. *Vegetatio* **71**, 157–173.

Zaman, A. U., and Khan, M. A. (1992). The role of buried viable seeds in saline desert plant community. *Bangladesh J. Bot.* **21**, 1–10.

Zammit, C., and Westoby, M. (1988). Pre-dispersal seed losses, and the survival of seeds and seedlings of two serotinous *Banksia* shrubs in burnt and unburnt heath. *J. Ecol.* **76**, 200–214.

Zammit, C. A., and Zedler, P. H. (1988). The influence of dominant shrubs, fire, and time since fires on soil seed banks in mixed chaparral. *Vegetatio* **75**, 175–187.

Zammit, C., and Zedler, P. H. (1994). Organisation of the soil seed bank in mixed chaparral. *Vegetatio* **111**, 1–16.

Zanin, G., Berti, A., and Zuin, M. C. (1989). Estimation du stock semencier d'un sol labouré ou en semis direct. *Weed Res.* **29**, 407–417.

Zhang, J., and Maun, M. A. (1994). Potential for seed bank formation in seven Great Lakes sand dune species. *Amer. J. Bot.* **81**, 387–394.

Zimmergren, D. (1980). The dynamics of seed banks in an area of sandy soil in southern Sweden. *Bot. Notiser* **133**, 633–641.

Zorner, P. S., Zimdahl, R. L. and Schweizer, E. E. (1984a). Sources of viable seed loss in buried dormant and non-dormant populations of wild oat (*Avena fatua* L.) seed in Colorado. *Weed Res.* **24**, 143–150.

Zorner, P. S., Zimdahl, R. L. and Schweizer, E. E. (1984b). Effect of depth and duration of seed burial on kochia (*Kochia scoparia*). *Weed Sci.* **32**, 602–607.

8

Causes of Within-Species Variations in Seed Dormancy and Germination Characteristics

I. PURPOSE

In many plant species, seeds collected in various locations or at different times in the same place and given identical dormancy breaking and germination test conditions germinate to different percentages. The purpose of this chapter is to explore the reasons for variation in seed dormancy breaking and germination requirements within a species. Genetics and the environment of the mother plant during the time of seed maturation are two of the most important factors controlling this variation; however, they do not operate independently. Thus, much research often is required to determine the relative importance of each of these two factors in the variation in seed dormancy and germination of a particular species.

Chapter 8 covers (1) the inheritance of seed dormancy and dormancy breaking and germination requirements, (2) variations in germination responses of seeds from different populations of the same species, and (3) environmental factors known to influence the mother plant, thereby causing changes in the germination characteristics of the progeny (seeds). Also, the kinds of changes that can occur in seeds as they develop under varying environmental conditions will be considered. Because seed polymorphism and cleistogamous/chasmogamous seeds represent an interplay of genetics, environment, and genetics × environmental factors, they are included in this chapter along with a discussion of their consequences on seedling recruitment.

II. INHERITANCE OF SEED DORMANCY

A. Species with Heritable Seed Dormancy

Seed dormancy has been shown to have a hereditary component in a number of species (Table 8.1), many of

which are crops or serious weeds. Studies on the genetics of seed dormancy in economically important species have resulted in the development of new varieties that either have, or lack, seed dormancy. For example, hybridization studies have been done to add physiological dormancy to *Arachis hypogaea* seeds (Stokes and Hull, 1930) and to remove physical dormancy from seeds of *Lupinus angustifolius* (Forbes and Wells, 1968), *Melilotus alba* (White and Stevenson, 1948), and *Glycine max* (Kilen and Hartwig, 1978). It should be realized, however, that physical dormancy can be a highly desirable trait, and certain varieties of legumes and cotton have been developed with physically dormant seeds.

B. Types of Inherited Seed Dormancy

With the exception of *Eschscholtzia californica* and one or more species of *Papaver,* all the entries in Table 8.1 have either physical or physiological dormancy. Because the seeds with physiological dormancy listed in Table 8.1 become nondormant during either a warm or a short period of cold stratification, they have nondeep physiological dormancy. Seeds of *E. californica* and *P. dubium* have underdeveloped embryos (Martin, 1946), and thus have either morphological or morphophysiological dormancy (see Chapter 3).

C. Selection of Pure Lines

By using selection for a number of generations, it has been possible for plant breeders to develop dormant and/or nondormant lines of many species. In *Sinapis arvensis* (Witcombe and Whittington, 1972; Garbutt and Witcombe, 1986), *Melilotus alba* (White and Stevenson, 1948), *Lupinus* spp. (Gladstones, 1970), and *Eleusine* spp. (Hilu and de Wet, 1980), lines with nondormant

TABLE 8.1 Species in Which Seed Dormancy Has a Hereditary Component[a]

Species	Reference
Arabidopsis thaliana	Karssen *et al.* (1989)
Arachis hypogaea	Stokes and Hull (1930)
Avena fatua	Naylor and Jana (1976), Jana and Naylor (1980), Adkins *et al.* (1986)
Beta vulgaris	Battle and Whittington (1971)
Brassica oleracea	Hodgkin and Hegarty (1978)
Collinsia verna	Kalisz (1986)
Eleusine spp.	Hilu and de Wet (1980)
Eschscholtzia californica[b]	Cook (1962)
Digitaria milanjiana	Hacker *et al.* (1984)
Galega orientalis[c]	Sain (1948)
Gilia capitata	Grant (1949)
Glycine max[c]	Kilen and Hartwig (1978)
Gossypium hirsutum[c]	Lee (1975)
Helianthus bolanderi	Olivieri and Jain (1978)
Hordeum vulgare	Boyd *et al.* (1971), Buraas and Skinnes (1984)
Lactuca sativa	Eenink (1981)
Lolium perenne	Haywood and Breese (1966)
Lotus corniculatus[c]	Sain (1948)
Lupinus angustifolius[c]	Forbes and Wells (1968)
Lupinus spp.[c]	Gladstones (1970)
Lycopersicon spp.	Whittington *et al.* (1965)
Medicago spp.[c]	Sain (1948)
Melilotus alba[c]	White and Stevenson (1948)
Oldenlandia corymbosa	Do Cao *et al.* (1978)
Oryza sativa	Chang and Li (1991)
Papaver spp.[b]	Harper and McNaughton (1960)
Petunia hybrida	Girard (1990)
Phaseolus vulgaris[c]	Lebedeff (1947)
Raphanus raphanistrum	Cheam (1986)
Senecio vulgaris	Kadereit (1984)
Sinapis arvensis	Garbutt and Witcombe (1986)
Solanum tuberosum	Simmonds (1964)
Sorghum vulgare	Gritton and Atkins (1961)
Stylosanthes humilis[c]	Cameron (1965)
Trifolium hirtum[c]	Jain (1982)
T. pratense[c]	Sain (1948)
T. repens[c]	Sain (1948)
T. subterraneum[c]	Morley (1958)
Triticum sp.	Gfeller and Svejda (1960)
Vicia cracca[c]	Sain (1948)
V. sativa[c]	Donnelly *et al.* (1972)
Vicia spp.[c]	Elkins *et al.* (1966)

[a] Seeds have physiological dormancy, unless otherwise indicated.
[b] Morphological or morphophysiological dormancy.
[c] Physical dormancy.

seeds were selected, whereas in *Arachis hypogaea* (Stokes and Hull, 1930) lines with dormant seeds were developed. In *Solanum tuberosum* (Simmonds, 1964), *Avena fatua* (Naylor and Jana, 1976), and *Oldenlandia corymbosa* (Do Cao *et al.*, 1978), both dormant and nondormant lines have been developed.

D. Crosses Within a Species

In genetic studies of seed dormancy, dormant and nondormant lines or varieties are crossed, and the germination responses of the progeny are determined. Seed dormancy in *Triticum* spp. is controlled by multiple genes that also control seed color (Gfeller and Svejda, 1960). Seed coat impermeability in *Lupinus angustifolius* is controlled by a single pair of dominant genes, and seed coat permeability in at least two cultivated varieties is determined by the recessive gene pair (Forbes and Wells, 1968). *Vicia* spp. have a two-gene inheritance system, and seeds with *aabb* have impermeable seed coats. The *A* gene acts as a simple dominant for impermeable seeds, whereas the *B* gene is dominant for permeable seeds when the A locus is homozygous recessive (*aa*) (Donnelly *et al.*, 1972). Seed impermeability in *Gossypium hisutum* largely is controlled by a single allelic pair, although a second allelic pair has some influence (Lee, 1975). In *Glycine max,* as few as three major genes seem to control seed permeability (Kilen and Hartwig, 1978).

Various kinds of crosses showed that seed dormancy is determined by a single gene in *Lactuca sativa* (Eenink, 1981) and by one major gene in *Senecio vulgaris* ssp. *denticulatus* (Kadereit, 1984). Hybridization studies of varieties of *Avena fatua* and *A. sativa* revealed that seed dormancy is a recessive trait (Garber and Quisenberry, 1923). Crosses and reciprocal backcrosses with dormant and nondormant pure lines of *A. fatua* showed that the time course of germination (rate of dormancy break and germination at 20°C) was controlled by three loci (Jana *et al.*, 1988). Seed dormancy in *Cucumis sativus* var. *hardwickii* is controlled by three to seven genes, and it may be conditioned by recessive genes (Staub *et al.*, 1989).

Crosses do not necessarily have to be made between lines with dormant and nondormant seeds to learn something about seed dormancy. If the level of seed dormancy in the parental lines is known, information on the dormancy occurring in the F_1, F_2, and F_3 generations provides insight into its inheritance. For example, results from the F_1, F_2, and F_3 generations of a cross between the Riso 1508 and Nordlys lines of *Hordeum vulgare* showed that seed dormancy is controlled by several recessive genes (Buraas and Skinnes, 1984).

E. Maternal, Paternal, and Embryonic Genotypes

The embryo and endosperm in seeds represent a new combination of genetic material, whereas the seed coat has the same genetic composition as the maternal parent. The genetic component of the endosperm varies, depending on the type of megasporogenesis in the species. As presented in most general botany textbooks, the *Polygonum*-type of embryo sac development results in the formation of a 3n endosperm when fertilization occurs, with one set of chromosomes coming from the paternal parent and two from the maternal parent. In the *Lilium*-type of embryo sac development, a 5n endosperm is formed when fertilization occurs, with one set of chromosomes coming from the paternal and four from the maternal parent. It should be noted, however, that Haig (1990) illustrated 18 different types of embryo sac development, including the *Polygonum*- and *Lilium*-types.

Hybridization studies involving dormant and nondormant pure lines or varieties of a species make it possible to determine if seed dormancy is controlled by the genotype of the maternal parent, the paternal parent, the embryo, or some combination of the three. Both a maternal and an embryonic component have been identified in seed dormancy of *Sinapis arvensis.* The maternal component is controlled by a single locus with two alleles, but the number of genes involved in the embryonic component of dormancy has not been determined (Garbutt and Witcombe, 1986). The genotype of the maternal parent has a significant effect on germination of the cluster of seeds in *Beta vulgaris,* whereas the paternal parent does not. However, the genetic make-up of the embryo has some influence on the germination rate (Battle and Whittington, 1971). Seeds of *Brassica oleracea* (Hodgkin and Hegarty, 1978) exhibit variation in germination percentages and rates that largely are attributed to genetics. Thus, a knowledge of the genetics of the maternal grandparent is very useful in predicting the germination behavior of *B. oleracea* seeds. The maternal parent in *Papaver* spp. plays an important role in controlling dormancy, but there also is direct embryonic determination (Harper and McNaughton, 1960). The genotype of the *Trifolium subterraneum* embryo is important in determining seed dormancy (Morley, 1958).

Dormancy in seeds of *Eschscholtzia californica* appears to be maternally inherited, but the pollen parent also has an effect (Cook, 1962). The inability of *Lactuca sativa* seeds to germinate in darkness at 25°C is controlled by a gene from the pollen parent (Rideau *et al.,* 1976). In crosses between dormant and nondormant lines of *Petunia hybrida,* the lack of germination in darkness at 25°C was controlled by the paternal parent, and

sensitivity of the seeds to light and gibberellic acid (GA) was controlled by the maternal parent (Girard, 1990).

F. Maternal Influence

In general, maternal influence is the effect that the maternal parent has on the offspring, and in this case we are interested in its effect on germination characteristics of the seeds. The mother plant may affect its seeds by one or more mechanisms: genetics, non-Mendelian inheritance, and/or through interactions with the environment (preconditioning). Some authors include all these mechanisms in discussions of the effects of the maternal parent on seeds (e.g., Roach and Wulff, 1987), but for the sake of increased clarity we have chosen to use a more restricted definition. In this chapter, maternal influence means information that is passed to the offspring in a more or less non-Mendelian fashion.

One way in which the maternal parent can influence offspring is through extrachromosomal or cytoplasmic inheritance, i.e., genetic information in the cytoplasm of the egg is passed to the offspring at the time of fertilization. For example, in *Linum usitatissimum* seed size is determined partly by cytoplasmic inheritance, and a small-seeded maternal parent has a stronger effect than a large-seeded one (Smith and Fitzsimmons, 1965). Cytoplasmic inheritance also may be important in determining seed size in *Oryza sativa* (Chandraratna and Sakai, 1960).

Maternal imprinting has been shown to occur in *Secale cereale,* when two B chromosomes occur in the maternal parent. DNA is replicated later during interphase in B than in regular (A) chromosomes (Jones and Rees, 1969), and transmission of B chromosomes occurs in a non-Mendelian fashion (Puertas *et al.,* 1989). If a mother plant has two B chromosomes, its offspring can be recognized even if they do not have any B chromosomes (Puertas *et al.,* 1989).

A second way in which information is passed from the mother to the offspring is via chemicals produced by the mother. This mechanism of maternal influence has been demonstrated in a number of animal systems and could be one explanation for maternal determination of dormancy in seeds. In the meal moth (*Ephestia* sp.), a hormone-like substance called kynurenin is involved in pigment synthesis in the skin of the larval stages. A larva with an *AA* genotype produces kynurenin and is black, whereas one with an *aa* genotype lacks kynurenin and is colorless. However, individuals with the *aa* genotype may have some pigmentation when they are young, but it eventually fades. Pigmentation in young *aa* individuals is explained by the fact that the mother had at least one *A* allele, which allowed the production of kynurenin. Thus, some kynurenin was in the cytoplasm of the egg, and it was enough to cause pigment development in the larva (see Strickberger, 1968).

The maternal influence in seed dormancy also may be the result of chemicals produced by the mother plant. Battle and Whittington (1971) suggested that the maternal influence in seed dormancy in *Beta vulgaris* is due to the formation of germination inhibitors by maternal tissues and that these compounds are transferred to the seeds. Because the genotype of the embryo in *Trifolium subterraneum* seeds is very important in determining dormancy, Morley (1958) hypothesized that a dormancy-causing chemical was produced in the embryo from a substrate provided by the mother plant or that an inhibitor was deposited in the seeds by the mother plant. In seeds such as those of *Trifolium subterraneum* (Morley, 1958) and *Brassica oleracea* (Hodgkin and Hegarty, 1978), where the maternal genotype has a strong influence on dormancy, it has been observed that the environment in which the seeds mature may have an effect on dormancy. Thus, the maternal effect can be modified.

G. Ploidy Level, B Chromosomes, and Isochromosomes

In a few species, the total number of chromosomes in the cells of the embryo makes a difference in germination responses. For example, somatic cells in *Nicandra physaloides* normally have $2n = 20$; however, two of these chromosomes are isochromosomes (i.e., a chromosome with identical genetic material on each side of the centromere). Because the isochromosomes fail to pair at meiosis, gametes can be produced that are missing the isochromosome. Deficient pollen apparently dies, but a deficient egg ($n = 9$) sometimes is fertilized, resulting in an embryo with a chromosome number of $2n = 19$. Seeds with a normal chromosome number ($2n = 20$) in the embryo germinate immediately after sowing, whereas those with a deficient chromosome number ($2n = 19$) exhibit a long delay before germinating (Darlington and Janaki-Ammal, 1945).

The Chihuahuan Desert ecotype of *Larrea divaricata* has a chromosome number of $2n = 26$, and seeds have a rapid germination rate. However, the Sonoran Desert ecotype has a chromosome number of $2n = 52$ and a slow germination rate (Yang, 1967; Yang and Lowe, 1968). Seeds of *Mimulus guttatus* from tetraploid ($n = 28$) and aneuploid ($n = 15$) populations germinate over a wider range of temperatures than those collected from diploid populations (Vickery, 1967). Seeds from tetraploid plants of *Lolium perenne* germinated at a slower rate than those from diploid plants when seeds were sown in mixtures, but rates were the same when seeds

were sown separately (Norrington-Davies and Harries, 1977). Germination characteristics of three allotetraploid species of *Clarkia* (*C. delicata, C. similis,* and *C. rhomboidea*) were compared to those of the putative diploid progenitors of each species, using a number of different constant temperatures and water potentials. Germination responses of *Clarkia delicata* and *C. similis* were similar to those of their diploid ancestors, whereas germination patterns of *C. rhomboidea* were quite different from those of the diploid parents (Smith-Huerta, 1984).

In the Alps of Switzerland and Italy, the diploid *Festuca pratensis* occurs in the woodland zone and below, whereas the tetraploid *F. pratensis* var. *apennina* occurs in the woodland zone and above on the mountains. Seeds of *F. pratensis* are nondormant and germinate after dispersal in mid-August, whereas those of *F. pratensis* var. *apennina* are dormant and require cold stratification for dormancy break. Consequently, the tetraploid does not germinate until spring (Tyler *et al.,* 1978). Seeds from common garden-grown (one generation) tetraploid and diploid plants of *Matricaria perforata* had similar optimum and maximum germination temperatures. However, seeds from tetraploids plants germinated 28–39% higher at suboptimal temperatures than those from diploids (Thomas *et al.,* 1994). Seeds from *Digitaria milanjiana* plants of different polyploid levels grown (from seeds) in a greenhouse had different levels of germination: hexaploids, 16%; tetraploids, 26%; and diploids, 38% (Hacker, 1988).

Another way in which the number of chromosomes may vary is by the presence of B chromosomes. The presence of B chromosomes in *Secale cereale* increased seed weight and delayed germination (Moss, 1966). Further, straw weight, plant weight, and tiller number are higher in plants of *S. cereale* with even numbers of B chromosomes than in those with an odd number (Jones and Rees, 1969). The germination rate of *Allium porrum* seeds increased with an increase in the number of B chromosomes (Vosa, 1966), but germination acceleration was only 1 day (Gray and Thomas, 1985). B chromosomes were more common in early- and drought-germinated seedlings of *A. schoenoprasum* than in late- and nondrought-germinated ones; however, the number of B chromosomes per cell did not increase (Plowman and Bougourd, 1994).

H. Inbreeding Depression

One consequence of inbreeding is an increase in the frequency of homozygous genotypes, some of which may be deleterious. Thus, inbreeding may cause a decline in the fitness of a species, and this reduction is called inbreeding depression (van Treuren *et al.,* 1993).

Inbreeding depression has received considerable attention, especially as it relates to the population sizes of rare species.

The degree of inbreeding depression is determined in plant species by comparing various attributes of fitness, including seed germination, of the progeny resulting from selfing vs outcrossing. Sometimes, actual selfing and outcrossing experiments are done (e.g., Scurfield, 1954; Ramsey and Vaughton, 1996), but in some studies selfing and outcrossing are inferred based on the size of the population (e.g., Menges, 1991; Heschel and Paige, 1995). That is, selfing is thought to be more common in small than in large populations and outcrossing to be more common in large than in small populations. Progeny from crosses between *Chamaecrista fasciculata* plants in the same neighborhood had a lower fitness than those from crosses between plants growing in different neighborhoods (Fenster, 1991). Seeds produced by female flowers of *Sidalcea oregana* ssp. *spicata* (Ashman, 1992) and *Eritrichum aretioides* (Puterbaugh *et al.,* 1997) germinated to higher percentages than those from hermaphroditic flowers. Also, seeds from female flowers of *E. aretioides* were larger than those from hermaphroditic flowers. Puterbaugh *et al.* (1997) suggested that the relatively high germination and large size of seeds from female (vs hermaphroditic) flowers are related to the fact that female flowers are outcrossed, whereas hermaphroditic flowers may be self-pollinated.

Inbreeding depression causes a reduction in (1) the number of viable seeds produced (Scurfield, 1954; Vander Kloet, 1984; Pettersson, 1992; van Treuren *et al.,* 1993; Lesica, 1993; Norman *et al.,* 1995; Ramsey and Vaughton, 1996), (2) the size of seeds (Vander Kloet, 1984; Pettersson, 1992; Dahlgaard and Warncke, 1995; Heschel and Paige, 1995; Norman *et al.,* 1995), (3) germination percentages (Vander Kloet, 1984; Fenster, 1991; Menges, 1991; Pettersson, 1992; Lesica, 1993; Carr and Dudash, 1995; Dahlgaard and Warncke, 1995; Heschel and Paige, 1995; Norman *et al.,* 1995; Hauser and Loeschcke, 1996; Ramsey and Vaughton, 1996), and (4) rates of germination (Biere, 1991a; Herrera, 1991; Hauser and Loeschcke, 1996; Ramsey and Vaughton, 1996). Inbreeding had no influence on the seed germination of *Sidalcea oregana* ssp. *spicata* (Ashman, 1992), and distance between mating plants of *Hymenoxys herbacea* had no effect on seed germination percentages (Moran-Palma and Snow, 1997).

III. INHERITANCE OF DORMANCY BREAKING AND GERMINATION CHARACTERISTICS

A. Degree of Dormancy

One of the clearest examples of the inheritance of degree of dormancy in freshly matured seeds is found in *Avena fatua*. In 10 genetically pure lines of this species, seed dormancy ranged from 0 to nearly 100%. Further, the lines with low dormancy levels required only a short period of time for seed development and seeds were heavy and few in number. In contrast, lines with high dormancy levels required a relatively long period of time for seed production and seeds were light in weight and many in number (Adkins *et al.,* 1986).

Another example of the degree of dormancy having a genetic component is found in atrazine-resistant and -susceptible biotypes of *Amaranthus* spp. Unfortunately, from the farmer's point of view the degree of dormancy is greater in resistant than in susceptible biotypes. Thus, herbicide-resistant plants produce seeds that potentially can form long-lived seed banks (Weaver and Thomas, 1986). However, after 20 months of burial, the viability of seeds of sulfonylurea-resistant and -susceptible biotypes of *Lactuca serriola* had decreased to 20% or less (Alcocer-Ruthling *et al.,* 1992).

Stimulatory effects of GA on the germination of *Lesquerella fendleri* seeds varied among 16 maternal sibships (Evans *et al.,* 1996). Germination of seeds of the 16 sibships ranged from 21 to 98% with 0 g liter^{-1} GA and from about 64 to 100% with 1 g liter^{-1} GA. In a greenhouse study, GA increased seed germination, as compared to controls, of 16 sibships by 2–35%, depending on the sibship. In a growth chamber study, GA increased seed germination in 14 sibships by 7–44%, but it decreased germination of two sibships by 3–7%. Thus, seeds of various maternal sibships varied in the degree of dormancy and sensitivity to GA.

Indirect evidence that degree of seed dormancy is a heritable trait is seen in the domestication of some plants. Humans have exerted strong selection pressure in their quest for certain kinds of seeds, i.e., large nutritious seeds that are easy to eat and readily germinable when sown. Thus, loss of dormancy may be one of the consequences if some seeds from the previous year's crop are planted year after year, and this would explain why seeds of various cultivars have less dormancy than those of their weedy relatives (Harlan *et al.,* 1973).

One way in which dormancy may have been lost (or decreased) was a reduction in the thickness of the seed coat. Paleoethnobotanical studies on *Chenopodium berlandieri* subsp. *jonesianum,* a premaize domesticate in eastern North America, have shown that crop and weed forms cooccurred (Smith and Cowan, 1987; Germillion, 1993). Seeds of the weed form had relatively thick (40–80 μm) seed coats, whereas those of the domesticated form had relatively thin (<20 μm) seed coats (Smith, 1989). About 50,000 carbonized thin-coated *C. berlandieri* seeds dated 1975±55 YBP have been found in remains of a basket in Russell Cave in Alabama, providing evidence that humans were storing seeds with thin seed coats, some of which probably were planted in the spring (Smith, 1984). It has been suggested that seeds with thin seed coats were less dormant, and thus they would have germinated faster than those with thick seed coats (Smith, 1984, 1989). Thin seed coats would offer less mechanical resistance to growth of the embryo (Chapter 3). However, seed coat thickness was not correlated with the degree of dormancy in seeds of *C. album* (Karssen, 1970). It seems logical that another reason for human selection of seeds with thin seed coats is that they were easier to prepare for consumption than those with thick coats.

B. Rate of Dormancy Loss

The length of the seed dormancy period in *Avena fatua* is controlled by several to many genes. However, the expression of genotypes causing long-term dormancy is sensitive to environmental conditions experienced by the mother plant as seeds develop. High (e.g., 28/22°C) temperatures and drought stress reduced the level of dormancy, whereas low (e.g., 15/10°C) temperatures and nonlimiting soil moisture did not (Naylor, 1983). The actual rate of dormancy loss in *A. fatua* seeds is controlled by three genes, and two of them influence the rate of dormancy loss at different periods after seed maturation (Jana *et al.*, 1979). Garber and Quinsenberry (1923) made crosses between *A. fatua* and *A. sativa* and showed that delayed germination was recessive. Johnson (1935) suggested that dormancy in these species was determined by a three-locus system, with the highest degree of dormancy occurring in the triple-recessive genotype. Genetic differences also exist in pure lines of *A. fatua* with respect to the influence of the maturation environment on the duration of dormancy. For example, high temperatures (Sawhney and Naylor, 1979) and water stress (Sawhney and Naylor, 1982) during seed maturation reduced the length of the dormancy period in some pure lines.

A theory of multigenic inheritance has been proposed in seeds of *Arachis* sp. for the inheritance of the rest period, which can last as long as 2 years (Hull, 1937). However, inheritance of the rate of softening of the impermeable seeds of *Phaseolus vulgaris* involves only a few genes (Lebedeff, 1947). The number of freshly matured seeds of *Trifolium subterraneum* with impermeable seed coats is strain specific (Quinlivan, 1965); however, strains vary in the percentage of impermeable seeds, depending on the geographical location of the site where plants are grown (Quinlivan and Millington, 1962).

C. Germination Requirements

Responses of seeds to various environmental factors have a genetic basis. Studies on the germination require-

ments of *Avena fatua* and *A. ludoviciana* and their hybrids have resulted in hypotheses concerning the number of genes involved in the temperature requirement for germination. The lack of germination of *A. fatua* at 5°C possibly is controlled by three recessive loci, and lack of germination of *A. ludoviciana* at 18°C may be controlled by a single recessive gene (Whittington *et al.*, 1970).

Germination percentages and rates were determined for seeds of six accessions of *Lycopersicon* over a range of temperatures. Germination responses were then quantified using a lognormal model for accelerated failure analysis. The model detected different genotypes based on the response times of seeds to temperature (Scott and Jones, 1985). Genotypes of the apomictic *Erigeron annuus* also exhibited significant differences in mean time to germination at various temperatures (Stratton, 1991).

To determine the genetic basis of the light requirement for the germination in seeds of *Nicotiana* sp., reciprocal crosses were made between light-requiring and light-insensitive strains. The offspring (F_1's) were like the maternal parent, demonstrating a strong maternal influence on dormancy. In fact, the effect of the maternal parent could be observed to some extent in the F_2's (Honing, 1930). Both parents of *N. tabacum* contribute to the light sensitivity of seeds, but the maternal was greater than the paternal contribution (Kasperbauer, 1968).

Freshly matured seeds of three nondormant lines of *Avena fatua* germinate at higher percentages in darkness than in red, far-red, blue, or white light. However, the degree of germination inhibition varies with the genetic line and kind of light (Hou and Simpson, 1991). Seeds of each of three dormant lines treated with GA to overcome dormancy germinated to about the same percentage in red or far-red light as they did in darkness.

Crosses were made between paraquat-resistant and -sensitive biotypes of *Erigeron canadensis,* and seeds of the F_1's were checked for sensitivity to the herbicide. Because three-fourths of the F_2 seeds were resistant to paraquat at a 10^{-3} M concentration, Yamasue *et al.* (1992) concluded that resistance is controlled by a single gene. Seeds from cultivars of barley (Niazi *et al.*, 1987, 1992) and genetic lines of *Haynaldoticum sardoum* (probably a hybrid between *Haynaldia villosa* and *Triticum durum*) (Cremonini *et al.*, 1992) may differ in their ability to germinate at increased salinity.

D. Germination Speed

Quantitative genetic studies showed that time (date) of seedling emergence in *Cleome serrulata* had a heritability of 1.23 (Farris, 1988). Because a heritability of

1.00 means progeny are exactly like the parents (Dobzhansky, 1970), Farris (1988) suggested that there were maternal effects or a higher level of additive genetic variance among dams than among sires to account for the >1.00 value for heritability in *C. serrulata*. Although *Lolium multiflorum* did not respond to selection for increased germination speed, it responded to selection for a later germination time (Nelson, 1980). Various cultivars of *Phaseolus vulgaris* differ in speed of germination at low (11°C) temperatures (Otubo *et al.*, 1996). However, time of germination in *Papaver dubium* had zero heritability (Arthur *et al.*, 1973).

E. Storability

Although seed viability during long-term dry storage is not exactly within the realm of seed germination ecology, it should be noted that storability is in part genetically determined. For more information and references on this subject, see Priestley (1986). Also, Halloin (1986) has reviewed the literature related to selecting seeds that are resistant to deterioration.

IV. VARIATION IN GERMINATION RESPONSES OF SEEDS FROM DIFFERENT POPULATIONS OF THE SAME SPECIES

A. Types of Variation

Species from a wide range of plant families, life cycle types, and plant communities exhibit differences in germination characteristics of seeds collected in different locations. Depending on the species, germination responses vary with latitude, elevation, soil moisture, soil nutrients, temperature, kind and density of plant cover, and degree of habitat disturbance of the sites where the seeds matured (Table 8.2). However, seeds of *Spartina alterniflora* (Seneca, 1974) collected at various latitudes and those of *Matricaria perforata* from across Canada (Thomas *et al.*, 1994) showed no differences in germination requirements. Scarified seeds of *Prosopis glandulosa* from throughout the southwestern United States germinated to near 100% over a range of temperatures (Peacock and McMillan, 1965). Seeds of *Capsella bursa-pastoris* from an altitudinal transect in the Alps showed no correlation between germination responses and collection site (Neuffer and Bartelheim, 1989). Altitude had no effect on the production of impermeable seeds of *Medicago* sp., but individual plants produced seeds varying in permeability from year to year. Further, in the same year, permeability varied from stem to stem on the same plant (Lute, 1928).

Germination characteristics of seeds collected from a number of sites can vary in many ways, but one of the most common is the degree of dormancy, as reflected by germination percentages of fresh seeds. Examples of species exhibiting habitat variation in degree of dormancy include *Abutilon theophrasti* (Warwick and Black, 1986), *Alopecurus myosuroides* (Naylor and Abdalla, 1982), *Amaranthus retroflexus* (McWilliams *et al.*, 1968), *Anthoxanthum odoratum* (Platenkamp, 1991), *Artemisia tridentata* (Meyer *et al.*, 1990), *Avena barbata, A. fatua* (Paterson *et al.*, 1976), *Betula alleghaniensis* (Wearstler and Barnes, 1977), *Bromus tectorum* (Meyer *et al.*, 1997), *Campanula punctata* (Inoue and Washitani, 1989), *Carex canescens* (Schutz and Milberg, 1997), *C. sempervirens* (Georges and Lazare, 1983), *Cedrus deodara* (Thapliyal and Gupta, 1980), *Chenopodium bonushenricus* (Dorne, 1973), *Digitaria milanjiana* (Hacker, 1984), *Eucalyptus pauciflora* (Beardsell and Mullett, 1984), *Euphorbia geniculata* (Kigel *et al.*, 1992), *Medicago sativa* (Dexter, 1955), *Nothofagus obliqua* (Donoso, 1979), *Picea mariana* (Morgenstern, 1969), *Picea* spp. (Roche, 1969), *Schoenus nigricans* (Bocchieri *et al.*, 1987), *Silene dioica* (Thompson, 1975), *Stellaria media* (Salisbury, 1974), *Taeniatherum asperum* (McKell *et al.*, 1962), *Themeda triandra* (Baxter *et al.*, 1993), *Phlox drummondii* (Schwaegerle and Bazzaz, 1987), *Pinus leucodermis* (Borghetti *et al.*, 1989), *Plantago lanceolata* (van Groenendael, 1986), *Poa annua* (Naylor and Abdalla, 1982; Standifer and Wilson, 1988), and *Senecio vulgaris* (Ren and Abbott, 1991). Achenes (cyselas) from different plants of *Onopordum acanthium* in the same populations of this species had variable germination (Perez-Garcia, 1993).

Population variations in germination rates of freshly matured seeds have been found in *Betula alleghaniensis* (Wearstler and Barnes, 1977), *Campanula punctata* (Inoue and Washitani, 1989), *Carex sempervirens* (Georges and Lazare, 1983), *Chenopodium bonus-henricus* (Dorne, 1973b), *Kochia indica* (Khatri *et al.*, 1991), *Leucochrysum albicans* (Gfeller and Kirkpatrick, 1994), *Liquidambar styraciflua* (Wilcox, 1968; Winstead, 1971), *Picea mariana* (Morgenstern, 1969), *Pseudoroegneria* (=*Agropyron*) *spicata* (Kitchen and Monsen, 1994), *Silene dioica* (Thompson, 1975), *Sonchus arvensis* (Pegtel, 1972), and *Taeniantherum asperum* (McKell *et al.*, 1962). Further, seeds of *Artemisia tridentata, Chrysothamnus nauseosus, Penstemon eatonii, P. palmeri,* and *Purshia tridentata* collected from sites with warm winters germinated faster under snowpack in the field than those collected from sites with cold winters (Meyer, 1990).

Environmental conditions required for germination also can vary between populations of a species. Seeds may differ in their sensitivity to substrate moisture (Lindauer and Quinn, 1972; Pegtel, 1972; Groves *et al.*, 1982; Sonaike and Okusanya, 1987; Fady, 1992), as well as to

Species	Collection sites vary with respect to	Reference	Species	Collection sites vary with respect to	Reference
Abies cephalonica	Latitude, precipitation	Fady (1992)	*C. maculosa*	Substrate	Nolan and Upadhyaya (1988)
Abutilon theophrasti	Latitude	Warwick and Black (1986)	*Ceratoides lanata*	Elevation	Moyer and Lang (1976)
Acacia nilotica	Latitude, longitude	Mathur et al. (1984)	*Chenopodium bonus-henricus*	Elevation	Dorne (1973b, 1981)
Acer negundo	Latitude	Williams and Winstead (1972)	*Chrysothamnus nauseosus*	Elevation	Meyer et al. (1990)
Adenocarpus decorticans	Elevation	Trillo and Carro (1993)	*Corchorus aestuans*	Soil moisture	Chawan and Sen (1973)
Agrostemma githago	Associated species	Thompson (1973)	*Coriaria fuscifolia*	Elevation	Smith (1975)
Alopecurus myosuroides	Elevation, habitat type	Naylor and Abdulla (1982)	*Dactylis glomerata*	Air temperature	Pannangpetch and Bean (1984)
Amaranthus powellii	Soil moisture, nutrients	Frost and Cavers (1975)		Latitude, longitude	Junttilla (1977), Probert et al. (1985a,c)
A. retroflexus	Latitude, moisture	McWilliams et al. (1968)	*Dactyloctenium aegyptium*	Salinity	Okusanya and Sonaike (1995)
Ammophila breviligulata	Latitude	Seneca and Cooper (1971)	*Dalbergia sissoo*	Longitude	Bangarwa et al. (1991)
Ampelodesmos mauritanicus	Latitude, longitude	Onnis et al. (1993)	*Danthonia sericea*	Soil moisture	Lindauer and Quinn (1972)
Anthoxanthum odoratum	Soil moisture	Platenkamp (1991)	*Digitaria milanjiana*	Soil moisture	Hacker (1984)
Artemisia tridentata	Elevation	Meyer (1990), Meyer et al. (1990), Meyer and Monsen (1991), Young et al. (1991)	*Dioscorea tokoro*	Latitude	Okagami and Kawai (1983)
			Echeveria venezuelensis	Elevation	Smith (1975)
Astragalus granatensis	Elevation	Trillo and Carro (1993)	*Echinochloa crus-galli*	Latitude, longitude, kind of crop	Honek and Martinkova (1996), Martinkova and Honek (1995a,b)
Atriplex confertifolia	Temperature	Sanderson et al. (1990)			
Avena barbata	Latitude	Paterson et al. (1976)	*Emex australis*	Latitude	Panetta and Randall (1993)
A. fatua	Latitude	Paterson et al. (1976)			
Betula alleghaniensis	Latitude	Wearstler and Barnes (1977)	*Eurotia lanata*	Soil chemistry	Workman and West (1967)
B. papyrifera	Longitude, latitude	Bevington (1986)	*Eucalyptus camaldulensis*	Salinity	Sands (1981)
B. pubescens	Latitude	Danchenko et al. (1977)	*E. pauciflora*	Elevation	Beardsell and Mullett (1984)
B. verrucosa	Latitude	Danchenko et al. (1977)	*Euphorbia geniculata*	Elevation	Kigel et al. (1992)
Campanula punctata	Latitude	Inoue and Washitani (1989)	*E. thymifolia*	Soil nutrients	Ramakrishnan (1965)
Capsella bursa-pastoris	Latitude, longitude	Stillwell and Sweet (1975)	*Festuca indigesta*	Elevation	Angosto and Matilla (1994)
Cardamine sp.	Elevation	Thurling (1966)	*F. pratensis*	Elevation	Tyler et al. (1978)
Castilleja fissifolia	Elevation	Smith (1975)	*Grayia brandegei*	Elevation	Meyer and Pendleton (1990)
Cedrus deodara	Longitude	Thapliyal and Gupta (1980)	*Hordeum spontaneum*	Humidity	Gutterman and Nevo (1994)
Centaurea diffusa	Substrate	Nolan and Upadhyaya (1988)	*Kochia indica*	Soil chemistry	Khatri et al. (1991)

(continues)

TABLE 8.2—*Continued*

Species	Collection sites vary with respect to	Reference	Species	Collection sites vary with respect to	Reference
Larrea tridentata	Latitude	McGee and Marshall (1993)	*Poa annua*	Soil moisture, mowing	Wu *et al.* (1987)
Lepidium lasiocarpum	Precipitation	Philippi (1993)		Elevation	Naylor and Abdalla (1982)
L. perfoliatum	Salinity	Choudhuri (1968)		Latitude	Standifer and Wilson (1988)
Leucochrysum albicans	Elevation	Gilfedder and Kirkpatrick (1994)	*P. trivialis*	Light quality	Hilton *et al.* (1984)
Leymus cinereus	Elevation	Meyer *et al.* (1995)	*Prunus serotina*	Latitude	Pitcher (1984)
Linum perenne	Elevation	Meyer and Kitchen (1994)	*Pseudoroegneria spicata*	Latitude, longitude	Kitchen and Monsen (1994)
Liquidambar styraciflua	Latitude, elevation	Wilcox (1968) Winstead (1971)	*Purshia tridentata*	Elevation	Meyer (1990)
Lychnis flos-cuculi	Latitude	Thompson (1970)	*Reynoutria japonica*	Elevation	Mariko *et al.* (1993)
Luffa aegyptiaca	Soil moisture	Sonaike and Okusanya (1987)	*Rottboellia exaltata*	Associated crops	Pamplona and Mercado (1981)
Medicago sativa	Elevation	Dexter (1955)	*Rubus* spp.	Latitude	Nybom (1980)
Melaleuca ericifolia	Salinity	Ladiges *et al.* (1981)	*Ruppia maritima*	Latitude	Koch and Dawes (1991)
M. thyoides	Salinity	van der Moezel and Bell (1987)	*Salvia columbariae*	Temperature, precipitation	Capon *et al.* (1978)
Mimulus guttatus	Elevation	Vickery (1983)	*Schoenus nigricans*	Elevation	Bocchieri *et al.* (1987)
Nemophila menziesii	Soil moisture, temp.	Cruden (1974)			
Nothofagus obliqua	Latitude, elevation	Donoso (1979)	*Senecio vulgaris*	Latitude	Ren and Abbott (1991)
Oldenlandia corymbosa	Latitude	Attims (1972b)	*Silene acaulis*	Substrate	Bianco and Bulard (1976)
Penstemon ambiguus	Elevation	Meyer *et al.* (1995)	*S. dioica*	Latitude	Thompson (1975)
P. eatonii	Elevation	Meyer (1990, 1992)	*S. otites*	Latitude	Thompson (1970)
P. pachyphyllus	Elevation	Meyer *et al.* (1995)	*Solanum ptycanthum*	Latitude	Hermanutz and Weaver (1991)
P. palmeri	Elevation	Meyer (1990)	*Sonchus arvensis*	Degree of disturbance	Pegtel (1972)
P. rostriflorus	Elevation	Meyer *et al.* (1995)			
P. utahensis	Elevation	Meyer *et al.* (1995)	*Sorbus aucuparia*	Elevation	Barclay and Crawford (1984)
Phlox drummondii	Plant cover, soil nutrients	Schwaegerle and Bazzaz (1987)	*Taeniatherum asperum*	Precipitation	McKell *et al.* (1962)
Picea abies	Latitude	Heide (1974)			
P. excelsa	Soil moisture	Youngberg (1952)	*Themeda australis*	Precipitation	Groves *et al.* (1982)
P. glauca	Latitude	Fraser (1971)	*T. triandra*	Elevation	Baxter *et al.* (1993)
P. mariana	Latitude	Morgenstern (1969)	*Thymelaea hirsuta*	Latitude	El-Keblawy *et al.* (1996)
Picea spp.	Elevation	Roche (1969)			
Pinus brutia	Latitude	Skorodilis and Thanos (1995)	*Tridax procumbens*	Soil nutrients	Ramakushnan and Jain (1965)
P. contorta	Elevation	Haais and Thrupp (1931)	*Triticum* sp.	Latitude	Pammel (1898)
			Tsuga canadensis	Latitude	Olson *et al.* (1959)
P. silvestris	Soil moisture	Youngberg (1952)	*T. heterophylla*	Latitude, longitude	Campbell and Ritland (1982)
P. strobus	Latitude, mean January temperature	Fowler and Dwight (1964)			
			Typha spp.	Latitude	McNaughton (1966)
Plantago lanceolata	Soil moisture	van Groenendael (1986)	*Uniola paniculata*	Latitude	Seneca (1972)

pH, calcium (Ramakrishnan and Jain, 1965), and salinity (Sands, 1981; Sonaike and Okusanya, 1987; van der Moezel and Bell, 1987; Okusanya and Sonaike, 1991).

Species with populations whose seeds vary in temperature requirements for germination include *Acer negundo* (Williams and Winstead, 1972), *Alopecurus myosuroides* (Naylor and Abdalla, 1982), *Artemisia tridentata* (Young *et al.*, 1991), *Betula papyrifera* (Bevington, 1986), *B. pubescens*, *B. verrucosa* (Danchenko *et al.*, 1977), *Cardamine* sp. (Thurling, 1966), *Cenchrus ciliaris* (Hacker and Ratcliff, 1989), *Ceratoides lanata* (Moyer and Lang, 1976), *Dactylis glomerata* (Pannangpetch and Bean, 1984), *Liquidambar styraciflua* (Winstead, 1971), *Luffa aegyptiaca* (Sonaike and Okusanya, 1987), *Nemophilia menziesii* (Cruden, 1974), *Picea abies* (Heide, 1974), *Pinus brutia* (Skordilis and Thanos, 1995), *Plantago lanceolata* (van Groenendael, 1986), *Poa annua* (Naylor and Abdalla, 1982; Wu *et al.*, 1987), *Solanum ptycanthum* (Hermanutz and Weaver, 1991), *Sonchus arvensis* (Pegtel, 1972), *Tsuga canadensis* (Olson *et al.*, 1959), *Typha* spp. (McNaughton, 1966), and *Silene dioica* (Thompson, 1975). Seeds also vary in their light:dark requirements for germination (Lindauer and Quinn, 1972; Gramshaw, 1976; Probert *et al.*, 1985a; Sonaike and Okusanya, 1987) and in their sensitivity to photoperiod (Naylor and Abdalla, 1982; Olson *et al.*, 1959), illuminance (Georges and Lazare, 1983; Probert *et al.*, 1985a; van Groenendael, 1986), and light quality (Hilton *et al.*, 1984). Because of agamospermy (i.e., production of asexual seeds), the genus *Taraxacum* is composed of numerous microspecies. Thus, each microspecies of *Taraxacum* grows in a different type of microhabitat rather than a single species growing in a variety of habitats. In the Netherlands, a study of 11 microspecies of *Taraxacum* revealed differences in germination rates and in optimum temperatures and light requirements for germination (van Loenhoud and Duyts, 1981).

Populations of a species may differ in the amount of cold stratification required to overcome dormancy of the seeds (Fowler and Dwight, 1964; Williams and Winstead, 1972; Dorne, 1973b; Tyler *et al.*, 1978; Donoso, 1979; Okagami and Kawai, 1983; Barclay and Crawford, 1984; Bevington, 1986; Meyer *et al.*, 1989; Meyer and Monsen, 1991; Meyer and Kitchen, 1994; Meyer *et al.*, 1995). However, seeds of *Carex canescens* from various locations in Sweden and Germany did not differ in the amount of cold stratification required (1 month) to break dormancy, although seeds did not have the same degree of dormancy (Schuetz and Milberg, 1997). Rates of afterripening of seeds held in dry storage can vary (Paterson *et al.*, 1976; Junttila, 1977; Pamplona and Mercado, 1981; Bocchieri *et al.*, 1987; Hacker and Ratcliff, 1989; Meyer *et al.*, 1997). The rate of induction

into secondary dormancy (Inoue and Washitani, 1989) and the proportion of seeds entering secondary dormancy (Panetta and Randall, 1993) also can vary.

B. Common Garden Studies

The variation in germination responses of plant species occurring over wide geographical areas and/or occupying habitats that differ with respect to environmental factors raises an important question: What is the reason for the variation? One answer is that the variation is genetically based. According to Turesson (1922a), natural selection occurs in the various sites where a species grows, which results in local genotypic differentiation. He proposed the word "ecotype" as the name for the genotypic product that develops in a particular habitat. The result of ecotypic differentiation is that variation within a species is correlated with habitat differences.

Early studies of ecotypic differentiation involved a common garden where plants from many places were brought to a single locality and grown for 1 to many years (Turesson, 1922b). In subsequent studies, however, plants were reciprocally transplanted between various habitats so that a number of common gardens was created (Clausen *et al.*, 1939, 1940). The first studies were concerned with morphological traits, but subsequently attention has been given to physiological characteristics and the search for physiological ecotypes (see review by Heslop-Harrison, 1964).

Some work has been done with germination characteristics of seeds produced by plants grown in common gardens. In these studies, seeds were collected from various sites and sown in common gardens; apparently no attempts were made to prevent cross pollination. After the resulting plants matured, seeds were obtained from them, and their germination requirements were determined. Seeds of *Silene inflata* and *Alyssoides utriculatum* collected from plants grown in a common garden at 280 m above sea level in France differed genotypically in seed size and weight and phenotypically in their germination behavior (Dorne, 1973). Seeds of *Lolium rigidum* collected from plants grown in two common gardens in Western Australia differed in their ability to germinate in darkness (Gramshaw, 1976). After being grown from seeds in a heated and in a cooled greenhouse, seeds of *Themeda australis* collected from eight populations in Australia and New Guinea exhibited differences in the degree of dormancy (Groves *et al.*, 1982). However, in a common garden experiment, also done in Australia, 11 selections of *Stylosanthes humilis* differed very little in the production of impermeable seeds (Cameron, 1967).

Plants grown from seeds of *Raphanus raphanistrum* collected at two sites were allowed to produce seeds in a common garden at a third site. The pattern of dormancy break was the same whether seeds were produced at the original sites or in the common garden (Cheam, 1986). Seeds of *Cenchrus ciliaris* (an obligate apomict) were collected from high (1200–1400 mm), medium (700–800 mm), low (400–500 mm), and very low (25–220 mm) annual rainfall sites in Africa, and plants from these seeds were allowed to produce seeds in a common garden in southeastern Queensland, Australia. High temperature treatments increased germination in all accessions, but germination percentages increased with an increase in precipitation at the original sites (Hacker and Ratcliff, 1989). Seeds of *Agrostemma githago* collected from 14 sites in Europe differed in the maximum and minimum temperatures for 50% germination after 4 days (Thompson, 1973). Seeds collected from plants (derived from seeds) from the 14 populations grown in greenhouses at Kew in England also exhibited differences in maximum and minimum temperatures for 50% germination after 4 days.

Seeds were collected in five populations of *Artemisia tridentata* in western Nevada, and seedlings grown from seeds from each site were transplanted to each of the five sites, resulting in five reciprocal transplant gardens. Seeds of *A. tridentata* from the "garden" at each site had lower germination percentages than those collected from native plants at each site (Young and Evans, 1989). Seeds of four ecotypes of *Digitaria milanjiana* were placed in nylon mesh bags on the soil surface in three contrasting sites in Australia for 41 weeks. Dormancy loss and germination characteristics of each ecotype were adaptive at some sites but not at others (Hacker *et al.*, 1984).

A second reason for variation in plants is that environmental conditions in 1 year may influence the growth of progeny the next year. This rather amazing fact was discovered by plant breeders who were developing new varieties of cereal crops (wheat, barley, oats, and corn). In the development of new varieties, it is important to be sure that plants will perform well regardless of where they are grown. Consequently, seeds produced in one site are sown in other areas and the seed production (yield) of the resulting plants is evaluated. Thus, common garden (field) studies are done throughout the anticipated range of use of the new variety. In a number of studies, plants resulting from seeds produced in one area had higher yields than those from seeds produced in some other area(s) (Lyon, 1902; Quinby *et al.*, 1962; McFadden, 1963). Furthermore, effects of the environment of maturation sometimes could be observed beyond the first generation. For example, the origin of *Triticum aestivum* (red wheat) seeds accounted for differences in yield in the second, but not in the third, generation of plants grown at the common site (Quinby

et al., 1962). It should be noted, however, that in some cases geographic origin of the seeds had no effect on the yield of a variety (McNeal *et al.*, 1960; Sarquis *et al.*, 1961).

Geographic origin also was found to influence other plant characteristics such as cold resistance in wheat (*Triticum* sp.) plants (Suneson and Peltier, 1936). Also, resistance to smut (*Ustilago levis* and *U. avenae*) varied in seeds of an oat variety (*Avena* sp.) grown at the same site in different years or at different sites in the same year (Tervet, 1944). Smut (*Tilletia tritici* and *T. laevis*) production in *Triticum* sp. varied from about 13 to 39–40% when seeds produced at different localities where grown at a common site (Holton and Herald, 1936).

The early common garden (field) studies with cereal crops contributed to the realization that the conditions under which a plant matures could have strong effects on the performance of the next generation of plants. Eventually the term "preconditioning" was used to describe this phenomenon. Although seed germination requirements per se were not a main focus of the common field studies of cereals, it does not take much imagination to see that preconditioning might also have an effect on seed germination. In one of the few common garden studies done on preconditioning effects on germination characteristics, Nelson *et al.* (1970) found that seeds of the annual grass *Taeniatherum asperum* produced at a warm, dry location were less dormant than those produced under cool, moist conditions.

To determine whether the variation in germination responses found in many species is due to genetics or preconditioning, it is recommended that seeds used in germination studies be collected from plants that have been grown in common gardens for one to several generations (Nelson *et al.*, 1970; Baskin and Baskin, 1973). However, Quinn and Colosi (1977) think that seeds should be used after only a single generation in a common garden. Their idea is that preconditioning effects would be lost when seeds mature in a common garden, and the remaining variation in germination responses would be due to genetic differences.

Studies on plants in general suggest that differences in germination responses may be due to an interaction between genetics and the environment. For example, some populations of a species may have genetic variability that is not expressed in the natural habitat, but it is expressed in a common garden (Heslop-Harrison, 1959a). Thus, some genetic differences seen in the common garden might not be expressed in the habitat due to environmental factors in the habitat. Further, common gardens may make it difficult to determine if plants have the genetic capacity to react adaptively to certain environments because various factors may not be present in the common garden (Heslop-Harrison, 1964). It has been suggested that populations of a species may differ genetically with respect to their sensitivity to preconditioning effects (Quinn and Colosi, 1977). Perhaps some information on the interaction between genetics and the environment in controlling variation in germination responses could be obtained by using seeds produced by species grown in reciprocal transplant gardens.

V. ENVIRONMENTAL FACTORS THAT CAUSE PRECONDITIONING EFFECTS

When the idea of preconditioning became associated with seed germination, much attention immediately was given to identifying factors that could cause changes in germination responses in the subsequent generation(s).

A. Carbon Dioxide Levels

Scarified seeds from *Ipomoea purpurea* grown at a low CO_2 concentration (350 ppmv) germinated to higher percentages than those from plants grown at a high concentration (700 ppmv). However, seeds of *Ipomoea* spp. and *Cassia* spp. grown at low and high concentrations were not significantly different (Farnsworth and Bazzaz, 1995).

B. Competition

Seeds from *Polygonum pensylvanicum* plants grown in competition with *Zea mays* come out of dormancy faster at 2°C than those from plants grown without competition. Also, scarification of seeds from *P. pensylvanicum* plants grown without competition resulted in higher (42 vs. 6%) germination than those from plants with competition (Jordon *et al.*, 1982). Seeds from *Nemophila menziesii* plants grown in competition with plants of *Bromus diandrus* exhibited a reduction in germination percentages and rates (Plantenkamp and Shaw, 1993).

C. Day Length

Short vs. long day lengths during the time seeds are being formed may cause differences in germination characteristics of mature seeds (Table 8.3). Seeds from plants of *Aegilops kotschyi* (Wurzburger and Koller, 1976), *Chenopodium amaranticolor* (Lona, 1947), *C. album* (Wentland, 1965; Karssen, 1970), *C. polyspermum* (Jacques, 1968; Pourrat and Jacques, 1975), *Diplotaxis harra* (Evenari, 1965), *Ononis sicula* (Evenari *et al.*, 1966), *Portulaca oleracea* (Gutterman, 1974), *Schismus arabicus* (Gutterman, 1996), and *Trigonella arabica* (Gutterman, 1978) grown under short days germinated

TABLE 8.3 Species Exhibiting Variations in Germination Characteristics When the Mother Plant Is Grown under Different Day Lengths

Species	Reference
Aegilops kotschyi	Wurzburger and Koller (1976)
A. ovata	Datta *et al.* (1972)
Amaranthus retroflexus	Kigel *et al.* (1977, 1979)
Avena fatua	Richardson (1979)
Beta vulgaris	Heide *et al.* (1976)
Cheiridopsis aurea	Gutterman (1991)
Chenopodium amaranticolor	Lona (1947)
C. album	Wentland (1965), Karssen (1970)
C. polyspermum	Jacques (1968), Pourrat and Jacques (1975)
Cucumis prophetarum	Gutterman and Porath (1975), Gutterman (1992a)
C. sativum	Gutterman and Porth (1975)
Diplotaxis harra	Evenari (1965)
Juttadinteria proximus	Gutterman (1991)
Lactuca scariola	Gutterman *et al.* (1975)
Ononis sicula	Evenari *et al.* (1966)
Plantago lanceolata	Case *et al.* (1996)
Polypogon monspeliensis	Gutterman (1982)
Portulaca oleracea	Gutterman (1974)
Rottboellia exaltata	Heslop-Harrison (1959b)
Schismus arabicus	Gutterman (1996)
Trigonella arabica	Gutterman (1978)

to lower percentages than those from plants grown under long days. A higher percentage of the seed coats of freshly matured seeds of *Ononis sicula* was permeable in seeds produced under short days than in those produced under long days. Also, seeds of this species produced under short days imbibed water and germinated faster than those produced under long days (Evenari *et al.*, 1966). Only the basal seed (of the two seeds produced in each spikelet) of *Aegilops kotschyi* is sensitive to day length during seed development (Wurzburger and Koller, 1976). Seeds of *Avena fatua* produced under 12-hr days germinated to 49%, whereas those produced under 18-hr days germinated to 66% (Richardson, 1979). Seeds produced by *Beta vulgaris* plants grown under 16-hr days showed a reduction in the ability to germinate (Heide *et al.*, 1976).

Other aspects of germination may be affected by the length of day to which the mother plant was exposed. Seeds from *Amaranthus retroflexus* plants grown under short days had higher germination percentages in darkness, greater response to cold stratification, and greater sensitivity to short light exposures than those from plants grown under long days (Kigel *et al.*, 1977). Seeds from *Rottboellia exaltata* plants grown under short days required a shorter period for loss of dormancy and germinated over a wider range of conditions than those from plants grown under long days (Heslop-Harrison, 1959). Seeds from *Lactuca scariola* plants shifted from long to short days germinated faster in both light and darkness than those in long days (Gutterman *et al.*, 1975).

Plants do not have to be exposed to a certain day length during their whole growth cycle for differences to occur in germination characteristics of the progeny (seeds). The sensitive period ranges from immediately after anthesis in *Aegilops kotschyi* (Wurzberger and Koller, 1976) to 8 days after flowering in *Chenopodium polyspermum* (Pourrat and Jacques, 1975) to the last 8 days of the maturation period in *Portulaca oleracea* (Gutterman, 1974) and *Ononis sicula* (Gutterman and Evenari, 1972). Seeds inside the ripe fruits of *Cucumis prophetarum* and *C. sativus* are hydrated and sensitive to photoperiod, and those from fruits exposed to short (8 hr) days germinated to higher percentages than those from fruits exposed to long (20 hr) days (Gutterman and Porath, 1975). In later studies, however, seeds from *C. prophetarum* fruits exposed to 8-hr days germinated to lower percentages than those from fruits exposed to 11.5-, 13-, 15-, or 18-hr days (Gutterman, 1992a).

D. Fungi

Seeds from *Panicum dichotomiflorum* plants infected with the smut *Ustilago destruens* had a lower percentage of viability and were less dormant than those produced by noninfected plants (Govinthasamy and Cavers, 1995). Application of foliar fungicides had no effect on germination of the seeds produced by wheat or triticale (Naylor, 1993).

E. Herbicides

Freshly matured seeds of *Galium aparine* collected in plots sprayed with MCPA (a phenoxyacetic acid) or 2,4-D were less dormant than those from controls (Aberg, 1956). Seeds from *Chenopodium album* plants treated with 2,4-D (at a rate of 1.1 kg ha^{-1}) were more dormant than those from controls, but both sets of seeds germinated equally well after cold stratification. Seeds from *C. album* plants treated with dalapon (4.5 kg ha^{-1}) were less dormant before or after cold stratification than those from controls (Fawcett and Slife, 1978a).

Seeds of *Setaria faberi* from 2,4-D-treated and control plants required cold stratification for germination. However, after overwintering in the field, seeds from 2,4-D-

treated plants germinated to significantly higher percentages than those from controls (Fawcett and Slife, 1978a). Seeds of *Amaranthus retroflexus* from dalapon-treated plants maintained a higher rate of viability while buried in soil in the field over winter than did those from control plants. Also, in spring a higher percentage of viable seeds from treated than from nontreated plants germinated (Fawcett and Slife, 1978a). Seeds from *Datura stramonium* plants treated with dalapon were less dormant than those from control plants (Fawcett and Slife, 1978a).

Seeds from *Avena fatua* plants sprayed with glyphosate at rates of 0.44, 0.88, and 1.76 kg a.i. ha^{-1} 10 days after anthesis germinated to significantly lower percentages than those from control plants. Seeds from plants sprayed at a rate of 0.44 kg a.i. ha^{-1} 15 days after anthesis germinated to the same degree (97%) as those of controls (98%), whereas seeds from plants sprayed with 0.88 and 1.76 kg a.i. ha^{-1} germinated to only 89 and 34%, respectively (Shuma *et al.*, 1995).

F. Hormones

When plants of an inbred strain of *Avena fatua* that normally produce seeds with physiological dormancy were treated with 0, 10, 100, and 1000 ppm GA, the resulting seeds germinated to 0, 30, 25, and 50%, respectively (Black and Naylor, 1959). Seeds from plants of *Lactuca scariola* treated with 10 mg l^{-1} of GA$_3$ showed an increase in dark germination compared to seeds from control plants; however, the germination of seeds from plants treated with 100 mg l^{-1} GA$_3$ was less than that of controls. In addition, the germination of seeds from plants treated with 20 mg l^{-1} abscisic acid (ABA) was promoted in darkness but not in light. Germination was inhibited in seeds collected 16 to 21 days after *L. scariola* plants were treated with 2-chloroethyl trimethylammonium chloride (CCC), but it was promoted in those collected 85 to 125 days after the treatment (Gutterman *et al.*, 1975). CCC was found in mature seeds collected from plants of *Phaseolus vulgaris* treated with this compound, but it was not present in immature seeds. GA$_3$ levels in immature seeds from control and GA$_3$-treated plants were the same, but GA$_3$ levels in mature seeds were seven times greater than those from nontreated plants (Felippe and Dale, 1968).

Seeds from plants of *Euphorbia pulcherrima* sprayed with 4-chlorophenoxyacetic acid (4-CL) germinated to 70–80%, whereas those from control plants germinated to 90–95%. Also, the germination rate of seeds from treated plants was quite slow (Stewart, 1961). Plants of *Salsola komarovii* treated with ABA produced mostly short-winged fruits, whereas those of controls produced a mixture of short- and long-winged fruits. Seeds in short-winged fruits came out of dormancy at a much slower rate in dry laboratory storage than those in long-winged fruits (Takeno and Yamaguchi, 1991).

G. Length of Growing Season

Length of the spring-growing season has an effect on the percentage of seeds of the winter annual *Trifolium subterraneum* that are impermeable to water. In Australian studies, abundant spring soil moisture prolonged the growing season and increased the percentage of seeds with impermeable seed coats (Aiken, 1939; Quinlivan, 1965). However, the environmental (photoperiod, temperature, moisture) and/or plant (age, size) factor(s) responsible for the increase in seed permeability was not identified.

H. Light Quality

Arabidopsis thaliana was one of the first species in which the spectral quality of light during seed development was shown to influence germination requirements of the progeny. Nondormant seeds from *A. thaliana* plants grown under cool, white fluorescent lights (deficient in far red), cool, white fluorescent plus incandescent lights, and incandescent lights (much far-red) germinated to 45, 12, and 0%, respectively, in darkness (McCullough and Shropshire, 1970). Seeds from plants grown under incandescent lights had a relatively high amount of the inactive, red-absorbing (Pi) form of phytochrome, and thus required light for germination. Seeds from plants grown under fluorescent lights, however, would have had a relative high amount of the active far-red-absorbing form of phytochrome (Pfr) and thus germinated in darkness (Hayes and Klein, 1974). Immature seeds were sensitive to light quality and remained sensitive up to 1 day before full maturation. Dehydration caused seeds to stop responding to light quality (Hayes and Klein, 1974).

The hydrated seeds inside fruits of *Cucumis prophetarum* and *C. sativus* not only remained sensitive to day length after the fruits matured, but they also responded to light quality, exhibiting the normal photoreversibility of phytochrome when alternately exposed to red and far-red light (Gutterman and Porath, 1975). Seeds from *Portulaca oleracea* plants exposed to red- and far-red light during seed production differed in germination percentages. However, seeds from plants exposed to red light did not germinate in darkness, suggesting that germination was mediated by some means other than phytochrome (Gutterman, 1974).

There is variation in the red:far-red (R:FR) ratio of light under which seeds of *Piper auritum* mature on the rainforest floor (Orozco-Segovia *et al.,* 1993). Seeds that matured at a R:FR ratio of 0.26 germinated to about 2, 5, and 35% at a R:FR ratio of 0.26, 0.40, and 0.86, respectively, in the laboratory; those maturing under a ratio of 0.40 germinated to 26, 71, and 91%, respectively; and those maturing under a ratio of 0.86 germinated to 5, 32, and 89%, respectively.

I. Mineral Nutrition

The mineral nutrients available to the mother plant can affect germination of the resulting seeds (Table 8.4). Increased calcium decreased the germination of *Trifolium subterraneum,* whereas increased potassium increased germination (James and Bancroft, 1951). Seeds from molybdenum-deficient plants of *Triticum aestivum* were less dormant than those from molybdenum-treated plants (Cairs and Kritzinger, 1993). Fertilizing plants of *Lathyrus ochrus* with a mixture of superphosphate, ammonium sulfate, and muriate of potash decreased the percentage of impermeable seeds and caused an

TABLE 8.4 Species Exhibiting Variations in Germination Responses When Mother Plants Produce Seeds under Different Levels of Mineral Nutrition

Species	Reference
Avena fatua	Thurston (1951)
A. ludoviciana	Thurston (1951)
Capsicum sp.	Harrington (1960)
Chenopodium album	Fawcett and Slife (1978b)
Daucus sp.	Harrington (1960)
Lactuca sativa	Thompson (1937)
Lathyrus ochrus	Gavrielit-Gelmond (1957)
Matthiola incana	Semeniuk (1964)
Nicotiana tabacum	Thomas and Raper (1979)
Trifolium alexandrinum	El Bagoury and Niyazi (1973)
T. subterraneum	James and Bancroft (1951)
Triplasis purpurea	Cheplick (1996)
Triticale	Naylor (1993)
Triticum sp.	Fox and Albrecht (1957)
Picea excelsa	Youngberg (1952)
Pinus resinosa	Youngberg (1952)
Pisum sp.	Peterson and Berger (1950)
Plantago major	Miao *et al.* (1991)
Rumex acetosella	Escarre and Houssard (1988)
Sorghum bicolor	Benech-Arnold (1993)
Triticum aestivum	Cairns and Kritzinger (1993)
Triticum sp.	Fox and Albrecht (1957)
Vicia faba	El Bagoury (1975)
Zea mays	Roy and Everett (1963)

increased rate of dormancy loss in seeds with physical dormancy (Gavrielit-Gelmond, 1957). Low percentages of permeable seeds were produced by *Vicia faba* plants fertilized with potassium sulfate and ammonium sulfate. Ammonium sulfate alone, however, caused a high percentage of impermeable seeds to be formed (El Bagoury and Niyazi, 1973). Nonfertilized plants of *Vicia faba* had a lower percentage of seeds with permeable seed coats than those from plants fertilized with potassium sulfate at a rate of 476 kg ha^{-1}. However, plants treated with ammonium sulfate (476 kg ha^{-1}) had fewer impermeable seeds than controls (El Bagoury, 1975).

A variety of fertilizer treatments has caused changes in germination characteristics of seeds with physiological dormancy; the most common response is a change in germination percentages. Fertilizers most commonly used in agriculture are mixtures of nitrogen (N), phosphorus (P), and potassium (K), and when NPK was applied to *Lactuca sativa* (Thompson, 1937) and *Zea mays* (Roy and Everett, 1963), germination percentages of the progeny increased. Seeds from plants of *Matthiola incana* treated with high and low levels of NPK germinated to 67 and 88%, respectively (Semeniuk, 1964). A nutrient pulse (120 ml of full-strength Hogland's solution) (size of pot was not stated) applied during the fruit-maturation stage of *Plantago major* plants in two consecutive years decreased the germination percentage of seeds tested in the absence of a *Poa* canopy, but increased it for those tested under light filtered though green *Poa* leaves (Miao et al., 1991).

Additions of individual elements to the soil in which plants are growing have been shown to have effects on germination of the progeny. Seeds from *Chenopodium album* plants treated with nitrogen (as ammonium nitrate) were less dormant than those from controls (Fawcett and Slife, 1978b), and those from *Triticum* sp. plants grown at high levels of nitrogen had higher rates and percentages of germination than those from nontreated plants. In dry years, however, seeds from *Triticum* sp. plants grown at increased nitrogen levels had the same or lower germination rates and percentages as those from nontreated plants (Fox and Albrecht, 1957). Nitrogen treatments of *Nicotiana tabacum* plants reduced the time for, and enhanced uniformity of, seed germination (Thomas and Raper, 1979). Nitrogen treatments increased seed size and germination percentages and rates in triticale (Naylor, 1993).

Seeds from *Daucus carota* plants deficient in potassium, and those from plants of *D. carota* and *Capsicum* sp. deficient in calcium, germinated to lower percentages than those from control plants. Germination of seeds from nitrogen- and potassium-deficient plants of *D. carota*, *Lactuca sativa*, and *Capsicum* sp. did not differ from that of controls (Harrington, 1960). Seeds from *Pisum sativum* plants grown in nutrient cultures

low (5 ppm) in magnesium germinated more slowly and to lower percentages than those from plants grown in cultures high (50 ppm) in magnesium (Peterson and Berger, 1950). Percentage viable seeds from plants of *Avena fatua* receiving 0, 0.8, and 2.4 g of manganese sulfate per pot (each pot held 2 kg of soil) was 28, 44, and 67%, respectively, and dormant seeds (as a percentage of viable seeds) was 64, 89, and 96%, respectively (Thurston, 1951).

J. Physiological Age of Plants

Differences in germination percentages have been observed in seeds collected from plants of various ages. Seeds from *Pinus monticola* trees 10–17 years of age germinated to 39%, whereas those from trees greater than 21 years old germinated to about 55% (Olson, 1932). Seeds from a young seed tree of *Cryptomeria japonica* began and finished germinating sooner than those from an old seed tree (Goo, 1948). Seeds from plants of *Amaranthus retroflexus* showed significant decreases in germination percentages as the age at which plants were induced to flower by a short-day treatment was increased from 6 to 15 days (Kigel et al., 1979).

Evidence that the age of the mother plant has an effect on the germination characteristics of the progeny also comes from studies in which seeds have been collected at different times from the same plants grown under uniform conditions. Seeds collected at weekly intervals during the period of seed production of *Lactuca sativa* plants grown in a greenhouse increased with the age of plants after the second harvest (Thompson, 1937). Seeds collected in July from *Spergularia mariana* plants grown under waterlogged, wet, and dry conditions in a greenhouse were 92–96% dormant, whereas those collected in August were only 18–20% dormant (Okusanya and Ungar, 1983). Seeds of *Oldenlandia corymbosa* produced in a growth chamber under a 16-hr daily photoperiod during July germinated to 80–90%, whereas those produced under the same conditions in August–October germinated to only 1–15% (Attims, 1972a). Under long days, old plants of *Ononis sicula* produced brown or green seeds, whereas young plants produced yellow seeds (Gutterman, 1980–1981b).

K. Position on Mother Plant

Seeds developing at different positions on the mother plant may not have the same dormancy-breaking and/or germination requirements (Table 8.5). One explanation for these differences is that resources are not allocated equally to all seeds. Thus, some seeds are larger than others, and large seeds may have different dormancy-breaking and/or germination requirements than small

TABLE 8.5 Species in Which the Position on the Mother Plant Where Seeds Develop Causes Differences in Germination Characteristics

Species	Reference
Aegilops neglecta	Maranon (1987)
A. ovata	Datta *et al.* (1970)
Aellenia autrani	Negbi and Tamari (1963), Werker and Many (1974)
Aethionema carneum	Zohary and Fahn (1950)
Agrostis curtisii	Gonzalez-Rabanal *et al.* (1994)
Apium graveolens	Thomas *et al.* (1978, 1979)
Atriplex lindleyi	Drennen *et al.* (1993)
Avena fatua	Raju and Ramaswamy (1983)
Avenula marginata	Gonzalez-Rabanal *et al.* (1994)
Bidens pilosa[a]	Forsyth and Brown (1982)
Cakile spp.	Maun and Payne (1989)
Cenchrus longispinus	Twentyman (1974)
Cotula coronopifolia[a]	van der Toorn and ten Hove (1982)
Dalbergia sissoo	Sahai (1994)
Daucus carota	Thomas *et al.* (1978), Gray (1979)
Dimorphotheca pluvialis[a]	Correns (1906)
Emex spinosa[a]	Evenari *et al.* (1977), Weiss (1980)
Emilia sonchifolia[a]	Marks and Akosin (1984)
Fezia pterocarpa	Hernandez B. and Clemente M. (1977)
Foeniculum vulgare	Thomas (1994)
Gossypium sp.	Enileev and Solov'ev (1960)
Grindelia lanceolata[a]	Baskin and Baskin (1979)
G. squarrosa[a]	McDonough (1975)
Gymnarrhena micrantha[a]	Koller and Roth (1964)
Hemizonia increscens[a]	Tanowitz *et al.* (1987)
Heterosperma pinnatum[a]	Venable *et al.* (1987)
Heterotheca grandiflora[a]	Flint and Palmblad (1978)
H. latifolia[a]	Baskin and Baskin (1976b), Venable (1985a)
Mesembryanthemum nodiflorum	Gutterman (1980–1981a, 1994b)
Platystemon californicus	Hannan (1980)
Petroselinum crispum	Thomas (1996)
Pseudarrhenatherum longifolium	Gonzalez-Rabanal *et al.* (1994)
Pteranthus dichotomus	Gutterman (1980–1981b)
Relhania genistaefolia[a]	Levyns (1935)
Rumex spp.	Cavers and Harper (1966)
Salicornia europaea	Ungar (1979)
S. patula	Grouzis *et al.* (1976)
Salsola komarovii	Yamaguchi *et al.* (1990)
Senecio jacobaea[a]	McEvoy (1984)
Synedrella nodiflora[a]	Ernst (1906), Manilala and Unni (1978), Marks and Akosin (1984)
Trifolium subterraneum	Halloran and Collins (1974), Taylor and Palmer (1979)
Xanthium canadense[a]	Crocker (1906)

[a] Member of the Asteraceae.

ones (Datta *et al.*, 1970; Halloran and Collins, 1974). Another possibility for differences in the level of dormancy is that seeds produced at one position (e.g., at the base of an inflorescence) developed under different environmental conditions than those produced at another position (e.g., at the top of an inflorescence). Also, the physiological age of the mother plant may vary when seeds are produced in different positions.

In some species, the position of seed development on the mother plant does not cause differences in dormancy. Seeds (achenes) produced by disc and ligulate (ray) flowers in capitula (heads) of two species of the plant family Asteraceae, *Picris echioides* (Sorensen, 1978) and *Tridax procumbens* (Marks and Akosim, 1984), differ morphologically, but their dormancies are the same (Sorensen, 1978). Also, both disc and ray achenes of *Senecio jacobaea* are nondormant (Baker-Kratz and Maguire, 1984). As discussed later, however, the position of seed development on the mother plant has effects on the dormancy and germination characteristics of the progeny of many species.

1. Different Parts of the Same Inflorescence

One of the best examples of how the position within an inflorescence influences germination of the progeny is found in a number of Asteraceae with both disc (central) and ray (peripheral) flowers in the head (Table 8.5). In these species, disc and ray flowers produce achenes that may be morphologically different, and, with the exception of *Emilia sonchifolia*, disc achenes are less dormant than those produced by ray flowers. In *E. sonchifolia*, ray achenes are less dormant than disc achenes. Varying results have been obtained for *Synedrella nodiflora*. At 26°C and a 12-hr daily photoperiod, disc achenes of *S. nodiflora* germinated to higher percentages than ray achenes (Manilal and Unni, 1978). However, at 27, 30, and 35°C in continuous light and at 27 and 30°C in darkness, ray achenes of *S. nodiflora* germinated to higher percentages than disc achenes (Marks and Akosim, 1984). Further, Ernst (1906) found that disc achenes germinated to higher percentages than ray achenes in low light and in darkness. Although ray and disc achenes of *Cotula coronopifolia* germinated to nearly 100%, ray achenes have a slower rate of germination than disc achenes (van der Toorn and ten Hove, 1982).

Other examples of variation in germination response by seeds produced in different parts of the same inflorescence include *Aegilops ovata* (Datta *et al.*, 1970), *A. neglecta, A. geniculata, A. triuncialis* (Maranon, 1987), *Agrostis curtisii, Avenula marginata,* and *Pseudarrhenatherum longifolium* (Gonzalez-Rabanal *et al.*, 1994). In these grasses, the basal seed (caryopsis) in a spikelet is larger and less dormant than the upper one. Also, in the grass *Avena fatua*, the primary seed in each spikelet matures sooner than the secondary one and is nondormant; the secondary seed is dormant (Raju and Ramaswamy, 1983).

Fruits borne at distal positions on branches of *Salsola komarovii* have longer wings and seeds with faster afterripening rates than those produced at proximal positions (Yamaguchi *et al.*, 1990). Fruits of *Aethionema careum* borne at the top of the main axis and on lateral racemes tend to be indehiscent and single-seeded, whereas those in other positions usually are dehiscent and have more than one seed. Seeds from indehiscent fruits appear to be more dormant than those from dehiscent fruits, as initiation of their germination was delayed 15 days. Fruits of *Aellenia autrani* produced at the distal portions of branches have thin, narrow wings and seeds with green, nondormant embryos, whereas fruits borne at basal positions have thick, wide wings and seeds with yellow, dormant embryos (Werker and Many, 1974).

Seeds in capsules produced along the main axis of the inflorescence and at proximal ends of inflorescence branches of *Oldenlandia corymbosa* are more dormant than those produced in capsules at distal ends of inflorescence branches (Do Cao *et al.*, 1978). The dispersal unit of *Pteranthus dichotomus* is an inflorescence with one fruit (pseudocarp) at its base, two fruits at the second order of branches, and four fruits at the third order. The number of orders is determined by moisture availability for plant growth. If only first-order pseudocarps are produced, they are nondormant, but if second-order pseudocarps develop, the first-order ones are dormant. If three orders of pseudocarps are produced, the third-order (outermost) ones are the least dormant (Evenari *et al.*, 1982).

Seeds produced by the central flower in a nodal segment of *Salicornia europaea* were heavier, less dormant, and germinated to higher percentages at 1% (0.17 *M*) NaCl than those produced by lateral flowers (Ungar, 1979). Central seeds in *S. patula* inflorescences did not require cold stratification or light for germination and germinated to 65% at 0.34 *M* (2.0%) NaCl at 25/15°C. However, lateral seeds required cold stratification and light for high germination percentages and germinated to 15% at 0.34 *M* NaCl at 25/15°C (Grouzis *et al.*, 1976; Berger, 1985).

Sometimes in the same inflorescence, but often in separate reduced inflorescences, some plants produce flowers that do not open; they are called cleistogamous or hidden flowers. By way of comparison, flowers that open normally are called chasmogamous flowers. Cleistogamous flowers may be borne on the aerial stem or they may be subterranean. Thus, cleistogamous and chasmogamous flowers may be produced at different positions on the mother plant. All discussion of germina-

tion of seeds from cleistogamous flowers is delayed until Section VIII,D.

2. Inflorescences in Different Positions on the Plant

Seeds of *Apium graveolens* (Apiaceae) from primary and secondary umbels germinated to higher percentages at 18°C in darkness than those from quaternary umbels. Also, seeds from quaternary umbels were less responsive to $GA_{4/7}$ and had a lower maximum temperature limit for germination than those from primary and secondary umbels (Thomas *et al.*, 1978, 1979). Plants of *Daucus carota* have a primary umbel and many secondary umbels, which are numbered from 1 to 8, starting at the top of the stem just below the primary umbel. Germination percentages of *D. carota* seeds harvested in early July from primary umbels were lower than those of seeds collected from umbels 1 and 5 and equal to those of seeds from umbel 8. However, seeds from the primary umbel collected in early July had a higher germination rate than those from lower umbels (Gray, 1979). Seeds from umbels 2 and 3 germinated to lower percentages than those from primary umbels, although they were the same size and weight as those from primary umbels. Large seeds were less dormant than small ones (Jacobsohn and Globerson, 1980).

3. Flowers at Different Heights on the Stem

Seeds from fruits (bolls) of *Gossypium* sp. produced at low and middle nodes were less dormant than those from high nodes on the stem. The number of seeds per boll increased with an increase in the height of the node at which they were produced (Enileev and Solov'ev, 1960).

4. Different Positions in a Bur

Crocker (1906) reported that the upper seed in the two-seeded dispersal unit (bur) of *Xanthium canadense* (=*X. strumarium* var. *canadense*) does not germinate until after the lower one has germinated. However, 9 of 51 (18%) burs of *X. strumarium* that germinated after receiving 12 weeks of cold stratification produced two seedlings (Baskin *et al.*, unpublished results).

The large seeds in burs of *Trifolium subterraneum* become permeable before the small ones, but the position of the burs on the lateral stem has some influence on the time when seeds become nondormant (Halloran and Collins, 1974; Taylor and Palmer, 1979). Burs of *Cenchrus longispinus* contain one to three spikelets, and each has a fertile floret that forms one seed. The seed in the central spikelet is larger than those in the lateral spikelets and comes out of dormancy in dry storage much faster than lateral seeds (Twentyman, 1974).

5. Different Positions in a Fruit

Seeds formed in the outer portion of the capsules of *Mesembryanthemum nodiflorum* (Aizoaceae) are the first to be dispersed and are less dormant than those produced in the interior of the capsule (Gutterman, 1980–1981a). Fruits of *Platystemon californicus* (Papaveraceae) are composed of multiple carpels, and seeds are formed within the carpels as well as in a central cavity to the inside of the carpels. Seeds formed in the carpels are dispersed as single-seeded joints of the carpels and require 8 months for dormancy break, but those from the central chamber of the fruit have no pericarp tissue attached to them and require only 4 months for dormancy break. Thus, seeds formed in the carpels are much more dormant than those produced in the chamber (Hannan, 1980).

Fruits of *Fezia pterocarpa* (Brassicaceae) have a dehiscent lower and an indehiscent upper part. At 12 and 16°C in light and at 21 and 26°C in darkness, seeds from the indehiscent portion of the fruit germinated to higher percentages than those from the dehiscent portion (Hernandez B. and Clemente M., 1977). In *Cakile* spp., also in the Brassicaceae, both the lower and the upper segments of the fruits are indehiscent. At maturity, the upper part is shed from the plant but the lower part of the fruit remains attached. Seeds from the two fruit segments of *C. edentula* var. *lacustris* germinated equally well in light and darkness at 20/10°C. However, seeds from lower segments germinated to higher percentages at 25/15 and 15/5°C in darkness than those from upper segments (Maun and Payne, 1989). There were no differences in germination percentages, sex expression, or vigor of seeds removed from apical, central, and peduncle regions of *Carica papaya* fruits (Sao Jose and Cunha, 1990).

L. Premature Defoliation

Seeds from *Gossypium hirsutum* plants defoliated completely by hand germinated to significantly lower percentages at 30/20°C than those from plants with 0 or 50% of the leaves removed (Minton *et al.*, 1979). However, the germination of *Datura ferox* seeds from plants with one-half the leaf area removed germinated to the same percentages as those from plants with no leaves removed (Sanchez *et al.*, 1981). Also, the leaf harvest of *Nicotiana tabacum* plants had no effect on germination (Thomas and Raper, 1979).

M. Soil Moisture

Depending on the species, water stress during the time of seed development causes increases or decreases in germination percentages of the seeds. Among species with physiological dormancy, water stress decreased the dormancy of seeds of *Amaranthus retroflexus* (Chadoeuf-Hannel and Barralis, 1982), *Avena fatua* (Sexsmith, 1969; Peters, 1982b; Sawhney and Naylor, 1982), *Datura ferox* (Sanchez *et al.*, 1981), *Sorghum bicolor* (Benech Arnold, 1993), and *S. halepense* (Benech Arnold, 1993; Benech Arnold *et al.*, 1992). However, low soil moisture increased the dormancy of seeds of *Arachis hypogaea* (Pallas *et al.*, 1977), *Salsola komarovii* (Yamaguchi *et al.*, 1990), and *Spergularia mariana* (Okusanya and Ungar, 1983).

In seeds with physical dormancy, water stress decreased the dormancy of *Gossypium* sp. seeds (Carver, 1936) and decreased the length of the dormancy period in those of *Erodium brachypodium* (Stamp, 1990). However, water stress increased the dormancy in seeds of *Cassia sophera* (Datta and Sen, 1981), *Lathyrus ochrus* (Gavrielit-Gelmond, 1957), and *Sesbania bispinosa* (Sharma *et al.*, 1978).

N. Solar Irradiance

To test the effects of differences in the total amount of sunlight under which plants are grown on germination of the resulting seeds, plants of *Amaranthus retroflexus* were grown under short (8) and long (16) days in both full sunlight and under screens transmitting only 27% of full sunlight. Seeds from shaded short-day plants germinated to higher percentages in darkness and were more responsive to cold stratification than those from nonshaded short-day plants. However, seeds from shaded long-day plants germinated to lower percentages in darkness and were less sensitive to cold stratification treatments than those from nonshaded, long-day plants (Kigel *et al.*, 1977). Seeds from plants of *Datura ferox* grown under a neutral density black plastic shade cloth transmitting 50% of the incident solar radiation were less dormant than those from plants grown in full sun (maximum irradiance incident on upper leaves of control plants was 1.2 cal cm^{-2} min^{-1}) (Sanchez *et al.*, 1981).

Seeds from *Abutilon theophrasti* plants grown under 0, 30, and 60% shade exhibited 77, 63, and 62% dormancy, respectively, in 1984 and 78, 65, and 64% dormancy, respectively, in 1985 (Bello *et al.*, 1995). Seeds from *Plantago lanceolata* plants exposed to reduced levels of irradiance under the canopy of grasses growing in a lawn were more sensitive to light than those from plants grown in full sun (Wulff *et al.*, 1994). Seeds from *Polygonum persicaria* plants grown under 100 and 8% full sun germinated to 68 and 52%, respectively (Sultan, 1996).

O. Temperature

Early evidence that temperature has a preconditioning effect came from studies in which temperature records were kept during the time of seed development under field conditions. Dormancy decreased with an increase in mean temperatures 10 or 30 days prior to seed harvest for seeds from *Lactuca sativa* plants grown at sites in Arizona and southern California (Harrington and Thompson, 1952). Dormancy decreased in seeds of *Rosa* spp. with an increase in mean temperatures 30 days prior to harvesting (VonAbrams and Hand, 1956). Seeds of *Triticum aestivum* produced in dry, warm years were less dormant than those matured in cool, wet summers (Belderok, 1961). Likewise, seeds of *Artemisia rhodantha* produced in warm years germinated to higher rates and percentages than those produced in cool years (Nosova, 1981).

Another approach to studying temperature as a preconditioning factor has been to grow plants at various temperatures in growth chambers and to compare germination responses of the resulting seeds. Thus, we now have a long list of species whose seeds vary in their germination characteristics when they are produced at different temperatures in either the field or in controlled environments (Table 8.6).

The most common result obtained when plants are grown at specific temperature regimes is that seeds produced by plants at high temperatures have higher germination percentages and/or rates than those produced at low temperatures (Stearns, 1960; Grant Lipp and Ballard, 1963; Datta *et al.*, 1972; Wiesner and Grabe, 1972; Boyce *et al.*, 1976; Wurzburger and Koller, 1976; Akpan and Bean, 1977; van der Vegte, 1978; Lexander, 1980; Peters, 1982a; Chadoeuf-Hannel and Barralis, 1983; Gray and Steckel, 1984; Somody *et al.*, 1984; Alexander and Wulff, 1985; Probert *et al.*, 1985b). However, seeds from *Syringa reflexa* plants grown at 18, 21, and 24°C were more dormant than those from plants grown at 15°C (Junttila, 1971). Also, seeds of *Lactuca sativa* produced at 26°C were more dormant, especially when incubated in darkness, than those matured at 20 or 23°C (Koller, 1962). Seeds from most *Themeda australis* plants grown at 33/28°C were significantly more dormant than those produced at 21/16°C. However, depending on the original collection site, seeds from some *T. australis* plants grown at 33/28 and 21/16°C did not differ in the degree of dormancy, whereas those from plants at the Tantangara site were less dormant when grown at 33/28 than at 21/16°C (Groves *et al.*, 1982).

The time during the reproductive cycle of a plant when temperature treatments are applied can be important in determining the germination requirements of the progeny. Exposure of *Hordeum vulgare* plants to

TABLE 8.6 Species Exhibiting Variations in Germination Responses When Mother Plants Produce Seeds under Different Temperatures

Species	Reference
Aegilops kotschyi	Wurzburger and Koller (1976)
A. ovata	Datta *et al.* (1972)
Allium cepa	Gray and Steckel (1984)
Amaranthus retroflexus	Kigel *et al.* (1977)
Anagallis arvensis	Grant Lipp and Ballard (1963)
Artemisia rhodantha	Nosova (1981)
Avena fatua	Sexsmith (1969), Peters (1982a), Somody *et al.* (1984), Sawhney *et al.* (1985), Sawhney (1989)
A. sterilis	Somody *et al.* (1984)
Beta vulgaris	Heide *et al.* (1976), Lexander (1980)
Cucumis anguria	Noronha *et al.* (1976)
Dactylis glomerata	Probert *et al.* (1985b)
Festuca arundinacea	Boyce *et al.* (1976)
Hordeum vulgare	Khan and Laude (1969)
Lactuca sativa	Harrington and Thompson (1952), Koller (1962)
Lolium multiflorum	Akpan and Bean (1977)
L. perenne	Akpan and Bean (1977)
Lolium sp.	Wiesner and Grabe (1972)
Medicago sativa	Dotzenko *et al.* (1967)
Mollugo verticillata	Attims (1972b)
Nicotiana tabacum	Thomas and Raper (1975)
Pennisetum typhoides	Mohamed *et al.* (1985)
Plantago lanceolata	Alexander and Wulff (1985)
Rosa spp.	VonAbrams and Hand (1956)
Stellaria media	van der Vegte (1978)
Syringa reflexa	Junttila (1971)
Themeda australis	Groves *et al.* (1982)
Trifolium subteraneum	Taylor and Palmer (1979)
Triticum aestivum	Belderok (1961)

54°C (heat stress) for 7–10 days after awns first appeared in the young spikes increased seed dormancy, whereas heat stress 3 weeks after awn emergence decreased dormancy (Khan and Laude, 1969). Exposure of detached culms of the Gulf cultivar of *Lolium* sp. to 15°C during anthesis reduced dormancy, but exposure to this temperature during seed ripening increased it (Wiesner & Grabe, 1972). Low pretransplant and high posttransplant temperatures in *Nicotiana tabacum* increased seed dormancy. Thus, for maximum germination of the progeny, *N. tabacum* plants need warm pretransplant and moderately warm or cool posttransplant temperatures (Thomas & Raper, 1975). The time when temperature has a preconditioning effect in *Avena fatua* is prior to anthesis (Sawhney *et al.*, 1985). Regardless of the maturation temperature, however, embryos excised from seeds of dormant lines of *A. fatua* germinated (Sawhney, 1989).

Seeds of *Cucumis anguria* developed at a relatively low mean temperature (18.4°C) required darkness to germinate at 25°C, whereas those developed at a relatively high mean temperature (24.2°C) germinated equally well in light and darkness at 25°C (Noronha *et al.*, 1976).

P. Time

Variations in germination occur in seeds of some species collected at the same site year after year (Harniss and McDonough, 1976; Young and Evans, 1977; Baskin and Baskin, 1982; Caron *et al.*, 1993; Johnston *et al.*, 1994) or at different times during a single growing season (Gutterman, 1992b, 1994a; Hume, 1994; Baskin and Baskin, 1995). Although the factor(s) responsible for differences in germination has not been determined, temperature, rainfall, and day length have been suggested (Fenner, 1992). Seeds collected from plants of *Fragaria vesca* during the fruiting period showed decreases in the length of time required for dormancy break (Bunning, 1954). Differences in dormancy-breaking requirement of these seeds may have been due to the increased age of the mother plant and/or changes in environmental conditions under which seeds matured.

Q. Ultraviolet Light

Seeds from *Dimorphotheca sinuata* plants grown under enhanced UV-B radiation (simulating a 20% atmospheric ozone depletion) germinated to a significantly lower percentage than those from plants grown under ambient UV-B radiation (Musil, 1996). Viability tests were not conducted on ungerminated seeds.

VI. SEED CHANGES AS A RESULT OF THE PRECONDITIONING ENVIRONMENT

In addition to differences in germination percentages and rates, the growth of plants under varying environmental conditions may cause changes in seed characteristics such as size, structure, and types and amounts of stored compounds. These changes may be related to seed germination and may help explain how certain environmental factors cause variation in germination responses.

A. Chemical Compounds

Gutterman (1973, 1974) suggested that the environment indirectly controls germination of the progeny via the kinds and amounts of compounds transferred to the seeds. He proposed that (1) the environmental conditions under which plants are grown can affect their chemical composition, (2) plants transfer compounds to developing seeds in proportion to the amounts present in the plant's tissues, and (3) compounds transported to seeds interact with phytochrome and thus influence germination.

A number of environmental factors have been shown to influence the chemical composition of seeds. Seeds produced by mycorrhizal plants of *Avena fatua* contain more phosphorus than those produced by nonmycorrhizal plants (Lu and Koide, 1991). The concentration of GA_{20} increased in cotyledons of *Pisum sativum* seeds produced under short days (Ingram and Browning, 1979), and the amount of amylose starch in the endosperm of *Zea mays* grains increased in response to low temperatures at the sites where the plants grew (Fergason and Zuber, 1962). Temperatures during grain development in *Hordeum vulgare* influenced the amount of α-amylase produced in endosperm halves in either the presence or the absence of GA. Endosperm in Himalaya barley seeds produced at 25/20°C did not require exogenous GA for the production of α-amylase, whereas endosperm from seeds produced at 19/14°C required GA for production of α-amylase (Nicholls, 1982). The barley-endosperm bioassay for gibberellins is sensitive to the variety of seeds used as well as to location where the seeds were produced (Jackson, 1971).

Compounds that may play a role in controlling germination also vary in response to varying environmental conditions during seed production. The iodine number of oil stored in seeds of *Linum usitatissimum* was 20 points higher in a relatively cool than in an excessively hot summer (Painter *et al.*, 1944). The combined effects of locality, soil type, and season could cause a two-fold increase in the amount of total protein in the seeds of *Triticum* sp.; however, the gliadin class of proteins was changed little by environmental conditions during seed development (Lee and Ronalds, 1967). Oil content in kernels of *Zea mays* varies with their position in the ear, with those in a mid-ear position having the highest oil content (Lambert *et al.*, 1967). The amount of oil in *Carthamus tintorius* seeds increased 7% in response to defoliation (Urie *et al.*, 1968).

B. Seed Size, Color, and Shape

In many species, the environment during the time of development causes seeds of different sizes (Table

8.7), colors, and shapes to be produced. The various factors known to cause these changes will be discussed next.

1. Mineral Nutrition

The addition of nitrogen to the soil in which plants are growing has had mixed results on seed size. Ammonium nitrate applied at a rate of 16.9 kg N ha^{-1} week^{-1} increased seed size in *Desmodium uncinatum* (Gibson & Humphreys, 1973), whereas increasing nitrogen from 0 to 80 kg ha^{-1} did not cause any change in weight of *Lolium perenne* seeds (Hebblethwaite, 1977). However, decreasing nitrogen resulted in a decrease in weight of *Glycine max* seeds (Streeter, 1978).

In other studies, plants have been fertilized with mixtures of mineral nutrients, and again the results vary with the species. A nutrient solution containing calcium nitrate, potassium phosphate, and magnesium sulfate added to the soil in which *Senecio sylvaticus* plants were growing increased seed weight; however, it had no effect on weight of *Chamaenerion angustifolium* seeds (van Andel and Vera, 1977). A dry fertilizer containing nitrogen, phosphorus, and potassium (in 6–10–4 proportions) was applied to soil around plants of *Asclepias* spp. Seed weight increased for *A. syriaca* but not for *A. verticillata* (Willson and Price, 1980). Plants of *Abutilon theophrasti* were fertilized with a solution containing nitrogen, phosphorus, and potassium at high (0.03 g N, 0.03g K, and 0.06 g P per plant), moderate (one-half of high), low (one-fourth of high), and very low (one-eight of high) concentrations. Seed weight increased from 6.6 to 8.8 mg with an increase in nutrient levels from very low to high (Parrish and Bazzaz, 1985). Whether grown in soil or hydroponically, an increase in phosphorus supply for plants of *A. theophrasti* increased seed weight (Lewis and Koide, 1990). Plants of *Senecio vulgaris* were fertilized with high, medium, and low levels of NPK (20–20–20), and the resulting seeds decreased in size with a decrease in nutrients (Aarssen and Burton, 1990).

2. Pollination Thoroughness and Seed Number

If the energy resources of the mother plant are limited, it makes sense that the number of seeds produced by the plant would have an effect on seed size, and vice versa. That is, with an increase (or decrease) in seed number, seed size would decrease (or increase). Is this always the case? McGinley *et al.* (1987) concluded that variable progeny sizes may be favored in a spatially and temporally variable environment if the density-dependent fitness loss of progeny is great and if parents control dispersal to suitable habitats. In a model concerning the relationship between size and number of progeny, McGinley and Charnov (1988) predicted that (1) optimal seed size is positively correlated with the amount of carbon and nitrogen available for seed production and (2) seed size and seed nitrogen content are negatively correlated.

Early studies on crop species (see review by Kidd and West, 1918) indicated that seed weight increased as number of seeds maturing per plant decreased. More recently, it has been shown that an increase in seed number caused a decline in seed weight of *Ipomopsis aggregata* (Wolf *et al.*, 1986), *Primula farinosa* (Baker *et al.*, 1994), *Raphanus raphanistrum* (Stanton, 1984; Stanton *et al.*, 1987), and *Rubus chamaemorus* (Agren, 1989). Weight of *mucuna andreana* seeds did not increase significantly with an increase in number of seeds per pod from one to five (Janzen, 1977). *Enterolobium cyclocarpum* seeds from pods with 1–5 seeds were 1.5–3 times heavier than those from pods with 7–16 seeds (Janzen, 1982).

An increase in the pollination level resulted in more fertilized ovules, an increase in seed abortion, and a decrease in seed weight of *Cirsium arvense* (Lalonde and Roitberg, 1989). Naturally pollinated flowers of self-compatible clones of *Clintonia borealis* set more seeds than those of self-incompatible clones, and seeds from self-compatible clones weighed less than those from self-incompatible clones (Galen and Weger, 1986). However, the addition of pollen to flowers of self-incompatible clones of *C. borealis* caused more, but also heavier, seeds to be formed. Fruit number and seed weight were lower for plants of *Clarkia timbloriensis* grown from seeds from self-pollinated than for those grown from outcrossed plants (Holtsford and Ellstrand, 1990). In *Prunella vulgaris,* there was a positive relationship between seeds per flower and biomass per seed (Winn and Werner, 1987), and seed size in *Primula maritima* increased with an increase in seed number per inflorescence (Baker *et al.*, 1994). In *Lupinus texensis,* no correlation was found between seed weight and total number of seeds per plant (Schaal, 1980), and in *Phytolacca americana* seed weight was independent of the number of seeds per fruit (Byrne and Mazer, 1990).

3. Position

Seeds produced in different parts of the same inflorescence may differ in weight. Central seeds in heads of *Tragopogon dubius* weighed less than the peripheral ones (McGinley, 1989), and basal seeds in spikelets of *Aegilops* spp. were heavier than those produced in more distal positions (Datta *et al.*, 1970; Maranon, 1989). In the grasses *Phalaris tuberosa* (Whalley *et al.*, 1966) and *Sorghum bicolor* (Hamilton *et al.*, 1982), seeds that matured (last) on the basal part of the panicle weighed

TABLE 8.7   Species Whose Seeds Vary in Size as a Consequence of the Environment in Which
the Mother Plant Was Growing

Species	Environmental factor	Reference
Abutilon theophrasti	Nutrients	Parrish and Bazzaz (1985)
	Shading	Bello *et al.* (1995)
Aegilops spp.	Position	Datta *et al.* (1972), Maranon (1989)
Amaranthus blitoides	Season	Cavers and Steele (1984)
A. retroflexus	Temperature, day length	Chadoeuf-Hannel and Barralis (1982), Schimpf (1977)
Apium graveolens	Position	Thomas *et al.* (1979)
Arctium minus	Defoliation	Reed and Stephenson (1972)
Asclepias spp.	Nutrition, light, defoliation	Willson and Price (1980)
Aster acuminatus	Position	Pitelka *et al.* (1983)
Atriplex heterosperma	Season	Frankton and Bassett (1968)
Beta vulgaris	Temperature	Wood *et al.* (1980)
Bilderdykia convolvulus	Herbicide	Andersson (1994b)
Brassica napus	Position	Clarke (1979)
	Predation	Williams and Free (1979)
Cakile spp.	Position	Maun and Payne (1989)
Carthamus tinctorius	Defoliation	Urie *et al.* (1968)
Catalpa speciosa	Defoliation	Stephenson (1980)
Cenchrus longispinus	Position	Twentyman (1974)
Chenopodium album[a]	Day length, season	Williams and Harper (1965), Cavers and Steele (1984)
C. rubrum	Day length	Cook (1975)
Cicer arietinum	Position	Sheldrake and Saxena (1979)
Cirsium arvense	Seed number	Lalonde and Roitberg (1989)
Cladrastis sp.	Seed number, position	Harris (1917)
Clintonia borealis	Seed number	Galen and Weger (1986)
Desmodium paniculatum	Position, temperature, moisture, seed number, photoperiod	Wulff (1986a)
D. uncinatum	Nutrition	Gibson and Humpheys (1973)
Diplotaxis tenuifolia	Season	Cavers and Steele (1984)
Enterolobium cyclocarpum	Seed number	Janzen (1982)
Erodium brachycarpum	Moisture	Stamp (1990)
Fedia spp.[a]	Position	Mathez and Xenda de Enrech (1985a)
Festuca arundinacea	Temperature	Bean (1971)
F. pratensis	Temperature	Akpan and Bean (1977)
Galium spurium	Herbicide	Andersson (1994b)
Glycine max	Moisture	Ramseur *et al.* (1984), Meckel *et al.* (1984)
	Nutrition	Streeter (1978)
	Position	Heindl and Brun (1984)
	Predation	Smith and Bass (1972)
	Season	Egli *et al.* (1978)
	Temperature	Egli and Wardlaw (1980)
Gymnocladus dioicus	Defoliation	Janzen (1976)
Halogeton glomeratus[a]	Day length	Williams (1960)
Heracleum lanatum	Defoliation	Hendrix (1984a)
Hordeum vulgare	Position	Giles (1990)

(continues)

TABLE 8.7—*Continued*

Species	Environmental factor	Reference
Impatiens capensis	Position	Waller (1982)
Ipomopsis aggregata	Seed number	Wolf *et al.* (1986)
Juglans major	Moisture	Stromberg and Patten (1990)
Lactuca sativa	Season	Soffer and Smith (1974)
	Temperature	Drew and Brocklehurst (1990)
Leontodon hispidus	Season	Fuller *et al.* (1983)
Lolium spp.	Temperature	Akpan and Bean (1977)
Lomatium grayi	Position	Thompson (1984)
L. salmoniflorum	Season	Thompson and Pellmyr (1989)
Lupinus albus	Moisture	Withers and Forde (1979)
L. texensis	Position	Schaal (1980)
Melilotus alba	Season	Cavers and Steele (1984)
Myosotis arvensis	Herbicide	Andersson (1994a)
Nemophila menziesii	Competition	Plantenkamp and Shaw (1993)
Onopordum acanthium	Season	Cavers and Steele (1984)
Pastinaca sativa	Position	Hendrix (1979)
Pennisetum typhoides	Temperature	Mohamed *et al.* (1985)
Petroselinum crispum	Position	Thomas (1996)
Phaseolus sp.	Position	Harris (1915)
Pinus longaeva	Tree age	Connor and Lanner (1991)
Pisum sativum	Temperature	Lambert and Linck (1958)
Plantago aristata	Temperature	Stearns (1960)
P. lanceolata	Temperature	Alexander and Wulff (1985)
	CO$_2$ enrichment	Wulff and Alexander (1985)
Poa pratensis	Temperature	Maun *et al.* (1969)
Polygonum spp.	Season	Hammerton and Jalloq (1970)
Populus deltoides	Position	Hardin (1984)
Raphanus raphanistrum	Seed number	Stanton *et al.* (1987)
	Position, seed number	Stanton (1984a)
Rubus chamaemorus	Seed number, defoliation	Agren (1989)
Rumex crispus	Predation, position	Maun and Cavers (1971)
Rumex spp.	Position	Cavers and Harper (1966)
	Defoliation	Bentley *et al.* (1980)
Salicornia patula	Position	Berger (1985)
Scabiosa columbaria	Season	Rorison (1972)
Senecio jacobaea	Defoliation	Crawley and Nachapong (1985)
S. sylvaticus	Nutrients	van Andel and Vera (1977)
S. vulgaris	Nutrients	Aarssen and Burton (1990)
Sesbania macrocarpa	Defoliation	Marshall *et al.* (1986)
Sorghum bicolor	Predation, position	Hamilton *et al.* (1982)
Staphylea sp.	Seed number	Harris (1917)
Strophostyles helvola	Position	Yanful and Maun (1996)
Thlaspi arvense	Herbicide	Andersson (1994b)
Tragopogon dubius	Position, season	McGinley (1989b)

(continues)

TABLE 8.7—*Continued*

Species	Environmental factor	Reference
Triticum aestivum	Light, moisture	Brocklehurst *et al.* (1978)
	Competition, irradiance	Martinez-Carrasco and Thorne (1979)
	Temperature	Campbell *et al.* (1981)
Triticum sp.	Irradiance	Jenner (1979)
	Position	Rawson and Evans (1970)
	Predation	Brenner and Rawson (1978)
	Temperature, irradiance	Wardlaw (1970)
	Temperature	Ford *et al.* (1976)
Verbascum blattaria	Season	Cavers and Steele (1984)
Zea mays	Position	Lambert *et al.* (1967)

a Seeds also vary in shape and/or color.

less than those at the top. In *Zea mays,* the heaviest seeds are at the base of the ear and the lightest ones are at the top (Lambert *et al.,* 1967).

Position in the inflorescence can influence the shape of seeds or dispersal units. For example, in genera of Asteraceae with both ray and disc flowers, the resulting achenes frequently can be distinguished by their shape, and these morphologies are correlated with differences in germination characteristics (Table 8.5). Also, the indehiscent fruits of *Fedia* spp. (Valerianaceae) have three different shapes depending on where they are produced in the inflorescence (Mathez and Xena de Enrech, 1985a).

Capsules from the base of catkins on one tree of *Populus deltoides* had heavier seeds than those from the tip of catkins. However, on a second tree, seeds in capsules from the top and bottom of catkins were not significantly different in weight (Hardin, 1984). Seeds of *Rumex crispus* (Cavers and Harper, 1966; Maun and Cavers, 1971) and *R. obtusifolius* (Cavers and Harper, 1966) from the lower portion of the panicle weighed more than those from the upper branches of the panicle. Seeds of *Brassica napus* were heavier in fruits produced on the bottom half than on the top half of the racemes (Clarke, 1979), but seed size was independent of position in the inflorescence of *Phytolacca rivinoides* (Byrne and Mazer, 1990).

Variation in seed size also occurs in some species with inflorescences borne at different positions on the mother plant. A good example of this is seen in the Apiaceae, where new umbels may be produced throughout the growing season. In *Lomatium grayi,* 16% of the variance in seed mass within individual plants was explained by differences between umbels (Thompson, 1984). In *Pastinaca sativa,* seeds in tertiary umbels were significantly smaller than those in primary or secondary umbels (Hendrix, 1979). In a later study, Hendrix (1984b) found that seeds from secondary and tertiary umbels were 73 and 50%, respectively, as heavy as those from primary umbels.

Inflorescence position also influences seed size in families other than the Apiaceae. Weights of disc and ray achenes from the same head of *Aster acuminatus* often are very similar, but those of achenes (disc and ray combined) from different heads on the same plant usually differ significantly (Pitelka *et al.,* 1983). Seeds produced in the primary racemes of *Desmodium uncinatum* are heavier than those formed in secondary and tertiary racemes (Gibson and Humphreys, 1973).

The node at which flowers are borne may result in different sizes of seeds. The weight of *Gossypium* sp. (Enileev and Solov'ev, 1960) and *Cicer arietinum* (Sheldrake and Saxena, 1979) seeds decreased with an increase in height of the node. However, the weight of *Glycine max* seeds was greater in the middle half than at the top or bottom one-fourth of the stem (Heindl and Brun, 1984).

Position of a seed in the fruit influences seed size in some species. In three varieties of *Phaseolus* sp. (greenbeans) (Harris, 1915) and in *Desmodium paniculatum* (Wulff, 1986a), seed weight increased with an increase in the distance from the base of the pod. In contrast, seed weight in pods of *Cladrastis* sp. containing two to four seeds decreased from the base toward the style (Harris, 1917). The basal seed in *Lupinus texensis* pods weighed less than the others; however, weights of seeds in positions 2–4 (proceeding toward the style) were not significantly different from each other (Schaal, 1980). The basal seed in pods of *Phaseolus coccineus* weighed less than the others (Rocha and Stephenson, 1990), whereas the central seed in *Strophostyles helvola* pods was heavier than those at the proximal or distal ends (Yanful and Maun, 1996).

Capsules of *Plantago coronopus* potentially can produce five seeds; four large ones in the lower part and a small one at the apex above the other four. The four lower seeds are dispersed when the capsule lid is shed, but the fifth seed is retained within the lid. The four lower seeds produce a conspicuous mucilage sheath within 2 min after touching a moist surface, but the upper seed does not produce as much mucilage (Dowling, 1933).

Seeds in the upper dehiscent part of *Cakile edentula* var. *lacustris, C. edentula*, var. *edentula*, and *C. maritima* fruits are heavier than those in the lower indehiscent part (Maun and Payne, 1989). *Raphanus raphanistrum* seeds near the base or in the middle of the fruit are heavier than those near the style (Stanton, 1984). Distal seeds are smaller than those in other positions because distal ovules are not fertilized until after those in the other parts of the fruit have been fertilized. Thus, distal seeds do not have as much time for development as the others (Mazer *et al.,* 1986).

Position of a seed in the fruit may determine its shape. *Stellaria media* seeds formed at the base of proximal capsules frequently have wings, whereas the more distal ones in the capsule do not have wings. Further, capsules borne at distal parts of the raceme are more likely not to have any winged seeds than those in proximal positions (Sterk, 1969).

4. Removal of Plant Parts

Various investigators have removed leaves, stems, inflorescences, flower buds, and young seeds and determined the effects on the size of the seeds produced by the mother plant. In defoliation studies, leaves and/ or stems are removed, whereas in predation studies, flowers, fruits, and/or seeds are removed. The reason for removing plant parts is to determine the effect damage to the mother plant by herbivores or predators has on changes in seed size.

Defoliation had no effect on seed size in *Rubus chamaemorus* (Agren, 1989), and in *Sesbania vesicaria* the size of seeds from plants with one-half or three-fourths of the leaves removed was not significantly different than controls (Marshall *et al.,* 1986). Defoliation of *Desmodium paniculatum* plants grown at 23/17°C did not change seed weights, but defoliation of plants grown at 29/23°C increased seed weight (Wulff, 1986a). Weights of seeds from partially defoliated plants of *Lotus corniculatus* were not significantly different than those of controls, but seeds from fertilized plants were significantly heavier than those from partially defoliated and nondefoliated (nonfertilized) plants (Stephenson, 1984). In a number of species, including *Arctium minus* (Reed and Stephenson, 1972), *Asclepias* spp. (Willson and Price, 1980), *Carthamus tinctorius* (Urie *et al.,* 1968), *Catalpa speciosa* (Stephenson, 1980), *Gymnocladus dioicus* (Janzen, 1976), *Heracleum* spp. (Bentley *et al.,* 1980),

Senecio jacobaea (Crawley and Nachapong, 1985), and *Sesbania macrocarpa* (Marshall *et al.,* 1986), defoliation caused a reduction in seed weight.

Predation, or pod removal, in *Glycine max* (Smith and Bass, 1972), spikelet removal in *Sorghum bicolor* (Hamilton *et al.,* 1982), and flower removal in *Rumex crispus* (Maun and Cavers, 1971) increased seed weight. Compensation also occurred in *Triticum* sp., and when developing seeds were removed the remaining ones were heavier than those produced by control plants (Bremmer and Rawson, 1978). Up to 60% of the fruits of *Brassica napus* had to be removed before plants showed a reduction in yield. However, if 80% of the fruits were removed, seed number per plant and seed weight were reduced (Williams and Free, 1979). The seed weight of *Zea mays* (B73 × MO17 and white popcorn) increased in response to removal of part of the ear, but this depended on genotype and on the time after silking when part of the ear was removed (Kiniry *et al.,* 1990).

Plants of *Mirabilis hirsuta* attacked by larvae of *Heliodines nyctaginella* (Lepidoptera), which ate flower buds, axillary buds, and leaves, produced fewer seeds than controls. Also, seeds from predated plants weighed less than those sprayed with an insecticide (Kinsman and Platt, 1984). In contrast, plants of *Banksia spinulosa* var. *neoanglica* sprayed with an insecticide showed no change in seed weight, although the number of seeds produced per plant increased 114% (Vaughton, 1990).

An indirect method of reduction in number and/or size of plant parts is the use of sublethal doses of herbicides, which stunt plant growth. Seed size was reduced when *Thlaspi arvense, Bilderdykia convolvulus,* and *Galium spurium* plants were treated with MCPA (a phenoxyacetic acid) or tribenuron-methyl (Andersson, 1994b). However, seed size was increased when plants of *Myosotis arvensis* were treated with MCPA (Andersson, 1994a).

5. Season

In a variety of species, including *Lactuca sativa* (Soffer and Smith, 1974), *Avena fatua* (Raju and Ramaswamy, 1983), *Leontodon hispidus* (Fuller *et al.,* 1983), *Amaranthus* spp., *Chenopodium album, Diplotaxis tenuifolia, Melilotus alba, Onopordum acanthium, Verbascum* spp. (Cavers and Steel, 1984), *Lomatium salmoniflorum* (Thompson and Pellmyr, 1989), *Panicum miliaceum* (Kane and Cavers, 1992), and *Scabiosa columbaria* (Rorison, 1972), seeds produced early in the growing season were heavier than those produced late in the growing season. In contrast, *Atriplex heterosperma* produces small, black seeds in early September and large, brown ones in mid-October (Frankton and Bassett, 1968). In some cultivars of *Glycine max,* early and late seed-filling periods do not result in size differences of the seeds. However, in certain cultivars with indeterminate growth, seeds that mature late in the season may be smaller than those that develop early in the season (Egli *et al.,* 1978). The time of development does not influence seed size in *Rubus chamaemorus* (Agren, 1989) or *Cajanus cajan* (Sheldrake and Narayanan, 1979).

Time of seed sowing, and thus time of seed production, may affect seed weight. The proportion of trigonous (relative to biconvex) seeds decreased with the advancement of the growing season (March, April, or May) when seeds of the summer annual *Polygonum persicaria* were sown. Month of sowing did not cause a decrease in the weight of trigonous seeds, but it caused a decrease in the weight of biconvex seeds. Month of sowing decreased the weight of *P. lapathifolium* seeds, but it did not cause changes in seed shape (Hammerton and Jalloq, 1970).

6. Soil Moisture

Seed weight was reduced when plants of *Lupinus albus* (Withers and Forde, 1979), *Triticum aestivum* (Brocklehurst *et al.,* 1978), *Glycine max* (cv. Williams) (Meckel *et al.,* 1984), and *Erodium brachycarpum* (Stamp, 1990) were subjected to drought stress. Nut weights of *Juglans major* in riparian habitats in central Arizona were correlated with spring rain. Large nuts were produced in wet years and small ones in dry years; only the large nuts were viable (Stromberg and Patten, 1990).

In contrast, plants of *Amaranthus retroflexus* subjected to continuous or temporary water stress in a greenhouse produced heavier seeds than those not receiving any drought treatment (Chadoeuf-Hannel and Barralis, 1982). Further, in eastern and central North America, the seed weight of *A. retroflexus* is greater in dry than in moist environments (Schimpf, 1977). Irrigation of 'Braxton' soybeans (*Glycine max*) caused an increase in the number of seeds, but seed weight was not increased. In fact, seed weight was greater for nonirrigated than for irrigated plants (Ramseur *et al.,* 1984).

7. Solar Irradiance and Day Length

Shading plants of *Triticum aestivum* during the endosperm cell expansion phase of seed development decreased starch accumulation, which in turn decreased seed weight (Brocklehurst *et al.,* 1978). In another study on wheat, shading for only 5–10 days at the start of anthesis slowed the accumulation of starch and lowered

final seed weight (Jenner, 1979). These results for wheat were confirmed by Martinez-Carrasco and Thorne (1979), who also found that thinning wheat plants just before anthesis increased final seed weight. Shading reduced the weight of *Asclepias verticillata* seeds but not those of *A. syriaca* (Willson and Price 1980); however, *Rubus chamaemorus* produced larger seeds in the shade than in full sunlight (Agren, 1989).

Chenopodium rubrum plants exposed to 15-hr days produced seeds that weighed less than those produced by plants exposed to 12-hr days. The time of maximum sensitivity to daylength was 6–12 days after the start of floral induction (Cook, 1975). In response to the shortening days of mid-summer, plants of *Halogeton glomeratus* produced brown seeds that weighed less than the black ones formed after mid-August (Williams, 1960). Williams and Harper (1965) suggested that day length may play a role in the development of the three types of seeds in *Chenopodium album*: (1) brown, nondormant seeds; (2) black, reticulate seeds that became nondormant with nitrate treatments but not with cold stratification; and (3) black, smooth seeds that became nondormant with nitrate treatments and partly nondormant with cold stratification.

In *Plantago lanceolata,* the effect of reduced solar irradiance during seed maturation on the subsequent germination of the progeny depended on (1) fertilizer levels during the maturation period and (2) the light regime used during the germination tests. Thus, ". . . expression of maternal genotype and environmental effects can depend on the conditions experienced by offspring . . ." (Schmitt *et al.,* 1992). Achenes from plants of *Polygonum persicaria* grown under 8 and 100% full sun weighed 1.2–1.3 and 1.5–1.7 mg, respectively, and the percentage pericarp [= (mg pericarp/mg achene × 100)] was 40.7–44.7 and 48.0–54.7, respectively (Sultan, 1996).

8. Temperature

Plants of various species have been grown over a range of temperatures, and sizes of the resulting seeds have been analyzed. Plants of *Aegilops ovata* (Datta *et al.,* 1972), *Desmodium paniculatum* (Wulff, 1986a), *Festuca arundinacea* (Bean, 1971), *F. pratensis* (Akpan and Bean, 1977), *Lactuca sativa* (Drew and Brocklehurst, 1990), *Lolium* spp. (Akpan and Bean, 1977), *Pennisetum typhoides* (Mohamed *et al.,* 1985), *Plantago aristata* (Stearns, 1960), *P. lanceolata* (Alexander and Wulff, 1985; Lacey, 1996), *Pisum* sp. (Lambert and Linck, 1958), *Poa pratensis* (Maun *et al.,* 1969), and *Triticum* sp. (Ford *et al.,* 1976; Campbell *et al.,* 1981) produced heavier seeds at low than at high temperatures. Temperatures immediately following anthesis of *Triticum* sp.

influenced the size of the resulting seeds. The heaviest seeds were produced if plants were exposed to high temperatures the first 10 days after anthesis and to low temperatures from 15 to 25 days after anthesis (Wardlaw, 1970).

Not all species produce larger seeds at low than at high temperatures. Plants of *Beta vulgaris* grown at high temperatures had a low seed yield because of a reduction in the number and size of fruits. However, *B. vulgaris* seeds formed at high temperatures were heavier than those formed at low temperatures (Wood *et al.*, 1980). Plants of *Glycine max* grown at 18/13 and at 33/28°C produced significantly smaller seeds than those grown at 24/19, 27/22, and 30/25°C; seed size was influenced by temperatures during flowering and pod set (Egli and Wardlaw, 1980). Temperature made little difference in the seed weight of *Stylosanthes humilis* plants grown over a range of temperatures, except for plants at a constant temperature of 10°C, where almost no seeds were produced (Skerman and Humphreys, 1973).

C. Structure

The environment experienced by the mother plant can cause structural changes in seeds, and in most cases these changes occur in the seed coats or other structures covering the embryo. High temperatures caused a significant increase in the mechanical resistance of the endosperm in seeds of *Syringa vulgaris* (Junttila, 1973). In contrast, seeds of *Lactuca sativa* produced at low temperatures were larger and had thicker pericarps than those matured at high temperatures. Pericarps, but not endosperm, from the low-temperature seeds had more mechanical resistance to embryo growth than those from high-temperature seeds (Drew and Brocklehurst, 1990).

Seeds of *Chenopodium polyspermum* matured under long days had low germination and thick seed coats, whereas those produced under short days had high germination and thin seed coats (Pourrat and Jacques, 1975). Also, the coats of *C. amaranticolor* seeds produced under long days were thicker than those formed under short days (Lona, 1947), and coats of *C. bonushenricus* seeds from high elevations were thicker and contained more polyphenols than those of seeds from low elevations (Dorne, 1981). Seed coats of *C. album* seeds produced under a daily 18-, 8-, or 8-hr plus 1-hr photoperiod (dark interruption) were 37, 14, and 15 μm thick, respectively. However, seeds of *C. album* produced under the 8-hr photoperiod were nondormant, and most of those produced under an 18- or 8-hr plus 1-hr photoperiod were dormant (Karssen, 1970). Thus, seed coat thickness is not correlated with dormancy. Also, seed dormancy in *Amaranthus retroflexus* was not

correlated with thickness of the seed coat (Kigel *et al.*, 1977).

As seen in Section V,C, day length can be important in the development of permeable/nonpermeable seed coats, with long days promoting the formation of impermeable seed coats (e.g., Quinlivan, 1965). Two studies on the effects of photoperiod on seed structure indicate that cuticles produced under long days are thicker and better developed than those formed under short days. In both *Ononis sicula* (Gutterman and Heydecker, 1973) and *Trigonella arabica* (Gutterman, 1978), yellow seeds with well-developed, thickened cuticles were formed under long days. Green seeds with less developed cuticle and brown seeds with no cuticle were produced under short-day conditions.

Plants of *Polygonum pensylvanicum* grown in competition with *Zea mays* produced seeds that were more dormant and had more lipid bodies in the epidermal cells of the embryo than those from plants grown without competition. Further, seeds from *P. pensylvanicum* plants grown with competition had pores in their walls, whereas those produced by competition-free plants did not (Jordan *et al.*, 1982).

D. Heritable Changes

Growth of *Linum usitatissimum* plants in soils treated with nitrogen, potassium, and phosphorus or with nitrogen and phosphorus resulted in offspring (plants) that were two to three times heavier (large) than the parents, whereas growth of plants in soils treated with only nitrogen and potassium resulted in offspring about half the size (small) of the parents. Nitrogen or phosphorus alone resulted in large and small types of offspring, respectively. These large and small plant types remained stable for six generations, and in reciprocal crosses and grafting experiments they remained genetically distinct (Durrant, 1958, 1962). Durrant (1962) called these genetically distinct forms of flax "genotrophs." In addition to plant height, other characters, including hairiness of capsule septa, amounts of nuclear DNA, ribosomal RNA content, and differences in peroxidases, esterase, and acid phosphatase isozymes, were changed by the fertilizer treatments (see review by Cullis, 1977). Bussey and Fields (1974) and Cullis (1977) constructed models showing how changes could have occurred in the DNA molecule after these fertilizer treatments were given. Fertilizer treatments also induced changes in flowering time and in final plant height in *Nicotiana rustica,* and the changes remained stable through two generations (Hill, 1965).

Although the heritable changes observed in flax and tobacco plants are not directly related to germination, it is possible that heritable changes in germination characteristics could be induced by some environmental conditions. The nutrient levels of the maternal environment resulted in different seed germination characteristics of *Plantago major* in the third generation (Miao *et al.*, 1991), and temperature-induced effects persisted into the third generation of *P. lanceolata* seeds (Case *et al.*, 1996). Selection among maternal genotypes may be one way in which effects of the maternal environment persist for more than one generation. Case *et al.* (1996) concluded that ". . . heritable GPT [grandparental temperature] effects arise from gametophytic selection or genomic modification."

VII. SEED POLYMORPHISM

The word "polymorphism" means many morphologies, so "seed polymorphism" means many sizes, shapes, or colors of seeds. However, it does not necessarily refer to the continuous range of variations in sizes, shapes, or colors of seeds found on an individual plant or in a population of a species. According to Harper *et al.* (1970), seed polymorphism is the production of two or more distinctly different types of seeds by a species. Furthermore, these types of seeds differ in their dormancy breaking and germination requirements.

A. Somatic Polymorphism

Seeds produced on the same plant may vary in size, shape, or color, and these variations may be associated with different germination responses. As seen in Section VI,B seed size, shape, and color frequently are changed by the conditions under which the mother plant is grown. Thus, the preconditioning environment is one of the major causes of seed polymorphism. Of all the preconditioning factors causing seed polymorphism, season and position on the mother plant seem to be the most likely to result in polymorphic seeds being produced on the same plant. Numerous species have seeds that are polymorphic with respect to size because of position on the mother plant or season of the year in which they were produced (Table 8.7).

In addition to those species producing polymorphic seeds on the same plant as a result of environmental factors, a number of other species produce two or more types of seeds on the same plant, but the cause is unknown (Table 8.8). The species listed in Table 8.8 have true seed polymorphism because the morphs or types of seeds differ in their germination characteristics. Depending on the species, the types vary with respect to (1) germination percentages and rates of freshly matured seeds (Kadman-Zahavi, 1955; Ellis and Ilnicki, 1968; Maurya and Ambrasht, 1973; Wulff, 1973; Nobs and

TABLE 8.8 Species with Dimorphic Seeds on the Same Plant

Species	Distinguishing characteristic	Reference
Aegilops geniculata	Color	Onnis *et al.* (1995)
Alysicarpus monilifer	Color	Maurya and Ambasht (1973)
Anthemis arvensis[a]	Size	Ellis and Ilnicki (1968)
Atriplex dimorphostegia	Shape	Koller (1957)
A. hortensis	Size, color	Nobs and Hagar (1974)
A. inflata	Color	Beadle (1952)
A. patula hastata	Color	Ungar (1971)
A. rosea	Color	Kadman-Zahavi (1955)
A. triangularis	Size	Khan and Ungar (1984b)
Bidens gardneri[a]	Shape	Felippe (1990)
B. odorata[a]	Size, shape	Corkidi *et al.* (1991)
B. pilosa[a]	Size	Rocha (1996)
Calluna vulgaris	Size, smoothness	*Helsper and Klerken (1984)*
Ceratocapnos heterocarpa	Shape	Ruiz de Clavijo (1994)
Crepis sancta[a]	Size, pappus	Imbert *et al.* (1996)
Dimorphotheca pluvialis[a]	Shape	Musil (1994)
Eremocarpus setigerus	Color	Cook (1972)
Grindelia lanceolata[a]	Shape	Baskin and Baskin (1979)
Hedypnois rhagadioloides[a]	Size, shape, pappus	Kigel (1992)
Hemizonia increscens[a]	Shape, size	Tanowitz *et al.* (1987)
Heterotheca grandiflora[a]	Shape, size	Flint and Palmbald (1978)
H. pinnatum[a]	Shape	Venable *et al.* (1995)
Hyptis suaveolens	Size	Wulff (1973)
Salsola komarovii	Color, wing size	Takeno and Yamaguchi (1991)
S. volkensii	Color	Negbi and Tamari (1963)
Senecio jacobaea[a]	Shape, size	McEvoy (1984)
Tragopogon dubius[a]	Size	Maxwell *et al.* (1994)

[a] Member of the Asteraceae.

Hagar, 1974; Takeno and Yamaguchi, 1991; Rocha, 1996), (2) light:dark requirements for germination (Beadle, 1952; Koller, 1957; Ungar, 1971; Negbi and Tamari, 1963; Corkidi *et al.*, 1991; Ruiz de Clavijo, 1994), or (3) maximum salinity at which seeds will germinate (Khan and Ungar. 1984a).

Dimorphic seeds also are found in *Dactyloctenium*

aegyptiam (Datta and Baser, 1978), *Hypochoeris glabra* (Baker and O'Dowd, 1982), and a number of other species, but studies are needed to determine if the types differ in their germination characteristics.

B. Genetic Polymorphism

Some species have two or more types of seeds, but these are produced on different plants that may grow in the same or separate populations. Both *Soliva valdiviana* and *S. pterosperma* have a range of achene sizes and shapes, and seven categories of shapes were described in a local population in New Zealand (Lovell *et al.,* 1986). Papillate and nonpapillate seeds are produced by separate plants of *Spergula arvensis,* which may grow together or in separate sites (New, 1959; Wagner, 1986). Some plants of *Senecio vulgaris* produce heads with disc and ray flowers (radiate morph), others produce only disc flowers (nonradiate morph), and some have short ray flowers (intermediate morph). These three morphs grow in mixed populations, and comparative studies of life history variations (Abbott, 1986) and germination behavior (Abbott *et al.,* 1988) of radiate and nonradiate morphs have been done in Britain. Three forms of *Valerianella umbilicata* are distinguished on the basis of fruit characteristics, which may occur together or in separate populations (Ware, 1983).

Nine forms of *Ipomoea pes-tigridis* are recognized on the basis of various morphological traits of the plants and by size, weight, color, and dormancy and germination characteristics of the seeds (Bhati and Sen, 1978). Two forms (*rubra* and *flava*) are recognized for *Trianthema portulacastrum,* and seeds of *flava* are more dormant than those of *rubra* (Mohammed and Sen, 1990). The branching growth form of *Glycine soja* has a higher percentage of dormant (impermeable) seeds than the twining growth form (Ohara and Shimamoto, 1994).

In some cases, the plants producing different kinds of seeds grow in separate populations. For example, populations of *Tephrosia purpurea* may have either short or long leaves, and seeds from plants with short or long leaves vary in their dimensions and in the length of the acid scarification treatment required to make the seed coats permeable (Mishra and Sen, 1986). Seeds from various populations of *Trianthema triquetra* differ in weight, size, color, and ability to germinate in different concentrations of NaCl or $MgSO_4$ (Mohammed and Sen, 1988). However, seeds from the green and purple biotypes of *Amaranthus hybridus* growing in different locations germinated to the same percentages at 25, 30, and 35°C (Maluf and Martins, 1991).

Many species produce polymorphic seeds on different plants under experimental conditions, with plants under varying conditions producing different seed morphs. The most common seed variation is seed size. These size differences can be the result of the environmental factors to which mother plants are exposed during seed development, and they are discussed in detail in Section VI.

C. Dimorphic Plants Whose Seeds Have Different Germination Responses

Some species have two or more types of plants. Seeds produced by the plants do not differ in size or shape, but they differ in germination responses. For example, the red biotype of *Trianthema portulacastrum* has red flowers, whereas the green biotype has white flowers. Seeds from the red biotype are less dormant and germinate faster than those from the green biotype (Rao and Reddy, 1982). Three types of plants are distinguished in *Calotropis procera* on the basis of the coronal scales in the flowers. Seeds from these three types of plants differ in degree of dormancy and in responses to high germination temperatures, pH, and soil moisture (Amritphale *et al.,* 1984). Seeds of the spring-germinating ecotype of *Galium spurium* var. *echinospermon* require cold stratification for loss of dormancy, whereas those of the autumn-germinating ecotype require warm stratification (Masuda and Washitani, 1992). Seeds from triazine-resistant plants of *Polygonum lapathifolium* germinate at lower temperatures than do those from sensitive plants (Gasquez *et al.,* 1981).

D. Mass Seed Collections and Seed Polymorphism

In some studies of seed polymorphism, seeds are collected from a large number of plants at a population site and pooled. Thus, it is impossible to tell if different types of seeds came from the same or different plants. For example, seeds of *Corynephorus canescens, Spergula vernalis,* and *Androsace septentrionalis* were collected and then sieved or measured to separate them into size classes. In all three species, small seeds germinated 1–2 days sooner than large ones, but large seeds eventually germinated to higher percentages than small ones (Symonides, 1978). Mass collections of *Hyptis suaveolens* seeds were sorted into three weight classes, and those in the small category required 10 times more photons of red light for 100% germination than those in the large category (Wulff, 1985).

Seeds of *Atriplex triangularis* were sorted into three size classes based on diameter. With an increase in temperature and salinity, the germination rate of seeds increased with an increase in seed size (Khan and Ungar, 1984b). Light brown (type I), dark brown–black (type II), and dark brown–black white-mottled (type III) seeds separated from a mass collection of *Rhynchosia capitata* required 60, 80, and 90 min of acid (concentrated H_2SO_4) scarification, respectively, for 100% germination (Sharma *et al.*, 1978). Pale violet and cream-colored seeds separated from *Capparis decidua* collections required 60 and 45 min of acid (concentrated H_2SO_4) scarification, respectively, for 33.3% germination (Paul and Sen, 1987). Yellow and yellow-mottled seeds of *Indigofera hochstetteri* required 5 and 10 min of acid (conconcentrated H_2SO_4) scarification, respectively, for 60% germination (Mishra and Sen, 1984). Class I (did not pass through a sieve with 1190-μm^2 holes; weight = 0.7 mg), class II (>1000 to <1190 μm^2; 1.3 mg), and class III (<1000 μm^2; 1.5 mg) seeds of *Heliocarpus popayanensis* germinated to 21, 72, and 76%, respectively, at 30°C (Gil, 1989). Large (3.3 mm in length), medium (2.8 mm), and small (2.4 mm) seeds from a mass collection of *Eurotia lanata* tested at 7.2, 10.6, 13.3, and 24.4/15.6°C exhibited an increase in germination at each temperature regime with an increase in seed size (Springfield, 1973).

Seeds of three varieties of *Trifolium subterraneum* collected by hand and separated into large and small size classes with sieves had slightly more large than small seeds. A higher percentage of the large than of the small seeds was permeable (Grant Lipp and Ballard, 1964). Pooled seeds from many capsules of *Nicotiana tabacum* were separated into weight classes of 76, 65, and 50 mg/1000 seeds. Germination percentages and rates in light and darkness at 20°C increased with seed weight (Kasperbauer and Sutton, 1977).

Seeds in a mass collection of *Isanthus brachiatus* were not obviously different with regard to size or color; however, they exhibited physiological differences. One group (15–35%) of seeds germinated after one cold stratification treatment, but the others required two, three, or more cold stratification treatments before they would germinate. Further, the cold stratification treatments had to be separated by several months of exposure to simulated summer temperatures before they were effective (Baskin and Baskin, 1975a). More research is needed to fully understand the germination ecology of species in which seed polymorphism is known only from mass seed collections.

E. Causes of Seed Polymorphism

Environment, genetics and environment × genotype interactions control the production of seeds with different sizes, shapes, and colors. The big question for each species becomes: What is the relative importance of each factor in determining seed variation? As discussed earlier, many environmental factors that act on the mother plant during seed development can cause seed polymorphism. The most common variation, however, seems to be in seed size, which has been documented in numerous species (Table 8.7). Thus, further discussion of the role of environmental factors in seed polymorphism is not needed, and the following material relates to the genetics of seed polymorphism.

In a number of species, polymorphism of seeds or indehiscent fruits can be attributed to heredity (Table 8.9). Seed coat characters in *Spergula arvensis* (New, 1959), fruit morphology in *Pletritis brachystemon* (Ganders *et al.*, 1977), and terminal fruit shape in *Fedia* spp. (Mathez and Xena de Enrech, 1985b) are determined by a single locus with two alleles. Crosses between pure or true types (Trow, 1912), varieties (Smith and Fitzsimmons, 1965; Voight *et al.*, 1966), clones (Hunt and Miller, 1965), genotypes (Antonovics and Schmitt, 1986), forms (Webb, 1986), strains (Murphy and Frey, 1962), inbred lines (Tedin, 1925; Leng, 1949), and cultivars (Williams and McGibbon, 1980) have demonstrated that seed size and morphology are inherited.

Narrow-sense heritability of seed size has been determined for several populations of *Capsella bursa-pastoris* (Hurka and Benneweg, 1979), for subsamples of a single population of *Lupinus texensis* (Schaal, 1980) and *Raphanus raphanistrum* (Stanton, 1984a), and Waller (1982) used data from sibling groups to derive an estimate of 26% for the combined genetic and environmental influence on seed size in *Impatiens capensis*. Fruit polymorphism in *Microseris* strain B87, which was derived from a cross between *M. pygmaea* and *M. bigelovii*, involves two, maybe three, linkage groups of genes. Two morphological markers are genetically linked to one of the genes, and the three combinations of genotypes result in recognizably different phenotypes (Backmann *et al.*, 1984).

In five genotypes of *Oenothera biennis*, progeny seed weight was positively correlated with that of the parents, and the weight of seeds of each genotype was about the same regardless of the level of competition. These data indicate that genetics rather than environmental factors control seed size (Kromer and Gross, 1987). Near-isogenic lines differing in seed size have been developed in *Glycine max* (Edwards and Hartwig, 1971), and seed size has been increased in some species as a result of several cycles of recurrent selection (Christie and Kalton, 1960; Draper and Wilsie, 1965; Fehr and Weber, 1968; Wright, 1976).

Other evidence of genetic effects on seed polymorphism comes from studies of the amount of genetic variation in populations of a species. Significant genetic variation for dimensions of the achenes was found

TABLE 8.9 Species in Which Seed (or Dispersal Unit)
Polymorphism Has a Genetic Component

Species	Distinguishing seed character	Reference
Agropyron intermedium	Size	Hunt and Miller (1965)
Anthoxanthum ordoratum	Size	Antonovics and Schmitt (1986)
Avena sativa	Size	Murphy and Frey (1962)
Bromus inermis	Size	Christie and Kalton (1960)
Camelina sativa	Size	Tedin (1925)
Capsella bursa-pastoris	Size	Hurka and Benneweg (1979)
Collinsia verna	Size	Kalisz (1989)
Crepis tectorum	Size	Andersson (1990)
Fedia spp.	Morphology	Mathez and Xena de Enrech (1985b)
Geranium carolinianum	Size	Roach and Wulff (1987)
Glycine max	Size	Fehr and Weber (1968), Edwards and Hartwig (1971)
Heterosperma pinnatum	Morphology	Venable and Burquez (1989)
Impatiens capensis	Size	Waller (1982)
Lespedeza cuneata	Size	Bates and Henson (1995)
Linum usitatissimum	Size	Smith and Fitzsimmons (1965)
Lotus corniculatus	Size	Draper and Wilsie (1965)
Lupinus albus	Size	Williams and McGibbon (1980)
L. mutabilis	Size	Williams and McGibbon (1980)
L. texensis	Size	Schaal (1980)
Microseris spp.	Morphology	Bachmann et al. (1984)
Oenothera biennis	Size	Kromer and Gross (1987)
Panicum antidotale	Size	Wright (1976)
Plectritis brachystemon	Morphology	Ganders et al. (1977)
Raphanus raphanistrum	Size	Stanton (1984a, 1985)
Senecio vulgaris	Morphology	Trow (1912)
Soliva sessilis	Morphology	Webb (1986)
Sorghum vulgare	Size	Voight et al. (1966)
Spergula arvensis	Morphology	New (1959)
Zea mays	Size	Leng (1949)

within and among six populations of *Heterosperma pinnatum* (Venable and Burquez, 1989), and isozyme patterns showed that plants of *Veronica peregrina* at the edge of vernal pools in California had significantly more variation than those in the center of the pools (Keeler, 1978). Although plants of *V. peregrina* growing in the center of the pools had larger seeds than those growing at the edges (Linhart, 1974), no attempt was made to correlate seed polymorphism with electrophoretic bands.

The heritability of seed size varies with the species: *Lupinus texensis*, 0.09 (Schaal, 1980); *Capsella bursa-pastoris*, 0.05–0.79 (Hurka and Benneweg, 1979); *Lespedeza cuneata*, 0.75–0.91 (Bates and Henson, 1955); and *Glycine max*, 0.93 (Fehr and Weber, 1968). However, in eight populations of *Plantago lanceolata*, no significant amount of heritability for seed weight was found (Primack and Antonovics, 1981).

Even in species with a significant amount of heritability for seed polymorphism, the environment also may be important in determining the final size and shape of seeds. A case in point is *Anthoxanthum odoratum* studied by Antonovics and Schmitt (1986). In this grass, about 17% of the variance in seed weight was due to maternal genetics, 3% to paternal genetics and 80% to environmental effects. Because the maternal parent × replicate (cross) effect was high for seed weight (16.5%), the environment has a strong effect on the mother plant during the time of seed development.

In addition to *Anthoxanthum odoratum* (Antonovics and Schmitt, 1986), paternal genetic effects on seed size have been found in *Crepis tectorum* (Andersson, 1990) and *Zea mays* (Leng, 1949). Heavier seeds of *Z. mays* result when eggs are fertilized by sperm carried by fast growing pollen tubes than when they are fertilized by sperm in slow-growing pollen tubes (Mulcahy, 1971).

F. Consequences on Seedling Recruitment

Seed polymorphism can be important in the recruitment of new individuals into the population. This is because different sizes and shapes of seeds (or fruits) can differ in percentages, rates, and time of germination, and the resulting seedlings can vary in size, vigor, and thus survivorship. Other aspects of reproductive biology, such as competitive ability, adult plant size, and yield, also may be influenced by seed polymorphism, but they are beyond the scope of this book.

1. Seed Size

Thompson (1990) found that the nuclear DNA content in 131 herbaceous angiosperms was positively correlated with seed size and negatively correlated with

TABLE 8.10   Effect of Seed Size on Germination under Natural or Simulated Habitat Temperatures

Species	Reference	Species	Reference
Large seeds germinate to higher percentages than small ones		*Raphanus raphanistrum*	Stanton (1984b)
		Sesbania macrocarpa	Marshall (1986)
Aegilops spp.	Maranon (1989)	*S. vesicaria*	Marshall (1986)
Anthoxanthum odoratum	Roach (1987)	*Striga asiatica*	Bebawi *et al.* (1984)
Asclepias syriaca	Morse and Schmitt (1985)	*Xanthium strumarium*	Zimmerman and Weis (1983)
Aster acuminatus	Pitelka *et al.* (1983)		
A. pilosus	Prinzie and Chmielewski (1994)	**Small seeds germinate to higher percentages than large ones**	
Beta vulgaris	Wood *et al.* (1980)	*Anthemis arvensis*	Ellis and Ilnicki (1968)
Bidens odorata	Corkidi *et al.* (1991)	*Arctium minus*	Reed and Stephenson (1972)
Cercocarpus montanus	Piatt (1973)	*Erodium brachycarpum*	Stamp (1990)
Crepis tectorum	Andersson (1990)	*Festuca pratensis*	Akpan and Bean (1977)
Cyperus spp.	Thullen and Keeley (1979)	*Glycine max*	Hopper *et al.* (1979), Edwards and
Desmodium paniculatum	Wulff (1986b)		Hartwig (1971)
Festuca pratensis	Akpan and Bean (1977)	*Lolium* spp.	Akpan and Bean (1977)
Garcinia mangostana	Hume and Cobin (1946)	*Medicago sativa*	Cooper *et al.* (1979)
Gossypium sp.	Enileev and Solov'ev (1960)	*Oryza sativa*	Krishnasamy and Seshu (1989)
Helianthus annuus	Radford (1977)	*Pennisetum typhoides*	Mohamed *et al.* (1985)
Leucaena leucocephala	Pathak *et al.* (1973)	*Rumex crispus*	Maun and Cavers (1971)
Leymus arenarius	Greipsson and Davy (1995)	*Sesbania drummondii*	Marshall (1986)
Linum usitatissimum	Harper and Obeid (1967)		
Lolium perenne	Naylor (1980)	**Germination is independent of seed size**	
Ludwigia leptocarpa	Dolan (1984)	*Dactylis glomerata*	Bretagnolle *et al.* (1995)
Mirabilis hirsuta	Weis (1982)	*Diplotaxis erucoides*	Perez-Garcia *et al.* (1995)
Panicum maximum	Mejia *et al.* (1978)	*D. virgata*	Perez-Garcia *et al.* (1995)
Pennisetum typhoides	Mohamed *et al.* (1985)	*Medicago sativa*	Beveridge and Wilsie (1959)
Pinus jeffreyi	Munns (1921)	*Oenothera biennis*	Gross and Kromer (1986)
Pogogyne abramsii	Zammit and Zedler (1990)	*Rumex obtusifolius*	Cideciyan and Malloch (1982)
Prunella vulgaris	Winn (1988)	*Zea mays*	Eagles and Hardacre (1979)
Quercus spp.	Tripathi and Khan 1990)		

the mean temperatures at which seeds of a given species would germinate. However, under natural or simulated habitat temperatures, large seeds of many species germinate to higher percentages than small ones (Table 8.10). However, small seeds of some species germinate to higher percentages than large ones. In other species, germination is independent of seed size (Table 8.10).

Salt tolerance for germination of *Atriplex triangularis* seeds increases with seed size (Khan and Ungar, 1984), and in *Hyptis suaveolens* the light requirement decreases with seed size (Wulff, 1973). The radicle emerges first when large seeds of *Taraxacum hamatiforme* germinate, but the cotyledons emerge first when small ones germinate (Mogie *et al.*, 1990). Seed size is important in determining the soil depth from which seedlings can emerge, with seedlings from large seeds being able to emerge from greater depths than those from small seeds (Strickler and Wassom, 1963; Harper and Obeid, 1967; Radford, 1977). Although no data for seed weights were presented, germination decreased as the number of seeds in burs of *Cenchrus ciliaris* increased (presumably decrease in seed size) from one to four (Hacker and Ratcliff, 1989). Depending on the year and type of seed (curved or straight), defoliation of *Arctium minus* reduced seed weight and increased germination percentages of cold stratified and of nonstratified seeds (Reed and Stephenson, 1972). Small seeds of *Zea mays* subjected to water stress in mannitol solutions germinated to higher percentages than large ones (Muchena and Grogan, 1977).

Large seeds of *Bidens odorata* (Corkidi *et al.*, 1991), *Impatiens capensis* (Waller, 1985), *Mirabilis hirsuta* (Weis, 1982), *Panicum virgatum* (Zhang and Maun, 1991), *Peltophorum inerme* (Anoliefo and Gill, 1992), and *Xanthium strumarium* (Zimmerman and Weis, 1983) germinated faster than small ones. However, germination was faster in small than in large seeds of *Cakile edentula* (Zhang, 1993), *Leymus arenarius* (Greipsson and Davy, 1995), *Rumex obtusifolius* (Cideciyan and Malloch, 1982), and *Solidago flexicaulis* (Chmielewski et al., 1989). Small seeds of *Erodium brachycarpum* became permeable faster than large ones; therefore, they are likely to germinate the year in which they are produced, with germination of large seeds being delayed until a subsequent year (Stamp, 1990). Seeds of *Desmodium paniculatum* (Wulff, 1986a), *Pennisetum typhoides* (Mohamed et al., 1985), *Lolium* spp., and *Festuca pratensis* (Akpan and Bean, 1977) produced at high temperatures weighed less and had higher germination rates and percentages than those produced at low temperatures.

Very small seeds of *Glycine max* were slower to germinate than small, medium, and large ones when incubated on wet paper towels; however, very small seeds had the fastest germination rate when seeds of all sizes were incubated on wet sand (Hopper et al., 1979). At soil moisture levels of 22.5, 25, 27.5, and 30%, small- and medium-sized seeds of *G. max* emerged from soil more rapidly than large ones (Edwards and Hartwig, 1971). Large and small seeds from hermaphroditic and male sterile flowers of *Plantago coronopus* have similar germination percentages at water potentials of −4 and −0.6 MPa, but small seeds from both types of flowers germinated faster than large ones (Schat, 1982).

Large seeds of *Quercus* spp. (Tripathi and Khan, 1990) and *Ludwigia leptocarpa* (Dolan, 1984) had earlier emergence dates than small ones, which probably was related to faster germination rates in large than in small seeds. However, in *Lolium perenne* (Naylor, 1980) and *Raphanus raphanistrum* (Stanton, 1984b), seed size and time to emergence were not correlated. Seed weight and mating system affected the emergence date for *Collinsia verna*. Outcrossed seeds were heavier and emerged earlier than selfed seeds (Kalisz, 1989). Small seeds of *Crepis tectorum* were less likely to germinate in autumn than large ones (Andersson, 1996).

2. Seed Morphology

Germination differences also occur in species whose seeds are dimorphic with respect to shape. At 20°C, dispersal units of *Atriplex dimorphostegia* germinated to 32 and 68% in light and darkness, respectively, whereas humped dispersal units germinated to only 6 and 38%, respectively. However, at 26°C, the germination of both types was inhibited by light and promoted by darkness (Koller, 1957). Seven categories of achene shapes have been described in *Soliva valdiviana* and *S. pterosperma*, and the hierarchy of germination rates in these seven categories was 1 to 3 > 4 to 6 > 7 (Lovell et al., 1986). In England, papillate seeds of *Spergula arvensis* germinated to higher percentages than the nonpapillate ones at 21°C, whereas the reverse was true at 13°C (New,

1958). In California, however, germination responses of *S. arvensis* seeds to temperature were not closely correlated with seed morphology (Wagner, 1988). A higher percentage of *Valerianella umbilicata* forma *intermedia* seeds (fruits) afterripened during the summer dormancy-breaking period and germinated in light and darkness in autumn as compared to those of *V. umbilicata* forma *patellaria* (Baskin and Baskin, 1976a).

Disc achenes of *Heterotheca latifolia* are less dormant than ray achenes, and thus the former have the potential to germinate and establish seedlings first (Venable and Levin, 1985a). Further, survival is higher in seedlings derived from disc than in those from ray achenes (Venable and Levin, 1985b). In *H. pinnatum,* disc achenes become nondormant faster than ray achenes, and thus they germinate sooner; however, early germination may result in high percentages of seedling death (Venable et al., 1987). Achenes from nonradiate plants of *Senecio vulgaris* are less dormant than those from radiate plants; consequently, more seedlings are established from nonradiate than from radiate achenes in autumn. However, the reverse is true in spring (Abbott et al., 1988). Ray achenes of *Galinsoga parviflora* have higher germination percentages and rates than disc achenes. Seedlings from ray achenes had higher survival rates than those from disc achenes at low and at medium levels of nitrogen, phosphorus, and potassium, whereas seedlings from disc achenes had higher survival rates at high levels of nitrogen, phosphorus, and potassium (Rai and Tripathi, 1987).

3. Seed Color

Seeds of different colors may have different germination requirements. Light-colored seeds of *Atriplex inflata* (Beadle, 1952) and *A. patula* var. *hastata* (Ungar, 1971) were less dormant than dark-colored ones, and brown seeds of *A. rosea* (Kadman-Zahavi, 1955) and *A. hortensis* (Nobs and Hagar, 1974) were less dormant than black ones. However, freshly matured brown seeds of *Alysicarpus monilifer* germinated to higher percentages than yellow ones (Maurya and Ambasht, 1973), and freshly matured brown seeds of *Senna obtusifolia* germinated to higher percentages than green ones (Nan, 1992). Freshly matured, green-colored dispersal units of *Salsola volkensii* germinated to high percentages in light and darkness, but yellow ones germinated to only low percentages in light (Negbi and Tamari, 1963). Green-colored fruits of *Aellenia autrani* were less dormant at 2 months of age than yellow ones, but after 10 months of dry storage, the germination of yellow fruits increased. The germination of green fruits declined after 10 months due to a loss of viability (Negbi and Tamari, 1963).

4. Seedling Size

In many species, large seeds give rise to seedlings that are larger than those produced by small seeds (Table 8.11). This makes good sense because a large seed has greater a store of reserve energy that can be used by the seedling in the early stages of growth than does a small one. Thus, seedlings from large seeds may have greater initial growth rates than those from small seeds. However, seedlings from small seeds of *Plantago aristata* produced at 21.1°C grew faster than those from large seeds produced at 15.6°C (Stearns, 1960). Also, *Pinus elliottii* seedlings from medium-sized seeds had a higher growth rate than those from either small or large seeds (Shoulders, 1961). Although relative growth rates of plants from different sizes of seeds of *Raphanus raphanistrum* were the same, plants from large seeds were three times larger than those from small seeds (Choe *et al.*, 1988). Thus, differences in adult plant sizes in this species are attributed to differences in the amount of

food reserves stored in the seeds (Stanton, 1985). Plants from different sizes of *Trifolium subterraneum* (Black, 1956) and of *Desmodium paniculatum* (Wulff, 1986b) seeds also grew at about the same rate, but these studies were terminated after only 31 and 14 days, respectively. Seed size and seedling vigor relationships were not significant in *Medicago sativa* or *Onobrychis viciaefolia* (Carleton and Cooper, 1972). Seed size can have an effect on seedling growth in noncompetitive but not in competitive cover (Gross, 1984).

It has been observed that the survivorship of seedlings from large seeds may be higher than that of seedlings from small seeds. Seedling survivorship increased in *Garcinia mangostana* with each increase in seed weight up to 1.3 g, after which seed size made no difference (Hume and Cobin, 1946). Regardless of cone size, seedlings of *Pinus elliottii* from large seeds had higher percentages of survival than those from medium-sized seeds. (Survivorship of seedlings from small seeds was not determined.) Large seeds from small cones (7.5 cm long) produced seedlings with higher percentages of survival than did large or small seeds from cones greater than 7.5 cm long (Langdon, 1958). Seedlings from large seeds of *Asclepias syriaca* had higher percentages of survival than those from small seeds (Morse and Schmitt, 1985). However, in a field study of seedling survival in *Desmodium paniculatum*, a higher percentage of seedlings from small than from large seeds was alive at the end of the growing season in both a path and a forest habitat (Wulff, 1986b). Seeds produced by plants of *Senecio vulgaris* grown at low nutrient levels were smaller than those produced by plants grown at high nutrient levels. Further, seedlings from small seeds were smaller than those from large seeds, and seedlings from small seeds survived longer in nutrient-poor soil than those from large seeds (Aarssen and Burton, 1990). Seed (nutlet) size in *Lithospermum caroliniense* had no effect on seedling survival (Weller, 1985).

The consequences of seed polymorphism on the fitness of species have not gone unnoticed by theoretical biologists, and much work has been done on various aspects of the evolutionary ecology of species that produce polymorphic seeds. Mechanisms for the evolution of seed polymorphism have been discussed (Westoby, 1981) and modeled (Silvertown, 1984; Venable, 1985b), and the role of this characteristic in the evolution of plant life histories has been considered (Silvertown, 1981; Venable, 1984). Multiple germination responses are viewed as a bet-hedging strategy (Westoby, 1981; Silvertown, 1985) or as an adaptation to reduce risk in a variable environment (Venable and Brown, 1988). Also, much thought has been given to optimal seed sizes for maximum fitness (Capinera, 1979; McGinley, 1989).

TABLE 8.11   Species In Which Large Seeds Give Rise to Larger Seedlings Than Do Small Seeds

Species	Reference
Abutilon theophrasti	Wulff and Bazzaz (1992)
Apium graveolens	Thomas *et al.* (1979)
Arabidopsis thaliana	Krannitz *et al.* (1991)
Asclepias syriaca	Morse and Schmitt (1985)
Atriplex triangularis	Ellison (1987)
Brassica spp.	Major (1977)
Buchloe dactyloides	Kneebone (1960)
Coreopsis lanceolata	Banovetz and Scheiner (1994)
Dactylis glomerata	Bretagnolle *et al.* (1995)
Desmodium paniculatum	Wulff (1986b)
Garcinia mangostana	Hume and Cobin (1946)
Hordeum vulgare	Boyd *et al.* (1971)
Linum usitatissimum	Harper and Obeid (1967)
Lithospermum caroliniense	Weller (1985)
Lotus corniculatus	Twamley (1967)
Medicago sativa	Beveridge and Wilsie (1959)
Oenothera biennis	Kromer and Gross (1987)
Oryza sativa	Krishnasamy and Seshu (1989)
Panicum miliaceum	Moore and Cavers (1985)
Pinus elliottii	Langdon (1958)
Raphanus raphanistrum	Stanton (1984b, 1985)
Sabal palmetto	Moegenburg (1996)
Trifolium subterraneum	Black (1956, 1957)
Triticum aestivum	Evans and Bhatt (1977)
Virola surinamensis	Howe and Richter (1982)

VIII. CLEISTOGAMY AND SEED GERMINATION ECOLOGY

A. Cleistogamous Flowers

Cleistogamy means the formation of flowers that do not open and which produce viable seeds as a result of autogamy. In contrast, flowers that open are called chasmogamous (CH) flowers and have various types of breeding systems, depending on the taxon. In a review of cleistogamy, Lord (1981) divided cleistogamous (CL) species into four categories. (1) *Preanthesis cleistogamy* refers to species in which pollination occurs during the bud stage, but the flower eventually opens. (2) In species with *pseudocleistogamy,* closed and open flowers are morphologically similar; however, those that fail to open are self-pollinated. Environmental stress may be the reason some flowers do not open, and Uphof (1938) listed a number of environmental factors, including flooding, drought, low and high temperatures, shading, and burial, that promote the formation of closed flowers. However, he did not say if the CL and CH flowers in the various species were morphologically similar or different. (3) *Complete cleistogamy* means that a species only produces CL flowers; this occurs in the orchids *Bulbophyllum dischidiifolium, Dendrobium chryseum, D. cleistogamum,* and *D. clausum* (Uphof, 1938) and in the grasses *Sporobolus subinclusus, Tetrapogon spathaceus* (Uphof, 1938), *Pheidochloa* sp., and *Briza* spp. (Connor, 1979). (4) In *true cleistogamy,* CL and CH flowers are morphologically different; these dimorphic flower are found in at least 29 families (Table 8.12). CL flowers are modified CH flowers with a reduced corolla and androecium (Lord, 1981), and the small size of CL flowers is due to a reduction in cell sizes and numbers (Mayers and Lord, 1984). In some species, e.g., *Commeliantia pringlei* (Parks, 1935), CH, CL, and transitional flowers (which are morphologically intermediate between CH and CL flowers) are formed.

Schoen and Lloyd (1984) have developed a basic model for the evolution of cleistogamy that predicts self-pollination in obligately closed flowers with small corollas and androecia. The idea is that resources saved when reduced flower parts are formed are used by the mother plant to make more and/or heavier seeds. CL flowers have a high pollen to ovule ratio, which results in "saved resources" for production of seeds.

Species with dimorphic flowers vary in position on the plant where CL flowers are formed. CH and CL flowers can be in the same (e.g., Wilken, 1982) or in different inflorescence(s) on the same plant. In the latter situation, inflorescences of CL flowers may be borne on the upper portion of the plant, where they are easily seen (e.g., Harlan, 1945), or they may be at ground level and hidden by leaves (Connor, 1979). These hidden or clandestine inflorescences occur in at least 10 genera of grasses (Campbell *et al.,* 1983) and are called cleistogenes. According to Chase (1918), cleistogenes have one to several florets (flowers), and the glumes are reduced or absent, depending on the species.

Some species are amphicarpic, which means that CH flowers are produced on aerial stems and CL flowers on subterranean stems (Table 8.13). However, plants of *Amphicarpaea bracteata* produce subterranean CL flowers and both aerial CL and CH flowers (Schively, 1904; Allard, 1932). Amphicarpy has been reported in *Trianophiles solitaria* (Cyperaceae), but the CL flowers are at ground level rather than being covered by soil (Haines and Lye, 1977). Thus, the so-called amphicarpic flowers in *T. solitaria* are very similar to the cleistogenes described in the Poaceae.

B. Causes of Cleistogamy

The occurrence of CL flowers in a species is, at least in part, genetically controlled. From 50 to 102 genes control the production of CH flowers in *Lespedeza cuneata,* and heritability of this character is 36% (Bates and Henson, 1955). Further, the ratio of CH to CL flowers (seeds) in *L. cuneata* has a heritability of 82.7% (Cope and Moll, 1969). Broad-sense heritability of CL flowers on *Danthonia spicata* plants was estimated to be 52.6% in the field and 71.6% in the greenhouse (Clay, 1982). Cleistogamy is genetically controlled in *Ruellia tuberosa, R. nudiflora,* and *R. brittoniana* (Long, 1977) and has a genetic component in *Lamium amplexicaule* (Correns, 1926, 1930). Cleistogamy was present in hybrids between *Antirrhinum glutinosum* and *A. majus* and involves several genes, some of which act as modifiers (Mather and Vines, 1951). Three different types of closed flowers have been identified in *Carthamus*

TABLE 8.12 Families with Dimorphic (Cleistogamous and Chasmogamous) Flowers[a]

Acanthaceae	Gesneriaceae	Poaceae
Asteraceae	Hydrocharitaceae	Polemoniaceae
Balsaminaceae	Juncaceae	Polygalaceae
Boraginaceae	Lamiaceae	Polygonaceae
Brassiaceae	Lentibulariaceae	Pontederiaceae
Campanulaceae	Malpighiaceae	Rubiaceae
Cistaceae	Malvaceae	Scrophulariaceae
Commelinaceae	Nyctaginaceae	Solanaceae
Fabaceae	Oxalidaceae	Violaceae
Gentianaceae	Plantaginaceae	

[a] From Lord (1981)

TABLE 8.13 Species That Produce Subterranean,
Cleistogamous and Aerial Chasmogamous
Flowers (Amphicarpy)

Species	Reference
Amphicarpaea bracteata[a]	Schively (1904), Allard (1932), Schnee and Waller (1986)
Amphicarpum purshii	Cheplick and Quinn (1982)
Cardamine chenopodifolia	Cheplick (1983)
Catananche lutea	Plitmann (1973)
Chloris chloridea	Connor (1979)
Commelina benghalensis	Maheshwari and Maheshwari (1955)
C. forskalaei	Maheshwari and Maheshwari (1955)
C. virginica	Uphof (1934)
Commelinantia pringlei	Parks (1935)
Emex spinosa	Evenari et al. (1977), Weiss (1980)
Fleurya podocarpa	Engler (1895)
Gymnarrhena micrantha	Koller and Roth (1964)
Lathyrus amphicarpos	Mattatia (1977b)
L. ciliolatus	Mattatia (1977b)
L. setifolius var. amphicarpos	Mattatia (1977b)
Paspalum amphicarpum	Connor (1979)
Pisum fulvum	Mattatia (1977a)
Polygala polygama	Shaw (1904)
Vicia sativa subsp. amphicarpa	Plitmann (1973)

[a] Produces both CH of CL aerial flowers

tinctorius, and each is controlled by a single gene (Dille and Knowles, 1975). The inheritance of cleistogamy is sex linked in section novorbis of Plantago, and individuals with hermaphroditic CL flowers produce offspring with CL flowers only (Schurhoff, 1924). Cleistogamy occurs in sorghum when the genes *py* and *gx* are present (Ayyangar and Ponnaiya, 1939).

CH and CL lines have been developed in Salpiglossis sinuata through selfing and selection. The CH line produces only CH flowers, whereas the CL line has both CL and CH flowers. In a series of crosses between the two lines, it was determined that CH plants are homozygous recessive (cc) and that CL plants are either homozygous dominant (CC) or heterozygous (Cc). Homozygous plants had a higher percentage of CL flowers than heterozygous plants, especially while plants were young. As plants age, the percentage of CL flowers increases in heterozygous plants. Modifier genes also may be involved in cleistogamy of this species, as excessively old chasmogamous (cc) plants sometimes produce CL flowers (Lee et al., 1976).

Environmental factors alone, or in combination with

genetics, also are important in determining the presence of CL flowers. As noted by Lord (1981), when species with the potential to produce dimorphic flowers are grown under natural conditions, both CH and CL flowers are formed at some time during the flowering season. For example, in Ruellia tuberosa, a tropical species, flowers are cleistogamic in the vernal, chasmogamic in the aestival, and cleistogamic in the serotinal period (Long, 1977). Viola odorata produced CH flowers in England from late September to March, semi-CL flowers in April and May, CL flowers from May to September, and semi-CL flowers in mid-September (Madge, 1929). In Tennessee, Viola egglestonii produced CH flowers from March to May and CL flowers from May to early November (Baskin and Baskin, 1975b). Changes in environmental conditions during the course of the flowering period may explain the switch in the type of flower produced.

Photoperiod is one of the most important environmental factors causing plants to stop forming one type of flower and to start producing another. Long days promote the formation of CL flowers and short days promote the production of CH flowers in a variety of species, including Viola cunninghamii (Holdsworth, 1966), V. palustris (Borgstrom, 1939; Evans, 1956), V. odorata (Borgstrom, 1939; Cooper and Watson, 1952; Mayers and Lord, 1983), V. biflora, V. canina, V. hirta, V. mirabilis, V. riviniana, V. silvestris (Borgstrom, 1939), and Bromus unioloides (Langer and Wilson, 1956). In contrast, short days promote formation of CL flowers, and long days promote formation of CH flowers in various species, including Falcata comosa (= Amphicarpaea bracteata) (Allard, 1932), Botriochloa decipiens (Heslop-Harrison, 1961), and Lamium amplexicaule (Lord, 1982). Day length could be the reason Mirabilis nyctaginea plants produce CH flowers during summer and CL flowers in autumn (Cruden, 1973).

In Impatiens biflora (= I. capensis) and I. pallida, increasing illuminance increased the proportion of CH flowers, whereas decreasing illuminance increased the percentage of CL flowers (Schemske, 1978; Waller, 1980). If subterranean CL flower buds of Commelinantia pringlei are placed in light, the reduced corolla opens, but the stamens and pistil are like those found in CL flowers (Parks, 1935). Subterranean flower buds of Commelina forskalaei exposed to direct sunlight produce CH flowers (Maheshwari and Maheshwari, 1955), and young stolons of Vicia sativa subsp. amphicarpa exposed to light produce CH flowers rather than CL flowers (Plitmann, 1973).

Water stress decreased CH and increased CL flowers in Stipa leucotricha (Brown, 1952), Impatiens capensis (Waller, 1980), and Collomia grandiflora (Minter and Lord, 1983). Treatment with abscisic acid also decreased CH and increased CL flower production in C. grand-

iflora. When drought-stressed plants of *C. grandiflora* were sprayed with GA_3, CH flowers were formed with corolla sizes intermediate between those of normal CH and CL flowers (Minter and Lord, 1983).

Increased plant density in natural populations of *Lithospermum caroliniense* was correlated strongly with an increase in the percentage of cleistogamic plants (Levin, 1972). Plant injury increased the production of CL flowers in *Impatiens capensis* (Waller, 1980), and grazing stimulated the production of cleistogams in *Stipa leucotricha* (Dyksterhuis, 1945) and *Danthonia* spp. (Clay, 1983a). CL flowers are formed near the base of the cymose inflorescence of *Collomia grandiflora*, whereas CH flowers are numerous in the mid and top portions of the inflorescence (Ellstrand *et al.*, 1984).

The formation of subterranean CL flowers (and fruits) in *Gymnarrhena micrantha* follows a "pessimistic strategy," i.e., the number of fruits is independent of plant weight (Zeide, 1978). However, the production of CH flowers (and fruits) in this species follows an "optimistic strategy," i.e., the number of fruits increases with an increase in plant weight.

C. Seed Size

Subterranean CL seeds are heavier than aerial CH seeds in a number of species, including *Cardamine chenopodifolia* (Cheplick and Quinn, 1982; Cheplick, 1989), *Catananche lutea* (Ruiz de Clavijo, 1995), *Emex spinosa* (Evenari *et al.*, 1977), *Gymnarrhena micrantha* (Koller and Roth, 1964), and *Vicia sativa* subsp. *amphicarpa* (Plitmann, 1973). In *Commelina forskalaei*, the subterranean CL fruit forms a single seed that is about the size of the largest seed in the two-seeded aerial CH fruit. Subterranean fruits of *C. benghalensis* are three-seeded, and the seed in the indehiscent part of the fruit is larger than the two seeds in the dehiscent portion. Aerial CH fruits are five-seeded, and the seed in the indehiscent part of the fruit is larger than the four in the dehiscent portion (Maheshwari and Maheshwari, 1955; Maheshwari, 1962). In *Amphicarpaea bracteata*, subterranean CL seeds are larger than the aerial CL and CH ones (Schnee and Waller, 1986).

Seeds from CH flowers are larger than those from CL flowers in *Danthonia compressa*, *D. spicata*, *D. sericea* (Clay, 1983a), and *Impatiens capensis* (Schemske, 1978; Waller 1982), but seeds from CL flowers of *Viola sororia* are larger than those from CH flowers (Solbrig, 1981). CL and CH seeds are the same size in *Collomia grandiflora* (Wilken, 1982) and in some populations of *I. capensis* (Schemske, 1978; Waller, 1982). In some habitats, CL seeds of *D. spicata* may be the same size (Weatherwax, 1928) or even larger (Chase, 1918) than CH seeds.

Various factors can influence the weight of CL and CH seeds. In some populations of *Impatiens capensis* (Waller, 1982) and in *Collomia grandiflora* (Wilken, 1982), CL and CH seeds weighed about the same. However, if the number of seeds in either CL or CH fruits decreased, seed weight increased (Wilken, 1982). The weight of CL seeds of *I. capensis* increased with an increase in the height on the stem on which seeds were produced and with proximity to the main stem (Waller, 1982). Weights of subterranean CL and aerial CH seeds of *Amphicarpum purshii* increased when plants were treated with a fertilizer containing nitrogen, phosphorus, and potassium (Cheplick, 1989). Finally, it should be noted that the weight of CL seeds in *I. capensis* (Waller, 1982) and *Lespedeza cuneata* (Bates and Henson, 1955) has a genetic component.

D. Seed Germination

Because seeds from CL and CH flowers are produced under different environmental conditions and/or at different positions on the mother plant, it is reasonable to expect them to differ in their germination responses. This expectation is fulfilled in many species, and we will begin our survey of these differences by discussing species with amphicarpy.

Seeds from subterranean CL flowers of *Emex spinosa* growing in Israel were more dormant than those from aerial CH flowers. Scarification broke the dormancy of CL seeds, and scarification or leaching improved germination percentages of CH seeds. Scarified CL seeds germinated to higher percentages in darkness than in light over a range of constant temperatures of 5 to 25°C, and scarified CH seeds germinated equally well at all temperatures in light and darkness. Leached CH seeds germinated to higher percentages in darkness than in light at temperatures from 5 to 20°C, but the reverse was true at temperatures of 25 to 35°C (Evenari *et al.*, 1977). In Australia, CL seeds of *E. spinosa* were less dormant than CH ones, and dormancy was overcome in both types by scarification followed by the addition of gibberellin and kinetin. Optimum temperatures for the germination of nondormant seeds were from 18/13 to 30/25°C for CL seeds and from 18/13 to 27/22°C for CH seeds. CH seeds required light to germinate to high percentages, but CL seeds did not (Weis, 1980).

Subterranean CL and aerial CH seeds of *Gymnarrhena micrantha* germinated to higher percentages in light than in darkness. In light, both types of seeds germinated to 80% or more at constant temperatures of 5 to 20°C; however, in darkness, only CH seeds germinated to ≥80% but only at 5 and 10°C (Koller and Roth, 1964). Subterranean CL seeds of *Amphicarpum purshii* were less dormant than aerial CH seeds, and cold stratification at 4°C improved germination of both types. However, even after 90 days of cold stratification, CL seeds germinated to higher percentages than CH seeds at 25/15, 30/20, and 35/20°C (McNamara and Quinn, 1977).

Large and small subterranean CL and large and small aerial CH seeds of *Commelina benghalensis* germinated to 90, 33, 20, and 0%, respectively. Scarification caused 97–100% of each of the four types of seeds to germinate at a 12/12-hr photoperiod. However, seeds germinated to 77, 66, 82, and 82%, respectively, in continuous light and to 28, 7, 55, and 66%, respectively, in continuous darkness (Walker and Evenson, 1985b). Large CH seeds of *C. benghalensis* were more responsive to heat, acid, and chlorox treatments than small CH seeds (Kim *et al.*, 1990), and CH seeds of this species were more responsive to scarification than CL seeds (Budd *et al.*, 1979)

Freshly matured subterranean CL seeds of *Amphicarpaea bracteata* germinated to 95–100%, whereas aerial CL and CH seeds germinated to only 20 and 15%, respectively (Schnee and Waller, 1986). In another study, subterranean CL, aerial CL, and CH seeds germinated to 86, 7, and 9%, respectively (Trapp and Hendrix, 1988). It appears that the subterranean CL seeds are nondormant, whereas many of the aerial CL and CH seeds have impermeable seed coats (physical dormancy).

Seeds from CL and CH flowers of nonamphicarpic species may differ in their germination responses. CL and CH seeds of *Danthonia unispicata* (Dobrenz and Beetle, 1966) and *Impatiens capensis* (Waller, 1984; Antlfinger, 1986) germinated to about the same percentages, whereas CL seeds of *Danthonia spicata* (Dobrenz and Beetle 1966) and *Dichanthelium clandestinum* (Bell and Quinn, 1985) germinated to higher rates and percentages than CH seeds. CL seeds of *Stipa leucothricha* generally germinated to higher percentages than CH ones at 20/10, 25/15, and 30/20°C at 0 and −0.25 MPa; however, 20/10°C and 0 MPa were the optimum germination conditions for both types of seeds (Call and Spoonts, 1989). Although CL seeds of *Viola egglestonii* germinated to higher percentages than CH seeds under natural temperatures, both types had the same germination phenology (Baskin and Baskin, 1975). In a burial study of CL and CH seeds of *V. rafinesquii*, CH seeds entered dormancy in early winter, whereas CL seeds did not enter dormancy until late winter–early spring (Baskin and Baskin, unpublished results). Solbrig (1981) apparently used only CL seeds in his germination studies of *Viola sororia*, in which he found genetic differences between families in germination rates and percentages.

Seeds from CL flowers of *Sporobolus neglectus, S. ozarkanus,* and *S. vaginiflorus* are either large or small, depending on position on the mother plant. Each short branch of the inflorescence has two pedicellate spikelets, and the distal one is larger than the proximal one. The large distal seeds became nondormant during 12 weeks of cold stratification, whereas only about half the small seeds were nondormant after 12 weeks at 5°C. However, dormancy was broken in all small seeds after 24 weeks of cold stratification (McGregor, 1990).

E. Consequences on Seedling Recruitment

In some species, differences in the sizes of young seedlings are related to whether the seeds were produced by CL or CH flowers. However, in many of these species, close observations reveal that one seed type is larger than the other, which may be the real explanation for the difference in sizes of the seedlings. In *Lespedeza cuneata* (Cope, 1966) and *Impatiens capensis* (Waller, 1984, 1985), CH seeds are larger than CL ones, and seedlings from CH seeds are larger than those from CL seeds. Further, plants from CH seeds of *I. capensis* had higher percentages of survival and were competitively superior to those from CL seeds (Waller, 1984). In greenhouse studies, however, the growth of CH and CL seedlings was similar (Antlfinger, 1986).

In Indiana, there were no consistent differences in sizes of CL or CH seeds of *Danthonia spicata,* and they germinated to similar percentages. Further, plants monitored from seedling to flowering stages did not differ in their appearance or vigor (Weatherwax, 1928). In contrast, in North Carolina, CL seeds of *D. spicata* were larger than CH ones, and in a seedling survival experiment conducted in North Carolina seedlings from CL seeds were larger than those from CH seeds (Clay, 1983b). In a later study carried out in North Carolina, seedling differences did not persist until the adult stage of the life cycle (Clay and Antonovics, 1985a).

In amphicarpic species such as *Amphicarpum purshii* (McNamara and Quinn, 1977; Cheplick and Quinn, 1982), *Cardamine chenopodifolia* (Cheplick, 1983), *Emex spinosa* (Evenari *et al.*, 1977), and *Gymnarrhena micrantha* (Koller and Roth, 1964), large subterranean CL seeds produce larger seedlings than small aerial CH seeds, and large seedlings have a greater chance of survival than small ones. Seedlings from subterranean CL seeds of *Amphicarpaea bracteata* were many times heavier than those from aerial CL or CH seeds (Trapp and Hendrix, 1988). Subterranean achenes of *Catananche lutea* produced larger seedlings than aerial ones, but seedlings from aerial achenes had higher growth rates than those from subterranean ones (Ruiz de Clavijo, 1995).

Plants from CL seeds of *Amphicarpum purshii* (McNamara and Quinn, 1977; Cheplick and Quinn, 1982), *Cardamine chenopodifolia* (Cheplick, 1983), and

Emex spinosa (Evenari *et al.*, 1977) flowered sooner than those from CH seeds. Plants from aerial CH seeds of *Commelina benghalensis* flowered sooner than those from subterranean CL seeds, and the plants from both CL and CH seeds produced more CH than CL seeds (Walker and Everson, 1985a). CL and CH seeds of *Calathea micans* are the same size, and the germination (and seedling size) of seeds sown in boxes of soil did not differ. In the field, more CH than CL seedlings became established in the understory, whereas there was some decrease in CH (but not CL) establishment in gaps (Le Corff, 1996).

CL and CH seeds of *Dichanthelium clandestinum* are about the same size (Campbell *et al.*, 1983), but germination rates and percentages of CL seeds were greater than those of CH seeds (Bell and Quinn, 1985). Plants from each seed type grown in monocultures in a greenhouse showed no difference in fitness; however, in a low density mixture (260 seedlings m^{-2}), plants from CH seeds were more vigorous than those from CL seeds. At densities of 650 and 1300 seedlings m^{-2}, however, there were no differences in the fitness of plants from the two seed types (Bell and Quinn, 1985).

Because seed size can make a big difference in seedling vigor, it is difficult to determine if there are genetic differences in the fitness of seedlings derived from CL and CH seeds. In one of the first attempts to answer this question, Clay and Antonovics (1985a) studied survival, flowering, fecundity, and net reproductive rates of *Danthonia spicata* for 3 years but found no overall significant differences between plants of CL and CH origin. In a subsequent study, genetic differences among offspring from CL and CH flowers from one maternal parent were examined using ramets. The results suggested that progeny from CL flowers were less variable than those from CH flowers (Clay and Antonovics, 1985b).

In an extension of their basic model on the evolution of cleistogamy, Schoen and Lloyd (1984) created the complex habitat model dealing with the evolution of cleistogamy in heterogenous parent environments. In this model, the environment varies in time and space, and plants respond by producing either CL or CH flowers. The most-fit phenotype is one that produces both CH and CL flowers, either at the same or at different times during the growing season. In a near and far dispersal model, Schoen and Lloyd (1984) consider the production of two seed (or dispersal) types by the same mother plant. In this model, they examined the evolution of cleistogamy in situations where CH and CL seeds have different patterns of dispersal, with seeds from CL flowers being more restricted in distance of dispersal than those from CH flowers. Under relatively poor environmental conditions for growth, the greatest fitness results from economically produced CL seeds that are more successful in germinating and becoming established close to the mother plants. However, if environmental conditions are highly favorable for growth, the greatest fitness results from an increased production of CH seeds that are dispersed away from the mother plant, where there would be less competition among the seedlings.

References

Aarssen, L. W., and Burton, S. M. (1990). Maternal effects at four levels in *Senecio vulgaris* (Asteraceae) grown on a soil nutrient gradient. *Am. J. Bot.* **77**, 1231–1240.

Abbott, R. J. (1986). Life history variation associated with the polymorphism for capitulum type and outcrossing rate in *Senecio vulgaris* L. *Heredity* **56**, 381–391.

Abbott, R. J., Horrill, J. C., and Noble, G. D. G. (1988). Germination behaviour of the radiate and non-radiate morphs of groundsel, *Senecio vulgaris* L. *Heredity* **60**, 15–20.

Aberg, E. (1956). Weed control research and development in Sweden. *Proc. 3rd. Brit. Weed Control Conf.* **1**, 141–164.

Adkins, S. W., Loewen, M., and Symons, S. J. (1986). Variation within pure lines of wild oats (*Avena fatua*) in relation to degree of primary dormancy. *Weed Sci.* **34**, 859–864.

Agren, J. (1989). Seed size and number in *Rubus chamaemorus*: Between-habitat variation, and effects of defoliation and supplemental pollination. *J. Ecol.* **77**, 1080–1092.

Aitken, Y. (1939). The problem of hard seeds in subterranean clover. *Proc. Roy. Soc. Vict.* **51**, 187–213.

Alcocer-Ruthling, M., Thill, D. C., and Shafii, B. (1992). Seed biology of sulfonylurea-resistant and -susceptible biotypes of prickly lettuce (*Lactuca serriola*). *Weed Tech.* **6**, 858–864.

Alexander, H. M., and Wulff, R. D. (1985). Experimental ecological genetics in *Plantago*. X. The effects of maternal temperature on seed and seedling characters in *P. lanceolata*. *J. Ecol.* **73**, 271–282.

Akpan, E. E. J., and Bean, E. W. (1977). The effects of temperature upon seed development in three species of forage grasses. *Ann. Bot.* **41**, 689–695.

Allard, H. A. (1932). Flowering behavior of the hog peanut in response to length of day. *J. Agri. Res.* **44**, 127–137.

Amritphale, D., Gupta, J. C., and Iyengar, S. (1984). Germination polymorphism in sympatric populations of *Calotropis procera*. *Oikos* **42**, 220–224.

Andersson, L. (1994a). Seed production and seed weight of six weed species treated with MCPA. *Swedish J. Agric. Res.* **24**, 95–100.

Andersson, L. (1994b). Effects of MCPA and tribenuron-methyl on seed production and seed size of annual weeds. *Swedish J. Agric. Res.* **24**, 49–56.

Andersson, S. (1990). Paternal effects on seed size in a population of *Crepis tectorum* (Asteraceae). *Oikos* **59**, 3–8.

Andersson, S. (1996). Seed size as a determinant of germination rate in *Crepis tectorum* (Asteraceae): Evidence from a seed burial experiment. *Can. J. Bot.* **74**, 568–572.

Angosto, T., and Matilla, A. J. (1994). Modifications in seeds of *Festuca indigesta* from two different altitudinal habitats. *Seed Sci. Technol.* **22**, 319–328.

Anoliefo, G. O., and Gill, L. S. (1992). Seed germination of *Peltophorum inerme* (Roxb.) Llanos (Leguminosae). *Int. J. Trop. Agric.* **10**, 246–253.

Antlfinger, A. E. (1986). Field germination and seedling growth of CH and CL progeny of *Impatiens capensis* (Balsaminaceae). *Am. J. Bot.* **73**, 1267–1273.

Antonovics, J., and Schmitt, J. (1986). Paternal and maternal effects on propagule size in *Anthoxanthum odoratum. Oecologia* **69**, 277–282.

Arthur, A. E., Gale, J. S., and Lawrence, M. J. (1973). Variation in wild populations of *Papaver dubium.* VII. Germination time. *Heredity* **30**, 189–197.

Ashman, T.-L. (1992). The relative importance of inbreeding and maternal sex in determining progeny fitness in *Sidalcea oregana* ssp. *spicata*, a gynodioecious plant. *Evolution* **46**, 1862–1874.

Attims, Y. (1972a). Influence de l'age physiologique de la plante mere sur la dormance des graines d'*Oldenlandia corybosa* L. (Rubiacees). *C. R. Acad. Sci. Paris Ser D.* **275**, 1613–1616.

Attims, Y. (1972b). Influence des conditions de culture de la plante mere sur l'entree en dormance des semences. *Bull. Group Etude Ryth. Biol.* **4**, 75–79.

Ayyangar, G. N. R., and Ponnaiya, B. W. X. (1939). Cleistogamy and its inheritance in sorghum. *Curr. Sci.* **8**, 418–419.

Bachmann, K., Chambers, K. L., and Price, H. J. (1984). Genetic components of heterocarpy in *Microseris* hybrid B87 (Asteraceae, Lactuceae). *Plant Syst. Evol.* **148**, 149–164.

Baker, G. A., and O'Dowd, D. J. (1982). Effects of parent plant density on the production of achene types in the annual *Hypochoeris glabra. J. Ecol.* **70**, 201–215.

Baker, K., Richards, A. J., and Tremayne, M. (1994). Fitness constraints on flower number, seed number and seed size in the dimorphic species *Primula farinosa* L. and *Armeria maritima* (Miller) Willd. *New Phytol.* **128**, 563–570.

Baker-Kratz, A. L., and Maguire, J. D. (1984). Germination and drymatter accumulation in dimorphic achenes of tansy ragwort(*Senecio jacobaea*). *Weed Sci.* **32**, 539–545.

Bangarwa, K. S., Singh, V. P., and Tomer, R. P. S. (1995). Progeny testing for seed quality parameters in *Dalbergia sissoo* Roxb. *Seed Sci. Technol.* **23**, 253–257.

Banovetz, S. J., and Scheiner, S. M. (1994). The effects of seed mass on the seed ecology of *Coreopsis lanceolata. Am. Midl. Nat.* **131**, 65–74.

Barclay, A. M., and Crawford, R. M. M. (1984). Seedling emergence in the rowan (*Sorbus aucuparia*) from an altitudinal gradient. *J. Ecol.* **72**, 627–636.

Baskin, J. M., and Baskin, C. C. (1973). Plant population differences in dormancy and germination characteristics of seeds: Heredity or environment? *Am. Midl. Nat.* **90**, 493–498.

Baskin, J. M., and Baskin, C. C. (1975a). Seed dormancy in *Isanthus brachiatus* (Labiatae). *Am. J. Bot.* **62**, 623–627.

Baskin, J. M., and Baskin, C. C. (1975b). Observations on the ecology of the cedar glade endemic *Viola egglestonii. Am. Midl. Nat.* **93**, 320–329.

Baskin, J. M., and Baskin, C. C. (1976a). Germination ecology of winter annuals: *Valerianella umbilicata* f. *patellaria* and f. *intermedia. J. Tennessee Acad. Sci.* **51**, 138–141.

Baskin, J. M., and Baskin, C. C. (1976b). Germination dimorphism in *Heterotheca subaxillaris* var. *subaxillaris* [=*H. latifolia*]. *Bull. Torrey Bot. Club* **103**, 201–206.

Baskin, J. M., and Baskin, C. C. (1979). Studies on the autecology and population biology of the monocarpic perennial *Grindelia lanceolata. Am. Midl. Nat.* **102**, 290–299.

Baskin, J. M., and Baskin, C. C. (1982). Comparative germination responses of the two varieties of *Arenaria patula. Trans. Kentucky Acad. Sci.* **43**, 50–54.

Baskin, J. M., and Baskin, C. C. (1995). Variation in the annual dormancy cycle in buried seeds of the weedy winter annual *Viola arvensis. Weed Res.* **35**, 353–362.

Bates, R. P., and Henson, P. R. (1955). Studies of inheritance in *Lespedeza cuneata* Don. *Agron. J.* **47**, 503–507.

Battle, J. P., and Whittington, W. J. (1971). Genetic variability in time to germination of sugar-beet clusters. *J. Agric. Sci., Camb.* **76**, 27–32.

Baxter, B. J. M., van Staden, J., and Granger, J. E. (1993). Seed germination response to temperature, in two altitudinally separate populations of the perennial grass *Themeda triandra. S. Afr. J. Sci.* **89**, 141–144.

Beadle, N. C. W. (1952). Studies in halophytes. I. The germination of the seed and establishment of the seedlings of five species of *Atriplex* in Australia. *Ecology* **33**, 49–62.

Bean, E. W. (1971). Temperature effects upon inflorescence and seed development in tall fescue (*Festuca arundinacea* Schreb.). *Ann. Bot.* **35**, 891–897.

Beardsell, D., and Mullett, J. (1984). Seed germination of *Eucalyptus pauciflora* Sieb. ex Spreng. from low and high altitude populations in Victoria. *Aust. J. Bot.* **32**, 475–480.

Bebawi, F. F., Eplee, R. E. and Norris, R. S. (1984). Effects of seed size and weight on witchweed (*Striga asiatica*) seed germination, emergence, and host-parasitization. *Weed Sci.* **32**, 202–205.

Belderok, B. (1961). Studies on dormancy in wheat. *Proc. Int. Seed Test. Assoc.* **26**, 697–760.

Bell, T. J., and Quinn, J. A. (1985). Relative importance of chasmogamously and cleistogamously derived seeds of *Dichanthelium clandestinum* (L.) Gould. *Bot. Gaz.* **146**, 252–258.

Bello, I. A., Owen, M. D. K., and Hatterman-Valenti, H. M. (1995). Effect of shade on velvetleaf (*Abutilon theophrasti*) growth, seed production, and dormancy. *Weed Technol.* **9**, 452–455.

Benech Arnold, R. (1993). Maternal effects on dormancy in *Sorghum* species. *In* "Basic and Applied Aspects of Seed Biology" (D. Come and F. Corbineau, eds.), pp. 607–614. Fourth International Workshop on Seeds, Universite Pierre et Marie Curie, Paris, France.

Benech Arnold, R. L., Fenner, M., and Edwards, P. J. (1992). Changes in dormancy level in *Sorghum halepense* seeds induced by water stress during seed development. *Funct. Ecol.* **6**, 596–605.

Bentley, S., Whittaker, J. B., and Malloch, A. J. C. (1980). Field experiments on the effects of grazing by a chrysomelid beetle (*Gastrophysa viridula*) on seed production and quality in *Rumex obtusifolius* and *Rumex crispus. J. Ecol.* **68**, 671–674.

Berger, A. (1985). Seed dimorphism and germination behaviour in *Salicornia patula. Vegetatio* **61**, 137–143.

Beveridge, J. L., and Wilsie, C. P. (1959). Influence of depth of planting, seed size, and variety on emergence and seedling vigor in alfalfa. *Agron. J.* **51**, 731–734.

Bevington, J. (1986). Geographic differences in the seed germination of paper birch (*Betula papyrifera*). *Am. J. Bot.* **73**, 564–573.

Bhati, P., and Sen, D. N. (1978). Adaptive polymorphism in *Ipomoea pes-tigridis* (Convolvulaceae), a common rainy season weed of the Indian arid zone. *Plant Syst. Evol.* **129**, 111–117.

Bianco, J., and Bulard, C. (1976). Physiologie de la germination de *Silene acaulis* L.) Jacq. ssp. *exscapa* (All.) J. Braun et ssp. *longiscapa* (Kern.) Hayeck. *Trav. Sci. Parc Nat. Vanoise* **7**, 107–115.

Biere, A. (1991a). Parental effects in *Lychnis flos-cuculi.* I. Seed size, germination and seedling performance in a controlled environment. *J. Evol. Biol.* **3**, 447–465.

Biere, A. (1991b). Parental effects on *Lychnis flos-cuculi.* II. Selection on time of emergence and seedling performance in the field. *J. Evol. Biol.* **3**, 467–486.

Black, J. N. (1956). The influence of seed size and depth of sowing on pre-emergence and early vegetative growth of subterranean clover (*Trifolium subterraneum* L.). *Aust. J. Agric. Res.* **7**, 98–109.

Black, J. N. (1957). The early vegetative growth of three strains of subterranean clover (*Trifolium subterraneum* L.) in relation to size of seed. *Aust. J. Agric. Res.* **8**, 1–14.

Black, M., and Naylor, J. M. (1959). Prevention of the onset of seed dormancy by gibberellic acid. *Nature* **184**, 468–469.

Bocchieri, E., Floris, G., and Mulas, B. (1987). Dormance, germination et contenu proteique des semences de deux populations de *Schoenus nigricans* cueillies a des altitudes differentes. *Can. J. Bot.* **65**, 617–621.

Borghetti, M., Vendramin, G. G., Giannini, R., and Schettino, A. (1989). Effects of stratification, temperature and light on germination of *Pinus leudoermis. Acta Oecol.* **10**, 45–56.

Borgstrom, G. (1939). Formation of cleistogamic and chasmogamic flowers in wild violets as a photoperiodic response. *Nature* **144**, 514–515.

Boyce, K. G., Cole, D. F., and Chilcote, D. O. (1976). Effect of temperature and dormancy on germination of tall fescue. *Crop Sci.* **16**, 15–18.

Boyd, W. J. R., Gordon, A. G., and LaCroix, L. J. (1971). Seed size, germination resistance and seedling vigor in barley. *Can. J. Plant Sci.* **51**, 93–99.

Bremmer, P. M., and Rawson, H. M. (1978). The weights of individual grains of the wheat ear in relation to their growth potential, the supply of assimilate and interaction between grains. *Aust. J. Plant Physiol.* **5**, 61–72.

Bretagnolle, F., Thompson, J. D., and Lumaret, R. (1995). The influence of seed size variation on seed germination and seedling vigour in diploid and tetraploid *Dactylis glomerata* L. *Ann. Bot.* **76**, 607–615.

Brocklehurst, P. A., Moss, J. P., and Williams, W. (1978). Effects of irradiance and water supply on grain development in wheat. *Ann. Appl. Biol.* **90**, 265–276.

Brown, W. V. (1952). The relation of soil moisture to cleistogamy in *Stipa leucotricha. Bot. Gaz.* **113**, 438–443.

Budd, G. D., Thomas, P. E. L., and Allison, J. C. S. (1979). Vegetative regeneration, depth of germination and seed dormancy in *Commelina benghalensis* L. *Rhodesia J. Agri. Res.* **17**, 151–153.

Bunning, E. (1954). Parental age and germinative characters of the seeds. *Ann. N. Y. Acad. Sci.* **57**, 484–487.

Buraas, T., and Skinnes, H. (1984). Genetic investigations on seed dormancy in barley. *Hereditas* **101**, 235–244.

Bussey, H., and Fields, M. A. (1974). A model for stably inherited environmentally induced changes in plants. *Nature* **251**, 708–710.

Byrne, M., and Mazer, S. J. (1990). The effect of position on fruit characteristics, and relationships among components of yield in *Phytolacca rivinoides* (Phytolaccaceae). *Biotropica* **22**, 353–365.

Cairns, A. L. P., and Kritzinger, J. H. (1993). The effect of molybdenum on seed dormancy in wheat. *In* "Basic and Applied Aspects of Seed Biology" (D. Come and F. Corbineau, eds.), Vol. 2, pp. 665–669. Fourth International Workshop on Seeds, Universite Pierre et Marie Curie, Paris, France.

Call, C. A., and Spoonts, B. O. (1989). Characterization and germination of chasmogamous and basal axillary cleistogamous florets of Texas winter grass. *J. Range Manage.* **42**, 51–55.

Cameron, D. F. (1965). Variation in flowering time and in some growth characteristics of Townsville lucerne (*Stylosanthes humilis*). *Aust. J. Exp. Agric. Anim. Husb.* **5**, 49–51.

Cameron, D. F. (1967). Hardseededness and seed dormancy of Townsville lucerne (*Stylosanthes humilis*) selections. *Aust. J. Exp. Agric. Anim. Husb.* **7**, 237–240.

Campbell, C. A., Davidson, H. R., and Winkleman, G. E. (1981). Effect of nitrogen, temperature, growth stage and duration of moisture stress on yield components and protein content of Manitou spring wheat. *Can. J. Plant Sci.* **61**, 549–563.

Campbell, C. S., Quinn, J. A., Cheplick, G. P., and Bell, T. J. (1983). Cleistogamy in grasses. *Annu. Rev. Ecol. Syst.* **14**, 411–441.

Campbell, R. K., and Ritland, S. M. (1982). Regulation of seed germination timing by moist chilling in western hemlock. *New Phytol.* **92**, 173–182.

Capinera, J. L. (1979). Qualitative variation in plants and insects: Effect of propagule size on ecological plasticity. *Amer. Nat.* **114**, 350–361.

Capon, B., Maxwell, G. L., and Smith, P. H. (1978). Germination responses to temperature pretreatment of seeds from ten populations of *Salvia columbariae* in the San Gabriel Mountains and Mojave Desert, California. *Aliso* **9**, 365–373.

Carleton, A. E., and Cooper, C. S. (1972). Seed size effects upon seedling vigor of three forage legumes. *Crop Sci.* **12**, 183–186.

Caron, G. E., Wang, B. S. P., and Schooley, H. O. (1993). Variation in *Picea glauca* seed germination associated with the year of cone collection. *Can. J. For. Res.* **23**, 1306–1313.

Carr, D. E., and Dudash, M. R. (1995). Inbreeding depression under a competitive regime in *Mimulus guttatus*: Consequences for potential male and female function. *Heredity* **75**, 437–445.

Case, A. L., Lacey, E. P., and Hopkins, R. G. (1996). Parental effects in *Plantago lanceolata* L. II. Manipulation of grandparental temperature and parental flowering time. *Heredity* **76**, 287–295.

Cavers, P. B., and Harper, J. L. (1966). Germination polymorphism in *Rumex crispus* and *Rumex obtusifolius. J. Ecol.* **54**, 369–382.

Cavers, P. B., and Steel, M. G. (1984). Patterns of change in seed weight over time on individual plants. *Am. Nat.* **124**, 324–335.

Carver, W. A. (1936). The effect of certain environmental factors on the development of cotton seed, germinating ability, and resultant yield of cotton. *Proc. Florida Acad. Sci.* **1**, 150. [Abstract]

Chadoeuf-Hannel, R., and Barralis, G. (1982). Influence de differents regimes hydriques sur la croissance vegetative, le poids et al germination des graines d'une mauvaise herbe cultivee en serre: *Amaranthus retroflexus* L. *Agronomie* **2**, 835–841.

Chadoeuf-Hannel, R., and Barralis, G. (1983). Evolution de l'aptitude a germer des graines d'*Amaranthus retroflexus* recoltees dans differentes conditions, au cours de leur conservation. *Weed Res.* **23**, 109–117.

Chandraratna, M. F., and Sakai, K.-I. (1960). A biometrical analysis of matroclinous inheritance of grain weight in rice. *Heredity* **14**, 365–373.

Chang, T.-T., and Li, C.-C. (1991). Genetics and breeding. *In* "Rice" (B. S. Luh ed.), Vol. I, pp. 23–101. Van Nostrand Reinhold, New York.

Chase, A. (1918). Axillary cleistogenes in some American grasses. *Am. J. Bot.* **5**, 254–258.

Chawan, D. D., and Sen, D. N. (1973). Diversity in germination behaviour and high temperature tolerance in the seeds of *Corchorus aestuans* Linn. *Ann. Arid Zone* **12**, 23–32.

Cheam, A. H. (1986). Seed production and seed dormancy in wild radish (*Raphanus raphanistrum* L.) and some possibilities for improving control. *Weed Res.* **26**, 405–413.

Cheplick, G. P. (1983). Differences between plants arising from aerial and subterranean seeds in the amphicarpic annual *Cardamine chenopodifolia* (Cruciferae). *Bull. Torrey Bot. Club* **110**, 442–448.

Cheplick, G. P. (1989). Nutrient availability, dimorphic seed production, and reproductive allocation in the annual grass *Amphicarpum purshii. Can. J. Bot.* **67**, 2514–2521.

Cheplick, G. P. (1996). Cleistogamy and seed heteromorphism in *Triplasis purpurea* (Poaceae). *Bull. Torrey Bot. Club* **123**, 25–33.

Cheplick, G. P., and Clay, K. (1989). Convergent evolution of cleistogamy and seed heteromorphism in two perennial grasses. *Evol. Trends Plants* **3**, 127–136.

Cheplick, G. P., and Quinn, J. A. (1982). *Amphicarpum purshii* and the "pessimistic strategy" in amphicarpic annuals with subterranean fruit. *Oecologia* **52**, 327–332.

Chmielewski, J. G, Semple, J. C., Burr, L. M., and Hawthorn,

W. R. (1989). Comparison of achene characteristics within and among diploid and tetraploid clones of *Solidago flexicaulis* and their significance in germination and resource allocation studies. *Can. J. Bot.* **67,** 1821–1832.

Choe, H. S., Chu C., Koch, G., Gorham, J., and Mooney, H. A. (1988). Seed weight and seed resources in relation to plant growth rate. *Oecologia* **76,** 158–159.

Choudhuri, G. N. (1968). Effect of soil salinity on germination and survival of some steppe plants in Washington. *Ecology* **49,** 465–471.

Christie, B. R., and Kalton, R. R. (1960). Recurrent selection for seed weight in bromegrass, *Bromus inermis* Leyss. *Agron. J.* **52,** 575–578.

Cideciyan, M. A., and Malloch, A. J. C. (1982). Effects of seed size on the germination, growth and competitive ability of *Rumex crispus* and *Rumex obtusifolius*. *J. Ecol.* **70,** 227–232.

Clarke, J. M. (1979). Intra-plant variation in number of seeds per pod and seed weight in *Brassica napus* 'Tower.' *Can. J. Plant Sci.* **59,** 959–962.

Clausen, J., Keck, D. D., and Hiesey, W. M. (1939). The concept of species based on experiment. *Am. J. Bot.* **26,** 103–106.

Clausen, J., Keck, D. D., and Hiesey, W. M. (1940). Experimental studies on the nature of species. I. Effect of varied environments on western North American plants. *Carnegie Inst. Wash. Pub.* 520.

Clay, K. (1982). Environmental and genetic determinants of cleistogamy in a natural population of the grass *Danthonia spicata*. *Evolution* **36,** 734–741.

Clay, K. (1983a). Variation in the degree of cleistogamy within and among species of the grass *Danthonia*. *Am. J. Bot.* **70,** 835–843.

Clay, K. (1983b). The differential establishment of seedlings from chasmogamous and cleistogamous flowers in natural populations of the grass *Danthonia spicata* (L.) Beauv. *Oecologia* **57,** 183–188.

Clay, K., and Antonovics, J. (1985a). Demographic genetics of the grass *Danthonia spicata*: Success of progeny from chasmogamous and cleistogamous flowers. *Evolution* **39,** 205–210.

Clay, K., and Antonovics, J. (1985b). Quantitative variation of progeny from chasmogamous and cleistogamous flowers in the grass *Danthonia spicata*. *Evolution* **39,** 335–348.

Connor, H. E. (1979). Breeding systems in the grasses: A survey. *New Zeal. J. Bot.* **17,** 547–574.

Connor, K. F., and Lander, R. M. (1991). Effects of tree age on pollen, seed, and seedling characteristics in Great Basin bristlecone pine. *Bot. Gaz.* **152,** 107–113.

Cook, A. D. (1972). Polymorphic and continuous variation in the seeds of dove weed, *Eremocarpus setigerus* (Hook.) Benth. *Am. Midl. Nat.* **87,** 366–376.

Cook, R. E. (1975). The photoinductive control of seed weight in *Chenopodium rubrum* L. *Am. J. Bot.* **62,** 427–431.

Cook, S. A. (1962). Genetic system, variation, and adaptations in *Eschscholzia californica*. *Evolution* **16,** 278–299.

Cooper, C. C., and Watson, D. P. (1952). Influence of daylength and temperature on the growth of greenhouse violets. *Proc. Am. Soc. Hort. Sci.* **59,** 549–553.

Cooper, C. S., Ditterline, R. L., and Welty, L. E. (1979). Seed size and rate effects upon stand density and yield of alfalfa. *Agron. J.* **71,** 83–85.

Cope, W. A. (1966). Growth rate and yield in sericea lespedeza in relation to seed size and outcrossing. *Crop Sci.* **6,** 566–568.

Cope, W. A., and Moll, R. H. (1969). Inheritance of yield, forage quality and seed characteristics in sericea lespedeza. *Crop Sci.* **9,** 467–470.

Corkidi, L., Rincon, E., and Vazquez-Yanes, C. (1991). Effects of light and temperature on germination of heteromorphic achenes of *Bidens odorata* (Asteraceae). *Can. J. Bot.* **69,** 574–579.

Correns, C. (1906). Das keimen der beiderlei fruchte der *Dimorphotheca pluvialis*. *Ber. Deutsch. Bot. Ges.* **24,** 173–176.

Correns, C. (1926). Genetische untersuchungen an *Lamium amplexicaule* L. I. *Biol. Zbl.* **46,** 65–79.

Correns, C. (1930). Genetische Üntersuchungen an *Lamium amplexicaule* L. IV. *Biol. Zbl.* **50,** 7–19.

Crawley, M. J., and Nachapong, N. (1985). The establishment of seedlings from primary and regrowth seeds of ragwort (*Senecio jacobaea*). *J. Ecol.* **73,** 255–261.

Cremonini, R., Lombardi, T., Stefani, A., and Onnis, A. (1992). x*Haynaldoticum sardoum* Meletti et Onnis: Effects of salinity and temperature on germination and early growth. *Cytobios* **72,** 83–92.

Crocker, W. (1906). Role of seed coats in delayed germination. *Bot. Gaz.* **42,** 265–291.

Cruden, R. W. (1973). Reproductive biology of weedy and cultivated *Mirabilis* (Nyctaginaceae). *Am. J. Bot.* **60,** 802–809.

Cruden, R. W. (1974). The adaptive nature of seed germination in *Nemophila menziesii* Aggr. *Ecology* **55,** 1295–1305.

Cullis, C. A. (1977). Molecular aspects of the environmental induction of heritable changes in flax. *Heredity* **38,** 129–154.

Dahlgaard, J., and Warncke, E. (1995). Seed set and germination in crosses within and between two geographically isolated small populations of *Saxifraga hirculus* in Denmark. *Nord J. Bot.* **15,** 337–341.

Danchenko, A. M., Markvart, V. R., and Shul'ga, L. V. (1977). Effect of origin of seed of drooping and white birch on its germination under various temperature regimes. *Soviet J. Ecol.* **8,** 74–76.

Darlington, C. D., and Janaki-Ammal, E. K. (1945). Adaptive isochromosomes in *Nicandra*. *Ann. Bot.* **9,** 267–281.

Datta, S. C., and Basu, R. (1978). Germination responses of seeds of *Dactyloctenium aegyptium*. In "Environmental Physiology and Ecology of Plants" (D. N. Sen and R. P. Bansal, eds.), pp. 235–248. B. Singh and M. P. Singh Publ., Dehra Dun, India.

Datta, S. C., Evenari, M., and Gutterman, Y. (1970). The heteroblasty of *Aegilops ovata* L. *Israel J. Bot.* **19,** 463–483.

Datta, S. C., Gutterman, Y., and Evenari, M. (1972). The influence of the origin of the mother plant on yield and germination of their caryopses in *Aegilops ovata*. *Planta* **105,** 155–164.

Datta, S. C., and Sen, S. (1981). Effect of the environment of the mother plants of *Cassia sophera* var. *purpurea* on the germination of their seeds. *Acta. Bot. Acad. Sci. Hung.* **27,** 319–323.

Dexter, S. T. (1955). Alfalfa seedling emergence from seed lots varying in origin and hard seed content. *Agron. J.* **47,** 357–361.

Dille, J. E., and Knowles, P. F. (1975). Histology and inheritance of the closed flower in *Carthamus tinctorius* (Compositae). *Am. J. Bot.* **62,** 209–215.

Dobrenz, A. K., and Beetle, A. A. (1966). Cleistogenes in *Danthonia*. *J. Range Manage.* **19,** 292–296.

Dobzhansky, T. (1970). "Genetics of the Evolutionary Process." Columbia University Press, New York.

Do Cao, T., Attims, Y., Corbineau, F., and Come, D. (1978). Germination des graines produites par les plantes de deux lignees d'*Oldenlandia corymbosa* L. (Rubiacees) cultivees dan des conditions controlees. *Physiol. Veg.* **16,** 521–531.

Dolan, R. W. (1984). The effect of seed size and maternal source on individual size in a population of *Ludwigia leptocarpa* (Onagraceae). *Am. J. Bot.* **71,** 1302–1307.

Donnelly, E. D., Watson, J. E., and McGuire, J. A. (1972). Inheritance of hard seed in *Vicia*. *J. Hered.* **63,** 361–365.

Donoso, C. (1979). Genecological differentiation in *Nothofagus obliqua* (Mirb.) Oerst. in Chile. *For. Ecol. Manage.* **2,** 53–66.

Dorne, A.-J. (1973a). Germination de deux especes a large distribution altitudinale: *Silene inflata* et *Alyssoides utriculatum*. Influence, sur le comportement germinatif, de la culture des porte-graines a basse altitude. *Phyton (Buenos Aires)* **31,** 25–39.

Dorne, A.-J. (1973b). Influence de l'altitude de recolte sur la capacite de germination des semences du *Chenopodium bonus- henricus* L. *C. R. Acad. Sci. Paris. Ser. D.* **277,** 305–308.

Dorne, A.-J. (1981). Variation in seed germination inhibition of *Chenopodium bonus-henricus* in relation to altitude of plant growth. *Can. J. Bot.* **59,** 1893–1901.

Dotzenko, A. D., Cooper, G. S., Dobrenz, A. K., Laude, H. M., Messengale, M. A., and Feltner, K. C. (1967). Temperature stress on growth and seed characteristics of grasses and legumes. *Colorado St. Univ. Agric. Exp. Sta. Tech. Bull.* 97.

Dowling, R. E. (1933). The reproduction of *Plantago coronopus*: An example of morphological and biological seed dimorphism. *Ann. Bot.* **47,** 861–872.

Draper, A. D., and Wilsie, C. P. (1965). Recurrent selection for seed size in birdsfoot trefoil, *Lotus corniculatus* L. *Crop Sci.* **5,** 313–315.

Drennan, P. M., van Staden, J., and Mtingane, B. M. (1993). Factors affecting germination of the dimorphic seeds of *Atriplex lindleyi*. *In* "Basic and Applied Aspects of Seed Biology" (D. Come and F. Corbineau, eds.), Fourth International Workshop on Seeds, Universite Pierre et Marie Curie, Paris, France.

Drew, R. L. K., and Brocklehurst, P. A. (1990). Effects of temperature of mother-plant environment on yield and germination of seeds of lettuce (*Lactuca sativa*). *Ann. Bot.* **66,** 63–71.

Durrant, A. (1958). Environmental conditioning of flax. *Nature* **181,** 928–929.

Durrant, A. (1962). The environmental induction of heritable change in *Linum. Heredity* **17,** 27–61.

Dyksterhuis, E. J. (1945). Axillary cleistogenes in *Stipa leucotricha* and their role in nature. *Ecology* **26,** 195–199.

Eagles, H. A., and Hardacre, A. K. (1979). Genetic variation in maize (*Zea mays* L.) for germination and emergence at 10°C. *Euphytica* **28,** 287–295.

Edwards, C. J., Jr., and Hartwig, E. E. (1971). Effect of seed size upon rate of germination in soybeans. *Agron. J.* **63,** 429–430.

Eenink, A. H. (1981). Research on the inheritance of seed dormancy in lettuce (*Lactuca sativa* L.) and selection for non-dormancy. *Euphytica* **30,** 371–380.

Egli, D. B., Leggett, J. E., and Wood, J. M. (1978). Influence of soybean seed size and position on the rate and duration of filling. *Agron. J.* **70,** 127–130.

Egli, D. B., and Wardlaw, I. F. (1980). Temperature response of seed growth characteristics of soybeans. *Agron. J.* **72,** 560–564.

El Bagoury, O. H. (1975). Effect of different fertilisers on the germination and hard seed percentage of broad bean seed (*Vicia faba*). *Seed Sci. Technol.* **3,** 569–574.

El Bagoury, O. H., and Niyazi, M. A. (1973). Effect of different fertilisers on the germination and hard seed percentage of Egyptian clover seeds (*Trifolium alexandrinum* L.). *Seed Sci. Technol.* **1,** 773–779.

El-Keblawy, A. A., Shaltout, K. H., Lovett Doust, J., and Lovett Doust, L. (1996). Maternal effects on progeny in *Thymelaea hirsuta. New Phytol.* **132,** 77–85.

Elkins, D. M., Hoveland, C. S., and Donnelly, E. D. (1966). Germination of *Vicia* species and interspecific lines as affected by temperature cycles. *Crop Sci.* **6,** 45–48.

Ellis, J. F., and Ilnicki, R. D. (1968). Seed dormancy in corn chamomile. *Weed Sci.* **16,** 111–113.

Ellison, A. M. (1987). Effect of seed dimorphism on the density-dependent dynamics of experimental populations of *Atriplex triangularis* (Chenopodiaceae). *Am. J. Bot.* **74,** 1280–1288.

Ellstrand, N. C., Lord, E. M., and Eckard, K. J. (1984). The inflorescence as a metapopulation of flowers: Position-dependent differ-ences in function and form in the cleistogamous species *Collomia grandiflora* Dougl. ex Lindl. (Polemoniaceae). *Bot. Gaz.* **145,** 329–333.

Engler, A. (1895). Uber amphicarpie bei *Fleurya podocarpa* Wedd., nebst einigen allgemeinen bemerkungen uber die erscheinung der amphicarpie und geocarpie. *Deut. Akad. Wiss. Berlin* **5,** 57–66.

Enileev, Kh. Kh., and Solov'ev, V. P. (1960). A study of the causes of diversity in germination behavior of cotton seeds. *Fiziol. Rast.* **7,** 20–24.

Ernst, A. (1906). Das keimen der dimorphen frchtchen von *Synedrella nodiflora* (L.) Grtn. *Ber. Deut. Bot. Ges.* **24,** 450–458.

Escarre, J., and Houssard, C. (1988). Aptitudes germinatives comparees de graines de *Rumex acetosella* issues de populations correspondant a des stades distincts d'une succession postculturale. *Can. J. Bot.* **66,** 1381–1390.

Evans, A. S., Mitchell, R. J., and Cabin, R. J. (1996). Morphological side effects of using gibberellic acid to induce germination: Consequences for the study of seed dormancy. *Am. J. Bot.* **83,** 543–549.

Evans, L. E., and Bhatt, G. M. (1977). Influence of seed size, protein content and cultivar on early seedling vigor in wheat. *Can. J. Plant Sci.* **57,** 929–935.

Evans, L. T. (1956). Chasmogamous flowering in *Viola palustris* L. *Nature* **178,** 1301.

Evenari, M. (1965). Physiology of seed dormancy, after-ripening and germination. *Proc. Int. Seed Test. Assoc.* **30,** 49–71.

Evenari, M., Kadouri, A., and Gutterman, Y. (1977). Ecophysiological investigations on the amphicarpy of *Emex spinosa* (L.) Campd. *Flora* **166,** 223–238.

Evenari, M., Koller, D., and Gutterman, Y. (1966). Effects of the environment of the mother plant on germination by control of seed-coat permeability to water in *Ononis sicula* Guss. *Aust. J. Biol. Sci.* **19,** 1007–1016.

Evenari, M., Shanan, L., and Tadmor, N. (1982). "The Negev: The Challenge of a Desert." Harvard Univ. Press, Cambridge, MA.

Fady, B. (1992). Effect of osmotic stress on germination and radicle growth in five provenances of *Abies cephalonica* Loud. *Acta. Oecol.* **13,** 67–79.

Farnsworth, E. J., and Bazzaz, F. A. (1995). Inter- and intrageneric differences in growth, reproduction, and fitness of nine herbaceous annual species grown in elevated CO_2 environments. *Oecologia* **104,** 454–466.

Farris, M. A. (1988). Quantitative genetic variation and natural selection in *Cleome serrulata* growing along a mild soil moisture gradient. *Can. J. Bot.* **66,** 1870–1876.

Fawcett, R. S., and Slife, F. W. (1978a). Effects of 2,4-D and dalapon on weed seed production and dormancy. *Weed Sci.* **26,** 543–547.

Fawcett, R. S., and Slife, F. W. (1978b). Effects of field applications of nitrate on weed seed germination and dormancy. *Weed Sci.* **26,** 594–596.

Fehr, W. R., and Weber, C. R. (1968). Mass selection by seed size and specific gravity in soybean populations. *Crop Sci.* **8,** 551–554.

Felippe, G. M. (1990). Germinacao de *Bidens gardneri* Baker, uma planta anual dos cerrados. *Hoehnea* **17,** 7–11.

Felippe, G. M., and Dale, J. E. (1968). Effects of CCC and gibberellic acid on the progeny of treated plants. *Planta* **80,** 344–348.

Fenner, M. (1992). Environmental influences on seed size and composition. *Horticult. Rev.* **13,** 183–213.

Fergason, V. L., and Zuber, M. S. (1962). Influence of environment on amylose content of maize endosperm. *Crop Sci.* **2,** 209–211.

Fenster, C. B. (1991). Gene flow in *Chamaesrista fasciculata* (Leguminosae). II. Gene establishment. *Evolution* **45,** 410–422.

Flint, S. D., and Palmblad, I. G. (1978). Germination dimorphism and developmental flexibility in the ruderal weed *Heterotheca grandiflora. Oecologia* **36,** 33–43.

Forbes, I., and Wells, H. D. (1968). Hard and soft seededness in blue lupine, *Lupinus angustifolius* L.: Inheritance and phenotype classification. *Crop Sci.* **8,** 195–197.

Ford, M. A., Pearman, I., and Thorne, G. N. (1976). Effects of variation in ear temperature on growth and yield of spring wheat. *Ann. Appl. Biol.* **82,** 317–333.

Forsyth, C., and Brown, N. A. C. (1982). Germination of the dimorphic fruits of *Bidens pilosa* L. *New Phytol.* **90,** 151–164.

Fowler, D. P., and Dwight, T. W. (1964). Provenance differences in the stratification requirements of white pine. *Can. J. Bot.* **42,** 669–675.

Fox, R. L., and Albrecht, W. A. (1957). Soil fertility and the quality of seeds. *Missouri Agri. Exp. Sta. Res. Bull.* 619.

Frankton, C., and Bassett, I. J. (1968). The genus *Atriplex* (Chenopodiaceae) in Canada. I. Three introduced species: *A. heterosperma, A. oblongifolia,* and *A. hortensis. Can. J. Bot.* **46,** 1309–1313.

Fraser, J. W. (1971). Cardinal temperatures for germination of six provences of white spruce seeds. *Dept. Fish. For. Can. For. Serv. Publ. No.* 1290.

Frost, R. A., and Cavers, P. B. (1975). The ecology of pigweeds (*Amaranthus*) in Ontario. I. Interspecific and intraspecific variation in seed germination among local collections of *A. powellii* and *A. retroflexus. Can. J. Bot.* **53,** 1276–1284.

Fuller, W., Hance, C. E., and Hutchings, M. J. (1983). Within-season fluctuations in mean fruit weight in *Leontodon hispidus* L. *Ann. Bot.* **51,** 545–549.

Galen, C., and Weger, H. G. (1986). Re-evaluating the significance of correlations between seed number and size: Evidence from a natural population of the lily, *Clintonia borealis. Am. J. Bot.* **73,** 346–352.

Ganders, F. R., Carey, K., and Griffiths, A. J. F. (1977). Outcrossing rates in natural populations of *Plectritis brachystemon* (Valerianaceae). *Can. J. Bot.* **55,** 2070–2074.

Garber, R. J., and Quisenberry, K. S. (1923). Delayed germination and the origin of false wild oats. *J. Hered.* **14,** 267–274.

Garbutt, K., and Witcombe, J. R. (1986). The inheritance of seed dormancy in *Sinapis arvensis* L. *Heredity* **56,** 25–31.

Gasquez, J., Darmency, H., and Compoint, J. P. (1981). Comparaison de la germination et de la croissance de biotypes sensibles et resistants aux triazines chez quatre especes de mauvaises herbes. *Weed Res.* **21,** 219–225.

Gavrielit-Gelmond, H. (1957). The effect of environment and agrotechnical methods on dormancy in *Lathyrus ochrus* seeds. *Proc. Int. Seed Test. Assoc.* **22,** 276–281.

Georges, F., and Lazare, J.-J. (1983). Contribution a l'etude ecologique du complexe orophile *Carex sempervirens* (Cyperaceae): etude experimentale de la germination de populations pyreneennes. *Can. J. Bot.* **61,** 135–141.

Gfeller, F., and Svejda, F. (1960). Inheritance of post-harvest seed dormancy and kernel colour in spring wheat lines. *Can. J. Plant Sci.* **40,** 1–6.

Gibson, T. A., and Humphreys, L. R. (1973). The influence of nitrogen nutrition of *Desmodium uncinatum* on seed production. *Aust. J. Agric. Res.* **24,** 667–676.

Gil, R. H. (1989). Tamano de la semilla y effecto de la temperatura en la germinacion de *Heliocarpus popayanensis* H.B.K. *Rev. For. Venezolana* **33,** 21–42.

Giles, B. E. (1990). The effects of variation in seed size on growth and reproduction in the wild barley *Hordeum vulgare* ssp. *spontaneum. Heredity* **64,** 239–250.

Gilfedder, L., and Kirkpatrick, J. B. (1994). Genecological variation in the germination, growth and morphology of four populations of a Tasmanian endangered perennial daisy, *Leucochrysum albicans. Aust. J. Bot.* **42,** 431–440.

Girard, J. (1990). Study of the inheritance of seed primary dormancy

and the ability to enter secondary dormancy in *Petunia*: Influence of temperature, light and gibberellic acid on dormancy. *Plant Cell Environ.* **13,** 827–832.

Gladstones, J. S. (1970). Lupins as crop plants. *Field Crop Abst.* **23,** 123–148.

Gonzalez-Rabanal, R., Casal, M., and Trabaud, L. (1994). Effects of high temperatures, ash and seed position in the inflorescence on the germination of three Spanish grasses. *J. Veg. Sci.* **5,** 289–294.

Goo, M. (1948). Effects of individual seed weight and seed coat on the germination of the seed collected from the young and old mother trees in *Cryptomeria japonica* D. Don. *Bull. Tokyo Univ. For.* **36,** 1–10.

Govinthasamy, T., and Cavers, P. B. (1995). The effects of smut (*Ustilago destruens*) on seed production, dormancy, and viability in fall panicum (*Panicum dichotomiflorum*). *Can J. Bot.* **73,** 1628–1634.

Gramshaw, D. (1976). Temperature/light interactions and the effect of seed source on germination of annual ryegrass (*Lolium rigidum* Gaud.) seeds. *Aust. J. Agric. Res.* **27,** 779–786.

Grant, V. (1949). Seed germination in *Gilia capitata* and its relatives. *Madrono* **10,** 87–93.

Grant Lipp, A. E., and Ballard, L. A. T. (1963). Germination patterns shown by the light-sensitive seed of *Anagallis arvensis. Aust. J. Biol. Sci.* **16,** 572–584.

Grant Lipp, A. E., and Ballard, L. A. T. (1964). The interrelation of dormancy, size, and hardness in seed of *Trifolium subterraneum* L. *Aust. J. Agric. Res.* **15,** 215–222.

Gray, C. T., and Thomas, S. M. (1985). Germination and B chromosomes in *Allium porrum* L. *J. Plant Physiol.* **121,** 281–285.

Gray, D. (1979). The germination response to temperature of carrot seeds from different umbels and times of harvest of the seed crop. *Seed Sci. Technol.* **7,** 169–178.

Gray, D., and Steckel, J. R. A. (1984). Viability of onion (*Allium cepa*) seed as influenced by temperature during seed growth. *Ann. Appl. Biol.* **104,** 375–382.

Greipsson, S., and Davy, A. J. (1995). Seed mass and germination behaviour in populations of the dune-building grass *Leymus arenarius. Ann. Bot.* **76,** 493–501.

Gremillion, K. J. (1993). Crop and weed in prehistoric eastern North America: The *Chenopodium* example. *Am. Antiq.* **58,** 496–509.

Gritton, E. T., and Atkins, R. E. (1961). Germination of sorghum seed as affected by dormancy. *Agron. J.* **55,** 169–174.

Gross, K. L. (1984). Effects of seed size and growth form on seedling establishment of six monocarpic perennial plants. *J. Ecol.* **72,** 369–387.

Gross, K. L., and Kromer, M. L. (1986). Seed weight effects on growth and reproduction in *Oenothera biennis* L. *Bull. Torrey Bot. Club* **113,** 252–258.

Grouzis, M., Berger, A., and Heim, G. (1976). Polymorphisme et germination des graines chez trois especes annuelles du genre *Salicornia. Oecol. Plant.* **11,** 41–53.

Groves, R. H., Hagon, M. W., and Ramakrishnan, P. S. (1982). Dormancy and germination of seed of eight populations of *Themeda australis. Aust. J. Bot.* **30,** 373–386.

Gutterman, Y. (1973). Differences in the progeny due to daylength and hormone treatment of the mother plant. *In* "Seed Ecology" (W. Heydecker, ed.), pp. 59–80. Pennsylvania State Univ. Press, University Park.

Gutterman, Y. (1974). The influence of the photoperiodic regime and red-far red light treatments of *Portulaca oleracea* L. plants on the germinability of their seeds. *Oecologia* **17,** 27–38.

Gutterman, Y. (1978). Seed coat permeability as a function of photoperiodical treatments of the mother plants during seed maturation in the desert annual plant: *Trigonella arabica*, del. *J. Arid Env.* **1**, 141–144.

Gutterman, Y. (1980–1981a). Annual rhythm and position effect in the germinability of *Mesembryanthemum nodiflorum*. *Israel J. Bot.* **29**, 93–97.

Gutterman, Y. (1980–1981b). Influences on seed germinability: Phenotypic maternal effects during seed maturation. *Israel J. Bot.* **29**, 105–117.

Gutterman, Y. (1982). Phenotypic maternal effects of photoperiod on seed germination. *In* "The Physiology and Biochemistry of Seed Development, Dormancy and Germination" (A. A. Khan, ed.), pp. 67–79. Elsevier Biomedical Press, Amsterdam.

Gutterman, Y. (1991). Comparative germination of seeds, matured during winter or summer, of some bi-seasonal flowering perennial desert Aiozaceae. *J. Arid Environ.* **21**, 283–291.

Gutterman, Y. (1992a). Influences of daylength and red or far-red light during the storage of ripe *Cucumis prophetarum* fruits, on seed germination in light. *J. Arid Environ.* **23**, 443–449.

Gutterman, Y. (1992b). Maturation dates affecting the germinability of *Lactuca serriola* L. achenes collected from a natural population in the Negev Desert highlands. Germination under constant temperatures. *J. Arid Environ.* **22**, 353–362.

Gutterman, Y. (1994a). Germinability under natural temperatures of *Lactuca serriola* L. achenes matured and collected on different dates from a natural population in the Negev Desert highlands. *J. Arid Environ.* **28**, 117–127.

Gutterman, Y. (1994b). Long-term seed position influences on seed germinability of the desert annual, *Mesembryanthemum nodiflorum* L. *Israel J. Plant Sci.* **42**, 197–205.

Gutterman, Y. (1996). Effect of day length during plant development and caryopsis maturation on flowering and germination, in addition to temperature during dry storage and light during wetting, of *Schismus arabicus* (Poaceae) in the Negev Desert, Israel. *J. Arid Environ.* **33**, 439–448.

Gutterman, Y., and Evenari, M. (1972). The influence of day length on seed coat colour, an index of water permeability, of the desert annual *Ononis sicula* Guss. *J. Ecol.* **60**, 713–719.

Gutterman, Y., and Heydecker, W. (1973). Studies of the surfaces of desert plant seeds. I. Effect of day length upon maturation of the seedcoat of *Ononis sicula* Guss. *Ann. Bot.* **37**, 1049–1050.

Gutterman, Y., and Nevo, E. (1994). Temperatures and ecological-genetic differentiation affecting the germination of *Hordeum spontaneum* caryopses harvested from three populations: The Negev Desert and opposing slopes on Mediterranean Mount Carmel. *Israel J. Plant Sci.* **42**, 183–195.

Gutterman, Y., and Porath, D. (1975). Influences of photoperiodism and light treatments during fruits storage on the phytochrome and on the germination of *Cucumis prophetarum* L. and *Cucumis sativus* L. seeds. *Oecologia* **18**, 37–43.

Gutterman, Y., Thomas, T. H., and Heydecker, W. (1975). Effect on the progeny of applying different day length and hormone treatments to parent plants of *Lactuca scariola*. *Physiol. Plant.* **34**, 30–38.

Haasis, F. W., and Thrupp, A. C. (1931). Temperature relations of lodgepole-pine seed germination. *Ecology* **12**, 728–744.

Hacker, J. B. (1984). Genetic variation in seed dormancy in *Digitaria milanjiana* in relation to rainfall at the collection site. *J. Appl. Ecol.* **21**, 947–959.

Hacker, J. B. (1988). Polyploid distribution and seed dormancy in relation to provenance rainfall in the *Digitaria milanjiana* complex. *Aust. J. Bot.* **36**, 693–700.

Hacker, J. B., Andrew M. H., McIvor, J. G., and Mott, J. J. (1984). Evaluation in contrasting climates of dormancy characteristics of seed of *Digitaria milanjiana*. *J. Appl. Ecol.* **21**, 961–969.

Hacker, J. B., and Ratcliff, D. (1989). Seed dormancy and factors controlling dormancy breakdown in buffel grass accessions from contrasting provenances. *J. Appl. Ecol.* **26**, 201–212.

Haig, D. (1990). New perspectives on the angiosperm female gametophyte. *Bot. Rev.* **56**, 236–274.

Haines, R. W., and Lye, K. A. (1977). Studies in African Cyperaceae. XV. Amphicarpy and spikelet structure in *Trianoptiles solitaria*. *Bot. Notiser* **130**, 235–240.

Halloin, J. M. (1986). Seed improvement through genetic resistance to pathogenesis. *In* "Physiological–Pathological Interactions Affecting Seed Deterioration," pp. 77–95. Crop Science Society of America Spec. Pub. No. 12.

Halloran, G. M., and Collins, W. J. (1974). Physiological predetermination of the order of hardseededness breakdown in subterranean clover (*Trifolium subterraneum* L.). *Ann. Bot.* **38**, 1039–1044.

Hamilton, R. I., Subramanian, B., Reddy, M. N., and Rao, C. H. (1982). Compensation in grain yield components in a panicle of rainfed sorghum. *Ann. Appl. Biol.* **101**, 119–125.

Hammerton, J. L., and Jalloq, M. C. (1970). Studies on weed species of the genus *Polygonum*. VI. Environmental effects on seed weight, seed polymorphism and germination behaviour in *P. lapathifolium* and *P. persicaria*. *Weed Res.* **10**, 204–217.

Hannan, G. L. (1980). Heteromericarpy and dual seed germination modes in *Platystemon californicus* (Papaveraceae). *Madrono* **27**, 163–170.

Hardin, E. D. (1984). Variation in seed weight, number per capsule and germination in *Populus deltoides* Bartr. trees in southeastern Ohio. *Am. Midl. Nat.* **112**, 29–34.

Harlan, J. R. (1945). Cleistogamy and chasmogamy in *Bromus carinatus* Hook. and Arn. *Am. J. Bot.* **32**, 66–72.

Harlan, J. R., de Wet, J. M., and Price, E. G. (1973). Comparative evolution of cereals. *Evolution* **27**, 311–325.

Harniss, R. O., and McDonough, W. T. (1976). Yearly variation in germination in three subspecies of big sagebrush. *J. Range Manage.* **29**, 167–168.

Harper, J. L., Lovell, P. H., and Moore, K. G. (1970). The shapes and sizes of seeds. *Annu. Rev. Ecol. Syst.* **1**, 327–356.

Harper, J. L., and McNaughton, I. H. (1960). The inheritance of dormancy in inter- and intraspecific hybrids of *Papaver*. *Heredity* **15**, 315–320.

Harper, J. L., and Obeid, M. (1967). Influence of seed size and depth of sowing on the establishment and growth of varieties of fiber and oil seed flax. *Crop Sci.* **7**, 527–532.

Harrington, J. F., and Thompson, R. C. (1952). Effect of variety and area of production on subsequent germination of lettuce seed at high temperatures. *Proc. Am. Soc. Hort. Sci.* **59**, 445–450.

Harrington, J. F. (1960). Germination of seeds from carrot, lettuce, and pepper plants grown under severe nutrient deficiencies. *Hilgardia* **30**, 219–235.

Harris, J. A. (1915). The influence of position in the pod upon the weight of the bean seed. *Am. Nat.* **49**, 44–47.

Harris, J. A. (1917). The weight of seeds as related to their number and position. *Torreya* **17**, 180–182.

Hauser, T. P., and Loeschcke, V. (1996). Drought stress and inbreeding depression in *Lychnis flos-cuculi* (Caryophyllaceae). *Evolution* **50**, 1119–1126.

Hayes, R. G., and Klein. W. H. (1974). Spectral quality influence of light during development of *Arabidopsis thaliana* plants in regulating seed germination. *Plant Cell Physiol.* **15**, 643–653.

Hayward, M. D., and Breese, E. L. (1966). The genetic organization of natural populations of *Lolium perenne*. I. Seed and seedling characters. *Heredity* **21**, 287–304.

Hebblethwaite, P. D. (1977). Irrigation and nitrogen studies in S.23

ryegrass grown for seed. 1. Growth, development, seed yield components and seed yield. *J. Agric. Sci. Camb.* **88**, 605–614.

Heide, O. M. (1974). Growth and dormancy in Norway spruce ecotypes. II. After-effects of photoperiod and temperature on growth and development in subsequent years. *Physiol. Plant.* **31**, 131–139.

Heide, O. M., Junttila, O., and Samuelsen, R. T. (1976). Seed germination and bolting in red beet as affected by parent plant environment. *Physiol. Plant.* **36**, 343–349.

Heindl, J. C., and Brun, W. A. (1984). Patterns of reproductive abscission, seed yield, and yield components in soybean. *Crop Sci.* **24**, 542–544.

Helsper, H. P. G., and Klerken, G. A. M. (1984). Germination of *Calluna vulgaris* (L.) Hull *in vitro* under different pH-conditions. *Acta Bot. Neerl.* **33**, 347–353.

Hendrix, S. D. (1979). Compensatory reproduction in a biennial herb following insect defloration. *Oecologia* **42**, 107–118.

Hendrix, S. D. (1984a). Reactions of *Heracleum lanatum* to floral herbivory by *Depressaria pastinacella*. *Ecology* **65**, 191–197.

Hendrix, S. D. (1984b). Variation in seed weight and its effects on germination in *Pastinaca sativa* L. (Umbelliferae). *Am. J. Bot.* **71**, 795–802.

Hermanutz, L. A., and Weaver, S. E. (1991). Variability in temperature-dependent germination in eastern black nightshade (*Solanum ptycanthum*). *Can. J. Bot.* **69**, 1463–1470.

Hernandez, J. E. B., and Clemente, M. M. (1977). Significado ecologico de la heterocarpia en diez especies de la tribu Brassiceae. El case de *Fezia pterocarpa* Pitard. *Ann. Inst. Bot. Cavanilles* **34**, 279–302.

Herrera, J. (1991). The reproductive biology of a riparian Mediterranean shrub, *Nerium oleander* L. (Apocynaceae). *Bot. J. Linn. Soc.* **106**, 147–172.

Heschel, M. S., and K. N. Paige. (1995). Inbreeding depression, environmental stress, and population size variation in scarlet gilia (*Ipomopsis aggregata*). *Conserv. Biol.* **9**, 126–133.

Heslop-Harrison, J. (1959a). Variability and environment. *Evolution* **13**, 145–147.

Heslop-Harrison, J. (1959b). Photoperiodic effects on sexuality, breeding system and seed germinability in *Rottboellia exaltata*. *Proc. IX Int. Bot. Congr.* **2**, 162–163. [Abstract]

Heslop-Harrison J. (1961). The function of the glume pit and the control of cleistogamy in *Botriochloa decipiens* (Hack.) C. E. Hubbard. *Phytomorphology* **11**, 378–383.

Heslop-Harrison, J. (1964). Forty years of genecology. *Adv. Ecol. Res.* **2**, 159–247.

Hill, J. (1965). Environmental induction of heritable changes in *Nicotiana rustica*. *Nature* **207**, 732–734.

Hilton, J. R., Froud-Williams, R. J., and Dixon, J. (1984). A relationship between phytochrome photoequilibrium and germination of seeds of *Poa trivialis* L. from contrasting habitats. *New Phytol.* **97**, 375–379.

Hilu, K. W., and de Wet, J. M. J. (1980). Effect of artificial selection on grain dormancy in *Eleusine* (Gramineae). *Syst. Bot.* **5**, 54–60.

Hodgkin, T., and Hegarty, T. W. (1978). Genetically determined variation in seed germination and field emergence of *Brassica oleracea*. *Ann. Appl. Biol.* **88**, 407–413.

Holdsworth, M. (1966). The cleistogamy of *Viola cunninghamii*. *Trans. Roy. Soc. New Zeal.* **3**, 169–174.

Holton, C. S., and Herald, F. D. (1936). Studies on the control and other aspects of bunt of wheat. *Washington Agric. Exp. Sta. Bull. No.* 339.

Holtsford, T. P., and Ellstrand, N. C. (1990). Inbreeding effects in *Clarkia tembloriensis* (Onagraceae) populations with different natural outcrossing rates. *Evolution* **44**, 2031–2046.

Honek, A., and Martinkova, Z. (1996). Geographic variation in seed dormancy among populations of *Echinochloa crus-galli*. *Oecologia* **108**, 419–423.

Honing, J. A. (1930). Nucleus and plasma in the heredity of the need of light for germination in *Nicotiana* seeds. *Genetica* **12**, 441–468.

Hopper, N. W., Overholt, J. R., and Martin, J. R. (1979). Effect of cultivar, temperature and seed size on the germination and emergence of soya beans (*Glycine max* (L.) Merr.). *Ann. Bot.* **44**, 301–308.

Hou, J. Q., and Simpson, G. M. (1991). Effects of prolonged light on germination of six lines of wild oat (*Avena fatua*). *Can. J. Bot.* **69**, 1414–1417.

Howe, H. F., and Richter, W. M. (1982). Effects of seed size on seedling size in *Virola surinamensis;* a within and between tree analysis. *Oecologia* **53**, 347–351.

Hull, F. H. (1937). Inheritance of rest period of seeds and certain characters in the peanut. *Florida Agri. Exp. Sta. Bull. No.* 314.

Hume, E. P., and Cobin, M. (1946). Relation of seed size to germination and early growth of mangosteen. *Proc. Am. Hort. Soc.* **48**, 298–302.

Hume, L. (1994). Maternal environment effects on plant growth and germination of two strains of *Thlaspi arvense* L. *Int. J. Plant Sci.* **155**, 180–186.

Hunt, O. J., and Miller, D. G. (1965). Coleoptile length, seed size, and emergence in intermediate wheatgrass (*Agropyron intermedium* (Host) Beauv.). *Agron. J.* **57**, 192–195.

Hurka, H., and Benneweg, M. (1979). Patterns of seed size variation in populations of the common weed *Capsella bursa-pastoris* (Brassicaceae). *Biol. Zbl.* **98**, 699–709.

Imbert, E., Escarre, J., and Lepart, J. (1996). Achene dimorphism and among-population variation in *Crepis sancta* (Asteraceae). *Int. J. Plant Sci.* **157**, 309–315.

Ingram, T. J., and Browning, G. (1979). Influence of photoperiod on seed development in the genetic line of peas G2 and its relation to changes in endogenous gibberellins measured by combined gas chromatography–mass spectrometry. *Planta* **146**, 423–432.

Inoue, K., and Washitani, I. (1989). Geographical variation in thermal germination responses in *Campanula punctata* Lam. *Plant Species Biol.* **4**, 69–74.

Jackson, D. I. (1971). Factors affecting response in the barley-endosperm bioassay for gibberellins. *J. Exp. Bot.* **22**, 613–619.

Jacobsohn, R., and Globerson, D. (1980). *Daucus carota* (carrot) seed quality: I. Effects of seed size on germination, emergence and plant growth under subtropical conditions. II. The importance of the primary umbel in carrot-seed production. *In* "Seed Production" (P. D. Hebblethwaite, ed.), pp. 637–646. Butterworths, London/Boston.

Jacques, R. (1968). Action de la lumiere par l'intermediaire du phytochrome sur la germination, la croissance et le developpement de *Chenopodium polyspermum* L. (I). *Physiol. Veg.* **6**, 137–164.

Jain, S. K. (1982). Variation and adaptive role of seed dormancy in some annual grassland species. *Bot. Gaz.* **143**, 101–106.

James, E., and Bancroft, T. A. (1951). The use of half-plants in a balanced incomplete block in investigating the effect of calcium, phosphorus, and potassium, at two levels each, on the production of hard seed in crimson clover, *Trifolium incarnatum*. *Agron. J.* **43**, 96–98.

Jana, A., Acharya, S. N., and Naylor, J. M. (1979). Dormancy studies in seed of *Avena fatua*. 10. On the inheritance of germination behaviour. *Can. J. Bot.* **57**, 1663–1667.

Jana, A., and Naylor, J. M. (1980). Dormancy studies in seed of *Avena fatua*. 11. Heritability for seed dormancy. *Can. J. Bot.* **58**, 91–93.

Jana, S., Upadhyaya, M. K., and Acharya, S. N. (1988). Genetic basis of dormancy and differential response to sodium azide in *Avena fatua* seeds. *Can. J. Bot.* **66**, 635–641.

Janzen, D. H. (1976). Effect of defoliation on fruit-bearing branches of the Kentucky coffee tree, *Gymnocladus dioicus* (Leguminosae). *Am. Midl. Nat.* **95**, 475–478.

Janzen, D. H. (1977). Variation in seed size within a crop of a Costa Rican *Mucuna andreana* (Leguminosae). *Am. J. Bot.* **64**, 347–349.

Janzen, D. H. (1982). Variation in average seed size and fruit seediness in a fruit crop of Guanacaste tree (Leguminosae: *Entrolobium cyclocarpum*). *Am. J. Bot.* **69**, 1169–1178.

Jenner, C. F. (1979). Grain-filling in wheat plants shaded for brief periods after anthesis. *Aust. J. Plant Physiol.* **6**, 629–641.

Johnson, L. P. V. (1935). The inheritance of delayed germination in hybrids of *Avena fatua* and *Avena sativa*. *Can. J. Bot.* **13**, 367–387.

Johnston, M., Fernandez, G., and Olivares, A. (1994). Capacidad germinativa en especies de la pradera annual Mediterranea. II. Efecto del ano de produccion de semillas. *Phyton (Buenos Aires)* **55**, 59–69.

Jones, R. N., and Rees, H. (1969). An anomalous variation due to B chromosomes in rye. *Heredity* **24**, 265–271.

Jordan, J. L., Staniforth, D. W., and Jordan, C. M. (1982). Parental stress and prechilling effects of Pennsylvania smartweed (*Polygonum pensylvanicum*) achenes. *Weed Sci.* **30**, 243–248.

Junttila, O. (1971). Effect of mother plant temperature on seed development and germination in *Syringa reflexa* Schneid. *Meld. Norges Landbruk. Shogsk.* **50**, 1–16.

Junttila, O. (1973). Seeds and embryo germination in *Syringa vulgaris* and *S. reflexa* as affected by temperature during seed development. *Physiol. Plant.* **29**, 264–268.

Junttila, O. (1977). Dormancy in dispersal units of various *Dactylis glomerata* populations. *Seed Sci. Technol.* **5**, 463–471.

Kadereit, J. W. (1984). Studies on the biology of *Senecio vulgaris* L. ssp. *denticulatus* (O. F. Muell.) P. D. Sell. *New Phytol.* **97**, 681–689.

Kadman-Zahavi, A. (1955). Notes on the germination of *Atriplex rosea*. *Bull. Res. Counc. Israel. Sect. D.* **4**, 375–378.

Kalisz, S. (1986). Variable selection on the timing of germination in *Collinsia verna* (Scrophulariaceae). *Evolution* **40**, 479–491.

Kalisz, S. (1989). Fitness consequences of mating system, seed weight, and emergence date in a winter annual, *Collinsia verna*. *Evolution* **43**, 1263–1271.

Kane, M., and Cavers, P. B. (1992). Patterns of weed weight distribution and germination with time in a weedy biotype of proso millet (*Panicum miliaceum*). *Can. J. Bot.* **70**, 562–567.

Karssen, C. M. (1970). The light promoted germination of the seeds of *Chenopodium album* L. III. Effect of the photoperiod during growth and development of the plants on the dormancy of the produced seeds. *Acta Bot. Neerl.* **19**, 81–94.

Karssen, C. M., Zagorski, S., Kepczynski, J., and Groot, S. P. C. (1989). Key role for endogenous gibberellins in the control of seed germination. *Ann. Bot.* **63**, 71–80.

Kasperbauer, M. J. (1968). Dark-germination of reciprocal hybrid seed from light-requiring and -indifferent *Nicotiana tabacum*. *Physiol. Plant.* **21**, 1308–1311.

Kasperbauer, M. J., and Sutton, T. G. (1977). Influence of seed weight on germination, growth, and development of tobacco. *Agron. J.* **69**, 1000–1002.

Keeler, K. H. (1978). Intra-population differentiation in annual plants. II. Electrophoretic variation in *Veronica peregrina*. *Evolution* **32**, 638–645.

Khan, M. A., and Ungar, I. A. (1984a). The effect of salinity and temperature on the germination of polymorphic seeds and growth of *Atriplex triangularis* Willd. *Am. J. Bot.* **71**, 481–489.

Khan, M. A., and Ungar, I. A. (1984b). Seed polymorphism and germination responses to salinity stress in *Atriplex triangularis* Willd. *Bot. Gaz.* **145**, 487–494.

Khan, R. A., and Laude, H. M. (1969). Influence of heat stress during

seed maturation on germinability of barley seed at harvest. *Crop Sci.* **9**, 55–58.

Khatri, R., Sethi, V., and Kaushik, A. (1991). Inter-population variations of *Kochia indica* during germination under different stresses. *Ann. Bot.* **67**, 413–415.

Kidd, F., and West, C. (1918). Physiological pre-determination: The influence of the physiological condition of the seed upon the course of subsequent growth and upon the yield. II. Review of literature. *Ann. Appl. Biol.* **5**, 112–142.

Kigel, J. (1992). Diaspore heteromorphism and germination in populations of the ephemeral *Hedypnois rhagadioloides* (L.) F. W. Schmidt (Asteraceae) inhabiting a geographic range of increasing aridity. *Acta Oecol.* **13**, 45–53.

Kigel, J., Gibly, A., and Negbi, M. (1979). Seed germination in *Amaranthus retroflexus* L. as affected by the photoperiod and age during flower induction of the parent plants. *J. Exp. Bot.* **30**, 997–1002.

Kigel, J., Lior, E., and Rubin, B. (1992). Biology of reproduction in the summer annual weed *Euphorbia geniculata* Ortega. *Weed Res.* **32**, 317–328.

Kigel, J., Ofir, M., and Koller, D. (1977). Control of the germination responses of *Amaranthus retroflexus* L. seeds by their parental photothermal environment. *J. Exp. Bot.* **28**, 1125–1136.

Kilen, T. C., and Hartwig, E. E. (1978). An inheritance study of impermeable seed in soybeans. *Field Crops Res.* **1**, 65–70.

Kim, S. Y., de Datta, S. K., and Mercado, B. L. (1990). The effect of chemical and heat treatments on germination of *Commelina benghalensis* L. aerial seeds. *Weed Res.* **30**, 109–116.

Kiniry, J. R., Wood, C. A., Spanel, D. A., and Bockholt, A. J. (1990). Seed weight response to decreased seed number in maize. *Agron. J.* **54**, 98–102.

Kinsman, S., and Platt, W. J. (1984). The impact of a herbivore upon *Mirabilis hirsuta*. *Oecologia* **65**, 2–6.

Kitchen, S. G., and Monsen, S. B. (1994). Germination rate and emergence success in bluebunch wheatgrass. *J. Range. Manage.* **47**, 145–150.

Kneebone, W. R. (1960). Size of caryopses in buffalograss (*Buchloe dactyloides* (Nutt.) Engelm.) as related to their germination and longevity. *Agron. J.* **52**, 553.

Koch, E. W., and Dawes, C. J. (1991). Influence of salinity and temperature on the germination of *Ruppia maritima* L. from the North Atlantic and Gulf of Mexico. *Aquat. Bot.* **40**, 387–391.

Koller, D. (1957). Germination-regulating mechanisms in some desert seeds. IV. *Atriplex dimorphostegia* Kar. et Kir. *Ecology* **38**, 1–13.

Koller, D. (1962). Preconditioning of germination in lettuce at time of fruit ripening. *Am. J. Bot.* **49**, 841–844.

Koller, D., and Roth, N. (1964). Studies on the ecological and physiological significance of amphicarpy in *Gymnarrhena micrantha* (Compositae). *Am. J. Bot.* **51**, 26–35.

Krannitz, P. G., Aarssen, L. W., and Dow, J. M. (1991). The effect of genetically based differences in seed size on seedling survival in *Arabidopsis thaliana* (Brassicaceae). *Am. J. Bot.* **78**, 446–450.

Krishnasamy, V., and Seshu, D. V. (1989). Seed germination rate and associated characters in rice. *Crop Sci.* **29**, 904–908.

Kromer, M., and Gross, K. L. (1987). Seed mass, genotype, and density effects on growth and yield of *Oenothera biennis* L. *Oecologia* **73**, 207–212.

Lacey, E. P. (1996). Parental effects in *Plantago lanceolata* L. I.: A growth chamber experiment to examine pre- and postzygotic temperature effects. *Evolution* **50**, 865–878.

Ladiges, P. Y., Foord, P. C., and Willis, R. J. (1981). Salinity and waterlogging tolerance of some populations of *Melaleuca ericifolia* Smith. *Aust. J. Ecol.* **6**, 203–215.

Lalonde, R. G., and Roitberg, B. D. (1989). Resource limitation and offspring size and number trade-offs in *Cirsium arvense* (Asteraceae). *Am. J. Bot.* **76**, 1107–1113.

Lambert, R. G., and Linck, A. J. (1958). Effects of high temperature on yield of peas. *Plant Physiol.* **33**, 347–350.

Lambert, R. J., Alexander, D. E., and Rodgers, R. C. (1967). Effect of kernel position on oil content in corn (*Zea mays* L.). *Crop Sci.* **7**, 143–144.

Langdon, O. G. (1958). Cone and seed size of south Florida slash pine and their effects on seedling size and survival. *J. For.* **56**, 122–127.

Langer, R. H. M., and Wilson, D. (1965). Environmental control of cleistogamy in prairie grass (*Bromus unioloides* H.B.K.). *New Phytol.* **64**, 80–85.

Lebedeff, G. A. (1947). Studies on the inheritance of hard seeds in beans. *J. Agric. Res.* **74**, 205–215.

Le Corff, J. (1996). Establishment of chasmogamous and cleistogamous seedlings of an ant-dispersed understory herb, *Calathea micans* (Marantaceae). *Am. J. Bot.* **83**, 155–161.

Lee, J. A. (1975). Inheritance of hard seed in cotton. *Crop Sci.* **15**, 149–152.

Lee, J. W., and Ronalds, J. A. (1967). Effect of environment on wheat gliadin. *Nature* **213**, 844–846.

Leng, E. R. (1949). Direct effect of pollen parent on kernel size in dent corn. *Agron. J.* **41**, 555–558.

Lesica, P. (1993). Loss of fitness resulting from pollinator exclusion in *Silene spaldingii* (Caryophyllaceae). *Madrono* **40**, 193–201.

Levin, D. A. (1972). Plant density, cleistogamy, and self-fertilization in natural populations of *Lithospermum caroliniense*. *Am. J. Bot.* **59**, 71–77.

Levyns, M. R. (1935). Germination in some South African seeds. *S. Afr. J. Bot.* **1**, 161–170.

Lewis, J. D., and Koide, R. T. (1990). Phosphorus supply, mycorrhizal infection and plant offspring vigor. *Funct. Ecol.* **4**, 695–702.

Lexander, K. (1980). Seed composition in connection with germination and bolting of *Beta vulgaris* L. (sugar beet). *In* "Seed Production" (P. D. Hebblethwaite, ed.), pp. 271–291. Butterworths, London.

Lindauer, L. L., and Quinn, J. A. (1972). Germination ecology of *Danthonia sericea* populations. *Am. J. Bot.* **59**, 942–951.

Linhart, Y. B. (1974). Intra-population differentiation in annual plants. I. *Veronica peregrina* L. raised under non-competitive conditions. *Evolution* **28**, 232–243.

Lona, F. (1947). L'influenza delle condizioni esterne durante l'embriogenese in *Chenopodium amaranticolor* Coste et Reyn. sulle qualita germinative dei semi e sul vigore delle plantule che ne derivano. Lavori di Bot. G. Gola Jubilee, Padova, pp. 324–352.

Long, R. W. (1977). Artificial induction of obligate cleistogamy in species-hybrids in *Ruellia* (Acanthaceae). *Bull. Torrey Bot. Club* **104**, 53–56.

Lord, E. M. (1981). Cleistogamy: A tool for the study of floral morphogenesis, function and evolution. *Bot. Rev.* **47**, 421–449.

Lord, E. M. (1982). Effect of daylength on open flower production in the cleistogamous species *Lamium amplexicaule* L. *Ann. Bot.* **49**, 261–263.

Lovell, P. H., Maxwell, C. D., and Jacob, N. (1986). Variation in cypsela morphology in *Soliva valdiviana* and *S. pterosperma* (Anthemideae, Asteraceae) in a local population in Auckland, New Zealand. *New Zeal. J. Bot.* **24**, 657–664.

Lu, X., and Koide, R. T. (1991). *Avena fatua* L. seed and seedling nutrient dynamics as influenced by mycorrhizal infection of the maternal generation. *Plant, Cell Environ.* **14**, 931–939.

Lute, A. M. (1928). Impermeable seed of alfalfa. *Colorada Agric. Exp. Sta. Bull.* 326.

Lyon, T. L. (1902). The importance of considering previous environment in conducting variety tests. *Proc. Soc. Prom. Agric. Sci.* **23**, 70–73.

Lyon, T. L. (1907). Modifications in cereal crops induced by changes in their environment. *Proc. Soc. Prom. Agric. Sci.* **28**, 144–163.

Madge, M. A. P. (1929). Spermatogenesis and fertilization in the cleistogamous flower of *Viola odorata*, var. *praecox*, Gregory. *Ann. Bot.* **43**, 543–577.

Maheshwari, J. K. (1962). Cleistogamy in angiosperms. *In* "Proceedings of the Summer School of Botany" (P. Maheshwari, B. M. Johri, and I. K. Vasil, eds.), pp. 145–155. Ministry of Scientific Research and Cultural Affairs, Government of India, New Delhi.

Maheshwari, P., and Maheshwari, J. K. (1955). Floral dimorphism in *Commelina forskalaei* Vahl and *C. benghalensis* L. *Phytomorphology* **5**, 413–422.

Major, D. J. (1977). Influence of seed size on yield and yield components of rape. *Agron. J.* **69**, 541–543.

Maluf, A. M., and Martins, P. S. (1991). Germinacao de sementes de *Amaranthus hybridus* L. e *A. viridis* L. *Rev. Brasil. Biol.* **51**, 417–425.

Manilal, K. S., and Unni, P. N. (1978). Performance variation with relation to morphological evolution in the dimorphic seeds of *Synedrella nodiflora* Gaertn. *In* "Environmental Physiology and Ecology of Plants" (D. N. Sen and R. P. Bansal, eds.), pp. 229–234. Bishen Singh and Mahendra Pal Singh Publishers, Dehra Dun, India.

Maranon, T. (1987). Ecologia del polimorfismo somatico de semillas y la sinaptospermia en *Aegilops neglecta* Req. ex Bertol. *Anales J. Bot. Madrid* **44**, 97–107.

Maranon, T. (1989). Variation in seed size and germination in three *Aegilops* species. *Seed Sci. Technol.* **17**, 583–588.

Mariko, S., Koizumi, H., Suzuki, J.-I., and Furukawa, A. (1993). Altitudinal variations in germination and growth responses of *Reynoutria japonica* populations on Mt. Fuji to a controlled thermal environment. *Ecol. Res.* **8**, 27–34.

Marks, M. K., and Akosim, C. (1984). Achene dimorphism and germination in three composite weeds. *Trop. Agric.* **61**, 69–73.

Marshall, D. L. (1986). Effect of seed size on seedling success in three species of *Sesbania* (Fabaceae). *Am. J. Bot.* **73**, 457–464.

Marshall, D. L., Levin, D. A., and Fowler, N. L. (1986). Plasticity of yield components in response to stress in *Sesbania macrocarpa* and *Sesbania vesicaria* (Leguminosae). *Am. Nat.* **127**, 508–521.

Martin, A. C. (1946). The comparative internal morphology of seeds. *Am. Midl. Nat.* **36**, 513–660.

Martinez-Carrasco, R., and Thorne, G. N. (1979). Physiological factors limiting grain size in wheat. *J. Exp. Bot.* **30**, 669–679.

Martinkova, Z., and Honek, A. (1995a). Termination of dormancy in caryopses of barnyard grass, *Echinochloa crus-galli*. *Ochr. Rostl.* **31**, 11–17.

Martinkova, Z., and Honek, A. (1995b). Seasonal changes in the occurrence of seed dormancy in caryopses of the barnyard grass, *Echinochloa crus-galli*. *Ochr. Rostl.* **31**, 249–256.

Masuda, M., and Washitani, I. (1992). Differentiation of spring emerging and autumn emerging ecotypes in *Galium spurium* L. var. *echinospermon*. *Oecologia* **89**, 42–46.

Mather, K., and Vines, A. (1951). Species crosses in *Antirrhinum*. II. Cleistogamy in the derivatives of *A. majus* × *A. glutinosum*. *Heredity* **5**, 195–214. + 1 plate.

Mathez, J., and Xena de Enrech, N. (1985a). Heterocarpy, fruit polymorphism and discriminating dissemination in the genus *Fedia* (Valerianaceae). *In* "Genetic Differentiation and Dispersal in Plants" (P. Jacquard, G. Heim, and J. Antonovics eds.), pp. 431–441. Springer-Verlag, Berlin.

Mathez, J., and Xena de Enrech, N. (1985b). Le polymorphisme genetique de la morphologie des fruits du genre *Fedia* Gaertn. (Valerianaceae). I. Determination du mecanisme de controle genetique chez les especes *F. cornucopiae* (L.) Gaertn. et *F. graciliflora* Fisch. and Meyer. *Candollea* **40**, 425–434.

Mathur, R. S., Sharma, K. K., and Rawat, M. M. S. (1984). Germination behaviour of various provenances of *Acacia nilotica* ssp. *indica*. *Indian For.* **110**, 435–449.

Mattatia, J. (1977a). Amphicarpy and variability in *Pisum fulvum*. *Bot. Notiser* **130**, 27–34.

Mattatia, J. (1977b). The amphicarpic species *Lathyrus ciliolatus*. *Bot. Notiser* **129**, 437–444.

Maun, M. A., Canode, C. L., and Teare, I. D. (1969). Influence of temperature during anthesis on seed set in *Poa pratensis* L. *Crop Sci.* **9**, 210–212.

Maun, M. A., and Cavers, P. B. (1971). Seed production and dormancy in *Rumex crispus*. II. The effects of removal of various proportions of flowers at anthesis. *Can. J. Bot.* **49**, 1841–1848.

Maun, M. A., and Payne, A. M. (1989). Fruit and seed polymorphism and its relation to seedling growth in the genus *Cakile*. *Can. J. Bot.* **67**, 2743–2750.

Maurya, A. N., and Ambasht, R. S. (1973). Significance of seed dimorphism in *Alysicarpus monilifer* DC. *J. Ecol.* **61**, 213–217.

Maxwell, C. D., Zobel, A., and Woodfine, D. (1994). Somatic polymorphism in the achenes of *Tragopogon dubius*. *Can. J. Bot.* **72**, 1282–1288.

Mayers, A. M., and Lord, E. M. (1983). Comparative flower development in the cleistogamous species *Viola odorata*. I. A growth rate study. *Am. J. Bot.* **70**, 1548–1555.

Mayers, A. M., and Lord, E. M. (1984). Comparative flower development in the cleistogamous species *Viola odorata*. III. A histological study. *Bot. Gaz.* **145**, 83–91.

Mazer, S. J., Snow, A. A., and Stanton, M. L. (1986). Fertilization dynamics and parental effects upon fruit development in *Raphanus raphanistrum*: Consequences for seed size variation. *Am. J. Bot.* **73**, 500–511.

McCullough, J. M., and Shropshire, W., Jr. (1970). Physiological predetermination of germination responses in *Arabidopsis thaliana* (L.) Heynh. *Plant Cell Physiol.* **11**, 139–148.

McDonough, W. T. (1975). Germination polymorphism in *Grindelia squarrosa* (Pursh) Dunal. *Northw. Sci.* **49**, 190–200.

McEvoy, P. B. (1984). Dormancy and dispersal in dimorphic achenes of tansy ragwort, *Senecio jacobaea* L. (Compositae). *Oecologia* **61**, 160–168.

McFadden, A. D. (1963). Effect of seed source on comparative test results in barley. *Can. J. Plant Sci.* **43**, 295–300.

McGee, K. P., and Marshall, D. L. (1993). Effects of variable moisture availability on seed germination in three populations of *Larrea tridentata*. *Am. Midl. Nat.* **130**, 75–82.

McGinley, M. A. (1989a). The influence of a positive correlation between clutch size and offspring fitness on the optimal offspring size. *Evol. Ecol.* **3**, 150–156.

McGinley, M. A. (1989b). Within and among plant variation in seed mass and pappus size in *Tragopogon dubius*. *Can. J. Bot.* **67**, 1298–1304.

McGinley, M. A., and Charnov, E. L. (1988). Multiple resources and the optimal balance between size and number of offspring. *Evol. Ecol.* **2**, 77–84.

McGinley, M. A., Temme, D. H., and Geber, M. A. (1987). Parental investment in offspring in variable environments: Theoretical and empirical considerations. *Am. Nat.* **130**, 370–398.

McGregor, R. L. (1990). Seed dormancy and germination in the annual cleistogamous species of *Sporobolus* (Poaceae). *Trans. Kansas Acad. Sci.* **93**, 8–11.

McKell, C. M., Robison, J. P., and Major, J. (1962). Ecotypic variation in medusahead, an introduced annual grass. *Ecology* **43**, 686–698.

McNamara, J., and Quinn, J. A. (1977). Resource allocation and reproduction in populations of *Amphicarpum purshii* (Gramineae). *Am. J. Bot.* **64**, 17–23.

McNaughton, S. J. (1966). Ecotype function in the *Typha* community-type. *Ecol. Monogr.* **36**, 297–325.

McNeal, F. H., Berg, M. A., Dubbs, A. L., Krall, J. L., Baldridge, D. E., and Hartman, G. P. (1960). The evaluation of spring wheat seed from different sources. *Agron. J.* **52**, 303–304.

McWilliams, E. L., Landers, Q. L., and Mahlstede, J. P. (1968). Variation in seed weight and germination in populations of *Amaranthus retroflexus* L. *Ecology* **49**, 290–296.

Meckel, L., Egli, D. B., Phillips, R. E., Radcliffe, D., and Leggett, J. E. (1984). Effect of moisture stress on seed growth in soybeans. *Agron. J.* **76**, 647–650.

Mejia P. V., Romero M. C., and Lotero C. J. (1978). Factores que afectan la germinacion y el vigor de la semilla del pasto guinea (*Panicum maximum* Jacq.). *Rev. Inst. Colombia Agropec.* **13**, 69–76.

Menges, E. S. (1991). Seed germination percentage increases with population size in a fragmented prairie species. *Conserv. Biol.* **5**, 158–164.

Meyer, S. E. (1990). Seed source differences in germination under snowpack in northern Utah. *In* "Proc. Fifth Billings Symposium on Disturbed Land Reclamation," Vol. 1, pp. 184–191. Montana State University Reclamation Research Unit, Bozeman.

Meyer, S. E. (1992). Habitat correlated variation in firecracker penstemon (*Penstemon eatonii* Gray: Scrophulariaceae) seed germination response. *Bull. Torrey Bot. Club* **119**, 268–279.

Meyer, S. E., Allen, P. S., and Beckstead, J. (1997). Seed germination regulation in *Bromus tectorum* (Poaceae) and its ecological significance. *Oikos* **78**, 475–485.

Meyer, S. E., Beckstead, J., Allen, P. S., and Pullman, H. (1995). Germination ecophysiology of *Leymus cinereus* (Poaceae). *Int. J. Plant Sci.* **156**, 206–215.

Meyer, S. E., and Kitchen, S. G. (1994). Life history variation in blue flax (*Linum perenne*: Linaceae): Seed germination phenology. *Am. J. Bot.* **81**, 528–535.

Meyer, S. E., Kitchen, S. G., and Carlson, S. L. (1995). Seed germination timing patterns in intermountain *Penstemon* (Scrophulariaceae). *Am. J. Bot.* **82**, 377–389.

Meyer, S. E., and Monsen, S. B. (1991). Habitat-correlated variation in mountain big sagebrush (*Artemisia tridentata* ssp. *vaseyana*) seed germination patterns. *Ecology* **72**, 739–742.

Meyer, S. E., Monsen, S. B., and McArthur, E. D. (1990). Germination response of *Artemisia tridentata* (Asteraceae) to light and chill: Patterns of between-population variation. *Bot. Gaz.* **151**, 176–183.

Meyer, S. E., McArthur, E. D., and Jorgensen, G. L. (1989). Variation in germination response to temperature in rubber rabbitbrush (*Chrysothamnus nauseosus*: Asteraceae) and its ecological implications. *Am. J. Bot.* **76**, 981–991.

Meyer, S. E., and Pendleton, R. L. (1990). Seed germination biology of spineless hopsage: Between-population differences in dormancy and response to temperature. *In* "Proceedings–Symposium on Cheatgrass Invasion, Shrub Die-off, and Other Aspects of Shrub Biology and Management" (E. D. McArthur, E. M. Romnex, S. D. Smith, and P. T. Tueller, compilers), pp. 187–192. USDA Forest Serv. Gen. Tech. Rep. INT-276.

Miao, S. L., Bazzaz, F. A., and Primack, R. B. (1991). Persistence of maternal nutrient effects in *Plantago major*: The third generation. *Ecology* **72**, 1634–1642.

Minter, T. C., and Lord, E. M. (1983). Effects of water stress, abscisic acid, and gibberellic acid on flower production and differentiation in the cleistogamous species *Collomia grandiflora* Dougl. ex Lindl. (Polemoniaceae). *Am. J. Bot.* **70**, 618–624.

Minton, E. B., Wanjura, D. F., and Bilbro, J. D. (1979). Effects of premature defoliation and plant kill on germination of cottonseed. *Agron. J.* **71**, 659–661.

Mishra, R. K., and Sen, D. N. (1984). A report on germination behaviour in dimorphic seeds of *Indigofera hochstetteri* Baker, in Indian arid zone. *Curr. Sci.* **53**, 380–381.

Mishra, R. K., and Sen, D. N. (1986). Polymorphism in populations of *Tephrosia purpurea* Pers. in Indian Desert ecosystem. *Flora* **178**, 183–190.

Moegenburg, S. M. (1996). *Sabal palmetto* seed size: Causes of variation, choices of predators, and consequences for seedlings. *Oecologia* **106**, 539–543.

Mogie, M., Latham, J. R., and Warman, E. A. (1990). Genotype-independent aspects of seed ecology in *Taraxacum*. *Oikos* **59**, 175–182.

Mohamed, H. A., Clark, J. A., and Ong, C. K. (1985). The influence of temperature during seed development on the germination characteristics of millet seeds. *Plant Cell Environ.* **8**, 361–362.

Mohammed, S., and Sen, D. N. (1988). A report on polymorphic seeds in halophytes 1. *Trianthema triquetra* L. in Indian Desert. *Curr. Sci.* **57**, 616–617.

Mohammed, S., and Sen, D. N. (1990). Biology and ecophysiology of *Trianthema portulacastrum* L. (Molluginaceae) in arid ecosystem. *Folia Geobot. Phytotax.* **25**, 145–158.

Moore, D. R. J., and Cavers, P. J. (1985). A comparison of seedling vigour in crop and weed biotypes of proso millet (*Panicum miliaceum*). *Can. J. Bot.* **63**, 1659–1663.

Morgenstern, E. K. (1969). Genetic variation in seedlings of *Picea mariana* (Mill.) BSP. *Silvae Genet.* **18**, 151–161.

Morse, D. H., and Schmitt, J. (1985). Propagule size, dispersal ability, and seedling performance in *Asclepias syriaca*. *Oecologia* **67**, 372–379

Moss, J. P. (1966). The adaptive significance of B-chromosomes in rye. *In* "Chromosomes Today" (C. D. Darlington and K. R. Lewis, eds.), Vol. 1, pp. 15–23. Oliver and Boyd, Edinburgh.

Moyer, J. L., and Lang, R. L. (1976). Variable germination response to temperature for different sources of winterfat seed. *J. Range Manage.* **29**, 320–321.

Moran-Palma, P., and Snow, A. A. (1997). The effect of interplant distance on mating success in federally-threatened, self-incompatible *Hymenoxys herbacea* = *H. acaulis* var. *glabra* (Asteraceae). *Am. J. Bot.* **84**, 233–238.

Morley, F. H. W. (1958). The inheritance and ecological significance of seed dormancy in subterranean clover (*Trifolium subterraneum* L.). *Aust. J. Biol. Sci.* **11**, 261–274.

Muchena, S. C., and Grogan, C. O. (1977). Effects of seed size on germination of corn (*Zea mays*) under simulated water stress conditions. *Can. J. Plant Sci.* **57**, 921–923.

Mulcahy, D. L. (1971). A correlation between gametophytic and sporophytic characteristics in *Zea mays* L. *Science* **171**, 1155–1156.

Munns, E. N. (1921). Effect of location of seed upon germination. *Bot. Gaz.* **72**, 256–260.

Murphy, C. F., and Frey, K. J. (1962). Inheritance and heritability of seed weight and its components in oats. *Crop Sci.* **2**, 509–512.

Musil, C. F. (1994). Ultraviolet-B irradiation of seeds affects photochemical and reproductive performance of the arid-environment ephemeral *Dimorphotheca pluvialis*. *Environ. Exp. Bot.* **34**, 371–378.

Musil, C. F. (1996). Accumulated effect of elevated ultraviolet-B radiation over multiple generations of the arid-environment annual *Dimorphotheca sinuata* DC. (Asteraceae). *Plant, Cell Environ.* **19**, 1017–1027.

Nan, X. (1992). "Comparison of Some Aspects of the Ecological Life History of an Annual and a Perennial Species of *Senna* (Leguminosae: Section *Chamaefistula*), with Particular Reference to Seed Dormancy." M.S. thesis, University of Kentucky, Lexington.

Naylor, J. M. (1983). Studies on the genetic control of some physiological processes in seeds. *Can. J. Bot.* **61**, 3561–3567.

Naylor, J. M., and Jana, S. (1976). Genetic adaptation for seed dormancy in *Avena fatua*. *Can. J. Bot.* **54**, 306–312.

Naylor, R. E. L. (1980). Effects of seed size and emergence time on subsequent growth of perennial ryegrass. *New Phytol.* **84**, 313–318.

Naylor, R. E. L. (1993). The effect of parent plant nutrition on seed size, viability and vigour and on germination of wheat and triticale at different temperatures. *Ann. Appl. Biol.* **123**, 379–390.

Naylor, R. E. L., and Abdalla, A. F. (1982). Variation in germination behaviour. *Seed Sci. Technol.* **10**, 67–76.

Negbi, M., and B. Tamari. (1963). Germination of chlorophyllous and achlorophyllous seeds of *Salsola volkensii* and *Aellenia autrani*. *Israel J. Bot.* **12**, 124–135.

Nelson, J. R., Harris, G. A., and Goebel, C. J. (1970). Genetic vs. environmentally induced variation in medusahead (*Taeniatherum asperum* [Simonkai] Nevski. *Ecology* **51**, 526–529.

Nelson, L. R. (1980). Recurrent selection for improved rate of germination in annual ryegrass. *Crop Sci.* **20**, 219–221.

Neuffer, B., and Bartelheim, S. (1989). Gen-ecology of *Capsella bursa-pastoris* from an altitudinal transect in the alps. *Oecologia* **76**, 521–527.

New, J. (1958). A population study of *Spergula arvensis*. I. Two clines and their significance. *Ann. Bot.* **22**, 457–477.

New, J. (1959). A population study of *Spergula arvensis*. II. Genetics and breeding behaviour. *Ann. Bot.* **23**, 23–33.

Niazi, M. L. K., Mahmood, K., and Malik, K. A. (1987). Salt tolerance studies in different cultivars of barley (*Hordeum vulgare* L.). *Pakistan J. Bot.* **19**, 17–27.

Niazi, M. L. K., Mahmood, K., Mujtaba, S. M., and Malik, K. A. (1992). Salinity tolerance in different cultivars of barely (*Hordeum vulgare* L.). *Biol. Plant.* **34**, 465–469.

Nicholls, P. B. (1982). Influence of temperature during grain growth and ripening of barley on the subsequent response to exogenous gibberellic acid. *Aust. J. Plant Physiol.* **9**, 373–383.

Nobs, M. A., and Hagar, W. G. (1974). Analysis of germination and flowering rates of dimorphic seeds from *Atriplex hortensis*. *Carnegie Inst. Wash. Yrbk.* **73**, 860–864.

Nolan, D. G., and Upadhyaya, M. K. (1988). Primary seed dormancy in diffuse and spotted knapweed. *Can. J. Plant Sci.* **68**, 775–783.

Norman, J. K., Sakai, A. K., Weller, S. G., and Dawson, T. E. (1995). Inbreeding depression in morphological and physiological traits of *Schiedea lydgatei* (Caryophyllaceae) in two environments. *Evolution* **49**, 297–306.

Noronha, A., Vicente, M., and Felippe, G. M. (1976). Effect of storage and growth conditions on photoblasticity of seeds of *Cucumis anguria*. *Hoehnea* **6**, 7–10.

Norrington-Davies, J., and Harries, J. H. (1977). Competition studies in diploid and tetraploid varieties of *Lolium perenne*. I. The influence of density and proportion of sowing. *J. Agric. Sci. Camb.* **88**, 405–410.

Nosova, L. I. (1981). Germination and germinability of seeds of *Artemisia rhodantha* Rupr. *Soviet J. Ecol.* **12**, 25–30.

Nybom, H. (1980). Germination in Swedish blackberries (*Rubus* L. subgen. *Rubus*). *Bot. Notiser* **133**, 619–631.

Ohara, M., and Shimamoto, Y. (1994). Some ecological and demographic characteristics of two growth forms of wild soybean (*Glycine soja*). *Can. J. Bot.* **72**, 486–492.

Okagami, N., and Kawai, M. (1983). Dormancy in *Dioscorea*: Range, duration and timing of high-temperature treatment in germination inhibition of *D. tokoro* seeds. *Plant Cell Physiol.* **24**, 509–515.

Okusanya, O. T., and Sonaike, A. A. (1991). Germination behaviour of

Dactyloctenium aegyptium from two localities in Nigeria. *Physiol. Plant.* **81**, 489–494.

Okusanya, O. T., and Ungar, I. A. (1983). The effects of time of seed production on the germination response of *Spergularia marina*. *Physiol. Plant.* **59**, 335–342.

Olivieri, A. M., and Jain, S. K. (1978). Effects of temperature and light variations on seed germination in sunflower (*Helianthus*) species. *Weed Sci.* **26**, 277–280.

Olson, D. S. (1932). Germinative capacity of seed produced from young trees. *J. For.* **30**, 871.

Olson, J. S., Stearns, F. W., and Nienstaedt, H. (1959). Eastern hemlock seeds and seedlings response to photoperiod and temperature. *Connecticut Agric. Exp. Sta. Bull. No. 620.*

Onnis, A., Bertacchi, A., Lombardi, T., and Stefani, A. (1995). Morphology and germination of yellow and brown caryopses of *Aegilops geniculata* Roth (Gramineae) population from Italy. *Giorn. Bot. Ital.* **129**, 813–821.

Onnis, A., Stefani, A., Arduini, I., Bertacchi, A., and Lombardi, T. (1993). Germination ecophysiology of Italian Gramineae species. *In* "Fourth International Workshop on Seeds: Basic and Applied Aspects of Seed Biology" (D. Come and F. Corbineau, eds.), Vol. 2, pp. 443–448. Universite Pierre et Marie Curie, Paris.

Orozco-Segovia, A., Sanchez-Coronado, M. E., and Vazquez-Yanes, C. 1993). Effect of maternal light environment on seed germination in *Piper auritum*. *Funct. Ecol.* **7**, 395–402.

Otubo, S. T., Ramalho, M. A. P., Abreu, A. F. B., and dos Santos, J. B. (1996). Genetic control of low temperature tolerance in germination of the common bean (*Phaseolus vulgaris* L.). *Euphytica* **89**, 313–317.

Painter, E. P., Nesbitt, L. L., and Stoa, T. E. (1944). The influence of seasonal conditions on soil formation and changes in the iodine number during growth of flaxseed. *Agron. J.* **36**, 204–213.

Pallas, J. E., Jr., Stansell, J. R., and Bruce, R. R. (1977). Peanut seed germination as related to soil water regime during pod development. *Agron. J.* **69**, 381–383.

Pammel, L. H. (1898). Some germination studies of cereals. *Proc. Soc. Prom Agric. Sci.* **19**, 194–203.

Pamplona, P. P., and Mercado, B. L. (1981). Ecotypes of *Rottboellia exaltata* L.f. in the Philippines. I. Characteristics and dormancy of seeds. *Philippine Agric.* **64**, 59–66.

Panetta, F. D., and Randall, R. P. (1993). Variation between *Emex australis* populations in seed dormancy/non-dormancy cycles. *Aust. J. Ecol.* **18**, 275–280.

Pannangpetch, K., and Bean, E. W. (1984). Effects of temperature on germination in populations of *Dactylis glomerata* from NW Spain and central Italy. *Ann. Bot.* **53**, 633–639.

Parks, M. (1935). Embryo sac development and cleistogamy in *Commelinantia pringlei*. *Bull. Torrey Bot. Club* **62**, 91–104 + 2 plates.

Parrish, J. A. D., and Bazzaz, F. A. (1985). Nutrient content of *Abutilon theophrasti* seeds and the competitive ability of the resulting plants. *Oecologia* **65**, 247–251.

Paterson, J. G., Goodchild, N. A., and Boyd, W. J. R. (1976). Effect of storage temperature, storage duration and germination temperature on the dormancy of seed of *Avena fatua* L. and *Avena barbata* Pott ex Link. *Aust. J. Agric. Res.* **27**, 373–379.

Pathak, P. S., Debroy, R., and Rai, P. (1973). Autecology of *Leucaena leucocephala* (Lam.) De Wit. I. Seed polymorphism and germination. *Trop. Ecol.* **15**, 1–10.

Paul, M. S., and Sen, D. N. (1987). A new report on dimorphism in seeds of *Capparis decidua* (Forsk.) Edgew. in Indian Desert. *Curr. Sci.* **56**, 1017–1019.

Peacock, J. T., and McMillan, C. (1965). Ecotypic differentiation in *Prosopis* (mesquite). *Ecology* **46**, 35–51.

Pegtel, D. M. (1972). Effects of temperature and moisture on the germination of two ecotypes of *Sonchus arvensis* L. *Acta Bot. Neerl.* **21**, 48–53.

Perez-Garcia, F. (1993). Germination behaviour of *Onopordum acanthium* L. (Asteraceae): Effect of the origin of the cypsela. *In* "Fourth International Workshop on Seeds: Basic and Applied Aspects of Seed Biology" (D. Come and F. Corbineau eds.), pp. 417–422. Universite Pierre et Marie Curie, Paris.

Perez-Garcia, F., Iriondo, J. M., and Martinez-Laborde, J. B. (1995). Germination behaviour in seeds of *Diplotaxis erucoides* and *D. virgata*. *Weed Res.* **35**, 495–502.

Peters, N. C. B. (1982a). The dormancy of wild oat seed (*Avena fatua* L.) from plants grown under various temperature and soil moisture conditions. *Weed Res.* **22**, 205–212.

Peters, N. C. B. (1982b). Production and dormancy of wild oat (*Avena fatua*) seed from plants grown under soil water stress. *Ann. Appl. Biol.* **100**, 189–196.

Peterson, A. E., and Berger, K. C. (1950). Effect of magnesium on the quality and yield of canning peas. *Proc. Am. Soil Sci. Soc.* **15**, 205–208.

Pettersson, M. W. (1992). Advantages of being a specialist female in nodioecious *Silene vulgaris* s. l. (Caryophyllaceae). *Am. J. Bot.* **79**, 1389–1395.

Philippi, T. (1993). Bet-hedging germination of desert annuals: Variation among populations and maternal effects in *Lepidium lasiocarpum*. *Am. Nat.* **142**, 488–507.

Piatt, J. R. (1973). Seed size affects germination of true mountain mahogany. *J. Range Manage.* **26**, 231–232.

Pitcher, J. A. (1984). Geographic variation patterns in seed and nursery characteristics of black cherry. *USDA For. Serv. Res. Paper SO–208.*

Pitelka, L. F., Thayer, M. E., and Hansen, S. B. (1983). Variation in achene weight in *Aster acuminatus*. *Can. J. Bot.* **61**, 1415–1420.

Platenkamp, G. A. J. (1991). Phenotypic plasticity and population differentiation in seeds and seedlings of the grass *Anthoxanthum odoratum*. *Oecologia* **88**, 515–520.

Platenkamp, G. A. J., and Shaw, R. G. (1993). Environmental and genetic maternal effects on seed characters in *Nemophila menziesii*. *Evolution* **47**, 540–555.

Plitmann, U. (1973). Biological flora of Israel. 4. *Vicia sativa* subsp. *amphicarpa* (Dorth.) Aschers. and Graebn. *Israel J. Bot.* **22**, 178–194.

Plowman, A. B., and Bougourd, S. M. (1994). Selectively advantageous effects of B chromosomes on germination behaviour in *Allium schoenoprasum* L. *Heredity* **72**, 587–593.

Pourrat, Y., and Jacques, R. (1975). The influence of photoperiodic conditions received by the mother plant on morphological and physiological characteristics of *Chenopodium polyspermum* L. seeds. *Plant Sci. Letts.* **4**, 273–279.

Priestley, D. A. (1986). "Seed Ageing: Implications for Seed Storage and Persistence in the Soil." Comstock Publishing Association, Cornell Univ. Press, Ithaca, NY.

Primack, R. B., and Antonovics, J. (1981). Experimental ecological genetics in *Plantago*. V. Components of seed yield in the ribwort plantain *Plantago lanceolata* L. *Evolution* **35**, 1069–1079.

Prinzie, T. P., and Chmielewski, J. G. (1994). Significance of achene characteristics and within-achene resource allocation in the germination strategy of tetraploid *Aster pilosus* var. *pilosus* (Asteraceae). *Am. J. Bot.* **81**, 259–264.

Probert, R. J., Smith, R. D., and Birch, P. (1985a). Germination responses to light and alternating temperatures in European populations of *Dactylis glomerata* L. I. Variability in relation to origin. *New Phytol.* **99**, 305–316.

Probert, R. J., Smith, R. D., and Birch, P. (1985b). Germination responses to light and alternating temperatures in European populations of *Dactylis glomerata* L. II. The genetic and environmental components of germination. *New Phytol.* **99**, 317–322.

Probert, R. J., Smith, R. D., and Birch, P. (1985c). Germination responses to light and alternating temperatures in European populations of *Dactylis glomerata* L. IV. The effects of storage. *New Phytol.* **101**, 521–529.

Puertas, M. J., Jimenez, M. M., Romera, F., Vega, J. M., and Diez, M. (1990). Maternal imprinting effect on B chromosome transmission in rye. *Heredity* **64**, 197–204.

Puterbaugh, M. N., Wied, A., and Galen, C. (1997). The functional ecology of gynodioecy in *Eritrichum aretioides* (Boraginaceae), the alpine forget-me-not. *Am. J. Bot.* **84**, 393–400.

Quinby, J. R., Reitz, L. P., and Laude, H. H. (1962). Effect of source of seed on productivity of hard red winter wheat. *Crop Sci.* **2**, 201–203.

Quinlivan, B. J. (1965). The influence of the growing season and the following dry season on the hardseededness of subterranean clover in different environments. *Aust. J. Agric. Res.* **16**, 277–291.

Quinlivan, B. J. (1966). The relationship between temperature fluctuations and the softening of hard seeds of some legume species. *Aust. J. Agric. Res.* **17**, 625–631.

Quinlivan, B. J., and Millington, A. J. (1962). The effect of a Mediterranean summer environment on the permeability of hard seeds of subterranean clover. *Aust. J. Agric. Res.* **13**, 377–387.

Quinn, J. A., and Colosi, J. C. (1977). Separating genotype from environment in germination ecology studies. *Am. Midl. Nat.* **97**, 484–489.

Radford, B. J. (1977). Influence of size of achenes sown and depth of sowing on growth and yield of dryland oilseed sunflowers (*Helianthus annuus*) on the Darling Downs. *Aust. J. Exp. Agri. Anim. Husb.* **17**, 489–494.

Raju, M. V. S., and Ramaswamy, S. N. (1983). Studies on the inflorescence of wild oats (*Avena fatua*). *Can. J. Bot.* **61**, 74–78.

Rai, J. P. N., and Tripathi, R. S. (1987). Germination and plant survival and growth of *Galinsoga parviflora* Cav. as related to food and energy content of its ray- and disc-achenes. *Acta Oecol.* **8**, 155–165.

Ramakrishnan, P. S. (1965). Studies on edaphic ecotypes in *Euphorbia thymifolia* L. II. Seed germination. *J. Ecol.* **53**, 157–162.

Ramakrishnan, P. S., and Jain, R. S. (1965). Germinability of the seeds of the edaphic ecotypes in *Tridax procumbens* L. *Trop. Ecol.* **6**, 53–55.

Ramseur, E. L., Quisenberry, V. L., Wallace, S. U., and Palmer, J. H. (1984). Yield and yield components of 'Braxton' soybeans as influenced by irrigation and intrarow spacing. *Agron. J.* **76**, 442–446.

Ramsey, M., and Vaughton, G. (1996). Inbreeding depression and pollinator availability in a partially self-fertile perennial herb *Blandfordia grandiflora* (Liliaceae). *Oikos* **76**, 465–474.

Rao, P. N., and Reddy, B. V. N. (1982). Comparative germination profiles of the seeds in the biotypes of *Trianthema portulacastrum* L. *Indian Bot. Reptr.* **1**, 23–26.

Rawson, H. M., and Evans, L. T. (1970). The pattern of grain growth within the ear of wheat. *Aust. J. Biol. Sci.* **23**, 753–764.

Reed, F. C., and Stephenson, S. N. (1972). Factors affecting seed number and size in burdock, *Arctium minus* Schk. *Michigan Acad.* **5**, 449–455.

Ren, Z., and Abbott, R. J. (1991). Seed dormancy in Mediterranean *Senecio vulgaris* L. *New Phytol.* **117**, 673–678.

Richardson, S. G. (1979). Factors influencing the development of primary dormancy in wild oat seeds. *Can. J. Plant Sci.* **59**, 777–784.

Rideau, M., Monin, J., Dommergues, P., and Cornu, A. (1976). Etude sur les mecanismes hereditaires de la dormance des akenes de *Lactuca sativa* L. *C. R. Acad. Sci. Paris Ser. D* **283**, 769–772.

Roach, D. A. (1987). Variation in seed and seedling size in *Anthoxanthum odoratum*. *Am. Midl. Nat.* **117**, 258–264.

Roach, D. A., and Wulff, R. D. (1987). Maternal effects in plants. *Annu. Rev. Ecol. Syst.* **18**, 209–235.

Rocha, O. J. (1996). The effects of achene heteromorphism on the dispersal capacity of *Bidens pilosa* L. *Int. J. Plant Sci.* **157**, 316–322.

Rocha, O. J., and Stephenson, A. G. (1990). Effect of ovule position on seed production, seed weight, and progeny performance in *Phaseolus coccineus* L. (Leguminosae). *Am. J. Bot.* **77**, 1320–1329.

Roche, L. (1969). A genecological study of the genus *Picea* in British Columbia. *New Phytol.* **68**, 505–554.

Rorison, I. H. (1972). Seed ecology: Present and future. *In* "Seed Ecology" (W. Heydecker, ed.), pp. 497–519. Pennsylvania State Univ. Press, University Park.

Roy, N. N., and Everett, H. L. (1963). Seed production, fertility levels, and cold test germination in corn. *Crop Sci.* **3**, 273–275.

Ruiz de Clavijo, E. (1994). Heterocarpy and seed polymorphism in *Ceratocapnos heterocarpa* (Fumariaceae). *Int. J. Plant Sci.* **155**, 196–202.

Ruiz de Clavijo, E. (1995). The ecological significance of fruit heteromorphism in the amphicarpic species *Catananche lutea* (Asteraceae). *Int. J. Plant Sci.* **156**, 824–833.

Sahai, K. (1994). Studies on seed position and their effect on germination and seedling survival in *Dalbergia sissoo* Roxb. *Indian For.* **120**, 464–465.

Sain, S. S. (1948). Inheritance of hard seeds in perennial leguminous forage. *Herb. Abst.* **43**, 206.

Salisbury, E. J., Sir. (1974). The variation in the reproductive organs of *Stellaria media* (*sensu stricto*) and allied species with special regard to their relative frequency and prevalent models of pollination. *Proc. Roy. Soc. Lond. B.* **185**, 331–342.

Sanchez, R. A., Eyherabide, G., and de Miguel, L. (1981). The influence of irradiance and water deficit during fruit development on seed dormancy in *Datura ferox* L. *Weed Res.* **21**, 127–132.

Sanderson, S. C., Stutz, H. C., and McArthur, E. D. (1990). Geographic differentiation in *Atriplex confertifolia*. *Amer. J. Bot.* **77**, 490–498.

Sands, R. (1981). Salt resistance in *Eucalyptus camaldulensis* Dehn. from three different seed sources. *Aust. For. Res.* **11**, 93–100.

Sao Jose, A. R., and Cunha, R. J. P. (1990). Influence of seed position in the fruit cavity on germination percentage, sex expression and seedling vigour of *Carica papaya* L. *Ecology Abstr.* **16**, 61.

Sarquis, A. V., Fischer, B. B., Parsons, F. G., and Miller, M. D. (1961). Geographic origin of barley seed. *California Agric.* **15**(4), 3.

Sawhney, R. (1989). Temperature control of dormancy and germination in embryos isolated from seeds of dormant and nondormant lines of wild oats (*Avena fatua*). *Can. J. Bot.* **67**, 128–134.

Sawhney, R., and Naylor, J. M. (1979). Dormancy studies in seed of *Avena fatua*. 9. Demonstration of genetic variability affecting the response to temperature during seed development. *Can. J. Bot.* **57**, 59–63.

Sawhney, R., and Naylor, J. M. (1982). Dormancy studies in seed of *Avena fatua*. 13. Influence of drought stress during seed development on duration of seed dormancy. *Can. J. Bot.* **60**, 1016–1020.

Sawhney, R., Quick, W. A., and Hsiao, A. I. (1985). The effect of temperature during parental vegetative growth on seed germination of wild oats (*Avena fatua* L.). *Ann. Bot.* **55**, 25–28.

Schaal, B. A. (1980). Reproductive capacity and seed size in *Lupinus texensis*. *Am. J. Bot.* **67**, 703–709.

Schat, H. (1981). Seed polymorphism and germination ecology of *Plantago coronopus* L. *Acta Oecol.* **2**, 367–380.

Schemske, D. W. (1978). Evolution of reproductive characteristics in *Impatiens* (Balsaminaceae): The significance of cleistogamy and chasmogamy. *Ecology* **59**, 596–613.

Schimpf, D. J. (1977). Seed weight of *Amaranthus retroflexus* in relation to moisture and length of growing season. *Ecology* **58**, 450–453.

Schively, A. F. (1904). Recent observation on *Amphicarpaea monoica*. *Contrib. Bot. Lab. Univ. Pennsylvania* **2**, 20–30.

Schmitt, J., Niles, J., and Wulff, R. D. (1992). Norms of reaction on seed traits to maternal environments in *Plantago lanceolata*. *Am. Nat.* **139**, 451–466.

Schnee, B. K., and Waller, D. M. (1986). Reproductive behavior of *Amphicarpaea bracteata* (Leguminosae), an amphicarpic annual. *Am. J. Bot.* **73**, 376–386.

Schoen, D. J., and Lloyd, D. G. (1984). The selection of cleistogamy and heteromorphic diaspores. *Biol. J. Linnean Soc.* **23**, 303–322.

Schuetz, W., and Milberg, P. (1997). Seed dormancy in *Carex canescens*: Regional differences and ecological consequences. *Oikos* **78**, 420–428.

Schurhoff, P. N. (1924). Die geschlechtsbegrenzte vererbung der kleistogamie bei *Plantago* Sect. *Novorbis*. *Ber. Deut. Botan. Gesell.* **42**, 311–321.

Schwaegerle, K. E., and Bazzaz, F. A. (1987). Differentiation among nine populations of *Phlox*: Response to environmental gradients. *Ecology* **68**, 54–64.

Scott, S. J., and Jones, R. A. (1985). Quantifying seed germination responses to low temperatures: Variation among *Lycopersicon* spp. *Environ. Exp. Bot.* **25**, 129–137.

Scurfield, G. (1954). Biological flora of the British Isles. *Deschampsia flexuosa* (L.) Trin. *J. Ecol.* **42**, 225–233.

Semeniuk, P. (1964). Effect of various levels of nitrogen, phosphorus, and potassium on seed production and germination of *Matthiola incana*. *Bot. Gaz.* **125**, 62–65.

Seneca, E. D. (1972). Germination and seedling response of Atlantic and Gulf coasts populations of *Uniola paniculata*. *Am. J. Bot.* **59**, 290–296.

Seneca, E. D. (1974). Germination and seedling response of Atlantic and Gulf coast populations of *Spartina alterniflora*. *Am. J. Bot.* **61**, 947–956.

Seneca, E. D., and Cooper, A. W. (1971). Germination and seedling response to temperature, daylength, and salinity by *Ammophila breviligulata* from Michigan and North Carolina. *Bot. Gaz.* **132**, 203–215.

Sexsmith, J. J. (1969). Dormancy of wild oat seed produced under various temperature and moisture conditions. *Weed Sci.* **17**, 405–407.

Sharma, M. M., Sharma, N. K., and Sen, D. N. (1978). A new report on differential seed coat dormancy in *Sesbania bispinosa* (Jacq.) W. F. Wight. *Folia Geobot. Phytotax.* **13**, 95–98.

Sharma, N. K., Sharma, M. M., and Sen, D. N. (1978). Seed perpetuation in *Rhynchosia capitata* DC. *Biol. Plant.* **20**, 225–228.

Shaw, C. H. (1904). The comparative structure of the flowers in *Polygala polygama* and *Polygala pauciflora* with a review of cleistogamy. *Contrib. Bot. Lab. Univ. Pennsylvania* **2**, 122–149.

Sheldrake, A. R., and Narayanan, A. (1979). Comparisons of earlier- and later-formed pods of pigeonpeas (*Cajanus cajan* (L.) Millsp.). *Ann. Bot.* **43**, 459–466.

Sheldrake, A. R., and Saxena, N. P. (1979). Comparisons of earlier- and later-formed pods of chickpeas (*Cicer arietinum* L.). *Ann. Bot.* **43**, 467–473.

Shoulders, E. (1961). Effect of seed size on germination, growth, and survival of slash pine. *J. For.* **59**, 363–365.

Shuma, J. M., Quick, W. A., Raju, M. V. S., and Hsiao, A. I. (1995). Germination of seeds from plants of *Avena fatua* L. treated with glyphosate. *Weed Res.* **35**, 249–255.

Silvertown, J. W. (1981). Seed size, life span, and germination date as coadapted features of plant life history. *Am. Nat.* **118**, 860–864.

Silvertown, J. W. (1984). Phenotypic variety in seed germination behavior: The ontogeny and evolution of somatic polymorphism in seeds. *Am. Nat.* **124**, 1–16.

Silvertown, J. W. (1985). When plants play the field. *In* "Evolution: Essays in Honour of John Maynard Smith" (P. J. Greenwood, P. H. Harvey, and M. Slatkin, eds.), pp. 143–153. Cambridge Univ. Press, Cambridge UK.

Simmonds, N. W. (1964). The genetics of seed and tuber dormancy in the cultivated potatoes. *Heredity* **19**, 489–504.

Skerman, R. H., and Humphreys, L. R. (1973). Effect of temperature during flowering on seed formation of *Stylosanthes humilis*. *Aust. J. Agric. Res.* **24**, 317–324.

Skordilis, A., and Thanos, C. A. (1995). Seed stratification and germination strategy in the Mediterranean pines *Pinus brutia* and *P. halepensis*. *Seed Sci. Res.* **5**, 151–160.

Smith, A. P. (1975). Altitudinal seed ecotypes in the Venezuelan Andes. *Am. Midl. Nat.* **94**, 247–250.

Smith, B. D. (1984). *Chenopodium* as a prehistoric domesticate in eastern North America: Evidence from Russell Cave, Alabama. *Science* **226**, 165–167.

Smith, B. D. (1989). Origins of agriculture in eastern North America. *Science* **246**, 1566–1570.

Smith, B. D., and Cowan, C. W. (1987). Domesticated *Chenopodium* in prehistoric eastern North America: New accelerator dates from eastern Kentucky. *Am. Antiq.* **52**, 355–357.

Smith, R. H., and Bass, M. H. (1972). Relationship of artificial pod removal to soybean yields. *J. Econ. Ent.* **65**, 606–608.

Smith, W. E., and Fitzsimmons, J. E. (1964). Maternal inheritance of seed weight in flax (*Linum usitatissimum* L.). *Can. J. Genet. Cytol.* **6**, 244. [Abstract]

Smith, W. E., and Fitzsimmons, J. E. (1965). Maternal inheritance of seed weight in flax. *Can. J. Genet. Cytol.* **7**, 658–662.

Smith-Huerta, N. L. (1984). Seed germination in related diploid and allotetraploid *Clarkia* species. *Bot. Gaz.* **145**, 246–252.

Soffer, H., and O. E. Smith. (1974). Studies on lettuce seed quality. III. Relationships between flowering pattern, seed yield, and seed quality. *J. Am. Soc. Hort. Sci.* **99**, 114–117.

Solbrig, O. T. (1981). Studies on the population biology of the genus *Viola*. II. The effect of plant size on fitness in *Viola sororia*. *Evolution* **35**, 1080–1093.

Somody, C. N., Nalewaja, J. D., and Miller, S. E. (1984). The response of wild oat (*Avena fatua*) and *Avena sterilis* accessions to photoperiod and temperature. *Weed Sci.* **32**, 206–213.

Sonaike, A. A., and Okusanya, O. T. (1987). Germination behaviour of *Luffa aegyptiaca* seeds from 19 localities in Nigeria. *Seed Sci. Technol.* **15**, 741–750.

Sorensen, A. E. (1978). Somatic polymorphism and seed dispersal. *Nature* **276**, 174–175.

Springfield, H. W. (1973). Larger seeds of winterfat germinate better. *J. Range Manage.* **26**, 153–154.

Staub, J. E., Globerson, D., and Genizi, A. (1989). Inheritance of seed dormancy in *Cucumis sativus* var. *hardwickii* (Royle) Alef. *Theor. Appl. Genet.* **78**, 143–151.

Stamp, N. E. (1990). Production and effect of seed size in a grassland annual (*Erodium brachycarpum*, Geraniaceae). *Am. J. Bot.* **77**, 874–882.

Standifer, L. C., and Wilson, P. W. (1988). Dormancy studies in three populations of *Poa annua* L. seeds. *Weed Res.* **28**, 359–363.

Stanton, M. L. (1984a). Developmental and genetic sources of seed weight variation in *Raphanus raphanistrum* L. (Brassicaceae). *Am. J. Bot.* **71**, 1090–1098.

Stanton, M. L. (1984b). Seed variation in wild radish: Effect of seed size on components of seedling and adult fitness. *Ecology* **65**, 1105–1112.

Stanton, M. L. (1985). Seed size and emergence time within a stand of wild radish (*Raphanus raphanistrum* L.): The establishment of a fitness hierarchy. *Oecologia* **67**, 524–531.

Stanton, M. L., Bereczky, J. K., and Hasbrouck, H. D. (1987). Pollination thoroughness and maternal yield regulation in wild radish, *Raphanus raphanistrum* (Brassicaceae). *Oecologia* **74**, 68–76.

Stearns, F. (1960). Effects of seed environment during maturation on seedling growth. *Ecology* **41**, 221–222.

Stephenson, A. G. (1980). Fruit set, herbivory, fruit reduction, and the fruiting strategy of *Catalpa speciosa* (Bignoniaceae). *Ecology* **61**, 57–64.

Stephenson, A. G. (1984). The regulation of maternal investment in an indeterminate flowering plant (*Lotus corniculatus*). *Ecology* **65**, 113–121.

Sterk, A. A. (1969). Biosystematic studies of *Spergularia media* and *S. marina* in the Netherlands. I. The morphological variability of *S. media. Acta Bot. Neerl.* **18**, 325–338.

Stewart, R. N. (1961). Effect on poisetttia progeny of applications of 4-chlorophenoxyacetic acid to young fruit on parent plant. *Bot. Gaz.* **123**, 43–46.

Stickler, F. C., and Wassom, C. E. (1963). Emergence and seedling vigor of birdsfoot trefoil as affected by planting depth, seed size, and variety. *Agron. J.* **55**, 78.

Stillwell, E. K., and Sweet, R. D. (1975). Germination, growth and flowering of shepherdspurse ecotypes. *Proc. Northe. Weed. Sci. Soc.* **29**, 148–153.

Stokes, W. E., and Hull, F. H. (1930). Peanut breeding. *Agron. J.* **22**, 1004–1019.

Stratton, D. A. (1991). Life history variation within populations of an asexual plant, *Erigeron annuus* (Asteraceae). *Am. J. Bot.* **78**, 723–728.

Streeter, J. G. (1978). Effect of N starvation of soybean plants at various stages of growth on seed yield and N concentration of plant parts at maturity. *Agron. J.* **70**, 74–76.

Strickberger, M. W. (1968). "Genetics." The Macmillan Company, New York.

Stromberg, J. C., and Patten, D. T. (1990). Variation in seed size of a southwestern riparian tree, Arizona walnut (*Juglans major*). *Am. Midl. Nat.* **124**, 269–277.

Sultan, S. E. (1996). Phenotypic plasticity for offspring traits in *Polygonum persicaria. Ecology* **77**, 1791–1807.

Suneson, C. A., and Peltier, G. L. (1936). Effect of source, quality, and condition of seed upon the cold resistance of winter wheats. *Agron. J.* **28**, 687–693.

Symonides, E. (1978). Effect of seed size, density and depth of sowing on the germination and survival of psammophyte seedlings. *Ekol. Pol.* **26**, 123–139.

Takeno, K., and Yamaguchi, H. (1991). Diversity in seed germination behavior in relation to heterocapry in *Salsola komarovii* Iljin. *Bot. Mag. Tokyo* **104**, 207–215.

Tanowitz, B. D., Salopek, P. F., and Mahall, B. E. (1987). Differential germination of ray and disc achenes in *Hemizonia increscens* (Asteraceae). *Am. J. Bot.* **74**, 303–312.

Taylor, G. B., and Palmer, M. J. (1979). The effect of some environmental conditions on seed development and hard-seededness in subterranean clover (*Trifolium subterraneum* L.). *Aust. J. Agric. Res.* **30**, 65–76.

Tedin, O. (1925). Vererbung, variation und systematik in der gattung *Camelina. Hereditas* **6**, 275–386.

Tervet, I. W. (1944). The relation of seed quality to the development of smut in oats. *Phytopathology* **34**, 106–115.

Thapliyal, R. C., and Gupta, B. N. (1980). Effect of seed source and stratification on the germination of deodar seed. *Seed Sci. Technol.* **8**, 145–150.

Thomas, A. G., Lefkovitch, L. P., Woo, S. L., Bowes, G. G., and Peschken, D. P. (1994). Effect of temperature on germination within and between diploid and tetraploid populations of *Matricaria perforata* Merat. *Weed Res.* **34**, 187–198.

Thomas, J. F., and Raper, C. D., Jr. (1975). Seed germinability as affected by the environmental temperature on the mother plant. *Tobacco Sci.* **19**, 104–106.

Thomas, J. F., and Raper, C. D., Jr. (1979). Germinability of tobacco seed as affected by culture of the mother plant. *Agron. J.* **71**, 694–695.

Thomas, T. H. (1994). Responses of florence fennel (*Foeniculum vulgare azoricum*) seeds to light, temperature and gibberellin $A_{4/7}$. *Plant Growth Regul.* **14**, 139–143.

Thomas, T. H. (1996). Relationships between position on the parent plant and germination characteristics of seeds of parsley (*Petroselinum crispum* Nym.). *Plant Growth Regul.* **18**, 175–181.

Thomas, T. H., Biddington, N. L., and O'Toole, D. F. (1979). Relationship between position on the parent plant and dormancy characteristics of seeds of three cultivars of celery (*Apium graveolens*). *Physiol. Plant.* **45**, 492–496.

Thomas, T. H., Gray, D., and Biddington, N. L. (1978). The influence of the position of the seed on the mother plant on seed and seedling performance. *Acta Hort.* **83**, 57–66.

Thompson, J. N. (1984). Variation among individual seed masses in *Lomatium grayi* (Umbelliferae) under controlled conditions: Magnitude and partitioning of the variance. *Ecology* **65**, 626–631.

Thompson, J. N., and Pellmyr, P. (1989). Origins of variance in seed number and mass: Interaction of sex expression and herbivory in *Lomatium salmoniflorum. Oecologia* **79**, 395–402.

Thompson, K. (1990). Genome size, seed size and germination temperature in herbaceous Angiosperms. *Evol. Trends Plants* **4**, 113–116.

Thompson, P. A. (1970). Characterization of the germination response to temperature of species and ecotypes. *Nature* **225**, 827–831.

Thompson, P. A. (1973). Effects of cultivation on the germination character of the corn cockle (*Agrostemma githago* L.). *Ann. Bot.* **37**, 133–154.

Thompson, P. A. (1975). Characterization of the germination responses of *Silene dioica* (L.) Clairv., populations from Europe. *Ann. Bot.* **39**, 1–19.

Thompson, R. C. (1937). The germination of lettuce seed as affected by nutrition of the plant and the physiological age of the plant. *Proc. Am. Soc. Hort. Sci.* **35**, 599–600.

Thullen, R. J., and Kelley, P. E. (1979). Seed production and germination in *Cyperus esculentus* and *C. rotundus. Weed Sci.* **27**, 502–505.

Thurling, N. (1966). Population differentiation in Australian *Cardamine*. III. Variation in germination response. *Aust. J. Bot.* **14**, 189–194.

Thurston, J. M. (1951). A comparison of the growths of wild and of cultivated oats in manganese-deficient soils. *Ann. Appl. Biol.* **38**, 289–302.

Trapp, E. J., and Hendrix, S. D. (1988). Consequences of a mixed reproductive system in the hog peanut, *Amphicarpaea bracteata*, (Fabaceae). *Oecologia* **75**, 285–290.

Trillo, T. A., and Carro, A. J. M. (1993). Germination, seed-coat structure and protein patterns of seeds from *Adenocarpus decorticans* and *Astragalus granatensis* growing at different altitudes. *Seed Sci. Technol.* **21**, 317–326.

Tripathi, R. S., and Khan, M. L. (1990). Effects of seed weight and microsite characteristics on germination and seedling fitness in two species of *Quercus* in a subtropical wet hill forest. *Oikos* **57**, 289–296.

Trow, A. H. (1912). On the inheritance of certain characters in the common groundsel—*Senecio vulgaris*, Linn.—and its segregates. *J. Genet.* **2**, 239–276.

Turesson, G. (1922a). The species and the variety as ecological units. *Hereditas* **3**, 100–113.

Turesson, G. (1922b). The genotypical response of the plant species to the habitat. *Hereditas* **3**, 211–350.

Twamley, B. E. (1967). Seed size and seedling vigor in birdsfoot trefoil. *Can. J. Plant Sci.* **47**, 603–609.

Twentyman, J. D. (1974). Environmental control of dormancy and germination in the seeds of *Cenchrus longispinus* (Hack.) Fern. *Weed Res.* **14**, 1–11.

Tyler, B., Borrill, M., and Chorlton, K. (1978). Studies in *Festuca*. X. Observations on germination and seedling cold tolerance in diploid *Festuca pratensis* and tetraploid *F. pratensis* var. *apennina* in relation to their altitudinal distribution. *J. Appl. Ecol.* **15**, 219–226.

Ungar, I. A. (1971). *Atriplex patula* var. *hastata* seed dimorphism. *Rhodora* **73**, 548–551.

Ungar, I. A. (1979). Seed dimorphism in *Salicornia europaea* L. *Bot. Gaz.* **140**, 102–108.

Uphof, J. C. Th. (1938). Cleistogamic flowers. *Bot. Rev.* **4**, 21–49.

Urie, A. L., Leininger, L. N., and Zimmer, D. E. (1968). Effects of degree and time of defoliation on yield and related attributes of safflower. *Crop Sci.* **8**, 747–750.

van Andel, J., and Vera, F. (1977). Reproductive allocation in *Senecio sylvaticus* and *Chamaenerion angustifolium* in relation to mineral nutrition. *J. Ecol.* **65**, 747–758.

Vander Kloet, S. P. (1984). Effects of pollen donors on seed production, seed weight, germination and seedling vigor in *Vaccinium corymbosum* L. *Am. Midl. Nat.* **112**, 392–396.

van der Moezel, P. G., and Bell, D. T. (1987). The effect of salinity on the germination of some Western Australian *Eucalyptus* and *Melaleuca* species. *Seed Sci. Technol.* **15**, 239–246.

van der Toorn, J., and ten Hove, H. J. (1982). On the ecology of *Cotula coronopifolia* L. and *Ranunculus sceleratus* L. II. Experiments on germination, seed longevity, and seedling survival. *Acta Oecol.* **3**, 409–418.

van der Vegte, F. W. (1978). Population differentiation and germination ecology in *Stellaria media* (L.) Vill. *Oecologia* **37**, 231–245.

van Groenendael, J. M. (1986). Life history characteristics of two ecotypes of *Plantago lanceolata* L. *Acta Bot. Neerl.* **35**, 71–86.

van Loenhoud, P. J., and Duyts, H. (1981). A comparative study of the germination ecology of some microspecies of *Taraxacum* Wigg. *Acta Bot. Neerl.* **30**, 161–182.

van Treuren, R., Bijlsma, R., Ouborg, N. J., and van Delden, W. (1993). The significance of genetic erosion in the process of extinction. IV. Inbreeding depression and heterosis effects caused by selfing and outcrossing in *Scabiosa columbaria*. *Evolution* **47**, 1669–1680.

Vaughton, G. (1990). Predation by insects limits seed production in *Banksia spinulosa* var. *neoanglica* (Proteaceae). *Aust. J. Bot.* **38**, 335–340.

Venable, D. l. (1984). Using intraspecific variation to study the ecological significance and evolution of plant life-histories. *In* "Perspectives on Plant Population Ecology" (R. Dirzo and J. Sarukhan, eds.), pp. 166–187. Sinauer, Sunderland, MA.

Venable, D. L. (1985a). Ecology of achene dimorphism in *Heterotheca latifolia*. III. Consequences of varied water availability. *J. Ecol.* **73**, 757–763.

Venable, D. L. (1985b). The evolutionary ecology of seed heteromorphism. *Am. Nat.* **126**, 577–595.

Venable, D. L., and Brown, J. S. (1988). The selective interactions of dispersal, dormancy, and seed size as adaptations for reducing risk in variable environments. *Am. Nat.* **131**, 360–384.

Venable, D. L., and Burquez M. A. (1989). Quantitative genetics of size, shape, life-history, and fruit characteristics of the seed-heteromorphic composite *Heterosperma pinnatum*. I. Variation within and among populations. *Evolution* **43**, 113–124.

Venable, D. L., Burquez, A., Corral, G., Morales, E., and Espinosa,

F. (1987). The ecology of seed heteromorphism in *Heterosperma pinnatum* in central Mexico. *Ecology* **68**, 65–76.

Venable, D. L., Dyreson, E., and Morales, E. (1995). Population dynamic consequences and evolution of seed traits of *Heterosperma pinnatum* (Asteraceae). *Am. J. Bot.* **82**, 410–420.

Venable, D. L., and Levin, D. A. (1985a). Ecology of achene dimorphism in *Heterotheca latifolia*. I. Achene structure, germination and dispersal. *J. Ecol.* **73**, 133–145.

Venable, D. L., and Levin, D. A. (1985b). Ecology of achene dimorphism in *Heterotheca latifolia*. II. Demographic variation within populations. *J. Ecol.* **73**, 743–755.

Vickery, R. K., Jr. (1967). Ranges of temperature tolerance for germination of *Mimulus* seeds from diverse populations. *Ecology* **48**, 647–653.

Vickery, R. K., Jr. (1983). Plasticity and polymorphism in seed germination of *Mimulus guttatus* (Scrophulariaceae). *Great Basin Nat.* **43**, 470–474.

Voigt, R. L., Gardner, C. O., and Webster, O. J. (1966). Inheritance of seed size in sorghum, *Sorghum vulgare* Pers. *Crop Sci.* **6**, 582–586.

VonAbrams, G. J., and Hand, M. E. (1956). Seed dormancy in *Rosa* as a function of climate. *Am. J. Bot.* **43**, 7–12.

Vosa, C. G. (1966). Seed germination and B-chromosomes in the leek (*Allium porrum*). *In* "Chromosomes Today" (C. D. Darlington and K. R. Lewis, eds.), Vol. 1, pp. 24–27. Oliver and Boyd, Edinburgh/London.

Wagner, L. K. (1986). Variation in seed-coat morph ratios in *Spergula arvensis* L. *Bull. Torrey Bot. Club* **113**, 28–35.

Wagner, L. K. (1988). Germination and seedling emergence in *Spergula arvensis*. *Am. J. Bot.* **75**, 465–475.

Walker, S. R., and Evenson, J. P. (1985a). Biology of *Commelina benghalensis* L. in south-eastern Queensland. 1. Growth, development and seed production. *Weed Res.* **25**, 239–244.

Walker, S. R., and Evenson, J. P. (1985b). Biology of *Commelina benghalensis* L. in south-eastern Queensland. 2. Seed dormancy, germination and emergence. *Weed Res.* **25**, 245–250.

Waller, D. M. (1980). Environmental determinations of outcrossing in *Impatiens capensis* (Balsaminaceae). *Evolution* **34**, 747–761.

Waller, D. M. (1982). Factors influencing seed weights in *Impatiens capensis* (Balsaminaceae). *Am. J. Bot.* **69**, 1470–1475.

Waller, D. M. (1984). Differences in fitness between seedlings derived from cleistogamous and chasmogamous flowers in *Impatiens capensis*. *Evolution* **38**, 427–440.

Waller, D. M. (1985). The genesis of size hierarchies in seedling populations of *Impatiens capensis* Meerb. *New Phytol.* **100**, 243–260.

Wardlaw, I. F. (1970). The early stages of grain development in wheat: Response to light and temperature in a single variety. *Aust. J. Biol. Sci.* **23**, 765–774.

Ware, D. M. E. (1983). Genetic fruit polymorphism in North American *Valerianella* (Valerianaceae) and its taxonomic implications. *Syst. Bot.* **8**, 33–34.

Warwick, S. I., and Black, L. D. (1986). Genecological variation in recently established populations of *Abutilon theophrasti* (velvetleaf). *Can. J. Bot.* **64**, 1632–1643.

Wearstler, K. W., Jr., and Barnes, B. V. (1977). Genetic diversity of yellow birch seedlings in Michigan. *Can. J. Bot.* **55**, 2778–2788.

Weatherwax, P. (1928). Cleistogamy in two species of *Danthonia*. *Bot. Gaz.* **85**, 104–109.

Weaver, S. E., and Thomas, A. G. (1986). Germination responses to temperature of atrazine-resistant and -susceptible biotypes of two pigweed (*Amaranthus*) species. *Weed Sci.* **34**, 865–870.

Webb, C. J. (1986). Variation in achene morphology and its implications for taxonomy in *Soliva* subgenus *Soliva* (Anthemideae, Asteraceae). *New Zeal. J. Bot.* **24**, 665–669.

Weis, I. M. (1982). The effects of propagule size on germination and seedling growth in *Mirabilis hirsuta*. *Can. J. Bot.* **60**, 1868–1874.

Weiss, P. W. (1980). Germination, reproduction and interference in the amphicarpic annual *Emex spinosa* (L.) Campd. *Oecologia* **45**, 244–251.

Weller, S. G. (1985). Establishment of *Lithospermum caroliniense* on sand dunes: The role of nutlet mass. *Ecology* **66**, 1893–1901.

Wentland, M. J. (1965). "The Effect of Photoperiod on the Seed Dormancy of *Chenopodium album*." Ph.D thesis, Univ. of Wisconsin, Madison.

Werke, E., and T. Many. (1974). Heterocarpy and its ontogeny in *Aellenia autrani* (Post) Zoh. Light- and electron-microscope study. *Israel J. Bot.* **23**, 132–144.

Westoby, M. (1981). How diversified seed germination behavior is selected. *Am. Nat.* **118**, 882–885.

Whalley, R. D. B., McKell, C. M., and Green, L. R. (1966). Seed physical characteristics and germination of hardinggrass (*Phalaris tuberosa* var. *stenoptera* (Hack.) Hitch.) *J. Range Manage.* **19**, 129–132.

White, W. J., and Stevenson, T. W. (1948). Permeable seeded strains of sweet clover (*Melilotus alba*), their development and nature. *Sci. Agri.* **28**, 206–222.

Whittington, W. J., Childs, J. D., Hartridge, J. M., and How, J. (1965). Analysis of variation in the rates of germination and early seedling growth in tomato. *Ann. Bot.* **29**, 59–71.

Whittington, W. J., Hillman, J., Gatenby, S. M., Hooper, B. E., and White, J. C. (1970). Light and temperature effects on the germination of wild oats. *Heredity* **25**, 641–650.

Wiesner, L. E., and Grabe, D. F. (1972). Effect of temperature preconditioning and cultivar on ryegrass (*Lolium* sp.) seed dormancy. *Crop Sci.* **12**, 760–764.

Wilcox, J. R. (1968). Sweetgum seed stratification requirements related to winter climate at seed source. *For. Sci.* **14**, 16–19.

Williams, M. C. (1960). Biochemical analyses, germination, and production of black and brown seeds of *Halogeton glomeratus*. *Weeds* **8**, 452–461.

Williams, I. H., and Free, J. B. (1979). Compensation of oil-seed rape (*Brassica napus* L.) plants after damage to their buds and pods. *J. Agric. Sci. Camb.* **92**, 53–59.

Williams, J. T., and Harper, J. L. (1965). Seed polymorphism and germination. I. The influence of nitrates and low temperatures on the germination of *Chenopodium album*. *Weed Res.* **5**, 141–150.

Williams, R. D., and Winstead, J. E. (1972). Population variation in seed germination and stratification of *Acer negundo* L. *Trans. Ky. Acad. Sci.* **33**, 43–48.

Williams, W., and McGibbon, R. (1980). Genetic control of seed weight and seed oil in *Lupinus mutabilis*. *Z. Pflanzenzuchtg.* **84**, 329–334.

Wilken, D. H. (1982). The balance between chasmogamy and cleistogamy in *Collomia grandiflora* (Polemoniaceae). *Am. J. Bot.* **69**, 1326–1333.

Willson, M. F., and Price, P. W. (1980). Resource limitation of fruit and seed production in some *Asclepias* species. *Can. J. Bot.* **58**, 2229–2233.

Winn, A. A. (1988). Ecological and evolutionary consequences of seed size in *Prunella vulgaris*. *Ecology* **69**, 1537–1544.

Winn, A. A., and Werner, P. A. (1987). Regulation of seed yield within and among populations of *Prunella vulgaris*. *Ecology* **68**, 1224–1233.

Winstead, J. E. (1971). Populational differences in seed germination and stratification requirements of sweetgum. *For. Sci.* **17**, 34–36.

Witcombe, J. R., and Whittington, W. J. (1972). The effects of selection for reduced dormancy in charlock (*Sinapis arvensis*). *Heredity* **29**, 37–49.

Withers, N. J., and Forde, B. J. (1979). Effects of water stress on *Lupinus albus*. *New Zeal. J. Agric. Res.* **22**, 463–474.

Wolf, L. L., Hainsworth, F. R., Mercier, T., and Benjamin, R. (1986). Seed size variation and pollinator uncertainty in *Ipomopsis aggregata* (Polemoniaceae). *J. Ecol.* **74**, 361–371.

Wood, D. W., Scott, R. K., and Longden, P. C. (1980). The effects of mother plant temperature on seed quality in *Beta vulgaris* L. (sugar beet). *In* "Seed Production" (P. D. Hebblethwaite, ed.), pp. 257–270. Butterworths, London/Boston.

Workman, J. P., and West, N. E. (1967). Germination of *Eurotia lanata* in relation to temperature and salinity. *Ecology* **48**, 659–661.

Wright, L. N. (1976). Recurrent selection for shifting gene frequency of seed weight in *Panicum antidotale* Retz. *Crop Sci.* **16**, 647–649.

Wu, L., Till-Bottraud, I., and Torres, A. (1987). Genetic differentiation in temperature-enforced seed dormancy among golf course populations of *Poa annua* L. *New Phytol.* **107**, 623–631.

Wulff, R. (1973). Intrapopulational variation in the germination of seeds in *Hyptis suaveolens*. *Ecology* **54**, 646–649.

Wulff, R. D. (1985). Germination of seeds of different sizes in *Hyptis suaveolens*: The response to irradiance and mixed red-far-red sources. *Can. J. Bot.* **63**, 885–888.

Wulff, R. D. (1986a). Seed size variation in *Desmodium paniculatum*. I. Factors affecting seed size. *J. Ecol.* **74**, 87–97.

Wulff, R. D. (1986b). Seed size variation in *Desmodium paniculatum*. II. Effects on seedling growth and physiological performance. *J. Ecol.* **74**, 99–114.

Wulff, R. D., and Alexander, H. M. (1985). Intraspecific variation in the response of CO_2 enrichment in seeds and seedlings of *Plantago lanceolata* L. *Oecologia* **66**, 458–460.

Wulff, R. D., and Bazzaz, F. A. (1992). Effect of the parental nutrient regime on growth of the progeny in *Abutilon theophrasti* (Malvaceae). *Am. J. Bot.* **79**, 1102–1107.

Wulff, R. D., Caceres, A., and Schmitt, J. (1994). Seed and seedling responses to maternal and offspring environments in *Plantago lanceolata*. *Funct. Ecol.* **8**, 763–769.

Wurzburger, J., and Koller, D. (1976). Differential effects of the parental photothermal environment on development of dormancy in caryopses of *Aegilops kotschyi*. *J. Exp. Bot.* **27**, 43–48.

Yamaguchi, H., Ichihara, K., Takeno, K., Hori, Y., and Saito, T. (1990). Diversities in morphological characteristics and seed germination behavior in fruits of *Salsola komarovii* Iljin. *Bot. Mag. Tokyo* **103**, 177–190.

Yamasue, Y., Kamiyama, K., Hanioka, Y., and Kusanagi, T. (1992). Paraquat resistance and its inheritance in seed germination of the foliar-resistant biotypes of *Erigeron canadensis* L. and *E. sumatrensis* Retz. *Pest. Biochem. Physiol.* **44**, 21–27.

Yanful, M., and Maun, M. A. (1996). Spatial distribution and seed mass variation of *Strophostyles helvola* along Lake Erie. *Can. J. Bot.* **74**, 1313–1321.

Yang, T. W. (1967). Ecotypic variation in *Larrea divaricata*. *Am. J. Bot.* **54**, 1041–1044.

Yang, T. W., and Lowe, C. H. (1968). Chromosome variation in ecotypes of *Larrea divaricata* in the North American desert. *Madrono* **19**, 161–192.

Young, J. A., and Evans, R. A. (1977). Squirreltail seed germination. *J. Range Manage.* **30**, 33–36.

Young, J. A., and Evans, R. A. (1989). Reciprocal common garden studies of the germination of seeds of big sagebrush (*Artemisia tridentata*). *Weed Sci.* **37**, 319–325.

Young, J. A., Palmquist, D. E., and Evans, R. A. (1991). Temperature profiles for germination of big sagebrush seeds from native stands. *J. Range Manage.* **44**, 385–390.

Youngberg, C. T. (1952). Effect of soil fertility on the physical and chemical properties of tree seed. *J. For.* **50**, 850–852.

Zammit, C., and Zedler, P. H. (1990). Seed yield, seed size and germination behaviour in the annual *Pogogyne abramsii*. *Oecologia* **84,** 24–28.

Zeide, B. (1978). Reproductive behavior of plants in time. *Am. Nat.* **112,** 636–639.

Zhang, J. (1993). Seed dimorphism in relation to germination and growth of *Cakile edentula*. *Can. J. Bot.* **71,** 1231–1235.

Zhang, J., and Maun, M. A. (1991). Establishment and growth of *Panicum virgatum* L. seedlings on a Lake Erie sand dune. *Bull. Torrey Bot. Club* **118,** 141–153.

Zimmerman, J. K., and Weis, I. M. (1983). Fruit size variation and its effects on germination and seedling growth in *Xanthium strumarium*. *Can. J. Bot.* **61,** 2309–2315.

Zohary, M., and Fahn, A. (1950). On the heterocapry of *Aethionema*. *Palestine J. Bot.* **5,** 28–31.

9

A Geographical Perspective on Germination Ecology: Tropical and Subtropical Zones

I. PURPOSE

The next objective of this book is to survey what is known about the germination ecology of plants growing in various types of vegetation throughout the world. To facilitate organization of the mass of literature, Walter's (1979) classification system of the vegetation zones (Fig. 9.1) has been adopted as a general outline. Walter divided vegetation into two broad categories: (1) tropical and subtropical zones and (2) temperate and arctic zones. Accordingly, this chapter covers the germination ecology of plants in tropical and subtropical zones, whereas Chapter 10 treats that of species in temperate and arctic zones.

Under tropical and subtropical zones, Walter listed (1) evergreen rain forests of lowlands and mountainsides; (2) semievergreen and deciduous forests; (3) dry woodlands, natural savannas, and grasslands; and (4) hot semideserts and deserts, located poleward to latitudes of 35°N and 35°S. This chapter covers the germination ecology of plants in each of these four vegetation zones, as well as that of species on tropical mountains. For each type of vegetation, the discussion includes a summary of available information on the germination of trees, shrubs, vines, and herbaceous species. Weeds also are considered, and attention is given to special biotic and abiotic factors influencing germination.

II. EVERGREEN RAIN FORESTS

Tropical evergreen rain forests (=tropical rain forests) occur in northern South America and extend northward to southern Mexico and southward on the east coast of Brazil to the Tropic of Capricorn (Fig. 9.1). In Africa, these forests occur on the Guinea Coast

in the Congo Basin and in eastern Madagascar (=Malagasy), and in Asia they are found in western India, Malaysia, Indonesia, the Philippines, and New Guinea. They also occur in northern and eastern Australia. Tropical rain forests mostly occur between 10°N and 10°S, and mean daily temperature is about 25–27°C (Walter, 1973). Rainfall is high and, as noted by Walter (1979), "... a month with less than 100 mm of rain is considered to be relatively dry." Short dry seasons sometimes occur, but leaves are not shed (Walter, 1979).

Tropical rain forests are dominated by phanerophytes (trees) and are famous for their high number of tree species. The tree canopy is layered into an upper, middle, and lower story, with the latter two being so dense that little light reaches the forest floor. Shrubs and herbs are present, but they are much less numerous than the lianas and epiphytes in the trees (Walter, 1979).

A. Climax Trees

Trees in tropical rainforests have been divided into two groups: nonpioneer (climax) and pioneer species. Within each group, there are small, medium, and large trees (Swaine and Whitmore, 1988). Seed germination requirements are an important criterion for distinguishing climax and pioneer trees. Seeds of climax species can germinate (and the seedlings become established) in dim light on the forest floor, whereas those of pioneer species require high light associated with a gap in the canopy for germination and seedling establishment (Swaine and Whitmore, 1988; Whitmore, 1989). Although seeds of some climax species germinate in gaps and those of some pioneer species germinate in shade, shade and gaps generally are optimal for the germina-

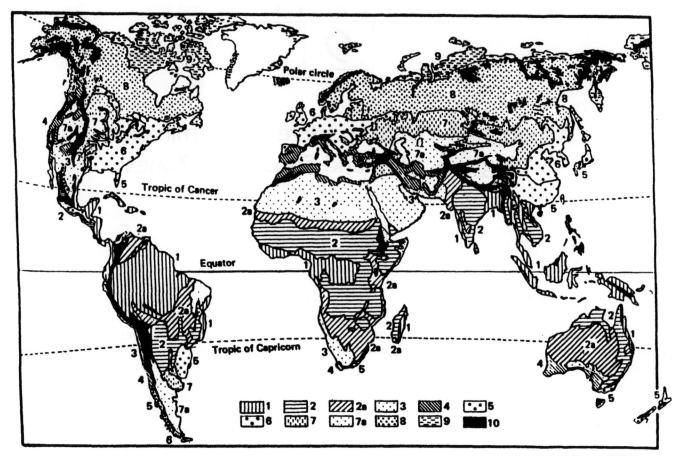

FIGURE 9.1   Walter's (1979) map of the vegetation of the world. I. Tropical and subtropical zones: 1, evergreen rain forests of lowlands and mountainsides (cloud forests); 2, semievergreen and deciduous forests; 2a, dry woodlands, natural savannas, or grasslands; and 3, hot semideserts and deserts, poleward up to a latitude of 35°C. (II) Temperate and arctic zones: 4, sclerophyllous woodlands with winter rain; 5, moist warm temperature woodlands; 6, deciduous (nemoral) forests; 7, steppes; 7a, semideserts and deserts with cold winters; 8, boreal coniferous forests; 9, tundra; and 10, mountains.

tion of climax and pioneer species, respectively (Raich and Khoon, 1990).

Seed size is another important difference between nonpioneer and pioneer species. Foster and Janson (1985) found that seed mass was significantly higher in 14 species of trees that become established under the forest canopy or in small gaps than in 22 species that require large gaps for establishment. The seed mass of shade-tolerant species in rain forests of the Malay Peninsula decreased with relative size of the plant: tall trees > woody climbers > small trees > shrubs > herbs (Metcalfe and Grubb, 1995). In a comparison of small trees in these same rain forests, Metcalfe and Grubb (1995) found that the seed mass of shade-tolerant species generally was larger than that of shade-intolerant ones. However, some shade-tolerant species had small seeds, and some shade-intolerant ones had large seeds. The large seeds of shade-tolerant species are thought to be of adaptive value because they contain relatively large amounts of food reserves that can be used for seedling establishment in dim light on the forest floor (Foster, 1986).

Although environmental conditions associated with shade on the forest floor are required for the germination of some seeds, these conditions may be favorable for the growth of seedlings and juveniles. Many canopy, as well as emergent, tree species require a gap in the canopy, and thus an increase in light, before young individuals can grow to maturity (Pickett, 1983). Seedlings of some climax trees are very tolerant of shade and may enter a state of dormancy until light is increased due to formation of a gap (Duke, 1969). The most shade-tolerant species require only a relatively small break in the canopy for seedling growth to be stimulated, and the resulting trees produce dense, dark wood that is quite valuable as lumber. Seedlings of other climax trees are less tolerant of shade and require fairly large gaps for seedling growth to be stimulated; these trees produce low-density, pale wood (Whitmore, 1989).

1. Types of Seed Dormancy

Relatively few studies have been done on the types of dormancy per se in seeds of rain forest trees. However, large numbers of seedlings can carpet the ground of rain forests shortly after seed dispersal (e.g., Barnard, 1956; Richards, 1957; Whitmore, 1975); these observations have prompted studies on germination rates of freshly matured seeds. Seeds were collected at maturity from trees, or picked up from the ground immediately following dispersal, and sown on a moist substrate in boxes or petri dishes under shade outdoors (Ng, 1973) or on a laboratory bench (Ng, 1973; Yap, 1981). Then, daily or weekly counts were made of the number of newly germinated seedlings. Extensive studies on the time required for the germination of indigenous trees have been conducted in Malaysia (Garrard, 1955; Ng, 1973, 1978, 1980; Ng and Asri, 1979; Wyatt-Smith, 1964a–f; Yap, 1981), but some work also has been done in Nigeria (Jones, 1956).

Ng (1978) divided tree species into three categories, according to germination rates: (1) rapid, all seeds germinated within 12 weeks; (2) intermediate, germination started before 12 weeks but was not completed by 12 weeks; and (3) delayed, germination started after 12 weeks. Species with rapid germination form carpets of seedlings near the mother plant, and no viable seeds remain at the end of the short germination period. Species with intermediate and delayed rates of seed germination do not produce masses of seedlings on the forest floor; instead, a few seeds germinate each month over a long period of time. Seeds of these species obviously remain viable longer than those with rapid germination (Ng, 1978). Ng (1978, 1980) suggested that species with rapid germination had no dormancy or that it was lost in a short period of time. Species with intermediate and delayed germination had differential dormancy; therefore, seeds became nondormant and germinated at varying times (Ng, 1978).

It is apparent that while seeds of some species studied by Ng were nondormant, others were dormant and required a dormancy-breaking period before they would germinate. The question is, where do we draw the line between slow to germinate and seed dormancy? In the rain forest habitat, seeds would be exposed to high temperatures between the time of seed dispersal and germination. As shown in Chapter 3, high temperatures promote dormancy loss in some seeds with nondeep physiological dormancy; thus, seeds of some rain forest trees could have nondeep physiological dormancy. Because some seeds with nondeep physiological dormancy

begin to come out of dormancy after 4 weeks of exposure to high temperatures (Chapter 3), we suggest that seeds of rain forest trees requiring more than 4 weeks to initiate germination are dormant.

By combining information on (1) time required for the initiation of germination and (2) characteristics of embryos and seeds in various plant families, the type of dormancy in seeds of various rainforest trees can be inferred (Table 9.1). Because several weeks may elapse between the time when the first and last seeds of a species germinate, the time when germination first started was used in our determinations of dormancy. If seeds of a species began to germinate before they were 4 weeks old, the species was considered to have nondormant seeds, but if seeds required longer than 4 weeks, the species had dormant seeds. If a species had dormant seeds and it belonged to a family known to have fully developed embryos and permeable seed coats, it was listed as having physiological dormancy. If a species had dormant seeds and it belonged to a family with fully developed embryos and impermeable seed or fruit coats, it was listed as having physical dormancy. Species belonging to families with underdeveloped embryos were listed as having morphological dormancy if seeds began to germinate before they were 4 weeks old and as having morphophysiologically dormancy if seeds required 4 or more weeks for the initiation of germination. It should be noted, however, that germination studies have been done for some seeds, and thus it was not necessary to infer the type of dormancy.

About 62% of the species listed in Table 9.1 have nondormant seeds, but the length of time required for the initiation of germination is variable. Seeds of *Pithecellobium arboreum, Inga* sp. (Flores and Mora, 1984), *Dryobalanops aromatica* (Sasaki *et al.,* 1979), and *Hura* sp. (Duke, 1969) can germinate before they are dispersed, and some members of the Dipterocarpaceae start to germinate within 5 days or less following dispersal (Yap, 1981). Some nondormant seeds, even in the Dipterocarpaceae, require 14–28 days for the initiation of germination (Ng, 1980).

Physiological dormancy is the most common type of dormancy in seeds of rain forest trees, being present in about 24% of the species listed in Table 9.1. In many genera, however, some species have seeds with physiological dormancy, whereas others have nondormant seeds. Because the loss of physiological dormancy occurs while seeds are exposed to high temperature, we conclude that seeds have nondeep physiological dormancy. Length of the high-temperature, dormancy-breaking treatment varies with the species. Seeds of *Phoebe tonduzii* (Flores *et al.,* 1985) begin to germinate about 1 month after sowing, but those of *Styrax benzoin* require 7–9 months (Kiew, 1982); those of *Sacroglottis*

TABLE 9.1 Types of Seed Dormancy in Nonpioneer Evergreen Rain Forest Trees

Species	Family	Type of dormancy[a,b]	Reference
Adenanthera bicolor	Fabaceae	PY*	Ng (1973)
A. pavonina	Fabaceae	PY*, ND*	Garrard (1955), Ng (1978)
Actinorhytis calapparia	Arecaceae	MPD*	Manokaran (1979)
Adinandra acuminata	Theaceae	ND*	Ng (1978)
Agathis dammara	Araucariaceae	ND*	Wyatt-Smith (1964e), Ng (1980)
Alangium ebenaceum	Alangiaceae	ND*	Ng (1973, 1978)
Alstonia angustiloba	Apocynaceae	ND*	Ng (1973, 1978)
A. scholaris	Apocynaceae	ND*	Beniwal and Singh (1989)
Amoora malaccensis	Meliaceae	ND*	Ng (1978)
Anacardium spruceanum	Anacardiaceae	ND*	Alencar and Magalhaes (1979)
Anisophyllea corneri	Rhizophoraceae	PD*	Ng (1980)
A. disticha	Rhizophoraceae	PD*	Ng (1973, 1978)
A. grandis	Rhizophoraceae	PD*	Ng (1978)
A. griffithii	Rhizophoraceae	PD*	Ng (1978)
Anisoptera laevis	Dipterocarpaceae	ND*	Ng and Asri (1979)
Antidesma neurocarpum	Euphorbiaceae	PD*	Ng (1978)
Aphania paucijuga	Sapindaceae	ND*	Ng (1980)
Aporusa arborea	Euphorbiaceae	ND*	Ng and Asri (1979)
A. aurea	Euphorbiaceae	ND*	Ng and Asri (1979)
Araucaria angustifolia	Araucariaceae	ND*	Tompsett (1984)
A. araucana	Araucariaceae	ND*	Tompsett 1984
A. hunsteinii	Araucariaceae	ND*	Tompsett (1984)
Areca latiloba	Arecaceae	MPD*	Manokaran (1979)
Aromadendron elegans	Magnoliaceae	MD*	Ng (1978)
Arthrophyllum ovalifolium	Araliaceae	MD*	Ng (1978)
Artocarpus elasticus	Moraceae	ND*	Ng (1973, 1978)
A. gomezianus	Moraceae	PD*	Ng (1978)
A. heterophyllus	Moraceae	ND*	Ng (1973)
A. integer	Moraceae	PD*	Ng (1973, 1978)
A. integrifolia	Moraceae	ND*	Beniwal and Singh (1989)
A. lanceifolius	Moraceae	ND*	Wyatt-Smith (1964e), Ng (1973, 1978)
A. nitidus	Moraceae	ND*	Ng 1980
A. rigidus	Moraceae	ND*	Ng and Asri (1979)
Arytera littoralis	Sapindaceae	ND*	Ng (1978)
Baccaurea griffithii	Euphorbiaceae	ND*	Ng (1978)
B. motleyana	Euphorbiaceae	ND*	Ng (1978)
B. pyriformis	Euphorbiaceae	ND*	Ng and Asri (1979)
B. racemosa	Euphorbiaceae	ND*	Ng and Asri (1979)
Balanocarpus heimii	Dipterocarpaceae	ND*	Ng (1973)
Barringtonia macrostachya	Lecythidaceae	PD*	Ng (1978, 1980)
B. scortechinii	Lecythidaceae	PD*	Ng (1980)
Bhesa robusta	Celastraceae	ND*	Ng (1978)
Blumeodendron tokbrai	Euphorbiaceae	PD*	Ng and Asri (1979)
Bombax ellipticum	Bombacaceae	ND*	Ng and Asri (1979)

(continues)

TABLE 9.1—*Continued*

Species	Family	Type of dormancy[a,b]	Reference
B. valetonii	Bombacaceae	ND*	Ng (1980)
Bouea macrophylla	Anacardiaceae	ND*	Ng (1978)
Bridelia penangiana	Euphorbiaceae	ND*	Ng and Asri (1979)
Brosimum lactescens	Moraceae	ND*	Gonzalez (1991)
Brownlowia helferiana	Tiliaceae	ND*	Ng and Asri (1979)
Brucea javanica	Simaroubaceae	ND*	Ng and Asri (1979)
Calophyllum ferrugineum	Clusiaceae	ND*	Ng (1980)
C. teysmanii	Clusiaceae	ND*	Ng (1980)
Campsiandra comosa	Fabaceae	PY*	da Silva *et al.* (1989)
Canarium littorale	Burseraceae	ND*, PD*	Ng (1973, 1980)
C. megalanthum	Burseraceae	ND*	Ng and Asri (1979)
C. patentinervium	Burseraceae	PD*	Wyatt-Smith, (1964b), Ng (1973)
C. pseudosumatranum	Burseraceae	PD*	Ng (1973, 1978)
C. schweingurthii	Burseraceae	PD*	Gilbert (1952)
C. strictum	Burseraceae	PD*	Beniwal and Singh (1989)
Carallia brachiata	Rhizophoraceae	PD*	Ng and Asri (1979)
Carapa guianensis	Meliaceae	PD*, ND*	Alencar and Magalhaes (1979), McHargue and Hartshorn (1983)
Cariniana micrantha	Lecythidaceae	ND	Alencar and Magalhaes (1979)
Caryocar villosum	Caryocaraceae	PD*	Alencar and Magalhaes (1979)
Caryota mites	Arecaceae	MPD*	Raich and Khoon (1990)
Cassia negrensis	Fabaceae	ND	da Silva *et al.* (1989)
C. nodosa	Fabaceae	ND*	Beniwal and Singh (1989)
Castanopsis foxworthyi	Fagaceae	PD*	Ng and Asri (1979)
C. javanica	Fagaceae	PD*	Ng (1980)
Castilla elastica	Moraceae	ND*	Gonzalez (1991)
Cedrela odorata	Meliaceae	ND*	Ng and Asri (1979), Gonzalez (1991)
Champereia manillana	Opiliaceae	PD*	Ng (1980)
Cheilosa malayana	Euphorbiaceae	ND*	Ng (1973, 1978)
Chisocheton divergens	Meliaceae	PD*	Ng (1980)
C. macrothyrsus	Meliaceae	PD*	Ng (1978)
Chlorocardium rodiei	Lauraceae	PD*	Ter Steege *et al.* (1994)
Clausena excavata	Rutaceae	ND*	Ng (1980)
Clerodendrum inerme	Verbenaceae	ND*	Ng (1980)
C. villosum	Verbenaceae	ND*	Ng (1980)
Cocos nucifera	Arecaceae	MPD*	Harries (1992)
Cola acuminata	Sterculiaceae	PY*	Oladokun (1985)
C. grisciflora	Sterculiaceae	PY*	Gilbert (1952)
Conopharyngia penduliflora	Apocynaceae	PD*	Jones (1956)
Corypha umbraculifera	Arecaceae	MPD*	Manokaran (1979)
Couepia longipendula	Lecythidaceae	ND*	Alencar and Magalhaes (1979)
Crossonephelis penangensis	Sapindaceae	ND*	Ng (1980)
Croton argyratum	Euphorbiaceae	ND*	Ng and Asri (1979)
Crudia curtisii	Fabaceae	PY*	Ng (1973, 1978, 1980)
C. pubescens	Fabaceae	PY*	da Silva *et al.* (1989)

(continues)

TABLE 9.1—*Continued*

Species	Family	Type of dormancy[a,b]	Reference
Cyathocalyx pruniferus	Annonaceae	MD*	Ng (1978)
Dacrydium comosum	Podocarpaceae	MPD*	Ng and Asri (1979)
Dacryodes costata	Burseraceae	ND*	Ng (1973, 1978)
D. kingii	Burseraceae	PD*	Ng (1973, 1978)
D. laxa	Burseraceae	ND*	Ng and Asri (1979)
D. rostrata	Burseraceae	ND*	Ng (1978)
D. rugosa	Burseraceae	ND*	Ng and Asri (1979)
Decaspermum fruticosum	Myrtaceae	PD*	Ng (1980)
Dendropanax arboreus	Araliaceae	MPD*	Gonzalez (1991)
Dialium maingayi	Fabaceae	PY	Ng (1988), Sasaki (1980a)
Dillenia grandifolia	Dilleniaceae	MPD*	Ng (1978)
D. suffructicosa	Dilleniaceae	MPD*	Ng (1973)
D. sumatrana	Dilleniaceae	MPD*	Ng (1980)
Dinizia excelsa	Fabaceae	PY	Vastano *et al.* (1983)
Diospyros andamanica	Ebenaceae	ND*	Ng and Asri (1979)
D. argentea	Ebenaceae	ND*	Ng (1980)
D. confertiflora	Ebenaceae	ND*	Ng (1980)
D. diepenhorstii	Ebenaceae	ND*	Ng (1973, 1978)
D. ismailii	Ebenaceae	ND*	Ng (1978)
D. maingayi	Ebenaceae	PD*	Ng (1978)
D. pendula	Ebenaceae	PD*	Ng (1978)
D. pilosantera	Ebenaceae	ND*	Ng (1978)
D. retrofracta	Ebenaceae	ND*	Ng and Asri (1979)
D. sumatrana	Ebenaceae	ND*	Ng (1973, 1978)
Dipterocarpus baudii	Dipterocarpaceae	ND*	Ng (1973), Yap (1981)
D. cornutus	Dipterocarpaceae	ND*	Ng and Asri (1979)
D. grandiflorus	Dipterocarpaceae	ND*	Ng (1973)
D. kunstleri	Dipterocarpaceae	ND*	Ng and Asri (1979)
D. oblongifolius	Dipterocarpaceae	ND*	Ng (1973), Yap (1981)
D. taudii	Dipterocarpaceae	ND*	Ng (1980)
D. turbinatus	Dipterocarpaceae	ND*	Tompsett (1987)
Dipteryx odorata	Fabaceae	PY*	Alencar and Magalhaes (1979)
D. panamensis	Fabaceae	ND	Flores (1992)
Dracontomelom mangiferum	Anacardiaceae	PY*	Ng (1973)
Dryobalanops aromatica	Dipterocarpaceae	ND	Sasaki *et al.* (1979), Itoh (1995)
D. lanceolata	Dipterocarpaceae	ND	Itoh (1995)
D. oblongifolia	Dipterocarpaceae	ND*	Ng (1973), Yap (1981)
Drypetes gossweileri	Euphorbiaceae	PD*	Gilbert (1952)
D. kikir	Euphorbiaceae	PD*	Ng and Asri (1979)
D. longifolia	Euphorbiaceae	PD*	Ng and Asri (1979)
Durio carinatus	Bombacaceae	ND*	Ng (1978)
D. graveolens	Bombacaceae	ND*	Ng and Asri (1979)
D. griffithii	Bombacaceae	ND*	Ng (1980)
D. lowianus	Bombacaceae	ND*	Ng and Asri (1979)

(continues)

TABLE 9.1—*Continued*

Species	Family	Type of dormancy[a,b]	Reference
D. oxleyanus	Bombacaceae	ND*	Ng and Asri (1979)
D. zibethinus	Bombacaceae	ND*	Ng (1978)
Dyera costulata	Apocynaceae	PD*	Wyatt-Smith, (1964b), Ng (1973)
Dysoxylum angustifolium	Meliaceae	ND*	Ng (1978)
D. arborescens	Meliaceae	PD*	Ng (1978)
D. cauliflorum	Meliaceae	PD*	Ng (1973, 1978, 1980)
Elaeocarpus floribundus	Elaeocarpaceae	PD*	Ng (1978, 1980), Beniwal and Singh (1989)
E. petiolatus	Elaeocarpaceae	PD*	Ng (1978)
E. stipularis	Elaeocarpaceae	PD*	Ng (1978)
Elateriospermum tapos	Euphorbiaceae	ND*	Ng (1973, 1978)
Endospermum malaccense	Euphorbiaceae	PD*	Ng and Asri (1979)
Eremoluma williamii	Sapotaceae	PD*	Alencar and Magalhaes (1979)
Erioglossum edule	Sapindaceae	ND*	Ng (1978)
Erythrina caffra	Fabaceae	PY	Nkang (1993)
Erythroxylum cuneatum	Erythroxylaceae	PD*	Ng and Asri (1979)
Eugeissona tristis	Arecaceae	MPD*	Manokaran (1979)
Eugenia claviflora	Myrtaceae	ND*	Ng (1973, 1978)
E. graeme-andersoniae	Myrtaceae	ND*	Ng (1978)
E. glabra	Myrtaceae	PD*	Ng (1973)
E. grandis	Myrtaceae	ND*	Ng (1973, 1978)
E. malaccensis	Myrtaceae	ND*	Ng and Asri (1979)
E. microcalyx	Myrtaceae	PD*	Ng (1978)
E. operculata	Myrtaceae	ND*	Ng (1978)
E. polyantha	Myrtaceae	ND*	Ng (1973, 1978)
Euphoria longan	Sapindaceae	ND*	Chin et al. (1984)
Eurycoma longifolia	Simaroubaceae	PD*	Ng (1980)
Euterpe edulis	Arecaceae	MD*	Mullette et al. (1981)
E. globosa	Arecaceae	MPD*	Bannister (1970)
Fagraea fragrans	Longaniaceae	ND*	Garrard (1955), Ng (1978)
Fahrenheitia pendula	Euphorbiaceae	ND*	Ng (1973)
Ficus benjamina	Moraceae	ND*	Ng (1973, 1978)
Flindersia brayleyana	Rutaceae	ND	Wick (1974)
Fordia lanceolata	Fabaceae	ND*, PY*	Ng (1980)
Garcinia atroviridis	Clusiaceae	ND*	Ng and Asri (1979)
G. mangostana	Clusiaceae	ND*	Ng (1973)
G. nigrolineata	Clusiaceae	PD*	Ng (1973, 1980)
G. opaca	Clusiaceae	PD*	Ng (1978)
G. parvifolia	Clusiaceae	PD*	Ng (1978, 1980)
G. scortechinii	Clusiaceae	PD*	Ng and Asri (1979)
Gardenia carinata	Rubiaceae	ND*	Ng (1973)
G. tubifera	Rubiaceae	ND*	Ng (1980)
Genipa americana	Rubiaceae	ND*	Gonzalez (1991)
Gironniera nervosa	Ulmaceae	PD*	Ng and Asri (1979)
Glochidion obscurum	Euphorbiaceae	ND*	Ng (1978)
G. sericeum	Euphorbiaceae	ND*	Ng (1978)

(continues)

TABLE 9.1—*Continued*

Species	Family	Type of dormancy[a,b]	Reference
Goethalsia meiantha	Tiliaceae	PY*	Gonzalez (1991)
Goniothalamus tortilipetalus	Annonaceae	MPD*	Ng and Asri (1979)
Gonystylus bancanus	Thymelaeaceae	ND*	Wyatt-Smith (1964f), Ng (1980)
G. confusus	Thymelaeaceae	ND*	Ng (1980)
G. maingayi	Thymelaeaceae	ND*	Ng (1980)
Grewia blattaefolia	Tiliaceae	PY*	Ng (1978)
G. laurifolia	Tiliaceae	PY*	Ng (1973, 1978)
Grossonephelis penangensis	Sapindaceae	ND*	Ng (1978)
Guarea cedrata	Meliaceae	PD*	Gilbert (1952)
Guilfoylia monostylis	Simaroubaceae	ND	Nkang and Chandler (1986), Nkang (1993)
Guioa pleuropteris	Sapindaceae	ND*	Ng (1978)
Gymnacranthera eugeniifolia	Myristicaceae	MPD*	Ng and Asri (1979)
Harpullia confusa	Sapindaceae	ND*	Ng (1978)
Heliciopsis velutina	Proteaceae	ND*	Ng and Asri (1979)
Heritiera simplicifolia	Sterculiaceae	ND*	Ng (1973, 1978)
Hevea spp.	Euphorbiaceae	ND	Barrueto *et al.* 1986
Hevea brasilensis	Euphorbiaceae	ND*	Beniwal and Singh (1989)
Hibiscus tiliaceus	Malvaceae	ND*, PY*	Ng and Asri (1979)
Hopea beccariana	Dipterocarpaceae	ND*	Ng and Asri (1979)
H. dyeri	Dipterocarpaceae	ND*	Ng (1978)
H. ferrea	Dipterocarpaceae	ND*	Wyatt-Smith (1964c)
H. helferi	Dipterocarpaceae	ND*	Ng (1973, 1978)
H. nervosa	Dipterocarpaceae	ND*	Ng (1973, 1978)
H. nutans	Dipterocarpaceae	ND*	Wyatt-Smith (1964d), Ng (1973, 1978)
H. odorata	Dipterocarpaceae	ND*	Ng (1973), Yap (1981)
H. sangal	Dipterocarpaceae	ND*	Ng and Asri (1979)
H. subalata	Dipterocarpaceae	ND*	Ng (1978)
H. wightiana	Dipterocarpaceae	ND*	Ng (1973)
Horsfieldia brachiata	Myristicaceae	MPD*	Ng (1973, 1978)
Hunteria corymbosa	Apocynaceae	ND*	Ng (1973)
H. zeylanica	Apocynaceae	ND*	Ng (1978)
Hura sp.	Euphorbiaceae	ND	Duke (1969)
Hydnocarpus woodii	Flacourtiaceae	PD*	Ng (1978, 1980)
Hyeronima alchorneoides	Euphorbiaceae	ND	Gonzalez (1991, 1992)
Hymenaea courbaril	Fabaceae	PY	Flores and Benavides (1990)
		ND*	Alencar and Magalhaes (1979)
Inga sp.	Fabaceae	ND	Flores and Mora (1984)
Intsia glauca	Fabaceae	ND*	Wyatt-Smith (1964e)
I. palembanica	Fabaceae	ND*, PY	Whitmore (1975), Sasaki, 1980a, Sasaki and Ng (1981)
I. platyclados	Fabaceae	ND*	Wyatt-Smith (1943e)
I. speciosa	Fabaceae	ND*	Wyatt-Smith (1964e)
Irvingia malayana	Ixonanthaceae	PD*	Ng (1973, 1978)
Ixonanthes icosandra	Ixonanthaceae	PD*	Ng (1978)
Jacaranda copaia	Bignoniaceae	ND*	Gonzalez (1991)

(continues)

TABLE 9.1—*Continued*

Species	Family	Type of dormancy[a,b]	Reference
Knema curtisii	Myristicaceae	MPD*	Ng (1978)
K. furfuracea	Myristicaceae	MPD*	Ng (1978)
K. laurina	Myristicaceae	MPD*	Ng (1973, 1978)
K. malayana	Myristicaceae	MPD*	Ng (1973)
K. scortechinii	Myristicaceae	MPD*	Ng (1980)
K. stenophylla	Myristicaceae	MPD*	Ng (1978)
Koompassia excelsa	Fabaceae	ND*	Ng (1980)
K. malaccensis	Fabaceae	ND*	Wyatt-Smith (1964b), Ng (1973, 1978)
Laetia procera	Flacourtiaceae	ND*	Gonzalez (1991)
Lagerstroemia speciosa	Lythraceae	ND*	Ng (1973, 1978)
Lansium domesticum	Meliaceae	ND*	Ng (1973, 1978)
Lecythia ampla	Lecythidaceae	PD*	Gonzalez (1991)
Leea indica	Leeaceae	ND*	Ng (1978)
Leptospermum flavescens	Myrtaceae	ND*	Ng (1973, 1978)
Licaria limbosa	Lauraceae	PD*	Flores *et al.* (1985)
Litchi chinensis	Sapindaceae	ND*	Ray and Sharma (1987)
Lithocarpus elegans	Fagaceae	PD*	Ng and Asri (1979)
L. ewyckii	Fagaceae	PD*	Ng (1978)
Litsea castanea	Lauraceae	PD*	Ng (1973, 1978)
Livistona kingiana	Arecaceae	MPD*	Manokaran (1979)
L. speciosa	Arecaceae	MPD*	Manokaran (1979)
Macaranga triloba	Euphorbiaceae	ND*	Ng and Asri (1979)
Machaerium inundatum	Fabaceae	ND, PY*	da Silva *et al.* (1989)
Macrolobium acaciifolium	Fabaceae	ND	da Silva *et al.,* (1989)
Madhuca butyracea	Sapotaceae	ND*	Beniwal and Singh (1989)
M. utilis	Sapotaceae	ND*	Ng (1978)
Maesopsis eminii	Rhamnaceae	. ND*	Mugasha and Msanga (1987)
Mallotus philippinensis	Euphorbiaceae	PD*	Ng (1980)
Mangifera indica	Anacardiaceae	ND	Corbineau *et al.* (1986)
Manilkara zapota	Sapotaceae	ND*	Garrard (1955)
Mastixia pentandra	Cornaceae	PD*	Ng (1978)
Melia azedarach	Meliaceae	PD*	Ng and Asri (1979)
Melecylon cantleyi	Memecylaceae	PD*	Ng and Asri (1979)
Mesua ferrea	Clusiaceae	ND*	Ng (1978)
Mezzettia leptopoda	Annonaceae	MPD*	Ng (1973, 1980)
Micromelium minutum	Rutaceae	ND*	Ng (1978)
Milletia atropurpurea	Fabaceae	PY*	Ng (1973, 1978)
Mimusops elengi	Sapotaceae	ND*	Ng (1978), Beniwal and Singh (1989)
Minquartia guianensis	Olacaceae	PD*	Gonzalez (1991)
Mitrephora maingayi	Annonaceae	MPD*	Ng and Asri (1979)
Monocarpia marginalis	Annonaceae	MPD*	Ng (1978, 1980)
Morinda citrifolia	Rubiaceae	PD*	Ng (1978)
Muntingia calabura	Elaeocarpaceae	ND*	Garrard (1955), Ng (1973)
Murraya paniculatum	Rutaceae	ND*	Ng (1978)

(continues)

TABLE 9.1—*Continued*

Species	Family	Type of dormancy[a,b]	Reference
Myristica crassa	Myristicaceae	MPD*	Ng (1978)
M. malaccensis	Myristicaceae	MPD*	Ng (1973, 1978)
Nauclea maingayi	Rubiaceae	ND*	Ng and Asri (1979)
N. subdita	Rubiaceae	ND*	Ng and Asri (1979)
Nectandra laurel	Lauraceae	PD*	Flores *et al.* (1985)
N. membranacea	Lauraceae	PD*	Gonzalez (1991)
N. plebosa	Lauraceae	PD*	Flores *et al.* (1985)
N. sinuata	Lauraceae	PD*	Flores *et al.* (1985)
Nenga pumila	Arecaceae	MPD*	Manokaran (1979)
Neobalanocarpus heimii	Dipterocarpaceae	ND*	Yap (1981)
Neoscortechinia kingii	Euphorbiaceae	ND*	Ng and Asri (1979)
Nephelium costatum	Sapindaceae	ND*	Ng and Asri (1979)
N. eriopetalum	Sapindaceae	ND*	Ng and Asri (1979)
N. glabrum	Sapindaceae	ND*	Ng (1978)
N. lappaceum	Sapindaceae	ND*	Ng (1973, 1978)
N. malaiense	Sapindaceae	ND*	Garrard (1955), Ng (1978)
N. wallichii	Sapindaceae	ND*	Ng and Asri (1979)
Ocotea cernua	Lauraceae	PD*	Flores *et al.* (1985)
O. novogratensis	Myristicaceae	MD*	Gonzalez (1991)
O. rodiaei	Lauraceae	PD*	Fanshawe (1974), Richards, (1957)
O. veraguensis	Lauraceae	PD*	Flores *et al.* (1985)
Ormosia venosa	Fabaceae	ND*	Ng (1973, 1978)
Oroxylon indicum	Bignoniaceae	ND*	Ng (1973, 1978)
Osmelia maingayi	Flacourtiaceae	ND*	Ng and Asri (1979)
Palaquium gutta	Sapotaceae	ND*	Ng and Asri (1979)
P. maingayi	Sapotaceae	ND*	Ng and Asri (1979)
Parashorea densiflora	Dipterocarpaceae	ND*	Ng (1973), Yap (1981)
P. lucida	Dipterocarpaceae	ND*	Wyatt-Smith (1964c), Ng (1980)
Parastemon urophyllus	Rosaceae	ND*	Ng and Asri (1979)
Parartocarpus bracteatus	Moraceae	ND*	Ng (1980)
Parkia javanica	Fabaceae	PY, ND*	Ng (1978), Longman and Jenik, (1987), Sasaki, (1980a)
P. pendula	Fabaceae	PY*	Alencar and Magalhaes (1979)
P. speciosa	Fabaceae	ND*	Ng (1978)
Paropsia vareciformis	Passifloraceae	ND*	Ng (1973, 1978)
Payena lucida	Sapotaceae	ND*	Ng (1973, 1978)
Pellacalyx saccardianus	Rhizophoraceae	PD*	Ng (1978)
Peltogyne prancei	Fabaceae	PY*	da Silva *et al.* (1989)
Peltophorum pterocarpum	Fabaceae	PY*, ND*	Ng (1978)
Pentaspadon officinalis	Anacardiaceae	ND*	Wyatt-Smith (1964e), Ng (1980)
Persea americana	Lauraceae	ND*	Flores *et al.* (1985)
P. caerulea	Lauraceae	PD*	Flores *et al.* (1985)
Phaeanthus opthalamicus	Annonaceae	MPD*	Ng and Asri (1979)
Phoebe brenesii	Lauraceae	ND*	Flores *et al.* (1985)
P. mexicana	Lauraceae	ND*	Flores *et al.* (1985)

(*continues*)

TABLE 9.1—*Continued*

Species	Family	Type of dormancy[a,b]	Reference
P. tonduzii	Lauraceae	ND*	Flores *et al.* (1985)
Phyllanthus acidus	Euphorbiaceae	ND*	Ng and Asri (1979)
P. emblica	Euphorbiaceae	PD*	Ng (1980)
		ND*	Beniwal and Singh (1989)
Pimelodendron griffithianum	Euphorbiaceae	ND*	Ng and Asri (1979)
Pinanga insignis	Arecaceae	MPD*	Manokaran (1979)
P. malaiana	Arecaceae	MPD*	Manokaran (1979)
Pithecellobium arboreum	Fabaceae	ND	Flores and Mora (1984)
P. clypearia	Fabaceae	ND*	Ng (1978)
P. ellipticum	Fabaceae	ND*	Ng (1978)
P. globosum	Fabaceae	ND*	Garrard, (1955)
P. macradenium	Fabaceae	ND*	Gonzalez (1991)
P. microcarpum	Fabaceae	ND*	Ng and Asri (1979)
P. pahangense	Fabaceae	ND*	Ng (1978)
Pittosporum ferrugineum	Pittosporaceae	MPD*	Ng (1980)
Planchonella glabra	Sapotaceae	ND*	Ng (1978)
P. maingayi	Sapotaceae	PD*	Ng (1978)
Planchonia grandis	Lecythidaceae	ND*	Ng (1980)
Platea latifolia	Icacinaceae	PD*	Ng (1978)
Podocarpus imbricatus	Podocarpaceae	MD*	Ng and Asri (1979)
P. neriifolius	Podocarpaceae	MD*	Ng (1978)
Polyalthia cinnamomea	Annonaceae	MPD*	Ng (1980)
P. glauca	Annonaceae	MPD*	Ng (1978)
P. jenkensii	Annonaceae	MPD*	Ng (1980)
		MD*	Beniwal and Singh (1989)
P. sclerophylla	Annonaceae	MPD*	Ng (1980)
Pometia pinnata	Sapindaceae	ND*	Ng (1978)
Pouteria malaccensis	Sapotaceae	ND*	Ng and Asri (1979)
Pritchardia minor	Arecaceae	MPD*	Manokaran (1979)
Prunus arborea	Rosaceae	PD*	Ng (1978, 1980)
Pseudobombax septenatum	Bombacaceae	ND*	Gonzalez (1991)
Pternandra coerulescens	Melastomataceae	ND*	Metcalfe (1996)
P. echinata	Melastomataceae	ND*	Metcalfe (1996)
Pterocarpus amazonicus	Fabaceae	ND	da Silva *et al.* (1989)
P. indicus	Fabaceae	ND*, PY*	Ng (1978)
P. soyauxii	Fabaceae	ND*	Gilbert (1952)
Pterocymbium javanicum	Sterculiaceae	ND*	Ng (1973, 1978)
Pterospermum javanicum	Sterculiaceae	ND*	Ng (1978)
Pterygota alata	Sterculiaceae	ND*, PY*	Ng (1978, 1980)
Ptychopyxis caput-medusae	Euphorbiaceae	ND*	Ng (1980)
P. costata	Euphorbiaceae	ND*	Ng and Asri (1979)
Pyrenaria acuminata	Theaceae	PD*	Ng (1980)
Randia acuminata	Rubiaceae	PD*	Jones (1956)
R. scortechinii	Rubiaceae	PD*	Ng (1978)

(continues)

TABLE 9.1—*Continued*

Species	Family	Type of dormancy[a,b]	Reference
Ricinodendron africanum	Euphorbiaceae	PD*	Jones (1956)
Rinorea anguifera	Violaceae	ND*	Ng (1978)
R. sclerocarpa	Violaceae	ND*	Ng and Asri (1979)
Rollinia microsepala	Annonaceae	MD*	Gonzalez (1991)
Sacoglottis trichogyna	Humiriaceae	PD*	Hartshorn, (1978)
Salacca conferta	Arecaceae	MPD*	Manokaran (1979)
Salmalia malabarica	Bombacaceae	PY*	Chatterji and Sen (1964)
Sandoricum koetjape	Meliaceae	ND*	Ng (1978)
Santiria griffithii	Burseraceae	PD*	Ng (1980)
S. laevigata	Burseraceae	PD*	Ng (1973)
S. nana	Burseraceae	PD*	Ng (1973)
S. oblongifolia	Burseraceae	ND*	Ng (1978)
S. rubiginosa	Burseraceae	PD*	Ng (1973)
Sapium baccatum	Euphorbiaceae	ND*	Beniwal and Singh (1989)
Sarcosperma paniculatum	Sarcospermataceae	PD*	Ng (1973, 1978)
Sarcotheca griffithii	Oxalidaceae	ND*	Ng (1980)
Saurauia roxburghii	Actinidiaceae	ND*	Ng (1978)
Scaphium marcopodum	Sterculiaceae	ND*	Ng and Asri (1979)
Scaphocalyx spathacea	Flacourtiaceae	ND*	Ng and Asri (1979)
Schima wallichii	Theaceae	ND*	Ng (1980), Beniwal and Singh (1989)
Schleronema micranthum	Bombacaceae	PY*	Alencar and Magalhaes (1979)
Scorodocarpus borneensis	Olacaceae	PD*	Ng (1980)
S. zeukeri	Olacaceae	ND*	Gilbert (1952)
Scutinanthe brunnea	Burseraceae	ND*	Ng and Asri (1979)
Serianthes dilmyi	Fabaceae	ND*	Ng (1980)
Shorea acuminata	Dipterocarpaceae	ND*	Ng and Asri (1979), Yap (1981)
S. agami	Dipterocarpaceae	ND*	Ng (1980)
S. argentifolia	Dipterocarpaceae	ND*	Ng (1980)
S. assamica	Dipterocarpaceae	ND*	Ng (1978), Yap (1981)
S. bracteolata	Dipterocarpaceae	ND*	Yap (1981)
S. curtisii	Dipterocarpaceae	ND*	Wyatt-Smith (1964e), Ng (1980), Yap (1981)
S. glauca	Dipterocarpaceae	ND*	Ng and Asri (1979)
S. kunstleri	Dipterocarpaceae	ND*	Ng (1980)
S. laevis	Dipterocarpaceae	ND*	Wyatt-Smith (1964e), Ng (1980)
S. lepidota	Dipterocarpaceae	ND*	Ng and Asri (1979)
S. leprosula	Dipterocarpaceae	ND*	Ng (1973, 1978)
S. macrophylla	Dipterocarpaceae	ND*	Ng (1973), Yap (1981)
S. martiniana	Dipterocarpaceae	ND*	Ng (1980)
S. materialis	Dipterocarpaceae	ND*	Wyatt-Smith (1964d), Ng (1980), Yap (1981)
S. maximum	Dipterocarpaceae	ND*	Ng (1973, 1978)
S. mecistopteryx	Dipterocarpaceae	ND*	Ng (1980)
S. multiflora	Dipterocarpaceae	ND*	Ng (1980)
S. ovalis	Dipterocarpaceae	ND*	Ng (1978)
S. palembanica	Dipterocarpaceae	ND*	Ng (1980)

(*continues*)

TABLE 9.1—*Continued*

Species	Family	Type of dormancy[a,b]	Reference
S. parvifolia	Dipterocarpaceae	ND*	Ng (1973, 1978)
S. pauciflora	Dipterocarpaceae	ND*	Ng and Asri (1979)
S. platyclados	Dipterocarpaceae	ND*	Ng (1978), Yap (1981)
S. resina-nigra	Dipterocarpaceae	ND*	Ng and Asri (1979)
S. resinosa	Dipterocarpaceae	ND*	Ng (1980)
S. robusta	Dipterocarpaceae	ND*	Purohit *et al.* (1982)
S. roxburghii	Dipterocarpaceae	ND*	Yap (1981)
S. rugosa	Dipterocarpaceae	ND*	Wyatt-Smith (1964f)
S. scabberima	Dipterocarpaceae	ND*	Ng and Asri (1979)
S. seminis	Dipterocarpaceae	ND*	Ng (1980)
S. singkawang	Dipterocarpaceae	ND*	Ng (1973, 1978)
S. stenoptera	Dipterocarpaceae	ND*	Ng (1980)
S. sumatrana	Dipterocarpaceae	ND*	Ng (1978), Yap (1981)
S. talura	Dipterocarpaceae	ND*	Wyatt-Smith (1964c), Ng (1978)
S. xanthophylla	Dipterocarpaceae	ND*	Ng and Asri (1979)
Simarouba amara	Simaroubaceae	ND*	Gonzalez (1991)
Sindora coriacea	Fabaceae	PY	Sasaki (1980a)
S. echinocalyx	Fabaceae	PY*	Ng (1978, 1988)
Sloanea javanica	Elaeocarpaceae	ND*	Ng and Asri (1979)
Spathodea campanulata	Bignonaceae	ND*	Garrard (1955), Ng (1973)
Sterculia foetida	Sterculiaceae	ND*	Ng (1980)
S. parviflora	Sterculiaceae	ND*	Ng (1973, 1978)
Stereospermum fimbriatum	Bignoniaceae	ND*	Ng and Asri (1979)
Strombosia javanica	Olacaceae	PD*	Ng (1973, 1978)
Stryphnodendron excelsum	Fabaceae	PY	Gonzalez (1991), Flores (1992)
S. guianense	Fabaceae	PY*	Alencar and Magalhaes (1979)
Swartzia polyphylla	Fabaceae	ND, PY*	da Silva *et al.* (1989)
Swietenia macrophylla	Meliaceae	ND*	Ng (1973), Beniwal and Singh, (1989)
Symingtonia populnea	Hamamelidaceae	ND*	Ng and Asri (1979)
Symphonia globulifera	Clusiaceae	ND*	Corbineau and Come (1988)
Synsepalum subcordatum	Sapotaceae	PD*	Gilbert (1952)
Styrax benzoin	Styracaceae	PD	Ng (1980), Kiew (1982)
Tabebuia argentea	Bignoniaceae	ND*	Ng and Asri (1979)
Tachigalia paniculata	Fabaceae	ND	da Silva *et al.* (1989)
Talauma betongensis	Magnoliaceae	MD*	Ng and Asri (1979)
Teijsmanniodendron coriaceum	Verbenaceae	ND*	Ng (1980)
T. pteropodum	Verbenaceae	ND*	Ng and Asri (1979)
Terminalia bellirica	Combretaceae	PD*	Ng (1973)
T. phellocarpa	Combretaceae	PD*	Ng (1978)
T. subspathulata	Combretaceae	ND*	Ng (1973, 1978)
T. superba	Combretaceae	PD*	Khasa (1992)
Tetrameles nudiflora	Passifloraceae	ND*	Beniwal and Singh (1989)
Tetramerista glabra	Tetrameristaceae	PD*	Wyatt-Smith (1964f), Ng (1978, 1980)
Theobroma cacao	Sterculiaceae	ND	King and Roberts (1982)

(continues)

TABLE 9.1—*Continued*

Species	Family	Type of dormancy[a,b]	Reference
Tonna australis	Meliaceae	ND	Walters (1974)
T. sureni	Meliaceae	ND	Noraini *et al.* (1994)
Trichilia dregeana	Meliaceae	ND*	Choinski (1990)
Trigonistrum hypoleucum	Trigoniaceae	ND*	Ng (1978)
Trigonostemon elegantissimus	Euphorbiaceae	ND*	Ng (1978)
Triomma malaccensis	Burseraceae	ND*	Ng (1978)
Tristania merguensis	Myrtaceae	ND*	Ng (1978)
Upuna borneensis	Dipterocarpaceae	ND*	Ng and Asri (1979)
Vatairea guianensis	Fabaceae	ND	da Silva *et al.* (1989)
Vatica cinerea	Dipterocarpaceae	ND*	Ng and Asri (1979)
V. lowii	Dipterocarpaceae	ND*	Ng (1978)
V. stapfiana	Dipterocarpaceae	ND*	Ng (1973, 1978)
V. umbonata	Dipterocarpaceae	ND*	Ng (1980), Yap (1981)
V. wallichii	Dipterocarpaceae	ND*	Ng (1980)
Virola koschnyi	Myristicaceae	MD	Flores (1992)
V. surinamensis	Myristicaceae	MD*	Cungha *et al.* (1995)
Vitex pubescens	Verbenaceae	PD*	Ng (1973, 1978)
Vochysia hondurensis	Vochysiaceae	ND*	Gonzalez (1991)
Walsura neurodes	Meliaceae	PD*	Ng (1978)
Xanthophyllum affine	Polygalaceae	ND*	Ng (1980)
X. amoenum	Polygalaceae	ND*	Ng and Asri (1979)
X. griffithii	Polygalaceae	ND*	Ng (1973, 1978)
X. intermedium	Polygalaceae	ND*	Ng and Asri, (1979)
X. obscurum	Polygalaceae	ND*	Ng (1973, 1978)
X. palembanicum	Polygalaceae	ND*	Ng (1980)
X. rufum	Polygalaceae	ND*	Ng and Asri (1979)
X. stipitatum	Polygalaceae	ND*	Ng and Asri (1979)
Xylopia caudata	Annonaceae	MPD*	Ng (1980)
X. xericophylla	Annonaceae	MPD*	Gonzalez (1991)
X. malayana	Annonaceae	MPD*	Ng and Asri (1979)
Zanthoxylum mayanum	Rutaceae	PD*	Gonzalez (1991)
Ziziphus jujuba	Rhamnaceae	ND*	Ng and Asri (1979)

[a] ND, nondormant: 0–4 weeks of incubation before initiation of germination; PD, physiological dormancy: more than 4 weeks incubation before initiation of germination and belonging to a family whose seeds have fully developed embryos and permeable seed coats; PY, physical dormancy: family has impermeable seed coats; MD, morphological dormancy: family has underdeveloped embryos, and germination was initiated in 0–4 weeks; and MPD, morphophysiological dormancy: family has underdeveloped embryos, and germination was initiated after 4 or more weeks of incubation.

[b] An asterisk means that the type of dormancy is inferred from information on time required for the initiation of germination and from characteristics of seeds in that family.

trichogyna require 18–24 months (Hartshorn, 1978). If the woody endocarp is removed from the drupes of *S. benzoin,* seeds germinate immediately (Kiew, 1982). Thus, in the field, a 7- to 9-month dormancy-breaking period is required before the embryo gains enough growth potential to crack open the endocarp.

Grushvitzky (1967) mentioned a number of plant families, including the Magnoliaceae, Degeneriaceae, Winteraceae, Lactoridaceae, Canellaceae, Annonaceae, Myristicaceae, Amborellaceae, Monimiaceae, and Lardizabalaceae, whose seeds have underdeveloped embryos. Further, some members of these families are rain forest trees; thus, it is expected that morphological or morphophysiological dormancy occurs in seeds of some species. It appears that about 2% of the species listed in Table 9.1 have morphological dormancy and 8% have morphophysiological dormancy. The Arecaceae (palms) have underdeveloped embryos (Martin, 1946), thus both morphological and morphophysiological dormancy may be present in this family. Depending on the species, seeds of palms may require 4–125 weeks for the initiation of germination (Koebernik, 1971; Manokaran, 1979). In a study of 241 species of palms, Koebernik (1971) found that seeds required from 1.6 to 125 weeks for germination. However, it is unclear if these numbers refer to days required for the initiation of germination or days required for all seeds of a species to germinate. The palms studied by Koebernik (1971) are not included in Table 9.1 because it is not possible to infer the type of dormancy.

Cola acuminata belongs to the Sterculiaceae, a family known to have impermeable seed coats but not underdeveloped embryos. However, seeds (nuts) of *C. acuminata* in most weight classes germinated to higher percentages following an unspecified period of dry storage than freshly collected ones, and the embryo in stored nuts was morphologically better developed than in fresh ones (Oladokun, 1985). de Vogel (1980) reported that mature seeds of *Buchanania* (Anacardiaceae) and *Gironniera* (Ulmaceae) appeared to be hollow, but they had underdeveloped embryos that elongated before seeds germinated. However, neither the Anacardiaceae nor the Ulmaceae is known to have underdeveloped embryos at the time of seed dispersal; thus, additional studies on tropical members of these families are needed.

Rain forest trees belong to plant families, including Anacardiaceae, Bombacaceae, Fabaceae, Malvaceae, Rhamnaceae, Sapindaceae, Sterculiaceae, and Tiliaceae, known to have seeds (or fruits) with impermeable coats (Table 3.5). Further, seeds of some members of these families require more than 4 weeks for the initiation of germination; therefore, we conclude that seeds of about 6% of the species listed in Table 9.1 have physical dormancy.

Seeds of the legume *Koompassia malaccensis* have thin seedcoats that are permeable to water (Sasaki, 1980a); this appears to be true for seeds of *K. excelsa* (Ng, 1980). Nondormant seeds also are found in some members of other families known to have physical dormancy (Table 9.1), e.g., Anacardiaceae, *Bouea, Mangifera,* and *Pentaspadon;* Bombacaceae, *Bombax and Durio;* Fabaceae, *Dipteryx, Fordia, Inga, Intsia, Ormosia, Parkia, Peltophorum, Pithecellobium,* and *Serianthes;* Malvaceae, *Hibiscus;* Rhamnaceae, *Ziziphus;* Sapindaceae, *Aphania, Arytera, Erioglossum, Euphoria, Guioa,*

Harpullia, Litchi, Madhuca, Nephelium, and *Pometia;* and Sterculiaceae, *Heritiera, Pterocymbium, Pterospermum, Pterygota, Scaphium,* and *Sterculia* (Table 9.1).

In *Pterygota* (Sterculiaceae) and *Intsia, Fordia,* and *Parkia* (Fabaceae), many members of a genus have nondormant seeds, but others have seeds with physical dormancy. Further, in *Litchi* (Sapindaceae), seeds that dry below a 20% moisture content develop impermeable seed coats (Ray and Sharma, 1987). Physical dormancy occurs in *Cola* (Sterculiaceae); *Crudia, Dialium, Hymenaea, Milletia, Sindora* (Fabaceae); *Grewia* (Tiliaceae); and *Dracontomelom* (Anacardiaceae). Thus, some genera in the Anacardiaceae, Fabaceae, Sterculiaceae, and Tiliaceae may have species with nondormant seeds, whereas others have seeds with physical dormancy.

The mechanism whereby physical dormancy is broken in the rain forest environment is unknown, but in the laboratory it is broken by concentrated sulfuric acid, hot water, and/or mechanical scarification (Sasaki, 1980a; Sasaki and Ng, 1981). Seeds of the legumes *Intsia palembanica* (Sasaki, 1980a; Sasaki and Ng, 1981) and *Sindora coriacea* (Sasaki, 1980a) can be made permeable when the strophiole is ruptured by scraping. Seeds of *Hymenaea courbaril* became permeable and began to germinate after 20–30 days of incubation on moist (field capacity) soil at 23°C (Flores and Benavides, 1990). Hatano and Tanaka (1979) reported that seeds of *Tectona grandis* (Verbenaceae) and *Terminalia ivorensis* (Combretaceae) have impermeable seed coats, as well as immature embryos. However, neither of these characteristics has been reported in seeds of other members of these two families.

2. Recalcitrant Seeds

The moisture content of recalcitrant seeds at the time of maturation is 30–70%, but it varies among species (Chapter 3) and even within the same species (Puchet and Vazquez-Yanes, 1987). Because metabolism is continuous in recalcitrant seeds (Lin and Chen, 1995), they lose viability if the moisture content drops below a certain critical level before germination occurs (Table 9.2). Embryos appear to be less tolerant of dehydration than the whole seed. For example, seeds of *Araucaria angustifolia* die when the moisture content decreases to 25% (Tompsett, 1984), but isolated embryos die when the moisture content decreases to less than about 40% (Espindola *et al.,* 1993). Orthodox seeds, however, have a low moisture content (15–20%) at maturity and can tolerate being dried to less than 5% without a loss of viability. In a survey of fresh weights of recalcitrant and orthodox seeds, Chin *et al.* (1989) found that recalcitrant seeds may be 10–77 times heavier than orthodox ones.

TABLE 9.2 Critical Moisture Content for Maintaining Viability
of Recalcitrant Seeds of Rain Forest Trees

Species	Moisture content (%)	Reference
Araucaria angustifolia	25	Tompsett (1984)
A. araucana	20	Tompsett (1984)
A. hunsteinii	14	Tompsett (1982)
Artocarpus heterophyllus	43	Chin *et al.* (1984)
A. integer	38	Hor *et al.* (1990)
Dipterocarpus obtusifolius	45	Tompsett (1987)
D. turbinatus	45	Tompsett (1987)
Durio zibethinus	26	Hor *et al.* (1990)
Euphoria longan	18	Chin *et al.* (1984)
Garcinia mangostana	24	Normah *et al.* (1997)
G. motleyana	<35	Normah *et al.* (1997)
Hevea brasiliensis	15–20	Chin *et al.* (1981)
Hopea odorata	17, 24	Tang and Tamari (1973), Corbineau and Come (1988)
H. helferi	19	Tang and Tamari (1973)
Inga, seven species	20–30	Pritchard *et al.* (1995)
Litchi chinensis	20	Ray and Sharma (1987)
Mangifera indica	25–30	Corbineau *et al.* (1986)
Nephelium lappaceum	13, 27	Chin *et al.* (1984), Hor *et al.* (1990)
N. malaiense	25	Chin *et al.* (1984)
Shorea acuminata	17	Chin *et al.* (1984)
S. platyclados	20	Tang (1971)
S. robusta	18	Purohit *et al.* (1982)
S. roxburghii	17	Corbineau and Come (1988)
S. talura	20	Sasaki (1980b)
Symphonia globulifera	37	Corbineau and Come (1988)
Theobroma cacao	36.7	King and Roberts (1982)
Trichilia dregeana	30	Choinski (1990)
Virola surinamensis	12	Cunha *et al.* (1995)

However, embryos are 15 times heavier in orthodox than in recalcitrant seeds. About 90 species have been listed in the literature as having recalcitrant seeds, but careful testing has shown that some of them, e.g., Coffea spp., *Citrus* spp. (Mumford and Grout, 1979; Usberti and Felippe, 1980b; Chin et al., 1984), *Elaeis guineensis* (Grout et al., 1983), and *Manihot esculenta* (Ellis et al., 1981) have orthodox rather than recalcitrant seeds. Seeds of some species do not fit well into either the orthodox or the recalcitrant category. For example, germination percentages were not decreased by a dehydration pretreatment that reduced the moisture content of *Coffea* (cultivar Number 39 from Tanzania) seeds to 9.4–10.5%, but they were decreased by about 12% after a dehydration pretreatment that reduced the moisture content to 6% (Ellis et al., 1990). Coffee cultivars vary in seed sensitivity to dehydration and temperature tolerance. Whereas orthodox seeds can be stored at −18°C at a moisture content of 5%, those of coffee are more likely to live if stored at 15°C at a moisture content of 10%. Thus, coffee seeds tolerate lower moisture contents and temperatures than typical tropical recalcitrant seeds, but they are not as tolerant of low moisture contents and low temperatures as orthodox seeds; they are intermediate between orthodox and recalcitrant (Ellis et al., 1990).

Seeds of *Carica papaya* also do not fit the definition of either orthodox or recalcitrant seeds (Ellis et al., 1991). Seeds at a moisture content of 7.9–9.4% germinated to 90% or more after 12 months of storage at 15°C. However, storage temperatures of 0 or −20°C (regardless of seed moisture content) or seed moisture contents of 4.2–5.9% (at 15, 0, or −20°C) reduced germination to 60% or less.

Recalcitrant seeds have been reported from a variety of species including deciduous temperate trees (USDA, 1948), aquatic grasses (Probert and Longley, 1989), mangroves (Farrant et al., 1985), trees of semievergreen tropical forests (Garwood and Lighton, 1990), and nonpioneer trees of tropical rain forests (Whitmore, 1975), including those requiring large gaps for seedling establishment (Johnson and Morales, 1972). The number of climax tree species in tropical rain forests with recalcitrant seeds is unknown, but Chin et al. (1989) suggested that recalcitrant seeds are more common in rain forests than in any other type of community. Rapid germination and short periods of seed viability, which are attributes of recalcitrant seeds, have been mentioned by various authors in discussions of rain forest species in Malaysia (Holttum, 1953; Ng, 1973, 1978), southeast Asia (Whitmore, 1975), Africa (Choinski, 1990), and Mexico (Moreno, 1977). Thus, recalcitrant seeds seem to be a characteristic of rain forests, regardless of geographical location.

Recalcitrant seeds have a high moisture content at the time of dispersal, but in some cases this may not be a sufficient level of hydration to promote germination. Seeds of *Shorea curtisii* germinated in 10 days if they were on wet litter but germinated in 6 days if they were soaked by frequent rains. The moisture content at the time of germination was 90–160% of that of a newly dispersed seed (Burgess unpublished data in Whitmore, 1975). The germination of *S. roxburghii, S. almon,* and *S. robusta* seeds declined when the moisture content dropped below 34, 40, and 40%, respectively (Tompsett, 1985). Thus, lack of rain following seed dispersal could prevent germination and result in eventual death of the seed.

Attempts to use species with recalcitrant seeds for reforestation in rain forest areas have resulted in the accumulation of some information on tolerances of seeds to drying and various temperatures. Seeds differ greatly in their ability to survive drying (Farrant *et al.,* 1988), with some *Dipterocarpus* spp. losing viability at 45% moisture and *Nephelium lappaceum* dying at 13% (Table 9.2). Moisture loss decreases with a decrease in storage temperature (Purohit *et al.,* 1982), but seeds differ in their tolerance to low temperatures. Seeds of *Shorea curtisii* (Tang, 1971), *Parashorea densiflora* (Sasaki, 1980b; Yap, 1981), *Shorea platyclados* (Tang and Tamari, 1973), and *Theobroma cacao* (Ibanez, 1968) are killed at 4°C, those of *S. talura* (Purohit *et al.,* 1982) and *Dryobalanops aromatica* (Jensen, 1971) at 5°C, and those of *Theobroma cacao* at 8°C (Boroughs and Hunter, 1963). The cotyledons are the site of cold damage in seeds of *T. cacao* (Ibanez, 1965, 1968). Seeds of *S. resina-nigra, S. multiflora, S. faguetiana,* and *S. hopefolia* lived for only about 1 month at 4°C (Sasaki, 1980b). Seeds of *S. robusta, S. roxburghii,* and *S. almon* lost viability at 6 and 11°C, whereas those of *S. robusta* and *S. almon* died at 16°C (Tompsett, 1985). However, seeds of *S. talura* survived at 4°C for 5 months if the moisture level of the seeds did not drop below 20% (Sasaki, 1980b). Research is being done to find ways to cryopreserve excised embryos of recalcitrant seeds (Becwar *et al.,* 1983; Hor *et al.* 1990; Maycock and Berjak, 1993; Chandel *et al.,* 1995). Also, about 53% of *Litchi chinensis* seeds pelletized with 4% Na-alginate and stored at 10°C were alive after 80 days, whereas only 12% of nonpelletized seeds were viable after 40 days; all nonpelletized seeds were dead after 80 days (Xu-Ping *et al.,* 1993). One problem with trying to store recalcitrant seeds is that they may germinate in closed containers, even on dry filter paper. Treatment with ABA and tetcyclacis or only tetcyclacis prevented the germination of *Hopea odorata* seeds stored dry at 15°C for 26 days, and 72 and 78% of the seeds germinated, respectively, when placed on a moist substrate (Maury-Lechon *et al.,* 1993).

3. Seed Longevity

It is assumed that recalcitrant seeds do not live very long, even on the forest floor; however, longevity studies under natural conditions have not been done. The closest thing we have to longevity studies of climax tree species is Ng's (1973, 1978, 1980) work in Malaysia, where most seeds were sown in soil-filled boxes buried in the ground (Ng, 1973). Seeds of 208 of the 335 species included in his study completed germination in 12 weeks, and all ungerminated seeds of the 208 species were nonviable; the proportion of these 208 species with recalcitrant seeds is unknown. Of the 127 species that did not complete germination in 12 weeks, 68 stopped germinating after 11–20 weeks, 23 after 21–30 weeks, 17 after 31–40 weeks, 5 after 41–50 weeks, and 9 after 51–100 weeks, 1 after 100–125 weeks, and 4 after 124–153 weeks (Ng, 1980).

In a model of the rate of loss of species from a seed bank in a Malaysian rain forest, Ng (1988) predicted that seeds of 50% of the 335 species would lose viability by the sixth week. All seeds of 75% of the species would be dead by week 12, and seeds of only 5% of the species would be viable after 30 weeks.

Seeds of *Canarium patinervium* lose viability after 31–53 days, those of *Dyera costulata* after 24–130 days, and those of *Koompassia malaccensis* after 13–57 days (Wyatt-Smith, 1964b; Poore, 1968). Seeds of *Khaya ivorensis* live for only a few days, and those of *Lovoa klaineana* are "short-lived" (Jones, 1956). It should be noted, however, that storage conditions (field or laboratory?) for seeds of these five species were not stated. In contrast to the relatively short lives of seeds of many species, seeds of *Anisophyllea* sp., *Barringtonia* sp., and *Scorodacarpus* sp. live for many months or years on the forest floor. These seeds are quite large, fully permeable, and apparently remain imbibed (Ng, 1988).

Seed banks or reserves in climax or mature rain forests are a potential source of information on the longevity of seeds of nonpioneer species. However, most of the seeds germinating in soil samples collected in primary or old-growth rain forests in Thailand (Cheke *et al.,* 1979), Malaya (Symington, 1932), Sabah (Chin, 1973), the Ivory Coast (de Rouw and van Oers, 1988), Ghana (Hall and Swaine, 1980), Nigeria (Keay, 1960), Australia (Hopkins and Graham, 1983, 1984), French Guiana (de Foresta and Prevost, 1986), Costa Rica (Lawton and Putz, 1988), Panama (Williams-Linera, 1990), and Mexico (Guevara S. and Gomez-Pompa, 1972) were from species characteristic of secondary forests or pioneer species. For example, in Australia only 0.9% (Hopkins and Graham, 1983) and in the Ivory Coast <1% (de Rouw and van Oers, 1988) of the seedlings that appeared in soil samples collected in primary forests were from nonpioneer tree species. In Ghana, 10% of the total species list and only 1% of all seedlings were from nonpioneer tree species (Hall and Swaine, 1980). The age of seeds of nonpioneer species in these soil samples from primary forests is unknown; the seeds might be several years old or freshly dispersed. Certainly, some rain forest species flower and produce seeds throughout the year (Sasaki *et al.*, 1979). Regardless of the age of the seeds of nonpioneer species that germinated in soil samples, the small number of species represented and the small number of seedlings per species are strong suggestions that seeds of most nonpioneer trees are short-lived.

4. Requirements for Germination

Studies on the environmental conditions required for the germination of seeds of climax trees are limited in number. In a few cases, the germination of seeds sown in shaded and nonshaded conditions has been compared. Seeds of *Dacryodes excelsa* (Quarterman, 1970), *Araucaria hunsteinii* (Havel, 1971), and *Anthocepalus cadamba* (Gonzalez and Grijpma, 1968) germinated to higher percentages in shade than in the sun. However, seeds of *Dryobalanops aromatica* and *D. lanceolata* germinated to 90–100% in both shade (under tree canopy) and sun (gaps in the tree canopy), but the seedling survival of both species was higher in gaps than in the shade (Itoh, 1995).

Six species from rain forests in the Congo germinated to higher percentages in shade than in sun, but one (*Antrocaryon micraster*) germinated to higher percentages in sun than in shade (Gilbert, 1952). In a Malaysian dipterocarp forest, more seeds of 22 species of trees germinated in shade of the forest than in a clearing, whereas more seeds of 12 species germinated in open sites than in the shade (Raich and Khoon, 1990). Seeds of *Pternandra coerulescens* germinated to 60, 70, and 0% under 6% daylight (red:far-red ratio of 0.7), 3% green shade (red:far-red ratio of 0.2), and darkness, respectively, those of *Pternandra echinata* germinated to 70, 55, and 5%, respectively, those of *Gymnotroches axillaris* to 10, 10, and 15%, respectively, and those of

Pellacalyx saccardianus to 85, 90, and 90%, respectively, (Metcalfe, 1996).

Optimal temperatures for the germination of *Araucaria angustifolia* (Espindola *et al.*, 1993), *Hopea odorata*, *Shorea roxburghii*, and *Symphonia globulifera* (Corbineau and Come, 1988) seeds were 25 or 30°C. Seeds of *S. roxburghii* germinated at 5°C, but those of the other two species germinated poorly or not at all at 5°C (Corbineau and Come, 1988). In another study, seeds of *S. roxburghii* did not germinate at 2, 6, and 11°C, but they germinated to 77% at 16°C. Seeds of *S. almon* and *S. robusta* also failed to germinate at 2, 6, and 11°C, but germinated to 32 and 63%, respectively, at 16°C (Tompsett, 1985). Seeds of *Hopea helferi* germinated to 93–99% at constant temperatures of 5, 10, 15, 20, 25, 30, and 35°C and at an alternating temperature regime of 25/28°C in 3 to 6 days, but at 40 and 42°C they germinated to only 59 and 52%, respectively (Tang and Tamari, 1973). *Dryobalanops aromatica* seeds germinated to 100% in 7 days at ambient laboratory temperatures (ca. 30°C) in open plastic boxes and to 80% in closed boxes (Jensen, 1971).

Seeds of *Litchi chinensis* germinated over a temperature range from 5 to 35°C, with 30°C being optimal, and those of *Euphoria longan* germinated at temperatures from 5 to 25°C, with 25°C being optimal (Xia *et al.*, 1992). The optimum germination temperature for *Elaeis guineensis* seeds was 38°C (Ferwerda, 1956) and was 25–35°C for *Citrus limonia;* constant temperatures of 15 and 20°C and alternating temperatures of 25/10°C reduced germination rates, but not percentages, of *C. limonia* seeds (Usberti and Felippe, 1980a). Seeds of *Mangifera indica* germinated at temperatures between 5 and 35°C, but optimum temperatures were 25 and 30°C (Corbineau and Come, 1988).

Because seeds of many rain forest species are recalcitrant, it is logical that high soil moisture conditions would be required for germination. Thus, germination may be prevented in gaps due to increased temperatures (Raich and Khoon, 1990) and, consequently, a decrease in water availability for germination. For example, seeds of *Chlorocardium rodiei* did not germinate to high percentages in large gaps in rain forests of Guyana unless they were buried (Ter Steege *et al.*, 1994). *Pithecellobium arboreum* seeds failed to germinate when placed on the soil surface and died unless it rained within 48 hr (Flores and Mora, 1984; Vazquez-Yanes and Toledo, 1989). Litter on the soil surface (Flores and Mora, 1984) and/or shading from vines and other plants (Savage, 1992) appears to promote germination, probably by reducing the loss of soil moisture. The presence of wings on seeds of *Dryobalanops aromatica*, *Shorea leprosula*, and *S. parvifolia* delayed them from reaching 60% germination by 1–5 days. Apparently, seeds with wings do

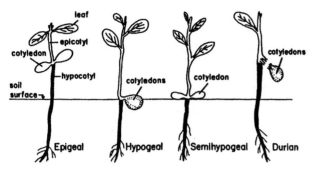

FIGURE 9.2 Types of germination reported by Ng (1978) in rain forest trees. Drawn from photographs in Ng, 1978.

not make good contact with the soil, thus water is not readily absorbed (Tang and Tamari, 1973).

5. Types of Seedlings

In his studies on seeds and seedlings of Malayan rain forest trees, Ng (1978) distinguished four types of germination: epigeal, hypogeal, semihypogeal, and Durian (Fig. 9.2). In epigeal germination, the cotyledons are carried above the soil surface by the elongating hypocotyl, and they expand and become photosynthetic in most species. In hypogeal germination, the hypocotyl does not elongate above the soil surface, and the cotyledons that remain inside the seed are at or slightly below the soil surface. In semihypogeal germination, the cotyledons emerge from the seed, but they remain on the soil surface and usually are nonphotosynthetic. In Durian germination, the cotyledons are lifted about the soil by elongation of the hypocotyl, but they remain inside the seed. Durian germination is not known outside the tropics and is found mainly in the genera *Durio, Rhizophora,* and *Dipterocarpus* (Ng, 1978). In Ng's (1978) study, 64, 18, 10, and 8% of the 180 species had epigeal, hypogeal, semihypogeal, and Durian germination, respectively.

Interest in the reproductive phase of the life cycle of tropical species has led to the publication of several volumes of drawings of tree seedlings (e.g., Duke, 1969; de Vogel, 1980). de Vogel (1980) described 16 different types of seedlings among the 150 woody taxa he studied in Malaysia, as compared to the 4 described by Ng (1978). de Vogel (1980) used developmental characteristics of juveniles as well as attributes of the cotyledons and hypocotyl in his classification system.

B. Pioneer Trees

1. Soil Seed Reserves

It is well known that pioneer trees become established following the removal or disturbance of rain for-

ests. However, it is not always clear if pioneer trees result from seeds present in the soil or from those dispersed to the site following the removal of climax species. Seeds of pioneer trees germinated in soil samples collected in primary (Chin, 1973; Enright, 1985; de Rouw and van Oers, 1988; Guevara S. and Gomez-Pompa, 1972; de Foresta and Prevost, 1986; Lawton and Putz, 1988; Cheke *et al.,* 1979; Hopkins and Graham, 1984; Hall and Swaine, 1980) and in secondary forests (Guevara S. and Gomez-Pompa, 1972; de Foresta and Prevost, 1986; Hopkins and Graham, 1984). However, as discussed in Chapter 7, it is difficult to know if seeds in soil samples are components of persistent, transient, or persistent plus transient seed banks. Thus, seeds of pioneer trees that germinate in soil samples may be many months or years in age or they may have been dispersed only recently.

Although many species of pioneer trees have a definite fruiting season (Whitmore, 1983), others produce seeds throughout the year. Therefore, regardless of when soil samples are collected, they might contain newly dispersed seeds. For example, *Cecropia sciadophylla* and *C. obtusa* produce seeds all year (Holthuijzen and Boerboom, 1982), and fruits are eaten by bats and small birds, which serve as seed dispersal agents to gaps and the surrounding forest (Charles-Dominique, 1986). The number of *Cecropia* seeds in the soil of an undisturbed rain forest in Surinam decreased during the long rainy season and increased during the dry season (Holthuijzen and Boerboom, 1982).

Field plots established in a 1-yr and 13-year-old gap and in a mature secondary forest in Mexico were sown with seeds of *Cecropia obtusifolia* and protected to prevent newly dispersed seeds from reaching the soil surface. Only about 5% of the seeds were alive after 8 months, and survivorship was less in the 1-year-old gap than at the other two sites (Martinez-Ramos and Alvarez-Buylla, 1986).

Cheke *et al.* (1979) determined annual seed rain and number of seeds in the soil of undisturbed rain forests in northern Thailand. Seeds of pioneer species were abundant in the soil, and their numbers exceeded the annual seed rain, indicating that a persistent seed bank of pioneer species existed in the rain forest. Martinez-Ramos and Alvarez-Buylla (1986) found that *Cecropia obtusifolia* seeds were dispersed to gaps in a Mexican rain forest at a rate of 7.2 seeds m^{-2} $year^{-1}$. However, due to low seed survival, only a small fraction of the seeds that potentially could germinate in a gap each year is from the seed bank.

There is some evidence that seeds, at least of a few species, could form persistent seed banks. Seeds of *Cecropia sciadophylla* and *C. obtusa* buried in Surinam in unglazed earthenware pots closed by lids showed no decline in viability after 62 and 48 months, respectively (Holthuijzen and Boerboom, 1982). A high percentage of *Carica papaya, Solanum diphyllum,* and *Trema micrantha* seeds buried in nylon net bags in Mexico for 2 years remained viable, but only 1–20% of those of *Myriocarpa longipes, Piper auritum,* and *P. hispidum* were alive after 2 years (Perez-Nasser and Vazquez-Yanes, 1986). Seeds of *Belotia campbellii* exhumed after 400 days of burial in Mexico showed an increase in germination percentages (Perez-Nasser and Vazquez-Yanes, 1986), and some seeds of *Piper umbellatum* were still viable after 3 years of burial at Los Tuxtlas in Mexico (Orozco-Segovia and Vazquez-Yanes, 1990).

2. Seed Dormancy

Freshly matured seeds of some pioneer species, e.g., *Anthocephalus cadamba* (Grijpma, 1967), *Vochysia ferruginea* (Gonzalez, 1991), and *Urera caracasana* (Orozco-Segovia *et al.,* 1987) are nondormant and begin to germinate within 1–3 weeks following dispersal. However, seeds of many species do not germinate within 4 weeks following dispersal. Because embryos are fully developed and seed coats are permeable to water, we conclude that the seeds have physiological dormancy.

The time required for loss of physiological dormancy varies with the species. Seeds of *Macaranga tanarius* sown outdoors in Malaysia germinated in 4–38 weeks (Ng, 1978), but those of *Trema guineensis* were still dormant after 24 weeks of dry storage at room temperature (Vazquez-Yanes, 1977). After 40 weeks, some dormancy loss had occurred in seeds of *T. quineensis,* and they germinated to 40%. Gibberellic acid and dry storage at 2°C were effective in overcoming dormancy in seeds of this species (Vazquez-Yanes, 1977). Fresh seeds of *Solanum erianthum, Trema orientalis,* and *Musanga cecropioides* did not germinate (Hall and Swaine, 1981), indicating that they were dormant, but the length of the dormancy-breaking period was not determined.

Seeds of some pioneer trees such as *Ochroma lagopus* (Vazquez-Yanes, 1974), *Belotia campbellii,* and *Heliocarpus donnell-smithii* (Vazquez-Yanes, 1981) have physical dormancy. As discussed in Chapter 6, seeds of *H. donnell-smithii* and *O. lagopus* germinate only in light gaps where the amplitude of the daily temperature fluctuations is sufficient to cause seeds to become permeable (Vazquez-Yanes, 1981; Vazquez-Yanes and Orozco-Segovia, 1982a). The physical dormancy of *O. lagopus* seeds can be broken by heating them to 80°C or by pouring boiling water over them (Vazquez-Yanes, 1974). These heat treatments simulate the effects of fire and help explain why this species appears after an area has been cleared and burned. Heat treatments at 60, 80, and 94°C were ineffective in promoting the germina-

tion of *B. campbellii* seeds, but after a year of dry storage at room temperatures they germinated to 54% (Vazquez-Yanes, 1981).

3. Germination Requirements

It seems reasonable that seeds of pioneer trees require special environmental conditions for dormancy break and/or germination; otherwise, seed reserves would be exhausted by *in situ* germination before a gap was created. In fact, one of the "generalizations for secondary succession" proposed by Budowski (1963) is that seeds of pioneer species remain viable, but ungerminated, in the soil until some event such as removal of forest vegetation triggers germination. The presence of physiological or physical dormancy would prevent germination. However, seeds of some pioneer species are nondormant when they are freshly matured, and those with dormancy eventually become nondormant. Thus, mechanisms that prevent the germination of nondormant seeds are important in maintaining seed banks of pioneer species.

Light quantity and quality are important ecological factors in regulating germination in nondormant seeds of many species and have been a primary research focus in trying to understand what controls the timing of germination in nondormant seeds of pioneer species. Shaded vs open (nonshaded, i.e. a gap) conditions influence the number of seedlings that emerge in soil samples containing seeds of pioneer species. Seeds of 80 pioneer species germinated in soil samples collected in Ghanian forests and placed in boxes set in full sun, whereas seeds of only 22 species germinated in boxes kept under an open black polythene tent in a mango grove. When the shaded boxes were moved to the sun, seeds of 38 additional species germinated (Hall and Swaine, 1980). Soil samples collected in the Ivory Coast produced more seedlings of pioneer trees in samples placed in the sun than in the shade (de Rouw and van Oers, 1988). However, secondary species with long life spans such as *Musanga cecropioides, Margaritia discoides,* and *Ficus sur* germinated to some extent in the shade. Large numbers of *Cecropia peltata* and *Didymopanax morototoni* seeds germinated in soil samples from a rain forest in Puerto Rico set in the sun, but few germinated in those placed in the shade (Bell 1970).

Seeds of the pioneer species *Sapium baccatum* and *Vitex pinnata* and of the nonnative pioneer *Gmelina arborea* sown in boxes of soil in Malaysia failed to germinate in the shade of a forest canopy (Mohamad and Ng, 1982). However, seeds sown in boxes in the sun or in boxes moved to the sun after 6 months of shading germinated to 60–80%. Seeds of *Alstonia macrophylla, Melastoma malabothricum,* and *Trema tomentosa* germinated to 66–88% in a gap but to only 1–3% under the shade of the forest canopy in Malaysia; however, germination increased when seeds were moved from the shade to the sun (Raich and Khoon, 1990). Seeds of 20 species (mostly herbaceous weeds) germinated in soil samples from a secondary forest in Costa Rica placed in full sun, but seeds of only 4 species germinated in samples placed in 50% full sun (Di Stefano and Chaverri, 1992).

Light has been shown to be required for the germination of some species under laboratory conditions. Seeds of *Cecropia sciadophylla* germinated to about 80% in light but to 0% in darkness at a constant temperature of 20°C (Holthuijzen and Boerboom, 1982). At an alternating temperature regime of 30/20°C, seeds of this species germinated to about 85 and 10% in light and darkness, respectively. Seeds of *C. obtusifolia* germinated to 77 and 0% in light and darkness, respectively, at 36/26°C, to 80 and 0% at 26°C, and to 18 and 0% at 36°C (Vazquez-Yanes, 1979). Seeds of *Chlorophora excelsa* germinated to 45–50% in both light and darkness at 30/20°C but to only about 37 and 3% in light and darkness, respectively, at 30°C (Chamshama and Downs, 1982). *Heliocarpus appendiculatus, Cecropia obtusifolia,* and *Piper auritum* seeds covered by one layer of forest litter (dead, moist, brown leaves) germinated to about 25, 15, and 27%, respectively, and those covered by three layers of leaves germinated to 10, 5, and 0%, respectively (Vazquez-Yanes and Orozco-Segovia, 1992). However, uncovered seeds germinated to about 55, 70, and 95%, respectively, and those covered with one piece of wet paper towel germinated to 40, 73, and 93%, respectively. Seeds of *Muntingia calabura* germinated to 44 and 0% in continuous white light and darkness, respectively, at 25°C (Laura *et al.* 1994).

Light quality in a clearing or gap vs that on the forest floor influences the germination of some pioneer species. Unfiltered light reaching the soil surface in a forest clearing has a relatively high red : far-red ratio, but light filtered through green leaves in a forest has a low red : far-red ratio (Stoutjesdijk, 1972). *Urera caracasana* seeds in petri dishes were placed along a transect from a gap in the forest to well within the forest and kept there for 13 days, and then they were incubated in darkness at 25°C for 15 days (Orozco-Segovia *et al.,* 1987). Seeds exposed to light with a red : far-red ratio of 0.2 (beneath forest canopy) or 0.5 did not germinate, but those exposed to light with a red : far-red ratio of 0.9 or 1.0 (gap) germinated to about 70%. Seeds of this species moved daily from beneath the forest canopy to the center of the gap for 1 min to 8 hr germinated to >50% only in the 8-hr exposure.

Seeds of *Cecropia obtusifolia* (Vazquez-Yanes, 1979; Vazquez-Yanes and Orozco-Segovia, 1986) and *Piper auritum* (Vazquez-Yanes and Smith, 1982; Orozco-Segovia and Vazquez-Yanes, 1989) germinated to 0–5% when placed on the soil surface in a Mexican rain forest, but germinated to 80–90% in diffuse light (shaded by cloth) outside the rain forest. Thus, the low red : far-red ratio in the forest, rather than the low irradiance, inhibited germination. Sunflecks in the forest had no effect on the germination of *Piper auritum* seeds (Orozco-Segovia *et al.,* 1993). Seeds of *Cecropia obtusifolia* and *Piper auritum* germinated to 5 and 0%, respectively, following exposures to white light for 120 min per day for 7 days, but germination in both species exceeded 80% after seeds were given 1 day of a 12-hr photoperiod (Vazquez-Yanes and Orozco-Segovia, 1987b, 1990). Laboratory studies confirmed that red stimulates and far-red light inhibits the germination of *C. obtusifolia* (Vazquez-Yanes and Smith, 1982; Vazquez-Yanes and Orozco-Segovia, 1986, 1990) and *P. auritum* (Vazquez-Yanes and Smith, 1982). The germination of *P. auritum* seeds was promoted by 13 days of a 12-hr daily photoperiod of only 0.005 μmol m^{-2} sec^{-1} of red light (Orozco-Segovia *et al.,* 1993).

Light filtered through dead leaves of six canopy species of a Mexican rain forest inhibited the germination of *Cecropia obtusifolia, Piper aequale, P. auritum,* and *P. hispidum* seeds (Vazquez-Yanes *et al.,* 1990). The red : far-red ratio above litter on the forest floor was 0.28, under dry litter 0.10 to 0.16, and under wet litter 0.17 to 0.25. Thus, litter may play a significant role in preventing the germination of seeds on the soil surface (Vazquez-Yanes *et al.,* 1990).

Responses of seeds of pioneer trees to light may dictate the size and type of gap in which they will germinate. In general, high-light requiring seeds germinate only in large gaps (Hartshorn, 1978; Denslow, 1980; Murray, 1986), and they do so the first few months following creation of the gap. After the gap has been colonized, shading by vegetation prevents seeds of some species from germinating (Murray, 1986). Seeds detect formation of a light gap by an increase in red light at the soil surface. The minimum gap size for 50% germination of *Cecropia obtusifolia* seeds was one in which seeds were unshaded for 4–5 hr each day (Vazquez-Yanes and Orozco-Segovia, 1990).

Seeds of *P. auritum* buried for 2 and 4 months required longer exposures to white light for germination than freshly matured ones (Orozco-Segovia and Vazquez-Yanes, 1989). Freshly matured seeds germinated to 60% after a 5-min exposure to white light, whereas only 20% of the seeds that had been buried for 120 days germinated after a 1440-min exposure to white light (Vazquez-Yanes and Orozco-Segovia, 1987a).

Although soil and air temperatures increase in gaps (Raich and Khoon, 1990), little attention has been given to the fact that seeds in gaps are exposed to higher temperatures than those under the forest canopy. What are the effects of increased temperatures on the germination of seeds of pioneer trees?

4. Effects of Fire

Fire frequently is used as a tool to help remove tropical forests from the land and it can have important consequences on seeds of pioneer species. When a 10- to 12-year-old secondary forest in Brazil was cut and burned, surface temperatures were 100–150°C for several hours. Temperatures at 2 cm were 95–125°C, whereas those at 5 cm were 45–105°C. All seeds of 15 species buried at 2 cm died, and only 20–30% of the seeds of five species buried at 5 cm survived and germinated. Most seeds buried at 10 cm and all of those buried at 20 cm survived (Brinkman and Vieira, 1971). While slash was being burned following the cutting of a tropical forest in Costa Rica, temperatures at the soil surface were about 200°C, and mean soil temperatures at 1 and 3 cm were about 100 and <38°C, respectively. These high temperatures killed 52% of the seeds in the soil, and 27% of the species were no longer represented in the seed bank (Ewel et al., 1981). Soil samples collected in secondary forests in Australia contained seeds of 39 species, and seeds of 38 species failed to germinate after the soil was subjected to a moist (soil at field capacity and wrapped in aluminum foil) heat treatment of 100°C for 1 hr; after this treatment, one seed of Acacia mangium germinated (Hopkins and Graham, 1984). Moist heat at 60°C for 1 hr destroyed seeds of 4 species, significantly reduced germination in 8 species, had no effect on seeds of 4 species, and enhanced germination of 1 species (A. mangium). Soil samples collected from a logged area before and after a fire in the Ivory Coast showed that about one half of the seeds were killed by fire. However, germination rates of seeds surviving the fire were increased (de Rouw and van Oers, 1988).

Available evidence indicates that a large percentage of the seeds, especially those in the top few centimeters of soil, is killed when a rain forest is cleared and the area is burned. Revegetation, or development of the secondary forest, is from seeds covered by 3–5 cm of soil or from those dispersed to the site following fire. Seed release appears to be enhanced by fire in some pioneer trees such as Ochroma lagopus (Aikman, 1955).

C. Dormancy and Germination of Shrubs

Seeds of Celastrus monospermoides (Ng and Asri, 1979), Clidemia discolor, C. japurensis (Ellison et al. 1993), Cordia subcordata (Garrard, 1955), Gaultheria leucocarpa (Ng, 1978), G. malayana (Ng, 1980), Leandra consimiles, L. dichotoma, Melastoma malabathricum (Metcalfe, 1996), Miconia barbinervis (Ellison et al., 1993), M. multispicata (Gonzalez, 1991), M. nervosa, M. simplex (Ellison et al., 1993), Plumeria obtusa, P. rubra (Chow and Tjondronegoro, 1978), Pternandra coerulescens, P. echinata (Metcalfe, 1996), Urophyllum hirsutum, and U. streptopodium (Metcalfe, 1996) are nondormant. However, seeds of Asystasia gangetica (Akamine, 1947), Clidemia crenulata, C. densiflora, Miconia centrodesma, M. dorsiloba, M. gracilis, M. grayumii, Ossaea macrophylla, O. micrantha (Ellison et al., 1993), and Rubus moluccanus (Ng and Asri, 1979) appear to have physiological dormancy, whereas those of Abroma augusta (Ng, 1980) have physical dormancy.

Light is important in controlling the seed germination of some, but not all, shrubs. Seeds of Verbesina greenmanii germinated to 6, 11, and 53% in a mature forest, secondary forest, and under artificial shade outside the forest, respectively (Vazquez-Yanes and Orozco-Segovia, 1982b), and those of Piper aequlae, germinated to 26 and 90% inside and outside a rain forest (Orozco-Segovia and Vazquez-Yanes, 1989). Soil samples placed in a clearing in Puerto Rico eventually had 121 seedlings of Psychotria berteriana, while those placed in the forest had only three (Bell, 1970). Thus, V. greenmanii, P. aequlae and P. berteriana are gap species.

Seeds of Piper hispidum, germinated equally well in the shade of mature and secondary forests and in artificial shade outside the forest (Vazquez-Yanes and Orozco-Segovia, 1982c; Orozco-Segovia and Vazquez-Yanes, 1989). No seedlings of Palicourea riparia appeared in soil samples placed in a clearing in a rain forest in Puerto Rico, but 83 emerged in samples placed in the forest (Bell, 1970). Two to 3 months were required before germination was initiated in P. riparia seeds sown in summer and winter in open and closed canopy areas in Puerto Rico (Lebron, 1979). Seeds in both summer and winter sowings germinated to higher percentages in open than in closed canopy areas. Urophyllum hirsutum seeds germinated to 95, 70, and 0% under 6% daylight (red : far-red ratio of 0.7), 3% green shade (red : far-red ratio of 0.2), and darkness, respectively, those of U. streptopodium to 85, 90, and 10%, respectively, and those of Pternandra echinata to 70, 55, and 5%, respectively (Metcalfe, 1996).

Seeds of Urera caracasana incubated in white light at 35/25°C did not begin to germinate until after 15 days, whereas those at 25°C did not start germinating until after 120 days (Vazquez-Yanes and Orozco-Segovia, 1987a). As seeds of Asystasia gangetica came out of dormancy during 135 days of dry storage at room temperature, they lost the alternating temperature requirement for germination (Akamine, 1947). Seeds of Plum-

eria obtusa and *P. rubra* germinated to 62–92% at 25 and 30°C in 4–9 days (Chow and Tjondronegoro, 1978).

D. Dormancy and Germination of Bamboos

Seeds of *Schizostachyum zollingeri* collected from fruiting plants in the northern part of the Malay Peninsula were nondormant and germinated to 65–70% within 5 days in light at 30°C. Germination ceased after 23 days, at which time 85% of the seeds had germinated. Seeds stored dry at 12–14°C for 60 days germinated to only 58%, indicating a decrease in viability and/or increase in dormancy (Wong, 1981). Seeds of *Dendrocalamus giganteus* also were nondormant, and germination was completed within 30 days (Beniwal and Singh, 1989).

E. Dormancy and Germination of Vines (Lianas)

Various species of climbers germinated in soil samples collected in a secondary forest in the Ivory Coast, and more seeds of most species germinated in samples placed in the sun than in those placed in the shade (de Rouw and van Oers, 1988). Seeds of *Ficus pertusa* and *F. tuerckheimii* required light for germination, and seedlings in the field were found on *Ficus* stems, on branches of a variety of trees, and on the ground. However, more seeds germinated on wet filter paper in petri dishes than on the various substrates in the habitat, indicating that high moisture conditions were required for germination (Titus *et al.*, 1990). Seeds of *Ficus chartacea* germinated to 95, 50, and 65% in 6% daylight (red : far-red ratio of 0.7), 3% green shade (red : far-red ratio of 0.2), and darkness, respectively; those of *F. fistulosa* to 100, 0, and 50%, respectively; and those of *F. grossularioides* to 100, 15, and 75%, respectively (Metcalfe, 1996).

Seeds (achenes) of the subgenus *Urostigma* of *Ficus* have a sticky mucilaginous coat; when this material dries, seeds become attached to the surface of tree branches, leaves, soil, rocks, and so on. Seeds placed on sterilized soil failed to germinate, whereas those placed on unsterilized soil germinated to 100%. Apparently, germination was inhibited until soil bacteria digested the mucilage (Ramirez B., 1976). In *Ficus religiosa*, another member of the subgenus *Urostigma*, seeds germinated equally well on a variety of substrates, including soil and sterile filter paper at high temperatures (30°C) in light. Seeds of this species required continuous hydration for germination, and those removed from a wet surface lost most of their imbibitional water in 1 hr (Galil and Meiri, 1981).

Calamus manan is a vine or rattan, and its seeds have underdeveloped embryos. Further, the seeds are recalcitrant and will die if their moisture content drops below 40–45%. Seeds germinate within 30 days if the sarcotesta is removed, but 2–4 months are required if it is left intact (Mori *et al.*, 1980; Dransfield and Manokaran, 1993). Thus, successful germination in the field requires a long period of favorable moisture conditions, during which time embryo growth occurs. We conclude that seeds of this species have some type of morphophysiological dormancy.

Four or more weeks were required for the germination of seeds of *Calamus manan* (3–15 weeks), *C. palustris* (6), *C. scipionum* (>4), *C. simplicifolius* (7–8), *C. subinermis* (ca. 4), *C. tumidus* (6–31), and *Daemonoropus margaritae* (7–8); thus, they have morphophysiological dormancy. In contrast, seeds of *Calamus tetradactylus, C. ovoideus* and *C. wailong* germinated in 2–3 weeks and thus have morphological dormancy (Dransfield and Manokaran, 1993).

F. Dormancy and Germination of Herbaceous Species

1. Forest Floor

A variety of herbaceous species grow on the rain forest floor (e.g., Burtt, 1976; Hall and Swain, 1981; Kiew, 1988), but little is known about their seed germination. In fact, newly germinated seedlings are seen infrequently. However, Burtt (1976) observed seedlings of *Monophyllaea* sp. and *Epithema* sp. in rocky forests on limestone, and Kiew (1988) found seedlings of *Begonia sinuata* and *B. phoeniogramma* on the forest floor. Many species have minute seeds that are dispersed by rain drops, and those of *Balanophora* sp. are so small (7 μg) they may be dispersed by beetles that visit the inflorescences (Kiew, 1988). In contrast, many forest herbs produce fleshy fruits that are dispersed by mammals and birds (Ridley, 1930), and ants disperse seeds of some species (Horvitz and Schemske, 1986, 1994).

High germination percentages were obtained for fresh seeds of *Begonia* sp. under greenhouse conditions (Burtt, 1976) and for those of *Nepsera aquatica* in light at ca. 25°C (Ellison *et al.*, 1993). Seeds of *Loxocarpus incana* and *Sonerila rudis* will germinate in the capsules if moisture is high (Kiew, 1988). Thus, we conclude that seeds of these four species are nondormant. Seeds of *Dianella densifolius* were dormant and required a 3-month dormancy-breaking period on a moist substrate before they would germinate. If seeds were stored dry, viability decreased with time, and after 3 months all of them were dead (Garrard, 1955). Because *D. densifolius* belongs to the Liliaceae, which has underdeveloped embryos, seeds of this species may have morphophysiological dormancy. Seeds of *Musa balbisiana* require 2–6 weeks for germination, but scarification decreased germination time under laboratory conditions to 6–10 days. However, when scarified seeds of *M. balbisiana* were sown on soil, they were attacked by fungi and killed (Stotzky *et al.,* 1962). In another study, scarified seeds of *M. balbisiana* germinated to 21 and 59% on nonsterile soil and sterile medium, respectively (Bhat *et al.,* 1994).

Seeds of *Calathea ovandensis* are dormant at maturity near the end of a wet season in Mexican rain forests and do not germinate until the beginning of the next wet season (Horvitz and Schemske, 1994). Seeds of *C. ovandensis* germinated to 0.4% under canopy shade, whereas those in a forest gap germinated to 73%; seeds in the shade were viable (Horvitz and Schemske, 1994). In both Ivory Coast (de Rouw and van Oers, 1988) and Ghana (Hall and Swaine, 1981), soil samples collected in mature forests and placed in clearings eventually had more seedlings of herbaceous species than those placed inside the forest. Thus, small light gaps may play a role in the establishment of some herbaceous species.

2. Disturbance Species

Seeds of some species growing in sites where rain forests once occurred are nondormant, whereas those of others have physiological dormancy (Table 9.3). Freshly matured achenes of *Tithonia rotundifolia* germinated to 0% in light at 25°C, but after 12 weeks of dry laboratory storage they germinated to about 70% (Upfold and van Staden, 1990). Dormancy in seeds of *Oldenlandia corymbosa* was broken by placing them on a wet substrate in darkness at 5, 10, 15, 20, and 25°C for 7–9 days (Corbineau and Come, 1980/81).

High temperatures are required for the germination of *Oldenlandia corymbosa* (Corbineau and Come, 1980–1981), *Phytolacca icosandra* (Edmisten, 1970), *Celosia cristata* (Okusanya, 1980), and *Alternanthera pungens* (Kaul, 1972), and alternating temperature regimes are needed for high germination percentages in seeds of *Dactyloctenium aegyptium* (Longman, 1969) and *Chromolaena odorata* (= *Eupatorium odoratum*) (Erasmus and van Staden, 1986). Seeds of many species require light for germination (Longman, 1969; Okusanya, 1980; Marks and Nwachuku, 1986; Erasmus and van Staden, 1986), and light quality, as modified by green leaves, may play a role in determining whether seeds germinate. Germination did not occur in *Piper umbellatum* seeds incubated under a forest canopy (Orozco-Segovia and Vazquez-Yanes, 1989), and it increased in seeds of *Phytolacca rivinoides, Witheringia solanacea,* and *W. coccoloboides* with an increase in the size of forest gaps (Murray, 1988). However, seedlings of *Panicum laxum* become established in the shade of shrubs, indicating that light filtered through green leaves did not inhibit germination (Cole, 1977).

Weeds, including *Emilia sonchifolia, Oldenlandia affinis,* and *Phyllanthus odontadenius,* produce seeds during both wet and dry seasons in Nigeria, and others, including *Schwenckia americana, Dissotis rotundifolia,* and *Scoparia dulcis,* are green all year but produce seeds only in the wet season. However, 15 annuals complete their life cycles when the dry season begins, and seeds do not germinate until the first rains of the wet season moisten the soil, usually in March (Marks, 1983). Seeds of *Borreria ocymoides* and *Platostoma africanum* that failed to germinate in the field in March (at the beginning of the wet season in Nigeria) did not germinate later in the wet season, although experiments showed that they were nondormant (Marks, 1983). Flooding promotes the germination of seeds of *Sphaeranthus indicus* (Shetty, 1967) and *Panicum laxum* (Cole, 1977), but inhibits the germination of *Alternanthera pungens* (Kaul, 1972).

About 75% of the *Mariscus alternifolius* and 50% of the *M. flavelliformis* and *Solenostemon monostachyus* seeds remained viable during a 15-month burial study in Nigeria. About 10% of the seeds of *Paspalum orbiculare, Eupatorium odoratum, Digitaria ciliaris, Eragrostis gangetica, Tridax procumbens, Borreria ocymoides, Ageratum conyzoides,* and *Triplotaxis stellulifera* and 0% of the seeds of *Bidens pilosa* and *Emilia sonchifolia* were alive after 15 months (Marks and Nwachuku, 1986). Regardless of when seeds of *Solenostemon monostachyus, Mariscus alternifolius,* and *Eupatorium odoratum* were exhumed during a 15-month period of burial, they germinated to 80–100% in light. Exhumed seeds of *Tridax procumbens* and *Digitaria ciliaris* showed increases in germination in darkness with the approach of the rainy season and decreases with the beginning of the dry season. Seeds of *Mariscus flabelliformis, Cyperus sphacelatus,* and *Eragrostis gangetica* were alternately dormant and nondormant, but they required light for germination when they were nondormant (Marks and Nwachuku, 1986).

G. Dormancy and Germination of Epiphytes

A true epiphyte is defined as a plant that uses another plant for physical support during its whole life, but it removes no nutrients from the phloem of its "host." Hemiepiphytes, however, are rooted at some time during their lives, either when they are seedlings or after they mature. Primary hemiepiphytes, e.g., Ficus and Clusia spp., become established in the canopy, and roots grow down to the ground (Benzing, 1990). In elfin forests of Monteverde, Costa Rica, epiphytes that are deposited on the ground when host trees break become rooted in soil and thus are an important means of establishment of plants in gaps (Lawton and Putz, 1988).

TABLE 9.3 Types of Seed Dormancy in Weedy Species That Invade Agricultural Soils
in Regions Where Evergreen Rain Forests Have Been Cut

Species	Family	Type of dormancy[a]	Reference
Ageratum conyzoides	Asteraceae	ND	Markes (1983)
Bidens pilosa	Asteraceae	ND	Marks and Nwachuku (1986)
Borreria ocymoides	Rubiaceae	PD	Marks (1983), Marks and Nwachuku (1986)
Cyperus sphacelatus	Cyperaceae	ND	Marks and Nwachuku (1986)
Digitaria ciliaris	Poaceae	ND, PD	Marks and Nwachuku (1986)
Emilia sonchifolia	Asteraceae	ND	Marks (1983)
Eragrostis gangetica	Poaceae	PD, ND	Marks and Nwachuku (1986)
Eupatorium odoratum	Asteraceae	ND	Marks and Nwachuku (1986)
Lindernia diffusa	Scrophulariaceae	PD	Marks (1983)
Mariscus alternifolius	Cyperaceae	ND	Marks and Nwachuku (1986)
M. flabelliformis	Cyperaceae	ND, PD	Marks and Nwachuku (1986)
Mitracarpus scaber	Rubiaceae	PD	Marks (1983)
Mollugo nudicaulis	Aizoaceae	PD	Marks (1983)
Oldenlandia corymbosa	Rubiaceae	PD	Corbineau and Come (1980–1981)
Otomeria guineensis	Rubiaceae	PD	Marks (1983)
Paspalum orbiculare	Poaceae	PD	Marks and Nwachuku (1986)
Phyllantus odontadenius	Euphorbiaceae	ND	Marks (1983)
Platostoma africanum	Lamiaceae	PD	Marks (1983)
Schwenckia americana	Solanaceae	PD	Marks (1983)
Solenostemon monostachyus	Lamiaceae	ND	Marks (1983)
Tithonia rotundifolia	Asteraceae	PD	Unfold and Van Staden (1990)
Tridax procumbens	Asteraceae	ND	Marks and Nwachuku (1986)
Triplotaxis stellulifera	Asteraceae	ND	Marks and Nwachuku (1986)

[a] Inferred from available germination data and characteristics of seeds in that family. ND, nondormant; PD, physiological dormancy.

Seeds of secondary hemiepiphytes germinate on the ground, and the plant climbs and becomes attached to a tree in the canopy. Then, the lower stems and roots decay, leaving the plant unattached to the soil (Benzing, 1990).

A recent survey of vascular epiphytes reveals 23,456 species in 876 genera and 84 families (Kress, 1986). The Orchidaceae has the most epiphytes, with 13,951 species in 440 genera. However, the Araceae has 1349 epiphytes in 13 genera, and the Bromeliaceae has 1144 epiphytes in 26 genera. Epiphytes are "typically most abundant in tropical montane forests characterized by daily precipitation throughout the year" (Madison, 1977). The greatest number of epiphytes is found in the neotropics, and the smallest number in tropical Africa; tropical Asia has an intermediate number (Madison, 1977). Some epiphytes occur in dry sites such as the cactus/shrub forests in Mexico and Peru (Benzing, 1990). Temperatures below freezing limit the geographical distribution of most epiphytes; therefore, with the exception of lichens and bryophytes, relatively few species are found outside the tropics (Benzing, 1990).

Seeds in about three-fourths the genera of epiphytes are less than 1 mm in length, and they are dispersed by wind or by the splashing of rain drops. Other species have winged or plumed seeds, and still others have fleshy fruits with animal-dispersed seeds. Regardless of the dispersal mechanism, seed size rarely exceeds 2 mm (Madison, 1977). Madison (1977) noted several advantages of small seeds in epiphytes: (1) potential habitat is showered by seeds, thus increasing the possibility that a few will fall into microsites suitable for germination and growth of seedlings; (2) small seeds are more likely to slip into cracks in tree bark, where soil and humus have accumulated than large ones; and (3) the large surface-to-volume ratio ensures the rapid imbibition of water.

Not much is known about the germination ecophysiology of epiphyte seeds (Benzing, 1990). Madison (1977) says that most epiphytes have nondormant seeds and that vivipary is found in fleshy fruited species, especially in the Araceae, Gesneriaceae, and Cactaceae. However, Benzing (1990) doubts that vivipary is as common as suggested by Madison. Germination has been studied in a few epiphytic orchids (see Chapter 11).

Seeds of the epiphytic climbers, *Hoya* spp., are nondormant and germinate as soon as they imbibe water (Rintz, 1978). Also, seeds of *Ficus stupenda* sown on various types of substrates on dipterocarp trees in a rain forest canopy in Indonesian Borneo germinated within 18 days (Laman, 1995), indicating that they were nondormant. The highest germination (41%) for *F. stupenda* seeds was on rotting wood in knotholes and crotches of broken branches, and the lowest (9%) was on bark (Laman, 1995).

Seeds of *Puya berteroniana* (Downs, 1956) and 17 other species in the Bromeliaceae require light for germination (Downs and Piringer, 1958), but those of 6 species germinate in darkness (Downs and Piringer, 1958). Germination was stimulated in light-requiring species by two 15-min light exposures per day. Optimum temperatures for the germination of *P. berteroniana* seeds were 16 and 20°C (Downs and Piringer, 1958; Downs, 1964), whereas those for *Aechmea nudicaulis, Billbergia elegans, Neoregelia concentrica, Vriesia scalaris,* and *Wiffrockia superba* seeds were 20 and/or 25°C (Downs, 1964). *Alchmea coelestis* seeds germinated equally well at 20, 25, and 30°C (Downs, 1964). At maturity in autumn, seeds of *Tillandsia usneoides* (Bromeliaceae) germinated readily at room temperatures, but they did not germinate on tree branches in southern Georgia until late May and early June (Guard and Henry, 1968).

H. Special Biotic Factors

Seed dispersal, especially as it relates to the coevolution of plants and animals, has become an important field of research in the tropics, and some massive pieces of work have been produced (e.g., Roth, 1987). The question of interest to us here is what are the effects of a dispersal agent(s), especially in the case of zoochory, on the germination ecology of seeds? For the most part, we know very little about the effects, if any, of animals on the subsequent germination of seeds in rain forests. Therefore, it is important that we do not make the assumption that the removal of seeds from one site to another or the passage through an animal's digestive system automatically increases the chance that a seed will germinate. In fact, the new site could be less favorable for germination than the old. Secondary dispersal of *Ficus hondurensis* seeds by ants results in some seeds being carried below ground. Light is required for germi-

nation, thus being below ground would prevent seeds from germinating (Roberts and Heithaus, 1986).

Seeds of *Calathea microcephala* and *C. ovandensis* moved to moist sites by ants immediately following dispersal germinated to higher percentages than those moved to moist sites some time after dispersal. If seeds were moved immediately to moist sites, aril removal had no effect on germination. However, if there was a delay before ants carried seeds to a moist site, seeds with the aril removed germinated to 76%, whereas those with the aril left intact germinated to only 44% (Horvitz, 1981). Seeds of *C. ovandensis* are dormant at maturity and do not begin to germinate for 270–300 days (Horvitz and Schemske, 1986).

Secondary dispersal of *Astrocaryum paramaca* seeds by the rodents *Proechimys* spp. resulted in some of them being cached under litter (Forget, 1991a), which could improve moisture conditions for germination. The germination of *Moronobea coccinea* seeds is enhanced by rodent burial (Forget, 1991b). Another consequence of secondary seed dispersal is that seeds could be removed some distance away from the mother tree, thereby decreasing the chances of seed predation (e.g., Hart, 1995).

Dispersal and germination of seeds of the liana *Strychnos aculeata* in west African rain forests appear to be dependent on elephants (Martin, 1991). Plants of this species produce cannon ball-like fruits that have a thick (7 mm) hard outer wall. According to Martin, elephants are the only animals in the region with jaws capable of crushing and eating these fruits. Further, seeds and seedlings of *S. aculeata* are found in elephant dung, demonstrating that fruits are eaten by elephants and that seeds are dispersed by them. More recent studies in Zaire have identified seven additional species (*Antrocaryon nannanii, Autranella congolensis, Klainedoxa gabonensis, Mammea africana, Omphalocarpum mortehani, Tetrapleura tetraptera,* and *Treculia africana*) with large, dull-colored, difficult-to-open fruits (i.e., covered with fibrous or hard pulp) that only an elephant could eat. A survey of dung piles of many kinds of animals in the region revealed that seeds of the seven species were found only in elephant dung (Yumoto and Maruhashi, 1995). No studies have been undertaken to determine how long it would take the seeds to be "released" from the fruits without an elephant eating them.

Seeds of *Cecropia obtusifolia* that passed through the gut of a tayra (*Eira barbara*) and two spider monkeys (*Ateles geogroyii*) gained the ability to germinate in far-red light, which means they lost the ability to respond to light quality differences in gaps vs. nongaps (Vazquez-Yanes and Orozco-Segovia, 1986). Bat-dispersed seeds, however, were still light sensitive (Vazquez-Yanes and Orozco-Segovia, 1986). Seeds of some species that have passed through an animal's digestive system show no changes in germination requirements. For example, seeds of *Ziziphus cinnamomum* collected from the feces of spider monkeys and capuchins had the same germination characteristics as those cleaned manually (Zhang and Wang, 1995). Also, germination percentages of *Dieffenbachia longispatha*, *Ipomoea trifida*, *Ficus glabrata*, *Cecropia* sp., and *Solanum ochraceo-ferrugineum* seeds removed from feces of the turtle *Rhinoclemmys funerea* were not significantly different from those of seeds that had not passed through the digestive system (Moll and Jansen, 1995). However, in other studies, passage through a digestive system results in the enhancement of germination. For example, germination percentages and rates of *Chrysophyllum* spp., *Tabernaemontana* spp., *Monodora myristica*, *Mimusops bagshawei*, *Pseudospondias microcarpa*, *Aframomum* spp., *Cordia millenii*, *Uvariopsis congensis*, *Ficus* spp., and *Cordia abyssinica* seeds increased after they passed through the gut of a chimpanzee (Wrangham *et al.*, 1994). Germination rates of *Clidemia densiflora* and *Miconia affinis* seeds increased, those of *M. simplex* decreased, and those of *Conostegia subcrustulata* remained the same after passing through a bird (Ellison *et al.*, 1993).

III. TROPICAL MOUNTAINS

The presence of mountains means that there are attitudinal vegetation belts in tropical regions (10°N to 10°S) ranging from evergreen rain forests to alpine communities (Walter, 1971). Some mountains such as the Andes in South America and the Ruwenzori Massif in Africa even have perennial snow on their peaks (Eyre, 1963). Tropical mountains of a sufficient height to have an alpine zone occur in South America, Africa, New Guinea, Borneo, Java, Sumatra, and Hawaii (Smith and Young, 1987).

Low elevation slopes on the windward side of mountains, where rainfall and humidity are continuously high, have evergreen rain forests; however, they may have savannas or deserts on the lee side (Eyre, 1963; Walter, 1971). Submontane, montane, subalpine, and alpine zones occur on tropical mountains. Temperatures decrease with an increase in altitude, whereas rainfall increases up to the montane or cloud zone (the wettest part of a mountain) and then decreases above the cloud zone. Due to high moisture levels, montane forests have an abundance of epiphytes, including ferns, lycopods, lianas, tree ferns, and bamboos, depending on location of the mountain. Precipitation decreases at about 1500–3000 m, which is above the cloud zone, and indicates the beginning of subalpine forests characterized by trees

with evergreen xeromorphic leaves. The upper subalpine forest in some mountains may be a stunted or elfin woodland that gives way to a shrub zone, but in other mountains it is adjacent to grasslands (Eyre, 1963). The alpine or paramos zone is encountered at about 3500–4000 m, and at this altitude night temperatures can drop below freezing (Walter, 1971).

A. Dormancy and Germination of Montane Species

Seeds of *Eucalyptus sieberi* from southeastern Australia were nondormant, and they germinated to higher percentages at a soil matric potential of −0.001 MPa than at 0 or −0.003 MPa (Bachelard, 1985). Optimum germination conditions were a high ($> -5 \times 10^{-6}$ MPa) matric water potential and a low ($<5 \times 10^{-9}$ MPa) vapor pressure deficit (Gibson and Bachelard, 1986). Wet–dry treatments prior to the time soil moisture became nonlimiting for the germination of *E. sieberi* seeds decreased the time for germination from 120 to 64 hr (Gibson and Bachelard, 1986).

Germination percentages of seeds of the forest trees *Dacryodes excelsa* and *Prestoea montana* in the montane of Puerto Rico were increased by high light (halogen light tubes), high temperatures, and the presence of litter, but were decreased by additional water (i.e., natural rainfall, but seeds were sown in/on soil in plastic pots with no drainage holes) (Everham *et al.*, 1996). Germination of the forest tree *Guarea guidonia* was increased by high light and the presence of litter, but was decreased by high temperatures and additional water. Germination of seeds of the disturbance species *Cecropia schreberiana* was increased by high light and high temperatures, but was decreased by the presence of litter and additional water. High temperatures increased germination of the disturbance species *Schefflera morototoni*, whereas additional water decreased it; other factors were not tested.

Seeds of the tree *Metrosideros polymorpha* from Hawaiian montane rain forests were nondormant at maturity and germinated to 96, 98, 90, and 10% under white, red, and far-red light and continuous darkness, respectively (Drake, 1993). Germination occurred over a range of constant temperatures from 10 to 34°C, but maximum germination percentages were at 22°C. Optimum photosynthetic photon flux density for the germination of *M. polymorpha* seeds at 20°C was 133 μmol m^{-2} sec^{-1} (Burton, 1982). Seed rain in the Hawaiian montane forests was 5461 filled *M. polymorpha* seeds m^{-2} year^{-1}, with 72% of the annual seed rain occurring in December–January (Drake, 1993). However, the density of seed rain into new, cool, volcanic lava flows decreased to 363, 137, 37, 25, and 20 m^{-2} at 25, 50, 100, 150, and

250 m, respectively, from the edge of the forest (Drake, 1992). Germination assays for soil seed banks in various seasons of the year suggested that the species forms only a transient seed bank (Drake, 1993).

Seeds of *Tetramolopium arenarium, T. consanguineum,* and *T. lepidotum* from dry montane shrublands in Hawaii germinated to a maximum of about 47, 46, and 57%, respectively, after 30 days of incubation in Colorado (presumably in a greenhouse) (Falkner *et al.,* 1997). However, age and storage conditions of seeds were not given; thus, it is impossible to know whether seeds were dormant at maturity. In general, germination percentages of the three species of *Tetramolopium* were increased by a decrease in light, an increase in soil moisture, and the presence of litter (Falkner *et al.,* 1997). Seeds of *Silene lanceolata,* also from dry montane shrublands in Hawaii, were mostly nondormant, but afterripening in dry storage increased the rate of germination (Halward and Shaw, 1996).

After 10 weeks of dry storage at room temperatures ($20 \pm 2°C$), seeds of *Myrica faya,* an introduced weedy tree in the montane rain forests of the Hawaiian Islands, germinated to 81% in an open-air greenhouse; freshly matured seeds were not tested (Walker, 1990). After 78 weeks of dry storage, germination declined to 29%, indicating that many seeds had lost viability or reentered dormancy. Germination of seeds that had passed through the digestive system of birds was not significantly different from that of seeds that had remained undispersed on the trees for an equivalent period of time. When seeds were placed under shade screens that reduced sunlight by 0, 30, 55, 63, 80, and 96%, maximum germination (75%) occurred under the 63% screen (Walker, 1990). *Myrica faya* litter inhibited the germination of *Metrosideros polymorpha* seeds in field experiments, but seeds of *M. polymorpha* germinated under *M. faya* trees if litter was removed. In greenhouse studies, leachates from *M. faya* litter did not inhibit the germination of *M. polymorpha* seeds (Walker and Vitousek, 1991).

Observations have been made on the germination phenology of *Chamaedorea bartlingiana* seeds in a Venezuelan cloud forest. Major peaks of germination occurred in September–December, near the end of major rains (Ataroff and Schwarzkopf, 1992); however, no information seems to be available on seed dormancy and/or germination requirements.

Pinus roxburghii in montane forests in India had nondormant seeds that germinated in 5–15 days at room temperatures (Beniwal and Singh, 1989). No relationship between initial seed weight and germination percentages was found in this species (Chauhan and Raina, 1980).

Data for various species in the montane zone in Ethiopia allow us to infer that seeds of the trees *Juniperus procera* and *Podocarpus gracilior* have physiological and morphophysiological dormancy, respectively (Teketay, 1993a). The shrubs *Bersama abyssinica, Ekebergia capensis,* and *Myrsine africana* from montane forests in Ethiopia have seeds with physiological dormancy, and *Abutilon longicuspe, Dodonea angustifolia, Rhus glutinosa, Senna didymobotrya,* and *S. bicapsularis* have seeds with physical dormancy (Teketay, 1993a, 1996b). The woody climbers *Clematis hirsuta* and *C. simensis* have morphophysiological dormancy (Teketay, 1993a). Seeds of the annual legumes *Trifolium multinerve, T. quartinianum, T. rueppellianum, T. steudneri,* and *T. tembense* have physical dormancy (Dauro *et al.,* 1997). Seeds buried in pots of soil in the field during the dry season were given three wet (watered each day for 20 days)/dry (no water for 30 days) cycles. Some seeds of each species germinated during each wet period, but germination increased with each successive wet period. Under natural conditions, germination occurs in response to rainfall, mostly in the long wet season from June to September (Dauro *et al.,* 1997).

B. Dormancy and Germination of Alpine Species

Tropical alpine areas are famous for the occurrence of giant rosette plants, a characteristic not shared with alpine areas in temperate latitudes (Rundel *et al.,* 1994). Ecological studies have been conducted on some of these "rosette trees," including *Argyroxiphium* in Hawaii, *Espeletia, Lupinus,* and *Puya* in South America; and *Dendrosenecio* and *Lobelia* in Africa (Smith and Young, 1987); however, not much is known about the germination of tropical alpine rosette species, and even less is known about the nonrosette alpine species.

Seeds of *Argyroxiphium macrocephalum* incubated in darkness germinated to 27% at a 25/8°C thermoperiod and to 14% at a constant 25°C; thiourea, potassium nitrate, kinetin, or gibberellic acid (GA) treatments did not improve germination (Siegel *et al.,* 1970). Incubation of seeds in continuous darkness may be one explanation for the low germination percentages in this study. More studies need to be done to determine the dormancy state of freshly matured seeds.

Freshly matured seeds of *Espeletia timotensis* collected in Venezuela were nondormant and germinated to 76–81% in light and to 45–67% in darkness at constant temperatures of 5, 10, 13, 16, and 19°C (Guariguata and Azocar, 1988). Fifty-one percent of the seeds placed in nylon bags and buried in soil in the habitat were still alive after 1 year, indicating that this species has the capacity to form a persistent seed bank. Further, naturally dispersed seeds (23.5 m^{-2}) were recovered from

soil in study plots at the end of the rainy season (Guariguata and Azocar, 1988), which presumably indicated
the end of the germination season.

Freshly matured seeds from open-pollinated flowers
of *Lobelia telekii* and *L. keniensis* germinated to 63 and
89%, respectively, at 25°C, whereas those from caged
flowers germinated to 33 and 62%, respectively
(Young, 1982).

IV. SEMIEVERGREEN TROPICAL FORESTS

Walter (1979) combined semievergreen and deciduous tropical forests (Fig. 9.1), but he noted that semievergreen forests are intermediate between tropical evergreen rain forests and tropical deciduous forests. In the
western hemisphere, semievergreen tropical forests are
found in Venezuela, western British Guiana, northeastern Colombia, the southern rim of the Amazon Basin,
southeastern Brazil, southern Panama, Trinidad, and
eastcentral Cuba. In the eastern hemisphere, they occur
in northeastern India, Burma, Thailand, Indochina,
eastern Java, and northern Australia (Eyre, 1963). For
the most part, semievergreen tropical forests do not
occur in Africa, although dry evergreen forests are
found in Nigeria and in west Africa to the north of the
tropical evergreen rain forests (Eyre, 1963). Semievergreen tropical forests have deciduous trees in the upper
story and evergreen trees in the lower story (Walter,
1979). According to Beard (1944), from one-third to
two-thirds of the trees in the upper story of a semievergreen tropical forest are deciduous, whereas over two-
third of those in the upper story of a deciduous (seasonal) forest are deciduous.

The absolute amount of precipitation per year in
areas supporting semievergreen tropical forests is less
than that in tropical evergreen rain forests, and there
is a definite dry season. Barro Colorado Island in Panama has semievergreen tropical forests (Knight, 1975;
Walter, 1971), and its annual rainfall varies from 190 to
360 cm, with little rain (18–27 cm) falling during the
dry season, which extends from mid-December to early
May. Annual mean maximum and minimum temperatures on the forest floor on Barro Colorado Island are
28.0 and 22.1°C, respectively (Croat, 1978).

A. Germination Phenology

Garwood (1982, 1983, 1986a) found that seed dormancy is very common in the woody dicots of Barro
Colorado Island. Field observations on 187 species indicated that essentially all germination occurs during the
rainy season (Garwood, 1982, 1983). Germination begins soon after the first rains of the wet season, but the

peak does not occur until after about 1 month. Although
75% of the species germinated in the first 3 months of
the wet season, germination of the other 25% was delayed until the last 5 months. Canopy and pioneer trees
and lianas germinate early in the wet season, whereas
the germination of understory species and shade-
tolerant trees occurs throughout the wet season (Garwood, 1983). Further, more seeds germinate in light
gaps than in the shade of the forest canopy, and seeds
germinate earlier in the wet season in the open than in
the shade (Garwood, 1986a). Both wind- and animal-
dispersed seeds have a peak of germination early in the
wet season. More animal-dispersed seeds germinate in
the mid- to late portion of the wet season than do wind-
dispersed seeds (Garwood, 1983).

Observations on 118 species revealed that 84, 62, and
76% of those whose seeds were dispersed in the dry,
early wet, and late wet seasons, respectively, germinated
during the wet season. Thus, seeds of about 75% of the
species can delay germination, indicating that they are
dormant at maturity (Garwood, 1983). Garwood's
(1983) studies on 157 species showed that the mean
length of time germination is delayed ranged from 2 to
370 days, and she distinguished three groups of species:

1. Rapid-rainy: seeds are dispersed during the wet
season and have a short period of dormancy. Thus, seed
dispersal and germination occur during the wet season.
Forty percent of the 157 species fit into this category;
half germinated in less than 2 weeks, whereas the others
required 2–16 weeks.

2. Delayed-rainy: seeds are dispersed during the wet
season and have a long period of dormancy, i.e., they
do not germinate until the beginning of the next wet
season. Eighteen percent of the species have this dormancy syndrome.

3. Intermediate-dry: seeds are dispersed in the dry
season and have a dormancy period of an intermediate
length; they germinate as soon as it starts to rain again.
Forty-two percent of the species are of this type. Garwood (1983) classified the dormancy periods as rapid,
intermediate, or delayed depending on whether seeds
required <2, 2–16, or >16 weeks, respectively, to germinate. As the period of time between dispersal and the
start of the wet season decreased, less time was required
for the germination of seeds of species in the delayed-
rainy and intermediate-dry groups (Garwood, 1983).

B. Dormancy and Germination of Trees

Few studies have been done to determine the types
of dormancy found in seeds of species growing in semievergreen tropical forests. However, the type of dormancy in seeds of many species can be inferred by com-

bining available information on germination (especially germination phenology) and characteristics of seeds in the family to which the species in question belongs.

Seeds of some semievergreen tropical forest trees are nondormant (Table 9.4), and those of *Gustavia superba, Protium tenuifolium, P. pamamense,* and *Randia armata* are recalcitrant (Garwood and Lighton, 1990). Members of the Annonaceae and Myristicaceae have seeds with underdeveloped embryos (Table 3.3) and require more than 4 weeks for germination (Garwood, 1983). Thus, it appears that these two families have seeds with morphophysiological dormancy.

Seed dormancy in trees belonging to the Anacardiaceae, Bombacaceae, Fabaceae, Malvaceae, Sapindaceae, Sterculiaceae, and Tiliaceae is attributed to impermeable seed coats (Table 9.4). Hot water and/or sulfuric acid treatments increased germination percentages of *Apeiba membranacea, A. tibourboue* (Tiliaceae), *Luehea seemannii* (Tiliaceae), *Trichospermum mexicanum* (Tiliaceae), and *Guazuma ulmnifolia* (Sterculiaceae), and seeds germinated equally well in light and darkness (Acuna and Garwood, 1987). Seeds of *Peltophorum inerme* (Fabaceae) germinated to 100 and 80% after mechanical scarification and hot water treatments, respectively, and they germinated equally well in light and darkness (Anoliefo and Gill, 1992).

Our knowledge of the environmental conditions required for the germination of nondormant seeds of trees is not extensive. Obviously, the onset of the wet season stimulates germination, and Garwood (1983) recorded "the first surge of emergence" within 10 days after some light rains. In the subcanopy tree *Faramea occidentalis,* a short moist period (an early rain) before the dry season ended causes failure of seedling emergence, possibly due to germination and the subsequent death of seedlings (Schupp, 1990).

Because more seeds germinated in natural light gaps than under the forest canopy (Garwood, 1986a), light quantity and/or quality may play a role in controlling germination. Many more seedlings of pioneer trees were found on Barro Colorado Island (Panama) in soil that had been disturbed by falling trees than in nondisturbed soil. The fact that large treefall gaps are colonized more frequently by pioneer trees than are small ones (Putz, 1983) indicates that light may be important in stimulating germination. Interestingly, Brokaw (1986) noted that treefalls are more common during the wet season, thus moisture would be nonlimiting when soil disturbance supposedly exposed light-requiring seeds to light. Seedlings of pioneer tree species appear during the first year after a gap is formed, but germination is not necessarily restricted to the first year. On Barro Colorado Island, *Treme micrantha* germinated in gaps in years 1 and 2 (following creation of the gap); those of *Cecropia*

insignis in years 1, 2, and 3; and those of *Miconia argentea* in years 1, 2, 3, and 4 (Brokaw, 1987).

In many trees whose seeds are wind dispersed, seeds germinate to higher percentages in the shade than in full sun, perhaps due to drying in the sun (Augspurger, 1984). Seeds of the pioneer tree *Gustavia superba* germinated to high percentages in the forest or at the edge of a gap, regardless of whether they were on the soil surface or buried 1–2 cm deep. In a gap, however, buried seeds of this species germinated to higher percentages than nonburied ones (Sork, 1985). Thus, burial by animals such as the agoutia (Sork, 1985; Forget, 1992) potentially would increase germination. Seeds of *Tachigalia versicolor,* a monocarpic canopy tree, germinated to higher percentages in shade than in light gaps (Kitajima and Augspurger, 1989).

Position of seeds within the wind-dispersed, indehiscent fruits of *Platypodium elegans* (Leguminosae) influences germination. In fruits with two seeds, the distal one usually germinates before the proximal one. However, if the proximal seed germinates first, the distal one is unlikely to germinate (Augspurger, 1986).

The optimum alternating temperature regime for germination of the nondormant seeds of *Pterocarpus macrocarpus* was 30/25°C (Liengsiri and Hellum, 1988). Seeds from five localities in Thailand germinated to ≥80% at constant temperatures of 24 to 33°C, and those from one locality germinated >80% at 22 to 33°C. At an osmotic potential of −0.5 MPa, seeds of *Eucalyptus pilularis* and *E. maculata* germinated to about 50% at ca. 100% relative humidity (RH), but they germinated to 0% at 57–78% RH (Bachelard, 1985). Thus, seeds tolerated some moisture stress if the RH was high.

After the fruits of *Hymenaea courbaril* are opened, usually by monkeys, seeds remain covered by pulp, and this material is removed by fungus-culturing ants, *Mycocepurus goeldii* (Oliveira *et al.,* 1995). If the pulp is not removed by ants, it becomes badly infected with fungi and may be one reason for the loss of seed viability. Seeds cleaned completely by ants, those cleaned partially by ants, and those not cleaned at all (pulp badly infected with fungi) germinated to 68, 53, and 18%, respectively (Oliveria *et al.,* 1995).

C. Dormancy and Germination of Bamboos

Seeds of *Bambusa tulda* were nondormant and germinated to 92% in 9–30 days at room temperatures (Beniwal and Singh, 1989).

D. Dormancy and Germination of Shrubs

Seeds of many shrubs are nondormant (Table 9.5), and those of *Pentagonia macrophylla, Piper reticulatum,*

TABLE 9.4 Types of Seed Dormancy In Semievergreen Forest Trees

Species	Family	Type of dormancy[a,b]	Reference
Aegle marmelos	Rutaceae	PD*	Beniwal and Singh (1989)
Albizia falcataria	Fabaceae	PY	Yap and Wong (1983)
Alchornea costaricensis	Euphorbiaceae	ND*	Garwood (1983)
Allophylus psilospermus	Sapindaceae	PY*	Garwood (1983)
Alseis blackiana	Rubiaceae	ND*	Garwood (1983)
Amoora wallichii	Meliaceae	ND*	Beniwal and Singh (1989)
Anacardium excelsum	Anacardiaceae	ND*	Garwood (1983)
Annona spraguei	Annonaceae	MPD*	Garwood (1983)
Apeiba membranacea	Tiliaceae	PY*, ND*	Acuna and Garwood (1987)
A. tibourbou	Tiliaceae	PY*, ND*	Acuna and Garwood (1987)
Araucaria angustifolia	Araucariaceae	PD	Ferreira and Handro (1979); Aquila and Ferreira (1984)
Artocarpus chaplasha	Moraceae	ND*	Beniwal and Singh (1989)
A. lakoocha	Moraceae	ND*	Beniwal and Singh (1989)
Bombacopsis sessilis	Bombacaceae	ND*	Garwood (1983)
Calophyllum longifolium	Clusiaceae	PD*	Garwood (1983)
Capparis frondosa	Capparidaceae	ND*	Garwood (1983)
Casearia arborea	Flacourtiaceae	PD*	Garwood (1983)
Cavanillesia platanifolia	Bombacaceae	ND*	Garwood (1983)
Ceiba pentandra	Bombacaceae	PY*, ND*	Garwood (1983)
Cephalotaxus griffithii	Cephalotaxaceae	MPD*	Beniwal and Singh (1989)
Chukrasia velutina	Meliaceae	ND*	Beniwal *et al.* (1989)
Cordia bicolor	Boraginaceae	ND*	Garwood (1983)
Coussapoa magnifolia[c]	Moraceae	ND*	Garwood (1983)
Coussarea curvigemmia	Rubiaceae	PD*	Garwood (1983)
Cupania sylvatica	Sapindaceae	ND*	Garwood (1983)
Desmopsis panamensis	Annonaceae	MPD*	Garwood (1983)
Eugenia nesiotica	Myrtaceae	PD*	Garwood (1983)
E. oerstedeana	Myrtaceae	PD*	Garwood (1983)
Eucalyptus maculata	Myrtaceae	ND*	Bachelard (1985)
E. pilularis	Myrtaceae	ND*	Bachelard (1985)
Faramea occidentalis	Rubiaceae	PD*	Garwood (1983)
Ficus costaricana	Moraceae	ND*	Garwood (1983)
F. insipida	Moraceae	ND*	Garwood (1983)
Gmelina arborea	Verbenaceae	ND	Yap and Wong (1983); Beniwal and Singh (1989)
Guapira standleyanum	Nyctaginaceae	ND*	Garwood (1983)
Guarea glabra	Meliaceae	ND*	Garwood (1983)
Guatteria dumetorum	Annonaceae	MPD*	Garwood (1983)
Guazuma ulmnifolia	Sterculiaceae	PY	Acuna and Garwood (1987)
Hampea appendiculata	Malvaceae	PY*, ND*	Garwood (1983)
Havetiopsis flexilis[c]	Clusiaceae	ND*	Garwood (1983)
Heisteria concinna	Olacaceae	PD*	Garwood (1983)
Hirtella triandra	Chrysobalanaceae	PD*	Garwood (1983)
Hyeronima laxiflora	Euphorbiaceae	ND*	Garwood (1983)
Hymenaea courbaril	Fabaceae	PY*	Garwood (1983)
Lacistema aggregatum	Lacistemaceae	ND*	Garwood (1983)

(continues)

TABLE 9.4—*Continued*

Species	Family	Type of dormancy[a,b]	Reference
Lacmellea panamensis	Apocynaceae	PD*	Garwood (1983)
Laetia thamnia	Flacourtiaceae	ND*	Garwood (1983)
Lafoensia punicifolia	Lythraceae	ND*	Garwood (1983)
Licania platypus	Chrysobalanaceae	ND*	Garwood (1983)
Lindackeria laurina	Flacourtiaceae	PD*	Garwood (1983)
Levistona jenkensii	Arecaceae	MPD*	Beniwal and Singh (1989)
Lonchocarpus velutinus	Fabaceae	PY*, ND*	Garwood (1983)
Luehea seemannii	Tiliaceae	PY, ND	Acuna and Garwood (1987)
Maquira costaricana	Moraceae	ND*	Garwood (1983)
Myrica fosteri	Myrtaceae	ND*	Garwood (1983)
Ochroma pyramidale	Bombacaceae	PY	Acuna and Garwood (1987)
Ormosia coccinea	Fabaceae	PY	Garwood (1986)
O. macrocalyx	Fabaceae	PY	Garwood (1986)
Parkia roxburghii	Fabaceae	ND*	Beniwal and Singh (1989)
Peltophorum inerme	Fabaceae	PY	Anoliefo and Gill (1992)
Pinus merkusii	Pinaceae	ND*	Beniwal and Singh (1989)
Poulsenia armata	Moraceae	ND*	Garwood (1983)
Pouteria stipitata	Sapotaceae	PD*	Garwood (1983)
P. unilocularis	Sapotaceae	ND	Garwood (1983)
Prioria copaifera	Fabaceae	ND	Janzen (1969)
Protium tenuifolium	Burseraceae	ND*	Garwood (1983)
Pterocarpus macrocarpus	Fabaceae	ND	Liengsiri and Hellum (1988)
P. rohrii	Fabaceae	PY*, ND*	Garwood (1983)
Pterospermum acerifolium	Sterculiaceae	ND*	Beniwal and Singh (1989)
Quararibea asterolepis	Bombacaceae	PY*, ND*	Garwood (1983)
Randia armata	Rubiaceae	ND*	Garwood (1983)
Rheedia acuminata	Clusiaceae	PD*	Garwood (1983)
Sapium caudatum	Euphorbiaceae	ND*	Garwood (1983)
Sloanea terniflora	Elaeocarpaceae	ND*	Garwood (1983)
Sorocea affinis	Moraceae	ND*	Garwood (1983)
Spondias mombin	Anacardiaceae	PY*	Garwood (1983)
S. radlkoferi	Anacardiaceae	PY*	Garwood (1983)
Sterculia apetala	Sterculiaceae	PY*, ND*	Garwood (1983)
Stereospermum chelonoldes	Bignoniaceae	ND*	Beniwal and Singh (1989)
Swartzia simplex	Fabaceae	PY*	Garwood (1983)
Swietenia macrophylla	Meliaceae	ND	Alvarenga and Flores (1988); Gerhardt (1996)
Symphonia globulifera	Clusiaceae	ND*	Garwood (1983)
Tabebuia guayacan	Bignoniaceae	ND*	Garwood (1983)
Tabernaemontana arborea	Apocynaceae	ND*	Garwood (1983)
Tachigalia versicolor	Fabaceae	PY*	Kitajima and Augspurger (1989)
Tectona grandis	Verbenaceae	PD*, ND*	Yap and Wong (1983)
Tocoyena pittieri	Rubiaceae	PD*	Garwood (1983)
Trema micrantha	Ulmaceae	PD*	Garwood (1983)
Trichilia cipo	Meliaceae	ND*	Garwood (1983)
T. emetica	Meliaceae	PD*	Mahgembe and Msanga (1988)

(*continues*)

TABLE 9.4—*Continued*

Species	Family	Type of dormancy[a,b]	Reference
Trichospermum mexicanum	Tiliaceae	PY	Acuna and Garwood (1987)
Tristania conferta	Myrtaceae	ND	Petteys (1974)
Turpinia occidentalis	Staphyleaceae	PD*	Garwood (1983)
Unonopsis pittieri	Annonaceae	MPD*	Garwood (1983)
Vantanea occidentalis	Humiriaceae	PD*	Garwood (1983)
Virola sebifera	Myristicaceae	MPD*	Garwood (1983)
V. surinamensis	Myristicaceae	MPD*	Garwood (1983)
Xylopia frutescens	Annonaceae	MPD*	Garwood (1983)
Zanthoxylum belizense	Rutaceae	PD*	Garwood (1983)
Z. setulosum	Rutaceae	PD*	Garwood (1983)
Zuelania guidonia	Flacourtiaceae	ND*	Garwood (1983)

[a] ND, nondormant; PD, physiological dormancy; PY, physical dormancy; MD, morphological dormancy; MPD, morphophysiological dormancy.
[b] An ansterisk means that the type of dormancy is inferred from available information on germination and on characteristics of seeds in that family.
[c] Species is a hemiepiphytic tree.

Psychotria marginata, and *P. limonensis* are recalcitrant (Garwood and Lighton, 1990). Seeds of several shrubs have physiological or physical dormancy, but a few have morphophysiological dormancy (Table 9.5).

The low number of species with physical dormancy in Table 9.5 may mean that few semievergreen tropical forest shrubs in the various families known to have seeds with impermeable seed coats have been investigated. However, shrubs in semievergreen tropical forests belonging to families known to have physical dormancy may be nondormant (i.e., seed coats are permeable to water). For example, the mean time for germination of *Cassia fruticosa, Pithecellobium rufescens* (Fabaceae), and *Herrania purpurea* (Sterculiaceae) seeds on Barro Colorado Island was 4.7, 29.6, and 11.1 days, respectively (Garwood, 1983), thus we consider the seeds to be nondormant. Studies are needed to determine if other shrubs in the Fabaceae and Sterculiaceae in semievergreen tropical forests have nondormant seeds or if they have physical dormancy.

Seeds of the shrub *Hybanthus prunifolius* germinated to a higher percentage and at a faster rate in sun than in semishade or shade (Augspurger, 1979).

E. Dormancy and Germination of Lianas

Available information allows us to infer that seeds of many lianas are nondormant, but those of some species have physiological, physical, or morphophysiological dormancy (Table 9.6). Seeds of the lianas *Connarus turczaninowii* and *Maripa panamensis* are nondormant and also recalcitrant (Garwood and Lighton, 1990). Lit-

tle is known about the dormancy breaking and germination requirements of seeds of lianas in semievergreen tropical forests. Seeds of *Cryptostegia grandiflora* stored dry at room temperatures for 6 months germinated to 52, 92, 95, 84, 20, and 0% at 15, 20, 25, 30, 35, and 40°C, respectively (Grice, 1996).

F. Dormancy and Germination of Herbaceous Species

Little information is available on the seed germination of herbaceous species growing in semievergreen tropical forests; however, some studies have been done on species growing in areas where the forests have been cut (Table 9.7). Most of these species are weeds (see Holm *et al.,* 1979) and may not be native to the semievergreen forest region per se; however, they can complete their life cycles under the climatic conditions where these forests occur.

Seeds of many herbaceous species are nondormant, but those of others are dormant. Because members of the Iridaceae have underdeveloped embryos (Martin, 1946) and seeds of *Orthrosanthus chimboracensis* germinated without being given special dormancy-breaking treatments (van Rooden *et al.,* 1970), we conclude that they have morphological dormancy. Physiological dormancy is the most common type of seed dormancy, but some species have seeds with physical dormancy.

Physical dormancy was broken in seeds of *Uraria picta* by acid scarification or dry heat treatments (Okusanya *et al.,* 1992), and seeds germinated to higher per-

TABLE 9.5   Types of Seed Dormancy in Semievergreen Forest Shrubs

Species	Family	Type of dormancy[a,b]	Reference
Bertiera guianensis	Rubiaceae	PD*	Garwood (1983)
Cassia fruticosa	Fabaceae	ND*	Garwood (1983)
Cephaelis ipecacuanha	Rubiaceae	PD*	Garwood (1983)
C. lasiocalyx	Boraginaceae	ND*	Garwood (1983)
Hamelia axillaris	Rubiaceae	ND*	Garwood (1983)
H. patens	Rubiaceae	ND*	Garwood (1983)
Hasseltia floribunda	Flacourtiaceae	ND*	Garwood (1983)
Herrania purpurea	Sterculiaceae	ND*	Garwood (1983)
Lycianthes maxonii	Solanaceae	ND*	Garwood (1983)
Mabea occidentalis	Euphorbiaceae	ND*	Garwood (1983)
Margaritaria nobilis	Euphorbiaceae	PD*	Garwood (1983)
Miconia nervosa	Melastomataceae	ND*	Garwood (1983)
Mouriri myrtilloides	Melastomataceae	PD*	Garwood (1983, 1986b)
Neea amplifolia	Nyctaginaceae	ND*	Garwood (1983)
Palicourea guianensis	Rubiaceae	PD*	Garwood (1983)
Parathesis microcalyx	Myrsinaceae	PD*	Garwood (1983)
Pithcellobium rufescens	Fabaceae	ND*	Garwood (1983)
Posoqueria latifolia	Rubiaceae	ND*	Garwood (1983)
Psychotria acuminata	Rubiaceae	PD*	Garwood (1983)
P. chagrensis	Rubiaceae	PD*	Garwood (1983)
P. deflexa	Rubiaceae	PD*	Garwood (1983)
P. emetica	Rubiaceae	PD*	Garwood (1983)
P. furcata	Rubiaceae	PD*	Garwood (1983)
P. grandis	Rubiaceae	PD*	Garwood (1983)
P. horizontalis	Rubiaceae	PD*	Garwood (1983, 1986b)
P. limonensis	Rubiaceae	PD*	Garwood (1983)
P. marginata	Rubiaceae	PD*	Garwood (1983)
P. pittieri	Rubiaceae	PD*	Garwood (1983)
P. racemosa	Rubiaceae	PD*	Garwood (1983)
Rinorea sylvatica	Violaceae	ND*	Garwood (1983)
Thevetia ahouai	Apocynaceae	PD*	Garwood (1983)
Tovomitopsis nicaraguensis	Clusiaceae	ND*	Garwood (1983)
Turnera panamensis	Turneraceae	ND*	Garwood (1983)
Vernonia patens	Asteraceae	ND*	Garwood (1979)
Ziziphus mauritiana	Rhamnaceae	PY*	Grice (1996)

[a] ND, nondormant; PD, physiological dormancy; PY, physical dormancy.
[b] An asterisk means that the type of dormancy is inferred from available information on germination and on characteristics of seeds in that family.

centages in light than in darkness. Alternating temperature regimes of 31/21 and 31/15°C promoted the germination of nondormant seeds of *U. picta*, but 21/15°C inhibited it (Okusanya *et al.*, 1991).

Nondormant seeds of *Corchorus capsularis* and *C. olitorius* germinated to higher percentages when soil moisture was at field capacity than at 80, 60, 40, or 20%

of field capacity; they did not germinate under waterlogged conditions (Newaz and Nazrul-Islam, 1991). Seeds of *C. olitorius* germinated in 50% sea water, and they required light for germination; the optimum temperature was 31°C (Okusanya, 1979). Seeds of *Stevia rebaudiana* germinated to higher percentages in light than in darkness (Felippe *et al.*, 1971). Seeds of *Phyto-*

TABLE 9.6 Types of Seed Dormancy in Semievergreen Forest Lianas[b]

Species	Family	Type of dormancy[a,b]	Reference
Acacia hayesii	Fabaceae	ND*	Garwood (1983)
Adenopodia polystachya	Fabaceae	PY*	Garwood (1983)
Amphilophium paniculatum	Bignoniaceae	PD*	Garwood (1983)
Anthodon panamense	Hippocrateaceae	PD*	Garwood (1983)
Calamus floribundus	Arecaceae	MPD*	Beniwal and Singh (1989)
C. tenuis	Arecaceae	MPD*	Beniwal and Singh (1989)
Callichlamys latifolia	Bignoniaceae	PD*	Garwood (1983)
Ceratophytum tetragonolobum	Bignoniaceae	ND*	Garwood (1983)
Clitoria javitensis	Fabaceae	ND*	Garwood (1983)
Combretum decandrum	Combretaceae	ND*	Garwood (1983)
C. fruticosum	Combretaceae	ND*	Garwood (1983)
Connarus panamensis	Connaraceae	ND*	Garwood (1983)
C. turczaninowii	Connaraceae	ND*	Garwood (1983)
Cydista aequinoctalis	Bignoniaceae	PD*	Garwood (1983)
Davilla nitida	Dilleniaceae	MPD*	Garwood (1983)
Doliocarpus dentatus	Dilleniaceae	MPD*	Garwood (1983)
D. major	Dilleniaceae	MPD*	Garwood (1983)
D. olivaceus	Dilleniaceae	MPD*	Garwood (1983)
Hippocratea volubilis	Hippocrateaceae	PD*	Garwood (1983)
Machaerium microphyllum	Fabaceae	ND*, PY*	Garwood (1983)
Maripa panamensis	Convolvulaceae	ND*	Garwood (1983)
Mascagnia hippocrateoides	Malpighiaceae	ND*	Garwood (1983)
M. nervosa	Malpighiaceae	ND*	Garwood (1983)
Mendoncia littoralis	Acanthaceae	PD*	Garwood (1983)
Odontadenia macrantha	Apocynaceae	PD*	Garwood (1983)
Paullinia fibrigera	Sapindaceae	PY*, ND*	Garwood (1983)
P. turbacensis	Sapindaceae	PY*, ND*	Garwood (1983)
Phryganocydia corymbosa	Bignonaceae	ND*	Garwood (1983)
Pithecoctenium crucigerum	Bignonaceae	ND*	Garwood (1983)
Prionostemma aspera	Hippocrateaceae	PD*	Garwood (1983)
Rhynchosia pyramidalis	Fabaceae	PY*	Garwood (1983)
Thinouia myriantha	Sapindaceae	PY*	Garwood (1983)
Tynnanthus croatianus	Bignoniaceae	ND*	Garwood (1983)

[a]MPD, morphophysiological dormancy; ND, nondormant; PD, physiological dormancy; PY, physical dormancy.
[b]An asterisk means that the type of dormancy is inferred from available information on germination and on characteristics of seeds in that family.

lacca rivinoides placed in a forest clearing in Panama germinated to 80–90%, whereas those placed 2–10 m into the forest did not germinate (Williams-Linera, 1990).

V. TROPICAL DECIDUOUS FORESTS

Tropical deciduous forests are located to the north and south of tropical evergreen rainforests in South America, on the Pacific side of Central America, and on islands in the Caribbean Sea. These forests also are found in India, southeast Asia, and northern and eastern Australia. Walter's map (Fig. 9.1) shows semievergreen/tropical deciduous forests in Africa to the north and south of the equatorial evergreen rain forests. However, the trees in these African deciduous forests are more widely spaced than those in tropical deciduous forests in the Americas, Asia, and Australia, and the communities are not clearly stratified (Eyre, 1963). Regions with

TABLE 9.7   Types of Seed Dormancy in Herbaceous Species Growing in Areas
Where Tropical Semievergreen Forests Have Been Cut

Species	Family	Type of dormancy[a,b]	Reference
Ageratum conyzoides[c]	Asteraceae	ND*	Van Rooden *et al.* (1970)
Amaranthus dubius[c]	Amaranthaceae	PD	Van Rooden *et al.* (1970)
Calotropis procera[c]	Asclepiadaceae	ND*	Van Rooden *et al.* (1970)
Chaptalia nutans[c]	Asteraceae	ND*	Garwood (1979)
Chloris inflata[c]	Poaceae	ND	Van Rooden *et al.* (1970)
Corchorus capsularis[c]	Tiliaceae	PY*	Newaz and Nazrul-Islam (1991)
C. olitorius[c]	Tiliaceae	PY*	Newaz and Nazrul-Islam (1991), Okusanya (1979)
Echinochloa colona[c]	Poaceae	PD	Van Rooden *et al.* (1970)
Eleusine indica[c]	Poaceae	PD	Van Rooden *et al.* (1970)
Euphorbia hirta[c]	Euphorbiaceae	PD*, ND*	Van Rooden *et al.* (1970)
Flaveria trinerva[c]	Asteraceae	PD	Van Rooden *et al.* (1970)
Galinsoga parviflora[c]	Asteraceae	ND*	Van Rooden *et al.* (1970)
Hyptis suaveolens[c]	Lamiaceae	PD	Van Rooden *et al.* (1970)
Impatiens sultani[c]	Balsaminaceae	ND*	Van Rooden *et al.* (1970)
Ipomoea coccinea[c]	Convolvulaceae	PY*	Van Rooden *et al.* (1970)
Leonurus sibiricus[c]	Lamiaceae	PD*, ND*	Van Rooden *et al.* (1970)
Leptochloa filiformis[c]	Poaceae	PD*, ND*	Van Rooden *et al.* (1970)
Melinis minutiflora[c]	Poaceae	ND*	Van Rooden *et al.* (1970)
Mentzelia aspera[c]	Loasaceae	PD	Van Rooden *et al.* (1970)
Merremia aegyptia[c]	Convolvulaceae	PY*	Van Rooden *et al.* (1970)
Orthrosanthus chimboracensis[c]	Iridaceae	MD*	Van Rooden *et al.* (1970)
Pennisetum pedicellatum[c]	Poaceae	PD*	Afolayan and Olugbami (1993)
P. setosum[c]	Poaceae	ND	Van Rooden *et al.* (1970)
Portulaca oleracea[c]	Portulacaceae	ND*	Van Rooden *et al.* (1970)
Pseudoelephantopus spicatus[c]	Asteraceae	ND*	Garwood (1979)
Ruellia tuberosa[c]	Acanthaceae	ND*	Van Rooden *et al.* (1970)
Setaria pallidefusca[c]	Poaceae	PD*	Afolayan and Olugbami (1993)
Sida acuta[c]	Malvaceae	PY*, ND*	Garwood (1979)
Stevia rebaudiana	Asteraceae	ND*	Felippe *et al.* (1991)
Synedrella nodiflora	Asteraceae	ND*	Van Rooden *et al.* (1970)
Tibouchina longifolia[c]	Melastomataceae	ND*	Van Rooden *et al.* (1970)
Tridax procumbens[c]	Asteraceae	PD*, ND*	Van Rooden *et al.* (1970)
Uraria picta[c]	Fabaceae	PY	Okusanya *et al.* (1992)
Verbesina caracasana[c]	Asteraceae	PD*	Van Rooden *et al.* (1970)
V. gigantea	Asteraceae	ND*	Garwood (1979)
Vernonia camescems	Asteraceae	ND*	Garwood (1979)
Witheringia solanacea	Scrophulariaceae	ND*	Garwood (1979)

[a]MD, morphological dormancy; ND, nondormant; PD, physiological dormancy; PY, physical dormancy.
[b]An asterisk means that the type of dormancy is inferred from available information on germination and on characteristics of seeds in that family.
[c]Listed as a weed by Holm *et al.* (1979).

tropical deciduous forests have distinct wet and dry seasons each year, with the wet season occurring during summer. Thus, plants are exposed to hot, wet summers and relatively cool, dry winters. With an increase in the distance from the equator, the length of the cool season increases and the amount of annual rainfall decreases (Walter, 1973).

Trees are the dominant life form in tropical deciduous forests, and they occur in two strata. Over two-thirds of the trees in the upper story are deciduous (Beard, 1944), whereas most of those in the lower story are evergreen (Richards, 1957; Eyre, 1963). The forests have few lianas and epiphytes, and herbaceous species (including grasses) usually are not abundant. The bamboo *Dendrocalamus strictus* is an important understory species in Burma, and bamboos and shrubs are important in Java (Eyre, 1963). However, in southeast Asia in soils that are either quite shallow or underlain by a hard pan, forests have an upper story but lack a lower one. Forests with only an upper story have many shrubs, grasses, and herbs, and epiphytes grow on the trees (Eyre, 1963). Trees are so widely spaced in Africa that tall grasses dominate the understory (Eyre, 1963).

A phenological study in a tropical deciduous forest in Costa Rica revealed that 75% of 113 trees lost their leaves during the dry season, and the peak of new leaf production was at the start of the wet season. Fifty-nine species flowered during the dry season, 35 at the beginning of the wet season in May and June, and 13 during both seasons. Seeds of 75 species matured during the dry season, but those of 19 species matured in the wet season; 7 species matured throughout the year. Most species whose seeds matured in the wet season had fleshy fruits, but about half the species whose seeds matured in the dry season had dry fruits and the other half fleshy fruits (Frankie *et al.*, 1974).

Trees in tropical deciduous forests in Ghana lost their leaves during the dry season and produced new ones during the wet season. Most trees (78%) flowered and produced fruits during the wet season, although a few (11%) flowered in the dry season or during both seasons (11%). More fleshy-fruited species produced seeds in the wet than in the dry season, and more dry-fruited species produced seeds in the dry than in the wet season (Lieberman, 1982). Rainfall peaks occurred in Pinkwae, Ghana, in May–June and October–November (autumn), and seed germination began in October and peaked in June–July at the end of the rainy season (Lieberman and Li, 1992). Although soil moisture conditions in tropical evergreen rain forests are suitable for seedling establishment most, or all, of the year (Richards, 1957), they are favorable for seedling establish-

ment in tropical deciduous forests only during the wet season (Lieberman and Li, 1992).

A. Dormancy and Germination of Trees

Seeds of some trees in tropical deciduous forests are nondormant, but those of most species are dormant (Table 9.8). The only known types of seed dormancy in these forests are physiological and physical, with physical being the most common of the two. Seeds of the subcanopy tree *Gliricidia sepium* mature at the end of the dry season in Costa Rica and are nondormant, germinating within 48–96 hr after the imbibition of water (Flores and Rivera, 1985). Seeds of the pioneer tree *Muntingia calabura* mature in Costa Rica early in the wet season and are nondormant; however, they require the high light and temperature conditions found in canopy gaps for germination (Fleming *et al.*, 1985). Freshly matured seeds of *Ficus bengalensis* germinated within 10 days, but those that passed through the digestive system of a bird had a higher germination rate and percentage than the controls (Midya and Brachmachary, 1991).

Seeds of *Diospyros melanoxylon*, a common tree in tropical deciduous forests in India, mature in April–June, which is during the hot, dry summer season, but they do not begin to germinate until early in the wet season, which lasts from June to October. If seeds do not germinate during the wet season, their germination is delayed until the second wet season following dispersal (Ghosh *et al.*, 1976). In contrast, seeds of *Mesua ferrea* are dispersed during the dry season in India but do not germinate until the subsequent wet season has passed and the next dry season has started (Richards, 1957).

Seeds of *Toona ciliata* germinated within 2 weeks after sowing, but germination percentages varied with the microsite: shade > mulch > open (Beniwal *et al.*, 1990b). Seven weeks after sowing, the (apparently) physiologically dormant seeds of *Adina cordifolia* had the highest germination (15%) under mulch: mulch > shade > open (Beniwal *et al.*, 1990a). Time to germination after water becomes nonlimiting varies with the species, e.g., *Cedrela odorata* seeds in 6 or 7 days and those of *Brosium alicastrum* in 17–21 days (Blain and Kellman, 1991). A 5-day drought treatment after sowing increased the time required for the germination of *Cedrela odorata* seeds to 17 days (Blain and Kellman, 1991). Seeds of *Tecoma stans* germinated to 9 and 0% at an osmotic potential (created with polyethylene glycol) of −1.0 and −1.5 MPa, respectively; however, they not only survived such treatments but germination rates and percentages were increased when seeds were transferred to distilled water (Cordero and Di Stefano, 1991).

TABLE 9.8 Types of Seed Dormancy in Tropical Deciduous Forest Trees

Species	Family	Type of dormancy[a,b]	Reference
Acacia catechu	Fabaceae	PY	Athaya (1990)
A. cochliacantha	Fabaceae	PY	Cervantes *et al.* (1996)
A. leucophaea	Fabaceae	PY	Athaya (1990), Zodape (1991)
A. mangium	Fabaceae	PY	Yap and Wong (1983)
A. nilotica	Fabaceae	PY	Zodape (1991), Ginwal *et al.* (1995)
A. pennatula	Fabaceae	PY	Cervantes *et al.* (1996)
Acrocarpus fraxiniifolius	Fabaceae	PY*	Beniwal and Singh (1989)
Adina cordifolia	Rubiaceae	PD*	Beniwal *et al.* (1990b)
Ailanthus excelsa	Simaroubaceae	ND*	Beniwal and Singh (1989)
A. integrifolia	Simaroubaceae	PD*	Beniwal and Singh (1989)
Albizia arunachalensis	Fabaceae	PY*	Beniwal and Singh (1989)
A. chinesis	Fabaceae	PY*	Beniwal and Singh (1989)
A. falcataria	Fabaceae	PY*	Beniwal and Singh (1989)
A. lebbek	Fabaceae	PY	Zodape (1991), Teketay (1996a)
A. lucidior	Fabaceae	PY*	Beniwal and Singh (1989)
A. procera	Fabaceae	PY	Athaya (1990)
Alnus nepalensis	Betulaceae	ND*	Beniwal and Singh (1989)
Altingia excelsa	Hamamelidaceae	PD*	Beniwal and Singh (1989)
Anacardium occidentale	Anacardiaceae	PY*	Beniwal and Singh (1989)
Anogeissus pendula	Combretaceae	PD*	Athaya (1985, 1990)
Anthocephalus chinensis	Naucleaceae	PD*	Beniwal and Singh (1989)
Aquilaria agallocha	Thymelaeaceae	ND*	Beniwal and Singh (1989)
Azadirachta indica	Meliaceae	ND*	Beniwal and Singh (1989)
Bauhinia purpurea	Fabaceae	PY*	Beniwal and Singh (1989)
B. racemosa	Fabaceae	PY	Zodape (1991)
B. variegata	Fabaceae	PY	Athaya (1990)
Bischofia javanica	Euphorbiaceae	ND*	Beniwal and Singh (1989)
Bridelia retusa	Euphorbiaceae	PD*	Beniwal and Singh (1989)
Brosimum alicastrum	Moraceae	ND*	Blain and Kellman (1991)
Cassia fistula	Fabaceae	PY	Athaya (1990), Todaria and Negi (1992)
C. glauca	Fabaceae	PY	Todaria and Negi (1992)
C. grandis	Fabaceae	PY	Flores *et al.* (1986)
C. javanica	Fabaceae	PY	Todaria and Negi (1992)
C. nodosa	Fabaceae	PY	Todaria and Negi (1992)
C. siema	Fabaceae	PY	Todaria and Negi (1992)
Casuarina equisetifolia	Casuarinaceae	ND*	Beniwal and Singh (1989)
Cedrela odorata	Meliaceae	ND	Blain and Kellman (1991)
Cordia elaeagnoides	Boraginaceae	PD*	van Groenendael *et al.* (1996)
Dalbergia assamica	Fabaceae	PY*	Beniwal and Singh (1989)
D. latifolia	Fabaceae	PY*	Beniwal and Singh (1989)
D. sissoo	Fabaceae	PY*	Beniwal and Singh (1989)
Dichrostachys cinerea	Fabaceae	PY	Zodape (1991), Bell and van Staden (1993)
Diospyros melanoxylon	Ebenaceae	ND*	Ghosh *et al.* (1976)
Emblica officinalis	Euphorbiaceae	PD*	Athaya (1990)
Enterolobium cyclocarpum	Fabaceae	PY	Janzen (1981a), Blain and Kellman (1991)

(continues)

TABLE 9.8—*Continued*

Species	Family	Type of dormancy[a,b]	Reference
Ficus bengalensis	Moraceae	ND*	Midya and Brahmachary (1991)
F. lushnathiana	Moraceae	ND	Figueiredo and Perin (1995)
F. sur	Moraceae	PD*	Teketay (1993a)
Gliricidia sepium	Fabaceae	ND	Flores and Rivera (1985)
Haematoxylon campachianum	Fabaceae	PY	Zodape (1991)
Kydia glabrescens	Malvaceae	PY*	Beniwal and Singh (1989)
Lagerstroemia parviflora	Lythraceae	PD*	Athaya (1990)
L. speciosa	Lythraceae	PD*	Beniwal and Singh (1989)
Leucaena esculenta	Fabaceae	PY	Cervantes *et al.* (1996)
L. leucocephala	Fabaceae	PY	Zodape (1991)
L. macrophylla	Fabaceae	PY	Cervantes *et al.* (1996)
Lysiloma divaricata	Fabaceae	PY	Cervantes *et al.* (1996)
Mitragyna parvifolia	Rubiaceae	PD*	Athaya (1990)
Muntingia calabura	Elaeocarpaceae	ND	Fleming *et al.* (1985)
Parkinsonia aculeata	Fabaceae	PY	Zodape (1991)
Peltophorum pterocarpum	Fabaceae	PY	Zodape (1991)
Pongamia pinnata	Fabaceae	PY	Athaya (1990)
Salmalia malabarica	Bombacaceae	PY*	Athaya (1990)
Santalum album	Santalaceae	PD*	Beniwal and Singh (1989)
Shorea assamica	Dipterocarpaceae	ND*	Beniwal and Singh (1989)
S. robusta	Dipterocarpaceae	ND*	Beniwal and Singh (1989)
Spondias pinnata	Anacardiaceae	PY*	Beniwal and Singh (1989)
S. purpurea	Anacardiaceae	PY*, ND*	Mandujano *et al.* (1994)
Sterculia villosa	Sterculiaceae	ND*	Beniwal and Singh (1989)
Swietenia macrophylla	Meliaceae	ND	Gerhardt (1996)
Tecoma stans	Bignoniaceae	ND	Pelton (1964)
Tectona grandis	Verbenaceae	PD*	Yadav (1992)
		ND*	Beniwal and Singh (1989)
Terminalia arjuna	Combretaceae	PD*	Athaya (1990)
T. bellirica	Combretaceae	PD*	Athaya (1990), Bhardwaj and Chakraborty (1994)
T. chebula	Combretaceae	PD*	Bhardwaj and Chakraborty (1994)
T. citrina	Combretaceae	PD*	Beniwal and Singh (1989)
T. tomentosa	Combretaceae	PD*	Negi and Todaria (1995)
Toona ciliata	Meliaceae	ND*	Beniwal *et al.* (1990b)
Zanthoxylum limonella	Rutaceae	PD*	Beniwal and Singh (1989)

[a]ND, nondormant; PD, physiological dormancy; PY, physical dormancy.
[b]An asterisk means that the type of dormancy is inferred from available information on germination and on characteristics of seeds in that family.

Physical dormancy in seeds of tropical deciduous forest trees has been broken by mechancial (Flores *et al.,* 1986; Athaya, 1990; Blain and Kellman, 1991) and acid (Zodape, 1991) scarification. Seeds of *Enterolobium cyclocarpum* became permeable after soaking in hot water (45°C) or being exposed to heat of a compost pile (45.8°C) or from a fire (Hunter, 1989).

B. Dormancy and Germination of Shrubs

Seeds of *Triumfetta rhomboidia* mature in India in November and December and are dispersed during the cool, dry season in December and January (winter) (Chaudhary, 1976). Although seeds are dormant at maturity, they become nondormant during the dry season and begin to germinate when rains wet the soil in July and August. Seeds came out of dormancy during dry laboratory storage. They germinated in both light and darkness, and higher percentages were obtained at 30°C than at 20 or 40°C. Achenes of *Brachylaena huillensis* (Kigomo *et al.,* 1994) and caryopses of the bamboo *Melocanna bambusoides* (Stapf, 1904) are nondormant.

C. Dormancy and Germination of Herbaceous Species

Herbaceous species are not an important component of tropical deciduous forests (Eyre, 1963); therefore, it is not surprising that there seems to be an almost complete lack of information on the germination ecology of native forest herbs. However, many forests have been cleared for agriculture and other uses, and studies have been done on seed germination of the herbaceous species, mostly weeds, now growing in these areas (Table 9.9).

Some (about 15%) herbaceous species have nondormant seeds, but most have dormant seeds (Table 9.9). Physiological dormancy is the most common type of dormancy, but physical and morphophysiological dormancy also are present. Lack of moisture during the hot, dry season promotes loss of dormancy, and thus germination percentages and rates increase. In fact, some species, e.g., *Oxygonum sinuatum,* do not come out of dormancy if they are kept on a moist substrate (Popay, 1975). However, some seeds either do not come out of dormancy during the first (second or third) dry season after dispersal or they reenter dormancy during the wet season. Consequently, seeds may not germinate until the second or some subsequent wet season (Popay, 1975). A 7-month period of afterripening in dry storage at room temperatures increased the germination of *Achyranthes aspera* utricles from 0 to 50%; however, when fresh utricles were washed in running water for 24 hr they germinated to 100% (Pandya and Pathak, 1980).

Seeds of most species germinated to higher percentages in light than in darkness (Popay, 1974; Ahlawat and Dagar, 1980; Felippe and Polo, 1983); however, those of *Ricinus communis* germinated to higher percentages in darkness than in light (Felippe and Polo, 1983). Seeds of *Tridax procumbens, Amaranthus graecizans,* and *Portulaca oleracea* germinated to higher percentages in light if they previously were incubated in darkness than if incubated continuously in light (Popay, 1974). Continuous light reduced the germination of *Asphodelus tenuifolius* seeds (Pandya and Reddy, 1990), and maximum germination of *Parthenium hysterophorus* seeds was obtained at a 10-hr daily photoperiod (Pandey and Dubey, 1988). The ability of *Hyptis suaveolens* seeds to germinate in darkness increased during loss of dormancy (Felippe *et al.,* 1983).

Because dormancy loss occurs during the dry season, large numbers of seeds can germinate following the first rains of the wet season. Thus, the germination of many species becomes synchronized (Popay, 1975). Flushes of seedlings appeared 5–8 days after a heavy rain in Kenya. In some species (e.g., *Oxygonum sinuatum* and *Rhynchelytrum repens*), a large number of seeds germinated in response to the first rains, a smaller number after the second rains, and few thereafter, regardless of the number of rains. In other species, such as *Bidens pilosa, Boerhaavia diffusa, B. erecta, Tridax procumbens,* and *Euphorbia geniculata,* three or four peaks of germination occur during the wet season (Popay, 1976).

After a 5-month period of afterripening at room temperatures (15–20°C), seeds of *Parthenium hysterophorus* from northern India germinated to higher percentages than those from southern India. Seeds from northern and southern locations germinated equally well at 25/20, 30/15, and 30/20°C, but those from the north germinated to higher percentages at 25/10 and 30/25°C than did those from the south (Pandey and Dubey, 1988).

D. Dormancy and Germination of Epiphytes

Freshly matured seeds of the hemiepiphytic *Ficus aurea* were nondormant, and they germinated to 90% in about 11 days at constant temperatures of 25–30°C (Swagel *et al.,* 1997). Seeds germinated to 86, 85, and 80% at water potentials (mannital solutions) of −0.1, −0.25, and −0.5 MPa, respectively, after 20 days, but they germinated to 33, 0, and 0% at −1.0, −1.5, and −2.0 MPa, respectively. Water potentials did not drop below −1.0 to −1.1 MPa in humus at the base of leaves in the canopy of palm trees at Hummingbird Cay, in the Bahamas. The continuous presence of water in humus on palm trees may explain why palms are the only host for *F. aurea* in the seasonally dry climate of the Bahamas (Swagel *et al.,* 1997).

E. Chlorophyllous Embryos

Janzen (1982a) surveyed seeds of 74 species of native shrubs, trees, and lianas in a tropical deciduous forest in Costa Rica to determine if embryos were chlorophyllous or achlorophyllous; 55% of the species had chlorophyllous embryos. Seeds with chlorophyllous embryos were produced in flat, strap-shaped or cylindrical fruits or in isolated spherical fruits that contained only one to four small- or medium-sized seeds. Seeds with achlorophyllous embryos were produced in fruits with seeds packed in layers or in fruits of species such as *Ardisia revoluta, Hirtella racemosa,* or *Aphelandia deppeana* that develop in the shade.

TABLE 9.9 Types of Seed Dormancy in Herbaceous Species Growing in Areas
Where Tropical Deciduous Forests Have Been Cut

Species	Family	Type of dormancy[a,b]	Reference
Achyranthes aspera[c]	Amaranthaceae	PD	Pandya and Pathak (1980)
Ageratum conyzoides[c]	Asteraceae	PD	Popay (1974)
Amaranthus deflexus[c]	Amaranthaceae	PD[d]	Felippe and Polo (1983)
A. graecizans[c]	Amaranthaceae	PD[d]	Popay (1973)
A. hybridus[c]	Amaranthaceae	PD	Felippe and Polo (1983)
Asphodelus tenuifolius[c]	Liliaceae	MPD*	Pandya and Reddy (1990)
Athroisma stuhlmanii[c]	Asteraceae	PD	Popay (1974)
Bidens biternata[c]	Asteraceae	PD	Ahlawat and Dagar (1980)
B. pilosa[c]	Asteraceae	PD[d]	Popay (1973), Felippe and Polo (1983)
Boerhaavia diffusa[c]	Nyctaginaceae	PD[d]	Popay (1973)
B. erecta[c]	Nyctaginaceae	PD	Popay (1973)
Cassia patellaria	Fabaceae	PY	Felippe and Polo (1983)
Cenchrus echinatus[c]	Poaceae	ND	Felippe and Polo (1983)
Cleome hirta[c]	Cleomaceae	PD	Popay (1974)
C. monophylla[c]	Cleomaceae	PD	Popay (1973)
Commelina benghalensis[c]	Commelinaceae	PD[d]	Popay (1973)
Crotalaria lanceolata[c]	Fabaceae	PY	Felippe and Polo (1983)
C. mucronata[c]	Fabaceae	PY[d]	Felippe and Polo (1983)
Dactyloctenium aegyptium[c]	Poaceae	PD	Popay (1973)
Datura stramonium[c]	Solanaceae	PD	Popay (1973)
Digitaria sanguinalis[c]	Poaceae	ND	Felippe and Polo (1983)
D. velutina[c]	Poaceae	PD	Popay (1974)
Echinochloa colonum[c]	Poaceae	PD[d]	Popay (1973)
Eleusine indica[c]	Poaceae	PD[d]	Popay (1973)
Emilia sonchifolia[c]	Asteraceae	PD[d]	Felippe and Polo (1983)
Eragrostis cilianensis[c]	Poaceae	PD	Popay (1974)
Eriochloa nubica	Poaceae	PD	Popay (1974)
Eupatorium pauciflorum	Asteraceae	PD*	Felippe and Polo (1983)
Euphorbia geniculata[c]	Euphorbiaceae	ND	Popay (1973)
Galinsoga parviflora[c]	Asteraceae	ND	Popay (1974)
Glycine wightii	Fabaceae	PY	Felippe and Polo (1983)

(continues)

The meaning of the seemingly large number of chlorophyllous embryos in tropical deciduous forest species is a mystery. Because similar surveys have not been made in other types of vegetation, we do not know if this number of species with chlorophyllous embryos is unusual or not.

F. Special Biotic Factor: Seed Predation

The primary consequence of seed predation is that seeds are killed. Thus, seed predators have little influence on germination per se, but they may have a huge impact on the seed biology of a species (Louda, 1982). Because of the research efforts of Daniel H. Janzen in Central America, more is known about seed predation in tropical deciduous forests than in any other type of forest.

The primary period of seed destruction for most species is while seeds are still attached to the mother plant, but some are eaten after dispersal. Janzen (1980) collected fruits and/or seeds from 975 species of dicots in Costa Rican tropical deciduous forests and found that 100 of them had predated seeds. Sixty-three species belonged to the Fabaceae, 11 to the Convolvulaceae,

TABLE 9.9—*Continued*

Species	Family	Type of dormancy[a,b]	Reference
Heliotropium subulatum	Borginaceae	PD	Popay (1973)
Hyptis suaveolens[c]	Lamiaceae	ND	Felippe *et al.* (1983)
Indigofera suffruticosa[c]	Fabaceae	PY	Felippe and Polo (1983)
Ipomoea acuminata[c]	Convolvulaceae	PY[d]	Felippe and Polo (1983)
I. coccinea[c]	Convolvulaceae	PY	Felippe and Polo (1983)
I. cynanchifolia[c]	Convolvulaceae	PY	Felippi and Polo (1983)
Lagascea mollis[c]	Asteraceae	PD[d]	Popay (1973)
Launaea cornuta	Asteraceae	PD[d]	Popay (1973)
Leonotis nepetaefolia[c]	Lamiaceae	PD	Felippe and Polo (1983)
Malvastrum coromandelianum[c]	Malvaceae	PY	Felippe and Polo (1983)
Mimosa invisa[c]	Fabaceae	PY	Felippe and Polo (1983)
Mitracarpus hirtus	Rubiaceae	ND	Felippe and Polo (1983)
Oxygonum sinuatum[c]	Polygonaceae	PD	Popay (1973, 1975)
Phyllanthus corcovadensis[c]	Euphorbiaceae	ND	Felippe and Polo (1983)
Porophyllum ruderale[c]	Asteraceae	ND	Felippe and Polo (1983)
Portulaca oleracea[c]	Portulacaceae	PD[d]	Popay (1973)
Rhynchelytrum repens[c]	Poaceae	PD[d]	Popay (1973)
Ricinus communis[c]	Euphorbiaceae	PD	Felippe and Polo (1983)
Rottboellia exaltata[c]	Poaceae	PD	Popay (1973)
Sida cordifolia[c]	Malvaceae	PY	Felippe and Polo (1983)
S. rhombifolia[c]	Malvaceae	PY	Felippe and Polo (1983)
S. spinosa[c]	Malvaceae	PY	Felippe and Polo (1983)
Solanum americanum	Solanaceae	PD	Felippe and Polo (1983)
Tagetes minuta[c]	Asteraceae	ND	Felippe and Polo (1983)
Tribulus terrestris[c]	Zygophyllaceae	PD	Popay (1973)
Trichodesma zeylanicum[c]	Boraginaceae	PD	Popay (1974)
Tridax procumbens[c]	Asteraceae	PD[d]	Popay (1973)
Wissadula subpeltata	Malvaceae	PY	Felippe and Polo (1983)

[a]MPD, morphophysiological dormancy; ND, nondormant; PD, physiological dormancy; PY, physical dormancy.

[b]An asterisk means that the type of dormancy is inferred from available information on germination and on characteristics of seeds in that family.

[c]Listed as a weed by Holm *et al.* (1979).

[d]15–50% of the seeds are nondormant.

and 36 to 16 other plant families. In all, 110 species of beetles were reared from predated seeds, and 75% of them were host specific. Beetles belonged to the Bruchidae, Curculionidae, and Cerambycidae, with the Bruchidae having the most species and thus killing the most seeds (Janzen, 1980). Predispersal predation by bruchid beetles has been studied in detail in a number of plants, including *Cassia grandis* (Janzen, 1971a) *C. biflora* (Silander, 1978), and *Guazuma ulmifolia* (Janzen, 1975).

Seeds of many legumes contain toxic nitrogenous compounds such as alkaloids and nonprotein amino acids (Janzen, 1969). One of the toxic nonprotein amino acids that has received much attention is L-canavanine,

a structural analog of L-arginine. This compound is distributed widely in the Papilionoideae (a subfamily of the Fabaceae) and has been studied in detail in seeds of the legume *Dioclea megacarpa* (Rosenthal, 1977). *Dioclea megacarpa* grows in tropical deciduous forests in Costa Rica, and its seeds are eaten by the bruchid beetle *Caryedes brasiliensis* (Janzen, 1980). Although seeds of *D. megacarpa* contain 8–12% canavanine by dry weight (Rosenthal *et al.*, 1977), larvae of *C. brasiliensis* feed exclusively on seeds of this legume (Rosenthal *et al.*, 1976). There are two reasons why larvae of *C. brasiliensis* are not killed by canavanine. (1) Arginyl transfer RNA synthetase in the larvae attaches to arginine but not to canavanine. Thus, the production of aberrant canavanine-containing proteins is avoided; such proteins are disfunctional (Rosenthal, 1976). Many insects are killed by canavanine because their arginyl transfer RNA synthetase cannot discriminate between arginine and canavanine; consequently, canavanine becomes incorporated into proteins, making them nonfunctional (Rosenthal et al., 1976; Rosenthal, 1983, 1988). (2) Larvae of C. brasiliensis break down canavanine and use nitrogen from it to construct various protein amino acids, like the plant D. megacarpa does (Rosenthal et al., 1977, 1982; Rosenthal, 1983). Arginase breaks canavanine into canaline and urea, and urease converts urea into carbon dioxide and ammonia. The larve have an enzyme that breaks the toxic canaline into homoserine and ammonia. Some ammonia, which also is toxic, is excreted, but some is added to glutamic acid to form glutamine, an important protein amino acid.

In addition to members of the Bruchidae, Curculionidae, and Cerambycidae, other insects, including a chalcid wasp, a bostrychid beetle, sucking bugs, weevils, and agaonid wasps (Janzen, 1979), destroy seeds in tropical deciduous forests in Costa Rica. Also, parrots (Janzen, 1980), squirrels (Janzen, 1971b), and tapiers (Janzen, 1981b) eat seeds.

When animals eat seeds, some, or all, of them may pass through the digestive system unharmed; thus, the animal serves as a dispersal agent. However, after seeds have been dispersed by an animal (or after they simply have fallen to the ground), they can be attacked and eaten. Seeds of Sterculia apetala were dispersed by gravity, squirrels, and monkeys. If seeds of this species fell beneath the mother plant, they were predated by cottonstrainer bugs (Dysdercus fasciatus). However, if seeds were dispersed 30–60 m from S. apetala trees by animals, the bugs did not find them (Janzen, 1972). Seeds of Cecropia peltata, Chlorophora tinctoria, and Piper amalago placed on the ground in a tropical deciduous forest in Costa Rica were removed by ants (especially Pheidole sp.) and rodents (primarily Liomys salvini) (Perry and Fleming, 1980). These data indicate that ants and rodents are important predators of small seeds after they are dispersed. Seeds of Enterolobium cyclocarpum were removed from piles of horse dung in Guanacaste, Costa Rica, by rodents (Liomys salvini and Sigmodon hispidus). Seeds were less likely to be eaten if dung piles were in grassy openings than in the forest and if they contained a small rather than a large number of seeds (Janzen, 1982b).

The main animal dispersers of Acacia faresiana seeds in Guanacaste are deer, horses, and ctenosaur lizards. Many unharmed seeds of A. faresiana were deposited in dung piles of these animals, but those on the surface of the piles were predated by the bruchid beetle Stator vachelliae (Traveset, 1990). Seeds of Andira inermis were dispersed from the mother tree by bats. However, after seeds were on the ground they were subject to predation by Cleogonus sp. weevils. Rates of predation were high under seed-bearing trees of A. inermis, intermediate under bat roosts, and low for seeds dropped as bats flew to their roosts (Janzen et al., 1976).

VI. TROPICAL DRY WOODLANDS, NATURAL SAVANNAS, AND GRASSLANDS

Walter (1973) defined savannas as "... homogenous grasslands upon which woody plants are more or less evenly distributed." If only a few scattered trees are present among grasses, the community may be a grass savanna, but if many trees are present it could be called a tree savanna (Bourliere and Hadley, 1983). Thus, dry woodlands and grasslands are a variation on the savanna theme. Savannas are found in Central and South America, Africa, India, southeast Asia, and Australia. Trees may be either evergreen or deciduous, and grasses are tall (up to 300–360 cm) and luxuriant, e.g., elephant grass (Pennisetum spp.) in savannas north of the equator in Africa, or short, sparse, and xerophilous, e.g., the Kalahari region in southern Africa (Eyre, 1963). The most famous savanna tree probably is the baobab (Adonsonia digitata), which actually grows throughout tropical Africa, not just in savannas (Steentoft, 1988).

Like tropical deciduous forests, savannas have a summer-wet and a winter-dry season. The dry season in savannas may last for 2.5–5 months in a "wet savanna" or 7.5–10 months in a "dry savanna" (Walter, 1979); however, total annual rainfall is important in determining the kind of vegetation present. On sandy loam soils in southwestern Africa, short tufted grasses grow in areas with 100 mm of rainfall (Walter, 1979). Tall grasses occur in areas with 200 mm of rainfall, and grasses and scattered shrubs (shrub savanna) are found in regions with 300 mm. Regions with 400 mm of rainfall support grasses, shrubs, and scattered trees (tree savanna). Lianas and epiphytes are rare in savannas (Sarmiento and Monasterio, 1983).

Edaphic savannas occur in Central Brazil (Sarmiento, 1983), near the Orinoco River in Colombia (Blydenstein, 1967), and in Venezuela (Meyers, 1933), Trinidad, British Honduras, Surinam (Eyre, 1963), and Africa (Walter, 1979). A hard lateritic crust (arecife) is formed in savanna soils that slows, but does not completely prevent, the drainage of rain water. Grasses have only the limited water supplies in the soil above the hard pan during the dry season, but tree roots that have grown through breaks in the arecife reach ample water supplies below it (Walter, 1979).

TABLE 9.10 Fifteen Phenological Patterns for Plants in Tropical Savannas[a]

Number	Time of carbon assimilation	Evergreen[b]	Life cycle[c]	Flowering time
1	All year	−	P	Beginning of wet season
2	All year	−	P	Middle of wet season
3	All year	−	P	Dry season
4	All year	+	P	All year
5	All year	−	P	Opportunistic
6	Seasonal	+	P	Beginning of wet season
7	Seasonal	+	P	Middle of wet season
8	Seasonal	+	P	Dry season
9	Seasonal	−	P	Beginning of wet season
10	Seasonal	−	P	Middle of wet season
11	Seasonal	−	P	Dry season
12	Seasonal	−	A	Beginning of wet season
13	Seasonal	−	A	Middle of wet season
14	Seasonal	−	A	Dry season
15	Seasonal	−	A	Opportunistic

[a]Described by Samiento and Monasterio (1983).
[b]−, no; +, yes.
[c]P, perennial; A, annual.

Sarmiento and Monasterio (1983) described 15 different phenological patterns for plants in tropical savannas (Table 9.10). Flowering (and presumably seed set) occurs in eight of the types during the wet season, four in the dry season, one all year, and two at variable times during the year, whenever conditions become favorable for growth. Evergreen trees, such as *Curatella americana* and *Byrsonima crassifolia* in the Venezuelan Llanos, may not flower until the dry season (Monasterio and Sarmiento, 1976), because anaerobic conditions due to the high water table under the arecife in the wet season are inhibitory (Foldats and Rutkis, 1975). Dominant grasses and sedges in the Venezuelan Llanos, as well as many of the ephemeral herbs, flower and fruit in the wet season (Monasterio and Sarmiento, 1976). The number of flowering species increases greatly in Nigerian savannas with the onset of the wet season and peaks from mid-October to mid-November, when most grasses flower; very few species flower during the winter dry season (Hopkins, 1968). Most herbaceous species, including grasses, in India flower and produce seeds in the wet season, and their shoots are dry by October (Misra, 1983).

A. Dormancy and Germination of Trees

The germination of savanna trees has received little research attention until relatively recently, because it was thought that trees reproduced primarily by vegetative means and that establishment from seeds was a rare event (Sarmiento and Monasterio, 1983). However, seedlings of 50 woody species were found in the cerrados

(savannas) of Brazil (Labouriau *et al.*, 1964), and 64% of the seedlings of *Kielmeyera coriacea* that germinated in the cerrados of central Brazil were alive after 5 years (Oliveira and Silva, 1993).

Seeds of some savanna trees are nondormant (Table 9.11), and those of *Hancornia speciosa* from the cerrados of Brazil are recalcitrant (Oliveira and Valio, 1992). Fresh seeds of *H. speciosa* have a high (54.7%) moisture content, and they lose viability if it drops below 25%. Oliveira and Valio (1992) speculated that the species survives in a seasonally dry habitat because some seeds are matured (and they germinate) during the wet season. Also, some seeds dispersed in the dry season may become buried before they dry to the critical moisture level.

Seeds of most savanna trees are dormant, and physical dormancy is the most common type (Table 9.11). Morphophysiological and physiological dormancies are present in seeds of some species, but little is known about their dormancy-breaking requirements. Physical dormancy has been broken by acid (Gill and Bamidele, 1981; Hashim, 1990; Choinski and Tuohy, 1991) and mechanical (Kariuki and Powell, 1988) scarification, boiling water (Brown and van Booysen, 1969; Hussain and Ilahi, 1988a), and dry heat (Brown and van Booysen, 1969; Gill *et al.*, 1981; Mucunguzi and Oryem-Origa, 1996). Further, the germination of *Dipteryx alata* (Melhem, 1975) and *Stryphnodendron barbadetimam* (Barradas and Handro, 1974) increased following a period of dry storage.

Little is known about the loss of physical dormancy in the natural habitat; however, seeds of *Acacia modesta*

TABLE 9.11 Types of Seed Dormancy in Trees of Tropical Dry Woodlands,
Natural Savannas, and Grasslands

Species	Family	Type of dormancy[a,b]	Reference
Acacia abyssinica	Fabaceae	PY*	Teketay (1993a)
A. albida	Fabaceae	PY	Hashim (1990), Hauser (1994)
A. aneura	Fabaceae	PY*	Jurado and Westoby (1992)
A. estrophiolata	Fabaceae	PY*	Jurado and Westoby (1992)
A. farnesiana	Fabaceae	PY*	Jurado and Westoby (1992)
A. gerrardii	Fabaceae	PY	Mucunguzi and Oryem-Origg (1996)
A. karro	Fabaceae	PY	Choinski and Tuohy (1991)
A. modesta	Fabaceae	PY	Hussain and Ilahi (1988a)
A. polyacantha	Fabaceae	PY	Hashim (1990)
A. seyal	Fabaceae	PY	Hashim (1990), Teketay (1996b)
A. sieberiana	Fabaceae	PY	Hashim (1990), Teketay (1996b)
A. tortilis	Fabaceae	PY	Choinski and Tuohy (1991), Teketay (1996b)
A. xanthophloea	Fabaceae	PY	Kariuki and Powell (1988)
Adansonia digitata	Bombacaceae	PY	Muthana *et al.* (1980)
Albizia amara	Fabaceae	PY	Hashim (1990)
Annona coriacea	Annonaceae	MPD*	Ferri (1959), Rizzini (1965)
Aspiodosperma tomentosum	Apocynaceae	ND*	Ferri (1959)
Bombax gracilipes	Bombacaceae	ND*	Ferri (1959)
B. martianum	Bombacaceae	ND*	de Melo *et al.* (1979)
B. tomentosum	Bombacaceae	ND*	de Melo *et al.* (1979)
Bowdichia virgilioides	Fabaceae	PY	Rizzini (1965)
Brachystegia spiciformis	Fabaceae	PY, ND	Kariuki and Powell (1988), Ernst (1988)
Caryocar brasiliense	Caryocaraceae	PD*	Rizzini (1965)
Casearia sylvestris	Flacourtiaceae	PD*	Rizzini (1965)
Cassia mystacicarpa	Fabaceae	ND*[c]	Rizzini (1965)
C. sieberiana	Fabaceae	PY	Gill *et al.* (1982)
Colophospermum mopane	Fabaceae	PY*	Choinski and Tuohy (1991)
Combretum apiculatum	Combretaceae	ND	Choinski and Tuohy (1991)
Copaifera oblongifolia	Fabaceae	ND*[c]	Rizzini (1965)
Croton macrostachyus	Euphorbiaceae	PD*	Teketay (1993a)
Dalbergia violacea	Fabaceae	ND*[c]	Arasaki and Felippe (1987)
Delonix regia	Fabaceae	PY	Gill *et al.* (1981)
Dialium guineense	Fabaceae	PY	Gill and Bamidele (1981)
Dimorphandra mollis	Fabaceae	PY*	Rizzini (1965)
Dipteryx alata	Fabaceae	PY	Melhem (1975)
Entada abyssinica	Fabaceae	PY	Teketay (1966a)
Enterolobium ellipticum	Fabaceae	PY*	Rizzini (1965)
Erythrina lysistemon	Fabaceae	PY	Teketay (1996a)

(continues)

TABLE 9.11—*Continued*

Species	Family	Type of dormancy[a,b]	Reference
E. mulungu	Fabaceae	PY*	Rizzini (1965)
Eucalyptus camaldulensis	Myrtaceae	ND	Yap and Wong (1983)
E. microtheca	Myrtaceae	ND*, PD*	Ismail *et al.* (1986)
Grevillea pteridifolia	Proteaceae	ND	Bahuguna and Lal (1994)
Hancornia speciosa	Apocynaceae	ND	Oliveira and Valio (1992)
Kielmeyera coriacea	Clusiaceae	ND*	de Melo *et al.* (1979), Oliveira and Silva (1993)
K. speciosa	Clusiaceae	ND*	Oliveira and Silva (1993)
Magonia pubescens	Sapindaceae	PY*, ND*	Joly *et al.* (1980)
Maytenus arbutifolia	Celastraceae	PD*	Teketay (1993a)
Maesa lanceolata	Myrsinaceae	PD*	Teketay (1993a)
Mimosa clausenii	Fabaceae	ND*c	Rizzini (1965)
M. laticifera	Fabaceae	ND*c	Rizzini (1965)
Olea africana	Oleaceae	MPD*	Teketay (1993a)
Parkia auriculata	Fabaceae	PY	Coutinho and Struffaldi (1971)
P. clapertoniana	Fabaceae	ND, PY	Gill and Bamidele (1981), Etejere *et al.* (1982)
Pinus caribaea	Pinaceae	ND	Yap and Wong (1983)
Piptadenia communis	Fabaceae	ND*c	Rizzini (1965)
P. falcata	Fabaceae	ND*, PY*	de Melo *et al.* (1979)
Pittosporum phylliraeoides	Pittosporaceae	MPD*	Jurado and Westoby (1992)
Plathymenia foliolosa	Fabaceae	ND*c	Rizzini (1965)
P. reticulata	Fabaceae	ND*c	Rizzini (1965)
Pouteria torta	Sapotaceae	PD*	Rizzini (1965)
Prosopis africana	Fabaceae	PY	Eze and Orole (1987), Hashim (1990)
Pterodon pubescens	Fabaceae	ND*, PY*	de Melo *et al.* (1979)
Qualea grandiflora	Vochysiaceae	PD*	de Melo *et al.* (1979)
Rapanea guianensis	Myrsinaceae	PD*	Joly and Felippe (1979a)
Reptonia buxifolia	Sapotaceae	PD*	Ilahi and Hussain (1988)
Sclerolobium aureum	Fabaceae	ND*c	Rizzini (1965)
S. paniculatum	Fabaceae	ND*c	Rizzini (1965)
Stryphnodendron adstringens	Fabaceae	ND*	Ferri (1959)
S. barbadetimam	Fabaceae	PY, ND	Barradas and Handro (1974)
Terminalia argentea	Combretaceae	PD*	Rizzini (1965)
Thieleodoxa lanceolata	Rubiaceae	PD*	Rizzini (1965)
Trachylobium verrucosum	Fabaceae	PY	Kariuki and Powell (1988)
Vochysia thyrsoidea	Vochysiaceae	PD*	Rizzini, 1965
Zeyhera digitalis	Bignoniaceae	ND*	Joly and Felippe (1979b)

[a]MD, morphological dormancy; MPD, morphophysiological dormancy; ND, nondormant; PD, physiological dormancy; PY, physical dormancy.
[b]An asterisk means that the type of dormancy is inferred from available information on germination and on characteristics of seeds in that family.
[c]It seems logical that freshly matured seeds of this species would have impermeable seed coats and thus physical dormancy. However, because no dormancy-breaking treatment was reported, we must assume that seeds were nondormant.

germinate during the monsoon (wet) season in Pakistan (Hussain and Ilahi, 1988a), indicating that dormancy was broken during the hot, dry season. Also, 167 seeds m^{-2} of *A. sieberiana* germinated after fires in savannas in Uganda (Sabiiti and Wein, 1987). Heat from fires can make seeds of *A. sieberiana* permeable and kill bruchid beetle larvae inside them (Sabiiti and Wein, 1987). Seeds of *Acacia albida* with bruchid beetle exit holes in them germinated if placed on a moist substrate before they were attacked by other insects or fungi (Hauser, 1994). Thus, biotic factors may play a role in the germination of some seeds.

Several trees listed in Table 9.11 belong to the Fabaceae, but they are listed as having nondormant seeds. It seems logical that seeds of these species would have physical dormancy, but the researchers did not mention any dormancy-breaking treatments. Thus, we have to ask if (1) the seeds were collected before the seed coats became impermeable, (2) the seed coats were scarified (or heat treated) to make them permeable before seeds were sown, or (3) the seed coats never become impermeable, as in seeds of the legumes *Pentaclethra* and *Inga* (Janzen, 1969).

Temperature and light requirements for germination have been determined for nondormant seeds of several species of trees. In a review of germination studies on cerrado species, Felippe and Silva (1984) found that constant temperatures of 25–30°C usually were optimal; germination percentages began to decline above 35°C and below 20°C. Alternating temperature regimes with a daily maximum of 25–35°C and a minimum of 15–20°C promoted high germination percentages. In *Magnoia pubescens*, a daily maximum temperature of 25°C combined with a minimum temperature of 5 or 10°C stimulated the same percentage (90) of seeds to germinate as an alternating temperature regime of 30/25°C (Joly et al., 1980).

Optimum seed germination temperatures for other trees are: *Acacia karoo*, 30°C (Choinski and Tuohy, 1991); *A. modesta*, 35°C (Hussain and Ilahi, 1988a); *A. tortilis*, 25°C (Choinski and Tuohy, 1991); *Andira humilis*, 35°C (Rizzini, 1970); *Colophospermum mopane*, 25°C (Choinski and Tuohy, 1991); *Combretum apiculatum*, 30°C (Choinski and Tuohy, 1991); *Dipteryx alata*, 30–35°C (Melhem, 1975); *Entada abyssinica*, 30°C (Teketay, 1996a); *Hancornia speciosa*, 25°C (Oliveira and Valio, 1992); *Rapanea guianensis*, 25/5°C (Joly and Felippe, 1979a); *Reptonia buxifolia*, 30°C (Ilahi and Hussain, 1988); and *Stryphnodendron barbadetimam*, 26–34°C (Barradas and Handro, 1974).

Seeds of various species, including *Dipteryx alata* (Melhem, 1975), *Magnoia pubescens* (Joly et al., 1980), *Zornia reticulata* (Felippe and Silva, 1984), *Dalbergia violacea* (Arasaki and Felippe, 1987), *Rapanea guianensis* (Joly and Felippe, 1979a), *Stryphnodendron barbadetimam* (Barradas and Handro, 1974), and *Gomphrena macrocephala* (Felippe and Silva, 1984), germinated equally well in light and darkness at temperatures of 25–30°C. However, seeds of *Z. reticulata* germinated faster in darkness than in light at 25°C (Felippe and Silva, 1984), and those of *D. violacea* germinated faster in darkness than in light at constant temperatures of 15–30°C (Arasaki and Felippe, 1987). Seeds of *Zeyhera digitalis*, however, germinated to higher percentages in darkness than in light at both 25 and 30°C (Joly and Felippe, 1979b).

Germination of tree seeds at the beginning of the wet season (Hussain and Ilahi, 1988a; Oliveira and Silva, 1993) emphasizes the importance of information on the sensitivity of nondormant seeds to moisture stress in understanding the timing of germination. Seeds of the African savanna trees *Colophospermum mopane* and *Combretum apiculatum* germinated at a water potential of −1.03 MPa, whereas the lowest water potential for the germination of *Acacia tortilis* and *A. karro* seeds was −0.51 MPa (Choinski and Tuohy, 1991). A 24-hr water imbibition pretreatment increased the germination of *C. mopane*, *A. tortilis*, and *A. karro* seeds at −0.51 MPa from 29 to 100%, 3 to 50%, and 0 to 90%, respectively. Water imbibition did not increase stress tolerance in *C. apiculatum* seeds (Choinski and Tuohy, 1991). More studies need to be done on the responses of nondormant seeds to varying water potentials.

B. Dormancy and Germination of Shrubs

Although some species have nondormant seeds, many of them have dormant seeds (Table 9.12). Both physiological and physical dormancies are present, but physical dormancy is the most common. Because few of the seeds with physiological dormancy have been studied in detail, little is known about their dormancy-breaking requirements. Seeds of *Bassia birchii* germinate in spring in western New South Wales, Australia, after being exposed to relatively low temperatures during winter (Auld, 1976).

Seeds (drupes) of the undershrub *Andira humilis* from Brazilian cerrados are dormant at maturity. Scarification of the water-permeable endocarps of this species stimulated germination within a few days; however, incubation at a constant temperature of 35°C promoted dormancy loss and germination (Rizzini, 1970). The drupe-like fruits of *Eromophila* have a hard woody endocarp and may contain more than one seed (Richmond and Chinnock, 1994). Optimum temperature regimes for the germination of *Eromophila* seeds were 25/5 and 25/11°C (Richmond and Chinnock, 1994). Although Richmond and Chinnock (1994) speculated that the fruits have physical dormancy, imbibition of scarified and nonscarified fruits has not been compared. Thus, available information indicates that the fruits probably have physiological dormancy. Richmond and Ghisalberti (1994) found water-soluble aromatic glycosides in the fruit walls, which inhibited germination. Thus, this may be a good genus in which to investigate the role of chemical inhibitors in the germination of physiologically dormant vs nondormant seeds (see Chapter 3).

The density of *Miconia albicans* and *Clidemia sericea* seedlings peaked the second and third years following savanna fires in Belize. However, increased germination

TABLE 9.12   Types of Seed Dormancy in Shrubs of Tropical Dry Woodlands,
Natural Savannas, and Grasslands

Species	Family	Type of dormancy[a,b]	Reference
Acacia brevispica	Fabaceae	PY	Teketay (1996a)
A. caven	Fabaceae	PY	Bahuguna and Lal (1993)
A. kempeana	Fabaceae	PY*	Jurado and Westoby (1992)
A. ligulata	Fabaceae	PY*	Jurado and Westoby (1992)
A. oerfta	Fabaceae	PY	Teketay (1996a)
A. senegal	Fabaceae	ND	Danthu *et al.* (1992)
A. seyal	Fabaceae	PY	Teketay (1996a)
A. tetragonophylla	Fabaceae	PY*	Jurado and Westoby (1992)
A. victoriae	Fabaceae	PY*	Jurado and Westoby (1992)
Albizia anthelmintica	Fabaceae	PY	Hashim (1990)
Andira humilis	Fabaceae	PY	Rizzini (1970)
Anemopaegma arvense	Bignoniaceae	ND*	Rizzini (1965)
Astronium fraxinifolium	Anacardiaceae	PY*	Rizzini (1965), de Melo *et al.* (1979)
A. urunduna	Anacardiaceae	PY*, ND*	de Melo *et al.* (1979)
Atalaya hemiglauca	Sapindaceae	PY*	Jurado and Westoby (1992)
Bassia birchii	Chenopodiaceae	PD*	Auld (1976)
Bauhinia rufescens	Fabaceae	PY	Hashim (1990)
Cadia purpurea	Fabaceae	PY	Teketay (1996a)
Calpurnia aurea	Fabaceae	PY*	Teketay (1993a)
Cassia artemisioides	Fabaceae	PY*	Jurado and Westoby (1992)
C. helmsii	Fabaceae	PY*	Jurado and Westoby (1992)
C. nemophila	Fabaceae	PY*	Jurado and Westoby (1992)
Chrysophyllum soboliferum	Sapotaceae	PD*	Rizzini (1965)
Clidemia sericea	Melastomataceae	ND	Miyanishi and Kellman (1986)
Combretum bauchinse	Cambretaceae	ND	Onyekwelu (1990)

(continues)

was not attributed to direct effects of fire on seeds but to an increase in light and bare mineral soil (Miyanishi and Kellman, 1986). Hot water broke physical dormancy in seeds of *Acacia caven* (Bahuguna and Lal, 1993), and acid scarification broke it in those of *A. senegal* (Danthu *et al.,* 1992); however, little is known about how physical dormancy is broken in seeds of these species in the field.

C. Dormancy and Germination of Bamboos

Intact caryopses of *Actinocladum verticillatum,* which is endemic to the Brazilian cerrados, were dormant at maturity. When the pericarp was removed by hand from the caryopses, seeds germinated to about 39 and 13% in light and darkness, respectively (Felippe and Filgueiras, 1986). Thus, caryopses of *A. verticillatum* appear to have physiological dormancy; dormancy-breaking requirements are unknown.

D. Dormancy and Germination of Herbaceous Species

A few species have nondormant seeds or seeds with morphophysiological dormancy, but most have seeds with either physical or physiological dormancy (Table 9.13). Although some species are listed as having nondormant seeds, further studies may show that freshly matured seeds of additional species also are nondormant. For example, Jurado and Westoby (1992) did an extensive study of 105 species in the central Australian arid zone, and from their results it appears that 23 of them have nondormant seeds. However, these authors stored the seeds for 3 months at room temperatures (except for a 24-hr drying period at 50°C) before they tested them for germination. Thus, seeds of some species may have been dormant initially and afterripened during dry storage before germination tests were con-

TABLE 9.12—*Continued*

Species	Family	Type of dormancy[a,b]	Reference
Dalbergia melanoxylon	Fabaceae	PY	Hashim (1990)
Dodonea angustifolia	Sapindaceae	PY*	Teketay (1993a)
D. viscosa	Sapindaceae	PY*	Qadir and Lodhi (1971), Hussain *et al.* (1991)
Enchylaena tomentosa	Chenopodiaceae	PD*	Jurado and Westoby (1992)
Eremophila gilesii	Myoporaceae	PD*	Richmond and Chinnock (1994)
E. maculata	Myoporaceae	PD*	Richmond and Ghisalberti (1994)
E. racemosa	Myoporaceae	PD*	Richmond and Ghisalberti (1994)
Erythroxylum ovalifolium	Erythroxylaceae	ND*	Fialho and Furtado (1993)
Esenbeckia pumila	Rutaceae	ND*	Rizzini (1965)
Indigofera rothii	Fabaceae	PY*	Teketay (1993a)
Leucaena leucocephala	Fabaceae	PY	Teketay (1996a)
Maireana scleroptera	Chenopodiaceae	PD*	Jurado and Westoby (1992)
Miconia albicans	Melastomataceae	ND	Miyanishi and Kellman (1986)
Parinarium obtusifolium	Rosaceae	PD*	Rizzini (1965)
Parkinsonia aculeata	Fabaceae	PY	Teketay (1996a)
Peschiera affinis	Apocynaceae	PD*	Rizzini (1965)
Piliostigma reticulatum	Fabaceae	PY	Hashim (1990)
Sclerolaena bicornis	Chenopodiaceae	PD*	Jurado and Westoby (1992)
Senna multiglandulosa	·Fabaceae	PY	Teketay (1996b)
S. occidentalis	Fabaceae	PY	Teketay (1996b)
S. septemtrionalis	Fabaceae	PY	Teketay (1996b)
Spiranthera odoratissima	Rutaceae	PD*	Rizzini (1965)
Stryphnodendron confertum	Fabaceae	ND*[c]	Rizzini (1965)
Ventilago viminalis	Rhamnaceae	PY*	Jurado and Westoby (1992)

[a]ND, nondormant; PD, physiological dormancy; PY, physical dormancy.

[b]An asterisk means that the type of dormancy is inferred from available information on germination and on characteristics of seeds in that family.

[c]It seems logical that fresh matured seeds of this species would have impermeable seed coats and thus physical dormancy. However, because no dormancy-breaking treatment was reported, we must assume that seeds were nondormant.

ducted. Species from their study were included in Table 9.13 only if (1) germination was ≤20% for the 3-month-old seeds and (2) seeds germinated to ≥50% following scarification or removal of dispersal structures.

Nondormant seeds may germinate in the same wet season in which they are produced. For example, seeds of *Sporobolus cubensis, Leptocoryphium lanatum,* and *Elyonurus adustus* are produced in the Venezuelan Llanos early in the wet season and germinate in that same season, as soon as they are dispersed (Silva and Ataroff, 1985). In contrast, seeds of *Andropogon semiberbis, Trachypogon plumosus,* and *Axonopus canescens* are produced in one wet season, but do not germinate until the beginning of the following wet season (Silva and Ataroff, 1985). These seeds have physiological dormancy at maturity and come out of dormancy during the dry season (Silva and Ataroff, 1985). Seeds of *Cynodon*

dactylon were nondormant at maturity in savannas of Botswana, but they were not dispersed until much of the dry season had passed. Thus, seeds were prevented from germinating until the wet season began (Ernst *et al.*, 1991; Veenendaal and Ernst, 1991).

Freshly matured seeds of numerous species have physiological dormancy, which is broken during the dry season (Andrew and Mott, 1983; Ernst *et al.*, 1991; Veenendaal and Ernst, 1991). Seeds of *Themeda australis, Chrysopogon fallox, Sorghum plumosum,* and *S. stipoideum* came out of dormancy during dry storage in the laboratory at a daily thermoperiod of 60/20°C and on the soil surface in the field in Australia (Mott, 1978). Seeds of all species, except possibly *T. australis,* came out of dormancy at 25°C, but rates of dormancy loss were slower at 25°C than at 60/22°C or in the field. GA_3 promoted germination, but cold stratification did not

TABLE 9.13 Dormancy Type and Optimum Temperature and Light: Dark (L:D) Requirements for Germination of Seeds of Herbaceous Species of Tropical Dry Woodlands, Natural Savannas, and Grasslands

Species[a]	Family	Optimum germination temperature (°C)	L:D requirement[b]	Reference
Morphophysiological dormancy				
Bulbine bulbosa*	Liliaceae	20/10	L = D	Willis and Groves (1991)
Nondormant				
Combretum bauchiense	Combretaceae	—[c]	—	Onyekwelu (1990)
Cynodon dactylon[d]	Poaceae	—	—	Veenendaal and Ernst (1991)
Digitaria ciliaris*,[d]	Poaceae	20–35	—	Torssell and McKeon (1976), Watt and Whalley (1982)
Elyonurus adustus	Poaceae	—	—	Silva and Ataroff (1985)
Leptocoryphium lanatum	Poaceae	—	—	Silva and Ataroff (1985)
Sporobolus cubensis	Poaceae	—	—	Silva and Ataroff (1985)
Physical dormancy				
Cassia alata[d]	Fabaceae	30	L = D	Gill et al. (1982)
C. hirsuta[d]	Fabaceae	30	L = D	Gill et al. (1982)
C. multijuga	Fabaceae	30	L = D	Gill et al. (1982)
C. nodosa	Fabaceae	30	L = D	Gill et al. (1982)
C. occidentalis[d]	Fabaceae	30	L = D	Gill et al. (1982)
C. spectabilis	Fabaceae	30	L = D	Gill et al. (1982)
Convolvulus remotus*	Convolvulaceae	12–28	—	Jurado and Westoby (1992)
Evolvulus alsinoides*,[d]	Convolvulaceae	12–28	—	Jurado and Westoby (1992)
Glycine tomentella	Fabaceae	35/25	—	Silcock et al. (1990)
Ipomoea lonchophylla	Convolvulaceae	35/25	—	Silcock et al. (1990)
Lotus cruentus*	Fabaceae	28	—	Jurado and Westoby (1992)
Malvastrum americanum*,[d]	Malvaceae	28	—	Jurado and Westoby (1992)
Muelleranthus trifoliolatus	Fabaceae	35/25	—	Silcock et al. (1990)
Psoralea eriantha	Fabaceae	35/25	—	Silcock et al. (1990)
P. patens*	Fabaceae	28	—	Jurado and Westoby (1992)
Rhynchosia minima[d]	Fabaceae	35/25	—	Silcock et al. (1990)
Stylosanthes humilis	Fabaceae	25	—	Torssell and McKeon (1976)
Swainsona canescens*	Fabaceae	12–28	—	Jurado and Westoby (1992)
Physiological dormancy				
Actinocladum verticillatum	Poaceae	25	L > D	Felippe and Filgueiras (1986)
Andrographis paniculata[d]	Acanthaceae	30	—	Chaudhary (1975)
Andropogon semiberbis	Poaceae	—	—	Silva and Ataroff (1985)
A. gayanus	Poaceae	30–35	L = D	Felippe et al. (1983)
Aristida armata	Poaceae	30/25	—	Brown (1982)
A. congesta[d]	Poaceae	—	—	Veenendaal and Ernst (1991)
A. latifolia*,[d]	Poaceae	20	—	Jurado and Westoby (1992)
A. ramosa	Poaceae	30	—	Lodge and Whalley (1981)
A. stipitata[a]	Poaceae	—	—	Veenendaal and Ernst (1991)
Astrebla lappacea	Poaceae	35/25	—	Silcock et al. (1990)
A. pectinata	Poaceae	12–28	—	Jurado and Westoby (1992)
Atriplex muelleri	Chenopodiaceae	35/25	—	Silcock et al. (1990)
Axonopus canescens	Poaceae	—	—	Silva and Ataroff (1985)
Bothriochloa ewartiana	Poaceae	35/25	—	Silcock et al. (1990)
B. macra	Poaceae	25	L = D	Hagon (1976), Lodge and Whalley (1981)
Brachiaria ruziziensis	Poaceae	25	L = D	Renard and Capelle (1976)
Cenchrus ciliaris[d]	Poaceae	—	—	Hacker (1989)
Chloris scariosa*	Poaceae	20	—	Jurado and Westoby (1992)
C. truncata[d]	Poaceae	20–30	—	Lodge and Whalley (1981)
C. virgata[d]	Poaceae	32	L > D	Mott (1978)
Chrysopogon fallox	Poaceae	32	L > D	Mott (1978)
C. latifolius	Poaceae	32	L > D	Mott (1978)
Cymbopogon jwarancusa	Poaceae	25	—	Ayaz et al. (1992)
C. obtectus	Poaceae	12–28	—	Jurado and Westoby (1992)
Dactyloctenium radulans[d]	Poaceae	35/25	—	Silcock et al. (1990)

(continues)

TABLE 9.13—*Continued*

Species[a]	Family	Optimum germination temperature (°C)	L:D requirement[b]	Reference
Danthonia linkii	Poaceae	25	—	Lodge and Whalley (1981)
Dichanthium sericeum	Poaceae	20	—	Lodge and Whalley (1981), Silcock *et al.* (1990)
Digitaria ammophila	Poaceae	35/25	—	Silcock *et al.* (1990)
D. brownii*	Poaceae	28	—	Jurado and Westoby (1992)
D. diminuta	Poaceae	35/25	—	Silcock *et al.* (1990)
Diplachne fusca*,[d]	Poaceae	12–28	—	Jurado and Westboy (1992)
Echinolaena inflexa*	Poaceae	25	—	Klink (1996)
Enneapogon avenaceus*	Poaceae	12	—	Jurado and Westoby (1992)
E. pallidus*	Poaceae	20	—	Jurado and Westoby (1992)
E. polyphyllus*	Poaceae	16–21	—	Ross (1976)
Enteropogon ramosus*	Poaceae	20, 28	—	Jurado and Westoby (1992)
Eragrostis dielsii*	Poaceae	20	—	Jurado and Westoby (1992)
E. eriopoda	Poaceae	42	—	Ross (1976)
		35/25	—	Silcock *et al.* (1990)
E. leptocarpa*	Poaceae	28	—	Jurado and Westoby (1992)
E. leptostachya[d]	Poaceae	20	—	Lodge and Whalley (1981)
E. setifolia	Poaceae	35/25	—	Silcock *et al.* (1990)
E. superba[d]	Poaceae	—	—	Veenendaal and Ernst (1991)
Eremophila gilesii*	Myoporaceae	25	—	Burrows (1971)
Eriachne mucronata*	Poaceae	35/25	—	Silcock *et al.* (1990)
Erianthus griffithii*	Poaceae	30	—	Ilahi *et al.* (1987)
Erymophyllum ramosum*	Asteraceae	10–20	L > D	Plummer and Bell (1995)
Eulalia trispicata	Poaceae	25	L > D	Pemadasa and Amarasinghe (1982)
E. fulva	Poaceae	35/25	—	Silcock *et al.* (1990)
Eupatorium adenophorum[d]	Asteraceae	25	L	Auld and Martin (1975)
Haloragis odontocarpa	Haloragidaceae	35/25	—	Silcock *et al.* (1990)
Harpagophytum procumbens[d]	Pedaliaceae	—	—	Ernst *et al.* (1988)
H. zeypheri	Pedaliaceae	—	—	Ernst *et al.* (1988)
Helichrysum apiculatum	Asteraceae	20/10	L = D	Willis and Groves (1991)
Helipterum albicans	Asteraceae	20/10	L = D	Willis and Groves (1991)
H. floribundum*	Asteraceae	12	—	Jurado and Westoby (1992)
Heteropogon contortus	Poaceae	30	L = D	Tothill (1977)
Hyptis suaveolens[d]	Lamiaceae	25–30	L	Wulff and Medina (1971), Felippe *et al.* (1983)
Imperata cylindrica[d]	Poaceae	30	L > D	Lock and Melburn (1971), Dickens and Moore (1974)
Iseilema vaginiflorum	Poaceae	35/25	—	Silcock *et al.* (1990)
Lepidium muelleri-ferdinandi*	Brassicaceae	16	—	Ross (1976)
		28	—	Jurado and Westoby (1992)
Leptorhynchos squamatus	Asteraceae	20/10	L = D	Willis and Groves (1991)
Mollugo cerviana*,[d]	Aizoaceae	51	—	Jurado and Westoby (1992)
Monachather paradoxa	Poaceae	35/25	—	Silcock *et al.* (1990)
Panicum decompositum	Poaceae	35/25	—	Silcock *et al.* (1990)
Paspalidium rarum	Poaceae	35/25	—	Silcock *et al.* (1990)
Pennisetum polystachyon[d]	Poaceae	25	L = D	Pemadasa and Amarasinghe, 1982
Portulaca filifolia*	Portulacaceae	20	—	Jurado and Westoby (1992)
Rapistrum rugosum[d]	Brassicaceae	10–25	D > L	Cousens *et al.* (1994)
Rhodanthe floribunda*	Asteraceae	20	L > D	Plummer and Bell (1995)
Rhynchelytrum repens[d]	Poaceae	—	—	Vennendaal and Ernst (1991)
Rottboellia exaltata[d]	Poaceae	25	—	Thomas and Allison (1975)
Salsola kali*,[d]	Chenopodiaceae	20	—	Jurado and Westoby (1992)
Schizachyrium tenerum*	Poaceae	25	—	Klink (1996)
Schmidtia pappophoroides	Poaceae	—	—	Veenendaal and Ernst (1991)
Schoenia cassiniana*	Asteraceae	10–20	L > D	Plummer and Bell (1995)
S. filifolia*	Asteraceae	5–20	L = D	Plummer and Bell (1995)
Sorghum intrans	Poaceae	—	—	Andrew and Mott (1983)

(continues)

TABLE 9.13—*Continued*

Species[a]	Family	Optimum germination temperature (°C)	L:D requirement[b]	Reference
S. plumosum[d]	Poaceae	32	L = D	Mott (1978)
S. stipoideum	Poaceae	32	L > D	Mott (1978)
Spergula arvensis*,[d]	Caryophyllaceae	April	—	Birch (1957)
Sporobolus actinocladus	Poaceae	35/25	—	Silcock *et al.* (1990)
S. caroli*	Poaceae	28	—	Jurado and Westoby (1992)
S. elongatus	Poaceae	25	—	Lodge and Whalley (1981)
Stipa bigeniculata	Poaceae	15/5	L = D	Hagon (1976)
S. variabilis	Poaceae	20	—	Lodge and Whalley (1981)
Streptoglossa odora*	Asteraceae	20	—	Jurado and Westoby (1992)
Stylidium graminifolium	Stylidiaceae	20/10	L > D	Willis and Groves (1991)
Themeda australis[d]	Poaceae	20/10	L = D	Hagon (1976), Mott (1978)
T. tremula	Poaceae	25	L = D	Pemadasa and Amarasinghe (1982)
T. triandra[d]	Poaceae	20–25	L > D	Lock and Milburn (1971)
Thyridolepis mitchelliana	Poaceae	35/25	—	Silcock *et al.* (1990)
Trachypogon plumosus	Poaceae	—	—	Silva and Ataroff (1985)
Tragus australianus*	Poaceae	28	—	Jurado and Westoby (1992)
T. berternianus[d]	Poaceae	30/25	—	Ernst and Tolsma (1988)
Tribulus terrestris[d]	Zygophyllaceae	—	—	Ernst and Tolsma (1988)
Urochloa mosambicensis[d]	Poaceae	30/15	—	Harty (1972)
Vernonia galamensis*	Asteraceae	25	L = D	Teketay (1993b)
Vittadinia muelleri	Asteraceae	25/15	L > D	Willis and Groves (1991)
Wahlenbergia stricta	Campanulaceae	15/5	D > L	Willis and Groves (1991)

[a] An asterisk means that the type of dormancy is inferred from available information on germination and on characteristics of seeds in that family.

[b] L, light required for germination; D, darkness required for germination; L > D, seeds germinated to a higher percentage in light than in darkness; D > L, seeds germinated to a higher percentage in darkness than in light; L = D, seeds germinated equally well in light and darkness.

[c] Data not available.

[d] Listed as a weed by Holm *et al.* (1979).

(Mott, 1978). Seeds of *Hyptis suaveolens* came out of dormancy during dry storage at 20°C (Wulff and Medina, 1971).

Seeds of various species came out of dormancy during dry storage at room temperatures: *Aristida congesta, A. stipitata, Chloris virgata, Eragrostis rigidor, E. superba, Heteropogon contortus, Rhynchelytrum repens, Schmidtia pappophoroides,* and *Tragus berteronianus* (Ernst and Tolsma, 1988; Ernst *et al.,* 1991; Veenendaal and Ernst, 1991). Germination percentages of *Aristida ramosa, Bothriochloa macera,* and *Sporobolus elongatus* seeds increased following dry storage at 12–27°C (Lodge and Whalley, 1981). Dormancy was reduced during dry storage at 70°C in seeds of *Aristida armata* (Brown, 1982); 60°C, *Cenchrus ciliaris* (Hacker, 1989); 35°C, *Rapistrum rugosum* (Cousens *et al.,* 1994); 45°C, *Rottboellia exaltata* (Thomas and Allison, 1975); 39/16°C, *Stipa bigeniculata* (Hagon, 1976); and 62/24°C, *Themeda australis* (Hagon, 1976). Cold stratification at 2–4°C reduced dormancy in seeds of *S. bigeniculata, T. australis* (Hagon, 1976), and *A. armata* (Brown, 1982), but induced dormancy in those of *C. ciliaris* (Hacker, 1989) and *R. rugosum* (Cousens *et al.,* 1994).

Seeds of the herbaceous perennials *Harpagophytum procumbens* and *H. zeyheri* from the Kalahari savanna region in Africa had physiological dormancy at maturity, but all attempts to break it, including treatment with GA_3, resulted in only 19% germination. Viable seeds of both species were found in the field in fruits that had been detached from the mother plants for 3 years (Ernst *et al.,* 1988).

The consequence of seeds coming out of dormancy during the dry season is that they are nondormant at the beginning of the wet season. Thus, germination occurs as soon as soil moisture becomes nonlimiting (Silcock, 1973; Chaudhary, 1975; Torsell and McKeon, 1976; Silva and Ataroff, 1985; Jurado and Westoby, 1992; Plummer and Bell, 1995). Seeds of the annual grasses *Tragus berteronianus* and *Urochloa panicoides* and the annual/perennial grass *Aristida congesta* germinate in savannas of Botswana mostly at the beginning and/or in the middle of the wet season. However, the actual time of germination is determined by the date and amount of rainfall. Small rains of 10–12 mm do not stimulate germination, but a large one of 30 mm or more does. Following a heavy rain, seedlings are seen within 3 days, and they continue to appear for 7–10 days. Rains early in the wet season will stimulate part of the seeds to germinate, and then subsequent rains cause others to germinate, resulting in two or three peaks of germination (Veenendaal, 1991). If rains do not come until the third month of the wet season, a good soaking of the soil will cause all the seeds to germinate. In western Victoria, Australia, seeds of *Rutidosis leptorrhynchoides* begin to germinate 8 days after the first autumn rains fall (Morgan, 1995).

Annuals in the Venezuelan Llanos have been divided into two categories: long- and short- (ephemeral) cycle species. Germination of long-cycle annuals, including *Aeschynomene hystrix, Pectis carthusianorum, P. ciliaris, Tephrosia tenella,* and *Borreria suaveolens,* occurs 1–2 months after the beginning of the wet season, and plants live for 6 to 7 months, maturing their seeds early in the dry season (Monasterio and Sarmiento, 1976). Seeds of short-cycle annuals, including *Andropogon brevifolius, Aristida capillacea, Borreria ocimoides, Diectomis fastigiata, Digitaria fragilis, Eragrostis maypurensis, Gymnopogon foliosus, Hyptis suaveolens, Microchloa indica,* and *Polycarpaea corymbosa,* germinate about 3 months after the onset of the wet season, and plants live for only a few weeks, maturing their seeds in the wet season (Monasterio and Sarmiento, 1976). Herbaceous species flower near, or at, the end of the wet season in savannas of northern Cameroon (Seghieri *et al.,* 1995); thus, seeds mature after soil moisture becomes limiting for germination. However, woody species flower during the beginning, middle, or end of the wet season, during the middle of the dry season, or at the end of the dry season through the beginning of the wet season, depending on the species (Seghieri *et al.,* 1995). Thus, seeds of some woody species mature in the wet season, when soil moisture is nonlimiting for germination, but those of others mature in the dry season, when soil moisture stress prevents germination.

The wetting and drying cycles seeds receive prior to the time soil moisture becomes nonlimiting for germination influences germination percentages (Thomas and Allison, 1975; Pemadasa and Amarasinghe, 1982) and rates (McKeon, 1984). Also, the amount and frequency of water applications (Grant, 1975), the presence of mulch and shade (Grant, 1975), and thus duration of water availability (Watt and Whalley, 1982) influences germination. In Sahel rangelands, the first small showers at the beginning of the wet season stimulate seeds of fast-germinating species to germinate, but those of slow-germinating species do not germinate until after there have been several rains. In some years, dry periods following the first rains kill seedlings of the fast-germinating species. Thus, timing of rains, speed of germination, and drought tolerance of seedlings are important in determining which species become established at a given site each year (Elberse and Breman, 1990). The level of moisture stress reducing germination from 100 to 50% varies: *Themeda australis,* −0.03 MPa; *Danthonia* sp., −0.06 MPa; *Bothriochloa macra,* −0.07 MPa; and *Stipa bigeniculata,* −0.09 MPa (Hagon and Chan, 1977).

Water caused long hairs on achenes of *Helichrysum cassinianum* to spread out on the soil surface and mucilage to form a covering layer on achenes of *Helipterum craspedioides* (Mott, 1974). The removal of hairs or mucilage decreased the germination of achenes sown on the soil surface under simulated light rains.

At water potentials slightly below those at which germination occurs, seeds of several grasses, including *Astrebla lappacea, Dichanthium sericeum, D. linkii, Chloris gayana, C. truncata, Setaria porphyrantha, Panicum coloratum, P. decompositum,* and *P. maximum* (Watt, 1974, 1978, 1982; Lambert *et al.,* 1990), imbibe and the coleorhiza and/or coleoptile emerges. However, the radicle and/or shoot does not break through the coleorhiza and/or coleoptile, respectively; this partially germinated state is called hydropedesis (Watt, 1974). Some hydropedetic seeds can be dried and then rehydrated without a loss of viability, and sometimes there is an increase in the germination rate (Watt, 1978). The ability of hydropedetic seeds to tolerate dehydration varies with the species (Watt, 1982). Watt (1978) considered the increased rate of germination of rehydrated hydropedetic seeds to be an advantage for rapid germination in habitats where the soil surface layers dry rapidly after a rain, e.g., the cracking black earths in New South Wales, Australia.

In addition to appropriate levels of soil moisture, seeds must be exposed to proper (species-dependent) temperature and light: dark conditions before they will germinate. The mean optimum germination temperature for species in Table 9.13 is 24.7 ± 0.6°C (mean ± SE). In 15 of the 33 species in which light: dark requirements for germination have been determined, light is either required for germination or seeds germinate to higher percentages in light than in darkness. Seeds of 16 species germinated equally well in light and darkness, and those of 2 species germinated to higher percentages in darkness than in light.

Seeds of *Hyptis suaveolens* are positively photoblastic (Wulff and Medina, 1971; Felippe *et al.,* 1983), and germination is controlled by the phytochrome system (Wulff and Medina, 1971). Small seeds of this species require 10 times more energy for 100% germination at 30°C under red light than large ones. Thus, large seeds require a shorter exposure to light to promote germination than small ones (Wulff, 1985). Seeds of *H. suaveolens* can germinate at temperatures from 10 to 40, but 25–30°C seems to be optimal (Felippe *et al.,* 1983).

E. Special Factor: Fire

It is not uncommon for savannas with a good grass cover to burn each year. These fires may be of natural origin, but many are started intentionally by humans (Gillon, 1983; Coutinho, 1982). One effect of fire is that it may stimulate an increase in flowering and thus seed production in shrubs and herbaceous species (Coutinho, 1982). Fire also creates favorable conditions for germination by removing litter and consequently increasing soil temperatures (Lacey *et al.*, 1982), increasing the amplitude of daily temperature fluctuations (Continho, 1990), and exposing bare mineral soil (Miyanishi and Kellman, 1986).

Heat from fires potentially can cause seeds with physical dormancy to become nondormant. For example, in Australian savannas ground fires promoted the germination of *Cassia eremophila* and *Dodonaea viscosa* seeds (Hodgkinson, 1991). However, Trollope (1982) speculates that fire in South African savannas promotes the germination of *Acacia karroo* seeds only when the fuel is very dry, fire is intense, and rains come after the burn. *Acacia gerrardii* seeds on the soil surface in Uganda were destroyed by fire, and the survival and germination of seeds after fire increased with the depth of burial in soil; seeds of *A. sieberiana* were more tolerant of high temperatures than those of *A. gerrardii* (Mucunguzi and Oryem-Origa, 1996).

Another possible effect of fire is that the smoke might stimulate germination. The germination of *Themeda triandra* seeds was promoted by plant-derived smoke, and sensitivity increased with the degree of seed imbibition. Also, aqueous smoke extracts increased germination at suboptimal and optimum germination temperatures, but ash and aqueous ash extracts had no effect on germination (Baxter *et al.*, 1994).

Seeds with physiological dormancy, as well as those with physical dormancy, can be killed by fire. Seeds of 12 of 13 species (including nine legumes) subjected to the heat of a fire in Venezuelan Llanos germinated to lower percentages than those not exposed to heat from a fire. The heat-treated seeds did not germinate because they had been killed (Boscan, 1967). Fires in savanna woodlands of the northern territory, Australia, killed 38% of the seeds of the annual grass *Sorghum intrans* (Watkinson *et al.*, 1989). Disseminules of the grasses *Cenchrus ciliaris, Enneapogon cenchroides, Eragrostis rigidor, Panicum maximum, Tragus berteronianus,* and *Urochloa trichopus* sown on the soil surface were destroyed completely by a savanna fire in Botswana (Ernst, 1991). Disseminules of *Setaria verticillata* survived the fire, but they germinated 4 months later than the controls. The ratio of disseminule mass to caryopsis mass was 1.38 for *S. verticillata,* but it ranged from 1.51 to 2.14 for the other species, implying that dispersal units with a high ratio were more likely to be killed in a fire than those with a low ratio (Ernst, 1991). Fire destroyed seeds of *Aristida junciformis, Heteropogon contortus, Panicum maximum, Sporobolus africanus,* and *Themeda triandra* placed on the soil surface in South Africa. However, when seeds were buried at 7–9 mm, heat from a fire resulted in a decrease (*T. triandra*), an increase (*H. contortus*), and no change (all other species) in germination percentages (Zacharias *et al.,* 1988). In Queensland, Australia, fire stimulated many more seeds to germinate than did mowing and removal of the cutoff plant material (Shaw, 1957).

Although temperatures at ground level during a fire may be quite high, e.g., 300°C (Miranda *et al.,* 1993) and 540°C (Hopkins, 1965), those beneath the soil surface remain relatively low (Gillon, 1983). During a fire in Brazilian cerrados, temperatures of 74, 47, 33, and 25°C were recorded at depths of 0, 1, 2, and 5 cm, respectively (Coutinho, 1990). Temperatures of 58, 37, and 32°C were recorded in *Themeda triandra* habitats in Australia, at depths of 0, 1, and 5 cm, respectively (Lock and Milburn, 1971).

Heat tolerance of seeds of various species has been determined under laboratory conditions. Seeds of *Gliricidia sepium, Securidaca diversifolia,* and *Marsdonia macrophylla* were killed by 5 min at 90°C, but 5 min at 150°C was required to kill those of *Copaifera officinalis* and *Bowdichia virgilioides* (Boscan, 1967). A dry heat treatment of 2 min at 116°C killed seeds of *Cenchrus ciliaris, Enneapogon cenchroides, Eragrostis rigidior, Setaria verticillata, Tragus berteronianus,* and *Urochloa trichopus,* and 3 min at 100°C killed seeds of all these species, except *T. berteronianus* (Ernst, 1991). After 2 min at 100°C, germination percentages of *E. cenchroides, E. rigidior, S. verticillata, T. berteronianus,* and *U. trichopus* seeds decreased due to the induction of dormancy (Ernst, 1991). The germination of *Heteropogon contortus* seeds was not reduced by a 1-min exposure to 50, 100, or 150°C, but no seeds germinated after a 1-min exposure to 200°C (Tothill, 1977). Seeds of *H. contortus* germinated equally well in burned and unburned field plots in southeastern Queensland, Australia (Tothill, 1977). Fire in *H. contortus* habitats removes litter on the soil surface and results in an increase in soil temperatures, which promote the germination of buried seeds of this species (Tothill, 1969).

Because soil temperatures, even those near the surface, remain relatively low when a fire sweeps over a savanna, seeds buried in the soil would not be killed (Gillon, 1983). Seeds of some species, e.g., *Dodonaea attenuata,* are carried below the soil surface by ants (Harrington and Driver, 1995). Further, dispersal units of some grasses have morphological features, including a sharply pointed callus, tufts of barbed hairs (at the base), and a hygroscopically active awn (Fig. 9.3), that increase their chances of becoming buried. A rigid, passive awn also is present in some species (Peart, 1979). The hygroscopically-active awn is straight when it is wet and bent when it is dry (Lock and Milburn, 1971). Diurnal fluctuations in RH of only 30% can cause the awn of *Themeda triandra* to move (Sindel *et al.*, 1993). Alternate wetting and drying (and consequently bending of the awn) cause the dispersal unit to move along the soil surface until it lodges in a crevice, under litter, or in some other favorable site for germination. Also, the hygroscopic awn pushes the dispersal unit firmly into the soil, and the hairs at the base of the unit prevent it from sliding back out of the soil (Murbach, 1900; Peart, 1979). A mean depth of self-burial of *T. triandra* seeds is 11 mm, thus seeds become buried deep enough to be protected from heat damage during fires (Lock and Milburn, 1971). Seeds of *T. triandra* sown at 0, 1, 2, and 3 cm with awns attached or detached gave maximum germination percentages at 1 cm (Sindel *et al.* 1993).

Various genera of savanna grasses, including *Danthonia, Dichelachne, Heteropogon, Schizachyrium, Stipa* (Peart, 1979), and *Themeda* (Lock and Milburn, 1971), have a hygroscopically active awn; however, not all grass dispersal units that become buried in savanna soils have hygroscopically active awns. Some grasses, such as *Cymbopogon, Digitaria, Entolasia, Eragrostis, Panicum*, and *Paspalidium*, are unawned, but some of their dispersal units become buried 2–5 cm deep in the soil; others remain on the surface (Peart, 1984). In passively awned grasses, such as *Aristida* and *Microlaena*, awns play a role in orientating the dispersal unit so that the callus is pointed into the soil. This orientation results in an increased chance of successful seedling establishment (Peart, 1981, 1984).

Cryptogeal germination is an adaptation to fire of newly germinated seedlings of some species in savannas. Although seeds may germinate on the soil surface, the fused cotyledon stalks and/or pseudoradicle elongate(s) and push(es) the radicle and/or plumule below ground (Jackson, 1974; Clarkson and Clifford, 1987; Onyekwelu, 1990). Thus, the shoot originates from beneath the soil surface, and in the event of a fire the apical meristem is protected. This type of germination occurs in a number of savanna shrubs and trees in the families Combretaceae, Sapotaceae, Rubiaceae, Ochnaceae, Fabaceae (Jackson, 1974), and Thymelaeaceae (Clarkson and Clifford, 1987).

VII. HOT SEMIDESERTS AND DESERTS

Hot semideserts and deserts are found on all continents, except Europe, and they extend poleward to about 35°N and S. This type of vegetation is found in southwestern North America, western South America, northern and southwestern Africa, southwestern Asia, and southern Australia (Walter, 1973). Hot deserts and arid shrublands also occur in coastal southwestern Malagasy (Evenari, 1985).

Rainfall in hot semideserts and deserts ranges from 0 to 600 mm per year and is highly variable both quantitatively and seasonally. Fog is a feature of the Namib, Peruvian–Chilean, Baja California, and Malagasy deserts, and dew is a source of water for organisms living in the Namib, Kalahari, Karroo, Baja California, Negev, and Malagasy deserts (Evenari, 1985). Mean annual air temperatures range from 20 to 25°C, except in Malagasy and the Sahel Desert, where they are 25 to 28°C (Evenari, 1985). Solar radiation is intense (710–1050 kJ cm^{-2} year^{-1}), and evaporation from "class A" pans is 2000–4000 mm year^{-1}, thus greatly exceeding annual rainfall (Evenari, 1985).

A diversity of land forms is found in deserts, including

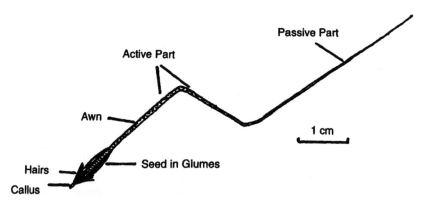

FIGURE 9.3 Dispersal unit of *Themeda triandra*, showing the sharply pointed callus, tufts of hairs, and hygroscopic awn. From Lock and Melburn (1971).

river beds that are dry most of the time, valleys, plateaus, basins or playas, sand dunes, and alluvial fans (Walter, 1979). Soils are poorly developed with little organic matter and nitrogen, and nonsandy soils may be saline. Calcium carbonate and gypsum are abundant and may form hardpans (Evenari, 1985). Depending on the desert, plants may grow in cracks in rocks, among gravel, in almost pure sand, or in heavy clays (Walter, 1979).

Shrubs (or suffrutescent chamaephytes) are the dominant life form in hot semideserts and deserts, with species showing a decrease in height with an increase in aridity (Shmida, 1985). Trees are present, but they may grow to a height of only a few meters and be restricted to edges of river beds (called wadies or arroyos) that are dry except during periods of flooding that follow rains, where their roots can reach ground water (Evenari, 1985; Rauh, 1985; Le Hourerou, 1986). Some herbaceous perennials in deserts have subterranean bulbs, tubers, or root stocks that store water, whereas others have root systems that grow deep enough to reach ground water (Eyre, 1963). Perennial grasses such as *Stipa, Panicum,* and *Aristida* may be abundant (Shmida, 1985), and annual species whose seeds germinate following rains are found in deserts of southwestern North America (MacMahon and Wagner, 1985), Peru–Chile (Rauh, 1985), the Middle East (Orshan, 1986), the Arabian Peninsula (Abd El Rahman, 1986), India (Gupta, 1986), northern Africa (Le Hourerou, 1986; Ayyad and Ghabbour, 1986), Malagasy (Rauh, 1986), Australia (Williams and Calaby, 1985), and Namibia (Walter, 1986). Annuals do not occur in deserts of Argentina (Mares *et al.,* 1985), and they have a minor role in those of Malagasy (Rauh, 1986).

Succulent plants are found in all hot semideserts and deserts, except those in Algeria and Australia (McCleary, 1968). Species of *Agave, Sedum,* and *Echeveria* and members of the Cactaceae are succulent in North American deserts (Eyre, 1963; MacMahon, 1979), whereas the main succulents in South American deserts are cacti belonging to the genera *Opuntia* and *Cereus* (Eyre, 1963). Cacti do not occur in the Old World, with the exception of the epiphytic *Rhipsalis* sp. in Africa (Shmida, 1985), and their ecological niche is filled by members of other families such as the Euphorbiaceae and Didiereaceae (Werger, 1986; Rauh, 1986). The greatest diversity of succulents seems to be in southern Africa in the Karroo-Namib Desert region, where many genera have succulent leaves, stems, stem–bases, and/or roots. Here, large and small trees, large, small, and dwarf shrubs, hemicryptophytic herbs, true geophytes, and annuals can be succulent (Werger, 1986). The semideserts of Malagasy also are famous for their numerous succulent species (Rauh, 1986).

A. Dormancy and Germination of Trees

A number of trees or subtrees grows in hot semideserts and deserts (*e.g.,* Leistner, 1979; MacMahon, 1979; Rauh, 1985, 1986; Shmida, 1985; Williams and Calaby, 1985; Abd El Rahman, 1986; Ayyad and Ghabbour, 1986; Le Hourerou, 1986; Orshan, 1986; Walter, 1986); however, seed germination studies have been done for only a relatively few species (Table 9.14). One reason for such a short list of trees is that many species are described in taxonomic manuals as "shrubs or small trees," in which case they have been included under shrubs.

Physical dormancy appears to be the most common type of dormancy found in seeds of desert trees. However, because of their economic value, legumes may have received more research attention than other kinds of trees. Desert trees belonging to families such as the Myricaceae, Didiereaceae, Apocynaceae, Canotiaceae, Rosaceae, Zygophyllaceae, Capparidaceae, and Ebenaceae could have seeds with physiological dormancy. Members of the Salicaceae and Platanaceae have nondormant seeds, which may be short-lived (Siegel and Brock, 1990). Thus, it is not surprising that trees in these two families grow in moist riparian habitats in deserts (Siegel and Brock, 1990).

Seeds of the date palm, *Phoenix dactylifera,* probably have underdeveloped embryos since they belong to the Arecaceae (Martin, 1946). Because radicle emergence occurs in 10–14 days in seeds of *P. dactylifera* incubated at 25 or 27°C (Khudairi, 1958), we conclude that these seeds have morphological dormancy. Studies on rates of embryo growth in seeds of this species would be very informative.

Physical dormancy is broken by mechanical or acid scarification (Khan *et al.,* 1984; Hoffman *et al.,* 1989) and by various heat treatments (Cavanagh, 1980). Seeds of *Sapindus drummondii* have a combination of physical and physiological dormancy, and a 60-min acid scarification treatment followed by 90 days of cold stratification resulted in high germination percentages (Munson, 1984).

One of the consequences of physical dormancy is that seeds of many desert trees may be able to live for long periods of time in the soil. For example, seeds of *Prosopis julifora* var. *velutina* were alive after 10 years of burial in the Sonoran Desert in jars of moist soil (Tschirley and Martin, 1960). Although seeds with physical dormancy may be long-lived in the desert, little is known about how their seed coats become permeable in nature.

Went (1955) observed leguminous trees growing in dry washes in deserts and concluded that their seeds were made permeable during periods of flooding, i.e.,

TABLE 9.14 Dormancy Type and Optimum Temperature and Light : Dark Requirement for Germination of Seeds of Trees of Hot Semideserts and Deserts[a]

Species	Family	Opt. germ. Temp. (°C)[b]	L:D[c]	Reference
Morphological dormancy				
Phoenix dactylifera	Arecaceae	25–27	—[d]	Khudairi (1958)
Nondormant				
Eucalyptus camaldulensis	Myrtaceae	—	—	Edgar (1977)
Platanus wrightii	Platanaceae	27	—	Siegel and Brock (1990)
Populus fremontii	Salicaceae	16–21	—	Siegel and Brock (1990)
Salix gooddingii	Salicaceae	27	—	Siegel and Brock (1990)
Physical dormancy				
Acacia constricta	Fabaceae	26–31	—	Cox et al. (1993)
A. erioloba	Fabaceae	30/15	—	Hoffman et al. (1989)
A. farnesiana	Fabaceae	30	L = D	Scifres (1974)
A. senegal	Fabaceae	30	—	Kaul and Manohar (1966)
Cercidium floridum	Fabaceae	21	L > D	Poole (1958); Scott et al. (1962)
C. microphyllum*	Fabaceae	28–32	L > D	Poole (1958)
Olneya tesota	Fabaceae	22–27	L > D	Poole (1958)
Prosopis glandulosa	Fabaceae	27	—	Scifres and Brock (1972)
P. juliflora	Fabaceae	25–35	—	Khan et al. (1984)
P. velutina	Fabaceae	29	—	Scifres and Brock (1969)
		26–31	—	Cox et al. (1993)
Physical and physiological dormancy				
Sapindus drummondii	Sapindaceae	—	—	Munson (1984)

[a] An asterisk (*) means that type of dormancy is inferred from available information on germination and on characteristics of seeds in that family.

[b] In some studies only one germination temperature was used. If that temperature resulted in high germination percentages, it is listed for the species.

[c] Light or dark requirements for germination: L, light was required for germination; D, darkness was required for germination; L > D, seeds germinated to a higher percentage in light than in darkness; D > L, seeds germinated to a higher percentage in darkness than in light; L = D, seeds germinated equally well in light and darkness.

[d] Data not available.

after a rain in the desert, washes fill with rapidly flowing water and the movement of sand and stones ground (mechanically scarified) seeds with impermeable seed coats, thereby breaking their dormancy. Although this idea frequently is mentioned in discussions of germination of desert seeds (e.g., Gutterman, 1993, 1994), it has not been tested in either the field or the laboratory.

Some attention has been given to the effects of an animal's digestive system on the breaking of physical dormancy in seeds of legumes that invade grazing lands. However, only 13 and 3% of Prosopis velutina seeds that passed through sheep and cattle, respectively, germinated, and only 6 and 1% of Acacia constricta seeds, respectively, germinated (Cox et al., 1993). Further, predation of A. erioloba seeds by bruchid beetles did not enhance germination (Hoffman et al., 1989).

Fires conceivably could play a role in breaking physical dormancy in seeds of some trees. Seeds of Prosopis velutina and Acacia constricta on the soil surface were killed by fire, but due to the activity of Merriam's kanga-

roo rats many seeds were buried at depths of 2–4 cm, where they were protected from extreme temperatures (Cox et al., 1993). During summer, however, seeds buried at 2–4 cm were subjected to natural soil temperatures that were high enough to stimulate germination. Thus, buried seeds of both species probably germinate in August (summer), when rainfall peaks and soil temperatures are about 30°C (Cox et al., 1993).

Although little is known about how physical dormancy is broken in seeds of most trees under desert conditions, it seems logical that the long dry, hot periods in deserts could result in a disruption of the strophiole. If physical dormancy is broken during the dry period, seeds would be nondormant and thus could germinate immediately after the beginning of a wet period. For example, Shreve (1917) observed germinating seeds of Cercidium microphylla in the Sonoran Desert during the first few days following a heavy rain in summer.

Optimum germination temperatures of nondormant seeds of trees range from 21 to 30°C (Table 9.14), and in the few species that have been studied germination was higher in light than in darkness. However, seeds of *Prosopis glandulosa* buried at depths of 0.5–1.5 cm germinated to 80% at soil temperatures of 27°C. Maximum seedling emergence under natural temperature regimes occurred after maximum daily temperatures in the surface 2.5 cm of soil were about 25°C for 4 days (Scifres and Brock, 1972). Bowers (1994) found that the minimum rainfall required to trigger the germination of *Cercidium microphylla* seeds was only 17 mm, but germination did not occur if temperatures dropped below 20°C.

Seeds of *Prosopis juliflora* germinated to 70% at an alternating temperature regime of 30/20°C and with no moisture stress (Scifres and Brock, 1969). Germination was reduced to 30% at −0.81 MPa, and no seeds germinated at −1.62 MPa. Seeds of *P. juliflora* germinated to 13, 81, and 25% at 38, 29, and 21°C, respectively, at an osmotic potential of −0.81 MPa. Thus, seeds of *P. juliflora* are more tolerant of moisture stress at an optimum germination temperature than they are at temperatures above and below it (Scifres and Brock, 1969). At a moisture stress of −0.8 MPa, seeds of *P. velutina* germinated to about 20%, but those of *Populus fremontii*, *Salix gooddingii*, and *Platanus wrightii* germinated to 3% or less (Siegel and Brock, 1990). Seeds of *Phoenix dactylifera* germinated to 20% on a substrate moistened with 1.8% NaCl (−1.4 MPa) (Khudairi, 1958), and those of *Prosopis juliflora* were not affected by −0.6 MPa (Khan *et al.*, 1987) or −0.8 MPa (Khan *et al.*, 1984) NaCl.

B. Dormancy and Germination of Shrubs

Physical and physiological dormancies are about equally important among desert shrubs, but a few species have nondormant seeds (Table 9.15). Many of the shrubs whose seeds have physical dormancy belong to the Fabaceae, but this type of dormancy also is found in shrubs belonging to the Convolvulaceae, Malvaceae, Sapindaceae, and Rhamnaceae. Shrubs whose seeds have physiological dormancy belong to a diversity of plant families.

Physical dormancy has been broken in seeds of shrubs by acid (Everitt, 1983a,b,c; Vora, 1989; Mehta and Sen, 1991) and mechanical (Chatterji and Mohnot, 1967; Whisenant and Ueckert, 1982; Everitt, 1983a; Mehta and Sen, 1991) scarification. The impermeable endocarps of *Ziziphus nummularia* fruits were broken with a hammer to release the nondormant seeds, and soaking in water increased the germination of freed seeds (Hussain *et al.*, 1993). Nonscarified seeds of *Eysenhardtia texana* (Whisenant and Ueckert, 1982) and *Acacia berlandieri* (Everitt, 1983a) germinated to 75 and

77%, respectively, after they had been stored dry at room temperatures for some unspecified period of time. Thus, freshly matured seeds of both species may have been physically dormant and seed coats became permeable during dry storage (e.g., Egley, 1976).

Physiological dormancy in seeds of some shrubs is broken while seeds are stored dry at room temperatures. Seeds of *Atriplex obovata* germinated to 42% after 100 days of dry storage at 23°C (Edgar and Springfield, 1977), and those of *Parthenium argentatum* (Emparan and Tysdal, 1957) and *Withania somnifera* (Hussain and Ilahi, 1988b) germinated to about 60% after 84 and 540 days, respectively, of dry storage at room temperatures. About half the seeds of *Welwitschia mirabilis* germinated after a 3-year period of dry storage at room temperatures (Pearson, 1910). Freshly matured seeds of *Zygophyllum dumosum* germinated to 58 and 51% in light at 20 and 26°C, respectively, but after 9.2 months of dry storage at room temperatures they germinated to 84 and 72%, respectively (Koller, 1955). A dry heat treatment at 50°C for 12 or 16 hr increased the germination of *Celtis pallida* seeds (Fulbright *et al.*, 1995). A 2-week cold stratification treatment increased the germination of *C. pallida* (Fulbright *et al.*, 1986a) and *Ehretia anacua* (Fulbright *et al.*, 1986b) seeds.

In a dormancy-breaking study conducted in the deserts of Qatar, seeds of *Zygophyllum qatarense* from nonhalophytic and halophytic sites were placed in fine-mesh nylon bags in June, returned to their respective habitats, and covered with plant debris and fine sand to a depth of <0.5 cm (Ismail and El-Ghazaly, 1990). At monthly intervals for 9 months, seeds were exhumed and tested for germination at 22/10°C. Seeds slowly came out of physiological dormancy, and by the following March those from both sites germinated to about

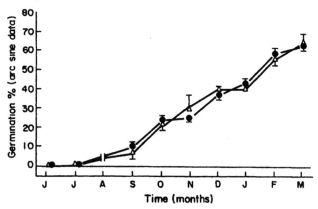

FIGURE 9.4 Germination percentages (mean ± SE) of seeds of *Zygophyllum qatarense* from nonhalophytic (●) and halophytic (△) sites that were buried in sand in their respective habitats and then exhumed at monthly intervals and tested at 23/10°C at an 11-hr daily photoperiod. From Ismail and El-Ghazaly (1990), with permission.

TABLE 9.15   Dormancy Type and Optimum Temperatures and Light:Dark
Requirements for Germination of Seeds of Shrubs of Hot Semideserts and Deserts[a]

Species	Family	Temp. (°C)[b]	L:D[c]	Reference
Nondormant				
Calotropsis procera	Asclepiadaceae	30	L > D	Sen and Chatterji (1965, 1968); Sen (1969)
Capparis decidua	Capparidaceae	28–30	L	Qaiser and Qadir (1971)
Datura alba*	Solanaceae	room	—[d]	Qadir and Lodhi (1971)
Diospyros texana	Ebenaceae	23–38	L = D	Everitt (1984), Van Auken and Bush (1992)
Ericameria austrotexana*	Asteraceae	20–25	L > D	Mayeux (1982)
Leptadenia pyrotechnica*	Asclepiadaceae	30–35	L = D	Sen and Chatterji (1968), Sen (1968, 1969)
Salazaria mexicana*	Lamiaceae	20	—	Kay et al. (1984)
Salsola glabrescens	Chenopodiaceae	20	L = D	Henrici (1935, 1939)
Salvadora oleoides	Salvadoraceae	30	L = D	Sen and Chawan (1969)
S. persica	Salvadoraceae	35	L = D	Sen and Chawan (1969)
Physical dormancy				
Abrus precatorius*	Fabaceae	room	L > D	Sen and Chatterji (1968)
Abutilon indicum	Malvaceae	room	—	Qadir and Lodhi (1971)
Acacia berlandieri*	Fabaceae	25	L = D	Everitt (1983a)
A. rigidula	Fabaceae	25	L = D	Everitt (1983a)
A. schaffneri	Fabaceae	20	L = D	Everitt (1983b)
A. smallii	Fabaceae	g.h.[e]	—	Vora (1989)
A. wrightii	Fabaceae	g.h.[e]	—	Vora (1989)
Cassia armata	Fabaceae	15	—	Kay et al. (1984)
C. italica	Fabaceae	28	—	Mehta and Sen (1991)
C. occidentalis	Fabaceae	room	—	Qadir and Lodhi (1971)
Condalia hookeri*	Rhamnaceae	g.h.[e]	—	Vora (1989)
C. microphylla	Rhamnaceae	30/20	L = D	Pelaez et al. (1996)
Convolvulus lanatus	Convolvulaceae	20–30	L = D	Koller and Cohen (1959)
C. negevensis	Convolvulaceae	20–30	L = D	Koller and Cohen (1959)
C. secundus	Convolvulaceae	20–30	L = D	Koller and Cohen (1959)
Coursetia axillarix*	Fabaceae	g.h.[e]	—	Vora (1989)
Crotalaria burhia*	Fabaceae	room	—	Bansal and Sen (1981)
Erythrina flabelliformis	Fabaceae	37/20, 23/7	—	Conn and Snyder-Conn (1981)
E. herbacea*	Fabaceae	g.h.[e]	—	Vora (1989)
Hermannia linearifolia*	Sterculiaceae	20	—	Henrici (1935, 1939)
Hibiscus micranthus*	Malvaceae	room	L = D	Sen and Chatterji (1968)
Indigofera tinctoria*	Fabaceae	room	—	Sen and Chatterji (1968)
Leucaena pulverulenta	Fabaceae	g.h.[e]	—	Vora (1989)
L. retusa	Fabaceae	30	L = D	Whisenant and Ueckert (1982)
Lotononis divaricata*	Fabaceae	20	L > D	Henrici (1939)
Mimosa hamata	Fabaceae	35	—	Chatterji and Mohnot (1967)
Parkinsonia aculeata	Fabaceae	25	L = D	Everitt (1983b)
Pavonia arabica	Fabaceae	room	L	Sen and Chatterji (1968)
Pithcellobium flexicaule	Fabaceae	g.h.[e]	—	Vora (1989)
P. pallens	Fabaceae	g.h.[e]	—	Vora (1989)
Prosopis stephaniana	Fabaceae	27–30	—	Khudairi (1956)
Sapindus drummondii	Sapindaceae	g.h.[e]	—	Vora (1989)
Sesbania drummondii*	Fabaceae	g.h.[e]	—	Vora (1989)
Sophora secundiflora	Fabaceae	25	L = D	Everitt (1983c)
Ziziphus nummularia	Rhamnaceae	20	—	Hussain et al. (1993)
Z. obtusifolia	Rhamnaceae	25	—	Speer and Wright (1981)
Physiological dormancy				
Acamplopappus sphaerocephalus*	Asteraceae	20	—	Kay et al. (1984)
Achillea fragrantissima*	Asteraceae	—	—	Shalaby and Youssef (1967)
Aerva persica*	Amaranthaceae	room	L > D	Sen and Chatterji (1968)
A. pseudo-tomentosa*	Amaranthaceae	room	—	Qadir and Lodhi (1971)
Ambrosia dumosa*	Asteraceae	15	—	Kay et al. (1984)

(continues)

TABLE 9.15—*Continued*

Species	Family	Opt. germ. Temp. (°C)[b]	L:D[c]	Reference
*Anisacanthus wrightii**	Acanthaceae	35	L = D	Whisenant and Ueckert (1981)
*Artemisia monosperma**	Asteraceae	15	L	Koller *et al.* (1964)
Atriplex canescens	Chenopodiaceae	15, 21	—	Graves *et al.* (1975), Kay *et al.* (1984)
A. capensis	Chenopodiaceae	20	L = D	Henrici (1935, 1939)
A. obovata	Chenopodiaceae	23	L > D	Edgar and Springfield (1977)
A. polycarpa	Chenopodiaceae	24/16	—	Chatterton and McKell (1969), Graves et al. (1975)
A. repanda	Chenopodiaceae	—	—	Fernandez H. (1978)
*Celtis pallida**	Ulmaceae	25, 30	L > D	Fulbright *et al.* (1986a, 1995)
*Ceratoides lanata**	Chenopodiaceae	10	—	Kay *et al.* (1984)
*Cryptostegia grandiflora**	Asclepidaceae	25–30	D > L	Sen (1968, 1969)
Ehretia anacua	Boraginaceae	30	—	Fulbright *et al.* (1986b)
*Ephedra nevadensis**	Ephedraceae	21	—	Graves *et al.* (1975)
*E. viridis**	Ephedraceae	15	—	Kay *et al.* (1984)
*Encelia virginensis**	Asteraceae	20	—	Kay *et al.* (1984)
*Eriogonum fasciculatum**	Polygonaceae	15	—	Kay *et al.* (1984)
Euryops multifidus	Asteraceae	20	—	Henrici (1935)
*Fouquieria splendens**	Fouquieriaceae	20–25	L = D	Freeman (1973a)
Franseria dumosa	Asteraceae	21	—	Graves et al. (1975)
*Grayia spinosa**	Chenopodiaceae	15	—	Kay *et al.* (1984)
*Halopeplis perfoliata**	Chenopodiaceae	5–35	—	Mahmoud *et al.* (1983)
*Haloxylon articulatum**	Chenopodiaceae	15–18	L > D	Sankary and Barbour (1972)
*H. salicornicum**	Chenopodiaceae	22	L > D	Choi et al. (1976), Kaul *et al.* (1990)
*Hymenoclea salsola**	Asteraceae	21	—	Graves et al. (1975)
*Isocoma drummondii**	Asteraceae	20–25	L > D	Mayeux and Scifres (1973)
*Isomeris arborea**	Cleomaceae	15	—	Kay *et al.* (1984)
*Larrea divaricata**	Zygophyllaceae	23	D > L	Barbour (1968), Rivera and Freeman (1979)
*Lepidospartum squamatum**	Asteraceae	21	—	Graves *et al.* (1975)
*Limonium axillare**	Plumbaginaceae	5–30	—	Mahmoud *et al.* (1983)
*Lycium andersonii**	Solanaceae	20	—	Kay *et al.* (1984)
*L. cooperi**	Solanaceae	20	—	Kay *et al.* (1984)
Osteospermum muricatum	Asteraceae	20	—	Henrici (1935)
*Parthenium argentatum**	Asteraceae	30/20	L > D	Benedict and Robinson (1946)
Pentzia sphaerocephala	Asteraceae	20	—	Henrici (1935, 1939)
Phymaspermum parvifolium	Asteraceae	20	—	Henrici (1935, 1939)
*Porlieria angustifolia**	Zygophyllaceae	25	L = D	Everitt (1983a)
Pteronia glauca	Chenopodiaceae	20	—	Henrici (1935)
*Salvia aegyptica**	Lamiaceae	room	L	Sen and Chatterji (1968)
Selago speciosa	Chenopodiaceae	20	—	Henrici (1935)
Tetragonia arbuscula	Aizoaceae	20	L = D	Henrici (1935, 1939)
Tripteris leptoloba	Asteraceae	20	—	Henrici (1935)
T. pachypteris	Asteraceae	20	—	Henrici (1935)
Welwitschia mirabilis	Welwitchiaceae	room	—	Pearson (1910)
Withania somnifera	Solanaceae	25	—	Hussain and Ilahi (1988)
Zygophyllum coccineum	Zygophyllaceae	25	D > L	Batanouny and Ziegler (1971b)
*Z. dumosum**	Zygophyllaceae	20	D > L	Koller (1955)
Z. gatarense	Zygophyllaceae	28/15	L = D	Ismail (1990), Ismail and El-Ghazaly (1990)

[a] An asterisk (*) means that type of dormancy is inferred from available information on germination and on characteristics of seeds in that family.

[b] In some studies, only one germination temperature was used. If that temperature resulted in high germination percentages, it is listed for the species.

[c] Light or dark requirements for germination: L, light was required for germination; D, darkness was required for germination; L > D, seeds germinated to a higher percentage in light than in darkness; D > L, seeds germinated to a higher percentage in darkness than in light; L = D, seeds germinated equally well in light and darkness.

[d] Data not available.

[e] g.h., greenhouse.

65% (Fig. 9.4). Germination usually occurs in the field from December to March, when rains not only wet the soil but lower the salinity (Ismail and El-Ghazaly, 1990).

Light:dark and temperature requirements for germination have been determined for a number of shrubs (Table 9.15). Seeds of only a few shrubs, including *Artemisia monosperma* (Koller *et al.*, 1964), *Capparis decidua* (Qaiser and Qadir, 1971), *Pavonia arabica* (Sen and Chatterji, 1968), and *Salvia aegyptica* (Sen and Chatterji, 1968), have an absolute light requirement for germination. Seeds of many shrubs germinate equally well in light and darkness, and those of some species germinate to higher percentages in darkness than in light. Although Kaul and Shankar (1988) reported that seeds of *Haloxylon salicornicum* germinated equally well in light and darkness, other studies (Choi *et al.*, 1976; Kaul *et al.*, 1990) have shown that they germinated to higher percentages in light than in darkness. It should be noted, however, that Kaul and Shankar removed seeds from dark cabinets on alternate days to check germination; thus, seeds may have received enough light to promote germination. Seeds of some shrubs, including *Erythrina flabelliformis* (Conn and Snyder-Conn, 1981), *Ziziphus nummularia* (Hussain *et al.*, 1993), and *Withania somnifera* (Hussain and Ilahi, 1988b), germinate when covered by soil, indicating that they can germinate in darkness. In fact, germination is sometimes improved by burial due to decreased moisture stress (Conn and Snyder-Conn, 1981).

A certain minimum amount of precipitation is required to stimulate germination in the desert habitat. A rainfall of 5 mm caused seeds of the undershrub *Achillea fragrantissima* to germinate, but additional precipitation was required for seedling survival (Shalaby and Youssef, 1967). A rainfall of 20 mm was required for the germination of seeds of *Arthocnemon glaucum* (Hammouda and Bakr, 1969), 17 mm for *Encelia farinosa*, and 25 mm for *Fouquieria splendens* (Bowers, 1994). Even if the minimum amount of water is present, germination does not occur unless temperatures and light:dark conditions are nonlimiting.

Temperature requirements for 60–100% germination range from 15 to 35°C, with temperatures of about 20–25°C being suitable for most species (Table 9.15). Seeds of *Larrea tridentata* germinated to 2–20% at temperatures (40/19, 43/22, and 46/25°C) simulating those in the Chihuahuan Desert (United States) during summer and to 40% at 23°C. Seeds of *L. tridentata* failed to germinate in moist Chihuahuan Desert soil until five consecutive relatively cool days occurred in September (Rivera and Freeman, 1979). Seeds of *L. tridentata* and *Ambrosia dumosa* germinated in the Mojave Desert after rain in late August, when minimum and maximum air temperatures were 15–20 and 29–34°C, respectively, and in Sep-

tember, when temperatures were 12–18 and 26–32°C, respectively (Ackerman, 1979). Seeds of *Celtis pallida* germinated to higher percentages at temperatures (25 and 30°C) simulating those under *Prosopis glandulosa* trees in May and June, when germination occurs in the field, than at temperatures (28–30°C) simulating those in spaces between trees (Fulbright *et al.*, 1995).

Optimum germination temperatures for seeds of *Encelia farinosa* (Kay *et al.*, 1984) are lower than those of *Fouquieria splendens* (Freeman, 1973a). Thus, in the Sonoran Desert in Arizona seeds of *E. farinosa* germinate in response to rainfall in the cool season, but not when temperatures are near freezing (Bowers, 1994). Seeds of *F. splendens* germinate in response to rains in summer (Shreve, 1917) and autumn (Bowers, 1994). Scarified seeds of *Erythrina flabelliformis,* another Sonoran Desert legume, incubated at simulated January (16.7/1.6°C), March (23.0/7.0°C), and July (37.0/20.0°C) habitat temperatures, germinated to 75% at July and March temperatures and to 0% at January temperatures (Conn and Snyder-Conn, 1981). Many seeds of shrubs, succulents, and forbs germinated in the arid shrublands of southern Karroo in South Africa following rains in autumn and early winter, when temperatures were decreasing, but none germinated in response to rains in summer when temperatures were at their yearly maximum (Milton, 1995).

The germination of *Fouquieria splendens* seeds decreased from 70% at a water potential of −0.51 MPa to 10% at −1.01 MPa (Freeman, 1973a), and that of *Diospyros texana* seeds decreased from about 95 to 45% at 0 and −0.6 MPa, respectively (Everitt, 1984). A water potential of −0.20 MPa decreased the germination of *Larrea tridentata* seeds from about 90 to 40%, and at a water potential of −1.01 MPa no seeds germinated (Barbour, 1968). Soil must be moist (field capacity or −0.03 MPa) for 2 continuous days before seeds of *Zygophyllum dumosum* start to germinate, and the highest germination percentages require 4 continuous days on moist soil. Germination of this species was inhibited at low (−0.10 to −10.00 MPa) and high (−0.002 to −0.0398 MPa) soil water potentials (Agami, 1986). A mannitol solution with an osmotic potential of −2.43 MPa inhibited the germination of *Z. coccineum* seeds, whereas at −2.02 MPa about 25% of them germinated (Batanouny and Ziegler, 1971a). Permeable seeds of *Eysenhardtia texana* germinated to 45, 5, and 0% in polyethylene glycol (PEG 6000) solutions at −0.40, −0.81 and −1.21 MPa, respectively, whereas those of *Leucaena retusa* germinated to 30, 5, and 0%, respectively (Whisenant and Ueckert, 1982).

A PEG solution of −0.4 MPa reduced the germination of *Acacia rigidula* seeds from 88 to 53%, that of *Porlieria angustifolia* from 66 to 35% (Everitt, 1983a), and that of *Anisacanthus wrightii* from about 98 to 58% (Whisenant and Ueckert, 1981). Seeds of *Isocoma coronopifolia* germinated to 49 and 0% in PEG solutions at −0.61 and −1.21 MPa, respectively, whereas those of *I. drummondii* germinated to 59 and 0%, respectively (Mayeux and Scifres, 1978). Seeds of *Haloxylon salicornicum* did not germinate in light or darkness at a moisture stress of −1.52 MPa when the convex side of the seed was down; however, there was some germination (percentage not given) at this moisture stress in light if the flat side was down (Kaul *et al.*, 1990).

Desert shrubs vary in their ability to germinate in the presence of moderate to high salt concentrations. The germination of *Isocoma drummondii* seeds was reduced from 80% at 500 ppm (−0.04 MPa) NaCl to 10% at 15,000 ppm (−1.17 MPa) NaCl (Mayeux and Scifres, 1978), and the germination of *Diospyros texana* seeds was reduced 60% at 31 ppm (−0.002 MPa) NaCl (Everitt, 1984). Seeds of *Fouquieria splendens* germinated in a 7080 ppm (−0.55 MPa) NaCl solution, indicating that the species is moderately salt tolerant (Freeman, 1973a). At 10,000 ppm (−0.78 MPa) NaCl, the germination of permeable seeds of *Acacia rigidula* was not reduced, but the germination of *Porlieria angustifolia* was reduced from 67 to 34% (Everitt, 1983a).

The germination of *Zygophyllum coccineum* seeds was reduced from 85 to 55% when the osmotic potential due to the addition of NaCl decreased from −0.40 to −0.81 MPa or when the osmotic potential due to the addition of Na_2SO_4 decreased from −0.20 to −0.61 MPa (Batanouny and Ziegler, 1971a). The germination of *Z. qatarense* seeds was reduced from about 50 to 0% when the osmotic potential due to the addition of NaCl decreased from −0.4 to −0.8 MPa (Ismail, 1990). The germination of *Zygophyllum dumosa* seeds was reduced from 75–80% to about 55% when the osmotic potential decreased due to an increase in NaCl (from −0.33 to −1.28 MPa) or Na_2SO_4 (from −0.26 to −0.92 MPa) (Coudret, 1971). Fruits of *Z. dumosum* contain a water-soluble inhibitor that decreases the ability of seeds to germinate in the presence of NaCl solutions. For example, seeds germinated to 55% in a −0.3 MPa NaCl solution, but none germinated when the inhibitor (1.0 g/ml) was added to the salt solution (Lerner *et al.*, 1959).

Seeds of *Halopeplis perfoliata* and *Limonium axillare* germinated to 96 and 85%, respectively, in 9225.5 ppm (−0.72 MPa) NaCl at 32/10°C (Mahmoud *et al.*, 1983). The germination of *Larrea tridentata* seeds was reduced from 70% at 500 ppm (−0.04 MPa) to 0% at 10,000 ppm (−0.78 MPa) NaCl (Barbour, 1968). However, seeds of *Haloxylon articulatum* germinated to 28% at 40,000 ppm (−3.12 MPa) NaCl (Sankary and Barbour, 1972). In the presence of chlorine ions, magnesium ions were less

inhibitory to the germination of *Z. album* seeds than those of sodium (Coudret, 1971).

Because most desert soils are alkaline (Evenari, 1985), there has been some interest in determining the pH requirements for the germination of desert shrubs. Seeds of *Larrea tridentata* germinated equally well at pH's of 7, 8, 9, and 10 (Barbour, 1968). In another study, however, seeds of this species showed no significant decreases in germination at pH's of 4.5, 7, or 8, but germination was reduced greatly at pH's of 9 and 10 (Lajtha *et al.*, 1987). Seeds of *Anisacanthus wrightii* germinated to 80–100% at pH's of 4 to 8 and to 60% at a pH of 9 (Whisenant and Ueckert, 1981), and those of *Diospyros texana* germinated to 93–97% at pH's of 4 to 11 (Everitt, 1984). The germination of *Acacia rigidula* and *Porlieria angustifolia* was not reduced at pH's of 3 to 11, but it was reduced at 2 and 12 (Everitt, 1983a). When tested over a range of pH's, seeds of *Fouquieria splendens* germinated to 60% or more only at pH 7.5 (Freeman, 1973a), and those of *Ziziphus nummularia* germinated to 50% or more only at pH 7 (Hussain *et al.*, 1993). Seeds of *Isocoma drummondii* germinated to 80–86% at pH's of 3 to 11, to 16% at a pH of 1, and to 73% at a pH of 12 (Mayeux and Scifres, 1978). Permeable seeds of *Eysenhardtia texana* germinated equally well at pH values from 5 to 9, but those of *Leucaena retusa* germinated to higher percentages at 7 to 9 than at 5 and 6 (Whisenant and Ueckert, 1982).

Germination is improved in some shrubs if seeds are leached. Soaking 60-month-old seeds of *Atriplex repanda* for 8 hr before sowing increased germination from 5 to 30%; additional soaking up to 40 hr did not improve germination (Johnston B. and Fernandez H., 1979). Removal of leachate from the vicinity of *A. polycarpa* seeds increased germination from 4 to 15% (Cornelius and Hylton, 1969). Seeds of *Zygophyllum dumosum* failed to germinate in the presence of fruit coats, indicating that a germination inhibitor was leached from them. Subsequent germination studies showed that the leachate from fruit coats inhibited the germination of lettuce seeds (Koller, 1955).

Leachates from fruit coats of *Larrea tridentata* did not inhibit the germination of *L. tridentata* seeds, whereas aqueous extracts from leaves and twigs of this species inhibited the germination of *Bouteloua eriopoda*, but not that of *Muhlenbergia porteri* or *L. tridentata* seeds (Knipe and Herbel, 1966). Activated carbon, which should absorb inhibitors, improved the germination of hulled and unhulled seeds of *L. tridentata, A. polycarpa*, and *Hymenoclea salsola* (Graves *et al.*, 1975). Leaching of *Zygophyllum coccineum* seeds promoted germination in light at 25°C and in darkness at 10°C; however, the leachate inhibited germination in light at 25°C (Batanouny and Ziegler, 1971b). Leaching significantly improved germination of the dispersal units (achene plus two staminate florets and a bract) of *Parthenium argentatum* (Naqvi and Hanson, 1982). Removal of the florets and bract, which contained seven phenolic compounds that inhibit germination, increased germination even more than leaching of the whole dispersal unit. Seeds of *Zygophyllum qatarense* leached in running tap water overnight did not germinate, indicating that something other than inhibitors per se prevented germination (Ismail, 1990).

C. Dormancy and Germination of Perennial Succulents

Temperature requirements for seed germination of some perennial succulents have been determined (Table 9.16), but little is known about dormancy. One reason for the lack of information on dormancy is that seeds usually are stored dry at room temperatures for long periods of time before germination studies are initiated. Thus, seeds with physiological dormancy could have lost

TABLE 9.16 Dormancy Type and Optimum Temperature and Light:Dark (L:D) Requirements for Germination of Seeds of Succulent Perennials in Hot Semideserts and Deserts[a]

Species	Family	Optimum germination temperature (°C)[b]	L:D[c]	Reference
Nondormant				
Beaucarnea gracilis	Nolinaceae	Room	—[e]	Cardel *et al.* (1997)
Euphorbia caducifolia	Agavaceae	25–30	L > D	Sen and Chatterji (1966)
Physiological dormancy				
Agave lecheguilla[*,d]	Agavaceae	25–30	L = D	Freeman (1973b)
A. parryi[*,e]	Agavaceae	20–23	L = D	Freeman *et al.* (1977)
Carnegiea gigantea[*,d]	Cactaceae	25	L	Alcorn and Kurtz (1959), McDonough (1964)
Cereus griseus	Cactaceae	25	L	Williams and Arias (1978)
Coryphanta gladispina[*]	Cactaceae	20–35	—	Zimmer (1969)
Echinocactus grusonii[*]	Cactaceae	25–35	—	Zimmer (1969)
Ferocactus histrix[*,d]	Cactaceae	24	L	del Castillo (1986)
F. glaucescens[*]	Cactaceae	15–30	—	Zimmer (1969)
Hamatocactus setispinus[*,d]	Cactaceae	26	—	Fearn (1974)
H. sinuatus[*,d]	Cactaceae	24	—	Fearn (1974)
Helianthocereus pasacana[*,d]	Cactaceae	15–22	—	Fearn (1974)
Lemaireocereus thurberi[*,d]	Cactaceae	25	L > D	McDonough (1964)
Lophocereus schottii[*,d]	Cactaceae	23	—	Fearn (1974)
Mammillaria fuauxiana[*]	Cactaceae	20–35	—	Zimmer (1969)
M. ingens[*]	Cactaceae	15–25	—	Zimmer (1969)
M. longimamma[*]	Cactaceae	20	—	Zimmer (1969)
M. potosina[*]	Cactaceae	20	—	Zimmer (1969)
M. durispina[*]	Cactaceae	20	—	Zimmer (1969)
Melocactus caesius	Cactaceae	25–41	—	Arias and Lemus (1984)
Opuntia discata[*]	Cactaceae	30	—	Potter *et al.* (1984)
O. edwardsii[*]	Cactaceae	30	—	Potter *et al.* (1984)
O. lindheimeri[*]	Cactaceae	30	—	Potter *et al.* (1984)
Rebutia xanthocarpa[*,d]	Cactaceae	17–30	—	Fearn (1974)
Yucca baccata[*,d]	Agavaceae	20–25	D > L	McCleary and Wagner (1973)
Y. brevifolia[*,d]	Agavaceae	20–24	D > L	McCleary and Wagner (1973)
Y. elata[*,d]	Agavaceae	20–25	D > L	McCleary and Wagner (1973)
Y. schidigera[*,d]	Agavaceae	23	—	Keeley and Meyers (1985)
Y. torreyi[*,d]	Agavaceae	23	—	Keeley and Meyers (1985)
Y. whipplei[*,d]	Agavaceae	25	—	Keeley and Tufenkian (1984)

[a] An asterisk means that the type of dormancy is inferred from available information on germination and on characteristics of seeds in that family.

[b] In some studies, only one germination temperature was used. If that temperature resulted in high germination percentages, it is listed for the species.

[c] L, light required for germination; D, darkness required for germination; L > D, seeds germinated to a higher percentage in light than in darkness; D > L, seeds germinated to a higher precentage in darkness than in light; and L = D, seeds germinated equally well in light and darkness.

[d] Seeds were stored dry at room temperatures for an unspecified period of time (or for 3 or more months) before germination studies were initiated.

[e] Data not available.

their dormancy before they were tested for germination. For example, seeds of *Cereus griseus* were dormant at maturity, but they came out of dormancy in dry laboratory storage within 8 weeks (Williams and Arias, 1978).

Optimum temperatures for seed germination of perennial succulents are 20–25°C (Table 9.16). Temperatures of 35–40°C inhibited the germination of seeds of *Agave parryi* (Freeman, 1975; Freeman *et al.,* 1977), *A. lecheguilla* (Freeman *et al.,* 1977; Freeman, 1973b), *Yucca whipplei* (Keeley and Tufenkian, 1984), *Euphorbia caducifolia* (Sen and Chatterji, 1966), and *Carnegiea gigantea* (Alcorn and Kurtz, 1959). Some seeds of *Lemaireocereus thurberi* germinated at 40°C in darkness, but none germinated at 45°C (McDonough, 1964). On a thermal gradient bar, the maximum and minimum temperatures for the germination of seeds of various cacti were: *Helianthocereus pasacana* (32.1, 12.3), *Rebutia xanthocarpa* (22.8, 11.5), *Parodia chysacanthion* (28.3, 11.3), *Carnegiea gigantea* (27.5, 16.3), *Lophocereus schottii* (28.1, 17.9), *Frailea pumila* (39.6, 11.5), and *Gymnocalycium spegazzinii* (38, 14.8°C) (Fearn, 1974). No *Carnegiea gigantea* or *Lemaireocereus thurberi* seeds (McDonough, 1964) germinated at 15°C, but *Agave parryi* (Freeman, 1975) and *Yucca elata* (McCleary and Wagner, 1973) seeds germinated to 5–20% at 10°C. However, no seeds of *Y. baccata* or *Y. brevifolia* germinated at 10°C (McCleary and Wagner, 1973).

The temperature requirements for the seed germination of most succulents suggest that germination occurs in deserts during the relatively cool season. In the Chihuahuan Desert, rains in September did not stimulate any germination of *Agave lecheguilla;* however, seeds germinated during a moist period in winter (Freeman *et al.,* 1977). Freeman *et al.* (1977) suggested that the inability of *A. lecheguilla* seeds to germinate at 40°C would prevent them from germinating following a summer shower, but germination may be possible in summer if there is a period with several rainy, cloudy days and moderate temperatures. Seeds of *Carnegiea gigantea* and *Echinocactus wislizeni* germinated in the Sonoran Desert following heavy rains in summer (Shreve, 1917). Seeds germinated in the Namib Desert of Namibia and southern Angola (Africa) in response to rainfall that occurred after temperatures had decreased in winter (von Willert *et al.,* 1992).

Seeds of *Melocactus caesius* (Arias and Lemus, 1984), *Cereus griseus* (Williams and Arias, 1978), and *Carnegiea gigantea* (Alcorn and Kurtz, 1959; McDonough, 1964) have an absolute light requirement for germination, but those of other species do not (Table 9.16). In fact, seeds of *Yucca baccata, Y. brevifolia,* and *Y. elata* germinated to higher percentages in darkness than in light (McCleary and Wagner, 1973). GA (1000 ppm)

stimulated 9% of the seeds of *C. gigantea* to germinate in darkness (McDonough, 1964).

Because seeds of succulents dispersed on the surface of desert soils would be subjected to high temperatures, high temperature pretreatments have been tried as a means of increasing germination percentages. Germination percentages of *Yucca baccata, Y. brevifolia, Y. elata,* and *Y. whipplei* seeds were increased significantly after a 5-min exposure of dry seeds to 90°C; however, 2 hr at 90°C and 5 min at 100, 110, and 120°C decreased germination percentages (Keeley and Meyers, 1985). A 5-min exposure to 110 and 130°C significantly reduced the germination of seeds of five subspecies of *Y. whipplei* (Keeley and Tufenkian, 1984).

Little work has been done on the effects of water stress or pH on the germination of succulents. Seeds of *Agave lecheguilla* germinated to about 95% at water potentials of 0, −0.20, and −0.51 MPa but to 63% at −1.01 MPa (Freeman, 1973b). Seeds of *A. parryi* germinated to about 83, 77, 67, and 5% at water potentials of 0, −0.20, −0.51, and −1.01 MPa, respectively (Freeman, 1975). Seeds of the cacti *Stenocereus thurberi, Pachycereus pecten-aboriginum,* and *Ferocactus peninsulae* survived hydration (3, 6, 12, 48, 72, or 80 hr)/dehydration (4, 14, 70, 100, 120, 130, 181, or 182 days) cycles, and mean time to germination (but not germination percentage) was increased (Dubrovsky, 1996). The optimum pH for the germination of *A. lecheguilla* seeds was 6.15; germination was decreased at pH's of 7.30 and 7.85 (Freeman, 1973b). Seeds of *A. parryi* germinated to the same percentages (values not given) at pH 7.0, 7.5, 8.0, and 8.5 after 8 days, but the rate of germination was greater at 7.5 than at 7.0 and 8.0 (Freeman, 1975).

The germination of *Cereus griseus* seeds increased from 8 to 63% when freshly matured seeds were leached with running water for 3 days (Williams and Arias, 1978). Concentrated aqueous extracts from stems of *Prosopis juliflora* inhibited the germination of *Euphorbia caducifolia* seeds, and dilute ones promoted it. It was concluded that the leachates from *P. juliflora* would never be concentrated enough in nature to have any effect on the germination of *E. caducifolia* seeds (Sen and Chawan, 1970).

D. Dormancy and Germination of Herbaceous Perennials

A few herbaceous perennials have nondormant seeds, but most of them have seeds with some type of dormancy (Table 9.17). Because (1) members of the Liliaceae have underdeveloped embryos (Martin, 1946) and (2) seeds of *Allium, Bellevalia,* and *Tulipa* were stored for several months before studies were initiated (Boeken and Gutterman, 1990; Gutterman *et al.,* 1995),

TABLE 9.17   Dormancy Type and Optimum Temperature and Light:Dark (L:D) Requirements
for Germination of Seeds of Herbaceous Perennials of Hot Semideserts and Deserts[a]

Species	Family	Optimum germination temperature (°C)[b]	L:D[c]	Reference
Nondormant[d]				
Achyranthes aspera[e]	Amaranthaceae	30	L = D	Khan et al. (1984)
Baccharis neglecta	Asteraceae	30/15	L	Baskin et al. (unpublished results)
Dipteracanthus patulus*	Acanthaceae	Room	L > D	Sen and Chatterji (1968)
Eupatorium havanense	Asteraceae	15/6	L	Baskin et al. (unpublished results)
Gymnosperma glutinosum	Asteraceae	25/15	L	Baskin et al. (unpublished results)
Morphophysiological dormancy				
Allium rothii*	Liliaceae	15	D > L	Gutterman et al. (1995)
A. truncatum*	Liliaceae	15	D > L	Gutterman et al. (1995)
Bellevalia desertorum*	Liliaceae	10–15	L = D	Boeken and Gutterman (1990)
B. eigii*	Liliaceae	10–15	L = D	Boeker and Gutterman (1990)
Tulipa systola*	Liliaceae	10–15	L = D	Boeker and Gutterman (1990)
Morphological dormancy				
Aloe barbadense*,[e]	Liliaceae	Room	—[d]	Sen and Chatterji (1968)
Physical dormancy[d]				
Cassia holosericea	Fabaceae	30	L = D	Khan et al. (1984)
Clitoria ternata[e]	Fabaceae	35	L > D	Mullick and Chatterji (1967)
Convolvulus microphyllus*	Convolvulaceae	Room	—	Sen and Chatterji (1968)
Corchorus depressus*	Tiliaceae	Room	—	Sen and Chatterji (1968)
Crotalaria medicaginea[e]	Fabaceae	28	—	Bansal and Sen (1981)
Indigofera linnaei[e]	Fabaceae	20	—	Kumar and Chaudhary (1991)
I. miniata	Fabaceae	30/20	—	Kissock and Haferkamp (1983)
Medicago pudica	Fabaceae	—	—	Mishra et al. (1990)
M. sativa[e]	Fabaceae	20–35	L = D	Hammouda and Bakr (1969)
Sida veronicaefolia[e]	Malvaceae	Room	—	Chaudhary and Singh (1991a)
Tephrosia uniflora*,[e]	Fabaceae	Room	—	Sen and Chatterji (1968)
Physical and physiological dormancy				
Alysicarpus monilifer[e]	Fabaceae	—	—	Mishra et al. (1990)
Physiological dormancy				
Achillea santolina[e]	Asteraceae	10	L > D	Hammouda and Bakr (1969)
Andropogon gayanus	Poaceae	31/22	—	Elberse and Breman (1989)
Aster ericoides[e]	Asteraceae	25/15	—	Baskin et al. (unpublished results)
Boerhavia diffusa[e]	Nyctaginaceae	Room	—	Sen and Chatterji (1968)
Brickellia dentata	Asteraceae	35/20	L = D	Baskin et al. (unpublished results)
Caylusea hexagyna	Resedaceae	35	L > D	Hammouda and Bakr (1969)
Chaptalia nutans[e]	Asteraceae	20/10	—	Baskin et al. (1994)
Cleome arabica	Cleomaceae	30	L = D	Hammouda and Bakr (1969)
Echinops spinosissimus	Asteraceae	20	D > L	Hammouda and Bakr (1969)
Engelmannia pinnatifida*	Asteraceae	30/20	—	Kissock and Haferkamp (1983)
Erigeron modestus	Asteraceae	15/6	—	Baskin et al. (1994)
Fagonia arabica	Zygophyllaceae	30	L = D	Hammouda and Bakr (1969)
Gaillardia multiceps	Asteraceae	—	—	Secor and Farhadnejad (1978)
G. pinnatifida	Asteraceae	—	—	Secor and Farhadnejad (1978)
G. suavis	Asteraceae	25/15	—	Baskin et al. (1994)
Glossocardia setosa	Asteraceae	Room	L > D	Sen and Chatterji (1968)
Hymenopappus scabiosaeus	Asteraceae	20/10	—	Baskin et al. (1994)
Hymenoxys scaposa	Asteraceae	20/10, 25/15	—	Baskin et al. (1994)
Leptotherium senegalense	Poaceae	31/22	—	Elberse and Breman (1989)
Nicotiana trigonophylla[e]	Solanaceae	18–35	L	Wells (1959)
Nothosaerus brachiata	Acanthaceae	Room	L > D	Sen and Chatterji (1968)
Onopordum alexandrinum	Asteraceae	30–35	L = D	Hammouda and Bakr (1969)
Oryzopsis miliacea*,[e]	Poaceae	30/20	L > D	Koller and Negbi (1959)
Panicum turgidum*	Poaceae	20	D > L	Koller and Roth (1963)

(continues)

TABLE 9.17—*Continued*

Species	Family	Optimum germination temperature (°C)[b]	L:D[c]	Reference
Peganum harmala[*,e]	Zygophyllaceae	30	L = D	Hammouda and Bakr (1969), Hussain and Nasrin (1985)
Peristrophe bicalyculate[e]	Acanthaceae	30	L = D	Khan *et al.* (1984)
Pergularia daemia	Asclepiadaceae	30	D > L	Sen (1968)
Pinaropappus roseus	Asteraceae	20/10	—	Baskin *et al.* (1994)
Solanum surattense[*,e]	Solanaceae	Room	L	Sen and Chatterji (1968)
Stipa speciosa[*]	Poaceae	15	—	Young and Evans (1980)
Varilla texana[*]	Asteraceae	28	—	Pluenneke and Joham (1972)
Viguiera dentata[e]	Asteraceae	25/15	L	Baskin *et al.* (unpublished results)
Zygophyllum simplex	Zygophyllaceae	30	L > D	Hammouda and Bakr (1969)

[a]An asterisk means that the type of dormancy is inferred from available information on germination and on characteristics of seeds in that family.
[b]In some studies, only the germination temperature was used. If that temperature resulted in high germination percentages, it is listed for the species.
[c]L, light required for germination; D, darkness required for germination; L > D, seeds germinated to a higher percentage in light than in darkness; D > L, seeds germinated to a higher percentage in darkness than in light; and L = D, seeds germinated equally well in light and darkness.
[d]Data not available.
[e]Listed as a weed in Holm *et al.* (1979).

we assume that they have morphophysiological dormancy. However, studies with freshly matured seeds may demonstrate that they have morphological dormancy, like those of *Aloe bardadense* (Sen and Chatterji, 1968).

Herbaceous perennials belonging to the Convolvulaceae, Fabaceae, Malvaceae, and Tiliaceae have seeds with physical dormancy (Table 9.17), which is broken by mechanical or acid scarification (Hammouda and Bakr, 1969; Kissock and Haferkamp, 1983; Khan *et al.*, 1984). Also, seeds of some species such as *Corchorus tridens* become permeable during many months of dry storage (Sen and Chatterji, 1968). Because seeds of *Alysicarpus monilifer* (Fabaceae) have dormant embryos and impermeable seed coats, a period of dry storage and scarification are required for germination (Mishra *et al.*, 1990).

Physiological dormancy is very common among desert herbaceous perennials (Table 9.17). Dormancy in seeds of *Andropogon gayanus* was broken by 10 weeks of exposure to daily temperature regimes of 50/20°C, and it was broken in seeds of *Leptotherium senegalense* by 24 hr of washing in running tap water (Elberse and Breman, 1989). Seeds (caryopses) of *Oryzopsis miliacea* also have a water-soluble germination inhibitor that can be removed by leaching with simulated rainfall (Koller and Negbi, 1959). Seeds of *Caylusea hexagyna* (Hammouda and Bakr, 1969) and *Peristrophe bicalyculata* (Khan *et al.*, 1984) became nondormant during dry storage, and GA_3 increased the germination of *Varilla texana* seeds from 11 to 65% (Pluenneke and Joham, 1972).

Achenes of the spring-maturing Asteraceae *Chaptalia nutans, Erigeron modestus, Gaillardia suavis, Hymenopappus scabiosaeus, Hymenoxys scaposa,* and *Pinaropappus roseus* subjected to simulated habitat temperatures during summer germinated to 65–100% at simulated habitat temperatures in autumn (Baskin *et al.*, 1994). Thus, achenes come out of dormancy in the field during summer, but germination is delayed until temperatures decrease and the soil moisture becomes nonlimiting in autumn. Achenes of the autumn-maturing Asteraceae *Baccharis neglecta, Eupatorium havavense,* and *Gymnosperma glutinosum* were nondormant at maturity and germinated to 95–98% at simulated autumn temperatures (Baskin *et al.*, unpublished results). At the time of maturation in autumn, achenes of *Aster ericoides, Brickellia dentata,* and *Viguiera dentata* germinated to 87, 96, and 44% at simulated habitat temperatures, respectively. Thus, if the soil is moist at the time of achene dispersal, germination would occur immediately. Dormancy loss occurred in achenes of *V. dentata* sown on wet sand at 15/6, 20/10, 25/15, and 30/15°C; consequently, exposure to winter habitat temperatures (15/6, 20/10°C) would promote germination during winter, several months before the onset of high temperatures and drought conditions in spring (Baskin *et al.*, unpublished results).

Optimum germination temperatures for seeds of herbaceous perennials range from 10°C in *Achillea santolina* (Hammouda and Bakr, 1969) to 35°C in *Clitoria ternata* (Mullick and Chatterji, 1967) and *Caylusea hexagyna* (Hammouda and Bakr, 1969). The mean (±SE) of all the temperatures in Table 9.17 is 22.7 ± 1.0°C.

Data for light:dark requirements for germination are available for 31 species. Seeds of 6 species require light for germination, 8 germinate to higher percentages in light than in darkness, 12 germinate equally well in light and darkness, and 5 germinate to higher percentages in darkness than in light (Table 9.17).

Seeds of *Caylusea hexagyna, Zygophyllum simplex, Achillea santolina, Peganum harmola,* and *Echinops spinosissimus* germinated to 0% after 5 mm of rainfall; to 0, 30, 10, 18, and 72%, respectively, after 10 mm; and to 7, 64, 51, 14, and 56%, respectively, after 20 mm (Hammouda and Bakr, 1969). *Panicum turgidum* caryopses with the flat side in contact with water germinate to higher percentages than those with the convex side touching the wet surface (Koller and Roth, 1963). A pit and mound topography, resulting from the digging activities of porcupines, in the Negev Desert (Israel) causes runoff water to accumulate in depressions. Thus, a favorable microhabitat is created for germination and seedling establishment of many species (Gutterman, 1982b; Gutterman and Herr, 1981).

Osmotic potentials (MPa) of NaCl solutions that reduce germination from 80–100% to 50% are: *Achyranthes aspera,* −0.7 (Khan *et al.,* 1984); *Peristrophe bicalyculata,* −0.3 (Khan *et al.,* 1984); *Achillea santolina,* −1.13; *Caylusea hexagyna,* −0.75; *Peganum harmala,* −0.38; and *Zygophyllum simplex,* −0.75 (Hammouda and Bakr, 1969). A 0.5% (−0.41 MPa) NaCl solution reduced the germination of *Allium rothii* seeds in darkness at 20°C from 91 to 21% and that of *A. truncatum* from 92 to 25% (Gutterman *et al.,* 1995).

Seeds of *Gaillardia multiceps,* which is restricted to areas where gypsum is exposed, germinated to 60% in water and to 91% in 1000 ppm $CaSO_4$ at a pH of 7.8 (Secor and Farhadnejad, 1978). Magnesium salts inhibited the germination of *Eragrostis antidotale, E. curvula, E. lehmanniana,* and *E. superba* seeds more than calcium salts, and magnesium chloride was more inhibitory than magnesium sulfate. Seeds of *E. superba* and *E. curvula* were more tolerant of increased osmotic potentials than those of the other two species (Ryan *et al.,* 1975).

E. Dormancy and Germination of Vines

Seeds of *Citrullus colocynthis* collected in Israel did not germinate in either darkness or light if seed coats were intact, but decoated seeds germinated to 98 and 11% in darkness and light, respectively. Nonscarified seeds imbibed water; therefore, they did not have physical dormancy. Piercing the seed coat did not promote germination, and no inhibitors were found (Koller *et al.,* 1963). Seeds of *C. colocynthis* collected in Iraq germinated to 85 and 15% in darkness and light, respectively, without disruption of the seed coat (Datta and

Chakravarty, 1962). After the seed coat was removed, seeds collected in India germinated to 70–100 and 20% in darkness and light, respectively (Bhandri and Sen, 1975).

Seeds of *Citrullus colocynthis* have physiological dormancy that decreases during dry storage (El Hajein, 1992). Mechanical restriction of the seed coats and a water-soluble germination inhibitor enhance physiological dormancy of the embryo in delaying germination (El Hajein and Neville, 1992).

F. Annual Species

1. Types of Dormancy and Dormancy-Breaking Requirements

Although freshly matured seeds of a few annuals are nondormant, those of most species have some type of dormancy (Table 9.18). Members of the Liliaceae and Apiaceae probably have morphophysiological dormancy, but studies have been done on only one species, *Spermolepis echinata* (Apiaceae). Embryos in freshly matured seeds of *S. echinata* are 0.42 mm in length and must grow to 0.75 mm before seeds can germinate. However, embryo growth does not occur until seeds have come out of physiological dormancy during summer (Baskin *et al.,* unpublished results). Thus, seeds of *S. echinata* have nondeep, simple morphophysiological dormancy (see Chapter 5).

Annuals belonging to the Convolvulaceae, Fabaceae, and Malvaceae have seeds with physical dormancy (Table 9.18). Also, desert annuals belonging to the Geraniaceae (e.g., Nasir, 1983), a family known to have seeds with impermeable seed coats (Table 3.5), may have physical dormancy. Mechanical or acid scarification (Hammouda and Bakr, 1969; Bansal and Sen, 1981; Elberse and Breman, 1989; Chaudhary and Singh, 1991a) and heat (60–80°C) treatments (Hammouda and Bakr, 1969; Elberse and Breman, 1989) cause seeds with physical dormancy to become permeable. However, factors controlling the loss of physical dormancy in the field have received little research attention. The rapid germination of *Astragalus coronopus, A. hispidulus* (Loria and Noy-Meir, 1979–1980), *A. tribuloides* (Gutterman, 1986; Gutterman and Evenari, 1994), *Erodium bryoniifolium* (Loria and Noy-Meir, 1979–1980), *E. cicutarium* (Tevis, 1958b; Johnson *et al.,* 1978; Juhren *et al.,* 1956), *E. crassifolium* (Gutterman and Evenari, 1994), *E. texanum* (Tevis, 1958b), *Indigofera cordifolia, I. hochesttetri* (Aziz and Khan, 1993), *Malva aegyptia* (Gutterman 1986; Gutterman and Evenari, 1994), *Malvastrum exile* (Tevis, 1958b), *Medicago laciniata* (Gutterman, 1986), *Tephrosia strigosa* (Aziz and Khan, 1994), *Trigonella arabica* (Gutterman and Evenari,

TABLE 9.18   Dormancy Type and Optimum Temperature and Light:Dark (L:D) Requirements
for Germination of Seeds of Annuals of Hot Semideserts and Deserts[e]

Species	Family[a]	Optimum germination temperature (°C)[b]	L:D[c]	Reference
Nondormant[d]				
Eriogonum abertianum	Polygonaceae	15/6	L > D	Baskin et al. (1993)
Halopeplis amplexicaulis	Chenopodiaceae	20	L > D	Tremblin and Binet (1982)
Ocium basilicum[e]	Lamiaceae	38/27	L	Ismail et al. (1990)
Physalis philadelphica*	Solanaceae	30–34	L > D	Madrid et al. (1989)
Morphophysiological dormancy				
Asphodelus tenuifolius*,[e]	Liliaceae	20–25	L = D	Hammouda and Bakr (1969)
Daucus pusillus*,[e]	Apiaceae	30/15	—[d]	Barton (1936)
Eryngium creticum*,[e]	Apiaceae	15	D > L	Hammouda and Bakr (1969)
Spermolepias echinata	Apiaceae	20/10	L > D	Baskin et al. (unpublished results)
Physical dormancy				
Alysicarpus ovalifolius[e]	Fabaceae	31/22	—	Elberse and Breman (1989)
A. vaginalis[e]	Fabaceae	28	—	Bansal and Sen (1981)
Astragalus hamosus[e]	Fabaceae	30–35	—	Hammouda and Bakr (1969)
Cassia tora[e]	Fabaceae	31/22	—	Elberse and Breman (1989)
Corchorus tridens[e]	Tiliaceae	Room	—	Sen and Chatterji (1968)
Hippocrepis cyclocarpa	Fabaceae	20–25	—	Hammouda and Bakr (1969)
Indigofera astragalina	Fabaceae	31/22	—	Elberse and Breman (1989)
I. cordifolia	Fabaceae	28	—	Bansal and Sen (1981)
I. hochstetteri	Fabaceae	—	—	Mishra and Sen (1984)
I. linifolia	Fabaceae	30	—	Kumar and Chaudhary (1991)
I. prieuriana	Fabaceae	31/22	—	Elberse and Breman (1989)
Ipomoea pes-tigridis[e]	Convolvulaceae	30	—	Bhati and Sen (1978)
Malva parviflora[e]	Malvaceae	20–25	—	Hammouda and Bakr (1969)
Rhynchosia capitata	Fabaceae	30	—	N. Sharma et al. (1978)
Scorpiurus sulcata[e]	Fabaceae	20–25	—	Hammouda and Bakr (1969)
Sesbania bispinosa	Fabaceae	25	—	M. Sharma et al. (1978)
Sida acuta[e]	Malvaceae	Room	—	Chaudhary and Singh (1991a)
Tephrosia hamiltonii*	Fabaceae	Room	—	Sen and Chatterji (1968)
T. strigosa*	Fabaceae	Summer	—	Aziz and Khan (1994)
Trifolium resupinatum[e]	Fabaceae	20–25	—	Hammouda and Bakr (1969)
Trigonella maritima	Fabaceae	20–25	—	Hammouda and Bakr (1969)
Zornia glochidiata	Fabaceae	31/22	—	Elberse and Breman (1989)
Physiological dormancy				
Allenia autrani	Chenopodiaceae	5–15	L > D	Negbi and Tamari (1963)
Amaranthus caudatus[e]	Amaranthaceae	35	L = D	Gutterman et al. (1992)
Aristida contorta	Poaceae	30	L	Mott (1972)
Atriplex dimorphostegia	Chenopodiaceae	26	D > L	Koller (1957)
Borreria articularis[e]	Rubiaceae	26	L > D	Kasera and Sen (1987)
Bromus rubens[e]	Poaceae	20–25	L = D	Hammouda and Bakr (1969)
Carrichtera annua*	Brassicaceae	15	L > D	Gutterman (1990a)
Cenchrus biflorus[e]	Poaceae	35	—	Kumar et al. (1971)
Chloris pilosa	Poaceae	31/22	—	Elberse and Breman (1989)
Cleome papillosa*	Cleomaceae	Room	—	Sen and Chatterji (1968)
Coreopsis bigelovii	Asteraceae	20	—	Capon and Van Asdall (1967)
Cryptantha pectocarya	Boraginaceae	15/6	L	Baskin et al. (unpublished results)
Cucumis callosus*,[e]	Curcubitaceae	30	D > L	Bansal and Sen (1978)
Dactyloctenium aegyptium[e]	Poaceae	30	—	Kumar et al. (1971)
Dicoma tomentosa	Asteraceae	Room	L > D	Sen and Chatterji (1968)
Digera muricata[e]	Amaranthaceae	28	—	Bansal and Sen (1981)
Echium sericeum	Boraginaceae	30–35	D > L	Hammouda and Bakr (1969)
Eriastrum diffusum	Polemoniaceae	15/6	L > D	Baskin et al. (1993)
Eriophyllum wallacei	Asteraceae	20	—	Capon and Van Asdall (1967)
Euphorbia dracunculoides*,[e]	Euphorbiaceae	Room	—	Sen and Chatterji (1968)
E. geniculata[e]	Euphorbiaceae	25	—	Kigel et al. (1992)

(continues)

TABLE 9.18—*Continued*

Species	Family[a]	Optimum germination temperature (°C)[b]	L:D[c]	Reference
E. polycarpa	Euphorbiaceae	20	—	Capon and Van Asdall (1967)
*Fagonia cretica**	Zygophyllaceae	Room	—	Sen and Chatterji (1968)
Gaillardia pulchella[e]	Asteraceae	20/10	—	Baskin *et al.* (1992)
Geraea canescens	Asteraceae	20	—	Capon and Van Asdall (1967)
Glinus lotoides[*,e]	Aizoaceae	Room	—	Sen and Chatterji (1968)
Gynandropsis gynandra[*,e]	Cleomaceae	Room	—	Sen and Chatterji (1968)
Helichrysum cassinianum	Asteraceae	15–20	L	Mott (1972)
Helipterum craspedioides	Asteraceae	15	L = D	Mott (1972), Mott and McComb (1975)
Hordeum spontaneum[e]	Poaceae	20	D > L	Gutterman and Nevo (1994), Gutterman *et al.* (1996)
Hymenoxys linearifolia	Asteraceae	25/15	—	Baskin *et al.* (1992)
H. odorata[*,e]	Asteraceae	20	L = D	Whisenant and Ueckert (1982)
Krigia gracilis	Asteraceae	20/10	—	Baskin *et al.* (1992)
Lepidium lasiocarpum	Brassicaceae	25	—	Barton (1936), Capon and Van Asdall (1967)
Lindheimera texana	Asteraceae	28/15	—	Baskin *et al.* (1995)
Mesembryanthemum nodiflorum	Aizoaceae	35/15	L > D	Gutterman (1980, 1981, 1990a)
Plantago albicans	Plantaginaceae	20–24	L = D	Hammouda and Bakr (1969)
P. coronopus	Plantaginaceae	20	L > D	Gutterman and Shem-Tov (1996)
P. ovata	Plantaginaceae	15	L = D	Hammouda and Bakr (1969)
P. purshii[e]	Plantaginaceae	20/10	L	Baskin *et al.* (unpublished results)
P. notata	Plantaginaceae	20–25	L = D	Hammouda and Bakr (1969)
Polypogon monspeliensis[*,e]	Poaceae	Room	L = D	Sen and Chatterji (1968)
Portulaca oleracea[e]	Portulacaceae	30	L > D	Chaudhary and Sinha (1990, 1991a,b)
P. quadrifida[e]	Portulacaceae	30	L > D	Chaudhary and Sinha (1990, 1991a,b)
Pyrrhopappus multicaulis	Asteraceae	15/6	—	Baskin *et al.* (1992)
Salsola volkensii	Chenopodiaceae	15–20	L	Negbi and Tamari (1963)
Salvia columbariae	Lamiaceae	20	—	Capon and Van Asdall (1967)
Schismus arabicus	Poaceae	20	D > L	Gutterman (1996a)
Sisymbrium altissium[e]	Brassicaceae	20	—	Capon and Van Asdall (1967)
Spergularia diandra	Caryophyllaceae	15	L > D	Gutterman (1994)
Streptanthus arizonicus	Brassicaceae	30/15	L > D	Barton (1936), Capon Van Asdall (1967)
Vaccaria pyramidata[*,e]	Caryophyllaceae	Room	L	Sen and Chatterji (1968)

[a] An asterisk means that the type of dormancy is inferred from available information on germination and on characteristics of seeds in that family.

[b] In some germination studies, only one germination temperature was used. If that temperature resulted in high germination percentages, it is listed for the species.

[c] L, light required for germination; D, darkness required for germination; L > D, seeds germinated to a higher percentage in light than in darkness; D > L, seeds germinated to a higher percentage in darkness than in light; and L = D, seeds germinated equally well in light and darkness.

[d] Data not available.

[e] Listed as a weed in Holm *et al.* (1979).

1994), and *T. stellata* (Loria and Noy-Meir, 1979–1980) seeds following rains in the field strongly suggests that environmental factors (e.g., high temperatures and/or lack of moisture) during the hot, dry season in the desert play a role in breaking physical dormancy.

Many desert annuals, belonging to a diversity of plant families, have seeds with physiological dormancy (Table 9.18), which is broken during the dry, hot season. Therefore, seeds germinate at the beginning of the hot, wet (monsoon) season (e.g., Aziz and Khan, 1993, 1994) and/or the cool, wet season (e.g., Went and Westergaard, 1949; Loria and Noy-Meir, 1979–1980; Burk,

1982). In many cases, scientists working on germination of desert annuals may not have realized that freshly matured seeds of the species they were studying had physiological dormancy because they stored the seeds for several months at room temperature before starting their studies (e.g., Barton, 1936; Negbi and Tamari, 1963; Kumar *et al.*, 1971; Mott, 1972; Kasera and Sen, 1987; Chaudhary and Sinha, 1990). Because seeds with non-deep physiological dormancy frequently come out of dormancy (afterripen) during dry storage at room temperatures, they may have been nondormant by the time germination tests were initiated.

Temperature requirements for the loss of physiological dormancy in seeds of desert annuals are not well understood. Seeds of many species come out of dormancy during storage at room temperatures (e.g., Barton, 1936; Hammouda and Bakr, 1969), but those of others come out of dormancy during dry storage at 70/15°C (Mott, 1972), 50/20°C (Elberse and Breman, 1989), 50°C (Capon and van Asdall, 1967), 40°C (Chaudhary and Sinha, 1990), and 35–40°C (Gutterman and Nevo, 1994). The minimum temperatures at which dormancy loss will occur are not known. Do seeds of any species require cold stratification? Also, the effects of alternate wetting and drying and of various RH levels on the loss of physiological dormancy in seeds of desert annuals have not been determined.

2. Germination Requirements

Because annuals germinate in either the hot, wet or the cool, wet season (depending on the desert), it is not surprising that nondormant seeds of one group of species have high (25–35°C) and those of the other have low (15–20°C) optimum germination temperatures (Table 9.18). For a given species, soil temperatures may be either too high or too low, e.g., temperatures of 4–10°C as well as 25–30°C inhibit the germination of *Hordeum bulbosum* (Ellern and Tadmor, 1956). One consequence of unfavorable germination temperatures is a reduction in the germination rate (Elberse and Breman, 1989), thus the soil dries before seeds can germinate (Ellen and Tadmor, 1956).

Seeds of *Ocimum basilicum* require alternating temperatures for germination (Ismail *et al.*, 1990), and those of *Hyoscyamus muticus* and *Eremobium microcarpa* germinate to higher percentages at alternating than at constant temperatures (Hammouda and Bakr, 1969). Seeds of *Amaranthus caudatus* vary in their response to temperature, depending on the light : dark conditions (Gutterman *et al.*, 1992). Seeds of *A. caudatus* germinated to 100% in light and darkness at 35°C, but continuous light was increasingly inhibitory with a decrease in temperature.

Light : dark requirements for germination have been determined for only a small fraction of desert annuals. Light promotes germination of most species, and seeds of a few (*Aristida contorta, Helichrysum cassinianum, Ocium basilicum, Salsola volkensii,* and *Vaccaria pyramidata*) have an absolute light requirement for germination (Table 9.18). Darkness is not required for seed germination in any species, but seeds of *Atriplex dimorphostegia, Cucumis callosus, Echium sericeum, Eryngium creticum,* and *Hordeum spontaneum* germinate to higher percentages in darkness than in light (Table 9.18).

A little work has been done on the effects of pH and of salinity on the germination of desert annuals. Germination of *Hymenoxys odorata* seeds increased from 91 to 97% with an increase in pH from 5 to 9 (Whisenant and Ueckert, 1982). Seeds of *Dactyloctenium aegyptium* germinated to about 12% and those of *Cenchrus biflorus* to 28% at 0.2 M (−0.91 MPa) NaCl (Kumar *et al.*, 1971). Seeds of *Bromus rubens* germinated to 90% in 0.17 M (−0.76 MPa) NaCl and to 0% at 0.33 M (−1.5 MPa) NaCl (Hammouda and Bakr, 1969).

Because germination of desert annuals may be followed by wonderful displays of flowers, much attention has been given to the amount of rainfall required to promote germination. For example, after 5 mm of rain, no seeds of *Plantago albicans, P. notata, Bromus rubens,* or *Echium sericeum* germinated; after 10 mm they germinated to 32, 16, 70, and 10%, respectively; and after 20 mm to 40, 68, 50, and 50%, respectively (Hammouda and Bakr, 1969). Seeds of *Spergularia diandra* (Gutterman, 1994b) and *Schismus arabicus* (Gutterman, 1992) germinated after 10 mm of rain, and those of *Pectis angustifolia, Lappula redowskii,* and *Lepidium lasiocarpum* germinated after 15 mm (Freas and Kemp, 1983).

Studies have been done in various deserts, attempting to correlate the amount of precipitation with the number and kinds (species) whose seeds germinate (Table 9.19). These studies reveal that temperatures at the time of rainfall are important in controlling the timing of germination of desert annuals. If rains come during autumn and winter, when temperatures are relatively low, seeds of winter annuals germinate; however, if rains come during summer, when temperatures are high, seeds of summer annuals germinate (Went, 1948, 1949, 1955; Juhren *et al.*, 1956; Mott, 1972; Burk, 1982; Kemp, 1983). Winter annuals germinated in soil samples collected in the Mojave Desert and incubated at 18/13°C, and summer annuals germinated in those placed at 27/26°C (Went, 1949). Even during autumn and winter, the timing of rainfall may determine which species of winter annuals germinate. Winter annual members of the Polemoniaceae, Hydrophyllaceae, Polygonaceae, Fabaceae, and Onagraceae germinated in the Mojave Desert in response to rains in September and October, whereas members of the Brassicaceae and Boraginaceae germinated in December (Beatley, 1974). Irrigation of Mojave (Johnson *et al.*, 1978) and Sonoran (Tevis, 1958a) desert sites in summer stimulated the germination of summer but not winter annuals, and natural rainfall in autumn–winter promoted the germination of winter annuals in both deserts. Seeds of 28 of 31 winter annuals germinated in the Negev Desert (Israel) in response to irrigation during summer that kept the soil continuously moist for a long period of time (Gutterman, 1986).

TABLE 9.19   Studies in Which Germination Was Observed to Follow Rainfall (or Irrigation) in Hot Semideserts or Deserts

Desert/country	Amount (mm) of rainfall	Month or season	No. of species germinated	Reference
Chihuahuan/United States	Wet	Summer	15	Kemp (1983)
	Wet	Winter	23	
Egypt	Wet	December–February	12	Hammouda and Bakr (1969)
Middle East	Wet	—[a]	1	Datta (1961)
Mojave/United States	90	June	3	Johnson et al. (1978)
	40	August	7	
Mojave/United States	51	September	44	Beatley (1967, 1974)
	41	September	15	
	21	September	14	
	89	March–June	58	
Mojave/United States	75	September–February	45	Went and Westergaard (1949)
Mojave/United States	51	January–March	11	Tevis (1958b)
Mojave/United States	187	August–September	14	Juhren et al. (1956)
Mojave/United States	95	February	5	Juhren et al. (1956)
Negev/Israel	31	November, December	14	Loria and Noy-Meir (1979–1980)
Negev/Israel	Kept wet	Summer	22	Gutterman (1986)
Negev/Israel	150+	August 1 and 30	22	Gutterman and Evenari (1994)
Sonoran/United States	100	July, August, September	3	Tevis (1958a)
	100	October	8	Tevis (1958a)
	100	November	7	Tevis (1958a)
	100	December	10	Tevis (1958a)
Sonoran/United States	140	December–February	16	Burk (1982)
	950	August	2	Burk (1982)
	180	September–October	12	Burk (1982)
Southern Australia	—[b]	Autumn–winter	4	Noble and Crisp (1979–1980)
Western Australia	70	Winter	—	Mott (1972)

[a] Information not given.
[b] Data not available.

The physiological state of seeds is an important aspect in the control of timing of germination. In a study of two annuals (*Eriogonum abertianum* and *Eriastrum diffusum*) in the Chihuahuan Desert, seeds were buried in soil in the field, and at various times for 16 months they were exhumed and tested for germination in light and darkness at 15/6, 20/10, 25/15, 30/15, and 35/20°C. Seeds of *E. diffusum* were dormant at maturity in May, and by October 65 and 38% of them germinated in light at 15/6 and 20/10°C, respectively (Fig. 9.5). Most of the nondormant seeds reentered dormancy during winter, but by the following October 92 and 58% of them germinated in light and darkness, respectively. Thus, like obligate winter annuals of temperate regions (Baskin and Baskin, 1985), most seeds of *E. diffusum* have an annual dormancy/nondormancy cycle.

Seeds of *Eriogonum abertianum* were nondormant at maturity in autumn and germinated to 87% in light at 15/6 and 20/10°C (Fig. 9.6). Maximum germination of seeds in darkness (37%) occurred in January/February. Seeds entered conditional dormancy during winter, and germination in light at 20/10°C decreased to 6%. Seeds

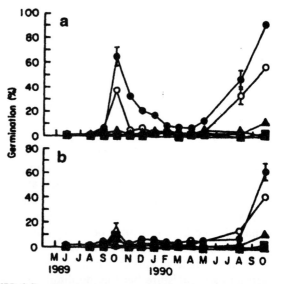

FIGURE 9.5   Germination percentages (mean ± SE, if ≥ 5%) of seeds of *Eriastrum diffusum* seeds exhumed after 0–16 months of burial in soil in the Chihuahuan Desert in Arizona and incubated at (a) a 14-hr daily photoperiod and (b) in continuous darkness. From Baskin et al. (1993), with permission.

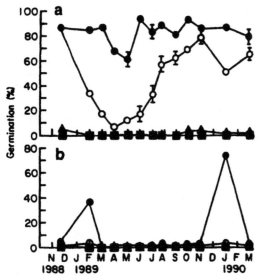

FIGURE 9.6 Germination percentages (mean ± SE, if ≥ 5%) of *Eriogonum abertianum* seeds exhumed after 0–16 months of burial in soil in the Chihuahuan Desert in Arizona and incubated at (a) a 14-hr daily photoperiod and (b) in continuous darkness. From Baskin *et al.* (1993), with permission.

were nondormant by the following autumn. Thus, like facultative winter annuals (Baskin and Baskin, 1980), seeds exhibited an annual conditional dormancy/nondormancy cycle.

Low temperatures are required for the germination of nondormant seeds of both *Eriastrum diffusum* and *Eriogonum abertianum;* therefore, germination is not possible in the summer, regardless of the amount of rain. *Eriastrum diffusum* has a shorter germination season than *E. abertianum* because most *E. diffusum* seeds that fail to germinate in autumn enter dormancy during winter. However, seeds of *E. abertianum* lose the ability to germinate at high (20/10) but not at low (15/6°C) temperatures during winter. Thus, *E. abertianum* seeds can germinate at low temperatures in autumn, winter, or spring.

Most populations of *Euphorbia geniculata* in Israel had nondormant seeds, and buried seeds did not enter dormancy. Exhumed seeds germinated to high percentages at 27/22°C, but no germination occurred at temperatures lower than 10 or 15°C, depending on the source of seed stock (Kigel *et al.,* 1992). Thus, seeds did not germinate in the field in winter because temperatures were below those required for germination. Seeds germinated in the field in spring, when temperatures increased and soil moisture became nonlimiting. No annual dormancy cycles were found in buried seeds of *E. geniculata.* In fact, because seeds germinated while they were buried, this species has a transient seed bank only (Kigel *et al.,* 1992).

One explanation for the germination of seeds following a rain storm, in addition to there being enough water present for the seeds to become fully imbibed, is that germination inhibitors are leached from the seeds. Thus, the amount of rain required to remove the inhibitors is thought to equal the amount of moisture required for seedling establishment and eventual maturation of the plant (Went, 1955). Although leaching has been shown to stimulate germination in the Mohave Desert (Went, 1949), no germination inhibitors have been isolated from seeds of species growing in this desert. In Israel, an unknown inhibitor has been found in the fruit bracts of *Atriplex dimorphostegia* (Kadman-Zahavi, 1955), and the inhibitor monoepoxylignanolide occurs in caryopses of *Aegilops ovata* (Lavie *et al.,* 1974).

Much remains to be learned about the role of inhibitors in the germination ecology of desert species and they need to be evaluated in relation to the type of dormancy present in seeds. An enhancement of germination following the leaching of seeds does not necessarily prove that an inhibitor has been removed. Seeds require several days of continuous imbibition for a dormancy break to occur, or germination rates may be slow the first time seeds are imbibed at the end of a long hot, dry season. Thus, if seeds are leached for a period of time, things other than the removal of inhibitors may be happening that result in germination. Experiments in which seeds are allowed to imbibe, but where no leaching is allowed to occur, would be very informative.

Another aspect of rainfall is that water dilutes the concentration of salts, which may inhibit germination via osmotic and/or toxic effects. In fact, the germination of nonhalophytes as well as halophytes is promoted when rainwater reduces salinity (Tremblin and Binet, 1982; Gutterman, 1980–1981, 1982; Mohammed and Sen, 1990).

Some species of desert annual grasses, including *Avena geniculata, A. barbata, A. sterilis, A. wiestii,* and *Stipa capensis,* have hygroscopic awns (Gutterman, 1992). These awns twist as they dry and become straight when they are wet; consequently, dispersal units move along the soil until they lodge in a depression. Also, awns play a role in the burial of dispersal units (Gutterman, 1992).

3. Germination Models

The unpredictability of rain in the desert habitat has served as a stimulus for some biologists to make models for the germination of desert annuals. Cohen's (1966) model for optimization of reproduction in a randomly varying environment has received much attention from seed ecologists:

$$S_{t+1} = S_t - S_t G - D(S_t - S_t G) + GY_t S_t,$$

where S is the number of seeds present; G is the fraction of seeds germinating each year; D is the fraction of seeds decaying (dying) each year; Y is the average number of seeds produced per germinated seedling, which depends on environmental conditions and is assumed to be independent of population density; and t is time.

The number of seeds of a particular species that germinates each year (G) increases with the probability that there will be enough rain for maturation of the plants and decreases as the probability of inadequate rainfall for seed production increases.

Venable and Lawlor's (1980) model considers the effects of dispersal and of predictive dormancy on the optimal germination fraction:

$$\lambda_{total} = D^1[u(GS + 1 - G) + v(1 - g) + w] + (1 - D^1) \\ (GS + 1 - G)^u (1 - g)^v,$$

where D is the proportion of newly produced seeds that disperses to a new site, D^1 is the fraction of all seeds that disperses (may include some seeds produced in previous years, but unlikely), G is the fraction of seeds germinating, S is the seed set of a successful plant, u is the probability of germination and reproduction, v is the probability of germination and death, w is the amount of predictive dormancy, and λ is the finite rate of increase per unit of time (e.g., year^{-1}).

As the number of seeds that become dispersed in space increases, the number of dormant seeds decreases. In those species with dimorphic seeds, reproduction is at a maximum if the seeds with the best dispersal mechanism are nondormant and those that are not so easily dispersed are dormant. One of the questions addressed with Venable and Lawlor's (1980) model is how many seeds should remain ungerminated (to protect against inaccurate predictions of the environment) in years when the environment is predicted to be good for subsequent seed set? That is, what kind of seed reserve should be left at the end of the germination season to ensure that the species does not become extinct at that population site if the weather, namely rainfall, becomes so unfavorable that all the seedlings/juveniles die before they have time to produce seeds? If environmental conditions that stimulate germination are not the same as those that permit seed set, lack of germination when the environment is predicted to be unfavorable for seed set has little or no effect on the optimal number of seeds with innate dormancy, i.e., those that would remain in the seed reserve. However, if the environmental conditions that prevent germination also are unsuitable for seed set, the optimal number of seeds with innate dor-

mancy increases as germination is increasingly controlled by predicted environmental favorability.

There are various reasons why seeds of a desert annual might not germinate. (1) Even after seeds have been subjected to dormancy-breaking treatments, such as exposure to summer temperatures, some of them may remain dormant (Freas and Kemp, 1983). (2) If light-requiring seeds are buried, and thus are in darkness, they cannot germinate. (3) In a species such as *Eriastrum diffusum,* if rains do not come when seeds are capable of germinating at late autumn–winter temperatures, seeds reenter dormancy. (4) Field studies in Arizona have revealed that high densities of seedlings can prevent additional seeds of desert annuals from germinating (Inouye, 1980).

A model to predict the timing of germination of ephemeral species in arid regions of South Australia was based on soil moisture data, soil surface temperatures, and germination phenology data; no equation was given (Noble and Crisp, 1979–1980). When soil moisture in the top 10 cm of soil exceed a threshold value of 5% by volume, degree-day units (figured on basing daily mean air temperature $>8°C$) began to accumulate. Germination occurred when accumulated degree-day units were greater than 75°C. The model accurately predicted the majority of times when germination occurred in the field and indicated that ephemeral species in the habitat have similar germination requirements. It only predicted the timing of germination of nondormant seeds and did not consider changes in dormancy states during the year.

As more information becomes available on the season-to-season requirements for the germination of seeds of desert annuals, it seems likely that additional germination models will be produced. In the two species (*Eriastrum diffusum* and *Eriogonum abertianum*) that have been studied, two different dormancy patterns were found. The annual dormancy/nondormancy cycle in *E. diffusum* is like the pattern found in annual species that occur in habitats where the favorable period for germination and completion of the life cycle is predictable (Baskin and Baskin, 1980, 1983a, 1984a). However, the annual dormancy/conditional dormancy cycle in *E. abertianum* is like the pattern found in annual species that occur in habitats where the favorable period for germination and seed maturation is unpredictable (Baskin and Baskin, 1977, 1978, 1981, 1983b,c, 1984b, 1986; Baskin *et al.,* 1987; Roberts and Boddrell, 1983; Roberts and Lockett, 1978; Roberts and Neilson, 1982a,b). Models for both types of desert annuals might be helpful in looking at possible evolutionary origins and relationships of the *Eriastrum* and *Eriogonum* types of patterns.

G. Special Factor: Water as It Relates to Dispersal

Adaptations for long-distance dispersal are uncommon in desert plants, and some species have adaptations that hinder seeds from being dispersed very far from the mother plant; these adaptations are called antitelechory (Ellner and Shmida, 1981). Ellner and Shmida (1981) list the major selective pressures for antitelechory as (1) regulating the within-season timing of germination, (2) spreading dispersal and germination over several years, (3) protection from predators, (4) anchorage against surface runoff, and (5) enhancing water uptake by seeds and seedlings. Some of these selective pressures for antitelechory are related to the water dispersal of seeds from the dead mother plant.

Within-season timing of germination is regulated in a number of species by rainfall that not only wets the soil for germination but is the dispersal agent for seeds. Seed dispersal by rain is called ombrohydrochory (Table 9.20). In species with ombrohydrochory, seeds are retained on the mother plant after they mature, and their dispersal is triggered by rain. Depending on the species, rain causes fruits to separate from plants; capitula to fall apart; capsules, pods, or capitula to open; bracts,

sepals, pedicels, or umbel rays to open; or capsules to explode (Gutterman, 1990b). The number of seeds released in some species is directly correlated with the amount of rainfall (Ellner and Shmida, 1981), and when the rain stops the structures close, preventing further dispersal until it rains again (Gutterman, 1981, 1982a). Thus, dispersal of seeds produced by a particular plant is spread over time.

The storage of reserve seeds on the mother plant in ombrohydrochorous species means that they are less likely to be eaten by seed predators than are those lying on the soil surface and that they are not swept away in flash floods (Ellner and Shmida, 1981). Seeds of some species, including *Anastatica hierochuntica* (Gutterman, 1990b), *Artemisia sieberi* (Gutterman, 1996b), *Blepharis ciliaris* (Gutterman and Witztum, 1977), *B. persica* (Gutterman *et al.*, 1973), *Carrictera annua* (Gutterman, 1990b), *Diplotaxis harra, Helianthemum vesicarum* (Gutterman, 1996b), *Plantago coronopus* (Gutterman, 1990b; Gutterman and Shem-Tov, 1996), *P. fastigiata* (Mauldin, 1943) and *Reboudia pinnata* (Gutterman, 1983), become covered with a layer of mucilage as soon as they come in contact with a wet substrate. Mucilage restricts the diffusion of oxygen to seeds under flooded conditions, which prevents germination if they are under

TABLE 9.20   Desert Annuals Whose Seeds are Dispersed by Rain (Ombrohydrochory)[a]

Species	Family	Reference
Aizoon hispanicum	Aizoaceae	Gutterman (1981)
Ammi visnaga	Apiaceae	Gutterman (1990b)
Anastatica hierochuntica	Brassicaceae	Friedman *et al.* (1978)
Anthemis pesudocotula	Asteraceae	Gutterman (1990b)
Asteriscus pygmaeus	Asteraceae	Gutterman and Ginott (1994a)
Astragalus tribuloides	Fabaceae	Gutterman (1990)
Blepharis ciliaris	Acanthaceae	Gutterman (1982a)
Carrichtera annua	Brassicaceae	Gutterman (1981, 1982a, 1994a)
Cichorium pumilum	Asteraceae	Gutterman (1990b)
Filago contracta	Asteraceae	Gutterman (1990b)
Gymnarrhena micrantha	Asteraceae	Gutterman (1982a)
Lepidium aucheri	Brassicaceae	Gutterman (1990b)
Mesembryanthemum nodiflorum	Aizoaceae	Gutterman (1981)
Pallinis spinosa	Asteraceae	Gutterman (1990b)
Phagnalon rupestre	Asteraceae	Gutterman (1990b)
Plantago coronopus	Plantaginaceae	Gutterman (1982a)
Reboudia pinnata	Brassicaceae	Gutterman (1990b)
Salvia horminum	Lamiaceae	Gutterman (1990b)
Trigonella stellata	Fabaceae	Gutterman (1982a)

[a] Only one species is listed per genus.

water (Gutterman *et al.*, 1973). Mucilage also anchors or attaches seeds to the soil (Gutterman, 1996b), delays drying of soil under the seed (Gutterman *et al.*, 1973), and prevents desiccation of the seed (Ismail *et al.*, 1990).

References

Abd El Rahman, A. A. (1986). The deserts of the Arabian Peninsula. *In* "Ecosystems of the World" (M. Evenari, I. Noy-Meir, and D. W. Goodall, eds.), Vol. 12B, pp. 29–54. Elsevier, Amsterdam.

Ackerman, T. (1979). Germination and survival of perennial plant species in the Mojave Desert. *Southw. Nat.* **24**, 399–408.

Acuna, P. I., and Garwood, N. C. (1987). Efecto de la luz y de la escarificacion en la germinacion de las semillas de cinco especies de arboles tropicales secundarios. *Rev. Biol. Trop.* **35**, 203–207.

Afolayan, A. J., and Olugbami, S. S. (1993). Seed germination and emergence of *Setaria pallidefusca* and *Pennisetum pedicellatum* (Cyperales: Poaceae) in Nigeria. *Rev. Biol. Trop.* **41**, 23–26.

Agami, M. (1986). The effects of different soil water potentials, temperature and salinity on germination of seeds of the desert shrub *Zygophyllum dumosum. Physiol. Plant.* **67**, 305–309.

Ahlawat, A. S., and Dagar, J. C. (1980). Effect of different pH, light qualities and some growth regulators on seed germination of *Bidens biternata* (Lour.) Merr. and Sherff. *Indian For.* **106**, 617–620.

Aikman, J. M. (1955). The ecology of balsa (*Ochroma lagopus* Swartz) in Ecuador. *Proc. Iowa Acad. Sci.* **62**, 245–252.

Akamine, E. K. (1947). Germination of *Asystasia gangetica* L. seed with special reference to the effect of age on the temperature requirement for germination. *Plant Physiol.* **22**, 603–607.

Alcorn, S. M., and Kurtz, E. B., Jr. (1959). Some factors affecting the germination of seed of the saguaro cactus (*Carnegiea gigantea*). *Am. J. Bot.* **46**, 526–529.

Alencar, J. da Cruz, and Magalhaes, L. M. S. (1979). Poder germinativo de sementes de doze especies florestais da regiao de Manaus. I. *Acta Amazonica* **9**, 411–418.

Alvarenga, S. (1988). Morfologia y germinacion de la semilla de caoba, *Swietenia macrophylla* King (Meliaceae). *Rev. Biol. Trop.* **36**, 261–267.

Aminuddin, M., and Ng, F. S. P. (1982). Influence of light on germination of *Pinus caribaea, Gmelina arborea, Sapium baccatum* and *Vitex pinnata. Malaysian For.* **45**, 62–68.

Andrew, M. H., and Mott, J. J. (1983). Annuals with transient seed banks: The population biology of indigenous *Sorghum* species of tropical north-west Australia. *Aust. J. Ecol.* **8**, 265–276.

Anoliefo, G. O., and Gill, L. S. (1992). Seed germination of *Peltophorum inerme* (Roxb.) Llanos (Leguminosae). *Int. J. Trop. Agric.* **10**, 246–253.

Aquila, M. E. A., and Ferreira, A. G. (1983). Germinacao de sementes escarificadas de *Araucaria angustifolia* em solo. *Cien. Cult.* **36**, 1583–1589.

Arasaki, F. R., and Felippe, G. M. (1987). Germinacao de *Dalbergia violacea*, uma especie dos cerrados. *Rev. Brasil. Biol.* **47**, 457–463.

Arias, I., and Lemus, L. (1984). Interaction of light, temperature and plant hormones in the germination of seeds of *Melocactus caesius* Went. (Cactaceae). *Acta Cien. Venezolana* **35**, 151–155.

Ataroff, M., and Schwarzkopf, T. (1992). Leaf production, reproductive patterns, field germination and seedling survival in *Chamaedorea bartlingiana*, a dioecious understory palm. *Oecologia* **92**, 250–256.

Athaya, C. D. (1985). Ecological studies of some forest tree seeds. I. Seed morphology. *Indian J. For.* **8**, 33–36.

Athaya, C. D. (1990). Seed dormancy studies of some forest tree seeds. *In* "International Symposium on Environmental Influences on Seed and Germination Mechanism: Recent Advances in Research and Technology" (D. N. Sen, S. Mohammed, P. K. Kasera, and T. P. Thomas, eds.), pp. 52–53. University of Jodhpur, Jodhpur, India. [Abstract]

Augspurger, C. K. (1979). Irregular rain cues and the germination and seedling survival of a Panamanian shrub (*Hybanthus prunifolius*). *Oecologia* **44**, 53–59.

Augspurger, C. K. (1984). Light requirements of neotropical tree seedlings: A comparative study of growth and survival. *J. Ecol.* **72**, 777–795.

Augspurger, C. K. (1986). Double- and single-seeded indehiscent legumes of *Platypodium elegans*: Consequences for wind dispersal and seedling growth and survival. *Biotropica* **18**, 45–50.

Auld, B. A. (1976). The biology of *Bassia birchii* (F. Muell.) F. Muell. *Weed Res.* **16**, 323–330.

Auld, B. A., and Martin, P. M. (1975). The autecology of *Eupatorium adenophorum* Spreng. in Australia. *Weed Res.* **15**, 27–31.

Ayaz, S., Hussain, F., and Ilahi, I. (1992). Germination strategy of *Cymbopogon jwarancusa* (Jones) Schult. *J. Sci. Technol. Univ. Peshawar* **16**, 59–64.

Ayyad, M. A., and Ghabbour, S. I. (1986). Hot deserts of Egypt and the Sudan. *In* "Ecosystems of the World" (M. Evenari, I. Noy-Meir, and D. W. Goodall, eds.), Vol. 12B. pp. 149–202. Elsevier, Amsterdam.

Aziz, S., and Khan, M. A. (1993). Survivorship patterns of some desert plants. *Pakistan J. Bot.* **25**, 67–72.

Aziz, S., and Khan, M. A. (1994). Life history strategies of *Tephrosia strigosa* Willd.—a desert summer annual. *Bangladesh J. Bot.* **23**, 139–146.

Bachelard, E. P. (1985). Effects of soil moisture stress on the growth of seedlings of three eucalypt species. I. Seed germination. *Aust. For. Res.* **15**, 103–114.

Bahuguna, V. K., and Lal, P. (1993). Introduction trials on *Acacia caven* at nursery stage germination behaviour and growth of the seedlings under Dehra Dun conditions. *Indian For.* **119**, 905–910.

Bahuguna, V. K., and Lal, P. (1994). Introduction trials on *Grevillea pteridifolia* (syn. *G. banksii*) germination behaviour of seeds at nursery stage under Dehra Dun climatic conditions. *Indian For.* **120**, 213–219.

Bannister, B. A. (1970). Ecological life cycle of *Euterpe globosa* Gaertn. *In* "A Tropical Rain Forest" (H. T. Odum, ed.), pp. B299-B314. U.S. Atomic Energy Commission, Oak Ridge, TN.

Bansal, R. P., and Sen, D. N. (1978). Contributions to the ecology and seed germination of *Cucumis callosus. Folia Geobot. Phytotax.* **13**, 225–233.

Bansal, R. P., and Sen, D. N. (1981). Differential germination behaviour in seeds of the Indian arid zone. *Folia Geobot. Phytotax.* **16**, 317–330.

Barbour, M. G. (1968). Germination requirements of the desert shrub *Larrea divaricata. Ecology* **49**, 915–923.

Barnard, R. C. (1956). Recruitment, survival and growth of timber-tree seedlings in natural tropical rain forest. *Malaysian For.* **19**, 156–161.

Barradas, M. M., and Handro, W. (1974). Algumas observacoes sobre a germinacao da semente do barbatimao, *Stryphnodendron barbadetimam* (Vell.) Mart. (Leguminosae-Mimosoideae). *Bol. Bot. Univ. Sao Paula* **2**, 139–150.

Barrueto, L. P., Da P. Pereira, I., and Neves, M. A. (1986). Influencia da maturacao fisiologica e do periodo entre a coleta e o inicio do armazenamento, sobre a viabilidade da semente de seringueira. *Turrialba* **36**, 65–75.

Barton, L. V. (1936). Germination of some desert seeds. *Contrib. Boyce Thompson Inst.* **8**, 7–11.

Baskin, C. C., Baskin, J. M., and Van Auken, O. W. (1992). Germination response patterns to temperature during afterripening of achenes of four Texas winter annual Asteraceae. *Can. J. Bot.* **70,** 2354–2358.

Baskin, C. C., Baskin, J. M., and Van Auken, O. W. (1994). Germination response patterns during dormancy loss in achenes of six perennial Asteraceae from Texas, USA. *Plant Species Biol.* **9,** 113–117.

Baskin, C. C., Baskin, J. M., and Van Auken, O. W. (1995). Temperature requirements for dormancy break and germination in achenes of the winter annual *Lindheimera texana* (Asteraceae). *Southw. Nat.* **40,** 268–272.

Baskin, C. C., Chesson, P. L., and Baskin, J. M. (1993). Annual seed dormancy cycles in two desert winter annuals. *J. Ecol.* **81,** 551–556.

Baskin, J. M., and Baskin, C. C. (1977). Role of temperature in the germination ecology of three summer annual weeds. *Oecologia* **30,** 377–382.

Baskin, J. M., and Baskin, C. C. (1978). Seasonal changes in the germination responses of *Cyperus inflexus* seeds to temperature and their ecological significance. *Bot. Gaz.* **139,** 231–235.

Baskin, J. M., and Baskin, C. C. (1980). Ecophysiology of secondary dormancy in seeds of *Ambrosia artemisiifolia. Ecology* **61,** 475–480.

Baskin, J. M., and Baskin, C. C. (1981). Seasonal changes in the germination responses of buried *Lamium amplexicaule* seeds. *Weed Res.* **21,** 299–306.

Baskin, J. M., and Baskin, C. C. (1983a). Seasonal changes in the germination responses of buried seeds of *Arabidopsis thaliana* and ecological interpretation. *Bot. Gaz.* **144,** 540–543.

Baskin, J. M., and Baskin, C. C. (1983b). Seasonal changes in the germination responses of fall panicum to temperature and light. *Can. J. Plant Sci.* **63,** 973–979.

Baskin, J. M., and Baskin, C. C. (1983c). The germination ecology of *Veronica arvensis. J. Ecol.* **71,** 57–68.

Baskin, J. M., and Baskin, C. C. (1984a). Role of temperature in regulating timing of germination in soil seed reserves of *Lamium purpureum. Weed Res.* **24,** 341–349.

Baskin, J. M., and Baskin, C. C. (1984b). Effect of temperature during burial on dormant and nondormant seeds of *Lamium amplexicaule* and ecological implications. *Weed Res.* **24,** 333–339.

Baskin, J. M., and Baskin, C. C. (1985). The annual dormancy cycle in buried weed seeds: A continuum. *BioScience* **35,** 492–498.

Baskin, J. M., and Baskin, C. C. (1986). Seasonal changes in the germination responses of buried witchgrass (*Panicum capillare*) seeds. *Weed Sci.* **34,** 22–24.

Baskin, J. M., Baskin, C. C., and McCormick, J. F. (1987). Seasonal changes in germination responses of buried seeds of *Portulaca smallii. Bull. Torrey Bot. Club* **114,** 169–172.

Batanouny, K. H., and Ziegler, H. (1971a). Eco-physiological studies on desert plants. II. Germination of *Zygophyllum coccineum* L. seeds under different conditions. *Oecologia* **8,** 52–63.

Batanouny, K. H., and Ziegler, H. (1971b). Eco-physiological studies on desert plants. V. Influence of soaking and redrying on germination of *Zygophyllum coccineum* seeds and the possible contribution of an inhibitor to the effect. *Oecologia* **8,** 209–217.

Baxter, B. J. M., van Staden, J., Granger, J. E., and Brown, N. A. C. (1994). Plant-derived smoke and smoke extracts stimulate seed germination of the fire-climax grass *Themeda triandra. Environ. Exp. Bot.* **34,** 217–223.

Beard, J. S. (1944). Climax vegetation in tropical America. *Ecology* **25,** 127–158.

Beatley, J. C. (1967). Survival of winter annuals in the northern Mojave desert. *Ecology* **48,** 745–750.

Beatley, J. C. (1974). Phenological events and their environmental triggers in Mojave desert ecosystems. *Ecology* **55,** 856–863.

Becwar, M. R., Stanwood, P. C., and Leonhardt, K. W. (1983). Dehy-dration effects on freezing characteristics and survival in liquid nitrogen of desiccation-tolerant and desiccation-sensitive seeds. *J. Am. Soc. Hort. Sci.* **108,** 613–618.

Bell, C. R. (1970). Seed distribution and germination experiment. *In* "A Tropical Rain Forest" (H. T. Odum, ed.), pp. D177–D182. U.S. Atomic Energy Commission, Oak Ridge, TN.

Bell, W. E., and van Staden, J. (1993). Seed structure and germination of *Dichrostachys cinerea. S. Afr. J. Bot.* **59,** 9–13.

Benedict, H. M., and Robinson, J. (1946). Studies on germination of guayule seed. *U.S. Dept. Agric. Tech. Bull. No.* 921.

Beniwal, B. S., Dhawan, V. K., and Joshi, S. R. (1989). Effect of shade and mulch on germination of *Chukrasia velutina,* Roemer. *Indian For.* **115,** 869–874.

Beniwal, B. S., Dhawan, V. K., and Joshi, S. R. (1990a). Effect of shade and mulch on germination of *Toona ciliata,* Roem. *Indian For.* **116,** 942–945.

Beniwal, B. S., Joshi, S. R., and Dhawan, V. K. (1990b). Effect of shade and mulch on germination of *Adina cordifolia* Hook. *Indian For.* **116,** 202–205.

Beniwal, B. S., and Singh, N. B. (1989). Observations on flowering, fruiting and germination behaviours of some useful forest plants of Arunachal Pradesh. *Indian For.* **115,** 216–227.

Benzing, D. H. (1990). "Vascular Epiphytes: General Biology and Related Biota." Cambridge Univ. Press, Cambridge, UK.

Bhandri, M. C., and Sen, D. N. (1975). Ecology of desert plants and observations on their seedlings. *Acta Agron. Acad. Sci. Hung.* **24,** 411–416.

Bhardwaj, S. D., and Chakraborty, A. K. (1994). Studies on time of seed collection, sowing and presowing seed treatments of *Terminalia bellirica* Roxb. and *Terminalia chebula* Retz. *Indian. For.* **120,** 430–439.

Bhat, S. R., Bhat, K. V., and Chandel, K. P. S. (1994). Studies on germination and cryopreservation of *Musa balbisiana* seed. *Seed Sci. Technol.* **22,** 637–640.

Bhati, P., and Sen, D. N. (1978). Temperature responses of seeds in *Ipomoea pes-tigridia* L. *Biol. Plant.* **20,** 221–224.

Birch, W. R. (1957). The seeding and germination of some Kenya weeds. *J. Ecol.* **45,** 85–91.

Blain, D., and Kellman, M. (1991). The effect of water supply on tree seed germination and seedling survival in a tropical seasonal forest in Veracruz, Mexico. *J. Trop. Ecol.* **7,** 69–83.

Blydenstein, J. (1967). Tropical savanna vegetation of the llanos of Colombia. *Ecology* **48,** 1–15.

Boeken, B., and Gutterman, Y. (1990). The effect of temperature on seed germination in three common bulbous plants of different habitats in the central Negev desert of Israel. *J. Arid Environ.* **18,** 175–184.

Boroughs, H., and Hunter, J. R. (1963). The effect of temperature on the germination of cacao seeds. *Proc. Am. Soc. Hort. Sci.* **82,** 222–224.

Boscan, V. C. G. (1967). Efectos del fuego sobre la reproduccion de algunas plantas de las llanos de Venezuela. *Bol. Soc. Venezolana Cien. Nat.* **27,** 70–103.

Bourliere, F., and Hadley, M. (1983). Present-day savannas: An overview. *In* "Ecosystems of the World" (F. Bouliere, ed.), pp. 1–17. Elsevier, Amsterdam.

Bowers, J. E. (1994). Natural conditions for seedling emergence of three woody species in the northern Sonoran Desert. *Madrono* **41,** 73–84.

Brinkmann, W. L. F., and Vieira, A. N. (1971). The effect of burning on germination of seeds at different soil depths of various tropical tree species. *Turrialba* **21,** 77–82.

Brokaw, N. V. L. (1986). Seed dispersal, gap colonization, and the case of *Cecropia insignis. In* "Frugivores and Seed Dispersal" (A. Estrada and T. H. Fleming, eds.), pp. 323–331. Junk, Dordrecht.

Brokaw, N. V. L. (1987). Gap-phase regeneration of three pioneer tree species in a tropical forest. *J. Ecol.* **75**, 9–19.

Brown, N. A. C., and Von Booysen, P. D. (1969). Seed coat impermeability in several *Acacia* species. *Agroplantae* **1**, 51–60.

Brown, R. F. (1982). Seed dormancy in *Aristida armata. Aust. J. Bot.* **30**, 67–73.

Budowski, G. (1963). Forest succession in tropical lowlands. *Turrialba* **13**, 42–44.

Burk, J. H. (1982). Phenology, germination, and survival of desert ephemerals in Deep Canyon, Riverside County, California. *Madrono* **29**, 154–163.

Burrows, W. H. (1971). A study of the phenology and germination of *Eremophila gilesii* in semiarid Queensland. *U.S. Dept. Agric. Misc. Publ. 1271.* Paper No. 14, pp. 150–159.

Burton, P. J. (1982). The effect of temperature and light on *Metrosideros polymorpha* seed germination. *Pacific Sci.* **36**, 229–240.

Burtt, B. L. (1976). Notes on rain-forest herbs. *Gard. Bull. Singapore* **29**, 73–80.

Capon, B., and Van Asdall, W. (1967). Heat pre-treatment as a means of increasing germination of desert annual seeds. *Ecology* **48**, 305–306.

Cardel, Y., Rico-Gray, V., Garcia-Franco, J. G., and Thien, L. B. (1997). Ecological status of *Beaucarnea gracilis*, an endemic species of the semiarid Tehuacan Valley, Mexico. *Conserv. Biol.* **11**, 367–374.

Cavanagh, A. K. (1980). A review of some aspects of the germination of acacias. *Proc. Roy. Soc. Vict.* **91**, 161–180.

Cervantes, V., Carabias, J., and Vazquez-Yanes, C. (1996). Seed germination of woody legumes from deciduous tropical forest of southern Mexico. *For. Ecol. Manage.* **82**, 171–184.

Chamshama, S. A. O., and Downs, R. J. (1982). Germination behaviour of *Chlorophora excelsa*, (Welw.) Benth. & Hook and *Podocarpus usambarensis*, Pilger. *Indian For.* **108**, 397–401.

Chandel, K. P. S., Chaudhury, R., Radhamani, J., and Malik, S. K. (1995). Desiccation and freezing sensitivity in recalcitrant seeds of tea, cocoa and Jackfruit. *Ann. Bot.* **76**, 443–450.

Charles-Dominique, P. (1986). Inter-relations between frugivorous vertebrates and pioneer plants: *Cecropia*, birds and bats in French Guyana. *In* "Frugivores and Seed Dispersal" (A. Estrada, and T. H. Fleming, eds.), pp. 119–135. Junk, Dordrecht.

Chatterji, U. N., and Mohnot, K. (1967). Thermo-physiological investigations on the imbibitions and germination of seeds of certain arid zone plants. *Acta Agron. Acad. Sci. Hung.* **16**, 7–16.

Chatterji, U. N., and Sen, D. N. (1964). On the eco-physiology of *Salmalia malabarica* (DC.) Scho. and Endl. seeds. *Sci. Cult.* **30**, 598–600.

Chatterton, N. J., and McKell, C. M. (1969). Time of collection and storage in relation to germination of desert saltbush seed. *J. Range Manage.* **22**, 355–356.

Chaudhary, R. L. (1975). Ecological observations on *Andrographis paniculata* Nees. *Botanique* **6**, 103–108.

Chaudhary, R. L. (1976). Seed germination of *Triumfetta rhomboidia* Jacq. Enum. *Botanique* **7**, 9–14.

Chaudhary, S. K., and Sinha, R. P. (1990). Dormancy and quantification of germination in *Portulaca quadrifida* and *P. oleracea. In* "International Symposium on environmental influences on seed and germination mechanism: Recent Advances in Research and Technology." (D. N. Sen, S. Mohammed, P. K. Kasera, and T. P. Thomas, eds.). pp. 59–60. University of Jodhpur, Jodhpur, India. [Abstract]

Chaudhary, S. K., and Sinha, R. P. (1991a). Effect of burial on germination of *Portulaca* species. *Geobios* **18**, 274–276.

Chaudhary, S. K., and Sinha, R. P. (1991b). Role of light quality in seed germination of *Portulaca* spp. *Proc. Natl. Acad. Sci. India* **61**(B) II, 221–222.

Chaudhary, S. K., and Singh, R. P. (1991a). Effect of scarification on germination of *Sida* spp. *Biojournal* **3**, 143–146.

Chaudhary, S. K., and Sinha, R. P. (1991b). Dormancy and quantification of germination in *Portulaca quadrifida* L. and *P. oleracea* L. *In* "Proc. Int. Seed Symposium" (D. N. Sen, and S. Mohammed, eds.), pp. 133–137. Jodhpur, India.

Chauhan, P. S., and Raina, V. (1980). Effect of seed weight on germination and growth of chir pine (*Pinus roxburghii* Sargent). *Indian For.* **106**, 53–59.

Cheke, A. S., Nanakorn, W., and Yankoses, C. (1979). Dormancy and dispersal of seeds of secondary forest species under the canopy of a primary tropical rain forest in northern Thailand. *Biotropica* **11**, 88–95.

Chin, H. F., Aziz, M., Ang, B. B., and Hamzah, S. (1981). The effect of moisture and temperature on the ultrastructure and viability of seeds of *Hevea brasiliensis. Seed Sci. Technol.* **9**, 411–422.

Chin, H. F., Hor, Y. L., and Mohd Lassim, M. B. (1984). Identification of recalcitrant seeds. *Seed Sci. Technol.* **12**, 429–436.

Chin, H. F., Krishnapillay, B., and Stanwood, P. C. (1989). Seed moisture: Recalcitrant vs. orthodox seeds. Crop Science Society of America Special Publication No. 14, pp. 15–22.

Chin, L. T. (1973). Occurrence of seeds in virgin forest top soil with particular reference to secondary species in Sabah. *Malaysian For.* **36**, 185–193.

Choi, M. A., Al-Ani, T. A., and Charchafchy, F. (1976). Germinability and seedling vigor of *Haloxylon salicornicum* as affected by storage and seed size. *J. Range Manage.* **29**, 60–62.

Choinski, J. S., Jr. (1990). Aspects of viability and post-germinative growth in seeds of the tropical tree, *Trichilia dregeana* Sonder. *Ann. Bot.* **66**, 437–442.

Choinski, J. S., Jr., and Tuohy, J. M. (1991). Effect of water potential and temperature on the germination of four species of African savanna trees. *Ann. Bot.* **68**, 227–233.

Chow, K. H., and Tjondronegoro, P. D. (1978). Fruit development, seed germination and seedling growth of frangipani (*Plumeria* L., Apocynaceae). *Malayan Nat. J.* **32**, 157–171.

Clarkson, J. R., and Clifford, H. T. (1987). Germination of *Jedda multicaulis* J. R. Clarkson (Thymelaeaceae). An example of cryptogeal germination in the Australian flora. *Aust. J. Bot.* **35**, 715–720.

Cohen, D. (1966). Optimizing reproduction in a randomly varying environment. *J. Theor. Biol.* **12**, 119–129.

Cole, N. H. A. (1977). Effect of light, temperature, and flooding on seed germination of the neotropical *Panicum laxum* Sw. *Biotropica* **9**, 191–194.

Conn, J. S., and Snyder-Conn, E. K. (1981). The relationship of the rock outcrop microhabitat to germination, water relations, and phenology of *Erythrina flabelliformis* (Fabaceae) in southern Arizona. *Southw. Nat.* **25**, 443–451.

Corbineau, F., and Come, D. (1980–1981). Some particularities of the germination of *Oldenlandia corymbosa* L. seeds (tropical Rubiaceae). *Israel J. Bot.* **29**, 157–167.

Corbineau, F., and Come, D. (1988). Storage of recalcitrant seeds of four tropical species. *Seed Sci. Technol.* **16**, 97–103.

Corbineau, F., Kante, M., and Come, D. (1986). Seed germination and seedling development in the mango (*Mangifera indica* L.). *Tree Physiol.* **1**, 151–160.

Cordero, S. R. A., and Di Stefano G., J. F. (1991). Efecto del estres osmotico sobre la germinacion de semillas de *Tecoma stans* (Bignoniaceae). *Rev. Biol. Trop.* **39**, 107–110.

Cornelius, D. R., and Hylton, L. O. (1969). Influence of temperature and leachate on germination of *Atriplex polycarpa.* *Agron. J.* **61,** 209–211.

Coudret, A. (1971). Action de differents sels sur le deroulement de la germination et sur l'évolution du phenomene respiratoire chez les graines de *Zygophyllum album* L. *Bull. Soc. Bot. France* **118,** 471–480.

Cousens, R., Armas, G., and Baweja, R. (1994). Germination of *Rapistrum rugosum* (L.) All. from New South Wales, Australia. *Weed Res.* **34,** 127–135.

Coutinho, L. M. (1982). Ecological effects of fire in Brazilian cerrado. *In* "Ecology of Tropical Savannas"(B. J. Huntley and B. H. Walker, eds.), pp. 273–291. Springer-Verlag, Berlin.

Coutinho, L. M. (1990). Fire in the ecology of the Brazilian cerrado. *In* "Fire in the Tropical Biota" (J. G. Goldammer, ed.), pp. 82–105. Springer-Verlag, Berlin.

Coutinho, L. M., and Struffaldi, Y. (1971). Observacoes sobre a germinacao das sementes e o crescimento das plantulas de uma leguminosa da mata amazonica de igapo (*Parkia auriculata* Spruce Mss.). *Phyton (Buenos Aires)* **28,** 149–159.

Cox, J. R., de Alba-Avila, A., Rice, R. W., and Cox, J. N. (1993). Biological and physical factors influencing *Acacia constricta* and *Prosopis velutina* establishment in the Sonoran Desert. *J. Range Mange.* **46,** 43–48.

Croat, T. B. (1978). "Flora of Barro Colorado Island." Stanford Univ. Press, Stanford, CA.

Cunha, R., Eira, M. T. S., and Rita, I. (1995). Germination and desiccation studies on wild nutmeg seed (*Virola surinamensis*). *Seed Sci. Technol.* **23,** 43–49.

Danthu, P., Roussel, J., Dia, M., and Sarr, A. (1992). Effect of different pretreatments on the germination of *Acacia senegal* seeds. *Seed Sci. Technol.* **20,** 111–117.

da Silva, M. F., Goldman, G. H., Magalhaes, F. M., and Moreira, F. W. (1989). Germinacao natural de 10 leguminosas arboreas da Amazonia—I. *Manaus Inst. Nac. Pesquisas Amazonica* **18,** 9–26.

Datta, S. C. (1961). Germination studies on the seeds of two desert plants. *Indian Agric.* **5,** 111–112.

Datta, S. C., and Chakravarty, H. L. (1962). Germination studies on the seeds of *Citrullus.* *Indian Agric.* **6,** 220–222.

Dauro, D., Mohamed-Saleem, M. A., and Gintzburger, G. (1997). Recruitment and survival of native annual *Trifolium* species in the highlands of Ethiopia. *Afr. J. Ecol.* **35,** 1–9.

de Foresta H., and Prevost, M.-F. (1986). Vegetation pionniere et graines du sol en foret Guyanaise. *Biotropica* **18,** 279–286.

del Castillo, R. F. (1986). Semillas, germinacion y establecimiento de *Ferocactus histrix. Cactaceas y Suculentas Mexicanas* **31,** 5–11.

de Melo, J., Ribeiro, J. F., and de F. Lima, V. L. G. (1979). Germinacao de sementes de algumas especies arboreas nativas do cerrado. *Revta. Brasil. Sem.* **1,** 8–12.

Denslow, J. S. (1980). Gap partitioning among tropical rainforest trees. *Biotropica* **12** (Suppl.), 47–95.

de Rouw, A., and van Oers, C. (1988). Seeds in a rainforest soil and their relation to shifting cultivation in the Ivory Coast. *Weed Res.* **28,** 373–381.

de Vogel, E. F. (1980). "Seedlings of Dicotyledons." Centre for Agricultural Publishing and Documentation, Wageningen.

Dickens, R., and Moore, G. M. (1974). Effects of light, temperature, KNO₃, and storage on germination on Cogongrass. *Agron. J.* **66,** 187–188.

Di Stefano G., J. F., and Chaverri, L. G. (1992). Potencial de germinacion de semillas en un bosque secundario premontano en San Pedro de Montes de Oca, Costa Rica. *Rev. Biol. Trop.* **40,** 7–10.

Downs, R. J. (1956). Seeds of *Puya berteroniana* require light for germination. *Bromeliad Soc. Bull.* **6,** 67–68.

Downs, R. J. (1964). Photocontrol of germination of seeds of the Bromeliaceae. *Phyton (Buenos Aires)* **21,** 1–6.

Downs, R. J., and Piringer, A. A. (1958). Seed germination in the bromeliaceae. *Bromeliad Soc. Bull.* **8,** 36–38.

Drake, D. R. (1992). Seed dispersal of *Metrosideros polymorpha* (Myrtaceae): A pioneer tree of Hawaiian lava flows. *Am. J. Bot.* **79,** 1224–1228.

Drake, D. R. (1993). Germination requirements of *Metrosideros polymorpha,* the dominant tree of Hawaiian lava flows and rain forests. *Biotropica* **25,** 461–467.

Dransfield, J., and Manokaran, N. (eds.). (1993). "Plant Resources of southeast Asia." Pudoc Scientific Publishers, Wageningen.

Dubrovsky, J. G. (1996). Seed hydration memory in Sonoran Desert cacti and its ecological implication. *Am. J. Bot.* **83,** 624–632.

Duke, J. A. (1969). On tropical tree seedlings. I. Seeds, seedlings, systems, and systematics. *Ann. Missouri Bot. Gard.* **56,** 125–161.

Edgar, J. G. (1977). Effects of moisture stress on germination of *Eucalyptus camaldulensis* Dehnh. and *E. regnans* F. Muell. *Aust. For. Res.* **7,** 241–245.

Edgar, R. L., and Springfield, H. W. (1977). Germination characteristics of broadscale: A possible saline-alkaline site stabilizer. *J. Range Manage.* **30,** 296–298.

Edmisten, J. (1970). Studies of *Phytolacca icosandra. In* "A Tropical Rain Forest" (H. T. Odum, ed.), pp. D183–D188. U.S. Atomic Energy Commission, Oak Ridge, TN.

Egley, G. H. (1976). Germination of developing prickly sida seeds. *Weed Sci.* **24,** 239–243.

Elberse, W. Th., and Breman, H. (1989). Germination and establishment of Sahelian rangeland species. I. Seed properties. *Oecologia* **80,** 477–484.

Elberse, W. Th., and Breman, H. (1990). Germination and establishment of Sahelian rangeland species. II. Effects of water availability. *Oecologia* **85,** 32–40.

El Hajzein, B. (1992). Etude de la germination de *Citrullus colocynthis* (L.) Schrad. II. Mise en evidence d'une dormance embryonnaire relative. *Rev. Cytol. Biol. Veget. Bot.* **15,** 175–181.

El Hajzein, B., and Neville, P. (1992). Etude de la germination de *Citrullus colocynthis* (L.) Schrad. I. L'influence inhibitrice des teguments seminaux. *Rev. Cytol. Biol. Veget. Bot.* **15,** 151–174.

Ellern, S. J., and Tadmor, N. H. (1956). Germination of range plant seeds at fixed temperatures. *J. Range. Manage.* **19,** 341–345.

Ellis, R. H., Hong, T. D., and Roberts, E. H. (1981). The influence of desiccation on cassava seed germination and longevity. *Ann. Bot.* **47,** 173–175.

Ellis, R. H., Hong, T. D., and Roberts, E. H. (1990). An intermediate category of seed storage behaviour? I. Coffee. *J. Exp. Bot.* **41,** 1167–1174.

Ellis, R. H., Hong, T. D., and Roberts, E. H. (1991). Effect of storage temperature and moisture on the germination of papaya seeds. *Seed Sci. Res.* **1,** 69–72.

Ellison, A. M., Denslow, J. S., Loiselle, B. A., and Brenes M., D. (1993). Seed and seedling ecology of Neotropical Melastomataceae. *Ecology* **74,** 1733–1749.

Ellner, S., and Shmida, A. (1981). Why are adaptations for long-range seed dispersal rare in desert plants? *Oecologia* **51,** 133–144.

Emparan, P. R., and Tysdal, H. M. (1957). The effect of light and other factors on breaking the dormancy of guayule seed. *Agron. J.* **49,** 15–19.

Enright, N. (1985). Existence of a soil seed bank under rainforest in New Guinea. *Aust. J. Ecol.* **10,** 67–71.

Erasmus, D. J., and van Staden, J. (1986). Germination of *Chromolaena odorata* (L.) K.&R. achenes: Effect of temperature, imbibition and light. *Weed Res.* **26,** 75–81.

Ernst, W. H. O. (1988). Seed and seedling ecology of *Brachystegia spiciformis*, a predominant tree component in Miombo Woodlands in south central Africa. *For. Ecol. Manage.* **25**, 195–210.

Ernst, W. H. O. (1991). Fire, dry heat and germination of savanna grasses in Botswana. *In* "Modern Ecology: Basic and Applied Aspects" (G. Esser and D. Overdieck, eds.), pp. 349–361. Elsevier, Amsterdam.

Ernst, W. H. O., Kuiters, A. T., and Tolsma, D. J. (1991). Dormancy of annual and perennial grasses from a savanna of southeastern Botswana. *Acta Oecol.* **12**, 727–739.

Ernst, W. H. O., Tietema, T., Veenendaal, E. M., and Masene, R. (1988). Dormancy, germination and seedling growth of two Kalaharian perennials of the genus *Harpagophytum* (Pedaliaceae). *J. Trop. Ecol.* **4**, 185–198.

Ernst, W. H. O., and Tolsma, D. J. (1988). Dormancy and germination of semi-arid annual plant species, *Tragus berteronianus* and *Tribulus terrestris. Flora* **181**, 243–251.

Espindola, L. S., Corbineau, F., and Come, D. (1993). Early events occurring during dehydration of recalcitrant embryos of *Araucaria angustifolia* seeds. *In* "Fourth International Workshop on Seeds: Basic and Applied Aspects of Seed Biology" (D. Come and F. Corbineau, eds.), pp. 873–878. Universite Pierre et Marie Curie, Paris.

Etejere, E. O., Fawole, M. O., and Sani, A. (1982). Studies on the seed germination of *Parkia clapertoniana. Turrialba* **32**, 181–185.

Evenari, M. (1985). The desert environment. *In* "Ecosystems of the World" (M. Evenari, I. Noy-Meir, and D. W. Goodall, eds.), Vol. 12A, pp. 1–22. Elsevier, Amsterdam.

Everham, E. M., III, Myster, R. W., and VanDeGenachte, E. (1996). Effects of light, moisture, temperature, and litter on the regeneration of five tree species in the tropical montane wet forest of Puerto Rica. *Am. J. Bot.* **83**, 1063–1068.

Everitt, J. H. (1983a). Seed germination characteristics of three woody plant species from south Texas. *J. Range Manage.* **36**, 246–249.

Everitt, J. H. (1983b). Seed germination characteristics of two woody legumes (retama and twisted acacia) from south Texas. *J. Range. Manage.* **36**, 411–414.

Everitt, J. H. (1983c). Germination of mescal bean (*Sophora secundiflora*) seeds. *Southwest. Nat.* **28**, 437–443.

Everitt, J. H. (1984). Germination of Texas persimmon seed. *J. Range. Manage.* **37**, 189–192.

Ewel, J., Berish, C., Brown, B., Price, N., and Raich, J. (1981). Slash and burn impacts on a Costa Rican wet forest site. *Ecology* **62**, 816–829.

Eyre, S. R. (1963). "Vegetation and Soils." Aldine Publishing Company, Chicago.

Eze, J. M. O., and Orole, B. C. (1987). Germination of the seeds *Prosopis africanum. Nigerian J. For.* **17**, 12–17.

Falkner, M. B., Laven, R. D., and Aplet, G. H. (1997). Experiments on germination and early growth of three rare and endemic species of Hawaiian *Tetramolopium* (Asteraceae). *Biol. Conserv.* **80**, 39–47.

Fanshawe, D. B. (1947). Studies of the trees of British Guiana. II. Greenheart (*Ocotea rodiaei* [Schomb.] Mez.). *Trop. Woods* **92**, 25–40.

Farrant, J. M., Berjak, P., and Pammenter, N. W. (1985). The effect of drying rate on viability retention of recalcitrant propagules of *Avicennia marina. S. Afr. J. Bot.* **51**, 432–438.

Farrant, J. M., Pammenter, N. W., and Berjak, P. (1988). Recalcitrance: A current assessment. *Seed Sci. Technol.* **16**, 155–166.

Fearn, B. (1974). An investigation into the effect of temperature on the seed germination of nine species of cacti using thermal gradient bars. *Cactus Succ. J.* **46**, 215–219.

Felippe, G. M., and Filgueiras, T. S. (1986). Germination of *Actinocladum verticillatum* (Ness) McClure ex Soderstrom, a bamboo from the Brazilian "cerrado" vegetation: Short communication. *Hoehnea* **13**, 95–100.

Felippe, G. M., Lucas, N. M. C., Behar, L., and Oliveira, M. A. C. (1971). Observacoes a respeito da germinacao de *Stevia rebaudiana* Bert. *Hoehnea* **1**, 81–93.

Felippe, G. M., and Polo, M. (1983). Germinacao de ervas invasoras: Efeito de luz e escarificacao. *Revta. Brasil. Bot.* **6**, 55–60.

Felippe, G. M., Polo, M., Cardosa, V. J. M., and Figeiredo-Ribeiro, R. C. L. (1983). Geminacao da unidade de dispersao de erva invasora *Hyptis suaveolens. An. Sem. Reg. Ecol.* **3**, 245–261.

Felippe, G. M., and Silva, J. C. S. (1984). Estudos de germinacao em especies do cerrado. *Revta. Brasil Bot.* **7**, 157–163.

Felippe, G. M., Silva, J. C. S., and Cardoso, V. J. M. (1983). Germination studies in *Andropogon gayanus* Kunth. *Revta. Brasil. Bot.* **6**, 41–48.

Fernandez, H. G. (1978). Influencia de la edad en la germinacion de *Atriplex repanda. Phyton (Buenos Aires)* **36**, 111–115.

Ferreira, A. G., and Handro, W. (1979). Aspects of seed germination in *Araucaria angustifolia* (Bert.) O. Ktze. *Revta. Brasil. Bot.* **2**, 7–13.

Ferri, M. G. (1959). Aspects of the soil-water-plant relationships in connexion with some Brazilian types of vegetation. *In* "Proceedings of the Abidjan UNESCO and the Commission for Technical Co-operation in Africa south of the Sahara. Symposium, UNESCO: Tropical Soils and Vegetation," pp. 103–109.

Ferwerda, J. D. (1956). Germination of oil palm seeds. *Trop. Agric. Trin.* **33**, 51–66.

Fialho, R. F., and Furtado, A. L. S. (1993). Germination of *Erythroxylum ovalifolium* (Erythroxylaceae) seeds within the terrestrial bromeliad *Neoregelia cruenta. Biotropica* **25**, 359–362.

Figueiredo, R. A. de, and Perin, E. (1995). Germination ecology of *Ficus luschnathiana* drupelets after bird and bat ingestion. *Acta Oecol.* **16**, 71–75.

Fleming, T. H., Williams, C. F., Bonaccorso, F. J., and Herbst, L. H. (1985). Phenology, seed dispersal, and colonization in *Muntingia calabura*, a neotropical pioneer tree. *Am. J. Bot.* **72**, 383–391.

Flores, E. M. (1992). "Trees and Seeds from the Neotropics." Vol 1. No. 1. *Dipteryx panamensis, Stryphnodendron excelsum* and *Virola koschnyi.* Museo National de Costa Rica, Depto. Historia Natural, Herbario Nacional de Costa Rica, San Jose, Costa Rica.

Flores, E. M., and Benavides, C. E. (1990). Germinacion y morfologia de la plantula de *Hymenaea courbaril* L. (Caesalpiniaceae). *Rev. Biol. Trop.* **38**, 91–98.

Flores, E. M., Fournier O., L. A., and Garcia, E. G. (1985). Morfologia y demografia de la germinacion en lauraceas de Costa Rica. *Rev. Biol. Trop.* **33**, 163–170.

Flores, E. M., and Mora, B. (1984). Germination and seedling growth of *Pithecellobium arboreum* Urban. *Turrialba* **34**, 485–488.

Flores, E. M., and Rivera, D. I. (1985). Germinacion y desarrollo de la plantula de *Gliricidia sepium* (Jacq.) Steud (Papilionaceae). *Rev. Biol. Trop.* **33**, 157–161.

Flores, E. M., Rivera, D. I., and Vasquez, N. M. (1986). Germinacion y desarrollo de la plantula de *Cassia grandis* L. (Caesalpinioideae). *Rev. Biol. Trop.* **34**, 289–296.

Foldats, E., and Rutkis, E. (1975). Ecological studies of chaparro (*Curatella americana* L.) and manteco (*Byrsonima crassifolia* H.B.K.) In Venezuela. *J. Biogeogr.* **2**, 159–178.

Forget, P.-M. (1991a). Scatterhoarding of *Astrocaryum paramaca* by *Proechimys* in French Guiana: Comparison with *Myoprocta exilis. Trop. Ecol.* **32**, 155–167.

Forget, P.-M. (1991b). Comparative recruitment patterns of two non-pioneer canopy tree species in French Guiana. *Oecologia* **85**, 434–439.

Forget, P.-M. (1992). Seed removal and seed fate in *Gustavia superba* (Lecythidaceae). *Biotropica* **24**, 408–414.

Foster, S. A. (1986). On the adaptive value of large seeds for tropical moist forest trees: A review and synthesis. *Bot. Rev.* **52**, 260–299.

Foster, S. A., and Janson, C. H. (1985). The relationship between seed size and establishment conditions in tropical woody plants. *Ecology* **66**, 773–780.

Frankie, G. W., Baker, H. G., and Opler, P. A. (1974). Comparative phenological studies of trees in tropical wet and dry forests in the lowlands of Costa Rica. *J. Ecol.* **62**, 881–919.

Freas, K. E., and Kemp, P. R. (1983). Some relationships between environmental reliability and seed dormancy in desert annual plants. *J. Ecol.* **71**, 211–217.

Freeman, C. E. (1973a). Germination responses of a Texas population of ocotillo (*Fouquieria splendens* Engelm.) to constant temperature, water stress, pH and salinity. *Am. Midl. Nat.* **89**, 252–256.

Freeman, C. E. (1973b). Some germination responses of lecheguilla (*Agave lecheguilla* Torr.). *Southw. Nat.* **18**, 125–134.

Freeman, C. E. (1975). Germination responses of a New Mexico population of Parry agave (*Agave parryi* Engelm. var. *parryi*) to constant temperature, water stress, and pH. *Southw. Nat.* **20**, 69–74.

Freeman, C. E., Tiffany, R. S., and Reid, W. H. (1977). Germination responses of *Agave lecheguilla*, *A. parryi*, and *Fouquieria splendens*. *Southw. Nat.* **22**, 195–204.

Friedman, J., Gunderman, N., and Ellis, M. (1978). Water response of the hygrochastic skeletons of the true rose of Jericho (*Anastatica hierochuntica* L.). *Oecologia* **32**, 289–301.

Fulbright, T. E., Flenniken, K. S., and Waggerman, G. L. (1986a). Enhancing germination of spiny hackberry seeds. *J. Range. Manage.* **39**, 552–554.

Fulbright, T. E., Flenniken, K. S., and Waggerman, G. L. (1986b). Methods of enhancing germination of anacua seeds. *J. Range. Manage.* **39**, 450–453.

Fulbright, T. E., Kuti, J. O., and Tipton, A. R. (1995). Effects of nurse-plant canopy temperatures on shrub seed germination and seedling growth. *Acta Oecol.* **16**, 621–632.

Galil, J., and Meiri, L. (1981). Drupelet germination in *Ficus religiosa* L. *Israel J. Bot.* **30**, 41–47.

Garrard, A. (1955). The germination and longevity of seeds in an equatorial climate. *Gard. Bull. (Singapore)* **14**, 534–545.

Garwood, N. C. (1979). "Seed Germination in a Seasonal Tropical Forest in Panama." Ph.D thesis, Univ. Chicago, Chicago.

Garwood, N. C. (1982). Seasonal rhythm of seed germination in a semideciduous tropical forest. *In* "The Ecology of a Tropical Forest: Seasonal Rhythms and Long-Term Changes" (E. G. Leigh, Jr., A. S. Rand, and D. M. Windson, eds.), pp. 173–185. Smithsonian Institution Press, Washington, DC.

Garwood, N. C. (1983). Seed germination in a seasonal tropical forest in Panama: A community study. *Ecol. Monogr.* **53**, 159–181.

Garwood, N. C. (1986a). Constraints on the timing of seed germination in a tropical forest. *In* "Frugivores and Seed Dispersal" (A. Estrada, and T. H. Fleming, eds.), pp. 347–355. Junk, Dordrecht.

Garwood, N. C. (1986b). Effects of acid and hot water pretreatments and seed burial on the germination of tropical moist forest seeds. *Turrialba* **36**, 479–484.

Garwood, N. C., and Lighton, J. R. B. (1990). Physiological ecology of seed respiration in some tropical species. *New Phytol.* **115**, 549–558.

Gerhardt, K. (1996). Germination and development of sown mahogany (*Swietenia macrophylla* King) in secondary tropical dry forest habitats in Costa Rica. *J. Trop. Ecol.* **12**, 275–289.

Ghosh, R. C., Mathur, N. K., and Singh, R. P. (1976). *Diospyros melanoxylon:* Its problems and cultivation. *Indian For.* **102**, 326–336.

Gibson, A., and Bachelard, E. P. (1986). Germination of *Eucalyptus sieberi,* L. Johnson seeds. I. Response to substrate and atmospheric moisture. *Tree Physiol.* **1**, 57–65.

Gilbert, G. (1952). Contribution a la biologie des essences forestieres Congolaises. *Bull. Soc. Roy. Bot. Belgique* **84**, 289–296.

Gill, L. S., and Bamidele, J. F. (1981). Seed morphology, germination and cytology of three savanna trees of Nigeria. *Nigerian J. For.* **11**, 16–23.

Gill, L. S., Hasaini, S. W. H., and Musa, A. H. (1982). Germination biology of some *Cassia* L. species (Leguminosae). *Legume Res.* **5**, 97–104.

Gill, L. S., Hasaini, S. W. H., and Olumekun, V. O. (1981). Germination, seedling and cytology of *Delonix regia* (Boj. ex Hook.) Raf. (Leguminosae). *Legume Res.* **4**, 51–55.

Gillon, D. (1983). The fire problem in tropical savannas. *In* "Ecosystems of the World" (F. Bourliere, ed.), Vol. 13, pp. 617–641. Elsevier, Amsterdam.

Ginwal, H. S., Rawat, P. S., Gera, M., Gera, N., and Srivastava, R. L. (1995). Study on the pattern of seed germination of various subspecies cum provenances of *Acacia nilotica* Willd. ex. Del. under nursery conditions. *Indian For.* **121**, 29–38.

Gonzalez J., E. (1991). Recoleccion y germinacion de semillas de 26 especies arboreas del bosque humedo tropical. *Rev. Biol. Trop.* **39**, 47–51.

Gonzalez J., E. (1992). Humedad y germinacion de semillas de *Hyeronima alchorneoides* (Euphorbiaceae). *Rev. Biol. Trop.* **40**, 139–141.

Gonzalez, M. C., and Grijpma, P. (1968). Germinacion y supervivencia al repique de *Anthocephalus cadamba* Miq. *Turrialba* **18**, 409–415.

Grant, P. J. (1975). Some factors affecting germination and emergence of sub-tropical pasture legumes in Rhodesia. *Proc. Grassl. Soc. S. Afr.* **10**, 65–71.

Graves, W. L., Kay, B. L., and Williams, W. A. (1975). Seed treatment of Mojave desert shrubs. *Agron. J.* **67**, 773–777.

Grice, A. C. (1996). Seed production, dispersal and germination in *Cryptostegia grandiflora* and *Ziziphus mauritiana,* two invasive shrubs in tropical woodlands in northern Australia. *Aust. J. Ecol.* **21**, 324–331.

Grijpma, P. (1967). *Anthocephalus cadamba,* a versatile, fast growing industrial tree species for the tropics. *Turrialba* **17**, 321–329.

Grout, B. W. W., Shelton, K., and Pritchard, H. W. (1983). Orthodox behaviour of oil palm seed and cryopreservation of the excised embryo for genetic conservation. *Ann. Bot.* **52**, 381–384.

Grushvitzky, I. V. (1967). After-ripening of seeds of primitive tribes of angiosperms, conditions and peculiarities. *In* "Physiologie, Okologie und Biochemie der Keimung" (H. Borris, ed.), Vol. 1, pp. 320–335. Ernst-Moritz-Arndt Universitat, Greifswald.

Guard, A. T., and Henry, M. (1968). Reproduction of Spanish moss, *Tillandsia usneoides* L., by seeds. *Bull. Torrey Bot. Club* **95**, 327–330.

Guariguata, M. R., and Azocar, A. (1988). Seed bank dynamics and germination ecology in *Espeletia timotensis* (Compositae), an Andean giant rosette. *Biotropica* **20**, 54–59.

Guevara S., S. and Gomez-Pompa, A. (1972). Seeds from surface soils in a tropical region of Veracruz, Mexico. *J. Arnold Arbor.* **53**, 312–335.

Gupta, R. K. (1986). The Thar desert. *In* "Ecosystems of the World" (M. Evenari, I. Noy-Meir, and D. W. Goodall, eds.), Vol. 12B, pp. 55–99. Elsevier, Amsterdam.

Gutterman, Y. (1980–1981). Annual rhythm and position effect in the germinability of *Mesembryanthemum nodiflorum*. *Israel J. Bot.* **29**, 93–97.

Gutterman, Y. (1981). Influence of quantity and date of rain on the dispersal and germination mechanisms, phenology, development and seed germinability in desert annual plants, and on the life cycle of geophytes and hemicryptophytes in the Negev desert. *In*

"Developments in Arid Zone Ecology and Environmental Quality." (H. Shuval, ed.), pp. 35–42. Balaban ISS, Philadelphia, PA.

Gutterman, Y. (1982a). Survival mechanisms of desert winter annual plants in the Negev highlands of Israel. *Sci. Rev. Arid Zone Res.* **1**, 249–283.

Gutterman, Y. (1982b). Observations on the feeding habits of the Indian crested porcupine (*Hystrix indica*) and the distribution of some hemicryptophytes and geophytes in the Negev desert highlands. *J. Arid. Environ.* **5**, 261–268.

Gutterman, Y. (1983). Mass germination of plants under desert conditions: Effects of environmental factors during seed maturation, dispersal, germination and establishment of desert annual and perennial plants in the Negev highlands, Israel. *In* "Developments in Ecology and Environmental Quality" (H. I. Shuval, ed.), pp. 1–10. Balaban ISS, Philadelphia, PA.

Gutterman, Y. (1986). Are plants which germinate and develop during winter in the Negev desert highlands of Israel, winter annuals? *In* "Environmental Quality and Ecosystem Stability" (A. Dubinsky, and Y. Steinberger, eds.), Vol. III A/B, pp. 135–144. Bar-Ilan Univ. Press, Ramat-Gan, Israel.

Gutterman, Y. (1990a). Do germination mechanisms differ in plants originating in deserts receiving winter or summer rain? *Israel J. Bot.* **39**, 355–372.

Gutterman, Y. (1990b). Seed dispersal by rain (ombrohydrochory) in some of the flowering desert plants in the deserts of Israel and the Sinai Peninsula. *Mitt. Inst. Allg. Bot. Hamburg* **23**, 841–852.

Gutterman, Y. (1992). Ecophysiology of Negev upland annual grasses. *In* "Desertified grasslands: Their Biology and Management" (G. P. Chapman, ed.), pp. 145–162. Linnean Society Symposium Series, No. 13. Academic Press, London.

Gutterman, Y. (1993). "Seed Germination in Desert Plants: Adaptations of Desert Organisms." Springer-Verlag, Berlin.

Gutterman, Y. (1994a). In memoria—Michael Evenari and his desert. Seed dispersal and germination strategies of *Spergularia diandra* compared with some other desert annual plants inhabiting the Negev Desert of Israel. *Israel J. Plant Sci.* **42**, 261–274.

Gutterman, Y. (1994b). Strategies of seed dispersal and germination in plants inhabiting deserts. *Bot. Rev.* **60**, 373–425.

Gutterman, Y. (1996a). Temperatures during storage, light and wetting affecting caryopses germinability of *Schismus arabicus*, a common desert annual grass. *J. Arid. Environ.* **33**, 73–85.

Gutterman, Y. (1996b). Some ecological aspects of plant species with mucilaginous seed coats inhabiting the Negev Desert of Israel. *In* "Preservation of Our World in the Wake of Change" (Y. Steinberger, ed), Vol. VI A/B, pp. 492–496. ISEEQS, Israel.

Gutterman, Y., Corbineau, F., and Come, D. (1992). Interrelated effects of temperature, light and oxygen on *Amaranthus caudatus* L. seed germination. *Weed Res.* **32**, 111–117.

Gutterman, Y., Corbineau, F., and Come, D. (1996). Dormancy of *Hordeum spontaneum* caryopses from a population on the Negev Desert highlands. *J. Arid. Environ.* **33**, 337–345.

Gutterman, Y., and Evenari, M. (1994). The influences of amounts and distribution of irrigation during the hot and dry season on emergence and survival of some desert winter annual plants in the Negev Desert. *Israel J. Plant Sci.* **42**, 1–14.

Gutterman, Y., and Ginott, S. (1994). Long-term protected 'seed bank' in dry inflorescences of *Asteriscus pygmaeus;* achene dispersal mechanism and germination. *J. Arid Environ.* **26**, 149–163.

Gutterman, Y., and Herr, N. (1981). Influences of porcupine (*Hystrix indica*) activity on the slopes of the northern Negev Mountains: Germination and vegetation renewal in different geomorphological types and slope directions. *Oecologia* **51**, 332–334.

Gutterman, Y., Kamenetsky, R., and Van Rooyen, M. (1995). A comparative study of seed germination of two *Allium* species from different habitats in the Negev Desert highlands. *J. Arid. Environ.* **29**, 305–315.

Gutterman, Y., and Nevo, E. (1994). Temperatures and ecological-genetic differentiation affecting the germination of *Hordeum spontaneum* caryopses harvested from three populations: The Negev Desert and opposing slopes on Mediterranean Mount Carmel. *Israel J. Plant Sci.* **42**, 183–195.

Gutterman, Y., and Shem-Tov, S. (1996). Structure and function of the mucilaginous seed coats of *Plantago coronopus* inhabiting the Negev Desert of Israel. *Israel J. Plant Sci.* **44**, 125–133.

Gutterman, Y., and Witztum, A. (1977). The movement of integumentary hairs in *Blepharis ciliaris* (L.) Burtt. *Bot. Gaz.* **138**, 29–34.

Gutterman, Y., Witztum, A., and Heydecker, W. (1973). Studies on the surfaces of desert plant seeds. II. Ecological adaptations of the seeds of *Blepharis persica*. *Ann. Bot.* **37**, 1051–1055.

Hacker, J. B. (1989). The potential for buffel grass renewal from seed in 16-year-old buffel grass-siratro pastures in southeast Queensland. *J. Appl. Ecol.* **26**, 213–222.

Hagon, M. W. (1976). Germination and dormancy of *Themeda australis, Danthonia* spp., *Stipa bigeniculata* and *Bothriochloa macra*. *Aust. J. Bot.* **24**, 319–327.

Hagon, M. W., and Chan, C. W. (1977). The effects of moisture stress on the germination of some Australian native grass seeds. *Aust. J. Exp. Agric. Anim. Husb.* **17**, 86–89.

Hall, J. B., and Swain, M. D. (1980). Seed stocks in Ghanaian forest soils. *Biotropica* **12**, 256–263.

Hall, J. B., and Swaine, M. D. (1981). Distribution and ecology of vascular plants in a tropical rain forest: Forest vegetation in Ghana. Junk, The Hague.

Halward, T., and Shaw, R. (1996). Germination requirements and conservation of an endangered Hawaiian plant species (*Silene lanceolata*). *Nat. Areas J.* **16**, 335–343.

Hammouda, M. A., and Bakr, Z. Y. (1969). Some aspects of germination of desert seeds. *Phyton (Austria)* **13**, 183–201.

Harries, H. C. (1992). Biogeography of the coconut *Cocus nucifera* L. *Principes* **36**, 155–162.

Harrington, G. N., and Driver, M. A. (1995). The effect of fire and ants on the seed-bank of a shrub in a semi-arid grassland. *Aust. J. Ecol.* **20**, 538–547.

Hart, T. B. (1995). Seed, seedling and sub-canopy survival in mono-dominant and mixed forests of the Ituri Forest, Africa. *J. Trop. Ecol.* **11**, 443–459.

Hartshorn, G. S. (1978). Tree falls and tropical forest dynamics. Pp. 617–638. *In* "Tropical Trees as Living Systems" (P. B. Tomlinson, and M. H. Zimmermann, eds.), pp. 617–638. Cambridge Univ. Press, New York.

Harty, R. L. (1972). Germination requirements and dormancy effects in seed of *Urochloa mosambicensis*. *Trop. Grassl.* **6**, 17–24.

Hashim, I. M. (1990). Abundance, seed pod nutritional characteristics, and seed germination of leguminous trees in South Kordofan, Sudan. *J. Range Manage.* **43**, 333–335.

Hatano, K., and Y. Tanaka. (1979). A review of papers published in the Proceedings of the IUFRO International Symposium on the physiology of seed germination. *Seed Sci. Technol.* **7**, 47–56.

Hauser, T. P. (1994). Germination, predation and dispersal of *Acacia albida* seeds. *Oikos* **71**, 421–426.

Havel, J. J. (1971). The *Araucaria* forests of New Guinea and their regenerative capacity. *J. Ecol.* **59**, 203–214.

Henrici, M. (1935). Germination of Karroo bush seeds. *S. Afr. J. Sci.* **32**, 223–234.

Henrici, M. (1939). Germination of Karroo bush seeds. Part II. *S. Afr. J. Sci.* **36**, 212–219.

Hodgkinson, K. C. (1991). Shrub recruitment response to intensity and season of fire in a semi-arid woodland. *J. Appl. Ecol.* **28**, 60–70.

Hoffman, M. T., Cowling, R. M., Douie, C., and Pierce, S. M. (1989). Seed predation and germination of *Acacia erioloba* in the Kuiseb River Valley, Namib Desert. *S. Afr. J. Bot.* **55**, 103–106.

Holthuijzen, A. M. A., and Boerboom, J. H. A. (1982). The *Cecropia* seedbank in the Surinam lowland rain forest. *Biotropica* **14**, 62–68.

Holttum, R. E. (1953). Evolutionary trends in an equatorial climate. *Symp. Soc. Exp. Biol.* **7**, 159–173.

Holm, L., Pancho, J. V., Herberger, J. P., and Plucknett, D. L. (1979). "A Geographical Atlas of World Weeds." Wiley, New York.

Hopkins, B. (1965). Observations on savanna burning in the Olokemeji Forest Reserve, Nigeria. *J. Appl. Ecol.* **2**, 367–381.

Hopkins, B. (1968). Vegetation of the Olokemeji Forest Reserve, Nigeria. V. The vegetation on the savanna site with special reference to its seasonal changes. *J. Ecol.* **56**, 97–115.

Hopkins, M. S., and Graham, A. W. (1983). The species composition of soil seed banks beneath lowland tropical rainforests in North Queensland, Australia. *Biotropica* **15**, 90–99.

Hopkins, M. S., and Graham, A. W. (1984). Viable soil seed banks in disturbed lowland tropical rainforest sites in North Queensland. *Aust. J. Ecol.* **9**, 71–79.

Hor, Y. L., Stanwood, P. C., and Chin, H. F. (1990). Effects of dehydration on freezing characteristics and survival in liquid nitrogen of three recalcitrant seeds. *Pertanika* **13**, 309–314.

Horvitz, C. C. (1981). Analysis of how ant behaviors affect germination in a tropical myrmecochore *Calathea microcephala* (P. & E.) Koernicke (Marantaceae): Microsite selection and aril removal by neotropical ants, *Odontomachus, Pachycondyla*, and *Solenopsis* (Formicidae). *Oecologia* **51**, 47–52.

Horvitz, C. C., and Schemske, D. W. (1986). Seed dispersal of a neotropical myrmecochore: Variation in removal rates and dispersal distance. *Biotropica* **18**, 319–323.

Horvitz, C. C., and Schemske, D. W. (1994). Effects of dispersers, gaps, and predators on dormancy and seedling emergence in a tropical herb. *Ecology* **75**, 1949–1958.

Hunter, J. R. (1989). Seed dispersal and germination of *Enterolobium cyclocarpum* (Jacq.) Griseb. (Leguminosae: Mimosoideae): Are megafauna necessary? *J. Biogeogr.* **16**, 369–378.

Hussain, F., and Ilahi, I. (1988a). Effect of different physico-chemical factors on the germination of *Acacia modesta* Wall. *Sarhad J. Agric.* **4**, 209–217.

Hussain, F., and Ilahi, I. (1988b). Germination behaviour of *Withania somnifera* (L.) Dunal. *Hamdard* **31**, 20–30.

Hussain, F., and Nasrin, R. (1985). Germination study on the seeds of *Peganum harmala*. *Pakistan J. Agric. Res.* **6**, 113–118.

Hussain, F., Shaukat, S., Ilahi, I., and Qureshi, M. Z. (1991). Note on the germination behaviour of *Dodonaea viscosa* (Linn.) Jacq. *Sci. Khyber* **4**, 45–49.

Hussain, F., Shaukat, S., Ilahi, I., and Qureshi, M. Z. (1993). Germination promotion of *Ziziphus nummularia*. *Hamdard* **36**, 46–55.

Ibanez, M. L. (1965). Estudios sobre el mecanismo de la sensibilidad al frio en semillas de cacao. *Turrialba* **15**, 194–198.

Ibanez, M. L. (1968). The effect of cold on tropical seeds. *Turrialba* **18**, 73–75.

Ilahi, I., and F. Hussain. (1988). Germination improvement of *Reptonia buxifolia* (Falc.) A.DC. with various physicochemical factors. *Pakistan J. For.* **38**, 69–78.

Ilahi, I., Siddiqi, N., and Hussain, F. (1987). Effects of physico-chemical factors on the germination and seedling growth of *Erianthus griffithii* (Munro) HK.f. *Sarhad J. Agric.* **3**, 349–363.

Inouye, R. S. (1980). Density-dependent germination response by seeds of desert annuals. *Oecologia* **46**, 235–238.

Ismail, A. M. A. (1990). Germination ecophysiology in populations of *Zygophyllum qatarense* Hadidi from contrasting habitats: Effect

of temperature, salinity and growth regulators with special reference to fusicoccin. *J. Arid Environ.* **18**, 185–194.

Ismail, A. M. A., and El-Ghazaly, G. A. (1990). Phenological studies on *Zygophyllum qatarense* Hadidi from contrasting habitats. *J. Arid. Environ.* **18**, 195–205.

Ismail, A. M. A., Khalifa, F. M., and Babiker, A. G. T. (1990). Age, environmental factors, germination and seedling emergence of *Ocimum basilicum* L. *Qatar Univ. Sci. Bull.* **10**, 155–166.

Ismail, A. M. A., Obeid, M., and Ahmed, A. E. (1986). Studies on the biology of *Eucalyptus microtheca* F. Muell. in the Gezira, Sudan. I. The behaviour of the seed. *Qatar Univ. Sci. Bull.* **6**, 123–135.

Itoh, A. (1995). Effects of forest floor environment on germination and seedling establishment of two Bornean rainforest emergent species. *J. Trop. Ecol.* **11**, 517–527.

Jackson, G. (1974). Cryptogeal germination and other seedling adaptations to the burning of vegetation in savanna regions: The origin of the pyrophytic habit. *New Phytol.* **73**, 771–780.

Janzen, D. H. (1969). Seed-eaters versus seed size, number, toxicity and dispersal. *Evolution* **23**, 1–27.

Janzen, D. H. (1971a). Escape of *Cassia grandis* L. beans from predators in time and space. *Ecology* **52**, 965–979.

Janzen, D. H. (1971b). Escape of juvenile *Dioclea megacarpa* (Leguminosae) vines from predators in a deciduous tropical forest. *Am. Nat.* **105**, 97–112.

Janzen, D. H. (1972). Escape in space by *Sterculia apetala* seeds from the bug *Dysdercus fasciatus* in a Costa Rican deciduous forest. *Ecology* **53**, 350–361.

Janzen, D. H. (1975). Intra- and interhabitat variations in *Guazuma ulmifolia* (Sterculiaceae) seed predation by *Amblycerus cistelinus* (Bruchidae) in Costa Rica. *Ecology* **56**, 1009–1013.

Janzen, D. H. (1979). How many babies do figs pay for babies? *Biotropica* **11**, 48–50.

Janzen, D. H. (1980). Specificity of seed-attacking beetles in a Costa Rican deciduous forest. *J. Ecol.* **68**, 929–952.

Janzen, D. H. (1981a). *Enterolobium cyclocarpum* seed passage rate and survival in horses, Costa Rican Pleistocene seed dispersal agents. *Ecology* **62**, 593–601.

Janzen, D. H. (1981b). Digestive seed predation by a Costa Rican Baird's tapir. *Biotropica* **13**, 59–63.

Janzen, D. H. (1982a). Ecological distribution of chlorophyllous developing embryos among perennial plants in a tropical deciduous forest. *Biotropica* **14**, 232–236.

Janzen, D. H. (1982b). Removal of seeds from horse dung by tropical rodents: Influence of habitat and amount of dung. *Ecology* **63**, 1887–1900.

Janzen, D. H., Miller, G. A., Hackforth-Jones, J., Pond, C. M., Hooper, K., and Janos, D. P. (1976). Two Costa Rican bat-generated seed shadows of *Andira inermis* (Leguminosae). *Ecology* **57**, 1068–1075.

Jensen, L. A. (1971). Observations on the viability of Borneo camphor *Dryobalanops aromatica* Gaertn. *Proc. Int. Seed Test. Assoc.* **36**, 141–146.

Johnson, H. B., Vasek, F. C., and Yonkers, T. (1978). Residual effects of summer irrigation on Mojave desert annuals. *Bull. South. California Acad. Sci.* **77**, 95–108.

Johnston B., M., and Fernandez H., G. (1979). Rol del pericarpio de *Atriplex repanda* en la germinacion. I. Efecto del lavado do los frutos en agua. *Phyton (Buenos Aires)* **37**, 145–151.

Johnson, P., and Morales, R. (1972). A review of *Cordia alliodora* (Ruiz & Pav.) Oken. *Turrialba* **22**, 210–220.

Joly, C. A. and Felippe, G. M. (1979a). Dormencia das sementes de *Rapanea guianensis* Aubl. *Revta. Brasil. Bot.* **2**, 1–6.

Joly, C. A., and Felippe, G. M. (1979b). Germinacao e fenologia de *Zeyhera digitalis* (Vell.) Hoehne. *Hoehnea* **8**, 35–40.

Joly, C. A., Felippe, G. M., Dietrich, S. M. C., and Campos-Takaki, G. M. (1980). Physiology of germination and seed gel analysis in two populations of *Magonia pubescens* St. Hil. *Revta. Brasil. Bot.* **3**, 1–9.

Jones, E. W. (1956). Ecological studies on the rain forest of southern Nigeria. IV. The plateau forest of the Okomu Forest Reserve. Part II. The reproduction and history of the forest. *J. Ecol.* **44**, 83–117.

Juhren, M., Went, F. W., and Phillips, E. (1956). Ecology of desert plants. IV. Combined field and laboratory work on germination of annuals in the Joshua Tree National Monument, California. *Ecology* **37**, 318–330.

Jurado, E., and Westoby, M. (1992). Germination biology of selected central Australian plants. *Aust. J. Ecol.* **17**, 341–348.

Kadman-Zahavi, A. (1955). Notes on the germination of *Atriplex rosea*. *Bull. Res. Counc. Israel* **4D**, 375–378.

Kariuki, E. M., and Powell, G. R. (1988). Pretreatment and germination of seeds of three leguminous tree species indigenous to Kenya. *Seed Sci. Technol.* **16**, 477–487.

Kasera, P. K., and Sen, D. N. (1987). Effect of different environmental factors and growth regulators on seed germination of *Borreria articularis* (Linn.) F. N. Will. *Biovigyanam* **13**, 112–116.

Kaul, A. (1972). Studies on seed germination of *Alternanthera pungens* H.B.K. in relation to its success in nature. *Trop. Ecol.* **13**, 104–109.

Kaul, A., Shankar, V., and Kumar, S. (1990). Effect of moisture stress, salinity and light conditions on the germination of *Haloxylon salicornicum*. "International Symposium on Environmental Influences on Seed and Germination Mechanism: Recent Advances in Research and Technology" (D. N. Sen, S. Mohammed, P. K. Kasera, and T. P. Thomas, eds.), pp. 95–96. University of Jodhpur, Jodhpur, India. [Abstract]

Kaul, A., and Shankar, V. (1988). Ecology of seed germination of the chenopod shrub *Haloxylon salicornicum*. *Trop. Ecol.* **29**, 110–115.

Kaul, R. N., and Manohar, M. S. (1966). Germination studies on arid zone tree seeds. I. *Acacia senegal* Willd. *Indian For.* **92**, 499–503.

Kay, B. L., Pergler, C. C., and Graves, W. L. (1984). Storage of seed of Mojave desert shrubs. *J. Seed Technol.* **9**, 20–28.

Keay, R. W. J. (1960). Seeds in forest soils. *Nigerian For. Inform. Bull.* (New Series) No. 4, pp. 1–4.

Keeley, J. E., and Tufenkian, D. A. (1984). Garden comparison of germination and seedling growth of *Yucca whipplei* subspecies (Agavaceae). *Madrono* **31**, 24–29.

Keeley, J. E., and Meyers, A. (1985). Effect of heat on seed germination of southwestern *Yucca* species. *Southw. Nat.* **30**, 303–304.

Kellman, M. C. (1974). The viable weed seed content of some tropical agricultural soils. *J. Appl. Ecol.* **11**, 669–678.

Kellman, M. (1980). Geographic patterning in tropical weed communities and early secondary successions. *Biotropica* **12**(Suppl.), 34–39.

Kemp, P. R. (1983). Phenological patterns of Chihuahuan desert plants in relation to the timing of water availability. *J. Ecol.* **71**, 427–436.

Khan, D., Ahmad, R., and Ismail, S. (1987). Germination, growth and ion regulation in *Prosopis juliflora* (Swartz) DC. under saline conditions. *Pakistan J. Bot.* **19**, 131–138.

Khan, D., Shaukat, S. S., and Faheemuddin, M. (1984). Germination studies of certain desert plants. *Pakistan J. Bot.* **16**, 231–254.

Khasa, P. D. (1992). Scarification of limba seeds with hot water, bleach, and acid. *Tree Planter's Notes* **43**(4), 150–152.

Khudairi. A. K. (1956). Breaking the dormancy of *Prosopis* seeds. *Physiol. Plant.* **9**, 452–461.

Khudairi, A. K. (1958). Studies on the germination of date-palm seeds. The effect of sodium chloride. *Physiol. Plant.* **11**, 16–22.

Kiew, R. (1982). Germination and seedling survival in kemenyan, *Styrax benzoin*. *Malaysian For.* **45**, 69–80.

Kiew, R. (1988). Herbaceous flowering plants. *In* "Key Environments: Malaysia," pp. 56–76. (Earl of Cranbrook, ed.). Pergamon Press, Oxford.

Kigel, J., Lior, E., Zamir, L., and Rubin, R. (1992). Biology of reproduction in the summer annual weed *Euphorbia geniculata* Ortega. *Weed Res.* **32**, 317–328.

Kigomo, B. N., Woodell, S. R., and Savill, P. S. (1994). Phenological patterns and some aspects of reproductive biology of *Brachylaena huillensis* O. Hoffm. *Afr. J. Ecol.* **32**, 296–307.

King, M. W., and Roberts, E. H. (1982). The imbibed storage of cocoa (*Theobroma cacao*) seeds. *Seed Sci. Technol.* **10**, 535–540.

Kissock, D. C., and Haferkamp, M. R. (1983). Presowing seed treatment and temperature effects on germination of *Engelmannia pinnatifida* and *Indigofera miniata* var. *leptosepala*. *J. Range Manage.* **36**, 94–97.

Kitajima, K., and Augspurger, C. K. (1989). Seeds and seedling ecology of a monocarpic tropical tree, *Tachigalia versicolor*. *Ecology* **70**, 1102–1114.

Klink, C. A. (1996). Germination and seedling establishment of two native and one invading African grass species in the Brazilian cerrado. *J. Trop. Ecol.* **12**, 139–147.

Knight, D. H. (1975). A phytosociological analysis of species-rich tropical forest on Barro Colorado Island, Panama. *Ecol. Monogr.* **45**, 259–284.

Knipe, D., and Herbel, C. H. (1966). Germination and growth of some semidesert grassland species treated with aqueous extract from creosotebush. *Ecology* **47**, 775–781.

Koebernik, J. (1971). Germination of palm seed. *Int. Palm Soc.* **15**, 134–137.

Koller, D. (1955). Germination-regulating mechanisms in some desert seeds. II. *Zygophyllum dumosum* Boiss. *Bull. Res. Counc. Israel* **4D**, 381–387.

Koller, D. (1957). Germination-regulating mechanisms in some desert seeds. IV. *Atriplex dimorphostegia* Kar. et Kir. *Ecology* **38**, 1–13.

Koller, D., and Negbi, M. (1959). The regulation of germination in *Oryzopsis miliacea*. *Ecology* **40**, 20–36.

Koller, D., and Cohen, D. (1959). Germination-regulating mechanisms in some desert seeds. VI. *Convolvulus lanatus* Vahl, *Convolvulus negevensis* Zoh. and *Convolvulus secundus* Desr. *Bull. Res. Counc. Israel* **7D**, 175–180.

Koller, D., Poljakoff-Mayber, A., Berg, A., and Diskin, T. (1963). Germination-regulating mechanisms in *Citrullus colocynthis*. *Am. J. Bot.* **50**, 597–603.

Koller, D., and Roth, N. (1963). Germination-regulating mechanisms in some desert seeds. VII. *Panicum turgidum* (Gramineae). *Israel J. Bot.* **12**, 64–73.

Koller, D., Sachs, M., and Negbi, M. (1964). Germination-regulating mechanisms in some desert seeds. VIII. *Artemisia monosperma*. *Plant Cell Physiol.* **5**, 85–100.

Kress, W. J. (1986). The systematic distribution of vascular epiphytes: An update. *Selbyana* **9**, 2–22.

Kumar, A., Joshi, M. C., and Babu, V. R. (1971). Some factors influencing the germination of seeds in two desert grasses. *Trop. Ecol.* **12**, 202–208.

Kumar, B., and Chaudhary, S. K. (1991). Effect of temperature on the germination of *Indigofera linnaei* Ali and *I. linifolia* Retz. *Biojournal* **3**, 235–238.

Labouriau, L. G., Valio, I. F. M., and Heringer, E. P. (1964). Sobre o sistema reprodutivo de plantas dos cerrados-I. *Anais Acad. Brasileira Cien.* **36**, 449–464.

Lacey, C. J., Walker, J., and Noble, I. R. (1982). Fire in Australian tropical savannas. *In* "Ecology of Tropical Savannas" (B. J. Huntley and B. H. Walker, eds.), pp. 246–272. Springer-Verlag, Berlin.

Lajtha, K., Weishampel, J., and Schlesinger, W. H. (1987). Phosphorus and pH tolerances in the germination of the desert shrub *Larrea tridentata* (Zygophyllaceae). *Madrono* **34**, 63–68.

Laman, T. G. (1995). *Ficus stupenda* germination and seedling establishment in a Bornean rain forest canopy. *Ecology* **76**, 2617–2626.

Lambert, F. J., Bower, M., Whalley, R. D. B., Andrews, A. C., and Bellotti, W. D. (1990). The effects of soil moisture and planting depth on emergence and seedling morphology of *Astrebla lappacea* (Lindl.) Domin. *Aust. J. Agric. Res.* **41**, 367–376.

Laura, V. A., de Alvarenga, A. A., and de F. Arrigoni, M. (1994). Effects of growth regulators, temperature, light, storage and other factors on the *Muntingia calabura* L. seed germination. *Seed Sci. Technol.* **22**, 573–579.

Lavie, D., Levy, E. C., Cohen, A., Evenari, M., and Gutterman, Y. (1974). New germination inhibitor from *Aegilops ovata* L. *Nature* **249**, 388.

Lawton, R. O., and Putz, F. E. (1988). Natural disturbance and gap-phase regeneration in a wind-exposed tropical cloud forest. *Ecology* **69**, 764–777.

Lebron, M. L. (1979). An autecological study of *Palicourea riparia* Bentham as related to rain forest disturbance in Puerto Rico. *Oecologia* **42**, 31–46.

Le Houerou, H. N. (1986). The desert and arid zones of northern Africa. *In* "Ecosystems of the World" (M. Evenari, I. Noy-Meir, and D. W. Goodall, eds.), pp. 101–147. Elsevier, Amsterdam.

Leistner, O. A. (1979). Southern Africa. *In* "Arid-Land Ecosystems: Structure, Functioning and Management" (D. W. Goodall, and R. A. Perry, eds.), Vol. 1, pp. 109–143. Cambridge Univ. Press, Cambridge, UK.

Lerner, H. R., Meyer, A. M., and Evenari, M. (1959). The nature of the germination inhibitors present in dispersal units of *Zygophyllum dumosum* and *Trigonella arabica*. *Physiol. Plant.* **12**, 245–250.

Lieberman, D. (1982). Seasonality and phenology in a dry tropical forest in Ghana. *J. Ecol.* **70**, 791–806.

Lieberman, D., and Li, M. (1992). Seedling recruitment patterns in a tropical dry forest in Ghana. *J. Veg. Sci.* **3**, 375–382.

Liengsiri, C., and Hellum, A. K. (1988). Effects of temperature on seed germination in *Pterocarpus macrocarpus*. *J. Seed Technol.* **12**, 66–75.

Lin, T.-P., and Chen, M.-H. (1995). Biochemical characteristics associated with the development of the desiccation-sensitive seeds of *Machilus thunbergii* Sieb. & Zucc. *Ann. Bot.* **76**, 381–387.

Lock, J. M., and Milburn, T. R. (1971). The seed biology of *Themeda triandra* Forsk. in relation to fire. *In* "The Scientific Management of Animal and Plant Communities for Conservation" (E. Duffey, and A. Watt, eds.), pp. 337–349. British Ecol. Soc. Symposium No. 11. Blackwell Scientific Publications, Oxford.

Lodge, G. M., and Whalley, R. D. B. (1981). Establishment of warm- and cool-season native perennial grasses on the north-west slopes of New South Wales. I. Dormancy and germination. *Aust. J. Bot.* **29**, 111–119.

Longman, K. A. (1969). The dormancy and survival of plants in the humid tropics. *Symp. Soc. Exp. Biol.* **23**, 471–488.

Longman, K. A., and Jenik, J. (1987). "Tropical Forest and Its Environment." 2nd Ed. Longman Scientific and Technical. Co-published with Wiley, New York.

Loria, M., and Noy-Meir, I. (1979–1980). Dynamics of some annual populations in a desert loess plain. *Israel J. Bot.* **28**, 211–225.

Louda, S. M. (1982). Limitation of the recruitment of the shrub *Haplopappus squarrosus* (Asteraceae) by flower- and seed-feeding insects. *J. Ecol.* **70**, 43–53.

MacMahon, J. A. (1979). North American deserts: Their floral and faunal components. *In* "Arid-Land Ecosystems: Structure, Functioning and Management"(D. W. Goodall, and R. A. Perry, eds.), Vol. 1, pp. 21–82. Cambridge Univ. Press, Cambridge, UK.

MacMahon, J. A., and Wagner, F. H. (1985). The Mojave, Sonoran and Chihuahuan deserts of North America. *In* "Ecosystems of the World" (M. Evenari, I. Noy-Meir, and D. W. Goodall, eds.), Vol. 12A, pp. 105–202. Elsevier, Amsterdam.

Madison, M. (1977). Vascular epiphytes: Their systematic occurrence and salient features. *Selbyana* **2**, 1–13.

Madrid, R. R., Caligaris, L. E. G., and Rincon, E. (1989). Algunos aspectos de la ecofisiologia de la germinacion en *Physalis philadelphica*. *Acta Bot. Mexicana* **7**, 33–41.

Mahgembe, J. A., and Msanga, H. P. (1988). Effect of physical scarification and gibberellic acid treatments in germination of *Trichilia emetica* seed. *Int. Tree Crops J.* **5**, 163–177.

Mahmoud, A., El Sheikh, A. M., and Abdul Baset, S. (1983). Germination of two halophytes: *Halopeplis perfoliata* and *Limonium axillare* from Saudi Arabia. *J. Arid Environ.* **6**, 87–98.

Mandujano, S., Gallina, S., and Bullock, S. H. (1994). Frugivory and dispersal of *Spondias purpurea* (Anacardiaceae) in a tropical deciduous forest in Mexico. *Rev. Biol. Trop.* **42**, 107–114.

Manokaran, N. (1979). Germination of Malaysian palms. *Malaysian For.* **42**, 50–52.

Mares, M. A., Morello, J., and Goldstein, G. (1985). The Monte desert and other subtropical, semi-arid biomes of Argentina, with comments on their relation to North American arid areas. *In* "Ecosystems of the World" (M. Evenari, I. Noy-Meir, and D. W. Goodall, eds.), Vol. 12A, pp. 203–237. Elsevier, Amsterdam.

Marks, M. K. (1983). Timing of seedling emergence and reproduction in some tropical dicotyledonous weeds. *Weed Res.* **23**, 325–332.

Marks, M. K., and Nwachuku, A. C. (1986). Seed-bank characteristics in a group of tropical weeds. *Weed Res.* **26**, 151–157.

Martin, A. C. (1946). The comparative internal morphology of seeds. *Am. Midl. Nat.* **36**, 513–660.

Martin, C. (1991). "The Rainforests of West Africa: Ecology–Threats–Conservation." Birkhauser Verlag, Basel. [English translation by Linda Tsardakas]

Martinez-Ramos, M., and Alvarez-Buylla, E. (1986). Seed dispersal, gap dynamics and tree recruitment: The case of *Cecropia obtusifolia* at Los Tuxtlas, Mexico. *In* "Frugivores and Seed Dispersal" (A. Estrada, and T. H. Fleming, eds.), pp. 332–346. Junk, Dordercht.

Mauldin, M. P. (1943). The rapid germination of a species with a mucilaginous seed coat, *Plantago fastigiata* Morris. *Agron. J.* **35**, 1023–1027.

Maury-Lechon, G., Dorffling, K., Trung, T. V., and Phong, N. T. (1993). Trials to delay germination in *Hopea odorata*, Vietnamese Dipterocarpaceae. *In* "Fourth International Workshop on Seeds: Basic and Applied Aspects of Seed Biology" (D. Come, and F. Corbineau, eds.), pp. 857–862. Universite Pierre et Marie Curie, Paris.

Maycock, D. J., and Berjak, P. (1993). Cryostorage of somatic embryos of coffee. *In* "Fourth International Workshop on Seeds:. Basic and Applied Aspects of Seed Biology" (D. Come, and F. Corbineau, eds), pp. 879–884. Universite Pierre et Marie Curie, Paris.

Mayeux, H. S., Jr. (1982). Germination of false broomweed (*Ericameria austrotexana*) seed. *Weed Sci.* **30**, 597–601.

Mayeux, H. S., Jr. and Scifres, C. J. (1978). Germination of goldenweed seed. *J. Range Manage.* **31**, 371–374.

McDonough, W. T. (1964). Germination responses of *Carnegiea gigantea* and *Lemaireocereus thurberi*. *Ecology* **45**, 155–159.

McCleary, J. A. (1968). The biology of desert plants. *In* "Desert Biology" (G. W. Brown, Jr, ed.), pp. 141–194. Academic Press, New York.

McCleary, J. A., and Wagner, K. A. (1973). Comparative germination and early growth studies of six species of the genus *Yucca*. *Am. Midl. Nat.* **90**, 503–508.

McHargue, L. A., and Hartshorn, G. S. (1983). Seed and seedling ecology of *Carapa guianensis. Turrialba* **33**, 399–404.

McKeon, G. M. (1984). Field changes in germination requirements: Effect of natural rainfall on potential germination speed and light requirement of *Stylosanthes humilis, Stylosanthes hamata* and *Digitaria ciliaris. Aust. J. Agric. Res.* **35**, 807–819.

Mehta, M., and Sen, D. N. (1991). Seed germination studies of *Cassia italica* (Mill.) Lamix-ex Andress in Indian Desert. *Ann. Arid Zone* **30**, 71–72.

Melhem, T. S. (1975). Fisiologia da germinacao das sementes de *Dipteryx alata* Vog. (Leguminosae-Lotoideae). *Hoehnea* **5**, 59–90.

Metcalfe, D. J. (1996). Germination of small-seeded tropical rain forest plants exposed to different spectral compositions. *Can. J. Bot.* **74**, 516–520.

Metcalfe, D., J., and Grubb, P. J. (1995). Seed mass and light requirements for regeneration in southeast Asian rain forest. *Can. J. Bot.* **73**, 817–826.

Meyers, J. G. (1933). Notes on the vegetation of the Venezuelan llanos. *J. Ecol.* **21**, 333–349.

Midya, S., and Brahmachary, R. L. (1991). The effect of birds upon germination of banyan (*Ficus bengalensis*) seeds. *J. Trop. Ecol.* **7**, 537–538.

Milton, S. J. (1995). Spatial and temporal patterns in the emergence and survival of seedlings in arid Karoo shrubland. *J. Appl. Ecol.* **32**, 145–156.

Miranda, A. C., Miranda, H. S., de F. O. Dias, I., and de S. Dias, B. F. (1993). Soil and air temperatures during prescribed cerrado fires in central Brazil. *J. Trop. Ecol.* **9**, 313–320.

Mishra, R. K., and Sen, D. N. (1984). A report on germination behaviour in dimorphic seeds of *Indigofera hochstetteri* Baker, in Indian arid zone. *Curr. Sci.* **53**, 380–381.

Mishra, R. K., Shukla, P., Choudhury, P. C., Ghosh, A., and Sengupta, K. (1990). *In* "International Symposium on Environmental Influences on Seed and Germination Mechanism: Recent Advances in Research and Technology" (D. N. Sen, S. Mohammed, P. K. Kasera, and T. P. Thomas, eds.), pp. 22–23 University of Jodhpur, Jodhpur, India. [Abstract]

Misra, R. (1983). Indian savannas. *In* "Ecosystems of the World (F. Bourliere, ed.), Vol. 13, pp. 151–166. Elsevier, Amsterdam.

Miyanishi, K., and Kellman, M. M. (1986). The role of fire in recruitment of two neotropical savanna shrubs, *Miconia albicans* and *Clidemia sericea. Biotropica* **18**, 224–230.

Mohamad, A. B., and Ng, F. S. P. (1982). Influence of light on germination of *Pinus caribaea, Gmelina arborea, Sapium baccatum* and *Vitex pinnata. Malaysian For.* **45**, 62–68.

Mohammed, S., and Sen, D. N. (1990). Seed germination studies of halophytes in Indian arid zone. *In* "International Symposium on Environmental Influences on Seed and Germination Mechanism: Recent Advances in Research and Technology" (D. N. Sen, S. Mohammed, P., K. Kasera, and T. P. Thomas, eds.), pp. 96–98. University of Jodhpur, Jodhpur, India. [Abstract]

Moll, D., and Jansen, K. P. (1995). Evidence for a role in seed dispersal by two tropical herbivorous turtles. *Biotropica* **27**, 121–127.

Monasterio, M., and Sarmiento, G. (1976). Phenological strategies of plant species in the tropical savanna and the semi-deciduous forest of the Venezuelan llanos. *J. Biogeogr.* **3**, 325–356.

Moreno, P. (1977). Latencia y viabilidad de semillas de arboles tropicales. *Interciencia* **2**, 298–302.

Morgan, J. W. (1995). Ecological studies of the endangered *Rutidosis leptorrhynchoides.* II. Patterns of seedling emergence and survival in a native grassland. *Aust. J. Bot.* **43**, 13–24.

Mori, T., Rahman, Z. H. A., and Tan, C. H. (1980). Germination and storage of rotan manau (*Calamus manan*) seeds. *Malaysian For.* **43**, 44–55.

Mott, J. J. (1972). Germination studies on some annual species from an arid region of Western Australia. *J. Ecol.* **60**, 293–304.

Mott, J. J. (1974). Factors affecting seed germination in three annual species from an arid region of Western Australia. *J. Ecol.* **62**, 699–709.

Mott, J. J. (1978). Dormancy and germination in five native grass species from savannah woodland communities of the Northern Territory. *Aust. J. Bot.* **26**, 621–631.

Mott, J. J., and McComb, A. J. (1975). Embryo dormancy and light requirements in the germination of *Helipterum craspedioides* (Compositae), an arid-zone annual. *Ann. Bot.* **39**, 1071–1075.

Mucunguzi, P., and Oryem-Origa, H. (1996). Effects of heat and fire on the germination of *Acacia sieberiana* D.C. and *Acacia gerrardii* Benth. in Uganda. *J. Trop. Ecol.* **12**, 1–10.

Mugasha, A. G., and Msanga, H. P. (1987). *Maesopsis eminii* seed coat impermeability is not the cause of sporadic and prolonged seed germination. *For. Ecol. Manage.* **22**, 301–305.

Mullett, J. H., Beardsell, D. V., and King, H. M. (1981). The effect of seed treatment on the germination and early growth of *Euterpe edulis* (family Palmae). *Sci. Hort.* **15**, 239–244.

Mullick, P., and Chatterji, U. N. (1967). Eco-physiological studies on seed germination: Germination experiments with the seeds of *Clitoria ternatea* Linn. *Trop. Ecol.* **8**, 117–125.

Mumford, P. M., and Grout, B. W. W. (1979). Desiccation and low temperature (−196°C) tolerance of *Citrus limon* seed. *Seed Sci. Technol.* **7**, 407–410.

Munson, R. H. (1984). Germination of western soapberry as affected by scarification and stratification. *HortScience* **19**, 712–713.

Murbach, L. (1900). Note on the mechanics of the seed-burying awns of *Stipa avenacea. Bot. Gaz.* **30**, 113–117.

Murray, K. G. (1986). Consequences of seed dispersal for gap-dependent plants: Relationships between seed shadows, germination requirements, and forest dynamic processes. *In* "Frugivores and Seed Dispersal" (A. Estrada, and T. H. Fleming, eds.), pp. 187–198. Junk, Dordrecht.

Murray, K. G. (1988). Avian seed dispersal of three neotropical gap-dependent plants. *Ecol. Monogr.* **58**, 271–298.

Muthana, K. D., Gyanchand, and Arora, G. D. (1980). Silvical studies on indigenous and exotic tree species in Rajasthan desert. *In* "Arid Zone Research and Development" (H. S. Mann, ed.), pp. 339–343. Scientific Publishers, Jodhpur, India.

Naqvi, H. H., and Hanson, G. P. (1982). Germination and growth inhibitors in guayule (*Parthenium argentatum* Gray) chaff and their possible influence in seed dormancy. *Am. J. Bot.* **69**, 985–989.

Nasir, Y. J., and Ali, S. I. (1983). Flora of Pakistan. Geraniaceae. No. 149. Pakistan Agricultural Research Council, Islamabad.

Negbi, M., and Tamari, B. (1963). Germination of chlorophyllous and achlorophyllous seeds of *Salsola volkensii* and *Aellenia autrani. Israel J. Bot.* **12**, 124–135.

Negi, A. K., and Todaria, N. P. (1995). Pre-treatment methods to improve germination in *Terminalia tomentosa* Wight & Arn. *Seed Sci. Technol.* **23**, 245–248.

Newaz, K. M. N., and Nazrul-Islam, A. K. M. (1991). The influence of moisture regimes and soil types on germination and establishment of two species of jute. *In* "Proc. Int. Seed Symposium" (D. N. Sen, and S. Mohammed, eds.), pp. 123–128. Jodhpur, India.

Ng, F. S. P. (1973). Germination of fresh seeds of Malaysian trees. *Malaysian For.* **36**, 54–65.

Ng, F. S. P. (1978). Strategies of establishment in Malayan forest trees. *In* "Tropical Trees as Living Systems" (P. B. Tomlinson, and M. H. Zimmermann, eds.), pp. 129–162. Cambridge Univ. Press, Cambridge, UK.

Ng, F. S. P. (1980). Germination ecology of Malaysian woody plants. *Malaysian For.* **43**, 406–438.

Ng, F. S. P. (1988). Forest tree biology. *In* "Key Environments: Malaysia" (Earl of Cranbrook, ed.), pp. 102–125. Pergamon Press, Oxford.

Ng, F. S. P., and Asri, N. S. (1979). Germination of fresh seeds of Malaysian trees IV. *Malaysian For.* **42,** 221–224.

Nkang, A. (1993). Time, temperature and germinability in rainforest seeds with orthodox and recalcitrant viability characteristics. *In* "Fourth International Workshop on Seeds: Basic and Applied Aspects of Seed Biology" (D. Come, and F. Corbineau, eds.), pp. 411–416. Universite Pierre et Marie Curie, Paris.

Nkang, A., and Chandler, G. (1986). Changes during embryogenesis in rainforest seeds with orthodox and recalcitrant viability characteristics. *J. Plant Physiol.* **126,** 243–256.

Noble, I. R., and Crisp, M. D. (1979–1980). Germination and growth models of short-lived grass and forb populations based on long term photo-point data at Koonamore, South America. *Israel J. Bot.* **28,** 195–210.

Noraini, M. T., Sainul, H. B., and Jamilah, M. S. (1994). Seed storage and germination of two Malaysian *Toona* species (Meliaceae). *Malayan Nat. J.* **48,** 83–87.

Normah, M. N., Ramiya, S. D., and Gintangga, M. (1997). Desiccation sensitivity of recalcitrant seeds: A study on tropical fruit species. *Seed Sci. Res.* **7,** 179–183.

Okusanya, O. T. (1979). Quantitative analysis of the effects of photoperiod, temperature, salinity and soil types on the germination and growth of *Corchorus olitorius. Oikos* **33,** 444–450.

Okusanya, O. T. (1980). Germination and growth of *Celosia cristata* L., under various light and temperature regimes. *Am. J. Bot.* **67,** 854–858.

Okusanya, O. T., Lakanmi, O. O., and Oyesiku, O. O. (1991). Germination ecology of the woody herb *Uraria picta,* from southern Nigeria. *J. Trop. Ecol.* **7,** 139–146.

Okusanya, O. T., Oyesiku, O. O., and Lakanmi, O. (1992). Seed dormancy in *Uraria picta. Nigerian J. Bot.* **5,** 209–218.

Oladokun, A. O. (1985). Germination studies on *Cola acuminata* (P. Beauv.) (Schott and Endlicher). *Turrialba* **35,** 109–115.

Oliveira, L. M. Q., and Valio, I. F. M. (1992). Effects of moisture content on germination of seeds of *Hancornia speciosa* Gom. (Apocynaceae). *Ann. Bot.* **69,** 1–5.

Oliveira, P. E., and Silva, J. C. S. (1993). Reproductive biology of two species of *Kielmeyera* (Guttiferae) in the cerrados of central Brazil. *J. Trop. Ecol.* **9,** 67–79.

Oliveira, P. S., Galetti, M., Pedroni, F., and Morellato, L. P. (1995). Seed Cleaning by *Mycocepurus goeldii* ants (Attini) facilitates germination in *Hymenaea courbaril* (Caesalpiniaceae). *Biotropica* **27,** 519–522.

Onyekwelu, S. S. C. (1990). Germination, seedling morphology and establishment of *Combretum bauchiense* Hutch. & Dalz. (Combretaceae). *Bot. J. Linnean Soc.* **103,** 133–138.

Orozco-Segovia, A., Sanchez-Coronado, M. E., and Vazquez-Yanes, C. (1993). Light environment and phytochrome-controlled germination in *Piper auritum. Funct. Ecol.* **7,** 585–590.

Orozco-Segovia, A., and Vazquez-Yanes, C. (1989). Light effect on seed germination in *Piper* L. *Acta Oecol.* **10,** 123–146.

Orozco-Segovia, A., and Vazquez-Yanes, C. (1990). Effect of moisture on longevity in seeds of some rain forest species. *Biotropica* **22,** 215–216.

Orozco-Segovia, A., and Vazquez-Yanes, C. (1993). Light environment and phytochrome-controlled germination in *Piper auritum. Funct. Ecol.* **7,** 585–590.

Orozco-Segovia, A., Vazquez-Yanes, C., and Coates-Estrada, R. (1987). Ecophysiological characteristics of the seed of the tropical forest pioneer *Urera caracasana* (Urticaceae). *Tree Physiol.* **3,** 375–386.

Orshan, G. (1986). The deserts of the Middle East. *In* "Ecosystems of the World" (M. Evenari, I. Noy-Meir, and D. W. Goodall, eds.), Vol. 12B, pp. 1–28. Elsevier, Amsterdam.

Pandey, H. N., and Dubey, S. K. (1988). Achene germination of *Parthenium hysterophorus* L.: Effects of light, temperature, provenance and achene size. *Weed Res.* **28,** 185–190.

Pandya, S. M., and Pathak, V. S. (1980). Seed dormancy imposed by covering structures in *Achyranthes aspera* Linn. *Geobios* **7,** 74–76.

Pandya, S. M., and Reddy, V. M. (1990). Germination responses of the seeds of jungle onion (*Asphodelus tenuifolius*), a weed. *In* "International Symposium on Environmental Influences on Seed and Germination Mechanism: Recent Advances in Research and Technology" (D. N. Sen, S. Mohammed, P. K. Kasera, and T. P. Thomas, eds.), pp. 72–74. University of Jodhpur, Jodhpur, India. [Abstract]

Pearson, H. H. W. (1910). On the embryo of *Welwitschia. Ann. Bot.* **24,** 759–766 + 1 plate.

Peart, M. H. (1979). Experiments on the biological significance of the morphology of seed-dispersal units in grasses. *J. Ecol.* **67,** 843–863.

Peart, M. H. (1981). Further experiments on the biological significance of the morphology of seed-dispersal units in grasses. *J. Ecol.* **69,** 425–436.

Peart, M. H. (1984). The effects of morphology, orientation and position of grass diaspores on seedling survival. *J. Ecol.* **72,** 437–453.

Pelaez, D. V., Boo, R. M., and Elia, O. R. (1996). The germination and seedling survival of *Condalia microphylla* Cav. in Argentina. *J. Arid Environ.* **32,** 173–179.

Pelton, J. (1964). A survey of the ecology of *Tecoma stans. Butler Univ. Bot. Stud.* **14,** 53–88.

Pemadasa, M. A., and Amarasinghe, L. (1982). The ecology of a montane grassland in Sri Lanka. *J. Ecol.* **70,** 483–490.

Perez-Nasser, N., and Vazquez-Yanes, C. (1986). Longevity of buried seeds from some tropical rain forest trees and shrubs of Veracruz, Mexico. *Malaysian For.* **49,** 352–356.

Perry, A. E., and Fleming, T. H. (1980). Ant and rodent predation on small, animal-dispersed seeds in a dry tropical forest. *Brenesia* **17,** 11–22.

Petteys, E. Q. P. (1974). *Tristania conferta* R. Br. Brushbox. *In* "Seeds of Woody Plants in the United States" (C. S. Schopmeyer, tech. coord.), pp. 817–818. USDA Forest Service.

Pickett, S. T. A. (1983). Differential adaptation of tropical tree species to canopy gaps and its role in community dynamics. *Trop. Ecol.* **24,** 68–84.

Pluenneke, R. H., and Joham, H. E. (1972). Studies on the germination of saladillo (*Varilla texana* Gray). *Phyton (Buenos Aires)* **30,** 141–145.

Plummer, J. A., and Bell, D. T. (1995). The effect of temperature, light and gibberellic acid (GA₃) on the germination of Australian everlasting daisies (Asteraceae, Tribe Inuleae). *Aust. J. Bot.* **43,** 93–102.

Poole, F. N. (1958). "Seed Germination Requirements of Four Desert Tree Species." M.S. thesis, Univ. of Arizona, Tucson.

Poore, M. E. D. (1968). Studies in Malaysian rain forest. I. The forest on Triassic sediments in Jengka Forest Reserve. *J. Ecol.* **56,** 143–196.

Popay, A. I. (1973). Germination and dormancy in the seeds of certain east African weed species. *In* "Proc. 4th Asian-Pacific Weed Sci. Soc. Conf." Rotorua, New Zealand, pp. 77–81.

Popay, A. I. (1974). Investigations into the behaviour of the seeds of some tropical weeds. I. Laboratory germination tests. *East Afr. Agric. For. J.* **40,** 31–43.

Popay, A. I. (1975). Investigations into the behaviour of the seeds of some tropical weeds. II. Dry soil storage and seasonal germination. *East Afr. Agric. For. J.* **40,** 408–415.

Popay, A. I. (1976). Investigations into the behaviour of the seeds of some tropical weeds. III. Patterns of emergence. *East Afr. Agric. For. J.* **41,** 304–312.

Potter, R. L., Petersen, J. L., and Ueckert, D. N. (1984). Germination responses of *Opuntia* spp. to temperature, scarification, and other seed treatments. *Weed Sci.* **32,** 106–110.

Pritchard, H. W., Haye, A. J., Wright, W. J., and Steadman, K. J. (1995). A comparative study of seed viability in *Inga* species: Desiccation tolerance in relation to the physical characteristics and chemical composition of the embryo. *Seed Sci. Technol.* **23,** 85–100.

Probert, R. J., and Longley, P. L. (1989). Recalcitrant seed storage physiology in three aquatic grasses (*Zizania palustris, Spartina anglica* and *Porteresia coarctata*). *Ann. Bot.* **63,** 53–63.

Puchet, C. E., and Vazquez-Yanes, C. (1987). Heteromorfismo criptico en las semillas recalcitrantes de tres especies arboreas de al selva tropical humeda de Veracruz, Mexico. *Phytologia* **62,** 100–106.

Purohit, A. N., Sharma, M. M., and Thapliyal, R. C. (1982). Effect of storage temperatures on the viability of sal (*Shorea robusta*) and talura (*Shorea talura*) seed. *For. Sci.* **28,** 526–530.

Putz, F. E. (1983). Treefall pits and mounds, buried seeds, and the importance of soil disturbance to pioneer tress on Barro Colorado Island, Panama. *Ecology* **64,** 1069–1074.

Qadir, S. A., and Lodhi, N. (1971). Germination behaviour of seeds of some common shrubs. *J. Sci. Univ. Karachi* **1,** 84–97.

Qaiser, M., and Qadir, S. A. (1971). A contribution to the autecology of *Capparis decidua* (Forssk.) Edgew. *Pakistan J. Bot.* **3,** 37–60.

Quarterman, E. (1970). Germination of seeds of certain tropical species. In: "A Tropical Rain Forest" (H. T. Odum, ed.), pp. D173–D175. U. S. Atomic Energy Commission, Oak Ridge, TN.

Raich, J. W., and Khoon, G. W. (1990). Effects of canopy openings on tree seed germination in a Malaysian dipterocarp forest. *J. Trop. Ecol.* **6,** 203–217.

Ramirez, W. B. (1976). Germination of seeds of new world *Urostigma* (*Ficus*) and of *Morus rubra* L. (Moraceae). *Rev. Biol. Trop.* **24,** 1–6.

Rauh, W. (1985). The Peruvian-Chilean deserts. *In* "Ecosystems of the World" (M. Evenari, I. Noy-Meir, and D. W. Goodall, eds.), Vol. 12A, pp. 239–267. Elsevier, Amsterdam.

Rauh, W. (1986). The arid region of Madagascar. *In* "Ecosystems of the World" (M. Evenari, I. Noy-Meir, and D. W. Goodall eds.), Vol. 12B, pp. 361–377. Elsevier, Amsterdam.

Ray, P. K., and Sharma, S. B. (1987). Growth, maturity, germination and storage of litchi seeds. *Sci. Hort.* **33,** 213–221.

Renard, C., and Capelle, P. (1976). Seed germination in ruzizi grass (*Brachiaria ruziziensis* Germain & Evrard). *Aust. J. Bot.* **24,** 437–446.

Richards, P. W. (1957). "The Tropical Rainforest: An Ecological Study." Univ. Press, Cambridge, UK.

Richmond, G. S., and Chinnock, R. J. (1994). Seed germination of the Australian desert shrub *Eremophila* (Myoporaceae). *Bot. Rev.* **60,** 483–503.

Richmond, G. S., and Ghisalberti, E. L. (1994). Seed dormancy and germination mechanisms in *Eremophila* (Myoporaceae). *Aust. J. Bot.* **42,** 705–715.

Ridley, H. N. (1930). "The Dispersal of Plants throughout the World." L. Reeve and Company, Ashford, Kent, England.

Rintz, R. E. (1978). The Peninsular Malaysian species of *Hoya* (Asclepiadaceae). *Malayan Nat. J.* **30,** 467–522.

Rivera, R. L., and Freeman, C. E. (1979). The effects of some alternating temperatures on germination of creosotebush (*Larrea tridentata*[D.C.] Cov.: Zygophyllaceae). *Southw. Nat.* **24,** 683–714.

Rizzini, C. T. (1965). Experimental studies on seedling development of cerrado woody plants. *Ann. Missouri Bot. Gard.* **52,** 410–426.

Rizzini, C. T. (1970). Inibidores de germinacao e crescimento em *Andira humilis* Benth. *An. Acad. Brasileira Cienc.* **42**(Suppl.), 329–366.

Roberts, H. A., and Boddrell, J. E. (1983). Field emergence and temperature requirements for germination in *Solanum sarrachoides* Sendt. *Weed Res.* **23,** 247–252.

Roberts, H. A., and Lockett, P. W. (1978). Seed dormancy and field emergence in *Solanum nigrum* L. *Weed Res.* **18,** 231–241.

Roberts, H. A., and Neilson, J. E. (1982a). Role of temperature in the seasonal dormancy of seeds of *Veronica hederifolia* L. *New Phytol.* **90,** 745–749.

Roberts, H. A., and Neilson, J. E. (1982b). Seasonal changes in the temperature requirements for germination of buried seeds of *Aphanes arvensis* L. *New Phytol.* **92,** 159–166.

Roberts, J. T., and Heithaus, E. R. (1986). Ants rearrange the vertebrate-generated seed shadow of a neotropical fig tree. *Ecology* **67,** 1046–1051.

Rosenthal, G. A. (1977). The biological effects and mode of action of L-canavanine, a structural analogue of L-arginine. *Quart. Rev. Biol.* **52,** 155–178.

Rosenthal, G. A. (1983). A seed-eating beetle's adaptations to a poisonous seed. *Sci. Amer.* **249,** 164–171.

Rosenthal, G. A. (1988). The protective action of a higher plant toxic product. *BioScience* **38,** 104–109.

Rosenthal, G. A., Dahlman, D. L., and Janzen, D. H. (1976). A novel means for dealing with L-canavanine, a toxic metabolite. *Science* **192,** 256–258.

Rosenthal, G. A., Hughes, C. G. and Janzen, D. H. (1982). L-Canavanine, a dietary nitrogen source for the seed predator *Caryedes brasiliensis* (Bruchidae). *Science* **217,** 353–355.

Rosenthal, G. A., Janzen, D. H., and Dahlman, D. L. (1977). Degradation and detoxification of canavanine by a specialized seed predator. *Science* **196,** 658–660.

Ross, M. A. (1976). The effects of temperature on germination and early growth of three plant species indigenous to central Australia. *Aust. J. Ecol.* **1,** 259–263.

Roth, I. (1987). "Stratification of a Tropical Forest as Seen in Dispersal Types." Junk, Dordrecht.

Rundel, P. W., Smith, A. P., and Meinzer, F. C. (eds.) (1994). "Tropical Alpine Environments: Plant Form and Function." Cambridge Univ. Press, Cambridge, MA.

Ryan, J., Miyamoto, S., and Stroehlein, J. L. (1975). Salt and specific ion effects on germination of four grasses. *J. Range Manage.* **28,** 61–64.

Sabiiti, E. N., and Wein, R. W. (1987). Fire and *Acacia* seeds: A hypothesis of colonization success. *J. Ecol.* **74,** 937–946.

Sankary, M. N., and Barbour, M. G. (1972). Autecology of *Haloxylon articulatum* in Syria. *J. Ecol.* **60,** 697–711.

Sarmiento, G. (1983). The savannas of tropical America. *In* "Ecosystems of the World" (F. Bourliere, ed.), Vol. 13, pp. 245–288. Elsevier, Amsterdam.

Sarmiento, G. and Monasterio, M. (1983). Life forms and phenology. *In:* "Ecosystems of the World" (F. Bourliere, ed.), Vol. 13, pp. 79–108. Elsevier, Amsterdam.

Sasaki, S. (1980a). Storage and germination of some Malaysian legume seeds. *Malaysian For.* **43,** 161–165.

Sasaki, S. (1980b). Storage and germination of dipterocarp seeds. *Malaysian For.* **43,** 290–308.

Sasaki, S., and Ng, F. S. P. (1981). Physiological studies on germination and seedling development in *Intsia palembanica* (Merbau). *Malaysian For.* **44,** 43–59.

Sasaki, S., Tan, C. H., and Zulfatah, H. A. R. (1979). Some observations on unusual flowering and fruiting of dipterocarps. *Malaysian For.* **42,** 38–45.

Savage, M. (1992). Germination of forest species under an anthropogenic vine mosaic in western Samoa. *Biotropica* **24**, 460–462.

Schupp, E. W. (1990). Annual variation in seedfall, postdispersal predation, and recruitment of a neotropical tree. *Ecology* **71**, 504–515.

Scifres, C. J. (1974). Salient aspects of huisache seed germination. *Southw. Nat.* **18**, 383–392.

Scifres, C. J., and Brock, J. H. (1969). Moisture-temperature interrelations in germination and early seedling development in mesquite. *J. Range Manage.* **22**, 334–337.

Scifres, C. J., and Brock, J. H. (1972). Emergence of honey mesquite seedlings relative to planting depth and soil temperature. *J. Range Manage.* **25**, 217–219.

Scott, F. M., Bystrom, B. G., and Bowler, E. (1962). *Cercidium floridum* seed coat, light and electron microscopic study. *Am. J. Bot.* **49**, 821–833.

Scotter, D. R. (1970). Soil temperatures under grass fires. *Aust. J. Soil Res.* **8**, 273–279.

Secor, J. B., and Farhadnejad, D. O. (1978). The seed germination ecology of three species of *Gaillardia* that occur in the gypsumland areas of eastern New Mexico. *Southw. Nat.* **23**, 181–186.

Seghieri, J., Floret, Ch., and Pontanier, R. (1995). Plant phenology in relation to water availability: Herbaceous and woody species in the savannas of northern Cameroon. *J. Trop. Ecol.* **11**, 237–254.

Sen, D. N. (1968). Ecology of desert plants and observations on their seedlings. II. Germination behaviour of seeds in Asclepiadaceae. *Osterr. Bot. Z.* **115**, 18–27.

Sen, N. (1969). Action of light in the germination of seeds and seedling growth in some Asclepiadaceae. *Acta Bot. Acad. Scien. Hung.* **15**, 327–335.

Sen, D. N., and Chatterji, U. N. (1965). Ecological studies on *Calotropis procera* (Ait.) R. Br. Australian Arid-Zone Research Conference, Alice Springs, Northern Territory, September 1965. pp. C-25–26.

Sen, D. N., and Chatterji, U. N. (1966). Eco-physiological observations on *Euphorbia caducifolia* Haines. *Sci. Cult.* **32**, 317–319.

Sen, D. N., and Chatterji, U. N. (1968). Ecology of desert plants and observations on their seedlings. I. Germination behaviour of seeds. *Bull. Bot. Soc. Bengal* **22**, 251–258.

Sen, D. N., and Chawan, D. D. (1969). Search for supplementary useful plants in Indian desert and their ecology. 1. *Salvadora persica* Linn. and *S. oleoides* Decne. *Indian For.* **95**, 681–688.

Sen, D. N., and Chawan, D. D. (1970). Ecology of desert plants and observations on their seedlings. III. The influence of aqueous extracts of *Prosopis juliflora* DC. on *Euphorbia caducifolia* Haines. *Vegetatio* **21**, 277–298.

Shalaby, A. F., and Youssef, M. M. (1967). Contribution to the autecology of *Achillea fragantissima* (Forssk.) Sch. Bip. with reference to its oil content. *Act. Agron. Acad. Scien. Hunga.* **16**, 375–382.

Sharma, M. M., Sharma, N. K., and Sen, D. N. (1978). A new report on differential seed coat dormancy in *Sesbania bispinosa* (Joaq.) W. F. Wight. *Folia Geobot. Phytotax.* **13**, 95–98.

Sharma, N. K., Sharma, M. M., and Sen, D. N. (1978). Seed perpetuation in *Rhynchosia capitata* DC. *Biol. Plant.* **20**, 225–228.

Shaw, N. H. (1957). Bunch spear grass dominance in burnt pastures in south-eastern Queensland. *Aust. J. Agric. Res.* **8**, 325–334.

Shetty, M. S. (1967). Germination and seedling establishment of *Sphaeranthus indicus* Linn. in relation to soil moisture. *Trop. Ecol.* **8**, 138–143.

Shmida, A. (1985). Biogeography of the desert flora. *In* "Ecosystems of the World" (M. Evenari, I. Noy-Meir, and D. W. Goodall, eds.) Vol. 12A, pp. 23–77. Elsevier, Amsterdam.

Shreve, F. (1917). The establishment of desert perennials. *J. Ecol.* **5**, 210–216.

Siegel, R. S., and Brock, J. H. (1990). Germination requirements of key Southwestern woody riparian species. *Desert Plants* **10** (1), 3–8.

Siegel, S. M., Carroll, P., Corn, C., and Speitel, T. (1970). Experimental studies on the Hawaiian silverswords (*Argyroxiphium* spp.): Some preliminary notes on germination. *Bot. Gaz.* **131**, 277–280.

Silander, J. A., Jr. (1978). Density-dependent control of reproductive success in *Cassia biflora*. *Biotropica* **10**, 292–296.

Silcock, R. G. (1973). Germination responses of native plant seeds to rainfall in south-west Queensland. *Trop. Grassl.* **7**, 99–104.

Silcock, R. G., Williams, L. M., and Smith, F. T. (1990). Quality and storage characteristics of the seeds of important native pasture species in south-west Queensland. *Aust. Rangel. J.* **12**, 14–20.

Silva, J. F., and Ataroff, M. (1985). Phenology, seed crop and germination of coexisting grass species from a tropical savanna in western Venezuela. *Acta Oecol.* **6**, 41–51.

Sindel, B. M., Davidson, S. J., Kilby, M. J., and Groves, R. H. (1993). Germination and establishment of *Themeda triandra* (kangaroo grass) as affected by soil and seed characteristics. *Aust. J. Bot.* **41**, 105–117.

Smith, A. P., and Young, T. P. (1987). Tropical alpine plant ecology. *Annu. Rev. Ecol. Syst.* **18**, 137–158.

Sork, V. L. (1985). Germination response in a large-seeded neotropical tree species, *Gustavia superba* (Lecythidaceae). *Biotropica* **17**, 130–136.

Speer, E. R., and Wright, H. A. (1981). Germination requirements of lotebush (*Ziziphus obtusifolia* var. *obtusifolia*). *J. Range Manage.* **34**, 365–368.

Stapf, O. (1904). On the fruit of *Melocanna bambusoides*, Trin., an endospermless, viviparous genus of Bambuseae. *Trans. Linnean Soc. Lond. Bot.* **6**, 401–425 + 3 plates.

Steentoft, M. (1988). "Flowering Plants in West Africa." Cambridge Univ. Press, Cambridge, UK.

Stotzky, G., Cox, E. A., and Goos, R. D. (1962). Seed germination studies in *Musa*. I. Scarification and aseptic germination of *Musa balbisiana*. *Am. J. Bot.* **49**, 515–520.

Stoutjesdijk, Ph. (1972). A note on the spectral transmission of light by tropical rainforest. *Acta Bot. Neerl.* **21**, 346–350.

Swagel, E. N., van H. Bernhard, A., and Ellmore, G. S. (1997). Substrate water potential constraints on germination of the strangler fig *Ficus aurea* (Moraceae). *Am. J. Bot.* **84**, 716–722.

Swaine, M. D., and Whitmore, T. C. (1988). On the definition of ecological species groups in tropical rain forests. *Vegetatio* **75**, 81–86.

Symington, C. F. (1932). The study of secondary growth on rain forest sites in Malaya. *Malaysian For.* **2**, 107–117.

Tang, H. T. (1971). Preliminary tests on the storage and collection of some *Shorea* species seeds. *Malaysian For.* **34**, 84–93.

Tang, H. T., and Tamari, C. (1973). Seed description and storage tests of some dipterocarps. *Malaysian For.* **36**, 38–53.

Teketay, D. (1993a). Problems associated with raising trees from seeds. The Ethiopian experience. *In* "Restoration of Tropical Forest Ecosystems" (H. Lieth and M. Lohmann, eds.), pp. 91–100. Kluwer Academic Publishers, Dordrecht.

Teketay, D. (1993b). Germination ecology of *Vernonia galamensis* (Cass.) Less. var. *ethiopica* M. G. Gilbert, a new industrial oilseed crop. *Trop. Ecol.* **34**, 64–74.

Teketay, D. (1996a). Germination ecology of twelve indigenous and eight exotic multipurpose leguminous species from Ethiopia. *For. Ecol. Manage.* **80**, 209–223.

Teketay, D. (1996b). The effect of different pre-sowing seed treatments, temperature and light on the germination of five *Senna* species from Ethiopia. *New For.* **11**, 155–171.

Ter Steege, H., Bokdam, C., Boland, M., Dobbelsteen, J., and Verburg, I. (1994). The effects of man made gaps on germination, early survival, and morphology of *Chlorocardium rodiei* seedlings in Guyana. *J. Trop. Ecol.* **10**, 245–260.

Tevis, L., Jr. (1958a). Germination and growth of ephemerals induced by sprinkling a sandy desert. *Ecology* **39**, 681–688.

Tevis, L., Jr. (1958b). A population of desert ephemerals germinated by less than one inch of rain. *Ecology* **39**, 688–695.

Thomas, P. E. L., and Allison, J. C. S. (1975). Seed dormancy and germination in *Rottboellia exaltata*. *J. Agric. Sci. Cambridge* **85**, 129–134.

Titus, J. H., Holbrook, M., and Putz, F. E. (1990). Seed germination and seedling distribution of *Ficus pertusa* and *F. tuerckheimii:* Are strangler figs autotoxic? *Biotropica* **22**, 425–428.

Todaria, N. P., and Negi, A. K. (1992). Pretreatment of some Indian *Cassia* seeds to improve their germination. *Seed Sci. Technol.* **20**, 583–588.

Tompsett, P. B. (1982). The effect of desiccation on the longevity of seeds of *Araucaria hunsteinii* and *A. cunninghamii*. *Ann. Bot.* **50**, 693–704.

Tompsett, P. B. (1984). Desiccation studies in relation to the storage of *Araucaria* seed. *Ann. Appl. Biol.* **105**, 581–586.

Tompsett, P. B. (1985). The influence of moisture content and storage temperature on the viability of *Shorea almon, Shorea robusta,* and *Shorea roxburghii* seed. *Can. J. For. Res.* **15**, 1074–1079.

Tompsett, P. B. (1987). Desiccation and storage studies on *Dipterocarpus* seeds. *Ann. Appl. Biol.* **110**, 371–379.

Torssell, B. W. R., and McKeon, G. M. (1976). Germination effects on pasture composition in a dry monsoonal climate. *J. Appl. Ecol.* **13**, 593–603.

Tothill, J. C. (1969). Soil temperatures and seed burial in relation to the performance of *Heteropogon contortus* and *Themeda australis* in burnt native woodland pastures in eastern Queensland. *Aust. J. Bot.* **17**, 269–275.

Tothill, J. C. (1977). Seed germination studies with *Heteropogon contortus*. *Aust. J. Ecol.* **2**, 477–484.

Traveset, A. (1990). Post-dispersal predation of *Acacia farnesiana* seeds by *Stator vachelliae* (Bruchidae) in Central America. *Oecologia* **84**, 506–512.

Tremblin, G., and Binet, P. (1982). Installation d'*Halopeplis amplexicaulis* (Vahl) Ung. dans une sebkha algerienne. *Acta Oecol.* **3**, 373–379.

Trollope, W. S. W. (1982). Ecological effects of fire in South African savannas. *In* "Ecology of Tropical Savannas" (B. J. Huntley and B. H. Walker, eds.), pp. 292–306. Springer-Verlag, Berlin.

Tschirley, F. H, and Martin, S. C. (1960). Germination and longevity of velvet mesquite seed in soil. *J. Range Manage.* **13**, 94–97.

United States Department of Agriculture, Forest Service. (1948). C. S. Schopmeyer (Tech. Coord.). Seeds of woody plants in the United States. *Agriculture Handbook No. 450*.

Upfold, S. J., and van Staden, J. (1990). The germination characteristics of *Tithonia rotundifolia*. *Ann. Bot.* **66**, 57–62.

Usberti, R., and Felippe, G. M. (1980a). Alguns aspectos da germinacao de sementes de limao-cravo (*Citrus limonia,* Osb.): Efeito deleterio da reidratacao lenta. *Cien. Cult.* **32**, 1094–1098.

Usberti, R., and Felippe, G. M. (1980b). Viabilidade de sementes de *Citrus limonia* Osb. com baixo teor de umidade, armazenadas em diferentes temperaturas. *Pesq. Agropec. Bras. (Brasilia),* **15**, 393–397.

Van Auken, O. W., and Bush, J. K. (1992). *Diospyros texana* Scheele (Ebenaceae) seed germination and seedling light requirements. *Texas J. Sci.* **44**, 167–174.

Van Groenendael, J. M., Bullock, S. H., and Perez-Jimenez, L. A., (1996). Aspects of the population biology of the gregarious tree *Cordia elaeagnoides* in Mexican tropical deciduous forest. *J. Trop. Ecol.* **12**, 11–24.

van Rooden, J., Akkermans, L. M. A., and van der Veen, R. (1970). A study on photoblastism in seeds of some tropical weeds. *Acta Bot. Neerl.* **19**, 257–264.

Vastano, B., Jr., Barbosa, A. P., and Goncalves, A. N. (1983). Tratamentos pre-germinativos de sementes de especies florestais Amazonicas. I. Angelim pedra (*Dinizia excelsa* Ducke–Leguminosae, Mimosoideae). *Acta Amazonica* **134**, 413–419.

Vazquez-Yanes, C. (1974). Studies on the germination of seeds of *Ochroma lagopus* Swartz. *Turrialba* **24**, 176–179.

Vazquez-Yanes, C. (1977). Germination of a pioneer tree (*Trema guineensis* Ficalho) from equatorial Africa. *Turrialba* **27**, 301–302.

Vazquez-Yanes, C. (1979). Notas sobre la ecofisiologia de la germinacion de *Cecropia obtusifolia* Bertol. *Turrialba* **29**, 147–149.

Vazqueq-Yanes, C. (1981). Germinacion do dos especies de Tiliaceas arboreas de la vegetacion secundaria tropical: *Belotia campbellii* y *Heliocarpus donnell-smithii*. *Turrialba* **31**, 81–83.

Vazquez-Yanes, C., and Orozco-Segovia, A. (1982a). Seed germination of a tropical rain forest pioneer tree (*Heliocarpus donnellsmithii*) in response to diurnal fluctuation of temperature. *Physiol. Plant* **56**, 295–298.

Vazquez-Yanes, C., and Orozco-Segovia, A. (1982b). Longevidad, latencia y germinacion de las semillas de *Verbesina greenmanii:* Efecto de la calidad de la luz. *Turrialba* **32**, 457–462.

Vazquez-Yanes, C., and Orozco-Segovia, A. (1982c). Germination of the seeds of a tropical rain forest shrub, *Piper hispidum* Sw. (Piperaceae) under different light qualities. *Phyton (Buenos Aires)* **42**, 143–149.

Vazquez-Yanes, C., and Orozco-Segovia, A. (1986). Dispersal of seeds by animals: Effect on light controlled dormancy in *Cecropia obtusifolia*. *In* "Frugivores and Seed Dispersal" (A. Estrada, and T. H. Fleming, eds.), pp. 71–82. Junk, Dordrecht.

Vazquez-Yanes, C., and Orozco-Segovia, A. (1987a). Fisiologia ecologica de semillas en las Estacion de Biologia Tropical "Los Tuxtlas," Veracruz, Mexico. *Rev. Biol. Trop.* **35**(Suppl.), 85–96.

Vazquez-Yanes, C., and Orozco-Segovia, A. (1987b). Light gap detection by the photoblastic seeds of *Cecropia obtusifolia* and *Piper auritum,* two tropical rain forest trees. *Biol. Plant.* **29**, 234–236.

Vazquez-Yanes, C., and Orozco-Segovia, A. (1990). Ecological significance of light controlled seed germination in two contrasting tropical habitats. *Oecologia* **83**, 171–175.

Vazquez-Yanes, C. and Orozco-Segovia, A. (1992). Effects of litter from a tropical rainforest on tree seed germination and establishment under controlled conditions. *Tree Physiol.* **11**, 391–400.

Vazquez-Yanes, C., Orozco-Segovia, A., Rincon, E., Sanchez-Coronado, M. D., Huante, P., Toledo, J. R., and Barradas, V. L. (1990). Light beneath the litter in a tropical forest: Effect on seed germination. *Ecology* **71**, 1952–1958.

Vazquez-Yanes, C., and Smith, H. (1982). Phytochrome control of seed germination in the tropical rain forest pioneer trees *Cecropia obtusifolia* and *Piper auritum* and its ecological significance. *New Phytol.* **92**, 477–485.

Vazquez-Yanes, C., and Toledo, J. R. (1989). El almacenamiento de semillas en las conservacion de especies vegetales. Problemas y applicaciones. *Bull. Soc. Bot. Mexico* **49**, 61–69.

Veenendaal, E. M. (1991). "Adaptive Strategies of Grasses in a Semiarid Savanna in Botswana." Doctoral thesis, Vrije Universiteit, Amsterdam.

Veenendaal, E. M., and Ernst, W. H. O. (1991). Dormancy patterns in accessions of caryopses from savanna grass species in south eastern Botswana. *Acta Bot. Neerl.* **40**, 297–309.

Venable, D. L., and Lawlor, L. (1980). Delayed germination and dispersal in desert annuals: Escape in space and time. *Oecologia* **46**, 272–282.

von Willert, D. J., Eller, B. M., Werger, M. J. A., Brinckmann, E., and Ihlenfeldt, H.-D. (1992). "Life Strategies of Succulents in Deserts with Special Reference to the Namib Desert." Cambridge Univ. Press, Cambridge, UK.

Vora, R. S. (1989). Seed germination characteristics of selected native plants of the lower Rio Grande Valley, Texas. *J. Range Manage.* **42,** 36–40.

Walker, L. R. (1990). Germination of an invading tree species (*Myrica faya*) in Hawaii. *Biotropica* **22,** 140–145.

Walker, L. R., and Vitousek, P. M. (1991). An invader alters germination and growth of a native dominant tree in Hawaii. *Ecology* **72,** 1449–1455.

Walter, H. (1971). "Ecology of Tropical and Subtropical Vegetation." Translated by D. Mueller-Dombois. Oliver and Boyd, Edinburgh.

Walter, H. (1973). "Vegetation of the Earth in Relation to Climate and the Eco-physiological Conditions." Translated from the second German edition by Joy Wieser. Springer-Verlag, Berlin.

Walter, H. (1979). "Vegetation of the Earth and Ecological Systems of the Geo-biosphere," 2nd Ed. Translated from the third, revised German edition by Joy Wieser. Springer-Verlag, Berlin.

Walter, H. (1986). The Namib desert. *In* "Ecosystems of the World" (M. Evenari, I. Noy-Meir, and D. W. Goodall, eds.), Vol. 12B, pp. 245–282. Elsevier, Amsterdam.

Walters, G. A. (1974). *Toona australis* Harms. Australian toon. *In* "Seeds of Woody Plants in the United States" (C. S. Schopmeyer, tech. coord.), pp. 813–814. USDA Forest Service, Agriculture Handbook No. 450.

Watkinson, A. R., Lonsdale, W. M., and Andrew, M. H. (1989). Modelling the population dynamics of an annual plant *Sorghum intrans* in the wet-dry tropics. *J. Ecol.* **77,** 162–181.

Watt, L. A. (1974). The effect of water potential on the germiation behaviour of several warm season grass species, with special reference to cracking black clay soils. *J. Soil Conserv. New S. Wales* **30,** 28–41.

Watt, L. A. (1978). Some characteristics of the germination of Queensland blue grass on cracking black earths. *Aust. J. Agric. Res.* **29,** 1147–1155.

Watt, L. A. (1982). Germination characteristics of several grass species as affected by limiting water potentials imposed through a cracking black clay soil. *Aust. J. Agric. Res.* **33,** 223–231.

Watt, L. A., and Whalley, R. D. B. (1982). Establishment of small-seeded perennial grasses on black clay soils in north-western New South Wales. *Aust. J. Bot.* **30,** 611–623.

Wells, P. V. (1959). An ecological investigation of two desert tobaccos. *Ecology* **50,** 626–644.

Went, F. W. (1948). Ecology of desert plants. I. Observations on germination in the Joshua Tree National Monument, California. *Ecology* **29,** 242–253.

Went, F. W. (1949). Ecology of desert plants. II. The effect of rain and temperature on germination and growth. *Ecology* **30,** 1–13.

Went, F. W. (1955). The ecology of desert plants. *Sci. Am.* **192**(4), 68–75.

Went, F. W., and Westergaard, M. (1949). Ecology of desert plants. III. Development of plants in the Death Valley National Monument, California. *Ecology* **30,** 26–38.

Werger, M. J. A. (1986). The Karoo and southern Kalahari. *In* "Ecosystems of the World" (M. Evenari, I. Noy-Meir, and D. W. Goodall, eds.), pp. 283–359. Elsevier, Amsterdam.

Whisenant, S. G., and Ueckert, D. N. (1981). Germination responses of *Anisacanthus wrightii* (Torr.) Gray (Acanthaceae) to selected environmental variables. *Southw. Nat.* **26,** 379–384.

Whisenant, S. G., and Ueckert, D. N. (1982). Factors influencing bitterweed seed germination. *J. Range Manage.* **35,** 243–245.

Whisenant, S. G., and Ueckert, D. N. (1982). Germination responses

of *Eysendardtia texana* and *Leucaena retusa. J. Range Manage.* **35,** 748–750.

Whitmore, T. C. (1975). "Tropical Rain Forests of the Far East." Clarendon Press, Oxford.

Whitmore, T. C. (1983). Secondary succession from seed in tropical rain forests. *For. Abst.* **44,** 767–779.

Whitmore, T. C. (1989). Canopy gaps and the two major groups of forest trees. *Ecology* **70,** 536–538.

Wick, H. L. (1974). *Flindersia brayleyana* F. Muell. Queensland-maple. *In* "Seeds of Woody Plants in the United States" (C. S. Schopmeyer, tech. coord.), pp. 409–410. USDA Forest Service, Agriculture Handbook No. 450.

Williams-Linera, G. (1990). Origin and early development of forest edge vegetation in Panama. *Biotropica* **22,** 235–241.

Williams, O. B., and Calaby, J. H. (1985). The hot deserts of Australia. *In* "Ecosystems of the World" (M. Evenari, I. Noy-Meir, and D. W. Goodall, eds.), Vol. 12A, pp. 269–312. Elsevier, Amsterdam.

Williams, P. M., and Arias, I. (1978). Physio-ecological studies of plant species from the arid and semi-arid regions of Venezuela. I. The role of endogenous inhibitors in the germination of seeds of *Cereus griseus* (Haw.) Br. & R. (Cactaceae). *Acta Cient. Venezolana* **29,** 93–97.

Willis, A. J., and Groves, R. H. (1991). Temperature and light effects on the germination of seven native forbs. *Aust. J. Bot.* **39,** 219–228.

Wong, K. M. (1981). Flowering, fruiting and germination of the bamboo *Schizostachyum zollingeri* in Perlis. *Malaysian For.* **44,** 453–463.

Wrangham, R. W., Chapman, C. A., and Chapman, L. J. (1994). Seed dispersal by forest chimpanzees in Uganda. *J. Trop. Ecol.* **10,** 355–368.

Wulff, R. D. (1985). Germination of seeds of different sizes in *Hyptis suaveolens:* The response to irradiance and mixed red-far-red sources. *Can. J. Bot.* **63,** 885–888.

Wulff, R., and Medina, E. (1971). Germination of seeds in *Hyptis suaveolens* Poit. *Plant Cell Physiol.* **12,** 567–579.

Wyatt-Smith, J. (1964a). Kapur forest. "Malayan Forest Records," Chapter 9.

Wyatt-Smith, J. (1964b). Lowland forests poor in red meranti and keruing species. "Malayan Forest Records," Chapter 10.

Wyatt-Smith, J. (1964c). White meranti-gerutu and schima-bamboo forests. "Malayan Forest Records," Chapter 12.

Wyatt-Smith, J. (1964d). 'Heath' forest. "Malayan Forest Records," Chapter 13.

Wyatt-Smith, J. (1964e). Hill forest. "Malayan Forest Records," Chapter 14.

Wyatt-Smith, J. (1964f). Swamp forest. "Malayan Forest Records," Chapter 15.

Xia, Q. H., Chen, R. Z., and Fu, J. R. (1992). Effects of desiccation, temperature and other factors on the germination of lychee (*Litchi chinensis* Sonn.) and longan (*Euphoria longan* Steud.) seeds. *Seed Sci. Technol.* **20,** 119–127.

Xu-Ping, Y., Jian, F., Shu-Jun, S., Chang-Kao, S., Xiu-Ying, X., and Jun-Bo, Z. (1993). Viability of *Litchi chinensis* seeds stored in Na-alginate pelletes. *In* "Fourth International Workshop on Seeds: Basic and Applied Aspects of Seed Biology" (D. Come and F. Corbineau eds.), pp. 863–865. Universite Pierre et Marie Curie, Paris.

Yadav, J. P. (1992). Pre-treatment of teak seed to enhance germination. *Indian For.* **118,** 260–264.

Yap, S. K. (1981). Collection, germination and storage of dipterocarp seeds. *Malaysian For.* **44,** 281–300.

Yap, S. K., and Wong, S. M. (1983). Seed biology of *Acacia mangium, Albizia falcataria, Eucalyptus* spp., *Gmelina arborea, Maesopsis eminii, Pinus caribaea* and *Tectona grandis. Malaysian For.* **46**, 26–45.

Young, J. A., and Evans, R. A. (1980). Germination of desert needlegrass. *J. Seed Technol.* **1**, 40–46.

Young, T. P. (1982). Bird visitation, seed-set, and germination rates in two species of *Lobelia* on Mount Kenya. *Ecology* **63**, 1983–1986.

Yumoto, T., and Maruhashi, T. (1995). Seed-dispersal by elephants in a tropical rain forest in Kahuzi-Beiga National Park, Zaire. *Biotropica* **27**, 526–529.

Zacharias, P. J. K., Tainton, N. M., and Oberholster, C. (1988). The effect of fire on germination in five common veld grasses. *J. Grassl. Soc. S. Afr.* **5**, 229–230.

Zhang, S.-Y. and Wang, L.-X. (1995). Fruit consumption and seed dispersal of *Ziziphus cinnamomum* (Rhamnaceae) by two sympatric primates (*Cebus apella* and *Ateles paniscus*) in French Guiana. *Biotropica* **27**, 397–401.

Zimmer, K. (1969). Effects of temperature on germination of cactus seeds. VI. Germination of some species of Mexican origin. *Biol. Abstr.* **50**, 9975.

Zodape, S. T. (1991). The improvement of germination of some forest species by acid scarification. *Indian For.* **117**, 61–65.

10

A Geographical Perspective on Germination
Ecology: Temperate and Arctic Zones

I. PURPOSE

Seed germination ecology of species growing in temperate and arctic regions of the world will be covered in this chapter. According to Walter (Fig. 9.1), the temperate–arctic zone includes (1) sclerophyllous woodlands with winter rain, (2) moist warm temperature woodlands, (3) deciduous (nemoral) forests, (4) steppes, (5) semideserts and deserts with cold winter, (7) boreal coniferous zone, (8) tundra, and (9) mountains. For each type of vegetation, the discussion includes a summary of available information on germination of trees, shrubs, vines, and herbaceous species. Weeds also are considered, and attention is given to special biotic and abiotic factors influencing germination.

II. SCLEROPHYLLOUS WOODLANDS WITH WINTER RAIN

This type of vegetation is characterized by dense thickets of woody plants with evergreen sclerophyllous leaves and is found in five widely separated parts of the world: the United States (California), Chile, the Mediterranean Basin, South Africa, and southern and southwestern Australia (Walter, 1979; di Castri, 1981; Barbour and Billings, 1988). The name varies, depending on the country; it is called chaparral in the United States, matorral in Chile and Spain, maquis in France and Israel, macchia in Italy, renosterveld in South Africa, and mallee in Australia (di Castri, 1981). With a decrease in precipitation, the space between shrubs increases and/or drought-deciduous shrubs increase; the community is called coastal sage in the United States, jaral in Chile, batha in Israel, and phrygana in Greece (di Castri, 1981). A low evergreen scrub

formation develops in response to increased human disturbance and is called garrigue in southern France and gariga in Italy. Evergreen sclerophyllous woodlands are replaced by heathlands in sites with low soil nutrients; these heathlands are called fynbos in South Africa and landes in France (di Castri, 1981). Tomaselli (1981) proposed that local names be discontinued and that evergreen sclerophyllous woodlands be referred to as matorral, using high, middle, and low matorral to indicate vegetation structure.

Matorral vegetation extends from the equator to 45°N and 37°30'S and is restricted to regions with a Mediterranean climate (di Castri, 1981), i.e., summers are hot and dry and winters are mild and moist. Annual precipitation may range from a low of 275–300 mm to a high of 900 mm, with 65% of it falling in winter (Aschmann, 1973). Summer rain is uncommon, especially in Europe (di Castri, 1981) and Australia (Parsons, 1981). Mean temperatures for 1 month during winter must be below 15°C (Aschmann, 1973), but if the average minimum temperature of the coldest month is ≤0°C, development of matorral is not extensive (Quezel, 1981).

Shrubs dominate matorral vegetation (di Castri, 1981) and grow to heights of 15–30 cm (subshrubs) or 5–8 m (high shrubs) (Floret et al., 1989). Some of the major species in mallee vegetation of Australia, including *Eucalyptus diversifolia, E. incrassata, E. behriana, E. socialis,* and *E. oleosa,* grow to heights of 5–10 m (Specht, 1981). Species such as *Erica arborea* in France (Floret et al., 1989), *Arbutus andrachne* and *Quercus calliprinos* in Israel (Orshan, 1989), *Cassine peragua* in South Africa (Le Roux et al., 1989), and *Luma chequen* and *Escallonia pulverulenta* in Chile (Montenegro et al., 1989) are described as shrubs or small trees. True trees (10 m or taller) include *Pinus halepensis, P. pinaster,* and *Quercus ilex* in the Mediterranean Basin (Floret et *al.,* 1989); *Kiggelaria africana, Olea capensis,* and *Virgilia oroboides* in South Africa (Le Roux *et al.,* 1989); *Maytenus boaria* in Chile (Montenegro *et al.,* 1989); and *Pinus sabiniana, P. coulteri,* and *Quercus wislizenii* in the United States (Hanes, 1981). Woody vines, including *Smilax, Lonicera, Clematis* (Walter, 1979), *Chironia* (Le Roux *et al.,* 1989), *Hardenbergia, Billardiera, Comesperma, Muehlenbeckia* (Specht, 1981), *Rubia, Tamus, Prasium* (Naveh, 1973), and *Aristolochia* (Naveh, 1975), also occur in matorral vegetation.

Various kinds of herbaceous perennials are found in the matorral of Chile (Rundel, 1981), the United States (Sweeney, 1956), South Africa (Le Roux *et al.,* 1989), the Mediterranean Basin (Walter, 1979), and Australia (Parsons, 1981). Annuals also occur in matorral vegetation, but indigenous species are not very important in Chile (Rundel, 1981) or South Africa (Eyre, 1963; Martin, 1966); however, a number of introduced spring-flowering annuals are found in disturbed sites in Chile (Rundel, 1981) and South Africa (Boucher and Molle, 1981).

Phenological studies in France (Floret *et al.,* 1989), Israel (Orshan, 1989), South Africa (Le Roux *et al.,* 1989), and Chile (Montenegro *et al.,* 1989) indicate that spring is the most important and summer the second most important flowering season. Spring is the most important flowering season in matorral vegetation of Australia and the United States, with winter and summer, respectively, being the second most important flowering season (Specht, 1988).

A. Dormancy and Germination of Trees

Seeds of most matorral trees are nondormant at maturity, but those of a few species have physical or physiological dormancy (Table 10.1). Many trees with nondormant seeds belong to the Proteaceae, Myrtaceae, or Pinaceae, and seeds may be retained in woody capsules or cones for long periods of time (Chapter 7). At the time of release, the nondormant seeds germinate on a variety of substrates if soil moisture is nonlimiting (Dean *et al.,* 1986; Thanos and Skordilis, 1987; Bell *et al.,* 1993). For example, seeds of mallee eucalyptus in Australia (*Eucalyptus diversifolia, E. incrassata,* and *E. oleosa*) were nondormant and germinated to 76–100% on calcareous, loamy, or siliceous sand at 25°C in light (Parsons, 1968).

Little is known about the germination of matorral trees with impermeable seed/fruit coats. Seedlings of *Rhus coriaria* have been found following fires in the habitat (Izhaki *et al.,* 1992), suggesting that high temperatures played a role in breaking physical dormancy of the fruits. Seeds of *Acacia melanoxylon* germinated after immersion in boiling water for 1 min (Bell and Bellairs, 1992).

Seeds of *Pinus sabiniana* and *P. coulteri* required 60–120 and 21–90 days of cold stratification, respectively, to come out of dormancy (Krugman and Jenkinson, 1974), and those of *Quercus wislizenii* germinated to 75% after receiving only 30–60 days of cold stratification (Olson, 1974a). Seeds of *Castanea sativa* germinated in winter (Bacilieri *et al.,* 1994), indicating that short periods of cold stratification may be necessary to overcome physiological dormancy.

The optimum germination temperature for seeds of matorral trees is about 21°C (Table 10.1). Thus, seeds would germinate during the cool season, when soil moisture is likely to be nonlimiting.

Data on light:dark requirements for germination are available for only six species (Table 10.1). Seeds of three species germinated to higher percentages in light than in darkness, and those of three species germinated equally well in light and darkness. Seeds of *Eucalyptus marginata* and *E. calophylla* germinated equally well in light and darkness at 13°C, but they germinated to higher percentages in darkness than in white light (86 μmol m^{-2} sec^{-1}) at 15°C (Bell, 1994). However, seeds of *E. marginata* germinated equally well in darkness and at wavelengths ranging from 430 to 720 nm, and those of *E. calophylla* germinated equally well in light and darkness at wavelengths of 430–520 nm (Rokich and Bell, 1995).

Seeds of *Pinus halepensis* and *P. brutia* collected in Greece germinated to about 80% in darkness at 20°C, but to only 30% or less in darkness at 25°C (Thanos and Skordilis, 1987). Far-red light inhibited germination and caused seeds of these two species to lose their ability to germinate in darkness; this effect was partly reversed by a red light treatment. Seeds of *Quercus ilex* and *Q. pubescens* germinated to higher percentages in the leaf-filtered light of *Q. ilex* coppices than in clearings (Bran *et al.,* 1990).

Seeds of *Eucalyptus incrassata* and *E. oleosa* sown in the field in April (autumn) germinated from May to July, the period of the year with high soil moisture (Parsons, 1968). However, seeds sown in September failed to germinate, presumably due to dry soil. Seeds of *E. camaldulensis,* which grows on floodplains in the eucalyptus shrublands, germinated to 68, 42, and 2% at substrate water potentials of 0, −0.2, and −1.0 MPa, respectively (Edgar, 1977). Some (percentages not given) seeds of *E. baxteri, E. incrassata,* and *E. oleosa* germinated at −0.55 MPa, and those of the latter two species germinated at −1.05 MPa (Facelli and Ladd, 1996). Litter (which would retard the loss of soil moisture but reduce light) promoted the germination of *E. baxteri* seeds, but it had no effect on the germination

TABLE 10.1   Dormancy in Seeds of Trees of Matorral Vegetation

Species	Family[a]	Temperature[b]	L:D[c]	Reference
Nondormant				
Banksia grandis[d]	Prote.	13	L > D	Bell (1994)
Eucalpytus calophylla[d]	Myrt.	13	L = D	Bell (1994)
E. camaldulensis	Myrt.	35	—[e]	Edgar (1977)
E. diversicolor	Myrt.	25	—	Devillez (1977)
E. diversifolia	Myrt.	25	—	Parsons (1968)
E. gomphocephala	Myrt.	20–25	—	Devillez (1977)
E. incrassata	Myrt.	25	—	Parsons (1968)
E. lehmanii	Myrt.	—	—	Dean *et al.* (1986)
E. marginata[d]	Myrt.	13	L = D	Bell (1994)
E. oleosa	Myrt.	15, 25	—	Parsons (1969), Bell and Bellairs (1992)
E. rudis	Myrt.	20, 25	—	Bell and Bellairs (1992)
E. wandoo	Myrt.	15/5	—	Bell and Bellaris (1992)
Leptospermum laevigatum	Myrt.	—	—	Dean *et al.* (1986)
Hakea gibbosa	Prote.	—	—	Dean *et al.* (1986)
H. sericea	Prote.	—	—	Dean *et al.* (1986)
H. suaveolens	Prote.	—	—	Dean *et al.* (1986)
Pinus brutia	Pin.	20	L > D	Shafig (1979), Thanos and Skordilis (1987)
P. halepensis	Pin.	20	L = D	Thanos and Skordilis (1987)
Quercus pinaster	Fag.	Winter	—	Bacilieri *et al.* (1994)
Q. pubescens	Fag.	Winter	—	Bacilieri *et al.* (1994)
Physical dormancy				
Acacia melanoxylon	Fab.	25	—	Bell and Bellaris (1992)
Rhus coriaria[d]	Aacardi	—	—	Izhaki *et al.* (1992)
Physiological dormancy				
Castanea sativa	Fag.	Winter	—	Bacilieri *et al.* (1994)
Pinus coulteri	Pin.	30/20	—	Krugman and Jenkinson (1974)
P. sabiniana	Pin.	22–26	L > D	Krugman and Jenkinson (1974)
Quercus wislizenii	Fag.	30/20	—	Olson (1974a)

[a]Add "aceae" to complete spelling of family name.

[b]Optimum germination temperature (°C) or test temperature that resulted in a high germination percentage.

[c]Light or dark requirements for germination: L, light required for germination; D, darkness required for germination; L > D, seeds germinated to a higher percentage in light than in darkness; D > L, seeds germinated to a higher percentage in darkness than in light; and L = D, seeds germinated equally well in light and darkness.

[d]Type of dormancy inferred.

[e]Data not available.

of *E. incrassata* or *E. oleosa* seeds. Further, seeds of *E. baxteri* germinated to higher percentages in darkness than in light, whereas there was no difference between germination in light and darkness in seeds of *E. incrassata* or *E. oleosa* (Facelli and Ladd, 1996). Seedling emergence from *E. salmonophloia* seeds on the soil surface and at depths of 2, 5, 10, and 20 mm was 84, 73, 55, 15, and 1%, respectively (Yates *et al.*, 1996), indicating that many seeds can germinate in darkness (especially at 2 and 5 mm depths).

B. Dormancy and Germination of Shrubs

Freshly-matured seeds of some matorral shrubs belonging to families such as the Proteaceae and Myrta-

ceae are nondormant (Table 10.2). Seeds of these species can be retained in woody capsules on the mother plant for many years. Fruits open after fire, thereby releasing the nondormant seeds (Siddiqi *et al.*, 1976; Bellairs and Bell, 1990). The germination rate (but not percentage) decreased in *Protea neriifolia* seeds stored dry at room temperatures for 4 months but not in those left in fruits on plants in the field (Le Maitre and Botha, 1991).

Nondormant seeds of *Myrtus communis* incubated at 25°C germinated to 100% after a single 9-hr light exposure and to 44% in continuous darkness (Ozturk *et al.*, 1983). Seeds from cross- and self-pollinated flowers of *Nerium oleander* germinated to near 100% at room temperature (12–20°C) and light conditions within 20 days (Herrera, 1991). Ray achenes of *Relhania genis-*

TABLE 10.2   Dormancy in Seeds of Shrubs of Matorral Vegetation

Species	Family[a]	Temperature[a]	L:D[a]	Reference
Nondormant				
Arbutus unedo	Eric.	20/15	L = D	Mesleard and Lepart (1991)
Banksia aspleniifolia	Prote.	20	—[b]	Siddiqi *et al.* (1976)
B. ericifolia	Prote.	20	—	Siddiqi *et al.* (1976)
B. hookeriana	Prote.	15	—	Enright *et al.* (1996)
B. serratifolia	Prote.	20	—	Siddiqi *et al.* (1976)
Felicia echinata[c]	Aster.	20/10	L = D	Pierce and Molly (1994)
Ficus carica	Mor.	30/20	L > D	Lisci and Pacini (1994)
Hakea amplexicaulis	Prote.	13	L > D	Bell (1994)
Kunzea recurva	Myrt.	15/5	—	Bell and Bellairs (1992)
Launaéa arborescens	Aster.	15–30	L > D	Schutz and Milberg (unpublished results)
Metalasia muricata[c]	Aster.	20/10	L > D	Pierce and Moll (1994)
Myrtus communis[c]	Myrt.	30/27	L > D	Ozturk *et al.* (1983)
Nerium oleander	Apocyn.	Room?	—	Herrera (1991)
Protea neriifolia	Prote.	10/5	—	Le Maitre and Botha (1991)
Xanthorrhoea australis[c]	Xanthorrhoe.	Autumn, 20/12	D > L	Gill and Ingwersen (1976), Curtis (1996)
X. gracilis[c]	Xanthorrhoe.	13	L = D	Bell (1994)
X. preissii[c]	Xanthorrhoe.	13	L = D	Bell (1994)
Morphophysiological dormancy				
Dendromecon rigida[c]	Papaver.	Cool season	—	Bullock (1989)
Garrya flavescens[c]	Garry.	—	—	Keeley (1987)
Hibbertia amplexicaulis[c]	Dilleni.	—	—	Schatral (1995)
H. commutata[c]	Dilleni.	—	—	Schatral (1995)
H. huegelii[c]	Dilleni.	—	—	Schatral (1995)
H. hypericoides[c]	Dilleni.	—	—	Schatral (1995, 1996)
H. montana[c]	Dilleni.	—	—	Bell and Bellairs (1992)
Patersonia occidentalis[c]	Irid.	—	—	Bell and Bellairs (1992)
Romneya trichocalyx[c]	Papaver.	23	—	Keeley and Keeley (1987)
Xanthosia atkinsoniana[c]	Api.	—	—	Bell and Bellairs (1992)
Physical dormancy				
Acacia browniana	Fab.	15/5	—	Bell and Bellairs (1992)
A. drummondii	Fab.	13	L = D	Bell (1994)
A. latericola	Fab.	10	—	Bell and Bellairs (1992)
A. myritifolia	Fab.	23	—	Clemens *et al.* (1977)
A. pulchella	Fab.	13	L = D	Bell (1994)
A. urophylla	Fab.	15	—	Bell and Bellairs (1992)
Bossiaea ornata	Fab.	15	—	Bell and Bellairs (1992)
Ceanothus leudodermis[c]	Rhamn.	23	L = D	Kelley (1987)
C. megacarpus[c]	Rhamn.	23	L = D	Kelley (1987)
C. oliganthus[c]	Rhamn.	23	D > L	Kelley (1987)
C. purpureus	Rhamn.	g.h.[d]	—	Quick (1935)
Chorizema ilicifolium	Fab.	15	L = D	Rokich and Bell (1995)
Cistus albidus	Cist.	17	—	Vuillemin and Bulard (1981)
C. incanus	Cist.	15	L = D	Thanos and Georghiou (1988)
C. ladanifer	Cist.	—	—	Alonso *et al.* (1992)
C. laurifolius	Cist.	—	—	Alonso *et al.* (1992)
C. monspeliensis	Cist.	17	—	Vuillemin and Bulard (1981)
C. psilosepalus	Cist.	20	—	Gonzalez-Rabanal and Casal (1995)
C. salvifolius	Cist.	20	L = D	Thanos and Georghiou (1988)
Cytisus scoparius	Fab.	22/17	—	Tarrega *et al.* (1992)
Daviesia cordata	Fab.	18	L = D	Bell (1994)
Fremontodendron californicus[c]	Sterculi.	—	—	Keeley (1987)
F. decumbens	Sterculi.	l.h.[e]	—	Boyd and Serafini (1992)
Genista florida	Fab.	22/17	—	Tarrega *et al.* (1992)
Gompholobium knightianum	Fab.	13	D > L	Bell (1994)
Halimium alyssoides	Cist.	20	—	Gonzalez-Rabanal and Casal (1995)
Helianthemum apenninum[c]	Cist.	Autumn	L = D	Martin *et al.* (1995)
H. hirtum[c]	Cist.	Autumn	L = D	Martin *et al.* (1995)

(continues)

TABLE 10.2—*Continued*

Species	Family[a]	Temperature[a]	L:D[a]	Reference
H. polygonoides	Cist.	25/15	—	Perez-Garcia *et al.* (1995)
H. squamatum	Cist.	25/15	—	Perez-Garcia *et al.* (1995)
Hovea chorizemifolia	Fab.	13	L = D	Bell (1994)
Lotus scoparius[c]	Fab.	23	L = D	Keeley (1987)
Malochothamnus fasciculatus[c]	Malv.	23	L = D	Keeley (1987)
Malosoma laurina[c]	Anacardi.	23	L = D	Keeley (1987)
Phylica ericoides	Rhamn.	20/10	—	Kilian and Cowling (1992)
Rhamnus californica[c]	Rhamn.	23	D > L	Keeley (1987)
R. crocea[c]	Rhamn.	23	L > D	Keeley (1987)
Rhus integrifolia	Anacardi.	Room?	—	Young (1972)
R. ovata	Anacardi.	g.h.	—	Stone and Juhren (1951)
R. trilobata[c]	Anacardi.	23	L > D	Keeley (1987)
Toxicodendron diversifolia[c]	Anacardi.	23	L > D	Keeley (1987)
Tuberaria guttata	Cist.	20	—	Gonzalez-Rabanal and Casal (1995)
Viminaria juncea[c]	Fab.	—	—	Bell and Bellairs (1992)
Physical and physiological dormancy				
Ceanothus cordulatus	Rhamn.	g.h	—	Quick (1959)
C. crassifolius	Rhamn.	g.h.	—	Quick (1959)
C. cuneatus	Rhamn.	g.h.	—	Quick (1935)
C. divaricatus	Rhamn.	g.h.	—	Quick (1935)
C. integerrimus	Rhamn.	g.h.	—	Quick (1935)
C. sanguineus	Rhamn.	g.h.	—	Gratkowski (1973)
C. sorediatus	Rhamn.	g.h.	—	Quick (1935)
Physiological dormancy				
Agathosma apiculata[c]	Rut.	20/10	L = D	Pierce and Moll (1994)
A. betulina[c]	Rut.	18/10	—	Blommaert (1972)
A. stenopetala[c]	Rut.	20/10	L = D	Pierce and Moll (1994)
Anigozanthos manglesii	Haemodor.	22	L = D	Sukhvibul and Considine (1994)
Anthospermum aethiopecum[c,f]	Rubi.	Room	—	Levyns (1935)
Astartea fascicularis[c]	Myrt.	—	—	Bell and Bellairs (1992)
Arctostaphylos glandulosa[c]	Eric.	—	—	Keeley (1987)
A. patula[c]	Eric.	—	—	Keeley (1987)
Artemisia californica[c]	Aster.	23	L	Keeley (1987)
Baeckea camphorosmae[c]	Myrt.	—	—	Bell and Bellaris (1992)
Calothamnus rupestris[c]	Myrt.	13	L = D	Bell (1994)
Coridothymus capitatus[c]	Lami.	15	L = D	Thanos *et al.* (1995)
Elytropappus rhinocerotis	Aster.	Room	—	Levyns (1935)
Erica andevalensis	Eric.	24–28	—	Aparicio (1995)
E. arborea[c]	Eric.	20/15	L = D	Mesleard and Lepart (1991)
E. junonia	Eric.	23/15	—	Small and Garner (1990)
Haplopappus squarrosa[c]	Aster.	23	L > D	Keeley (1987)
Helichrysum stoechas[c]	Aster.	Autumn	L = D	Martin *et al.* (1995)
Leucadendron daphnoides	Prote.	20/10	L = D	Brown and van Staden (1973)
L. tinctum	Prote.	20/10	—	Brown and Dix (1985)
Leucospermum cordifolium[c]	Prote.	Winter	—	Brits and van Niekerk (1986)
Melaleuca preissiana[c]	Myrt.	15	—	Bell and Bellairs (1992)
Muraltica muricata[c]	Polygal.	30/15	—	Pierce and Moll (1994)
Origanum dictamnus[c]	Lami.	20/15	L = D	Thanos and Doussi (1995)
Passserina paleacea	Thymelae.	20/10	—	Kilian and Cowling (1992)
P. vulgaris[c]	Thymelae.	20/10	—	Pierce and Moll (1994)
Pericalymma ellipticum[c]	Myrt.	—	—	Bell and Bellairs (1992)
Protea compacta	Prote.	20/10	L = D	Brown and van Staden (1973)
P. magnifica	Prote.	20	—	Deall and Brown (1981)
Relhania genistaefolia[f]	Aster.	Room	—	Levyns (1935)
Salvia apeana[c]	Lami.	—	—	Keeley (1987)
S. fruticosa[c]	Lami.	20/15	L > D	Thanos and Doussi (1995)
S. mellifera	Lami.	23/13	L > D	Keeley (1986)
S. pomifera[c]	Lami.	20/15	L > D	Thanos and Doussi (1995)
Sarcopoterium spinosum	Ros.	20	—	Litav and Orshan (1971)

(continues)

TABLE 10.2—*Continued*

Species	Family[a]	Temperature[a]	L:D[a]	Reference
Satureja thymbra[c]	Lami.	10–20	L > D	Thanos *et al.* (1995)
Sideritis syriaca[c]	Lami.	20/15	L > D	Thanos and Doussi (1995)
Thryptomene calycina	Myrt.	20	L = D	Beardsell *et al.* (1993)
Thymelaea hirsuta	Thymelae.	15	—	Shaltout and El-Shourbagy (1989)
Viguiera laciniata[c]	Aster.	23	L > D	Keeley (1987)

[a]See notes for Table 10.1.
[b]Data not available.
[c]Type of dormancy inferred.
[d]Greenhouse.
[e]Lath house.
[f]Seeds were stored dry at room temperatures for an unspecified period of time; thus, they may have had physiological dormancy, which was broken during afterripening.

taefolia were nondormant and germinated to 72% at room temperatures in Rondelbosch, South Africa, in November and December, but most of the disc achenes were dormant and germinated to only 17% (Levyns, 1935).

No matorral shrubs are known to have seeds with morphological dormancy, but some have seeds with morphophysiological dormancy (Table 10.2). No studies have been done on the dormancy-breaking and germination requirements of these seeds.

Many matorral shrubs have seeds (or fruits) with physical dormancy (Table 10.2). Physical dormancy can be broken by mechanical scarification, immersion in hot water (Clemens *et al.*, 1977), exposure to temperatures ranging from 50 to 150°C for 1–15 min (Tarrega and Trabaud, 1992; Gonzalez-Rabanal and Casal, 1995), and fire (Stone and Juhren, 1951). The appearance of *Cistus laurifolius* and *C. ladanifer* seedlings in experimental plots following fire (Alonso *et al.*, 1992) suggests that heat from the fire made the seeds permeable.

A few matorral shrubs, especially some members of the genus *Ceanothus*, have both physical and physiological dormancy. Quick (1935) obtained high germination percentages in various *Ceanothus* species by giving seeds hot water treatments to break physical dormancy, followed by cold stratification for 3 months to break physiological dormancy.

Many matorral shrubs have seeds with physiological dormancy (Table 10.2). Various shrubs belong to the Myrtaceae and Proteaceae; therefore, their seeds may be held for long periods of time in indehiscent woody fruits. However, unlike seeds of matorral trees (and some shrubs) in the Myrtaceae and Proteaceae, seeds of many shrubs in these families are dormant when released from the fruits (e.g., Deall and Brown, 1981; Beardsell *et al.*, 1993). Seeds of *Protea magnifica* required 28 days of cold stratification to germinate to 92% (Deall and Brown, 1981), and cold stratification

promoted germination in seeds of *P. compacta* and *Leucadendron daphnoides* (Brown and van Staden, 1973). Further, seeds of *Thryptomene calycina* germinated within 50 days at 20°C (Beadsell *et al.*, 1993), indicating that warm stratification may be required for dormancy loss. Thus, seeds of matorral shrubs with aerial seed banks can be dormant at the time of dispersal, and germination is delayed until dormancy is broken.

Physiological dormancy in seeds of matorral shrubs not belonging to the Proteaceae or Myrtaceae is broken by cold stratification (Small and Garner, 1980; Mesleard and Lepart, 1991) or by dry storage at room temperatures (Levyns, 1935; Litav and Orshan, 1971; Sukhvibul and Considine, 1994; Thanos *et al.*, 1995). Seeds of *Erica andevalensis* receiving 0 and 20 days of cold stratification germinated to 21 and 86%, respectively (Aparicio, 1995). Cold stratification decreased the germination of *Anigozanthos manglesii* seeds, whereas a period of dry storage increased it (Sukhvibul and Considine, 1994).

Seeds of *Anthospermum aethiopicum* were reported to be nondormant (Levyns, 1935), but they were not tested until after 5 months of dry storage at room temperatures, during which time seeds could have afterripened. Freshly matured seeds of *Leucospermum cordifolium* germinated to only about 5% when sown in the field in January (summer) in South Africa; however, they germinated to 20–40% when sown during the relatively cool months of May to September (Brits and van Niekerk, 1986). It is not known if freshly matured seeds were dormant or if they required low temperatures for germination.

In the Corsican maquis, essentially all seeds of *Arbutus unedo* and about 60% of those of *Erica arborea* were nondormant at maturity, and seeds of both species germinated equally well in light and darkness (Mesleard and Lepart, 1991). Stratification was more effective than desiccation in overcoming dormancy in the dormant seeds of *E. arborea*. Seeds of *A. unedo* were dispersed in winter and germinated in spring, leaving no ungerminated viable seeds in the soil. Seeds of *E. arborea* were dispersed at the beginning of summer, when environmental conditions are not optimal for seedling establishment. Although a soil seed bank is formed in this species, it is not long-lived (Mesleard and Lepart, 1991).

The mean optimum germination temperature for matorral shrubs is about 19°C (Table 10.2); thus, relatively low temperatures associated with the winter-wet season would favor germination of most species. The optimum temperature for the germination of *Satureja thymba* seeds, as well as the maximum temperature at which they could germinate, increased during afterripening in dry storage (Thanos *et al.*, 1995).

Data on light : dark requirements for germination are available for 49 shrubs (Table 10.2). Seeds of only one (2%) shrub, *Artemisia californica* (Keeley, 1987), require light for germination; those of 29 (59%) species germinate equally well in light and darkness, 15 (30%) germinate to higher percentages in light than in darkness, and 4 (8%) germinate to higher percentages in darkness than in light. Thus, some seeds of most species can germinate if they are covered by soil and/or litter, but increased germination percentages are to be expected if seeds are in light rather than in darkness.

C. Dormancy and Germination of Vines

Seeds of the legumes *Hardenbergia comptoniana*, *Kennedia coccinea*, and *K. prostrata* have physical dormancy, which was broken by immersion in boiling water (Bell, 1994). Optimum germination temperature of permeable seeds was 13, 18, and 18°C, respectively, and seeds of all three species germinated to higher percentages in darkness than in light.

D. Dormancy and Germination of Herbaceous Species

Nondormancy has not been reported in freshly matured seeds of herbaceous matorral species. Morphological, morphophysiological, and physical dormancy are represented among herbaceous matorral species, but most species have seeds with physiological dormancy (Table 10.3).

Seeds of *Anemone coronaria* have morphological dormancy, and embryos grow immediately if seeds are placed on a moist substrate in darkness at temperatures of 10–20°C (Bullowa *et al.*, 1975). Three species (*Anthropodium minus*, *Papaver californicum*, and *Watsonia fourcadei*) are listed as having morphophysiological dormancy because (1) they belong to families with underdeveloped embryos and (2) freshly matured seeds do not germinate. However, little is known about the dormancy-breaking and germination requirements of these seeds. Gibberellic acid (GA₃) and a 1-week cold stratification treatment increased germination percentages of *Watsonia fourcadei* seeds (Esterhuizen *et al.*, 1986).

Only a few herbaceous species listed in Table 10.3 have seeds with physical dormancy, but others belonging to the Cistaceae, Convolvulaceae, Fabaceae, and Malvaceae probably have this type. Physical dormancy in seeds of herbaceous matorral species has been broken by mechanical scarification, acid scarification, immersion in hot water (Perez-Garcia *et al.*, 1995), or dry heat treatments (Christensen and Muller, 1975). Seeds of *Onobrychis peduncularis* germinated to 83% at 15°C, after 1 year of dry storage at 5°C (Perez-Garcia *et al.*, 1995). Because freshly matured seeds of this species were not tested, it is unknown if physical dormancy was broken during dry storage at 5°C or if seeds were permeable to water at maturity. The appearance of numerous newly germinated seedlings of *Astragalus didymocarpus*, *Lotus micranthus*, *L. subpinnatus*, *Trifolium ciliolatum*, and *T. microcephalum* following fires in California chaparral (Stocking, 1966) suggests that heat from fire broke physical dormancy in seeds of these legumes.

Most herbaceous matorral species have seeds with physiological dormancy (Table 10.3). Although a relatively few species have been studied in detail, it appears that dormancy loss readily occurs while seeds are stored dry at temperatures of 20°C or higher (Gramshaw, 1972; van de Venter and Small, 1974; Hagon and Simmons, 1978; Ren and Abbott, 1991; Espigares and Peco, 1993). Seeds of many species probably come out of dormancy in the field during summer, when the habitat is dry and hot, and nondormant seeds germinate in autumn or winter, when it is moist and cool.

Dormancy was broken in *Amsinckia hispida* seeds by a 2-week cold stratification period (Connor, 1965). Thus, the dormancy break in some matorral species may not occur until the beginning of the cool season, when soil moisture is adequate for seeds to imbibe, which would explain why seeds of *A. hispida* germinate throughout the autumn (Connor, 1965). Dormancy loss occurred in *Senecio vulgaris* seeds during dry storage at ≥15°C, but it also took place during 2 weeks of cold stratification (Ren and Abbott, 1991). Thus, if the dormancy break was not completed in seeds of this species at the end of summer, it could occur during the low-temperature period of autumn and winter.

The mean optimum germination temperature for seeds of herbaceous matorral species is about 20°C (Table 10.3). However, timing of germination is controlled by an interaction between the temperature require-

TABLE 10.3 Dormancy in Seeds of Herbaceous Species of Matorral Vegetation

Species	Family[a]	Temperature[a]	L:D[a]	Reference
Morphological dormancy				
Anemone coronaria[b]	Ranuncul.	10–20	D > L	Bullowa et al. (1975)
Morphophysiological dormancy				
Anthropodium minus[c]	Lili.	20/10	D	Morgan and Lunt (1994)
Papaver californicum[c]	Papaver.	23	—[d]	Keeley and Keeley (1987)
Watsonia fourcadei[c]	Irid.	10	L > D	Esterhuizen et al. (1986)
Physical dormancy				
Calystegia cyclostegius	Convolvul.	Winter	—	Christensen and Muller (1975)
Dichondra repens[b,c]	Convolvul.	20	L = D	Morgan and Lunt (1994)
Lavatera oblongifolia	Malv.	15	—	Perez-Garcia et al. (1995)
Lotus salsuginosus[c]	Fab.	23	L = D	Keeley (1984), Keeley et al. (1985)
Onobrychis peduncularis	Fab.	15	—	Perez-Garcia et al. (1995)
Physical and physiological dormancy				
Trifolium subterraneum	Fab.	25/15	—	Quinlivan and Nicol (1971)
Physiological dormancy				
Alyssum akamasicum[e]	Brassic.	10	D > L	Kadis and Georghiou (1993)
Amsinckia hispida[b]	Boragin.	12–13	—	Connor (1965)
Camissonia californica[c]	Onagr.	23	—	Keeley and Keeley (1987)
Caryotophora skiatophytoides[c]	Azizo.	20/10	—	Hickey and van Jaarsveld (1995)
Clarkia rhomboidea[c]	Onagr.	27/18	—	Sweeney (1956)
Collinsia parryi[c]	Scrophulari.	23	—	Keeley and Keeley (1987)
Craspedia variabilis[c]	Aster.	20/10	L = D	Morgan and Lunt (1994)
Emex australis[b]	Polygon.	30/20	L > D	Hogan and Simmons (1978)
E. spinosa[b]	Polygon.	30/20	L > D	Hogan and Simmons (1978)
Emmenanthe penduliflora[c]	Hydrophyll.	27/18	—	Sweeney (1956)
Gillia capitata[c]	Polemoni.	27/18	—	Sweeney (1956)
G. gilioides[c]	Polemoni.	27/18	—	Sweeney (1956)
Helichrysum scorpioides[c]	Aster.	20	L = D	Morgan and Lunt (1994)
Hordeum leporinum[b,c]	Po.	25	—	Cocks and Donald (1973)
Lagenifera gracilis[c]	Aster.	20	L = D	Morgan and Lunt (1994)
Leptorhynchos linearis[c]	Aster.	20/10	D > L	Morgan and Lunt (1994)
Lolium rigidum[b,c]	Po.	20–25, 24/12	L > D	Gramshaw (1972), Cocks and Donald (1973)
Lomandra drummondii	Dasypogon.	15	—	Plummer et al. (1995)

(continues)

ments for germination of nondormant seeds of a species and temperatures in the habitat when autumn rains come. For example, in pastures dominated by annual species in central Spain, germination in one group of species was promoted by low (15/10°C) temperatures and that of another group by relatively high (20/15°C) temperatures; species in a third group were indifferent. Thus, if rains came early, autumn pastures were dominated by a different set of species than if they came relatively late (Espigares and Peco, 1993). If rains were late, temperatures were too low for rapid germination, but the long period of favorable soil moisture allowed a high percentage of the seeds to germinate (Gramshaw and Stern, 1977).

Data on light : dark requirements for germination are available for 17 herbaceous matorral species (Table 10.3). Seeds of one species require light for germination, six germinate to higher percentages in light than in darkness, six germinate equally well in light and darkness,

three germinate to higher percentages in darkness than in light, and one requires darkness for germination. Seeds of some species, e.g., Lolium rigidum (Gramshaw and Stern, 1977), germinate to higher percentages in light than in darkness in the laboratory. However, seeds of this species buried at depths of 0.2–2.5 cm in the field germinated to higher percentages than those sown on the soil surface (Smith, 1968; Gramshaw, 1972). Buried seeds may be exposed to higher moisture conditions than those on the soil surface. The ability of seeds to germinate beneath the soil surface may ensure a higher percentage of seedling survival due to increased moisture availability than if seeds germinate on the surface (Rokich and Bell, 1995).

Seeds of Marah oreganus (Cucurbitaceae) have cryptogeal germination, which means the fused cotyledon bases cover the epicotyl. During growth of the cotyledon bases, the epicotyl and radicle are pushed into the soil (Schlising, 1969). The cotyledonary petiole tube

TABLE 10.3 (*Continued*)

Species	Family[a]	Temperature[a]	L:D[a]	Reference
L. sonderi	Dasypogon.	20	—	Plummer *et al.* (1995)
Marah oregans[c]	Cucurbit.	Autumn	—	Schlising (1969)
Mentzelia dispersa[c]	Loas.	27/18	—	Sweeney (1956)
Mimulus bolanderi[c]	Scrophulari.	27/18	—	Sweeney (1956)
Oenothera micrantha[c]	Onagr.	27/18	—	Sweeney (1956)
Origanum cordifolium[e]	Lami.	10	L = D	Kadis and Georghiou (1993)
Penstemon spectabilis[c]	Scrophulari.	23	—	Keeley and Keeley (1987)
Phacelia heterophylla[c]	Hydrophyll.	27/18	—	Sweeney (1956)
Rafinesquia californica[c]	Aster.	23	—	Keeley and Keeley (1987)
Saphesia flaccida[c]	Azizo.	20/10	—	Hickey and van Jaarsveld (1995)
Senecio vulgaris[b]	Aster.	9–13	L	Hilton (1983), Ren and Abbott (1991)
Silene coeli-rosea[c]	Caryophyll.	15–20	—	Thompson (1970)
S. fabaria[c]	Caryophyll.	20	—	Thompson (1970)
S. gigantea[c]	Caryophyll.	19	—	Thompson (1970)
S. italica[c]	Caryophyll.	17	—	Thompson (1970)
S. multinervia[c]	Caryophyll.	23	—	Keeley and Keeley (1987)
S. secundiflora[c]	Caryophyll.	19	—	Thompson (1970)
Skiatophyllum tripolium[c]	Azizo.	20/10	—	Hickey and van Jaarsveld (1995)
Solenogyne domenii[c]	Aster.	20	L > D	Morgan and Lunt (1994)
Strelitzia juncea[c]	Strelitzi.[f]	25	—	van de Venter and Small (1974)
S. reginae[c]	Strelitzi.[f]	25	—	van de Venter and Small (1974)
Velleia trinvervis[c]	Goodeni.	15	—	Bell and Bellairs (1992)
Veronica plebeia[c]	Scrophulari.	20	L > D	Morgan and Lunt (1994)

[a]See notes for Table 10-1.

[b]Listed as a weed in Holm *et al.* (1979)

[c]Type of dormancy inferred.

[d]Data not available.

[e]Seeds were stored dry at room temperatures for an unspecified period of time during which physiologically dormancy seeds could have after-ripened.

[f]The Strelitziaceae has been considered as a subfamily (Strelitzioideae) of Musaceae (Lawrence, 1951), which is known to have seeds with impermeable seed coats (Table 3.5). Thus, members of the Strelitziaceae could have physical dormancy.

may be 5–25 cm long, thus the radicle and epicotyl are placed well beneath the soil surface, where they are protected from summer drought and fire.

Low amounts of rainfall in early autumn may wet seeds at or near the soil surface, but they are insufficient to promote germination. In fact, seeds may be subjected to several wetting and drying cycles before soil moisture is adequate for germination. These moisture cycles may increase germination rates of seeds when adequate rainfall for germination finally occurs (Cocks and Donald, 1973). Also, in the fynbos of South Africa, rain causes the hygroscopic fruits of *Bergeranthus scapigerus* and *Dorotheanthus bellidiformis* to open, thereby releasing seeds (Lockyer, 1930).

E. Special Factor: Fire

Matorral vegetation is highly prone to burning (di Castri, 1981; Trabaud, 1981), and various adaptations to fire have evolved (Naveh, 1975). A remarkable consequence of fire in matorral vegetation is that it frequently is followed by the appearance of thousands of seedlings.

Increased germination after fires has been documented in (1) the United States (Quick, 1959; Hanes and Jones, 1967; Vogl and Schorr, 1972; Hanes, 1972); (2) Chile (Trabaud, 1981); (3) the Mediterranean region (Naveh, 1973, 1975; Harif, 1978; Godron *et al.*, 1981; Margaris, 1981; Carreira *et al.*, 1992; Calvo *et al.*, 1992; Izhaki *et al.*, 1992); (4) South Africa (Bouche, 1981; Moll and Gubb, 1981); and (5) Australia (Specht *et al.*, 1958; Parsons, 1981). Increased germination after fires occurs in annuals (Horton and Kraebel, 1955; Specht *et al.*, 1958; Mills, 1986), perennials (Naveh, 1975), shrubs (Juhren, 1966; Moreno and Oechel, 1991), and trees (Trabaud, 1981). However, fire may increase, decrease, or have no effect on the germination of individual species (Davis *et al.*, 1989). The number of annuals is greatest the first or second year after fire, depending on the season of the year in which fire occurs, and then it subsequently decreases. Perennials increase as annuals decrease (Calvo *et al.*, 1991, 1992).

Fire may influence germination via the production of heat, ash, charate, and volatile chemicals and/or changes in environmental conditions of microsites. The effects of these various factors on the germination of matorral species will be surveyed.

1. Heat Effects on Germination

Temperatures on the soil surface during fires in matorral vegetation may reach 500°C (Martin, 1966) to 600°C (Sweeney, 1956), but those recorded at depths of 2 cm may be only 50–225°C (Sweeney, 1956; Martin, 1966; Davis et al., 1989; Bradstock et al., 1992). Maximum soil temperatures during autumn fires in semiarid mallee shrublands in central New South Wales, Australia, varied depending on depth and type of fuel (Bradstock et al., 1992). In general, the maximum soil temperature during a fire decreased with soil depth, but the duration of elevated temperatures increased with soil depth.

Exposure to fire temperatures of 460°C increased the germination of *Adenostoma fasciculatum, Cryptantha muricata, Phacelia brachyloba, Camissonia intermedia,* and *Streptanthus heterophyllus* seeds under field conditions in California, but temperatures of 570°C decreased germination of these species (Moreno and Oechel, 1991). (Increased fire intensity and temperatures were obtained by adding cut shrubs to plots prior to burning.) Maximum germination of *Ceanothus greggii* seeds occurred after exposure to a fire temperature of 570°C, and maximum germination of *Lotus strigosus* seeds occurred after exposure to 620°C (Moreno and Oechel, 1991). However, under laboratory conditions, temperatures simulating those on the soil surface during fires are lethal to seeds (Chapter 6); some seeds even lose viability at 80–120°C (Keeley et al., 1985; Musil, 1991). Thus, some seeds in the study of Moreno and Oechel (1991) may have been covered with soil, and consequently they were not exposed to lethal temperatures. Seeds of *Cistus albidus, C. monspeliensis,* and *C. salvifolius* under litter on the soil surface were killed by fire under laboratory conditions, whereas those buried at soil depths of 2–2.5 cm survived and germinated to 81, 71, and 51%, respectively (Trabaud and Oustric, 1989).

Seeds of *Ceanothus greggii* and *C. crassifolius,* which reproduce by seeds following a fire, have an activation energy (kcal/mol) of 20.3 and 19.1, respectively, whereas those of *C. leudodermis* and *C. spinosus,* which reproduce by sprouts following a fire, have an activation energy of 45.5 and 42.4, respectively (Barro and Poth, 1988). Further, the maximum death rate (percentage killed/min at a given temperature) for seeds of the four species was *C. greggii,* 29.1; *C. crassifolius,* 25.0; *C. leucodermis,* 63.3; and *C. spinosus,* 57.9. Thus, seeds with a low activation energy are less likely to be killed by high temperatures than those with a high activation energy.

Depending on the species, dry heat treatments may

increase, decrease, or have no effect on germination percentages (Table 10.4). Many species exhibiting increased germination after exposure to 80–120°C belong to families with physical dormancy. Thus, high temperatures may cause seed or fruit coats to become permeable to water (Margaris, 1981). Several species showing no response to heat treatments belong to families with physical dormancy; thus, intensity and/or duration of heat treatments may have been insufficient to overcome dormancy. Some species exhibiting an increase in germination following heat treatments have seeds with permeable seed coats (physiological dormancy); heat treatment might accelerate afterripening. A lack of response of seeds with physiological dormancy to heat treatments could mean that environmental conditions such as cold stratification are required to break dormancy. A decrease in germination following heat treatments could be due to the induction of dormancy by high temperatures or to loss of viability.

Soil temperatures after fires in matorral of Chile remained high for many hours, and seeds of seven species buried at 5 cm (next to the temperature probe) apparently were killed (Munoz and Fuentes, 1989). Thus, these researchers concluded that fires would not promote seed germination of shrubs. However, low-intensity fires in California chaparral resulted in higher densities of annual and subshrub seedlings than high-intensity fires (Tyler, 1995). In fire-prone habitats, frequent low-intensity fires may promote higher germination percentages of seeds in the soil than infrequent high-intensity fires.

Heat from fires is important in causing the woody, seed-storing structures of serotinous species to open and release seeds (Chapter 7). Serotiny is found in a number of matorral genera: *Protea, Leucadendron, Erica,* and *Helipterum* in South Africa (Bond, 1985; Lamont et al., 1991); *Banksia, Hakea, Petrophile, Melaleuca, Callitris, Leptospermum, Kunzea, Eucalyptus, Casuarina,* and *Callistemon* in Australia (Parsons, 1981; Lamont et al., 1991); and *Pinus, Erica, Cupressus,* and *Tetraclinis* in the Mediterranean region (Naveh, 1975; Le Hourerou, 1981; Lamont et al., 1991). Thus, fires in matorral vegetation may be followed by the appearance of many seedlings due to seed dispersal (e.g., Wellington and Noble, 1985).

2. Chemical Stimulants

de Lange and Boucher (1990) found that smoke from burning a mixture of fresh and dry plant materials collected in a South African fynbos community stimulated 128 seeds of the fynbos shrublet *Audouinia capitata* to germinate in field plots, whereas no germination occurred in nontreated plots. Smoke was held over soil

TABLE 10.4 Effects of Dry Heat Treatments on Germination of Seeds of Matorral Species

Species	Family[a]	Temperature[b]	D[c]	Reference
Increased germination[e]				
Acacia myrtifolia[d]	Fab.	80	120	Auld and O'Connell (1991)
Agathosma betulina	Rut.	80	20	Blommaert (1972)
Aotus ericoides[d]	Fab.	80	60	Auld and O'Connell (1991)
Calluna vulgaris	Eric.	110	5	Gonzalez-Rabanal and Casal (1995)
Calopsis impolita	Restion.	120	6	Musil and de Witt (1991)
Camissonia hirtella	Onagr.	150	5	Keeley *et al.* (1985)
Ceanothus crassifolius[d]	Rhamn.	100	60	Christensen and Muller (1975)
C. divaricatus[d]	Rhamn.	116	5	Wright (1931)
C. megacarpus[d]	Rhamn.	100	5	Hadley (1961)
Cistus albidus[d]	Cist.	100	5	Vuillemin and Bulard (1981)
C. incanus[d]	Cist.	120	1.5	Aronne and Mazzoleni (1989)
C. monspeliensis[d]	Cist.	100	5	Vuillemin and Bulard (1981)
C. psilosepalus[d]	Cist.	110	5	Gonzalez-Rabanal and Casal (1995)
Colliguaya odorifera	Euphorbi.	100	5	Munoz and Fuentes (1989)
Conostylis setosa	Haemodor.	105	2	Bell *et al.* (1987)
Cytisus scoparius[d]	Fab.	130	1	Tarrega *et al.* (1992)
Daboecia cantabrica	Eric.	110	5	Conzalez-Rabanal and Casal (1995)
Erica ciliaris	Eric.	80	5	Gonzalez-Rabanal and Casal (1995)
E. umbellata	Eric.	110	5	Gonzalez-Rabanal and Casal (1995)
Eriodictyon crassifolium	Hydrophyll.	120	5	Keeley (1987)
Fremontodendron decumbens[d]	Sterculi.	100	5	Boyd and Serafini (1992)
Genista florida[d]	Fab.	150	1	Tarrega *et al.* (1992)
Halimium alyssoides[d]	Cist.	110	5	Gonzalez-Rabanal and Casal (1995)
Hardenbergia violaceae[d]	Fab.	90	70	Auld and O'Connell (1991)
Lotus scoparius[d]	Fab.	100	60	Christensen and Muller (1975)
Metalasia muricata	Aster.	100	6	Musil (1991)
Muelenbeckia hastulata	Polygon.	100	5	Munoz and Fuentes (1989)
Peumus boldus	Monimi.	100	5	Munoz and Fuentes (1989)
Phylica ericoides[d]	Rhamn.	100	5	Kilian and Cowling (1992)
P. stipularis[d]	Rhamn.	100	3	Musil (1991)
Podalyria calyptrata	Fab.	100	1	Jeffery *et al.* (1988)
Rhus laurina[d]	Anacardi.	93	5	Wright (1931)
R. trilobata[d]	Anacardi.	120	5	Keeley (1987)
Saliva apiana	Lami.	70	60	Keeley (1987)
Schinus polygamus	Anacardi.	100	5	Munoz and Fuentes (1989)
Stylidium soboliferum	Stylidi.	100	30	Enright *et al.* (1997)
Thermopsis macrophylla[d]	Fab.	80	5	Borchert (1989)
Trevoa trinervis[d]	Rhamn.	100	5	Munoz and Fuentes (1989)
Trymalium spathulatum[d]	Rhamn.	105	2	Bell *et al.* (1987)
T. ledifolium[d]	Rhamn.	105	2	Bell *et al.* (1987)
Tuberaria guttata[d]	Cist.	110	5	Gonzalez-Rabanal and Casal (1995)
Decreased germination[f]				
Agrostis curtisii	Po.	80	5	Gonzalez-Rabanal and Casal (1995)
Arbutus unedo	Eric.	120	10	Mesleard and Lepart (1991)
Calochortus splendens	Lili.	120	5	Keeley and Keeley (1987)
Comarostaphylis diversifolia	Eric.	70	1	Keeley (1987)
Cordylanthus filifolius	Scrophulari.	120	5	Keeley and Keeley (1987)
Delphinium cardinale	Ranuncul.	120	5	Keeley and Keeley (1987)
D. parryi	Ranuncul.	120	5	Keeley and Keeley (1987)
Dichelostemma pulchella	Alli.	120	5	Keeley and Keeley (1987)
Erica arborea	Eric.	120	10	Mesleard and Lepart (1991)
Heterotheca grandiflora	Aster.	120	5	Keeley and Keeley (1987)

(continues)

TABLE 10.4—*Continued*

Species	Family[a]	Temperature[b]	D[c]	Reference
Madia gracilis	Aster.	120	5	Keeley and Keeley (1987)
Marah macrocarpus	Cucurbit.	120	5	Keeley *et al.* (1985)
Paeonia californica	Paeoni.	120	5	Keeley *et al.* (1985)
Penstemon heterophyllus	Scrophulari.	120	5	Keeley and Keeley (1987)
Perezia microcephala	Aster.	120	5	Keeley and Keeley (1987)
Pinus halepensis	Pin.	110	5	Martinez-Sanchez *et al.* (1995)
P. pinaster	Pin.	150	5	Martinez-Sanchez *et al.* (1995)
Pterostegia drymarioides	Polygon.	120	5	Keeley and Keeley (1987)
Quillaja boldus	Ros.	100	5	Munoz and Fuentes (1989)
Stipa coronata	Po.	120	5	Keeley *et al.* (1985)
Zigadenus fremontii	Lili.	120	5	Keeley *et al.* (1985)
Germination not changed[g]				
Agathosma apiculata	Rut.	100	29	Pierce and Moll (1994)
A. conostephioides	Rut.	100	30	Enright *et al.* (1997)
A. stenopetala	Rut.	100	20	Pierce and Moll (1994)
Agrostis curtisii	Po.	110	5	Gonzalez-Rabanal and Casal (1995)
A. delicatula	Po.	110	5	Gonzalez-Rabanal and Casal (1995)
Astroloma conostephioides	Epacrid.	100	30	Enright *et al.* (1997)
Avenula marginata	Po.	110	5	Gonzalez-Rabanal and Casal (1995)
Barbarea americana	Brassic.	120	5	Sweeney (1956)
Clarkia spp.	Onagr.	120	5	Keeley and Keeley (1987)
Colletia spinosa[d]	Rhamn.	100	5	Munoz and Fuentes (1989)
Cryptantha muricata	Boragin.	90	5	Sweeney (1956)
Eriophyllum confertiflorum	Aster.	100	60	Christensen and Muller (1975)
Felicia echinata	Aster.	100	20	Pierce and Moll (1994)
Festuca reflexa	Po.	90	5	Sweeney (1956)
Galium parisiense	Rubi.	120	5	Keeley and Keeley (1987)
Helianthemum scoparium[d]	Cist.	120	5	Kelley *et al.* (1985)
Hypericum concinnum	Clusi.	90	5	Sweeney (1956)
Laxmannia orientalis	Rubi.	100	30	Enright *et al.* (1997)
Leucopogon glacialis	Epacrid.	100	30	Enright *et al.* (1997)
Lotus scoparius[d]	Fab.	100	5	Keeley (1987)
Metalasia muricata	Aster.	100	20	Pierce and Moll (1994)
Microseris linearifolia	Aster.	120	5	Keeley and Keeley (1987)
Mimulus californicus	Scrophulari.	100	5	Sweeney (1956)
Muraltia squarrosa	Aster.	100	20	Pierce and Moll (1994)
Passerina paleacea	Thymelae.	100	5	Kilian and Cowling (1992)
P. vulgaris	Thymelae.	100	5	Pierce and Moll (1994)
Ranunculus californicus	Ranuncul.	100	5	Sweeney (1956)
Rhamnus californica[d]	Rhamn.	160	5	Wright (1931)
Rhus integrifolia[d]	Anacardi.	70	1	Keeley (1987)
Solanum douglasii	Solan.	120	5	Keeley and Keeley (1987)
Staberoha distachya	Restion.	120	3	Musil and de Witt (1991)
Stipa lepida	Po.	120	5	Keeley and Keeley (1987)
Virgilia oroboides	Fab.	80	60	Jeffrey *et al.* (1988)

[a]See notes for Table 10.1.
[b]Treatment temperature (°C).
[c]Duration (min) of treatment.
[d]Species belongs to a family with physical dormancy.
[e]Germination of heat-treated seeds increased by 25% or more.
[f]Germination of heat-treated seeds decreased by 25% or more.
[g]Germination of heated and nonheated seeds varied less than 25%.

and seeds in treatment plots for 30 min by use of a plastic tent. Further, an extract, obtained by bubbling smoke through water, stimulated germination. The chemical(s) responsible for the enhancement of germination was not identified, but ethylene was eliminated as a possibility (de Lange and Boucher, 1990).

The work of de Lange and Boucher (1990) stimulated other scientists to investigate the effects of smoke on dormancy break and germination, and a number of matorral species have been identified whose seeds are stimulated by smoke (Table 10.5). However, smoke does not promote seed germination in all species, including some *Erica* spp. (Brown *et al.*, 1993), *Calopsis impolita*, *Phylica pubescens* (Brown, 1993a), *Caryotophora skiatophytoides*, and *Saphesia flaccida* (Hickey and van Jaarsveld, 1995). Smoke from 27 species collected in montane grasslands of Drakensberg, South Africa, failed to stimulate the germination of *Themeda triandra* seeds (Baxter *et al.*, 1995).

Seeds of the fynbos shrub *Erica hebecalyx* were treated with ethylene and ammonia, since these compounds can be produced during fires (van de Venter and Esterhuizen, 1988). Germination increased 10 and 15% after dry seeds of this species were exposed to 1% ethylene or 50 ppm ammonia, respectively. Dry heat (96.5°C for 3 min) increased germination by 14%.

Twelve compounds have been identified in aqueous extracts of smoke from burning *Passerina vulgaris* and/or *Themeda triandra* plant material (van Staden *et al.*, 1995). However, the stimulatory compound(s) has not been found. A high concentration (1:1 dilution) of aqueous smoke extracts from burning leaves of *Acacia mearnsii*, *Eucalyptus grandis*, *Hypoxis colchicifolia*, *Pinus patula*, and *Themeda triandra* and from burning tissue paper inhibited the germination of light-sensitive seeds of Grand Rapids lettuce, but a low concentration (1:40 dilution) promoted germination (Jagar *et al.*, 1996). The germination stimulant(s) was produced if leaves of *T. triandra* were heated to 160–200°C but not burned. Heated agar and cellulose also produced the stimulant(s), but heated starch, glucose, galactose, and glucuronic acid did not.

Ashes are a product of fires, and they have been placed on, or near, seeds to determine their effects on germination. Sweeney (1956) treated seeds of 16 California chaparral herbaceous species with ashes obtained by burning excelsior and observed a 5% increase in germination of only one species, *Oenothera micrantha* var. *hirtella*. Germination was decreased 5–20% in 11 of the species and was not changed in 4 of them. *Adenostoma fasciculatum* ash placed on the soil surface under mature *A. fasciculatum* plants in California did not promote seeds of any species to germinate (Christensen and Muller, 1975). The germination of *Pinus halepensis*

seeds sown on top of a 1- or 5-cm-deep layer of ashes (derived from *P. halepensis* twigs and needles) was reduced by 72 and 94%, respectively, whereas that of *Cistus salviifolius* seeds was reduced by 74 and 99%, respectively (Ne'eman *et al.*, 1993). A suspension of ash in water reduced the germination of nondormant seeds of *Agrostis curtisii*, *Calluna vulgaris*, *Cistus psilosepalus*, *Daboecia cantabrica*, *Erica ciliaris*, *E. umbellata*, *Halimium alyssoides*, and *Tuberaria guttata* (Gonzalez-Rabanal and Casal, 1995). An ash solution increased the germination of *A. curtisii* and *Pseudarrhenatherum longifolium* seeds produced in apical, but not in basal, portions of the inflorescence (Gonzales-Rabanal *et al.*, 1994).

Another product of fire is partially burned (charred) pieces of plant material, especially stems. Germination of the California chaparral annual *Emmenanthe penduliflora* was stimulated if seeds were placed on potting soil, but not on filter paper, adjacent to charred stems of the shrub *Adenostoma fasciculatum* (Wicklow, 1977). Presumably, some chemical(s) was formed or released when stems were charred, which promoted germination. Unburned and ashed stems had no effect on germination. Jones and Schlesinger (1980) obtained similar results with charred *A. fasciculatum* stems, except *E. penduliflora* seeds germinated on both soil and filter paper.

Subsequently, charate or powdered charred wood has been shown to stimulate seed germination in a large number of matorral species (Table 10.6). However, charate has no stimulatory effects on seeds of some species, and it decreases germination in others (Table 10.7). A lack of response to charate in species whose seeds have physical dormancy may mean that the seed/fruit coats were impermeable when the charate treatment was applied.

Charred wood from various shrubs, including nonchaparral species, produces stimulatory compounds (Keeley and Pizzorno, 1986). Further, the water-soluble stimulatory substance(s) is produced when wood is heated at 175°C for 30 min; thus, it is not necessary to actually burn the wood. Keeley and Pizzorno (1986) found that the compound comes from heated hemicellulose, suggesting that an oligosaccharin-type molecule is formed when the hemicellulose molecule xylan or other hemicellulose molecules that have glucuronic acid side chains are heated. The stimulatory compound(s) in charred wood can be extracted by using acetone, hexane, or methane (Keeley, 1991). Identification of the compound(s) in charate that stimulates seeds to germinate will be very exciting.

Not only does the compound(s) in charred wood that promotes germination need to be identified, but information is needed on how it works physiologically. Does charate substitute for light? Whereas nontreated seeds

TABLE 10.5   Matorral Species in Which Seed Germination Is Stimulated by Smoke
and/or Smoke Extracts

Species	Family[a]	Germination[b]	Reference
Actinostrobus acuminatus	Cupress.	70	Dixon *et al.* (1995)
Andersonia lehmanniana	Epacrid.	10	Dixon *et al.* (1995)
Anigozanthos bicolor	Haemodor.	6	Dixon *et al.* (1995)
A. humilis	Haemodor.	13	Dixon *et al.* (1995)
A. manglesii	Haemodor.	28	Dixon *et al.* (1995)
Audouinia capitata	Bruni.	11	de Lange and Boucher (1990)
Billardiera bicolor	Pittospor.	46	Dixon *et al.* (1995)
Burchardia umbellata	Colchic.	19	Dixon *et al.* (1995)
Caulanthus hetereophyllus	Brassic.	—[c]	Fotheringham *et al.* (1995)
Chaenactis glabriuscula	Aster.	—	Fotheringham *et al.* (1995)
Codonocarpus cotinifolius	Gyrostemon.	17	Dixon *et al.* (1995)
Conospermum triplinervium	Prote.	12	Dixon *et al.* (1995)
Conostylis neocymosa	Haemodor.	80	Dixon *et al.* (1995)
C. setosa	Haemodor.	50	Dixon *et al.* (1995)
Emmenanthe penduliflora	Hydrophyll.	—	Fotheringham *et al.* (1995)
Erica 26 species	Eric.		Brown *et al.* (1993a)
Geleznowia verrucosa	Rut.	70	Dixon *et al.* (1995)
Grevillea wilsonii	Prote.	70	Dixon *et al.* (1995)
Guillenia lasiophylla	Brassic.	—	Fotheringham *et al.* (1995)
Gyrostemon ramulosus	Gyrostemon.	10	Dixon *et al.* (1995)
Hypocalymma angustifolium	Myrt.	26	Dixon *et al.* (1995)
Lechenaultia biloba	Goodeni.	40	Dixon *et al.* (1995)
L. macrantha	Goodeni.	23	Dixon *et al.* (1995)
Lysinema ciliatum	Prote.	44	Dixon *et al.* (1995)
Metalasia densa	Aster.	13	Brown (1993a)
Neurachne alopecuroidea	Po.	40	Dixon *et al.* (1995)
Othonna quinquedentata	Aster.	19	Brown (1993a)
Patersonia occidentalis	Irid.	30	Dixon *et al.* (1995)
Petrophile drummondii	Prote.	65	Dixon *et al.* (1995)
Pimelea spectabilis	Thymelae.	25	Dixon *et al.* (1995)
P. sylvestris	Thymelae.	30	Dixon *et al.* (1995)
Restio praeacutus	Restion.	18	Brown (1993a)
R. similis	Restion.	10	Brown (1993a)
Romneya coulteri	Papaver.	99	Fotheringham *et al.* (1995)
Salvia columbariae	Lami.	—	Fotheringham *et al.* (1995)
Scaevola caliptera	Goodeni.	15	Dixon *et al.* (1995)
Senecio grandiflorus	Aster.	19	Brown (1993a)
Sphenotoma capitatum	Prote.	45	Dixon *et al.* (1995)
Spyridium globosum	Rhamn.	14	Dixon *et al.* (1995)
Staberoha disticha	Restion.	22	Brown (1993a)
Stirlingia latifolia	Prote.	42	Dixon *et al.* (1995)
Stylidium soboliferum	Stylidi.	23	Enright *et al.* (1997)
Syncarpha eximia	Prote.	61	Brown (1993a)
S. vestita	Prote.	76	Brown (1993b)
Tetratheca hirsuta	Tremandr.	12	Dixon *et al.* (1995)
Thamnochortus pellucidus	Restion.	24	Brown (1993a)
Thysanotus multiflorus	Antheric.	30	Dixon *et al.* (1995)
Verticordia densiflora	Myrt.	67	Dixon *et al.* (1995)
Widdringtonia cupressoides	Cupress.	13	Brown (1993a)

[a]See notes for Table 10.1.
[b]Percentage increase in germination compared to control.
[c]Information not given.

TABLE 10.6 Matorral Species in Which Seed Germination Is Stimulated by the Presence of Charred Wood or Aqueous Extracts of It

Species	Family[a]	Germination[b]	Reference
Adenostoma fasciculatum	Ros.	7	Keeley (1987)
Agoseris heterophylla	Aster.	30	Keeley and Keeley (1987)
Antirrhinum coulterianum	Scrophulari.	40	Keeley and Keeley (1987)
A. kelloggi	Scrophulari.	24	Keeley and Keeley (1987)
Arctostaphylos glandulosa	Eric.	5	Keeley (1987)
Artemisia californica	Aster.	5	Keeley (1987)
Burcharida umbellata	Colchi.	24	Bell *et al.* (1987)
Camissonia california	Onagr.	46	Keeley and Keeley (1987)
Chaenactis artemisiaefolia	Aster.	14	Keeley *et al.* (1985)
Clarkia epilobioides	Onagr.	33	Keeley and Keeley (1987)
C. purpurea	Onagr.	32	Keeley and Keeley (1987)
Clematis lasiantha	Ranuncul.	47	Keeley (1987)
Collinsia parryi	Scrophulari.	53	Keeley and Keeley (1987)
Cryptantha intermedia	Boragin.	19	Keeley and Keeley (1987)
C. muricata	Boragin.	43	Keeley *et al.* (1985)
Daucus pusillus	Api.	12	Keeley and Keeley (1987)
Dryandra carduacea	Prote.	13	Bell *et al.* (1987)
D. fraseri	Prote.	15	Bell *et al.* (1987)
Emmenanthe penduliflora	Hydrophyll.	48, 31	Wicklow (1977), Jones and Schlesinger (1980)
Eriodictylon crassifolium	Hydrophyll.	27	Keeley (1987)
Eriophyllum confertiflorum	Aster.	48	Keeley *et al.* (1985)
Festuca megalura	Po.	12	Keeley *et al.* (1985)
Fremontodendron decumbens	Papaver.	14	Boyd and Serafini (1992)
Galium angustifolium	Rubi.	21	Keeley and Keeley (1987)
G. parisiense	Rubi.	15	Keeley and Keeley (1987)
Garrya flavescens	Garry.	61	Keeley (1987)
Gilia australis	Polemoni.	49	Keeley and Keeley (1987)
G. capitata	Polemoni.	75	Keeley and Keeley (1987)
Gnaphalium californica	Aster.	33	Keeley and Keeley (1987)
Lepidium nitidum	Brassic.	20	Keeley and Keeley (1987)
Malacothrix clevelandii	Aster.	26	Keeley and Keeley (1987)
Mimulus aurantiacus	Scrophulari.	6	Keeley (1987)
Papaver californicum	Papaver.	89	Keeley and Keeley (1987)
Penstemon centranthifolius	Scrophulari.	16	Keeley and Keeley (1987)
P. heterophyllus	Scrophulari.	20	Keeley and Keeley (1987)
P. spectabilis	Scrophulari.	61	Keeley and Keeley (1987)
Phacelia cicutaria	Hydrophyll.	27	Keeley *et al.* (1985)
P. fremontii	Hydrophyll.	15	Keeley *et al.* (1985)
P. minor	Hydrophyll.	13	Keeley and Keeley (1987)
Rafinesquia californica	Aster.	51	Keeley and Keeley (1987)
Rhus trilobata	Anacardi.	27	Keeley (1987)
Romneya coulteri	Papaver.	40	Keeley and Keeley (1987)
R. trichocalyx	Papaver.	24	Keeley (1987)
Salvia columbariae	Lami.	43	Keeley (1984)
S. mellifera	Lami.	21	Keeley (1987)
Silene multinervia	Caryophyll.	38	Keeley and Keeley (1987)
Stephanomeria virgata	Aster.	11	Keeley and Keeley (1987)
Syncarpha vestita	Aster.	82	Brown (1993)
Streptanthus heterophyllus	Brassic.	24	Keeley and Keeley (1987)
Toxicodendron diversifolia	Anacardi.	10	Keeley (1987)
Trymalium ledifolium	Rhamn.	11	Bell *et al.* (1987)

[a]See notes for Table 10.1.
[b]Percentage increase in germination as compared to control.

TABLE 10.7   Matorral Species in Which Seed Germination Was Not Stimulated by the Presence of Charred Wood or Aqueous Extracts of It[a]

Species	Effect on germination[b]	Species	Effect on germination[b]
Agathosma apiculata	0	*Hypocalymma angustifolium*	0
Agrostocrinum scabrum	0	*H. robusta*	−
Allium praecox	0	*Isopogon dubius*	0
Anigozanthus manglesii	0	*Keckiella antirrhinoides*	0
Antirrhinum nuttallianum	−	*K. cordifolia*	0
Apiastrum angustifolium	0	*K. ternata*	−
Arctostaphylos patula	0	*Lactuca serriola*	0
Avena barbata	0	*Leschenaultia biloba*	0
Billardiera bicolor	0	*Lomatium dasycarpum*	−
B. floribunda	0	*Lonicera subspicata*	−
Bloomeria crocea	0	*Madia gracilis*	0
Boronia fastigiata	0	*Malosoma laurina*	−
Brassica nigra	−	*Marah macrocarpus*	0
Camissonia hirtella	−	*Metalasia muricata*	−
Ceanothus cuneatus	0	*Microseris linearifolia*	0
C. megacarpus	−	*Muraltia squarrosa*	−
Cercocarpus betuloides	0	*Paeonia californica*	−
Chorizanthe fimbriata	0	*Passerina paleacea*	0
Clarkia unguiculata	0	*Patersonia sericea*	0
Clematis pubescens	0	*Perezia microcephala*	−
Comarostaphylis diversifolia	−	*Phacelia brachyloba*	0
Conostylis setosa	0	*P. grandiflora*	0
Cordylanthus filifolius	0	*P. parryi*	0
Delphinium cardinale	−	*P. viscida*	0
D. parryi	−	*Phylica ericoides*	−
Descurainia pinnata	0	*Phyllanthus calycinus*	0
Dianella revoluta	0	*Pimelea spectabilis*	0
Dicentra chrysantha	0	*Porophyllum gracile*	0
D. ochroleuca	0	*Pterostegis drymarioides*	−
Dichelostemma pulchella	−	*Quercus dumosa*	−
Dryandra formosa	−	*Rhus integrifolia*	−
D. nivea	0	*Salvia apiana*	−
D. polycephala	−	*Scrophularia californica*	−
D. sessilis	−	*Silene gallica*	−
Eriogonum fasciculatum	−	*Sisymbrium orientale*	−
Eucrypta chrysanthemifolia	0	*Solanum douglasii*	−
Felicia echinata	0	*Sollya heterophylla*	0
Fremontodendron californicum	0	*Stipa coronata*	−
Hakea lissocarpha	−	*Thysanotus multiflorus*	0
H. trifurcata	−	*Trymalium spathulatum*	0
Haplopappus squarrosus	−	*Vigueira laciniata*	0
Heteromeles arbutifolia	0	*Xanthorrhoea gracilis*	−
Heterotheca grandiflora	−	*X. preissii*	−
Hybanthus floribunda	0	*Zigadenus fremontii*	−

[a] From Keeley *et al.* (1985), Bell *et al.* (1987), Keeley and Keeley (1987), Keeley (1987), and Pierce and Moll (1994).
[b] 0, no effect; −, decreased.

of *Salvia mellifera* germinated to higher percentages in light than in darkness, charate-treated seeds germinated equally well in light and darkness (Keeley, 1986). However, charate-treated seeds of *Emmenanthe penduliflora* and *Eriophyllum confertiflorum* germinated to higher percentages in light than in darkness (Keeley and Pizzorno, 1986).

Evidence shows that stimulatory substances also are

formed when soil is heated. Extracts of soil heated to 195°C for 10 min stimulated the germination of seeds of *Emmenanthe penduliflora* and *Eriophyllum confertiflorum* (Keeley and Nitzberg, 1984). Are the stimulatory substances produced by heating soil the same as those found in charred wood?

Although compounds that stimulate seed germination may be isolated from smoke, charred wood, or heated soil in the laboratory, their role in germination ecology cannot be understood until they have been tested in the field, which is not an easy thing to do. The addition of ash and charate to burned and clipped plots in California chaparral had no effect on germination of the shrubs *Adenostoma fasciculatum* and *Ceanothus greggii*, whereas results for the herbaceous species *Phacelia brachyloba, Cryptantha muricata, Camissonia intermedia, Streptanthus heterophyllus,* and *Lotus strigosus* were mixed (Moreno and Oechel, 1991), i.e., germination was stimulated by the addition of ash and charate in one burning treatment and was decreased in another. The addition of ash and charate to clipped plots had no effect.

3. Changes in Microsites

A fire changed the red : far-red ratio at the soil surface from 0.3 to 1.1 in phrygana vegetation in Greece by removing green leaves, which caused the germination of seeds of the dwarf shrub *Sarcopoterium spinosum* to increase from 3 to 30% (Roy and Arianoutsou-Faraggitaki, 1985). Fire also would remove litter from the soil surface, consequently exposing seeds to light and perhaps stimulating germination. Seeds of dicots in soil samples collected from nonburned and recently burned (2 years) California chaparral germinated to higher percentages in light than in darkness. More monocot seeds in samples from nonburned chaparral germinated in light than in darkness, whereas those in samples from burned chaparral germinated to the same percentages in light and in darkness (Keeley, 1984). The germination of seeds of annuals, perennials, and subshrubs was not affected by the removal of shrubs in field studies in California chaparral (Tyler, 1995).

Another effect of removal of litter is an increase in soil surface temperatures, as well as an increase in the amplitude of daily temperature fluctuations. Summer soil temperatures in sclerophyll woodlands in Australia increased when litter was removed by fire, with 40°C being recorded at a depth of 45 mm and 60°C at a depth of 4 mm (Auld and Bradstock, 1996). These temperatures were high enough to break physical dormancy in seeds of legumes in the soil seed bank. Soil temperatures did not exceed 30°C in areas where the litter was undisturbed. Seeds of the South African fynbos shrubs *Leucospermum cordifolium* and *Serruria florida* require alternating temperature regimes for maximum germination. Soil temperature measurements indicate that temperature requirements for germination are fulfilled in the field in burned, but not in unburned, areas and that they are fulfilled in burned areas only during the cool months, from May to September (Brits, 1986). An increase in the amplitude of daily temperature fluctuations

may be responsible for breaking physical dormancy in some matorral species, but this has not been documented.

Fire may decrease (Debano and Conrad, 1978; Stock and Lewis, 1986) or increase (Christensen, 1973) the total nitrogen content of matorral soils. Nitrogen is added to soils in burned areas as ammonium and as organic nitrogen in the ash, and ammonium is converted to nitrate by bacteria, after the soil becomes moist in autumn (Christensen, 1973). The slight increase in soil pH that occurs after fires (Christensen, 1973; Sweeney, 1956) favors the growth of nitrifying and N-fixing bacteria (Sweeney, 1956). Because nitrate has been shown to stimulate germination (Chapter 4,V,F), increases in soil nitrate levels could have significant consequences on germination. Nitrates applied at the level found in burned field plots promoted the germination of *Emmenanthe penduliflora, Phacelia grandiflora,* and *Salvia mellifera* seeds, and ammonium ions promoted the germination of *P. grandiflora* and *S. mellifera* seeds (Thanos and Rundel, 1995). However, these authors found that the concentration of nitrogenous compounds in charate extracts was too low to account for the stimulatory effects of charate on seed germination in these three species.

Some ammonium is present in soils in unburned chaparral, but its nitrification is prevented, possibly by inhibitors in foliar leachates from shrubs or by the high lignin content in fallen leaves (Christensen, 1973). One consequence of fire may be the destruction of compounds that inhibit the growth of nitrifying bacteria. Soil microbes produce substances that inhibit germination (Kaminsky, 1981; Keeley and Nitzberg, 1984), in which case fire would destroy the compounds and/or microbes. Fire also could destroy allelopathic chemicals that prevent germination (McPherson and Muller, 1969).

F. Special Factor: Allelopathy

When metabolic compounds produced by one plant inhibit any physiological activity, including seed germination of other plants, the antagonistic chemical interaction is called allelopathy (see Muller, 1966). In trying to explain why so many seeds germinate in matorral vegetation after fires, researchers have devoted some attention to allelopathy. The hypothesis is that chemicals capable of inhibiting germination are leached out of living leaves of shrubs or leaf litter on the soil surface by rainwater, where they accumulate in the upper 1–3 cm of soil, preventing germination. A fire destroys leaves on shrubs, litter on the soil surface, and inhibitory chemicals in the surface layers of the soil. Thus, toxins are no longer present after a fire, and seeds can germinate as soon as the soil moisture becomes nonlimiting (Muller *et al.,* 1968; McPherson and Muller, 1969).

1. Leachates

Various studies have been done to determine the effects of leachates from plant material on the germination of species in the California chaparral. Seeds of five chaparral annuals incubated in the presence of water leachates of organic material collected from the soil surface showed no inhibition of germination (Sweeney, 1956). The germination of scarified seeds of *Ceanothus megacarpus* decreased 62% when fresh leaves of the shrub *Adenostoma fasciculatum* were added to petri dishes. Fallen dead leaves (duff) of *A. fasciculatum* had no effect on the germination of permeable seeds of *Ceanothus megacarpus,* but ashed duff decreased germination 35% (Hadley, 1961). The removal of crowns (or shoots) of *A. fasciculatum* plants from field plots without disturbing the soil resulted in the germination of 30 species of herbs and shrubs. The implication of these results is that germination inhibitors are produced by *A. fasciculatum* leaves, leached out by rainwater, and accumulated in the soil, where they prevent germination (Muller *et al.,* 1968; McPherson and Muller, 1969).

Experiments showed that soil moisture, light, and nutrients were adequate for the growth of herbs in stands of *Adenostoma fasciculatum* (McPherson and Muller, 1969). Natural rainwater leachates (raindrip) collected from beneath plants of *A. fasciculatum* in the field and concentrated to 10 times (10×) their original strength inhibited the germination of the native dicots *Calandrinia ciliata, Helianthemum scoparium,* and *Silene multinervia* and the introduced grass *Bromus rigidus* (McPherson and Muller, 1969). A 4× leachate of *A. fasciculatum* leaves also inhibited the germination of *H. scoparium* seeds (Keeley and Keeley, 1981). However, 4× leachates of *A. fasciculatum* promoted the germination of *Paeonia californica, Salvia columbariae, Convolvulus cyclostegius,* and *Descurainia pinnata* seeds, and 1× leachates promoted the germination of *Oenothera micrantha, Cryptantha muricata, Apiastrum angustifolium, D. pinnata, Brassica nigra, Phacelia cicutaria,* and *Phacelia grandiflora* seeds (Keeley and Keeley, 1981). Leachates from *Rhus integrifolia, Quercus dumosa,* and *A. fasciculatum* leaves did not inhibit the germination of nondormant seeds of *R. integrifolia* (Lloret and Zedler, 1991).

Much work needs to be done to elucidate the role of chemicals leached from the foliage of shrubs on the germination ecology of matorral species. What are the chemicals in leachates that inhibit or promote germination? If these chemicals are applied to soil in the field, will they influence germination? After chemicals are leached from plant material, do they remain stable in the soil? Can the inhibitory/stimulatory chemicals be isolated from soils in field plots? How does fire affect the stability and function of these chemicals?

Water-soluble germination inhibitors are found in leaf litter and in intact canopies of the California chaparral shrubs *Arctostaphylos glauca* and *A. glandulosa* (Muller *et al.,* 1968). Growth-inhibiting chemicals in leachates of foliar branches and leaf litter of *A. glandulosa* var. *zacaensis* included arbutin, hydroquinone, two unknown phenols, and tannic, gallic, chlorogenic, vanillic, *p*-hydroxybenzoic, and ferulic acids. One of the unknown phenols completely inhibited the germination of *Lactuca sativa.* When soil from under shrubs was analyzed, only four of the leachate compounds (ferulic, *p*-hydroxybenzoic, and vanillic acid, and one unknown phenol) were found, but three new compounds (*p*-coumaric, *o*-coumaric, and syringic acids) were present (Chou and Muller, 1972). Chou and Muller (1972) suggested that physical and biotic processes may result in changes in compounds after they enter the soil.

The toxicity of leachates from *Arctostaphylos glandulosa* var. *zacaensis* leaves, as determined by the radicle growth of *Bromus rigidus* seedlings, was increased when leaves were given heat treatments up to 160°C (Chou and Miller, 1972). Leachates from leaves subjected to temperatures greater than 160°C decreased in toxicity with each increase in temperature, with no toxicity in leachates from leaves subjected to 200°C. The toxicity of leachates of *A. glandulosa* leaves that had been heated decreased when temperatures were increased to 180°C and were not toxic after leaves were heated to 200°C (Chou, 1973). The high levels of toxic compounds found in chaparral soils immediately following a fire declined 40% after the first rainfall and 70% after the second one (Chou, 1973).

Suppression of the growth of herbaceous species, particularly grasses, in heaths of the Galician region of Spain has been attributed in part to inhibitory compounds leached from the shrub *Erica scoparia.* Leaves of *E. scoparia* placed in the bottom of petri dishes and covered with wet filter paper inhibited radicle growth in seedlings of *Trifolium pratense.* A chromatographic bioassay of *E. scoparia* leachate showed two zones of growth inhibition (R_f 0.10–0.40 and 0.60–0.80) and two zones of growth promotion (0.40–0.60 and 0.9–1.0). Seven phenolic compounds were identified in the two zones of growth inhibition: 2-hydroxyphenylacetic, *p*-coumaric, ferulic, gentisic, syringic, vanillic, and p-hydroxybenzoic acids (Ballester *et al.,* 1977). Effects of these compounds on seed germination need to be determined.

Plants of *Casuarina pusilla, Leptospermum myrinoides, Leucopogon ericoides,* and *L. virgatus* in the mallee of Australia are either stunted or absent beneath the canopies of *Eucalyptus baxteri* and *E. obliqua* (del Moral *et al.,* 1978). Water leachates of *E. baxteri* leaves contained gentisic and ellagic acids, and they inhibited the growth of *E. viminalis* seedlings. Foliar leachates of *E. obliqua,* which contained tannic, ellagic, and chlorogenic acids, also inhibited the growth of *E. viminalis.* Litter extracts of *E. baxteri* and *E. obliqua* inhibited the growth of *E. viminalis,* but those of *E. nitida* did not. Litter leachates of *E. baxteri* contained phenolic glycosides, tannins, and gentisic, gallic, caffeic, and ellagic acids, whereas litter leachates of *E. obliqua* contained chlorogenic, ellagic, and gallic acids (del Moral *et al.,* 1978). The effects, if any, of these chemicals on seed germination ecology were not studied.

The presence of toxic compounds in leachates of leaves of shrubs does not allow us to conclude that they inhibit seed germination. Many other environmental factors must be considered. For example, toxic compounds in leachates of naturalized trees of *Eucalyptus camaldulensis* in California did not inhibit the growth of annual herbs on sand but were inhibitory on poorly drained soils with a high colloidal content (del Moral and Muller, 1970). Also, chemicals in leachates may interact with other factors. Kaminsky (1981) determined the amounts of available phenolics in soil in mature stands of *Adenostoma fasciculatum* throughout the year, and at no time were the levels high enough to inhibit germination. Also, soils had a high affinity for phenolics, making it unlikely that these compounds would be absorbed by seeds. Soils contained phytotoxins that inhibited germination, but they were of microbial (not *A. fasciculatum*) origin. In some unknown way, the presence of *A. fasciculatum* shrubs promoted the production of phytotoxins by microbes (Kaminsky, 1981).

2. Volatiles

The contact between grasslands and soft chaparral on the coast of California is marked by bare zones extending 1–2 m out from the crowns of shrubs (Muller, 1966). Soft chaparral is dominated by *Salvia* spp., *Artemisia californica, Eriogonum fasciculatum,* and *Encelia californica* (Gray and Schlesinger, 1981). One explanation for the bare zone is that volatile compounds released by the shrubs inhibit the growth of herbaceous species. For example, seeds of the grasses *Bromus rigidus* and *Festuca megalura* failed to germinate in the presence of volatiles from homogenized leaves of *Salvia leucophylla* in 500 ml of atmosphere (Muller, 1966).

What volatiles are released by shrubs? The terpenes α-pinene, camphene, β-pinene, cineole, dipentene, and camphor were identified in the volatilization products

from the leaves of *Salvia leucophylla, S. mellifera,* and *S. apiana* (Muller and Muller, 1964). The root growth of *Cucumis sativus* seedlings was inhibited more by the presence of camphor and cineole in the atmosphere than by pinenes (Muller and Muller, 1964). Camphor and cineole were found in the air immediately surrounding plants of *S. leucophylla* and *S. mellifera* growing in the field or in a greenhouse (Muller, 1965). Because these two terpenes were highly soluble in hard paraffin, Muller (1965) suggested that they entered young seedlings by becoming dissolved in lipids of the cutin layer covering mesophyll cells.

The role of terpenes in seed germination ecology is not well understood. Not only are terpenes present in the atmosphere of aromatic shrubs, they also are absorbed by soil, especially dry soil in the field (Muller, 1966). The germination of *Bromus rigidus* was prevented when seeds were sown on soil that had been exposed to air recirculated over a mass of homogenized *Salvia* sp. leaves for 18 hr (Muller, 1966). Are terpenes adsorbed by soils in sufficient quantities to inhibit germination in the field? What happens to the terpene concentration when the soil is wet? After a rain of 37 mm, Went *et al.* (1952) observed many *Salvia mellifera* seedlings under *S. mellifera* and *Adenostoma fasciculatum* shrubs, but *A. fasciculatum* seedlings were more abundant under *A. fasciculatum* than under *S. mellifera* shrubs. What effect do terpenes have on soil microbes, especially those producing phytotoxins? Is lack of seed germination the only reason for bare zones around aromatic shrubs? Probably not. It has been suggested that seeds (Kelly and Parker, 1990) and small seedlings (Bartholomew, 1970) can be destroyed by animals. Further, seedlings are stunted by volatiles and subsequently killed by drought (Muller and del Moral, 1971). Shrub removal (by cutting) in maritime chaparral in California had no effect on the germination of seeds of shrubs, herbaceous perennials, or annuals, but it increased the survival of annuals. The exclusion of mammalian herbivores increased seedling survival of all types of plants, and fire increased the seed germination of shrubs and subshrubs but reduced the germination of annuals (Tyler, 1996).

G. Special Factor: Myrmecochory

Myrmecochory is a mutually beneficial plant–ant relationship in which plants (myrmecochores) produce dispersal units (diaspores) that are dispersed by ants. Seeds are the usual dispersal unit, but in some species mericarps, drupes, or nutlets are dispersed by ants (Berg, 1975). Ants are attracted to diaspores (hereafter called seeds) by the presence of an external food body called an elaiosome. The compound in the food body that attracts ants varies, and 1,2-diolein (Marshall *et al.,* 1979; Brew *et al.,* 1989), triolein, olein acid (Brew *et al.,* 1989), and perhaps amino acids (Brew *et al.,* 1989), stimulate collecting behavior. Ants carry seeds to their nests, thereby effecting dispersal, and they eat the elaiosome. Viable, intact seeds then are discarded, either on the soil surface or in subterranean middens (Berg, 1975; Buckley, 1982; Beattie, 1983; Bond and Slingsby, 1983).

Myrmecochorus plants are found in various kinds of habitats ranging from tropical to temperate forests to deserts (Sernander, 1906; Nordhagen, 1959; Buckley, 1982; Beattie, 1983). However, the greatest concentration of myrmecochores is in dry heath and sclerophyllous vegetation in Australia and South Africa (Table 10.8) with about 1500 species (24 families and 87 genera) in Australia (Berg, 1975) and 1300 species (29 families and 78 genera) in South Africa (Bond and Slingsby, 1983). Berg (1975) estimated that 300 species of myrmecochores occur in other parts of the world. Two myrmecochores, *Dendromecon rigida* (Berg, 1966; Bullock 1989) and *Fremontodendron decumbens* (Boyd, 1996), have been reported from the California chapar-

ral, and at least 10 are known from the garrigue of the Mediterranean region (Sernander 1906; Muller, 1933).

Although myrmecochores in Europe, North America, and Asia are herbaceous and grow in mesic habitats, those in xeric matorral vegetation are mostly woody shrubs (Nordhagen, 1959; Berg, 1966, 1975; Bond and Slingsby, 1983). Further, elaiosomes of myrmecochores in mesic sites are soft and collapse soon after seed dispersal, whereas those of myrmecochores in xeric habitats are firm and dry and can persist indefinitely (Berg, 1975). However, elaiosomes persisted on seeds of *Fremontodendron decumbens* for only 2 months; they were consumed by small invertebrates and decomposers (Boyd, 1996). Seed coats adjacent to the elaiosome of *F. decumbens* were three times thicker than elsewhere on the seed, "suggesting an adaptation to discourage ants from consuming seeds" (Boyd, 1996).

The evolutionary convergence of myrmecochory in matorral vegetation of South Africa and Australia has been studied (Milewski and Bond, 1982), and attention has been given to the fact that myrmecochores are more frequent on sites with infertile soil than on those with

TABLE 10.8   Plant Families with Myrmecochorous Species in Matorral Vegetation in Australia (Berg, 1975) and South Africa (Bond and Slingsby, 1983)

Family	Australia	South Africa	Family	Australia	South Africa
Aizoaceae	x	x	Mimosaceae	x	
Apiaceae	x	x	Menyanthaceae	x	x
Asteraceae		x	Papilionaceae	x	x
Balanophoraceae		x	Penaeaceae		x
Bruniaceae		x	Poaceae		x
Caesalpiniaceae	x		Polygalaceae	x	x
Celastraceae	x		Proteaceae		x
Cyperaceae	x	x	Restionaceae		x
Dilleniaceae	x		Retziaceae		x
Epacridaceae	x		Rhamnaceae	x	
Euphorbiaceae	x	x	Rosaceae		x
Fumariaceae		x	Rubiaceae	x	
Geissolomataceae		x	Rutaceae	x	x
Goodeniaceae	x		Santalaceae		x
Gyrostemonaceae	x		Sapindaceae	x	
Haemadoraceae		x	Stilbaceae		x
Hypoxidaceae		x	Sterculiaceae	x	
Iridaceae		x	Thymelaeaceae		x
Juncaceae	x		Tremandraceae	x	
Lamiaceae	x		Violaceae	x	x
Liliaceae	x	x	Zygophyllaceae		x
Malvaceae		x			

fertile soil (Rice and Westoby, 1981; Mossop, 1989; Westoby *et al.*, 1991). Various hypotheses have been developed to explain the selective advantage of myrmecochory, and these have been summarized by Beattie (1983). (1) Ant nests have more mineral nutrients than the surrounding soil, and these nutrients promote seedling establishment. (2) If a seed is inside an ant's nest, it is protected from being eaten by rodents, birds, beetles, and other animals. (3) Seeds of sympatric congeners are moved by ants, thereby reducing interspecific competition of seedlings for microsites. (4) Seeds are placed in nests below the soil where they are protected from fires that occur in the habitat. (5) Seedling survival for each plant species is enhanced at some optimum distance from the mother plant, thus the movement of seeds by ants promotes reproductive success.

Total N and available P were no higher in the soil adjacent to field-established seedlings of the Australian myrmecochores *Grevillea sericea, Eriostemon australasius, Dillwynia retorta, Actinotus minor, Bossiaea ensata,* and *Gompholobium glabratum* than they were in the soil around the nonmyrmecochores *Entolasia stricta, Casuarina distyla,* and *Banksia ericifolia* (Rice and Westoby, 1986). In the fynbos of South Africa, seedlings from ant-dispersed seeds of *Leucospermum conocarpodendron* emerged in sites with less total N and available P than those from seeds that were passively dispersed (Bond and Stock, 1989). Dispersal of Australian *Acacia* spp. whose seeds have arillate appendages may be via birds or ants. Bird-dispersed seeds fall onto nutrient-enriched soils beneath the canopies of trees where birds roost, whereas ant-dispersed seeds are placed in nutrient-enriched soil associated with ant mounds (Davidson and Morton, 1984).

Studies indicate that ants do not accomplish long-distance dispersal. The mean dispersal distance of *Acacia suaveolens* seeds by ants in Australia was 1.2 m, and maximum distances were 2 to 10 m (Andersen, 1988). The mean dispersal distance of *Leucospermum conocarpodendron* seeds in South Africa was 2.5 m, and the maximum distance was 9.8 m (Slingsby and Bond, 1985). Andersen (1988) argues that these dispersal distances are sufficient to be of benefit to the reproductive success of myrmecochores. However, thick clusters of seedlings may result when seeds germinate in ant nests (Slingsby and Bond, 1985; Brits, 1987), causing one to wonder about the consequences of seedling (sibling ?) competition on survival.

In addition to the possible selective advantages of myrmecochory, what are the consequences on seed germination ecology? The two most important effects of myrmecochory on seeds are (1) removal of the elaiosome and (2) movement away from the mother plant. Seeds of myrmecochores in matorral vegetation have

thick, hard seed coats that are smooth and shiny (Berg, 1975; Bond and Slingsby, 1983), and some species belong to families known to have physical dormancy (e.g., Fabaceae, Malvaceae, Rhamnaceae, Sapindaceae, and Sterculiaceae). Thus, removal of the elaiosome by ants probably has no effect on the seed coat. However, comparisons have not been made of seeds before and after they were handled by ants to determine if they differ with respect to imbibition of water. Seeds of *Leucospermum conocarpodendron* with and without the elaiosome germinated to 24 and 22%, respectively, when buried 30 mm deep in sand and watered regularly. Seeds with and without elaiosomes germinated to 12 and 2%, respectively, when sown on the soil surface in fynbos vegetation that was subsequently burned. The explanation given for the higher germination in seeds with than without the elaiosome is that before the fire occurred ants had moved seeds with the elaiosome into their nests, where they were protected (Slingsby and Bond, 1985). Thus, the survival of *L. conocarpodendron* in a fire-prone habitat may depend on seed burial by ants, and ants only bury seeds with an elaiosome (Slingsby and Bond, 1985). Burial by ants also is important for seed survival in the fynbos myrmecochore *Mimetes culcullatus* (Bond and Slingsby, 1984).

Unless a seed buried by ants eventually germinates and the seedling emerges, the genotype is lost from the population. The germination of buried seeds depends on whether conditions required for dormancy break and germination are fulfilled beneath the soil. For example, if seeds with physical dormancy are buried too deep heat from fires will not be sufficient to overcome dormancy (Drake, 1981). The critical depth for burial and subsequent germination of seeds with physical dormancy has not been determined. Buried nondormant seeds germinate only if temperatures are nonlimiting. Seeds of *Leucospermum cordifolium* will germinate at depths of 30–45 mm, and records showed that daily temperature fluctuations were optimal for germination at depths of 10, 20, and 40 mm in burned, but not in unburned, South African fynbos in winter (Brits, 1987). Can temperatures be favorable for germination at soil depths that are too great for seedling emergence? Germination of seeds while they are buried in soil indicates that they do not require light. However, it is not known if seeds of all myrmecochores will germinate in darkness, thus studies are needed on the light : dark requirements for germination. Do seeds of any myrmecochores require darkness for germination?

III. MOIST WARM TEMPERATURE WOODLANDS

Moist warm temperature woodlands occur on all five continents and on the islands of Japan, Tasmania, Malagasy, and New Zealand (Fig. 9.1). This vegetation type is found in North America along the Pacific coast in the northwestern United States and southwestern Canada and in scattered locations from Virginia to Florida in the United States. In Central America, it is found on mountains in Mexico and Guatemala, and in South America in southeastern Brazil and Chile (Walter, 1979). This vegetation zone forms a narrow band around the Mediterranean Sea and then extends across the Middle East to the western and southwestern slopes of the Himalayan Mountains and the southern Indian Hills (Walter, 1979; Ovington, 1983a; Gopal and Meher-Homji, 1983). In southeastern Asia, moist warm temperature woodlands are found in southern China, Vietnam, Thailand, and Burma (Walter, 1979; Ovington, 1983a). Moist warm temperature woodlands are found on the southern Cape coast of the Republic of South Africa, on mountains throughout the continent of Africa (Donald and Theron, 1983), and along the southeast coast of Australia (Walter, 1979; Ashton, 1981).

Although moist warm temperature woodlands occur in regions that may have some frost during winter, the mean daily minimum temperature for the coolest month is above 10°C (Walter, 1979). Rainfall varies from about 600 mm year^{-1} in Australia (Walter, 1979) to 4000 mm year^{-1} in parts of Chile (Veblen et al., 1983). Thus, some moist warm temperature woodlands have a Mediterranean climate with much of the rain falling in winter, whereas others have a humid climate with no significant drought at any time of the year.

Moist warm temperature woodlands are dominated by hardwood trees with evergreen leaves; therefore, they often are called temperate broad-leaved evergreen forests (Ovington, 1983a). Evergreen leaves are hard, waxy, and reduced in size in regions with a Mediterranean climate, and communities are called temperate broad-leaved sclerophyll forests. Evergreen leaves in humid regions, however, are not sclerophyllous, and the forests are referred to as temperate broad-leaved evergreen rain forests (Ovington, 1983a). Temperate broad-leaved sclerophyll forests are found mostly in the Northern Hemisphere and in Australia, whereas temperate broad-leaved evergreen rain forests occur in Asia, southern Japan, South America, Africa, Tasmania, and New Zealand (Ovington, 1983a).

Information concerning vegetation, floristics, and life forms of moist warm temperature woodlands has been compiled from various sources (Debazac, 1983; Donald and Theron, 1983; Gopal and Meher-Homji, 1983; Olson, 1983; Ovington and Pryor, 1983; Satoo, 1983; Veblen et al., 1977; Veblen et al., 1980, 1983; Wardle et al., 1983). According to these authors, the dominant plants in moist warm temperature woodlands are trees that grow to heights of 25–60 m; however, some are only 15–25 m tall. The Lauraceae is the only plant family that occurs (as trees) in moist warm temperature woodlands of Africa, Asia, Australia, India, the Mediterranean region, North America, New Zealand, and South America. Families with trees in four or more of these places include the Cunoniaceae (4) and Monimiaceae (4) in the southern hemisphere and the Cupressaceae (4), Fagaceae (6), Myrtaceae (6), Podocarpaceae (5), and Theaceae (4) in both the northern and the southern hemispheres. Many small trees, and sometimes tree ferns, form a subcanopy, under which there are shrubs and herbs. Moist warm temperature woodlands also have lianas and epiphytes.

A. Trees

1. Types of Dormancy

About half the trees listed in Table 10.9 have seeds that are nondormant at maturity. Seeds of some species with nondormant seeds, including *Dysoxylum spectabile* (Court and Mitchell, 1988) and *Machilus thunbergii* (Lin and Chen, 1995), are very susceptible to desiccation and thus may be recalcitrant. Desiccation reduced germination percentages and rates of *Griselinia littoralis* seeds, suggesting that they might be recalcitrant (Bannister et al., 1996). Further, the fleshy exocarp and fibrous endocarp protect the embryo in this species from desiccation.

A few trees have seeds with morphological or physical dormancy, but many have either morphophysiological or physiological dormancy. A few species with morphophysiological dormancy, including *Podocarpus henkelii* (Dodd and van Staden, 1981) and *Dacrycarpus dacrydioides* (Fountain et al., 1989), also have recalcitrant seeds. No seeds of *D. dacrydioides* germinated until after 110 days at natural (summer) habitat temperatures in New Zealand; most germinated 127–135 days after sowing (Fountain et al., 1989). However, fresh seeds of *Podocarpus henkelii* germinated to 68% after 160 days, scarified (fresh) seeds germinated to 100% in 22 days, and those that were cold stratified and then scarified germinated to 100% in 15 days (Dodd and van Staden, 1981). Thus, *P. henkelii* and *D. dacrydioides* seeds have to remain at high moisture levels while morphophysiological dormancy is broken or else they lose viability. Examples of other species with recalcitrant morphophysiologically dormant seeds are unknown. Seeds of *Ilex opaca* (Bonner, 1974a) and *Magnolia grandiflora* (Olson et al., 1974) in the southeastern United States have morphophysiological dormancy; however, they can be stored dry at 0–5°C, but not at room temperatures, for several years without loss of viability.

TABLE 10.9   Dormancy in Seeds of Trees of Moist Warm Temperature Woodlands

Species	Family[a]	Temperature[a]	L:D[a]	Reference
Nondormant				
Agathis australis	Araucari.	25	—[b]	Barton (1978)
Alnus nepalensis	Betul.	25	—	Thapliyal and Rawat (1991)
A. nitida	Betul.	20	—	Thapliyal and Rawat (1991)
A. rubra	Betul.	24/16	L > D	Schopmeyer (1974b), Bormann (1983)
Araucaria angustifolia	Araucari.	21–29	—	Walters (1974)
A. bidwillii	Araucari.	21–29	—	Walters (1974)
A. columnaris	Araucari.	30/20	—	Scowcroft (1988)
A. cunninghamii	Araucari.	21–29	—	Walters (1974)
A. heterophylla	Araucari.	21–29	—	Walters (1974)
Aristotelia serrata[c]	Elaeocarp.	Summer	L = D	Burrows (1995b)
Atherosperma moschatum	Atherospermat.	25	L = D	Read (1989)
Athrotaxis cupressoides	Taxodi.	17	L = D	Read (1989)
A. selaginoides	Taxodi.	17	L = D	Read (1989)
Banksia aemula	Prote.	28/10	—	Sonia and Heslehurst (1978)
B. integrifolia	Prote.	20	—	Sonia and Heslehurst (1978)
B. serrata	Prote.	20	—	Sonia and Heslehurst (1978)
Carya floridana	Jugland.	g.h.[d]	—	McCarthy and Bailey (1992)
Castanopsis chrysophylla	Fag.	—	—	Hubbard (1974a)
Casuarina glauca	Casurin.	30/20	—	Olson and Petteys (1974)
Cinnamomum cecidodaphne	Laur.	Room	—	Beniwal and Singh (1989)
Cordyline australis[c]	Agav.	Autumn	L = D	Burrows (1995b)
Cryptomeria japonica	Taxodi.	Room	—	Beniwal and Singh (1989)
Dysoxylum spectabile	Meli.	16	L = D	Court and Mitchell (1988)
Eucalyptus amygdalina	Myrt.	Autumn	—	Battaglia (1996)
E. botryoides	Myrt.	25	L = D	Clifford (1953), Devillez (1977)
E. maculata	Myrt.	25	L = D	Grose and Zimmer (1957)
E. microcorys	Myrt.	25	L = D	Clifford (1953), Devillez (1977)
E. rubida	Myrt.	25	L = D	Clifford (1953), Grose and Zimmer (1957)
E. tereticornis	Myrt.	25–35	—	Devillez (1971)
Eucryphia lucida	Eucryphi.	30/20	L = D	Hickey *et al.* (1982), Read (1989)
Ficus microcarpa	Mor.	Room?	L	Thrower (1986)
Fuchsia excorticata[c]	Onagr.	Summer	L > D	Burrows (1995a)
Griselinia littoralis[c]	Griselini.	Autumn	L > D	Burrows (1995a)
Lithocarpus densiflorus	Fag.	Autumn	—	Roy (1974b)
Litsea sp.	Laur.	Autumn	—	Wardle *et al.* (1983)
Machilus thunbergii	Laur.	22	—	Lin and Chen (1995)
Melicytus ramiflorus[c]	Viol.	Summer	L = D	Burrows (1995a)
Metrosideros umbellata	Myrt.	Room?	—	Wardle (1971)
Nothofagus cunninghamii[c]	Fag.	30/20	L = D	Hickey *et al.* (1982), Read (1989)
N. fusca	Fag.	Autumn–winter	—	Hocking (1931–1935)
N. gunnii	Fag.	25	L = D	Read (1989)
N. procera	Fag.	30/20	L = D	Shafiq (1979)
N. truncata	Fag.	Autumn	—	Wardle *et al.* (1983)
Picea polita	Pin.	30/20	L = D	Heit (1968a), Safford (1974)
Quercus lamellosa	Fag.	Room	—	Beniwal and Singh (1989)
Q. serrata	Fag.	Autumn	—	Matsuda (1985)
Sequoia sempervirens[c]	Taxodi.	20/30	—	Boe (1974)
Syzygium jambolana	Myrt.	Room	—	Beniwal and Singh (1989)
Terminalia myriocarpa	Combret.	Room	—	Beniwal and Singh (1989)
Weinmannia racemosa[c]	Cunoni.	Room	—	Wardle and MacRae (1966)
Morphological dormancy				
Manglietia caveana	Magnoli.	Room	—	Beniwal and Singh (1989)
M. insignis	Magnoli.	Room	—	Beniwal and Singh (1989)
Michelia champaca	Magnoli.	Room	—	Beniwal and Singh (1989)
M. doltsopa	Magnoli.	Room	—	Beniwal and Singh (1989)
M. montana	Magnoli.	Room	—	Beniwal and Singh (1989)
Talauma phellocarpa	Magnoli.	Room	—	Beniwal and Singh (1989)

(continues)

TABLE 10.9—*Continued*

Species	Family[a]	Temperature[a]	L:D[a]	Reference
Morphophysiological dormancy				
Carpodetus serratus[c]	Escalloni.	—	L	Burrows (1993)
Dacrydium cupressinum[c]	Podocarp.	Summer	—	Norton *et al.* (1988)
D. dacrycarpus[c]	Podocarp.	—	—	Fountain *et al.* (1989)
Hedycarya arborea[c]	Monimi.	Winter	L = D	Burrows (1995d)
Ilex opaca	Aquifoli.	Spring	—	Bonner (1974a), Hu *et al.* (1979)
Magnolia grandiflora	Magnoli.	Spring	—	Olson *et al.* (1974)
M. virginiana	Magnoli.	18	—	del Tredici (1981)
Manglietia azedarach	Magnoli.	Room	—	Beniwal and Singh (1989)
Phyllocladus aspleniifolius	Podocarp.	g.h.	—	Barker (1995)
Pittosporum eugenioides[c]	Pittospor.	—	—	Burrows (1993)
P. tenuifolium[c]	Pittospor.	—	L = D	Burrows (1993)
Podocarpus henkelii[c]	Podocarp.	25	—	Dodd and van Staden (1981)
Prumnopitys taxifolia[c]	Podocarp.	Winter	—	Burrows (1994a)
Pseudopanax arboreus[c]	Arali.	—	L > D	Burrows (1993)
P. crassifolius[c]	Arali.	25	—	Horrell *et al.* (1989), Burrows (1993)
P. ferox[c]	Arali.	19, 25	—	Horrell *et al.* (1989), Bannister and Bridgeman (1991)
P. simplex[c]	Arali.	25	—	Horrell *et al.* (1989)
Rhopalostylis sapida[c]	Arec.	Summer	L	Burrows (1995d)
Schefflera digitata[c]	Arali.	Winter	L > D	Burrows (1995b)
Physical Dormancy				
Alectryon excelsus[c]	Sapind.	Autumn–spring	L > D	Burrows (1993, 1996c)
Chordospartium muritai	Fab.	20–24	—	Williams *et al.* (1996)
C. stevensonii	Fab.	20–25	L = D	Conner and Conner (1988)
Hoheria angustifolia[c]	Malv.	Winter	L > D	Burrows (1996d)
Plagianthus regius[c]	Malv.	—	—	Burrows (1993)
Sophora microphylla	Fab.	—	—	Burrows (1989)
Physiological dormancy				
Abies firma	Pin.	30/20	—	Franklin (1974)
Acer campbellii	Acer.	Room	—	Beniwal and Singh (1989)
A. macrophyllum	Acer.	30/20	—	Olson and Gabriel (1974)
Arbutus menziesii	Eric.	21	—	Roy (1974a)
Atherosperma moschatum[c]	Antherospermat.	30/20	—	Hickey *et al.* (1982)
Beilschmiedia tawa[c]	Laur.	Autumn–spring	—	Knowles and Beveridge (1982)
Carpodetus serratus	Escalloni.	25	—	Horrell *et al.* (1989)
Cinnamomum camphora	Laur.	Room	—	Beniwal and Singh (1989)
Castanopsis indica	Fag.	Room	—	Beniwal and Singh (1989)
Corynocarpus laevigatus[c]	Corynocarp.	Autumn	L > D	Burrows (1996c), Bannister *et al.* (1996)
Elaeocarpus hookerianus	Elaeocarp.	25	—	Horrell *et al.* (1989)
Machilus globosa	Laur.	Room	—	Beniwal and Singh (1989)
Morus laevigata	Mor.	Room	—	Beniwal and Singh (1989)
Myoporum laetum[c]	Myopor.	Spring	L > D	Burrows (1993, 1996b)
Myrsine australis	Myrsin.	Winter–spring	L > D	Burrows (1996d)
Nothofagus obliqua	Fag.	30/20	L	Shafiq (1979)
Paratrophis microphylla[c]	Mor.	Autumn–winter	L > D	Burrows (1994b)
Pennantia corymbosa	Icacin.	25, Winter	L = D	Horrell *et al.* (1989), Burrows (1995b,d)
Persea borbonia	Laur.	g.h.	—	Nokes (1986)
Picea sitchensis	Pin.	22	L = D	Taylor and Wareing (1979),
		30/20	L > D	Gosling (1988), Gosling and Rigg (1990), Li *et al.* (1994)
Pseudotsuga menziesii	Pin.	25	L = D	Richardson (1959), Allen (1962), Devillez (1973)
Tsuga heterophylla	Pin.	20	—	Ching (1958), Edwards and Olson (1973)
Vitex lucens[c]	Verben.	Room	—	Godley (1971)

[a]See notes for Table 10.1.
[b]Data not available.
[c]Type of dormancy is inferred.
[d]Greenhouse.

Little is known about the breaking of morphophysiological dormancy in seeds of moist warm temperature woodland trees. Seeds of *Ilex opaca* have underdeveloped embryos (Hu *et al.*, 1979) and require cold stratification for germination (Bonner, 1974a). Seeds of *Magnolia grandiflora* have underdeveloped embryos, and a maximum of 35% of them will germinate after 1 month of drying at room temperatures (Evans, 1933). However, up to 90% germination can be obtained following 3–6 months of cold stratification (Olson *et al.*, 1974). Embryos in *M. virginiana* seeds grew after seeds that had been cold stratified at 2°C for 90 days were transferred to 18°C. Embryo growth began immediately, and after 32 days at 18°C the radicle had emerged (del Tredici, 1981). Seeds of *Pseudopanax crassifolius*, *P. arboreus*, and *P. ferox* germinated to only 49, 21, and 18%, respectively, after 3 months of cold stratification (Bannister and Bridgman, 1991). Perhaps seed germination in these species would be increased if a warm stratification pretreatment was given prior to cold stratification.

Although various trees in moist warm temperature woodlands belong to families, including the Anacardiaceae, Malvaceae, Fabaceae, Rhamnaceae, Sapindaceae, and Sterculiaceae (e.g., Ovington, 1983b), known to have seeds with physical dormancy, only a few species are listed under physical dormancy in Table 10.9. Is the low number of species with physical dormancy due to a lack of research or to the presence of nondormancy, instead of physical dormancy, in seeds of some species in these families? *Alectryon excelsus* is listed under physical dormancy because (1) it belongs to the Sapindaceae and (2) Burrows (1993) found that seeds of this species germinated over a 2- to 5-month period. However, Fountain and Outred (1991) suggested that seeds of this species are nondormant. Seeds of *Chordospartium stevensonii* (Fabaceae) have physical dormancy and germinated to 100% following mechanical or acid scarification (Conner and Conner, 1988). Nothing is known about the loss of physical dormancy under field conditions in moist warm temperature woodlands.

Physiological dormancy is broken in seeds of many trees by cold stratification. The cold stratification period required to break dormancy varies: *Persea borbonia*, 30 days (Nokes, 1986); *Abies firma*, 30–60 days (Franklin, 1974); *Acer macrophyllum*, 40–60 days (Olson and Gabriel, 1974); and *Arbutus menziesii*, 60 days (Roy, 1974a). A 4-week cold stratification treatment at 4°C increased the germination rate, but not percentage, of *Eucryphia lucida*, *Nothofagus gunnii*, and *Athrotaxis selaginoides* seeds (Read, 1989). Seeds of *Carpodetus serratus* germinated to high percentages in a greenhouse (25°C) after being outdoors during winter on South Island, New Zealand (Horrell *et al.*, 1989).

The radicle emerges from seeds of *Beilschmiedia tawa* (Knowles and Beveridge, 1982) and *Quercus serrata* (Matsuda, 1985) in autumn, but the epicotyl does not emerge until spring. It is unknown if cold stratification is required to break the dormancy of the epicotyl in seeds of these species. Although seeds of some trees such as *Pseudotsuga menziesii* (Allen, 1958) and *Tsuga heterophylla* (Edwards, 1973) will germinate at maturity, cold stratification increases germination rates. Also, 3 week of cold stratification lowered the temperature requirement for the germination of *P. menziesii* and *Picea sitchensis* seeds (Gosling, 1988).

Not all seeds with physiological dormancy require a cold stratification treatment for germination. Wet, fleshy drupes of *Vitex lucens* produced seedlings during a 2- to 2.5-month period of storage in a plastic bag at room temperatures (Godley, 1971), suggesting that dormancy was broken by high temperatures.

2. Germination Requirements

The mean optimum seed germination temperature of trees in moist warm temperature woodlands is about 24°C (Table 10.9); thus, seeds of most species sown under natural temperatures generally germinate in summer, autumn, or spring. The maximum germination of *Mallotus japonicus* seeds was obtained when night temperatures of 24–32°C alternated with day temperatures of 36°C; however, light filtered through green leaves had no effect on germination (Washitani and Takenaka, 1987). Because (1) more seeds germinated in forest gaps than under the canopy of trees and (2) light filtered through green leaves was not inhibitory to germination, these authors concluded that seeds of *M. japonicus* detect gaps via an increase in soil temperatures.

Light : dark requirements for germination have been determined for seeds of 37 species of trees. Seeds of 4 (11%) species germinate only in light, 11 (30%) germinate to a higher percentage in light than in darkness, and 22 (59%) germinate equally well in light and darkness. The germination of *Alnus rubra* seeds is inhibited by light filtered through green leaves (increased FR : R ratio) and is reduced as irradiance decreases (Bormann, 1983; Haeussler and Tappeiner, 1993).

Further studies on temperature, light : dark, and moisture requirements for germination may help explain the patterns of seedling establishment in forests. Seedlings of *Nothofagus cunninghamii* (Howard, 1973) and *Agathis australis* (Ecroyd, 1982) require old stumps and logs or litter-free mineral soil for high percentages of survival, whereas a covering of litter promotes the establishment of *Dysoxylum spectabile* seedlings (Court and Mitchell, 1988). Does darkness due to the presence of litter prevent the germination of *N. cunninghamii* and *A. australis* seeds or does litter inhibit seedling establishment? Can seeds of *D. spectabile* germinate in darkness and/or seedlings tolerate litter? Seeds of *Eucalyptus obliqua* germinate to higher percentages in darkness than in light, but litter apparently increases germination percentages due to the retention of soil moisture (Facelli and Ladd, 1996).

Scarified seeds of *Chordospartium stevensonii* germinated to 71% at a water potential of −1.5 MPa (Conner and Conner, 1988). Tolerances of seeds of other trees to decreased soil moisture need to be determined. Acorns of *Quercus leucotrichophora* germinated to higher percentages when they were parallel to the soil surface than when they were vertical, with the embryo pointed either up or downward (Lal and Karnataka, 1993). Is the position effect on germination due to water relations?

B. Shrubs

1. Types of Dormancy

Freshly matured seeds of some shrubs are nondormant, but most have dormant seeds, with physiological dormancy being the most common (Table 10.10).

Some moist warm temperature woodland shrubs, including *Pseudopanax* (Araliaceae), *Daphniphyllum* (Daphniphyllaceae), *Xymalos* (Monimiaceae), *Pittosporum* (Pittosporaceae), and *Drimys* (Winteraceae), belong to families whose seeds have underdeveloped embryos; thus, studies on seeds of these shrubs will, no doubt, reveal the presence of morphological or morphophysiological dormancy in additional species. Little is known about breaking morphophysiological dormancy in seeds of shrubs. Seeds of *Pseudowintera colorata* germinated during winter in New Zealand (Burrows, 1995d), suggesting that cold stratification played a role in dormancy break and embryo growth.

The physical dormancy of *Desmodium tiliaefolium* seeds was broken by immersing them in boiling water for 5 min or by mixing them in fresh cow dung for 3 days, presumably where temperatures were relatively high (Maithani *et al.,* 1991). Fruit coats of *Discaria toumatou* can be made permeable by acid scarification (Keogh and Bannister, 1992); however, permeable fruits also require cold stratification before germination occurs (Keogh and Bannister, 1994); GA$_3$ substitutes for cold stratification (Keogh and Bannister, 1992).

Dormancy-breaking requirements have not been investigated in most shrubs whose seeds have physiological dormancy. For many species, all we know is that a period longer than 30 days is required for germination after seeds are sown under natural temperature regimes. Physiological dormancy is broken in seeds of *Acer circi-*

natum by 90–189 days of cold stratification (Olson and Gabriel, 1974), and those of *Erica lusitanica* overwintered outdoors in New Zealand germinated to higher percentages than freshly matured ones (Mather and Williams, 1990).

2. Germination Requirements

When exposed to natural temperature changes, seeds of shrubs mostly germinate in spring, summer, or autumn (Table 10.10), and in the laboratory the optimum germination temperature ranges from 16–18°C in seeds of *Sambucus australia* (Bolli, 1994) to 30/20°C in those of *Acer circinatum* (Olson and Garbiel, 1974). Light : dark requirements have been determined for 20 species. Seeds of two (10%) require light for germination, 12 (60%) germinate to higher percentages in light than in darkness, 5 (25%) germinate equally well in light and darkness, and 1 (5%) germinates to higher percentages in darkness than in light (Table 10.10).

Freshly matured seeds of *Erica lusitanica* did not germinate in darkness, but those collected (from a second site) after overwintering in capsules on the mother plant in the field germinated to 57–92% in darkness (Mather and Williams, 1990). One implication of these data is that dormancy loss occurred while seeds overwintered in the field, and they gained the ability to germinate in darkness. However, because the two sets of seeds came from different sites, those from the second location may have been less dormant at maturity than those from the first one.

Exudates from *Melicytus ramiflorus* seeds inhibited germination of *M. ramiflorus, Coprosma robusta, Fuchsia excorticata,* and *Kunzea ericoides* seeds, but not that of *Sophora microphylla* seeds (Partridge and Wilson, 1990). These authors suggested that germination of *M. ramiflorus* would occur only when there was sufficient rainfall to remove leachate from the vicinity of the seeds. However, this idea has not been tested in the field.

C. Dormancy and Germination of Bamboos

Bamboos are found in the forest understorey, in gaps, and/or in sites where moist warm temperature woodlands have been destroyed in South America (Veblen *et al.,* 1983), India (Gopal and Meher-Homji, 1983), Africa (Donald and Theron, 1983), and Asia (Satoo, 1983). Seeds of *Dendrocalamus hamiltonii* were nondormant at maturity and required only 8–28 days of incubation at room temperature for germination to occur (Beniwal and Singh, 1989). However, most seeds of *Cephalostachyum capitatum* were dormant (probably

TABLE 10.10 Dormancy in Seeds of Shrubs of Moist Warm Temperature Woodlands

Species	Family[a]	Temperature[a]	L:D[a]	Reference
Nondormant				
Coprosma robusta[b]	Rubi.	Autumn	L > D	Burrows (1995b)
Coriaria arborea[b]	Coriari.	Summer	L = D	Burrows (1995c)
Kunzea ericoides[b]	Myrt.	Winter	L > D	Burrows (1996c)
Macropiper excelsum[b]	Piper.	Summer	L = D	Burrows (1995a)
Myrsine divaricata[b]	Myrsin.	Autumn	D > L	Burrows (1996b)
Morphological dormancy				
Sambucus australis	Caprifoli.	16–18	L > D	Bolli (1994)
Morphophysiological dormancy				
Ascarina lucida[b]	Chloranth.	Autumn	L = D	Burrows (1996e)
Pseudowintera colorata[b]	Winter.	Winter	L = D	Burrows (1995d)
Zamia floridana	Cycad.	29/21	—[c]	Dehgan and Schutzman (1983)
Z. furfuraceae	Cycad.	g.h.[d]	—	Smith (1978), Dehgan and Schutzman (1993)
Z. integrifolia	Cycad.	g.h.	—	Smith (1978)
Physical dormancy				
Desmodium tiliaefolium	Fab.	Room?	—	Maithani *et al.* (1991)
Dodonaea viscosa	Sapind.	Spring–summer	L > D	Burrows (1995d)
Physical and physiological dormancy				
Discaria toumatou[b]	Rhamn.	20	—	Keogh and Bannister (1992, 1994)
Physiological dormancy				
Acer circinatum	Acer.	30/20	—	Olson and Gabriel (1974)
Coprosma foetidissima[b]	Rubi.	Winter	L	Burrows (1996d)
C. grandifolia[b]	Rubi.	Autumn	L > D	Burrows (1996e)
C. lucida[b]	Rubi.	—	L	Burrows (1993)
Erica lusitanica	Eric.	25/18	L > D	Mather and Williams (1990)
Gaultheria shallon	Eric.	21	—	Dimock *et al.* (1974)
Leptospermum scoparium[b]	Myrt.	Spring	L > D	Burrell (1965), Mohan *et al.* (1984a,b)
Melicope simplex[b]	Rut.	Spring	L > D	Burrows (1993, 1996b)
Melicytus lanceolatus[b]	Viol.	Winter	L > D	Burrows (1996e)
Myrsine divaricata[b]	Myrsin.	—	L > D	Burrows (1993)
Myrtus obcordata[b]	Myrt.	Winter	L > D	Burrows (1995b)
Solanum laciniatum[b]	Solan.	Winter	L > D	Burrows (1993, 1996e)
Streblus heterophyllus	Mor.	Autumn–winter	L = D	Burrows (1995d)
Urtica ferox[b]	Urtic.	Spring	—	Burrows (1993, 1996b)

[a]See notes for Table 10.1.
[b]Type of dormancy inferred.
[c]Data not available.
[d]Greenhouse.

physiological dormancy), and only 25% of them had germinated after 31–90 days.

Seeds collected from 2-, 3-, 4-, 5-, and 6-year old culms of *Phyllastachys pubescens* in early February in Japan germinated to 40–80% at 18°C with no correlation with respect to culm age (Watanabe *et al.,* 1982). Freshly matured seeds were not tested for germination; therefore, it is not known if seeds were nondormant at maturity or if cold stratification received in the field broke physiological dormancy. Seeds of *Sinarundinaria fangiana* produced in 1983 germinated in field plots in China in the summers of 1984 and 1985 (Taylor and Zisheng, 1988). Seeds retrieved from the field in June, after snow melt, germinated to 65%, indicating that dormancy break occurred during winter. However, freshly-matured seeds were not tested for germination; there-

fore, they may have been nondormant at the time of maturation.

D. Dormancy and Germination of Lianas and Vines

Freshly matured seeds of *Ripogonum scandens* are nondormant, and the taxon belongs to the Liliaceae, which has underdeveloped embryos (Martin, 1946). Therefore, the species is listed under morphological dormancy in Table 10.11. Germination percentages of *R. scandens* decreased in proportion to the duration of presowing drying, with no seeds germinating after 21–22 days of drying (MacMillian, 1972). Seeds of *Calamus flagellum* (Beniwal and Singh, 1989) and *Clematis foetida* (Burrows, 1993, 1996a) are dormant, and the species

TABLE 10.11   Dormancy in Seeds of Lianas and Vines of Moist Warm
Temperature Woodland Areas

Species	Family[a]	Temperature[a]	L:D[a]	Reference
Morphological dormancy				
Ripogonum scandens	Smilac.	—[b]	L > D,	MacMillan (1972),
			L = D	Burrows (1993, 1996a)
Morphophysiological dormancy				
Calamus flagellum	Arec.	Room	—	Beniwal and Singh (1989)
Clematis foetida	Ranuncul.	—	L	Burrows (1993, 1996a)
Physical dormancy				
Calystegia japonica[c]	Convolvul.	12–24	—	Washitani and Masuda (1990)
C. tuguriorum	Convolvul.	—	—	Burrows (1989, 1996a)
Vigna angularis[c]	Fab.	12–36	—	Washitani and Masuda (1990)
Physiological dormancy				
Ampelopsis brevipedunculata	Vit.	28–36	—	Washitani and Masuda (1990)
Freycinetia baueriana	Pandan.	Spring	L	Burrows (1996d)
Morrenia odorata	Asclepiad.	25/20	L > D	Singh and Achhireddy (1984)
Muehlenbeckia australis[d]	Polygon.	—	L > D	Burrows (1993, 1996a)
Parsonsia heterophylla	Apocyn.	—	L > D	Burrows (1993, 1996a)
Passiflora mollissima[c]	Passiflor.	25	—	Williams and Buxton (1995)
P. pinnatistipula[c]	Passiflor.	25	—	Williams and Buxton (1995)
Rubus cissoides	Ros.	—	L > D	Burrows (1993, 1996a)
Tetrapathaea tetrandra	Passiflor.	—	L = D	Burrows (1993, 1996a)

[a]See notes for Table 10.1.
[b]Data not available.
[c]Type of dormancy is inferred.
[d]Some freshly matured seeds were nondormant, but most had physiological dormancy.

belong to families with underdeveloped embryos (see Table 3.3). Thus, both species are listed under morphophysiological dormancy in Table 10.11. Other lianas and vines in moist warm temperature woodlands, including *Popowia* (Annonaceae), *Hedera* (Araliaceae), *Boquila,* and *Stauntonia* (Lardizabalaceae) belonging to families with small embryos, may have morphological or morphophysiological dormancy.

Seeds of a few lianas and vines have physical dormancy, but essentially nothing is known about their dormancy-breaking and germination requirements. The list of lianas and vines with physical dormancy may be extended when seeds/fruits of more species in the Cistaceae, Fabaceae, and Rhamnaceae are investigated.

Physiological dormancy is found in lianas and vines of several families (Table 10.11). Cold stratification promoted the germination of seeds of *Ampelopsis brevipedunculata* (Washitani and Masuda, 1990), whereas those of *Morrenia odorata* germinated after a 5-month period of dry storage at 5°C (Singh and Achhireddy, 1984). Because freshly matured seeds of *M. odorata* were not tested, there is a possibility that they were nondormant before being stored dry at 5°C. Seeds of other species, e.g., *Passiflora mollissima* (Williams and Buxton, 1995), reach maximum germination only after a long period of incubation at relatively high temperatures, indicating that high temperatures are required for loss of dormancy.

Information on temperature and light : dark requirements for the germination of liana and vine seeds is sparse. Available data suggest that high temperatures (20–25°C) and light would promote seed germination, but further studies may reveal species whose nondormant seeds have requirements that differ from these.

E. Herbaceous Species

1. Types of Dormancy

About 20% of the herbaceous species studied thus far have seeds that are nondormant at maturity (Table 10.12). The most common types of dormancy are physiological and morphophysiological, but seeds of some species have morphological or physical dormancy.

Dormancy loss in seeds with morphophysiological dormancy is promoted by either cold stratification or dry storage at room temperatures, depending on the species. Cold stratification increased the germination of *Dioscorea* spp. (Okagami and Kawai, 1982; Okagami, 1986; Washitani and Masuda, 1990), *Polygonatum odoratum, Scilla scilloides,* and *Thalictrum simplex* seeds (Washitani and Masuda, 1990). Dry storage at 25°C promoted dormancy loss in seeds of *Arthropodium cirratum;* cold stratification at 2°C induced seeds of this species into secondary dormancy (Conner and Conner, 1988). No em-

TABLE 10.12 Dormancy in Seeds of Herbaceous Species of Moist Warm
Temperature Woodland Areas

Species	Family[a]	Temperature[a]	L:D[a]	Reference
Nondormant				
Achyranthes fauriei[b]	Amaranth.	28–16[c]	—[d]	Washitani and Masuda (1990)
Artemisia princeps	Aster.	6–24[c]	—	Washitani and Masuda (1990)
Cirsium pendulum	Aster.	20–24[c]	—	Washitani and Masuda (1990)
Coriaria plumosa[e]	Coriari.	Summer–autumn	L > D	Burrows (1995c)
C. sarmentosa[e]	Coriari.	Summer	L > D	Burrows (1995c)
Lactuca indica[e]	Aster.	36–24[c]	—	Washitani and Masuda (1990)
Lobelia boninensis	Campanul.	25/20	L	Mariko and Kachi (1995)
Metaplexis japonica[b]	Asclepid.	36–20[c]	—	Washitani and Masuda (1990)
Miscanthus sacchariflorus	Po.	36–32[c]	—	Washitani and Masuda (1990)
Physalis alkekengi[b]	Solan.	28–32[c]	—	Washitani and Masuda (1990)
Sambucus javanica	Caprifoli.	16–18	L > D	Bolli (1994)
Scoparia dulcis	Scrophulari.	30/20	L	Jain and Singh (1989)
Morphological dormancy				
Hosta albomarginata[e]	Lili.	20–28[c]	—	Washitani and Masuda (1990)
Morphophysiological dormancy				
Arthropodium cirratum[e]	Lili.	12–25	D > L	Conner and Conner (1988)
Dioscorea japonica[e]	Lili.	20, 24–28[c]	D > L	Okagami (1986), Washitani and Masuda (1990)
D. izuensis[e]	Lili.	20	—	Okagami (1986)
D. quinqueloba[e]	Lili.	14	L = D	Okagami and Kawai (1982)
D. septemloba[e]	Lili.	17	D > L	Okagami and Kawai (1982), Okagami (1986)
D. tenuipes[e]	Lili.	23	L = D	Okagami and Kawai (1982), Okagami (1986)
Polygonatum odoratum[e]	Lili.	16	—	Washitani and Masuda (1990)
Scilla scilloides[b,e]	Lili.	16–28[c]	—	Washitani and Masuda (1990)
Thalictrum simplex[b,e]	Ranuncul.	20–28[c]	—	Washitani and Masuda (1990)
Torilis japonica[b,e]	Api.	16–20[c]	—	Washitani and Masuda (1990)
Physical dormancy				
Amphicarpaea edgeworthii[b,e]	Fab.	36–28[c]	—	Washitani and Masuda (1990)
Glycine soja[b,e]	Fab.	36–22[c]	—	Washitani and Masuda (1990)
Physiological dormancy				
Agrimonia pilosa[b,e]	Ros.	20–32[c]	—	Washitani and Masuda (1990)
Amsonia elliptica[e]	Apocyn.	20	—	Washitani and Masuda (1990)
Cirsium japonicum[b]	Aster.	28	—	Washitani and Masuda (1990)

(continues)

bryo growth studies have been done for any species whose seeds have morphophysiological dormancy.

The dormancy-breaking requirements of herbaceous species with physical dormancy have not been investigated. However, germination of the legumes *Amphicarpaea edgeworthii* and *Glycine soja* seeds increased if they (1) were stored dry at 4°C for 4 month and then given 1 month of moist chilling or (2) allowed to overwinter in the field in Japan (Washitani and Masuda, 1990).

Cold stratification breaks physiological dormancy in seeds of many herbaceous species (Mondrus-Engle, 1981; Washitani and Kabaya, 1988; Washitani and Masuda, 1990). Cold stratification decreased the minimum temperature requirement for germination in seeds of 19 species from a moist tall grassland community in warm temperate Japan (Washitani and Masuda, 1990). Seeds of some species with physiological dormancy, especially weedy grasses, come out of dormancy during

dry storage at room temperatures (Popay, 1975). Freshly matured achenes of *Soliva valdiviana* and *S. pterosperma*, which have invaded areas that once supported moist warm temperature woodlands in New Zealand, germinated to about 80 and 60% in light, respectively. However, seeds of both species germinated to about 90% following 4 months of afterripening in dry storage at room temperatures (Johnson and Lovell, 1980). Seeds of the nonweedy species *Coriaria angustissima* sown under natural temperature conditions in New Zealand did not begin to germinate until the end of summer (Burrows, 1995c), indicating that exposure to high temperatures may break dormancy.

2. Germination Requirements

Mean (±SE) optimum germination temperature is 21.2 ± 1.0°C for herbaceous species, 20.0 ± 1.1°C for nonweedy species, and 21.6 ± 1.5°C for weedy species

TABLE 10.12—*Continued*

Species	Family[a]	Temperature[a]	L:D[a]	Reference
Codonopsis ussuriensis	Campanul.	12–16[c]	—	Washitani and Masuda (1990)
Coriaria angustissima[e]	Coriari.	Winter	L > D	Burrows (1995c)
Euphorbia adenochlora	Euphorbi.	8–16[c]	—	Washitani and Masuda (1990)
Galium spurium[b,e]	Rubi.	16–20[c]	—	Washitani and Masuda (1990)
Hordeum glaucum	Po.	15	—	Popay (1975)
H. hystrix	Po.	15, 20	—	Popay (1975)
H. leporinum[b]	Po.	10, 15	—	Popay (1975)
H. marinum[b]	Po.	10, 15	—	Popay (1975)
H. murinum[b]	Po.	10–20	—	Popay (1975)
Humulus japonicus[b]	Mor.	4–16[c]	—	Washitani and Masuda (1990)
Melothria japonica[b]	Cucurbit.	32–24[c]	—	Washitani and Masuda (1990)
Paederia scandens[b]	Rubi.	16–32[c]	—	Washitani and Masuda (1990)
Pennisetum alopecuroides[b]	Po.	16–20[c]	—	Washitani and Masuda (1990)
P. macrourum[b,e]	Po.	30	—	Harradine (1980)
Phragmites australis[b]	Po.	36–32[c]	—	Washitani and Masuda (1990)
Polygonum japonicum[b,e]	Polygon.	12, 36	—	Washitani and Masuda (1990)
P. longisetum[b]	Polygon.	16–24[c]	—	Washitani and Masuda (1990)
P. perfoliatum[b]	Polygon.	4–16[c]	—	Washitani and Masuda (1990)
P. sieboldii[b]	Polygon.	36–24[c]	—	Washitani and Masuda (1990)
Primula sieboldii[e]	Primul.	24/12	—	Washitani and Kabaya (1988)
Rubia akane[e]	Rubi.	8–20[c]	—	Washitani and Masuda (1990)
Rumex acetosa[b]	Polygon.	12	—	Washitani and Masuda (1990)
R. conglomeratus[b]	Polygon.	8–12[c]	—.	Washitani and Masuda (1990)
Soliva pterosperma	Aster.	25–29	L > D	Johnson and Lovell (1980)
S. valdiviana	Aster.	25–29	L > D	Johnson and Lovell (1980)
Stachys japonica[e]	Lami.	28	—	Washitani and Masuda (1990)
Stellaria aquatica[b]	Caryophyll.	4–12[c]	—	Washitani and Masuda (1990)
Zea perennis[e]	Po.	20	—	Mondras-Engle (1987)

[a]See notes for Table 10.1.
[b]Listed as a weed by Holm *et al.* (1979).
[c]Seeds were kept at the first constant temperature for 1–8 days before being exposed to the second constant temperature.
[d]Data not available.
[e]Type of dormancy is inferred.

(Table 10.12). Data on light:dark requirements are available for only 13 species. Seeds of two species require light for germination, 6 germinate to higher percentages in light than in darkness, 2 germinate equally well in light and darkness, and 3 germinate to higher percentages in darkness than in light. In Florida, seeds of *Chrysopsis floridana* germinated to higher percentages in field plots in which litter had been removed than in those in which litter was not removed (Lambert and Menges, 1996). Further, more seeds germinated in disturbed than in nondisturbed soil. These data imply that seeds would germinate to higher percentages in light than in darkness, but experiments to test this idea have not been done.

F. Dormancy and Germination of Epiphytes

Epiphytes of moist warm temperature woodlands include members of the Asteraceae, Bromeliaceae, Cam-

panulaceae, Cistaceae, Gesneriaceae, Grislineaceae, Lardizabalaceae, Liliaceae, Myrtaceae, Orchidaceae, Peperomiaceae, Pittosporaceae, and Rubiaceae (see references given earlier); however, few germination studies have been done. Seeds of *Tillandsia usneoides* germinated readily at room temperatures (Guard and Henry, 1968); thus, they may be nondormant.

By knowing something about the family to which an epiphyte belongs, a little can be inferred about the germination ecology of some species. Members of the Orchidaceae have seeds with undifferentiated embryos (Chapters 3 and 11), thus they have morphological dormancy. The Lardizabalaceae, Liliaceae, and Pittosporaceae have seeds with underdeveloped embryos (Table 3.3); therefore, they would have either morphological or morphophysiological dormancy. Because members of the Cistaceae have impermeable seed coats (Table 3.5), epiphytes belonging to this family may have physical dormancy. Seeds in the remaining families could be either nondormant or have physiological dormancy.

G. Special Factor: Myrmecochory

Forty-one myrmecochorous species in 20 genera and 16 families (Aristolochiaceae, Berberidaceae, Boraginaceae, Caryophyllaceae, Cyperaceae, Euphorbiaceae, Iridaceae, Juncaceae, Lamiaceae, Lardizabalaceae, Lauraceae, Moraceae, Papaveraceae, Polygalaceae, Scrophulariaceae, and Violaceae) have been investigated in the warm temperate zone of southern Japan (Nakanishi, 1988, 1994). Nakanishi (1988) notes that unlike myrmecochorous species of the cool temperate region of Japan, which mostly grow in forests, those in the warm temperate region usually grow " ... in open or semiopen habitats such as at the forest margins, in cracks in stone walls, in grasslands, and at the roadside." Most myrmecochores disperse their seeds in May, June, or July, but the dispersal season is extended until October, November, or December in a few of them. Further, seeds of *Humulus japonicus* and *Melampyrum roseum* are dispersed only from October–December (Nakanishi, 1988). Studies are needed on the effects of dispersal and handling by ants on seed germination.

IV. DECIDUOUS (NEMORAL) FORESTS

Walter's map (Fig. 9.1) shows temperate deciduous forests in eastern Asia, western and central Europe, eastern North America, and southern Chile. Thus, temperate deciduous forests are restricted to the northern hemisphere, except for those occurring in inland valleys and on leeward slopes of Patagonia in southern Chile (Eyre, 1963). In addition to the areas shown on Walter's map, temperate deciduous forests also occur in the Near East from western Turkey to eastern Iran and on the lower and middle elevations of the Caucasus Mountains in Caucasia (Rohring, 1991a,b).

Most plants in temperate deciduous forests have broad, thin, nonsclerophyllous leaves that are shed at the end of the growing season. Leaf shedding is an adaptation of plants for survival in a climate with temperatures below freezing part of the year. However, deciduous trees require a growing season of at least 120 days with a mean temperature above 10°C to produce enough photosynthate for maintenance and growth. If the growing season is less than 120 days, conifers replace deciduous trees (Walter, 1973). Another requirement for the growth of temperate deciduous trees is enough rainfall during the growing season for deciduous leaves to remain turgid and photosynthetically active.

Generally, temperate deciduous forests are not found in continental climates characterized by hot dry summers and cold winters. However, in both North America and Europe, a narrow belt of deciduous forests extends northwest and northeast, respectively, between boreal forests occurring on the northern edge and steppe vegetation on the southern edge (Eyre, 1963).

Temperate deciduous forests have one or two tree strata, a shrub stratum, and a herbaceous stratum (Walter, 1973). Trees, of course, dominate the plant community, and about 65 genera comprise the canopy in North America, Europe, and Asia and about 25 in southern Chile (Rohring, 1991c). A comparison of Rohring's (1991c) lists of important tree genera in the northern and southern hemisphere temperate deciduous forest shows no overlap. The only deciduous trees (seven species) in temperate deciduous forests of Chile belong to the genus *Nothofagus*. Other trees in southern hemisphere forests have evergreen leaves and frequently also occur in temperate broad-leaved evergreen forests (Schmaltz, 1991). Because there is an overlap between species in temperate deciduous and broad-leaved evergreen forests in the southern hemisphere, the discussion of germination of species in temperate deciduous forests will be restricted to temperate deciduous forests of the northern hemisphere. The germination ecology of species in temperate deciduous forests of Chile is included under moist warm temperature woodlands (see Section III).

A. Trees

1. Types of Dormancy

Some canopy (Table 10.13), subcanopy (Table 10.14), and successional (Table 10.15) trees of temperate deciduous forests have nondormant seeds, but they are more common among successional than among canopy or subcanopy trees; nondormant seeds are wind dispersed.

Morphophysiological dormancy (MPD) occurs in both canopy and subcanopy trees, but it is more common in canopy than in subcanopy species. The type of MPD in *Liriodendron tulipifera* seeds has not been determined; however, their cold stratification requirement for germination (Crocker, 1930) suggests that they have either intermediate or deep complex MPD. Based on data of Afanasiev (1937), seeds of *Magnolia acuminata* also appear to have some type of deep complex MPD. The germination requirements of *Fraxinus excelsior* (Villiers and Wareing, 1964) and *F. nigra* (Steinbauer, 1937) indicate that seeds of these species have deep simple MPD. Although Ives (1923) concluded that the rudimentary embryos of *Ilex opaca* were nondormant, Hu *et al.* (1979) showed that they were dormant. Thus, this species presumably has deep simple

TABLE 10.13　Seed Dormancy in Canopy Trees of Temperate Deciduous Forests

Species	Family[a]	Strat.[b]	Temperature[a]	L:D[a]	Reference
Nondormant					
Ulmus americana	Ulm.	0	30/20	L = D	McDermott (1953), Brinkman (1974d)
U. campestris	Ulm.	0	Autumn	—[c]	von Kirchner et al. (1931)
U. davidiana	Ulm.	0	Summer	L	Seiwa (1987)
U. japonica	Ulm.	0	30/20	—	Brinkman (1974d)
U. montana	Ulm.	0	Autumn	—	von Kirchner et al. (1931)
Morphophysiological dormancy					
Fraxinus excelsior	Ole.	365[d]	5	—	Villers and Wareing (1964)
F. nigra	Ole.	60–90[d]	30/20	—	Steinbauer (1937)
		126	20–21	—	Vanstone and LaCroix (1975)
Liriodendron tulipifera	Magnoli.	70	21	L = D	Crocker (1930), Bonner (1967)
Magnolia acuminata	Magnoli.	140	26/5, 26/15	—	Afanasiev (1937)
Physical and physiological dormancy					
Tilia americana	Tili.	90[e]	Spring	—	Spaeth (1932), Barton (1934)
T. cordata	Tili.	Winter[e]	Spring	—	Barton (1934)
T. europaea	Tili.	Winter[e]	Spring	—	Heit (1969b)
T. japonica	Tili.	Winter[e]	Spring	—	Heit (1969b)
T. platyphyllos	Tili.	Winter[e]	Spring	—	Barton (1934), Heit (1969b)
T. tomentosa	Tili.	Winter[e]	Spring	—	Heit (1969b)
Physiological dormancy					
Acer palmatum	Acer.	100–300	23	—	Toth and Garrett (1989)
A. platanoides	Acer.	105	20	—	Pinfield et al. (1974)
A. pseudoplatanus	Acer.	49–63	20	L = D	Pinfield and Stobart (1972)
A. saccharum	Acer.	72	5	—	Webb and Dumbroff (1969)
Aesculus glabra	Hippocastan.	120	30/20	—	Rudolf (1974a)
A. hippocastanum	Hippocastan.	56–84	16	—	Pritchard et al. (1996)
A. indica	Hippocastan.	Winter, 30	Spring, 30	—	Maithani et al. (1990), Bhagat et al. (1993b)
A. octandra	Hippocastan.	120	30/20	—	Rudolf (1974a)
Carya glaba	Jugland.	90–120	30/20	—	Bonner and Maisenhelder (1974a)
C. illinoinensis	Jugland.	0	30–35	—	van Staden and Dimalla (1976), Adams and
		60–90	g.h.[f]	—	Thielges (1978)
C. ovata	Jugland.	30–150	21	—	Barton (1936a)
C. tomentosa	Jugland.	90–150	30/20	—	Bonner and Maisenhelder (1974a)

(continues)

MPD, but more studies are needed to be sure. Seeds of *Kalopanax pictus* require warm plus cold stratification for germination (Rudolf, 1974e), thus they have either deep simple or nondeep complex MPD.

Subcanopy trees in the Fabaceae (*Cladrastis, Gymnocladus*), Rhamnaceae (*Hovenia*), and Sapindaceae (*Koelreuteria*) and successional trees in the Fabaceae (*Gleditsia* and *Robinia*) have seeds with physical dormancy (Tables 10.14 and 10.15). Physical dormancy also occurs in canopy trees in the Tiliaceae (*Tilia*); however, seeds of many *Tilia* spp. have physiological dormancy in addition to physical dormancy (Table 10.13). A combination of physical and physiological dormancy also occurs in the successional species *Cercis canadensis* and *Cotinus obovatus* (Table 10.15).

Physiological dormancy is the most important type of seed dormancy in canopy, subcanopy, and successional trees of temperate deciduous forests. These seeds require 1–4 months of cold stratification, depending on the species, to come out of dormancy. However, cold stratification is not very effective in breaking the dormancy in seeds of some species, including *Carpinus betulus* (Suszka, 1968), *C. caroliniana* (Rudolf and Phipps, 1974), or *Sorbus aria* (Devillez, 1979) unless they first receive a period of warm stratification. The physiological reason(s) why these seeds require warm plus cold stratification is unknown. Warm plus cold stratification is required for dormancy break in seeds with some types of MPD, but those of *Carpinus* and *Sorbus* have fully developed embryos (Martin, 1946).

An interesting variation on the physiological dormancy theme occurs in dispersal units (acorns) of the white oak section (subgenus *Lepidobalanus*) of the genus *Quercus*. The radicle is nondormant in freshly matured acorns of most white oaks, but the plumule (epicotyl) is dormant. Radicles emerge from acorns immediately following dispersal in autumn, and a root system develops in autumn and early winter (Olson,

TABLE 10.13—*Continued*

Species	Family[a]	Strat.[b]	Temperature[c]	L:D[a]	Reference
Celtis occidentalis	Ulm.	60–90	21	—	Barton (1939)
Fagus grandifolia	Fag.	90	30/20	—	Rudolf and Leak (1974)
F. sylvatica	Fag.	84	20	—	Frankland and Wareing (1966)
Fraxinus americana	Ole.	56–140	30/20	—	Steinbauer (1937), Bonner (1975)
F. pennsylvanica	Ole	90	25/18	—	Cram and Lindquist (1982)
Juglans cinerea	Jugland.	60–120	21	—	Barton (1936a)
J. nigra	Jugland.	60–120	21	—	Barton (1936a)
J. regia	Jugland.	57	24/22	—	Martin *et al.* (1969)
Liquidambar styraciflua	Hamamelid.	56	24,16, 32/24	L = D	Bonner and Farmer (1966), Bonner (1967)
		0–32	30/20	—	Wilcox (1968)
Nyssa sessiliflora	Nyss.	—	Room	—	Beniwal and Singh (1989)
Prunus padus	Ros.	210[d]	10–15	—	Suszka (1967)
Quercus alba	Fag.	Epicotyl	18/10	—	Korstian (1927), Farmer (1977)
Q. arkansana	Fag.	120	g.h.[f]	—	Wirges and Yeiser (1984)
Q. borealis	Fag.	90	34	—	Aikman (1934), Godman and Mattson (1980)
Q. falcata	Fag.	32, 60	30/20	—	Jones and Brown (1966), Bonner (1984)
Q. griffithii	Fag.	—	Room	—	Beniwal and Singh (1989)
Q. ilicifolia	Fag.	Epicotyl	—	—	Allen and Farmer (1977)
Q. incana	Fag.	—	Room	—	Beniwal and Singh (1989)
Q. lyrata	Fag.	Epicotyl	—	—	Bonner and Vozzo (1987)
Q. macrocarpa	Fag.	90	g.h.[f]	—	Aikman (1934)
Q. nigra	Fag.	28–49	30/20	—	Peterson (1983), Bonner (1984)
Q. petraea	Fag.	Epicotyl	—	—	Jones (1959)
Q. prinus	Fag.	Epicotyl	18/10	—	Korstian (1927), Farmer (1977)
Q. rubra	Fag.	84	25/15	—	Hopper *et al.* (1985)
Q. robur	Fag.	Epicotyl	—	—	Jones (1959)
Q. velutina	Fag.	90	g.h.[f]	—	Aikman (1934)
Tsuga canadensis	Pin.	0–120	26/5	L = D	Olson *et al.* (1959)

[a]See notes for Table 10.1.
[b]Cold stratification (days).
[c]Data not available.
[d]Seeds require warm stratification prior to cold stratification for germination.
[e]Seeds coats have to become permeable prior to cold stratification.
[f]Greenhouse.

1974b). Cold stratification is required to break dormancy of the shoot, which does not emerge until spring. Thus, white oaks whose acorns have nondormant radicles and dormant shoots are said to have epicotyl dormancy (Farmer, 1977). Low temperatures in late autumn–early winter actually induced dormancy in plumules of *Q. prinus* and *Q. alba* (Farmer, 1977).

Members of the black oak section (subgenus *Erythrobalanus*) of *Quercus* have seeds in which both the radicle and the epicotyl are dormant, and cold stratification is required for germination (Olson, 1974b). However, one black oak, *Q. ilicifolia,* has epicotyl dormancy, and after the radicle emerges, 6 weeks of cold stratification are required to break dormancy of the shoot (Allen and Farmer, 1977).

2. Germination Requirements

Temperatures at which seeds of trees of temperate deciduous forests will germinate range from 5/1°C in *Sorbus aucuparia* (Flemion, 1931) to 32/24°C in *Liquidambar styraciflua* (Bonner and Farmer, 1966). Although seeds of a few species such as *Acer pensylvanicum* (Wilson *et al.,* 1979), *Sorbus* spp. (Devillez, 1979), and *Juniperus virginiana* (Pack, 1921) require temperatures of 2–5°C for germination, those of other species germinate to high percentages at 20 to 30°C (Tables 10.13, 10.14, and 10.15). Cold stratification decreased the minimum temperature for the germination of *Betula populifolia* seeds from 32 or 32/15°C to 15°C (Joseph, 1929) and that of *Alnus glutinosa* seeds from 18 to 7°C (McVean, 1955).

Secondary dormancy has been induced in seeds of *Magnolia acuminata* by prolonged (>12 weeks) periods of cold stratification. However, seeds came out of dormancy again after a second period of cold stratification (Afanasiev, 1937). Secondary dormancy also may be induced in seeds, if they are transferred from low (cold stratification) to high (germination) temperatures before dormancy loss is completed. Seeds of *Rhodotypos kerrioides* (Flemion, 1933), *Cornus florida* (Davis,

TABLE 10.14 Dormancy in Seeds of Subcanopy Trees of Temperate Deciduous Forests

Species	Family[a]	Strat.[b]	Temperature[a]	L:D[a]	Reference
Nondormant					
Betula alnoides	Betul.	0	Room	—[c]	Beniwal and Singh (1989)
Ulmus rubra	Ulm.	0	30/20	—	Brinkman (1974d)
Morphophysiological dormancy					
Aralia spinosa	Arali.	30–60	21	—	Nokes (1986)
Ilex opaca	Aquifoli.	60[d]	—	—	Bonner (1974a)
Kalopanax pictus	Arali.	60–90[d]	Spring	—	Rudolf (1974e)
Magnolia globosa	Magnoli.	—	Room	—	Beniwal and Singh (1989)
Physical dormancy					
Cladrastis lutea	Fab.	0	Room?	—	Heit (1967), Frett and Dirr (1979)
Gymnocladus dioicus[f]	Fab.	0	g.h.[e]	—	Wieshuegel (1935), Yeiser (1983)
Hovenia dulcis	Rhamn.	0	Room?	—	Heit (1967)
Koelreuteria paniculata	Sapind.	0	30/20	—	Rudolf (1974f)
Physiological dormancy					
Acer campestre	Acer.	90	23	—	Toth and Garrett (1989)
A. ginnala	Acer.	60		D > L	Dumbroff and Webb (1970)
A. rubrum	Acer.	90	5–15	L = D	McDermott (1953)
		42–56	30/20	L > D	Farmer and Cunningham (1981), Wang and Haddon (1978)
A. pensylvanicum	Acer.	90–180	5	—	Wilson *et al.* (1979)
Ailanthus altissima	Simaroub.	60–80	30/20	—	Little (1974)
		0	30/20	—	Graves (1990)
Amelanchier canadensis	Ros.	90–120	21	—	Crocker (1930), Crocker and Barton (1931)
A. laevis	Ros.	120	20	—	Hilton *et al.* (1965)
Betula cylinrostachys	Betul.	—	Room	—	Beniwal and Singh (1989)
B. lenta	Betul.	28	15	—	Joseph (1929)
Carpinus betulus	Betul.	84–108[d]	20	—	Suszka (1968)
C. caroliniana	Betul.	180[d]	27/15	—	Rudolf and Phipps (1974)
Cornus florida	Corn.	100–130	15–27	—	Davis (1927)
C. kusa	Corn.	120	21	—	Barton (1939)
Diospyros virginiana	Eben.	60	21	—	Barton (1939)
Halesia carolina	Styrac.	60	21	—	Giersbach and Barton (1932)
Morus rubra	Mor.	90–120		—	Core (1974)
Nyssa sylvatica	Nyss.	60–90	21	—	Barton (1939)
Ostrya virginiana	Betul.	140	25/10	—	Schopmeyer and Leak (1974)
Prunus nepalensis	Ros.	—	Room	—	Beniwal and Singh (1989)
Sciadopitys verticillata	Taxodi.	90	21	—	Barton (1930)
Sorbus aria	Ros.	180[d]	2	L > D	Devillez (1979)
S. aucuparia	Ros.	180	2, 5/1	L > D	Rudolf (1936)
		113	8/2	D	Flemion (1931), Devillez (1979)
S. torminalis	Ros.	180	2	L > D	Devillez (1979)
Ulmus crassifolia	Ulm.	60–90		—	Brinkman (1974d)
U. serotina	Ulm.	60–90		—	Brinkman (1974d)

[a]See notes for Table 10.1.
[b]Cold stratification (days).
[c]Data not available.
[d]Seeds require warm prior to cold stratification for germination.
[e]Seeds were germinated in a greenhouse.
[f]Can sometimes be a canopy tree.

1927), and *Sorbus aucuparia* (Flemion, 1931) transferred from 0–5°C to 15–20°C before dormancy was broken entered secondary dormancy completely. Consequently, a second period of cold stratification was required before seeds would germinate. Nondormant (i.e., cold stratified) seeds of *Juniperus virginiana* were induced into dormancy if they were incubated at temperatures above 15°C (Pack, 1921). High temperatures apparently also induced nondormant seeds of *Malus* spp. into secondary dormancy (Harrington and Hite, 1923).

Light:dark requirements for germination have been determined for only a few tree species. Because foresters have propagated many trees by planting seeds outdoors

TABLE 10.15 Dormancy in Seeds of Successional Trees of Temperate Deciduous Forests

Species	Family[a]	Strat.[b]	Temperature[c]	L:D[a]	Reference
Nondormant					
Acer saccharinum	Acer.	0	30	—[c]	Jones (1920)
Pinus kesiya	Pin.	—	Room	—	Beniwal and Singh (1989)
Platanus occidentalis	Platan.	0	30/20	L > D	McDermott (1953), Bonner (1974b)
Populus ciliata	Salic.	0	20	—	Sah and Singh (1995)
P. deltoides	Salic.	0	32	—	Farmer and Bonner (1967)
P. grandidentata	Salic.	0	29–32	—	Faust (1936)
Salix alba	Salic.	0	20	—	van Splunder *et al.* (1995)
S. caroliniana	Salic.	0	—	L	Brinkman (1974c)
S. nigra	Salic.	0	30/20	L	Brinkman (1974c), van Splunder *et al.* (1995)
S. rigida	Salic.	0	22	L	Brinkman (1974c)
S. triandra	Salic.	0	25	—	van Splunder *et al.* (1995)
S. viminalis	Salic.	0	20	—	van Splunder *et al.* (1995)
Physical dormancy					
Cercis chinensis	Fab.	0	21	—	Barton (1947)
Gleditsia triacanthos[g]	Fab.	0	30/20	L = D	Heit (1967, 1968c)
Robinia pseudoacacia	Fab.	0	g.h.[d]	L = D	Chapman (1936), Wilson (1937)
			20		Roberts and Carpenter (1983), Sadhu and Kaul (1989)
Physical and physiological dormancy					
Cercis canadensis	Fab.	35–56[e]	21	—	Afanasiev (1944)
Cotinus obovatus	Anacardi.	60–80[e]	30/20	—	Rudolf (1974d), Nokes (1986)
Physiological dormancy					
Alnus glutinosa	Betul.	—	20–28	—	McVean (1955)
Betula nigra	Betul.	30–60	24–27	L > D	McDermott (1953), Nokes (1986)
B. populifolia	Betul.	55	32	—	Weiss (1926)
Juglans nigra	Jugland.	210–570	29/22	—	von Althen (1971)
Juniperus virginiana	Cupress.	100, 65	5, 21	—	Pack (1921), Crocker (1930)
Maclura pomifera	Mor.	30	30/20	—	Bonner and Ferguson (1974)
Malus coronaria	Ros.	120	10	—	Crossley (1974a)
M. ioensis	Ros.	60	30/20	—	Crossley (1974a)
M. floribunda	Ros.	60–120	—	—	Crossley (1974a)
Paulownia tomentosa	Scrophulari.	0	30/20	L	Bonner and Burton (1974)
		60	21	—	Carpenter and Smith (1981)
Pinus clausa	Pin.	28	25	—	Outcalt (1991)
P. echinata	Pin.	30–60	21	—	Barton (1928)
P. palustris	Pin.	30–60	21	L > D	Barton (1928), McLemore and Hansbrough (1970)
P. rigida	Pin.	30–90	21	—	Barton (1930)
P. strobus	Pin.	60	21	—	Barton (1930)
P. taeda	Pin.	30, 84	21/18	L > D	Belcher and Jones (1966), Biswas *et al.* (1972)
Prunus avium	Ros.	210[f]	15–20	—	Suszka (1967)
P. cornuta	Ros.	Winter	Spring	—	Bhagat *et al.* (1993a)
P. maackii	Ros.	60[d]	30/20	—	Morgenson (1986)
P. pensylvanica	Ros.	120	20	—	Hilton *et al.* (1965)
		60–90[d]	30/5	—	Laidlaw (1987)
P. serotina	Ros.	30–60	21	—	Barton (1928, 1939)
P. virginiana	Ros.	112–168	27/21	—	Lockley (1980)
Sassafras albidum	Laur.	120	30/20	—	Bonner and Maisenhelder (1974b)
Taxodium distichum	Taxodi.	112	21/18	—	Biswas *et al.* (1972)
Thuja occidentalis	Cupress.	30	21	—	Barton (1939)

[a]See notes for Table 10.1.
[b]Cold stratification (days).
[c]Data not available.
[d]Seeds were germinated in a greenhouse.
[e]Seeds were scarified and then cold stratified.
[f]Seeds require warm stratification prior to cold stratification for germination.
[g]Can sometimes be a canopy tree.

(Schopmeyer, 1974a), it can be assumed that seeds of most trees can germinate in darkness, i.e., beneath soil and/or litter. However, further studies may show that seeds of many species germinate to higher percentages in light than in darkness. Seeds of *Paulownia tomentosa* (Bonner and Burton, 1974) and *Salix* spp. (Brinkman, 1974c) required light for germination, and those of *Platanus occidentalis* and *Betula nigra* (McDermott, 1953) germinated to higher percentages in light than in darkness. Seeds of *Acer pseudoplatanus* (Pinfield and Stobart, 1972), *Ulmus americana* (McDermott, 1953), and *Alnus glutinosa* (McVean, 1955) germinated equally well in light and darkness. Seeds of *Acer ginnala* (Dumbroff and Webb, 1970) germinated to higher percentages in darkness than in light, and those of *Sorbus aucuparia* have been reported to require darkness (Flemion, 1931) for germination or to germinate to higher percentages in light than in darkness (Devillez, 1979).

Cold stratification may change the light requirement for germination. Whereas freshly matured seeds of *Tsuga canadensis* required a daily photoperiod of 8 or 12 hr for about 40% germination, stratified seeds germinated to 66–80% at 0, 8, 12, 14, 16, and 20 hr of light per day (Olson *et al.*, 1959). Freshly matured seeds of *Liquidambar styraciflua* germinated to 0, 0, and 3.4% after 0, 0.25, and 24 hr of light, respectively, but seeds stratified for 4 weeks germinated to 98.6, 98.6, and 100%, respectively (Bonner, 1967). Nonstratified seeds of *Acer rubrum* germinated equally well in light and darkness, whereas cold-stratified seeds germinated to higher percentages in light than in darkness (Wang and Haddon, 1978).

Seeds of some trees are recalcitrant and will die if their moisture content drops below a critical level, e.g., seeds of *Quercus nigra* (Bonner, 1996) and *Q. rubra* (Pritchard, 1991) die if the moisture content drops below 10–15% and 15–20%, respectively. Examples of trees with recalcitrant seeds include *Acer saccharum* (Jones, 1920), *Fagus, Quercus, Aesculus, Castanea, Carya, Corylus,* and *Juglans* (Holmes and Buszewicz, 1958). Seeds of these trees can be stored for several months on a moist substrate at low temperatures (Holmes and Buszewicz, 1958). Imbibed seeds of *Aesculus hippocastanum* come out of dormancy at 2–5°C, and started germinating at this temperature after 21–25 weeks (Pritchard, 1996). However, seeds imbibed at 16°C did not come out of dormancy and were viable after 3 years. Thus, seeds of some recalcitrant species can be stored for more than 1 year if they are imbibed at temperatures too high for dormancy loss to occur. Although seeds of *Populus, Salix, Ulmus, Betula,* and *Alnus* are not recalcitrant, they lose viability shortly after maturity unless they are (1) placed on a moist substrate at temperatures suitable for germination or (2) dried and placed at low tempera-

tures. Seeds of *Fraxinus,* most *Acer* spp., *Liriodendron, Prunus,* and *Pyrus* also can be dried and stored at low temperatures (Holmes and Buszewicz, 1958).

Seeds of many temperate deciduous forests trees are dispersed in late summer and/or autumn. Thus, both recalcitrant and orthodox seeds are subjected to ideal conditions (high moisture and low temperatures) in late autumn and winter for storage (of recalcitrant seeds) and breaking of dormancy. One reason seed banks are rare for many forest trees is that the seeds either germinate or die. Dry, low temperature storage is impossible on the forest floor, so any ungerminated seeds are likely to lose viability the following summer, when they are subjected to high moisture and high temperature conditions.

Soil moisture conditions are important in determining whether seeds germinate. Acorns of *Quercus pedunculata* and *Q. sessiliflora* will not germinate on moist soil unless the micropyle is pointed downward (Watt, 1919). Excessive evaporation from the radicle when the micropyle is turned upward probably prevents germination. An increase in moisture stress from 0 to -1.0 MPa decreased the germination percentage of *Populus deltoides* seeds incubated at 15, 21, 27, 32, and 38°C, and no germination occurred at -1.5 MPa except at 32 and 38°C (Farmer and Bonner, 1967). Cold-stratified seeds of *Liquidambar styraciflua* incubated at 30–38°C germinated to about 100% at moisture stresses up to -0.5 MPa and to 80% or more at -1.0 MPa; no germination occurred at -1.5 MPa (Bonner and Farmer, 1966). Nonstratified seeds germinated to 100% only at 0 and -0.1 MPa (Bonner and Farmer, 1966).

Soils in low-lying areas may be water-logged at the time of seed germination, but the effect of excess moisture varies with the species. Acorns of *Quercus pedunculata* and *Q. sessiliflora* did not germinate when immersed in water or when the water table was 2–3 cm below the soil surface (Watt, 1923). Seeds of *Nyssa aquatica* (Shunk, 1939), *N. sylvatica* (DeBell and Naylor, 1972), *Taxodium distichum,* and *Liquidambar styraciflua* (DuBarry, 1963) germinated to higher percentages under nonflooded than under flooded conditions. However, seeds of *Fraxinus pensylvanica, F. caroliniana, F. americana,* and *Platanus occidentalis* germinated to higher percentages in flooded than in nonflooded conditions (DuBarry, 1963).

B. Dormancy and Germination of Bamboos

Seeds of *Dendrocalamus sikkimensis* were nondormant and germinated within 9–31 days at room temperatures (Beniwal and Singh, 1989).

C. Shrubs

1. Types of Dormancy

Freshly matured seeds of a few shrubs, including *Kalmia* spp. (Jaynes, 1971), *Rhododendron maximum* (Olson, 1974a), *Viburnum nudum,* and *V. scabrellum* (Giersbach, 1937a), are nondormant (Table 10.16). However, because seeds do not mature until autumn, germination is prevented at the time of seed dispersal in late autumn or early winter because temperatures in the habitat are below those required for germination. Thus, germination is delayed until temperatures increase in spring. It is not known if cold stratification lowers the temperature requirement for germination in any of these species. If further investigations show that the temperature requirement for germination is lowered by cold stratification, then the freshly matured seeds are conditionally dormant, i.e., they have nondeep physiological dormancy.

Morphological dormancy appears to be rare in seeds of temperate deciduous forest shrubs, but several species have seeds with morphophysiological dormancy (Table 10.16). Cold stratification is required to break dormancy in seeds with MPD, but warm stratification also is required, especially for members of the Caprifoliaceae and Taxaceae. *Viburnum* spp. have seeds with deep simple epicotyl MPD and thus require high temperatures for radicle emergence and 2–3 months of cold stratification for the production of shoots (Giersbach, 1937a).

Seeds/fruits of some shrubs have physical dormancy (Table 10.16), but little is known about the loss of dormancy in the habitat. In seeds with both physical and physiological dormancy, physical dormancy presumably is broken during summer, and physiological dormancy is broken during winter. Therefore, seeds germinate in spring, e.g., *Rhus aromatica* (Li, unpublished results).

Seeds of many shrubs have physiological dormancy and require cold stratification for various periods of time, depending on the species, to become nondormant (Table 10.16). However, high temperature (60, 80, and 90°C) and humidity (100%) conditions break dormancy in seeds of *Kalmia hirsuta* (Jaynes, 1968).

2. Germination Requirements

Seeds of a few shrubs, including *Rhodotypos kerrioides* (Flemion, 1933), *Rosa multiflora* (Crocker and Barton, 1931), and *Symphoricarpos racemosus* (Flemion, 1934), germinate at 5°C, but those of most species germinate well at 20–30°C (Table 10.16). Thus, seeds of most species would come out of dormancy in response to cold stratification during winter and germinate at the beginning of the growing season. Light:dark requirements for germination have been determined for eight species. Seeds of two species required light, two germinated equally well in light and darkness, two germinated to higher percentages in light than darkness, and two germinated to higher percentages in darkness than in light.

D. Dormancy and Germination of Vines

Nondormancy and morphological dormancy are known in seeds of vines, but those of most species have either morphophysiological or physiological dormancy, with physiological dormancy being the most common type (Table 10.17). Physical dormancy has not been reported. Both morphophysiological and physiological dormancy in seeds of vines are broken by cold stratification, and seeds germinate at 20–30°C. Thus, dormancy is broken in response to cold stratification in the habitat during winter, and seeds germinate in spring. Seeds of *Vitis vinifera, Smilax glauca,* and *S. rotundifolia* germinate to higher percentages in light than in darkness, whereas those of *Dioscorea nipponica* and *Hedera helix* germinate equally well in light and darkness (Table 10.17). Light:dark requirements for seed germination of other vines need to be determined.

E. Dormancy and Germination of Herbaceous Woodland Species

Seeds of some woodland herbs are nondormant at maturity (Table 10.18) and thus germinate to 90–100% over a range of daily thermoperiods (Grime *et al.,* 1981; Baskin and Baskin, 1984c, 1985d). Seeds of these species either require light for germination, or they germinate to higher percentages in light than in darkness.

More studies need to be done to determine when seeds that are nondormant at maturity germinate in the forests. Seeds of *Campanula americana* dispersed in late summer–early autumn can germinate immediately; however, if seeds are dispersed in winter, germination is delayed until spring (Baskin and Baskin, 1984c). Achenes of *Geum canadense* dispersed as soon as they mature in early autumn (mid-September) can germinate immediately. Achenes dispersed in late autumn, winter, or spring can germinate in spring. If achenes remain undispersed during winter they lose the ability to germinate at high, but not at low, temperatures. Thus, achenes dispersed in spring can germinate in spring, but those dispersed in summer cannot germinate until the following autumn or spring (Baskin and Baskin, 1985d).

Morphological dormancy occurs in seeds of *Isopyrum biternatum* (Baskin and Baskin, 1986a), and it appears

TABLE 10.16 Dormancy in Seeds of Shrubs of Temperature Deciduous Forests

Species	Family[a]	Strat.[b]	Temperature[a]	L:D[a]	Reference
Nondormant					
Kalmia angustifolia	Eric.	0	22	—[c]	Jaynes (1971)
K. microphylla	Eric.	0	22	—	Jaynes (1971)
Rhododendron maximum	Eric.	0	26/21	L	Olson (1974a)
Woodfordia fruticosa	Lythr.	0	25/20	—	Bhagat et al. (1992b)
Morphological dormancy					
Viburnum nudum	Caprifoli.	0	26/18	—	Giersbach (1937a)
V. scabrellum	Caprifoli.	0	26/18	—	Giersbach (1937a)
Morphophysiological dormancy					
Aralia hispida	Arali.	90–120	21	—	Barton (1939)
Asimina triloba	Annon.	100	21	—	Barton (1939)
Berberis thunbergii	Berberid.	90	24/13	—	Davis (1927), Rudolf (1974b)
Lonicera maackii	Caprifoli.	42	25/15	L > D	Luken and Goessling (1995), Hidayati et al. (unpublished results)
L. oblongifolia	Caprifoli.	60–90[d]	30/20	—	Brinkman (1974a)
Sambucus canadensis	Caprifoli.	60[d]	20	—	Heit (1969b)
Symphoricarpos orbiculatus	Caprifoli.	150[d]	10	—	Flemion and Parker (1942)
S. racemosus	Caprifoli.	180[d]	5	—	Flemion (1934)
Taxus baccata	Tax.	60–120[d]	16/13	—	Heit (1969a)
T. cuspidata	Tax.	60–120[d]	16/13	—	Heit (1969a)
Viburnum acerifolium	Caprifoli.	60[d]	30/20	—	Giersbach (1937a)
V. dentatum	Caprifoli.	75[d]	30/20	—	Giersbach (1937a)
V. dilatatum	Caprifoli.	90[d]	30/20	—	Giersbach (1937a)
V. lentago	Caprifoli.	60–90[d]	30/20	—	Giersbach (1937a)
V. opulus	Caprifoli.	60–90[d]	30/20	—	Giersbach (1937a)
V. prunifolia	Caprifoli.	60–90[d]	30/20	—	Giersbach (1937a)
V. rufidulum	Caprifoli.	60–90[d]	30/20	—	Giersbach (1937a)
V. trilobum	Caprifoli.	60–90[d]	20	—	Knowles and Zalik (1958)
Physical dormancy					
Amorpha fruticosa	Fab.	0	30/20	—	Hutton and Porter (1937)
Dodonaea viscosa	Sapind.	0	20	—	Bhagat and Singh (1994a)
Grewia optiva	Tili.	0	Room?	—	Lata and Verma (1993)
Indigofera gerardiana	Fab.	0	20	—	Bhagat and Singh (1994a)
Rhus copallina	Anacardi.	0	20	—	Heit (1969b)
R. glabra	Anacardi.	0	20	—	Heit (1969b), Farmer et al. (1982)
R. typhina	Anacardi.	0	20	—	Heit (1969b), Farmer et al. (1982)
Physical and physiological dormancy					
Ceanothus americana	Rhamn.	60–90[e]	Spring	—	Nokes (1986)
Cotinus coggygria	Anacardi.	60[e]	Spring	—	Heit (1967, 1968b)
Rhus aromatica	Anacardi.	30[e]	Spring	—	Heit (1968b), Li (unpublished results)
R. trilobata	Anacardi.	40[e]	Spring	—	Heit (1968b)
Physiological dormancy					
Alnus rugosa	Betul.	60	24–27	L = D	McDermott (1953)
Arctostaphylos uva-ursi	Eric.	120[f]	10	—	Giersbach (1934a)
Bumelia lanuginosa	Sapot.	30–60[e]	30/20	—	Bonner and Schmidtling (1974), Nokes (1986)
Callicarpa americana	Verben.	60	Room?	—	Nokes (1986)
Chionanthus virginica	Ole.	60–90[d]	21	—	Nokes (1986)
Clethra alnifolia	Clethr.	30	Spring	—	Nokes (1986)
Cornus amomum	Corn.	84	30/10	D > L	Allen and Farmer (1977)
C. obliqua	Corn.	84	—	D > L	Allen and Farmer (1972)
C. stolonifera	Corn.	90–120	30/10	—	Heit (1968b)
Corylus avellana	Coryl.	84	20	—	Frankland and Wareing (1966)
Cotoneaster dielsiana	Ros.	120	10	—	Giersbach (1934b)
C. divaricata	Ros.	120[d]	10	—	Giersbach (1934b)
C. horizontalis	Ros.	120[d]	21	—	Giersbach (1934b)
C. zabelii	Ros.	120	10	—	Giersbach (1934b)

(continues)

TABLE 10.16—*Continued*

Species	Family[a]	Strat.[b]	Temperature[a]	L:D[a]	Reference
Crataegus cordata	Ros.	135	21	—	Flemion (1938)
C. coccinea	Ros.	135	21	—	Flemion (1938)
C. flava	Ros.	135[d]	21	—	Flemion (1938)
C. mollis	Ros.	96	21	L = D	Davis and Rose (1912)
C. punctata	Ros.	135[d]	21	—	Flemion (1938)
C. crus-galli	Ros.	135[d]	21	—	Flemion (1938)
C. rotundifolia	Ros.	135[d]	21	—	Flemion (1938)
Elaeagnus umbellata	Elaeagn.	90–112	30/10, 20/10	—	Heit (1968b), Fowler and Fowler (1987)
Erica tetralix	Eric.	Winter	22/12	L	Pons (1989)
Euonymus americana	Celastr.	139	21	—	Rudolf (1974i)
E. atropurpureus	Celastr.	60	30/20	—	Rudolf (1974i)
E. europaeus	Celastr.	90[d]	0–20	—	Nikolaeva (1969)
E. maackii	Celastr.	75[d]	0–20	—	Nikolaeva (1969)
E. sacrosacta	Celastr.	180	0–20	—	Nikolaeva (1969)
E. verrucosa	Celastr.	105[d]	0–20	—	Nikolaeva (1969)
Fothergilla major	Hamamelid.	87–159	27/21	—	Farmer (1978)
Gaultheria procumbens	Eric.	30–75	21	—	Barton (1939), Dimrock *et al.* (1974)
Hamamelis virginiana	Hamamelid.	120–180	Spring	—	Heit (1968b)
Kalmia cuneata	Eric.	56	21–27	—	Jaynes (1971)
K. hirsuta	Eric.	H[g]	22	—	Jaynes (1968)
K. latifolia	Eric.	56	21–27	—	Jaynes (1971)
Ligustrum spp.	Ole.	60	30/10	—	Heit (1968b)
Lindera benzoin	Laur.	90–120[d]	21	—	Schroeder (1935)
Myrica carolinensis	Myric.	120	21	—	Barton (1932)
M. pensylvanicum	Myric.	90–150[d]	30/10	—	Heit (1968b)
Prunus americana	Ros.	150	21	—	Giersbach and Crocker (1932)
P. fruticosa	Ros.	120[d]	30/20	—	Morgenson (1986)
P. virginiana	Ros.	120–160	25/10	—	Grisez (1974)
Ptelea isophylla	Rut.	61–120	21	—	Schroeder (1937)
P. serrata	Rut.	150	21	—	Schroeder (1937)
P. trifoliata	Rut.	181	17, 22/16	—	McLeod and Murphy (1977)
Pyrus arbutifolia	Ros.	90	21	—	Crocker and Barton (1931)
Rhamnus carolianiana	Rhamn.	30	7–10	—	Nokes (1986)
R. carthartica	Rhman.	40–60	20, 30/20	—	Heit (1968b)
R. frangula	Rhamn.	40–60	20, 30/20	—	Heit (1968b)
Rhodotypos kerrioides	Ros.	90[d]	5	—	Flemion (1933)
R. scandens	Ros.	90[d]	21	—	Flemion (1933)
Ribes cynosbati	Grossulari.	90	25/10	—	Fivaz (1931)
R. grossularia	Grossulari.	90–120	21	—	Barton (1939)
R. rotundifolium	Grossulari.	120	25/10	—	Fivaz (1931)
Rosa multiflora	Ros.	50–120	5, 21	—	Crocker and Barton (1931), Barton (1939)
R. rugosa	Ros.	56[d]	20	—	Svejda and Poapst (1972), Svejda (1972)
Rubus spp.[h]	Ros.	90–120[d]	23–21	L > D	Scott and Ink (1957), Scott and Draper (1967)
Rubus allegheniensis	Ros.	90[d]	30/20	—	Brinkman (1974b)
R. idaeus	Ros.	120	30/20, 20	—	Brinkman (1974b), Nesme (1985)
R. odoratus	Ros.	120	30/20	—	Brinkman (1974b)
Zanthoxylum americanum	Ruta.	120	30/20	—	Bonner (1974c)

[a]See notes for Table 10.1.
[b]Cold stratification (days).
[c]No data available.
[d]Warm prior to cold stratification promotes loss of dormancy.
[e]Seeds coats have to be made permeable before seeds can be cold stratified effectively.
[f]Acid scarification plus warm stratification prior to cold stratification increased germination.
[g]Heat treatments broke dormancy.
[h]Seeds of blackberries and black and red raspberries were studied, but names of species were not given.

TABLE 10.17 Dormancy in Seeds of Vines of Temperate Deciduous Forests

Species	Family[a]	Strat.[b]	Temperature[a]	L:D[a]	Reference
Nondormant					
Dioscorea nipponica	Dioscore.	0	20	L = D	Okagami and Kawai (1982)
Morphological dormancy					
Hedera helix	Arali.	0	29/6	L = D	Grime *et al.* (1981)
Morphophysiological dormancy					
Clematis montana	Ranuncul.	30	Room?	—[c]	Qadir and Lodhi (1971)
C. virginiana	Ranuncul.	60	30/20	—	Rudolf (1974c)
Lonicera hirsuta	Caprifoli.	60[d]	30/20	—	Brinkman (1974a)
Schisandra chinensis	Schisandr.	—	—	—	Nikolaeva (1969)
Smilax glauca	Smilac.	150	22	L > D	Pogge and Bearce (1989)
S. rotundifolia	Smilac.	150	22	L > D	Pogge and Bearce (1989)
Physiological dormancy					
Ampelopsis sp.	Vit.	30–60	Spring	—	Nokes (1986)
Brunnichia ovata	Polygon.	?[e]	35	—	Shaw *et al.* (1991)
Campsis radicans	Bignoni.	60	30/20	—	Bonner (1974d)
Celastrus scandens	Celastr.	90	30/20	—	Heit (1968b)
Mitchella repens	Rubi.	150–180	21	—	Brinkman and Erdmann (1974)
Parthenocissus quinquefolia	Vit.	60	30/20	—	Gill and Pogge (1974)
P. tricuspidata	Vit.	60	30/20	—	Gill and Pogge (1974)
Tamus communis	Dioscore.	Winter[d]	Spring	—	Burkill (1937)
Vitis aestivalis	Vit.	90	21	—	Flemion (1937)
V. bicolor	Vit.	135	21	—	Flemion (1937)
V. vinifera	Vit.	60	35/15	L > D	Pereira and Maeda (1986)

[a]See notes for Table 10.1.
[b]Cold stratification (days).
[c]Data not available.
[d]Seeds require warm stratification prior to cold stratification for germination.
[e]Seeds stored dry at 5°C for a nonspecified period of time.

to be the type found in seeds of *Heloniopsis orientalis,* a member of the Liliaceae, which has seeds with linear embryos (Martin, 1946). Seeds of *H. orientalis* are dispersed in mid- to late May (early summer), and germination is completed by early July (Takahashi, 1984). Thus, no dormancy-breaking treatment is required, and embryo growth and germination occur as soon as seeds are placed under suitable temperature and moisture conditions.

Morphophysiological dormancy is very common among woodland herbs (Table 10.18). Seeds of winter annuals [e.g., *Corydalis favula* (Baskin and Baskin, 1994) and *Chaerophyllum procumbens* (Baskin and Baskin, unpublished results)] require only warm stratification to break embryo dormancy. Thus, dormancy is broken in summer, and embryo growth and germination occur in autumn. Seeds of perennials and biennials (monocarpic perennials) with MPD may require warm stratification for loss of dormancy, but they all require cold stratification. Thus, germination occurs in spring, or is completed in spring, in those species whose seeds have deep simple epicotyl MPD. Because seeds of most species germinate to high percentages at low temperatures, they germinate in late winter or early spring in

the field. The optimum temperature regime for the germination of *Perideridia americana* seeds in the laboratory was 25/15°C, but the peak of germination for seeds exposed to natural temperatures was in late winter (early March) when maximum and minimum daily temperatures were 16.4 and 4.3°C, respectively (Baskin and Baskin, 1993). Thus, many seeds of this, and perhaps other, species may germinate in the field in late winter–early spring before temperatures have increased to the optimum for the species.

Light:dark requirements for germination have been determined for only nine species whose seeds have MPD (Table 10.18). Seeds of four species germinated to higher percentages in light than in darkness, four germinated equally well in light and darkness, and one germinated to a higher percentage in darkness than in light. Obviously, a general statement concerning the relative importance of light in the germination ecology of species with MPD should be delayed until data on additional species are available.

Physical dormancy is uncommon among seeds of herbaceous woodland species, and the only reports of it are in two species of *Geranium* (Table 10.18). Detailed studies have not been undertaken to determine how

TABLE 10.18 Dormancy in Seeds of Herbaceous Species of Temperate Deciduous Forests

Species	Family[a]	LC[b]	Tr[c]	Temperature[d]	L:D[d]	Reference
Nondormant						
Apropyron caninum	Po.	P		20/15	L > D	Grime *et al.* (1981)
Campanula americana	Campanul.	B,A		25/15	L > D	Baskin and Baskin (1984c)
Digitalis purpurea	Scrophulari.	B		20/15	L > D	Grime *et al.* (1981)
Epilobium montanum	Onagr.	P		20/15	L	Grime *et al.* (1981)
Geum canadense	Ros.	P		20/10	L	Baskin and Baskin (1985d)
Luzula sylvaticum	Junc.	P		20/15	L	Grime *et al.* (1981)
Mycelis muralis	Aster.	P		20/15	L > D	Grime *et al.* (1981)
Morphological dormancy						
Heloniopsis orientalis	Lili.	P[d]		15, 20	L > D	Takahashi (1984)
Isopyrum biternatum	Ranuncul.	P		15/6	—[e]	Baskin and Baskin (1986a)
Morphophysiological dormancy						
Actaea pachypoda	Ranuncul.	P	W + C	15/6	—	Baskin and Baskin (1988)
Allium burdickii	Lili.	P	W + C	25/15	—	Baskin and Baskin (1988)
A. tricoccum	Lili.	P	W + C	20/10	—	Baskin and Baskin (unpublished results)
A. ursinum	Lili.	P	W + C	10	—	Ernst (1979)
Anemone nemorosa	Ranuncul.	P	C	11–17	L = D	Grime *et al.* (1981), Shirreffs (1985)
Arisaema dracontium	Ar.	P	C + W + C	10–15	—	Pickett (1913)
A. triphyllum	Ar.	P	C + W + C	10–15	—	Pickett (1913)
Asarum canadense	Aristolochi.	P	W + C	20/10	—	Barton (1944), Baskin and Baskin (1986c)
Arum maculatum	Ar.	P	W? + C	—	—	Sowter (1949)
Caulophyllum thalictroides	Berberid.	P	C + W + C	5	—	Barton (1944)
Chaerophyllum procumbens	Api.	A	W	20/10	L > D	Baskin and Baskin (unpublished results)
Cimicufuga racemosa	Ranuncul.	P	W + C	15/6	—	Baskin and Baskin (1985a)
Clintonia borealis	Lili.	P	W? + C	20/10	—	Baskin and Baskin (1988, unpublished results)
Convallaria majalis	Lili.	P	C + W + C	5	—	Barton and Schroeder (1941)
Corydalis flavula	Fumari.	A	W	20/10	—	Baskin and Baskin (1994)
Cryptotaenia canadensis	Api.	P	C	15/6	L > D	Baskin and Baskin (1988a)
Delphinium tricorne	Ranuncul.	P	C	5	L > D	Baskin and Baskin (1994b)
Disporum lanuginosum	Lili.	P	W? + C	15/6	—	Baskin and Baskin (1988)
Eranthis hiemalis	Ranuncul.	P	W + C	20–25	—	Frost-Christensen (1974)
Erigenia bulbosa	Api.	P	W + C	15/6	—	Baskin and Baskin (1988, unpublished results)
Erythronium albidum	Lili.	P	W + C	15/6	—	Baskin and Baskin (1985b)
E. americanum	Lili.	P	W + C	15/6	—	Baskin and Baskin (1985b)
E. rostratum	Lili.	P	W + C	15/6	—	Baskin and Baskin (unpublished results)
Frasera caroliniensis	Gentian.	B	C[f]	5	—	Threadgill *et al.* (1981), Baskin and Baskin (1986b)
Hepatica acutiloba	Ranuncul.	P	W + C	15/6	—	Baskin and Baskin (1985a)
Hyacinthoides non-scripta	Lili.	P	W	11	—	Thompson and Cox (1978)
Hydrophyllum appendiculatum	Hydrophyll.	P	W + C	5	—	Baskin and Baskin (1985c)
H. macrophyllum	Hydrophyll.	P	W + C	15/6	—	Baskin and Baskin (1983a)
Jeffersonia diphylla	Berberid.	P	W + C	5	—	Baskin and Baskin (1989a)
J. dubia	Berberid.	P	W + C	—	—	Grushvitzky (1967)
Lilium canadense	Lili.	P	W + C	20	—	Barton (1936b)
L. japonicum	Lili.	P	W + C	20	—	Barton (1936b)
Myrrhis odorata	Api.	P	C	5	—	Lhotska (1977)
Osmorhiza claytonii	Api.	P	W + C	5	—	Baskin and Baskin (1991)
O. longistylis	Api.	P	W + C	15/6	—	Baskin and Baskin (1984a)
Paeonia suffruticosa	Paeoni.	P	W + C	13	—	Barton (1933)
Panax ginseng	Arali.	P	W + C	—	—	Nikolaeva (1969)

(continues)

TABLE 10.18—*Continued*

Species	Family[a]	LC[b]	Tr[c]	Temperature[d]	L:D[e]	Reference
P. quinquefolia	Arali.	P	W + C	5	—	Stoltz and Snyder (1985)
Perideridia americana	Api.	P	C	25/15	L = D	Baskin and Baskin (1993)
Polygonatum biflorum	Lili.	P	C + W + C	20/10	—	Baskin and Baskin (1988, unpublished results)
P. commutatum	Lili.	P	C + W + C	25	—	Barton (1944)
Ranunculus ficaria	Ranuncul.	P	W + C	11/9	L = D	Taylor and Markham (1978)
Sanguinaria canadensis	Papaver.	P	W + C, C + W + C	20	—	Barton (1944)
Sanicula canadensis	Api.	B	C	20/10	D > L	Baskin and Baskin (unpublished results)
Smilacina racemosa	Lili.	P	C + W + C	5	—	Barton and Schroeder (1941)
Stylophorum diphyllum	Papaver.	P	C	15/6	L > D	Baskin and Baskin (1984b)
Thaspium pinnatifidum	Api.	B	C	20/10	L = D	Baskin et al. (1992b)
Tricyrtis flava	Lili.	P[d]	C	10	—	Takahashi (1981)
T. macranthopsis	Lili.	P[d]	C	10	—	Takahashi (1981)
T. nana	Lili.	P[d]	C	10	—	Takahashi (1981)
T. ohsumiensis	Lili.	P[d]	C	10	—	Takahashi (1981)
T. perfoliata	Lili.	P[d]	C	10	—	Takahashi (1981)
T. erectum	Lili.	P	C + W + C	5	—	Barton (1944)
T. flexipes	Lili.	P	W + C	25/15	—	Baskin and Baskin (1988, unpublished results)
T. grandiflorum	Lili.	P	C + W + C	21	—	Barton (1994)
T. sessile	Lili.	P	W + C	20/10	—	Baskin and Baskin (1988, unpublished results)
Trollius ledebouri	Ranuncul.	P	—[f]	20	—	Hepher and Roberts (1995)
Uvularia perfoliata	Lili.	P	C + W + C	Spring	—	Whigham (1974)
Physical dormancy						
Geranium maculatum	Gerani.	P	C	25	—	Martin (1965)
G. robertianum	Gerani.	A-B	S	—	L = D	Slade and Causton (1979)
Physiological dormancy						
Alliaria petiolata	Brassic.	B	C	15/6	L = D	Baskin and Baskin (1992a)
Atropa belladonna	Solan.	P	C	20/15	L > D	Grime et al. (1981)
Brachypodium sylvaticum	Po.	P	W[h]	20/15	L > D	Grime et al. (1981)
Campanula latifolia	Campanul.	P	C	20/15	—	Grime et al. (1981)
Cardamine concatenata	Brassic.	P	W + C	5	—	Baskin and Baskin (1995)
Carex flacca	Cyper.	P	C	20?	L = D	Taylor (1956)
C. sylvatica	Cyper.	P	W[h]	20/15	L	Grime et al. (1981)
Collinsia verna	Scrophulari.	A	W	15/6	—	Baskin and Baskin (1983b)
Deschampsia flexuosa	Po.	P	W[h]	25	L = D	Scurfield (1954)
Diarrhena americana	Po.	P	C	20/10	L > D	Baskin and Baskin (1988)
Floerkea proserpinacoides	Limnanth.	A	W + C	5	—	Baskin et al. (1988)
Galeobdolon luteum	Lami.	P	C	5	—	Grime et al. (1981)
Galium aparine	Rubi.	A	W[h]	20/15	L = D	Grime et al. (1981)
Geum urbanum	Ros.	P	W[h]	16–20	—	Graves and Taylor (1988)
Impatiens capensis	Balsamin.	A	C	15	—	Leck (1979)
I. pallida	Balsamin.	A	C	15	—	Nozzolillo and Thie (1993)
I. parviflora	Balsamin.	A	C	20–25?	—	Coombe (1956)
Lactuca floridana	Aster.	B	C	30/15	L	Baskin and Baskin (1988)
Lamiastrum galeobdolon	Lami.	P	C	20/15	—	Grime et al. (1981), Packham (1983)
Melampyrum cristatum	Scrophulari.	A	C	25?	—	Horrill (1972)
Melica nutans	Po.	P	W[h]	20/15	L > D	Grime et al. (1981)
Mercurialis perennis	Euphorbi.	P	C?	—	—	Mukerji (1936)
Mertensia virginica	Boragin.	P	C	15/6	—	Baskin and Baskin (1988, unpublished results)
Milium effusum	Po.	P	C	16, 21	—	Thompson (1980)
Moehringia trinervia	Caryophyll.	A	W[h]	20/15	L > D	Grime et al. (1981)
Nemophilia aphylla	Hydrophyll.	A	W	15/6	L = D	Baskin et al. (1993a)
Oxalis acetosella	Oxalid.	P	C	20/15	—	Packham (1978)

(continues)

TABLE 10.18—*Continued*

Species	Family[a]	LC[b]	Tr[c]	Temperature[a]	L:D[a]	Reference
Phacelia bipinnatifida	Hydrophyll.	B	C	20/10	L = D	Baskin and Baskin (1988, unpublished results)
P. ranunculacea	Hydrophyll.	A	W	20/10	L > D	Baskin *et al.* (1993a)
Polemonium reptans	Polemoni.	P	W	20/10	L > D	Baskin and Baskin (1992b)
Polymnia canadensis	Aster.	B	C	25/15	L > D	Bender (1991)
Primula vulgaris	Primul.	P	W[h]	20/15	L	Grime *et al.* (1981)
Silene dioica	Caryophyll.	B-P	W	20–25	—	Thompson (1975)
Stachys sylvatica	Lami.	P	C	—	D > L	Slade and Causton (1979)
Teucrium scorodonia	Lami.	P	C	25?	L = D	Hutchinson (1968)
Tiarella cordifolia	Saxifrag.	P	C	30/15	L	Baskin and Baskin (1988, unpublished results)
Tovara virginiana	Polygon.	P	C	30/15	L > D	Baskin and Baskin (1988)
Uniola sessiliflora	Po.	P	C	40/20	D > L	Wolters (1970)
Urtica dioica	Urti.	P	W[h]	20/15	L	Grime *et al.* (1981)
Viola riviniana	Viol.	P	C	20/15	—	Grime *et al.* (1981)

[a]See notes for Table 10.1.
[b]Life cycle type: A, annual; B, biennial or monocarpic perennial, P, polycarpic perennial.
[c]Dormancy-breaking treatment: C, cold stratification; S, scarification; W, warm stratification.
[d]Seeds are presumed to have morphological dormancy.
[e]No data available.
[f]Seeds nondispersed in the field for 5–12 months required warm + cold stratification for germination.
[g]Seeds were given a gibberellic acid treatment rather than cold stratification.
[h]Seeds afterripened in dry storage.

dormancy is broken in seeds of these species, but Martin (1965) found that *G. maculatum* seeds germinated to 76% after they had been cold stratified at 5–10°C for 3 months. Ellenberg (1988) lists the legumes *Astragalus glycyphllos, Lathyrus linifolius,* and *Vicia sepium* as growing under broadleaved woodlands in central Europe; thus there may be a few other herbaceous woodland species with physical dormancy.

Physiological dormancy is almost as prevalent among woodland herbs as MPD and occurs in a great diversity of plant families. The Hydrophyllaceae is the only family represented in Table 10.18 that has some species with MPD and others with physiological dormancy. Some species with physiological dormancy, including *Galium aparine, Glechoma hederacea, Geum urbanum, Silene dioica,* and *Alliaria petiolata,* not only grow in forests, but are also found in forest edges (Ellenberg, 1988). A few species, including *A. petiolata, Atropa belladonna, G. aparine, G. hederacea,* and *Urtica dioica,* are listed as weeds by Holm *et al.* (1979).

Seeds of the winter annuals *Collinsia verna, Galium aparine, Moehringia trinervia, Nemophilia aphylla,* and *Phacelia ranunculacea* and those of the perennial *Polemonium reptans* require high summer temperatures for loss of dormancy (Table 10.18). However, seeds of summer annuals, biennials, and most perennials require cold stratification for loss of dormancy. The effects of cold stratification have not been tested on seeds of *Carex*

sylvatica, Deschampsia flexuosa, Geum urbanum, Primula vulgaris, and *Urtica dioica.* However, because (1) these species are perennials and (2) their seeds afterripen in dry storage at room temperatures, it is assumed that they have nondeep physiological dormancy that could be broken by short periods of cold stratification. Although cold stratification breaks dormancy in seeds of *Cardamine concatenata* and *Floerkea proserpinacoides,* a period of warm stratification prior to cold stratification enhances the effectiveness of cold stratification in breaking dormancy.

With the exception of winter annuals and the perennial *Polemonium reptans,* dormancy break in seeds of most herbaceous woodland species with physiological dormancy occurs in winter; therefore, seeds germinate in early spring. The temperature for maximum germination of nondormant seeds ranges from 5°C in the winter-germinating ephemeral *Floerkea proserpinacoides* (Baskin *et al.,* 1988) to 30/15°C in the perennials *Tiarella cordifolia* and *Tovara virginiana* (Baskin and Baskin, unpublished results). Light:dark requirements for seed germination have been determined for 23 species with physiological dormancy (Table 10.18). Five (22%) species require light for germination, 9 (39%) germinate to higher percentages in light than in darkness, 7 (30%) germinate equally well in light and darkness, and 2 (9%) germinate to higher percentages in darkness than in light.

F. Dormancy and Germination of Herbaceous Nonwoodland Species

Within the region of temperate deciduous forests, many hectares of land are not covered by forests. Rock outcrops, dunes, rivers banks, and edges of lakes are naturally devoid of trees, whereas other sites, including fields, gardens, pastures, vineyards, hay meadows, roadsides, railroad and power line right-of-ways, vacant city lots, and lawns, have been made treeless by human activities. Natural events such as landslides and storms also may remove trees from an area. Treeless habitats support a large number of herbaceous species that usually do not occur in forests; therefore, an understanding of seed germination ecology of species in the region of temperate deciduous forests must include species in nonforested habitats.

1. Weeds

A species is called a weed if it occurs in such an abundance where humans are growing plants for food, fiber, shelter, forage, or beauty that it causes a decrease in the growth of the desired species. Within a given geographical area, weeds may be either native or introduced. Also, a species may be a serious weed in one country and a rare species in another. Thus, sometimes it is difficult to decide whether a species should be called a weed. Weed biologists have given considerable attention to this problem, and Holm *et al.* (1979) conducted a worldwide survey to identify which species are weeds in various countries. In the following discussion of the germination ecology of herbaceous weeds of the temperate deciduous forest region, a species is included only if (1) it grows in one or more countries where temperate deciduous forests occur and (2) Holm *et al.* (1979) list it as a weed in their book, "A Geographical Atlas of World Weeds." A total of 220 weeds are included in our survey, but it should be noted that many of them can occur as weeds within the region of temperate grasslands. Further, some weeds discussed under temperate grasslands (Table 10.26) can occur as weeds in the temperate deciduous forest region. However, each weed is covered only once in our discussion, i.e., either under temperate deciduous forest or under grasslands.

Freshly matured seeds of 36 (16%) of the 220 weeds included in Table 10.19 are nondormant and can germinate to 90–100% at various temperatures, depending on the species. The optimum temperature for germination of the majority of these species is from 20 to 30°C, with alternating temperature regimes of 25/15 and 30/15°C being optimal for many of them.

Seeds of most species whose seeds are nondormant at maturity either require light for germination or they germinate to higher percentages in light than in darkness (Table 10.19). Seeds of *Bromus commutatus* (Froud-Williams and Chancellor, 1986) and *Tussilago farfara* (Baskin and Baskin, unpublished results) germinate equally well in light or darkness. Achenes of *Lactuca serriola* require light for germination at low (15/6°C) temperatures, but germinate to 100% in both light and darkness at high (35/20°C) temperatures (Baskin and Baskin, unpublished results). Achenes of *L. serriola* lost their light requirement for germination when buried in soil at natural temperatures (in England) from October to April (Marks and Prince, 1982).

Although seeds of a species may germinate to high percentages over a range of temperatures at the time of maturity, they still may have a small amount of innate dormancy. Thus, after seeds are given an appropriate dormancy-breaking treatment, germination percentages and/or the temperature range over which seeds will germinate may increase. For example, seeds of *Rumex crispus* collected on July 19, 1981 and placed on moist sand 3 days later germinated to 45, 77, 82, 90, and 85% at 15/6, 20/10, 25/15, 30/15, and 35/20°C, respectively (Baskin and Baskin, 1985f). However, seeds exhumed after 5 weeks of burial at natural temperatures in Kentucky germinated to 67, 98, 98, 97, and 98% at the five thermoperiods, respectively. The high germination percentages of freshly matured seeds of *R. crispus* might tempt one to say that the seeds are nondormant, but that would be a mistake because some seeds of this species undergo dormancy loss after maturation. Further studies may reveal that other species listed under "nondormant" in Table 10.19 have at least small percentages of freshly matured seeds with some physiological dormancy. Such seeds are in a late state of conditional dormancy.

Morphological dormancy is known in some weedy members of the Apiaceae (Table 10.19). Due to the relatively high temperature requirements for germination, seeds of these species can germinate if they are dispersed in summer or early autumn. However, if dispersal is delayed until mid- to late autumn or winter, seeds do not germinate until spring (Baskin and Baskin, 1979b, 1990a). Seeds of *Conium maculatum* retained on the mother plant during late autumn and winter enter physiological dormancy; consequently, they have morphophysiological dormancy. Physiological dormancy is broken during the following summer, and thus seeds of *C. maculatum* would have only morphological dormancy by the second autumn after maturity (Baskin and Baskin, 1990a).

Morphophysiological dormancy (MPD) is broken in seeds of weeds by either warm or cold stratification, depending on the species (Table 10.19). Seeds of winter annuals require only high temperatures for dormancy break and thus have nondeep simple MPD. Seeds of

TABLE 10.19   Dormancy in Seeds of 220 Weeds of Nonforested Areas (Excluding Those in Wet Habitats) within the Region of Temperate Deciduous Forests

Species	Family[a]	LC[b]	Tr[c]	Temperature[e]	L:D[d]	Reference
Nondormant						
Achillea millefolium[d]	Aster.	P		25/15	L > D	Bostock (1978)
Agropyron repens[d]	Po.	P		30/20	L > D	Williams (1971)
Amphiachyris dracunculoides	Aster.	wsA		30/15	L > D	Baskin and Baskin (1983e)
Arctium minus[d]	Aster.	B		30/15	L	Baskin and Baskin (1988, unpublished results)
Avena fatua[e]	Po.	wA		16–21	—[f]	Friesen and Shebeski (1961), Sawhney et al. (1986)
Bidens pilosa	Aster.	sA		25/20	L > D	Reddy and Singh (1992)
Briza media	Po.	P		22/12	L	Maas (1989)
Bromus commutatus[d]	Po.	wA		7–21	L = D	Froud-William and Chancellor (1986)
Carduus acanthoides[d]	Aster.	B-A		30/20	L > D	McCarty et al. (1969)
C. nutans[d]	Aster.	B-A		30/20	L > D	McCarty et al. (1969), McCarty and Scifres (1969)
Chrysanthemum leucanthemum[d]	Aster.	P		30/15	L > D	Povilaitis (1956), Baskin and Baskin (1988)
Coreopsis tinctoria[d]	Aster.	wA		25/15	L	Baskin et al. (unpublished results)
Cotula coronopifolia	Aster.	A-P		5, 30/15	—	van der Toorn and ten Hove (1982)
Crepis tectorum[d]	Aster.	wA		10–35	L > D	Anderson (1968), Najda et al. (1982)
Dipsacus sylvestris[d]	Dipsac.	B		30/20	—	Werner (1975a)
Eleusine indica[d]	Po.	sA		35/20	L > D	Toole and Toole (1940), Fulwider and Engel (1959)
Erigeron annuus	Aster.	wA		20	—	Hayashi and Numata (1967)
E. canadensis[d]	Aster.	wA		35/20	L > D	Baskin and Baskin (1988, unpublished results)
E. philadelphicus[d]	Aster.	wA		35/20	L > D	Baskin and Baskin (unpublished results)
E. sumatrensis	Aster.	wA		20	—	Hayashi and Numata (1967)
Galinsoga parviflora[d]	Aster.	sA		35/20	L	Baskin and Baskin (1981b)
Gnaphalium obtusifolium[d]	Aster.	B		20/10	L	Baskin and Baskin (unpublished results)
Helenium amarum[d]	Aster.	wsA		30/15	L > D	Baskin and Baskin (1973a)
Holcus lanatus[d]	Po.	P		20/16	L > D	Williams (1983), Grime et al. (1981)
Hypochoeris radicata[d]	Aster.	P		20/15	L	Grime et al. (1981)
Lactuca canadensis[d]	Aster.	B		30/15, 25/15	L	Baskin et al. (unpublished results)
L. serriola[d]	Aster.	A-B		20/10	L > D	Marks and Prince (1982), Baskin and Baskin (unpublished results)
Matricaria inodora	Aster.	A		15/10	L > D	Lonchamp and Gora (1980)
Oxalis corniculata[d]	Oxalid.	P		17	L	Holt (1987)
Phyllanthus urinaria	Euphorbi.	A		25	L	Wehtje et al. (1992)
Richardia scabra	Rubi.	A-P		30/20	L	Biswas et al. (1975)
Rumex obtusifolius[d]	Polygon.	P		30/20, 20	L > D	Steinbauer and Grigsby (1960)
Senecio vulgaris[d]	Aster.	wA		10	L > D	Popay and Roberts (1970)
Taraxacum officinale[d]	Aster.	P		20, 10, 10–18, 25	L > D	Mezynski and Cole (1974), Maguire and Overland (1959), Washitani (1984), Letchamo and Gosselin (1996)
T. platycarpum	Aster.	P		6–16	—	Washitani and Ogawa (1989)
Tussilago farfara	Aster.	P		25/15	L = D	Robocker (1977), Baskin and Baskin (unpublished results)
Morphological dormancy						
Cicuta maculata[d]	Api.	P		21/15	—	Mulligan and Munro (1981)
Conicum maculatum[d,g]	Api.	B		30/15	L > D	Baskin and Baskin (1990a)
Daucus carota	Api.	B	C[h]	30/15	L > D	Baskin and Baskin (1988, unpublished results)
Pastinaca sativa[d,g]	Api.	B		30/15	L > D	Baskin and Baskin (1979b)
Torilis japonica[d]	Api.	wA		20/10	L > D	Baskin and Baskin (1975a)
Morphophysiological dormancy						
Aegopodium podagaria	Api.	P	C	5	—	Grime et al. (1981)
Aethusa cynapium	Api.	sA	C	20/10	L	Martin (1946) Roberts and Boddrell (1985)
Anthriscus silvestris[d]	Api.	B	C	8/4	—	Lhotska (1978)

(continues)

TABLE 10.19—*Continued*

Species	Family[a]	LC[b]	Tr[c]	Temperature[a]	L:D[a]	Reference
A. vulgaris	Api.	wA	W	12/7 → 25/18	—	Lhotska (1978)
Apium leptophyllum	Api.	wA	W	15/6	L > D	Baskin *et al.* (unpublished results)
Bupleurum rotundifolium	Api.	wA	W	10, 15/6	D > L	Baskin and Baskin (1974a)
Chelidonium majus	Papaver.	P	C	20/15	—	Grime *et al.* (1981)
Fumaria officinalis[d]	Fumari.	sA	C	20	—	Jeffery and Nalewaja (1970, 1973)
Papaver dubium[d]	Papaver.	sA	C	5	—	Grime *et al.* (1981)
P. rhoeas[d]	Papaver.	wsA	C	20/10, 5	—	McNaughton and Harper (1964), Grime *et al.* (1981)
Ranunculus sceleratus[d]	Ranuncul.	wA	H	20/5	L > D	van der Toorn and ten Hove (1982), Lehmann (1909)
		wsA	C	21/11	L > D	Probert *et al.* (1989)
Physical dormancy						
Abutilon theophrasti[d]	Malv.	sA	S	15–40	—	Horowitz and Taylorson (1983)
Anoda cristata[d]	Malv.	sA	S	30	L = D	Solano *et al.* (1976)
Cardiospermum halicacabum[d]	Sapind.	sA	S	35	—	Johnston *et al.* (1979a)
Cassia marilandica[d]	Fab.	sA	S	30/15	L = D	Nan (1992)
C. obtusifolia[d]	Fab.	sA	S	21–33	L = D	Creel *et al.* (1968), Nan (1992)
Convolvulus arvensis[d]	Convolvul.	sA	S	35/20	L = D	Weaver and Riley (1982)
Crotolaria spectabilis	Fab.	sA	S	35	L = D	Egley (1979)
Cuscuta europea	Convolvul.	sA	C	27	—	Gaertner (1956)
Erodium cicutarium[d]	Gerani.	wA	S	20/15	—	Grime *et al.* (1981)
Geranium dissectum	Gerani.	wA	S	20/15	—	Grime *et al.* (1981)
G. molle	Gerani.	wA	S	20/15	—	Grime *et al.* (1981)
G. pratense	Gerani.	P	S	20/15	L > D	Grime *et al.* (1981)
Ipomoea hederacea[d]	Convolvul.	sA	S	25, 30	L = D	Gomes *et al.* (1978)
I. lacunosa[d]	Convovul.	sA	S	30	—	Gomes *et al.* (1978)
I. pandurata[d]	Convolvul.	P	S	25	—	Horak and Wax (1991)
I. purpurea[d]	Convolvul.	sA	S	35/20	L = D	Baskin and Baskin (unpublished results)
Jacquemontia tamnifolia	Convolvul.	sA	S	25	—	Eastin (1983)
Lathyrus latifolius	Fab.	P	S	20	—	Atwater (1980)
L. pratensis[d]	Fab.	P	S	20/15	L = D	Grime *et al.* (1981)
Lotus corniculata[d]	Fab.	P	S	20/15	—	Grime *et al.* (1981)
Malva rotundifolia[d]	Malv.	A-B	S	15–30	—	Makowski and Morrison (1989)
Medicago lupulina[d]	Fab.	A-P	S	20/15	L = D	Grime *et al.* (1981)
Melilotus alba[d]	Fab.	B	S	25?	—	Hamly (1932)
M. altissima	Fab.	B	S	20/15	L = D	Grime *et al.* (1981)
Sesbania exaltata[d]	Fab.	sA	S	35	L = D	Johnston *et al.* (1979b)
Sicyos angulatus[d]	Cucurbit.	sA	S	25	—	Mann *et al.* (1981)
Sida spinosa[d]	Malv.	sA	S	35/20	L = D	Baskin and Baskin (1984f)
Trifolium campestre[d]	Fab.	wA	S	20/15	L = D	Grime *et al.* (1981)
T. pratense[d]	Fab.	P	S	20/15	L = D	Grime *et al.* (1981)
T. repens[d]	Fab.	P	S	20/15	L = D	Grime *et al.* (1981)
Vicia cracca[d]	Fab.	P	S	20/15	—	Grime *et al.* (1981)
V. hirsuta	Fab.	wA	S	20/15	L = D	Grime *et al.* (1981)
V. sativa[d]	Fab.	wA	S	20/10	L = D	Grime *et al.* (1981)
Physical and physiological dormancy						
Geranium carolinianum[d]	Gerani.	wA	S	15/6	L = D	Baskin and Baskin (1974b)
Physiological dormancy						
Acalypha ostryaefolia[d]	Euphorbi.	sA	C[l]	35/20	L > D	Baskin and Baskin (unpublished results)
Aegilops cylindrica[d]	Po.	wA	W	10–20	L = D	Morrow *et al.* (1982)
Agrostemma githago[d]	Caryophyll.	wA	W	5–25	L > D	Thompson (1973), de Klerk and Smulders (1984)
Alopecurus myosuroides[d]	Po.	wA	W	25/15	L > D	Wellington and Hitchings (1965), Wallgren and Aamisepp (1977), Wallgren and Avholm (1978)
Amaranthus hybridus[l]	Amaranth.	sA	C[l]	35/20	L > D	Baskin and Baskin (1987a)
A. retroflexus	Amaranth.	sA	C	20	—	Hayashi and Numata (1967)
Ambrosia artemisiifolia[d]	Aster.	sA	C	30/15	L > D	Baskin and Baskin (1980)

(continues)

TABLE 10.19—*Continued*

Species	Family[a]	LC[b]	Tr[c]	Temperature[d]	L:D[e]	Reference
A. trifida[d]	Aster.	sA	C	10–24	—	Abul-Fatih and Bazzaz (1979)
Ampelamus albidus[d]	Asclepiad.	P	C	30/15, 35/20	L = D	Baskin and Baskin (unpublished results)
Anagallis arvensis[d]	Primul.	wA	W	9–12	L	Pandey (1969), Singh (1969)
Anthemis cotula[d]	Aster.	wA	W C?	20	—	Kay (1971), Gealy *et al.* (1985)
Apera spica-venti	Po.	wA	W	20/15	L > D	Wallgren and Aamisepp (1977), Wallgren and Avholm (1978)
Arabidopsis thaliana[d]	Brassic.	wA	W	10–20	L	Baskin and Baskin (1972a)
Arenaria serpyllifolia[d]	Caryophyll.	wA	W	15–20	L > D	Ratcliffe (1961)
Asclepias syriaca[d]	Asclepiad.	P	C	30/15	L = D	Evetts and Burnside (1972), Baskin and Baskin (1977b)
A. tuberosa[d]	Asclepiad.	P	C	30/15	L > D	Baskin and Baskin (unpublished results)
Aster pilosus[d]	Aster.	P	C	30/15	L	Baskin and Baskin (1979a)
Atriplex hastata[d]	Chenopodi.	sA	C	20/15	L > D	Grime *et al.* (1981)
A. patula[d]	Chenopodi.	sA	C	20/15	L > D	Grime *et al.* (1981)
Avena fatua[d,e]	Po.	wA	W	21,16	L > D	Hsiao and Simpson (1971)
Barbarea vulgaris[d]	Brassic.	A-B	W	20/10	L > D	Baskin and Baskin (1989b)
Berteroa incana[d]	Brassic.	A	?	25, 30/20	L = D	Toole *et al.* (1957)
Bidens cernua[d]	Aster.	sA	C	35/20	L = D	Hogue (1976), Baskin and Baskin (1988, unpublished results)
B. connata[d]	Aster.	sA	C	35/20	L = D	Baskin and Baskin (unpublished results)
Brassica arvensis[d]	Brassic.	wA	W	30/20	L > D	Lonchamp and Gora (1980)
Bromus japonicus[d]	Po.	wA	W	15/6	L = D	Baskin and Baskin (1981a)
B. secalinus[d]	Po.	wA	W	15	—	Steinbauer and Grigsby (1956)
Camelina microcarpa[d]	Brassic.	sA	C	35/20	L > D	Toole *et al.* (1957), Chepil (1946)
Capsella bursa-pastoris[d]	Brassic.	wA	W	20/10	L > D	Baskin and Baskin (1989c)
Cardamine hirsuta	Brassic.	wA	W	15–20	L > D	Ratcliffe (1961)
Carex nigra	Cyper.	P	C	20/15	L > D	Grime *et al.* (1981)
Cerastium viscosum[d]	Caryophyll.	wA	W	15/6	L	Baskin and Baskin (1988, unpublished results)
Chaenorrhinum minus[d]	Scrophulari.	sA	C	30/20	L > D	Lonchamp and Gora (1980)
Chenopodium album[d]	Chenopodi.	sA	C[i]	30/15	L > D	Baskin and Baskin (1977a, unpublished results)
C. bonus-henricus	Chenopodi.	sA	C	30	—	Herron (1953)
C. rubrum[d]	Chenopodi.	sA	C	20/15	L	Grime *et al.* (1981)
Cirsium arvense[d]	Aster.	P	C	30/20	L > D	Roberts and Chancellor (1979), Wilson (1979)
C. vulgare[d]	Aster.	B	W[h]	20	—	Michaux (1989)
Cynoglossum officinale[d]	Boragin.	B	C	5	D > L	van Breeman (1984)
Cyperus aristatus[d]	Cyper.	sA	C	30/15	L	Baskin and Baskin (1978)
C. esculentus[d]	Cyper.	P	C[k]	30, 20, 38/32	L > D	Justice and Whitehead (1947), Thullen and Keeley (1979)
C. odoratus[d]	Cyper.	P	C	35/20	L	Baskin *et al.* (1989)
C. rotundus[d]	Cyper.	P	W	30/20, 38/32	L > D	Justice (1956), Thullen and Keeley (1979)
Dactylis glomerata[d]	Po.	P	C	28/10	L > D	Sprague (1940), Grime *et al.* (1981)
Danthonia spicata[d]	Po.	P	C	25/10	—	Toole (1939)
Dianthus armeria[d]	Caryophyll.	B	W	30/15	L	Baskin and Baskin (1988, unpublished results)
Digitaria ischaemum[d]	Po.	sA	C	35/20	L > D	Baskin and Baskin (1988, unpublished results)
D. sanguinalis[d]	Po.	sA	C	35/20	L > D	Toole and Toole (1941)
Diodea teres[d]	Rubi.	sA	C	30/15	L > D	Baskin and Baskin (1988, unpublished results), Baird and Dickens (1991)
Echinochloa crus-galli[d]	Po.	sA	C[h]	30/10, 30	L = D	Watanabe and Hirokawa (1975), Barrett and Wilson (1983)
Echinocystis lobata[d]	Cucurbit.	sA	C	30/20	—	Choate (1940)
Echium vulgare[d]	Boragin.	B		30/20	L > D	van Breeman (1984), Baskin and Baskin (unpublished results)
Erechtites hieracifolia[d]	Aster.	sA	C	35/20	L > D	Baskin and Baskin (1996)

(continues)

TABLE 10.19—*Continued*

Species	Family[a]	LC[b]	Tr[c]	Temperature[a]	L:D[a]	Reference
Euphorbia nutans[d]	Euphorbi.	sA	C	35/20	L > D	Baskin and Baskin (unpublished results)
E. maculata[d]	Euphorbi.	sA	C	35/20	L > D	Baskin and Baskin (1979d)
Galium aparine[d]	Rubi.	wA	W	10–17	D > L	van der Weide (1993)
Helenium autumnale[d]	Aster.	P	C	30/15	L	Baskin and Baskin (unpublished results)
Helianthus annuus[d]	Aster.	sA	C	30/15	L > D	Baskin and Baskin (1988, unpublished results)
Hieracium aurantiacum	Aster.	P	C	22	—	Stergios (1976)
H. pratense	Aster.	P	C[h]	30/20	L > D	Panebianco and Willemsen (1976)
Holosteum umbellatum[d]	Caryophyll.	wA	W	10	L > D	Baskin and Baskin (1973b)
Hordeum pusillum[d]	Po.	wA	W	30/20	L = D	Fischer *et al.* (1982)
Hyoscyamus niger	Solan.	A-B	C	20	—	Hussain *et al.* (1984)
Kickxia spuria[d]	Apocyn.	sA	C	30/20	L > D	Lonchamp and Gora (1980)
Lamium amplexicaule[d]	Lami.	wA	W	15/6	L > D	Baskin and Baskin (1981c, 1984d)
L. purpureum[d]	Lami.	wA	W	15/6	L	Baskin and Baskin (1984e)
Lapsana communis[d]	Aster.	wA	W	15/10	L > D	Lonchamp and Gore (1980)
Lepidium virginicum[d]	Brassic.	wA	W	15/6	L > D	Toole *et al.* (1957), Baskin and Baskin (1988, unpublished results)
Leptochloa filiformis[d]	Po.	sA	C	35/20	L > D	Baskin *et al.* (1993b)
Linaria vulgaris[d]	Scrophulari.	P	C	20	—	Nadeau and King (1991)
Lithospermum arvense[d]	Boragin.	wA	W	15/6	L = D	Baskin and Baskin (1988, unpublished results)
Lobelia inflata[d]	Campanul.	A-B	C	30/15	L	Baskin and Baskin (1992c)
L. cardinalis[d]	Campanul.	P	C	30/15	L	Baskin *et al.* (unpublished results)
Lolium multiflorum[d]	Po.	wA	W	10/5	—	Young *et al.* (1975)
Lychnis alba[d]	Caryophyll.	P	W	30/15	L > D	Baskin and Baskin (1988, unpublished results)
Mikania scandens	Aster.	P	C	30/15	L = D	Baskin *et al.* (1993c)
Mollugo verticillata[d]	Aizo.	sA	C	40/25	L > D	Baskin and Baskin (1988, unpublished results)
Muhlenbergia scherberi[d]	Po.	P	C	35/20	L	Baskin and Baskin (1985e)
Nardus stricta[d]	Po.	P	C	25/20	L > D	Grime *et al.* (1981)
Odontites verna	Scrophulari.	A	C	20/15	—	Grime *et al.* (1981)
Oenothera biennis[d]	Onagr.	B	C	30/15	L = D	Baskin and Baskin (1994a)
Onopordum acanthium[d]	Aster.	B	C	30/20	L > D	Scifres and McCarty (1969)
Oryza sativa[d]	Po.	wA	W	30	—	Cohn and Hughes (1981)
Panicum capillare[d]	Po.	sA	C	30/15	L > D	Baskin and Baskin (1986d)
P. dichotomiflorum[d]	Po.	sA	C	35/20	L > D	Baskin and Baskin (1983c)
Physalis angulata[d]	Solan.	sA	C[i]	30/20	L = D	Thomson and Witt (1987)
P. virginiana[d]	Solan.	sA	C[i]	30/20	L = D	Thompson and Witt (1987)
Phytolacca americana[d]	Phytolacc.	P	C	35/20	L > D	Baskin and Baskin (unpublished results)
Plantago aristata[d]	Plantagin.	wA	W	30/20	L > D	Stearns (1955), Steinbauer and Grigsby (1957)
P. lanceolata[d]	Plantagin.	P	C[h]	30/20	L	Steinbauer and Grigsby (1957)
P. major[d]	Plantagin.	P	C[h]	30/20	L	Steinbauer and Grigsby (1957)
P. virginica[d]	Plantagin.	wA	W	20/10	L	Baskin *et al.* (unpublished results)
Poa annua[d]	Po.	wA	W	10, 15	—	Standifer and Wilson (1988)
P. pratense[d]	Po.	P	C[h]	10, 15	—	Sprague (1940)
Polygonum aviculare[d]	Polygon.	sA	C	35/20	L > D	Courtney (1968), Baskin and Baskin (1990b)
P. convolvulus[d]	Polygon.	sA	C	25	L = D	Justice (1941)
P. pensylvanicum[d]	Polygon.	sA	C	35/20	L > D	Baskin and Baskin (1987a, unpublished results)
P. lapathifolium[d]	Polygon.	sA	C	30/10	L = D	Watanabe and Hirokawa (1975)
Portulaca oleracea[d]	Portulac.	sA	W	35/20	L > D	Baskin and Baskin (1988b)
Potentilla recta[d]	Ros.	P	W	25/15	L	Baskin and Baskin (1990c)
Raphanus raphanistrum	Brassic.	A-B	W	30/20	D > L	Mekenian and Willensen (1975)
Rhinanthus crista-galli	Scrophulari.	sA	C	15	—	Vallance (1952)
Rumex acetosella[d]	Polygon.	wA	W	20/15	L = D	Grime *et al.* (1981)

(continues)

TABLE 10.19—*Continued*

Species	Family[a]	LC[b]	Tr[c]	Temperature[a]	L:D[a]	Reference
R. crispus[d]	Polygon.	P	W	30/15	L	Baskin and Baskin (1985f)
Saponaria officinalis[d]	Caryophyll.	P	C	30/20	—	Steinbauer and Grigsby (1956)
Senecio erucifolius	Aster.	P	C	25/5	—	Otzen and Doornbos (1980)
Setaria lutescens[d]	Po.	sA	C[h]	30/15	L > D	Norris and Schoner (1980), Baskin *et al.* (1996)
S. viridis[d]	Po.	sA	C	25	—	vanden Born (1971)
Silene antirrhina[d]	Caryophyll.	wA	W	15/6	L	Baskin and Baskin (1988, unpublished results)
S. noctiflora	Caryophyll.	wA	W	20	L = D	Povilaitis (1956)
Sisymbrium altissimum[d]	Brassic.	wA	W	20 → 35 → 20	L	Toole *et al.* (1957)
S. officinale[d]	Brassic.	wA	W	20 → 35 → 20	L > D	Toole *et al.* (1957)
Solanum carolinense[d]	Solan.	P	C[h]	30/20	L > D	Ilnicki *et al.* (1962)
S. dulcamara[d]	Solan.	P	C	25/10, 35/25	D > L	Roberts and Lockett (1977), Pegtel (1985)
S. nigrum[d]	Solan.	sA	C[h]	30/10	L > D	Roberts and Lockett (1978a), Givelberg *et al.* (1984)
S. ptycanthum	Solan.	—	C[m]	30/20	L	Hermanutz and Weaver (1991)
S. sarrachoides[d]	Solan.	sA	C[b]	30/20, 30	—	Roberts and Boddrell (1983)
Solidago altissima[d]	Aster.	P	C	30/15	L	Baskin *et al.* (1993c)
Sonchus arvensis[d]	Aster.	P	C[h]	30/20	L = D	Maguire and Overland (1959), Pegtel (1976)
Sorghum halepense[d]	Po.	P	C[h]	40/20[n]	—	Taylorson and McWhorter (1969)
Spergula arvensis[d]	Caryophyll.	wA	W	27/19	—	Wagner (1988)
Sporobolus vaginiflorus[d]	Po.	sA	C	20	—	Baskin and Caudle (1967)
Stellaria media[d]	Caryophyll.	wA	W	15/6	L > D	Baskin and Baskin (1988, unpublished results)
Thlaspi arvense[d]	Brassic.	wA	W	15/6	L > D	Baskin and Baskin (1989d)
T. perfoliatum[d]	Brassic.	wA	W	23/12	L > D	Baskin and Baskin (1979c)
Valerianella olitoria	Valerian.	wA	W	15/6	L > D	Baskin and Baskin (1988, unpublished results)
Verbascum thapsus[d]	Scrophulari.	B	C	35/20	L > D	Baskin and Baskin (1981d)
V. blattaria[d]	Scrophulari.	B	C	30/15	L > D	Baskin and Baskin (1981d)
Verbena officinalis	Scrophulari.	P	C	20/15	L > D	Grime *et al.* (1981)
Veronica arvensis[d]	Scrophulari.	wA	W	15/6	L	Baskin and Baskin (1983d)
V. hederifolia[d]	Scrophulari.	wA	W	10/4	—	Roberts and Lockett (1978b)
V. peregrina[d]	Scrophulari.	wA	W	20/10	L	Baskin and Baskin (1983f)
Viola arvensis[d]	Viol.	wA	W	15/6	L > D	J. Baskin and Baskin (1995)
Xanthium pensylvanicum[d]	Aster.	sA	C	23	L > D	Esashi *et al.* (1986)

[a]See notes for Table 10.1.

[b]Life cycle type: A, annual; wA, winter annual; sA, summer annual; wsA, winter and summer annual; B, biennial or monocarpic perennial; P, polycarpic perennial.

[c]Dormancy-breaking treatment: W, warm stratification; C, cold stratification.

[d]Species also is a weed in temperate grasslands.

[e]Genetically dormant and nondormant lines occur in *Avena fatua*.

[f]No data available.

[g]Some freshly matured seeds had MPD, which was broken during summer.

[h]Seeds afterripened in dry laboratory storage.

[i]Seeds afterripened to some extent at high temperatures.

[j]Incorrectly reported as *Amaranthus retroflexus*.

[k]Seeds afterripen equally well when stored dry at room temperature or moist at 2, 10, 20/10, and 20/2°C (Justice and Whitehead, 1947).

[l]Berries were allowed to dry for 2 months (at room temperature?) during which time seeds could have afterripened. Dormancy loss also would have occurred during cold stratification in the field.

[m]Berries were stored at 4°C for 5 months.

[n]Seeds were at 40°C for 2 hr and then returned to 20°C.

summer annuals, perennials, and biennials require cold stratification for loss of MPD and thus have deep complex MPD. Seeds of *Papaver rhoeas* germinate in spring and autumn, with most germination occurring in spring (Roberts and Feast, 1970). Grime *et al.* (1981) obtained 75% germination for *P. rhoeas* seeds while they were at 5°C, but McNaughton and Harper (1964) obtained 89 and 97% germination for seeds of this species kept at 30/10 and 20/10°C, respectively. Did the high or the low phases of these alternating temperature regimes break dormancy? More studies need to be done on this and other weeds whose seeds have underdeveloped embryos.

Light:dark requirements for germination have been studied in only four weeds whose seeds have MPD. Light is required for the germination of *Aethusa cynapium* seeds (Roberts and Boddrell, 1985), and it is required in the early, but not the late, stages of dormancy loss in seeds of *Apium leptophyllum* (Baskin *et al.*, unpublished results). Seeds of *Bupleurum rotundifolium* germinate to higher percentages in darkness than in light (Baskin and Baskin, 1974a), and those of *Ranunculus sceleratus* germinate to higher percentages in light than in darkness (van der Toorn and ten Hove, 1982; Probert *et al.*, 1989).

Holms *et al.* (1979) list 212 genera of weeds that belong to families known to have physical dormancy, and about 65 (31%) of them are represented by herbaceous species in the temperate deciduous forest regions of the world. Experimental studies on members of 21 genera with physical dormancy in the temperate deciduous forest region have shown that mechanical and/or acid scarification results in high germination percentages in both light and darkness (Table 10.19). The ability of permeable seeds to germinate in darkness helps explain why they will germinate during burial. In fact, seed burial may enhance germination and seedling establishment. Optimum seed-planting depths for maximum germination of various taxa of *Ipomoea* are 0.5 to 4 cm (Gomes *et al.*, 1978; Eastin, 1983; Horak and Wax, 1991), and maximum germination of *Cardiospermum halicacabum* occurred in seeds sown at depths of 1–3 cm (Johnston *et al.*, 1979a). A few seedlings emerged when scarified seeds of *C. halicacabum* (Johnston *et al.*, 1979a) and *Sesbania exaltata* (Johnston *et al.*, 1979b) were placed at a depth of 12 cm and of scarified seeds of *Sicyos angulatus* (Mann *et al.*, 1981) placed at a depth of 16 cm.

Seeds of 63% of the weeds in Table 10.19 have physiological dormancy and require either warm or cold stratification before they will germinate to the highest percentages possible for the species. Seeds of summer annuals and most perennials and biennials require cold stratification to come out of dormancy, whereas those of winter annuals require exposure to high temperatures to do so. A period of dry storage at room temperatures decreased the length of the cold stratification period

required to break dormancy in seeds of the summer annuals *Echinocystis lobata* (Choate, 1940) and *Digitaria* spp. (Toole and Toole, 1941). In general, if seeds come out of dormancy at high temperatures, optimum germination temperatures are low (15/6, 20/10°C), and if they come out of dormancy at low temperatures, optimum germination temperatures are high (25/15, 30/15, 35/20°C).

As discussed in Chapter 4, nondormant, imbibed seeds of winter annuals can be induced into secondary dormancy if they are exposed to low temperatures. This has been documented in a number of weeds, including *Alopecurus myosuroides* (Wellington and Hitchings, 1966), *Bromus japonicus* (Baskin and Baskin, 1981a), *Hordeum pusillum* (Fischer *et al.*, 1982), *Lamium purpureum* (Baskin and Baskin, 1984e), *Thlaspi arvense* (Baskin and Baskin, 1989d), *Torilis japonica* (Baskin and Baskin, 1975a), and *Raphanus raphanistrum* (Mekenian and Willemsen, 1975). However seeds that require low temperatures for loss of dormancy may be induced into secondary dormancy if nondormant seeds are exposed to high temperatures, e.g., seeds of the summer annuals *Polygonum aviculare* (Courtney, 1968) and *Ambrosia artemisiifolia* (Baskin and Baskin, 1980) enter dormancy in early summer when temperatures begin to increase.

Light:dark requirements for germination have been determined for 109 of the 133 weeds whose seeds have physiological dormancy (Table 10.19). Twenty-four (22%) species required light for germination, but none required darkness. Seeds of 19 (17%) species germinated equally well in light and darkness, 61 (57%) germinated to higher percentages in light than in darkness, and 4 (4%) germinated to higher percentages in darkness than in light. *Bromus commutatus* seeds germinated to higher percentages in darkness than in light as they came out of dormancy, but nondormant seeds germinated to higher percentages in light than in darkness (Froud-Williams and Chancellor, 1986). White light inhibited the germination of *Avena fatua* seeds at low (1 ml per dish) levels of water, but it promoted germination at high (2 ml per dish) levels of water (Hsiao and Simpson, 1971).

Seeds of *Galium aparine* germinate to higher percentages in darkness than in light (van der Weide, 1993); thus, it is not surprising that they germinate when buried in soil. The depth of burial interacts with the temperature in controlling the germination of *G. aparine* seeds. Whereas seeds at 1 cm germinated (= seedling emergence) to 90–100% at 8, 10, 15, 20, and 30°C, those at 3 cm germinated to this percentage only at 8, 10, and 15°C (van der Weide, 1993). Sixty-five percent of the seedlings of *G. aparine* emerged from a depth of 9 cm in soils with a low resistance (1.6 kg/cm^2), whereas only 25% emerged from 2 cm in soils with a high resistance (10.0 kg/cm^2) (van der Weide, 1993).

Seedlings of *Sonchus arvensis* (Pegtel, 1976) and *Onopordum acanthium* (Scifres and McCarty, 1969) emerged from 4 but not 5 cm, those of *Setaria viridis* (vanden Born, 1971) from 10 cm and to a limited degree (7%) from 12 cm, and those of *Ampelamus albidus* (Soteres and Murray, 1981) from 5 but not 7.5 cm. The optimum depth of planting for *Ambrosia trifida* was 2 cm, but 30% of the seedlings emerged from 16 cm (Abul-Fatih and Bazzaz, 1979).

Optimum pH values for germination vary with the species, but usually they are about neutral (Scifres and McCarty, 1969; Soteres and Murray, 1981). However, seeds of some species germinate to high percentages over a broad range of pH values: *Antemis cotula*, 4 to 7 (Gealey *et al.*, 1985); *Ampelamus albidus*, 4 to 10 (Evetts and Burnside, 1972); *Carduus nutans* and *C. acanthoides*, 3 to 9 (McCarty *et al.*, 1969); and *Richardia scabra*, 3 to 9 (Biswas *et al.*, 1975).

Optimum soil moisture conditions for the germination of seeds of most weeds (e.g., Wallgren and Aamisepp, 1977) are around field capacity. However, seeds of *Echinochloa crus-galli* variety *oryzicola* not only germinate under water, but 75% of the seedlings emerged from seeds flooded to a depth of 8 cm (Barrett and Wilson, 1983). This variety of barnyard grass will germinate in an oxygen-free environment (Kennedy *et al.*, 1980).

A number of studies have been done to determine the effects of moisture stress on the germination of weed seeds. Germination percentages were decreased from 75–100% to about 50%: *Cirsium arvense*, about −0.7 MPa (Wilson, 1979); *Setaria viridis*, between −0.4 and −0.8 MPa (Manthey and Nalewaja, 1987); *Ampelamus albidus*, between −0.51 and −0.71 MPa (Evetts and Burnside, 1972); *Ampelamus albidus*, more negative than −1.28 MPa (Soteres and Murray, 1981); *Physalis virginiana* and *P. angulata*, −0.8 MPa (Thomson and Witt, 1987); *Onopordum acanthium*, −1.22 MPa (Scifres and McCarty, 1969); *Carduus acanthoides* −8.1 MPa; and *C. nutans*, −1.52 MPa (McCarty *et al.*, 1969). Priming seeds of 60 species (in 21 different families) at 15°C for 14 days in PEG 6000 at −1.0 or −1.5 MPa enhanced germination percentages of 15 species and reduced it in 8 (Tallowin *et al.*, 1994).

2. Nonweeds

This section covers herbaceous species that (1) are native to the region of temperate deciduous forests, (2) grow in nonforested habitats in the region, and (3) are not listed as a weed by Holm *et al.* (1979). Species growing in bodies of water or in wet soil at their edges are covered in Chapter 11.

A comparison of weeds (Table 10.19) and nonweeds (Table 10.20) in this survey reveals that whereas seeds of 38 weeds are nondormant at maturity, those of only two nonweeds are nondormant. Although various kinds of seed dormancy are found in nonweeds, physiological dormancy is the most common.

Delphinium carolinianum is the only nonweed listed in Table 10.20 whose seeds have morphological dormancy. Seeds of this species mature in early summer and will germinate to near 100% at low (5, 15/6°C) temperatures within 3 to 4 weeks. Germination in the field is delayed until autumn when soils become continuously moist, and temperatures decline enough to be nonlimiting (Baskin and Baskin, unpublished results).

Morphophysiological dormancy (MPD) in seeds of the three species of winter annuals is broken during exposure to high summer temperatures (Table 10.20); consequently, seeds have nondeep simple MPD. Seeds of all the perennial species with MPD come out of dormancy during cold stratification. Thus, presumably, seeds of these species have deep complex MPD, but further studies may show that some of them have intermediate complex MPD. Three species (*Papaver somnifera*, *Silaum silaus*, and *Smyrnium olusatrum*) are listed under MPD because (1) they belong to families whose seeds may have underdeveloped embryos (Martin, 1946) and (2) seeds are dormant.

Eleven species, representing 10 genera and three families, have physical dormancy (Table 10.20). Scarified seeds germinate to 95–100% at temperatures of 20–30°C in light and darkness. However, little is known about the germination ecology of these species in the field except that fire promotes the germination of *Iliamna corei* seeds (Baskin and Baskin, 1997). Whereas only 1% of the *I. corei* seeds sown on soil in metal flats and exposed to natural seasonal temperature changes germinated, 13% of those subjected to heat of a fire germinated. Furthermore, additional seeds germinated when flats (and seeds) were subjected to heat of a fire in subsequent years, and after 5 years a total of 2 and 39% of seeds in nonburned and burned flats, respectively, had germinated. After 105 min in a drying oven set at 80–90°C, germination in nonburned and burned flats increased to 38 and 61%, respectively. In another experiment, seeds of *I. corei* given a dry heat treatment at 130, 140, and 150°C for 4 min germinated to 71, 79, and 71%, respectively (Baskin and Baskin, unpublished results).

Physiological dormancy occurs in nonweedy annuals, biennials (monocarpic perennials), and perennials (Table 10.20). Nonweedy winter annuals require high temperatures for loss of dormancy, and nondormant seeds generally germinate best at relatively low temperatures, e.g., 10/4, 15/6, or 20/10°C. However, the optimum germination temperature for nondormant seeds of the win-

TABLE 10.20   Seed Dormancy in Nonweedy Herbaceous Species Growing in Nonforested Areas within the Region of Temperate Deciduous Forests

Species	Family[a]	LC[b]	TR[c]	Temperature[a]	L:D[a]	Reference
Nondormant						
Primula prolifera	Primul.	P?		20/10	L > D	Thompson (1969)
P. smithiana	Primul.	P?		15	L > D	Thompson (1969)
Morphological dormancy						
Delphinium carolinianum	Ranuncul.	P		20/10	L = D	Baskin and Baskin (unpublished results)
Morphophysiological dormancy						
Chaerophyllum tainturieri	Api.	wA	W	20/10	L	Baskin and Baskin (1990d)
C. temulentum	Api.	P	C	5	—[d]	Grime *et al.* (1981)
Conopodium majus	Api.	P	C	5	—	Grime *et al.* (1981)
Hemerocallis sp.	Lili.	P	C	22	L = D	Giersbach and Voth (1957)
Heracleum mantegazzianum	Api.	P	C	Room	—	Tilley *et al.* (1996)
Hornungia petraea	Lili.	wA	W	20–15	L > D	Ratcliffe (1961)
Meconopsis cambricum	Papaver.	P	C	5	—	Schroeder and Barton (1939)
Myrrhis odorata	Api.	P	C	5	—	Grime *et al.* (1981)
Nothoscordum bivalve	Lili.	P	C	15/6	D > L	Baskin and Baskin (1979g)
Papaver somnifera[e]	Papaver.	wA	W	20/15	L > D	Grime *et al.* (1981)
Ptilimnium nuttallii	Api.	wA	W	20/10	L > D	Baskin *et al.* (unpublished results)
Silaum silaus[e]	Api.	P	C	5	—	Grime *et al.* (1981)
Smyrnium olusatrum[e]	Api.	P	C	5	—	Grime *et al.* (1981)
Tofieldia calyculata	Lili.	P	C	22/12	L = D	Maas (1989)
Trepocarpus aethusae	Api.	wA	W	20/10	L > D	Baskin *et al.* (unpublished results)
Trollius europaeus	Ranuncul.	P	C	22/12	D > L	Mass (1989),
				20/8	L = D	Milberg (1994a)
Zigadenus densus	Lili.	P	C	20/10	—	Baskin *et al.* (1993e)
Physical dormancy						
Anthyllis vulneraria	Fab.	P	S	20/15	L = D	Grime *et al.* (1981)
Astragalus tennesseensis	Fab.	P	S	25	—	Baskin and Quarterman (1969)
Baptisia australis	Fab.	P	S	25/10	—	Baskin and Baskin (1988, unpublished results)
Geranium lucidum	Gerani.	wA	S	20/15	L = D	Grime *et al.* (1981)
G. sanguineum	Gerani.	P	S	20/15	L > D	Grime *et al.* (1981)
Iliamna corei	Malv.	P	S	30/15	L = D	Baskin and Baskin (1997)
Lathyrus montanus	Fab.	P	S	20/15	L = D	Grime *et al.* (1981)
Lotus uliginosus	Fab.	P	S	20/15	L = D	Grime *et al.* (1981)
Pediomelum subacaule	Fab.	P	S	30	—	Baskin and Quarterman (1970)
Sida hermaphrodita	Malv.	P	S	35/20	L = D	Spooner *et al.* (1985), Baskin and Baskin (1988, unpublished results)
Trifolium stoloniferum	Fab.	P	S	30/15	—	Baskin and Baskin (unpublished results)
Physiological dormancy						
Agalinis fasciculata	Scrophulari.	sA	C	20/10	L	Baskin *et al.* (unpublished results)
Aira praecox	Po.	wA	W	15/6	L > D	Newman (1963)
Allium cernuum	Lili.	P	C	20/10	L = D	Baskin and Baskin (1988, unpublished results)
Alyssum alyssoides	Brassic.	wA	W	15/6	L > D	Baskin and Baskin (1974c)
A. calycinum	Brassic.	wA	W	23	—	Hajkova and Krehule (1972)
A. saxatile	Brassic.	P	C	25/15	L > D	Lhotska (1988)
Aphanes arvensis	Ros.	wA	W	4, 10/4	L > D	Roberts and Neilson (1982)
Arabis laevigata	Brassic.	B	C	25/15	L > D	Bloom *et al.* (1990)
A. shortii	Brassic.	P	C	15/6	L > D	Baskin *et al.* (unpublished results)
A. serotina	Brassic.	B	C	20/10	L > D	Baskin and Baskin (unpublished results)
Arenaria fontinalis	Caryophyll.	wA	W	15/6	L = D	Baskin and Baskin (1987b)
A. glabra	Caryophyll.	wA	W	15/6	L > D	Baskin and Baskin (1982a)
A. patula	Caryophyll.	wA	W	15	—	Caudle and Baskin (1968)
Aristida longespica	Po.	sA	C	35	—	Baskin and Caudle (1967)
Aster divaricatus	Aster.	P	C	20/10	L > D	Baskin *et al.* (1993c)
A. ptarmicoides	Aster.	P	C	20/10	L	Baskin and Baskin (1988, unpublished results)
Bidens laevis	Aster.	sA	C	35/20	L	Leck *et al.* (1994)
B. polylepis	Aster.	sA	C	30/15	L > D	Baskin *et al.* (1995a)
Boltonia decurrens	Aster.	P	C	30/15	L	Baskin and Baskin (1988, unpublished results)

(continues)

TABLE 10.20—*Continued*

Species	Family[a]	LC[b]	TR[c]	Temperature[e]	L:D[e]	Reference
B. diffusa	Aster.	P	C	30/15	L	Baskin *et al.* (unpublished results)
Carex panicea	Cyper.	P	C	20/15	—	Grime *et al.* (1981)
Carlina vulgaris	Aster.	B	C	22/12	L > D	van Tooren and Pons (1988)
Catapodium rigidum	Po.	wA	W	25	L = D	Clark (1974)
Cerastium nutans	Caryophyll.	wA	W	23/12	L > D	Baskin and Baskin (1988, unpublished results)
C. semidecandrum	Caryophyll.	wA	W	5	—	Rozijn and van Andel (1985)
Cyperus granitophilus	Cyper.	sA	C	25/15	L	Baskin *et al.* (unpublished results)
Delphinium alabamicum	Ranuncul.	P	C	5	—	Baskin and Baskin (unpublished results)
D. exaltatum	Ranuncul.	P	C	15/6	L > D	Baskin *et al.* (unpublished results)
Diamorpha cymosa	Crassul.	wA	W	15	L	Baskin and Baskin (1972b)
Draba muralis	Brassic.	wA-B	W	15–20	L = D	Ratcliffe (1961)
D. verna	Brassic.	wA	W	10	L	Baskin and Baskin (1970)
Echinacea simulata	Aster.	P	C	20/10	L > D	Baskin *et al.* (1993c)
E. tennesseensis	Aster.	P	C	20/10	L > D	Baskin *et al.* (1993c)
Elephantopus carolinianus	Aster.	P	C	35/20	—	Baskin and Baskin (unpublished results)
Eragrostis spectabilis	Po.	P	C	30	L = D	Baskin and Baskin (unpublished results)
Eupatorium fistulosum	Aster.	P	C	30/15	L > D	Baskin *et al.* (1993c)
Gentiana andrewsii	Gentian.	P	C	25/15	—	Thompson (1969a)
G. asclepiadea	Gentian.	P	C	22/12	L > D	Maas (1989)
G. crinita	Gentian.	B	C	24/16	L > D	Farmer (1978)
G. cruciata	Gentian.	P	C	25/15	—	Thompson (1969a)
G. quinquefolia	Gentian.	B	C	15/6	L	Baskin and Baskin (1988, unpublished results)
Gentianella campestris	Gentian.	B	C	20/8	D > L	Milberg (1994b)
Grindelia lanceolata	Aster.	B	C	23/12	L > D	Baskin and Baskin (1979e)
Helianthus atrorubens	Aster.	P	C	30/15	—	Baskin *et al.* (unpublished results)
Heliotropium tenellum	Boragin.	sA	C	15/6	—	Baskin and Baskin (1988, unpublished results)
Hypericum gentianoides	Clusi.	sA	C	20/10	L	Baskin and Baskin (unpublished results)
Impatiens glandulifera	Balsamin.	sA	C	20	—	Mumford (1988)
Isanthus brachiatus	Lami.	sA	C	30/15	L > D	Baskin and Baskin (1975b, unpublished results)
Iva annua	Aster.	sA	C	30/15	L > D	Baskin and Baskin (unpublished results)
Jamesianthus alabamensis	Aster.	P	C	35/20	L > D	Baskin and Baskin (unpublished results)
Krigia oppositifolia	Aster.	wA	W	25/15	L	Baskin *et al.* (1991)
Kuhnia eupatorioides	Aster.	P	C	30/15	L = D	Baskin *et al.* (1993c)
Lactuca floridana	Aster.	B	C	30/15	L = D	Baskin *et al.* (unpublished results)
Leavenworthia crassa	Brassic.	wA	W	15	—	Caudle and Baskin (1968)
L. exigua	Brassic.	wA	W	23/12	L > D	Baskin and Baskin (1972c)
L. stylosa	Brassic.	wA	W	20	L > D	Baskin and Baskin (1971a)
L. torulosa	Brassic.	wA	W	15	L > D	Baskin and Baskin (1971a)
L. uniflora	Brassic.	wA	W	23/12	L > D	Baskin and Baskin (1971a)
Lesquerella filiformis	Brassic.	wA	W	30/15	L > D	Baskin and Baskin (1988, unpublished results)
L. lescurii	Brassic.	wA	W	20/10	L > D	Baskin *et al.* (1992)
L. stonensis	Brassic.	wA	W	30/15	L > D	Baskin and Baskin (1990e)
Liatris squarrosa	Aster.	P	C	35/20	L > D	Baskin and Baskin (1989e)
Linum catharticum	Lin.	sA	C	15/5	L	van Tooren and Pons (1988)
Lobelia cardinalis	Campanul.	P	C	30/15	L	Baskin *et al.* (unpublished results)
L. gattingeri	Campanul.	wsA	W,C	20	L	Baskin and Baskin (1979f)
L. inflata	Campanul.	P	C	20/10	L	Baskin and Baskin (1992c)
Manfreda virginica	Agav.	P	C	28/18	L = D	Baskin and Baskin (1971c)
Melampyrum lineare	Scrophulari.	sA	C[f]	3	—	Cantlon *et al.* (1963)
M. pratense	Scrophulari.	wA	W[f]	10	—	Masselink (1980)
Molinia caerulea	Po.	P	C	22/12	L > D	Maas (1989)
Myosotis hispida	Boragin.	wA	W	23	—	Hajkova and Krehule (1972)
M. ramosissima	Boragin.	wA	W	5–10	—	Janssen (1973)
Panicum flexile	Po.	sA	C	30/15	L > D	Baskin and Baskin (1988, unpublished results)
Pedicularis lanceolata	Scrophulari.	P	C	20/10	L > D	Baskin *et al.* (unpublished results)
Penstemon hirsutus	Scrophulari.	P	C	20/10	—	Schroeder and Barton (1939)
Phacelia dubia	Hydrophyll.	wA	W	15	L > D	Baskin and Baskin (1971b)

(continues)

TABLE 10.20—*Continued*

Species	Family[a]	LC[b]	TR[c]	Temperature[a]	L:D[a]	Reference
Phleum arenarium	Po.	wA	W	5	—	Ernst (1981)
Pimpinella major	Api.	P	C	5	—	Grime *et al.* (1981)
Portulaca smallii	Portulac.	sA	C	30/15	L	Baskin *et al.* (1987)
Primula chungensis	Primul.	P?	C	25/15	L	Thompson (1969b)
P. farinosa	Primul.	P	C	22/12	L > D	Maas (1989)
P. japonica	Primul.	P	C	25/5	L > D	Thompson (1969)
P. veris	Primul.	P	C	5, 15	L	Grime *et al.* (1981), Milberg (1994a)
Ratibida pinnata	Aster.	P	C	35/20	L > D	Baskin *et al.* (1993c)
Ruellia humilis	Acanth.	P	C	35/20	L > D	Baskin and Baskin (1982b)
Sabatia angularis	Gentian.	B	C	30/15	L	Baskin and Baskin (unpublished results)
Saxifraga tridactylites	Saxifrag.	wA	W	15–20	L > D	Ratcliffe (1961)
Scutellaria parvula	Lami.	P	W	30/15	—	Baskin and Baskin (1982c)
Sedum pulchellum	Crassul.	wA	W	20/10	L > D	Caudle and Baskin (1968), Baskin and Baskin (1977c)
Silene regia	Caryophyll.	P	C	20/10	L > D	Baskin and Baskin (1988, unpublished results)
Spergula morisonii	Caryophyll.	wsA	W	22/12	L > D	Pons (1989)
Solidago shortii	Aster.	P	C	25/15	L > D	Buchele *et al.* (1991)
Specularia perfoliata	Campanul.	wA	W	15/6	L	Baskin *et al.* (unpublished results)
Succisa pratensis	Dipsac.	P	C	22/12	L > D	Maas (1989)
Talinum calcaricum	Portulac.	P	C	30/15	L	Baskin and Baskin (unpublished results)
T. calycinum	Portulac.	P	C	30/15	L	Baskin and Baskin (unpublished results)
T. mengesii	Portulac.	P	C	25/15	L	Baskin and Baskin (unpublished results)
T. parviflorum	Portulac.	P	C	30/15	L	Baskin and Baskin (unpublished results)
T. rugospermum	Portulac.	P	C	30/15	L	Baskin and Baskin (unpublished results)
T. teretifolium	Portulac.	P	C	30/15	L	Baskin and Baskin (unpublished results)
Teesdalia nudicaulis	Brassic.	wA	W	10	L > D	Newman (1963)
Trisetum flavescens	Po.	P	C[g]	15	L = D	Dixon (1995)
Valerianella carinata	Valerian.	wA	W	20/15	L = D	Grime *et al.* (1981)
V. radiata	Valerian.	wA	W	23/12	L = D	Baskin and Baskin (1988, unpublished results)
V. umbilicata	Valerian.	wA	W	30/16	L > D	Baskin and Baskin (1976)
Verbesina alternifolia	Aster.	P	C	30/15	L > D	Baskin *et al.* (1993c)
V. virginica	Aster.	P	C	25/15	L = D	Baskin *et al.* (unpublished results)
Veronica dillenii	Scrophulari.	wA	W	23	—	Hajkova and Krehule (1972)
Viola egglestonii	Viol.	P	C	20/10	—	Baskin and Baskin (1975c)
V. lutea	Viol.	P	C	5	—	Grime *et al.* (1981)
V. rafinesquii	Viol.	wA	W	20/10	L > D	Baskin and Baskin (1972d)

[a]See notes for Table 10.1.

[b]Life cycle type: wA, winter annual; sA, summer annual; wsA, winter and summer annual; B, biennial or monocarpic perennial; P, polycarpic perennial.

[c]Dormancy-breaking treatment: W, warm stratification; C, cold stratification.

[d]Data not available.

[e]Inferred to have morphophysiological dormancy.

[f]Seedlings require cold stratification to break epicotyl dormancy.

[g]Seeds afterripened in dry storage, exhibiting an increase in germination at low temperatures.

ter annuals *Lesquerella filiformis* (Baskin and Baskin, unpublished results) and *L. stonensis* (Baskin and Baskin, 1990e) is 30/15°C. Summer annuals require cold stratification for loss of dormancy, and nondormant seeds generally germinate best at high temperatures. However, the optimum germination temperature for nondormant seeds of *Heliotropium tenellum* is 15/6°C (Baskin and Baskin, unpublished results) and for *Impatiens glandulifera* 20°C (Mumford, 1988). Seeds of most winter and summer annuals either require light or germinate to higher percentages in light than in darkness.

Only a few winter annuals germinate equally well in light and darkness.

Seeds of biennials and of all perennials, except *Scutellaria parvula,* require cold stratification for loss of physiological dormancy. Seeds of *S. parvula* require high temperatures to come out of dormancy, thus they germinate in autumn. Seeds of biennials germinate over a range of temperatures, depending on the species, and either require light for germination or germinate to higher percentages in light than in darkness. Seeds of *Lactuca floridana,* however, germinate equally well in light and darkness (Baskin *et al.,* unpublished results). Optimum germination temperatures for perennials vary with the species and are either high (30/15, 35/20°C) or low (5, 15/6, 20/10°C). Five species germinate equally well in light and darkness, and 28 others either require light or germinate to higher percentages in light than in darkness. Seeds of *Nothoscordum bivalve* germinate to a higher percentage in darkness than in light. It should be noted that although nondormant seeds of some species germinate equally well in light and darkness at optimum temperatures, they may germinate to higher percentages in light than in darkness at nonoptimal germination temperatures (e.g., Baskin and Baskin, 1982b,c).

G. Special Factors

1. Myrmecochory

Beattie (1983) has compiled a list of plant genera with myrmecochorus species, and many of them are found in the region of temperate deciduous forests (Table 10.21). Myrmecochores are an important component of the herbaceous layer of forests (Buckley, 1982). For example, ants disperse seeds of about 30% of the spring-flowering herbaceous species in deciduous forests of eastern North America (Lanza *et al.,* 1992).

Seed preferences of ants are related to elaiosome mass and chemical composition, which vary with the plant species. The elaiosome mass of *Trillium grandiflorum, T. erectum,* and *T. undulatum* seeds was 3.50, 3.90, and 1.05 mg, respectively; total lipids were 1.1, 0.71, and 0.04 mg, respectively; and oleic acid was 34.4, 32.84, and 35.86%, respectively (Lanza *et al.,* 1992). Oleic acid was the most abundant fatty acid in elaiosomes of all three species, and linoleic acid was more abundant in elaiosomes of *T. erectum* and *T. grandiflorum* than in those of *T. undulatum.* Oleic acid stimulates ant attraction, and linoleic acid stimulates ant feeding behavior. Thus, ants are attracted to seeds of the three species of *Trillium,* but fewer seeds of *T. undulatum* are taken than those of *T. erectum* or *T. grandiflorum* (Lanza *et al.,* 1992).

Because ants move seeds from the sites where they were originally dispersed and eat the elaiosome, perhaps even scarifying the seed coat (e.g., Culver and Beattie, 1978), do ants enhance the germination of seeds? Seeds of *Viola papilionacea* collected from refuse piles outside ant nests germinated to 64%, whereas those not handled by ants germinated to 56% (Culver and Beattie, 1978). The germination of *Viola odorata* and *V. hirta* seeds was improved (but not significantly) by scarification and elaiosome removal by ants (Culver and Beattie, 1980). Removal of elaiosomes from seeds of *Sanguinaria*

TABLE 10.21 Plant Genera with Myrmecochorus Species Occurring in the Region of Temperate Deciduous Forests[a]

Genus	Family	Genus	Family
Acalypha	Euphorbiaceae	*Lamium*	Labiatae
Adonis	Ranunculaceae	*Leucojum*	Amaryllidaceae
Ajuga	Labiatae	*Luzula*	Juncaceae
Allium	Liliaceae	*Melampyrum*	Scrophulariaceae
Anchusa	Boraginaceae	*Melica*	Poaceae
Anemone	Ranunculaceae	*Mercurialis*	Euphorbiaceae
Arenaria	Caryophyllaceae	*Mertensia*	Boraginaceae
Asarum	Aristolochaceae	*Moehringia*	Caryophyllaceae
Ballota	Labiatae	*Myosotis*	Boraginaceae
Borago	Boraginaceae	*Nemophilia*	Hydrophyllaceae
Buxus	Buxaceae	*Nonea*	Boraginaceae
Carduus	Asteraceae	*Ornithogalum*	Liliaceae
Carex	Cyperaceae	*Parietaria*	Urticaceae
Centaurea	Asteraceae	*Pedicularis*	Scrophulariaceae
Chelidonium	Papaveraceae	*Phyteuma*	Campanulaceae
Chrysogonum	Asteraceae	*Polygala*	Polygalaceae
Cirsium	Asteraceae	*Potentilla*	Rosaceae
Claytonia	Portulacaceae	*Primula*	Primulaceae
Cleome	Cleomaceae	*Pulmonaria*	Boraginaceae
Corydalis	Fumariaceae	*Ranunculus*	Ranunculaceae
Croton	Euphorbiaceae	*Reseda*	Resedaceae
Cyclamen	Primulaceae	*Rhynchosia*	Fabaceae
Datura	Solanaceae	*Sanguinaria*	Papaveraceae
Delphinium	Ranunculaceae	*Scabiosa*	Dipsacaceae
Desmodium	Fabaceae	*Scilla*	Liliaceae
Dicentra	Fumariaceae	*Sclerolaena*	Chenopodiaceae
Dichromena	Cyperaceae	*Sieglingia*	Poaceae
Dischidia	Asclepiadaceae	*Stylophorum*	Papaveraceae
Epimedium	Berberidaceae	*Symphytum*	Boraginaceae
Euphorbia	Euphorbiaceae	*Thesium*	Santalaceae
Fumaria	Fumariaceae	*Tephrosia*	Fabaceae
Gagea	Liliaceae	*Teucrium*	Labiatae
Galanthus	Amaryllidaceae	*Trillium*	Liliaceae
Galeobdolon	Labiatae	*Triodia*	Poaceae
Helleborus	Ranunculaceae	*Ulex*	Fabaceae
Hepatica	Ranunculaceae	*Uvularia*	Liliaceae
Hibiscus	Malvaceae	*Veronica*	Scrophulariaceae
Hybanthus	Violaceae	*Viola*	Violaceae
Iris	Iridaceae	*Waldsteinia*	Rosaceae
Jeffersonia	Berberidaceae		

[a]From Beattie (1983).

canadensis with forceps significantly increased germination (Lobstein and Rockwood, 1993). However, this treatment did not improve the germination of *Asarum canadense, Jeffersonia diphylla, Viola striata,* or *V. pensylvanica* seeds (Rockwood and Blois, 1986; Lobstein and Rockwood, 1993), and results for *Dicentra cucullaria* were inconclusive. Seeds of *Mercurialis annua* germinated to 12% when the elaiosome was removed with a razor blade, but elaiosomeless seeds taken from ant nests in June germinated to 89% (Pacini, 1990). The length of time seeds had been in the nest was not given; thus, the author may have been comparing germination in freshly matured (i.e., those with hand-removed elaiosome) vs. warm- and/or cold-stratified (those from ant nests) seeds.

Secondary dispersal by ants may result in an escape from seed predation (Culver and Beattie, 1978; Turnbull and Culver, 1983) or from competition from equal-aged specific congeners (Handel, 1978). Also, seeds may be placed in a more favorable environment (nest) for germination and seedling establishment than they were originally. The nest environment increased the germination of *Viola odorata* and *V. hirta* seeds (Culver and Beattie, 1980), and emergence of *Corydalis aurea* was greater for seeds in nests than for those planted at a comparable depth in the vicinity of nests. Mapping studies showed that juveniles and seedlings of *Polygala vulgaris* and *Viola curtisii* are closely associated with ant nests richer in potassium, phosphate, and nitrate than the surrounding areas (Oostermeijer, 1989). Percentages of seedling survival were slightly higher in ant-planted than in hand-planted (control) seeds (Hanzawa *et al.,* 1988). Thus, it has been suggested that increased levels of potassium, phosphate, and nitrate in nests are partly responsible for higher seedling survival in nests than in surrounding areas (Beattie and Culver, 1983; Oostermeijer, 1989). In calcareous grasslands in England, *Arenaria serpyllifolia* (an annual), *Cerastium holosteoides, Helianthemum chamaecistus,* and *Thymus drucei* (perennial chamaephytes) were more abundant on ant mounds than in the surrounding vegetation. However, soils did not differ from those of the surrounding areas with regard to pH, calcium carbonate, or extractable phosphorus, but they were lower in organic matter and nitrogen than other soils (King, 1977).

2. Litter

Temperate deciduous forests are characterized by leaf senescence in autumn, and leaf-litter production ranges from 180 to 360 g^{-2} $year^{-1}$ (Rohring, 1991d). This mass of material deposited on the forest floor may have a great impact on the herbaceous flora. For example, few plants of the grasses *Holcus mollis, Poa trivialis,* or *Milium effusum* occur in sites in deciduous woodlands in the British Isles where the litter mass is more than 200 g m^{-2} (Sydes and Grime, 1981). However, the herbs *Anemone nemorosa, Endymion non-scriptus,* and *Galeobdolon luteum* and seedlings of the tree *Fraxinus excelsior* grow where the litter mass is 400 g m^{-2} or more. Does the absence of species such as *H. mollis, P. trivialis,* or *M. effusum* from areas with heavy litter indicate a lack of germination or seedling survival?

Litter inhibits germination in several species (some of which are not forest herbs), including *Arabis laevigata* (Bloom *et al.,* 1990), *Carduus nutans* (Hamrick and Lee, 1987), *Pinus pungens* (Williams *et al.,* 1990), *Ulmus davidiana* (Seiwa, 1997), and, indirectly, *Acer saccharum, Aster divaricatus, Eupatorium rugosum, Hieracium pratense, Hypericum punctatum,* and *Veronica officinalis* (Beatty and Sholes, 1988). Further, litter reduced the germination of *Pinus taeda* (Pomeroy, 1949), *Cichorium intybus* (Smith and Capelle, 1992), *Betula alleghaniensis* (Peterson and Facelli, 1992), *Juniperus virginiana, Cornus florida, Carya tomentosa,* and *Quercus rubra* (Myster, 1994) seeds, but not those of *Rhus typhina* (Peterson and Facelli, 1992). The germination of *Capsella bursa-pastoris* and *Senecio vulgaris* seeds decreased with an increase in *Poa annua* litter (Bergelson, 1990). In general, the presence of ground cover was more inhibitory for the germination of small than of large seeds (Reader, 1993). Rye mulch (rye plants killed by herbicide) decreased the germination of *Chenopodium album, Portulaca aleracea,* and *Digitaria sanguinalis* seeds in no-till corn fields but not that of *Amaranthus retroflexus* or *Taraxacum officinale* seeds (Mohler and Calloway, 1992). However, 2.5 cm of litter was required for successful germination and establishment of *Quercus montana* (Barrett, 1931), and removal of litter in treefall pits did not increase the germination of *Prenanthes altissima, Epipactis helleborine, Arisaema atrorubens,* or *Caulophyllum thalictroides* seeds (Beatty and Sholes, 1988), indicating that the presence of litter did not inhibit germination. Seeds of *Carduus acanthoides* germinated to higher percentages in gaps with litter than in those without it (Feldman *et al.,* 1994).

There are several ways in which litter could influence the germination ecology of a species. If litter decreases the amount of light reaching the soil surface, the germination of light-requiring seeds might be prevented. However, if seeds germinate to higher percentages in darkness than in light, litter might increase germination. To understand the effects of litter on germination, we need to know (1) the light:dark requirements for the germination of seeds in various stages of afterripening and (2) the quality and quantity of light reaching the soil surface. Further, litter from different species may not create the same light environment on the soil surface, e.g., light transmittance through *Solidago* litter > *Setaria* litter > *Quercus* litter (Facelli and Pickett, 1991a). Darkness or far-red light induced dormancy in seeds of *Ulmus davidiana*, thus explaining why they did not germinate under litter in the field (Seiwa, 1997). After dormant seeds were cold stratified, they germinated in white or red light.

The presence of litter influences soil temperatures and/or moisture levels (Facelli and Pickett, 1991a,b; Fowler, 1988; Williams *et al.*, 1990), but the effects of these changes on the germination of species in the temperate deciduous forest region have not been investigated; some studies have been done in temperate grasslands (see Section V,E,1). Litter sometimes forms a mechanical barrier that prevents seeds from reaching the soil surface (Koroleff, 1954; Cabiaux and Devillez, 1977; Hamrick and Lee, 1987). Thus, seeds are held up in/on the litter where insufficient moisture levels delay germination, and other factors such as pathogens may result in a loss of viability.

3. Heavy Metals

Depending on the species and the concentration, metals may stimulate, inhibit, or have no effect on germination. For example, a high copper concentration (160 μM) significantly reduced the germination of *Minuartia hirsuta*, *Silene compacta*, *Alyssum montanum*, and *Thlaspi ochroleucum* seeds, but a low copper concentration (8 μM) promoted the germination of *M. hirsuta*, *S. compacta*, and *A. montanum* seeds (Ouzounidou, 1995). Manganese at 1 and 10 mg l^{-1} stimulated the germination of *Hordeum vulgare* seeds, but the same concentration of lead reduced germination. In mixtures of manganese and lead, an increase of either lead or manganese reduced germination (Pathak *et al.*, 1987). Lead concentrations above 0.01% reduced the germination of *Lepidium sativum* and *Sinapis alba* seeds, and 0.5% inhibited it (Dilling, 1926).

The germination of *Miscanthus floridulus* and *M. transmorrisonensis* seeds decreased with an increase in concentrations (ranging from 200 to 1000 ppm) of lead, mercury, copper, and cadmium. Seeds were most sensitive to cadmium and least sensitive to lead (Hsu and Chou, 1992). An increase in aluminum concentration (from 0 to 25 ppm) reduced the germination of *Alopecurus pratensis*, *Deschampsia flexuosa*, *Festuca pratensis*, and *Lolium perenne* seeds (Hackett, 1964). At concentrations of 1, 5, and 10 mg/liter, chromium (as $K_2Cr_2O_7$) did not affect the germination of *Salvia sclarea* seeds, but radicle growth was inhibited (Corradi *et al.*, 1993).

Because ecotypes of various species are tolerant of metals such as copper, lead, zinc, nickel, chromium, or cobalt, it is possible that seeds produced by plants grow-ing in contaminated sites may germinate in higher concentrations of metals than those produced by plants growing in uncontaminated areas. However, as noted by Archambault and Winterhalder (1995), germinated-based metal tolerance tests usually are not done. These authors found that seeds produced by *Agrostis scabra* plants growing in contaminated sites in Sudbury, Ontario, Canada, had a significantly higher ability to germinate in the presence of metals in contaminated soils than seeds produced by plants growing in uncontaminated sites. Because low levels of copper stimulated the germination of seeds of *Alyssum montanum*, *Minuartia hirsuta*, and *Silene compacta* collected in copper-rich soils in Greece (Ouzounidou, 1995), it would be interesting to know if seeds of these species collected from plants growing in noncontaminated soils would be stimulated to germinate by low levels of copper.

4. Herbicides

Little information is available on the effects of herbicides on germination, especially noncultivated species. Depending on the species and the kind and concentration of herbicide, germination may be enhanced (Huffman and Jacoby, 1984), reduced (Morash and Freedman, 1989), or unaffected (Shaukat, 1974; Benjamini, 1986). Low concentrations of a particular herbicide may stimulate germination, but high concentrations inhibit germination (Shaukat, 1974; Huffman and Jacoby, 1984; Morash and Freedman, 1989). Studies are needed to determine if herbicides have the same effects on seed germination in the field that they do under laboratory/greenhouse conditions.

Tribenuron-methyl applied to growing plants of *Galium spurium* reduced the weight of seeds, and germination decreased with a decrease in seed weight (Andersson, 1996). In the same study, germination was reduced in seeds of *Fallopia convolvulus* when mother plants were treated with MCPA (4-chlor-2-methylphenoxyacetic acid), and it was reduced in those of *Thlaspi arvense* when mother plants were treated with tribenuron-methyl.

V. STEPPES OR GRASSLANDS OF THE TEMPERATE ZONES

Natural extratropical grasslands may be called prairies, steppes, veld, or pampas (Coupland, 1993), and they occur in North and South America, Europe, Asia, New Zealand, South Africa, and southeastern Australia (Eyre, 1963; Walter, 1973; Tainton and Walker, 1993; Moore, 1993) (see region 7 in Fig. 9.1). In general, temperate grasslands receive less rain than deciduous forests but more than deserts, and Ripley (1992) has determined that the average total annual precipitation for grasslands is 640 mm. Grasslands frequently occur in areas with semipermanent high atmospheric pressure and greater levels of solar radiation than other sites at the same latitude (Ripley, 1992). The continental location of many grasslands means that summers are hotter and winters colder than they are at the edges of land masses (Walter, 1973). Mean annual temperatures vary greatly, and in North America, for example, they range from about 3°C in Edmonton, Alberta, Canada, to 23°C in southern Texas (Ripley, 1992).

As the name "grassland" indicates, grasses are the dominant plants, and the systematic index of "Ecosystems of the World: Natural Grasslands" (Coupland, 1992b, 1993) lists 273 genera in the Poaceae. Sedges also are important, and they, along with grasses, are referred to as graminoides. A rich array of herbaceous, nongraminoides (forbs) occurs in many grasslands, and small shrubs and halfshrubs may be present (Coupland, 1992a). Theoretically, trees are not a part of climax grassland vegetation, but they may be present in ecotones with forests, along ravines, streams, and rivers, or scattered in a savanna-like fashion in some grasslands (Coupland, 1992a; Lavrenko and Karamysheva, 1993).

A. Dormancy and Germination of Trees

Many tree species found in grasslands also grow in adjacent forest communities, thus the germination ecology of these species is covered under the appropriate type of forest. However, some trees, such as *Populus deltoides* subsp. *monilifera*, are more common in grasslands than in other areas. Seeds of this species are nondormant at maturity; consequently, germination peaks at the time of seed dispersal in early summer (Shafroth *et al.*, 1995a). Seeds sown on moist soil germinated to maximum percentages in full sunlight (vs. shade) when the groundwater level was at 42 cm (vs 12, 20, 32, or 523 cm) below the soil surface (Shafroth *et al.*, 1995a).

B. Dormancy and Germination of Shrubs

About 30 genera of shrubs are important in grasslands (Coupland, 1992b, 1993). Most shrubs are members of the Asteraceae, Chenopodiaceae, Fabaceae, or Rosaceae, but some belong to the Caprifoliaceae, Lamiaceae, Lythraceae, Polygonaceae, Rhamnaceae, or Tamariaceae.

Genera of shrubs in the Asteraceae include *Agania, Asterothamnus, Artemisia, Baccharis, Eupatorium, Olgaea,* and *Vernonia*. Achenes of *Baccharis* required no pretreatment for germination at 30/20°C (Olson, 1974c), whereas those of *Artemisia* (Deitschman, 1974) required cold stratification to come out of physiological dor-

mancy. Shrubs in the Fabaceae (*Acacia, Amorpha, Calophaca, Caragana*) and Rhamnaceae (*Ceanothus*) have seeds (or fruits) with impermeable coats. Dietz and Slabaugh (1974) reported that the germination of *Caragana arborescens* seeds was increased by cold stratification at 5°C for 15 days. However, it is unclear if low temperatures caused seed coats to become permeable or if seeds also have physiological dormancy, which was broken by cold stratification.

It is assumed that shrubs in the Rosaceae (*Amygdalus, Elaeagnus, Cerasus, Cotoneaster, Margyricarpus, Potentilla,* and *Rosa*) and Chenopodiaceae (*Anabasis, Atriplex, Kochia, Nanophyton,* and *Salsola*) have seeds with physiological dormancy that is broken by cold stratification. Seeds of *Elaeagnus commutata* reached a maximum of 75% germination after 100 days of cold stratification at 5°C (Corns and Schraa, 1962), and those of *E. angustifolia* germinated to 60% after 84 days of cold stratification at 5°C (Hogue and LaCroix, 1970).

Shrubs or small trees in the genus *Tamarix* may be frequent in saline sites such as halophytic meadows (Ting-Cheng, 1993), coastal marshes (Smeins *et al.,* 1992), and along streams in desert grasslands (Schmutz *et al.,* 1992). Freshly matured seeds of *T. pentandra* germinated to 96, 95, 95, 95, 92, 87, 98, and 91% at constant temperatures of 19, 22, 27, 29, 33, 35, 41, and 43°C, respectively (Wilgus and Hamilton, 1962). At temperatures of 3–27°C, seeds retained high viability if stored in sealed containers with a desiccant, but they lost viability if stored at "moderate humidity" (Wilgus and Hamilton, 1962). Thus, in the field dispersed seeds probably would remain viable for only a few weeks.

C. Dormancy and Germination of Herbaceous Species

1. Forbs

None of the forbs studied thus far have seeds that are nondormant at maturity. Germination data for *Geum triflorum* (Sorensen and Holden, 1974), *Liatris aspera,* and *L. pycnostachya* (Salac and Hesse, 1975) seem to indicate that seeds of these species are nondormant at maturity. However, Sorensen and Holden (1974) stored seeds at room temperatures from autumn until January. Salac and Hesse (1975) do not say how long seeds were stored, nor do they give the storage conditions. Thus, seeds could have afterripened in these two studies before germination tests were initiated; the three species are listed under physiological dormancy in Table 10.22.

Seeds of a few species have underdeveloped embryos and require cold stratification for germination. Thus, these seeds have morphophysiological dormancy, but the type of MPD has not been determined. There is a

TABLE 10.22 Seed Dormancy in Forbs of Temperate Zone Steppes or Grasslands

Species	Family[a]	Tr[b]	D[c]	Temperature[e]	L:D[d]	Reference
Morphopysiological dormancy						
Anemone cylindrica	Ranuncul.	C	84	16–22	—[d]	Kis (1984)
A. patens	Ranuncul.	C	60	18–21	—	Greene and Curtis (1950)
Cicuta maculata	Api.	C	120	21	—	Sorensen and Holden (1974)
Eryngium yuccifolium	Api.	C	150	18–21	—	Greene and Curtis (1950)
Lilium philadelphicum	Lili.	C	150	18–21	—	Greene and Curtis (1950)
Zizia aptera	Api.	C	150	18–21	—	Greene and Curtis (1950)
Physical dormancy						
Amorpha canescens	Fab.	S		21	—	Sorensen and Holden (1974)
Astragalus bisulcatum	Fab.	S		22	—	Smreciu *et al.* (1988)
A. canadensis	Fab.	S		21	—	Sorensen and Holden (1974)
A. crassicarpus	Fab.	S		21	—	Sorensen and Holden (1974), Smreciu *et al.* (1988)
A. striatus	Fab.	S		22	—	Smreciu *et al.* (1988)
Callirhoe triangulata	Malv.	S		24	—	Voigt (1977)
Dalea purpureum	Fab.	S		Spring	—	Dickerson *et al.* (1981)
Desmanthus illinoensis	Fab.	S		Room?	—	Latting (1961)
D. velutinus	Fab.	S		30/20	—	Haferkamp *et al.* (1984)
Hippocrepis unisiliquosa	Fab.	S		15	—	Ehrman and Cocks (1996)
Lespedeza capitata	Fab.	S		30/20	—	Segelquist (1971)
L. stuevei	Fab.	S		Spring	—	Dickerson *et al.* (1981)
L. virginica	Fab.	S		30/20	—	Segelquist (1971)
Medicago blancheana	Fab.	S		15	—	Ehrman and Cocks (1996)
M. constricta	Fab.	S		15	—	Ehrman and Cocks (1996)
M. polymorpha	Fab.	S		15	—	Ehrman and Cocks (1996)
M. rigidula	Fab.	S		15	—	Ehrman and Cocks (1996)
M. rotata	Fab.	S		15	—	Ehrman and Cocks (1996)
Onobrychis crista-galli	Fab.	S		15	—	Ehrman and Cocks (1996)
Oxytropis monticola	Fab.	S		22	—	Smreciu *et al.* (1988)
O. sericea	Fab.	S		22	—	Smreciu *et al.* (1988)
Psoralea agrophylla	Fab.	S		24	L = D	Spessard (1988)
P. esculenta	Fab.	S		24	D > L	Spessard (1988)
Thermopsis rhombifolia	Fab.	S		22	—	Smreciu *et al.* (1988)
Trifolium argutum	Fab.	S		15	—	Ehrman and Cocks (1996)
T. ballatum	Fab.	S		15	—	Ehrman and Cocks (1996)
T. cherleri	Fab.	S		15	—	Ehrman and Cocks (1996)
T. pauciflorum	Fab.	S		15	—	Ehrman and Cocks (1996)
T. scutatum	Fab.	S		15	—	Ehrman and Cocks (1996)
T. spumosum	Fab.	S		15	—	Ehrman and Cocks (1996)
Trigonella monspeliaca	Fab.	S		15	—	Ehrman and Cocks (1996)
Physiological dormancy						
Agoseris cuspidata	Aster.	C	120	18–21	—	Greene and Curtis
Apocynum androsaemifolium	Apocyn.	C	60	18–21	—	Greene and Curtis (1950)
Artemisia frigida	Aster.	C[e]	Winter	10	L > D	Bai and Romo (1994), Bai *et al.* (1996)
Asclepias floridana	Asclepiad.	C	120	18–21	—	Greene and Curtis (1950)
A. meadii	Asclepiad.	C	70	—	—	Betz (1989)
A. ovalifolia	Asclepiad.	C	120	18–21	—	Greene and Curtis (1950)

(continues)

possibility that seeds of *Anemone cylindrica* and *A. patens* could have morphological rather than morphophysiological dormancy, as indicated in Table 10.22. Freshly matured seeds were not tested, but seeds of these two species germinated after being stored dry at room temperatures from autumn until January (Sorensen and Holden, 1974). Thus, seeds may have had only morphological dormancy, which required no dormancy-breaking treatment, or they may have had nondeep simple MPD, which was broken during dry storage at room temperatures.

With the exception of *Callirhoe triangulata* (Malvaceae), the species listed under physical dormancy in Table 10.22 are legumes. Scarification results in high percentages of germination at simulated spring or autumn temperatures, but little is known about light : dark

TABLE 10.22—*Continued*

Species	Family[a]	Tr[b]	D[c]	Temperature[e]	L:D[d]	Reference
A. sullivantii	Asclepiad.	C	150	18–21	—	Greene and Curtis (1950)
A. tuberosa	Asclepiad.	C	60	18–21	—	Greene and Curtis (1950)
A. viridiflora	Asclepiad.	C	30	18–21	—	Greene and Curtis (1950)
Aster novae-angliae	Aster.	C	56	16–21	—	Kis (1984)
Besseya wyomingensis	Scrophulari.	C	60–90	22	—	Smreciu *et al.* (1988)
Cacalia atriplicifolia	Aster.	C	150	18–21	—	Greene and Curtis (1950)
Coreopsis palmata	Aster.	C	60	24	—	Voigt (1977)
Coriospermum villosum	Chenopodi.	C	90	21	—	Tolstead (1941)
Dodecatheon meadia	Primul.	C	84	27/10	L	Turner and Quarterman (1968)
Echinacea angustifolia	Aster.	C	84	20/10	L > D	Baskin *et al.* (1992a)
E. pallida	Aster.	C	30	25	—	Albrecht and Smith-Jochum (1990)
Eriocoma hymenoides	Po.	C	90	21	—	Tolstead (1941)
Euphorbia corollata	Euphorbi.	C	60	24	—	Voigt (1977)
Gentiana andrewsii	Genti.	C	84	16–22	—	Kis (1984)
G. puberula	Genti.	C	90	18–21	—	Greene and Curtis (1950)
Grindelia squarrosa	Aster.	C	60–90	22, 25/20	L > D	Smreciu *et al.* (1988), McDonough (1975)
Helianthus maximillianii	Aster.	C	56	30/15	L = D	Baskin *et al.* (unpubl.)
H. mollis	Aster.	C	60	24	—	Voigt (1977)
H. petiolaris	Aster.	C	42	21	—	Tolstead (1941)
H. subrhombiodeus	Aster.	C	60–90	22	—	Smreciu *et al.* (1988)
Kuhnia eupatorioides	Aster.	C	60	18–21	—	Greene and Curtis (1950)
Lepidium densiflorum	Brassic.	W	—	21	—	Tolstead (1941)
Liatris aspera	Aster.	C	105	33/19	—	Salac and Hesse (1975)
L. cylindrica	Aster.	C	120	18–21	—	Greene and Curtis (1950)
L. ligulistylis	Aster.	C	150	18–21	—	Greene and Curtis (1950)
L. pycnostachya	Aster.	C	105	33/19	—	Salac and Hesse (1975)
Parthenium integrifolium	Aster.	C	150	18–21	—	Greene and Curtis (1950)
Penstemon angustifolius	Scrophulari.	C	90	21	—	Tolstead (1941)
P. grandiflorus	Scrophulari.	C	105	33/19	—	Salac and Hesse (1975)
Plantago purshii	Plantagin.	W	—	21	—	Tolstead (1941)
Potentilla arguta	Ros.	C	120	18–21	—	Greene and Curtis (1950)
Prenanthes racemosa	Aster.	C	150	18–21	—	Greene and Curtis (1950)
Ratibida columinifera	Aster.	C	90	21	—	Tolstead (1941)
R. pinnata	Aster.	C	90	18–21	—	Greene and Curtis (1950)
Rumex venosus	Polygon.	C	90	21	—	Tolstead (1941)
Silphium laciniatum	Aster.	C	150	18–21	—	Greene and Curtis (1950)
S. perfoliatum	Aster.	C	150	18–21	—	Greene and Curtis (1950)
S. terebinthinaceum	Aster.	C	150	18–21	—	Greene and Curtis (1950)
Solidago rigida	Aster.	C	60–90	22	—	Smreciu *et al.* (1988)

[a]See notes for Table 10.1.
[b]Treatment: C, cold stratification; S, scarification; W, high summer temperatures.
[c]Length of treatment (days).
[d]Data not available.
[e]Seeds were in vials buried in the field in Saskatoon, Saskatchewan, Canada, for 10 months.

requirements for germination. Also, much remains to be learned about the loss of physical dormancy under natural conditions. Voigt (1977) found that cold stratification increased the germination of *C. triangulata* seeds by 41% and that of *Lespedeza virginica* seeds by 39%.

Most temperate grassland forbs have physiological dormancy, which is broken by cold stratification (Table 10.22). Thus, seeds would come out of dormancy in response to cold stratification during winter and germinate in the habitat in spring. Newly germinated seedlings

of many species have been observed in spring (Clements and Weaver, 1924; Tolstead, 1941). Seeds of some forbs germinate following dry storage at low (2–5°C) (Wilson and McCarty, 1984; Halinar, 1981) or high (16, 20°C) temperatures (Blake, 1935; Salac and Hesse, 1975; Owens and Call, 1985) or following treatment with gibberellic acid (GA) (Kis, 1984; Chandler and Jan, 1985). These data indicate that seeds of many forbs probably have nondeep physiological dormancy.

The optimum germination temperature for seeds

after they come out of physiological dormancy is about 21°C, but few studies have defined temperature and light:dark requirements for the germination of forb seeds. One reason for the lack of research on specific temperature requirements for germination may be related to the fact that an important stimulus for learning how to germinate forb seeds has been to use the species in restoration projects, especially in the tallgrass prairies of the midcontinental United States (Sorensen and Holden, 1974). Because simulated spring temperatures promote high germination percentages of cold-stratified seeds, there has been little incentive to try other temperatures. Little is known about light:dark requirements of most species because (1) seeds were not tested in both light and darkness (Blake, 1935; Greene and Curtis, 1950; Salac and Hesse, 1975) or (2) they were incubated in darkness, but dishes were removed and examined (usually in room light) on a regular basis, thereby probably fulfilling any light requirement for germination the seeds might have had (e.g., Sorensen and Holden, 1974; Voigt, 1977; Smreciu et al., 1988).

A few forbs such as *Lepidium densiflorum* and *Plantago purshii* are winter annuals, and their seeds come out of dormancy during summer and germinate in autumn (Tolstead, 1941).

2. Graminoids

a. Dormancy

Although the term graminoids includes both grasses and sedges, little work has been done on the germination of sedges that grow in temperate grasslands. However, it is widely recognized that grasses in the grasslands, and grasses in general, have seed dormancy. In fact, Simpson (1990) gives references for seed dormancy studies on about 200 forage grasses and numerous weedy species in his book "Seed Dormancy in Grasses."

Because seeds of many grasses from temperate grasslands afterripen during dry storage and GA promotes germination, it appears that freshly matured seeds have nondeep physiological dormancy, i.e., embryos lack sufficient growth potential to overcome the inhibiting effects of the palea and lemma. Treatments that remove the inhibiting effects of covering layers (scarification) or increase the growth potential of the embryo (GA or KNO_3) result in germination. Thus, it is relatively easy to germinate seeds of grasses, using a variety of nonecological treatments (Table 10.23).

Seeds of most species, except those of winter annuals, are dispersed in summer/autumn and they lie on or in the soil during winter. Loss of dormancy occurs during winter and seeds germinate in spring. For spring-germinating grasses, the way to simulate dormancy-breaking conditions in the field is to give the seeds a period of cold stratification. However, the number of studies investigating the length of the cold stratification period required to break dormancy in seeds of temperate grasslands grasses is surprisingly low (Table 10.24). In general, grass seeds do not require extended periods of cold stratification for high germination percentages, which is another reason for concluding that they have nondeep physiological dormancy.

Some grasses, including *Bromus commutatus*, *Festuca octoflora* (Tolstead, 1941), *B. tectorum* (Hulbert, 1955), *Vulpia bromoides*, *V. myuros* (Dillon and Forcella, 1984), *B. mollis*, *B. diandrus*, *B. madrilensis*, *Avena barbata*, *Taeniatherum asperum*, *V. megalura*, and *Aira caryophyllea* (Young et al., 1981), are winter annuals, and dormancy break is promoted by high temperatures. Dormancy loss in *B. tectorum* was accelerated by placing dry seeds at constant temperatures of 40 and 50°C for 4–28 days (Thill et al., 1980). Seed dormancy in *B. tectorum* was at a minimum in the field in autumn (after the summer high temperature afterripening period), reached a peak in winter, and began to decline again in spring (Evans and Young, 1975). More information about the germination of annual weedy grasses will be given (see later).

b. Germination Requirements

Grasses can be divided into two categories based on the photosynthetic pathway: cool season species with the C_3 photosynthetic pathway and warm season species with the C_4 photosynthetic pathway. As the names indicate, the most favorable periods for growth are cool and warm seasons, respectively (Waller and Lewis, 1979). Using regression analysis and extrapolation, Jordan and Haferkamp (1989) determined that the temperatures at which the germination rate of cool and warm season range grasses approached zero were 3.7–6.3 and 7.8–13.7°C, respectively. The mean optimum germination temperature for cool and warm season grasses is 16.4 ± 0.6 and 25.5 ± 0.8°C, respectively (Table 10.25). (For an alternating temperature regime, the two numbers were averaged.) These temperatures imply that seeds of cool season species would germinate earlier in spring than those of warm season grasses.

If seeds of warm season species germinate later than cool season ones, seeds of warm season species should germinate under greater moisture stress than those of cool season species. Data in Table 10.25 indicate that this is true. Mean (±SE) moisture stress that reduced germination to about 50% was −0.64 ± −0.1 MPa and −1.02 ± −0.12 MPa for cool and warm season species, respectively.

The tolerance of moisture stress varies with temperature. For example, at −0.58 MPa, seeds of *Festuca altaica*

TABLE 10.23   Nonnatural Treatments That Promote Germination of Seeds of Grasses
in Temperate Zone Steppes or Grasslands

Species	% germination	Reference
Dry storage of spring-germinating seeds		
Andropogon gerardii	100	Coukos (1944)
A. hallii	90	Shaidaee et al. (1969)
Bouteloua curtipendula	95	Shaidaee et al. (1969)
B. gracilis	95	Shaidaee et al. (1969)
B. hirsuta	64	Tolstead (1941)
Calamovilfa longifolia	98	Maun (1981)
Elymus canadensis	66	Robocker et al. (1953)
E. virginicus	66	Robocker et al. (1953)
Eragrostis trichodes	32	Robocker et al. (1953)
Hilaria belangeri	92	Ralowicz and Mancino (1992)
Panicum virgatum	90	Shaidaee et al. (1969)
Poa sandbergii	95	Evans et al. (1977)
Schizachyrium scoparium	100	Coukos (1944)
Sorghastrum nutans	93	Coukos (1944)
Stipa viridula	25	Robocker et al. (1953)
Mechanical injury of caryopsis or covering layers		
Bothriochloa intermedia	77	Ahring et al. (1975)
B. ischaemum	76	Ahring et al. (1975)
Buchloe dactyloides	73	Thornton (1966), Ahring and Todd (1977)
Distichlis stricta	98	Sabo et al. (1979)
Eragrostis lehmanniana	75	Wright (1973)
Panicum virgatum	84	Sautter (1962), Zhang and Maun (1989)
Phalaris arundinacea	88	Junttila et al. (1978)
Schizachyrium scoparium	95	Hussain and Ilahi (1990)
Sporobolus contractus	67	Toole (1941)
S. cryptandrus	71	Toole (1941)
S. flexuosus	74	Toole (1941)
S. giganteus	59	Toole (1941)
Stipa viridula	96	Frank and Larson (1970)
Taeniatherum asperum[a]	96	Nelson and Wilson (1969)
Tripsacum dactyloides	86	Anderson (1985)
Gibberellic acid		
Andropogon gerardii	46	Kucera (1966)
Hilaria belangeri	80	Ralowicz et al. (1992)
Paspalum plicatulum	35	Fulbright and Flenniken (1988)
Schizachyrium scoparium	21	Svedarsky and Kucera (1970)
Sorghastrum nutans	25	Svedarsky and Kucera (1970)
Stipa columbiana	81	Young et al. (1990)
S. viridula	49–100	Fulbright et al. (1983)
Tripsacum dactyloides	82	Anderson (1985)
Heating pretreatments		
Buchloe dactyloides	58	Ahring and Todd (1977)
Eragrostis trichodes	100	Ahring et al. (1963)
Potassium nitrate		
Buchloe dactyloides	60	Wenger (1941)
Eragrostis trichodes	63	Ahring et al. (1963)
Panicum obtusum	56	Toole (1940)
Sporobolus aspera	39	Toole (1941)
S. contractus	47	Toole (1941)
S. cryptandrus	47	Toole (1941)
Stipa viridula	69	Wiesner and Kinch (1964)

[a]Awns were removed.

TABLE 10.24   Temperate Zone Steppe or Grassland Grasses in Which Physiological Dormancy
Has Been Broken by Cold Stratification

Species	Stratification (days)	Temperature[a]	L:D[a]	Reference
Andropogon gerardii	14	20	—[b]	Hsu *et al.* (1985)
Bouteloua hirsuta	90	21	—	Tolstead (1941)
Buchloe dactyloides	60	35/20	—	Pladeck (1940), Wenger (1941)
Elymus canadensis	Winter	Spring	—	Christiansen and Landers (1966)
Eragrostis spectabilis	90	21	—	Tolstead (1941)
E. trichodes	14	30/20	—	Ahring *et al.* (1963)
Leymus cinereus	6–12	20/10	—	Meyer *et al.* (1995)
Oryzopsis hymenoides	28	15/5	D > L	Clark and Bass (1970)
Panicum virgatum	14	30	—	Hsu *et al.* (1985)
Paspalum plicatulum	30	35/25	L > D	Fulbright and Flenniken (1988)
Setaria macrostachya	14	35/20	—	Toole (1940)
Sorghastrum nutans	4	28/18	L > D	Emal and Conard (1973)
Sporobolus aspera	14	35/20	—	Toole (1941)
S. contractus	28	35/20	—	Toole (1941)
S. cryptandrus	28	35/20	L > D	Toole (1941)
Stipa viridula	12	30/15	—	Rogler (1960)
Tripsacum dactyloides	42–56	30/20	—	Ahring and Frank (1968)

[a]See notes for Table 10.1.
[b]No data available.

germinated to higher percentages in both light and darkness at 10°C than at 20°C, whereas those of *Bromus inermis* germinated to higher percentages in both light and darkness at 20°C than at 10°C (Grilz *et al.*, 1994). Seeds of *Bouteloua eriopoda* and *Eragrostis chloromelas* germinated over a broader range of soil moisture conditions at relatively low (38–51°C) temperatures than at high (53–67°C) soil temperatures (Herbel and Sosebee, 1969).

One consequence of seeds being exposed to moisture stress is that the rate of germination may be increased at some time in the future, when soil moisture becomes nonlimiting. After 8 days of absorbing water vapor from air at 23°C and at water potentials of −0.2, −0.4, and −0.6 MPa, the germination of *Agropyron desertorum* seeds was speeded up by 10.5, 3, and 0.75 days, respectively (Wilson, 1973). Imbibition of seeds of this species at a low temperature (at favorable moisture conditions) also increased the rate of germination when the temperature was increased; 24 days at 2°C hastened germination by 10 days at 5°C (Wilson, 1973). Imbibing *Agropyron dasystachyum*, *Pseudoroegneria* (= *Agropyron*) *spicata*, *Poa canbyi*, *Poa sandbergii*, *Sitanion hystrix*, *Festuca ovina* and *Leymus* (= *Elymus*) *cinereus* seeds in PEG 8000 (−1.0 to −2.5 MPa) for 7 days at 25°C (= priming) increased germination rates by 4 to 8 days (Hardegree, 1994). Germination percentages for seeds of *Bouteloua curtipendula*, *Cenchrus ciliaris*, *Eragrostis lehmanniana*, and *Panicum coloratum* seeds primed at −1.5 to −7.7 MPa for 1 to 14 days were higher when

primed at high water potentials for a short period of time than when primed at low water potentials for a long period of time (Hardegree and Emmerich, 1992). There were no subsequent effects on the germination rate of seeds of *Bromus tectorum*, *Elymus elymoides*, *Lolium perenne*, *Poa annua*, or *Pseudoroegneria spicata* dried at an early or intermediate stage of germination. If seeds were dried slowly just prior to emergence of the radicle, there was no effect on the rate of germination following subsequent rehydration. However, the germination rate decreased following rehydration in seeds dried rapidly (−150 MPa) just prior to germination (Allen *et al.*, 1993).

Another cause of moisture stress can be the presence of salt(s). Seeds of *Sorghastrum nutans* from a nonsaline area incubated in NaCl solutions of 0%, 0.0099% (−0.0099 MPa), 0.0499% (−0.0499 MPa), 0.0999% (−0.0999 MPa), 0.497% (−0.497 MPa), and 0.990% (−0.990 MPa) germinated to 31, 65, 20, 20, 30, and 0%, respectively, whereas those from a saline area germinated to 50, 95, 40, 30, 0, and 0%, respectively (Sautter, 1962). A solution of Na_2SO_4 (simulating conditions on mine spoils) at −0.31 MPa reduced the germination of *Panicum virgatum*, *Sporobolus airoides*, *Schizachyrium scoparium*, and *Stipa viridula* seeds, whereas $MgSO_4$ (−0.24 MPa) and Na_2SO_4 + $MgSO_4$ (−0.26 MPa) reduced germination only in seeds of *Stipa viridula*. Neither Na_2SO_4 nor $MgSO_4$ reduced the germination of *Agropyron dasystachyum* or *Poa canbyi* seeds (Ries and Hofmann, 1983).

TABLE 10.25 Germination Requirements of Nondormant Seeds of Grasses in
Temperate Zone Steppes or Grasslands

Species	Ph[a]	Temperature[b]	L:D[b]	M[c]	Reference
Agropyron cristatum	3	13, 15/4	—[d]	—	Ashby and Hellmers (1955), McElgunn (1974)
A. desertorum	3	13, 20, 15/4	—	0.51	McGinnies (1960), McElgunn (1974)
A. elongatum	3	7			
A. inerme	3	20	—	0.51	McGinnes (1960)
A. intermedium	3	13, 20, 15/4	—	1.02	McGinnies (1960), McElgunn (1974)
A. riparium	3	10	—		McElgunn (1974)
A. smithii	3	27/16, 30/15, 18.5/10	L = D	0.1, 0.7	Knipe (1973), Toole (1976), Sabo *et al.* (1979)
A. trachycaulon	3	15/4	—	—	McElgunn (1974)
A. trichophorum	3	18, 20, 15/4	—	1.02	McGinnies (1960), Smoliak and Johnson (1968)
Andropogon gerardii	4	20	—	—	Hsu *et al.* (1985)
A. hallii	4	35/25	L = D	—	Stubbendieck and McCully (1976)
A. inerme	3	20	—	0.51	McGinnies (1960)
Bothriochloa caucasica	4	25	—	—	Hsu *et al.* (1985)
B. macra	4	30/20	L > D	0.25	Maze *et al.* (1993)
Bouteloua curtipendula	4	31/12, 23	L = D	>1.60	Sabo *et al.* (1979)
B. eriopoda	4	—	—	2.03	Knipe and Herbel (1960)
B. gracilis	4	27, 29.5/18, 35/27	L = D	0.30, >1.60	Ashby and Hellmers (1955), Bokhari *et al.* (1975), Sabo *et al.* (1979)
Bromus inermis	3	13, 18, 20	—	1.02	McGinnes (1960), Smoliak and Johnson (1968)
B. rubens	3	17/10	—	—	Ashby and Hellmers (1955)
Buchloe dactyloides	4	24/13	—	0.10	Bokhari *et al.* (1975)
Chloris truncata	4	22/12	L > D	0.25	Maze *et al.* (1993)
Danthonia parryi	3	18	—	—	Ashby and Hellmers (1955)
D. caespitosa	3	22/12	L > D	—	Maze *et al.* (1993)
Ehrharta calycina	3	23/10	—	—	Ashby and Hellmers (1955)
E. angustus	3	13, 27/16	—	—	McElgunn (1974)
E. junceus	3	7, 13, 20	—	0.04	McGinnes (1960), Smoliak and Johnson (1968), McElgunn (1974)
Elymus scabrus	3	30/20	L = D	0.5	Maze *et al.* (1993)
E. triticoides	3	30/17	—	—	Ashby and Hellmers (1955)
Enteropogon ramosus	4	35/20	L > D	0.75	Maze *et al.* (1993)
Eragrostis curvula	4	35/20, 30	L > D	0.75	Maze *et al.* (1993), Martin and Cox (1984)

(continues)

The chemical composition of salts may have effects on germination other than those associated with moisture stress per se. At −0.30 MPa, NaCl, CaCl$_2$, KCl, and MgCl$_2$ decreased the germination of *Sporobolus airoides* by 72, 83, 90, and 93%, respectively (Hyder and Yasmin, 1972). The inhibition of MgCl$_2$ partly was overcome by the addition of CaCl$_2$ or NaCl, with CaCl$_2$ being more effective, but the addition of KCl had little effect. Seeds of *Panicum antidotale, Eragrostis lehmanniana, E. superba,* and *E. curvula* also were more tolerant of Ca than they were of Mg salts; however, when subjected to NaCl, CaCl$_2$, MgCl$_2$, or Na$_2$SO$_4$ the response varied with the species (Ryan *et al.,* 1975).

The germination of buried seeds under various levels of moisture stress may be related to the ability of seedlings to penetrate the hard top layer or crust on the soil surface. Increasing the strength of simulated (wax) soil crusts in the absence of moisture stress decreased the percentage of seedling emergence for *Elymus cinereus, E. junceus, Agropyron trichophorum,* and *Bromus inermis,* but not for *Festuca arundinacea* and *A. elongatum* (Frelich *et al.,* 1973). For a soft soil crust, increasing moisture stress from 0 to −0.84 MPa did not reduce emergence of the six grasses, but with medium and hard crusts an increase in moisture stress to −0.84 MPa generally reduced emergence.

Although soil moisture stress plays a significant role in the germination ecology of seeds, too much water may reduce germination percentages. Seeds of *Oryzopsis hymenoides* (cultivar Nezpar) germinated to only 0–16% on paper or sand that was oversaturated with water; however, at moisture tensions of −0.005 to −0.10 MPa, germination increased to 40–76% (Blank and Young, 1992). However, seeds of the winter annual grasses *Tuctoria greenei* and *Orcuttia californica* that

TABLE 10.25—*Continued*

Species	Ph[a]	Temperature[b]	L:D[b]	M[c]	Reference
E. lehmanniana	4	27	—	1.11	Martin and Cox (1984), Knipe and Herbel (1960)
Festuca altaica	3	15, 20, 20/15	—	0.5, 0.81	Romo *et al.* (1991)
F. rubra	3	27	—	—	Ashby and Hellmers (1955)
F. scabrella	3	13	—	—	Ashby and Hellmers (1955)
Hilaria jamesii	4	29, 33/24	L = D	1.3	Sabo *et al.* (1979)
H. mutica	4	—	—	1.11	Knipe and Herbel (1960)
Koeleria cristata	3	18	—	—	Ashby and Hellmers (1955)
Melica imperfecta	3	23/10	—	—	Ashby and Hellmers (1955)
Muhlenbergia porteri	4	30/17	—	1.52	Ashby and Hellmers (1955), Knipe and Herbel (1960)
M. wrightii	4	12–34.5	L = D	1.3	Sabo *et al.* (1979)
Oryzopsis miliacea	3	30/17	—	—	Ashby and Hellmers (1955)
Panicum obtusum	4	35/20	—	—	Toole (1940)
P. virgatum	4	30	—	—	Hsu *et al.* (1985)
			L > D	—	Zhang and Maun (1989)
Paspalum plicatulum	4	35/20	L > D	—	Fulbright and Flenniken (1988)
Phalaris arundinacea	3	15/4	—	—	McElgunn (1974)
		26–32	L > D	—	Junttila *et al.* (1978)
Poa sandbergii	3	20/5, 20/15, 20/10	—	—	Romo *et al.* (1991)
P. scabrella	3	23/10, 17/10	—	—	Ashby and Hellmers (1955)
Schizachyrium scoparium	4	18	—	—	Ashby and Hellmers (1955)
		27/16.5	L = D	—	Sabo *et al.* (1979)
Setaria macrostachya	4	35/10	—	—	Toole (1940)
Sitanion hystrix	3	20/10	—	—	Young and Evans (1977)
Sorghastrum nutans	4	30/20	L > D	—	Hsu *et al.* (1985), Emal and Conard (1973)
Sporobolus contractus	4	26/15.5	L > D, L = D	1.0	Sabo *et al.* (1979)
S. cryptandrus	4	35/20	L > D	1.0	Sabo *et al.* (1979)
Stipa comata	3	27	—	—	Ashby and Hellmers (1955)
S. leucotricha	3	25–30	—	—	White and Van Auken (1996)
Stipa viridula	3	20, 20.15	D > L	—	Fulbright *et al.* (1983)
Tripsacum dactyloides	4	30/20	—	—	Ahring and Frank (1968)

[a]Cool C_3 (3) or warm C_4 (4) season species (data from Waller and Lewis, 1979).
[b]See notes for Table 10.1.
[c]Moisture stress ($-MPa$) that reduced germination from 75–100% to about 50%.
[d]Data not available.

grow in vernal pools in grasslands in California can germinate under water (Keeley, 1988).

Regardless of photosynthetic pathway, seeds of grasses, except those of *Stipa viridula*, germinate to higher percentages in light than in darkness, or they germinate equally well in light and darkness (Table 10.25). Seeds of *S. viridula* germinated to higher percentages in darkness than in light (Fulbright *et al.*, 1983). Unfortunately, data on light:dark requirements for germination are not available for most species of grasses, and nothing is known about the dormancy-breaking and germination requirements of sedges in grasslands.

Seeds of numerous grasses, including *Poa sandbergii* (Evans *et al.*, 1977), *Bromus carinatus, B. inermis, Agropyron trachycaulon, A. smithii, A. spicatum, Elymus glaucus, Arrhenatherum elatius, E. triticoides, Stipa arida, Festuca ovina* (Plummer, 1943), *B. tectorum*, and *Taeniatherum asperum* (Evans and Young, 1972),

germinate to higher percentages when covered with 0.6–1.0 cm of soil than when placed on the soil surface. However, these results do not necessarily prove that seeds of these species germinate to higher percentages in darkness than in light. Soil moisture conditions may have been more conducive for germination under the soil crust than on the surface. For example, Toole (1976) found that seeds of *A. smithii* germinated equally well in light and darkness. Thus, the higher germination percentage of buried than of nonburied seeds (Plummer, 1943) must be due to something other than light:dark requirements per se. Seeds of *P. bulbosa* germinated to a higher percentage on the soil surface than at a depth of 0.6 cm (Plummer, 1943); therefore, they may be inhibited by darkness. Another reason for the enhancement of germination of buried seeds may be a reduction in temperature, which could be very important in promoting germination of seeds of winter annuals such as *B. tectorum* and *T. asperum* (Evans and Young, 1972) in early autumn, when soil surface temperatures are above those required for germination.

D. Dormancy and Germination of Weeds

When perennial, indigenous graminoids and forbs of grasslands are removed, or their cover greatly reduced, a habitat is created for weedy species. The major cause of disturbance is human activities, but some destruction of vegetation occurs via natural abiotic and biotic events, e.g., Platt (1975). In this discussion, a species was considered to be a grassland weed if it was listed in the systematic index of "Ecosystems of the World: Natural Grasslands" (Coupland, 1992b, 1993) and/or the "Flora of the Great Plains" (Great Plains Flora Association, 1986), and it is listed as a weed by Holm *et al.* (1979).

Most weedy species are herbaceous, and many of them are annual forbs and grasses (Le Houerou, 1993a,b; Lavrenko and Karamysheva, 1993; Tainton and Walker, 1993; Moore, 1993; Coupland, 1992c; Heady *et al.*, 1992; Herlocker *et al.*, 1993). However, not every annual forb in grasslands behaves as a weed. The annual forb *Notiosciadium pampicola* (Apiaceae) is a rare species in the Rio de la Plata grasslands of South America (Soriano, 1992). Some weeds in temperate grassland regions are biennials (monocarpic perennials) or perennials. Thistles, many of which are biennials, belonging to the genera *Atractylis, Carduus, Carlina, Carthamus, Centaurea, Cirsium, Cousinia, Cynara, Galactites, Onopordum, Scolymus,* and *Silybum* (Le Houerou, 1993a; Soriano, 1992) can be serious pests, especially in overgrazed grasslands. Cool season perennial, adventive grasses, including *Agrostis alba, Festuca arundinacea, Phleum pratense, Poa compressa,* and *P. pratensis,* invade disturbed grasslands, and some perennial forbs with low palatability increase (and become weedy), when species with high palatability are removed by cattle (Kucera, 1992).

Freshly matured seeds of grassland weeds have either physical or physiological dormancy, but physiological dormancy is the most common type (Table 10.26). Weeds with physiological dormancy include summer annuals, biennials, and perennials whose seeds require cold stratification to come out of dormancy, and winter annuals and a few perennials whose seeds require exposure to high summer temperatures to come out of dormancy. The mean ($\pm$SE) optimum germination temperature was 22.7 $\pm$ 1.0°C for seeds coming out of dormancy during cold stratification and 17.0 $\pm$ 1.3°C for those coming out of dormancy at high temperatures. Regardless of dormancy-breaking requirements, most nondormant seeds germinate to higher percentages in light

than in darkness, or they germinate equally well in light and darkness. Only seeds of *Bromus tectorum* (Hulbert, 1955) and *Hordeum jubatum* (Banting, 1979) germinated to higher percentages in darkness than in light.

E. Special Factors

1. Litter

Annual aboveground production of dry matter in temperate grasslands ranges from 200–1000 g m^{-2} in tallgrass prairies (Kucera, 1992) to 1.4–154 g m^{-2} in desert grasslands and velds (Schmutz *et al.*, 1992; Tainton and Walker, 1993). Aboveground parts of most grassland plants senesce at the end of the growing season, and consequently much litter accumulates on the soil surface. The presence of litter may have beneficial or detrimental effects on seed germination, depending on the species.

Litter influences germination via modification of the microhabitat of seeds. The germination of *Bouteloua rigidiseta* and *Aristida longiseta* seeds was increased by a rocky soil surface and by the presence of a low cover of litter, which decreased desiccation (Fowler, 1986). Straw litter and gauze fabric reduced summer soil surface temperatures in desert grasslands in Arizona by 10°C and increased soil moisture by 3.2 and 0.6%, respectively. Only 11–200 seedlings m^{-2} of *Bouteloua filiformis, B. hirsuta, Aristida divaricata, A. glabrata,* and *Trichachne californica* were recorded on bare soil, whereas 346–2256 and 1467–4444 seedlings m^{-2} became established under straw and gauge, respectively (Glendening, 1942).

Most perennial-dominated grasslands have been replaced by annual-dominated grasslands in the western United States, and litter can be important in the establishment of annuals each year. In fact, management for maximum herbage yield of annual grasses in sites with annual rainfall of 250–1400 mm in California requires that 56–110 g m^{-2} of litter be left on the soil surface (Heady *et al.*, 1992). Litter decreased daytime temperatures of the surface layer of soil by 9°C in Nevada, but compared to bare soil, night temperatures increased by 5°C (Evans and Young, 1970). Litter also increased relative humidity at the soil surface by about 30% and delayed depletion of soil moisture (at the surface) by about 1 week. These modifications of the microenvironment by litter increase the germination of seeds of many annual grasses (Young and Evans, 1989).

Another aspect of the effect of litter on germination is that it prevents grass seeds from being blown away by the wind. In contrast, small seeds of dicots, such as those of the weeds *Sisymbrium altissimum* and *Salsola*

TABLE 10.26 Seed Dormancy of Weeds of Temperate Zone Steppes or Grassland Areas

Species	Family[a]	LC[b]	Tr[c]	Temperature[e]	L:D[a]	Reference
Physical dormancy						
Astragalus hamousus	Fab.	A	S	15	—[d]	Ehrman and Cocks (1996)
Coronilla scorpioides	Fab.	A	S	15	—	Ehrman and Cocks (1996)
Hymenocarpus circinnatus	Fab.	A	S	15	—	Ehrman and Cocks (1996)
Malva parviflora[e]	Malv.	wA	S[f]	—	—	Sumner and Cobb (1967)
Medicago minima	Fab.	A	S	15	—	Ehrman and Cocks (1996)
M. orbicularis	Fab.	A	S	15	—	Ehrman and Cocks (1996)
M. turbinata	Fab.	A	S	15	—	Ehrman and Cocks (1996)
Scorpiurus muricatus	Fab.	A	S	15	—	Ehrman and Cocks (1996)
Swainsona salsula	Fab.	P	S	—	—	Robocker et al. (1964)
Trifolium campestre	Fab.	A	S	15	—	Ehrman and Cocks (1996)
T. lappaceum	Fab.	A	S	15	—	Ehrman and Cocks (1996)
T. purpureum	Fab.	A	S	15	—	Ehrman and Cocks (1996)
T. resupinatum	Fab.	A	S	15	—	Ehrman and Cocks (1996)
T. scabrum	Fab.	A	S	15	—	Ehrman and Cocks (1996)
T. stellatum	Fab.	A	S	15	—	Ehrman and Cocks (1996)
T. tomentosum	Fab.	A	S	15	—	Ehrman and Cocks (1996)
Physiological dormancy						
Artemisia absinthium[e]	Aster.	P	C[g]	20/10 30/10	L > D	Maw et al. (1985)
A. frigida[e]	Aster.	P	C[h]	20/10	L > D	Wilson (1982)
Bromus tectorum[e]	Po.	wA	W	15	D > L	Hulbert (1955)
Cenchrus longispinus[e]	Po.	sA	C[h]	30/10	L = D	Boydston (1989)
Centaurea maculosa[e]	Aster.	B	C	20	—	Eddleman and Romo (1988)
Chenopodium ambrosioides[e]	Chenopodi.	sA	C	30/15	—	Herron (1953)
C. botyrs[e]	Chenopodi.	sA	C	30/15	—	Herron (1953)
Chondrilla juncea[e]	Aster.	P	W	11	L = D	Panetta (1988)
Crupina vulgaris[f]	Aster.	wA	W	15/4	—	Patterson and Mortensen (1985)
Cucurbita foetidissima	Cucurbit.	P	W?	25	—	Horak and Sweat (1994)
Descurainia sophia[e]	Brassic.	wA	W	30/10	L > D	Best (1977)
Eriochloa villosa	Po.	sA	C	20	—	Hatterman-Valenti et al. (1996)
Erodium botyrs	Gerani.	wA	W,S	15/10	—	Young et al. (1975b)
Euphorbia escula[e]	Euphorbi.	P	C[g]	30/20	—	Brown and Porter (1942), Selleck and Coupland (1954)
E. marginata[e]	Euphorbi.	sA	C	30/20	—	Heit (1942)
Fagopyrum tataricum	Polygon.	wA?	W	20	—	Vanden Born and Corns (1958)
Festuca octoflora	Po.	wA	W	20	L > D	Hylton and Bass (1991)
Gypsophila paniculata[i]	Caryophyll.	P	C?	20	L = D	Darwent and Coupland (1966)
Hordeum jubatum[e]	Po.	P	C[j]	27/15, 20/18	D > L	Banting (1979), Badger and Ungar (1989)
Hypericum perforatum[e]	Clusi.	P	W	15, 30/20	L > D	Campbell (1985)
Linaria dalmatica[e]	Scrophulari.	P	C	30/20	—	Maguire and Overland (1959)
Lolium persicum	Po.	wA	W	10–15	L > D	Banting and Gebhardt (1979)
Marrubium vulgare	Lami.	P	C	26, 30/15	—	Stritzke (1975), Lippai et al. (1996)
Salsola kali	Chenopodi.	sA	C[g]	31/21	—	Allen (1982)
Sorghum vulgare[e]	Po.	sA	C[g]	30/20	—	Burnside (1965)
Taeniatherum asperum	Po.	wA	W	10–15	—	Young et al. (1968)
Tribulus terrestris[e]	Zygophyll.	sA	W[g]	32, 35	—	Johnson (1932), Squires (1969)
Vulpia bromoides	Po.	wA	W	17–25	L > D	Dillon and Forcella (1984)
V. myuros[e]	Po.	wA	W	22	L > D	Dillon and Foercella (1984)

[a]See notes for Table 10.1.
[b]Life cycle type: A, annual; wA, winter annual; sA, summer annual; B, biennial or monocarpic perennial; P, polycarpic perennial.
[c]Dormancy-breaking treatment: W, high summer temperatures; C, cold stratification; S, scarification.
[d]Data not available.
[e]Species also is a weed in the deciduous forest region.
[f]Embryo has some physiological dormancy.
[g]Will afterripen in dry laboratory storage.
[h]Dry storage at 5°C.
[i]Not listed by Holm et al. (1979), but important in Canadian prairie provinces.
[j]Some freshly matured seeds will germinate to high percentage at limited conditions.

kali, lodge between soil particles and are not removed by wind (Evans and Young, 1970). The result is that bare areas lacking litter may become populated with annual forbs.

Grass seeds may be suspended in litter rather than falling to the soil surface, where they would become covered by litter. Seeds of *Poa sandbergii,* a native perennial bunch grass in western North America, will not germinate unless their bases are in contact with a wet substrate (Evans *et al.,* 1977); however, those of the annual nonnative grasses *Bromus rubens, B. rigidus, B. mollis, B. madritensis,* and *Taeniatherum asperum* germinate without the base touching a wet substrate (Young *et al.,* 1971), i.e., seeds of these annuals caught in litter can absorb enough water from the near-saturated air in spaces between pieces of litter to germinate (Young and Evans, 1989). Seeds of some grasses held in litter have a narrower temperature range for germination than those lying on a wet substrate. Seeds of *B. tectorum* and *B. mollis* suspended in simulated litter (layers of cotton and bronze gauze) germinated to 0% at 10 and 15°C, but germinated to 40–68% at 20 and 25°C. In contrast, seeds of these two species incubated on wet germination pads in petri dishes germinated to 92–100% at 10, 15, 20, and 25°C (Young *et al.,* 1971). Seeds of *B. rubens, B. rigidus, B. madritensis,* and *T. asperum* germinated over the range of temperatures in simulated litter, but germinated to higher percentages on a wet substrate than in the litter (Young *et al.,* 1971).

Litter did not increase seed germination of the perennial bunch grass *Stipa pulchra* or of the annuals *Lotus subpinnatus, Plantago erecta, Lasthenia californica, Microseris douglasii, Agoseris heterophylla, Vulpia microstachys, Bromus mollis,* and *Calycadenia multiglandulosa* in serpentine grasslands in California (Gulmon, 1992). However, leached litter stimulated seed germination of the annual *Trifolium albopurpureum* (Gulmon, 1992). The presence of litter in desert grasslands in Arizona, decreased the germination of *Eragrostis lehmanniana* to less than 10 seeds m^{-2}, whereas removal of litter by clipping and fire in July resulted in about 275 and 145 seedlings m^{-2}, respectively (Sumrall *et al.,* 1991). The increased amplitude of daily temperature fluctuations following the removal of litter promoted germination.

Seeds of *Festuca scabrella* covered by 0, 1.3, 2.5, and 3.8 cm of litter germinated to 77, 60, 34, and 16%, respectively; however, those covered by 1.3 cm of soil germinated to 81% (Johnston, 1961). Thus, inhibition of germination by litter was caused by something other than darkness. Seeds of *Dipsacus sylvestris* seeds sown in the field germinated to 32% if *Agropyron repens* litter was removed, but germinated to only 0.8% if litter was left in place (Werner, 1975b).

Dead *Poa pratensis* litter inhibited the germination of *Centaurea nigra, Dipsacus sylvestris,* and *Hypericum perforatum* seeds but not those of *Verbascum thapsus.* Leachates of *P. pratensis* litter inhibited the germination of *C. nigra* and *D. sylvestris* seeds (Bosy and Reader, 1995). Extracts from green leaves of *Bouteloua gracilis* and *Agropyron smithii* inhibited the germination of *B. gracilis, A. smithii,* and *Buchloe dactyloides* seeds more than those from dead leaf litter (Bokhari, 1978).

2. Fire

Fire is an important factor in grasslands, and it can be natural or anthropogenic in origin; lightening is the most common natural cause (Daubenmire, 1968; Collins and Wallace, 1990). Temperatures of fires vary, depending on the amount and kind of fuel, topography, and wind speed (Gibson *et al.,* 1990). The hottest place during a fire is 5–10 cm above the soil surface where temperatures of 200°C (Bailey and Anderson, 1980), 90–353°C (Stinson and Wright, 1969), 482°C (Bentley and Fenner, 1958), and 516°C (McKell *et al.,* 1962) have been recorded. Obviously, seeds in the litter or those attached to flowering stalks would be burned by fire and thus killed. Seeds of a few species, e.g., *Erodium cicutarium* (Stamp, 1984), actively work themselves down into the soil, where they are protected from the flames.

Soil is a good insulator and at depths of 2–3 cm, temperatures during fires may be only 90–120°C (Bentley and Fenner, 1958). Also, maximum soil temperatures are maintained for only short periods of time, e.g., less than 5 min (Bentley and Fenner, 1958). How much heat can seeds of forbs and graminoides tolerate before they are killed? Dry seeds of *Avena fatua, Bromus hordeaceus,* and *B. rigidus* were killed during a 5-min exposure to temperatures of 116–127, 127–138, and 93–104°C, respectively (Wright, 1931). However, *Lolium perenne* seeds germinated to about 40% after a 5-hr exposure to dry heat at 100°C (Crosier, 1956).

Imbibed seeds generally are more sensitive to high temperatures than dehydrated ones (Sweeney, 1956). Seeds of *Elymus caput-medusae* at moisture contents of 8.1, 11.0, and 15.4% germinated to 80, 33, and 3%, respectively, after a 2-min exposure to 180°C. Seeds of *Bromus mollis* at moisture contents of 9.0, 11.5, and 18.6% germinated to 82, 51, and 34%, respectively, after 2 min at 180°C (McKell *et al.,* 1962). From 90 to 100% of the seeds of *Aegilops triuncialis, B. mollis, B. rigidus, Festuca megalura,* and *Hordeum hystrix* buried in moist soil at, or near, field capacity died after being heated for 1 hr at 48 or 49°C (Laude, 1957). From 90 to 100% of *B. catharticus, Festuca arundinacea, Lolium perenne, Oryzopsis miliacea, Phalaris tuberosa,* and *Stipa cernua* seeds buried in moist soil died as a result of being heated for 6 hr at 47–53°C (Laude *et al.,* 1952).

More studies on the effects of heat from fires on seed viability of grassland species are needed. Such studies may help explain observed changes in the relative abundance of forbs and graminoids following fires (Kucera and Koelling, 1964). For example, seedlings of dicots were more abundant in nonburned than in burned plots in a dry year in a central Iowa tallgrass prairie, but there was no difference between burned and nonburned plots in a wet year (Glenn-Lewin *et al.,* 1990).

Another possible effect of heat from fires is that it breaks dormancy in species whose seeds have physical dormancy. This is an important consideration in the germination ecology of legumes and other species with impermeable seed coats in grasslands. As discussed in Chapter 6, heat treatments will break physical dormancy in seeds of many species. Seeds of *Lespedeza* spp. germinated to high percentages after 4 min of moist heat treatment at 60 or 70°C and dry heat at 70, 80, or 90°C (Cushwa *et al.,* 1968). However, seeds of *Lespedeza* spp. were killed by moist heat at 90°C and by dry heat at 100°C. Further insight into the germination ecology of grassland species with physical dormancy could be gained from determining seed temperatures (perhaps by coating them with temperature-sensitive paints) during fires in the habitat and monitoring subsequent rates and percentages of germination.

Fires have various effects on microsites of seeds. When litter is burned, increased solar radiation on the soil surface increases the temperature (Hulbert, 1969; Sharrow and Wright, 1977; Whisenant *et al.,* 1984). Dense litter (11–20 cm) resulting from 15 years without fires reduced soil temperatures 12–15°C in May in the loess hills of central Nebraska (Weaver and Rowland, 1952). Thus, germination may occur earlier in the growing season in burned than in nonburned areas. Also, removal of litter improves the light environment at the soil surface. The light level was 400 ft-c (4304 lux) 8 cm above the soil surface in a nonburned tallgrass prairie in Oklahoma, whereas it was 9000–10,000 ft-c (96,840–107,600 lux) in a burned portion of the prairie (Penfound and Kelting, 1950). Burned areas in grasslands may have reduced soil moisture compared to nonburned areas (Anderson, 1965; Hulbert, 1969; Sharrow and Wright, 1977; Whisenant *et al.,* 1984). Also, fire causes an increase in soil nitrate levels (White and Gartner, 1975; Sharrow and Wright, 1977; Hobbs and Schimel, 1984). Logically, a decreased soil moisture would reduce germination, and an increase in temperature, light, or nitrates might increase it. However, the direct effects of these microenvironmental changes on germination of seeds in burned grasslands are unknown.

VI. SEMIDESERTS AND DESERTS WITH COLD WINTERS

Temperate semideserts and deserts are found in western North America, southeastern South America (Patagonia), eastern Europe, and middle and central Asia (region 7a in Fig. 9.1). Cold winters, which may have some frozen precipitation, distinguish these temperate deserts from tropical arid regions; however, summers in temperate arid regions generally are hot (West, 1983a,b). Mean annual temperatures vary with location, and in Asia, for example, they range from 3.9°C at Dalandzadgad in the Gobi Desert (Walter and Box, 1983a) to 16.7°C at Kerki in the Turanian lowland (Walter and Box, 1983b).

Temperate semideserts receive 200–500 mm of precipitation per year, whereas deserts receive less than 200 mm (West, 1983a). Most of the precipitation falls in winter in Patagonia and the western parts of the North American and Eurasian temperate semideserts and deserts, whereas summer monsoonal rainfall increases in importance on the eastern edges of this arid region in North America and Eurasia (West, 1983c).

Shrubs are dominant plants in temperate semideserts and deserts (hereafter called cold deserts) (West, 1983a), but herbaceous perennials and annuals also may be present (Walter and Box, 1983c; West, 1983e). The proportion of herbaceous perennials, including grasses, increases with an increase in precipitation (West, 1983c). An interesting feature of cold deserts is the presence of salt lakes and pans with associated halophytic vegetation (West, 1983c; Walter and Box, 1983c; Soriano, 1983).

A. Dormancy and Germination of Shrubs

Cold deserts on each continent have a diversity of shrub species, and there is some overlap of genera. For example, *Artemisia, Atriplex,* and *Ephedra* are found in Eurasia and North America, *Lycium* occurs in Eurasia, North America, and South America, and *Berberis* is in North America and South America (Walter and Box, 1983d; West, 1983d,e; Soriano, 1983). The most important families of shrubs are the Chenopodiaceae and Asteraceae (West, 1983f).

Neither morphological nor morphophysiological dormancy has been reported for seeds of cold desert shrubs. However, further studies may reveal that seeds of shrubs/semishrubs/vines in the Ranunculaceae, such as *Clematis orientalis* and *C. songarica,* have morphological or morphophysiological dormancy. Physical dormancy also has not been reported in seeds of cold desert shrubs. However, it seems reasonable that some (all?) shrubs belonging to the Anacardiaceae (*Schinus*), Convolvulaceae (*Convolvulus*), and Fabaceae (*Adesmia, Ammodendron, Ammopiptanthus, Astragalus, Colutea, Lespedeza,* and *Oxytropis*) would have impermeable seed coats.

Physiological dormancy is the only type of dormancy reported thus far in seeds of cold desert shrubs (Table 10.27). Seeds come out of dormancy during cold stratification, but those of many species also become nondormant during dry storage at room temperatures. Further, dry storage also can decrease the length of the subsequent cold stratification period required to break dormancy (Meyer, 1989; Meyer and Monsen, 1989).

Because most of the precipitation in many cold des-

erts is in winter, it is expected that (1) nondormant seeds would germinate to high percentages at relatively low temperatures and (2) germination could occur in late winter. Seeds of *Atriplex confertifolia, Ceratoides lanata, Ephedra nevadensis,* and *Grayia spinosa* have been observed germinating between October and March, whenever a rainfall event exceeded 16 mm (Ackerman, 1979). Optimum temperatures for germination vary with the species and range from 3°C in *Artemisia nova* (Deitschman, 1974) to 30/15°C in *Chrysothamnus nauseosus* (Sabo *et al.,* 1979). The mean (±SE) optimum germination temperature for species listed in Table 10.27 is 14.8 ± 1.1°C, which is slightly less than that (16.4°C) of cool

TABLE 10.27   Seed Dormancy in Shrubs of Temperate Semideserts and Deserts

Species	Family[a]	Tr[b]	Temperature[a]	L:D[a]	M[c]	Reference
Physiological dormancy						
Artemisia macrocephala	Aster.	—[d]	14–17	—	—[e]	Sveshnikova (1948)
A. nova	Aster.	C	3	—	—	Deitschman (1974)
A. skorniakovii	Aster.	—[d]	14–17	—	—	Sveshnikova (1948)
A. tridentata	Aster.	C[f]	18.5	L = D, L > D	1.0	Weldon *et al.* (1959), Deitschman (1974), Sabo *et al.* (1979)
Atriplex canescens	Chenopodi.	C[f]	16–19	—	0.3	Springfield (1969, 1970), Stidham *et al.* (1980)
A. centralasiatica	Chenopodi.	—[d]	1.5–8.5	—	—	Sveshnikova (1948)
A. confertifolia	Chenopodi.	C?[g]	12, 16/12	L = D	0.4	Sabo *et al.* (1979), Foiles (1974)
A. gardneri	Chenopodi.	C	24/13	—	—	Ansley and Abernethy (1984)
A. lentiformis	Chenopodi.	?[h]	25/10	L = D	—	Young *et al.* (1980)
A. obovata	Chenopodi.	?[i]	20	L > D	—	Edgar and Springfield (1977)
A. patula	Chenopodi.	?[h]	5/25	L > D	—	Young *et al.* (1980)
A. polycarpa	Chenopodi.	?[d,g]	15	—	0.4	Chatterton and McKell (1969), Cornelius and Hylton (1969)
Ceratoides lanata	Chenopodi.	C[f,g,h]	10–27	—	0.7, 1.5	Springfield (1972a,b, 1968)
C. latens	Chenopodi.	C[f]	0–5/15–20	—	—	Dettori *et al.* (1984)
Chrysothamnus nauseosus	Aster.	C[g]	30/15	L = D, L > D	0.2, 0.4	Khan *et al.* (1987), Sabo *et al.* (1979)
Ephedra nevadensis	Ephedr.	—[h]	20	—	0.4	Young *et al.* (1977)
E. viridis	Ephedr.	—[h]	2/15 → 25	—	—	Young *et al.* (1977)
Grayia brandegei	Chenopodi.	C[f]	15	—	—	Pendleton and Meyer (1990)
G. spinosa	Chenopodi.	C	15/5	—	0.3	Wood *et al.* (1976), Shaw *et al.* (1994)
Gutierrezia sarothrae	Aster.	?[i]	16–21	—	0.61	Kruse (1970)
Kochia prostrata	Chenopodi.	—[g]	20	—	—	Young *et al.* (1981)
Mahonia fremontii	Berberid.	C	15/6	L = D	—	Baskin *et al.* (1993d)
Purshia glandulosa	Ros.	C	15	—	—	Young and Evans (1981)
P. tridentata	Ros.	C	2/15	—	—	Young and Evans (1976), Evans and Young (1977)
Sarcobatus vermiculatus	Chenopodi.	C	10, 30/20	L = D	0.7	Eddleman (1979), Sabo *et al.* (1979), Romo and Haferkamp (1987)
Zygophyllum rosovii	Zygophyll.	—[d]	1.5–8.5	—	—	Sveshnikova (1948)

[a]See notes for Table 10.1.
[b]Dormancy-breaking treatment: C, cold stratification.
[c]Moisture stress (−MPa) that reduced germination from 75–100% to about 50%.
[d]Fresh seeds germinated to high percentages.
[e]Data not available.
[f]Apparently, seeds will afterripen in dry storage at room temperatures.
[g]Stored dry at a low (2–5°C) temperature.
[h]Stored dry at room temperature.
[i]Storage conditions not given.

season grasses of temperate steppes and grasslands. Cold stratification decreased the minimum germination temperature for seeds of *C. nauseosus* (Meyer *et al.*, 1989), but increased it for those of *Sarcobatus vermiculatus* (Eddleman, 1979).

Information on light:dark requirements for the germination of seeds is limited (Table 10.27). However, seeds of two species germinated to higher percentages in light than in darkness, and four germinated equally well in light and darkness. Seeds of *Artemisia tridentata* and *Chrysothamnus nauseosus* may germinate to higher percentages in light than in darkness, but sometimes there is no difference between germination in light and in darkness. Seeds of *Atriplex gardneri* (Ansley and Abernethy, 1984), *A. canescens,* and *Ephedra viridis* (Williams *et al.,* 1974) germinated to high percentages when buried to a depth of 1 cm in soil, demonstrating their capacity to germinate in darkness.

Because (1) the wet season coincides with the cool season in many cold deserts and (2) seeds of many species germinate well at low temperatures, germination would occur at a time when soil moisture is high. Thus, it is expected that seeds might be quite sensitive to moisture stress. The mean (±SE) moisture stress that reduced seed germination of the shrubs in Table 10.27 from 75 to 100% to about 50% is -0.59 ± -0.12 MPa. For comparison, this value is about the same as that (-0.64 ± -0.1 MPa) for cool season grasses of temperate steppes and grasslands, but less negative than that (-1.02 ± -0.12 MPa) for warm season grasses.

Leachates from bracts enclosing fruits of *Grayia brandegei* inhibited the germination of debracted fruits. Much of the inhibition was due to the high osmotic potential of the leachate. However, germination inhibition was greater in the leachate than at a comparable osmotic potential in solutions of NaCl or mannitol, indicating that a chemical germination inhibitor also was present. The leachate decreased germination more in freshly matured than in nondormant fruits (Pendleton and Meyer, 1990).

B. Dormancy and Germination of Herbaceous Species

1. Perennials

A type of polycarpic perennial called "geophytes" is found in some cold deserts. These plants have underground storage structures such as bulbs and produce a short-lived shoot with flowers during the growing season. Geophytes are mostly absent in North and South American cold deserts (West, 1983d; Soriano, 1983), but they are a very important component of the flora/vegetation in Eurasia (Walter and Box, 1983c,e; Breckle, 1983), where there are about 20 genera of geo-

phytes (West, 1983f). Although little information seems to be available on the germination of geophytes, we can infer a little about the types of seed dormancy by knowing the families to which they belong. Morphophysiological dormancy may be very important, as eight genera are in the Liliaceae, one each in the Araceae, Fumariaceae, Leonticaceae, and Ranunculaceae, and two each in the Iridaceae and Apiaceae. All, or many, members of these families have underdeveloped embryos (Martin, 1946). *Tulipa tarda* (Liliaceae) has deep complex MPD (Nikolaeva, 1969). Geophytes in the Cyperaceae and Polygonaceae are expected to have physiological dormancy. None of the geophytes belong to families known to have physical dormancy.

Nongeophyte, herbaceous perennials in cold deserts belong to 64 genera in 23 families, with the Asteraceae (16 genera) and Poaceae (12 genera) being the most important (West, 1983f). No herbaceous perennials are known to have seeds that are nondormant at maturity, and morphological dormancy still has not been reported. However, further studies on species belonging to the Apiaceae, Iridaceae, and Liliaceae may reveal some perennials whose seeds have morphological dormancy or increase the number known to have MPD. Only three species have MPD (Table 10.28). Physical dormancy occurs in various perennials (Table 10.28), but it probably occurs in other genera, including *Psoralea, Sophora, Adesmia,* and *Convolvulus.* Scarified seeds of *Alhagi, Astragalus,* and *Sphaeralcea* germinated to high percentages (Table 10.28); however, those of *Sphaeralcea* germinated to low percentages (Page *et al.,* 1966), suggesting that the embryos might have some physiological dormancy.

Seeds of most cold desert herbaceous perennials have physiological dormancy, which is broken by cold stratification. However, seeds stored dry at (presumably) room temperatures in some studies eventually germinated to high percentages, indicating that afterripening occurred during dry storage. Seeds of *Penstemon palmeri* require high temperatures to come out of dormancy, and cold stratification induced about half of them into secondary dormancy (Kitchen and Meyer, 1992; Meyer and Kitchen, 1992). Seeds of this species behave like those of a facultative winter annual, germinating mostly in autumn but also in spring (Meyer and Kitchen, 1992).

The mean (±SE) optimum germination temperature for the herbaceous perennials listed in Table 10.28 is $16.7 \pm 1.4°C$. With the exception of a few species such as the C_4 species *Sporobolus airoides* and *Distichlis stricta,* the optimum temperature for germination is 20°C or less, or the optimum is an alternating regime with a night temperature of 15°C or less. Seeds of *S. airoides* and *D. stricta* germinate best at 35/20 and 40/36°C, respectively. The ecological implication of these data is

TABLE 10.28   Seed Dormancy in Herbaceous Perennials of Temperate Semideserts and Deserts

Species	Family[a]	Tr[b]	Temperature[a]	L:D[a]	M[c]	Reference
Morphophysiological dormancy						
Ferula ferulioides	Api.	C	4–10	—[d]	—	Nikolaeva (1969)
F. songorica	Api.	C	4–6	—	—	Nikolaeva (1969)
Ranunculus testiculatus	Ranuncul.	H?	5	—	—	Young *et al.* (1992)
Physical dormancy						
Alhagi pseudalhagi[e]	Fab.	S	25	—	—	Kerr *et al.* (1965)
Astragalus borodini	Fab.	—[f]	14–17	—	—	Sveshnikova (1948)
A. lentiginosus[e]	Fab.	S	13/7	L = D	0.64	Ziemkiewicz and Cronin (1981)
Medicago sativa	Fab.	—[f]	14–17	—	—	Sveshnikova (1948)
Oxytropis poncinsii	Fab.	—[f]	14–17	—	—	Sveshnikova (1948)
Sphaeralcea grossulariaefolia	Malv.	S[g]	22/15	L = D	—	Page *et al.* (1966)
Physiological dormancy						
Agropyron spicatum	Po.	—[h]	25/5	—	—	Young *et al.* (1981)
A. tenerum	Po.	—[f]	2–9	—	—	Sveshnikova (1948)
Atropis pamtrica	Po.	—[f]	2–9	—	—	Sveshnikova (1948)
Balsamorhiza sagitata	Aster.	C	10	—	—	Young and Evans (1979)
Bromus inermis	Po.	—[f]	14–17	—	—	Sveshnikova (1948), Goodwin *et al.* (1996)
B. tectorum	Po.	H	14–17	—	—	Sveshnikova (1948)
Distichlis stricta[e]	Po.	—[i]	40/36	L = D	1.6	Sabo *et al.* (1979)
Elymus cinereus	Po.	—[i]	20/15, 20	—	1200[j]	Evans and Young (1983), Roundy *et al.* (1985)
E. junceus	Po.	—[i]	20/15, 30/20	—	—	Young and Evans (1982)
Eriogonum umbellatum	Polygon.	—[k]	30/10, 35/15	—	—	Young (1989)
Festuca idahoensis	Po.	—[i]	20/15, 23/4	—	—	Young *et al.* (1981), Goodwin *et al.* (1955, 1996)
F. ovina[e]	Po.	—[i]	20/10, 20/15	—	—	Young *et al.* (1981)
Gypsophila capituliflora	Caryophyll.	—[f]	14–17	—	—	Sveshnikova (1948)
Hordeum turkestanicum	Po.	—[f]	2–9	—	—	Sveshnikova (1948)
Nicotiana attenuata	Solan.	C	18–35	L	0.5	Wells (1959)
Oryzopsis hymenoides	Po.	C	15/5	L > D	—	Clark and Bass (1970)
Penstemon cyananthus	Scrophulari.	C	20/10	—	—	Meyer and Kitchen (1994)
P. eatonii	Scrophulari.	C	15	—	—	Kitchen and Meyer (1991)
P. humilis	Scrophulari.	C	2	—	—	Kitchen and Meyer (1991)
P. palmeri	Scrophulari.	H	15	L > D	—	Kitchen and Meyer (1992), Meyer and Kitchen (1992)
P. rostriflorus	Scrophulari.	C	15	—	—	Kitchen and Meyer (1991)
P. sepalulus	Scrophulari.	C	15	—	—	Kitchen and Meyer (1991)
P. strictus	Scrophulari.	C	15	—	—	Allen and Meyer (1990)
Poa canbyi	Po.	—[i]	20/15	—	—	Young *et al.* (1981)
Sporobolus airoides[e]	Po.	C	35/20	—	0.81, 1.22	Toole (1941), Knipe (1971)
Wyethia amplexicaulis[e]	Aster.	C	25/10	—	—	Young and Evans (1979)

[a]See notes for Table 10.1.

[b]Dormancy-breaking treatment; C, cold stratification; H, high temperatures; S, scarification.

[c]Moisture stress (−MPa) that reduced germination from 75–100% to about 50%.

[d]Data not available.

[e]Listed as a weed in Holm *et al.* (1979).

[f]Stored dry (at room temperatures?) for several years before testing.

[g]67% germination was obtained after treatment with diethyl dioxide.

[h]Afterripened for 5 months in dry storage (at room temperatures?).

[i]Storage conditions not given, presumably stored dry at room temperatures, during which time afterripening occurred.

[j]Joules kg^{-1}.

[k]Stored dry at room temperatures.

that seeds of most species germinate during the cool season, but those of *S. airoides* and *D. stricta* germinate during the warm season.

Light:dark requirements for germination have been determined for only three species; seeds of all three germinate to higher percentages in light than in darkness (Table 10.28). Seeds of *S. airoides* and *D. stricta* can germinate under moderate moisture stress, which is consistent with the idea that they germinate during the warm season. Nondormant seeds of *Festuca idahoensis* germinated to 70–90%, depending on geographical origin, at 23/14°C at a soil water potential of −1.0 MPa

(Goodwin *et al.*, 1996). Seeds of perennials germinate in spring in middle Asian deserts if the soil is sufficiently moist; therefore, the number of seedlings fluctuates each year, depending on winter/spring precipitation (Walter and Box, 1983b).

2. Annuals

Cold desert annuals belong to more than 70 genera in 20 families, and the most important families, in decreasing order of number of species, are Brassicaceae, Asteraceae, Poaceae, Chenopodiaceae, Papilionaceae, Caryophyllaceae, Boraginaceae, and Polemoniaceae (West, 1983f). As expected, many species are winter annuals that germinate in autumn/winter, whenever the cool season rains begin (Walter and Box, 1983b; Soriano, 1983; West, 1983g). Most winter annuals complete their life cycle before early summer; however, a few species, such as *Diarthron vesiculosum* and *Aphanopleura capillifolia*, live all summer (Walter and Box, 1983b). Further, seeds of the annuals *Aristida adscensionis* and *Tribulus terrestris* germinate in spring, and plants behave as summer annuals (Walter and Box, 1983b). Walter and Box (1983b) reported 143 species of annuals in the Karakum Desert of middle Asia, and 98 of them were winter and 45 were summer annuals. Roots of some of these summer annuals penetrated 70 cm into the soil.

It is assumed that annuals in the (1) Apiaceae and Liliaceae have morphological and/or morphophysiological dormancy, (2) Fabaceae have physical dormancy, and (3) various other families have physiological dormancy. However, little research has been done on the germination ecology of cold desert annuals, except for a few weedy species (see later).

C. Dormancy and Germination of Weeds

Disturbance of the natural vegetation in cold deserts frequently is followed by the invasion of weeds, usually alien species. These invaders may be winter or summer annuals or perennials. Examples of weedy winter annuals, for which data are available, include the grass *Bromus tectorum* and various members of the Brassicaceae. Following a period of afterripening at actual or simulated summer temperatures, seeds of winter annuals germinate optimally at low autumn temperatures: *B. tectorum* at 15°C (Table 10.26) and *Descurainia pinnata*, *Lepidium perfoliatum*, and *Sisymbrium altissimum* at 5–20°C (Young *et al.*, 1970).

Summer annual weeds such as *Salsola kali* (= *S. iberica*) (Young and Evans, 1972b) and *Halogeton glomeratus* (Cronin, 1973) and perennials such as *Marrubium vulgare* (Young and Evans, 1986) and *Gypsophila*

paniculata (Table 10.26) require cold stratification for maximum germination. Optimum germination temperatures of nondormant seeds are relatively high: *S. kali*, 31/21°C (Allen, 1982); *H. glomeratus*, 22°C (Cronin, 1973); *M. vulgare*, 20°C (Young and Evans, 1986); and *G. paniculata*, 20°C (Darwent and Coupland, 1966). Nondormant seeds of *S. kali* var. *tenuifolia* germinate in the field when night temperatures are 0–5°C and day temperatures are 12–25°C (Evans and Young, 1972b). It should be noted that *H. glomeratus* plants produce both black and brown seeds, and black seeds are less dormant than brown ones (Williams, 1960). When buried in soil in the field, black seeds lost viability faster than brown ones (Robocker *et al.*, 1969).

Although the weeds just mentioned are only a small fraction of those occurring in cold deserts, they probably represent the basic types of dormancy-breaking and germination requirements found among weeds in these deserts, with the possible exception of physical dormancy. Weedy members of the Fabaceae and Geraniaceae, no doubt, would have physical dormancy.

D. Special Factor: Salinity

Rain water that accumulates in depressions or lakes and ground water that comes to the surface contain dissolved salts, especially sodium chloride but also sodium sulfate. Because many of the low areas in cold deserts have no external drainage, water remains in place until it seeps into the ground or evaporates. Salts accumulate as water evaporates, and a salt lake or salt pan results (Walter and Box, 1983a,c; West, 1983d). If a basin is deep and receives much runoff or spring water, the input of water exceeds evaporation, and a salt lake is formed. If evaporation is greater than the amount of water reaching the area, a salt pan is present.

Salinity is so high at the edges of salt lakes and in the center of salt pans that vascular plants are excluded. However, salinity decreases as elevation increases, and concentric bands of salt-tolerant plants, or halophytes occur, with the the most intolerant species in the low-saline areas. Cold desert halophytes belong to 41 genera in 10 families, with 23 belonging to the Chenopodiaceae. Other families with 2 or more genera are the Asteraceae, Poaceae, Plumbaginaceae, Tamaricaceae, and Zygophyleaceae (West, 1983f).

Seeds of halophytes usually germinate to higher percentages in distilled water than in NaCl solutions, and transferring seeds from these solutions to fresh water increases germination (Williams and Ungar, 1972; Rajput and Sen, 1990). The concentration of NaCl that inhibits germination (i.e., reduces germination from near 100% to 10% or less) varies with the species: *Kochia americana*, 6% (Clarke and West, 1969); *Suaeda de-*

pressa, 4% (Ungar, 1962); *Ceratoides lanata,* 3% (Workman and West, 1967); and *Atriplex polycarpa,* 2.3% (Chatterton and McKell, 1969). The seedling stage can be more tolerant of increased salinity than the germination stage of the life cycle (Choudhuri, 1968). Seedlings of *S. depressa* grew better at 1 than at 0% NaCl, when nitrogen was nonlimiting (Williams and Ungar, 1972).

Germination percentages at increased salinities may depend on the seed source and incubation temperatures. For example, in 2% NaCl seeds of *Ceratoides lanata* from Rush Valley, Utah, germinated to 16 and 14% at 12.8/1.7 and 29.4/18.3°C, respectively, whereas those from Cisco, Utah, germinated to 42 and 20%, respectively (Workman and West, 1967). Seeds collected from *Lepidium perfoliatum* plants growing in a saline area germinated to higher percentages in NaCl and PEG 400 solutions (at both −0.25 and −0.45 MPa) than those from nonsaline habitats (Choudhuri, 1968).

Salts can have toxic and/or osmotic effects on germination (Rajput and Sen, 1990; Romo and Haferkamp, 1987). However, in a natural saline habitat, sensitivity to the osmotic potential may prevent germination before ions reach a toxic level and be an important factor controlling the timing of germination. As pointed out by Romo and Haferkamp (1987), rainfall not only reduces osmotic stress, but it dilutes the concentration of ions, especially at the soil surface. Thus, osmotic/ionic conditions become suitable for germination following the winter rains, and germination occurs as soon as the temperature increases enough to become nonlimiting. See Chapter 11,VII for a more detailed discussion of the germination of halophytes.

VII. BOREAL CONIFEROUS AND TEMPERATE SUBALPINE ZONES

In the northern hemisphere, a circumpolar belt of coniferous forests, called the boreal forest or taiga, occurs to the north of deciduous forests, steppes, or cold deserts and extends to the treeline at the edge of the Arctic tundra (region 8 in Fig. 9.1). Boreal forests are found in North America as far north as the Brooks Range in Alaska, and in Eurasia they come within a few kilometers of the coast of the Arctic Ocean (Walter, 1973; Larsen, 1980).

The southern and northern limits of the boreal forest occur where the average daily temperature is more 10°C for 120 and 30 days per year, respectively, and the cold season is 6 and 8 months, respectively (Walter, 1973). However, the belt of a boreal forest does not have a uniform width, especially as it stretches across North America (Fig. 9.1), possibly due to the position of bodies of cold water such as the Hudson Bay. Precipitation

varies from 25 to 100 cm per year, but snow cover is not deep (Barbour *et al.,* 1980).

The dominant plants in boreal forests are evergreen conifers, with members of the genera *Abies, Larix, Picea,* and *Pinus* being very important. However, some broad-leaved, deciduous species belonging to the genera *Alnus, Betula, Populus,* and *Salix* also occur in the region (Larsen, 1980), especially near the transition with deciduous forests and/or in successional sites (Walter, 1973). In each genus, different species of trees occur on different continents (Larsen, 1980). The tree canopy in a mature conifer forest is a single layer, and large shrubs are widely spaced. The ground layer has many small shrubs, a few herbaceous species, and masses of mosses and lichens. Many members of the ground layer are circumpolar in distribution (Larsen, 1980).

A special feature of the boreal region is the presence of numerous lakes and various kinds of wetland communities (muskeg, swamp, fen, and bog), which are distinguished by water depth, concentrations of minerals in the water, presence of sphagnum vs. sedges, and shaded vs. nonshaded by trees (Larsen, 1980). Heathlands occur in the region of boreal forests, and they may represent an increase in the dominance of shrubs (especially members of the Ericaceae) following the removal of trees (Larsen, 1980).

Conifer forests extend south of the boreal forest region in mountains in temperate regions of the northern hemisphere. The zone on mountains where conifer forests have the same physiognomy as boreal forests (but different species of *Picea, Abies, Pinus,* and *Larix*) is called the subalpine. Subalpine forests are at moderate to high elevations, depending on latitude and direction of slope, and their upper boundary is the treeline. The subalpine zone generally is about 5°C warmer and has 20–50% more precipitation than the boreal forest region (Barbour *et al.,* 1980). The germination ecology of boreal species and those in subalpine zones of temperate regions of the northern hemisphere will be discussed together, but see Section IX for woodland and montane species.

Subalpine vegetation in temperate regions of the southern hemisphere is adjacent to temperate broad-leaved evergreen forests in many places (Donald and Theron, 1983; Veblen *et al.,* 1983; Ovington and Pryor, 1983). Forests are common in the subalpine of the southern hemisphere, but shrublands and grasslands also are present. Important trees in the subalpine of Australia are *Eucalyptus pauciflora* and *E. stellulata,* and in Tasmania subalpine forests are dominated by *E. coccifera, E. gunnii, Nothofagus gunnii, Athrotaxis cupressoides,* and *A. selaginoides* (Costin, 1981). Subalpine forests of New Zealand are dominated by *N. menziesii* and *N. solandri* (Wardle, 1991), and those in southern Chile are dominated by *N. betuloides*, with *N. pumilio* forming the treeline (Veblen *et al.*, 1983). However, certain areas of the subalpine zones in Australia, Tasmania, and Chile are dominated by shrubs and/or grasses and forbs. A subalpine zone occurs on some mountains in southern Africa, including the Drakensberg Mountains in Cape, Natal, and Lesotho, the Amatole Mountains in Cape, and Mount Mlanje in Malawi, and the vegetation is grassland or shrub communities (Killick, 1978).

A. Dormancy and Germination of Species in Boreal and North-Temperate Subalpine Zones

1. Trees

Some species have nondormant seeds, whereas others have seeds with physiological dormancy (Table 10.29). None of the species have seeds with underdeveloped embryos or impermeable seed coats; therefore, morphological, morphophysiological, or physical dormancy does not occur in trees of this region. Only 1 month or less of cold stratification is required for seeds of some species with physiological dormancy to come out of dormancy, but those of others such as *Pinus cembra* and *P. sibirica* require several months of cold stratification (Table 10.29).

Seeds of many trees germinate at high temperatures without first being cold stratified (Table 10.29). There are three possible explanations for this observation. (1) Seeds are nondormant at maturity, like those of *Salix arbusculoides* that germinate after dispersal in summer. In fact, seeds of *S. arbusculoides* can germinate within 24 hr if they fall on a wet substrate (Densmore and Zasada, 1983). The number of species in boreal and north-temperate subalpine forests with nondormant seeds is unknown.

(2) Seeds may be conditionally dormant at maturity and thus can germinate at high, but not at low, temperatures. Germination is prevented at the time of dispersal in autumn because habitat temperatures are above those required for germination. Cold stratification in late autumn and winter may lower the temperature requirements for germination, and seeds can germinate in spring at relatively low temperatures that were inhibitory the previous autumn. Whereas the minimum temperature for the germination of freshly harvested (airdried) seeds of *Betula papyrifera* was 20°C, it decreased to 15°C after 4 weeks of cold stratification (Joseph, 1929). It is suspected that the minimum temperature for germination is lowered in many species during winter because autumn-dispersed seeds germinate at low temperatures in early spring (Wahlenberg, 1924). Seeds of *Abies lasiocarpa* have been observed germinating in

late-persisting snow banks (Franklin and Krueger, 1968). Further, seeds of *A. nordmanniana* germinated at 2°C after 40 weeks of incubation (Edwards, 1962) and those of *A. fraseri* at 5°C after 30 weeks (Blazich and Hinesley, 1984). However, 3 weeks of cold stratification had no effect on the temperature requirement for the germination of *Pinus sylvestris* seeds (Gosling, 1988). Thus, cold stratification may not decrease the temperature requirement for germination in all species. More studies are needed on the effects of cold stratification on temperature requirements for seed germination. If cold stratification lowers the temperature requirement for seed germination, species should be listed under physiological dormancy instead of nondormant.

(3) Germination of tree seeds at high temperatures may, in part, reflect the germination testing protocol rather than an absolute high temperature requirement for germination per se. For example, the Association of Official Seed Analysts (1965) recommends that seeds of most conifers be tested at a 30/20°C daily thermoperiod. If seeds were tested over a range of temperatures, they might germinate to high (higher?) percentages at temperatures lower than 30/20°C. Nonstratified seeds of *Pinus sylvestris* germinated to 81% at 30 (8 hr light)/20 (16 hr dark)°C (Krugman and Jenkinson, 1974), but they germinated to over 90% at constant temperatures of 15, 20, 25, and 30°C in darkness (Gosling, 1988). Would nonstratified seeds of this species also germinate to >90% in light at 15, 20, 25, and 30°C?

Light:dark requirements for germination have been determined for only 19 species of boreal trees (Table 10.29). Seeds of 1 species required light for germination, 7 germinated equally well in light and darkness, 7 germinated to higher percentages in light than in darkness, and 4 germinated to higher percentages in darkness than in light. One reason for this lack of information on light:dark requirements for germination may be the fact that germination tests of conifers frequently are conducted using only an 8-hr daily photoperiod (Association Official Seed Analysts, 1965); thus, the relative importance of light and darkness for germination cannot be determined.

Nonstratified seeds of *Betula papyrifera* (Bevington and Hoyle, 1981), *Picea mariana* (Farmer *et al.*, 1984), and *Pinus contorta* (Li *et al.*, 1994) germinated to higher percentages in light than in darkness; however, stratified seeds of these species germinated equally well in light and darkness. Nonstratified seeds of *Larix occidentalis* and *Pinus monticola* germinated to higher percentages in light than in darkness, but stratified seeds germinated to higher percentages in darkness than in light (Li *et al.*, 1994). Studies on nonstratified seeds of *B. pubescens* and *B. verrucosa* indicated that they required light for germination (Junttila, 1976a); however, stratified seeds

TABLE 10.29   Seed Dormancy in Trees of Boreal and North-Temperate Subalpine Forests

Species	Family[a]	Strat.[b]	Temperature[a]	L:D[a]	Reference
Nondormant					
Abies nordmanniana	Pin.	0	30/20	—[c]	Franklin (1974)
Alnus incana	Pin.	0	30/20	—	Schopmeyer (1974b)
A. tenuifolia	Pin.	0	32/20	—	Schopmeyer (1974b)
Betula pubescens	Betul.	0	24	L	Junttila (1976a)
Larix dahurica	Pin.	0	30/20	—	Rudolf (1974g)
L. decidua	Pin.	0	30/20	—	Rudolf (1974g)
L. lyallii	Pin.	0	18	—	Rudolf (1974g)
Picea abies	Pin.	0	30/20	L > D	Safford (1974), Patten (1963)
P. asperata	Pin.	0	—	L = D	Heit (1968a)
P. engelmannii	Pin.	0	30/20	—	Heit (1968a), Safford (1974)
P. glauca	Pin.	0	30/20	L = D	Safford (1974), Li et al. (1994)
P. koyamai	Pin.	0	30/20	—	Heit (1968a), Safford (1974)
P. smithiana	Pin.	0	30/20	—	Safford (1974)
Pinus gerardiana	Pin.	0	Room	—	Beniwal and Singh (1989)
P. mugo	Pin.	0	30/20	—	Krugman and Jenkinson (1974)
P. nigra	Pin.	0	30/20	—	Krugman and Jenkinson (1974)
P. rubens	Pin.	0	24–26	L > D	Baldwin (1934)
P. smithiana	Pin.	0	30/20	—	Safford (1974)
P. sylvestris	Pin.	0	10–30	—	Krugman and Jenkinson (1974)
Populus balsamifera	Salic.	0	5–25	—	Zasada and Viereck (1975)
P. tremula	Salic.	0	30/20	—	Schreiner (1974)
P. tremuloides	Salic.	0	30/20, 5–25	—	Zasada and Viereck (1975), Fechner et al. (1981)
Salix arbusculoides	Salic.	0	15, 25	—	Densmore and Zasada (1983)
Physiological dormancy					
Abies balsamea	Pin.	28	30/20	—	Franklin (1974)
A. densa	Pin.	—	Room	—	Beniwal and Singh (1989)
A. firma	Pin.	60	25/13	L = D	Asakawa (1959)
A. fraseri	Pin.	112	30/20	L > D	Adkins et al. (1984), Blazick and Hinesley (1984)
A. lasiocarpa	Pin.	28	30/20	D > L	Franklin (1974), Li et al. (1994)
A. magnifica	Pin.	28	30/20	—	Franklin (1974)
A. mariesii	Pin.	42	25/13	—	Asakawa (1959)
A. mayriana	Pin.	42	25/13	L > D	Asakawa (1959)
A. sachalinensis	Pin.	30–60	30/20	—	Franklin (1974)
A. veitchii	Pin.	0–14	30/20	—	Franklin (1974)
Betula payrifera	Betul.	14	25	L = D	Bevington and Hoyle (1981)
Chamaecyparis nootkatensis	Cupress.	90	24	—	Pawuk (1993)
Larix kaempferi	Pin.	42	25/13	D > L	Asakawa (1959)
L. laricina	Pin.	30, 60	30/20	L > D	Rudolf (1974g), Farmer and Reinholt (1986)
L. leptolepis	Pin.	0–30	30/26	—	Rudolf (1974g)
L. occidentalis	Pin.	0–42	30/20	D > L	Rudolf (1974g), Li et al. (1994)
Picea bicolor	Pin.	Winter	Spring	—	Heit (1968a)
P. breweriana	Pin.	Winter	Spring	—	Heit (1968a)
P. glehnii	Pin.	21–42	30/20, 25	L > D	Asakawa (1959), Safford (1974)
P. jezoenis	Pin.	21	30/20	—	Safford (1974)
P. mariana	Pin.	24	20/10	L = D	Farmer et al. (1984)
P. polita	Pin.	42	25/13	L > D	Asakawa (1959)
Pinus albicaulis	Pin.	90–120	30/20	—	Krugman and Jenkinson (1974)
P. aristata	Pin.	0–30	30/20	L = D	Heit (1968c), Krugman and Jenkinson (1974)
P. banksiana	Pin.	0–7	30/20	—	Krugman and Jenkinson (1974)
P. cembra	Pin.	90–270	30/20	—	Krugman and Jenkinson (1974)
P. contorta	Pin.	21	24, 27	L = D	Haais and Thrupp (1931), Li et al. (1984)
P. flexilis	Pin.	60	20	—	Barton (1930)
P. pumila	Pin.	120–150	25	—	Krugman and Jenkinson (1974)
P. monticola	Pin.	60	20	D > L	Barton (1930), Gansel (1986), Li et al. (1994)
P. sibirica	Pin.	60–90	30/20	—	Krugman and Jenkinson (1974)
Tsuga mertensiana	Pin.	90	30/20	—	Ruth (1974)

[a]See notes for Table 10.1.
[b]Cold stratification (days).
[c]Data not available.

TABLE 10.30   Seed Dormancy in Shrubs of Forested and Nonforested Areas
in Boreal and North-Temperate Subalpine Regions.

Species	Family[a]	Strat.[b]	Temperature[a]	L:D[a]	Reference
Nondormant					
Calluna vulgaris	Eric.	0	10–28	L	Helsper and Klerken (1984), Grime *et al.* (1981)
Kalmia polifolia	Eric.	0	22	—[c]	Jaynes (1971)
Salix alaxensis	Salic.	0	5–25	—	Densmore and Zasada (1983)
S. barrattiana	Salic.	0	5–25	—	Densmore and Zasada (1993)
S. bebbiana	Salic.	0	5–25	—	Densmore and Zasada (1983)
S. chamissonis	Salic.	0	5–25	—	Densmore and Zasada (1983)
S. fuscescens	Salic.	0	5–25	—	Densmore and Zasada (1983)
S. hastata	Salic.	0	10–25[d]	—	Steshenko (1966), Densmore and Zasada (1983)
S. herbacea	Salic.	0	22–32	—	Junttila (1967b)
S. interior	Salic.	0	5–25	—	Densmore and Zasada (1983)
S. lanata	Salic.	0	5–25	—	Steshenko (1966), Densmore and Zasada (1983)
S. lasiandra	Salic.	0	10–25	—	Densmore and Zasada (1983)
S. monticola	Salic.	0	15–25	—	Densmore and Zasada (1983)
S. mrytillifolia	Salic.	0	15–25	—	Densmore and Zasada (1983)
S. novae-angliae	Salic.	0	5–25	—	Densmore and Zasada (1983)
S. planifolia	Salic.	0	5–25	—	Densmore and Zasada (1983)
S. scouleriana	Salic.	0	5–25	—	Densmore and Zasada (1983)
S. setchelliana	Salic.	0	5–25	—	Densmore and Zasada (1983)
Vaccinium caespitosum	Eric.	0	21	—	McLean (1967)
Morphophysiological dormancy					
Aralia nudicaulis	Arali.	60–70[e]	30/20	—	Blum (1974)
Sambucus racemosa	Caprilfoli.	90[e]	20/15	—	Brinkman (1974f)
Symphoricarpos albus	Caprifoli.	180[e]	30/20	—	Evans (1974)
Taxus canadensis	Tax.	60–120[e]	15/13	—	Rudolf (1974h)
Viburnum cassinoides	Caprifoli.	90[e]	30/20	—	Gill and Pogge (1974)
Physiological dormancy					
Alnus crispa	Betul.	60	30/20	—	Schopmeyer (1974b)
A. viridis	Betul.	14	20/10	—	Farmer *et al.* (1985)
Amelanchier alnifolia	Ros.	120	21	—	McLean (1967)
Arctostaphylos uva-ursi	Eric.	90[f]	21	—	McLean (1967)
Betula nana	Betul.	14	24	—	Junttila (1970)
Cornus canadensis	Corn.	71–112[g]	g.h.[h]	—	Nichols (1934)
C. stolonifera	Corn.	30	25/10	—	Acharya *et al.* (1992)
Corylus cornuta	Coryl.	60–120	30/20	—	Brinkman (1974e)
Diapensia lapponica	Diapensi.	71–112[g]	g.h.[h], 10–20	L	Nichols (1934), Densmore (1997)
Empetrum nigrum	Empetr.	60[f]	20/15	—	Grime *et al.* (1974)

(continues)

2. Shrubs

also may germinate in darkness. Cold stratification increased the germination of *Picea mariana* seeds in darkness at 20/10°C from about 30 to 100% (Farmer *et al.*, 1984), and this treatment increased the germination of *Larix laricina* seeds in darkness at low (15/5, 20/10°C) temperatures (Farmer and Reinholt, 1986).

2. Shrubs

This discussion includes shrubs growing in forests as well as those growing in nonforested sites such as peatland or heath communities. Some shrubs have seeds that are nondormant and a few have seeds with morphophysiological dormancy. The majority of seeds have physiological dormancy (Table 10.30).

Morphophysiological dormancy is found in seeds of *Aralia nudicaulis* (Blum, 1974), *Sambucus racemosa* (Brinkman, 1974f), *Symphoricarpos albus* (Evans, 1974), *Taxus canadensis* (Rudolf, 1974h), and *Viburnum cassinoides* (Gill and Pogge, 1974), but the type of MPD has not been determined for any of these species. However, since warm followed by cold stratification promotes germination in all of them, their seeds must have either some type of deep simple MPD or nondeep complex MPD.

Cold stratification is not very effective in breaking dormancy in seeds of *Empetrum nigrum* or *Rosa acicularis* unless they first receive warm stratification (Table 10.30). It is not clear why these seeds require a warm stratification pretreatment. Neither species belongs to

TABLE 10.30—*Continued*

Species	Family[a]	Strat.[b]	Temperature[c]	L:D[d]	Reference
Kalmia polifolia	Eric.	71–112[g]	g.h.	L	Nichols (1934)
Ledum palustre	Eric.	30	20, 25	L	Calmes and Zasada (1982)
Rhamnus alnifolia	Rhamn.	60	—	—	Hubbard (1974b)
Rhododendron lapponicum	Eric.	71–112[g]	g.h.[h]	—	Nichols (1934)
Ribes hudsonianum	Grossulari.	90–120	g.h.[h]	—	Pfister (1974)
R. lacustre	Grossulari.	120–200	g.h.[h]	—	Pfister (1974)
Rosa acicularis	Ros.	90[e]	25	—	Densmore and Zasada (1977)
Rubus chamaemorus	Ros.	270	18	L > D	Taylor (1971), Warr *et al.* (1979)
Salix arctica	Salic.	30	25	—	Densmore and Zasdada (1983)
S. brachycarpa	Salic.	30	25	—	Densmore and Zasada (1983)
S. glauca	Salic.	30	25	—	Densmore and Zasada (1983)
S. ovalifolia	Salic.	30	25	—	Densmore and Zasada (1983)
S. pentandra	Salic.	56	18–33	—	Junttila (1976b)
S. phlebophylla	Salic.	30	25	—	Densmore and Zasada (1983)
S. polaris	Salic.	56	21–33	—	Junttila (1976b), Densmore and Zasada (1982)
S. reticulata	Salic.	56	19–32	—	Junttila (1976b), Densmore and Zasada (1982)
S. rotundifolia	Salic.	30	25	—	Densmore and Zasada (1982)
Shepherdia canadensis	Elaeagn.	60	21	—	McLean (1967)
Sorbus amurensis	Ros.	123	3	—	Razumova (1985)
S. caschmiriana	Ros.	95	3	—	Razumova (1985)
S. gracilis	Ros.	100	3	—	Razumova (1985)
S. koehneana	Ros.	95	3	—	Razumova (1985)
S. serotina	Ros.	110	3	—	Razumova (1985)
Spirea beauverdiana	Ros.	30	15, 20, 25	L	Calmes and Zasada (1982)
S. glauca	Ros.	90	10–25	—	Zasada and Viereck (1975), Densmore and Zasada (1983)
Vaccinium angustifolium	Eric.	83	g.h.	—	Crossley (1974b)
V. myrtillus	Eric.	21	18–28	L	Ritchie (1956), Grime *et al.* (1981)
V. uliginosum	Eric.	30	20, 25	L > D	Calmes and Zasada (1982)
V. vitis-idaea	Eric.	Winter[i]	18–28	L	Ritchie (1955), Grime *et al.* (1981)

[a]See notes for Table 10.1.
[b]Cold stratification (days).
[c]Data not available.
[d]Seeds not tested at 5°C.
[e]Seeds require warm stratification prior to cold stratification for dormancy break.
[f]Seeds were acid scarified prior to cold stratification.
[g]Seeds were outdoors in Connecticut during winter.
[h]Seeds were germinated in a heated greenhouse.
[i]Seeds were outdoors in England until January.

a family known to have physical dormancy (Table 3.5) or underdeveloped embryos (Table 3.3). However, a warm stratification pretreatment could increase the effectiveness of cold stratification in overcoming physiological dormancy of the embryo, as it does in seeds of *Cardamine concatenata* (Baskin and Baskin, 1995) and *Floerkea proserpinacoides* (Baskin *et al.*, 1988).

With respect to germination ecology, two kinds of shrubby willows occur in boreal/subalpine regions. One group of willows has nondormant seeds that are dispersed in summer and germinate immediately over a range of temperatures, including those in the habitat in summer (Junttila, 1976b; Densmore and Zasada, 1983). The second group has conditionally dormant seeds that are dispersed in autumn and germinate over only a limited range of temperatures. Temperatures required

for the germination of newly dispersed seeds are higher than those in the habitat in autumn; therefore, germination is prevented. During cold stratification in late autumn and winter, the temperature requirement for germination is lowered, and by spring seeds germinate at low spring temperatures (Junttila, 1976; Densmore and Zasada, 1982; Zasada *et al.*, 1983).

Cold stratification also decreases the temperature requirement for germination in seeds of *Ledum palustre, Spirea beauverdiana,* and *Vaccinium uliginosum* (Calmes and Zasada, 1982). Although *V. caespitosum* is listed in Table 10.30 as having nondormant seeds, it would not be surprising if cold stratification lowered the temperature requirement for germination in seeds of this species, indicating presence of MPD.

Seeds of many shrubs have physiological dormancy

and require cold stratification before they can germinate at spring temperatures in spring. Species whose seeds require 60 days or less of cold stratification for germination, as well as those like *Ledum palustre*, *Vaccinium uliginosum*, and *Spirea beauverdiana* whose freshly matured seeds can germinate at some temperatures, have nondeep physiological dormancy. Seeds requiring 90 days or more of cold stratification to come out of dormancy require further studies to determine the type of physiological dormancy (see Chapter 3). GA$_3$ substitutes for cold stratification in overcoming dormancy of *Rhododendron lapponicum* seeds (Junttila, 1972); thus, seeds of this species may have nondeep physiological dormancy.

Light:dark requirements for seed germination have been determined for only nine shrubs (Table 10.30). Seeds of seven species require light for germination, whereas two germinate to higher percentages in light than in darkness. One reason for the lack of information on light:dark requirements is that germination tests frequently are conducted only at a daily photoperiod. Seeds of *Vaccinium myrtillus*, *V. vitis-idaea*, *V. uliginosum*, and *V. oxycoccos* germinated along a gradient from a forest to an open bog in southeast Sweden; however, the best conditions for seedling establishment were "small disturbances with high moisture and organic soil content" (Eriksson and Froborg, 1995). Seeds on the soil surface in such sites would receive full sunlight, implying that high irradiance was more conducive for germination than low irradiance or darkness. Seeds of *Ledum groenlandicum* and *L. palustre* also require full sunlight and high soil moisture for germination, as well as a pH of about 5.5 (Karlin and Bliss, 1983).

3. Herbaceous Species

Some herbaceous species in boreal and north-temperate subalpine regions have nondormant seeds, but most have seeds that are dormant at maturity, with morphophysiological dormancy and physiological dormancy being about equally important (Table 10.31). Freshly matured seeds of some herbaceous species germinate to high percentages, especially at relatively high temperatures, and thus are listed as nondormant (Table 10.31). However, studies are needed to determine whether cold stratification lowers the temperature requirement for germination in seeds of boreal/subalpine herbaceous species. The optimum germination temperature regime for nonstratified (fresh) and cold stratified (12 weeks) seeds of *Trientalis borealis* was 20/10°C (Baskin and Baskin, unpublished results). Thus, cold stratification did not lower the temperature requirement for germination. Seeds of *T. europaea* tested at 20°C germinated to 42% without cold stratification and to

56% following 18 weeks of cold stratification (Matthews, 1942). Thus, cold stratification increased the percentage of germination, but it is not known if the optimum germination temperature was decreased. Light:dark requirements have been determined for seeds of only five species whose seeds are nondormant at maturity. Seeds of three species required light for germination, and those of two germinated to higher percentages in light than in darkness.

Seeds of *Veratrum californicum* (Williams and Cronin, 1968), *Osmorhiza chilensis*, *O. occidentalis*, *Erythronium grandiflorum* (Baskin *et al.*, 1995), and *Heracleum lanatum* (= *H. sphondylium*) (Stokes, 1952) require only cold stratification for dormancy break and embryo growth, thus they appear to have nondeep complex MPD. The kind of MPD has not been determined for the other species listed under MPD in Table 10.31. However, because seeds of *Smilacina racemosa* (Table 5.7) have deep simple double MPD, this type of MPD may occur in seeds of *S. stellata*. Further, because seeds of *Actaea pachypoda* have deep simple epicotyl MPD (Table 5.5), this type may occur in seeds of *A. rubra*.

The temperature requirements for seed germination in some species with MPD are quite low (e.g., 1, 5, 5/1°C), and seeds of *Osmorhiza occidentalis* have been observed germinating under snow (Baskin *et al.*, 1995). Other species, however, germinate at temperatures (20/9, 22/17°C) that might occur in the habitat after snow melt. For example, 72% of the seeds of *Clintonia borealis* subjected to natural seasonal temperatures germinated over a 30-day period when mean daily maximum and minimum temperatures were 20.1 and 9.1°C, respectively (Baskin and Baskin, unpublished results). Light:dark requirements for seed germination have been determined for only six boreal/subalpine species with MPD (Table 10.31). Seeds of one species germinated equally well in light and darkness, whereas five germinated to higher percentages in light than in darkness.

Physical dormancy has been reported in seeds of several herbaceous species (Table 10.31). However, other genera of Fabaceae (*Anthyllis*, *Astragalus*, *Hedysarum*, *Lathyrus*, *Lotus*, *Medicago*, *Onobrychis*, *Oxytropis*, *Trifolium*, and *Vicia*) and Geraniaceae (*Erodium*) occur in boreal and north-temperate subalpine zones (Huxley, 1967; Larsen, 1980), and it is expected that their seeds also have physical dormancy. Little is known about the dormancy-breaking requirements of seeds with physical dormancy in boreal and north-temperate subalpine zones, except that heat from fires can make seeds of some species permeable (see Section VII,C,1). It would be informative to determine the effects of winter temperatures on dormancy break in these species.

Species in a variety of families in boreal/subalpine

TABLE 10.31   Seed Dormancy in Herbaceous Species of Forested and Nonforested Areas
in Boreal and North-Temperate Subalpine Regions

Species	Family[a]	Strat.[b]	Temperature[a]	L:D[a]	Reference
Nondormant seeds					
Androsace septentrionalis[c]	Primul.	0	32/22	—[d]	McDonough (1970)
Carex xerantica	Cyper.	0	22/17	—	McDonough (1970)
Epilobium hirsutum[c]	Onagr.	0	12–33	L	Grime *et al.* (1981)
Luzula pilosa	Junc.	0	9–23	L > D	Grime *et al.* (1981)
L. sylvatica	Junc.	0	6–20	L	Grime *et al.* (1981)
Lythrum salicaria[c]	Lythr.	0	19–37	L	Grime *et al.* (1981)
Mycelis muralis[c]	Aster.	0	10–24	L > D	Grime *et al.* (1981)
Rudbeckia occidentalis[c]	Aster.	0	22/17	—	McDonough (1969)
Rumex salicifolius[c]	Polygon.	0	32/22	—	McDonough (1970)
Trientalis borealis	Primul.	0	20/10	—	Baskin and Baskin (unpublished results)
Morphological dormancy					
Cymopterus acaulis	Api.	0	22/17	—	McDonough (1970)
Morphophysiological dormancy					
Actaea rubra[c]	Ranuncul.	112	22/17	—	McDonough (1969)
Clematis hirsutissima	Ranuncul.	112	22/17	—	McDonough (1969)
Caltha palustris[c]	Ranuncul.	71–112	g.h.[e]	—	Nichols (1934)
Clintonia borealis	Lili.	71–112	g.h.[e], 20/9	—	Nichols (1934), Baskin and Baskin (unpublished results)
Coptis asplenifolia	Ranuncul.	Winter	Spring	—	Tappeiner and Alaback (1989)
Delphinium barbeyi[c]	Ranuncul.	91	1	—	Williams and Cronin (1968)
D. occidentale	Ranuncul.	91	1	—	Williams and Cronin (1968)
Erythronium grandiflorum	Lili.	126	5	—	Baskin *et al.* (1995b)
Heracleum lanatum	Api.	112	22/17	—	McDonough (1969)
Linnaea borealis	Caprifoli.	60	21	—	McLean (1967)
Maianthemum canadense	Lili.	71–112	g.h.[e]	—	Nichols (1934)
Meconopsis betonicifolia	Papaver.	—	15/5	L > D	Thompson (1968)
M. dhwojii	Papaver.	—	15/5, 15	L > D	Thompson (1968)
M. gracilipes	Papaver.	—	15/5	L = D	Thompson (1968)
M. horridula	Papaver.	—	15/5	L > D	Thompson (1968)
M. napaulensis	Papaver.	—	25/15	L > D	Thompson (1968)
M. regia	Papaver.	—	15/5	L > D	Thompson (1968)
Osmorhiza chilensis	Api.	140	5/1	—	Baskin *et al.* (1995b)
O. occidentalis	Api.	140	1	—	Baskin *et al.* (1995b)
Smilacina stellata	Lili.	71–112	g.h.[e]	—	Nichols (1934)
Veratrum californicum[c]	Lili.	91	1	—	Williams and Cronin (1968)

(continues)

zones have physiological dormancy, and cold stratification is required to break it (Table 10.31). None require high temperatures for loss of dormancy. Seeds of some species germinate at temperatures near freezing (e.g., 1, 2°C), which probably means they can germinate in/under melting snow. It is hard to evaluate the temperature requirements for the germination of some species because seeds were tested only at relatively high temperatures (20/15, 20, 15°C). Although nondormant seeds can germinate at high temperatures, they also may be able to germinate at low temperatures that simulate those in the habitat in early spring. However, seeds of *Carex eburnea, Polygonum amphibium, P. viviparum, Potentilla rivalis, Ranunculus hyperboreus, Rorippa palustris,* and *Tanacetum vulgare* generally germinated to

higher percentages at 30/15°C than at 20/10°C (Hogenbirk and Wein, 1992). Light:dark requirements have been determined for only four species, and seeds of all of them germinated to higher percentages in light than in darkness.

Both availability of seeds and/or safe sites for germination and seedling establishment were important in the recruitment in plant populations in deciduous and coniferous woodlands in central Sweden (Eriksson and Ehrlen, 1992). Low numbers of seeds limited the recruitment of *Actaea spicata, Convallaria majalis,* and *Linnaea borealis* seeds, whereas both limited seed and microsite availability reduced the recruitment of *Maianthemum bifolium, Trientalis europaea, Sorbus aucuparia, Lathyrus montanus, Rubus*

TABLE 10.31—*Continued*

Species	Family[a]	Strat.[b]	Temperature[a]	L:D[a]	Reference
Physical dormancy					
Anthyllis vulneraria	Fab.	0	20	—	Granstrom and Schimmel (1993)
Genista anglica	Fab.	0	20	—	Mallik and Gimingham (1985)
Geranium bicknellii	Gerani.	0	20	—	Granstrom and Schimmel (1993)
G. bohemicum	Gerani.	0	20	—	Granstrom and Schimmel (1993)
G. lanuginosum	Gerani.	0	20	—	Granstrom and Schimmel (1993)
G. viscosissimum	Gerani.	0	22/17	—	McDonough (1969)
Lathyrus ochroleucus	Fab.	0	21	—	McLean (1967)
Lupinus arcticus	Fab.	0	21	—	McLean (1967)
Physiological dormancy					
Agoseris glauca	Po.	56	22/17	—	McDonough (1970)
Agropyron trachycaulon	Po.	56	17/12	—	McDonough (1970)
Carex egglestonii	Cyper.	30	24/18	—	Johnson *et al.* (1965)
C. hoodii	Cyper.	112	22/17	—	McDonough (1969)
C. sempervirens	Cyper.	40	20	L > D	Georges and Lazare (1983)
Circaea alpina	Onagr.	71–112	g.h.[e]	—	Nichols (1934)
Cirsium foliosum	Aster.	112	2	—	McDonough (1969)
Epilobium angustifolium[c]	Onagr.	30	21	—	McLean (1967)
Eupatorium purpureum	Aster.	71–112	g.h.[e]	—	Nichols (1934)
Frasera speciosa	Gentian.	112	2	—	McDonough (1969)
Gentiana pneumonanthe	Gentian.	56	20/10	—	Thompson (1969)
Mertensia ciliata	Boragin.	200	18	—	Pelton (1961)
Molinia caerulea	Po.	56	22/12	L > D	Pons (1989)
Oxalis acetosella[c]	Oxalid.	180	20/15	—	Grime *et al.* (1981)
Penstemon cyananthus	Scrophulari.	112	15	—	Kitchen and Meyer (1991)
P. subglaber	Scrophulari.	112	15	—	Kitchen and Meyer (1991)
Pedicularis parryi	Scrophulari.	56	22/17	—	McDonough (1970)
Polemonium foliosissimum	Polemoni.	—[f]	15	L > D	McDonough and Laycock (1975)
Senecio integerrimus	Aster.	112	2	—	McDonough (1969)
S. serra	Aster.	112	2	—	McDonough (1969)
Trientalis europaea	Primul.	126	20	L > D	Matthews 1942

[a]See notes for Table 10.1.
[b]Cold stratification (days).
[c]Listed as a weed in Holm *et al.* (1979).
[d]Data not available.
[e]Greenhouse.
[f]Seeds may have afterripened during dry storage at room temperatures.

saxatilis, and *Vaccinium myrtillus*. *Ribes alpinum*, *V. vitis-idaea*, *Paris quadrifolia*, *Frangula alnus*, and *V. opulus* were not seed limited, but the cause of low recruitment was not determined.

B. Dormancy and Germination of Species in South-Temperate Subalpine Zones

1. Trees

Seeds of *Eucalyptus delegatensis*, *E. regnans*, *E. fastigata*, and *E. glaucescens* have physiological dormancy and required 3–6 weeks of cold stratification for high germination percentages at 20, 25, 25, and 20°C, respectively (Krugman, 1974). However, the optimum germination temperature for *E. regnans* seeds was 15°C (Devillez, 1977), and it was between 15 and 20°C for those

of *E. delegatensis* (Battaglia, 1993). Cold stratification for 4 or 8 weeks greatly increased germination percentages of *E. niphophila* (8 weeks), *E. gigantea* (4 weeks), *E. pauciflora* (8 weeks) and *E. dives* (4 weeks) seeds at 24°C (Pryor, 1954). Seeds of *Eucalyptus pauciflora* sown at the treeline in southeastern Australia in autumn did not germinate until the following spring, after they had been cold stratified. Germination in spring did not occur until air temperatures were 6°C or above (Ferrar *et al.*, 1988). Seeds of both *E. delegatensis* and *E. amygdalina* sown at elevations of 350 and 540 m on the east coast of Tasmania germinated when mean daily temperatures were higher than 6–7°C (Battagalia, 1996). Newly germinated seedlings of *Nothofagus menziesii* (Wardle, 1967; Manson, 1974) and *N. solandri* (Wardle, 1970) have been observed in subalpine forests of New Zealand in spring, implying that cold stratification during winter broke seed dormancy. In fact, seeds of *N. menziesii* germinated to 90% after 8–12 weeks at 2–5°C (Wardle, 1967).

Seeds of *Eucalyptus regnans* require light for germination, but those of *E. pauciflora* do not (Clifford, 1953). Light shade promoted the germination of *Nothofagus solandri* seeds in the field in New Zealand, but heavy shade at elevations of 979 and 1333 m delayed germination, possibly due to a reduction in soil temperatures (Wardle, 1970). More seedlings of *N. menziesii* were found on raised bryophyte-covered sites than on litter-covered logs or soil (Allen, 1987). These observations suggest that litter inhibited germination, possibly by reducing light. Seeds of *Eucalyptus delegatensis* sown in Tasmania in autumn, winter, and summer germinated to higher percentages than those sown in spring, regardless of microsite (i.e., hillock, flat, or depression), except germination on a flat microsite was the same in winter and spring (Battaglia and Reid, 1993). Seeds germinated to higher percentages in autumn, winter, and summer in a depression than on a hillock or flat microsite.

Hoheria glabrata is a small tree in the Malvaceae, thus its seeds potentially have physical dormancy. Freshly matured seeds germinated sporadically, but after dry storage in a plastic bag at 4°C for 42 days, they germinated to 60% (Haase, 1987). Do freshly matured seeds of this species have impermeable seed coats, and, if so, do they become permeable at low temperatures?

Seeds of *Eucalyptus regnans* germinated to 50–80% at substrate water potentials of 0, −0.2, −0.4, and −0.6 MPa, but they germinated to only 25 and 5% at −0.8 and −1.0 MPa, respectively (Edgar, 1977).

2. Shrubs

Essentially nothing is known about the germination ecology of shrubs in the south-temperate subalpine zone, although shrub vegetation may be more important than forests in some areas (Wardle, 1991; Killick, 1978). Seeds of *Dracophyllum traversii* (Epacridaceae) germinated to 80% at room temperatures following 6–8 weeks of cold stratification at 4°C (Haase, 1986a). Haase (1986c) obtained 50% germination for *Olearia ilicifolia* (Asteraceae) achenes that were cold stratified for 3 days and then incubated at a 16-hr daily photoperiod at 20°C for 25 days. Because Haase (1986c) was careful to determine whether achenes were filled, we know that the ungerminated achenes were viable; they must have had physiological dormancy. Results for achenes of *O. arborescens*, *O. avicenniaefolia*, and *O. nummularifolia* were similar to those for achenes of *O. ilicifolia*.

Achenes of *Olearia colensoi* germinated in the field in spring, but attempts to break dormancy by cold stratification at 5°C for 35 days were unsuccessful (Wardle

et al., 1970). Achenes of *Senecio bennettii* germinated to 41% in light at 20°C; they did not germinate in darkness. The remaining 59% of the achenes were dormant and did not germinate after 8 weeks of cold stratification; however, they germinated following scarification (Haase, 1986b).

3. Herbaceous Species

Little information is available on the germination ecology of herbaceous species in south-temperate subalpine zones. After 2 months of dry storage at room temperatures, achenes of *Acaena caesiiglauca*, *A. fissistipula*, *A. glabra*, and *A. profundeincisa* (Rosaceae) from the subalpine zone in New Zealand germinated to 61, 61, 56, and 43%, respectively, at a 16-hr daily photoperiod at 20°C (Conner, 1987). The optimal germination temperature for all species, except *A. glabra*, was 20°C; for *A. glabra*, it was 17°C. Unfortunately, the effects of cold stratification were not determined.

C. Special Factors

1. Fire

Fire is a natural phenomenon in boreal and subalpine zones (Heinselman, 1970; Rowe and Scotter, 1973; Wright, 1974; Killick, 1978; Wardle, 1991) and is needed to maintain the diversity and stability of conifer forests (Habeck and Mutch, 1973; Zackrisson, 1977). The frequency of fire has been altered, however, by human activities, which has caused changes in plant communities. Increased fire frequency has contributed to the conversion of forest to shrub- or grass-dominated communities (Wardle, 1991), and in the subalpine of the Amatole Mountains in southern Africa it may have changed fynbos shrub vegetation to grassland (Killick, 1978).

Fire in boreal and subalpine communities has various effects on germination ecology. The most severe consequence of fire is the death of seeds, which may result from them actually being consumed by the flames (Legg *et al.,* 1992). Other seeds die because they are exposed to high temperatures. Treatments at 200–220°C for 5 min killed seeds of *Abies magnifica* and *Pseudotsuga menziesii*, while 220–240°C for 5 min killed those of *Pinus contorta* (Wright, 1931). Most seeds of *Calluna vulgaris* were killed after 60 sec at 160°C (Whittaker and Gimingham, 1962) or 2 min at 100°C (Mallik and Gimingham, 1985). Even seeds of *Geranium lanuginosum*, *G. bohemicum*, and *Anthyllis vulneraria*, which have physical dormancy, were killed during a 10-min treatment at 100°C (Granstrom and Schimmel, 1993). Seeds of *Betula pendula* and *Luzula pilosa* died after 10 min at 70°C, and those of *Picea abies* and *Pinus sylvestris* died after 10 min at 65°C (Granstrom and Schimmel, 1993).

Heat from fires can promote loss of dormancy in seeds with impermeable seed coats. Seeds of the legume *Anthyllis vulneraria* germinated to about 90% after 10 min at temperatures of 50–85°C (Granstrom and Schimmel, 1993), and those of the legume *Genista angelica* germinated to 96% after 2 min at 100°C (Mallik and Gimingham, 1985). Seeds of *Geranium bicknellii* germinated to 90% or higher after 10 min at 65–100°C, those of *G. bohemicum* after 10 min at 55–90°C, and those of *G. lanuginosum* after 10 min at 60–95°C (Granstrom and Schimmel, 1993). A short high-temperature pretreatment increased germination percentages of *Calluna vulgaris* (Whittaker and Gimingham, 1962), *Hypericum pulchrum*, *Vaccinium myrtillus*, *V. vitis-idaea* (Mallik and Gimingham, 1985), and *Abies magnifica* (Wright, 1931) seeds, although they have physiological dormancy.

Another effect of the heat from fires is that it could melt resins and cause serotinous cones to open, thereby releasing seeds. Examples of species in the boreal/subalpine zone with serotinous cones are *Pinus banksiana*, *P. contorta*, and *Picea mariana* (see Chapter 7,II).

Soil temperatures (Raison, 1979), pH, and conductivity (Smith, 1970) increase after a fire, and soils also may be drier after a fire than before due to a loss of organic matter (Sims, 1976). Removal of litter and increased temperatures further promote the loss of moisture from the soil surface. Another factor influencing the microenvironment of seeds is the presence of ash. Although ashes are a natural fertilizer (Stark, 1979), their presence can inhibit germination via high pH (10.6–12.5) and hydroxide and bicarbonate levels (Thomas and Wein, 1990). Thomas and Wein (1990) found that leaching of ash improved the germination of *Pinus banksiana*; however, fewer seeds germinated in leached *Populus tremuloides* ash than in leached conifer (*Picea mariana/ P. banksiana*) ash.

Although fire potentially could make microenvironmental conditions unsuitable for germination, seeds of some species germinate to higher percentages in burned than in nonburned sites. For example, seeds of *Betula papyrifera*, *Alnus crispa*, *Picea mariana*, *Salix alaxensis*, *S. scouleriana*, and *S. bebbiana* sown on unburned, scorched, lightly burned, moderately burned, and heavily burned plots in Alaska germinated only in the moderately and heavily burned plots (Zasada *et al.*, 1983). Further, more seeds of each species germinated in heavily burned than in moderately burned plots.

2. Acid Precipitation and Heavy Metals

Parts of the boreal forest region and subalpine conifer forests receive precipitation with a pH as low as 2.1

(Likens and Bormann, 1974). Some deciduous forests, especially those in eastern North America and northeastern Europe, also are being impacted by acidic rain and snow (Likens and Bormann, 1974). Although the optimum pH for germination of most species is slightly acidic (Chapter 2), increased acidity could be important in the germination ecology of species in heavily impacted areas. Germination of *Picea glauca*, *P. rubens*, *Abies balsamea*, and *Betula papyrifera* seeds was inhibited by simulated acid rain at pH 2.6 but not 3.6, 4.6, or 5.6, whereas germination of *Picea mariana*, *Pinus strobus*, *P. banksiana*, *P. resinosa*, *P. sylvestris*, *B. alleghaniensis*, and *Acer rubrum* seeds was not inhibited at pH 2.6, 3.6, 4.6, or 5.6 (Percy, 1986).

Seeds of *Picea rubens* germinated equally well at pH 3.4–6.2 (Moore and Gillette, 1988), those of *Pinus contorta* germinated to significantly higher percentages at pH 2.2 and 3.0 than in distilled water (pH 6.5), and those of *Picea glauca* and *Pinus banksiana* germinated equally well at pH 2.2 and 6.5. *Pinus banksiana* seeds germinated to a higher percentage at pH 3.0 than at pH 6.5, but those of *Picea mariana* germinated equally well at pH 3.0 and 6.5; the germination of *P. mariana* was reduced at pH 2.2 (Abouguendia and Redmann, 1979).

One effect of increased acidity is an increase in the availability of toxic metals. Nosko *et al.* (1988) tested seeds of *Picea glauca* in aluminum concentrations of 0, 50, 100, and 500 μM but found no decrease in germination. Seeds of *P. rubens*, *Abies balsamea*, *Betula alleghaniensis*, and *B. papyrifera* were not reduced by aluminum (10 or 100 mg liter^{-1}), cadmium (1.0 mg liter^{-1}), copper (5 or 10 mg liter^{-1}), lead (5 or 20 mg liter^{-1}), zinc (5 or 10 mg liter^{-1}), or by a pH of 3, 4, or 5 (Scherbatskoy *et al.*, 1987). Further, acidity and metals showed no interaction in controlling germination; however, seeds of *A. balsamea* and *B. alleghaniensis* germinated to higher percentages at pH 3 than at pH 4 or 5 (Scherbatskoy *et al.*, 1987). Nickel (10 mg liter^{-1}) increased the germination of *Acer ginnala* and *P. abies* seeds and decreased the germination of *B. papyrifera* and *Pinus banksiana* seeds (Heale and Ormrod, 1983). However, copper (20 mg liter^{-1}) increased the germination of *B. papyrifera* and *P. abies*, decreased germination of *A. ginnala*, but had little effect on *P. banksiana* seeds. Copper and nickel combined reduced the germination of all four species (Heale and Ormrod, 1983).

Some studies have been done on species in forests adjacent to boreal/subalpine forests. Species from the deciduous forest area, including *Pinus strobus*, *Juniperus virginiana*, and *Betula alleghaniensis*, germinated to higher percentages at pH 3.0 than at pH 5.7, whereas seeds of *Acer saccharum*, *Carya ovata*, and *Fagus grandifolia* germinated equally well at pH 3.0 and pH 5.7 (Lee and Weber, 1979). Raynal *et al.* (1982) found that seeds of *P. strobus* germinated to a higher percentage at pH 3.0 than at pH 5.6, those of *A. saccharum* and *Tsuga canadensis* germinated equally well at pH 3.0 and pH 5.6, and those of *B. lutea* germinated to a higher percentage at pH 5.6 than at pH 3.0. Seeds of *Pseudotsuga menziesii* from western North American montane and moist warm temperate forests germinated to higher percentages at pH 3.0 than at pH 5.7 (Lee and Weber, 1979); however, in another study, seeds germinated equally well (about 90%) at pH 5.6, pH 4.0, and pH 3.0 but to only about 30% at pH 2.0 (McColl and Johnson, 1983).

In conclusion, a pH of about 3.0 inhibits seed germination in some species, but it has no effect or promotes germination in others. Because acid rain may have a pH of 3.0–3.6 (see Raynal *et al.*, 1982), it is important to learn more about the effect of pH in this range on the germination of species with various life forms, not only in boreal/subalpine forests but also in other types of vegetation.

VIII. ARCTIC AND TEMPERATE-ZONE ALPINE TUNDRA

According to Walter (1973), tundra (region 10 in Fig. 9.1) is the northernmost vegetational zone in the eastern and western hemispheres and begins north of the boreal coniferous forest zone. Walter's map does not show tundra vegetation in the southern hemisphere; however, it is found on the southernmost tip of South America (Moore, 1983), ice-free regions of the Antarctic Peninsula and nearby islands (Longton, 1985), and periantarctic islands (Stonehouse, 1989).

The climate of the northern tundra region is characterized by short cool summers and long cold winters. The average temperature of the warmest summer month is less than 10°C, whereas temperatures of the coldest winter month range from −22 to −40°C, depending on location (Trewartha, 1968). With increases in latitude, solar radiation decreases, and finally a point is reached when much of the solar radiation is required to melt the snow cover. Thus, the growing season is only 40 days (Barry *et al.*, 1981). Precipitation also decreases northward, and yearly totals may be only 25–30 cm. The period of greatest precipitation for inland sites is summer and autumn, but some coastal areas may have more precipitation in autumn and winter than in summer (Trewartha, 1968). Because temperatures, and thus potential evaporation rates, are low, the soil is saturated with water in many areas during summer, although rainfall is low. Further, the presence of permafrost prevents water from seeping very deep into the soil; consequently, countless marshes and wet areas result (Walter,

1973). In contrast, at the northern limits of land on islands in the Arctic Ocean, very little moisture is available for plant growth, and the region is called a polar desert (Bliss, 1981a; Stonehouse, 1989).

In the southern hemisphere, the boundaries for tundra vegetation are the 10°C July isotherm, which follows the west coast of South America north for a short distance and then crosses the continent at about the Strait of Magellan, and the southern boundary is the northern limit of the Antarctic ice pack (Stonehouse, 1989). On the Antarctic Peninsula and neighboring islands, the mean temperature of the warmest month is −1 to 2°C, and that of the coldest month is rarely below −15°C. Annual precipitation is 20–100 cm. On the periantarctic or subantarctic islands, the mean temperature of the warmest month is 0–6°C, and that of the coldest month is rarely below −2°C (Stonehouse, 1989). These islands, as well as the tip of South America (Tierra del Fuego), have high rainfall (200–500 cm on Tierra del Fuego), are cloudy most of the summer, and are pounded by fierce westerly winds (Walter, 1973; Moore, 1983; Walton, 1985).

Polar deserts near the North Pole have mostly lichens and mosses, with less than 100 species of angiosperms, many of which grow as compact cushion plants (Stonehouse, 1989). Vegetation in the high Arctic tundra consists of sedge–moss–meadow communities, with an occasional low shrub, dwarf shrub heath, or cottongrass community. The low Arctic tundra just north of the boreal coniferous forests is characterized by low shrub–subshrub (0.4–0.6 m tall), tall shrub (2–5 m tall), cottongrass tussocks–subshrub, subshrub heath, sedge–moss, and sedge–grass–moss meadow communities (Bliss, 1981a).

The Antarctic Peninsula has two native angiosperms, *Deschampsia antarctica* (Poaceae) and *Colobanthus quitensis* (Caryophyllaceae), and various lichen, moss, and alga communities (Komarkova *et al.*, 1985; Longton, 1985). Vegetation on the subantarctic islands is grasslands, herb fields, bogs, swamps, fern brakes, and various bryophyte and lichen communities (Walton, 1985), and Tierra del Fuego has dwarf shrubs, cushion bog, graminoid bog, and tussock grassland communities (Moore, 1983).

High mountains in temperate regions of both the northern and the southern hemispheres have tundra vegetation at their peaks, thus the name alpine tundra. The elevation of alpine tundra varies with the latitude and mass of the mountain, slope, degree, and aspect; however, it is found above subalpine vegetation and below permanent snow fields. Alpine tundra vegetation receives higher daily levels of solar radiation than tundra vegetation in polar regions, and alpine tundra has a greater daily fluctuation in temperature and deeper snow cover in winter than arctic tundra (Trewartha,

1968; Barry *et al.,* 1981). Vegetation in alpine tundra regions consists of low shrub, subshrub, meadow, sedge-moss, cushion plant–lichen (Bliss, 1981b), bogs, heath, and grasslands (Killick, 1978).

A. Dormancy and Germination of Shrubs

Freshly matured seeds of most shrubs will germinate without first being cold stratified, thus they are listed as nondormant in Table 10.32. However, in view of the relatively high germination temperatures of seeds of some of these species, it would not be surprising if future studies showed that cold stratification reduced the temperature requirements for seed germination. If cold stratification lowers the temperature requirement for germination, the species should be listed under physiological dormancy.

Morphological and morphophysiological dormancy have not been reported in seeds of tundra shrubs, but some species have seeds with physiological dormancy (Table 10.32). However, cold stratification is not an absolute requirement for germination in seeds with physiological dormancy. That is, freshly matured seeds germinate, at least to a low percentage at one or more temperatures, but cold stratification increases germination percentages (e.g., Bannister, 1990) and/or lowers the temperature requirement for germination. For example, nonstratified seeds of *Chamaedaphne calyculata* germinated to 0, 10, and 21% at 10, 15, and 20°C, respectively, under long (22 hr light) days and to 0, 0, and 0%, respectively, under short (13-hr) days, whereas stratified seeds germinated to 0, 41, and 65%, respectively, under long days and to 0, 0, and 0%, respectively, under short days (Densmore, 1997). Freshly matured seeds of *Dryas octopetala,* which germinated to 48% after 4 days in light at 25°C, failed to germinate when sown outdoors in autumn in England (Elkington, 1971). However, seeds germinated the following spring, suggesting that cold stratification had lowered the temperature requirement for germination.

TABLE 10.32   Seed Dormancy in Shrubs and Subshrubs of Northern and Southern (*) Hemisphere Polar and Alpine Tundra Vegetation

Species	Family[a]	Strat.[b]	Temperature[a]	L:D[a]	Reference
Nondormant					
Cassiope tetragona	Eric.	0	22	—[c]	Bliss (1958)
Dryas integrifolia	Ros.	0	22	—	Bliss (1958)
*Gaultheria antipoda	Eric.	0	20	L > D	Bannister (1990)
*G. depressa	Eric.	0	20	L	Bannister (1990)
*Hebe cupressoides	Scrophulari.	0	25	—	Simpson (1976)
*H. diosmifolia	Scrophulari.	0	25	—	Simpson (1976)
*H. hulkeana	Scrophulari.	0	12	—	Simpson (1976)
*H. lavaudiana	Scrophulari.	0	12	—	Simpson (1976)
*H. pinguifolia	Scrophulari.	0	12	—	Simpson (1976)
*H. raoulii	Scrophulari.	0	15	—	Simpson (1976)
*H. salcifolia	Scrophulari.	0	25	—	Simpson (1976)
*H. speciosa	Scrophulari.	0	25	—	Simpson (1976)
*H. strictissima	Scrophulari.	0	25	—	Simpson (1976)
*H. traversii	Scrophulari.	0	15	—	Simpson (1976)
Potentilla fruticosa	Ros.	0	18	—	Bonde (1965a)
Salix phylicifolia	Salic.	0	22	—	Steshenko (1966)
S. planifolia	Salic.	0	22, 5–25	—	Densmore and Zasada (1983)
S. setchelliana	Salic.	0	21/10	—	Douglas (1995)
Physiological dormancy					
Chamaedaphne calyculata	Eric.	90	20	L	Densmore (1997)
Dryas octopetala	Ros.	0	20/10	—	Weilenmann (1981)
*Hebe coarctata	Scrophulari.	42	25	—	Simpson (1976)
*H. pauciramosa	Scrophulari.	28	25	—	Simpson (1976)
*H. decumbens	Scrophulari.	C?[d]	10	—	Simpson (1976)
Ledum decumbens	Eric.	90	15	L	Densmore (1997)
*Pernettya macrostigma	Eric.	26	20	—	Bannister (1990)

[a]See notes for Table 10.1.
[b]Cold stralification (days).
[c]Data not available.
[d]Seeds stored dry at room temperatures.

B. Dormancy and Germination of Herbaceous Species

Species whose seeds germinate to 50% or more without first being cold stratified are listed as nondormant in Table 10.33, but further studies may show that some of them have physiological dormancy. A big problem with the list of species with nondormant seeds is that seeds frequently were stored for long periods of time before germination studies were initiated. Depending on storage conditions and the species, considerable afterripening could have taken place before seeds were tested for germination. For example, seeds of *Melandrium furcatum, Arenaria rubella, Cerastium beeringianum, Hymenoxys acaulis* (Bonde, 1965b), *Alopecurus pratensis, Carex tripartita,* and *C. redowskiana* (Steshenko, 1966) stored dry at room temperature afterripened within 2–10 months and showed great increases in germination percentages; thus, these species are listed under physiological dormancy. Seeds of *Carex doenitzii* were stored dry at 4°C for 6 months and then given a 2-week cold stratification treatment at 5°C before germination tests were conducted; no nonstratified seeds were tested (Kibe and Masuzawa, 1994). Thus, it is not known if the high germination percentages were due to lack of dormancy in freshly matured seeds or if dormancy was lost prior to the time seeds were tested for germination. This species is not listed under any category in Table 10.33.

From data in Table 10.33, it appears that species with nondormant seeds require relatively high temperatures for germination; however, in many studies only one test temperature was used. Our understanding of the germination ecology of herbaceous species in tundra vegetation would be enhanced greatly by studies on freshly matured seeds, using a range of germination test temperatures and both light and darkness. Also, we need to know if cold stratification lowers the temperature requirement for germination and/or modifies light:dark requirements for germination. Cold stratification lowered the temperature requirement for seeds of *Saxifraga triscuspidata* tested under short (13 hr light) but not long (22 hr light) daily photoperiods (Densmore, 1997). Nonstratified seeds of this species germinated to 38, 93, 99, and 100% at 5, 10, 15, and 20°C, respectively, under long days and to 0, 6, 68, and 93%, respectively, under short days. However, after 88–93 days of cold stratification, seeds germinated to 33, 83, 96, and 85%, respectively, under long days and to 32, 75, 96, and 93%, respectively, under short days. Pelton (1956) incubated seeds only in darkness, and from his data one could conclude that seeds of some species are dormant. However, seeds may be capable of germinating over a range of temperatures if their light requirement was fulfilled,

e.g., *Epilobium.* Another problem with studies to determine light:dark requirements for germination is that some investigators have opened "dark-incubated" seeds during the course of the study to count seedlings and thus may have fulfilled the light requirement of the seeds. These questionable data have not been included in Table 10.33.

Seeds of *Achillea moschata, Hieracium staticifolium, Oxyria digyna,* and *Rumex scutatus* sown in autumn 1985 on the foreland of the Morteratsch Glacier in the Alps of Switzerland germinated in spring (Stocklin and Baumber, 1996). The lack of germination in autumn may have been due to the physiological dormancy of seeds and/or to temperatures of the habitat being too low for germination. For example, the optimum temperature of freshly matured, nondormant seeds of *O. digyna* is about 20°C (Table 10.33). Thus, low temperatures in the field in autumn near the glacier may have prevented the germination of *O. digyna* seeds.

A few herbaceous tundra species have seeds with morphological or morphophysiological dormancy. However, data for some species are very limited, and it is difficult to decide which type of dormancy seeds have. Seeds of *Tofieldia calyculata* have underdeveloped embryos, and they germinated when seeds were incubated in light at 23°C. Because 22% of the seeds germinated after only 5 days and 97% had germinated after 25 days (Bianco and Bulard, 1974), embryo growth apparently began as soon as seeds imbibed water. Thus, it is concluded that seeds have morphological dormancy. Seeds of *Papaver* also have underdeveloped embryos (Martin, 1946). However, because seeds of *P. radicatum* were stored for several (?) months at 15°C before being stored at −22°C for 3 months (Olson and Richards, 1979), there is no way to know if embryos in freshly matured seeds are dormant. If they were dormant, they could have afterripened while seeds were at 15°C, in which case seeds have nondeep simple morphophysiological dormancy. Until further studies are done, however, this species is listed under morphophysiological dormancy.

Seeds of *Caltha introloba,* which also have underdeveloped embryos, required 75 days before any seeds germinated, regardless of the incubation temperature. Cold stratification reduced the time of first germination to about 15 days. Also, cold stratification lowered the temperature requirement for germination and increased the germination percentage (Wardlaw *et al.,* 1989). Seeds of this species have MPD, but not all of them require cold stratification for germination. About 40% of the seeds incubated at 29.5°C germinated after 160 days (Wardlaw *et al.,* 1989). Thus, a portion of the seeds may have nondeep simple MPD (i.e., require only warm stratification for dormancy break), whereas others may

TABLE 10.33   Seed Dormancy in Herbaceous Species of Northern and Southern (*) Hemisphere Polar and Alpine Tundra Vegetation

Species	Family[a]	Strat.[b]	Temperature[a]	L:D[a]	Reference
Nondormant					
Achillea millefolium[c]	Aster.	0[d]	20	—[e]	Kaye (1997)
Androsace septentrionalis[c]	Primul.	0[f]	18	—	Bliss (1958), Bonde (1965a)
Agropyron latiglume	Po.	0[g]	22/15	L > D	Acharya (1989)
Antennaria alpina	Aster.	0[f]	18	—	Bonde (1965a)
A. parvifolia	Aster.	0[h]	18	—	Pelton (1956)
A. rosea	Aster.	0[h]	18	—	Pelton (1956)
Arabis lemmonii	Brassic.	0[i]	22	—	Bock (1976)
Arctagrostis latifolia	Po.	0	22	—	Bliss (1958)
Arenaria fendleri	Caryophyll.	0[f]	18	—	Bonde (1965a)
A. groenlandica	Caryophyll.	0	19–23	—	Marchand and Roach (1980)
A. obtusiloba	Caryophyll.	0[f]	18	—	Bliss (1958), Bonde (1965a)
Artemisia arctica	Aster.	0[f]	18	—	Bonde (1965a)
A. scopulorum	Aster.	0[f]	18	L > D	Bliss (1958), Bonde (1965a), Chambers *et al.* (1987)
A. tilesii	Aster.	0	22	—	Steshenko (1966)
Aster sibiricus	Aster.	0	22	—	Bliss (1958)
Besseya alpina	Scrophulari.	0[f]	18	—	Bonde (1965a)
Biscutella laevigata[c]	Brassic.	0[j]	20/10	—	Weilemann (1981)
Carduus defloratus	Aster.	0	20/10	—	Zuur-Isler (1982)
Carex phaeocephala	Cyper.	0	20	—	Kaye (1997)
Chionophila jamesii	Scrophulari.	0[f]	18	—	Bonde (1965a)
Chrysosplenium tetrandrum	Saxifrag.	0	15/5	—	Leck (1980)
Crepis nana	Aster.	0	22	—	Bliss (1958
Epilobium latifolium	Onagr.	0	22	—	Bliss (1958)
Erigeron pinnatisectus	Aster.	0[f]	22	—	Bliss (1958), Bonde (1965a)
E. purpuratus	Aster.	0	22	—	Bliss (1958)
E. subtrinervis	Aster.	0[d]	20	—	Kaye (1997)
Eriophorum vaginatum	Cyper.	0[k]	27/23, 30	L > D	Gartner *et al.* (1986) Wein and Mclean (1973)
Geum rossii	Ros.	0[i]	22	L = D	Bock (1976), Chambers *et al.* (1987)
G. turbinatum	Ros.	0[d]	30/25	L = D	Bliss (1958), Sayers and Ward (1966)
Haplopappus pygmaeus	Aster.	0[f]	18	—	Bonde (1965a)
Hymenoxys grandiflora	Aster.	0[f]	18	—	Bliss (1958), Bonde (1965a)
Juncus trifidus	Junc.	0	12–15	—	Marchand and Roach (1980)
Koenigia islandica	Polygon.	0[i]	15/5	L = D	Reynolds (1984)
Loiseleuria procumbens	Eric.	0[i]	23	L	Bianco and Pellegrin (1973)
Myosotis alpestris	Boragin.	0[j]	20/10	—	Weilenmann (1981)
Oxyria digyna	Polygon.	0	20, 22/18	L > D	Mooney and Billings (1961), Chabot and Billings (1972)
Parnassia palustris	Parnassi.	0	22	—	Bliss (1958)
Pedicularis groenlandica	Scrophulari	0[i]	22	—	Bock (1976)
Petasites frigidus	Aster.	0	22	—	Bliss (1958)
Phleum alpinum	Po.	0	22, 17/12	—	Bliss (1958), McDonough (1969)
Poa alpina	Po.	0[g]	22/15	D > L	Bliss (1958), Bock (1976), Acharya (1989)
P. arctica	Po.	0[i]	22	—	Bock (1976)
Rhododendron ferugineum	Eric.	0[i]	23	L	Bianco and Bulard (1974)
Rumex alpinus	Polygon.	0[m]	24	L	Bianco 1972
R. nivalis	Polygon.	0[j]	20/10	—	Weilenmann (1981)
Sagina linnaei	Caryophyll.	0[j]	20/10	—	Weilenmann (1981)
Saxifraga caespitosa	Saxifrag.	0	12	—	Bell and Bliss (1980)
S. cernua	Saxifrag.	0	12	—	Bell and Bliss (1980)
S. rhomboidea	Saxifrag.	0[f]	18	—	Bonde (1965a)
S. triscuspidata	Saxifrag.	0	15, 20	L > D	Densmore (1997)
Sedum lanceolatum	Crassul.	0[f]	18	—	Bonde (1965a)
S. stenopetalum	Crassul.	0[d]	25/10	L = D	Sayers and Ward (1966)
Senecio doronicum	Aster.	0	20/10	—	Zuur-Isler (1982)
S. mutabilis	Aster.	0[h]	18	—	Pelton (1956)

(*continues*)

TABLE 10.33—*Continued*

Species	Family[a]	Strat.[b]	Temperature[a]	L:D[a]	Reference
Sesleria coerulea	Po.	0[j]	20/10	—	Weilenmann (1981)
Soldanella alpina	Primul.	0	20/10	—	Zuur-Isler (1982)
S. pusilla	Primul.	0[j]	20/10	—	Weilenmann (1981)
Solidago alpestris	Aster.	0	20/10	—	Zuur-Isler (1982)
Taraxacum alpinum	Aster.	0[j]	Room	—	Schutz and Urbanska (1984)
Trisetum spicatum	Po.	0[d]	20/3, 25/20	L > D L = D	Pelton (1956), Bliss (1958), Sayers and Ward (1966), Clebsch and Billings (1976)
Veronica alpina	Scrophulari.	0[j]	20/10	—	Weilenmann (1981)
V. wormskjoldii	Scrophulari.	0[f]	18	—	Bonde (1965a)
Morphological dormancy					
Aquilegia pubescens	Ranuncul.	0[j]	27/23	—	Chabot and Billings (1972)
Ligusticum scoticum	Api.	0	20	L = D	Palin (1988)
Lloydia serotina	Lili.	0[f]	18	—	Bonde (1965a)
Pulsatilla slavica	Ranuncul.	0	25, 30/10	L = D	Lhotska and Moravcova (1989)
Ranunculus sabinei	Ranuncul.	0	12	—	Bell and Bliss (1980)
Tofieldia calyculata	Lili.	0[j]	23	L	Bianco and Bulard (1974)
Morphophysiological dormancy					
Anemone occidentalis	Ranuncul.	—[n]	20	—	Kaye (1997)
*Caltha introloba	Ranuncul.	130	8.9	—	Wardlaw *et al.* (1989)
Lomatium dissectum	Api.	330	18	—	Pelton (1956)
Papaver radicatum	Papaver.	0[o]	19–23	—	Olson and Richards (1979)
Physical dormancy					
Astragalus cottonii	Fab.	0	20	—	Kaye (1997)
A. subpolaris	Fab.	0	22	—	Steshenko (1966)
Geranium albiflorum	Gerani.	0	22	—	Steshenko (1966)
Hedysarum arcticum	Fab.	0	22	—	Steshenko (1966)
H. occidentale	Fab.	0	20	—	Kaye (1997)
Lupinus arcticus	Fab.	0	22	—	Bliss (1958)
L. latifolius	Fab.	0	20	—	Kaye (1997)
Oxytropis campestris	Fab.	0	22	—	Bliss (1958)
O. jacquinii	Fab.	0	20/10	—	Weilenmann (1981)
O. sordida	Fab.	0	22	—	Steshenko (1966)
O. viscida	Fab.	0	20	—	Kaye (1997)
Trifolium alpinum	Fab.	0	20/10	—	Weilenmann (1981)
T. dasyphyllum	Fab.	0	22	—	Bliss (1958)
T. nanum	Fab.	—[l]	22	—	Bock (1976)
T. parryi	Fab.	0	21–24	—	Blakenship and Smith (1967)
Physiological dormancy					
*Acaena magellanica	Ros.	35	20/10	L > D	Walton (1977)
*A. tenera	Ros.	28	20/10	L > D	Walton (1977)
Alopecurus pratensis	Po.	—[p]	g.h.[v]	—	Steshenko (1966)
Arenaria rubella	Caryophyll.	0[f]	18	—	Bonde (1965a)
Bartsia alpina	Scrophulari.	—[q]	20/10	—	Zuur-Isler (1982)
Bistorta bistortoides	Polygon.	—[o]	4	L > D	Allessio (1969)
Bromus sitchensis	Po.	0[n]	20	—	Kaye (1997)
Carex redowskiana	Cyper.	—[p]	g.h.[v]	—	Steshenko (1966)
C. tripartita	Cyper.	—[p]	g.h.[v]	—	Steshenko (1966)
*Celmisia angustifolia	Aster.	84	5	L > D	Scott (1975)
*C. coriacea	Aster.	84	5	L > D	Scott (1975)
*C. gracilenta	Aster.	—[r]	20	L > D	Scott (1975)
Cerastium berringianum	Cyper.	—[f]	18	—	Bonde (1965b)
C. latifolium	Caryophyll.	14	20/10	—	Zuur-Isler (1982)
Danthonia intermedia	Po.	—[n]	20	—	Kaye (1997)
Daphne striata	Thymelae.	—[q]	20/10	—	Zuur-Isler (1982)
Deschampsia atropurpurea	Po.	12	20	—	Kaye (1997)
D. caespitosa[c]	Po.	20 / 0[d,l,s]		L = D	Sayers and Ward (1966), Chambers *et al.* (1987), Kaye (1997)
Elymus glaucus	Po.	—[n]	20	—	Kaye (1997)

(continues)

TABLE 10.33—*Continued*

Species	Family[a]	Strat.[b]	Temperature[a]	L:D[a]	Reference
Erigeron compositus	Aster.	—[n]	20	—	Kaye (1997)
Erysimum arenicola	Brasic.	—[n]	20	—	Kaye (1997)
Festuca pratensis var. apennina	Po.	14	25	—	Linnington *et al.* (1979)
F. ovina[c]	Po.	—[i]	17/12	—	McDonough (1970), Harmer and Lee (1978)
Galium bifolium	Rubi	Winter	Spring	—	Pelton (1956)
Gentiana acaulis	Gentian.	60	30/15	—	Giersbach (1937b)
G. campestris	Gentian.	—[q]	20/10	—	Zuur-Isler (1982)
*G. corymbifera	Gentian.	56	5	—	Simpson and Webb (1980)
*G. gracifolia	Gentian.	56	10	—	Simpson and Webb (1980)
*G. patula	Gentian.	56	5–12	—	Simpson and Webb (1980)
G. punctata	Gentian.	28	25/15	—	Thompson (1969)
G. purpurea	Gentian.	56	20/10	—	Thompson (1969)
*G. saxosa	Gentian.	70–91	5	—	Simpson and Webb (1980)
G. septemfida	Gentian.	56	15/5	—	Thompson (1969)
Hymenoxys acaulis	Aster.	—[f]	18	—	Bonde (1965a)
Melandrium furcatum	Caryophyll.	—[f]	18	—	Bonde (1965a)
Minuartia verna	Caryophyll.	14	20/10	—	Zuur-Isler (1982)
Phlox diffusa	Polemoni.	12	20	—	Kaye (1997)
Poa incurva	Po.	6	20	—	Kaye (1997)
Polygonum confertiflorum	Polygon.	60	15/5	L > D	Reynolds (1984)
P. douglasii	Polygon.	60	15/5	L > D	Reynolds (1984)
Potentilla diversifolia[c]	Ros.	90[u]	18/4	L > D	Chambers *et al.* (1987)
Sibbaldia procumbens	Ros.	0[f,l]	18	L > D	Bonde (1965a), Chambers *et al.* (1987)
Silene acaulis	Caryophyll.	0[l]	23	L > D	Bliss (1958), Bianco and Bulard (1976)
Stipa occidentalis	Po.	12	20	—	Kaye (1997)
Thesium alpinum	Santal.	—[q]	20/10	—	Zuur-Isler (1982)
Viola calcarata	Viol.	—[q]	20/10	—	Zuur-Isler (1982)

[a]See notes for Table 10.1

[b]Cold stratification (days).

[c]Listed as a weed in Holm *et al.* (1979).

[d]Tested after a 3- to 6-week period of drying at room temperatures.

[e]Data not available.

[f]Seeds were stored at room temperatures or at 18°C from August to December.

[g]Seeds were 2 or more months old when tested; storage conditions not given.

[h]Seeds were stored at 18°C for an unspecified period of time before testing.

[i]Seeds were stored for an unspecified period of time (at room temperatures?) and presumably tested at room temperatures (22°C?).

[j]Seeds were stored dry for 2–5 years at 4°C.

[k]Fresh seeds were not tested. Seeds were tested after an unspecified period of time in (dry?) storage at 5°C.

[l]Cold stratification increased germination percentages.

[m]Seeds were stored at room temperatures for 42 months or longer.

[n]Seeds afterripened during dry storage at room temperatures.

[o]Seeds were stored for several months (or years?) at 15°C before being stored at −22°C for 3 months.

[p]Stored "for some time under laboratory conditions" (in Leningrad).

[q]Scarification and/or GA increased germination.

[r]Nonstratified seeds germinated to 80% after 70 days at 20°C.

[s]Dry storage at 1–2°C increased germination percentages.

[t]Length of cold stratification period not given; 2–3 months of dry storage also broke dormancy.

[u]Fresh seeds were not tested; presumed to be dormant.

[v]Greenhouse.

have deep complex MPD (i.e., require only cold stratification for dormancy break).

Physical dormancy occurs in members of the Fabaceae and Geraniaceae (Table 10.33), but other families known to have physical dormancy (Table 3.5) do not seem to be represented in tundra vegetation (e.g., Polunin, 1959; Costin et al., 1979). From the few data available on the germination of tundra species with physical dormancy, it appears that seed coats become permeable during dry storage at various temperatures. However, the effects of habitat temperatures on dormancy break and germination are unknown. Permeable seeds germinate to high percentages at about 20°C, but little is known about the minimum temperature requirements for germination. Thus, permeable seeds of tundra species with physical dormancy need to be tested over a range of temperatures. No information is available on light:dark requirements for germination.

Most tundra species with physiological dormancy require cold stratification to come out of dormancy (Table 10.33). Although seeds of Celmisia gracilenta (Scott, 1975) and Bistorta (=Polygonum) bistortoides (Allesio, 1969) afterripened at 20°C and at about 22°C (room temperature), respectively, they also probably would come out of dormancy during cold stratification. The optimum temperature for germination ranges from 5°C to 25/15°C, with seeds of many species germinating to high percentages at about 20°C. Cold-stratified seeds need to be tested over a range of temperatures to help gain an understanding of the minimum temperatures at which seeds can germinate. Germination at low temperatures, which would characterize the beginning of the short tundra growing season, would give seedlings the maximum period of time for establishment and growth before onset of winter. Light:dark requirements for seed germination have been determined for 12 species. Seeds of 11 species germinated to higher percentages in light than in darkness, and those of one species germinated equally well in light and darkness.

Data in Table 10.33 give the impression that more tundra species have nondormant seeds than have physiological dormancy. In fact, Amen (1966) stated that, "A requirement for pregermination chilling (stratification) or for after-ripening account for very little alpine seed dormancy." However, investigators have been unable to germinate seeds of numerous tundra species (Table 10.34), implying that seeds are dormant. Based on what is known about seed germination in the various families to which these species belong, we can infer the type of dormancy in these species. Thus, physiological dormancy is expected in most of them, but some should have morphophysiological or physical dormancy. For example, it seems logical that seeds of Cerastium arctium, Minuartia rubra (Caryophyllaceae) and Phippsia

algida (Poaceae) would have physiological dormancy. In fact, when seeds of these three species were sown in the field in the Northwest Territories, Canada, immediately after snow melt in late June, they germinated to a higher percentage the second than the first growing season (Bell and Bliss, 1980), indicating that cold stratification was required to break dormancy in many of them.

After failing to germinate seeds of some species, several authors have tried scarification to promote germination, which has worked with varying degrees of success (Amen and Bonde, 1964; Rochow, 1970; Amen, 1965; Bell and Amen, 1970; Weilenmann, 1981; Zuur-Isler, 1982). Although scarification tells us very little about germination ecology, it suggests that embryos lack sufficient growth potential for germination and that the seeds have physiological dormancy (Chapter 3). Much work remains to be done to elucidate the dormancy-breaking and germination requirements of herbaceous species in tundra vegetation.

C. Special Factor: Vivipary

An unusual feature of tundra vegetation is the presence of several species that reproduce asexually by formation of vegetative propagules in the inflorescence; this is called vivipary (see Chapter 11,VII,D). Callaghan and Emanuelsson (1985) suggest that vivipary is an adaptation of species to high stress or disturbance habitats and report that 20% of the vascular plant species in the fore-field vegetation of the Karsa Glacier in Swedish Lapland are viviparous. Vegetative propagules are small plantlets or bulbils, and their number per inflorescence varies, depending on species and habitat severity (Law et al., 1983; Clebsch and Billings, 1976). Bulbils are produced in some species such as Saxifraga cernua in the axils of leaves on the stem as well as in the rosette (Wehrmeister and Bonde, 1977).

Bulbils usually are nondormant and start to grow as soon as they come in contact with moist soil (Callaghan and Collins, 1981; Harmer and Lee, 1978). Plantlets of Festuca vivipara started to grow at maturity in autumn within a few days when placed at 3, 10, 15, and 20°C; however, growth was slower at 3°C than at the other temperatures (Harmer and Lee, 1978). Bulbils stored at 100 relative humidity at 3°C for 18 weeks grew rapidly when placed at 20°C (Harmer and Lee, 1978), indicating that they could initiate growth the following spring after overwintering under snow.

IX. MOUNTAINS

Walter's map (Fig. 9.1) shows mountains in temperate regions of the northern and southern hemispheres.

TABLE 10.34   Inferred Type of Seed Dormancy in Herbaceous Species of Northern and Southern (*)
Hemisphere Polar and Alpine Tundra Vegetation[a]

Species	Family[b]	Dormancy[c]	Reference
Achillea lanulosa	Aster.	PD	Bonde (1965a)
Agropyron scribneri	Po.	PD	Bock (1976)
Allium geyeri	Lili.	MPD	Bonde (1965a)
Alopecurus alpinus	Po.	PD	Eurola (1972)
Andromeda polifolia	Eric.	PD	Bliss (1958)
Anthoxanthum alpinum	Po.	PD	Zuur-Isler (1982)
A. odoratum	Po.	PD	Weilenmann (1981)
Anthyllis alpestris	Fab.	PY	Zuur-Isler (1982)
Arabis coerulea	Brassic.	PD	Weilenmann (1981)
Arctostaphylos alpina	Eric.	PD	Bliss (1958)
Aster paucicapitatus	Aster.	PD	Kaye (1997)
Caltha leptosepala	Ranuncul.	MPD	Bonde (1965a)
Calyptridium umbellatum	Portulac.	PD	Chabot and Billings (1972)
Campanula uniflora	Campanul.	PD	Bonde (1965a), Eurola (1972)
Cardamine bellididfolia	Brassic.	PD	Eurola (1972), Bell and Bliss (1980)
C. rotundifolia	Brassic.	PD	Bonde (1965a)
Carex albonigra	Cyper.	PD	Amen and Bonde (1964)
C. aquatilis	Cyper.	PD	Bliss (1958)
C. bigelowii	Cyper.	PD	Bliss (1958)
C. drummondiana	Cyper.	PD	Bliss (1958)
C. ebenea	Cyper.	PD	Amen and Bonde (1964)
C. ericetorum	Cyper.	PD	Zuur-Isler (1982)
C. firma	Cyper.	PD	Weilenmann (1981)
C. lachenalii	Cyper.	PD	Eurola (1972)
C. misandra	Cyper.	PD	Steshenko (1966), Eurola (1972)
C. mucronata	Cyper.	PD	Weilenmann (1981)
C. parviflora	Cyper.	PD	Weilenmann (1981)
C. paysonis	Cyper.	PD	Haggas et al. (1987)
C. scopulorum	Cyper.	PD	Bliss (1958)
C. sempervirens	Cyper.	PD	Weilenmann (1981), Zuur-Isler (1982)
Castilleja occidentalis	Scrophulari.	PD	Bonde (1965a), Crumley (1972)
C. nana	Scrophulari.	PD	Chabot and Billings (1972)
Cerastium arcticum	Caryophyll.	PD	Eurola (1972), Bell and Bliss (1980)
*Chionochloa rigida	Po.	PD	Mark (1965)
Cirsium hookerianum	Aster.	PD	Bonde (1965a)
Claytonia megarhiza	Portulac.	PD	Bonde (1965a)
Cochlearia officinalis	Brassic.	PD	Bell and Bliss (1980)
Cryptantha nubigena	Boragin.	PD	Chabot and Billings (1972)
Dodecatheon jeffreyi	Primul.	PD	Chabot and Billings (1972)
Draba alpina	Brassic.	PD	Eurola (1972)
D. aurea	Brassic.	PD	Bonde (1965a)
D. bellii	Brassic.	PD	Eurola (1972), Bell and Bliss (1980)
D. crassifolia	Brassic.	PD	Bliss (1958)
D. lemmonii	Brassic.	PD	Chabot and Billings (1972)

(continues)

TABLE 10.34—*Continued*

Species	Family[b]	Dormancy[c]	Reference
D. micropetala	Brassic.	PD	Eurola (1972)
D. nivalis	Brassic.	PD	Eurola (1972)
D. oblongata	Brassic.	PD	Eurola (1972)
D. subcapitata	Brassic.	PD	Eurola (1972)
Dupontia fischeri	Po.	PD	Eurola (1972)
Epilobium alpinum	Onagr.	PD	Bonde (1965a), Kaye (1997)
E. halleanum	Onagr.	PD	Pelton (1956)
E. obcordatum	Onagr.	PD	Chabot and Billings (1972)
Erigeron simplex	Aster.	PD	Bonde (1965a)
Eriophorum angustifolium	Cyper.	PD	Bliss (1958)
E. scheuchzeri	Cyper.	PD	Eurola (1972)
Erysimum nivale	Brassic.	PD	Bonde (1965a)
Festuca hyperborea	Po.	PD	Eurola (1972)
Gentiana clusii	Gentian.	PD	Weilenmann (1981)
G. kochiana	Gentian.	PD	Weilenmann (1981), Zuur-Isler (1982)
G. romanzovii	Gentian.	PD	Bonde (1965a), Crumley (1972)
Gnaphalium supinum	Aster.	PD	Weilenmann (1981)
Heuchera parvifolia	Saxifrag.	PD	Bonde (1965a)
Hierochloe alpina	Po.	PD	Eurola (1972)
Homogyne alpina	Aster.	PD	Weilenmann (1981), Zuur-Isler (1982)
Hulsea algida	Aster.	PD	Chabot and Billings (1972)
Hutchinsia alpina	Brassic.	PD	Weilenmann (1981)
Juncus biglumis	Junc.	PD	Eurola (1972)
Kalmia polifolia	Eric.	PD	Bliss (1958)
Leontodon hyoseroides	Aster.	PD	Zuur-Isler (1982)
Lewisia pygmaea	Portulac.	PD	Chabot and Billings (1972)
Ligusticum mutellina	Api.	MPD	Weilenmann (1981)
Luzula arctica	Junc.	PD	Eurola (1972)
L. confusa	Junc.	PD	Eurola (1972)
L. parviflora	Junc.	PD	Bell and Amen (1970)
L. spicata	Junc.	PD	Amen (1965, 1967), Bell and Amen (1970)
Melandrium affine	Caryophyll.	PD	Eurola (1972)
M. apetalum	Caryophyll.	PD	Steshenko (1966), Eurola (1972)
Menyanthes trifoliata	Menyanth.	PD	Bliss (1958)
Mertensia fusiformis	Boragin.	PD	Pelton (1956)
M. viridis	Boragin.	PD	Bonde (1965a)
Minuartia rubella	Caryophyll.	PD	Eurola (1972), Bell and Bliss (1980)
Orthocarpus imbricatus	Scrophulari.	PD	Kaye (1997)
Papaver dahlianum	Papaver.	MPD	Eurola (1972)
Pedicularis bracteosa	Scrophulari.	PD	Kaye (1997)
P. capitata	Scrophulari.	PD	Bliss (1958), Kaye (1997)
P. dasyantha	Scrophulari.	PD	Eurola (1972)
P. hirsuta	Scrophulari.	PD	Eurola (1972)
P. labradorica	Scrophulari.	PD	Bliss (1958)
P. lanata	Scrophulari.	PD	Bliss (1958)
P. parryi	Scrophulari.	PD	Bliss (1958)

(*continues*)

TABLE 10.34—*Continued*

Species	Family[b]	Dormancy[c]	Reference
Penstemon davidsonii	Scrophulari.	PD	Chabot and Billings (1972)
P. whippleanus	Scrophulari.	PD	Bonde (1965a)
Phacelia sericea	Hydrophyll.	PD	Bonde (1965a)
Phippsia algida	Po.	PD	Eurola (1972), Bell and Bliss (1980)
Poa alpigena	Po.	PD	Eurola (1972)
P. glauca	Po.	PD	Eurola (1972)
Polemonium boreale	Polemoni.	PD	Eurola (1972)
P. eximium	Polemoni.	PD	Chabot and Billings (1972)
P. viscosum	Polemoni.	PD	Bonde (1965a), Bliss (1958)
Potentilla aurea	Ros.	PD	Weilenmann (1981)
P. hyparctica	Ros.	PD	Eurola (1972), Bell and Bliss (1980)
P. pulchella	Ros.	PD	Eurola (1972)
Primula auricula	Primul.	PD	Weilenmann (1981)
P. integrifolia	Primul.	PD	Weilenmann (1981)
Pseudocymopteris montanus	Api.	MPD	Bock (1976)
Ranunculus grenierianus	Ranuncul.	MPD	Weilenmann (1981)
R. hyperboreus	Ranuncul.	MPD	Eurola (1972)
R. lapponicus	Ranuncul.	MPD	Bliss (1958)
R. montanus	Ranuncul.	MPD	Weilenmann (1981)
R. nivalis	Ranuncul.	MPD	Eurola (1972)
R. pygmaeus	Ranuncul.	MPD	Eurola (1972)
R. sulphureus	Ranuncul.	MPD	Steshenko (1966), Eurola (1972)
Sagina intermedia	Caryophyll.	PD	Eurola (1972)
Sambucus microbotrys	Caprifoli.	MPD	Pelton (1956)
Saxifraga bronchialis	Saxifrag.	PD	Bonde (1965a)
S. flagellaris	Saxifrag.	PD	Bonde (1965a), Eurola (1972)
S. hieraciifolia	Saxifrag.	PD	Eurola (1972)
S. hirculus	Saxifrag.	PD	Steshenko (1966), Eurola (1972)
S. hyperborea	Saxifrag.	PD	Eurola (1972)
S. nivalis	Saxifrag.	PD	Eurola (1972)
S. oppositifolia	Saxifrag.	PD	Eurola (1972)
S. punctata	Saxifrag.	PD	Bliss (1958)
S. rivularis	Saxifrag.	PD	Eurola (1972)
S. tolmiei	Saxifrag.	PD	Chabot and Billings (1972)
Sedum integrifolium	Crassul.	PD	Bonde (1965a)
S. rhodanthum	Crassul.	PD	Bonde (1965a)
Senecio congestus	Aster.	PD	Bliss (1958)
Silene willdenowii	Caryophyll.	PD	Zuur-Isler (1982)
Stellaria crassipes	Caryophyll.	PD	Eurola (1972)
S. weberi	Caryophyll.	PD	Bonde (1965a)
Taraxacum arcticum	Aster.	PD	Eurola (1972)
Thalictrum occidentale	Ranuncul.	MPD	Kaye (1997)
Thlaspi alpestre	Brassic.	PD	Bonde (1965a), Rochow (1970)
Valeriana capitata	Valerian.	PD	Steshenko (1966), Bock (1976)

[a]Seeds are dormant ($\leq 50\%$ germination), but the type of dormancy has not been determined.
[b]See notes for Table 10.1.
[c]MPD, morphophysiological dormancy; PD, physiological dormancy; PY, physical dormancy.

Because an increase in elevation corresponds to a decrease in temperature, vegetation on high slopes differs from that on the land adjacent to mountains. Vegetation is arranged in altitudinal belts with the number and width of these zones, depending on aspect, slope, wind and rainfall patterns, geographical location, and mass of the mountains. From top to bottom, the general vegetation zones on mountains are tundra, subalpine, montane, and woodland. This chapter already has covered the germination ecology of species growing in the alpine tundra (Section VIII) and subalpine (Section VII) zones; thus, the purpose of this section is to discuss the seed germination of plants in the montane and woodland zones.

In many parts of the temperate region, vegetation in the montane and woodland zones is the same as that found on land surrounding the mountains. In Australia (Costin, 1981), New Zealand (Wardle *et al.,* 1983), southern South America (Veblen *et al.,* 1983), southern Africa (Donald and Theron (1983), southern Japan, Taiwan, southeastern China (Satoo, 1983), and parts of the Himalaya Mountains (Gopal and Meher-Homji, 1983), temperate broad-leaved evergreen forests are found in the montane and woodland zones. (See Section III for germination studies.) In eastern North America (Barbour *et al.,* 1980), northern Japan (Maekawa, 1974), and parts of the Himalaya Mountains and the south Indian Hills (Gopal and Meher-Homji, 1983), the montane and woodland zones have deciduous forests. (See Section IV for germination studies.) In other places, such as middle Asia and western North America, the woodland and montane zones have semideserts and steppe grasslands (Walter and Box, 1983f; West, 1983d,e). (See Sections V and VI for germination studies.)

A. Montane

In western North America (Eyre, 1963) and in the mountains of southcentral Europe (Ellenberg, 1988), a distinctive formation, usually dominated by conifers, is found in the montane. Our purpose here is to survey the germination ecology of species occurring in this unique vegetation type.

1. Dormancy and Germination of Trees

Some montane trees have seeds that are nondormant at maturity, and the optimum germination temperature is about 23°C (Table 10.35). Light:dark requirements have been determined for seeds of only four species. Seeds of three species germinated to higher percentages in light than in darkness and those of one species germinated equally well in light and darkness.

Most montane trees have seeds with physiological dormancy, and cold stratification for 30–90 days is required for germination (Table 10.35). The mean optimum germination temperature for seeds is about 22°C. Light:dark requirements for germination have been determined for only seven species. Seeds of one species germinated to higher percentages in light than in darkness, those of five germinated equally well in light and darkness, and those of *Abies amabilis* germinated to higher percentages in darkness than in light (Table 10.35). Freshly matured seeds of *A. grandis, Pinus ponderosa,* and *Pseudotsuga menziesii* germinated to higher percentages in light than in darkness, but after cold stratification they germinated equally well in light and darkness (Li *et al.,* 1994). Seeds of *Cedrus deodara* (Chandra and Ram, 1980), *Picea smithiana,* and *A. pindrow* (Singh and Singh, 1990) germinated to 34–57% . while they were buried in soil or humus. However, no seeds of these three species were tested in light; therefore, the relative importance of light and darkness for the germination of seeds of these species is unknown.

The germination of *Picea smithiana* seeds increased with increase in seed weight (Singh *et al.,* 1990). The side of *Pseudotsuga menziesii* seeds adjacent to the cone scale during development is flatter than the one away from the scale. Seeds of this species germinate slower when the flat side touches a moist substrate than when it is turned away from the substrate (Sorensen and Campbell, 1981).

2. Dormancy and Germination of Shrubs

A diversity of shrubs occurs in montane forests (e.g., Elmore, 1976; Ellenberg, 1988), but information on dormancy-breaking and germination requirements of seeds is very limited. Nondormancy and morphophysiological and physical dormancies are represented among montane shrubs, but most species have seeds with physiological dormancy (Table 10.36). Relatively long periods of cold stratification are required to break physiological dormancy, and seeds of *Berberis repens* need warm followed by cold stratification (Rudolf, 1974b). The physiologically dormant seeds of *Ceanothus integerrimus* also have physical dormancy, thus seed coats must become permeable before seeds can be cold stratified. Seeds of *C. integerrimus* (Talley and Griffin, 1980) and *C. fendleri* (Vose and White, 1987) probably are made permeable when exposed to the heat of fires.

The optimum germination temperature for seeds of montane shrubs is about 22°C, but essentially nothing is known about the light:dark requirements for germination.

TABLE 10.35 Seed Dormancy in Trees of North-Temperate Conifer-Dominated Montane Forests

Species	Family[a]	Strat.[b]	Temperature[a]	L:D[a]	Reference
Nondormant					
Cupressus torulosa	Cupress.	0	20/9	L = D	Rao (1988)
Picea abies[c]	Pin.	0	30/20	L > D	Patten (1963), Safford (1974)
P. omorika	Pin.	0	30/20	—[d]	Heit (1968a), Safford (1974)
P. orientalis	Pin.	0	30/20	—	Heit (1968a), Safford (1974)
P. pungens	Pin.	0	30/20	—	Heit (1968a), Safford (1974)
Pinus contorta[c]	Pin.	0	24, 27	—	Haasis and Thrupp (1931)
P. gerardiana	Pin.	0	32/16	—	Sprackling (1976)
P. wallichiana	Pin.	0	Room	—	Beniwal and Singh (1989)
Quercus leucotrichophora	Fag.	0	25	—	Rao and Singh (1985)
Q. floribunda	Fag.	0	20/9	L > D	Rao (1988)
Q. roxburghii	Fag.	0	25	—	Rao and Singh (1985)
Sequoiadendron giganteum	Taxodi.	0	20, 23	—	Stark (1968), Fins (1981)
Thuja plicata	Cupress.	0	30/20	L > D	Li *et al.* (1994)
Physiological dormancy					
Abies alba	Pin.	21	20	—	Edwards (1962)
A. amabilis	Pin.	28	30/20	D > L	Franklin (1974), Li *et al.* (1994)
A. concolor	Pin.	0–60	30/20	—	Franklin (1974)
A. grandis	Pin.	20–40	30/20	L = D	Heit (1968c), Franklin (1974), Li *et al.* (1994)
A. homolepis	Pin.	30–60	30/20	—	Franklin (1974)
A. pindrow	Pin.	Winter	Spring	—	Singh and Singh (1984)
A. procera	Pin.	0–42	30/20	—	Franklin (1974)
Cedrus deodora	Pin.	—	Room	—	Beniwal and Singh (1989)
Cupressus torulosa	Cuppress.	—	Room	—	Beniwal and Singh (1989)
Juniperus recurva	Cupress.	—	Room	—	Beniwal and Singh (1989)
Larix griffithii	Pin.	—	Room	—	Beniwal and Singh (1989)
Libocedrus decurrens	Cupress.	60	20	—	Barton (1930)
Picea smithiana	Pin.	Winter	Summer	—	Singh and Singh (1984)
Pinus densiflora	Pin.	60–90	20	—	Barton (1930)
P. jeffreyi	Pin.	90	15	—	Stone (1957b)
P. koraiensis	Pin.	60–90	20	—	Barton (1930)
P. lambertiana	Pin.	90	5–25	—	Stone (1957a), Baron (1978)
P. leucodermis	Pin.	70	12, 16, 20	L = D	Borghetti *et al.* (1989)
P. ponderosa	Pin.	30	20	L = D	Barton (1930), Li *et al.* (1994)
P. thunbergii	Pin.	30–60	20	—	Barton (1930)
Pseudotsuga menziesii[e]	Pin.	0–120, 28	20, 30/20	L > D, L = D	Allen (1958, 1962), Richardson (1959), Li *et al.* (1994)
Tsuga dumosa	Pin.	—	Room	—	Beniwal and Singh (1989)
T. heterophylla[e]	Pin.	0–84	20	L = D	Ching (1958), Edwards (1973), Edwards and Olson (1973)

[a]See notes for Table 10.1.
[b]Cold stratification (days).
[c]Also in the subalpine.
[d]Data not available.
[e]Also in moist warm temperature woodlands.

3. Dormancy and Germination of Herbaceous Species

Germination studies have been done for only a small fraction of the montane herbaceous species. In about half of these species, seeds germinate to high percentages without first being cold stratified and thus are listed as nondormant in Table 10.37. However, it would be interesting to know if cold stratification would lower the temperature requirement for germination in these species. Several montane herbaceous species have physiological dormancy, which is broken by cold stratification. Thus, seeds come out of dormancy in the field in late autumn and winter and germinate in early spring. One species in the montane, *Penstemon eatonii*, has seeds with physiological dormancy, but populations of this species growing at low elevations in the warm desert have nondormant seeds (Meyer, 1992).

Physical dormancy is documented in only one species

TABLE 30.36 Seed Dormancy in Shrubs of North-Temperate Conifer-Dominated Montane Forests

Species	Family[a]	Strat.[b]	Temperature[a]	L:D[a]	Reference
Nondormant					
Salix exigua	Salic.	0	22	—[c]	Brinkman (1974c)
Morphophysiological dormancy					
Sambucus glauca	Caprifoli.	98	21	—	Brinkman (1974f)
Physical dormancy					
Indigofera gerardiana	Fab.		25	—	Bhagat et al. (1992)
Physical and physiological dormancy					
Ceanothus integerrimus	Rhamn.	60[d]	20–30	—	Heit (1970)
C. ledifolius	Rhamn.	60[d]	20–30	—	Heit (1970)
Physiological dormancy					
Acer glabrum	Acer.	180	10–15	—	Olson and Gabriel (1974)
Amelanchier utahensis	Ros.	45	10–30	L/D	Heit (1970)
Arctostaphylos patula	Eric.	120	30/20	—	Berg (1974), Carlson and Sharp (1975)
Crataegus rivularis	Ros.	84–112	—	—	Shaw (1984)
Berberis repens	Berberid.	196[e]	21	—	Rudolf (1974j)
Prunus virginiana	Ros.	120–160	—	—	Shaw (1984)
Rhamnus purshiana	Rhamn.	140	30/20	L > D	Radwan (1976)
Ribes aureum	Grossulari.	60	—	—	Shaw (1984)
R. montigenum	Grossulari.	200–300	g.h.[f]	—	Pfister (1974)
Rubus ellipticus	Ros.	C[g]	20	—	Bhagat and Singh (1995)

[a]See notes for Table 10.1
[b]Cold stratification (days).
[c]Data not available.
[d]Seeds were scarified or treated with hot water prior to cold stratification.
[e]Warm stratification pretreatment given prior to cold stratification.
[f]Germinated in a greenhouse.
[g]Seeds germinated in response to scarification treatments.

TABLE 10.37 Seed Dormancy in Herbaceous Species of North-Temperate Conifer-Dominated Montane Forests

Species	Family[a]	Strat.[b]	Temperature[a]	L:D[a]	Reference
Nondormant					
Agastache urticifolia	Lami.	0	24/2	L > D	McDonough (1976), Hoffman (1985)
Bromus ciliatus	Po.	0	24	L = D	Hoffman (1985)
Elymus glaucus[c]	Po.	0	24/10	L > D	Hoffman (1985)
Erigeron speciosa	Aster.	0	24/10	L > D	Hoffman (1985)
Helenium hoopesii[c]	Aster.	0	36/16	—[d]	Cox and Klett (1984)
Rudbeckia occidentalis	Aster.	0	30/15	L > D	McDonough (1969), Florez and McDonough (1974)
Solidago spathulata	Aster.	0	24/10	L > D	Hoffman (1985)
Veronica biloba	Scrophulari.	0	—	—	McDonough (1976)
Physical dormancy					
Oxytropis lambertii[c]	Fab.	0	36/16	—	Cox and Kleet (1984)
Physiological dormancy					
Penstemon eatonii	Scrophulari.	112	20/10	—	Meyer (1992)
P. leiophyllus	Scrophulari.	112	15	—	Kitchen and Meyer (1991)
P. rostriflorus	Scrophulari.	28–56	15	—	Kitchen and Meyer (1991)
P. sepalulus	Scrophulari.	84–112	15	—	Kitchen and Meyer (1991)
Phacelia compacta	Hydrophyll.	142	g.h.[e]	—	Quick (1947)
P. mutabilis	Hydrophyll.	142	g.h.	—	Quick (1947)

[a]See notes for Table 10.1.
[b]Cold stratification (days).
[c]Listed as a weed in Holm et al. (1979).
[d]Data not available.
[e]Seeds germinated in a greenhouse.

TABLE 10.38 Seed Dormancy in Trees of the Woodland Zone

Species	Family[a]	Strat.[b]	Temperature[a]	L:D[a]	Reference
Nondormant					
Pinus brutia	Pin.	0	30/20	—[c]	Krugman and Jenkinson (1974)
P. edulis	Pin.	0	30/20	—	Krugman and Jenkinson (1974)
P. halepensis	Pin.	0	30/20	—	Krugman and Jenkinson (1974)
P. quadrifolia	Pin.	0	30/20	—	Krugman and Jenkinson (1974)
Populus fremontii	Salic.	0	20	—	Shafroth *et al.* (1995b)
Quercus agrifolia	Fag.	0	30/20	—	Olson (1974b), Matsuda and McBride (1989)
Q. arizonica	Fag.	0	30/20	—	Nyandiga and McPherson (1992)
Q. douglasii	Fag.	0	30/20	—	Olson (1974b)
Q. dumosa	Fag.	0	30/20	—	Olson (1974b)
Q. emoryi	Fag.	0	30/20	—	Nyandiga and McPherson (1992)
Q. gambelii	Fag.	0	g.h.[d]	—	Sopp *et al.* (1977)
Physiological dormancy					
Juniperus ashei	Cupress.	120	30/20	—	Johnsen and Alexander (1974)
J. occidentalis	Cupress.	98–140	g.h.[d]	—	Young *et al.* (1988)
J. osteosperma	Cupress.	98	g.h.[d]	—	Young *et al.* (1988)
J. pinchotii	Cupress.	120	18	—	Smith *et al.* (1975)
J. scopulorum	Cupress.	120–150	30/20, 15	—	Johnsen and Alexander (1974), Benson (1976)
Pinus coulteri	Pin.	30–40	25/19, 20	—	Barton (1930), Johnstone and Clare (1931)
P. monophylla	Pin.	20–30	25/19	—	Johnstone and Clare (1931)
P. sabiniana	Pin.	45	25/19	—	Johnstone and Clare (1931)
Quercus agrifolia	Fag.	7–14	Room	—	Matsuda and McBride (1989)
Q. chrysolepis	Fag.	17–46	30/20	—	Olson (1974b), Matsuda and McBride (1989)
Q. wislizenii	Fag.	17–28	30/20	—	Olson (1974b), Matsuda and McBride (1989)

[a]See notes for Table 10.1.
[b]Cold stratification (days).
[c]Data not available.
[d]Greenhouse.

(Table 10.37). However, various genera of Fabaceae (*Astragalus, Lathyrus, Lotus, Lupinus,* and *Thermopsis*), Geraniaceae (*Geranium,*) and Malvaceae (*Iliamna* and *Sidalcea*) occur in the montane zone (Dixon, 1935; Ellenberg, 1988), and their seeds probably have physical dormancy. Because fire is an important factor in montane coniferous forests (Habeck and Mutch, 1973; Kilgore, 1973), it may be important in promoting the germination of some seeds with physical dormancy. The appearance of newly germinated seedlings of *Lupinus abramsii, L. cervinus* (Talley and Griffin, 1980), and *Iliamna rivularis* (Steele and Geier-Hayes, 1989) after forest fires suggests that seeds have physical dormancy that is broken by high temperatures.

Optimum germination temperatures range from 15°C in *Penstemon* spp. (Kitchen and Meyer, 1991) to 36/16°C in *Helenium hoopesii* (Cox and Klett, 1984). Light:dark requirements for germination have been determined for only six herbaceous species. Seeds of five species germinated to higher percentages in light than in darkness, and those of one germinated equally well in light and darkness (Table 10.37).

Fires in the montane zone may promote germination in species whose seeds have physical dormancy, but the heat may be detrimental to seeds lacking physical dormancy. Heat treatments for 15 min at 75°C killed all *Carex deweyana, Cirsium vulgare, Epilobium angustifolium, Gaultheria shallon, Rhododendron macrophyllum,* and *Senecio sylvaticus* seeds in wet soil, indicating that fires in the habitat could destroy seeds in the soil seed bank (Clark and Wilson, 1994).

Ant-dispersed herbaceous species, including *Claytonia lanceolata, Delphinium nelsonii, Mertensia fusiformis,* and *Viola nuttallii,* usually were not found growing on ant mounds in montane meadows in Colorado (Culver and Beattie, 1983). These authors speculate that ants either destroy the seeds or leave them too deep in the soil for seedlings to emerge.

B. Woodland (Colline)

The woodland zone is the lowest belt of vegetation on mountains, or it can be the vegetation covering low-lying hills in the western United States and in southern

TABLE 10.39 Seed Dormancy in Shrubs of the North-Temperate Woodland Zone

Species	Family[a]	Strat.[b]	Temperature[a]	L:D[a]	Reference
Nondormant					
Artemisia frigida	Aster.	0	24/14	L = D	Sabo *et al.* (1979)
Fallugia paradoxa	Ros.	0[c]	25/16, 20–25	L = D	Sabo *et al.* (1979), Veit and Van Auken (1993)
Menodora scabra	Ole.	0	24/17	L = D	Sabo *et al.* (1979)
Tamarix ramosissima	Tamaric.	0	20	—[d]	Shafroth *et al.* (1995)
Physical dormancy					
Robinia neomexicana	Fab.	0	36/16	—	Cox and Klett (1984)
Sphaeralcea incana	Malv.	0	24/17	L = D	Sabo *et al.* (1979)
Physical and physiological dormancy					
Rhus trilobata	Anacardi.	30[e], 90[f]	30/20	—	Heit (1970), Shaw (1984)
Physiological dormancy					
Cercocarpus ledifolius	Ros.	21	10	—	Young *et al.* (1978)
C. montanus	Ros.	28	22–27	L = D	Piatt (1973, 1976)
Cowania stansburiana	Ros.	30	30/10, 25	L = D	Heit (1970), Sabo *et al.* (1979)
Elaeagnus commutata	Elaeagn.	30–90	30/20	—	Morgenson (1990)
Purshia tridentata[g]	Ros.	80–90	30/10	—	Peterson (1953), McHenry and Jensen (1967)
Shepardia argentea	Elaeagn.	60–90	—	—	Shaw (1984) .

[a]See notes for Table 10.1.
[b]Cold stratification (days).
[c]Achenes were removed from the pericarp by rubbing fruits together or stored dry at room temperatures for 1 month; these treatments could have resulted in the loss of a cold stratification requirement for germination.
[d]Data not available.
[e]Seeds were acid scarified prior to cold stratification.
[f]Seeds required a hot water treatment to break physical dormancy followed by 30–90 days of cold stratification to break physiological dormancy.
[g]Also in temperate semideserts and deserts.

Europe. Trees may reach 20 m in height and may be dense or widely spaced, with shrubs and herbaceous species occurring between them. The most important genera of trees are *Pinus, Quercus,* and *Juniperus,* and some woodlands are dominated by *Pinus* and/or *Juniperus* and others by *Quercus.* Many woodland trees, as well as shrubs, also occur in adjacent matorral vegetation (Munz and Keck, 1949; Polunin and Walters, 1985).

1. Dormancy and Germination of Trees

About half the woodland zone trees have seeds that are nondormant at maturity, whereas others have seeds with physiological dormancy, which is broken by cold stratification (Table 10.38). None of the trees have seeds with underdeveloped embryos or impermeable seed coats. The mean optimum germination temperature is about 23°C; no information is available on light:dark requirements for seed germination.

2. Dormancy and Germination of Shrubs

Although many shrubs occur in the woodland zone, little is known about seed germination of most of them.

A few species have nondormant seeds, whereas others have seeds with physical and/or physiological dormancy (Table 10.39). Physiological dormancy is broken by cold stratification, but the environmental conditions resulting in a loss of physical dormancy have not been investigated. The mean optimum germination temperature is about 21°C, with optimum temperatures ranging from 10°C in *Cercocarpus ledifolius* (Young *et al.,* 1978) to 36/16°C in *Robinia neomexicana* (Cox and Klett, 1984). Light:dark requirements have been determined for only six species, and seeds of all of them germinated equally well in light and darkness.

3. Dormancy and Germination of Herbaceous Species

Essentially nothing is known about the germination ecology of most herbaceous species in the woodland zone. Seeds of *Penstemon cyanocaulis* (Meyer and Kitchen, 1994), *P. ambiguus, P. fremontii, P. leonardii, P. platyphyllus* (Kitchen and Meyer, 1991), and *Mirabilis multiflorus* (Cox and Klett, 1984) have physiological dormancy and require 56–112 days of cold stratification before they will germinate. Because the woodland zone often is adjacent to matorral vegetation, it seems reasonable that the woodland zone in some places would have a somewhat Mediterranean-type climate with winter rains. Therefore, it is expected that some herbaceous species might have seeds with physiological dormancy that is broken in summer, and that nondormant seeds would germinate with the onset of rainfall in autumn. Seeds of *Nemophila menziesii* (Cruden, 1974), and apparently those of *Phacelia vallicola, P. quickii,* and *P. lemmonii* (Quick, 1947), have physiological dormancy and afterripen during dry storage at room temperature. However, it is unknown whether seeds of these species come out of dormancy in the field during summer and germinate in autumn. Obviously, more studies are needed on the germination ecology of herbaceous species in the woodland zone.

References

Abouguendia, Z. M., and Redmann, R. E. (1979). Germination and early seedling growth of four conifers on acidic and alkaline substrates. *For. Sci.* **25**, 358–360.

Abul-Fatih, H. A., and Bazzaz, F. A. (1979). The biology of *Ambrosia trifida* L. II. Germination, emergence, growth and survival. *New Phytol.* **83**, 817–827.

Acharya, S. N. (1989). Germination response of two alpine grasses from the Rocky Mountains of Alberta. *Can. J. Plant Sci.* **69**, 1165–1177.

Acharya, S. N., Chu, C. B., Hermesh, R., and Schaalje, G. B. (1992). Factors affecting red-osier dogwood seed germination. *Can. J. Bot.* **70**, 1012–1016.

Ackerman, T. L. (1979). Germination and survival of perennial plant species in the Mojave Desert. *Southw. Nat.* **24**, 399–408.

Adams, J. C., and Thielges, B. A. (1978). Seed treatment for optimum pecan germination. *Tree Plant. Notes* **29**(3), 12–13, 35.

Adkins, C. R., Hinesley, L. E., and Blazich, F. A. (1984). Role of stratification, temperature, and light in Fraser fir germination. *Can. J. For. Res.* **14**, 88–93.

Afanasiev, M. (1937). A physiological study of dormancy in seed of *Magnolia acuminata. Cornell Univ. Agric. Exp. Sta. Mem.* 208.

Afanasiev, M. (1944). A study of dormancy and germination of seeds of *Cercis canadensis. J. Agric. Res.* **69**, 405–419.

Ahring, R. M., Dunn, N. L., Jr., and Harlan, J. R. (1963). Effect of various treatments in breaking seed dormancy in sand lovegrass, *Eragrostis trichodes* (Nutt.) Wood. *Crop Sci.* **3**, 131–133.

Ahring, R. M., Eastin, J. D., and Garrison, C. S. (1975). Seed appendages and germination of two Asiatic bluestems. *Agron. J.* **67**, 321–325.

Ahring, R. M., and Frank, H. (1968). Establishment of eastern gamagrass from seed and vegetative propagation. *J. Range Manage.* **21**, 27–30.

Ahring, R. M., and Todd, G. W. (1977). The bur enclosure of the caryopses of buffalograss as a factor affecting germination. *Agron. J.* **69**, 15–17.

Aikman, J. M. (1934). The effect of low temperature on the germination and survival of native oaks. *Proc. Iowa Acad. Sci.* **41**, 89–93.

Albrecht, M. L. and Smith-Jochum, C. (1990). Germination and establishment of *Echinacea* spp. (Compositae). *Wildflower* **3**(2), 6–11.

Allen, E. B. (1982). Germination and competition of *Salsola kali* with native C_3 and C_4 species under three temperature regimes. *Bull. Torrey Bot. Club* **109**, 39–46.

Allen, G. S. (1958). Factors affecting the viability and germination

behavior of coniferous seed. II. Cone and seed maturity, *Pseudotsuga menziesii* (Mirb.) Franco. *For. Chron.* **34**, 275–282.

Allen, G. S. (1962). Factors affecting the viability and germination behavior of coniferous seed. VI. Stratification and subsequent treatment, *Pseudotsuga menziesii* (Mirb.) Franco. *For. Chron.* **38**, 485–496.

Allen, P. S., Debaene, S. B. G. and Meyer, S. E. (1993). Regulation of grass seed germination under fluctuating moisture regimes. *In* "Fourth International Workshop on Seeds" (C. Come and F. Corbineau, eds.), Vol. 2, pp. 387–392. Universite Pierre et Marie Curie.

Allen, P. S., and Meyer, S. E. (1990). Temperature requirements for seed germination of three *Penstemon* species. *HortScience* **25**, 191–193.

Allen, R., and Farmer, R. E., Jr. (1972). Germination of silky dogwood. *J. Wildl. Manage.* **41**, 767–770.

Allen, R. and Farmer, R. E., Jr. (1977). Germination characteristics of bear oak. *South. J. Appl. For.* **1**, 19–20.

Allen, R. B. (1987). Ecology of *Nothofagus menziesii* in the Catlins Ecological Region, south-east Otago, New Zealand. II. Seedling establishment. *New Zeal. J. Bot.* **25**, 11–16.

Allessio, M. L. (1969). Variability in germination of *Bistorta bistortoides* (Pursh) Small. *Bull. Torrey Bot. Club* **96**, 673–689.

Alonso, I., Luis, E., and Tarrega, R. (1992). First phases of regeneration of *Cistus laurifolius* and *Cistus ladanifer* after burning and cutting in experimental plots. *Intl. J. Wildl. Fire* **2**, 7–14.

Amen, R. D. (1965). Seed dormancy in the alpine rush, *Luzula spicata* L. *Ecology* **46**, 361–364.

Amen, R. D. (1966). The extent and role of seed dormancy in alpine plants. *Quart. Rev. Biol.* **41**, 271–281.

Amen, R. D. (1967). The effects of gibberellic acid and scarification on the seed dormancy and germination in *Luzula spicata. Physiol. Plant.* **20**, 6–12.

Amen, R. D., and Bonde, E. K. (1964). Dormancy and germination in alpine *Carex* from the Colorado Front Range. *Ecology* **45**, 881–884.

Andersen, A. N. (1988). Dispersal distance as a benefit of myrmecochory. *Oecologia* **75**, 507–511.

Anderson, K. L. (1965). Time of burning as it affects soil moisture in an ordinary upland Bluestem prairie in the Flint Hills. *J. Range Manage.* **18**, 311–316.

Andersson, L. (1996). Characteristics of seeds and seedlings from weeds treated with sublethal herbicide doses. *Weed Res.* **36**, 55–64.

Anderson, R. C. (1985). Aspects of the germination ecology and biomass production of eastern gamagrass (*Tripsacum dactyloides* L.). *Bot. Gaz.* **146**, 353–364.

Anderson, R. N. (1968). "Germination and Establishment of Weeds for Experimental Purposes." W. F. Humphrey Press, Geneva, NY.

Ansley, R. J., and Abernethy, R. H. (1984). Seed pretreatments and their effects on field establishment of spring-seeded Gardner saltbush. *J. Range Manage.* **37**, 509–513.

Aparicio, A. (1995). Seed germination of *Erica andevalensis* Cabezudo and Rivera (Ericaceae), an endangered edaphic endemic in southwestern Spain. *Seed Sci. Technol.* **23**, 705–713.

Archambault, D. J., and Winterhalder, K. (1995). Metal tolerance in *Agrostis scabra* from the Sudbury, Ontario, area. *Can. J. Bot.* **73**, 766–775.

Aronne, G., and Mazzoleni, S. (1989). The effects of heat exposure on seeds of *Cistus incanus* L. and *Cistus monspeliensis* L. *Giorn. Bot. Ital.* **123**, 283–289

Asakawa, S. (1959). Germination behavior of several coniferous seeds. *J. Jap. For. Soc.* **41**, 430–435.

Aschmann, H. (1973). Distribution and peculiarity of Mediterranean ecosystems. *In* "Mediterranean Type Ecosystems, Origin and Structure" (F. di Castri and H. A. Mooney, eds.), pp. 11–19. Springer-Verlag, Berlin.

Ashby, W. C., and Hellmers, H. (1955). Temperature requirements for germination in relation to wild-land seeding. *J. Range Manage.* **8,** 80–83.

Ashton, D. H. (1981). Tall open-forests. *In* "Australian Vegetation" (R. H. Groves, ed.), pp. 121–151. Cambridge Univ. Press, Cambridge, UK.

Association of Official Seed Analysts. (1965). Rules for testing seeds. *Proc. Assoc. Offic. Seed Anal.* **54,** 1–112.

Atwater, B. R. (1980). Germination, dormancy and morphology of the seeds of herbaceous ornamental plants. *Seed Sci. Technol.* **8,** 523–573.

Auld, T. D., and Bradstock, R. A. (1996). Soil temperatures after the passage of a fire: Do they influence the germination of buried seeds? *Aust. J. Ecol.* **21,** 106–109.

Auld, T. D., and O'Connell, M. A. (1991). Predicting patterns of post-fire germination in 35 eastern Australian Fabaceae. *Aust. J. Ecol.* **16,** 53–70.

Bacilieri, R., Bourchet, M. A., Bran, D., Grandjanny, M., Maistre, M., Perret, P., and Romane, F. (1994). Natural germination as resilience component in Mediterranean coppice stands of *Castanea sativa* Mill. and *Quercus ilex* L. *Acta Oecol.* **15,** 417–429.

Badger, K. S., and Ungar, I. A. (1989). The effects of salinity and temperature on the germination of the inland halophyte *Hordeum jubatum. Can. J. Bot.* **68,** 1420–1425.

Bai, Y., and Romo, T. (1994). Germination of previously buried seeds of fringed sage (*Artemisia frigida*). *Weed Sci.* **42,** 390–397.

Bai, Y., Romo, J. T., and Hou, J. (1996). Phytochrome action in seed germination of fringed sage (*Artemisia frigida*). *Weed Sci.* **44,** 109–113.

Bailey, A. W., and Anderson, M. L. (1980). Fire temperatures in grass, shrub and aspen forest communities of central Alberta. *J. Range Manage.* **33,** 37–40.

Baird, J. H., and Dickens, R. (1991). Germination and emergence of Virginia buttonweed (*Diodia virginiana*). *Weed Sci.* **39,** 37–41.

Baldwin, H. I. (1934). Germination of red spruce. *Plant Physiol.* **9,** 491–532.

Ballester, A., Albo, J. M., and Vieitez, E. (1977). The allelopathic potential of *Erica scoparia* L. *Oecologia* **30,** 55–61.

Bannister, P. (1990). Seed germination in *Gaultheria antipoda, G. depressa,* and *Pernettya macrostigma. New Zeal. J. Bot.* **28,** 357–358.

Bannister, P., and Bridgman, J. (1991). Responses of seeds of three species of *Pseudopanax* to low temperature stratification, removal of fruit flesh, and application of gibberellic acid. *New Zeal. J. Bot.* **29,** 213–216.

Bannister, P., Bibby, T., and Jameson, P. E. (1996). An investigation of recalcitrance in seeds of three native New Zealand tree species. *New Zeal. J. Bot.* **34,** 583–590.

Banting, J. D. (1979). Germination, emergence and persistence of foxtail barley. *Can. J. Plant Sci.* **59,** 35–41.

Banting, J. D., and Gebhardt, J. P. (1979). Germination, afterripening, emergence, persistence and control of Persian darnel. *Can. J. Plant Sci.* **59,** 1037–1045

Barbour, M. G. and Billings, W. D. (eds.) (1988). "North American Terrestrial Vegetation." Cambridge Univ. Press, Cambridge, UK.

Barbour, M. G., Burk, J. H., and Pitts, W. D. (1980). "Terrestrial Plant Ecology." Benjamin-Cummings, Menlo Park, CA.

Barker, P. C. J. (1995). *Phyllocladus aspleniifolius*: Phenology, germination, and seedling survival. *New Zeal. J. Bot.* **33,** 325–337.

Baron, F. J. (1978). Moisture and temperature in relation to seed structure and germination of sugar pine (*Pinus lambertiana* Dougl.). *Am. J. Bot.* **65,** 804–810.

Barrett, L. I. (1931). Influence of forest litter on the germination and early survival of chestnut oak, *Quercus montana,* Willd. *Ecology* **12,** 476–484.

Barrett, S. C. H., and Wilson, B. F. (1983). Colonizing ability in the *Echinochloa crus-galli* complex (barnyard grass). II. Seed Biology. *Can. J. Bot.* **61,** 556–562.

Barro, S. C., and M. Poth. (1988). Seeding and sprouting *Ceanothus* species: Their germination responses to heat. *In* "Time Scales and Water Stress" (F. di Castri, Ch. Floret, S. Rambal, and J. Roy, eds.), pp. 155–158. Proc. 5th Int. Conf. Mediterranean Ecosystems, I.U.B.S., Paris.

Barry, R. G., Courtin, G. M., and Labine, C. (1981). Tundra climates. *In* "Tundra Ecosystems: A Comparative Analysis" (L. C. Bliss, O. W. Heal, and J. J. Moore eds.), pp. 81–114. Cambridge Univ. Press, Cambridge, UK.

Bartholomew, B. (1970). Bare zone between California shrub and grassland communities: The role of animals. *Science* **170,** 1210–1212.

Barton, I. L. (1978). Temperature and its effect on the germination and initial growth of kauri (*Agathis australis*). *New Zeal. J. Bot.* **3,** 327–331.

Barton, L. V. (1928). Hastening the germination of southern pine seeds. *Boyce Thomp. Inst. Plant Res. Prof. Paper* **1**(9), 58–69.

Barton, L. V. (1930). Hastening the germination of some coniferous seeds. *Am. J. Bot.* **17,** 88–115.

Barton, L. V. (1932). Germination of bayberry seeds. *Contrib. Boyce Thomp. Inst.* **4,** 19–25.

Barton, L. V. (1933). Seedling production in tree peony. *Contrib. Boyce Thomp. Inst.* **5,** 451–460.

Barton, L. V. (1934). Dormancy in *Tilia* seeds. *Contrib. Boyce Thomp. Inst.* **6,** 69–89.

Barton, L. V. (1936a). Seedling production in *Carya ovata* (Mill.) K. Koch, *Juglans cinerea* L., and *Juglans nigra* L. *Contrib. Boyce Thomp. Inst.* **8,** 1–6.

Barton, L. V. (1936b). Germination and seedling production in *Lilium* sp[p]. *Contrib. Boyce Thomp. Inst.* **8,** 297–309.

Barton, L. V. (1939). Experiments at Boyce Thompson Institute on germination and dormancy of seeds. *Sci. Hort.* **7,** 186–193.

Barton, L. V. (1944). Some seeds showing special dormancy. *Contrib. Boyce Thomp. Inst.* **13,** 259–271.

Barton, L. V. (1947). Special studies on seed coat impermeability. *Contrib. Boyce Thomp. Inst.* **14,** 355–362.

Barton, L. V., and Schroeder, E. M. (1941). Dormancy in seeds of *Convallaria majalis* L. and *Smilacina racemosa. Contrib. Boyce Thompson Inst.* **12,** 277–300.

Baskin, C. C., and Baskin, J. M. (1988). Germination ecophysiology of herbaceous plant species in a temperate region. *Am. J. Bot.* **75,** 286–305.

Baskin, C. C., and Baskin, J. M. (1994a). Germination requirements of *Oenothera biennis* seeds during burial under natural seasonal temperature cycles. *Can. J. Bot.* **72,** 779–782.

Baskin, C. C., and Baskin, J. M. (1994b). Deep complex morphophysiological dormancy in seeds of the mesic woodland herb *Delphinium tricorne* (Ranunculaceae). *Int. J. Plant Sci.* **155,** 738–743.

Baskin, C. C., and Baskin, J. M. (1995). Warm plus cold stratification requirement for dormancy break in seeds of the woodland herb *Cardamine concatenata* (Brassicaceae), and evolutionary implications. *Can. J. Bot.* **73,** 608–612.

Baskin, C. C., and Baskin, J. M. (1996). Role of temperature and light in the germination ecology of buried seeds of weedy species of disturbed forests. II. *Erechtites hieracifolia. Can. J. Bot.* **74,** 2002–2005.

Baskin, C. C., Baskin, J. M., and Chester, E. W. (1991). Temperature response pattern during afterripening of achenes of the winter annual *Krigia oppositifolia* (Asteraceae). *Plant Species Biol.* **6**, 111–115.

Baskin, C. C., Baskin, J. M., and Chester, E. W. (1993a). Seed germination ecology of two mesic woodland winter annuals, *Nemophila aphylla* and *Phacelia ranunculacea* (Hydrophyllaceae). *Bull. Torrey Bot. Club* **120**, 29–37.

Baskin, C. C., Baskin, J. M., and Chester, E. W. (1993b). Germination ecology of *Leptochloa panicoides*, a summer annual grass of seasonally dewatered mudflats. *Acta Oecol.* **14**, 693–704.

Baskin, C. C., Baskin, J. M., and Chester, E. W. (1995a). Role of temperature in the germination ecology of the summer annual *Bidens polylepis* Blake (Asteraceae). *Bull. Torrey Bot. Club* **122**, 275–281.

Baskin, C. C., Baskin, J. M., and Hoffman, G. R. (1992a). Seed dormancy in the prairie forb *Echinacea angustifolia* var. *angustifolia* (Asteraceae): Afterripening pattern during cold stratification. *Int. J. Plant Sci.* **153**, 239–243.

Baskin, C. C., Baskin, J. M., and Leck, M. A. (1993c). Afterripening pattern during cold stratification of achenes of ten perennial Asteraceae from eastern North America, and evolutionary implication. *Plant Species Biol.* **8**, 61–65.

Baskin, C. C., Baskin, J. M., and Meyer, S. E. (1993d). Seed dormancy in the Colorado Plateau shrub *Mahonia fremontii* (Berberidaceae) and its ecological and evolutionary implications. *Southw. Nat.* **38**, 91–99.

Baskin, C. C., Baskin, J. M., and McDearman, W. W. (1993e). Seed germination ecophysiology of two *Zigadenus* (Liliaceae) species. *Castanea* **58**, 45–53.

Baskin, C. C., Chester, E. W., and Baskin, J. M. (1992b). Deep complex morphophysiological dormancy in seeds of *Thaspium pinnatifidum* (Apiaceae). *Int. J. Plant Sci.* **153**, 565–571.

Baskin, C. C., Meyer, S. E., and Baskin, J. M. (1995b). Two types of morphophysiological dormancy in seeds of two genera (*Osmorhiza* and *Erythronium*) with an Arcto-Tertiary distribution pattern. *Am. J. Bot.* **82**, 293–298.

Baskin, C. C., Baskin, J. M., and El-Moursey, S. A. (1996). Seasonal changes in germination responses of buried seeds of the weedy summer annual grass *Setaria glauca*. *Weed Res.* **36**, 319–324.

Baskin, C. C. and Quarterman, E. (1969). Germination requirements of seeds of *Astragalus tennesseensis*. *Bull. Torrey Bot. Club* **96**, 315–321.

Baskin, J. M., and Baskin, C. C. (1970). Germination eco-physiology of *Draba verna*. *Bull. Torrey Bot. Club* **97**, 209–216.

Baskin, J. M., and Baskin, C. C. (1971a). Germination ecology and adaptation to habitat in *Leavenworthia* spp. (Cruciferae). *Am. Midl. Nat.* **85**, 22–35.

Baskin, J. M., and Baskin, C. C. (1971b). Germination ecology of *Phacelia dubia* var. *dubia* [interior] in Tennessee glades. *Am. J. Bot.* **58**, 98–104.

Baskin, J. M., and Baskin, C. C. (1971c). The ecological life history of *Agave virginica* L. in Tennessee cedar glades. *Am. Midl. Nat.* **86**, 449–462.

Baskin, J. M., and Baskin, C. C. (1972a). Ecological life cycle and physiological ecology of seed germination of *Arabidopsis thaliana*. *Can. J. Bot.* **50**, 353–360.

Baskin, J. M., and Baskin, C. C. (1972b). Germination characteristics of *Diamorpha cymosa* seeds and an ecological interpretation. *Oecologia* **10**, 17–28.

Baskin, J. M., and Baskin, C. C. (1972c). The ecological life cycle of the cedar glade endemic *Leavenworthia exigua* var. *exigua*. *Can. J. Bot.* **50**, 1711–1723.

Baskin, J. M., and Baskin, C. C. (1972d). Physiological ecology of germination of *Viola rafinesquii*. *Am. J. Bot.* **59**, 981–988.

Baskin, J. M., and Baskin, C. C. (1973a). Ecological life cycle of *Helenium amarum* in central Tennessee. *Bull. Torrey Bot. Club* **100**, 117–124.

Baskin, J. M., and Baskin, C. C. (1973b). Studies on the ecological life cycle of *Holosteum umbellatum*. *Bull. Torrey Bot. Club* **100**, 110–116.

Baskin, J. M., and Baskin, C. C. (1974a). Some aspects of the autecology of *Bupleurum rotundifolium* in Tennessee cedar glades. *J. Tennessee Acad. Sci.* **49**, 21–24.

Baskin, J. M., and Baskin, C. C. (1974b). Some eco-physiological aspects of seed dormancy in *Geranium carolinianum* L. from central Tennessee. *Oecologia* **16**, 209–219.

Baskin, J. M., and Baskin, C. C. (1974c). Germination and survival in a population of the winter annual *Alyssum alyssoides*. *Can. J. Bot.* **52**, 2439–2445.

Baskin, J. M., and Baskin, C. C. (1975a). Ecophysiology of seed dormancy and germination in *Torilis japonica* in relation to its life cycle strategy. *Bull. Torrey Bot. Club* **102**, 67–72.

Baskin, J. M., and Baskin, C. C. (1975b). Seed dormancy in *Isanthus brachiatus* (Labiatae). *Am. J. Bot.* **62**, 623–627.

Baskin, J. M., and Baskin, C. C. (1975c). Observations on the ecology of the cedar glade endemic *Viola egglestonii*. *Am. Midl. Nat.* **93**, 320–329.

Baskin, J. M., and Baskin, C. C. (1976). Germination ecology of winter annuals: *Valerianella umbilicata*, f. *patellaria* and f. *intermedia*. *J. Tennessee Acad. Sci.* **51**, 138–141.

Baskin, J. M., and Baskin, C. C. (1977a). Role of temperature in the germination ecology of three summer annual weeds. *Oecologia* **30**, 377–382.

Baskin, J. M., and Baskin, C. C. (1977b). Germination of common milkweed (*Asclepias syriaca* L.) seeds. *Bull. Torrey Bot. Club* **104**, 167–170.

Baskin, J. M., and Baskin, C. C. (1977c). Germination ecology of *Sedum pulchellum* Michx. (Crassulaceae). *Am. J. Bot.* **64**, 1242–1247.

Baskin, J. M., and Baskin, C. C. (1978). Seasonal changes in the germination response of *Cyperus inflexus* seeds to temperature and their ecological significance. *Bot. Gaz.* **139**, 231–235.

Baskin, J. M., and Baskin, C. C. (1979a). The germination strategy of oldfield aster (*Aster pilosus*). *Am. J. Bot.* **66**, 1–5.

Baskin, J. M., and Baskin, C. C. (1979b). Studies on the autecology and population biology of the weedy monocarpic perennial, *Pastinaca sativa*. *J. Ecol.* **67**, 601–610.

Baskin, J. M., and Baskin, C. C. (1979c). The ecological life cycle of *Thlaspi perfoliatum* and a comparison with published studies on *Thlaspi arvense*. *Weed Res.* **19**, 285–292.

Baskin, J. M., and Baskin, C. C. (1979d). Timing of seed germination in the weedy summer annual *Euphorbia supina* [maculata]. *Bartonia No.* **46**, 63–68.

Baskin, J. M., and Baskin, C. C. (1979e). Studies on the autecology and population biology of the monocarpic perennial *Grindelia lanceolata*. *Am. Midl. Nat.* **102**, 290–299.

Baskin, J. M., and Baskin, C. C. (1979f). The ecological life cycle of the cedar glade endemic *Lobelia gattingeri*. *Bull. Torrey Bot. Club* **106**, 176–181.

Baskin, J. M., and Baskin, C. C. (1979g). The ecological life cycle of *Nothoscordum bivalve* in Tennessee cedar glades. *Castanea* **44**, 193–202.

Baskin, J. M., and Baskin, C. C. (1980). Ecophysiology of secondary dormancy in seeds of *Ambrosia artemisiifolia*. *Ecology* **61**, 475–480.

Baskin, J. M., and Baskin, C. C. (1981a). Ecology of germination and flowering in the weedy winter annual grass *Bromus japonicus*. *J. Range Manage.* **34**, 369–372.

Baskin, J. M., and Baskin, C. C. (1981b). Temperature relations of seed germination and ecological implications in *Galinsoga par-viflora* and *G. quadriradiata. Bartonia No.* **48,** 12–18.

Baskin, J. M., and Baskin, C. C. (1981c). Seasonal changes in the germination responses of buried *Lamium amplexicaule* seeds. *Weed Res.* **21,** 299–306.

Baskin, J. M., and Baskin, C. C. (1981d). Seasonal changes in germination responses of buried seeds of *Verbascum thapsus* and *V. blattaria* and ecological implications. *Can. J. Bot.* **59,** 1769–1775.

Baskin, J. M., and Baskin, C. C. (1982a). Germination ecophysiology of *Arenaria glabra,* a winter annual of sandstone and granite outcrops of southeastern United States. *Am. J. Bot.* **69,** 973–978.

Baskin, J. M., and Baskin, C. C. (1982b). Temperature relations of seed germination in *Ruellia humilis,* and ecological implications. *Castanea* **47,** 119–131.

Baskin, J. M., and Baskin, C. C. (1982c). Ecological life cycle and temperature relations of seed germination and bud growth of *Scutellaria parvula. Bull. Torrey Bot. Club* **109,** 1–6.

Baskin, J. M., and Baskin, C. C. (1983a). Germination ecophysiology of eastern deciduous forest herbs: *Hydrophyllum macrophyllum. Am. Midl. Nat.* **109,** 63–71.

Baskin, J. M., and Baskin, C. C. (1983b). Germination ecology of *Collinsia verna,* a winter annual of rich deciduous woodlands. *Bull. Torrey Bot. Club* **110,** 311–315.

Baskin, J. M., and Baskin, C. C. (1983c). Seasonal changes in the germination responses of fall panicum to temperature and light. *Can. J. Plant Sci.* **63,** 973–979.

Baskin, J. M., and Baskin, C. C. (1983d). Germination ecology of *Veronica arvensis. J. Ecol.* **71,** 57–68.

Baskin, J. M., and Baskin, C. C. (1983e). Ecophysiology of seed germination and flowering in common broomweed, *Amphiachyris dracunculoides* (DC.) Nutt. *J. Range Manage.* **36,** 619–622.

Baskin, J. M., and Baskin, C. C. (1983f). Seasonal changes in the germination responses of seeds of *Veronica peregrina* during burial, and ecological implications. *Can. J. Bot.* **61,** 3332–3336.

Baskin, J. M., and Baskin, C. C. (1984a). Germination ecophysiology of the woodland herb *Osmorhiza longistylis* (Umbelliferae). *Am. J. Bot.* **71,** 687–692.

Baskin, J. M., and Baskin, C. C. (1984b). Germination ecophysiology of an eastern deciduous forest herb *Stylophorum diphyllum. Am. Midl. Nat.* **111,** 390–399.

Baskin, J. M., and Baskin, C. C. (1984c). The ecological life cycle of *Campanula americana* in northcentral Kentucky. *Bull. Torrey Bot. Club* **111,** 329–337.

Baskin, J. M., and Baskin, C. C. (1984d). Effect of temperature during burial on dormant and non-dormant seeds of *Lamium amplexicaule* L. and ecological implications. *Weed Res.* **24,** 333–339.

Baskin, J. M., and Baskin, C. C. (1984e). Role of temperature in regulating timing of germination in soil seed reserves of *Lamium purpureum* L. *Weed Res.* **24,** 341–349.

Baskin, J. M. and Baskin, C. C. (1984f). Environmental conditions required for germination of prickly sida (*Sida spinosa*). *Weed Sci.* **32,** 786–791.

Baskin, J. M., and Baskin, C. C. (1985a). Epicotyl dormancy in seeds of *Cimicifuga racemosa* and *Hepatica acutiloba. Bull. Torrey Bot. Club* **112,** 253–257.

Baskin, J. M., and Baskin, C. C. (1985b). Seed germination ecophysiology of the woodland spring geophyte *Erythronium albidum. Bot. Gaz.* **146,** 130–136.

Baskin, J. M., and Baskin, C. C. (1985c). Germination ecophysiology of *Hydrophyllum appendiculatum,* a mesic forest biennial. *Am. J. Bot.* **72,** 185–190.

Baskin, J. M., and Baskin, C. C. (1985d). Role of dispersal date and changes in physiological responses in controlling timing of

germination in achenes of *Geum canadense. Can. J. Bot.* **63,** 1654–1658.

Baskin, J. M., and Baskin, C. C. (1985e). Dormancy breaking and germination requirements of nimble will (*Muhlenbergia schreberi* Gmel.) seeds. *J. Range Manage.* **38,** 513–515.

Baskin, J. M., and Baskin, C. C. (1985f). Does seed dormancy play a role in the germination ecology of *Rumex crispus*? *Weed Sci.* **33,** 340–343.

Baskin, J. M., and Baskin, C. C. (1986a). Germination ecophysiology of the mesic deciduous forest herb *Isopyrum biternatum. Bot. Gaz.* **147,** 152–155.

Baskin, J. M., and Baskin, C. C. (1986b). Change in dormancy status of *Frasera caroliniensis* seeds during overwintering on parent plant. *Am. J. Bot.* **73,** 5–10.

Baskin, J. M., and Baskin, C. C. (1986c). Seed germination ecophysiology of the woodland herb *Asarum canadense. Am. Midl. Nat.* **116,** 132–139.

Baskin, J. M., and Baskin, C. C. (1986d). Seasonal changes in the germination responses of buried witchgrass (*Panicum capillare*) seeds. *Weed Sci.* **34,** 22–24.

Baskin, J. M., and Baskin, C. C. (1987a). Temperature requirements for after-ripening in buried seeds of four summer annual weeds. *Weed Res.* **27,** 385–389.

Baskin, J. M., and Baskin, C. C. (1987b). Seed germination and flowering requirements of the rare plant *Arenaria fontinalis* (Caryophyllaceae). *Castanea* **52,** 291–299.

Baskin, J. M., and Baskin, C. C. (1988a). The ecological life cycle of *Cryptotaenia canadensis* (L.) DC. (Umbelliferae), a woodland herb with monocarpic ramets. *Am. Midl. Nat.* **119,** 165–173.

Baskin, J. M., and Baskin, C. C. (1988b). Role of temperature in regulating the timing of germination in *Portulaca oleracea. Can. J. Bot.* **66,** 563–567.

Baskin, J. M., and Baskin, C. C. (1989a). Seed germination ecophysiology of *Jeffersonia diphylla,* a perennial herb of mesic deciduous forests. *Am. J. Bot.* **76,** 1073–1080.

Baskin, J. M., and Baskin, C. C. (1989b). Seasonal changes in the germination responses of buried seeds of *Barbarea vulgaris. Can. J. Bot.* **67,** 2131–2134.

Baskin, J. M., and Baskin, C. C. (1989c). Germination responses of buried seeds of *Capsella bursa-pastoris* exposed to seasonal temperature changes. *Weed Res.* **29,** 205–212.

Baskin, J. M., and Baskin, C. C. (1989d). Role of temperature in regulating timing of germination in soil seed reserves of *Thlaspi arvense* L. *Weed Res.* **29,** 317–326.

Baskin, J. M., and Baskin, C. C. (1989e). Ecophysiology of seed germination and flowering in *Liatris squarrosa. Bull. Torrey Bot. Club* **116,** 45–51.

Baskin, J. M., and Baskin, C. C. (1990a). Seed germination ecology of poison hemlock, *Conium maculatum. Can. J. Bot.* **68,** 2018–2024.

Baskin, J. M., and Baskin, C. C. (1990b). The role of light and alternating temperatures on germination of *Polygonum aviculare* seeds exhumed on various dates. *Weed Res.* **30,** 397–402.

Baskin, J. M., and Baskin, C. C. (1990c). Role of temperature and light in the germination ecology of buried seeds of *Potentilla recta. Ann. Appl. Biol.* **117,** 611–616.

Baskin, J. M., and Baskin, C. C. (1990d). Germination ecophysiology of seeds of the winter annual *Chaerophyllum tainturieri:* A new type of morphophysiological dormancy. *J. Ecol.* **78,** 993–1004.

Baskin, J. M., and Baskin, C. C. (1990e). Seed germination biology of the narrowly endemic species *Lesquerella stonensis* (Brassicaceae). *Plant Species Biol.* **5,** 205–213.

Baskin, J. M., and Baskin, C. C. (1991). Nondeep complex morphophysiological dormancy in seeds of *Osmorhiza claytonii* (Apiaceae). *Am. J. Bot.* **78,** 588–593.

Baskin, J. M. and Baskin, C. C. (1992a). Seed germination biology of the weedy biennial *Alliaria petiolata*. *Nat. Areas J.* **12**, 191–197.

Baskin, J. M., and Baskin, C. C. (1992b). Germination ecophysiology of the mesic deciduous forest herb *Polemonium reptans* var. *reptans* (Polemoniaceae). *Plant Species Biol.* **7**, 61–68.

Baskin, J. M., and Baskin, C. C. (1992c). Role of temperature and light in the germination ecology of buried seeds of weedy species of disturbed forests. I. *Lobelia inflata*. *Can. J. Bot.* **70**, 589–592.

Baskin, J. M., and Baskin, C. C. (1993). The ecological life cycle of *Perideridia americana* (Apiaceae). *Am. Midl. Nat.* **129**, 75–86.

Baskin, J. M., and Baskin, C. C. (1994). Nondeep simple morphophysiological dormancy in seeds of the mesic woodland winter annual *Corydalis flavula* (Fumariaceae). *Bull. Torrey Bot. Club* **121**, 40–46.

Baskin, J. M., and Baskin, C. C. (1995). Variation in the annual dormancy cycle in buried seeds of the weedy winter annual *Viola arvensis*. *Weed Res.* **35**, 353–362.

Baskin, J. M., and Baskin, C. C. (1997). Methods of breaking seed dormancy in the endangered species *Iliamna corei* (Malvaceae), with special reference to heating. *Nat. Areas J.* **17**, 313–323.

Baskin, J. M., Baskin, C. C., and Chester, E. W. (1992). Seed dormancy pattern and seed reserves as adaptations of the endemic winter annual *Lesquerella lescurii* (Brassicaceae) to its floodplain habitat. *Nat. Areas J.* **12**, 184–190.

Baskin, J. M., Baskin, C. C, and McCann, M. T. (1988). A contribution to the germination ecology of *Floerkea proserpinacoides* (Limnanthaceae). *Bot. Gaz.* **149**, 427–431.

Baskin, J. M., Baskin, C. C., and McCormick, J. F. (1987). Seasonal changes in the germination responses of buried seeds of *Portulaca smallii*. *Bull. Torrey Bot. Club* **114**, 169–172.

Baskin, J. M., Baskin, C. C., and Spooner, D. M. (1989). Role of temperature, light and date seeds were exhumed from soil on germination of four wetland perennials. *Aquat. Bot.* **35**, 387–394.

Baskin, J. M., and Caudle, C. (1967). Germination and dormancy in cedar glade plants. I. *Aristida longespica* and *Sporobolus vaginiflorus*. *J. Tennessee Acad. Sci.* **42**, 132–133.

Baskin, J. M., and Quarterman, E. (1970). Autecological studies of *Psoralea subacaulis*. *Am. Midl. Nat.* **84**, 376–397.

Battaglia, M. (1993). Seed germination physiology of *Eucalyptus delegatensis* R. T. Baker in Tasmania. *Aust. J. Bot.* **41**, 119–136.

Battaglia, M. (1996). Effects of seed dormancy and emergence time on the survival and early growth of *Eucalyptus delegatensis* and *E. amygdalina*. *Aust. J. Bot.* **44**, 123–137.

Battaglia, M., and Reid, J. B. (1993). The effect of microsite variation on seed germination and seedling survival of *Eucalyptus delegatensis*. *Aust. J. Bot.* **41**, 169–181.

Baxter, B. J. M., Granger, J. E., and van Staden, J. (1995). Plant-derived smoke and seed germination: Is all smoke good smoke? That is the burning question. *S. Afr. J. Bot.* **61**, 275–277.

Beardsell, D. V., Knox, R. B., and Williams, E. G. (1993). Germination of seeds from the fruits of *Thryptomene calycina* (Myrtaceae). *Aust. J. Bot.* **41**, 263–273.

Beattie, A. J. (1983). Distribution of ant-dispersed plants. *Sonderbd. Nauturwiss. Ver. Hamburg* **7**, 249–270.

Beattie, A. J., and Culver, D. C. (1983). The nest chemistry of two seed-dispersing ant species. *Oecologia* **56**, 99–103.

Beatty, S. W., and Sholes, O. D. V. (1988). Leaf litter effect on plant species composition of deciduous forest treefall pits. *Can. J. For. Res.* **18**, 553–559.

Belcher, E. W., Jr., and Jones, L. (1966). Influence of light, temperature, and stratification on loblolly and slash pine seed germination. *Proc. Assoc. Offic. Seed Anal.* **56**, 89–94.

Bell, D. T. (1994). Interaction of fire, temperature and light in the germination response of 16 species from the *Eucalyptus marginata*

forest of south-western Western Australia. *Aust. J. Bot.* **42**, 501–509.

Bell, D. T., and Bellairs, S. M. (1992). Effects of temperature on the germination of selected Australian native species used in the rehabilitation of bauxite mining disturbance in Western Australia. *Seed Sci. Technol.* **20**, 47–55.

Bell, D. T., Plummer, J. A., and Taylor, S. K. (1993). Seed germination ecology in southwestern Western Australia. *Bot. Rev.* **59**, 24–73.

Bell, D. T., Vlahos, S., and Watson, L. E. (1987). Stimulation of seed germination of understorey species of the northern jarrah forest of Western Australia. *Aust. J. Bot.* **35**, 593–599.

Bell, K. L., and Amen, R. D. (1970). Seed dormancy in *Luzula spicata* and *L. parviflora*. *Ecology* **51**, 492–496.

Bell, K. L., and Bliss, L. C. (1980). Plant reproduction in a high arctic environment. *Arctic Alpine Res.* **12**, 1–10.

Bellairs, S. M., and Bell, D. T. (1990). Canopy-borne seed store in three Western Australian communities. *Aust. J. Ecol.* **15**, 299–305.

Bender, M. H. (1991). "Contribution to the Basic and Theoretical Population Biology of Monocarpic Perennials ("Biennials")." Ph.D thesis, University of Kentucky, Lexington.

Beniwal, B. S., and Singh, N. B. (1989). Observations on flowering, fruiting and germination behaviours of some useful plants of Arunachal Pradesh. *Indian For.* **115**, 216–227.

Benjamini, L. (1986). Effect of carbofuran on seed germination and initial development of seven crops. *Phytoparasitica* **14**, 219–230

Benson, D. A. (1976). Stratification of *Juniperus scopulorum*. *Tree Plant. Notes* **27**(2), 11, 23.

Bentley, J. R., and Fenner, R. L. (1958). Soil temperatures during burning related to postfire seedbeds on woodland range. *J. For.* **56**, 737–740.

Berg, A. R. (1974). *Arctostaphylos* Adans. Manzanita. *In* "Seeds of Woody Plants in the United States" (C. S. Schopmeyer, Tech. Cord.), pp. 228–231. Forest Service USDA Agriculture Handbook No. 450.

Berg, R. Y. (1966). Seed dispersal of *Dendromecon*: Its ecologic, evolutionary, and taxonomic significance. *Amer. J. Bot.* **53**, 61–73.

Berg, R. Y. (1975). Myrmecochorous plants in Australia and their dispersal by ants. *Aust. J. Bot.* **23**, 475–508.

Bergelson, J. (1990). Life after death: Site pre-emption by the remains of *Poa annua*. *Ecology* **71**, 2157–2165.

Best, K. F. (1977). The biology of Canadian weeds. 22. *Descurainia sophia* (L.) Webb. *Can. J. Plant Sci.* **57**, 499–507.

Betz, R. F. (1989). Ecology of Mead's milkweed (*Asclepias meadii* Torrey). *In* "Prairie Pioneers: Ecology, History and Culture" (T. B. Bragg and J. Stubbendieck, eds.), pp. 187–191. Proc. 11th North Amer. Prairie Conf. University of Nebraska, Lincoln.

Bevington, J. M., and Hoyle, M. C. (1981). Phytochrome action during prechilling induced germination of *Betula papyrifera* Marsh. *Plant Physiol.* **67**, 705–710.

Bhagat, S., Singh, O., and Lalhal, J. S. (1993a). Preliminary studies on germination of birdcherry (*Prunus cornuta* seeds in the nursery. *Indian For.* **119**, 295–298.

Bhagat, S., Singh, O., and Singh, V. (1993b). Effect of seed weight on germination, survival and initial growth of horsechestnut (*Aesculus indica* Colebr.) in the nursery. *Indian For.* **119**, 627–629.

Bhagat, S., and Singh, V. (1994). Storage capacity of some temperate shrubs. *Indian For.* **120**, 258–261.

Bhagat, S., and Singh, V. (1995). Studies on effect of concentrated sulphuric acid treatment on germination of *Rubus ellipticus* seed. *Indian For.* **121**, 643–646.

Bhagat, S., Singh, V., and Singh, O. (1992a). Seed scarification requirement in *Indigofera gerardiana* Wall. *Indian For.* **118**, 429–431.

Bhagat, S., Singh, V., and Singh, O. (1992b). Studies on germination behaviour and longevity of *Woodfordia fruiticosa* Kurz seeds. *Indian For.* **118,** 797–799.

Bianco, J. (1972). Etude de la germination de *Rumex alpinus* L. *Travaux Scient. Parc Nat. Vanoise* **2,** 27–34.

Bianco, J., and Bulard, C. (1974). Etude de la germination des graines de *Rhododendron ferrugineum* L. et de *Tofieldia calyculata* (L.) Wahlnb. *Travaux Scient. Parc Nat. Vanoise* **5,** 121–130.

Bianco, J., and Bulard, C. (1976). Effet de la stratification sur la germination des graines de *Silene acaulis* (L.) Jacq. ssp *exscapa* (All.) J. Braun et ssp. *longiscapa* (Kern.) Hayek. *C. R. Acad. Sci. Paris* **283 D,** 1489–1491.

Bianco, J., and Pellegrin, M. C. (1973). Physiologie de la germination d'une plante alpine: *Loiseleuria procumbens* (L.) Desv. *Travaux Scient. Parc Nat. Vanoise* **3,** 43–51.

Biswas, P. K., Bell, P. D., Crayton, J. L., and Paul, K. B. (1975). Germination behavior of Florida pusley seeds. I. Effects of storage, light, temperature and planting depths on germination. *Weed Sci.* **23,** 400–403.

Biswas, P. K., Bonamy, P. A., and Paul, K. B. (1972). Germination promotion of loblolly pine and baldcypress seeds by stratification and chemical treatments. *Physiol. Plant.* **27,** 71–76.

Blake, A. K. (1935). Viability and germination of seeds and early life history of prairie plants. *Ecol. Monogr.* **5,** 405–460.

Blank, R. R., and Young, J. A. (1992). Influence of matric potential and substrate characteristics on germination of Nezpar Indian ricegrass. *J. Range Manage.* **45,** 205–209.

Blankenship, J. O., and Smith, D. R. (1967). Breaking seed dormancy in Parry's clover by acid treatment. *J. Range Manage.* **20,** 50.

Blazich, F. A., and Hinesley, L. E. (1984). Low temperature germination of Fraser fir seed. *Can J. For. Res.* **14,** 948–949.

Bliss, L. C. (1958). Seed germination in arctic and alpine species. *Arctic* **11,** 180–188.

Bliss, L. C. (1981a). North American and Scandinavian tundras and polar deserts. *In* "Tundra Ecosystems: A Comparative Analysis" (L. C. Bliss, O. W. Heal, and J. J. Moore, eds.), pp. 8–34. Cambridge Univ. Press, Cambridge, UK.

Bliss, L. C. (1981b). Summary. *In* "Tundra Ecosystems: A Comparative Analysis" (L. C. Bliss, O. W. Heal, and J. J. Moore, eds.), pp. 38–44. Cambridge Univ. Press, Cambridge, UK.

Blommaert, K. L. J. (1972). Buchu seed germination. *S. Afr. J. Bot.* **38,** 237–239.

Bloom, C. T., Baskin, C. C., and Baskin, J. M. (1990). Germination ecology of the facultative biennial *Arabis laevigata* variety *laevigata. Am. Midl. Nat.* **124,** 214–230.

Blum, B. M. (1974). *Aralia* L. Aralia. *In* "Seeds of Woody Plants in the United States" (C. S. Schopmeyer, Tech. Coord.), pp. 220–222. USDA Forest Service Agriculture Handbook No. 450.

Bock, J. H. (1976). The effects of increased snowpack on the phenology and seed germinability of selected alpine species. *In* "Ecological Impacts of Snowpack Augmentation in the San Juan Mountains, Colorado: Final Report, San Juan Ecology Project" (H. W. Steinhoff and J. D. Ives, eds.), pp. 265–280. Colorado State University, Fort Collins.

Boe, K. N. (1974). *Sequoia sempervirens* (D. Don.) Endl. Redwood. *In* "Seeds of Woody Plants in the United States" (C. S. Schopmeyer, Tech. Coord.), pp. 764–766. USDA Forest Service Agriculture Handbook No. 450.

Bokhari, U. G. (1978). Allelopathy among prairie grasses and its possible ecological significance. *Ann. Bot.* **42,** 127–136.

Bokhari, U. G., Singh, J. S., and Smith, F. M. (1975). Influence of temperature regimes and water stress on the germination of three range grasses and its possible ecological significance to a shortgrass prairie. *J. Appl. Ecol.* **12,** 153–163.

Bolli, R. (1994). Revision of the genus *Sambucus.* Dissertations Botanicae. Band 223. J. Cramer, Berlin.

Bond, W. J. (1985). Canopy-stored seed reserves (serotiny) in Cape Proteaceae. *S. Afr. J. Bot.* **51,** 181–186.

Bond, W. J., and Slingsby, P. (1983). Seed dispersal by ants in shrublands of the Cape Province and its evolutionary implications. *S. Afr. J. Sci.* **79,** 231–233.

Bond, W., and Slingsby, P. (1984). Collapse of an ant-plant mutualism: The Argentine ant (*Iridomyrmex humilis*) and myrmecochorous Proteaceae. *Ecology* **65,** 1031–1037.

Bond, W. J., and Stock, W. D. (1989). The costs of leaving home: Ants disperse myrmecochorous seeds to low nutrient sites. *Oecologia* **81,** 412–417.

Bonde, E. K. (1965a). Further studies on the germination of seeds of Colorado alpine plants. Univ. Colorado Stud. No. 18.

Bonde, E, K. (1965b). Studies on the germination of seeds of Colorado alpine plants. Univ. Colorado Stud. No. 14.

Bonner, F. T. (1967). Germination of sweetgum seed in response to light. *J. For.* **65,** 339.

Bonner, F. T. (1974a). *Ilex* L. Holly. *In* "Seeds of Woody Plants in the United States" (C. S. Schopmeyer, Tech. Coord.), pp. 450–453. USDA Forest Service Agriculture Handbook No. 450.

Bonner, F. T. (1974b). *Platanus* L. Sycamore. *In* "Seeds of Woody Plants in the United States" (C. S. Schopmeyer, Tech. Coord.), pp. 641–644. USDA Forest Service Agriculture Handbook No. 450.

Bonner, F. T. (1974c). *Xanthoxylum* L. Prickly-ash. *In* "Seeds of Woody Plants in the United States" (C. S. Schopmeyer, Tech. Coord.), pp. 859–861. USDA Forest Service Agriculture Handbook No. 450.

Bonner, F. T. (1974d). *Campsis radicans* (L.) Seem. Common trumpetcreeper. *In* "Seeds of Woody Plants in the United States" (C. S. Schopmeyer, Tech. Coord.), pp. 260–261. USDA Forest Service Agriculture Handbook No. 450.

Bonner, F. T. (1975). Germination temperatures and prechill treatments for white ash (*Fraxinus americana* L.). *Proc. Assoc. Offic. Seed Anal.* **65,** 60–65.

Bonner, F. T. (1984). Testing for seed quality in southern oaks. *South. For. Exp. Sta. Res. Note* SO-306.

Bonner, F. T. (1996). Responses to drying of recalcitrant seeds of *Quercus nigra* L. *Ann. Bot.* **78,** 181–187.

Bonner, F. T., and Burton, J. D. (1974). *Paulownia tomentosa* (Thunb.) Siev. & Zucc. Royal paulownia. *In* "Seeds of Woody Plants in the United States" (C. S. Schopmeyer, Tech. Coord.), pp. 572–573. USDA Forest Service Agriculture Handbook No. 450.

Bonner, F. T., and Farmer, R. E., Jr. (1966). Germination of sweetgum in response to temperature, moisture stress, and length of stratification. *For. Sci.* **12,** 40–43.

Bonner, F. T., and Ferguson, E. R. (1974). *Maclura pomifera* (Raf.) Schneid. Osage-orange. *In* "Seeds of Woody Plants in the United States" (C. S. Schopmeyer, Tech. Coord.), pp. 525–526. USDA Forest Service Agriculture Handbook No. 450.

Bonner, F. T., and Maisenhelder, L. C. (1974a). *Carya* Nutt. Hickory. *In* "Seeds of Woody Plants in the United States" (C. S. Schopmeyer, Tech. Coord.), pp. 269–272. USDA Forest Service Agriculture Handbook No. 450.

Bonner, F. T., and Maisenhelder, L. C. 1974b). *Sassafras albidum* (Nutt.) Nees. Sassafras. *In* "Seeds of Woody Plants in the United States" (C. S. Schopmeyer, Tech. Coord.), pp. 761–762. USDA Forest Service Agriculture Handbook No. 450.

Bonner, F. T., and Schmidtling, R. C. (1974). *Bumelia lanuginosa* (Michx.) Pers. Gum bumelia. *In* "Seeds of Woody Plants in the United States" (C. S. Schopmeyer, Tech. Coord.), pp. 258–259. USDA Forest Service Agriculture Handbook No. 450.

Bonner, F. T., and Vozzo, J. A. (1987). Seed biology and technology of *Quercus. South. For. Exp. Sta. Gen. Tech. Rep.* SO-66.

Borchert, M. (1989). Postfire demography of *Thermopsis macrophylla* H.A. var. *agnina* J. T. Howell (Fabaceae), a rare perennial herb in chaparral. *Am. Midl. Nat.* **122,** 120–132.

Borghetti, M., Vendramin, G. G., Giannini, R., and Schettino, A. (1989). Effects of stratification, temperature and light on germination of *Pinus leucodermis. Acta Oecol.* **10,** 45–56.

Bormann, B. T. (1983). Ecological implications of phytochrome-mediated seed germination in red alder. *For. Sci.* **29,** 734–738.

Bostock, S. J. (1978). Seed germination strategies of five perennial weeds. *Oecologia* **36,** 113–126.

Bosy, J. L., and Reader, R. J. (1995). Mechanisms underlying the suppression of forb seedling emergence by grass (*Poa pratensis*) litter. *Funct. Ecol.* **9,** 635–639.

Boucher, C. (1981). Autecological and population studies of *Orothamnus zeyheri* in the Cape of South Africa. *In* "The Biological Aspects of Rare Plant Conservation" (H. Synge, ed.), pp. 343–353. Wiley, Chichester, UK.

Boucher, C., and Moll, E. J. (1981). South African Mediterranean shrublands. *In* "Ecosystems of the World" (F. di Castri, D. W. Goodall, and R. L. Specht, eds.), Vol. 11, pp. 233–248. Elsevier, Amsterdam.

Boyd, R. S. (1996). Ant-mediated seed dispersal of the rare chaparral shrub *Fremontodendron decumbens* (Sterculiaceae). *Madrono* **43,** 299–315.

Boyd, R. S., and Serafini, L. L. (1992). Reproductive attrition in the rare chaparral shrub *Fremontodendron decumbens* Lloyd (Sterculiaceae). *Am. J. Bot.* **79,** 1264–1272.

Boydston, R. A. (1989). Germination and emergence of longspine sandbur (*Cenchrus longispinus*). *Weed Sci.* **37,** 63–67.

Bradstock, R. A., Auld, T. D., Ellis, M. E., and Cohn, J. S. (1992). Soil temperatures during bushfires in semi-arid, mallee shrublands. *Aust. J. Ecol.* **17,** 433–440.

Bran, D., Lobreaux, P., Maistre, M., Perret, P., and Romane, F. (1990). Germination of *Quercus ilex* and *Q. pubescens* in a *Q. ilex* coppice. *Vegetatio* **87,** 45–50.

Breckle, S. W. (1983). Temperate deserts and semi-deserts of Afghanistan and Iran. *In* "Ecosystems of the World" (N. E. West, ed.), Vol. 5, pp. 271–319. Elsevier, Amsterdam.

Brew, C. R., O'Dowd, D. J., and Rae, I. D. (1989). Seed dispersal by ants: Behaviour-releasing compounds in elaiosomes. *Oecologia* **80,** 490–497.

Brinkman, K. A. (1974a). *Lonicera* L. Honeysuckle. *In* "Seeds of Woody Plants in the United States" (C. S. Schopmeyer, Tech. Coord.), pp. 515–519. USDA Forest Service Agriculture Handbook No. 450.

Brinkman, K. A. (1974b). *Rubus* L. Blackberry, raspberry. *In* "Seeds of Wood Plants in the United States" (C. S. Schopmeyer, Tech. Coord.), pp. 738–743. USDA Forest Service Agriculture Handbook No. 450.

Brinkman, K. A. (1974c). *Salix* L. Willow. *In* "Seeds of Woody Plants in the United States" (C. S. Schopmeyer, Tech. Coord.), pp. 746–750. USDA Forest Service Agriculture Handbook No. 450.

Brinkman, K. A. (1974d). *Ulmus* L. Elm. *In* "Seeds of Woody Plants in the United States" (C. S. Schopmeyer, Tech. Coord.), pp. 829–834. USDA Forest Service Agriculture Handbook No. 450.

Brinkman, K. A. (1974e). *Corylus* L. Hazel, Filbert. *In* "Seeds of Woody Plants in the United States" (C. S. Schopmeyer, Tech. Coord.), pp. 343–345. USDA Forest Service Agriculture Handbook No. 450.

Brinkman, K. A. (1974f). *Sambucus* L. Elder. *In* "Seeds of Woody Plants in the United States" (C. S. Schopmeyer, Tech. Coord.), pp. 754–757. USDA Forest Service Agriculture Handbook No. 450.

Brinkman, K. A., and Erdmann, G. G. (1974). *Mitchella repens* L. Partridgeberry. *In* "Seeds of Woody Plants in the United States" (C. S. Schopmeyer, Tech. Coord.), p. 543. USDA Forest Service Agriculture Handbook No. 450.

Brits, G. J. (1986). Influence of fluctuating temperatures and H_2O_2 treatment on germination of *Leucospermum cordifolium* and *Serruria florida* (Proteaceae) seeds. *S. Afr. J. Bot.* **52,** 286–290.

Brits, G. J. (1987). Germination depth *vs.* temperature requirements in naturally dispersed seeds of *Leucospermum cordifolium* and *L. cuneiforme* (Proteaceae). *S. Afr. J. Bot.* **53,** 119–124.

Brits, G. J., and van Niekerk, M. N. (1986). Effects of air temperature, oxygenating treatments and low storage temperature on seasonal germination response of *Leucospermum cordifolium* (Proteaceae) seeds. *S. Afr. J. Bot.* **52,** 207–211.

Brown, E. O., and Porter, R. H. (1942). The viability and germination of seeds of *Convolvulus arvensis* L., and other perennial weeds. *Iowa Agric. Exp. Sta. Res. Bull.* **294,** 475–504.

Brown, N. A. C. (1993a). Promotion of germination of fynbos seeds by plant-derived smoke. *New Phytol.* **123,** 575–583.

Brown, N. A. C. (1993b). Seed germination in the fynbos fire ephemeral, *Syncarpha vestita* (L.) B. Nord is promoted by smoke, aqueous extracts of smoke and charred wood derived from burning the Ericoid-leaved shrub, *Passerina vulgaris* Thoday. *Int. J. Wildl. Fire* **3,** 203–206.

Brown, N. A. C., and Dix, L. (1985). Germination of the fruits of *Leucadendron tinctum. S. Afr. J. Bot.* **51,** 448–452.

Brown, N. A. C., Kotze, G., and Botha, P. A. (1993). The promotion of seed germination of Cape *Erica* species by plant-derived smoke. *Seed Sci. Technol.* **21,** 573–580.

Brown, N. A. C., and van Staden, J. (1973). The effect of scarification, leaching, light, stratification, oxygen and applied hormones on germination of *Protea compacta* R. Br. and *Leucadendron daphnoides* Meisn. *S. Afr. J. Bot.* **39,** 185–195.

Buchele, D. E., Baskin, J. M., and Baskin, C. C. (1991). Ecology of the endangered species *Solidago shortii*. III. Seed germination ecology. *Bull. Torrey Bot. Club* **118,** 288–291.

Buckley, R. C. (1982). Ant-plant interactions: A world review. *In* "Ant-Plant Interactions in Australia" (R. C. Buckley, ed.), pp. 111–141. Junk, The Hague.

Bullock, S. H. (1989). Life history and seed dispersal of the short-lived chaparral shrub *Dendromecon rigida* (Papaveraceae). *Am. J. Bot.* **76,** 1506–1517.

Bullowa, S., Negbi, M., and Ozeri, Y. (1975). Role of temperature, light and growth regulators in germination in *Anemone coronaria* L. *Aust. J. Plant Physiol.* **2,** 91–100.

Burkill, I. H. (1937). The life-cycle of *Tamus communis* L. *J. Bot.* **75,** 1–12, 33–43, 65–74.

Burnside, O. C. (1965). Seed and phenological studies with shattercane. *Univ. Nebraska Coll. Agric. Exp. Sta. Res. Bull.* 220.

Burrell, J. (1965). Ecology of *Leptospermum* in Otago. *New Zeal. J. Bot.* **3,** 3–16.

Burrows, C. J. (1989). Patterns of delayed germination in seeds. *New Zeal. Nat. Sci.* **16,** 13–19.

Burrows, C. J. (1993). Germination requirements of the seeds of native trees, shrubs and vines. *Canterbury Bot. Soc. J.* **27,** 42–48.

Burrows, C. J. (1994a). Germinating matai seeds: An inadvertent experiment. *Canterbury Bot. Soc. J.* **28,** 40–41.

Burrows, C. J. (1994b). Germination experiments with seeds from the native New Zealand woody plant flora. *In* "Seed Symposium: Seed Development and Germination" (P. Coolbear, C. A. Cornford, and K. M. Pollock, eds.), pp. 17–23. Proceedings of a joint New Zealand Society of Plant Physiologists and Agronomy Society of New Zealand Symposium on Seed Physiology held at Tauranga, August 1991. Agronomy Society of New Zealand Spec. Publ No. 9.

Burrows, C. J. (1995a). Germination behaviour of seeds of the New Zealand species *Fuchsia excorticata, Griselinia littoralis, Macropiper excelsum*, and *Melicytus ramiflorus. New Zeal. J. Bot.* **33**, 131–140.

Burrows, C. J. (1995b). Germination behaviour of the seeds of the New Zealand species *Aristotelia serrata, Coprosma robusta, Cordyline australis, Myrtus obcordata*, and *Schefflera digitata. New Zeal. J. Bot.* **33**, 257–264.

Burrows, C. J. (1995c). Germination behaviour of the seeds of four New Zealand species of *Coriaria* (Coriariaceae). *New Zeal. J. Bot.* **33**, 265–275.

Burrows, C. J. (1995d). Germination behaviour of the seeds of six New Zealand woody plant species. *New Zeal. J. Bot.* **33**, 365–377.

Burrows, C. J. (1996a). Germination behaviour of the seeds of seven New Zealand vine species. *New Zeal. J. Bot.* **34**, 93–102.

Burrows, C. J. (1996b). Germination behaviour of seeds of the New Zealand woody species *Melicope simplex, Myoporum laetum, Myrsine divaricata*, and *Urtica ferox. New Zeal. J. Bot.* **34**, 205–213.

Burrows, C. J. (1996c). Germination behaviour of the seeds of the New Zealand woody species *Alectryon excelsus, Corynocarpus laevigatus*, and *Kunzea ericoides. New Zeal. J. Bot.* **34**, 489–498.

Burrows, C. J. (1996d). Germination behaviour of the seeds of the New Zealand woody species *Coprosma foetidissima, Freycinetia baueriana, Hoheria angustifolia*, and *Myrsine australis. New Zeal. J. Bot.* **34**, 499–508.

Burrows, C. J. (1996e). Germination behaviour of seeds of the New Zealand woody species *Ascarina lucida, Coprosma grandifolia, Melicytus lanceolatus*, and *Solanum laciniatum. New Zeal. J. Bot.* **34**, 509–515.

Cabiaux, C., and Devillez, F. (1977). Etude de L'influence des facteurs du milieu sur la germination et la levee des plantules du bouleau pubescent. *Bull. Soc. Roy. Bot. Belgique* **110**, 96–112.

Callaghan, T. V., and Collins, N. J. (1981). Life cycles, population dynamics and the growth of tundra plants. *In* "Tundra Ecosystems: A Comparative Analysis" (L. C. Bliss, O. W. Heal, and J. J. Moore, eds.), pp. 257–284. Cambridge Univ. Press, Cambridge, UK.

Callaghan, T. V., and Emanuelsson, U. (1985). Population structure and processes of tundra plants and vegetation. *In* "The Population Structure of Vegetation" (J. White, ed.), pp. 399–439. Junk, Dordrecht.

Calmes, M. A., and Zasada, J. C. (1982). Some reproductive traits of four shrub species in the black spruce forest type of Alaska. *Can. Field-Nat.* **96**, 35–40.

Calvo, L., Tarrega, R., and Luis, E. (1991). Regeneration in *Quercus pyrenaica* ecosystems after surface fire. *Intl. J. Wildland Fire* **1**, 205–210.

Calvo, L., Tarrega, R., and Luis, E. (1992). The effect of human factors (cutting, burning and uprooting) on experimental heathland plots. *Pirineos* **140**, 15–27.

Campbell, M. H. (1985). Germination, emergence and seedling growth of *Hypericum perforatum* L. *Weed Res.* **25**, 259–266.

Cantlon, J. E., Curtis, E. J. C., and Malcolm, W. M. (1963). Studies of *Melampyrum lineare. Ecology* **44**, 466–474.

Carlson, J. R., and Sharp, W. C. (1975). Germination of high elevation manzanitas. *Tree Plant. Notes* **26**(3), 10–11, 25.

Carpenter, S. B., and Smith, N. D. (1981). Germination of *Paulownia* seeds in the presence and absence of light. *Tree Plant. Notes* **32**(4), 27–29.

Carreira, J. A., Sanchez-Vazquez, F., and Niell, F. X. (1992). Short-term and small-scale patterns of post-fire regeneration in a semi-arid dolomitic basin of southern Spain. *Acta Oecol.* **13**, 241–253.

Caudle, C., and Baskin, J. M. (1968). The germination pattern of three winter annuals. *Bull. Torrey Bot. Club* **95**, 331–335.

Chabot, B. F., and Billings, W. D. (1972). Origins and ecology of the Sierran alpine flora and vegetation. *Ecol. Monogr.* **42**, 163–199.

Chambers, J. C., MacMahon, J. A., and Brown, R. W. (1987). Germination characteristics of alpine grasses and forbs: A comparison of early and late seral dominants with reclamation potential. *Reclam. Reveg. Res.* **6**, 235–249.

Chandler, J. M., and Jan, C. C. (1985). Comparison of germination techniques for wild *Helianthus* seeds. *Crop Sci.* **25**, 356–358.

Chandra, J. P., and Ram, A. (1980). Studies on depth of sowing deodar (*Cedrus deodara*) seed. *Indian For.* **106**, 852–855.

Chapman, A. G. (1936). Scarification of black locust seed to increase and hasten germination. *J. For.* **34**, 66–74.

Chatterton, N. J., and McKell, C. M. (1969). *Atriplex polycarpa*. I. Germination and growth as affected by sodium chloride in water cultures. *Agron. J.* **61**, 448–450.

Chepil, W. S. (1946). Germination of weed seeds. I. Longevity, periodicity of germination, and vitality of seeds in cultivated soil. *Sci. Agric.* **26**, 307–346.

Ching, T. M. (1958). Some experiments on the optimum germination conditions for western hemlock (*Tsuga heterophylla* Sarg.). *J. For.* **56**, 277–279.

Choate, H. A. (1940). Dormancy and germination in seeds of *Echinocystis lobata. Am. J. Bot.* **27**, 156–160.

Chou, C.-H. (1973). The effect of fire on the California chaparral vegetation. *Bot. Bull. Acad. Sinica* **14**, 23–34.

Chou, C.-H., and Muller, C. H. (1972). Allelopathic mechanisms of *Arctostaphylos glandulosa* var. *zacaensis. Am. Midl. Nat.* **88**, 324–347.

Choudhuri, G. N. (1968). Effect of soil salinity on germination and survival of some steppe plants in Washington. *Ecology* **49**, 465–471.

Christiansen, P. A., and Landers, R. Q. (1966). Notes on prairie species in Iowa. I. Germination and establishment of several species. *Proc. Iowa Acad. Sci.* **73**, 51–59.

Christensen, N. L. (1973). Fire and the nitrogen cycle in California chaparral. *Science* **181**, 66–68.

Christensen, N. L., and Muller, C. H. (1975). Relative importance of factors controlling germination and seedling survival in *Adenostoma* chaparral. *Am. Midl. Nat.* **93**, 71–78.

Clark, D. C., and Bass, L. N. (1970). Germination experiments with seeds of Indian ricegrass, *Oryzopsis hymenoides* (Roem. and Schult.) Ricker. *Proc. Assoc. Offic. Seed Anal.* **60**, 226–239.

Clark, D. L., and Wilson, M. V. (1994). Heat-treatment effects on seed bank species of an old-growth Douglas-fir forest. *Northw. Sci.* **68**, 1–5.

Clark, S. C. (1974). Biological flora of the British Isles. *Catapodium rigidum* (L.) D. C. Hubbard. *J. Ecol.* **62**, 937–958.

Clarke, L. D., and West, N. E. (1969). Germination of *Kochia americana* in relation to salinity. *J. Range Manage.* **22**, 286–287.

Clebsch, E. E. C., and Billings, W. D. (1976). Seed germination and vivipary from a latitudinal series of populations of the arctic-alpine grass *Trisetum spicatum. Arctic Alpine Res.* **8**, 255–262.

Clemens, J., Jones, P. G., and Gilbert, N. H. (1977). Effect of seed treatments on germination in *Acacia. Aust. J. Bot.* **25**, 269–276.

Clements, F. E., and Weaver, J. E. (1924). Experimental vegetation. The relation of climaxes to climate. *Carnegie Inst. Washington Publ. No.* 355.

Clifford, H. T. (1953). A note on the germination of *Eucalyptus* seed. *Aust. For.* **17**, 17–20.

Cocks, P. S., and Donald, C. M. (1973). The germination and establishment of two annual pasture grasses (*Hordeum leporinum* Link and *Lolium rigidum* Gaud). *Aust. J. Agric. Res.* **24**, 1–10.

Cohn, M. A., and Hughes, J. A. (1981). Seed dormancy in red rice (*Oryza sativa*). I. Effect of temperature on dry-afterripening. *Weed Sci.* **29**, 402–404.

Collins, S. L., and Wallace, L. L. (eds.). (1990). "Fire in North American Tallgrass Prairies." Univ. Oklahoma Press, Norman.

Conner, A. J., and Conner, L. N. (1988). Germination and dormancy of *Arthropodium cirratum* seeds. *New Zeal. Nat. Sci.* **15**, 3–10.

Conner, L. N. (1987). Seed germination of five subalpine *Acaena* species. *New Zeal. J. Bot.* **25**, 1–4.

Conner, L. N., and Conner, A. J. (1988). Seed biology of *Chordospartium stevensonii*. *New Zeal. J. Bot.* **26**, 473–475.

Connor, D. J. (1965). Seed production and seed germination of *Amsinckia hispida*. *Aust. J. Exp. Agric. Anim. Husb.* **5**, 495–499.

Coombe, D. E. (1956). *Impatiens parviflora* DC. *J. Ecol.* **44**, 701–713.

Cornelius, D. R., and Hylton, L. O. (1969). Influence of temperature and leachate on germination of *Atriplex polycarpa*. *Agron. J.* **61**, 209–211.

Corner, E. J. H. (1976). "The Seeds of Dicotyledons," Vol. 1. Cambridge Univ. Press, Cambridge, UK.

Corns, W. G., and Schraa, R. J. (1962). Dormancy and germination of seeds of silverberry (*Elaeagnus commutata* Bernh.). *Can. J. Bot.* **40**, 1051–1055.

Core, E. L. (1974). Red mulberry. *Morus rubra* L. *In* "Shrubs and Vines for Northeastern Wildlife" (J. D. Gill, and W. M. Healy, eds.), pp. 106–107. USDA For. Serv. Gen. Tech. Rep. NE-9.

Corradi, M. G., Bianchi, A., and Albasini, A. (1993). Chromium toxicity in *Salvia sclarea*-I. Effects of hexavalent chromium on seed germination and seedling development. *Environ. Exp. Bot.* **33**, 405–413.

Costin, A. B. (1981). Alpine and sub-alpine vegetation. *In* "Australian Vegetation" (R. H. Groves, ed.), pp. 361–376. Cambridge Univ. Press, Cambridge, UK.

Costin, A. B., Gray, M., Totterdell, C. J., and Wimbush, D. J. (1979). "Kosciusko Alpine Flora." CSIRO/Collins, Australia.

Coukos, C. J. (1944). Seed dormancy and germination in some native grasses. *Agron. J.* **36**, 337–345.

Coupland, R. T. (1992a). Approach and generalizations. *In* "Ecosystems of the World" (R. T. Coupland, ed.), Vol. 8A, pp. 1–6. Elsevier, Amsterdam.

Coupland, R. T. (1992b). "Ecosystems of the World" (R. T. Coupland, ed.), Vols. 8A and 8B. Elsevier, Amsterdam.

Coupland, R. T. (1992c). Mixed prairie. *In* "Ecosystems of the World" (R. T. Coupland, ed.), Vol. 8A, pp. 151–182. Elsevier, Amsterdam.

Coupland, R. T. (ed.). (1992). "Ecosystems of the World," Vol. 8A. Elsevier Amsterdam.

Coupland, R. T. (ed.). (1993). Ecosystems of the World," Vol. 8B. Elsevier, Amsterdam.

Court, A. J., and Mitchell, N. D. (1988). The germination ecology of *Dysoxylum spectabile* (Meliaceae). *New Zeal. J. Bot.* **26**, 1–6.

Courtney, A. D. (1968). Seed dormancy and field emergence in *Polygonum aviculare*. *J. Appl. Ecol.* **5**, 675–684.

Cox, R. A., and Klett, J. E. (1984). Seed germination requirements of native Colorado plants for use in the landscape. *The Plant Propag.* **30**(2), 6–10.

Cram, W. H., and Lindquist, C. H. (1982). Germination of green ash is related to seed moisture content at harvest. *For. Sci.* **28**, 809–812.

Creel, J. M., Jr., Hoveland, C. S., and Buchanan, G. A. (1968). Germination, growth, and ecology of sicklepod. *Weed Sci.* **16**, 396–400.

Crocker, W. (1930). Harvesting, storage and stratification of seeds in relation to nursery practice. *Boyce Thomp. Inst. Plant Res. Prof. Pap.* **1**(15), 114–120.

Crocker, W., and Barton, L. V. (1931). After-ripening, germination, and storage of certain rosaceous seeds. *Contrib. Boyce Thomp. Inst.* **3**, 385–404.

Cronin, E. H. (1973). Pregermination treatments of black seed of halogeton. *Weed Sci.* **21**, 125–127.

Crosier, W. (1956). Longevity of seeds exposed to dry heat. *Proc. Assoc. Offic. Seed Anal.* **46**, 72–74.

Crossley, J. A. (1974a). *Malus* Mill. Apple. *In* "Seeds of Woody Plants in the United States" (C. S. Schopmeyer, Tech. Coord.), pp. 531–534. USDA Forest Service Agriculture Handbook No. 450.

Crossley, J. A. (1974b). *Vaccinium* L. Blueberry. *In* "Seeds of Woody Plants in the United States" (C. S. Schopmeyer, Tech. Coord.), pp. 840–843. USDA Forest Service Agriculture Handbook No. 450.

Cruden, R. W. (1974). The adaptive nature of seed germination in *Nemophila menziesii* Aggr. *Ecology* **55**, 1295–1305.

Crumley, R. E. (1972). Germination and dormancy responses of several alpine species. *Disser. Abst.* **32**, 3806-B.

Culver, D. C., and Beattie, A. J. (1978). Myrmecochory in *Viola*: Dynamics of seed-ant interactions in some West Virginia species. *J. Ecol.* **66**, 53–72.

Culver, D. C., and Beattie, A. J. (1980). The fate of *Viola* seeds dispersed by ants. *Am. J. Bot.* **67**, 710–714.

Culver, D. C., and Beattie, A. J. (1983). Effect of ant mounds on soil chemistry and vegetation patterns in a Colorado montane meadow. *Ecology* **64**, 485–492.

Curtis, N. P. (1996). Germination and seedling survival studies of *Xanthorrhoea australis* in the Warby Range State Park, northeastern Victoria, Australia. *Aust. J. Bot.* **44**, 635–647.

Cushwa, C. T., Martin, R. E., and Miller, R. L. (1968). The effects of fire on seed germination. *J. Range Manage.* **21**, 250–254.

Darwent, A. L., and Coupland, R. T. (1966). Life history of *Gypsophila paniculata*. *Weeds* **14**, 313–318.

Daubenmire, R. (1968). Ecology of fire in grasslands. *Adv. Ecol. Res.* **5**, 209–266.

Davidson, D. W., and Morton, S. R. (1984). Dispersal adaptations of some *Acacia* species in the Australian arid zone. *Ecology* **65**, 1038–1051.

Davis, F. W., Borchert, M. I., and Odion, D. C. (1989). Establishment of microscale vegetation pattern in maritime chaparral after fire. *Vegetatio* **84**, 53–67.

Davis, O. H. (1927). Germination and early growth of *Cornus florida*, *Sambucus canadensis*, and *Berberis thunbergii*. *Bot. Gaz.* **84**, 225–263.

Davis, W. E., and Rose, R. C. (1912). The effect of external conditions upon the after-ripening of the seeds of *Crataegus mollis*. *Bot. Gaz.* **54**, 49–62.

Dawson, J. (1988). "Forest Vines to Snow Tussocks." Victoria Univ. Press, Wellington, New Zealand.

Deall, G. B., and Brown, N. A. C. (1981). Seed germination in *Protea magnifica* Link. *S. Afr. J. Sci.* **77**, 175–176.

Dean, S. J., Holmes, P. M., and Weiss, P. W. (1986). Seed biology of invasive alien plants in South Africa and south west Africa/Namibia. *In* "The Ecology and Management of Biological Invasions in Southern Africa" (I. A. W. MacDonald, F. J. Kruger, and A. A. Ferrar, eds.), pp. 157–170. Oxford Univ Press, Cape Town.

Debano, L. F., and Conrad, C. E. (1978). The effect of fire on nutrients in a chaparral ecosystem. *Ecology* **59**, 489–497.

Debazac, E. F. (1983). Temperate broad-leaved evergreen forests of the Mediterranean region and Middle East. *In* "Ecosystems of the World" (J. D. Ovington, ed.), Vol. 10, pp. 107–123. Elsevier, Amsterdam.

DeBell, D. S., and Naylor, A. W. (1972). Some factors affecting germination of swamp tupelo seeds. *Ecology* **53**, 504–506.

Dehgan, B., and Johnson, C. R. (1983). Improved seed germination of *Zamia floridana* (*sensu lato*) with H_2SO_4 and GA_3. *Sci. Hort.* **19**, 357–361.

Dehgan, B., and Schutzman, B. (1983). Effect of H_2SO_4 and GA_3 on seed germination of *Zamia furfuracea*. *Hortscience* **18**, 371–372.

Deitschman, G. H. (1974). *Artemisia* L. Sagebrush. *In* "Seeds of

Woody Plants in the United States" (C. S. Schopmeyer, Tech. Coord.), USDA Forest Service Agriculture Handbook No. 450.

De Klerk, G. J., and Smulders, R. (1984). Protein synthesis in embryos of dormant and germinating *Agrostemma githago* L. seeds. *Plant Physiol.* **75**, 929–935.

de Lange, J. H., and Boucher, C. (1990). Autecological studies on *Audouinia capitata* (Bruniaceae). I. Plant-derived smoke as a seed germination cue. *S. Afr. J. Bot.* **56**, 700–703.

del Moral, R., and Muller, C. H. (1970). The allelopathic effects of *Eucalyptus camaldulensis. Am. Midl. Nat.* **83**, 254–282.

del Moral, R., Willis, R. J., and Ashton, D. H. (1978). Suppression of coastal heath vegetation by *Eucalyptus baxteri. Aust. J. Bot* **26**, 203–219.

Del Tredici, P. (1981). *Magnolia virginiana* in Massachusetts. *Arnoldia* **41**(2), 36–49.

Densmore, R. V. (1997). Effect of day length on germination of seeds collected in Alaska. *Am. J. Bot.* **84**, 274–278.

Densmore, R., and Zasada, J. C. (1977). Germination requirements of Alaskan *Rosa acicularis. Can. Field-Nat.* **91**, 58–62.

Densmore, R., and Zasada, J. C. (1983). Seed dispersal and dormancy patterns in northern willows: Ecological and evolutionary significance. *Can. J. Bot.* **61**, 3207–3216.

Dettori, M. L., Balliette, J. F., Young, J. A., and Evans, R. A. (1984). Temperature profiles for germination of two species of winterfat. *J. Range Manage.* **37**, 218–222.

Devillez, F. (1973). Temperatures d'incubation et structures semencieres agissant sur la germination du *Pseudotsuga menziesii* (Mirb.) Franco var. *menziesii. Seed Sci. Technol.* **1**, 749–758.

Devillez, F. (1977). Germination et chorologie de diverses especes du genre *Eucalyptus* L'Herit. *Doc. Phytosoc. N. S.* **1**, 71–83.

Devillez, F. (1979). Influence de la stratification sur les graines et les fruits de *Sorbus aria* (L.) Crantz, *S. aucuparia* L. et *S. torminalis* (L.) Crantz. *Bull. Acad. Roy. Belgique Cl. Sci.* **65**, 312–329.

di Castri, F. (1981). Mediterranean-type shrublands of the world. *In* "Ecosystems of the World" (F. di Castri, D. W. Goodall, and R. L. Specht, eds.), Vol. 11, pp. 1–52. Elsevier, Amsterdam.

Dickerson, J. A., Longren, W. G., and Hadle, E. K. (1981). Native forb seed production. *In* "The Prairie Peninsula: In the "Shadow" of Transeau" (R. L. Stuckey, and K. J. Reese, eds.), pp. 216–222. Proc. 6th North Amer. Prairie Conf., The Ohio State Univ., Columbus, *Ohio Biol. Surv. Biol. Notes* No. 15.

Dietz, D. R., and Slabaugh, P. E. (1974). *Caragana arborescens* Lam. Siberian peashrub. *In* "Seeds of Woody Plants in the United States" (C. S. Schopmeyer, Tech. Coord.), pp. 262–264. USDA Forest Service Agriculture Handbook No. 450.

Dilling, W. J. (1926). Influence of lead and the metallic ions of copper, zinc, thorium, beryllium and thallium on the germination of seeds. *J. Appl. Biol.* **13**, 160–167.

Dillon, S. P., and Forcella, F. (1984). Germination, emergence, vegetative growth and flowering of two silvergrasses, *Vulpia bromoides* (L.) S. F. Gray and *V. myuros* (L.) C. C. Gmel. *Aust. J. Bot* **32**, 165–175.

Dimock, E. J., II, Johnston, W. F., and Stein, W. I. (1974). *Gaultheria* L. Wintergreen. *In* "Seeds of Woody Plants in the United States" (C. S. Schopmeyer, Tech. Coord.), pp. 422–426. USDA Forest Service Agriculture Handbook No. 450.

Dixon, H. (1935). Ecological studies on the high plateaus of Utah. *Bot. Gaz.* **97**, 272–320.

Dixon, J. M. (1995). Biological Flora of the British Islea. *Trisetum flavescens* (L.) Beauv. (*T. pratense* Pers., *Avena flavenscens* L.). *J. Ecol.* **83**, 895–909.

Dixon, K. W., Roche, S., and Pate, J. S. (1995). The promotive effect of smoke derived from burnt native vegetation on seed germination of Western Australian plants. *Oecologia* **101**, 185–192.

Dodd, M. C., and van Staden, J. (1981). Germination and viability studies on the seeds of *Podocarpus henkelii* Stapf. *S. Afr. J. Sci.* **77**, 171–174.

Donald, D. G. M., and Theron, J. M. (1983). Temperate broad-leaved evergreen forests of Africa south of the Sahara. *In* "Ecosystems of the World" (J. D. Ovington, ed.), Vol. 10, pp. 135–168. Elsevier, Amsterdam.

Douglas, D. A. (1995). Seed germination, seedling demography, and growth of *Salix setchelliana* on glacial river gravel bars in Alaska. *Can. J. Bot.* **73**, 673–679.

Drake, W. E. (1981). Ant-seed interaction in dry sclerophyll forest on North Stradbroke Island, Queensland. *Aust. J. Bot.* **29**, 293–309.

DuBarry, A. P., Jr. (1963). Germination of bottomland tree seed while immersed in water. *J. For.* **61**, 225–226.

Dumbroff, E. B., and Webb, D. P. (1970). Factors influencing the stratification process in seeds of *Acer ginnala. Can. J. Bot.* **48**, 2009–2015.

Eastin, E. F. (1983). Smallflower morningglory (*Jacquemontia tamnifolia*) germination as influenced by scarification, temperature, and seeding depth. *Weed Sci.* **31**, 727–730.

Ecroyd, C. E. (1982). Biological flora of New Zealand. 8. *Agathis australis* (D. Don) Lindl. (Araucariaceae) Kauri. *New Zeal. J. Bot.* **20**, 17–36.

Eddleman, L. E. (1979). Germination in black greasewood (*Sarcobatus vermiculatus* (Hook.) Torr.). *Northw. Sci.* **53**, 289–294.

Eddleman, L. E., and Romo, J. T. (1988). Spotted knapweed germination response to stratification, temperature, and water stress. *Can. J. Bot.* **66**, 653–657.

Edgar, J. G. (1977). Effects of moisture stress on germination of *Eucalyptus camaldulensis* Dehnh. and *E. regnans* F. Muell. *Aust. For. Res.* **7**, 241–245.

Edgar, R. L., and Springfield, H. W. (1977). Germination characteristics of broadscale: A possible saline-alkaline site stabilizer. *J. Range Manage.* **30**, 296–298.

Edwards, D. G. W. (1973). Effects of stratification on western hemlock germination. *Can J. For. Res.* **3**, 522–527.

Edwards, D. G. W., and Olsen, P. E. (1973). A photoperiod response in germination of western hemlock seeds. *Can J. For. Res.* **3**, 146–148.

Edwards, G. (1962). The germination requirements of *Abies* species. *Proc. Int. Seed Test. Assoc.* **27**, 142–180.

Egley, G. H. (1979). Seed coat impermeability and germination of showy crotalaria (*Crotalaria spectabilis*) seeds. *Weed Sci.* **27**, 355–361.

Ehrman, T., and Cocks, P. S. (1996). Reproductive patterns in annual legume species on an aridity gradient. *Vegetatio* **122**, 47–59.

Elkington, T. T. (1971). Biological flora of the British Isles. *Dryas octopetala. J. Ecol.* **59**, 887–905.

Ellenberg, H. (1988). "Vegetation Ecology of Central Europe," 4th Ed. Cambridge Univ. Press, Cambridge, UK [English translation by G. K. Strutt].

Elmore, F. H. (1976). "Shrubs and Trees of the Southwest Uplands." Southwest Parks and Monuments Assoc., Globe, AZ.

Emal, J. G., and Conard, E. C. (1973). Seed dormancy and germination in Indiangrass as affected by light, chilling, and certain chemical treatments. *Agron. J.* **65**, 383–385.

Enright, N. J., Goldblum, D., Ata, P., and Ashton, D. H. (1997). The independent effects of heat, smoke and ash on emergence of seedlings from the soil seed bank of a healthy *Eucalyptus* woodland in Grampians (Gariwerd) National Park, western Victoria. *Aust. J. Ecol.* **22**, 81–88.

Enright, N. J., Lamont, B. B., and Marsula, R. (1996). Canopy seed bank dynamics and optimum fire regime for the highly serotinous shrub, *Banksia hookeriana. J. Ecol.* **84**, 9–17.

Eriksson, O., and Froborg, H. (1996). "Windows of opportunity" for recruitment in long-lived clonal plants: Experimental studies of seedling establishment in *Vaccinium* shrubs. *Can. J. Bot.* **74**, 1369–1374.

Ericksson, O., and Ehrlen, J. (1992). Seed and microsite limitation of recruitment in plant populations. *Oecologia* **91**, 360–364.

Ernst, W. H. O. (1979). Population biology of *Allium ursinum* in northern Germany. *J. Ecol.* **67**, 347–362.

Ernst, W. H. O. (1981). Ecological implication of fruit variability in *Phleum arenarium* L., an annual dune grass. *Flora* **171**, 387–398.

Esashi, Y., Fuwa, N., Kojima, K., and Hase, S. (1986). Light actions in the germination of cocklebur seeds. IV. Disappearance of red light-requirement for the germination of upper seeds subject to anoxia, chilling, cyanide or azide pretreatment. *J. Exp. Bot.* **37**, 1652–1662.

Espigares, T., and Peco, B. (1993). Mediterranean pasture dynamics: The role of germination. *J. Veg. Sci.* **4**, 189–194.

Esterhuizen, A. D., van de Venter, H. A., and Robbertse, P. J. (1986). A preliminary study on seed germination of *Watsonia fourcadei*. *S. Afr. J. Bot.* **52**, 221–225.

Eurola, S. (1972). Germination of seeds collected in Spitsbergen. *Ann. Bot. Fennici* **9**, 149–159.

Evans, C. R. (1933). Germination behavior of *Magnolia grandiflora*. *Bot. Gaz.* **94**, 729–753.

Evans, K. E. (1974). *Symphoricarpos* Duham. Snowberry. *In* "Seeds of Woody Plants in the United States" (C. S. Schopmeyer, Tech. Coord.), pp. 787–789. USDA Forest Service Agriculture Handbook No. 450.

Evans, R. A., and Young, J. A. (1970). Plant litter and establishment of alien annual weed species in rangeland communities. *Weed Sci.* **18**, 697–703.

Evans, R. A., and Young, J. A. (1972a). Microsite requirements for establishment of annual rangeland weeds. *Weed Sci.* **20**, 350–356.

Evans, R. A., and Young, J. A. (1972b). Germination and establishment of *Salsola* in relation to seedbed environment. II. Seed distribution, germination, and seedling growth of *Salsola* and microenvironmental monitoring of the seedbed. *Agron. J.* **64**, 219–224.

Evans, R. A., and Young, J. A. (1975). Enhancing germination of dormant seeds of downy brome. *Weed Sci.* **23**, 354–357.

Evans, R. A., and Young, J. A. (1977). Bitterbrush germination with constant and alternating temperatures. *J. Range Manage.* **30**, 30–32.

Evans, R. A., and Young, J. A. (1983). 'Magnar' basin wildrye: Germination in relation to temperature. *J. Range Manage.* **36**, 395–398.

Evans, R. A., Young, J. A., and Roundy, B. A. (1977). Seedbed requirements for germination of Sandberg bluegrass. *Agron. J.* **69**, 817–820.

Evetts, L. L., and Burnside, O. C. (1972). Germination and seedling development of common milkweed and other species. *Weed Sci.* **20**, 371–378.

Eyre, S. R. (1963). "Vegetation and Soils, A World Picture." Aldine, Chicago.

Facelli, J. M., and Ladd, B. (1996). Germination requirements and responses to leaf litter of four species of eucalypt. *Oecologia* **107**, 441–445.

Facelli, J. M., and Pickett, S. T. A. (1991a). Plant litter: Light interception and effects on an old-field plant community. *Ecology* **72**, 1024–1031.

Facelli, J. M., and Pickett, S. T. A. (1991b). Plant litter: Its dynamics and effects on plant community structure. *Bot. Rev.* **57**, 1–32.

Farmer, R. E., Jr. (1977). Epicotyl dormancy in white and chestnut oaks. *For. Sci.* **23**, 329–332.

Farmer, R. E., Jr. (1978). Germination of *Fothergilla major* seeds. *Plant Propag.* **24**(4), 12–14.

Farmer, R. E., Jr. (1978). Propagation of a southern Appalachian population of fringed gentian. *Bull. Torrey Bot. Club* **105**, 139–142.

Farmer, R. E., Jr., and Bonner, F. T. (1967). Germination and initial growth of eastern cottonwood as influenced by moisture stress, temperature, and storage. *Bot. Gaz.* **128**, 211–215.

Farmer, R. E., Jr., Charrette, P., Searle, I. E., and Tarjan, D. P. (1984). Interaction of light, temperature, and chilling in the germination of black spruce. *Can. J. For. Res.* **14**, 131–133.

Farmer, R. E., Jr., and Cunningham, M. (1981). Seed dormancy of red maple in east Tennessee. *For. Sci.* **27**, 446–448.

Farmer, R. E., Jr., Lockley, G. C., and Cunningham, M. (1982). Germination patterns of the sumacs, *Rhus glabra* and *Rhus coppalina*: Effects of scarification time, temperature and genotype. *Seed Sci. Technol.* **10**, 223–231.

Farmer, R. E., Jr., Maley, M. L., Stoehr, M. U., and Schnekenburger, F. (1985). Reproductive characteristics of green alder in northwestern Ontario. *Can. J. Bot.* **63**, 2243–2247.

Farmer, R. E., Jr., and Reinholt, R. W. (1986). Seed quality and germination characteristics of tamarack in northwestern Ontario. *Can. J. For. Res.* **16**, 680–683.

Faust, M. E. (1936). Germination of *Populus grandidentata* and *P. tremuloides*, with particular reference to oxygen consumption. *Bot. Gaz.* **97**, 808–821.

Fechner, G. H., Burr, K. E., and Myers, J. F. (1981). Effects of storage, temperature, and moisture stress on seed germination and early seedling development of trembling aspen. *Can. J. For. Res.* **11**, 718–722.

Feldman, S. R., Vesprini, J. L., and Lewis, J. P. (1994). Survival and establishment of *Carduus acanthoides* L. *Weed Res.* **34**, 265–273.

Ferrar, P. J., Cochrane, P. M., and Slatyer, R. O. (1988). Factors influencing germination and establishment of *Eucalyptus pauciflora* near the alpine tree line. *Tree Physiol.* **4**, 27–43.

Fins, L. (1981). Seed germination of giant sequoia. *Tree Plant. Notes* **32**(2), 3–8.

Fischer, M. L., Stritzke, J. F., and Ahring, R. M. (1982). Germination and emergence of little barley (*Hordeum pusillum*). *Weed Sci.* **30**, 624–628.

Fivaz, A. E. (1931). Longevity and germination of seeds of *Ribes*, particularly *R. rotundifolium*, under laboratory and natural conditions. USDA Tech. Bull. No. 261.

Flemion, F. (1931). After-ripening, germination, and vitality of seeds of *Sorbus aucuparia* L. *Contrib. Boyce Thomp. Inst.* **3**, 413–439.

Flemion, F. (1933). Physiological and chemical studies of after-ripening of *Rhodotypos kerrioides* seeds. *Contrib. Boyce Thomp. Inst.* **5**, 143–159.

Flemion, F. (1934). Physiological and chemical changes preceding and during the after-ripening of *Symphoricarpos racemosus* seeds. *Contrib. Boyce Thomp. Inst.* **6**, 91–102.

Flemion, F. (1937). After-ripening at 5°C favors germination of grape seeds. *Contrib. Boyce Thomp. Inst.* **9**, 7–15.

Flemion, F. (1938). Breaking the dormancy of seeds of *Crataegus* species. *Contrib. Boyce Thomp. Inst.* **9**, 409–423.

Flemion, F., and Parker, E. (1942). Germination studies of seeds of *Symphoricarpos orbiculatus*. *Contrib. Boyce Thomp. Inst.* **12**, 301–307.

Floret, Ch., Galan, M. J., Le Floc'h, E., LePrince, F., and Romane, R. (1989). France. *In* "Plant Pheno-morphological Studies in Mediterranean Type Ecosystems" (G. Orshan, ed.), pp. 9–97. Kluwer, Dordrecht.

Florez, A., and McDonough, W. T. (1974). Seed germination, and growth and development of *Rudbeckia occidentalis* Nutt. (western coneflower) on Aspen Ridge in Utah. *Am. Midl. Nat.* **91**, 160–169.

Foiles, M. W. (1974). *Atriplex*. Saltbush. *In* "Seeds of Woody Plants in the United States" (C. S. Schopmeyer, Tech. Coord.), pp. 240–243. USDA Forest Service Agriculture Handbook No. 450.

Fotheringham, C. J., Siu, E., and Keeley, J. E. (1995). Smoke stimulated germination in chaparral and coastal sage species of California. *Bull. Ecol. Soc. Am.* **76** (No. 3, Part 3), 328–329. [Abstract]

Fountain, D. W., Holdsworth, J. M., and Outred, H. A. (1989). The dispersal unit of *Dacrycarpus dacrydioides* (A. Rich.) de Lauvenfels (Podocarpaceae) and the significance of the fleshy receptacle. *Bot. J. Linn. Soc.* **99**, 197–207.

Fountain, D. W., and Outred, H. A. (1991). Germination requirements of New Zealand native plants: A review. *New Zeal. J. Bot.* **29**, 311–316.

Fowler, N. L. (1986). Microsite requirements for germination and establishment of three grass species. *Am. Midl. Nat.* **115**, 131–145.

Fowler, N. L. (1988). What is a safe site? Neighbor, litter, germination date, and patch effects. *Ecology* **69**, 947–961.

Fowler, L. J., and Fowler, D. K. (1987). Stratification and temperature requirements for germination of autumn olive (*Elaeagnus umbellata*) seed. *Tree Plant. Notes* **38**(1), 14–17.

Frank, A. B., and Larson, K. L. (1970). Influence of oxygen, sodium hypochlorite, and dehulling on germination of green needlegrass seed (*Stipa viridula* Trin.). *Crop Sci.* **10**, 679–683.

Franklin, B., and Wareing, P. F. (1966). Hormonal regulation of seed dormancy in hazel (*Corylus avellana* L.) and beech (*Fagus sylvatica* L.). *J. Exp. Bot.* **17**, 596–611.

Franklin, J. F. (1974). *Abies* Mill. Fir. *In* "Seeds of Woody Plants in the United States" (C. S. Schopmeyer, Tech. Coord.), pp. 168–183. USDA Forest Service Agriculture Handbook No. 450.

Franklin, J. F., and Krueger, K. W. (1968). Germination of true fir and mountain hemlock seed on snow. *J. For.* **66**, 416–417.

Frelich, J. R., Jensen, E. H., and Gifford, R. O. (1973). Effect of crust rigidity and osmotic potential on emergence of six grass species. *Agron. J.* **65**, 26–29.

Frett, J. L., and Dirr, M. A. (1979). Scarification and stratification requirements for seeds of *Cercis canadensis* L. (redbud), *Cladrastis lutea* (Michx. f.) C. Koch. (yellowwood), and *Gymnocladus dioicus* (L.) C. Koch. (Kentucky coffee tree). *Plant Propag.* **25**(2), 4–6.

Friesen, G., and Shebeski, L. H. (1961). The influence of temperature on the germination of wild oat seeds. *Weeds* **9**, 634–638.

Frost-Christensen, H. (1974). Embryo development in ripe seeds of *Eranthis hiemalis* and its relation to gibberellic acid. *Physiol. Plant.* **30**, 200–205.

Froud-Williams, R. J., and Chancellor, R. J. (1986). Dormancy and seed germination of *Bromus catharticus* and *Bromus commutatus*. *Seed Sci. Technol.* **14**, 439–450.

Fulbright, T. E., and Flenniken, K. S. (1988). Causes of dormancy in *Paspalum plicatulum* (Poaceae) seeds. *Southw. Nat.* **33**, 35–39.

Fulbright, T. E., Redente, E. F., and Wilson, A. M. (1983). Germination requirements of green needlegrass (*Stipa viridula* Trin.). *J. Range Manage.* **36**, 390–394.

Fulwider, J. R., and Engel, R. E. (1959). The effect of temperature and light on germination of seed of goosegrass, *Eleusine indica*. *Weeds* **7**, 359–361.

Gaertner, E. E. (1956). Dormancy in the seed of *Cuscuta europea*. *Ecology* **37**, 389.

Gansel, C. (1986). Comparison of seed stratification methods for western white pine. *In* "Proceedings: Combined Western Forest Nursery Council and Intermountain Nursery Association Meeting" (T. D. Landis, Tech. Coord.), pp. 19–22. USDA Forest Service Gen. Tech. Rep. RM-137.

Gartner, B. L., Chapin, F. S., III, and Shaver, G. R. (1986). Reproduction of *Eriophorum vaginatum* by seed in Alaskan tussock tundra. *J. Ecol.* **74**, 1–18.

Gealy, D. R., Young, F. L., and Morrow, L. A. (1985). Germination of mayweed (*Anthemis cotula*) achenes and seed. *Weed Sci.* **33**, 69–73.

Georges, F., and Lazare, J.-J. (1983). Contribution a l'etude ecologique du complexe orophile *Carex sempervirens* (Cyperaceae): Etude experimentale de la germination de populations phreneennes. *Can. J. Bot.* **61**, 135–141.

Gibson, D. J., Hartnett, D. C., and Merrill, G. L. S. (1990). Fire temperature heterogeneity in contrasting fire prone habitats: Kansas tallgrass prairie and Florida sandhill. *Bull. Torrey Bot. Club* **117**, 349–356.

Giersbach, J. (1934a). Germination and seedling production of *Arctostaphylos uva-ursi*. *Contrib. Boyce Thomp. Inst.* **9**, 71–78.

Giersbach, J. (1934b). After-ripening and germination of cotoneaster seeds. *Contrib. Boyce Thomp. Inst.* **6**, 323–338.

Giersbach, J. (1937a). Germination and seedling production of species of *Viburnum*. *Contrib. Boyce Thomp. Inst.* **9**, 79–90.

Giersbach, J. (1937b). Some factors affecting germination and growth of gentian. *Contrib. Boyce Thomp. Inst.* **9**, 91–103.

Giersbach, J., and Barton, L. V. (1932). Germination of seeds of the silver bell, *Halesia carolina*. *Contrib. Boyce Thomp. Inst.* **4**, 27–37.

Giersbach, J., and Crocker, W. (1932). Germination and storage of wild plum seeds. *Contrib. Boyce Thomp. Inst.* **4**, 39–51.

Giersbach, J., and Voth, P. D. (1957). On dormancy and seed germination in *Hemerocallis*. *Bot. Gaz.* **118**, 223–237.

Gill, A. M., and Ingwersen, F. (1976). Growth of *Xanthorrhoea australis* R.Br. in relation to fire. *J. Appl. Ecol.* **13**, 195–203.

Gill, J. D., and Pogge, F. L. (1974). *Parthenocissus* Planch. Creeper. *In* "Seeds of Woody Plants in the United States" (C. S. Schopmeyer, Tech. Coord.), pp. 568–571. USDA Forest Service Agriculture Handbook No. 450.

Givelberg, A., Horowitz, M., and Poljakoff-Mayber, A. (1984). Germination behaviour of *Solanum nigrum* seeds. *J. Exp. Bot.* **35**, 588–598.

Glendening, G. E. (1942). Germination and emergence of some native grasses in relation to litter cover and soil moisture. *Agron. J.* **34**, 797–804.

Glen-Lewin, D. C., Johnson, L. A., Jurik, T. W., Akey, A. Leoschke, M., and Rosburg, T. (1990). Fire in central North American grasslands: Vegetative reproduction, seed germination, and seedling establishment. *In* "Fire in North American Tallgrass Prairies" (S. L. Collins and L. L. Wallace, eds.), pp. 22–45, Univ. Oklahoma Press, Norman.

Godley, E. J. (1971). The fruit of *Vitex lucens* (Verbenaceae). *New Zeal. J. Bot.* **9**, 561–568.

Godman, R. M., and Mattson, G. A. (1980). Low temperatures optimum for field germination of northern red oak. *Tree Plant. Notes* **31**(2), 32–34.

Godron, M., Guillerm, J. L., Poissonet, J., Poissonet, P. Thiault, M., and Trabaud, L. (1981). Dynamics and management of vegetation. *In* "Ecosystems of the World" (F. di Castri, D. W. Goodall, and R. L. Specht, eds.), Vol. II, pp. 317–344. Elsevier, Amsterdam.

Gomes, L. F., Chandler, J. M., and Vaughan, C. E. (1978). Aspects of germination, emergence, and seed production of three *Ipomoea* taxa. *Weed Sci.* **26**, 245–248.

Gonzalez-Rabanal, F., and Casal, M. (1995). Effect of high temperatures and ash on germination of ten species from gorse shrubland. *Vegetatio* **116**, 123–131.

Gonzalez-Rabanal, R., Casal, M., and Trabaud, L. (1994). Effects of high temperatures, ash and seed position in the inflorescence on the germination of three Spanish grasses. *J. Veg. Sci.* **5**, 289–294.

Goodwin, J. R., Doescher, P. S., and Eddleman, L. E. (1995). After-ripening in *Festuca idahoensis* seeds: Adaptive dormancy and implications for restoration. *Restor. Ecol.* **3**, 137–142.

Goodwin, J. R., Doescher, P. S., and Eddleman, L. E. (1996). Germination of Idaho fescue and cheatgrass seeds from coexisting populations. *Northw. Sci.* **70**, 230–241.

Gopal, B., and Meher-Homji, V. M. (1983). Temperate broad-leaved evergreen forests of the Indian subcontinent. *In* "Ecosystems of the World" (J. D. Ovington, ed.), Vol. 10, pp. 125–134. Elsevier, Amsterdam.

Gosling, P. G. (1988). The effect of moist chilling on the subsequent germination of some temperate conifer seeds over a range of temperatures. *J. Seed Technol.* **12**, 90–98.

Gosling, P. G., and Rigg, P. (1990). The effect of moisture content and prechill duration on the efficiency of dormancy breakage in Sitka spruce (*Picea sitchensis*) seed. *Seed Sci. Technol.* **18**, 337–343.

Gramshaw, D. (1972). Germination of annual ryegrass seeds (*Lolium rigidum* Gaud.) as influenced by temperature, light, storage environment, and age. *Aust. J. Agri. Res.* **23**, 779–787.

Gramshaw, D., and Stern, W. R. (1977). Survival of annual ryegrass (*Lolium rigidum* Gaud.) in a Mediterranean type environment. I. Effect of summer grazing by sheep on seed numbers and seed germination in autumn. *Aust. J. Agri. Res.* **28**, 81–91.

Granstrom, A and Schimmel, J. (1993). Heat effects on seeds and rhizomes of a selection of boreal forest plants and potential reaction to fire. *Oceologia* **94**, 307–313.

Gratkowski, H. (1973). Pregermination treatments for redstem *Ceanothus* seeds. USDA For. Serv. Res. Paper PNW-156.

Graves, J. D., and Taylor, K. (1988). A comparative study of *Geum rivale* L., and *G. urbanum* L. to determine those factors controlling their altitudinal distribution. III. The response of germination to temperature. *New Phytol.* **110**, 391–397.

Graves, W. R. (1990). Stratification not required for tree-of-heaven seed germination. *Tree Plant. Notes* **41**(2), 10–12.

Gray, J. T., and Schlesinger, W. H. (1981). Biomass, production, and litterfall in the coastal sage scrub of southern California. *Am. J. Bot.* **68**, 24–33.

Great Plains Flora Association. (1986). "Flora of the Great Plains. Univ. Press Kansas, Lawrence.

Green, H. C., and Curtis, J. T. (1950). Germination studies of Wisconsin prairie plants. *Am. Midl. Nat.* **43**, 186–194.

Grilz, P. L., Romo, J. T., and Young, J. A. (1994). Comparative germination of smooth brome and plains rough fescue. *Prairie Nat.* **26**, 157–170.

Grime, J. P., Mason, G., Curtis, A. V., Rodman, J., Band, S. R., Mowforth, M. A. G., Neal, A. M., and Shaw, S. (1981). A comparative study of germination characteristics in a local flora. *J. Ecol.* **69**, 1017–1059.

Grisez, T. J. (1974). *Prunus* L. Cherry, peach, and plum. *In* "Seeds of Woody Plants in the United States" (C. S. Schopmeyer, Tech. Coord.), pp. 658–673. USDA Forest Service Agriculture Handbook No. 450.

Grose, R. J., and Zimmer, W. J. (1957). Preliminary laboratory studies on light requirements for the germination of some eucalypt seeds. *Aust. For.* **21**, 76–80.

Grushvitzky, I. V. (1967). After-ripening of seeds of primitive tribes of angiosperms: Conditions and peculiarities. *In* "Physiologie, Okologie und Biochemie der Keimung" (H. Borriss, ed.), Vol. 1, pp. 320–336. Ernst-Moritz-Arndt Universitat, Griefswald.

Guard, A. T., and Henry, M. (1968). Reproduction of Spanish moss, *Tillandsia usneoides* L., by seeds. *Bull. Torrey Bot. Club* **95**, 327–330.

Gulmon, S. L. (1992). Patterns of seed germination in Californian serpentine grassland species. *Oecologia* **89**, 27–31.

Haase, P. (1986a). An ecological study of the subalpine tree *Dracophyllum traversii* (Epacridaceae) at Arthur's Pass, South Island, New Zealand. *New Zeal. J. Bot.* **24**, 69–78.

Haase, P. (1986b). An ecological study of the subalpine shrub *Senecio*

bennettii (Compositae) at Arthur's Pass, South Island, New Zealand. *New Zeal. J. Bot.* **24**, 247–262.

Haase, P. (1986c). Phenology and productivity of *Olearia ilicifolia* (Compositae) at Arthur's Pass, South Island, New Zealand. *New Zeal. J. Bot.* **24**, 369–379.

Haase, P. (1987). Ecological studies on *Hoheria glabrata* (Malvaceae) at Arthur's Pass, South Island, New Zealand. *New Zeal. J. Bot.* **25**, 401–409.

Haasis, F. W., and Thrupp, A. C. (1931). Temperature relations of lodgepole-pine seed germination. *Ecology* **12**, 728–744.

Habeck, J. R., and Mutch, R. W. (1973). Fire-dependent forests in the northern Rocky Mountains. *J. Quat. Res.* **3**, 408–424.

Hackett, C. 1964. Ecological aspects of the nutrition of *Deschampsia flexuosa* (L.) Trin. I. The effect of aluminium, manganese and pH on germination. *J. Ecol.* **52**, 159–167.

Hadley, E. B. (1961). Influence of temperature and other factors on *Ceanothus megacarpus* seed germination. *Madrono* **16**, 132–138.

Haeussler, S., and Tappeiner, J. C., II (1993). Effect of the light environment on seed germination of red alder (*Alnus rubra*). *Can. J. For. Res.* **23**, 1487–1491.

Haferkamp, M. R., Kissock, D. C., and Webster, R. D. (1984). Impact of presowing seed treatments, temperature and seed coats on germination of velvet bundleflower. *J. Range Manage.* **37**, 185–188.

Haggas, L., R. W. Brown, and Johnston, R. S. (1987). Light requirement for seed germination of Payson sedge. *J. Range Manage.* **40**, 180–184.

Hagon, M. W., and Simmons, D. M. (1978). Seed dormancy of *Emex australis* and *E. spinosa*. *Aust. J. Agri. Res.* **29**, 565–575.

Hajkova, L., and Krekule, J. (1972). The developmental pattern in a group of therophytes. *Flora* **161**, 111–120.

Halinar, M. (1981). Germination studies, and purity determinations on native Wisconsin prairie seeds. *In* "The Prairie Peninsula: In the "Shadow of Transeau" (R. L. Stuckey and K. J. Reese, eds.), pp. 227–231. Proc. 6th North Amer. Prairie Conf., The Ohio State Univ., Columbus, *Ohio Biol. Surv. Biol. Notes* No. 15.

Hamly, D. H. (1932). Softening of the seeds of *Melilotus alba*. *Bot. Gaz.* **93**, 345–375.

Hamrick, J. L., and Lee, J. M. (1987). Effect of soil surface topography and litter cover on the germination, survival, and growth of musk thistle (*Carduus nutans*). *Am. J. Bot.* **74**, 451–457.

Handel, S. N. (1978). The competitive relationship of three woodland sedges and its bearing on the evolution of ant dispersal of *Carex pedunculata*. *Evolution* **32**, 151–163.

Hanes, T. L. (1972). Succession after fire in the chaparral of southern California. *Ecol. Monogr.* **41**, 27–52.

Hanes, T. L. (1981). California chaparral. *In* "Ecosystems of the World" (F. di Castri, D. W. Goodall and R. L. Specht, eds.), Vol. 11, pp. 139–174. Elsevier, Amsterdam.

Hanes, T. L., and Jones, H. W. (1967). Postfire chaparral succession in southern California. *Ecology* **48**, 259–264.

Hanzawa, F. M., Beattie, A. J., and Culver, D. C. (1988). Directed dispersal: Demographic analysis of an ant-seed mutualism. *Am. Nat.* **131**, 1–13.

Hardegree, S. P. (1994). Drying and storage effects on germination of primed grass seeds. *J. Range Manage.* **47**, 196–199.

Hardegree, S. P., and Emmerich, W. E. (1992). Effect of matric-priming duration and priming water potential on germination of four grasses. *J. Exp. Bot.* **43**, 233–238.

Harif, I. (1978). The effect of burning on the germination and development of the maquis plants. *Israel J. Bot.* **27**, 44. [Abstract]

Harmer, R., and Lee, J. A. (1978). The germination and viability of *Festuca vivipara* (L.) Sm. plantlets. *New Phytol.* **81**, 745–751.

Harradine, A. R. (1980). The biology of African feather grass (*Penni-*

setum macrourum Trin.) in Tasmania. I. Seedling establishment. *Weed Res.* **20**, 165–169.

Harrington, G. T., and Hite, B. C. (1923). After-ripening and germination of apple seeds. *J. Agric. Res.* **23**, 153–161.

Hatterman-Valenti, H., Bello, I. A., and Owen, M. D. K. (1996). Physiological basis of seed dormancy in woolly cupgrass (*Eriochloa villosa* [Thunb.] Kunth. *Weed Sci.* **44**, 87–90.

Hayashi, I., and Numata, M. (1967). Ecology of pioneer species of early stages in secondary succession L. *Bot. Mag. Tokyo* **80**, 11–22.

Heady, H. F., Bartolome, J. W., Pitt, M. D., Savelle, G. D., and Stroud, M. C. (1992). California prairie. *In* "Ecosystems of the World" (R. T. Coupland, ed.), Vol. 8A, pp. 313–335. Elsevier, Amsterdam.

Heale, E. L., and Ormrod, D. P. (1983). Effects of nickel and copper on seed germination and growth and development of seedlings of *Acer ginnala, Betula papyrifera, Picea abies* and *Pinus banksiana. Reclam. Reveg. Res.* **2**, 41–54.

Heinselman, M. L. (1970). The natural role of fire in the northern conifer forests. *Naturalist (Minneapolis)* **21**(4), 15–23.

Heit, C. E. (1942). Snow-on-the-mountain (*Euphorbia marginata*) seed data: A rapid method of detecting viability. *Proc. Assoc. Offic. Seed Anal.* **34**, 78–82.

Heit, C. E. (1967). Propagation from seed. 6. Hardseededness: A critical factor. *Am. Nurseryman* **125**(10), 10–12, 88–96.

Heit, C. E. (1968a). Propagation from seed. 13. Some western and exotic spruce species. *Am. Nurseryman* **127**(8), 12–13, 51–63.

Heit, C. E. (1968b). Propagation from seed. 15. Fall planting of shrub seeds for successful seedling production. *Am. Nurseryman* **128**(4), 8–10, 70–80.

Heit, C. E. (1968c). Thirty-five years' testing of tree and shrub seed. *J. For.* **66**, 632–634.

Heit, C. E. (1969a). Propagation from seeds. 18. Testing and growing seeds of popular taxus forms. *Am. Nurseryman* **129** (2), 10–11, 118–128.

Heit, C. E. (1969b). Propagation from seed. 7. Germinating six hardseeded groups. *Am. Nurseryman* **125**(12), 10–12, 37–41, 44–45.

Heit, C. E. (1970). Germinative characteristics and optimum testing methods for twelve western shrub species. *Proc. Assoc. Offic. Seed Anal.* **60**, 197–205.

Helsper, H. P. G., and Klerken, G. A. M. (1984). Germination of *Calluna vulgaris* (L.) Hull in vitro under different pH conditions. *Acta Bot. Neerl.* **33**, 347–353.

Hepher, A., and Roberts, J. A. (1985). The control of seed germination in *Trollius ledebouri*: The breaking of dormancy. *Planta* **166**, 314–320.

Herbel, C. H., and Sosebee, R. E. (1969). Moisture and temperature effects on emergence and initial growth of two range grasses. *Agron. J.* **61**, 628–631.

Herlocker, D. J., Dirschl, H. J., and Frame, G. (1993). Grassland of east Africa. *In* "Ecosystems of the World" (R. T. Coupland, ed.), Vol. 8B, pp. 221–264. Elsevier, Amsterdam.

Hermanutz, L. A., and Weaver, S. E. (1991). Variability in temperature-dependent germination in eastern black nightshade (*Solanum ptycanthum*). *Can. J. Bot.* **69**, 1463–1470.

Herrera, J. (1991). The reproductive biology of a riparian Mediterranean shrub, *Nerium oleander* L. (Apocynaceae). *Bot. J. Linnean Soc.* **106**, 147–172.

Herron, J. W. (1953). Study of seed production, seed identification, and seed germination of *Chenopodium* spp. *Cornell Univ. Agric. Exp. Sta. Mem.* 320.

Hickey, B., and van Jaarsveld, E. (1995). Propagation of three rare and endangered vygies. *Br. Cact. Succ. J.* **13**(4), 141–146.

Hickey, J. E., Blakesley, A. J., and Turner, B. (1982). Seedfall and germination of *Nothofagus cunninghamii* (Hook.) Oerst., *Eu-*

cryphia lucida (Labill.) Baill and *Atherosperma moschatum* Labill: Implications for regeneration practice. *Aust. For. Res.* **13**, 21–28.

Hilton, J. R. (1983). The influence of light on the germination of *Senecio vulgaris* L. *New Phytol.* **94**, 29–37.

Hilton, R. J., Jaswal, A. S., Teskey, B. J. E., and Barabas, B. (1965). Rest period studies on seeds of *Amelanchier, Prunus,* and *Sorbus. Can. J. Plant Sci.* **45**, 79–85.

Hobbs, N. T., and Schimel, D. S. (1984). Fire effects on nitrogen mineralization and fixation in mountain shrub and grassland communities. *J. Range Manage.* **37**, 402–405.

Hocking, G. H. (1931–35). A note on the germination of some native species. *New Zeal. Inst. For.* **3**, 225–227.

Hoffman, G. R. (1985). Germination of herbaceous plants common to aspen forests of western Colorado. *Bull. Torrey Bot. Club* **112**, 409–413.

Hogenbirk, J. C., and Wein, R. W. (1992). Temperature effects on seedling emergence from boreal wetland soils: Implications for climate change. *Aquat. Bot.* **42**, 361–373.

Hogue, E. J. (1976). Seed dormancy of nodding beggarticks (*Bidens cernua* L.). *Weed Sci.* **24**, 375–378.

Hogue, E. J., and LaCroix, L. J. (1970). Seed dormancy of Russian olive (*Elaeagnus angustifolia* L.). *J. Am. Soc. Hort. Sci.* **95**, 449–452.

Holm, L., Pancho, J. V., Herberger, J. P., and Plucknett, C. L. (1979). "A Geographical Atlas of World Weeds." Wiley, New York.

Holmes, G. D., and Buszewicz, G. (1958). The storage of seed of temperate forest tree species. *For. Abstr.* **19**, 313–322, 455–476.

Holt, J. S. (1987). Factors affecting germination in greenhouse-produced seeds of *Oxalis corniculata*, a perennial weed. *Am. J. Bot.* **74**, 429–436.

Hopper, G. M., Smith, D. W., and Parrish, D. J. (1985). Germination and seedling growth of northern red oak: Effects of stratification and pericarp removal. *For. Sci.* **31**, 31–39.

Horak, M. J., and Sweat, J. K. (1994). Germination, emergence, and seedling establishment of buffalo gourd (*Cucurbita foetidissima*). *Weed Sci.* **42**, 358–363.

Horak, M. J., and Wax, L. M. (1991). Germination and seedling development of bigroot morningglory (*Ipomoea pandurata*). *Weed Sci.* **39**, 390–396.

Horowitz, M., and Taylorson, R. B. (1983). Effect of high temperatures on imbibition, germination, and thermal death of velvetleaf (*Abutilon theophrasti*) seed. *Can. J. Bot.* **61**, 2269–2276.

Horrell, B. A., Jameson, P. A., and Bannister, P. (1989). Propagation and regulation of phase change in some New Zealand heteroblastic species. *Proc. Int. Plant Propag. Soc.* **39**, 268–274.

Horrill, A. D. (1972). Biological flora of the British Isles. *Melampyrum cristatum* L. *J. Ecol.* **60**, 235–244.

Horton, J. S., and Kraebel, C. J. (1955). Development of vegetation after fire in the chamise chaparral of southern California. *Ecology* **36**, 244–262.

Howard, T. M. (1973). Studies in the ecology of *Nothofagus cunninghamii* Oerst. *Aust. J. Bot.* **21**, 67–78.

Hsiao, A. I.-H., and Simpson, G. M. (1971). Dormancy studies in seed of *Avena fatua.* 7. The effects of light and variation in water regime on germination. *Can. J. Bot.* **49**, 1347–1357.

Hsu, F. H., and Chou, C.-H. (1992). Inhibitory effects of heavy metals on seed germination and seedling growth of *Miscanthus* species. *Bot. Bull. Acad. Sinica* **33**, 335–342.

Hsu, F. H., Nelson, C. J., and Matches, A. G. (1985). Temperature effects on germination of perennial warm-season forage grasses. *Crop Sci.* **25**, 215–220.

Hu, C.-Y., Rogalski, F., and Ward, C. (1979). Factors maintaining *Ilex* rudimentary embryos in the quiescent state and the ultrastructural changes during in vitro activation. *Bot. Gaz.* **140**, 272–279.

Hubbard, R. L. (1974a). *Castanopsis* (D. Don.) Spach. Chinkapin. *In*

"Seeds of Woody Plants in the United States" (C. S. Schopmeyer, Tech. Coord.), pp. 276–277. USDA Forest Service Agriculture Handbook No. 450.

Hubbard, R. L. (1974b). *Rhamnus* L. Buckthorn. *In* "Seeds of Woody Plants in the United States" (C. S. Schopmeyer, Tech. Coord.), pp. 704–708. USDA Forest Service Agriculture Handbook No. 450.

Huffman, A. H., and Jacoby, P. W., Jr. (1984). Effects of herbicides on germination and seedling development of three native grasses. *J. Range Manage.* **37,** 40–43.

Hulbert, L. C. (1955). Ecological studies of *Bromus tectorum* and other annual bromegrasses. *Ecol. Monogr.* **25,** 181–213.

Hulbert, L. C. (1969). Fire and litter effects in undisturbed bluestem prairie in Kansas. *Ecology* **50,** 874–877.

Hussain, F., and Ilahi, I. (1990). Germination study on some grasses. *Sci. Khyber* **3,** 209–215.

Hussain, F., Khanzada, G., and Haqvi, H. H. (1984). Preliminary studies on the germination of *Hyoscyamus niger* L., and *Hyoscyamus insanus* Stock under laboratory conditions. *J. Sci. Technol.* **8,** 1–3.

Hutchinson, T. C. (1968). Biological flora of the British Isles. *Teucrium scorodonia* L. *J. Ecol.* **56,** 901–911.

Hutton, M. E.-J., and Porter, R. H. (1937). Seed impermeability and viability of native and introduced species of Leguminosae. *Iowa St. Coll. J. Sci.* **12,** 5–24.

Huxley, A. (1967). "Mountain Flowers in Colour." Blandford Press, London.

Hyder, S. Z., and Yasmin, S. (1972). Salt tolerance and cation interaction in alkali sacaton at germination. *J. Range Manage.* **25,** 390–392.

Hylton, L. O., Jr., and Bass, L. N. (1961). Germination of sixweeks fescue. *Proc. Assoc. Off. Seed Anal.* **51,** 118–122.

Ilnicki, R. D., Tisdell, T. F., Fertig, S. N., and Furrer, A. H., Jr. (1962). Life history studies as related to weed control in the Northeast. 3. Horse nettle. *Univ. Rhode Island Agri. Exp. Sta. Bull.* 368.

Ives, S. A. (1923). Maturation and germination of seeds of *Ilex opaca*. *Bot. Gaz.* **76,** 60–77.

Izhaki, H. Lahav, and Ne'eman, F. (1992). Spatial distribution patterns of *Rhus coriaria* seedlings after fire in a Mediterranean pine forest. *Acta Oecol.* **13,** 279–289.

Jager, A. K., Light, M. E., and van Staden, J. (1996). Effects of source of plant material and temperature on the production of smoke extracts that promote germination of light-sensitive lettuce seeds. *Environ. Exp. Bot.* **36,** 421–429.

Jain, R., and Singh, M. (1989). Factors affecting goatweed (*Scoparia dulcis*) seed germination. *Weed Sci.* **37,** 766–770.

Janssen, J. G. M. (1973). Effects of light, temperature and seed age on the germination of the winter annuals *Veronica arvensis* L., and *Myosotis ramosissima* Rochel ex. Schult. *Oecologia* **12,** 141–146.

Jaynes, R. A. (1968). Breaking seed dormancy of *Kalmia hirsuta* with high temperatures. *Ecology* **49,** 1196–1198.

Jaynes, R. A. (1971). Seed germination of six *Kalmia* species. *J. Am. Soc. Hort. Sci.* **96,** 668–672.

Jeffery, D. J., Holmes, P. M., and Rebelo, A. G. (1988). Effects of dry heat on seed germination in selected indigenous and alien legume species in South Africa. *S. Afr. J. Bot.* **54,** 28–34.

Jeffery, L. S., and Nalewaja, J. D. (1970). Studies of achene dormancy in fumitory. *Weed Sci.* **18,** 345–348.

Jeffery, L. S., and Nalewaja, J. D. (1973). Changes in fumitory achenes during low temperature after-ripening. *Weed Sci.* **21,** 310–313.

Johnsen, T. N., Jr., and Alexander, R. A. (1974). *Juniperus* L. Juniper. *In* "Seeds of Woody Plants in the United States" (C. S. Schopmeyer, Tech. Coord.), pp. 460–469. USDA Forest Service Agriculture Handbook No. 450.

Johnson, C. D., and Lovell, P. H. (1980). Germination, establishment,

and spread of *Soliva valdiviana* (Compositae). *New Zeal. J. Bot.* **18,** 487–493.

Johnson, E. (1932). The Puncture vine in California. *Univ. California Agric. Exp. Sta. Bull.* 528.

Johnson, W. M., Blankenship, J. O., and Brown, G. R. (1965). Explorations in the germination of sedges. *U.S. Forest Serv. Res. Note* RM-51.

Johnston, A. (1961). Some factors affecting germination, emergence, and early growth of three range grasses. *Can. J. Plant Sci.* **41,** 41–70.

Johnston, G. R., and Clare, R. S. (1931). Hastening the germination of western pine seeds. *J. For.* **29,** 895–906.

Johnston, S. K., Murray, D. S., and Williams, J. C. (1979a). Germination and emergence of balloonvine (*Cardiospermum halicacabum*). *Weed Sci.* **27,** 74–76.

Johnston, S. K., Walker, R. H., and Murray, D. S. (1979b). Germination and emergence of hemp sesbania (*Sesbania exaltata*). *Weed Sci.* **27,** 290–293.

Jones, C. S., and Schlesinger, W. H. (1980). *Emmenanthe penduliflora* (Hydrophyllaceae): Further consideration of germination response. *Madrono* **27,** 122–125.

Jones, E. W. (1959). Biological flora of the British Isles. *Quercus* L. *J. Ecol.* **47,** 169–222.

Jones, H. A. (1920). Physiological study of maple seeds. *Bot. Gaz.* **69,** 127–152.

Jones, L., and Brown, C. L. (1966). Cause of slow germination in cherrybark and northern red oak. *Proc. Assoc. Offic. Seed Anal.* **56,** 83–88.

Jordan, G. L., and Haferkamp, M. R. (1989). Temperature responses and calculated heat units for germination of several range grasses and shrubs. *J. Range Manage.* **42,** 41–45.

Joseph, H. C. (1929). Germination and vitality of birch seeds. *Bot. Gaz.* **87,** 127–151.

Juhren, M. C. (1966). Ecological observations of *Cistus* in the Mediterranean vegetation. *For. Sci.* **12,** 415–426.

Junttila, O. (1970). Effects of stratification, gibberellic acid and germination temperature on the germination of *Betula nana*. *Physiol. Plant.* **23,** 425–433.

Junttila, O. (1972). Effect of gibberellic acid on dark and light germination at different temperatures of *Calluna*, *Ledum* and *Rhododendron* seeds. *Physiol. Plant.* **26,** 239–243.

Junttila, O. (1976a). Effects of red and far-red irradiation on seed germination in *Betula verrucosa* and *B. pubescens*. *Z. Pflanzenphysiol.* **80,** 426–435.

Junttila, O. (1976b). Seed germination and viability in five *Salix* species. *Astarte* **9,** 19–24.

Junttila, O., Landgraff, A., and Nilsen, A. J. (1978). Germination of *Phalaris* seeds. *Acta Hort.* **83,** 163–166.

Justice, O. L. (1941). A study of dormancy in seeds of *Polygonum*. *Cornell Univ. Agri. Exp. Sta. Mem.* 235.

Justice, O. L. (1956). Germination behavior in seeds of nutgrass (*Cyperus rotundus* L.). *Proc. Assoc. Offic. Seed Anal.* **46,** 67–71.

Justice, O. L., and Whitehead, M. D. (1947). Seed production, viability, and dormancy in the nutgrasses *Cyperus rotundus* and *C. esculentus*. *J. Agri. Res.* **73,** 303–318.

Kadis, C. C., and Georghiou, K. (1993). The germination physiology of the endangered plants of Cypress, *Alyssum akamasicum* and *Origanum cordifolium*. *In* "Fourth International Workshop on Seeds: Basic and Applied Aspects of Seed Biology" (D. Come and F. Corbineau, eds.), Vol. 2, pp. 461–465. Universite Pierre et Marie Curie, Paris.

Kaminsky, R. (1981). The microbial origin of the allelopathic potential of *Adenostoma fasciculatum*. *Ecol. Monogr.* **51,** 365–382.

Karlin, E. F., and Bliss, L. C. (1983). Germination ecology of *Ledum groenlandicum* and *Ledum palustre* ssp. *decumbens. Arctic Alpine Res.* **15,** 397–404.

Kay, Q. O. N. (1971). Biological flora of the British Isles. *Anthemis cotula* L. *J. Ecol.* **59,** 623–636.

Kaye, T. N. (1997). Seed dormancy in high elevation plants: Implications for ecology and restoration. "Conservation and Management of Native Plants and Fungi" (T. N. Kaye, A. Liston, R. M. Love, D. L. Luoma, R. J. Meinke, and M. V. Wilson, eds.), pp. 115–120. Native Plant Society of Oregon, Corvallis.

Keeley, J. E. (1984). Factors affecting germination of chaparral seeds. *Bull. South. California Acad. Sci.* **83,** 113–120.

Keeley, J. E. (1986). Seed germination patterns of *Salvia mellifera* in fire-prone environments. *Oecologia* **71,** 1–5.

Keeley, J. E. (1987). Role of fire in seed germination of woody taxa in California chaparral. *Ecology* **68,** 434–443.

Keeley, J. E. (1988). Anaerobiosis as a stimulus to germination in two vernal pool grasses. *Am. J. Bot.* **75,** 1086–1089.

Keeley, J. E. (1991). Seed germination and life history syndromes in the California chaparral. *Bot. Rev.* **57,** 81–116.

Keeley, J. E., and Keeley, S. C. (1987). Role of fire in the germination of chaparral herbs and suffrutescents. *Madrono* **34,** 240–249.

Keeley, J. E., Morton, B. A., Pedrosa, A., and Trotter, P. (1985). Role of allelopathy, heat and charred wood in the germination of chaparral herbs and suffrutescents. *J. Ecol.* **73,** 445–458.

Keeley, J. E., and Nitzberg, M. E. (1984). Role of charred wood in the germination of the chaparral herbs *Emmenanthe penduliflora* (Hydrophyllaceae) and *Eriophyllum confertiflorum* (Asteraceae). *Madrono* **31,** 208–218.

Keeley, S. C., and Keeley, J. E. (1981). The role of allelopathy, heat, and charred wood on the germination of chaparral herbs. *In* "Dynamics and Management of Mediterranean-Type Ecosystems" (C. E. Conrad and W. C. Oechel, eds.), pp. 128–134. USDA Forest Service Gen. Tech. Rep. PSW-58.

Keeley, S. C., and Pizzorno, M. (1986). Charred wood stimulated germination of two fire-following herbs of the California chaparral and the role of hemicellulose. *Am. J. Bot.* **73,** 1289–1297.

Kelly, V. R., and Parker, V. T. (1990). Seed bank survival and dynamics in sprouting and nonsprouting *Arctostaphylos* species. *Am. Midl. Nat.* **124,** 114–123.

Kennedy, R. A., Barrett, S. C. H., vander Zee, D., and Rumpho, M. E. (1980). Germination and seedling growth under anaerobic conditions in *Echinochloa crus-galli* (barnyard grass). *Plant Cell Environ.* **3,** 243–248.

Keogh, J. A., and Bannister, P. (1992). A method for inducing rapid germination in seed of *Discaria toumatou* Raoul. *New Zeal. J. Bot.* **30,** 113–116.

Keogh, J. A., and Bannister, P. (1994). Seed structure and germination in *Discaria toumatou* (Rhamnaceae). *Weed Res.* **34,** 481–490.

Kerr, H. D., Robocker, W. C., and Muzik, T. J. (1965). Characteristics and control of camelthorn. *Weeds* **13,** 156–163.

Khan, M. A., Sankhla, N., Weber, D. J., and McArthur, E. D. (1987). Seed germination characteristics of *Chrysothamnus nauseosus* ssp. *viridulus* (Astereae, Asteraceae). *Great Basin Nat.* **47,** 220–226.

Kibe, T., and Masuzawa, T. (1994). Seed germination and seedling growth of *Carex doenitzii* growing on alpine zone of Mt. Fuji. *J. Plant Res.* **107,** 23–27.

Kilgore, B. M. (1973). The ecological role of fire in Sierran conifer forests. Its application to national park management. *Quat. Res.* **3,** 496–513.

Kilian, D., and Cowling, R. M. (1992). Comparative seed biology and co-existence of two fynbos shrub species. *J. Veg. Sci.* **3,** 637–646.

Killick, D. J. B. (1978). The Afro-alpine region. *In* "Biogeography and Ecology of Southern Africa" (M. J. A. Werger, ed.), pp. 515–560. Junk, The Hague.

King, T. J. (1977). The plant ecology of ant-hills in calcareous grasslands. I. Patterns of species in relation to ant-hills in southern England. *J. Ecol.* **65,** 235–256.

Kis, B. (1984). Germination of prairie plants under ambient and controlled conditions. *Michigan Bot.* **23,** 93–95.

Kitchen, S. G., and Meyer, S. E. (1991). Seed germination of intermountain Penstemons as influenced by stratification and GA_3 treatments. *J. Environ. Hort.* **9,** 51–56.

Kitchen, S. G., and Meyer, S. E. (1992). Temperature-mediated changes in seed dormancy and light requirement for *Penstemon palmeri* (Scrophulariaceae). *Great Basin Nat.* **52,** 53–58.

Knipe, D., and Herbel, C. H. (1960). The effects of limited moisture on germination and initial growth of six grass species. *J. Range Manage.* **13,** 297–302.

Knipe, O. D. (1971). Effect of different osmotica on germination of alkali sacaton (*Sporobolus airoides* Torr.) at various moisture stresses. *Bot. Gaz.* **132,** 109–112.

Knipe, O. D. (1973). Western wheatgrass germination as related to temperature, light, and moisture stress. *J. Range Manage.* **26,** 68–69.

Knowles, B., and Beveridge, A. E. (1982). Biological flora of New Zealand. 9. *Beilschmiedia tawa* (A. Cunn.) Benth. et Hook. f. ex Kirk (Lauraceae) Tawa. *New Zeal. J. Bot.* **20,** 37–54.

Knowles, R. H., and Zalik, S. (1958). Effects of temperature treatment and of a native inhibitor of seed dormancy and of cotyledon removal on epicotyl growth in *Viburnum trilobum* Marsh. *Can. J. Bot.* **36,** 561–566.

Komarkova, V., Poncet, S., and Poncet, J. (1985). Two native Antarctic vascular plants, *Deschampsia antarctica* and *Colobanthus quitensis*: A new southernmost locality and other localities in the Antarctic Peninsula area. *Arctic Alpine Res.* **17,** 401–416.

Koroleff, A. (1954). Leaf litter as a killer. *J. For.* **52,** 178–182.

Korstian, C. F. (1927). Factors controlling germination and early survival in oaks. *Yale Univ. Sch. For. Bull. No.* 19.

Krugman, S. L. (1974). *Eucalyptus* L'Herit. Eucalyptus. *In* "Seeds of Woody Plants in the United States" (C. S. Schopmeyer, Tech. Coord.), pp. 384–392. USDA Forest Service Agriculture Handbook No. 450.

Krugman, S. L., and Jenkinson, J. L. (1974). *Pinus* L. Pine. *In* "Seeds of Woody Plants in the United States" (C. S. Schopmeyer, Tech. Coord.), pp. 598–638. USDA Forest Service Agriculture Handbook No. 450.

Kruse, W. H. (1970). Temperature and moisture stress affect germination of *Gutierrezia sarothrae. J. Range Manage.* **23,** 143–145.

Kucera, C. L. (1966). Some effects of gibberellic acid on grass seed germination. *Iowa St. J. Sci.* **41,** 137–143.

Kucera, C. L. (1992). Tall-grass prairie. *In* "Ecosystems of the World" (R. T. Coupland, ed.), Vol. 8A, pp. 227–268. Elsevier, Amsterdam.

Kucera, C. L., and Koelling, M. (1964). The influence of fire on composition of central Missouri prairie. *Am. Midl. Nat.* **72,** 142–147.

Laidlaw, T. F. (1987). Drastic temperature fluctuation: The key to efficient germination of pin cherry. *Tree Plant. Notes* **38**(3), 30–32.

Lal, P., and Karnataka, D. C. (1993). Effect of orientation of seed sowing and soil mixture on germination behaviour of *Quercus leucotrichophora* (syn. *Q. incana* Roxb.). *Indian For.* **119,** 122–125.

Lambert, B. B., and Menges, E. S. (1996). The effects of light, soil disturbance and presence of organic litter on the field germination and survival of the Florida goldenaster, *Chrysopsis floridana* Small. *Florida Scient.* **59,** 121–137.

Lamont, B. B., Le Maitre, D. C., Cowling, R. M., and Enright, N. J. (1991). Canopy seed storage in woody plants. *Bot. Rev.* **57,** 277–317.

Lanza, J., Schmitt, M. A., and Awad, A. B. (1992). Comparative chemistry of elaiosomes of three species of *Trillium*. *J. Chem. Ecol.* **18**, 209–221.

Larsen, J. A. (1980). "The Boreal Ecosystem." Academic Press, New York.

Lata, S., and Verma, K. R. (1993). Pre-sowing treatment of bhimal (*Grewia optiva* Drummond) seeds. *Indian For.* **119**, 135–138.

Latting, J. (1961). The biology of *Desmanthus illinoensis. Ecology* **42**, 487–493.

Laude, H. M. (1957). Comparative pre-emergence heat tolerance of some seeded grasses and of weeds. *Bot. Gaz.* **119**, 44–46.

Laude, R. M., Shrum, J. E., Jr., and Biehler, W. E. (1952). The effect of high soil temperatures on the seedling emergence of perennial grasses. *Agron. J.* **44**, 110–112.

Lavrenko, E. M., and Karamysheva, Z. V. (1993). Steppes of the former Soviet Union and Mongolia. *In* "Ecosystems of the World" (R. T. Coupland, ed.), Vol. 8B, pp. 3–59. Elsevier, Amsterdam.

Law, R., Cook, R. E. D., and Manlove, R. J. (1983). The ecology of flower and bulbil production in *Polygonum viviparum. Nord. J. Bot.* **3**, 559–565.

Lawrence, G. H. M. (1951). "Taxonomy of Vascular Plants." Macmillan, New York.

Leck, M. A. (1979). Germination behavior of *Impatiens capensis* Meerb. *Bartonia No.* **46**, 1–14.

Leck, M. A. (1980). Germination in Barrow, Alaska, tundra soil cores. *Arctic Alpine Res.* **12**, 343–349.

Leck, M. A., Baskin, C. C., and Baskin, J. M. (1994). Germination ecology of *Bidens laevis* (Asteraceae) from a tidal freshwater wetland. *Bull. Torrey Bot. Club* **121**, 230–239.

Lee, J. J., and Weber, D. E. (1979). The effect of simulated acid rain on seedling emergence and growth of eleven woody species. *For. Sci.* **25**, 393–398.

Legg, C. J., Maltby, E., and Proctor, M. C. F. (1992). The ecology of severe moorland fire on the north York moors: Seed distribution and seedling establishment of *Calluna vulgaris. J. Ecol.* **80**, 737–752.

Lehmann, E. (1909). Zur Keimungsphysiologie und -biologie von *Ranunculus scleratus* L. und einigen anderen Samen. *Ber. Deut. Bot. Gesell.* **27**, 476–494.

Le Houerou, H. N. (1981). Impact of man and his animals on Mediterranean vegetation. *In* "Ecosystems of the World" (F. di Castri, D. W. Goodall, and R. L. Specht, eds.), Vol. 11, pp. 479–521. Elsevier, Amsterdam.

Le Houerou, H. N. (1993a). Grazing lands of the Mediterranean Basin. *In* "Ecosystems of the World" (R. T. Coupland, ed.), Vol. 8B, pp. 171–196. Elsevier, Amsterdam.

Le Houerou, H. N. (1993b). Grasslands of the Sahel. *In* "Ecosystems of the World" (R. T. Coupland, ed.), Vol. 8B, pp. 197–220. Elsevier, Amsterdam.

Lei, S. A. (1997). Variation in germination response to temperature and water availability in blackbrush (*Coleogyne ramosissima*) and its ecological significance. *Great Basin Nat.* **57**, 172–177.

Le Maitre, D. C., and Botha, S. A. (1991). The effects of exposure to different environments on the viability of *Protea neriifolia* seeds. *S. Afr. J. Bot.* **57**, 226–228.

Le Roux, A., Perry, P., and Kyriacou, X. (1989). South Africa. *In* "Plant Pheno-morphological studies in Mediterranean Type Ecosystems," pp. 159–346. Kluwer, Dordrecht.

Letchamo, W., and Gosselin, A. (1996). Light, temperature and duration of storage govern the germination and emergence of *Taraxacum officinale* seed. *J. Hort. Sci.* **71**, 373–377.

Levyns, M. R. (1935). Germination in some South African seeds. *S. Afr. J. Bot.* **1**, 161–170.

Lhotska, M. (1977). Notes on the ecology of germination in *Myrrhis odorata. Folia Geobot. Phytotax.* **12**, 209–213.

Lhotska, M. (1978). Contribution to the ecology of germination of the synanthropic species of the family Daucaceae II. Genus *Anthriscus. Acta Bot. Slovaca Acad. Sci. Slovacae Ser. A.* **3**, 157–165.

Lhotska, M. (1988). The ecology of germination and reproduction of less frequent and vanishing species of the Czechoslovak flora. I. *Alyssum saxatile* L. *Folia Geobot. Phytotax.* **23**, 321–324.

Lhotska, M., and Moravcova, L. (1989). The ecology of germination and reproduction of less frequent and vanishing species of the Czechoslovak flora. II. *Pulsatilla slavica* Reuss. *Folia Geobot. Phytotax.* **24**, 211–214.

Li, X. J., Burton, P. J., and Leadem, C. L. (1994). Interactive effects of light and stratification on the germination of some British Columbia conifers. *Can J. Bot.* **72**, 1635–1646.

Likens, G. E., and Bormann, F. H. (1974). Acid rain: A serious regional environmental problem. *Science* **184**, 1176–1179.

Lin, T.-P., and Chen, M.-H. (1995). Biochemical characteristics associated with the development of the desiccation-sensitive seeds of *Machilus thunbergii* Sieb. & Zucc. *Ann. Bot.* **76**, 381–387.

Linnington, S., Bean, E. W., and Tyler, B. F. (1979). The effects of temperature upon seed germination in *Festuca pratensis* var. *apennina. J. Appl. Ecol.* **16**, 933–938.

Lippai, A., Smith, P. A., Price, T. V., Weiss, J., and Lloyd, C. J. (1996). Effects of temperature and water potential on germination of horehound (*Marrubium vulgare*) seeds from two Australian localities. *Weed Sci.* **44**, 91–99.

Lisci, M., and Pacini, E. (1994). Germination ecology of drupelets of the fig (*Ficus carica* L.). *Bot. J. Linn. Soc.* **114**, 133–146.

Litav, M., and Orshan, G. (1971). Biological flora of Israel. 1. *Sarcopoterium spinosum* (L.) Sp. *Israel J. Bot.* **20**, 48–64.

Little, S. (1974). *Ailanthus altissima* (Mill.) Swingle. Ailanthus. *In* "Seeds of Woody Plants in the United States" (C. S. Schopmeyer, Tech. Coord.), pp. 201–202. USDA Forest Service Agriculture Handbook No. 450.

Lloret, F., and Zedler, P. H. (1991). Recruitment pattern of *Rhus integrifolia* populations in periods between fire in chaparral. *J. Veg. Sci.* **2**, 217–230.

Lobstein, M. B., and Rockwood, L. L. (1993). Influence of elaiosome removal on germination in five ant-dispersed plant species. *Virginia J. Sci.* **44**, 59–72.

Lockley, G. C. (1980). Germination of chokecherry (*Prunus virginiana*) seeds. *Seed Sci. Technol.* **8**, 237–244.

Lockyer, S. (1930). Seed dispersal from hygroscopic *Mesembryanthemum* fruits: *Bergeranthus scapigerus* Schw., and *Dorotheanthus bellidiformis* N.E. Br., with a note on *Carpanthea pomeridiana* N.E. Br. *Ann. Bot.* **44**, 639–655.

Lonchamp, J. P., and Gora, M. (1980). Evolution de la faculte germinative de semences de mauvaises herbes au cours de leur conservation au sec. Inst. Nat. Recher. Agron., Dijon, France.

Longton, R. E. (1985). Terrestrial habitats: Vegetation. *In* "Key Environments: Antarctica" (W. N. Bonner and D. W. H. Walton, eds.), pp. 73–105. Pergamon Press, Oxford.

Luken, J. O., and Goessling, N. (1995). Seedling distribution and potential persistence of the exotic shrub *Lonicera maackii* in fragmented forests. *Am. Midl. Nat.* **133**, 124–130.

Maas, D. (1989). Germination characteristics of some plant species from calcareous fens in southern Germany and their implications for the seed bank. *Holarctic Ecol.* **12**, 337–344.

MacMillan, B. H. (1972). Biological flora of New Zealand. 7. *Ripogonum scandens* J. R. et G. Forst. (Smilacaceae) Supplejack, Kareao. *New Zeal. J. Bot.* **10**, 641–672.

Maekawa, F. (1974). Origin and characteristics of Japan's flora. *In* "The Flora and Vegetation of Japan" (M. Numata, ed), pp. 33–86. Elsevier, Amsterdam.

Maguire, J. D., and Overland, A. (1959). Laboratory germination of seeds of weedy and native plants. *Washington Agric. Exp. Sta. Circ. No.* 349.

Maithani, G. P., Bahuguna, V. K., and Lal, P. (1991). Seed germination behaviour of *Desmodium tiliaefolium* G. Donm: An important shrub species of Himalayas. *Indian For.* **117**, 593–595.

Maithani, G. P., Thapliyal, R. C., Bahuguna, V. K., and Sood, O. P. (1990). Enhancement of seed germination and seedling growth of *Aesculus indica* by stratification. *Indian For.* **116**, 577–579.

Makowski, R. M. D., and Morrison, I. N. (1989). The biology of Canadian weeds. 91. *Malva pusilla* Sm. (= *M. rotundifolia* L.). *Can J. Plant Sci.* **69**, 861–879.

Mallik, A. U., and Gimingham, C. H. (1985). Ecological effects of heather burning. II. Effects on seed germination and vegetative regeneration. *J. Ecol.* **73**, 633–644.

Mann, R. K., Rieck, C. E., and Witt, W. W. (1981). Germination and emergence of burcucumber (*Sicyos angulatus*). *Weed Sci.* **29**, 83–86.

Manson, B. R. (1974). The life history of silver beech (*Nothofagus menziesii*). *Proc. New Zeal. Ecol. Soc.* **21**, 27–31.

Manthey, D. R., and Nalewaja, J. D. (1987). Germination of two foxtail (*Setaria*) species. *Weed Technol.* **1**, 302–304.

Marchand, P. J., and Roach, D. A. (1980). Reproductive strategies of pioneering alpine species: Seed production, dispersal, and germination. *Arctic Alpine Res.* **12**, 137–146.

Margaris, N. S. (1981). Adaptive strategies in plants dominating Mediterranean-type ecosystems. *In* "Ecosystems of the World" (F. di Castri, D. W. Goodall, and R. L. Specht, eds.), Vol. 11, pp. 309–315. Elsevier, Amsterdam.

Mariko, S., and Kachi, N. (1995). Seed ecology of *Lobelia boninensis* Koidz. (Campanulaceae), an endemic species in the Bonin Islands (Japan). *Plant Species Biol.* **10**, 103–110.

Mark, A. F. (1965). Flowering, seeding, and seedling establishment of narrow-leaved snow tussock, *Chionochloa rigida. New Zeal. J. Bot.* **3**, 180–193.

Marks, M. K., and Prince, S. D. (1982). Seed physiology and seasonal emergence of wild lettuce *Lactuca serriola. Oikos* **38**, 242–249.

Marshall, D. L., Beattie, A. J., and Bollenbacher, W. E. (1979). Evidence for diglycerides as attractants in an ant-seed interaction. *J. Chem. Ecol.* **5**, 335–344.

Martin, A., Grzeskowiak, V., and Puech, S. (1995). Germination variability in three species in disturbed Mediterranean environments. *Acta Oecol.* **16**, 479–490.

Martin, A. C. (1946). The comparative internal morphology of seeds. *Am. Midl. Nat.* **36**, 513–660.

Martin, A. R. H. (1966). The plant ecology of the Grahamstown Nature Reserve. II. Some effects of burning. *S. Afr. J. Bot.* **32**, 1–39.

Martin, G. C., Mason, M. I. R., and Forde, H. I. (1969). Changes in endogenous growth substances in the embryos of *Juglans regia* during stratification. *J. Am. Soc. Hort. Sci.* **94**, 13–17.

Martin, M. C., Sr. (1965). An ecological life history of *Geranium maculatum. Am. Midl. Nat.* **73**, 111–149.

Martin, M. H., and Cox, J. R. (1984). Germination profiles of introduced lovegrasses at six constant temperatures. *J. Range Manage.* **37**, 507–509.

Martinez-Sanchez, J. J., Marin, A., Herranz, J. M., Ferrandis, P., and de las Heras, J. (1995). Effects of high temperatures on germination of *Pinus halepensis* Mill. and *P. pinaster* Aiton subsp. *pinaster* seeds in southeast Spain. *Vegetatio* **116**, 69–72.

Masselink, A. K. (1980). Germination and seed population dynamics in *Melampyrum pratense* L. *Acta Bot. Neerl.* **29**, 451–468.

Mather, L. J., and Williams, P. A. (1990). Phenology, seed ecology, and age structure of Spanish heath (*Erica lusitanica*) in Canterbury, New Zealand. *New Zeal. J. Bot.* **28**, 207–215.

Matsuda, K. (1985). Studies on the early phase of the regeneration

of a konara oak (*Quercus serrata* Thunb.) secondary forest. II. The establishment of current-year seedlings on the forest floor. *Jap. J. Ecol.* **35**, 145–152.

Matsuda, K., and McBride, J. R. (1989). Germination characteristics of selected California oak species. *Am. Midl. Nat.* **122**, 66–76.

Matthews, J. R. (1942). The germination of *Trientalis europaea. J. Bot.* **80**, 12–17.

Maun, M. A. (1981). Seed germination and seedling establishment of *Calamovilfa* on Lake Huron sand dunes. *Can. J. Bot.* **59**, 460–469.

Maw, M. G., Thomas, A. G., and Stahevitch, A. (1985). The biology of Canadian weeds. 66. *Artemisia absinthium* L. *Can. J. Plant Sci.* **65**, 389–400.

Maze, K. M., Koen, T. B., and Watt, L. A. (1993). Factors influencing the germination of six perennial grasses of central New South Wales. *Aust. J. Bot.* **41**, 79–90.

McCarthy, B. C., and Bailey, D. R. (1992). Seed germination and seedling establishment of *Carya floridana* (Sarg.) Small (Juglandaceae). *Bull. Torrey Bot. Club* **119**, 384–391.

McCarty, M. K., and Scifres, C. J. (1969). Life cycle studies with musk thistle. *Univ. Nebraska Coll. Agri. Res. Bull.* 230.

McCarty, M. K., Scifres, C. J., Smith, A. L., and Horst, G. L. (1969). Germination and early seedling development of musk and plumeless thistles. *Nebraska Agri. Exp. Sta. Res. Bull.* 229.

McColl, J. G., and Johnson, R. (1983). Effects of simulated acid rain on germination and early growth of Douglas-fir and ponderosa pine. *Plant Soil* **74**, 125–129.

McDermott, R. E. (1953). Light as a factor in the germination of some bottomland hardwood seeds. *J. For.* **51**, 203–204.

McDonough, W. T. (1969). Effective treatments for the induction of germination in mountain rangeland species. *Northw. Sci.* **43**, 18–22.

McDonough, W. T. (1970). Germination of 21 species collected from a high-elevation rangeland in Utah. *Am. Midl. Nat.* **84**, 551–554.

McDonough, W. T. (1975). Germination polymorphism in *Grindelia squarrosa* (Pursh) Dunal. *Northw. Sci.* **49**, 190–200.

McDonough, W. T. (1976). Germination of seeds treated with gibberellic acid and kinetin during stratification. *Phyton (Buenos Aires)* **34**, 41–44.

McDonough, W. T., and Laycock, W. A. (1975). Growth and development of *Polemonium foliosissimum* on high elevation rangeland in Utah. *Am. Midl. Nat.* **94**, 27–37.

McElgunn, J. D. (1974). Germination response of forage grasses to constant and alternating temperatures. *Can. J. Plant Sci.* **54**, 265–270.

McGinnies, W. J. (1960). Effects of moisture stress and temperature on germination of six range grasses. *Agron. J.* **52**, 159–162.

McHenry, W. B., and Jensen, L. A. (1967). Response of bitterbrush (*Purshia tridentata*) seed to certain germination methods. *Proc. Assoc. Offic. Seed Anal.* **57**, 89–95.

McKell, C. M., Wilson, A. M., and Kay, B. L. (1962). Effective burning of rangelands infested with medusahead. *Weeds* **10**, 125–131.

McLean, A. (1967). Germination of forest range species from southern British Columbia. *J. Range Manage.* **20**, 321–322.

McLemore, B. F., and Hansbrough, R. T. (1970). Influence of light on germination of *Pinus palustris* seeds. *Physiol. Plant.* **23**, 1–10.

McLeod, K. W., and Murphy, P. G. (1977). Germination ecology of *Ptelea trifoliata. Am. Midl. Nat.* **97**, 363–372.

McNaughton, I. H., and Harper, J. L. (1964). Biological Flora of the British Isles. *Papaver* L. *J. Ecol.* **52**, 767–793.

McPherson, J. K., and Muller, C. H. (1969). Allelopathic effects of *Adenostoma fasciculatum*, "Chamise," in the California chaparral. *Ecol. Monogr.* **39**, 177–198.

McVean, D. N. (1955). Ecology of *Alnus glutinosa* (L.) Gaertn. II. Seed distribution and germination. *J. Ecol.* **43**, 61–71.

Mekenian, M. R., and Willemsen, R. W. (1975). Germination characteristics of *Raphanus raphanistrum*. I. Laboratory studies. *Bull. Torrey Bot. Club* **102**, 243–252.

Mesleard, F., and Lepart, J. (1991). Germination and seedling dynamics of *Arbutus unedo* and *Erica arborea* on Corsica. *J. Veg. Sci.* **2**, 155–164.

Meyer, S. E. (1989). Warm pretreatment effects on antelope bitterbrush (*Purshia tridentata*) germination response to chilling. *Northw. Sci.* **63**, 146–153.

Meyer, S. E. (1992). Habitat correlated variation in firecracker penstemon (*Penstemon eatonii* Gray: Scrophulariaceae) seed germination response. *Bull. Torrey Bot. Club* **119**, 268–279.

Meyer, S. E., Beckstead, J., Allen, P. S., and Pullman, H. (1995). Germination ecophysiology of *Leymus cinereus* (Poaceae). *Int. J. Plant Sci.* **156**, 206–215.

Meyer, S. E., and Kitchen, S. G. (1992). Cyclic seed dormancy in the short-lived perennial *Penstemon palmeri*. *J. Ecol.* **80**, 115–122.

Meyer, S. E., and Kitchen, S. G. (1994). Habitat-correlated variation in seed germination response to chilling in *Penstemon* Section *Glabri* (Scrophulariaceae). *Am. Midl. Nat.* **132**, 349–365.

Meyer, S. E., McArthur, E. D., and Jorgensen, G. L. (1989). Variation in germination response to temperature in rubber rabbitbrush (*Chrysothamnus nauseosus*: Asteraceae) and its ecological implications. *Am. J. Bot.* **76**, 981–991.

Meyer, S. E., and Monsen, S. B. (1989). Seed germination biology of antelope bitterbrush (*Purshia tridentata*). *USDA For. Serv. Gen. Tech. Rep.* INT-256, 147–157.

Mezynski, P. R., and Cole, D. F. (1974). Germination of dandelion seed on a thermogradient plate. *Weed Sci.* **22**, 506–507.

Michaux, B. (1989). Reproductive and vegetative biology of *Cirsium vulgare* (Savi) Ten. (Compositae: Cynareae). *New Zeal. J. Bot.* **27**, 401–414.

Milberg, P. (1994a). Germination ecology of the polycarpic grassland perennials *Primula veris* and *Trollius europaeus Ecography* **17**, 3–8.

Milberg, P. (1994b). Germination ecology of the endangered grassland biennial *Gentianella campestris*. *Biol. Conserv.* **70**, 287–290.

Milewski, A. V., and Bond, W. J. (1982). Convergence of myrmecochory in Mediterranean Australia and South Africa. *In* "Ant-Plant Interactions in Australia" (R. C. Buckley, ed.), pp. 89–98. Junk, The Hague.

Mills, J. N. (1986). Herbivores and early postfire succession in southern California chaparral. *Ecology* **67**, 1637–1649.

Mohan, E., Mitchell, N., and Lovell, P. (1984a). Seasonal variation in seedfall and germination of *Leptospermum scoparium* (manuka). *New Zeal. J. Bot.* **22**, 103–107.

Mohan, E., Mitchell, N., and Lovell, P. (1984b). Environmental factors controlling germination of *Leptospermum scoparium* (manuka). *New Zeal. J. Bot.* **22**, 95–101.

Mohler, C. L., and Calloway, M. B. (1992). Effects of tillage and mulch on the emergence and survival of weeds in sweet corn. *J. Appl. Ecol.* **29**, 21–34.

Moll, E. J., and Gubb, A. A. (1981). Aspects of the ecology of *Staavia dodii* in the south western cape of South Africa. *In* "The Biological Aspects of Rare Plant Conservation" (H. Synge, ed.), pp. 331–342. Wiley, Chichester, UK.

Mondrus-Engle, M. (1981). Tetraploid perennial teosinte seed dormancy and germination. *J. Range Manage.* **34**, 59–61.

Montenegro, G., Avila, G., Aljaro, M. E., Osorio, R., and Gomez, M. (1989). Chile. *In* "Plant Pheno-morphological Studies in Mediterranean Type Ecosystems" (G. Orshan, ed.), pp. 347–387. Kluwer, Dordrecht.

Mooney, H. A., and Billings, W. D. (1961). Comparative physiological ecology of arctic and alpine populations of *Oxyria digyna*. *Ecol. Monogr.* **31**, 1–29.

Moore, A. M., and Gillette, A. (1988). Germination of red spruce and Fraser fir seeds exposed to simulated acid rain. *J. Elisha Mitchell Sci. Soc.* **104**, 137–140.

Moore, D. M. (1983). "Flora of Tierra del Fuego." Anthony Nelson, England and Missouri Botanical Garden, USA

Moore, R. M. (1993). Grasslands of Australia. *In* "Ecosystems of the World" (R. T. Coupland, ed.), Vol. 8B, pp. 315–360. Elsevier, Amsterdam.

Morash, R., and Freedman, B. (1989). The effects of several herbicides on the germination of seeds in the forest floor. *Can. J. For. Res.* **19**, 347–350.

Moreno, J. M., and Oechel, W. C. (1991). Fire intensity effects on germination of shrubs and herbs in southern California chaparral. *Ecology* **72**, 1993–2004.

Morgan, J. W., and Lunt, I. D. (1994). Germination characteristics of eight common grassland and woodland forbs. *Victorian Nat.* **111**, 10–17.

Morgenson, G. (1986). Seed stratification treatments for two hardy cherry species. *Tree Plant. Notes* **37**(3), 35–38.

Morgenson, G. (1990). Pregermination treatment and stratification of silverberry seed. *Tree Plant. Notes* **41**(1), 24–25.

Morrow, L. A., Young, F. L., and Flom, D. G. (1982). Seed germination and seedling emergence of jointed goatgrass (*Aegilops cylindrica*). *Weed Sci.* **30**, 395–398.

Mossop, M. K. (1989). Comparison of seed removal by ants in vegetation on fertile and infertile soils. *Aust. J. Ecol.* **14**, 367–374.

Mukerji, S. K. (1936). Contributions to the autecology of *Mercurialis perennis* L. *J. Ecol.* **24**, 38–81.

Muller, C. H. (1965). Inhibitory terpenes volatilized from *Salvia* shrubs. *Bull. Torrey Bot. Club* **92**, 38–45.

Muller, C. H. (1966). The role of chemical inhibition (allelopathy) in vegetational composition. *Bull. Torrey Bot. Club* **93**, 332–351.

Muller, C. H., and del Moral, R. (1971). Role of animals in suppression of herbs by shrubs. *Science* **173**, 462–463.

Muller, C. H., Hanawalt, R. B., and McPherson, J. K. (1968). Allelopathic control of herb growth in the fire cycle of California chaparral. *Bull. Torrey Bot. Club* **95**, 225–231.

Muller, P. (1933). Verbreitungsbiologie der Garigueflora. *Beih. Botan. Central.* **50**, 395–469.

Muller, W. H., and Muller, C. H. (1964). Volatile growth inhibitors produced by *Salvia* species. *Bull. Torrey Bot. Club* **91**, 327–330.

Mulligan, G. A., and Munro, D. B. (1981). The biology of Canadian weeds. 48. *Cicuta maculata* L., *C. douglasii* (DC.) Coult. & Rose and *C. virosa* L. *Can. J. Plant Sci.* **61**, 93–105.

Mumford, P. M. (1988). Alleviation and induction of dormancy by temperature in *Impatiens glandulifera* Royle. *New Phytol.* **109**, 107–110.

Munoz, M. R., and Fuentes, E. R. (1989). Does fire induce shrub germination in the Chilean matorral? *Oikos* **56**, 177–181.

Munz, P. A., and Keck, D. D. (1949). California plant communities. *El Aliso* **2**, 87–105.

Musil, C. F. (1991). Seed bank dynamics in sand plain lowland fynbos. *S. Afr. J. Bot.* **57**, 131–142.

Musil, C. F., and de Witt, D. M. (1991). Heat-stimulated germination in two Restionaceae species. *S. Afr. J. Bot.* **57**, 175–176.

Myster, R. W. (1994). Contrasting litter effects on old field tree germination and emergence. *Vegetatio* **114**, 169–174.

Nadeau, L. B., and King, J. R. (1991). Seed dispersal and seedling establishment of *Linaria vulgaris* Mill. *Can. J. Plant Sci.* **71**, 771–782.

Najda, H. G., Darwent, A. L., and Hamilton, G. (1982). The biology of Canadian weeds. 54. *Crepis tectorum* L. *Can. J. Plant Sci.* **62**, 473–481.

Nakanishi, H. (1988). Myrmecochores in warm-temperate zone of Japan. *Japan. J. Ecol.* **38**, 169–176.

Nakanishi, H. (1994). Myrmecochorous adaptations of *Corydalis* species (Papaveraceae) in southern Japan. *Ecol. Res.* **9**, 1–8.

Nan, X. (1992). "Comparison of Some Aspects of the Ecological Life History of an Annual and a Perennial Species of *Senna* (Leguminosae: Section *Chamaefistula*), with Particular Reference to Seed Dormancy." M.S. thesis, University of Kentucky, Lexington.

Naveh, Z. (1973). The ecology of fire in Israel. *Proc. Tall Timbers Fire Ecol. Conf.* **13**, 131–170.

Naveh, Z. (1975). The evolutionary significance of fire in the Mediterranean region. *Vegetatio* **29**, 199–208.

Naylor, J. M., and Simpson, G. M. (1961). Dormancy studies in seed of *Avena fatua*. 2. A gibberellin-sensitive inhibitory mechanism in the embryo. *Can. J. Bot.* **39**, 281–295.

Ne'eman, G., Meir, I., and Ne'eman, R. (1993). The influence of pine ash on the germination and early growth of *Pinus halepensis* Mill. and *Cistus salviifolius* L. *Water. Sci. Technol.* **27**, 525–532.

Nelson, J. R., and Wilson, A. M. (1969). Influence of age and awn removal on dormancy of medusahead seeds. *J. Range Manage.* **22**, 289–290.

Nesme, X. (1985). Respective effects of endocarp, testa and endosperm, and embryo on the germination of raspberry (*Rubus idaeus* L.) seeds. *Can. J. Plant Sci.* **65**, 125–130.

Newman, E. I. (1963). Factors controlling the germinate date of winter annuals. *J. Ecol.* **51**, 625–638.

Nichols, G. E. (1934). The influence of exposure to winter temperatures upon seed germination in various native American plants. *Ecology* **15**, 364–373.

Nikolaeva, M. G. (1969). Physiology of deep dormancy in seeds. Izdatel'stvo "Nauka," Leningrad. [Translated from Russian by Z. Shapiro, National Science Foundation, Washington, DC]

Nikolaeva, M. G. (1977). Factors controlling the seed dormancy pattern. *In* "The Physiology and Biochemistry of Seed Dormancy and Germination" (A. A. Khan, ed.), pp. 51–74. North-Holland, Amsterdam.

Nokes, J. (1986). "How to Grow Native Plants of Texas and the Southwest." Texas Monthly Press, Austin.

Nordhagen, R. (1959). Remarks on some new or little known myrmecochorous plants from North America and east Asia. *Bull. Res. Counc. Israel Sec. D.* **7**, 184–201.

Norris, R. F., and Schoner, C. A., Jr. (1980). Yellow foxtail (*Setaria lutescens*) biotype studies: Dormancy and germination. *Weed Sci.* **28**, 159–163.

Norton, D. A., Herbert, J. W., and Beveridge, A. E. (1988). The ecology of *Dacrydium cupressinum*: A review. *New Zeal. J. Bot.* **26**, 37–62.

Nosko, P, Brassard, P., Kramer, J. R., and Kershaw, K. A. (1988). The effect of aluminum on seed germination and early seedling establishment, growth, and respiration of white spruce (*Picea glauca*). *Can. J. Bot.* **66**, 2305–2310.

Nozzolillo, C., and Thie, I. (1983). Aspects of germination of *Impatiens capensis* Meerb., formae *capensis* and *immaculata*, and *I. pallida* Nutt. *Bull. Torrey Bot. Club* **110**, 335–344.

Nyandiga, C. O., and McPherson, G. R. (1992). Germination of two warm-temperate oaks, *Quercus emoryi* and *Quercus arizonica*. *Can. J. For. Sci.* **22**, 1395–1401.

Okagami, N. (1986). Dormancy in *Dioscorea*: Different temperature adaptation of seeds, bulbils and subterranean organs in relation to north-south distribution. *Bot. Mag. Tokyo* **99**, 15–27.

Okagami, N., and Kawai, M. (1982). Dormancy in *Dioscorea*: Differences of temperature responses in seed germination among six Japanese species. *Bot. Mag. Tokyo* **95**, 155–166.

Olson, A. R., and Richards, J. H. (1979). Temperature responses of germination in arctic poppy (*Papaver radicatum* Rottb). seeds. *Arctic Alpine Res.* **11**, 343–348.

Olson, D. F., Jr. (1974a). *Rhododendron* L. Rhododendron. *In* "Seeds of Woody Plants in the United States" (C. S. Schopmeyer, Tech. Coord.), pp. 709–712. USDA Forest Service Agriculture Handbook No. 450.

Olson, D. F., Jr. (1974b). *Quercus* L. Oak. *In* "Seeds of Woody Plants in the United States" (C. S. Schopmeyer, Tech. Coord.), pp. 692–703. USDA Forest Service Agriculture Handbook No. 450.

Olson, D. F., Jr. (1974c). *Baccharis* L. Baccharis. *In* "Seeds of Woody Plants in the United States" (C. J. Schopmeyer, Tech. Coord.), pp. 244–246. USDA Forest Service Agriculture Handbook No. 450.

Olson, D. F., Jr. (1983). Temperate broad-leaved evergreen forests of southeastern North America. *In* "Ecosystems of the World" (J. D. Ovington, ed.), Vol. 10, pp. 103–105. Elsevier, Amsterdam.

Olson, D. F., Jr., Barnes, R. L., and Jones, L. (1974). *Magnolia* L. Magnolia. *In* "Seeds of Woody Plants in the United States" (C. S. Schopmeyer, Tech. Coord.), pp. 527–530. USDA Forest Service Agriculture Handbook No. 450.

Olson, D. F., Jr., and Gabriel, W. J. (1974). *Acer* L. Maple. *In* "Seeds of Woody Plants in the United States" (C. S. Schopmeyer, Tech. Coord.), pp. 187–194. USDA Forest Service Agriculture Handbook No. 450.

Olson, D. F., Jr., and Petteys, E. Q. P. (1974). *Casuarina* L. Casuarina. *In* "Seeds of Woody Plants in the United States" (C. S. Schopmeyer, Tech. Coord.), pp. 278–280. USDA Forest Service Agriculture Handbook No. 450.

Olson, J. S., Stearns, F. W., and Nienstaedt, H. (1959). Eastern hemlock seeds and seedlings. Response to photoperiod and temperature. *Connecticut Agri. Exp. Sta. Bull.* 620.

Oostermeijer, J. G. B. (1989). Myrmecochory in *Polygala vulgaris* L., *Luzula campestris* (L.) DC. and *Viola curtisii* Forster in a Dutch dune area. *Oecologia* **78**, 302–311.

Orshan, G. (1989). Israel. *In* "Plant Pheno-morphological Studies in Mediterranean Type Ecosystems" (G. Orshan, ed.), pp. 99–157. Kluwer, Dordrecht.

Otzen, D., and G. Doornbos. (1980). Observations on germination and seedling establishment of *Senecio erucifolius* L. in relation to its northern boundary. *Acta Bot. Neerl.* **29**, 419–427.

Outcalt, K. W. (1991). Stratification increases germination of Ocala sand pine seed in dry soil. *Seed Sci. Technol* **19**, 511–517.

Ouzounidou, G. (1995). Effect of copper on germination and seedling growth of *Minuartia, Silene, Alyssum* and *Thlaspi. Biol. Plant.* **37**, 411–416.

Ovington, J. D. (1983a). Introduction. *In* "Ecosystems of the World" (J. D. Ovington, ed.) Vol. 10, pp. 1–4. Elsevier, Amsterdam.

Ovington, J. D. (ed.) (1983b). "Ecosystems of the World," Vol. 10. Elsevier, Amsterdam.

Ovington, J. D., and Pryor, L. D. (1983). Temperate broad-leaved evergreen forests of Australia. *In* "Ecosystems of the World" (J. D. Ovington, ed.), Vol. 10, pp. 73–101. Elsevier, Amsterdam.

Owens, D. W., and Call, C. A. (1985). Germination characteristics of *Helianthus maximiliana* Schrad. and *Simsia calva* (Engelm. & Gray) Gray. *J. Range Manage.* **38**, 336–339.

Ozturk, M., Secmen, O., Gork, G., Kondo, K., and Segawa, M. (1983). Ecological studies of macchia elements in Aegean region of Turkey. *Mem. Fac. Integrated Arts Sci. Hiroshima Univ. Ser. IV* **8**, 51–86.

Pacini, E. (1990). *Mercurialis annua* L. (Euphorbiaceae) seed interactions with the ant *Messor structor* (Latr.), Hymenoptera: Formicidia. *Acta. Bot. Neerl.* **39**, 253–262.

Pack, D. A. (1921). After-ripening and germination of *Juniperus* seeds. *Bot. Gaz.* **71**, 32–60.

Packham, J. R. (1978). Biological flora of the British Isles. *Oxalis acetosella* L. *J. Ecol.* **66**, 669–693.

Packham, J. R. 1983. Biological flora of the British Isles. *Lamiastrum galeobdolon* (L.) Ehrend. & Polatschek. *J. Ecol.* **71**, 975–997.

Page, R. J., Goodwin, D. L., and West, N. E. (1966). Germination requirements of scarlet globemallow. *J. Range Manage.* **9**, 145–146.

Palin, M. A. (1988). Biological flora of the British Isles. *Ligusticum scoticum* L. (*Haloscias scoticum* (L.) Fr.). *J. Ecol.* **76**, 889–902.

Pandey, S. B. (1969). Photocontrol of seed germination in *Anagallis arvensis* Linn. *Trop. Ecol.* **10**, 96–138.

Panebianco, R., and Willemsen, R. W. (1976). Seed germination of *Hieracium pratense*, a successional perennial. *Bot. Gaz.* **137**, 255–261.

Panetta, F. D. (1988). Factors determining seed persistence of *Chondrilla juncea* L. (skeleton weed) in southern Western Australia. *Aust. J. Ecol.* **13**, 211–224.

Parsons, R. F. (1968). An introduction to the regeneration of mallee eucalypts. *Proc. Roy. Soc. Vict.* **81**, 59–68.

Parsons, R. F. (1981). *Eucalyptus* scrubs and shrublands. *In* "Australian Vegetation" (R. H. Groves, ed.), pp. 227–252. Cambridge Univ. Press, Cambridge, UK.

Partridge, T. R., and Wilson, M. D. (1990). A germination inhibitor in the seeds of mahoe (*Melicytus ramiflorus*). *New Zeal. J. Bot.* **28**, 475–478.

Pathak, S. M., Mukhiya, Y. K., and Singh, V. P. (1987). Mercury, manganese interaction studies on barley: Germination and phytotoxicity. *Indian J. Plant Physiol.* **30**, 13–19.

Patten, D. T. (1963). Light and temperature influence on Engelmann spruce seed germination and subalpine forest advance. *Ecology* **44**, 817–818.

Patterson, D. T., and Mortensen, D. A. (1985). Effects of temperature and photoperiod on common crupina (*Crupina vulgaris*). *Weed Sci.* **33**, 333–339.

Pawuk, W. H. (1993). Germination of Alaska-cedar seed. *Tree Plant. Notes* **44**(1), 21–24.

Pegtel, D. M. (1976). "On the Ecology of Two Varieties of *Sonchus arvensis* L." Ph.D thesis, State University of Groningen, Groningen, The Netherlands.

Pegtel, D. M. (1985). Germination in populations of *Solanum dulcamara* L. from contrasting habitats. *New Phytol.* **100**, 671–679.

Pelton, J. (1956). A study of seed dormancy in eighteen species of high altitude Colorado plants. *Butler Univ. Stud. Bot.* **13**, 74–84.

Pelton, J. (1961). An investigation of the ecology of *Mertensia ciliata* in Colorado. *Ecology* **42**, 38–52.

Pendleton, R. L., and Meyer, S. E. (1990). Seed germination biology of spineless hopsage: Inhibition by bract leachate. *In* "Proceedings of the Symposium on Cheatgrass Invasion, Shrub Die-off, and Other Aspects of Shrub Biology and Management," pp. 181–186. USDA For. Serv. Gen. Tech. Rep. INT-276.

Penfound, W. T., and Kelting, R. W. (1950). Some effects of winter burning on a moderately grazed pasture. *Ecology* **31**, 554–560.

Percy, K. (1986). The effects of simulated acid rain on germinative capacity, growth and morphology of forest tree seedlings. *New Phytol.* **104**, 473–484.

Pereira, M. F. A., and Maeda, J. A. (1986). Environmental and endogenous control of germination of *Vitis vinifera* seeds. *Seed Sci. Technol.* **14**, 227–235.

Perez-Garcia, F., Iriondo, J. M., Gonzalez-Benito, M. E., Carnes, L. F., Tapia, J., Prieto, C., Plaza, R., and Perez, C. (1995). Germination studies in endemic plant species of the Iberian Peninsula. *Israel J. Plant Sci.* **43**, 239–247.

Peterson, C. J., and Facelli, J. M. (1992). Contrasting germination and seedling growth of *Betula alleghaniensis* and *Rhus typhina* subjected to various amounts and types of plant litter. *Am. J. Bot.* **79**, 1209–1216.

Peterson, J. K. (1983). Mechanisms involved in delayed germination of *Quercus nigra* L. seeds. *Ann. Bot.* **52**, 81–92.

Peterson, R. A. (1953). Comparative effect of seed treatments upon seedling emergence in seven browse species. *Ecology* **34**, 778–785.

Pfister, R. D. (1974). *Ribes* L. Currant, gooseberry. *In* "Seeds of Woody Plants in the United States" (C. S. Schopmeyer, Tech. Coord.), pp. 720–727. USDA Forest Service Agriculture Handbook No. 450.

Piatt, J. R. (1973). Seed size affects germination of true mountainmahogany. *J. Range Manage.* **26**, 231–232.

Piatt, J. R. (1976). Effects of water stress and temperature on germination of true mountainmahogany. *J. Range Manage.* **29**, 138–140.

Pickett, F. L. (1913). The germination of seeds of *Arisaema*. *Indiana Acad. Sci. Proc.* **1913**, 125–128.

Pierce, S. M., and Moll, E. J. (1994). Germination ecology of six shrubs in fire-prone Cape fynbos. *Vegetatio* **110**, 25–41.

Pinfield, N. J., Davies, H. V., and Stobart, A. K. (1974). Embryo dormancy in seeds of *Acer platanoides*. *Physiol. Plant.* **32**, 268–272.

Pinfield, N. J., and Stobart, A. K. (1972). Hormonal regulation of germination and early seedling development in *Acer pseudoplatanus* (L.). *Planta* **104**, 134–145.

Pladeck, M. M. (1940). The testing of buffalo grass "seed," *Buchloe dactyloides* Engelm. *Agron. J.* **32**, 486–494.

Platt, W. J. (1975). The colonization and formation of equilibrium plant species associations on badger distrubances in a tall-grass prairie. *Ecol. Monogr.* **45**, 285–305.

Plummer, A. P. (1943). The germination and early seedling development of twelve range grasses. *Agron. J.* **35**, 19–34.

Plummer, J. A., Crawford, A. D., and Taylor, S. K. (1995). Germination of *Lomandra sonderi* (Dasypogonaceae) promoted by pericarp removal and chemical stimulation of the embryo. *Aust. J. Bot.* **43**, 223–230.

Pogge, F. L., and Bearce, B. C. (1989). Germinating common and cat greenbriar. *Tree Plant. Notes* **40**(1), 34–37.

Polunin, N. (1959). "Circumpolar Arctic Flora." Clarendon Press, Oxford.

Polunin, O., and Walters, M. (1985). "A Guide to the Vegetation of Britain and Europe." Oxford Univ. Press, Oxford.

Pomeroy, K. B. (1949). The germination and initial establishment of loblolly pine under various surface soil conditions. *J. For.* **47**, 541–543.

Pons, T. L. (1989a). Dormancy and germination of *Calluna vulgaris* (L.) Hull and *Erica tetralix* L. seeds. *Acta Oecol.* **10**, 35–43.

Pons, T. L. (1989b). Dormancy, germination and mortality of seeds in heathland and inland sand dunes. *Acta Bot. Neerl.* **38**, 327–335.

Popay, A. I. (1975). Laboratory germination of barley grass. *Proc. New Zeal. Weed Pest Cont. Conf.* **28**, 7–11.

Popay, A. I., and Roberts, E. H. (1970). Factors involved in the dormancy and germination of *Capsella bursa-pastoris* (L.) Medik., and *Senecio vulgaris* L. *J. Ecol.* **58**, 103–122.

Povilaitis, B. (1956). Dormancy studies with seeds of various weed species. *Proc. Int. Seed Test. Assoc.* **21**, 87–111.

Pritchard, H. W. (1991). Water potential and embryonic axis viability in recalcitrant seeds of *Quercus rubra*. *Ann. Bot.* **67**, 43–49.

Pritchard, H. W., Tompsett, P. B., and Manger, K. R. (1996). Development of a thermal time model for the quantification of dormancy loss in *Aesculus hippocastanum* seeds. *Seed Sci. Res.* **6**, 127–135.

Probert, R. J., Dickie, J. B., and Hart, M. R. (1989). Analysis of the effect of cold stratification on the germination response to light and alternating temperatures using selected seed populations of *Ranunculus sceleratus* L. *J. Exp. Bot.* **40**, 293–301.

Pryor, L. D. (1954). Improved germination of some alpine eucalypts by stratification. *Aust. For.* **18**, 104–106.

Qadir, S. A., and Lodhi, N. (1971). Germination behaviour of seeds of some common shrubs. *J. Sci. Univ. Karachi* **1**, 84–92.

Quezel, P. (1981). Floristic composition and phytosociological structure of sclerophyllous matorral around the Mediterranean. *In* "Ecosystems of the World" (F. di Castri, D. W. Goodall, and R. L. Specht, eds.), Vol. 11, pp. 107–121. Elsevier, Amsterdam.

Quick, C. R. (1935). Notes on the germination of *Ceanothus* seeds. *Madrono* **3**, 135–140.

Quick, C. R. (1947). Germination of *Phacelia* seeds. *Madrono* **9**, 17–20.

Quick, C. R. (1959). *Ceanothus* seeds and seedlings on burns. *Madrono* **15**, 79–81.

Quinlivan, B. J., and Nicol, H. I. (1971). Embryo dormancy in subterranean clover seeds. I. Environmental control. *Aust. J. Agric. Res.* **22**, 599–606.

Radwan, M. A. (1976). Germination of cascara seed. *Tree Plant. Notes* **27**(2), 20–23.

Raison, R. J. (1979). Modification of the soil environment by vegetation fires, with particular reference to nitrogen transformations: A review. *Plant Soil* **51**, 73–108.

Rajput, P., and Sen, D. N. (1990). Effect of some salts on seed germination of *Atriplex argentina*. *In* "International Symposium on Environmental Influences on Seed and Germination Mechanism: Recent Advances in Research and Technology" (D. N. Sen, S. Mohammed, P. K. Kasera, and T. P. Thomas, eds.), pp. 99–100. Univ. Jodhpur, India. [Abstract]

Ralowicz, A. E., and Mancino, C. F. (1992). Afterripening in curly mesquite seeds. *J. Range Manage.* **45**, 85–87.

Ralowicz, A., Mancino, C., and Kopec, D. (1992). Technical note: Chemical enhancement of germination in curly mesquite seed. *J. Range Manage.* **45**, 507–508.

Rao, P. B. (1988). Effects of environmental factors on germination and seedling growth in *Quercus floribunda* and *Cupressus torulosa*, tree species of central Himalaya. *Ann. Bot.* **61**, 531–540.

Rao, P. B., and Singh, S. P. (1985). Response breadths on environmental gradients of germination and seedling growth in two dominant forest tree species of central Himalaya. *Ann. Bot.* **45**, 783–794.

Ratcliffe, D. (1961). Adaptation to habitat in a group of annual plants. *J. Ecol.* **49**, 187–203.

Raynal, D. J., Roman, J. R., and Eichenlaub, W. M. (1982). Response of tree seedlings to acid precipitation I. Effect of substrate acidity on seed germination. *Environ. Exp. Bot.* **22**, 377–383.

Razumova, M. V. (1985). The biology of seed germination in the species of the genus *Sorbus* (Rosaceae). *Bot. Zhur.* **72**, 77–83.

Read, J. (1989). Phenology and germination in some rainforest canopy species at Mt. Field National Park, Tasmania. *Papers Proc. Roy. Soc. Tasmania* **123**, 211–221.

Reader, R. J. (1993). Control of seedling emergence by ground cover and seed predation in relation to seed size for some old-field species. *J. Ecol.* **81**, 169–175.

Reddy, K. N., and Singh, M. (1992). Germination and emergence of hairy beggarticks (*Bidens pilosa*). *Weed Sci.* **40**, 195–199.

Ren, Z., and Abbott, R. J. (1991). Seed dormancy in Mediterranean *Senecio vulgaris* L. *New Phytol.* **117**, 673–678.

Reynolds, D. N. (1984). Alpine annual plants: Phenology, germination, photosynthesis, and growth of three Rocky Mountain species. *Ecology* **65**, 759–766.

Rice, B., and Westoby, M. (1981). Myrmecochory in sclerophyll vegetation of the West Head, New South Wales. *Aust. J. Ecol.* **6**, 291–298.

Rice, B., and Westoby, M. (1986). Evidence against the hypothesis that ant-dispersed seeds reach nutrient-enriched microsites. *Ecology* **67**, 1270–1274.

Richardson, S. D. (1959). Germination of Douglas-fir seed as affected by light, temperature, and gibberellic acid. *For. Sci.* **5**, 174–181.

Ries, R. E., and Hofmann, L. (1983). Effect of sodium and magnesium sulfate on forage seed germination. *J. Range Manage.* **36**, 658–662.

Ripley, E. A. (1992). Grassland climate. *In* "Ecosystems of the World" (R. T. Coupland, ed.), Vol. 8A, pp. 7–24. Elsevier, Amsterdam.

Ritchie, J. C. (1955). Biological flora of the British Isles. *Vaccinium vitis-idaea* L. *J. Ecol.* **43**, 701–708.

Ritchie, J. C. (1956). Biological flora of the British Isles. *Vaccinium myrtillus* L. *J. Ecol.* **44**, 291–299.

Roberts, D. R., and Carpenter, S. B. (1983). The influence of seed scarification and site preparation on establishment of black locust on surface-mined sites. *Tree Plant. Notes* **34**(3), 28–30.

Roberts, H. A., and Boddrell, J. E. (1983). Field emergence and temperature requirements for germination in *Solanum sarrachoides* Sendt. *Weed Res.* **23**, 247–252.

Roberts, H. A., and Boddrell, J. E. (1985). Temperature requirements for germination of buried seeds of *Aethusa cynapium* L. *Weed Res.* **25**, 267–274.

Roberts, H. A., and Chancellor, R. J. (1979). Periodicity of seedling emergence and achene survival in some species of *Carduus, Cirsium* and *Onopordum*. *J. Appl. Ecol.* **16**, 641–647.

Roberts, H. A., and Feast, P. M. (1970). Seasonal distribution of emergence in some annual weeds. *Exp. Hort.* **21**, 36–41.

Roberts, H. A., and Lockett, P. M. (1977). Temperature requirements for germination of dry-stored, cold-stored and buried seeds of *Solanum dulcamara* L. *New Phytol.* **79**, 505–510.

Roberts, H. A., and Lockett, P. M. (1978a). Seed dormancy and field emergence in *Solanum nigrum* L. *Weed Res.* **18**, 231–241.

Roberts, H. A., and Lockett, P. M. (1978b). Seed dormancy and periodicity of seedling emergence in *Veronica hederifolia* L. *Weed Res.* **18**, 41–48.

Roberts, H. A., and Neilson, J. E. (1982). Seasonal changes in the temperature requirements for germination of buried seeds of *Aphanes arvensis* L. *New Phytol.* **92**, 159–166.

Robocker, W. C. (1977). Germination of seeds of common yarrow (*Achillea millefolium*) and its herbicidal control. *Weed Sci.* **25**, 456–459.

Robocker, W. C., Curtis, J. T., and Ahlgren, H. L. (1953). Some factors affecting emergence and establishment of native grass seedlings in Wisconsin. *Ecology* **34**, 194–199.

Robocker, W. C., Kerr, H. D., and Burns, V. F. (1964). Characteristics and control of Swainsonpea. *Weeds* **12**, 189–191.

Robocker, W. C., Williams, M. C., Evans, R. A., and Torell, P. J. (1969). Effects of age, burial, and region on germination viability of halogeton seed. *Weed Sci.* **17**, 63–65.

Rochow, T. F. (1970). Ecological investigations of *Thlaspi alpestre* L. along an elevational gradient in the central Rocky Mountains. *Ecology* **51**, 649–656.

Rockwood, L. L., and Blois, M. C. (1986). Effects of elaiosome removal on germination of two ant-dispersed plants. *Am. J. Bot.* **73**, 675. [Abstract]

Rogler, G. A. (1960). Relation of seed dormancy of green needlegrass (*Stipa viridula* Trin.) to age and treatment. *Agron. J.* **52**, 467–469.

Rohring, E. (1991a). Introduction. *In* "Ecosystems of the World" (E. Rohrig and B. Ulrich, eds.), Vol. 7, pp. 1–5. Elsevier, Amsterdam.

Rohrig, E. (1991b). Deciduous forests of the Near East. *In* "Ecosystems of the World" (E. Rohrig and B. Ulrich, eds.), Vol. 7, pp. 527–537. Elsevier, Amsterdam.

Rohrig, E. (1991c). Floral composition and its evolutionary development. *In* "Ecosystems of the World" (E. Rohrig and B. Ulrich, eds.), Vol. 7, pp. 17–23. Elsevier, Amsterdam.

Rohrig, E. (1991d). Seasonality. *In* "Ecosystems of the World" (E.

Rohrig and B. Ulrich, eds.), Vol. 7, pp. 25–33. Elsevier, Amsterdam.

Rokich, D. P., and Bell, D. T. (1995). Light quality and intensity effects on the germination of species from the jarrah (*Eucalyptus marginata*) forest of Western Australia. *Aust. J. Bot.* **43**, 169–179.

Romo, J. T., Grilz, P. L., Bubar, C. J., and Young, J. A. (1991). Influences of temperature and water stress on germination of plains rough fescue. *J. Range Manage.* **44**, 75–81.

Romo, J. T., and Haferkamp, M. R. (1987). Effects of osmotic potential, potassium chloride, and sodium chloride on germination of greasewood (*Sarcobatus vermiculatus*). *Great Basin Nat.* **47**, 110–116.

Roundy, B. A., Young, J. A., and Evans, R. A. (1985). Germination of basin wildrye and tall wheatgrass in relation to osmotic and matric potential. *Agron. J.* **77**, 129–135.

Rowe, J. S., and Scotter, G. W. (1973). Fire in the boreal forest. *Quat. Res.* **3**, 444–464.

Roy, D. F. (1974a). *Arbutus menziesii* Pursh. Pacific madrone. *In* "Seeds of Woody Plants in the United States" (C. S. Schopmeyer, Tech. Coord.), pp. 226–227. USDA Forest Service Agriculture Handbook No. 450.

Roy, D. F. (1974b). *Lithocarpus densiflorus* (Hook. & Arn.) Rehd. Tanoak. *In* "Seeds of Woody Plants in the United States" (C. S. Schopmeyer, Tech. Coord.), pp. 512–514. USDA Forest Service Agriculture Handbook No. 450.

Roy, J., and Arianoutsou-Faraggitaki, M. (1985). Light quality as the environmental trigger for the germination of the fire-promoted species *Sarcopoterium spinosum* L. *Flora* **177**, 345–349.

Rozijn, N. A. M. G., and van Andel, J. (1985). Analysis of the germination syndrome of dune annuals. *Flora* **177**, 175–185.

Rudolf, P. O. 1936. Note on seed germination of European mountain ash. *J. For.* **34**, 533–534.

Rudolf, P. O. (1974a). *Aesculus* L. Buckeye, horsechestnut. *In* "Seeds of Woody Plants in the United States" (C. S. Schopmeyer, Tech. Coord.), pp. 195–199. USDA Forest Service Agriculture Handbook No. 450.

Rudolf, P. O. (1974b). *Berberis* L. Barberry, mahonia. *In* "Seeds of Woody Plants in the United States" (C. S. Schopmeyer, Tech. Coord.), pp. 247–251. USDA Forest Service Agriculture Handbook No. 450.

Rudolf, P. O. (1974c). *Clematis* L. Clematis. *In* "Seeds of Woody Plants in the United States" (C. S. Schopmeyer, Tech. Coord.), pp. 331–334. USDA Forest Service Agriculture Handbook No. 450.

Rudolf, P. O. (1974d). *Cotinus* Mill. Smoketree. *In* "Seeds of Woody Plants in the United States" (C. S. Schopmeyer, Tech. Coord.), pp. 346–348. USDA Forest Service Agriculture Handbook No. 450.

Rudolf, P. O. (1974e). *Kalopanax pictus* (Thunb.) Nakai. Kalopanax. *In* "Seeds of Woody Plants in the United States" (C. S. Schopmeyer, Tech. Coord.), pp. 472–473. USDA Forest Service Agriculture Handbook No. 450.

Rudolf, P. O. (1974f). *Koelreuteria paniculata* Laxm. Panicled golden raintree. *In* "Seeds of Woody Plants in the United States" (C. S. Schopmeyer, Tech. Coord.), pp. 474–475. USDA Forest Service Agriculture Handbook No. 450.

Rudolf, P. O. (1974g). *Larix* Mill. Larch. Rudolf, P. O. *In* "Seeds of Woody Plants in the United States" (C. S. Schopmeyer, Tech. Coord.), pp. 478–485. USDA Forest Service Agriculture Handbook No. 450.

Rudolf, P. O. (1974h). *Taxus* L. Yew. *In* "Seeds of Woody Plants in the United States" (C. S. Schopmeyer, Tech. Coord.), pp. 799–802. USDA Forest Service Agriculture Handbook No. 450.

Rudolf, P. O. (1974i). *Euonymus* L. Euonymus. *In* "Seeds of Woody Plants in the United States" (C. S. Schopmeyer, Tech. Coord.), pp. 393–397. USDA Forest Service Agriculture Handbook No. 450.

Rudolf, P. O. (1974j). *Berberis* L. Barberry, mahonia. *In* "Seeds of Woody Plants in the United States" (C. S. Schopmeyer, Tech. Coord.), pp. 247–251. USDA Forest Service Agriculture Handbook No. 450.

Rudolf, P. O., and Leak, W. B. (1974). *Fagus* L. Beech. *In* "Seeds of Woody Plants in the United States" (C. S. Schopmeyer, Tech. Coord.), pp. 401–405. USDA Forest Service Agriculture Handbook No. 450.

Rudolf, P. O., and Phipps, H. (1974). *Carpinus* L. Hornbeam. *In* "Seeds of Woody Plants in the United States" (C. S. Schopmeyer, Tech. Coord.), pp. 266–268. USDA Forest Service Agriculture Handbook No. 450.

Rundel, P. W. (1981). The matorral zone of central Chile. *In* "Ecosystems of the World" (F. di Castri, D. W. Goodall, and R. L. Specht, eds.), Vol. 11, pp. 175–201. Elsevier, Amsterdam

Ruth, R. H. (1974). *Tsuga* (Endl.) Carr. Hemlock. *In* "Seeds of Woody Plants in the United States" (C. S. Schopmeyer, Tech. Coord.), pp. 819–827. USDA Forest Service Agriculture Handbook No. 450.

Ryan, J., Miyamoto, S., and Stroehlein, J. L. (1975). Salt and specific ion effects on germination of four grasses. *J. Range Manage.* **28**, 61–64.

Sabo, D. G., Johnson, G. V., Martin, W. C., and Aldon, E. F. (1979). Germination requirements of 19 species of arid land plants. *USDA For. Serv. Res. Pap.* RM-210.

Sadhu, R. N., and Kaul, V. (1989). Seed-coat dormancy in *Robinia pseudo-acacia*. *Indian For.* **115**, 483–487.

Safford, L. O. (1974). *Picea* A. Dietr. Spruce. *In* "Seeds of Woody Plants in the United States" (C. S. Schopmeyer, Tech. Coord.), pp. 587–597. USDA Forest Service Agriculture Handbook No. 450.

Sah, V. K., and Singh, V. (1995). Effect of temperature and storage on seed germination in *Populus ciliata* Wall. ex Royle in Garhwal Himalaya. *Indian For.* **121**, 273–275.

Salac, S. S., and Hesse, M. C. (1975). Effects of storage and germination conditions on the germination of four species of wild flowers. *J. Am. Soc. Hort. Sci.* **100**, 359–361.

Satoo, T. (1983). Temperate broad-leaved evergreen forests of Japan. *In* "Ecosystems of the World" (J. D. Ovington, ed.). Vol. 10, pp. 169–189. Elsevier, Amsterdam.

Sautter, E. H. (1962). Germination of switchgrass. *J. Range Manage.* **15**, 108–110.

Sawhney, R., Hsiao, A. I. and Quick, W. A. (1986). The influence of diffused light and temperature on seed germination of three genetically nondormant lines of wild oats (*Avena fatua*) and its adaptive significance. *Can. J. Bot.* **64**, 1910–1915.

Sayers, R. L., and Ward, R. T. (1966). Germination responses in alpine species. *Bot. Gaz.* **127**, 11–16.

Schatral, A. (1995). The structure of the seed in some Western Australian species of the genus *Hibbertia* (Dilleniaceae). *Bot. J. Linn. Soc.* **119**, 257–263.

Schatral, A. (1996). Dormancy in seeds of *Hibbertia hypericoides* (Dilleniaceae). *Aust. J. Bot.* **44**, 213–222.

Scherbatskoy, T., Klein, R. M., and Badger, G. J. (1987). Germination responses of forest tree seed to acidity and metal ions. *Environ. Exp. Bot.* **27**, 157–164.

Schlising, R. A. (1969). Seedling morphology in marah (Cucurbitaceae) related to the California Mediterranean climate. *Am. J. Bot.* **56**, 552–561.

Schmaltz, J. (1991). Deciduous forests of southern South America. *In* "Ecosystems of the World" (E. Rohrig and B. Ulrich, eds.), Vol. 7, pp. 557–578. Elsevier, Amsterdam.

Schmutz, E. M., Smith, E. L., Ogden, P. R., Cox, M. L., Klemmedson, J. O., Norris, J. J., and Fierro, L. C. (1992). Desert grassland. *In* "Ecosystems of the World" (R. T. Coupland, ed.). Vol. 8A, pp. 337–362. Elsevier, Amsterdam.

Schopmeyer, C. S. (Tech. Coord.) (1974a). "Seeds of Woody Plants in the United States." USDA Forest Service Agriculture Handbook No. 450.

Schopmeyer, C. S. (1974b). *Alnus* B. Ehrh. Alder. *In* "Seeds of Woody Plants in the United States" (C. S. Schopmeyer, Tech. Coord.), pp. 206–211. USDA Forest Service Agriculture Handbook No. 450.

Schopmeyer, C. S., and Leak, W. B. (1974). *Ostrya virginiana* (Mill.) K. Koch. Eastern hophornbeam. *In* "Seeds of Woody Plants in the United States" (C. S. Schopmeyer, Tech. Coord.), pp. 564–565. USDA Forest Service Agriculture Handbook No. 450.

Schreiner, E. J. (1974). *Populus* L. Poplar. *In* "Seeds of Woody Plants in the United States." (C. S. Schopmeyer, Tech. Coord.), pp. 645–655. USDA Forest Service Agriculture Handbook No. 450.

Schroeder, E. M. (1935). Dormancy in seeds of *Benzoin aestivale* L. *Contrib. Boyce Thomp. Inst.* **7,** 411–419

Schroeder, E. M. (1937). Germination of fruits of *Ptelea* species. *Contrib. Boyce Thomp. Inst.* **8,** 355–359.

Schroeder, E. M. and Barton, L. V. (1939). Germination and growth of some rock garden plants. *Contrib. Boyce Thomp. Inst.* **10,** 235–255.

Schutz, M., and Urbanska, K. M. (1984). Germinating behaviour and growth potential in *Taraxacum alpinum* (2n = 32) from the Swiss Alps. *Ber. Geobot. Inst. ETH Stiftung Rubel* **51,** 118–131.

Schutz, W. (1995). Keimungsokologie von 5 horstbildendenden *Carex*-arten nasser standorte. *Verhandl. Gesell. Okol.* **24,** 155–160.

Scifres, C. J., and McCarty, M. K. (1969). Some factors affecting germination and seedling growth of Scotch thistle. *Univ. Nebraska Coll. Agri. Res. Bull.* 228.

Scott, D. (1975). Some germination requirements of *Celmisia* species. *New Zeal. J. Bot.* **13,** 653–664.

Scott, D. H., and Draper, A. D. (1967). Light in relation to seed germination of blueberries, strawberries and *Rubus. HortScience* **2,** 107–108.

Scott, D. H., and Ink, D. P. (1957). Treatment of *Rubus* seeds prior to after-ripening to improve germination. *J. Am. Soc. Hort. Sci.* **69,** 261–267.

Scowcroft, P. G. (1988). Germinability of Cook pine (*Araucaria columnaris*) seeds under different storage conditions. *Tree Plant. Notes* **39**(3), 17–25.

Scurfield, G. (1954). Biological flora of the British Isles. *Deschampsia flexuosa* (L.) Trin. *J. Ecol.* **42,** 225–233.

Segelquist, C. A. (1971). Moistening and heating improve germination of two legume species. *J. Range Manage.* **24,** 393–394.

Seiwa, K. (1997). Variable regeneration behaviour of *Ulmus davidiana* var. *japonica* in response to disturbance regime for risk spreading. *Seed Sci. Res.* **7,** 195–207.

Selleck, G. W., and Coupland, R. T. (1954). Effect of temperature on germination of seeds of leafy spurge. Canada Weed Committee, Western Section, Research Report, p. 96.

Sernander, R. (1906). "Entwurf einer monographie der Europaischen myrmekochoren." Almqvist & Wiksells Boktryckeri-A.-B., Uppsala.

Shafiq, Y. (1979). Some effects of light and temperature on the germination of *Pinus brutia, Nothofagus obliqua* and *Nothofagus procera* seeds. *Seed Sci. Technol.* **7,** 189–193.

Shafroth, P. B., Auble, G. T., and Scott, M. L. (1995). Germination and establishment of the native plains cottonwood (*Populus deltoides* Marshall subsp. *monilifera*) and the exotic Russian-olive (*Elaeagnus angustifolia* L.). *Conserv. Biol.* **9,** 1169–1175.

Shafroth, P. B., Friedman, J. M., and Ischinger, L. S. (1995). Effects of salinity on establishment of *Populus fremontii* (cottonwood) and *Tamarix ramosissima* (saltcedar) in southwestern United States. *Great Basin Nat.* **55,** 56–65.

Shaidaee, G., Dahl, B. E., and Hansen, R. M. (1969). Germination

and emergence of different age seeds of six grasses. *J. Range Manage.* **22,** 240–243.

Shaltout, K. H., and El-Shourbagy, M. N. (1989). Germination requirements and seedling growth of *Thymelaea hirsuta* (L.) Endl. *Flora* **183,** 429–436.

Sharrow, S. H., and Wright, H. A. (1977). Effects of fire, ash, and litter on soil nitrate, temperature, moisture and tobosagrass production in the Rolling Plains. *J. Range Manage.* **30,** 266–270.

Shaukat, S. S. (1974). The effects of simazine, atrazine and 2,4-D on germination and early seedling growth of *Oryza sativa. Pakistan J. Bot.* **6,** 141–149.

Shaw, D. R., Mack, R. E., and Smith, C. A. (1991). Redvine (*Brunnichia ovata*) germination and emergence. *Weed Sci.* **39,** 33–36.

Shaw, N. (1984). Producing bareroot seedlings of native shrubs. *In* "The Challenge of Producing Native Plants for the Intermountain Area" (P. M. Murphy, compiler), pp. 6–15. USDA For. Serv. Gen. Tech. Rep. INT-168.

Shaw, N. L., Haferkamp, M. R., and Hurd, E. G. (1994). Germination and seedling establishment of spiny hopsage in response to planting date and seedbed environment. *J. Range Manage.* **47,** 165–174.

Shirreffs, D. A. (1985). Biological flora of the British Isles. *Anemone nemorosa* L. *J. Ecol.* **73,** 1005–1020.

Shunk, I. V. (1939). Oxygen requirements for germination of seeds of *Nyssa aquatica*-tupelo gum. *Science* **90,** 565–566.

Siddiqi, M. Y., Myerscough, P. J., and Carolin, R. C. (1976). Studies in the ecology of coastal heath in New South Wales. *Aust. J. Ecol.* **1,** 175–183.

Simpson, G. M. (1990). "Seed Dormancy in Grasses." Cambridge Univ. Press, Cambridge, UK.

Simpson, M. J. A. (1976). Seeds, seed ripening, germination and viability in some species of *Hebe. Proc. New Zeal. Ecol. Soc.* **23,** 99–108.

Simpson, M. J. A., and Webb, C. J. (1980). Germination in some New Zealand species of *Gentiana*: A preliminary report. *New Zeal. J. Bot.* **18,** 495–501.

Sims, H. P. (1976). The effect of prescribed burning on some physical soil properties of jack pine sites in southeastern Manitoba. *Can. J. For. Res.* **6,** 58–68.

Singh, K. P. (1969). Seed dormancy and its control by germination inhibitor in *Anagallis arvensis* L. var. *caerulae* Gren et Godr. *Proc. Natl. Inst. Sci. India B* **35,** 161–171.

Singh, M., and Achhireddy, N. R. (1984). Germination ecology of milkweedvine (*Morrenia odorata*). *Weed Sci.* **32,** 781–785.

Singh, O., and Singh, V. (1990). Germination and growth of spruce and silver fir in relation to covering media. *Indian For.* **116,** 278–282.

Singh, V., Bhagat, S., and Singh, O. B. (1990). Effect of seed weight on germination and initial seedling growth in spruce (*Picea smithiana* Wall. Boiss). *Indian For.* **116,** 403–406.

Singh, V., and Singh, R. V. (1984). Seed dispersal, seed germination and seedling establishment in natural forests of silver fir and spruce. II. Seed germination and seedling establishment. *Indian For.* **110,** 632–639.

Slade, E. A. and Causton, D. R. (1979). The germination of some woodland herbaceous species under laboratory conditions: A multifactorial study. *New Phytol.* **83,** 549–557.

Slingsby, P., and Bond, W. J. (1985). The influence of ants on the dispersal distance and seedling recruitment of *Leucospermum conocarpodendron* (L.) Buek (Proteaceae). *S. Afr. J. Bot.* **51,** 30–34.

Small, J. G. C., and Garner, C. J. (1980). Gibberellin and stratification required for the germination of *Erica junonia*, an endangered species. *Z. Pflazenphysiol.* **99,** 179–182.

Smeins, F. E., Diamond, D. D., and Hanselka, C. W. (1992). Coastal prairie. *In* "Ecosystems of the World" (R. T. Coupland, ed.), Vol. 8A, pp. 269–290. Elsevier, Amsterdam.

Smith, D. F. (1968). The growth of barley grass (*Hordeum leporinum*) in annual pasture. 1. Germination and establishment in comparison with other annual pasture species. *Aust. J. Exp. Agric. Anim. Husb.* **8,** 478–483.

Smith, D. W. (1970). Concentrations of soil nutrients before and after fire. *Can. J. Soil Sci.* **50,** 17–29.

Smith, G. S. (1978). Seed scarification to speed germination of ornamental cycads (*Zamia* spp.). *HortScience* **13,** 436–438.

Smith, M., and Capelle, J. (1992). Effects of soil surface microtopography and litter cover on germination, growth and biomass production of chicory (*Cichorium intybus* L.). *Am. Midl. Nat.* **128,** 246–253.

Smith, M. A., Wright, H. A., and Schuster, J. L. (1975). Reproductive characteristics of redberry juniper. *J. Range Manage.* **28,** 126–128.

Smoliak, S., and Johnston, A. (1968). Germination and early growth of grasses at four root-zone temperatures. *Can. J. Plant Sci.* **48,** 119–127.

Smreciu, E. A., Currah, R. S., and Toop, E. (1988). Viability and germination of herbaceous perennial species native to southern Alberta grasslands. *Can. Field-Nat.* **102,** 31–38.

Solano, F., Schrader, J. W., and Coble, H. D. (1976). Germination, growth, and development of spurred anoda. *Weed Sci.* **24,** 574–578.

Sonia, L., and Heslehurst, M. R. (1978). Germination characteristics of some *Banksia* species. *Aust. J. Ecol.* **3,** 179–186.

Sopp, D. F., Salac, S. S., and Sutton, R. K. (1977). Germination of Gambel oak seed. *Tree Plant. Notes* **28**(2), 4–5.

Sorensen, F. C., and Campbell, R. K. (1981). Germination rate of Douglas-fir [*Pseudotsuga menziesii* (Mirb.) Franco] seeds affected by their orientation. *Ann. Bot.* **47,** 467–471.

Sorensen, J. T., and Holden, D. J. (1974). Germination of native prairie forb seeds. *J. Range Manage.* **27,** 123–126.

Soriano, A. (1983). Deserts and semi-deserts of Patagonia. *In* "Ecosystems of the World" (N. E. West, ed.), Vol. 5, pp. 423–460. Elsevier, Amsterdam.

Soriano, A. (1992). Rio de la Plata grasslands. *In* "Ecosystems of the World" (R. T. Coupland ed.), Vol. 8A, pp. 367–407. Elsevier, Amsterdam.

Soteres, J. K., and Murray, D. S. (1981). Germination and development of honeyvine milkweed (*Ampelamus albidus*) seed. *Weed Sci.* **29,** 625–628.

Sowter, F. A. (1949). Biological flora of the British Isles. *Arum maculatum* L. *J. Ecol.* **37,** 207–219.

Spaeth, J. N. (1932). Hastening germination of basswood seeds. *J. For.* **30,** 925–928.

Specht, R. L. (1981). Mallee ecosystems in southern Australia. *In* "Ecosystems of the World" (F. di Castri, D. W. Goodall, and R. L. Specht, eds.), Vol. 11, pp. 203–230. Elsevier, Amsterdam.

Specht, R. L. (ed.). (1988). "Mediterranean-Type Ecosystems." Kluwer, Dordrecht.

Specht, R. L., Rayson, P., and Jackman, M. E. (1958). Dark Island Heath (Ninety-mile Plain, South Australia). VI. Pyric succession: Changes in composition, coverage, dry weight, and mineral nutrient status. *Aust. J. Bot.* **6,** 59–88 + 4 plates.

Spessard, L. L. (1988). Seed-germination studies of *Psoralea esculenta* Pursh (indian turnip) and *Psoralea argophylla* Pursh (silver scurfpea). *Trans. Nebraska Acad. Sci.* **16,** 123–126.

Spooner, D. M., Cusick, A. W., Hall, G. F., and Baskin, J. M. (1985). Observations on the distribution and ecology of *Sida hermaphrodita* (L.) Rusby (Malvaceae). *Sida* **11,** 215–225.

Sprackling, J. A. (1976). Germination of *Pinus gerardiana* seeds following storage and stratification. *Tree Plant. Notes* **27**(3), 5–6, 22.

Sprague, V. G. (1940). Germination of freshly harvested seeds of several *Poa* species and of *Dactylis glomerata*. *J. Am. Soc. Agron.* **32,** 715–721.

Springfield, H. W. (1968). Germination of winterfat seeds under different moisture stresses and temperatures. *J. Range Manage.* **21,** 314–316.

Springfield, H. W. (1969). Temperatures for germination of fourwing saltbush. *J. Range Manage.* **22,** 49–50.

Springfield, H. W. (1970). Germination and establishment of fourwing saltbush in the Southwest. *USDA Forest Serv. Res. Pap.* RM-55.

Springfield, H. W. (1972a). Winterfat seeds undergo after-ripening. *J. Range Manage.* **25,** 479–480.

Springfield, H. W. (1972b). Optimum temperatures for germination of winterfat. *J. Range Manage.* **25,** 69–70.

Squires, V. R. (1969). Ecological factors contributing to the success of *Tribulus terrestris* L. as a weed in a winter rainfall environment in southern Australia. *Proc. Ecol. Soc. Aust.* **4,** 55–66.

Stamp, N. E. (1984). Self-burial behaviour of *Erodium cicutarium* seeds. *J. Ecol.* **72,** 611–620.

Standifer, L. C., and Wilson, P. W. (1988). A high temperature requirement for after ripening of imbibed dormant *Poa annua* L. seeds. *Weed Res.* **28,** 365–371.

Stark, N. (1968). Seed ecology of *Sequoiadendron giganteum*. *Madrono* **19,** 267–277.

Stark, N. (1979). Plant ash as a natural fertilizer. *Environ. Exp. Bot.* **19,** 59–68.

Stearns, F. (1955). The influence of light and temperature on germination and flowering of five species of *Plantago*. *N. Central Weed Cont. Conf.* **12,** 8.

Steele, R., and Geier-Hayes, K. (1989). The Douglas-fir/ninebark habitat type in central Idaho: Succession and management. *USDA For. Serv. Gen. Tech. Rep.* INT-252.

Steinbauer, G. P. (1937). Dormancy and germination of *Fraxinus* seeds. *Plant Physiol.* **12,** 813–824.

Steinbauer, G. P., and Grigsby, B. (1956). Some correlations between germination and dormancy of weed seeds in the laboratory and in the field. *Proc. N. Cent. Weed Cont. Conf.* **13,** 35–36.

Steinbauer, G. P., and Grigsby, B. (1957). Dormancy and germination characteristics of the seeds of four species of *Plantago*. *Proc. Assoc. Offic. Seed Anal. N. Am.* **47,** 158–164.

Steinbauer, G. P., and Grigsby, B. (1960). Dormancy and germination of the docks (*Rumex* spp.). *Proc. Assoc. Offic. Seed Anal.* **50,** 112–117.

Stergios, B. G. (1976). Achene production, dispersal, seed germination, and seedling establishment of *Hieracium aurantiacum* in an abandoned field community. *Can. J. Bot.* **54,** 1189–1197.

Steshenko, A. P. (1966). Biology of newly collected seeds of the high mountain zone plants of the Pamirs. *Bot. Zhur.* **48,** 965–978. [In Russian with English summary]

Stidham, N. D., Ahring, R. M., Powell, J., and Claypool, P. L. (1980). Chemical scarification, moist prechilling, and thiourea effects on germination of 18 shrub species. *J. Range Manage.* **33,** 115–118.

Stinson, K. J., and Wright, H. A. (1969). Temperatures of headfires in the southern mixed prairie of Texas. *J. Range Manage.* **22,** 169–174.

Stock, W. D., and Lewis, O. A. M. (1986). Soil nitrogen and the role of fire as a mineralizing agent in a South African coastal fynbos ecosystem. *J. Ecol.* **74,** 317–328.

Stocking, S. K. (1966). Influence of fire and sodium-calcium borate on chaparral vegetation. *Madrono* **18,** 193–224.

Stocklin, J., and Baumler, E. (1996). Seed rain, seedling establishment and clonal growth strategies on a glacier foreland. *J. Veg. Sci.* **7,** 45–56.

Stokes, P. (1952). A physiological study of embryo development in *Heracleum sphondylium* L. I. The effect of temperature on embryo development. *Ann. Bot.* **16,** 441–447.

Stoltz, L. P., and Snyder, J. C. (1985). Embryo growth and germination of American ginseng seed in response to stratification temperatures. *HortScience* **20,** 261–262.

Stone, E. C., and Juhren, G. (1951). The effect of fire on the germination of the seed of *Rhus ovata* Wats. *Am. J. Bot.* **38**, 368–372.

Stone, E. C. (1957a). Embryo dormancy and embryo vigor of sugar pine as affected by length of storage and storage temperatures. *For. Sci.* **3**, 357–371.

Stone, E. C. (1957b). Embryo dormancy of *Pinus jeffreyi* Murr. seed as affected by temperature, water uptake, stratification, and seed coat. *Plant Physiol.* **32**, 93–99.

Stonehouse, B. (1989). "Polar Ecology." Blackie, Glasgow/London.

Stritzke, J. F. (1975). Germination characteristics and chemical control of horehound. *J. Range Manage.* **28**, 225–226.

Stubbendieck, J., and McCully, W. G. (1976). Effect of temperature and photoperiod on germination and survival of sand bluestem. *J. Range Manage.* **29**, 206–208.

Sukhvibul, N., and Considine, J. A. (1994). Regulation of germination of seed of *Anigozanthos manglesii. Aust. J. Bot.* **42**, 191–203.

Sumner, D. C., and Cobb, R. D. (1967). Germination characteristics of cheeseweed (*Malva parviflora* L.) seeds. *Agron. J.* **59**, 207–208.

Sumrall, L. B., Roundy, B. A., Cox, J. R., and Winkel, V. K. (1991). Influence of canopy removal by burning or clipping on emergence of *Eragrostis lehmanniana* seedlings. *Int. J. Wildl. Fire* **1**, 35–40.

Suszka, B. (1967). Studia nad spoczynkiem i kielkowaniem nasion roznych gatunkow z rodzaju *Prunus* L. *Arb. Korn.* **12**, 221–282.

Suszka, B. (1968). Conditions for the breaking of dormancy and germination of hornbeam (*Carpinus betulus* L.) seeds. *Arb. Korn.* **13**, 167–190.

Svedarsky, D., and Kucera, C. L. (1970). Effects of gibberellic acid and post-harvest age on germination of prairie grasses. *Iowa St. J. Sci.* **44**, 513–518.

Svejda, F. (1972). Water uptake of rose achenes. *Can. J. Plant Sci.* **52**, 1043–1047.

Svejda, F. J., and Poapst, P. A. (1972). Effects of different afterripening treatments on germination and endogenous growth inhibitors in *Rosa rugosa. Can. J. Plant Sci.* **52**, 1049–1058.

Sveshnikova, M. (1948). On germination capacity of plants of the higher mountain deserts of Pamir. *Dok. Akad. Nauk SSSR* **61**, 925–927. [In Russian]

Sweeney, J. R. (1956). "Responses of Vegetation to Fire." Univ. California Press, Berkeley.

Sydes, C., and Grime, J. P. (1981). Effects of tree leaf litter on herbaceous vegetation in deciduous woodland. I. Field investigations. *J. Ecol.* **69**, 237–248.

Tainton, N. M., and Walker, B. H. (1993). Grasslands of southern Africa. *In* "Ecosystems of the World" (R. T. Coupland, ed.). Vol. 8B, pp. 265–290. Elsevier, Amsterdam.

Takahashi, H. (1981). Seed dormancy in *Tricyrtis* Sects. *flavae* and *brachycyrtis* (Liliaceae). *Sci. Rep. Fac. Educ. Gifu Univ.* (*Nat. Sci.*) **6**, 684–691.

Takahashi, H. (1984). Germination ecology of *Heloniopsis orientalis* (Liliaceae). *Sci. Rep. Fac. Educ. Gifu Univ.* (*Nat. Sci.*) **8**, 1–8.

Talley, S. N., and Griffin, J. R. (1980). Fire ecology of a montane pine forest, Junipero Serra Peak, California. *Madrono* **27**, 49–60.

Tallowin, J. R. B., Rook, A. J., and Brookman, S. K. E. (1994). The effects of osmotic pre-sowing treatment on laboratory germination in a range of wild flower species. *Ann. Appl. Biol.* **124**, 363–370.

Tappeiner II, J. C., and Alaback, P. B. (1989). Early establishment and vegetation growth of understory species in the western hemlock-Sitka spruce forests of southeast Alaska. *Can. J. Bot.* **67**, 318–326.

Tarrega, R., Calvo, L., and Trabaud, L. (1992). Effect of high temperatures on seed germination of two woody Leguminosae. *Vegetatio* **102**, 139–147.

Taylor, A. H., and Zisheng, Q. (1988). Regeneration from seed of *Sinarundinaria fangiana*, a bamboo, in the Wolong Giant Panda Reserve, Sichuan, China. *Am. J. Bot.* **75**, 1065–1073.

Taylor, F. J. (1956). Biological flora of the British Isles. *Carex flacca* Schreb. *J. Ecol.* **44**, 281–290.

Taylor, J. S., and Wareing, P. F. (1979). The effect of light on the endogenous levels of cytokinins and gibberellins in seeds of sitka spruce (*Picea sitchensis* Carriere). *Plant Cell Environ.* **2**, 173–179.

Taylor, K. (1971). Biological flora of the British Isles. *Rubus chamaemorus* L. *J. Ecol.* **59**, 293–306.

Taylor, K., and Markham, B. (1978). Biological flora of the British Isles. *Ranunculus ficaria* L. (*Ficaria verna* Huds.; *F. ranunculodies* Moench). *J. Ecol.* **66**, 1011–1031.

Taylorson, R. B., and McWhorter, C. G. (1969). Seed dormancy and germination in ecotypes of Johnsongrass. *Weed Sci.* **17**, 359–361.

Thanos, C. A., and Doussi, M. A. (1995). Ecophysiology of seed germination in endemic Labiates of Crete. *Israel J. Plant Sci.* **43**, 227–237.

Thanos, C. A., and Georghiou, K. (1988). Ecophysiology of fire-stimulated seed germination in *Cistus incanus* ssp. *creticus* (L.) Heywood and *C. salvifolius* L. *Plant Cell Environ.* **11**, 841–849.

Thanos, C. A., Kadis, C. C., and Skarou, F. (1995). Ecophysiology of germination in the aromatic plants thyme, savory and oregano (Labiatae). *Seed Sci. Res.* **5**, 161–170.

Thanos, C. A., and Rundel, P. W. (1995). Fire-followers in chaparral: Nitrogenous compounds trigger seed germination. *J. Ecol.* **83**, 207–216.

Thanos, C. A., and Skordilis, A. (1987). The effects of light, temperature and osmotic stress on the germination of *Pinus halepensis* and *P. brutia* seeds. *Seed Sci. Technol.* **15**, 163–174.

Thapliyal, R. C., and Rawat, M. M. S. (1991). Studies on germination and viability of seed of two species of Himalayan alders, *Alnus nitida* and *A. nepalensis. Indian For.* **117**, 256–261.

Thill, D. C., Schirman, R. D., and Appleby, A. P. (1980). Influence of afterripening temperature and endogenous rhythms on downy brome (*Bromus tectorum*) germination. *Weed Sci.* **28**, 321–323.

Thomas, P. A., and Wein, R. W. (1990). Jack pine establishment on ash from wood and organic soil. *Can. J. For. Res.* **20**, 1926–1932.

Thompson, P. A. (1968). Germination responses of *Meconopsis* species. *J. Roy. Hort. Soc.* **93**, 336–343.

Thompson, P. A. (1969a). Effects of after-ripening and chilling treatments on the germination of species of *Gentiana* at different temperatures. *J. Hort. Sci.* **44**, 343–358.

Thompson, P. A. (1969b). Some effects of light and temperature on the germination of some *Primula* species. *J. Hort. Sci.* **44**, 1–12.

Thompson, P. A. (1970). Germination of species of Caryophyllaceae in relation to their geographical distribution in Europe. *Ann. Bot.* **34**, 427–449.

Thompson, P. A. (1973). Effects of cultivation on the germination character of the corn cockle (*Agrostemma githago* L.). *Ann. Bot.* **37**, 133–154.

Thompson, P. A. (1975). Characterization of the germination responses of *Silene dioica* (L.) Clairv., populations from Europe. *Ann. Bot.* **39**, 1–19.

Thompson, P. A. (1980). Germination strategy of a woodland grass: *Milium effusum* L. *Ann. Bot.* **46**, 593–602.

Thompson, P. A., and Cox, S. A. (1978). Germination of the bluebell (*Hyacinthoides non-scripta* (L.) Chouard) in relation to its distribution and habitat. *Ann. Bot.* **42**, 51–62.

Thomson, C. E., and Witt, W. W. (1987). Germination of cutleaf groundcherry (*Physalis angulata*), smooth groundcherry (*Physalis virginiana*), and eastern black nightshade (*Solanum ptycanthum*). *Weed Sci.* **35**, 58–62.

Thornton, M. L. (1966). Seed dormancy in buffalograss (*Buchloe dactyloides*). *Proc. Assoc. Offic. Seed Anal.* **56**, 120–123.

Threadgill, P. F., Baskin, J. M., and Baskin, C. C. (1981). Dormancy in seeds of *Frasera caroliniensis* (Gentianaceae). *Am. J. Bot.* **68**, 80–86.

Thrower, S. L. (1986). Ecological records No. 5: Some observations concerning the conditions necessary for germination of Chinese banyan (*Ficus microcarpa* Linn. f.) seeds. *Mem. Hong Kong Nat. Hist. Soc.* **17**, 13–14.

Thullen, R. J., and Keeley, P. E. (1979). Seed production and germination in *Cyperus esculentus* and *C. rotundus*. *Weed Sci.* **27**, 502–505.

Tiley, G. E. D., Dodd, F. S., and Wade, P. M. (1996). Biological Flora of the British Isles. *Heracleum mantegazzianum* Sommier & Levier. *J. Ecol.* **84**, 297–319.

Ting-Cheng, Z. (1993). Grasslands of China. *In* "Ecosystems of the World" (R. T. Coupland, ed.), Vol. 8B, pp. 61–82. Elsevier, Amsterdam.

Tolstead, W. L. (1941). Germination habits of certain sand-hill plants in Nebraska. *Ecology* **22**, 393–397.

Tomaselli, R. (1981). Main physiognomic types and geographic distribution of shrub systems related to Mediterranean climates. *In* "Ecosystems of the World" (F. di Castri, D. W. Goodall, and R. L. Specht, eds.), Vol. 11, pp. 95–106. Elsevier, Amsterdam.

Toole, E. H., and Toole, V. K. (1940). Germination of seed of goosegrass, *Eleusine indica*. *J. Am. Soc. Agron.* **32**, 320–321.

Toole, E. H., and Toole, V. K. (1941). Progress of germination of seed of *Digitaria* as influenced by germination temperature and other factors. *J. Agric. Res.* **63**, 65–90.

Toole, E. H., Toole, V. K., Hendricks, S. B., and Borthwick, H. A. (1957). Effect of temperature on germination of light-sensitive seeds. *Proc. Int. Seed Test. Assoc.* **22**, 196–204.

Toole, V. K. (1939). Germination of the seeds of poverty grass, *Danthonia spicata*. *Agron. J.* **31**, 954–965.

Toole, V. K. (1940). Germination of seeds of vine-mesquite, *Panicum obtusum*, and plains bristle-grass, *Setaria macrostachya*. *Agron. J.* **32**, 503–512.

Toole, V. K. (1941). Factors affecting the germination of various dropseed grasses (*Sporobolus* spp.). *J. Agric. Res.* **62**, 691–715.

Toole, V. K. (1976). Light and temperature control of germination in *Agropyron smithii* seeds. *Plant Cell Physiol.* **17**, 1263–1272.

Toth, J., and Garrett, P. W. (1989). Optimum temperatures for stratification of several maple species. *Tree Plant. Notes* **40**(3), 9–12.

Trabaud, L. (1981). Man and fire: Impacts on Mediterranean vegetation. *In* "Ecosystems of the World" (F. di Castri, D. W. Goodall, and R. L. Specht, eds.), Vol. 11, pp. 523–537. Elsevier, Amsterdam.

Trabaud, L., and Oustric, J. (1989). Influence du feu sur la germination des semences de quatre especes ligneuses mediterraneennes a reproduction sexuee obligatoire. *Seed Sci. Technol.* **17**, 589–599.

Trewartha, G. T. (1968). "An Introduction to Climate," 4th Ed. McGraw-Hill, New York.

Turnbull, C. L., and Culver, D. C. (1983). The timing of seed dispersal in *Viola nuttallii*: Attraction of dispersers and avoidance of predators. *Oecologia* **59**, 360–365.

Turner, B. H., and Quarterman, E. (1968). Ecology of *Dodecatheon meadia* L. (Primulaceae) in Tennessee glades and woodland. *Ecology* **49**, 909–915.

Tyler, C. M. (1995). Factors contributing to postfire seedling establishment in chaparral: Direct and indirect effects of fire. *J. Ecol.* **83**, 1009–1020.

Tyler, C. M. (1996). Relative importance of factors contributing to postfire seedling establishment in maritime chaparral. *Ecology* **77**, 2182–2195.

Ungar, I. A. (1962). Influence of salinity on seed germination in succulent halophytes. *Ecology* **43**, 763–764.

Vallance, K. B. (1952). The germination of the seeds of *Rhinanthus crista-galli*. *Ann. Bot.* **16**, 409–420.

van Althen, F. W. (1971). Extended stratification assures prompt walnut germination. *For. Chron.* **47**, 349.

van Breemen, A. M. M. (1984). Comparative germination ecology of three short-lived monocarpic Boraginaceae. *Acta Bot. Neerl.* **33**, 283–305.

Vanden Born, W. H. (1971). Green foxtail: Seed dormancy, germination and growth. *Can. J. Plant Sci.* **51**, 53–59.

Vanden Born, W. H., and Corns, W. G. (1958). Studies on seed dormancy, plant development, and chemical control of tartary buckwheat (*Fagopyrum tataricum* (L.) Gaertn.). *Can. J. Plant Sci.* **38**, 357–366.

van de Venter, H. A., and Esterhuizen, A. D. (1988). The effect of factors associated with fire on seed germination of *Erica sessiliflora* and *E. hebecalyx* (Ericaceae). *S. Afr. J. Bot.* **54**, 301–304.

van de Venter, H. A., and Small, J. G. C. (1974). Dormancy in seeds of *Strelitzia* Ait. *S. Afr. J. Sci.* **70**, 216–217.

van der Toorn, J., and ten Hove, H. J. (1982). On the ecology of *Cotula coronopifolia* L. and *Ranunculus sceleratus* L. *Oecol. Plant.* **3**, 409–418.

van der Weide, R. Y. (1993). "Population Dynamics and Population Control of *Galium aparine* L." Ph.D thesis, Agricultural University, Wageningen, The Netherlands.

van Splunder, I. V., Coops, H., Voesenek, L. A. C. J., and Blom, C. W. P. M. (1995). Establishment of alluvial forest species in floodplains: The role of dispersal timing, germination characteristics and water level fluctuations. *Acta Bot. Neerl.* **44**, 269–278.

van Staden, J., and Dimalla, G. G. (1976). Regulation of germination of pecan, *Carya illinoensis*. *Z. Pflanzenphysiol.* **78**, 66–75.

van Staden, J., Drewes, F. E., and Jager, A. K. (1995). The search for germination stimulants in plant-derived smoke extracts. *S. Afr. J. Bot.* **61**, 260–263.

Vanstone, D. E., and LaCroix, L. J. (1975). Embryo immaturity and dormancy of black ash. *J. Am. Soc. Hort. Sci.* **100**, 630–632.

van Tooren, B. F., and Pons, T. L. (1988). Effects of temperature and light on the germination in chalk grassland species. *Funct. Ecol.* **2**, 303–310.

Veblen, T. T., Ashton, D. H., Schlegel, F. M., and Veblen, A. T. (1977). Distribution and dominance of species in the understorey of a mixed evergreen-deciduous *Nothofagus* forest in south-central Chile. *J. Ecol.* **65**, 815–830.

Veblen, T. T., Schlegel, F. M., and Escobar R., B. (1980). Structure and dynamics of old-growth *Nothofagus* forests in the Valdivian Andes, Chile. *J. Ecol.* **68**, 1–31.

Veblen, T. T., Schlegel, F. M., and Oltremari, J. V. (1983). Temperate broad-leaved evergreen forests of South America. *In* "Ecosystems of the World" (J. D. Ovington, ed.), Vol. 10, pp. 5–31. Elsevier, Amsterdam.

Veit, V., and Van Auken, O. W. (1993). Factors influencing the germination of seeds of *Fallugia paradoxa* (Rosaceae). *Texas J. Sci.* **45**, 325–333.

Villiers, T. A., and Wareing, P. F. (1964). Dormancy in fruits of *Fraxinus excelsior* L. *J. Exp. Bot.* **15**, 359–367.

Vogl, R. J., and Schorr, P. K. (1972). Fire and manzanita chaparral in the San Jacinto Mountains, California. *Ecology* **53**, 1179–1188.

Voigt, J. W. (1977). Seed germination of true prairie forbs. *J. Range Manage.* **30**, 439–441.

von Kirchner, O., Loew, E., and Schroter, C. (1931). Legensgeschichte der Blutenpflanzen Mitteleuropas. Verlagsbuchhandlung Eugen Uler, Stuttgart.

Vose, J. M., and White, A. S. (1987). Processes of understory seedling recruitment 1 year after prescribed fire in an Arizona ponderosa pine community. *Can. J. Bot.* **65**, 2280–2290.

Vuillemin, J., and Bulard, C. (1981). Ecophysiologie de la germination de *Cistus albidus* L. et *Cistus monspeliensis* L. *Natur. Monspeliensia Ser. Bot.* **46**, 1–11.

Wagner, L. K. (1988). Germination and seedling emergence in *Spergula arvensis*. *Am. J. Bot.* **75**, 465–475.

Wahlenberg, W. G. (1924). Fall sowing and delayed germination of western white pine seed. *J. Agric. Res.* **28**, 1127–1131.

Waller, S. S., and Lewis, J. K. (1979). Occurrence of C₃ and C₄ photosynthetic pathways in North American grasses. *J. Range Manage.* **32**, 12–28.

Wallgren, B. E., and Aamisepp, A. (1977). Biology and control of *Alopecurus myosuroides* Huds. and *Apera spica-venti* L. *In* "Proc. EWRS Symposium: Methods of Weed Control and Their Integration," pp. 229–241.

Wallgren, B., and Avholm, K. (1978). Dormancy and germination of *Apera spica-venti* L. and *Alopecurus myosuroides* Huds. seeds. *Swed. J. Agri. Res.* **8**, 11–15.

Walter, H. (1973). "Vegetation of the Earth: In Relation to Climate and the Eco-physiological Conditions." The English Universities Press Ltd., London.

Walter, H. (1979). "Vegetation of the Earth and Ecological Systems of the Geo-biosphere," 2nd Ed. Springer-Verlag, New York.

Walter, H., and Box, E. O. (1983a). The deserts of central Asia. *In* "Ecosystems of the World" (N. E. West, ed.), Vol. 5, pp. 193–236. Elsevier, Amsterdam.

Walter, H., and Box, E. O. (1983b). Middle Asian deserts. *In* "Ecosystems of the World" (N. E. West, ed.), Vol. 5, pp. 79–104. Elsevier, Amsterdam.

Walter, H., and Box, E. O. (1983c). Caspian lowland biome. *In* "Ecosystems of the World" (N. E. West, ed.), Vol. 5, pp. 9–41. Elsevier, Amsterdam.

Walter, H., and Box, E. O. (1983d). Semi-deserts and deserts of central Kazakhstan. *In* "Ecosystems of the World" (N. E. West, ed.), Vol. 5, pp. 43–78. Elsevier, Amsterdam.

Walter, H., and Box, E. O. (1983e). The Karakum Desert, an example of a well-studied eu-biome. *In* "Ecosystems of the World" (N. E. West, ed.), Vol. 5, pp. 105–159, Elsevier, Amsterdam.

Walter, H., and Box, E. O. (1983f). The orobiomes of middle Asia. *In* "Ecosystems of the World" (N. E. West, ed.), Vol. 5, pp. 161–191. Elsevier, Amsterdam.

Walters, G. A. (1974). *Araucaria* (Jussieu). Araucaria. *In* "Seeds of Woody Plants in the United States" (C. S. Schopmeyer, Tech. Coord.), pp. 223–225. USDA Forest Service Agriculture Handbook No. 450.

Walton, D. W. H. (1977). Studies on *Acaena* (Rosaceae). I. Seed germination, growth and establishment in *A. megellanica* (Lam.) Vahl and *A. tenera* Alboff. *Br. Antarct. Surv. Bull.* **45**, 29–40.

Walton, D. W. H. (1985). The sub-Antarctic islands. *In* "Key Environments: Antarctica" (W. N. Bonner and D. W. H. Walton. eds.), pp. 293–317. Pergamon Press, Oxford.

Wang, B. S. P., and Haddon, B. D. (1978). Germination of red maple seed. *Seed Sci. Technol.* **6**, 785–790.

Wardlaw, I. F., Moncur, M. W., and Totterdell, C. J. (1989). The growth and development of *Caltha introloba* F. Muell. II. The regulation of germination, growth and photosynthesis by temperature. *Aust. J. Bot.* **37**, 291–303.

Wardle, P., and MacRae, A. H. (1966). Biological flora of New Zealand. 1. *Weinmannia racemosa* Linn. f. (Cunoniaceae). Kamahi. *New Zeal. J. Bot.* **4**, 114–131.

Wardle, P. (1967). Biological flora of New Zealand. 2. *Nothofagus menziesii* (Hook. f) Oerst. (Fagaceae) silver beech. *New Zeal. J. Bot.* **5**, 276–302.

Wardle, P. (1970). The ecology of *Nothofagus solandri*. 3. Regeneration. *New Zeal. J. Bot.* **8**, 571–608.

Wardle, P. (1971). Biological flora of New Zealand. 6. *Metrosideros*

umbellata Cav. [syn. *M. lucida* (Forst.f.) A. Rich.] (Myrtaceae) southern rata. *New Zeal. J. Bot.* **9**, 645–671.

Wardle, P. (1991). "Vegetation of New Zealand." Cambridge Univ. Press, Cambridge, UK.

Wardle, P., Bulfin, M. J. A., and Dugdale, J. (1983). Temperate broadleaved evergreen forests of New Zealand. *In* "Ecosystems of the World" (J. D. Ovington, ed.), Vol. 10, pp. 33–71. Elsevier, Amsterdam.

Wardle, P., Field, T. R. O., and Spain, A. V. (1970). Biological flora of New Zealand. 5. *Olearia colensoi* Hook. f. (Compositae) leatherwood, tupari. *New Zeal. J. Bot.* **9**, 186–214.

Warr, H. J., Savory, D. R., and Bal, A. K. (1979). Germination studies of bakeapple (cloudberry) seeds. *Can. J. Plant Sci.* **59**, 69–74.

Washitani, I. (1984). Germination responses of a seed population of *Taraxacum officinale* Weber to constant temperatures including the supra-optimal range. *Plant Cell Environ* **7**, 655–659.

Washitani, I., and Kabaya, H. (1988). Germination responses to temperature responsible for the seedling emergence seasonality of *Primula sieboldii* E. Morren in its natural habitat. *Ecol. Res.* **3**, 9–20.

Washitani, I., and Masuda, M. (1990). A comparative study of the germination characteristics of seeds from a moist tall grassland community. *Funct. Ecol.* **4**, 543–557.

Washitani, I., and Ogawa, K. (1989). Germination responses of *Taraxacum platycarpum* seeds to temperature. *Plant Species Biol.* **4**, 123–130.

Washitani, I., and Takenaka, A. (1987). Gap-detecting mechanism in the seed germination of *Mallotus japonicus* (Thunb.) Muell. Arg., a common pioneer tree of secondary succession in temperate Japan. *Ecol. Res.* **2**, 191–201.

Watanabe, Y., and Hirokawa, F. (1975). Requirement of temperature conditions in germination of annual weed seeds and its relation to seasonal distribution of emergence in the field. *In* "Proc. 5th Asian-Pacific Weed Sci. Soc. Conf," pp. 38–41. Tokyo.

Watanabe, M., Ueda, K., Manabe, I., and Akai, T. (1982). Flowering, seeding, germination, and flowering periodicity of *Phyllostachys pubescens*. *J. Jap. For. Soc.* **64**, 107–111.

Watt, A. S. (1919). On the causes of failure of natural regeneration in British oakwoods. *J. Ecol.* **7**, 173–203.

Watt, A. S. (1923). On the ecology of British beechwoods with special reference to their regeneration. *J. Ecol.* **11**, 1–48.

Weaver, J. E., and Rowland, N. W. (1952). Effects of excessive natural mulch on development, yield, and structure of native grassland. *Bot. Gaz.* **114**, 1–19.

Weaver, S. E., and Riley, W. R. (1982). The biology of Canadian weeds. 53. *Convolvulus arvensis* L. *Can. J. Plant Sci.* **62**, 461–472.

Webb, D. P., and Dumbroff, E. B. (1969). Factors influencing the stratification process in seeds of *Acer saccharum*. *Can. J. Bot.* **47**, 1555–1563.

Wehrmeister, R. R., and Bonde, E. K. (1977). Comparative aspects of growth and reproductive biology in arctic and alpine populations of *Saxifraga cernua* L. *Arctic Alpine Res.* **9**, 401–406.

Wehtje, G. R, Gilliam, C. H., and Reeder, J. A. (1992). Germination and growth of leafflower (*Phyllanthus urinaria*) as affected by cultural conditions and herbicides. *Weed Technol.* **6**, 139–143.

Weilenmann, K. (1981). Importance of germination and first developmental phases of alpine taxa from various habitats. *Ber. Beobot. Inst. ETH, Stiftung Rubel* **48**, 68–119. [In German with English summary]

Wein, R. W., and MacLean, D. A. (1973). Cotton grass (*Eriophorum vaginatum*) germination requirements and colonizing potential in the Arctic. *Can. J. Bot.* **51**, 2509–2513.

Weiss, F. (1926). Seed germination in the gray birch, *Betula populifolia*. *Am. J. Bot.* **13**, 737–742.

Weldon, L. W., Bohmont, D. W., and Alley, H. P. (1959). The interrelation of three environmental factors affecting germination of sagebrush seed. *J. Range Manage.* **12**, 236–238.

Wellington, A. B., and Noble, I. R. (1985). Post-fire recruitment and mortality in a population of the mallee *Eucalyptus incrassata* in semi-arid, south-eastern Australia. *J. Ecol.* **73**, 645–656.

Wellington, P. S., and Hitchings, S. (1965). Germination and seedling establishment of blackgrass (*Alopecurus myosuroides* Huds.). *J. Natl. Inst. Agric. Bot.* **10**, 262–273.

Wellington, P. S., and Hitchings, S. (1966). Seed dormancy and the winter annual habit in blackgrass (*Alopecurus myosuroides* Huds.). *J. Natl. Inst. Agri. Bot.* **10**, 628–643.

Wells, P. V. (1959). An ecological investigation of two desert tabaccos. *Ecology* **40**, 626–644.

Wenger, L. E. (1941). Soaking buffalo grass (*Buchloe dactyloides*) seed to improve its germination. *Agron. J.* **33**, 135–141.

Went, F. W., Juhren, G., and Juhren, M. C. (1952). Fire and biotic factors affecting germination. *Ecology* **33**, 351–364.

Werner, P. A. (1975a). The biology of Canadian weeds. 12. *Dipsacus sylvestris* Huds. *Can. J. Plant Sci.* **55**, 783–794.

Werner, P. A. (1975b). The effects of plant litter on germination in teasel, *Dipsacus sylvestris* Huds. *Am. Midl. Nat.* **94**, 470–476.

West, N. E. (ed.) (1983a). Approach. *In* "Ecosystems of the World" Vol. 5, pp. 1–2. Elsevier, Amsterdam.

West, N. E. (1983b). Overview of North American temperate deserts and semi-deserts. *In* "Ecosystems of the World" (N. E. West, ed.), Vol. 5, pp. 321–330. Elsevier, Amsterdam.

West, N. E. (1983c). Comparisons and contrasts between the temperate deserts and semi-deserts of three continents. *In* "Ecosystems of the World" (N. E. West, ed.), Vol. 5, pp. 461–472. Elsevier, Amsterdam.

West, N. E. (1983d). Intermountain salt-desert shrubland. *In* "Ecosystems of the World" Vol. 5, pp. 375–397. Elsevier, Amsterdam.

West, N. E. (1983e). Great Basin-Colorado Plateau sagebrush semi-desert. *In* "Ecosystems of the World" (N. E. West, ed.), Vol. 5, pp. 331–349. Elsevier, Amsterdam.

West, N. E. (ed.) (1983f). "Ecosystems of the World" Vol. 5. Elsevier, Amsterdam.

West, N. E. (1983g). Western intermountain sagebrush steppe. *In* "Ecosystems of the World" (N. E. West, ed.), Vol. 5, pp. 351–374. Elsevier, Amsterdam.

Westoby, M., French, K., Hughes, L., Rice, B., and Rodgerson, L. (1991). Why do more plant species use ants for dispersal on infertile compared with fertile soils? *Aust. J. Ecol.* **16**, 445–455.

Whigham, D. (1974). An ecological life history study of *Uvularia perfoliata* L. *Am. Midl. Nat.* **91**, 343–359.

Whisenant, S. G., Scifres, C. J., and Ueckert, D. N. (1984). Soil water and temperature response to prescribed burning. *Great Basin Nat.* **44**, 558–562.

White, E. M., and Gartner, F. R. (1975). Immediate effects of prairie fire on the soil nitrate, ammonium, available phosphorus and total N contents. *Proc. South Dakota Acad. Sci.* **54**, 188–193.

White, J. J., and Van Auken, O. W. (1996). Germination, light requirements, and competitive interactions of *Stipa leucotricha* (Gramineae). *Southw. Nat.* **41**, 27–34.

Whittaker, E., and Gimingham, C. H. (1962). The effects of fire on regeneration of *Calluna vulgaris* (L.) Hull. from seed. *J. Ecol.* **50**, 815–822.

Wicklow, D. T. (1977). Germination response in *Emmenanthe penduliflora* (Hydrophyllaceae). *Ecology* **58**, 201–205.

Wiesehuegel, E. G. (1935). Germinating Kentucky coffee tree. *J. For.* **33**, 533–534.

Wiesner, L. E., and Kinch, R. C. (1964). Seed dormancy in green needlegrass. *Agron. J.* **56**, 371–373.

Wilcox, J. R. (1968). Sweetgum seed stratification requirements related to winter climate at seed source. *For. Sci.* **14**, 16–19.

Wilgus, F., and Hamilton, K. C. (1962). Germination of salt cedar seed. *Weeds* **10**, 332–333.

Williams, C. E., Lipscomb, M. V., Johnson, W. C., and Nilsen, E. T. (1990). Influence of leaf litter and soil moisture on early establishment of *Pinus pungens*. *Am. Midl. Nat.* **124**, 142–152.

Williams, E. D. (1971). Germination of seeds and emergence of seedlings of *Agropyron repens* (L.) Beauv. *Weed Res.* **11**, 171–181.

Williams, E. D. (1983). Effects of temperature fluctuation, red and far-red light and nitrate on seed germination of five grasses. *J. Appl. Ecol.* **20**, 923–935.

Williams, M. C. (1960). Biochemical analyses, germination, and production of black and brown seed of *Halogeton glomeratus*. *Weeds* **8**, 452–461.

Williams, M. C., and Cronin, E. H. (1968). Dormancy, longevity, and germination of seed of three larkspurs and western false hellebore. *Weeds* **16**, 381–384.

Williams, M. D., and Ungar, I. A. (1972). The effect of environmental parameters on the germination, growth, and development of *Suaeda depressa* (Pursh) Wats. *Am. J. Bot.* **59**, 912–918.

Williams, P. A., and Buxton, R. P. (1995). Aspects of the ecology of two species of *Passiflora* (*P. mollissima* (Kunth) L. Bailey and *P. pinnatistipula* Cav.) as weeds in South Island, New Zealand. *New Zeal. J. Bot.* **33**, 315–323.

Williams, P. A., Norton, D. A., and Nicholas, J. M. (1996). Germination and seedling growth of an endangered native broom, *Chordospartium muritai* A. W. Purdie (Fabaceae), found in Marlborough, South Island, New Zealand. *New Zeal. J. Bot.* **34**, 199–204.

Williams, W. A., Cook, O. D., and Kay, B. L. (1974). Germination of native desert shrubs. *California Agric.* **28**, 13.

Wilson, A. M. (1973). Responses of crested wheatgrass seeds to environment. *J. Range Manage.* **26**, 43–46.

Wilson, B. F., Hibbs, D. E., and Fischer, B. C. (1979). Seed dormancy in striped maple. *Can. J. For. Res.* **9**, 263–266.

Wilson, J. K. (1937). Scarification and germination of black locust seeds. *J. For.* **35**, 241–246.

Wilson, R. G., Jr. (1979). Germination and seedling development of Canada thistle. *Weed Sci.* **27**, 146–151.

Wilson, R. G., Jr. (1982). Germination and seedling development of fringed sagebrush (*Artemisia frigida*). *Weed Sci.* **30**, 102–105.

Wilson, R. G., Jr., and McCarty, M. K. (1984). Germination, and seedling and rosette development of Flodman thistle (*Cirsium flodmanii*). *Weed Sci.* **32**, 768–773.

Wirges, G., and Yeiser, J. (1984). Stratification and germination of Arkansas oak acorns. *Tree Plant. Notes* **35**(2), 36–38.

Wolters, G. L. (1970). Breaking dormancy of longleaf *Uniola* seeds. *J. Range Manage.* **23**, 178–180.

Wood, M. K., Knight, R. W., and Young, J. A. (1976). Spiny hopsage germination. *J. Range Manage.* **29**, 53–56.

Workman, J. P., and West, N. E. (1967). Germination of *Eurotia lanata* in relation to temperature and salinity. *Ecology* **48**, 659–661.

Wright, E. (1931). The effect of high temperatures on seed germination. *J. For.* **29**, 679–687.

Wright, H. E., Jr. (1974). Landscape development, forest, fires, and wilderness management. *Science* **186**, 487–495.

Wright, L. N. (1973). Seed dormancy, germination environment, and seed structure of Lehmann lovegrass, *Eragrostis lehmanniana* Nees. *Crop Sci.* **13**, 432–435.

Yates, C. J., Hobbs, R. J., and Bell, R. W. (1996). Factors limiting the recruitment of *Eucalyptus salmonophloia* in remnant woodlands. III. Conditions necessary for seed germination. *Aust. J. Bot.* **44**, 283–296.

Yeiser, J. L. (1983). Germinative pretreatments and seedcoat impermeability for the Kentucky coffeetree. *Tree Plant. Notes* **34**(2), 33–35.

Young, D. A. (1972). The reproductive biology of *Rhus integrifolia* and *Rhus ovata* (Anacardiaceae). *Evolution* **26**, 406–414.

Young, J. A. (1989). Germination of seeds of sulphur flower (*Eriogonum umbellatum* Torr.). *J. Seed Technol.* **13**, 31–38.

Young, J. A., Eckert, R. E., Jr., and Evans, R. A. (1981). Temperature profiles for germination of blue-bunch and beardless wheatgrasses. *J. Range Manage.* **34**, 84–89.

Young, J. A., Emmerich, F. L., and Patten, B. (1990). Germination of seeds of Columbia needlegrass. *J. Seed Technol.* **14**, 94–100.

Young, J. A., and Evans, R. A. (1972). Germination and establishment of *Salsola* in relation to seedbed environment. I. Temperature, afterripening, and moisture relations of *Salsola* seeds as determined by laboratory studies. *Agron. J.* **64**, 214–218.

Young, J. A., and Evans, R. A. (1976). Stratification of bitterbrush seeds. *J. Range Manage.* **29**, 421–425.

Young, J. A., and Evans, R. A. (1977). Squirreltail seed germination. *J. Range Manage.* **30**, 33–36.

Young, J. A., and Evans, R. A. (1979). Arrowleaf balsamroot and mules ear seed germination. *J. Range Manage.* **32**, 71–74.

Young, J. A., and Evans, R. A. (1981). Germination of seeds of antelope bitterbrush, desert bitterbrush, and cliff rose. USDA, *Agric. Res. Results, Western Series. No.* 17.

Young, J. A., and Evans, R. A. (1982). Temperature profiles for germination of cool season range grasses. USDA, *Agric. Res. Results Western Series No.* 27.

Young, J. A., and Evans, R. A. (1986). Germination of white horehound (*Marrubium vulgare*) seeds. *Weed Sci.* **34**, 266–270.

Young, J. A., and Evans, R. A. (1989). Seed production and germination dynamics in California annual grasslands. *In* "Grassland Structure and Function: California Annual Grassland" (L. F. Huenneke, and H. A. Mooney, eds.), pp. 39–45. Kluwer, Dordrecht.

Young, J. A., Evans, R. A., Budy, J. D., and Palmquist, D. E. (1988). Stratification of seeds of western and Utah juniper. *For. Sci.* **34**, 1059–1066.

Young, J. A., Evans, R. A., and Eckert, R. E., Jr. (1968). Germination of medusahead in response to temperature and afterripening. *Weed Sci.* **16**, 92–95.

Young, J. A., Evans, R. A., Eckert, R. E., Jr., and Ensign, R. D. (1981). Germination-temperature profiles for Idaho and sheep fescue and Canby bluegrass. *Agron. J.* **73**, 716–720.

Young, J. A., Evans, R. A., Gifford, R. O., and Eckert, R. E., Jr.

(1970). Germination characteristics of three species of Cruciferae. *Weed Sci.* **18**, 41–48.

Young, J. A., Evans, R. A., and Kay, B. L. (1971). Germination of caryopses of annual grasses in simulated litter. *Agron. J.* **63**, 551–555.

Young, J. A., Evans, R. A., and Kay, B. L. (1975a). Germination of Italian ryegrass seeds. *Agron. J.* **67**, 386–389.

Young, J. A., Evans, R. A., and Kay, B. L. (1975b). Dispersal and germination dynamics of broadleaf filaree, *Erodium botrys* (Cav.) Bertol. *Agron. J.* **67**, 54–57.

Young, J. A., Evans, R. A., and Kay, B. L. (1977). *Ephedra* seed germination. *Agron. J.* **69**, 209–211.

Young, J. A., Evans, R. A., and Neal, D. L. (1978). Treatment of curlleaf *Cercocarpus* seeds to enhance germination. *J. Wildl. Manage.* **42**, 614–620.

Young, J. A., Evans, R. A., Raguse, C. A., and Larson, J. R. (1981). Germinable seeds and periodicity of germination in annual grasslands. *Hilgardia* **49**(2), 1–37.

Young, J. A., Evans, R. A., Stevens, R., and Everett, R. L. (1981). Germination of *Kochia prostrata* seed. *Agron. J.* **73**, 957–961.

Young, J. A., Kay, B. L., George, H., and Evans, R. A. (1980). Germination of three species of *Atriplex. Agron. J.* **72**, 705–709.

Young, J. A., Martens, E., and West, N. E. (1992). Germination of bur buttercup seeds. *J. Range Manage.* **45**, 358–362.

Young, J. A., Wight, J. R., and Mowbray, J. E. (1993). Field stratification of antelope bitterbrush seeds. *J. Range Manage.* **46**, 325–330.

Zackrisson, O. (1977). Influence of forest fires on the north Swedish boreal forest. *Oikos* **29**, 22–32.

Zasada, J. C., Nordum, R. A., van Veldhuizen, R. M., and Teutsch, C. E. (1983). Artificial regeneration of trees and tall shrubs in experimentally burned upland black spruce/feather moss stands in Alaska. *Can. J. For. Res.* **13**, 903–913.

Zasada, J. C., and Viereck, L. A. (1975). The effect of temperature and stratification on germination in selected members of the Salicaceae in interior Alaska. *Can. J. For. Res.* **5**, 333–337.

Zhang, J., and Maun, M. A. (1989). Seed dormancy of *Panicum virgatum* L. on the shoreline sand dunes of Lake Erie. *Am. Midl. Nat.* **122**, 77–87.

Ziemkiewicz, P. F., and Cronin, E. H. (1981). Germination of seed of three varieties of spotted locoweed. *J. Range Manage.* **34**, 94–97.

Zuur-Isler, D. (1982). Germinating behaviour and early life phases of some species from alpine serpentine soils. *Ber. Beobot. Inst. ETH, Stiftung Rubel* **49**, 76–107. [In German with English summary]

11

Germination Ecology of Plants with Specialized Life Cycles and/or Habitats

I. PURPOSE

In Chapters 9 and 10, the germination ecology of plant species growing in the major vegetation regions of the world was surveyed; however, little attention was given specifically to parasites, saprophytes, orchids, carnivorous plants, aquatics, halophytes, or psammophytes. There are two reasons for devoting a chapter to the germination ecology of these plants. (1) Each of the seven groups has representatives in more than one vegetation zone; thus, a consideration of their germination ecology needs to be approached from a global perspective. (2) The groups have such unusual life cycles and/or unique habitat requirements that they have long attracted much attention from botanists, ecologists, and/or horticulturalists. The purposes of this chapter are to (1) survey the kinds of dormancy found in freshly matured seeds in each of the seven groups of plants and (2) discuss available data on the environmental conditions required to break dormancy and stimulate germination.

II. PARASITIC PLANTS

A parasite is an organism that obtains nutrients from another organism, called the host, usually to the latter's detriment. Parasitic flowering plants are divided into two groups: holoparasites and hemiparasites. Holoparasites lack chlorophyll and receive fixed carbon, water, and minerals from the host plant, whereas hemiparasites have chlorophyll and can fix carbon, but obtain much of their water and minerals from the host (Begon *et al.*, 1996).

Kuijt (1969) believes that parasitism has evolved at least eight different times in unrelated groups of dicots. None of the monocots are parasitic (Kuijt, 1969), and

only one gymnosperm (*Podocarpus ustus*) has been reported to be a parasite (Delaubenfels, 1959; Wilkinson and Musselman, 1979).

A. Holoparasites

1. Types of Seed Dormancy

Members of the Balanophoraceae, Orobanchaceae, Hydnoraceae, Lennoaceae, and Rafflesiaceae are holoparasites (Kuijt, 1969) and have seeds with undifferentiated embryos that consist of only a few cells (Table 3.4). Studies on the morphology/anatomy of germination in members of these families indicate that a radicle and cotyledons per se are never formed. The radicular pole of the embryo in seeds of *Cistanche tubulosa* and *Orobanche aegyptiaca* (Orobanchaceae) grows by cell division, emerges through the micropyle, and develops into a multicellular structure called the germ tube; the plumular pole of the embryo remains inside the seed coat (Fig. 11.1). Although the germ tube looks like a root, it is neither a root nor a root meristem (Rangaswamy, 1967). If seeds are sown on a nutrient medium, the germ tube enlarges at the tip and forms a tubercle, in which shoot buds are differentiated (Rangan, 1965; Rangaswamy, 1967). The tubercle also can give rise to roots (Kumar and Rangaswamy, 1977). If gibberellic acid (GA) (Rangaswamy, 1967) or kinetin (Kumar and Rangaswamy, 1977) is present in the medium, the radicular pole may give rise to roots, and the plumular pole produces a shoot. The addition of kinetin (2 ppm) or both kinetin (2 ppm) and 2,4-D (2 ppm) to a modified White's medium promoted the germination of *Aeginetia indica* seeds, and the resulting tubercles occasionally gave rise to roots (Chennaveeraiah *et al.*, 1971).

If a host root is present, the germ tube penetrates it

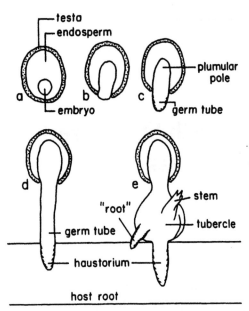

FIGURE 11.1   Germination morphology of *Orobanche* seeds in the presence of a host plant root. Drawn from information in Kadry and Tewfic (1956a), Rangaswamy (1967), and Kumar and Rangaswamy (1977).

and becomes the primary haustorium (Fig. 11.1). It is presumed that chemicals in exudates of host plant roots stimulate development of the haustorium; however, such compounds have not been identified (Parker and Riches, 1993). If a germinated seed of *Orobanche crenata* is more than 3 mm away from a host root, the germ tube does not reach the root and thus dies (Kadry and Tewfic, 1956b). The part of the germ tube remaining on the outside of the host root enlarges and forms a tubercle, from which the shoot and root-like structures are initiated (Garmon, 1890; Kadry and Tewfic, 1956a,b). When the root-like organs of *O. crenata* come in contact with roots of the host, they form secondary haustoria that enter the host roots (Kadry and Tewfic, 1956a). In the perennial *Conopholis americana*, tubercles live for a maximum of 13 years and reach diameters of about 50 mm (Baird and Riopel, 1986a). Tubercles develop a vascular cambium that is closely aligned with that of the host; thus, during the life of a tubercle additional xylem connections are made with the host each year (Baird and Riopel, 1986b).

In *Balanophora abbreviata* (Balanophoraceae), the tiny fruits are attached to newly formed roots of the legumes *Acacia suma* or *Pithecellobium dulce* by sticky, tubular extensions of the endosperm. Then, the undifferentiated embryo produces the germ tube or primary haustorium that penetrates the root and makes contact with vascular tissues of the host (Fig. 11.2). Cells on the lower side of the embryo divide and form a nodule that

gradually increases in size and breaks through the fruit pericarp. Lobes of the nodule give rise to flowering shoots (Arekal and Shivamurthy, 1976).

Little is known about seed germination in the Hydnoraceae, Lennoaceae, or Rafflesiaceae, but presumably the morphology is similar to that in the Orobanchaceae and Balanophoraceae. After haustorial connections are made with the host, species of Balanophoraceae, Hydnoraceae, Lennoaceae, and Orobanchaceae develop a plant body external to the host (Suessenguth, 1927; Kuijt, 1969). In the Rafflesiaceae, however, the parasite exists as an endophyte inside the body of the host, but flower buds eventually break through the host tissues and can be seen outside the host (Kuijt, 1969). Apparently, the endophyte grows throughout the body of the host. Seeds of the parasite *Cytinus hypocistis* (Rafflesiaceae) were sown adjacent to the roots of a host plant, and 4 years later an inflorescence of *C. hypocistis* appeared on the stem of the host plant (Heinricher, 1917).

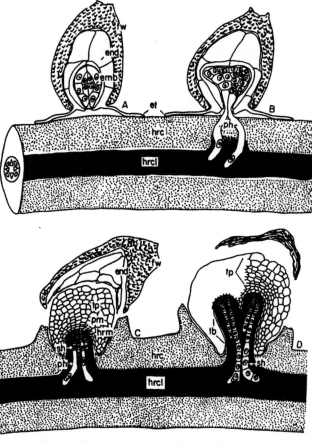

FIGURE 11.2   Germination morphology of *Balanophora abbreviata* (Balanophoraceae). emb, embryo; end, endosperm; et, endosperm tubule; fw, fruit wall; hrc, host root cortex; hrcl, host root vascular tissue; hrm, host root meristem; ph, primary haustorium; pm, parasite meristem; sh, secondary haustorium; tb, tuber bundle; and tp, tuber parenchyma. From Arekal and Shivamurthy (1976), with permission.

As described by Nikolaeva (1969), seeds with morphological dormancy (MD) have embryos that may consist of a few undifferentiated cells, but germination does not occur until embryo development is complete. Although embryos in seeds of holoparasites consist of a few undifferentiated cells only, meristem differentiation and organ formation take place during germination, not prior to it (Rangaswamy, 1967). Thus, seeds of holoparasites do not strictly comply with Nikolaeva's (1969) definition of morphological dormancy. However, because morphological development must precede formation of a plant, dormancy in seeds of holoparasites is a "morphological problem."

In addition to morphological underdevelopment, seeds of various Orobanchaceae require warm or cold stratification before the germ tube or primary haustorium emerges from the seed (French and Sherman, 1976; van Hezewijk et al., 1993). Because these treatments are the usual dormancy-breaking requirements of seeds with physiological dormancy (PD) (Chapter 3), seeds of some holoparasites appear to have morphophysiological dormancy (MPD).

Three pieces of evidence suggest that the type of physiological dormancy found in the Orobanchaceae (and perhaps other holoparasites) is nondeep PD. (1) Seeds of *Orobanche crenata* (Edwards, 1972; Al-Menoufi and Zaitoun, 1987) and *O. oxyloba* (Al-Menoufi and Zaitoun, 1987) germinated to higher percentages after several years of dry laboratory storage than they did when they were freshly matured. The ability of seeds to come out of dormancy during dry storage at room temperatures is a characteristic of nondeep PD (Chapter 3). Unfortunately, some studies on Orobanchaceae do not give the age of seeds used in germination experiments (e.g., Nash and Wilhelm, 1960; Jain and Foy, 1992; Nun and Mayer, 1993). Thus, we do not know how common an afterripening requirement is in members of this family. However, Musselman (1980) indicates that a period of afterripening is a necessary step for the germination of *Orobanche* seeds. (2) Another attribute of seeds with nondeep PD is that GA promotes germination (Chapter 3), and GA stimulated the germination of seeds of *Orobanche ramosa*, *O. ludoviciana* var. *cooperi* (Nash and Wilhelm, 1960), *O. racemosa* (Abu-Shakra et al., 1970), *O. crenata* (Pieterse, 1981), *O. aegyptiaca* (Kumar and Rangaswamy, 1977), and *Aeginetia indica* (French and Sherman, 1976). (3) If the seed coat is softened by calcium hypochlorite, seeds of *Orobanche* spp. germinate without being warm stratified (Cezard, 1973). Acid scarification (0.5 N H_2SO_4) increased the germination of *O. crenata* seeds by about 20% compared to nonscarified controls (Lopez-Granados and Garcia-Torres, 1996). Thus, the embryo of fresh seeds apparently lacks sufficient growth poten-

tial for the germ tube to break through the seed coat; this is a characteristic of nondeep PD (Chapter 3).

2. Dormancy-Breaking Requirements

Although embryos in seeds of the Balanophoraceae, Hydnoraceae, Lennoaceae, and Rafflesiaceae are undifferentiated, it is unknown if they also are physiologically dormant. The Orobanchaceae are the only holoparasites for which any information is available concerning physiological dormancy. This family has received attention because a number of species are serious pests on crop species, especially in regions with a Mediterranean climate. However, some important weedy *Orobanche* extend into the temperate and tropical zones (Cordas, 1973; Parker and Riches, 1993).

According to Musselman (1980), seeds of Orobanchaceae will not germinate unless afterripening is followed by a period of warm stratification, which is called "conditioning" (Musselman, 1980; Pieterse, 1981; van Hezewijk et al., 1993) or "preconditioning" (Jain and Foy, 1992) by various people who study the germination of parasitic plants. Because "preconditioning" has been used to describe how environmental factors acting on the mother plant influence germination (Chapter 8), we suggest that "conditioning" rather than "preconditioning" be used in reference to preparing seeds of parasitic species for germination. Information available on Orobanchaceae seeds that require conditioning prior to germination indicates that their germination ecology is somewhat similar to that of nonparasitic winter annuals. In fact, it seems reasonable to think of these parasites as winter annuals because annual Orobanchaceae in regions with a Mediterranean climate complete their life cycle during the cool wet season (Mandaville, 1990).

After seeds of nonparasitic winter annuals are placed at low temperatures (5, 15/6°C) in a refrigerator or incubators, no dormancy loss occurs in some species, but in others seeds will gain the ability to germinate at low temperatures only (Baskin and Baskin, 1986). With each increase in the dormancy-breaking temperature, the rate of dormancy loss and the maximum temperature for germination increase (Baskin and Baskin, 1984a, 1986). As seeds first begin to come out of dormancy in early summer, they will germinate at low temperatures, but with additional loss of dormancy during mid- to late summer the maximum temperature at which they germinate increases. Thus, if seeds are exposed to high summer temperatures (25/15, 30/15, 35/20°C) for 3–4 months they can germinate over a wide range of temperatures, including those that occur in the field in temperate regions during autumn.

Seeds of *Orobanche crenata* routinely are given a 2-week period of conditioning at 25°C (Pieterse, 1981) or

23°C (Pieterse, 1991) and those of *O. aegyptiaca* a 10-day period at 22–25°C (Jain and Foy, 1992) before germination studies are initiated. Seeds of *Aeginetia indica* required 40–45 days, or longer, to germinate at 20, 15, 30, or 26–28°C, even if they were given chemical or heat shock treatments (French and Sherman, 1976). Thus, before they will germinate, seeds of various Orobanchaceae require a period of conditioning at temperatures that break physiological dormancy in seeds of nonparasitic winter annuals. However, seeds of Orobanchaceae must be on a moist substrate for loss of physiological dormancy (French and Sherman, 1976; Pieterse, 1981; Jain and Foy, 1992), whereas those of nonparasitic winter annuals will come out of dormancy if stored dry at 25–30°C (Baskin and Baskin, 1983).

Seeds of *Orobanche aegyptiaca* exhibit an increase in respiration and synthesis of proteins during conditioning. If conditions are unsuitable for germination at the end of the conditioning period, rates of respiration and protein synthesis decline, and seeds remain viable for long periods of time. Seeds can be dried after a conditioning pretreatment with no loss of viability (Nun and Mayer, 1993). When conditions become nonlimiting for germination, a second peak of respiratory activity and protein synthesis occurs, and if the germ tube fails to reach a host root, the seed/seedling dies (Nun and Mayer, 1993).

Seeds of *Orobanche crenata* were placed on a wet substrate at 0, 5, 10, 15, 20, 25, 30, and 35°C for 0–8 weeks, after which they were tested for germination at 20°C. Seeds pretreated at 0 and 35°C failed to germinate at 20°C, whereas those pretreated at 5–30°C showed an increase in rate and percentage of germination at 20°C with each increase in temperature (van Hezewijk *et al.,* 1993). Unfortunately, only one germination temperature was used in this study. Thus, it is not known if the temperate range at which seeds would germinate increased or if seeds at higher temperatures gained the ability to germinate over a wider range of temperatures than those held at low temperatures. Seeds of *O. crenata* and *O. aegyptiaca* gave the highest germination percentages when they were pretreated on wet filter paper for 3 or 5 weeks at 18, 23, or 28°C and then tested for germination at the same temperatures (Kasasian, 1973).

Seeds of some nonparasitic winter annuals subjected to decreasing temperatures in late autumn and winter lose the ability to germinate at all temperatures (dormant seeds) (Baskin and Baskin, 1984b), whereas those of other species lose the ability to germinate at high but not at low temperatures (conditionally dormant) (Baskin and Baskin, 1984a). Depending on the species, an autumn thermoperiod of 20/10°C, which may be optimal for germination of nondormant seeds, causes seeds to lose the ability to germinate at high temperatures. Later, when temperatures decrease to 15/6°C, or lower,

seeds also lose the ability to germinate at low temperatures (Baskin and Baskin, 1984b).

The germination of *O. crenata* seeds declined after prolonged (>7 weeks) exposure to 10, 15, and 20°C, with the decline increasing with a decrease in temperature. The decrease in germination was not due to loss of viability, and it was concluded that seeds had entered secondary dormancy (van Hezewijk *et al.,* 1993). Although an 8-day cold stratification treatment broke dormancy in *Aeginetia indica* seeds, dormancy was reinduced in seeds kept at 3–5°C for 36 days or longer (French and Sherman, 1976).

Densities of up to 4 million *O. crenata* seeds m^{-2} have been recorded in Spain (Lopez-Granados and Garcia-Torres, 1993), and buried seeds can live for 10 or more years (Kadry and Tewfic, 1956b; Zahran, 1982). Thus, in view of the great difficulty in eliminating seeds from agricultural soils (ter Borg, 1987) and the detrimental effects of this genus on crops (Parker and Riches, 1993), it is imperative that we learn more about dormancy in buried seeds. Specifically, do seeds lose the ability to germinate when exposed to low temperatures in winter? Do they lose the ability to germinate only at high temperatures, and thus enter conditional dormancy?

As noted by Parker and Riches (1993), farmers have learned that a delay in the sowing date of crops such as faba beans and lentils in autumn until temperatures have declined results in a decrease in *Orobanche* infestations (Silim *et al.,* 1991; van Hezewijk *et al.,* 1987). As the season progresses from autumn to winter, temperatures decrease and so does the germination of *Orobanche* seeds. Temperatures may be below those required for germination and/or seeds may enter secondary dormancy. Seeds of *O. crenata* sown in a faba bean field in Cordoba, Spain, in October, November, December, and January germinated in each of these months, with the peak occurring in November or December. Germination of seeds in January of two different years, when minimum soil temperatures ranged from about 0 to −3°C and from 7 to 4°C, respectively (Mesa-Garcia and Garcia-Torres, 1986), may indicate that seeds had only entered conditional dormancy. Also, dormancy break occurs at 5°C in seeds of *O. crenata* (van Hezewijk *et al,* 1993); therefore, seeds that were dormant in autumn may have come out of dormancy during winter. Thus, germination may be possible in this species in early spring if soil moisture and host plants are available. However, van Hezewijk *et al.* (1987) suggest that maturing plants of faba beans or lentils are not infected in spring because their roots are too deep in the soil.

Jacobsohn *et al.* (1987) and Lopez-Granados and Garcia-Torres (1993) placed the tiny seeds of *Orobanche* in bags, buried them in soil in the field, and then exhumed and tested them, demonstrating that it is possible to monitor changes in dormancy states (or lack thereof) in buried *Orobanche* seeds throughout the year. However, van Hezewijk (1994) actually has monitored changes in dormancy states in seeds of *O. crenata*. She buried seeds in fine-mesh nylon bags at depths of 5, 15, and 45 cm in soil in Syria and at depths of 8 and 23 cm in Spain. Seeds were exhumed at 1.5- to 2-month intervals for 18 and 22 months, respectively, and were tested for germination at 20°C, with and without a warm stratification pretreatment of 11 days at 20°C. During the germination tests, seeds were treated with the synthetic germination stimulant GR24. Seeds exhibited seasonal changes in their ability to germinate at 20°C (Fig. 11.3), like those of a winter annual. In general, seeds germinated to high percentages from midsummer or autumn until early winter and to low percentages in midwinter, spring, and early summer. As noted by van Hezewijk (1994), use of only one test temperature did not allow her to determine if seeds were "truly dormant" or in conditional dormancy in spring. Regardless of the actual dormancy state, few seeds could germinate in spring when temperatures in the field are about 20°C. It is hoped that more test temperatures can be used in future studies to determine if seeds can germinate at temperatures lower than 20°C during winter.

In addition to the weedy *Orobanche* species that extend into the temperate zone from Mediterranean regions, there are indigenous species of *Orobanche* and other genera in the Orobanchaceae in the temperate zone. However, little is known about the germination ecology of Orobanchaceae in the temperate region. In fact, it is not known if the underdeveloped embryos have PD. It seems logical that physiological dormancy in embryos would be broken during cold stratification in winter and that seeds would germinate in spring. In *Orobanche cernua* and *O. ramosa*, which can be serious pests on summer crops such as sunflowers and tobacco in the temperate region (Parker and Riches, 1993), it would not be surprising to find at least some varieties whose seeds have the same dormancy-breaking and germination requirements as those of nonparasitic summer annuals. Seeds (6-month-old) collected from *O. crenata* plants growing in faba bean fields in southern Spain germinated to 85% after a 24-hr cold stratification treatment at 0°C (Lopez-Granados and Garcia-Torres, 1996). Certainly, some species require high, rather than low, temperatures for germination. Seeds of *O. cernua* germinate in Spain as soil temperatures increase in May–July (Garcia-Torres *et al.*, 1987; Castejon-Munoz *et al.*, 1993), and those of *O. ramosa* germinate in July and August, when soil temperatures are at their yearly maximum (Garcia-Torres *et al.*, 1987). An exciting study would be to bury seeds of these species in soil in temperate and Mediterranean regions and monitor their dormancy state throughout the year. Would seeds come out of dormancy during winter and enter secondary dormancy in late summer or early autumn, like those of the summer annual grass *Panicum capillare* (Baskin and Baskin, 1986)?

Seeds of *Epifagus virginiana*, a parasite on *Fagus grandifolia* trees in temperate deciduous forests of eastern North America (Musselman and Mann, 1978), germinated to 57% on soil following 150 days of (presumably dry) storage at 4°C (Williams and Zuck, 1986). Because seeds "in stoppered test tubes" probably were too dry for dormancy-breaking events to occur, the only type of dormancy in these seeds may be morphological underdevelopment of the embryo.

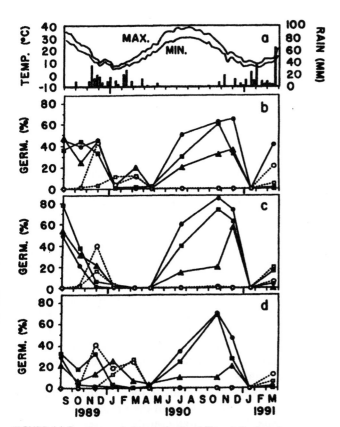

FIGURE 11.3 Seasonal changes in the ability of *Orobanche crenata* seeds to germinate at 20°C following various periods of burial at three soil depths in a field in Syria. (a) Mean daily maximum and minimum weekly temperatures at a soil depth of 10 cm in Syria and weekly rainfall (bars). Seeds were collected in (b) Syria, (c) Egypt, and (d) Spain and buried at depths of 5 (O ●), 15 (□ ■), or 45 (▲) cm in Syria. The immediate response to germination stimulant is represented by open symbols and dotted lines. The response to the germination stimulant after optimal conditioning is shown by closed symbols and solid lines. From van Hezewijk *et al.* (1994), with permission.

3. Germination Requirements

After dormancy-breaking treatments were completed, seeds of Orobanchaceae germinated to high percentages when incubated at the following temperatures: *Orobanche aegyptiaca,* 22–25°C (Jain and Foy, 1992); *O. crenata,* 22°C (Pieterse, 1991); 25°C (Pieterse, 1981); and 15–20° (van Hezewijk *et al,* 1991, 1993); *Aeginetica indica,* 25–30°C (French and Sherman, 1976); and *Epifagus virginiana,* 26–27°C (Williams and Zuck, 1986). The optimum germination temperature for seeds of *A. indica* was 25–30°C (French and Sherman, 1976), and for those of *O. crenata* it was 15–20°C (van Hezewijk *et al.,* 1991). Pieterse (1979) concluded that temperatures of 20–25°C were the best for germination studies of Orobanchaceae. Alternating temperatures of 25/15 and 10/5°C were neither better nor worse than constant temperatures for the germination of *O. crenata* seeds; however, more seeds germinated at 25/15 than at 10/5°C (van Hezewijk, 1994).

Because haustoria of the Orobanchaceae become attached to roots of the host plant, it makes sense that seeds of these parasites could germinate in darkness. Thus, germination studies often are done in darkness (Jain and Foy, 1992); in a few studies, comparisons of germination in light and darkness have been made. *Orobanche crenata* seeds germinated to higher percentages in darkness than in light (Hiron, 1973; Pieterse, 1981), whereas those of *Aeginetia indica* required darkness for germination (French and Sherman, 1976). Light inhibited the germination of old, but not new, seeds of *O. aegyptiaca* (Jain and Foy, 1987). However, seeds of *Epifagus virginiana* germinated to a higher percentage in light than in darkness (Williams and Zuck, 1986).

Studies using soil-filled plexiglass vessels buried in the field showed that seeds of *Orobanche* germinated to higher percentages on saturated than on dry (<40% water, dry weight basis) soil (Linke and Vogt, 1987). In addition, seeds germinated further from the host root and germ tubes were longer in wet than in dry soil.

When seeds of holoparasites (i.e., Orobanchaceae) are warm stratified and placed in darkness at appropriate temperatures, few, or no, seeds may germinate unless a host root or aqueous solutions of root exudates are present (Sunderland, 1960; Brown, 1965; Abu-Shakra *et al.,* 1970; Edwards, 1972; Hiron, 1973; Farah, 1987) for at least 24–48 hr (Hiron, 1973). This observation has prompted investigators to conclude that germination in the Balanophoraceae, Orobanchaceae, Rafflesiaceae, Hydnoraceae, and Lennoaceae requires a chemical stimulant from the host plant's roots (Stewart and Press, 1990).

Considerable effort has been put into trying to identify the chemicals that promote germination in the Orobanchaceae, especially the weedy species of *Orobanche.* However, isolation and identification of these stimulators are quite difficult, and we still do not know much

about what they are or how they work. Problems are many:

1. The amount of stimulatory compounds produced per root appears to be very low (Brown *et al.,* 1951a; Hauck *et al.,* 1992).
2. Molecules are unstable at high temperatures, under alkaline conditions (Brown *et al.,* 1951b), or in the presence of mineral acids (Mallet, 1973).
3. Root exudates that stimulate germination are mixtures of compounds (Brown *et al,* 1949).
4. Roots of different host species may produce different mixtures of stimulants (Sunderland, 1960; Abu-Shakra *et al.,* 1970).
5. Root exudates may contain germination inhibitors (Whitney, 1979).
6. Stimulants are produced for varying periods of time by roots of different species (Khalaf, 1992).
7. Seeds of *Orobanche* have endogenous germination stimulators (Hiron, 1973).
8. The amount of stimulant produced by host roots varies with the season in which a plant is grown (Lopez-Granados and Garcia-Torres, 1996).

Roots of some species produce chemicals that stimulate the germination of *Orobanche* seeds, but the plants producing them are not parasitized (Brown *et al.,* 1951a). For example, exudates of flax roots stimulated more seeds of *O. ramosa* to germinate than those of sorghum or tomato roots (Hameed *et al.,* 1973). However, flax was not parasitized, whereas sorghum and tomato were parasitized. Flax and various other species have been used to promote the germination of *Orobanche* seeds, and thus they are called "trap" or "catch" crops. However, some highly virulent subspecies of *O. aegyptiaca* also will parasitize flax (Kleifeld *et al.,* 1994).

Another problem in studies of germination stimulants could be seasonal variability in the sensitivity of *Orobanche* seeds to the compounds, i.e., if seeds are nondormant they may be very sensitive and germinate at very low concentrations of the stimulants, whereas conditionally dormant seeds may be insensitive and germinate only if concentrations are quite high. Also, the concentration of stimulants in root exudates could vary seasonally.

Attempts are being made to identify the germination stimulants in root exudates of the principal host species for *Orobanche.* A partially purified germination stimulant from faba bean root exudates possibly may be an unsaturated lactone (Mallet, 1973). Thin-layer chromatography (TLC) analysis of root exudates of broad bean, tomato, and flax suggests that the stimulatory compound has an unsaturated lactone, a carbonyl group, and C—C and C=C bonds. Further, the nuclear magnetic resonance (NMR) spectrum indicated an alycyclic structure or an aliphatic chain (Al-Menoufi *et al.,* 1987). To date, strigol, or its analog, is the most promising chemical to promote the germination of buried seeds of *Orobanche* in fields (Saghir, 1979); however, strigol has not been isolated from root exudates of *Orobanche* host plants.

Increased soil nutrients (Bischof and Kock, 1973) or the use of nitrogen fertilizers can cause a reduction of *Orobanche* infestations in crop species (Kasasian, 1973; Abu-Irmaileh, 1981; Jain and Foy, 1987). Thus, studies have been done to determine the effects of various nitrogenous compounds on the germination of *Orobanche* seeds. The presence of sodium nitrate during warm stratification of *O. aegyptiaca* seeds stimulated germination, but ammonium sulfate and urea inhibited it (Jain and Foy, 1987, 1992). Exposure to urea and ammonium sulfate during warm stratification also inhibited the germination of *O. crenata* (Pieterse, 1991). In contrast, van Hezewijk (1994) found that ammonium sulfate supplied during warm stratification did not inhibit germination, unless (1) a nitrification inhibitor was present and (2) germ tube growth was normal. Neither germination nor germ tube growth was inhibited by the presence of urea during warm stratification.

The presence of 8 mM ammonium sulfate (and a nitrification inhibitor) during conditioning reduced the germination of *Orobanche crenata* seeds from 46 to 26%, but 8 mM ammonium sulfate alone had no effect (van Hezewijk and Verkleij, 1996). Application of 4 mM ammonium sulfate during the germination phase reduced germination from 50 to 16%, and the addition of a nitrification inhibitor reduced it to 2%. Urea had limited effects on germination and germ tube growth, and when a urease inhibitor was added there was no effect. Nitrate did not inhibit germination or germ tube growth (van Hezewijk, 1994). van Hezewijk suggested that ammonia inhibits the germination of *O. crenata* seeds because they have a low ability to detoxify it. Ammonium nitrate applied during the germination phase decreased germination as the concentration increased from 0 to 100 ppm; however, root exudates had to be present for any germination to occur, even at 0 ppm ammonium nitrate (Abu-Irmaileh, 1994). Regardless of the ammonium nitrate concentration, the stimulatory effects on germination of root exudates decreased: flax > lentil > pepper > tomato > wheat (Abu-Irmaileh, 1994).

Although nitrogen fertilizers can cause a decrease in soil pH, it is unlikely that a change in pH accounts for a decrease in *Orobanche* infestations of crops (van Hezewijk, 1994). In fact, seeds of *O. crenata* germinated equally well between pH 5 and 8.5; however, germ tube length increased with an increase in pH. At pH 4, only 0.7% of the seeds germinated, but the length of germ tubes was not reduced (van Hezewijk, 1994).

Preemergence herbicides such as imazethapyr, imazapyr, or chlorsulfuron can provide good control of *Orobanche cernua* in *Helianthus annuus* fields if the soil is relatively dry when the compound is applied and crop seeds are sown (Garcia-Torres *et al.,* 1994). Some success has been achieved in controlling *O. ramosa* in tobacco using foliar sprays of sulfosate or imazaquin; glyphosate was effective in controlling *O. ramosa* but it damaged the tobacco leaves (Lolas, 1994).

B. A Holoparasite with Chlorophyll: *Cuscuta*

The genus *Cuscuta* usually is thought of as a holoparasite, but many species of this genus have a small amount of chlorophyll, especially in the seedling stage of the life cycle (Varrelman, 1937). Thus, plants are able to carry on limited photosynthesis (MacLeod, 1961; Pattee *et al.,* 1965; Baccarini, 1966). However, some species, such as *C. europaea,* lack chlorophyll (Machado and Zetsche, 1990), and high irradiance can reduce pigment levels in those with chlorophyll (Panda and Choudhury, 1992). In species with chlorophyll, the amount of chlorophyll is low, the number of thylakoids in the chloroplasts is reduced, and no typical grana are present (Baccarini, 1966; Machado and Zetsche, 1990). Thus, a very limited amount of photosynthesis is possible, but it is not enough to support the plant after carbohydrate reserves of the seed/seedling are exhausted (Baccarini, 1966; Dinelli *et al.,* 1993). For all practical purposes, *Cuscuta* is a holoparasite and depends on a host plant for photosynthate. In fact, *Cuscuta* is very efficient in removing assimilates (Tsivion, 1978; Dinelli *et al.,* 1993) and leaves little or nothing for developing seeds of the host (Wolswinkel, 1974).

Seeds of *Cuscuta* have long, coiled embryos (Martin, 1946) that are interpreted to be miniature seedlings (Rao and Rama Rao, 1990). These seedlings consist of a radicular end that lacks a meristem and a short stem with an apical meristem and scale-like leaves (Rao and Rama Rao, 1990). The coiled mature embryo, which lacks a true radicle, is one of the many characteristics listed by Johri (1987a) to justify placing *Cuscuta* in the Cuscutaceae rather than the Convolvulaceae.

The root does not grow at the time of germination and serves as a support structure for the shoot. After the shoot produces a haustorium and attaches to a host, the root dies (Rao and Rama Rao, 1990; Parker and Riches, 1993). The period of time a seedling can live before attaching to a host and the length that can be attained by the shoot prior to reaching a host plant vary with the species (Parker and Riches, 1993).

1. Types of Seed Dormancy

The main cause of seed dormancy in *Cuscuta* is impermeability of the seed coat. In a young developing seed, the seed coat consists of four layers: external epidermis, hypodermis, palisade, and parenchyma cells. Cells of the palisade layer are thick-walled and mature into macrosclereids (Hutchison and Ashton, 1979), accounting for physical dormancy of the seeds. Seeds of *C. pedicellata* have two palisade layers (Lyshede, 1992), whereas those of *C. campestris* (Hutchison and Ashton, 1979) and apparently many other species of *Cuscuta* have only one palisade layer (Corner, 1976).

Various techniques for breaking the seed coat are used by people who study the biology and control of *Cuscuta*. These methods include (1) abrasion with sand paper (Ashton and Santana, 1976; Hutchison and Ashton, 1979; Rao and Reddy, 1991), (2) pricking with a pin (Hutchison and Ashton, 1980; Rao and Reddy, 1991), and (3) acid scarification (Midgley, 1926; Tingey and Allred, 1961; Gaertner, 1950; Hassawy, 1973; Hutchison and Ashton, 1979; Rao and Reddy, 1991). Scarification results in high germination percentages of seeds of many *Cuscuta* species (Table 11.1), thus seeds have physical dormancy. However, seeds of some species also have physiological dormancy. Seeds of *C. approximata* have both physical and physiological dormancy, thus they do not germinate unless cold stratification follows scarification (Tingey and Allred, 1961; Allred and Tingey, 1964). Seeds of *C. epilinium* and *C. epithymum* did not germinate following acid scarification (Gaertner, 1950), indicating the presence of physiological dormancy. Do seeds of these species have both physical and physiological dormancy? Seeds of *C. europaea* do not require scarification and germinate following (and during) cold stratification (Gaertner, 1956). Did cold stratification make seeds of *C. europaea* permeable? Imbibition studies need to be done on seeds of *C. europaea*, *C. epilinium*, and *C. epithymum*.

TABLE 11.1 Species of *Cuscuta* Whose Seeds Germinate following Scarification, thus Indicating the Presence of Physical Dormancy

Species	% germination	Reference
Cuscuta californica	70	Gaertner (1950)
C. campestris	100	Hutchinson and Ashton (1979)
C. cephalanthi	100	Gaertner (1950)
C. chinensis	68, 91	Hassawy (1973), Rao and Reddy (1991)
C. compacta	60	Gaertner (1950)
C. glomerata	80	Gaertner (1950)
C. gronovii	88	Gaertner (1950)
C. indecora	100, 89	Gaertner (1950), Tingey and Allred (1961)
C. suaveolens	100	Gaertner (1950)

2. Dormancy-Breaking Requirements

Although dormancy in seeds of many *Cuscuta* species is broken by scarification in the laboratory, little is known about how seed coats become permeable under natural environmental conditions. However, seeds seem to have definite germination seasons in nature. Seeds of *C. campestris* sown in the field in Davis, California, in August germinated the following March and April (Hutchison and Ashton, 1980), indicating that winter conditions promoted loss of dormancy. Further, seeds exhumed after burial in the field in California from November to March and tested in darkness at 30°C germinated to 42%. Seeds stored dry at 3–8°C and in the field (in a jar) for 13 months germinated to 60 and 83%, respectively (Hutchison and Ashton, 1980). Thus, drying and/or exposure to low and/or high temperatures caused seed coats to become permeable. The appearance of *C. glomerata* in prairies in years when there has been a fire suggests that high temperatures of the fire may have caused seeds to become permeable (McCormac and Windus, 1993). Preliminary evidence suggests that freezing and thawing may not be very effective in making seeds of *Cuscuta* permeable. Midgley (1926) subjected seeds of a *Cuscuta* sp. to 20 cycles of freezing and thawing and only obtained 10% germination. Seeds of more species need to be tested for their response to freezing and thawing.

Physiological dormancy in seeds of *Cuscuta* spp. has been reported only in temperate regions and is broken by cold stratification during winter. The amount of cold stratification required to break physiological dormancy varies from 2–3 weeks in seeds of *C. approximata* (Tingey and Allred, 1961) to 3 months in those of *C. europaea* (Gaertner, 1956). It seems logical that some *Cuscuta* spp. in subtropical regions also could have physiological dormancy, either alone or in combination with physical dormancy, and that high temperatures would be required for dormancy break.

3. Germination Requirements

Seeds of *Cuscuta* do not require a chemical stimulus from the host plant to germinate (Rao and Reddy, 1991; Parker and Riches, 1993).

Temperature requirements for germination are relatively high, and 30/25 (Hassawy, 1973), 30 (Hutchison and Ashton, 1979), 22 (Tingey and Allred, 1961), and 38°C (Rao and Reddy, 1991) have been used for germination tests. Maximum and minimum temperatures for the germination of *C. campestris* seeds were 36 and 16°C, respectively, and optimum temperatures were 30–33°C (Allred and Tingey, 1964; Hutchison and Ashton, 1980). Maximum and minimum temperatures for the germination of *C. approximata* seeds were 35 and 2°C, respectively, with 16°C being optimal (Allred and Tingey, 1964). The optimum temperature for the germination of seeds of *C. trifolii, C. prodani,* and *C. tinei* was 20°C and for those of *C. campestris, C. pentagona, C. lupuliformis,* and *C. monogyna* 30–33°C (Stojanovic and Mijatovic, 1973). In field studies in California, seeds of *C. approximata* and *C. indecora* began to germinate when soil temperatures were about 4 and 11°C, respectively; those of *C. campestris* began germinating a few days later than those of *C. indecora* (temperatures were not given). Seeds of *C. approximata* stopped germinating by mid-May, but those of *C. indecora* and *C. campestris* continued to germinate into summer (Allred and Tingey, 1964).

Studies have not been done to determine the light:dark requirements for the germination of *Cuscuta* seeds, and some authors do not say if they incubated seeds in light or darkness e.g., Gaertner, 1956; Tingey and Allred, 1961; Allred and Tingey, 1964; Stojanovic and Mijatovic, 1973; Rao and Reddy, 1991). Seeds of *C. campestris* germinated to 100% in darkness (Hutchison and Ashton, 1979), and those of *C. chinensis* germinated to 68% when covered with 5 mm of soil (Hassawy, 1973). A soil depth of 10 mm reduced the germination of *C. trifolii* seeds from about 50 to 15% but increased the germination of *C. campestris* seeds from about 13 to 20% (Stojanovic and Mijatovic, 1973). Depending on the type of soil, seedlings of *C. campestris* emerged when scarified seeds were sown at depths of 25–80 mm, with maximum emergence occurring from depths of 3–5 mm (Hutchison and Ashton, 1980).

C. A Hemiparasitic Member of the Lauraceae: *Cassytha*

There are about 20 species in the genus *Cassytha* and the center of diversity is Australia (Parker and Riches, 1993). Also, several species occur in Africa, and *C. filiformis* is found throughout the New and Old World tropics (Kuijt, 1969). Plants of *Cassytha* are similar to those of *Cuscuta* and can be confused with them. A few roots develop on newly germinated seedling of *Cassytha,* but after attachment to a host they die (Kuijt, 1969). Stems and the scale-like leaves are green, but rates of photosynthesis can be quite low (de la Harpe *et al.,* 1981).

Seeds of Lauraceae have large, investing embryos and no endosperm (Martin, 1946). However, nothing is known about types of seed dormancy, dormancy-breaking requirements, or germination requirements of nondormant seeds of *Cassytha* spp. Because seeds have large, fully developed embryos (Sastri, 1962), they would not have morphological or morphophysiological

dormancy. The dispersal unit is a drupe with the endocarp containing a palisade layer of lignified cells (Corner, 1976); thus, there is a possibility that they have physical dormancy. Imbibition studies need to be done on scarified and nonscarified seeds to determine if the endocarp is permeable to water.

D. Rooted Hemiparasites Whose Seeds Require a Host Stimulus for Germination: *Striga* and *Alectra*

The hosts of *Striga* and *Alectra* (Scrophulariaceae) are grasses and legumes, respectively, and many species in these two genera are important economic plants (Parker and Riches, 1993). Plants of these hemiparasites have green leaves, but the amounts of chlorophyll present and the levels of photosynthetic activity are lower than those in nonparasitic plants (Parker and Riches, 1993). Further, respiration rates are high (Graves *et al.,* 1990, 1992); therefore, plants may have little, or no, net carbon gain from their own photosynthesis. Thus, much of the photosynthate for growth and seed production is derived from the host (Gouws *et al.,* 1980; Parker and Riches, 1993), and plants essentially behave as holoparasites.

Like members of the Orobanchaceae, *Striga* and *Alectra* have "dust" seeds. Seed lengths range from 0.2 to 1.0 mm and weights from 0.3 to 12.4 μg, with as many as 200–1800 seeds being produced per capsule (Hartman and Tanimonure, 1991; Parker and Riches, 1993). Some seeds live in the soil for at least 14 years (Bebawi *et al.,* 1984; Eplee and Westbrooks, 1990), and the mean number of seeds in some agricultural fields may be as high as 12 (Visser and Wentzel, 1980; Hartman and Tanimonure, 1991) or 17.9 (Smith and Webb, 1996) per 100 g of soil. Seeds have a heart-shaped embryo with two cotyledons and a radicular pole, and the embryo is surrounded by an endosperm consisting of one to several layers of cells (Nwoke and Okonkwo, 1978; Okonkwo and Nwoke, 1978).

In the germination of *Striga* and *Alectra,* the radicle pushes through the micropylar end of the seed, leaving the plumular pole inside the seed coats (Nwoke and Okonkwo, 1978; Okonkwo and Nwoke, 1978). Because food reserves in the endosperm obviously are very limited, growth of the radicle does not exceed 2–4 mm, depending on the species (Williams, 1960; Okonkwo, 1975; Okonkwo and Raghavan, 1982). In the presence of a germination stimulus, radicles of *S. gesnerioides* grew to a maximum length of about 2 mm in 11 days, whereas those of *A. vogelii* reached a maximum length of 3 mm in 8 days (Okonkwo and Raghavan, 1982). If the radicle fails to reach a host root by the time the food reserves are exhausted, the seedling dies (Williams, 1961b; Okonkwo, 1987).

If a radicle is 0–4 mm away from the host root, it may actually curve toward the root (Williams, 1960, 1961b). A chemical signal or stimulus is required for the radicle to start developing into a haustorium, but the compounds are not the same ones that promote germination (Stewart and Press, 1990; Press *et al.*, 1990). In fact, many phenolic compounds and various cytokinins will induce the formation of haustoria in *S. asiatica* and *S. hermonthica* (Timko *et al.*, 1989). A specific haustorium inducer for *Striga* has been isolated from root exudates of the host plant *Sorghum bicolor* and has been identified as 2,6-dimethoxy-*p*-benzoquinone (Chang and Lynn, 1986). The formation of haustoria is temperature and time dependent, occurring optimally at relatively high temperatures (27, 30°C) and limited to the first week following germination (Riopel *et al.*, 1990).

After the radicle of the parasite comes in contact with the appropriate chemical stimulus, growth stops and its tip swells and produces structures resembling root hairs (Williams, 1960, 1961a,b; Riopel and Baird, 1987; Press *et al.*, 1990) (Fig. 11.4). The hairs are important in the initial contact between the radicle and the host root (Nickert *et al.*, 1979). When the radicle joins a host plant root, it penetrates the cortex by dis-

solving cells and makes contact with protoxylem vessels of the host's stele. Thus, the radicle becomes the primary haustorium, and its presence stimulates cells of the pericycle of the host root to divide. Eventually, cells derived from the pericycle and the haustorium itself result in the formation of a large tuberous haustorium (Uttaman, 1950; Nwoke and Okonkwo, 1978; Okonkwo and Nwoke, 1978). In *Alectra vogelii*, some of the pericycle-derived cells in the haustorium give rise to roots (Nwoke and Okonkwo, 1978). After the xylem of the host and haustorium become connected, the plumule of the embryo gives rise to a shoot (Nwoke and Okonkwo, 1978; Okonkwo and Nwoke, 1978). Roots produced from the base of the shoot can develop into secondary haustoria (Okonkwo and Nwoke, 1975a), and a parasite may become attached to the same host root more than once or to different roots of the same host plant (Parker and Riches, 1993).

1. Types of Seed Dormancy

Conventional wisdom regarding the germination of *Striga* and *Alectra* is that seeds require dormancy-breaking treatments before germination is possible (e.g., Brown and Edwards, 1944; Musselman, 1980; Patterson, 1987; Dawoud and Sauerborn, 1994). However, no studies have specifically addressed the question of what kind of dormancy the seeds have.

Photographs of longitudinal thin sections of imbibed seeds of *Alectra vogelii* (Nwoke and Okonkwo, 1978) and *Striga gesnerioides* (Okonkwo and Nwoke, 1978) show that seed coats lack a palisade layer of macrosclereids and thus would be permeable to water. Thus, seeds do not have physical dormancy. The minute seeds of *Striga* and *Alectra* fit Martin's (1946) description of dwarf rather than micro seeds because they have differentiated embryos (Nwoke and Okonkwo, 1978; Okonkwo and Nwoke, 1978). The embryo in dwarf seeds is large relative to the total size of the seed, and little embryo growth is possible without rupturing the seed coat. Thus, morphological dormancy is not possible in dwarf seeds such as those of *Striga* and *Alectra.*

If seeds of *Striga* and *Alectra* do not have physical or morphological dormancy, that leaves only physiological dormancy (PD) as a possibility. As seen later, the dormancy-breaking treatments for seeds of those two genera (dry storage at room temperatures and warm stratification) also will break physiological dormancy in seeds of other species, especially those with nondeep PD. Also, seeds of *S. asiatica* germinate if they are acid scarified or if the seed coat is punctured at the radicle end (Egley, 1972). These data indicate that embryos lack sufficient growth potential to push through the seed coat and that seeds have physiological dormancy. Seeds

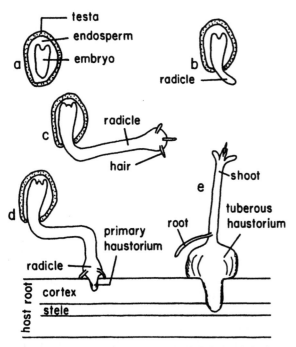

FIGURE 11.4 Germination morphology of *Striga* seeds in the presence of a host plant root. Drawn from information in Okonkwo and Nwoke (1975, 1978), Okonkwo and Raghavan (1982), and Riopel and Baird (1987).

treated with root exudates (germination stimulants) from cotton germinated to higher percentages under osmotic stress than those in controls (Egley, 1972). Thus, germination stimulants appear to increase the growth potential of the embryo; however, as seen later, they also may increase ethylene production in seeds.

2. Dormancy-Breaking Requirements

Seeds of *Striga* and *Alectra* require a period of afterripening or postharvest ripening followed by conditioning (warm stratification) before they will germinate (Botha, 1948; Okonkwo and Nwoke, 1975b; Musselman, 1980; Patterson, 1987). The afterripening requirement for germination varies with species and storage temperature. The afterripening phase was completed after 6–8 weeks in seeds of *S. asiatica* stored dry at 100% relative humidity (RH) at 31°C; however, after 20 weeks only 17% of the seeds were viable (Table 11.2). In contrast, afterripening was slower at 0, 5, and 24°C than at 31°C, but most of the seeds were viable after 20 weeks. Seeds of *S. hermonthica* stored dry at 0, 60, and 100% RH at both 15 and 35°C afterripened faster under 100% RH at 35°C than at the other conditions (Egbers *et al.*, 1991). However, the highest number of germinating seeds was from the 60% RH at 35°C treatment. Seeds at high temperature and high RH lost viability faster than those at low temperature and low RH. Thus, like seeds of winter annuals (Baskin and Baskin, 1986), those of *Striga* afterripen more completely at high than at low temperatures. Further, like seeds of most species that do not have physical dormancy (Chapter 7), high temperatures and high RHs promoted loss of viability of *Striga* seeds (Egbers *et al.*, 1991). Very little is known about *Alectra* seeds.

More attention needs to be given to the afterripening phase of *Striga* and *Alectra* seeds, not only to learn more about it, but to ensure that the results of different investigators are comparable. In a few studies (e.g., Brown and Edwards, 1944; Kust, 1963; Bebawi *et al.*, 1984; Pieterse *et al.*, 1993), freshly matured seeds were

tested for germination. With this baseline information, it then is possible to determine how much change in germination over time can be attributed to the afterripening phase of dormancy loss. As seeds age, (1) their sensitivity to conditioning treatments increases, (2) the ability of conditioned seeds to germinate in the absence of a host chemical stimulus increases, and (3) the ability of seeds to be conditioned at low (12°C) temperatures increases (Vallance, 1950). For example, seeds of *S. asiatica* held in dry storage at 23–24°C for 27 months required only 11 days of conditioning to germinate to 77%, whereas those stored dry for 12–16 months germinated to only 2–22% after 11 days of conditioning (Worsham, 1961).

In some studies (e.g., Patterson *et al.*, 1982; Hauck *et al.*, 1992), the age of seeds at the time they were used in experiments is not given; thus, there is no way to know if the afterripening phase was completed. In other studies (e.g., Egley, 1972; Okonkwo and Nwoke, 1975b; Bebawi *et al.*, 1986; Hsiao *et al.*, 1988a), seeds were 1 or more years old, and presumably the afterripening phase was complete when the studies began. Unless investigators use seeds that are in the same state of the afterripening phase, or those that have completed it, differences between studies may be due to variations in the physiological state of the seeds rather than to experimental treatments.

Afterripened seeds of *Striga* and *Alectra* fail to germinate even in the presence of host stimulants unless they receive a period of conditioning (Brown and Edwards, 1944; Visser, 1975). For this treatment, seeds are sown on a substrate moistened with water and placed in darkness at 20–30°C (Table 11.3). Is darkness required for conditioning? Not much information is available with which to answer this question, except for seeds of *S. asiatica*. Light given during the conditioning period did not prevent conditioning from occurring, but it could inhibit germination after the stimulant was applied (Worsham, 1987). Conditioned seeds exposed to light and then placed in darkness at 29°C for 1 day germinated

---

TABLE 11.2 Germination Percentages of *Striga asiatica* Seeds[a] and Percentage Viability after 20 Weeks[b]

Storage conditions		Weeks of storage							
Temperature (°C)	RH (%)	2	4	6	8	12	16	24	% viability after 20 weeks
0	100	3	8	16	30	38	39	43	94
5	50	3	7	25	50	60	68	77	96
24	50	3	15	25	48	67	72	88	95
31	100	3	64	88	92	52	17	0	17

[a] Seeds were incubated at 31°C following storage at 50 or 100% RH at 0, 5, 24, or 31°C for 2 to 24 weeks.
[b] Data from Kust (1963).

TABLE 11.3 Optimum Periods of Time and Temperatures for Conditioning Seeds of Alectra and Striga

Species	Conditioning		Optimum germination temperature (°C)	Reference
	Length (days)	Temperature (°C)		
Alectra vogelii	10	25 D[a]	25 D	Okonkwo and Nwoke (1975b)
	14	28 ?[b]	28 ?	Visser (1975)
	15	23 D	25/15 D	Dawoud and Sauerborn (1994)
	14	30 ?	30 ?	Botha (1948)
	9	35 ?	30 ?	Botha (1950)
A. sessiliflora	6	30 D	20 D	van der Merwe (1993)
Striga asiatica	15–30	23–24 D	32–33 D	Worsham (1961)
	10–12	31 ?	31 ?	Kust (1963)
	14–20	29 D	29 D	Egley (1972)
	14	20 D	30 D	Hsiao *et al.* (1988a)
	7	25 D	30 D	Hsiao *et al.* (1988a)
	5	30 D	30 D	Hsiao *et al.* (1988a)
	3	35 D	30 D	Hsiao *et al.* (1988a)
	6	35 D	35 D	Brown and Edwards (1944)
	20	22 D	35 D	Brown and Edwards (1944)
S. densiflora	—	—	35 D	Solomon (1953)
S. gesnerioides	10	25 D	30 D	Igbinnosa and Okonkwo (1992)
	21	33 ?	33 ?	Reid and Parker (1979)
S. hermonthica	14	25 ?	30 ?	Egbers *et al.* (1991)
	14	23 D	23 D	Pieterse (1991)
	21	30 ?	30 ?	Pieterse *et al.* (1993)
	21–28	25 ?	30 ?	Pieterse *et al.* (1993)
	10	23 D	30/20 D	Dawoud and Sauerborn (1994)
	15	22 ?	22 ?	Vallance (1950)
S. hermonthica	8	33 D	33 D	Bebawi *et al.* (1987)

[a] Seeds were in darkness.
[b] Light:dark conditions were not given.

to high percentages when the stimulant was applied. However, seeds exposed to light and then held in darkness for 1 day at 22, 16, or 4°C prior to the application of stimulant showed decreases in germination with decreases in temperature (Worsham and Egley, 1990). Red light given on days 7–12 of the conditioning period inhibited germination, but far-red light reversed this effect (Worsham and Egley, 1990). Thus, germination is controlled by the phytochrome system.

The optimum length of the conditioning period decreases with an increase in temperature (Table 11.3). Vallance (1950) found that germination was higher for *Striga hermonthica* seeds conditioned at 22°C than at 12, 27, or 32°C. Also, if seeds were conditioned for 3 days at 27 or 32°C and then moved to 22°C, subsequent germination was reduced, being equal to that of seeds conditioned only at high temperatures. However, some researchers have conditioned seeds of this species at 30 or 33°C (Table 11.3), apparently with good success.

Minimum temperatures for conditioning have been of interest to weed biologists because they might help determine the geographical distributions of these pests.

Seeds of *S. hermonthica* could be conditioned at 12°C (Vallance, 1950), and those of *S. asiatica* were conditioned at 10°C, but not at 5°C (Hsiao *et al.*, 1988a). Seeds of *S. asiatica* were conditioned and germinated (0.5%) at 12/12-hr daily thermoperiods of 20/14°C, but not at 17/11°C (Patterson *et al.*, 1982). The effect of conditioning *S. asiatica* seeds at a low temperature (15°C) could be overcome by increasing the concentration of the chemical stimulant (strigol) during germination (Hsiao *et al.*, 1988b). Further, seeds of *S. asiatica* survived freezing at −7 and at −15°C for 15 days (Patterson *et al.*, 1982). Patterson *et al.*, (1982) concluded that low temperatures were unlikely to prevent the northward spread of *S. asiatica* in temperate North America.

Germination percentage may drop to near zero if the conditioning period lasts for more than a few weeks (Brown and Edwards, 1944; Vallance, 1950; Bebawi *et al.*, 1986; Pieterse *et al.*, 1993). After 30–33 days of conditioning at 22°C, seeds of *S. hermonthica* germinated to only 5–10% (Vallance, 1950), and after 53 days at 23–24°C, seeds of *S. asiatica* germinated to 0% (Worsham, 1961). However, seeds of *Alectra vogelii* germinated to 84% after 40 days at 30°C (Botha, 1948).

One reason for low germination percentage after extended periods of conditioning could be loss of viability. Seeds of *S. hermonthica* died when conditioned for 2 weeks at 40°C or for 5 weeks at 35°C; however, 83% of them were alive (and germinated) after 5 weeks at 25°C (Pieterse *et al.,* 1993). Another reason for the reduction in the germination of seeds given extended periods of conditioning is that they may have entered secondary dormancy. Many seeds of *S. hermonthica* that failed to germinate after conditioning did so after drying followed by a second period of conditioning (Vallance, 1950), demonstrating that they were still alive.

Soil depth and type can influence conditioning under field conditions. Seeds of *S. hermonthica* placed on the soil surface and tested (using ethephon as the stimulant) at intervals of 5–80 days exhibited variation in germination ranging from 37 to 83%, whereas those buried at 5 cm germinated to 65–85% (Babiker *et al.,* 1987). Seeds buried at 15–30 cm germinated to 58–72% initially, but after 80 days germination decreased to 0–4%. Thus, depending on the microsite, seeds could be dried and conditioned many times before the environment became favorable for germination. However, in some microsites they may remain in a constant state of readiness to germinate. There is a great need for burial studies to monitor changes (or lack thereof) in buried seeds throughout the year. Seeds of *S. hermonthica* stored dry at room temperatures for 22 months exhibited cycles in their ability to respond to conditioning and chemical stimulation, reaching a maximum in summer and a minimum in winter (Mangnus *et al.,* 1992a).

The germination of seeds of *S. hermonthica* buried at depths of 5 and 20 cm in soil in Kenya increased following periods of high rainfall (Pieterse *et al.,* 1993), indicating that conditioning did not occur when the soil was dry. Although seeds of *Striga* and *Alectra* must be on a moist substrate for conditioning to occur, they can be conditioned at surprisingly low osmotic potentials. Seeds of *S. hermonthica* germinated to 80–95% after being conditioned for 10 days at osmotic potentials ranging from 0 to −1.5 MPa, but those of *Alectra vogelii* germinated to about 90, 89, 65, and 53% at 0, −0.3, −0.6, −0.9, −1.2, and −1.5 MPa, respectively (Dawoud and Sauerborn, 1994). However, after 20 days of conditioning under these osmotic potentials, seeds of *A. vogelii* germinated to 85–90%.

Seeds of *Alectra vogelii* require oxygen during the conditioning period (Botha, 1948), whereas those of *Striga asiatica* can be conditioned in low oxygen concentrations associated with flooding (Hsiao *et al.,* 1981). Seeds of *S. asiatica* immersed in water required 10 days of conditioning to germinate to 84% (Hsiao *et al.,* 1981). However, seeds immersed for 2 days and then dried for 2 days required only an additional 2 days of immersion

before being capable of germination to 85% (Hsiao *et al.,* 1987). The authors were not sure if the conditioning period was decreased due to drying per se or to the removal of leachates from the seeds' environment. The germination of *S. euphrasioides* seeds is promoted by washing them with water (Rangaswamy and Rangan, 1966), and those of *S. asiatica* germinated to higher percentages if the volume of water in which they were conditioned was increased (Pavlista *et al.,* 1979a). Seeds of *S. asiatica* produce unidentified, water-soluble germination inhibitors, but their effects can be overridden by germination stimulants produced by roots of host plants (Kust, 1966). Thus, it makes sense that washing seeds and/or keeping them in a large volume of water during conditioning would promote germination by removing inhibitors. In other studies, however, increasing the number of *S. asiatica* seeds up to 1000 seeds per 1.2 ml of water (Hsiao *et al.,* 1979) or 150–200 seeds per 6 ml of water (with five disks of glass fiber filter paper) (Mangus *et al.,* 1992a) increased germination. No attempt has been made to explain these results.

After the conditioning phase of dormancy break is completed, compounds in the exudates from roots of host plants promote germination (Worsham, 1987). However, the time required for conditioning and the amount of stimulant required to promote germination at the end of conditioning are increased (not decreased) if chemical stimulants are added during the conditioning period (Hsiao *et al.,* 1979, 1981, 1987, 1988b). Washing *S. asiatica* seeds conditioned in the chemical stimulant (strigol) increased germination when seeds subsequently were treated with strigol during germination (Hsiao *et al.,* 1979). Pavlista *et al.* (1979b) suggested that strigol slowed conditioning by retarding the leaching of germination inhibitors from the seeds. GA$_3$, kinetin, and zeatin inhibited conditioning of *S. asiatica* seeds, but their effects were overcome by increasing the length of the conditioning period or by increasing the concentration of stimulant (strigol) during germination (Hsiao *et al.* 1988b). Seeds conditioned in 50 or 100 m*M* sucrose or NaNO$_3$ were inhibited for 3 days, but the inhibition disappeared after 5 days (Hsiao *et al.,* 1988b).

3. Germination Requirements

This section will consider the environmental conditions necessary for the germination of *Striga* and *Alectra* seeds that are fully afterripened and conditioned.

a. Chemicals

Stimulation of seed germination by chemicals found in exudates of living roots of host, and even some nonhost, plants has received much research attention. The idea that a host plant might promote seed germination came from observations that parasitic plants appeared if seeds were sown in pots of soil containing a host plant, but not in those with no host plants (Brown, 1965). Saunders (1933) showed that water leached from pots of sand containing five plants of maize would stimulate the germination of *Striga lutea* (=*S. asiatica*) seeds and that the effect was not due to the acidity of the root exudate or to the mineral nutrition of the host. Brown and Edwards (1944) demonstrated that although some seeds of *S. asiatica* would germinate without chemical stimulation of a host plant, germination percentages were much higher when seeds where sown in the immediate vicinity of a living root of *Sorghum vulgare*. At a distance of 6 mm, only 1% of the seeds germinated. Seeds of *Alectra vogelii* germinated on 2% sucrose agar medium only if intact seeds of the host *Voandzeia subterranea* were present, and germination usually occurred in proximity of root tips (Groenewald *et al.*, 1979).

To better understand the effects of host stimulation on germination, various investigators (e.g., Brown and Edwards, 1944, 1946; Vallance, 1950) began preparing "standard solutions" of stimulants by growing a constant number of sorghum seedlings on floats in water in a closed container in darkness for a given period of time at a constant temperature. Water containing root exudates was used in germination studies. Standard solutions of exudates of living roots of *Gossypium hirsutum* (Egley, 1990), *Zea mays* (Worsham *et al.*, 1959; Yoshikawa *et al.*, 1978), and even *Coleus* (Egley, 1965) have been prepared and used to study the germination of *Striga*, whereas exudates from roots of various species, including *Vicia faba*, *Glycine hispida* (Botha, 1948), *Voandzeia subterranea*, *Vigna unguiculata*, *Phaseolus multiflorus*, and *Helianthus annuus*, have been used in studies of *Alectra vogelii* seeds (Visser, 1975). A continuous system for the recovery of root exudates from *Sorghum vulgare* (millet) and *V. unguiculata* has been developed (Muller *et al.*, 1993).

Various problems are encountered in using root exudates to stimulate seed germination. (1) Seed exudates are mixtures of compounds, including one or more germination stimulants (Brown *et al.*, 1952; Siame *et al.*, 1993) and inhibitors (Weerasuriya *et al.*, 1993). (2) Strains or variants of *Striga* spp. have evolved that vary in their sensitivity to root exudates. For example, seeds from some variants of the white-flowered form of *S. asiatica* respond to exudates of *Sorghum bicolor*, whereas others respond to those from *Pennisetum typhoides* (Lakshmi and Jayachandra, 1979). (3) Potential host plants vary in the amount of chemical stimulants produced, e.g., the stimulatory activity of exudates from *Striga*-resistant cultivars of sorghum is much less than that of susceptible cultivars (Weerasuriya *et al.*, 1993).

Tremendous efforts over a long period of time have been devoted to isolating and identifying stimulatory chemicals in root exudates (e.g., Brown *et al.*, 1949; Sunderland, 1960; Worsham *et al.*,1964; Visser and Botha, 1974; Visser, 1975; Visser *et al.*, 1987; Muller *et al.*, 1992; Siame *et al.*, 1993), and some success has occurred. The first germination stimulant of *Striga* seeds to be identified was isolated from root exudates of *Gossypium hirsutum* (cotton), a nonhost species, and it was given the trivial name "strigol" (Cook *et al.*, 1966). Strigol now has been found in root exudates of the host plants *Zea mays* (corn or maize), *Panicum miliaceum* (proso millet), and *Sorghum bicolor* (sorghum) (Siame *et al.*, 1993). Strigol was the major germination stimulant in exudates from corn and proso millet, but a minor one in those from sorghum roots (Siame *et al.*, 1993).

The chemical structure of strigol has been determined (Cook *et al.*, 1972) and synthesized (Heather *et al.*, 1974). Thus, strigol is available for germination studies. In addition, the stimulatory properties of precursors, analogs, derivatives, and fragments of strigol have been investigated (Pepperman *et al.*, 1982; Mangnus and Zwanenburg, 1992; Mangnus *et al.*, 1992b). One of the analogs (GR-7) stimulates the germination of seeds of *Orobanche ramosa*, *Striga hermonthica*, *S. asiatica* (Johnson *et al.*, 1976), and *Alectra sessiliflora* (van der Merwe, 1993), and GR-24 stimulates the germination of those of *S. hermonthica*, *S. forbesii*, *Alectra vogelii* (Jackson and Parker, 1991), *S. asiatica*, and *S. aspera* but not *S. densiflora* (Parker and Riches, 1993). Analogs GR-7 and GR-24 did not stimulate the germination of *S. gesnerioides* seeds (Igbinnosa and Okonkwo, 1992).

The germination of conditioned *Striga asiatica* seeds increases with increasing concentrations of the synthetic *dl*-strigol (Hsiao *et al.*, 1981, 1987), suggesting that this compound applied to cropland could help control *Striga* infestations. In a 21-day soil leaching experiment, sufficient amounts of surface-applied *dl*-strigol reached soil depths of 22.5–30 cm to stimulate most *S. asiatica* seeds to germinate. Further, *dl*-strigol was not degraded in the soil (Hsiao *et al.*, 1983). Synthetic analogs of strigol applied 6 weeks before sorghum seeds were sown in boxes of soil reduced the number of buried *S. asiatica* seeds by 65% (Johnson *et al.*, 1976). However, in other studies, synthetic analogs did not stimulate as much germination as did strigol, and they decomposed at low pH and high temperatures and in light (Mhehe, 1987). Also, the stimulant GR-7 applied to soil before seeds were conditioned did not increase germination (Babiker and Hamdoun, 1982). GR-7, GR-45, and GR-60 stimulated higher percentages of germination in petri dishes than in pots of soil in a greenhouse or in field plots; however, all compounds stimulated 1–15% germination of *S. asiatica* seeds buried at soil depths of 15 cm (Steven and Eplee, 1979). Much research remains to be done before strigol compounds can be used as effective control agents for infestation of *Striga* in crops.

The first host-derived germination stimulant (a benzoquinone) for *Striga* to be identified came from root exudates of sorghum (Chang *et al.*, 1986). This compound is called sorgoleone, and it occurs in hydrophobic droplets exuded from root hairs of sorghum (Netzly *et al.*, 1988). The unstable, reduced form (dihydroquinone) of sorgoleone stimulates germination, whereas the oxidized form does not (Chang *et al.*, 1986). Six sorgoleone compounds of various molecular weights have been found in sorghum root hair droplets, and all will stimulate the germination of *Striga* seeds (Housley *et al.*, 1987). Netzly *et al.* (1988) suggest that sorgoleones are allelochemicals that may inhibit the growth of some competitors of sorghum; these compounds inhibit the growth of lettuce (Netzly and Butler, 1986; Housley *et al.*, 1987) and pigweed seedlings (Housley *et al.*, 1987). Apparently, *Striga* has evolved to use these compounds for host recognition.

Another germination stimulant for *Striga,* called sorgolactone, also has been isolated from sorghum root exudates. This compound is related closely to $(+)-$strigol (Hauck *et al.*, 1992). A germination stimulant, called alectrol, for seeds of *Alectra vogellii* and *S. gesnerioides* has been identified from root exudates of the host plant *Vigna unguiculata*. Spectroscopic data indicate that alectrol is related closely to $(+)-$strigol (Muller *et al.*, 1992).

Many compounds, besides stimulants in roots exudates, will promote the germination of *Striga* and *Alectra* seeds, especially if they have been conditioned (Table 11.4). The most promising of these compounds to promote "suicidal seed germination" is ethylene gas (Egley and Dale, 1970), and its use resulted in a 90% reduction in the number of buried *S. asiatica* seeds in fields in North Carolina (Eplee, 1975). Ethephon (2-chloroethylphosphonic acid), which generates ethylene, did not promote the germination of nonconditioned *S. hermonthica* seeds buried in soil. However, ethephon added to soil after irrigation reduced the number of aerial shoots of *S. hermonthica* by 39–84% in field plots (Babiker and Hamdoun, 1983); presumably, seeds were conditioned after the soil was moistened. Another possible problem of promoting the germination of buried seeds is that CO_2 can inhibit ethylene-induced germination (Egley and Dale, 1970).

One consequence of treating seeds with strigol or its analogs, or even the defoliant thidiazuron, is an increase in ethylene production, which is accompanied by an increase in germination percentages (Jackson and Parker, 1991; Babiker *et al.*, 1993b; Logan and Stewart, 1995). A model has been proposed for the germination of *Striga* in which host-derived stimulants, such as strigol, cause the synthesis of ethylene to increase in conditioned seeds. Then, ethylene initiates the biochemical changes leading to germination (Babiker *et al.*, 1993a,b; Logan and Stewart, 1991, 1995). However, seeds of *S. forbesii* and *Alectra vogelii* are unresponsive to ethylene given alone, but they are stimulated to germinate by strigol (GR-24) (Jackson and Parker, 1991). Further, norbornadine, which blocks the production of ethylene, did not prevent the germination of seeds of *S. forbesii* treated with GR-24 (Jackson and Parker, 1991). These data suggest that the gas did not penetrate the seed coats of *S. forbesii* and *A. alectra* or that strigol does not promote germination via ethylene production. Thus, there may be more than one mechanism of dormancy release (Jackson and Parker, 1991).

The mystery concerning chemical stimulation of seed germination in *Striga* appears to be a long way from being solved. Incredibly, water from 22 of 29 different streams, ponds, and lakes in North Carolina and moist sections of freshly cut stems from 118 (57 families) of 163 plant species stimulated seeds of *S. asiatica* to germinate (Dale and Egley, 1971).

Much research has been conducted to try to find ways to control *Striga*. Under *in vitro* conditions, the herbicides chlorthal-dimethyl, dicamba, and pendimenthalin prevented the germination of seeds of *S. hermonthica* and *S. gesnerioides* that has been conditioned and treated with germination stimulants produced by *Sorghum bicolor* roots (Bagonneaud-Berthome *et al.*, 1995). However, dicamba, clopyralid, and linuron promoted the germination of *S. gesnerioides* seeds when exudates of *S. bicolor* roots were not present. Bagonneaud-Berthome *et al.* (1995) noted that little is known about how these compounds inhibit or promote the germination of *Striga* seeds. An isolate of the fungus *Fusarium oxysporum,* growing on sorghum straw that was mixed into pots of soil where seeds of *S. hermonthica* and sorghum were sown, prevented any plants of *S. hermonthica* from emerging (Ciotola *et al.*, 1995). In root chamber studies, the presence of fungus prevented the germination of *S. hermanthica* seeds, and seedlings that became attached to sorghum roots were killed (Ciotola *et al.*, 1995).

Nitrogen fertilizers applied to *Zea mays* (corn) fields in Natal caused a 93% reduction in the number of *Striga asiatica* plants at both high and low levels of phosphorus. In the absence of nitrogen, however, high levels of potassium led to an increase in *S. asiatica* (Farina *et al.*, 1985). Nitrogen also helped suppress *S. hermonthica* in sorghum fields in Sudan (Bebawi, 1987) and decreased the numbers of *S. asiatica* plants in pots of soil containing a sorghum plant (Yaduraju and Setty, 1979). Although the number of *S. hermonthica* plants decreased with an increase in nitrogen levels in corn fields in Kenya, the

TABLE 11.4   Compounds that Stimulate ≥50% Germination of *Striga* and *Alectra*
Seeds Previously Conditioned in Water

Species	Compound	Reference
Alectra vogelii	Sodium hypochlorite	Okonkwo and Nwoke (1975b)
	Calcium hypochlorite	Okonkwo and Nwoke (1975b)
Striga asiatica	Abscisic acid	Hsiao *et al.* (1988b)
	Benzaldehyde-6-purinylhydrazone	Worsham *et al.* (1959)
	6-Benzylaminopurine	Worsham *et al.* (1959)
	Cytokinins	Babiker *et al.* (1993a)
	Ethylene	Egley and Dale (1970), Eplee (1975)
	6-(2-Furfuryl)aminopurine (kinetin)	Worsham *et al.* (1959)
	GA₃ + kinetin	Yoshikawa *et al.* (1978)
	GA₃ + kinetin + NAA (1-napthaleneacetic acid)	Yoshikawa *et al.* (1978)
	Kinetin + NAA	Yoshikawa *et al.* (1978)
	6-(2-Phenethyl)aminopurine	Worsham *et al.* (1959)
	6-Phenylaminopurine	Worsham *et al.* (1959)
	15 Sesquiterpene lactones	Fischer *et al.* (1989, 1990)
	6-(2-Thenyl)aminopurine	Worsham *et al.* (1959)
	Thidiazuron	Babiker *et al.* (1992)
	Zeatin	Hsiao *et al.* (1988b)
S. hermonthica	Thidiazuron	Babiker *et al.* (1992), Logan and Stewart (1995)
	Ethylene	Bebawi and Eplee (1986)

impact of nitrogen varied between years and seemed to be related to the severity of infestation (Mumera and Below, 1993). In laboratory studies, urea inhibited the germination of *S. hermonthica* seeds, whereas sodium nitrate had no effect; the response to ammonium sulfate was variable (Pesch and Pieterse, 1982; Pieterse, 1991).

Conditioned seeds of *S. hermonthica* placed near sorghum roots germinated to higher percentages in 1 mM than in 3 mM ammonium nitrate. Also, the percentage of haustoria that became attached was lower and early growth of seedlings slower in 1 mM than in 3 mM ammonium nitrate (Cechin and Press, 1993). These authors suggest that increased nitrogen decreases the number of *S. hermonthica* plants by decreasing the amount of stimulant produced and/or released by sorghum roots.

b. Temperature

Most germination studies of *Striga* and *Alectra* seeds have been done at temperatures ranging from 22 to 35°C (Table 11.3). Seeds obviously will germinate at constant temperatures, but alternating temperatures are not inhibitory (Patterson *et al.,* 1982; Dawoud and Sauerborn, 1994). Constant temperatures of 30–35°C and daily thermoperiods of 32/26, 29/20, and 25/15°C are optimal for germination (Table 11.3; Patterson *et al.,* 1982; Patterson, 1987). Low germination percentages are obtained at simulated early spring temperatures such as 20/10, 20/14, and 15°C (Patterson *et al.,* 1982;

Hsiao *et al.,* 1988a; Dawoud and Sauerborn, 1994); thus, germination could be delayed in the field until temperatures increase in late spring. Such a delay would ensure that roots of host plants are actively growing at the time seeds of *Striga* and *Alectra* germinate.

c. Light and Darkness

Because seeds of *Striga* and *Alectra* germinate in the rhizosphere of the host plant, it makes sense that they would be able to germinate in darkness. However, do they actually require darkness for germination? This is a difficult question to answer because seeds usually are tested only in darkness (Table 11.3). Conditioned seeds of *S. asiatica* treated with root exudates of cotton germinated to higher percentages in darkness than in light. However, seeds that were acid scarified and those that had a hole punched in them at the radicle end germinated equally well in light and darkness (Egley, 1972). Light exposures of only 5 sec given 1–3 hr after seeds were treated with stimulants inhibited the germination of *S. asiatica* seeds (Worsham and Egley, 1990). Light exposures given 4 hr after seeds were treated with stimulants had no effect. Seeds of *S. angustifolia* (Rao *et al.,* 1985), *S. euphrasioides* (Kumar and Solomon, 1940; Rangaswamy and Rangan, 1966), and *S. densiflora* (Kumar and Solomon, 1940) germinated to higher percentages in light and in darkness. Studies need to be done on the light:dark requirements for the germination of seeds of other *Striga* species and on those of *Alectra* species.

d. Other Factors

Seeds of *Striga* are somewhat sensitive to pH. Seeds of *S. asiatica* conditioned at a pH of 7.2 or 6.6 germinated to higher percentages at low levels of host stimulant than those conditioned at pH 5.5 or 4.0. However, the optimum pH for conditioning and germination were 6.6 and 5.5, respectively (Worsham and Egley, 1990).

Seeds of *S. asiatica* conditioned and incubated while immersed in water germinated to higher percentages than those on the surface of moist filter paper (Worsham and Egley, 1990). Seeds in aerated water, or in water under near anaerobic conditions during both conditioning and germination, germinated to higher percentages than those moved from one condition to the other following the conditioning treatment (Worsham and Egley, 1990). That is, seeds germinated to high percentages in anaerobic or aerobic conditions, but a different oxygen concentration during conditioning than during germination reduced the percentage of seeds that germinated. Further, in the presence of cotton root exudates, seeds of *S. asiatica* germinated to 65–96% in oxygen, nitrogen, carbon dioxide, or air (Egley, 1972). Thus, if host plants are present, seeds of *S. asiatica* (and perhaps other *Striga* and *Alectra* species) should be capable of conditioning and germinating in soil under the low oxygen concentrations that could follow heavy rains or irrigation.

E. Rooted Hemiparasites Whose Seeds Do Not Require a Host Stimulus for Germination

This group of hemiparasites includes the genus *Krameria* in the Krameriaceae, 17 genera in the Santalaceae, 24 in the Scrophulariaceae (Kuijt, 1969), and 10 in the Olacaceae/Opiliaceae (Kuijt, 1969; Kubat, 1987). Plants of rooted hemiparasitic species have green leaves (Musselman and Mann, 1978), but species differ in their ability to grow without being attached to a host plant. In some species, plants grow equally well with or without being connected to a host (Musselman and Mann, 1979), but in other species, plants attached to a host are much healthier than those growing independently (Govier and Harper, 1964; Govier *et al.*, 1967; Chuang and Heckard, 1971; Klaren and van de Dijk, 1976). In some species, seedlings grow for several weeks and may produce several pairs of leaves. However, unless a haustorium becomes attached to a host plant root, the plant (1) dies (Musselman, 1969, 1972; Grelen and Mann, 1973; Oesau, 1973; Musselman and Mann, 1977a; Malcolm, 1966; Baskin *et al.*, 1991) or (2) remains stunted and never flowers (Cantlon *et al.*, 1963). Thus, some hemiparasites

are unable to complete their life cycle in the absence of a host plant. After a haustorium penetrates the root of a host plant, the hemiparasite obtains water, minerals, and organic compounds from the host (Govier and Harper, 1964; Malcolm, 1966; Govier *et al.*, 1967, 1968).

As noted by Parker and Riches (1993), seeds produced by hemiparasitic plants are much larger than those of *Orobanche* or *Striga*, ranging from 0.5 mm in length in *Castilleja* (Malcolm, 1966) to 5.0 mm in *Melampyrum* (Parker and Riches, 1993). The increase in seed size means an increase in food reserves, thus seedlings can become established without the radicle having to make contact with a host root. In fact, host root exudates are not required to stimulate germination in these species. (Musselman, 1969; Chancellor, 1973; Okonkwo and Nwoke, 1974; Musselman and Mann, 1977a,b; Campion-Bourget, 1983).

Although seedlings can become established in the absence of a host plant, formation of haustoria usually requires the presence of chemical stimulants. However, haustoria can be formed on roots of *Comandra* seedlings, even if host roots are not present (Piehl, 1965). Chemicals in exudates of host plant roots cause one or more roots on the hemiparasite plant to produce haustoria. In the Scrophulariaceae, all roots have the potential to give rise to haustoria (Musselman and Dickison, 1975). Haustoria are induced by a chemical(s) found in root or yeast exudates (Atsatt *et al.*, 1978; Riopel, 1979; Riopel and Musselman, 1979) as well as one leached from soya bean or flax seeds, cotton string, cotton, paper towels, tissue paper (Atsatt *et al.*, 1978), or gum tragacanth from the legume *Astragalus gummifer* (Steffens *et al.*, 1982). Identified compounds that will induce haustoria formation in *Agalinus*, a hemiparasitic member of the Scrophulariaceae, include xenognosin A and B from gum tragacanth (Lynn *et al.*, 1981) and soyasapogenol B from root exudates of *Lespedeza* (Steffens *et al.*, 1983).

1. Types of Seed Dormancy

Because seeds in the Krameriaceae, Olacaceae, Santalaceae, and Scrophulariaceae do not have a palisade layer of macroscelerids in the seed coat (Corner, 1976), they do not have physical dormancy. Seeds of the Krameriaceae have large investing embryos and no endosperm (Martin, 1946); therefore, any dormancy in this family would be of the physiological type. This is also true for members of the Scrophulariaceae, which, like *Striga* and *Alectra*, have dwarf seeds (Martin, 1946). Seeds of at least some members of the Santalaceae, including *Buckleya*, *Comandra*, *Pyrularia* (Martin, 1946), and *Santalum* (Rangaswamy and Rao, 1963), have underdeveloped linear embryos and thus potentially have morphological or morphophysiological dormancy. Because seeds of *Comandra* (Piehl, 1965), *Santalum* (Srimathi and Rao, 1969), and *Buckleya* (Musselman and Mann, 1979) require long periods of cold or warm stratification before they will germinate (Table 11.5), it is concluded that the embryos are physiologically dormant and that these seeds have MPD. Requirements for embryo growth and dormancy break have not been determined. Germination percentages and rates

of *Santalum album* seeds were increased by treatments with GA, whereas treatments with potassium nitrate, ammonium nitrate, sodium nitrite, and potassium nitrite had no effect (Nagaveni and Srimathi, 1980).

Little information is available on seed morphology and germination for members of the Olacaceae and Opiliaceae. However, these families belong to the order Santalales (Lawrence, 1951), and seeds of *Heisteria longipes* in the Olacaceae have rudimentary embryos (Kuijt, 1969). Thus, there is a possibility that seeds of

TABLE 11.5   Dormancy-Breaking and Germination Requirements in Seeds of Hemiparasitic Species Whose Seeds Do Not Require a Host Stimulus for Germination

Species	D[a]	Tr[b]	Days[c]	Temperature[d]	L:D[e]	Reference
Agalinis fasciculata[a]	PD	C	84	20/10	L	Baskin and Baskin (unpublished results)
Aureolaria grandiflora[f]	PD	C	150	21	L = D	Musselman (1969)
A. pectinata[f]	PD	C	84	20/10	L	Baskin and Baskin (unpublished results)
A. pedicularia[f]	PD	C	150	21	L = D	Musselman (1969)
A. patula[f]	PD	C	63	20/10	L	Baskin and Baskin (unpublished results)
A. virginica[f]	PD	C	70	20/10	L = D	King (1989)
Bartsia odontites[f]	PD	C	72	15	—[g]	Vallance (1951)
Buchnera hispida[f]	PD	W	50	25, 29/27	L	Okonkwo and Nwoke (1974), Nwoke and Okonkwo (1980)
Buckleya distichophylla	MPD?	C	120[h]	25?[i]	—	Musselman and Mann (1979)
Castilleja coccinea[f]	PD	C	42	20	L	Malcolm (1966)
Comandra californica	MPD?	C	Winter	25?[i]	—	Piehl (1965)
C. umbellata	MPD?	C	Winter	25?[i]	—	Piehl (1965)
C. pallida	MPD?	C	Winter	25?[i]	—	Piehl (1965)
Euphrasia spp.[f]	PD	C	84	g.h.[j]	—	Yeo (1961)
Krameria lanceolata	ND?			24	—	Musselman and Mann (1977b)
Macranthera flammea[f]	PD	C	150	25?[i]	—	Musselman (1972)
Melampyrum lineare[f]	PD	C[k]	60–80	3	—	Curtis and Cantlon (1963), Cantlon et al. (1963)
M. pratense[f]	PD	W	75	7.5	L = D	Masselink (1980)
Odontites verna[f]	PD	C	28	10–23	—	Chancellor (1973)
Pedicularis lanceolata[f]	PD	C	84	15/6	L > D	Baskin et al. (unpublished results)
Rhinanthus alectorolophus[f]	PD	C	35–70	5–15	—	ter Borg (1987)
R. angustifolius[f]	PD	C	70–105	20	—	ter Borg (1987)
R. aristatus[f]	PD	C	35–70	5–15	—	ter Borg (1987)
R. burnati[f]	PD	C	79	15–20	—	Campion-Bourget (1983)
R. crista-galli[f]	PD	C	119	2	—	Vallance (1952)
R. glaber[f]	PD	C	21, 70	25/15	—	Bakker et al. (1966)
R. mediterraneus[f]	PD	C	35–70	5–15	—	ter Borg (1987)
R. minor[f]	PD	C	70–105	20	—	ter Borg (1987)
Santalum album	MPD?	W	40[+]	24–27	—	Srimathi and Rao (1969)
Sopubia delphinifolia[f]	PD	C	4	24	L	Sahai and Shivanna (1985)
Tomanthera auriculata[f]	PD	C	84	15/6	L	Baskin et al. (1991)

[a] Type of dormancy: ND, nondormant; PD, physiological dormancy; MPD, morphophysioloigcal dormancy.

[b] Dormancy-breaking treatment: C, cold stratification; W, warm stratification.

[c] Length of dormany-breaking treatment.

[d] Optimum germination temperature.

[e] Light:dark requirements for germination: L, seeds require light for germination, L = D, seeds germinate equally well in light and darkness; L > D, seeds germinate to higher percentage in light than in darkness; D > L, seeds germinate to a higher percentage in darkness than in light; and D, seeds require darkness for germination.

[f] Species belongs to the Scrophulariaceae.

[g] Information not available.

[h] Seeds appear to have been stored dry at 3°C.

[i] Presumably at room temperature.

[j] Greenhouse.

[k] Some seeds required a 1-month period of moist storage at 20°C prior to cold stratification.

these families have underdeveloped embryos and MD or MPD.

2. Dormancy-Breaking Requirements

Data on dormancy-breaking requirements are available for species in 17 of the 55 genera of rooted hemiparasites, and 13 of these 17 genera are in the Scrophulariaceae. Thus, the picture we now have of germination of this group of parasitic plants may change considerably as more data become available, especially if studies are done on species of Olacaceae and Santalaceae from subtropical and tropical regions.

Physiological, as well as morphophysiological, dormancy in seeds of many rooted hemiparasites is broken by cold stratification (Table 11.5), and thus seeds would come out of dormancy during winter and germinate in spring. The length of the cold stratification period needed to overcome dormancy ranges from 28 days in seeds of *Odontites verna* (Chancellor, 1973) to 150 days in those of *Aureolaria* spp. (Musselman, 1969). In a few species, however, high temperatures of summer break seed dormancy, and germination occurs in autumn. The amount of warm stratification required to break dormancy ranges from 50 days in *Buchnera hispida* (Okonkwo and Nwoke, 1974) to 75 days in *Melampyrum pratense* (Masselink, 1980). Germination of a cohort of nonscarified seeds of *Santalum album* began after 40 days of warm stratification and was still in progress after 70 days, when the study was terminated (Srimathi and Rao, 1969). Thus, we do not know how much warm stratification is required for all seeds in a cohort of this species to come out of dormancy.

A short (30-day) warm stratification pretreatment prior to cold stratification increased germination in seeds of *Melampyrum lineare* (Curtis and Cantlon, 1963; Cantlon *et al.,* 1963); however, seeds appear to have a type of epicotyl dormancy. The radicle emerges from *M. lineare* seeds during cold stratification, but the epicotyl does not emerge if seeds are moved to temperatures, such as 15°C, which are favorable for shoot growth. Epicotyl dormancy persists until seeds (with the emerged radicle) are given an additional 20–30 days of cold stratification (Cantlon *et al.,* 1963). After 75 days of warm stratification, radicles emerged if *M. pratense* seeds were at low temperatures (7.5°C), after which seeds required cold stratification for 28 days at 5°C to break epicotyl dormancy (Masselink, 1980). Seeds of *M. arvense* (Oesau, 1973), *M. nemorosum*, and *M. sylvaticum* (Oesau, 1975) also germinate at low temperatures (2–10°C) and require an extended period of cold stratification for seedling development.

Cold stratification breaks dormancy in seeds of the annual species *Odontites verna*, but high (23, 33°C) temperatures induce nondormant seeds into secondary dormancy (Chancellor, 1973). Thus, if seeds remain viable during long periods of burial, they may exhibit annual dormancy/nondormancy cycles.

3. Germination Requirements

The most interesting thing about the temperature requirements for the germination of rooted hemiparasites is the low optimum temperatures (mean of about 15°C) for the germination of members of the Scrophulariaceae (Table 11.5). Unfortunately, the papers on *Buckleya* (Musselman and Mann, 1979) and *Comandra* (Piehl, 1965) in the Santalaceae do not state the test temperatures (presumably they were at room temperatures). However, seeds of *Santalum album* germinated at 24–27°C (Srimathi and Rao, 1969).

Information on the light : dark requirements for germination is limited to members of the Scrophulariaceae (Table 11.5). Seeds of a few species germinate equally well in light and darkness, and those of other species require light for germination. Seeds of *Buchnera hispida* require light for germination and will germinate at photoperiods of 0.25, 1, 4, 8, 12, and 24 hr of light per day (Nwoke and Okonkwo, 1980). Increases in the length of a dark incubation pretreatment increased the sensitivity of seeds of *B. hispida* to light and decreased the time for germination (Okonkwo and Nwoke, 1974; Nwoke and Okonkwo, 1980).

F. Nonrooted Hemiparasites: Mistletoes

Mistletoes are flowering plants in the families Loranthaceae (900 species, 65 genera) and Viscaceae (400 species, 7 genera) (Johri, 1987b). However, a few taxonomists have placed some of the mistletoes in a third family called the Eremolepidaceae (Calder, 1983), whereas other authorities include all mistletoes in the Loranthaceae (Kuijt, 1969). The diversity of mistletoes is greatest in tropical and subtropical regions, but genera such as *Arceuthobium, Phoradendron,* and *Viscum* are found in temperate regions. All genera of mistletoes, except *Arceuthobium,* are restricted to either the eastern or the western hemisphere; *Arceuthobium* is in both the Old and the New World but only in the northern hemisphere (Kuijt, 1969; Hawksworth, 1987).

Most mistletoes are rootless, hemiparasitic shrubs that grow on branches of trees (Kuijt, 1969). However, *Atkinsonia, Gaiadendron,* and *Nuytsia,* which are relatively primitive members of the Loranthaceae (Barlow, 1983), are rooted hemiparasitic shrubs or trees with haustorial connections to roots of host plants (Kuijt, 1969). Some mistletoes such as *Arceuthobium* (Hawksworth, 1961) and *Phoradendron* (Kuijt, 1989) form an extensive endophyte in the host plant's body, but stems of the hemiparasite are formed on the outside of the host; those of *Arceuthobium* are leafless (Hawksworth, 1961). The vegetative plant body of *Tristerix aphyllus* is entirely endophytic, and its host species are cacti, including *Trichocereus chilensis, Eulychnia acida* (Mauseth *et al.,* 1985), *Echinopsis chilensis,* and *E. skotsbergii* (del Rio *et al.,* 1995, 1996). Buds of *T. aphyllus* break through to the surface of the cactus plant during sexual reproduction, and flowering, seed set, and germination occur on the exterior of the host. After a *T. aphyllus* seed germinates, the radicle tip swells, forming a haustorial disk that adheres to the host. Filaments of cells grow from the haustorium into the cactus plant, and then the haustorium dies (Mauseth *et al.,* 1985).

Leaves and often the stems of mistletoes are green, and the chlorophyll content of leaf tissue sometimes can be comparable to that of the host tree (de la Harpe *et al.,* 1981; Elias, 1987; Parker and Riches, 1993). However, the amount of fixed carbon obtained from the host varies from an estimated 23–45% in *Viscum album* (Richter and Popp, 1992) to 62% in *Phoradendron juniperum* (Marshall and Ehleringer, 1990). The movement of photosynthate from the host to the mistletoe can be so great that it has devastating effects on the host (Room, 1971). However, fixed carbon can move from the mistletoe to the host (Rediske and Shea, 1961; Kuijt, 1964). Rootless mistletoe plants growing on branches of trees must obtain all of their water (Fisher, 1983) and minerals (Lamont, 1983a) from the host. In fact, mistletoes may transpire water at a greater rate than the host (Tocher *et al.,* 1984). Although shrubs or trees of *Atkinsonia, Nuytsia,* and *Gaiadendron,* which have true roots growing in soil, form haustorial connections with other species (Fineran and Hocking, 1983), little information is available on how much water, photosynthate, and minerals are derived from their host plants (Hocking and Fineran, 1983).

In mistletoes, the placenta has been drastically reduced to a small protuberance or mound of undifferentiated tissue at the base of the ovary called the mamelon. In some genera, the placenta is absent, and the ovary does not form a cavity (Bhandari and Vohra, 1983). Embryo sacs are formed in the mamelon or in the subdermal layers of the ovary if the mamelon is absent. In the Loranthaceae, tips of the embryo sacs grow to various lengths, depending on the genus, with some of them growing all the way up the style to the stigma. Each embryo sac gives rise to an endosperm, but if only one seed eventually matures in a fruit (which frequently happens), the various endosperms come together to form a composite tissue (Bhatnagar and Johri, 1983). Endosperm is absent in *Psittacanthus* (Loranthaceae),

but the suspensor remains as a cup-like structure around the radicular end of the embryo, and it apparently absorbs nutrients from the fruit wall and transfers them to the embryo (Kuijt, 1967). In *Amyema* (Lamont, 1982a) and *Lysiana* (Bhatnagar and Johri, 1983), both in the Loranthaceae, and in the Viscaceae (Bhandaria and Vohra, 1983), the endosperm is chlorophyllous.

The embryo in seeds of the Loranthaceae has two, sometimes three or more, cotyledons (Kuijt, 1967; Bhatnagar and Johri, 1983), which may be so compressed that the seedling looks like a monocot (Johri and Bajaj, 1965; Bajaj, 1967). In some Viscaceae, the cotyledons are very small and are either scarcely differentiated (Kuijt, 1960, 1969) or absent (Tainter, 1967). The radicular end of the embryo in most mistletoes does not have a root cap, but it does have a tunica-corpus organization with a single tunica layer (Kuijt, 1967). Some authors say that mistletoe embryos lack a true radicle and think that the radicular end of the embryo is part of the hypocotyl (e.g., Hawksworth, 1961; Salle, 1983; Bhatnagar and Johri, 1983). However, Kuijt (1969) thinks mistletoes have a radicle. In mistletoes whose "radicles" lack a root cap, the radicle elongates several centimeters, depending on the species, and cells in the tip (in the single tunica layer) give rise to a swollen structure called a holdfast (Fig. 11.5). The holdfast secures the mistletoe to the host, and cells in the holdfast give rise to the haustorium, which penetrates the host (Gill and Hawksworth, 1961). Radicles in *Atkinsonia, Nuytsia,* and *Gaiadendron* have a root cap (Kuijt, 1965), and apparently haustoria are formed only on secondary roots. The radi-

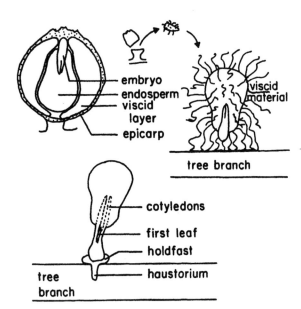

FIGURE 11.5 Germination morphology of a mistletoe. Drawn from information in Bajaj (1967), Kuijt (1969), and Calder (1983).

Instead, the root grows into the soil, and a leafy shoot develops above ground. After a period of growth, the underground part of the stem swells to form a tuber. Lateral roots growing from the tuber can give rise to haustoria (Kuijt, 1963). *Psittacanthus schiedeanus* has neither a root cap nor a tunica-corpus organization (Kuijt, 1967), and the primary haustorium arises from the side of the radicle (Kuijt, 1970).

In mature mistletoe seeds (Fig. 11.5), a small portion of the embryo, usually the radicular end, protrudes a little from the endosperm (Bajaj, 1967, Kuijt, 1969; Johri and Bhojwani, 1970, Reid, 1987). Because ovules are not formed and there is no seed coat, mistletoes do not have true seeds. The endosperm is covered by the fruit wall, and the fruit is called a pseudoberry, except in *Atkinsonia* and *Gaiadendron*, where it is a pseudodrupe (Bhatnagar and Johri, 1983; Bhandari and Vohra, 1983).

In the Loranthaceae, the fruit wall is divided into four zones: an outermost leathery epicarp, a viscid zone (mesocarp), a parenchymatous or fleshy zone, and a vascular zone (Bhatnagar and Johri, 1983). In the Viscaceae, the fruit wall is divided into three zones: a leathery epicarp traversed by vascular bundles, a viscid mesocarp, and a parenchymatous endocarp (Bhandari and Vohra, 1983). Thus, in the Loranthaceae the viscid zone is the outside of the vascular bundles, whereas in the Viscaceae it is to the inside of the vascular bundles (Kuijt, 1969).

A mucilaginous material called viscin is released when the epicarp is removed and cells of the viscid zone are disturbed. Then, the seed is surrounded by a jelly-like mass that is hygroscopic and imbibes even more water. Eventually, however, viscin dries, and seeds are "cemented" to the surface of tree branches or various other substrates (Hawksworth, 1961). The hardened viscin is thought to reduce water loss from the embryo and endosperm (Kuijt, 1960).

1. Dispersal of Seeds

Although some mistletoes seeds have been reported to germinate while they are enclosed by the epicarp (Mauseth *et al.*, 1985; Reid, 1987), we are unaware of any observations of the radicle actually penetrating the epicarp. Because the epicarp is tough and leathery, birds eat only the seeds (Liddy, 1983). Thus, removal of the epicarp to "free" the seeds is an important event in the life cycle of mistletoes. Seeds certainly do not have to pass through the digestive system of birds before they will germinate (Cannon, 1904; Lamont, 1982b, 1983b, Lamont and Perry, 1977).

In the dwarf mistletoes (*Arceuthobium*) and to a lesser degree in other members of the Viscaceae such as *Viscum japonicum* (Sahni, 1933), seeds are dispersed from the fruit by an explosive mechanism. The mucilaginous material associated with the viscin cells imbibes water and internal hydrostatic pressure increases as the fruit ripens (Hinds *et al.*, 1963). Finally, the pressure is so great that the epicarp comes apart at the dehiscent layer at the base of the fruit and seeds are shot into the air (Heinrichter, 1915). Seeds of *Arceuthobium* are dispersed up to 10 (Dowding, 1929) to 14 (Hudler and French, 1976) meters from the mother plant, whereas those of *V. japonicum* travel only 0.6 m (Sahni, 1933). Seed dispersal in *Arceuthobium* may (Hudler and French, 1976) or may not (Muir, 1977) be correlated with increasing daily temperatures in the habitat. After dispersal, seeds are washed by rain into crevices in trees bark or to the base of conifer needles (Hawksworth, 1965).

Some Loranthaceae such as *Nuytsia floribunda* have dry fruits, and mechanisms of dispersal are unknown (Kuijt, 1969). Most Loranthaceae, however, have juicy fruits. Fruits of *Loranthus celastroides* have a circular line of dehiscence on the upper end. If the fruit touches something as it falls from the host tree, the top opens like an operculum, and the viscid seed slides out of the fruit (McLuckie, 1923). For fruits of most Loranthaceae, birds are the primary means of seed dispersal. Depending on the mistletoe species, mature fruits can be white, yellow, orange, red, blue, purple, or black, and Kuijt (1969) has wondered what role the color preferences of birds has played in the evolution of mistletoes. Seeds of mistletoes are ground up and killed while they are in the gizzards of some species of birds. However, the digestive tract in other birds is modified in a way that seeds do not enter the gizzard. Thus, seeds remain in the digestive tract for less than an hour and emerge with much of the viscin material still attached to them (Kuijt, 1969). Depending on the bird species, seeds are regurgitated in pellets (Godschalk, 1983) or defecated, and birds sometimes rub their anus on branches to remove the sticky seeds (Liddy, 1983).

In South Africa, seeds of *Tapinanthus natalitius, T. leendertziae,* and *Viscum combreticola* are eaten and dispersed by the yellow-fronted tinkerbird (*Pogoniulus chrysoconus*). These mistletoes basically have the attributes of species with a model 1 type of dispersal strategy, i.e., plants produce large, nutritious fruits a few at a time throughout the year. Thus, food is always available for the bird, which feeds almost exclusively on fruits of these mistletoes (Godschalk, 1983).

In *Acacia papyrocarpa* woodlands in South Australia, the spiny-cheeked honeyeater (*Acanthagenys rufogularis*) and the mistletoe bird (*Dicaeum hirundinaceum*) eat seeds of the mistletoe *Amyema quandang* and deposit them on live branches. Based on the number of 2-month-old seedlings, the honeyeater seemed to be the most effective seed disperser (Reid, 1989). The mistletoe has some ripe fruits throughout the year, with the peak of fruiting occurring in summer. Low densities of mistletoe birds feed on mistletoe fruits all year, but they also eat insects. Honeyeaters feed on mistletoe fruits all year, but also eat mistletoe nectar during the winter flowering season. Thus, reproduction in this mistletoe may have evolved in response to flower pollination and seed dispersal by frugivorous birds (Reid, 1990). According to Reid (1991), frugivorous birds specializing on fruits of mistletoes have evolved in Australasia, the old World tropics, and the Neotropics.

Seeds of *Phoradendron robustissimum* are deposited on tree branches in Costa Rica by defecating birds. When seeds were sown on branches of the host tree *Sapium oligoneuron*, successful seedling establishment was more likely to occur on branches 10–14 mm in diameter than on those in any other size class (Sargent, 1995). Few seedlings were established on branches <5 mm in diameter because many of the branches died; the presence of *P. robustissimum* seeds apparently contributed to the death of small branches. No seedlings became established on branches larger than 80 mm in diameter. Bark thickness increased with an increase in branch diameter, and lack of seedling establishment may have been due to a lack of bark penetration by the haustorium (Sargent, 1995).

The Chilean mockingbird (*Mimus thenca*) is more likely to deposit seeds of the endophytic mistletoe *Tristerix aphyllus* on a host cactus (*Eulychnia acida* or *Echinopsis skottsbergii*) that already has been parasitized than on one that has not been parasitized by the mistletoe (del Rio *et al.*, 1996). Further, tall cacti are more likely to be parasitized than short ones (del Rio *et al.*, 1995).

2. Types of Seed Dormancy

Seeds of the Loranthaceae and Viscaceae have developed embryos of the linear type (Martin, 1946; Kuijt, 1969); consequently, embryos must grow before germination is possible. Seeds of *Loranthus celastroides* (McLuckie, 1923), *Tapinanthus bangwensis* (Room, 1973), *Amyema preissii* (Lamont and Perry, 1977), *Phoradendron juniperinum* (Dawson and Ehleringer, 1991), and *Tristerix aphyllus* (Mauseth *et al.*, 1985) germinate within 1–9 days after they are removed from the fruit. Seeds of *Arceuthobium vaginatum* germinate in the field in Colorado immediately following dispersal in autumn (Hawksworth, 1965). Thus, there is no delay in germination after conditions become nonlimiting, and we conclude that the only type of dormancy in seeds of these species is morphological dormancy.

There is evidence that seeds of some mistletoes have

morphophysiological dormancy. Although seeds of *Arceuthobium americanum* (Hawksworth, 1965; Muir, 1977) and *A. tsugense* (Carpenter *et al.*, 1979) are dispersed in autumn, they do not germinate until spring, after being subjected to cold stratification for a number of weeks. There are two possible reasons for a delay in germination: (1) Temperatures at the time of dispersal in autumn are too low for embryo growth and germination or (2) seeds have physiological dormancy in addition to MD and thus have MPD; therefore, cold stratification is required to break physiological dormancy. Seeds of *A. campylopodum* incubated at 13.3°C required 60 days to reach 64% germination (Scharpf and Parmeter, 1962); however, seeds stored dry at 1.5°C for 240 days germinated to about 53 and 56% after only 16 days of incubation at 17 and 19°C, respectively (Beckman and Roth, 1968). These studies indicate that the seeds had physiological dormancy, which was broken during the period of cold storage. Because seeds of *Arceuthobium* have the hygroscopic viscid material attached to them and a moisture content of about 35% (Scharpf, 1970), it is reasonable to think that their moisture content during storage at 1.5°C was high enough for them to be cold stratified and come out of physiological dormancy.

Freshly matured seeds of *Arceuthobium americanum*, *A. campylopodum*, *A. laricis*, and *A. douglasii* did not germinate during a 10-day incubation period in light at 22–23°C; however, after 2 weeks of dry storage at 5°C they germinated to 60–80% (Wicker, 1974). Further studies are needed on seeds of spring-germinating species of *Arceuthobium* to determine (1) rates of loss of the physiological dormancy in these seeds on a wet substrate vs "dry" storage and (2) if any embryo growth occurs during cold stratification or only after physiological dormancy is broken.

Another indication of MPD in mistletoe seeds comes from studies of growth responses of embryos on synthetic media. If embryos from seeds of *Phoradendron tomentosum* (Bajaj, 1970) and *Dendrophthoe falcata* (Bajaj, 1968) are enclosed by the endosperm, germination is delayed. Does this mean that the embryos of these seeds have physiological dormancy and, therefore, do not have enough growth potential to overcome the mechanical resistance of the endosperm? If this is true, the embryos would have MPD. Because the embryos covered by endosperm eventually grew, this may mean that physiological dormancy was broken during warm stratification.

3. Dormancy-Breaking Requirements

No special dormancy-breaking treatments are required in mistletoe seeds with only morphological dormancy. In fact, some embryo growth occurs while seeds are still in the fruit, but it stops before the radicle penetrates the epicarp (Lamont, 1982a). Lamont and Perry (1977) have proposed various reasons why radicles do not penetrate the epicarp:

1. The osmotic potential of the viscin material is great enough to inhibit germination.
2. Germination inhibitors occur in tissues of the fruit.
3. The epicarp prevents light from reaching the embryo.
4. The epicarp inhibits gas exchange.

Lamont and Perry (1977) and Lamont (1982a) have studied these possibilities with regard to the germination of *Amyema preissii*.

Seeds of *Amyema preissii* surrounded by viscin and placed in polyethylene glycol (PEG 20000) at an osmotic potential of −1.5 MPa germinated to near 100%; thus, the osmotic potential does not explain the lack of germination (Lamont and Perry, 1977). The epicarp of *A. preissii* fruits transmits up to 540 μmol m^{-2} sec^{-1}, 400–700 nm, of light; consequently, light was not limiting for germination (Lamont, 1982a). Small cuts in the epicarp of *A. preissii* fruits, as well as in those of *Viscum capense, V. rotundifolium,* and *V. continuum* (Lamont, 1982b), stimulated germination, thus apparently the germination inhibitor was a volatile compound. Seeds of *A. preissii* germinated in 100% O_2 or N_2, especially in light, but CO_2 was inhibitory (Lamont, 1982a).

Carbon dioxide in fruits of *Amyema preissii* was 26.6% by volume, which is 760 times greater than that of ambient air. At CO_2 concentrations of 0–5%, 100% of the seeds germinated, but at 20% CO_2 only 4% of them did so. Thus, the concentration of CO_2 in fruits of this species is high enough to prevent germination until the epicarp is removed by birds (Lamont, 1982a). It seems logical that the buildup of CO_2 would occur during fruit maturation and that the amount of embryo growth would be determined by how fast it accumulated. However, the amount of CO_2 was greater in green than in mature berries of *A. preissii* (Lamont, 1982a).

Many more studies are needed on the germination of mistletoe seeds from a range of habitats to determine the relative importance of MD and MPD and to better understand the dormancy-breaking requirements. Although 2 weeks of cold stratification will break the physiological dormancy in seeds of some *Arceuthobium* spp. that have MPD (Wicker, 1974), very little is known about other mistletoes whose seeds germinate in spring in temperate regions (e.g., Cannon, 1904). Do these seeds have MPD, and is cold stratification required to break the physiological dormancy? There is a good possibility that seeds of some mistletoes have MPD and that the physiological dormancy is broken by warm stratification. A good place to look for such seeds is in

tropical and/or subtropical regions, especially those with a seasonally dry climate. Seeds of *Phoradendron tomentosum* and *Dendrophthoe falcata,* whose embryos grow slowly when covered by endosperm (Bajaj, 1968, 1970), might be interesting to study.

4. Germination Requirements

Mistletoe seeds do not have to be in contact with a host plant to germinate (Yan, 1993; Clay *et al.,* 1985). In fact, seeds will germinate on a variety of substrates, including soil, fences, gates (McLuckie, 1923), dead twigs, thorns, leaves, other mistletoes (Cannon, 1904), and telephone wires (Room, 1973), as well as on branches of host and nonhost trees (Clay *et al.,* 1985). However, haustorial disks were more likely to form if seeds of *Phoradendron tomentosum* germinated on the same species of host tree that supported the maternal plant (Clay *et al.,* 1985).

Large seeds of *Phoradendron juniperinum* germinated to higher percentages than small ones, and seeds produced by old plants were heavier than those produced by young plants. For example, there was a twofold difference in the mass of *P. juniperinum* seeds produced by 4- and 14-year-old plants (Dawson and Ehleringer, 1991).

Data on temperature requirements for the germination of mistletoe seeds are very limited. However, in the few studies where nondormant seeds have been tested over a range of temperatures, some germination occurred from 5 to about 35°C (Beckman and Roth, 1968; Lamont, 1982a). The optimum germination temperature is known for a few species, including *Arceuthobium campylopodum,* 19°C (Beckman and Roth, 1968); *A. abietinum,* 13°C (Scharpf, 1970); *A. pusillum,* 15°C (Bonga and Chakraborty, 1968), *Amyema preissii,* 25°C (Lamont, 1982a); *Nuytsia floribunda,* 20°C (Lamont, 1983b); and *Viscum album,* 15–20°C (Salle, 1983).

Seeds of a number of mistletoes, including *Arceuthobium occidentale* (Scharpf, 1970), *A. campylopodum* (Scharf and Parmeter, 1962; Knutson, 1984), *A. tsugense* (Knutson, 1984), *Phoradendron flavescens* (Gardner, 1921), *Viscum articulatum, V. orientale* (Wiesner, 1894), *V. capense, V. rotundifolium, V. continuum* (Lamont, 1982b), *Loranthus repandus, L. pentandrus* (Wieser, 1894), and *L. europaeus* (Weisner, 1897), germinated to higher percentages in light than in darkness. Further, unshaded seeds of *Tapinanthus bangwensis* germinated to higher percentages and were more likely to become established than shaded ones (Room, 1973). Seeds of *Viscum cruciatum* (Glimcher, 1938) and *V. album* (Wiesner, 1897) will not germinate in darkness, whereas those of *Arceuthobium americanum, A. campylopodium, A. vaginatum* (Holmes *et al.,* 1968), *Nuytsia flori-*

bunda (Lamont, 1983b), and *Loranthus floribunda* (Wiesner, 1897) germinate equally well in light and darkness.

Rigby (1959) found that the color of the embryo and endosperm could be used as an indicator of whether seeds required light for germination. *Amyema pendula* seeds with white embryos and endosperm required light for germination after the pericarp was removed. Seeds of *A. linophylla* and *Lysiana exocarpi* with green embryos and red or green endosperm germinated in darkness, and seedlings grew in darkness. However, seedlings did not form a clamp (holdfast) until they were exposed to light. Seeds of *A. miquelii* with green embryos and a white endosperm germinated in darkness, but seedlings did not grow until they were exposed to light.

Mistletoe seeds vary in moisture requirements for germination. Seeds of *Amyema preissii* do not require either a moist substrate or a high RH for germination (Lamont and Perry, 1977), whereas those of *Loranthus celastroides* require a high RH but not a moist substrate (McLuckie 1923). Seeds of *Arceuthobium pusillum* on a moist substrate germinated to about 93% in 3 weeks, but those at a RH of 98% germinated to 82% in 6 months (Bonga, 1972).

III. MYCOHETEROTROPHIC PLANTS OTHER THAN ORCHIDS

Achlorophyllous angiosperms that are not parasitic on autotrophic (green) plants have been called "saprophytes," meaning that their source of carbon is from dead organic material in the habitat (Furman and Trappe, 1971). However, these colorless plants do not obtain nutrition directly from dead organic material; therefore, the name saprophyte is inappropriate. Instead, they obtain nutrition from fungi, and fungi obtain it from dead organic material and/or autotrophs. Thus, the so-called saprophytes are heterotrophic, and their ultimate source of carbon is from fungi. In fact, "saprophytes" are actually parasitic on fungi (Bakshi, 1959), and the term "mycoheterotrophic" clearly explains their nutritional status (Leake, 1994).

About 400 species of mycoheterotrophs are known, and they occur in 87 genera and 10 families, including the Burmanniaceae, Corsiaceae, Gentianaceae, Geosiridaceae, Monotropaceae, Orchidaceae, Petrosaviaceae, Polygalaceae, Pyrolaceae, and Triuridaceae (Groom, 1895; Furman and Trappe, 1971; Maas, 1979; Leake, 1994). According to Leake (1994), mycoheterotrophs have evolved independently many times in widely separated taxonomic groups, but they have many characteristics in common. (1) Plants frequently grow in densely shaded habitats with much leaf litter. (2) Thousands of seeds may be produced in a single fruit. (3) Seeds are dust-like, rarely exceeding 1 mm in the longest dimension, and are dispersed by wind; animals and water may be important dispersal agents in some species. (4) Embryos are not differentiated into hypocotyl, epicotyl, and cotyledon(s), and depending on the species they may have only a few cells. (5) Seeds have only a limited amount of endosperm, which, relative to the size of the embryo, is species specific.

Four forms of parasitism have evolved in angiosperms, and they are distinguished on the basis of how the parasite becomes associated with the host: Orobanchoid, Cuscutoid, Pyroloid (Monotropoid), and Orchidoid (Teryokhin and Nikiticheva, 1982). In all four forms, the embryo and endosperm of seeds have been reduced, and there has been a shift to progressively earlier stages of ontogenesis as to when the parasite makes contact with the host. In the Orobanchoid form of parasitism, the haustorium originates from the radicular pole of the embryo, and in the Cuscutoid form it originates from the terminal pole. However, in the Pyroloid and Orchidoid forms, an haustorium is not formed, and an association is established with fungi (Francke, 1934; Teryokhin and Nikiticheva, 1982). In the Pyroloid form, the sporophyte of the parasite arises from the radicular pole of the embryo, whereas in the Orchidoid form it arises from the epicotyl (Teryokhin and Nikiticheva, 1982).

Mycoheterotrophs in the Pyrolaceae and Monotropaceae have the Pyroloid form of parasitism, whereas those in the Orchidaceae have the Orchidoid form (Teryokhin and Nikiticheva, 1982). No information seems to be available on the form of parasitism for mycoheterotrophs in the Burmanniaceae, Corsiaceae, Gentianaceae, Geosiridaceae, Petrosaviaceae, Polygalaceae, or Triuridaceae.

Little is known about the germination of seeds of mycoheterotrophs with the Pyroloid form of parasitism, except that seeds of *Monotropa hypopithys* do not germinate unless the appropriate symbiont is present (Francke, 1934). Because (1) mycoheterotrophic and autotrophic Orchidaceae have the Orchidoid form of parasitism (Teryokhin and Nikiticheva, 1982; Leake, 1994) and (2) early seedling stages of both types of orchids are unable to photosynthesize (Harley and Smith, 1983), the germination of mycoheterotrophic and autotrophic orchids will be considered together in Section IV.

IV. ORCHIDS

The Orchidaceae is estimated to have about 19,500 species (Dressler, 1993), which grow in all kinds of habitats, except extreme deserts, from Alaska and northern Sweden south to Tierra del Fuego and Macquarie Island (Dressler, 1981). Orchids can be epiphytic, terrestrial, lithophytic, semiaquatic (but not aquatic), and subterranean (Withner, 1959; Arditti, 1979). Much variation exists in the family with respect to plant size, flower size and color, floral structure, leaf morphology, and seed shape (Arditti, 1979; Dressler, 1993). Seeds may be flattened, lenticular, globular, ovoid, winged, thread-like, balloon-like, or kernel-like (Dressler, 1993). Clifford and Smith (1969) found five basic shapes of seeds in 49 species from the orchid tribes Epidendreae and Neottieae (Fig. 11.6). Orchid seed coats are either glossy and crustose or loose and papery (Barthlott, 1975; Dressler, 1981, 1993), and walls of cells in the papery seed coats may be smooth or reticulated (Clifford and Smith, 1969; Barthlott, 1975; Healey *et al.,* 1980; Dressler, 1981, 1993).

One characteristic common to all orchids is that their seeds are tiny and appear dust-like to the naked eye. Seed lengths vary from 0.18 mm in *Oberonia iridifolia* to 3.85 mm in an *Epidendrum* hybrid (Clifford and Smith, 1969). See Arditti (1967) for a table summarizing various characteristics of seeds of 98 orchid taxa. Data on seed shape and size, shape of cells in the seed coat, and nature of any thickenings on the walls of these cells can be important in the classification of orchids, especially at generic, subtribal, and tribal levels (Clifford and Smith, 1969; Barthlott, 1975; Healey *et al.,* 1980; Dressler, 1993).

Weight of an individual orchid seed ranges from 0.3 μg in *Schomburgkia undulata* to 14 μg in *Galeola [lindleyana* (?)] (Koch and Schulz, 1975). The low weight is attributed to lack of an endosperm and to an embryo consisting of relatively few cells. Although double fertilization takes place in a number of species, the primary endosperm nucleus degenerates, and endosperm is not formed (Savina, 1974).

In addition to their low weight, seeds also have a high seed/embryo volume ratio (i.e., air space is present inside the seed coats). Thus, orchid seeds are very buoyant and can float on the surface of water (Stoutamire,

1983) or be carried by wind (Healey *et al.,* 1980). Four terrestrial species of orchids were among the first plants to recolonize Krakatoa after all plant life on the island was destroyed by volcanic activity in 1883. Because the nearest seed source was 400 km away, wind dispersal seems to be the most reasonable explanation for how orchids reached the island (Went, 1957).

With a decrease in the size and weight of orchid seeds, there has been an increase in the mean number of seeds produced per capsule, ranging from 376 in *Listera cordata* (Stoutamire, 1964) to 4,000,000 in *Cycnoches ventricosum* var. *chlorochilon* (Arditti, 1961). According to Teryokhin and Nikiticheva (1982), a reduction in seed size and an increase in the number of seeds occurred during the evolution of parasitic plants in response to the scattered distribution of host plants. Thus, as the number of seeds produced by parasitic plants increases, the possibility that some of them will reach a suitable host is enhanced.

A. Types of Seed Dormancy

According to Dressler (1993), "... orchids seem most closely related to something that would once have been classified in the Liliaceae or possibly the Amaryllidaceae." In present-day members of the Liliaceae and Amaryllidaceae, seeds have endosperm and differentiated (linear) embryos, whereas those of Orchidaceae lack endosperm and have an undifferentiated embryo (Martin, 1946). However, a rudimentary cotyledon occurs in the orchids *Arundina, Bletilla, Dendrochilum, Epidendrum, Polystachya, Sobralia,* and *Thunia* (Nishimura, 1991). These genera belong to the Epidendroideae, the most advanced subfamily of orchids (Dressler, 1993).

Absence of a radicle, and usually of a cotyledon, in orchid seeds means that much development and growth are required before a seedling is produced; thus, seeds of the Orchidaceae have morphological dormancy (see Chapter 3). During germination, the embryo swells until the seed coat is ruptured, but a radicle, and usually a cotyledon per se, are not produced. Cells in the embryo divide and produce a globular structure called a protocorm that is less than 1 mm in diameter (Weatherhead *et al.,* 1986). Some epidermal cells produce hairs, which may give rise to rhizoids. After several weeks, depending on species and growth conditions, the protocorm becomes polarized. The apical side gives rise to a leaf primordium, and a root primordium develops exogenously on the side of the protocorm (Fig. 11.7). When the leaf primordium originates, the protocorm becomes spindle-shaped (Curtis, 1943; Knudson, 1950; Stoutamire, 1964; Arditti, 1979; Linden, 1980).

Freshly matured seeds of some orchids germinate

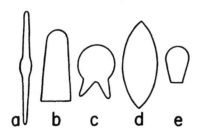

FIGURE 11.6 Basic shapes of seeds found in 49 orchid species belonging to the tribes Epidendreae and Neottieae. From Clifford and Smith (1969), with permission.

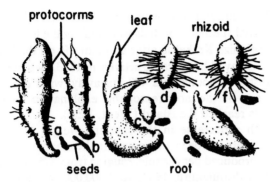

FIGURE 11.7 Seeds and protocorms of various orchids. (a) *Liparis loeselii*, 41 months old. (b) *Goodyera oblongifolia*, 12 months old. (c) *Habenaria ciliaris*, 14 months old. (d) *H. blephariglottis*, 14 months old. (e) *H. ciliaris* × *blephariglottis*, 14 months old. (f) *H. blephariglottis* × *psycodes*, 14 months old. From Stoutamire (1964), with permission.

immediately when they are placed on synthetic media (Curtis, 1936; Lindquist, 1958, 1965; Harvais and Hadley, 1967; Stoutamire, 1974), but those of other species do not germinate for many months (Curtis, 1936; Liddell, 1944; Knudson, 1950; Stoutamire, 1964; De Pauw et al., 1995). One reason for lack of germination may be that seeds have some type of dormancy in addition to morphological dormancy. Knudson (1950) said that, "Dormancy in orchid seeds is not uncommon. ... It is of interest that the dormancy is inherent in the embryo."

Physical dormancy is not known to occur in orchid seeds, and seeds of many species have been observed to imbibe water when placed on a moist substrate (Stoutamire, 1974; Light, 1992). Orchid seed coats are only one cell thick (Ramsbottom, 1922–1923), and they lack the water-impermeable palisade layers characteristic of seeds with physical dormancy. Lignified bands are found on the walls of cells in some orchid seed coats, but they do not completely cover the walls (Shushan, 1959). However, some genera such as *Vanilla, Selenipedium,* and *Galeola,* have sclerotic seed coats composed of lignified cells (Withner, 1955; Wirth and Withner, 1959). It would be interesting to do some imbibition curves for seeds of these genera. The suspensor projects through the micropyle in seeds of some orchids, e.g., Cattleya (Shushan, 1959), and it provides an entry point for water and fungi into the seed.

Various pieces of evidence suggest that seeds of some orchids have PD in addition to MD, thus they would have MPD. (1) Immature seeds of various species, including *Calanthe discolor* (Arditti, 1982), *Calypso bulbosa* (Arditti, 1982), *Cypripedium acaule* (Withner, 1953; Light, 1989), *C. reginae, C. calceolus* var. *parviflorum, C. candidum* (De Pauw and Remphrey, 1993), *Epipactis gigantea* (Arditti et al., 1981; Arditti, 1982),

Orchis mascula, O. morio (Linden, 1992), *Platanthera bifolia* (Linden, 1980, 1992), and *Vanilla* sp. (Withner, 1955) germinate to higher percentages than mature ones, thus (physiological?) dormancy develops during seed ripening (Stoutamire, 1964). (See Table 2.1 in Rasmussen, 1995.) (2) After seeds are placed on sterile culture media (Liddell, 1944) or in soil from the habitat (MacDougal and Dufrenoy, 1946), there may be a delay of several months before embryo growth occurs, indicating that PD must be broken before MD is broken. (3) Soaking seeds of *Dactylorhis maculata, Gymnadenia conopsea, Orchis morino* (Linden, 1980), and *O. purpurella* (Harvais and Hadley, 1967) in aqueous sodium hypochlorite increased germination. Although the exact effect of sodium hypochlorite on the seed coat is unknown, it seems possible that molecules were oxidized and the seed coat was weakened. If weakening of the seed coat promoted germination, this means the growth potential of the embryo was insufficient to overcome the restriction of the seed coat. Thus, the embryo has some PD. (4) Seeds of *Cypripedium reginae* (Ballard, 1987) and *Ponerorchis graminifolia* (Ichihashi, 1989) require cold stratification for germination. Further, the appearance of *Corallorhiza odontorhiza, Goodyera pubescens,* and *Galearis spectabilis* seedlings in late spring or early summer (Rasmussen and Whigham, 1993) suggests that cold stratification plays a role in the germination ecology of these species. Seeds of *Vanilla fragans* require a period of afterripening before they will germinate (Knudson, 1950). Because cold stratification and afterripening are known to break PD (see Chapter 3), it is assumed that seeds requiring these treatments for germination have PD. However, cold stratification decreased the germination of *Cypripedium calceolus, Epipactis atrorubens,* and *E. helleborine* seeds (Van Waes and Debergh, 1986). Does cold stratification induce PD in seeds of these species? (5) The germination of *Dendrobium nobile* seeds was promoted by GA (Miyazaki and Nagamatsu, 1965), indicating that they may have some type of nondeep or intermediate MPD.

Arditti *et al.* (1990) seem to doubt that PD plays a role in the germination of orchids because seeds of some species germinate in response to high rather than low temperatures (Nakamura, 1976). These authors apparently did not realize that some seeds with PD require high, whereas others require low, temperatures for loss of PD (see Chapter 3).

B. Dormancy-Breaking Requirements

For most orchids, it is not known if seeds have only MD or if they have MPD. If orchid seeds have only MD, they do not have dormancy-breaking requirements per se. However, the temperature and light:dark requirements for germination may be very exacting (see later), thus serving as mechanisms that control the timing of germination in nature. However, if orchid seeds have MPD, MD cannot be broken until (or after) environmental conditions are favorable for the breaking of PD. Thus, both dormancy-breaking and germination requirements would play a role in controlling the timing of germination of seeds with MPD.

It seems logical that seeds of many orchids in temperate regions would have MPD and that cold stratification would be required to break the PD part of MPD. Thus, germination would occur in spring, at the beginning of the growing season. Four weeks of cold stratification were required to break PD in seeds of *Ponerorchis graminifolia* (Ichihashi, 1989), and 9 weeks were required to break it in seeds of *Cypripedium reginae* (Ballard, 1987). (See Table 4.2 in Rasmussen, 1995 for length of cold stratification periods required to break dormancy in seeds of various species.) However, dormancy could be broken by warm stratification during summer in temperate regions, and seeds germinate in autumn at the beginning of the cool, moist season, as in winter annuals. This germination phenology has not been documented for orchids in the field in temperate regions. However, a warm stratification requirement for loss of PD may help explain why seeds of various species [e.g., *Calanthe discolor* and *C. izuinsularis* (Nishimura, 1982)] require several months of incubation at room temperature before they germinate. In subtropical and tropical regions with alternate wet and dry seasons, loss of dormancy during the dry season would permit germination to occur at the beginning of the wet season. A year of exposure to greenhouse temperatures was required to break the PD in seeds of *Vanilla fragans* (Knudson, 1950).

Rasmussen and Whigham (1993) have developed a technique for sowing orchid seeds in the habitat and then retrieving them. Thus, the technology now is available for monitoring germination phenology under natural conditions. Data on germination phenology will allow researchers to design experiments to better understand the germination ecology of orchids. (1) If seeds germinate soon after maturation, they probably have only MD. (2) If germination occurs in spring, seeds may require cold stratification for loss of PD. (3) If germination occurs in autumn, seeds may require exposure to high summer temperatures for loss of PD. It may be possible to modify the Rasmussen–Whigham technique so that large numbers of seeds could be stored in the habitat. Then, seeds could be retrieved and tested over a range of environmental conditions to determine temperature and light:dark requirements for germination at different times of the year.

C. Germination Requirements

1. Presence of Symbionts

The dogma of orchid seed germination until the early 1900s was that seeds would not germinate unless they were infected by a fungus (see Curtis, 1939 and Arditti, 1967 for reviews). However, Knudson (1922) germinated seeds of *Laelia* and *Cattleya* under sterile conditions in culture tubes on media containing sugars; thus, seeds germinated in the absence of fungi. Further, after seedlings produced four to five green leaves, they no longer required sugars and grew normally when transplanted to a compost of peat and sphagnum in flower pots. Knudson's technique for germinating seeds of *Laelia* and *Cattleya* in the absence of fungi has made it possible for researchers to learn much about the ecophysiology of seed germination in orchids.

When seeds of many orchids are placed on a moist substrate, the embryo swells, breaks through the seed coat, and produces a protocorm with epidermal hairs (Burgeff, 1959; Shushan, 1959; Harrison and Arditti, 1978; Hadley, 1982; Harley and Smith, 1983; Warcup, 1985; Smreciu and Currah, 1989). However, seedling development stops at the protocorm stage until an exogenous source of carbohydrates, usually sugars, is supplied (Harley and Smith, 1983; Harrison and Arditti, 1978) or until the protocorm becomes infected by symbiotic fungi (Warcup, 1985). Embryos of some orchids do not produce a protocorm unless exogenous sugars (Curtis, 1939; Downie, 1940; Manning and van Staden, 1987) or fungi (Knudson, 1925; Downie, 1940; Warcup, 1985; Smreciu and Currah, 1989) are present.

Orchid seeds lack endosperm; consequently, food reserves are stored in cells of the embryo. Cells in freshly matured seeds contain lipids and proteins but no starch (Knudson, 1925; Harrison, 1977; Arditti, 1979; Manning and van Staden, 1987; Rasmussen, 1990; Richardson *et al.*, 1992). However, embryos of some species contain a small amount of free sugars, including sucrose, glucose, maltose, arabinose, and rhamnose (Manning and van Staden, 1987). Ultrastructural and histochemical studies of the development and mobilization of food reserves in orchid seeds help us understand why exogenous carbohydrates are important in the germination of orchid seeds. Seeds of *Disa polygonoides* and *Disperis fanniniae* placed on culture media (Knudson C) with and without sucrose showed different patterns in the way food reserves in embryo cells were used (Manning and van Staden, 1987). In the presence of sucrose, protein breakdown was rapid in seeds of both species and began before lipid hydrolysis. Glyoxysomes (type of microbody containing enzymes that help convert lipids to sugars) were not present in embryo cells at maturity, but microbodies (possibly mitochondria) appeared in the vicinity of lipid bodies after seeds of both species were incubated for 4 days in the presence of sucrose. After microbodies became apparent, starch grains were formed, implying that lipids had been converted to starch. In the absence of sucrose, no hydrolysis of proteins or lipids took place in embryo cells of *D. polygonoides,* whereas protein, but not lipid, hydrolysis occurred in those of *D. fanniniae.* Further, in *D. fanniniae* the free sugars that were present in embryo cells at the time of maturation were converted to starch.

Lipid reserves in *Cattleya aurantiaca* seeds placed on Knudson C culture media with no sucrose were broken down slowly, but when sucrose was present they were broken down rapidly. However, glyoxysomes were not formed, even when sucrose was present, and mitochondria were associated with lipid bodies (Harrison, 1977). Further, acetate, a product of lipid hydrolysis, was not converted to sugar (Harrison and Arditti, 1978). Thus, it appears that seeds are biochemically incapable of converting fats to sugars. Arditti (1979) suggests that acetyl CoA, which is released from the lipid bodies, may be oxidized during the conversion of energy, via the Kreb's cycle in mitochondria, which might explain the close association of mitochondria with lipid bodies.

Orchid seeds can use their lipid reserves only slowly and not very efficiently, and they cannot use starch or cellulose (Arditti, 1979). Thus, after the protocorm is formed additional development does not occur until a source of sugar molecules becomes available. In the laboratory, sugars added to culture media are absorbed by the protocorm, but in nature free sugars are not readily available for absorption (Smith, 1966). Symbiont fungi are the main source of sugars in nature for protocorms of all orchids and of all stages in the life cycle of mycoheterotrophic orchids. Not only do the hyphae contain sugars, but they also contain enzymes that can break down starch in cells of the orchid (Burgeff, 1959; Purves and Hadley, 1976). In fact, after protocorms become infected with fungi, starch synthesis may be suppressed (Purves and Hadley, 1976).

The germination of orchid seeds in sterile culture containing starch, instead of sugars, is promoted by the addition of various genera of basidiomycetes, including *Armillaria, Ceratobasidium, Corticium, Fomes, Hypholoma, Marasmius, Pleurotus, Rhizoctonia, Sistotrema, Thanatephorus, Tulasnella,* and *Xerotus* (Cappelletti, 1935; Burgeff, 1936; Porter, 1942; Hadley, 1970b; Warcup, 1973; Smreciu and Currah, 1989). In addition, the phycomycetes *Phytophthora* and *Choanophora,* the ascomycetes *Ceratostomella, Pencillium,* and *Chaelomium* (Curtis, 1939), and some bacteria (Knudson, 1922; Wilkinson *et al.,* 1989) promote germination. The most common genus of fungi stimulating the germination of

orchid seeds and promoting the growth of protocorms and seedlings is *Rhizoctonia* (Arditti, 1979).

Seeds of an orchid species may be promoted to germinate by several different species of fungi (Curtis, 1939; Warcup, 1973, 1985; Masuhara and Katsuya, 1989), whereas the same species of fungi may stimulate seed germination in various species of orchids (Curtis, 1939; Hadley, 1970b; Clements *et al.,* 1986; Masuhara and Katsuya, 1989). Further, fungi isolated from orchid rhizomes may (Warcup, 1971; Masuhara and Katsuya, 1989; Muir, 1989) or may not (Curtis, 1939; Muir, 1989; Smreciu and Currah, 1989) promote the germination of seeds of that same species. Thus, it generally has been concluded that orchid seeds do not require a specific fungus for germination (Curtis, 1939; Hadley, 1970b; Arditti, 1979).

The question of whether some orchids are species specific with regard to fungal symbionts for germination has not been resolved completely (Arditti, 1979). However, some strong fungus–orchid associations do exist. In Australian terrestrial orchids, the fungus *Sebacina vermifera* has been isolated from roots of *Caladenia* spp. and the fungus *Tulasnella colospora* from roots of *Diuris* spp. (Warcup, 1971). Warcup (1973) found that seeds of two species of *Pterostylis* were stimulated to germinate only by the fungus *Ceratobasidium cornigerum,* whereas seeds of two species of *Diuris* were stimulated to germinate only by the fungus *Tulasnella calospora.* More than one species of the fungus *Tulansnella* stimulated seeds of two species of *Thelymitra* to germinate, but *C. cornigerum* was not very effective in promoting germination. Various isolates of *T. calospora* differ in their ability to stimulate seeds of *Diuris* spp. and *Thelymitra* spp. to germinate (Warcup, 1973). When seeds of *Spiranthes sinensis* were buried in the field in Japan, the resulting protocorms mostly were infected with *Rhizoctonia repens;* however, in the laboratory 30 other *Rhizoctonia* isolates induced seed germination. Thus, specificity was not the same during *in situ* and *in vitro* germination (Masuhara and Katsuya, 1994). Also, the fungus that stimulates seed germination may not promote seedling growth (Muir, 1989); thus, different fungi may be associated with different stages of the life cycle (Harley and Smith, 1983). Curtis (1939) suggested that species of fungi were correlated with the type of ecological habitat rather than with the species of orchid, i.e., if an orchid species grew in a restricted type of habitat, it would be associated with fewer fungi than if it grew in a variety of habitats.

Fungi infect the embryo of imbibed orchid seeds via the dead suspensor cells (Downie, 1943; Clements, 1988; Richardson *et al.,* 1992) or the protocorm via epidermal hairs or rhizoids (Harvais and Hadley, 1967; Williamson and Hadley, 1970; Rasmussen, 1990; Rasmussen *et al.,*

1990a). However, infection of a seed does not ensure that germination will occur (Smreciu and Currah, 1989) or that a symbiotic relationship will develop between the fungus and the orchid. Infected cells in the seed or protocorm may kill the fungus or the fungus may kill the seed or protocorm (Burgeff, 1959). Fungi may promote seed germination, but then become pathogenic and kill the protocorm (Knudson, 1925; Harvais and Hadley, 1967; Smreciu and Currah, 1989). Even a supposedly good symbiotic fungus sometimes kills protocorms (Harvais, 1974). However, fungi known to be plant pathogens can live in a mycorrhizal association with orchid seeds or protocorms without killing them (Knudson, 1925; Harvais and Hadley, 1967; Williamson and Hadley, 1970; Harvais, 1974). In culture media, successful establishment of an orchid–fungal association is enhanced if (1) the medium is high in starch and low in sugars (Knudson, 1925) and (2) introduction of the fungus occurs prior to formation of the protocorm (Harvais, 1974). Symbiotic relationships between *Orchis purpurella* protocorms and *Rhizoctonia solani* were stable at 11°C, but were unstable at 17 and 20°C and ended in parasitism of the protocorm (Harvais and Hadley, 1967).

If a mycorrhizal association develops between a protocorm and a species of fungi, hyphae are active in some, but not all, cells of the protocorm. Infected cells are located to the inside of the epidermal layer in the basal portion of the protocorm and are divided into two layers. The outermost layer of cells contains active hyphae, and the inner one contains hyphae that are collapsing and being digested. After hyphae are digested, new ones enter the cells and subsequently are digested (Knudson, 1925; Ramsbottom, 1922–1923; Burgeff, 1959; Harley and Smith, 1983).

When hyphae are broken, sugars, vitamins, amino acids, hormones, proteins, nucleic acids, and minerals are released into the orchid cell. What do orchids obtain from fungi that promote(s) seed germination and protocorm growth? The main research approach to try to answer this question has been to add chemicals to sterile culture media and determine their ability to stimulate germination and/or protocorm growth (Arditti, 1967).

Numerous kinds of soluble carbohydrates have been tested for their ability to promote the germination of orchid seeds (Arditti, 1979). Some compounds have little or no effect, others promote seed germination of many species, and a few stimulate seed germination of only a limited number of species (Arditti, 1967, 1979). Carbohydrates that stimulate germination include monosaccharides: fructose, glucose, mannose, ribose, and xylose; disaccharides: cellobiose, lactose, maltose, melibiose, sucrose, trehalose, and turanose; trisaccharides: melezitose and raffinose; tetrasaccharide: stachyose; sugar alcohols: arabinitol, ribitol, sorbitol, and xyli-

tol; and organic acids: malate and pyruvate (La Grade, 1929; Wynd, 1933a; Arditti, 1979). Sucrose, D-glucose, and D-fructose are the carbohydrates most likely to stimulate germination (Withner, 1959), whereas L-glucose and L-fructose inhibit germination (Harley and Smith, 1983). There is no doubt that orchid seeds infected by fungi receive a mixture of compounds, including trehalose and mannitol, that are common metabolic products of fungi (Smith, 1967; Harley and Smith, 1983). Trehalose stimulates germination (Smith, 1973; Purves and Hadley, 1976), but mannitol does not (Smith, 1967). However, mannitol can promote the growth of orchid seedlings (Ernst, 1967).

Other kinds of compounds that have been tested for their ability to stimulate germination and protocorm growth include various nitrogen sources (Curtis, 1947; Spoerl, 1948; Harvais, 1972; Harvais and Raitsakas, 1975; Mead and Bulard, 1975, 1979; Van Waes and Debergh, 1986), vitamins (Noggle and Wynd, 1943; Harvais, 1972; Mead and Bulard, 1975; Arditti and Harrison, 1977), hormones (Hadley and Harvais, 1968; Hadley, 1970a; Van Waes and Debergh, 1986; De Pauw *et al.*, 1995), and minerals (Wynd, 1933b,c; Harvais, 1973; Arditti, 1977). For a full account of the effects of various compounds on orchid seeds and protocorms, see Arditti (1967). If a suitable carbon source is present in the culture medium, vitamins or another compound(s) may promote seed germination and/or protocorm growth; however, no compound has been found that substitutes for sugar or symbiotic fungi (Arditti *et al.*, 1990).

Downie (1949) prepared water-soluble extracts from disintegrated hyphae of the endophytic fungus of *Goodyera repens* and found that they stimulated germination. However, in later studies, filtrates of fungi did not stimulate germination (Arditti *et al.*, 1981). Further, orchid seeds separated from fungi by dialysis tubing failed to germinate, but after the tubing was cut seeds germinated within 10 days (Clements, 1988). Thus, the stimulation of orchid seed germination is attributed to the actual presence of the fungi and not to some compound secreted from the hyphae. However, filtrates of culture media in which fungi have been growing and mycelial extracts can increase respiration rates of orchid protocorms (Blakeman *et al.*, 1976). Arditti *et al.* (1990) speculated that "... seeds require genetic input from the fungi before they can germinate." That is, in the evolution of orchid seeds, genetic information required for certain physiological functions such as photosynthesis and metabolism of lipids may have been lost as seed size was reduced. These impaired physiological processes could be restored by genetic information transferred from the fungus to the orchid via plasmids (Arditti *et al.*, 1990). However, photosynthesis does not occur in fungi, thus its initiation in fungus-infected protocorms may be due to something other than the transfer of a gene(s) for this process from the fungus to the orchid. Fungal infection of protocorm cells results in the production of new classes of DNA, but this could be the result of increased cell growth instead of a response to the fungi per se (Williamson and Hadley, 1969).

In orchid species whose mature plants have green, photosynthetically active leaves, newly produced protocorms may be colorless (even if exposed to light) or green, depending on the species (Curtis, 1936; Stoutamire, 1964; Arditti et al., 1981). In time, colorless protocorms may turn green if they are exposed to light and supplied with sugar (Knudson, 1924; Lindquist, 1965; Harrison, 1977; Harrison and Arditti, 1978; Oliva and Arditti, 1984). Even in the presence of sugar and light, however, it may take many months for colorless protocorms to turn green (Arditti et al., 1981).

A period of protocorm dormancy occurs in various orchids, including *Coeloglossum viride* (Hadley, 1970a), *Cypripedium* spp. (Withner, 1953), *Dactylorhiza purpurella* (Hadley, 1970a), *Goodyera pubescens* (Arditti, 1982), *Masdevallia nidifica* (Light, 1992), *Platanthera bifolia* (Hadley, 1970a), and *Spiranthes gracilis* (Curtis, 1936). Protocorms of *D. purpurella*, *C. viride*, and *P. bifolia* required 6–8 weeks of exposure to temperatures of about 3°C for loss of dormancy, and optimum temperatures for chlorophyll production were below 15°C (Hadley, 1970a). Thus, dormancy break of protocorms of these species would occur in temperate regions in winter, and conditions would be suitable for production and growth of leaves in early spring. Dormant protocorms of *M. nidifica* produced a leaf after about 1 year at 20°C (Light, 1992).

In the presence of sugar, nondormant protocorms placed in darkness can produce etiolated leaves (Knudson, 1924), but if sugar is absent, green protocorms in light may not produce leaves (Harrison and Arditti, 1978; Arditti et al., 1981). Thus, chlorophyll can be present, but apparently photosynthesis does not occur (Knudson, 1924). In the absence of sucrose, thylakoids in the chloroplasts are abnormal, and starch does not accumulate (Homes and Vanseveren-Van Espen, 1973). Protocorms of *Cattleya aurantiaca* required only about 30 days of growth on media with sucrose for 50% of them to form plantlets; plantlets were formed even if protocorms were transferred from sucrose-enriched to sucrose-free media after 30 days (Harrison and Arditti, 1978). During the time protocorms were on media with sucrose, chlorophyll and soluble protein contents, RuBP carboxylase activity, and photosynthetic rate increased. However, photosynthesis was correlated with levels of RuBP carboxylase, but not with those of chlorophyll (Harrison and Arditti, 1978).

2. Fungi and Physiological Dormancy

Seeds of tropical orchids, especially those of epiphytic species, are much easier to germinate under asymbiotic conditions on media containing sugars than are those of most temperate-zone species (Arditti et al., 1982). Problems and delays in the seed germination of temperate orchids may be due to the presence of physiological dormancy (Stoutamire, 1974; Reyburn, 1978). Two reasons can be suggested for the evolution of PD in seeds of orchids in temperate regions. (1) Depending on the species, climatic conditions are not equally favorable for seedling growth and survival throughout the year. Spring may be optimal for survival of the protocorms of some orchids, whereas autumn may be optimal for those of others. If PD is broken by cold stratification during winter, germination would occur in spring, but if PD is broken by warm stratification, germination would occur in autumn. Thus, the presence of PD ensures that seed germination occurs at the beginning of the favorable season for protocorm establishment. Rasmussen (1995) says, "The general impression is that most of the holoarctic species germinate in spring." (See Table 3.1 in Rasmussen, 1995.) (2) PD may prevent orchid seeds from germinating until the season of the year when the appropriate symbiotic fungus is active. If a seed germinated before the fungus was active, the young protocorm might exhaust the food reserves and die.

Little is known about the phenological life cycle of symbiotic fungi of orchids. However, seasonal changes in density have been documented for fungal communities; maximum hyphal densities can occur in spring, summer, summer and autumn, spring and autumn (England and Rice, 1957; Widden, 1981; Widden and Parkinson, 1973; Wright and Bollen, 1961), or at the end of the dry season (Moubasher and El-Dohlob, 1970). It is difficult to study seasonal variation of an individual species of fungi (Gams and Domsch, 1969; Dick, 1981), and only a few studies have shown seasonal fluctuations in fungal populations in soil at the genus (Mabee and Garner, 1974) or at the species (Moubasher and El-Dohlob, 1970) level. Bacteria in roots of the orchids *Thelymitra crinita* and *Lyperanthus nigricans* in Australia are at peak densities in midwinter, but levels of infection decline by late winter (Wilkinson et al., 1989). Studies comparing the germination phenology of orchid seeds with phenological life cycles of symbiotic fungi in both temperate and tropical regions would contribute greatly to our understanding of the germination ecology of orchids. In wet tropical regions, fungi may be active all year, which may be one reason why PD seems to be less common in seeds of tropical orchids than in those of the temperate zone.

In orchids whose seeds are infected by fungi growing

through the suspensor cells, germination per se appears to be stimulated by the fungi, but do these orchid seeds have PD? Does PD have to be broken before fungi can promote germination or can fungi break PD? Because fungi are the "host" in the orchid–fungus system, do fungi produce a substance(s) that promote(s) germination? Attempts to identify germination stimulators in culture media in which fungi have been growing have been unsuccessful (Arditti et al., 1981), but a germination promoter might be produced by hyphae after they enter orchid cells. Would strigol have any effect on the germination of orchid seeds? There also is a possibility that the presence of hyphae causes orchid cells themselves to produce stimulatory compounds. Orchids are known to produce a series of compounds collectively called phytoalexins in response to fungal or bacterial infections (Stoessl and Arditti, 1984). Perhaps some phytoalexins can stimulate germination.

3. Environmental Factors

The optimum temperature for germination of orchid seeds ranges from 10°C in *Goodyera repens* (Downie, 1949) to 32°C in *Vanilla fragans* (Knudson, 1950), with the optimum for many species being 23 or 25°C (Table 11.6). Since constant temperatures of 20–25°C have been used very successfully by orchid growers for seed germination and seedling growth (see Arditti, 1967), growers have not been very interested in testing the effects of alternating temperature regimes on seed germination. Thus, most of the available data on responses of orchid seeds to temperature cannot be extrapolated to field conditions.

Optimum temperatures in the field may promote the growth of symbiotic fungi, which in turn may stimulate seed germination. At the optimum temperature (23.6°C) in the laboratory, seeds of *Dactylorhiza majalis* germinated to 42 and 21% under symbiotic and asymbiotic conditions, respectively (Rasmussen et al., 1990b). Symbiotic and asymbiotic germination at 27.2°C was 21 and 4%, respectively, and at 21°C germination was 23 and 12%, respectively.

Seeds of many orchids require darkness for germination, but those of others either germinate in light or darkness or require light for germination (Table 11.6). Hadley (1982) noted that seeds of terrestrial orchids germinated in darkness under the soil surface, whereas those of epiphytic species germinated in light on aerial substrates. Curtis (1943) found protocorms of *Cypripedium parviflorum* at soil depths of 2–5 cm, and he speculated that moving water might be one mechanism whereby seeds were transported to these levels. A 10- to 14-day exposure to light promoted the germination of *Dactylorhiza majalis* seeds that subsequently were incubated in darkness (Rasmussen et al., 1990a). Thus, light received by seeds prior to the time they become buried in soil may promote germination.

When orchids are grown from seeds using culture media methods, seeds and protocorms frequently are kept in darkness for a period of time and then moved to light, where protocorms turn green and produce a leaf (Arditti, 1982). In terrestrial orchids whose seeds germinate in darkness beneath the soil surface, the protocorm is unlikely ever to be exposed to light unless there is a soil disturbance that results in it being brought to the surface. Four- and 8-week-old protocorms of *Cypripedium reginae* died when moved from darkness to light, but many 12-week-old protocorms survived in light and were healthy (Harvais, 1973). For most protocorms produced beneath the soil surface, food reserves supplied by symbiotic fungi are used to form a colorless leaf, which grows to the soil surface and turns green after exposure to sunlight.

Maximum germination in the achlorophyllous orchid *Galeola septentrionalis,* using the normal mixture of atmospheric gases (20.9% O_2, 0.03% CO_2, 79% N_2), occurred at a pressure of 0.18 MPa. Normal (sea level) atmospheric pressure is 0.101 MPa. Optimum concentrations of O_2 and CO_2 at 0.18 MPa were 5 and 8%, respectively (Nakamura et al., 1975). Thus, decreased O_2 and increased CO_2 levels in the soil burial environment of orchid seeds may be an important factor in their germination. However, information on the gaseous environment of buried orchid seeds is not available.

The optimum pH for seed germination and seedling development of orchids varies with the species and ranges from 4.5 to 8.0 (see Table XVI in Arditti, 1967). The optimum pH for the germination of *Cypripedium reginae* seeds incubated in light was 7.0, but it was 5.5–6.0 for those incubated in darkness (Reyburn, 1978). One explanation given for the stimulatory effects of fungi on orchid seed germination is that fungi lower the pH of the germination medium (La Garde, 1929).

The germination ecology of epiphytic orchids may be influenced by the presence of inhibitory substances in the bark of trees. Some species of epiphytic orchids in Mexican cloud forests occur on oak trees (*Quercus* spp.), whereas others do not. Oaks without orchids have compounds such as ellagic and other gallic acid derivatives and leucoanthocyanin tannins in their bark that inhibit seed germination and the development of protocorms, whereas those with orchids do not have these inhibitory compounds in their bark (Frei and Dodson, 1972).

V. CARNIVOROUS PLANTS

About 500 species of angiosperms have the capacity to trap and partially digest small animals (mainly in-

TABLE 11.6   Optimum Temperature (°C) and Light:Dark (L:D) Requirements for Seed
Germination of Various Species of Orchids

Species	°C	L:D[a]	Reference
Anacamptis pyramidalis	23	D	Van Waes and Debergh (1986)
Aplectrum hyemale	23	L > D	Oliva and Arditti (1984)
Calypso bulbosa	—[b]	D > L	Arditti *et al.* (1981)
Coeloglossum viride	23	D	Harvais and Hadley (1967)
Cypripedium californicum	23	D = L	Oliva and Arditti (1984)
C. reginae	25	D > L	Ballard (1987), Harvais (1973)
	23	D = L	Oliva and Arditti (1984)
Dactylorhiza maculata	23	D	Van Waes and Debergh (1986)
D. majalis	24	—	Rasmussen *et al.* (1990a,b)
D. sambucina	23	D	Van Waes and Debergh (1986)
Epipactis atrorubens	23	D	Van Waes and Debergh (1986)
E. gigantea	—	L = D	Arditti *et al.* (1981)
E. helleborine	23	D	Van Waes and Debergh (1986)
Goodyera oblongifolia	25	D = L	Arditti *et al.* (1981, 1982)
G. tesselata	25	L	Arditti *et al.* (1982)
Gymnadenia conopsea	23	D	Van Waes and Debergh (1986)
G. odoratissima	23	D	Van Waes and Debergh (1986)
Listera ovata	23	D	Van Waes and Debergh (1986)
Ophrys sphegodes	23	L > D	Mead and Bulard (1975)
Orchis laxiflora	23	L > D	Mead and Bulard (1975)
O. morino	23	D	Van Waes and Debergh (1986)
O. purpurella	23	D > L	Harvais and Hadley (1967)
Platanthera chlorantha	23	D	Van Waes and Debergh (1986)
P. saccata	—	L > D	Arditti *et al.* (1981)
Spiranthes gracilis	23	D	Oliva and Arditti (1984)
S. romanzoffiana	23	L	Oliva and Arditti (1984)
S. spiralis	23	D	Van Waes and Debergh (1986)
Vanilla fragans	32	D	Knudson (1950)

[a] See note for Table 11.5.

[b] Information not available.

sects); collectively, they are called carnivorous or insectivorous plants. These fascinating plants grow from arctic to tropical regions of the world (Lloyd, 1942) and occur in nine families and 19 genera: Sarraceniaceae (*Darlingtonia, Heliamphora,* and *Sarracenia*), Nepenthaceae (*Nepenthes*), Droseraceae (*Aldrovanda, Dionaea, Drosophyllum,* and *Drosera*), Byblidaceae (*Byblis*), Cephalotaceae (*Cephalotus*), Dioncophyllaceae (*Triphyophyllum*), Lentibulariaceae (*Biovularia, Genlisea, Pinguicula, Polypompholyx,* and *Utricularia*), Marytniaceae (*Ibicella*), and Bromeliaceae (*Brocchinia* and *Catopsis*) (Juniper *et al.,* 1989).

Carnivorous plants are characterized by a suite of characters that Juniper *et al.* (1989) call the carnivorous syndrome, which probably has evolved several times since the upper Cretaceous (Juniper *et al.,* 1989). The plants (1) attract insects (and sometimes other small animals), (2) have some way to prevent insects from escaping, (3) kill the "prey," (4) partially digest its body, and (5) absorb some of the molecules. These plants frequently are weak-rooted perennials that live in open,

well-lighted habitats, where competition from other plants is minimal. Many carnivorous plants are calcifuges and may be found growing on infertile sands or in shallow soil on granite outcrops, but some grow on wet, limestone cliffs or in marl bogs. Plants are tolerant of low levels of nutrients in the soil, temporary or permanent flooding (depending on the species), and some tolerate an occasional low-temperature fire (Juniper *et al.,* 1989).

A. Types of Seed Dormancy

Little specific information is available on seeds and embryos of carnivorous plants. Seed length of *Sarracenia* spp. ranges from 1.2 to 2.5 mm (McDaniel, 1971), and Martin (1946) shows a *Sarracenia purpurea* seed ca. 2.0 mm in length containing a 0.4-mm linear embryo. Because seeds of this species are dormant at maturity (Mandossian, 1966), we conclude that they have morphophysiological dormancy. More studies are needed on embryos of members of the Sarraceniaceae. Seeds of *Darlingtonia* (Sarraceniaceae) are physiologically dormant at maturity (Pietropaolo and Pietropaolo, 1986) and thus also may have MPD. However, because seeds of *Heliamphora* (Sarraceniaceae) appear to lack physiological dormancy at maturity (Pietropaolo and Pietropaolo, 1986), they may have only morphological dormancy.

The Lentibulariaceae produces numerous small seeds (Lawrence, 1951), and the embryo is not differentiated in *Utricularia* (Swamy and Mohan Ram, 1969). Thus, seeds of *Pinguicula* spp. and *Utricularia* spp. that "germinate" soon after sowing (Swamy and Mohan Ram, 1969; Pietropaolo and Pietropaolo, 1986) would have MD, and those with delayed germination (Pietropaolo and Pietropaolo, 1986) would have MPD. After 4 weeks of incubation, "... a prominent mound appeared at the plumular pole of the [*U. inflexa*] embryo, which caused bursting of the seed coat." This mound of cells gave rise to several structures called "cotyledonoides" and a shoot bud (Swamy and Mohan Ram, 1969).

At least some members of the Cephalotaceae, Droseraceae (Martin, 1946), and Nepenthaceae (Kaul, 1982) have dwarf seeds; these families are not known to have physical dormancy (Table 3.5). Seeds of *Dionaea* and some species of *Nepenthes* (Pietropaolo and Pietropaolo, 1986) and tropical species of *Drosera* (Vickery, 1933; Pietropaolo and Pietropaolo, 1986) will germinate if they are sown at maturity (Pietropaolo and Pietropaolo, 1986), indicating that they are nondormant or conditionally dormant. However, Smith (1931) had difficulty germinating seeds of *Dionaea* in a greenhouse. Freshly matured seeds of some species of *Nepenthes* (Corker, 1986), *Drosophyllum* (Swamy and Mohan Ram, 1967), *Cephalotus,* and temperate species of *Drosera* (Pietropaolo and Pietropaolo, 1986) do not germinate and thus would have PD. The germination of seeds of *Nepenthes gracilis* and *N. rafflesiana* placed on moist filter paper, soil, or sand in a laboratory in Singapore was delayed for 20 days, but then germinated to about 80% (Wee, 1978). In other studies, the germination of *N. gracilis* seeds was delayed about 21 (Green, 1967) and 30 (Ah-Lan and Prakash, 1973) days. These results may indicate that seeds of *N. gracilis* have nondeep PD, which was broken during the 20–30 days of incubation prior to the start of germination.

Swamy and Mohan Ram (1967) were not able to germinate nonscarified seeds of *Drosophyllum* on culture media, but excised embryos grew normally. Thus, it seems likely that the increased germination of scarified seeds of *Drosophyllum* is due to the low growth potential of the embryos, and thus PD.

No information is available on seed dormancy of carnivorous species belonging to the Dioncophyllaceae, Byblidaceae, Bromeliaceae, or Martyniaceae.

B. Longevity of Stored Seeds

Professional growers of carnivorous plants recommend that freshly matured seeds be sown immediately (either for germination per se or for cold stratification treatments) (e.g., Schnell, 1976). Otherwise, seeds should be placed in dry storage under refrigeration (Table 11.7). There is a good reason for suggesting that seeds be stored at low temperatures. Seeds of the tropical *Nepenthes gracilis* stored dry in glass tubes with corks at room temperatures in Singapore died within 1 month (Garrard, 1955), and those of this species stored at 30°C and 76–80% RH in air-tight containers died within 1 week (Ah-Lan and Prakash, 1973). About 25% of *Drosera aliciae* seeds (stored dry at room temperatures?) lost their ability to germinate after 1 year (Ferreira and Small, 1974), presumably due to loss of viability. Even at low temperatures, seeds of some genera such as *Dionaea* and *Cephalotus* remain viable for only a year or less (Pietropaolo and Pietropaolo, 1986).

C. Dormancy-Breaking Requirements

Cold stratification breaks dormancy in seeds of *Sarracenia purpurea* (Mandossian, 1966), *Pinguicula vulgaris,* and *Drosera rotundifolia* (Grime *et al.,* 1981), and it is recommended for those of *Darlingtonia, Cephalotus,* and temperate species of *Drosera, Pinguicula,* and *Utricularia* (Table 11.7). Seeds of *D. rotundifolia* mostly germinated in May in Sweden (Redbo-Torstensson, 1994), indicating that seeds came out of dormancy during winter. However, a few seeds germinated in September, suggesting that some dormancy loss occurred in summer. In tropical and subtropical areas, PD is lost during either dry storage or warm stratification. The maximum germination of *Nepenthes mirabilis* seeds was obtained after 2 months of dry storage at 25°C (Corker, 1986), whereas seeds of *Drosera aliciae* required a 24-day period of imbibition at 15/10°C before they started to germinate (Ferreira and Small, 1974).

D. Germination Requirements

Nondormant seeds of many carnivorous plants germinate at temperatures of about 20 to 30°C (Table 11.7). However, the optimum germination temperature for *Drosera aliciae* seeds was 10°C in continuous light and 15°C in continuous darkness. At a 14-hr daily photoperiod and at daily alternating temperature regimes of 15/10, 20/10, and 25/10°C, optimum germination (70%) of *D. aliciae* seeds occurred at 20/10°C (Ferreira and Small, 1974). Cold-stratified seeds of *Sarracenia purpurea* germinated to higher percentages in constant light than in

TABLE 11.7 Recommended Dormancy-Breaking and Germination Requirements for
Seeds of Carnivorous Plants[a]

Genus	Cold stratification required	Store at low temperatures to maintain viability	Optimum germination temperature (°C)	Sow on soil surface	Special treatments
Aldrovanda	?	?	?	?	?
Byblis	?	+[b]	21–32	?	Fire, hot water, or GA
Cephalotus	+	+[c]	21–32	+	None
Darlingtonia	+	+	21–29	+	None
Dionaea	−[d]	+	26.7–29.5	+	None
Drosera					
Pygmy	?	?	21–27	+	None
Temperate	+	?	21–29	+	None
Tropical	−	?	21–35	+	None
Tuberous	?	?	?	?	Fire
Drosophyllum	?	?	21	−	Scarification
Genlisea	?	?	?	?	?
Heliamphora	−	?	20–22	+	None
Nepenthes					
Lowland type	−	+	21–29	+	None
Highland type	−	+	10–21	+	None
Pinguicula					
Temperate	+	+	?	?	None
Subtropical	−	?	16–30	+	None
Sarracenia	+	+	21–29	+	None
Utricularia	+	+	?	+[e]	None

[a] Information from Pietropaolo and Pietropaolo (1986). +, yes; −, no.
[b] Seeds of *Byblis liniflora* must be stored 2 months before planting.
[c] Seeds do not store well.
[d] Sow seeds as soon as they mature.
[e] Seeds of aquatic species are sown on the surface of water.

darkness at 28°C, but they germinated equally well in light and darkness at 22°C (Mandossian, 1966).

Seeds of *Nepenthes mirabilis* (Corker, 1986) and *Pinguicula grandiflora* (Grime *et al.,* 1981) required light for germination, and those of *P. vulgaris* germinated to higher percentages in light than in darkness (Maas, 1989). Because many carnivorous plants have dwarf seeds, it is expected that they have a light requirement for germination. In fact, the general recommendation for growing plants from seeds is to sow them on the soil surface (Table 11.7).

Seeds of *Drosera aliciae* germinated to 60–98% at pH values ranging from 2.5 to 6.5, with the lowest germination occurring at pH 2.5 and the highest at pH 5.5 (Ferreira and Small, 1974). Seeds of *Utricularia juncea* and *U. cornuta* germinated to 98 and 87%, respectively, at pH 4.5–5.0, but only if seeds were in Moore's solution in flasks. They germinated to only 1–13% when placed on filter paper moistened with buffer or GA solutions at pH values of 4.5–5.0 (Kondo, 1971). Thus, either Moore's solution per se and/or the low levels of oxygen in the solution-containing flasks promoted germination.

VI. AQUATIC (NONSALINE) PLANTS

This discussion concerns the seed germination ecology of plants growing in, or in association with, fresh water that covers the soil surface to varying depths (*sensu* Riemer, 1984). Cook *et al.* (1974) list 71 families of angiosperms with genera that contain aquatic species. Aquatic plants grow in, or adjacent to, standing bodies of water such as lakes, peatlands, muskegs, mires, bogs, fens, ponds, reservoirs, swamps, moors, marshes, and lagoons and/or flowing bodies of water such as rivers, streams, and canals (Gore, 1983; Riemer, 1984). Four general types of aquatic plants are recognized (Riemer, 1984). (1) Floating, unattached plants have roots hanging in the body of water and leaves floating on its surface; these plants can be moved about by wind and/or currents. (2) Floating, attached plants have roots growing in soil at the bottom of the body of water, while their leaves float on the surface; some species also have submerged, as well as floating, leaves. (3) Submerged plants also are rooted in soil, but their leaves remain beneath the water's surface; flowers of some species may reach the surface. (4) Emergent plants are rooted in water, but their stems and leaves are mostly above the surface.

In habitats such as mudflats or the shores of lakes and rivers, seeds may not germinate until the water recedes, exposing wet soil. Such habitats in temperate regions often are flooded during the cool season, and the period of nonflooding coincides with the growing season (Voightlander and Poppe, 1989). Thus, these seasonally nonflooded habitats are favorable for the growth of many plant species, especially summer annuals (Salisbury, 1970; Webb et al., 1988; Chester, 1992). Seeds germinate after floodwaters recede, and plants flower and set seeds before sites are flooded in autumn or winter. In a year with normal amounts of precipitation during the growing season (and consequently no abnormal increases in water level), neither roots nor shoots of these plants are flooded during the growing season. However, such species are included with aquatics, as emergents, because their seeds are flooded for a portion of the year. Shrubs and trees growing in sites with soils that are waterlogged or flooded for part, or all, of the year are discussed in Chapters 9 and 10.

A. Types of Seed Dormancy

Freshly matured seeds of a number of aquatics germinate to high percentages without any pretreatments; therefore, the species are listed under "Nondormant" in Table 11.8. In many of these studies, however, seeds were tested only at one temperature regime. Thus, if a range of test temperatures is used in future studies, the results may show that fresh seeds of some of these species are conditionally dormant, i.e., have physiological dormancy (PD) rather than nondormancy.

Morphological dormancy (MD) occurs in emergent aquatics in the Apiaceae, Araceae, Iridaceae, and Menyanthaceae (Table 11.8). Studies on other aquatic members of these families and of those in the Amaryllidaceae, Cannaceae, Hydrophyllaceae, Liliaceae, and Ranunculaceae will, no doubt, reveal other species with MD and possibly some with morphophysiological dormancy (MPD).

A small group of aquatics has seeds with physical dormancy (Table 11.8). *Nelumbo* (Nelumbonaceae) is the most famous aquatic with physical dormancy (see Chapter 7,IV), but this characteristic also is found in seeds of aquatic members of the Convolvulaceae (*Ipomoea*) and Fabaceae (*Aeschynomene*, *Lotus*, and *Neptunia*). Seeds of the aquatic legumes *Sesbania bispinosa*, *S. hirtistyla*, and *S. sesban* (Cook et al., 1974) also probably have physical dormancy.

The most common type of dormancy in seeds of aquatic species is PD (Table 11.8), and it may be nondeep PD in seeds of many species. Species whose seeds come out of dormancy during warm stratification would have nondeep PD. Scarification of the permeable seed or fruit coat in some aquatics results in increased germination percentages (Table 11.9), indicating a low growth potential of the embryo. In nature, the growth potential of the embryo increases during warm or cold stratification treatments, depending on the species. Because scarified permeable seeds of the species germinate without cold or warm stratification, many of them may have nondeep PD. In some species, a dormancy-breaking treatment per se is not absolutely required for seed germination, but cold (Szmeja, 1987) or warm (Obeid and El Seed, 1976) stratification increases the rate of germination. Thus, seeds have nondeep PD. GA can stimulate the germination of seeds with nondeep PD (Chapter 3), and it promotes the germination of freshly matured seeds of *Zizania aquatica* (Cardwell et al., 1978), *Z. palustris* (Oelke and Albrecht, 1980), and *Sagittaria graminea* (Chabreck et al., 1983). Thus, seeds of these three species have nondeep PD.

B. Dormancy-Breaking Requirements

1. Temperature

Seeds with MPD require 60 or more days of cold stratification to germinate (Table 11.8). Because seeds are unlikely to receive more than about 200 days of cold stratification (0–10°C) during winter in any part of the world, it is doubtful that those requiring more than 200 days of cold stratification for germination have received the proper dormancy-breaking treatment. Seeds may require warm followed by cold stratification for loss of dormancy and/or embryo growth.

Seeds of aquatics with physical dormancy usually germinate in the laboratory in response to mechanical (Francho, 1986; Datta and Biswas, 1970) or chemical (Datta and Sinha-Roy, 1975; Ohga, 1926) scarification treatments that render the seed coat permeable to water. Essentially nothing is known about the mechanism whereby seeds become permeable in nature. However, a possible clue is provided by the response of *Neptunia oleracea* (Fabaceae) seeds to alternating temperature regimes. Only two cycles of high (60°C for 8 hr) and low (20°C for 16 hr) temperatures resulted in 100% germination of *N. oleracea* seeds (Sharma et al., 1984). In Bharatpur, India, this species grows in seasonal pools, and in October and November it produces seeds that fall into the water (Sharma et al., 1984). All the water evaporates from the pools during the dry season, and seeds are left on or at the surface of dried mud where day and night temperatures of 50 ± 5 and 22 ± 3°C (mean ± SE?), respectively, may occur. Pools begin to fill with water at the beginning of the wet season, and seeds germinate. Thus, loss of physical dormancy in some aquatics may depend on the alternating tempera-

TABLE 11.8   Habitat and Seed Dormancy-Breaking and Germination Requirements
of Aquatic Plants

Species	H[a]	Tr[b]	D[c]	T[d]	L:D[e]	Reference
Nondormant						
Baldellia ranunculoides	E			20	L = D	Forsberg (1966)
Cyperus papyrus	E			25	—[f]	Gaudet (1977)
Griffithella hookeriana	S			27	L	Vidyashankari and Mohan Ram (1987)
Hydrilla verticillata	S			23–28	L	Lal and Gopal (1993)
Lemna perpusilla	FU			26	L	Posner and Hillman (1962)
Limosella aquatica	E			Room	L	Salisbury (1967)
Ludwigia leptocarpa	E			32	—	Christy and Sharitz (1980)
Marathrum haenkeanum	S			22	L	Philbrick and Novelo (1994)
M. rubrum	S			22	L	Philbrick and Novelo (1994)
Nasturtium nasturtium-aquaticum	FA			19/15[g]	—	Muenscher (1936a)
Orontium aquaticum	S			19/15[g]	—	Muenscher (1936a)
Oserya coulteriana	S			22	L	Philbrick and Novelo (1994)
Petasites hybridus	E			20/15	L = D	Grime *et al.* (1981)
Podostemum ceratophyllum	S			23	—	Philbrick (1984)
Rumex hydrolapathum	E			20/15	L > D	Grime *et al.* (1981)
Saururus cernuus	E			25	—	Thien *et al.* (1994)
Tristicha trifaria	S			22	L	Philbrick and Novelo (1994)
Typha glauca	E			25/15	L > D	Galinato and van der Valk (1986)
T. latifolia	E			35, 25–27	L	Muenscher (1926a), Bonnewell *et al.* (1983), Gopal and Sharma (1983)
Vallisneria americana	S			19/15[g]	D > L	Muenscher (1936a), Kimber *et al.* (1995)
V. spiralis	S			30	L > D	Choudhuri (1966)
Vanroyenella plumosa	S			22	L	Philbrick and Novelo (1994)
Morphophysiological dormancy						
Acorus calamus	E	C	210	19/15[g]	—	Muenscher (1936a)
		C	270	30/20	—	Shipley and Parent (1991)
Angelica sylvestris	E	C	90	20/15	—	Grime *et al.* (1981)
Calla palustris	E	C	90	20/15	—	Grime *et al.* (1981)
Dulichium aurundinaceum	E	C	270	30/20	—	Shipley and Parent (1991)
Iris angustifolia	E	C	60	20–25	—	Coops and van der Velde (1995)
I. pseudacorus	E	C	360	20/15	—	Grime *et al.* (1981)
I. versicolor	E	C	270	30/20	—	Shipley and Parent (1991)
I. virginica	E	C	70	30/20	—	Morgan (1990)
Menyanthes trifoliata	E	C?	—	Room	L > D	Hewett (1964)
Peucedanum palustre	E	C	Winter	15–25	L > D	Meredith and Grubb (1993)
Physical dormancy						
Aeschynomene aspera	E	S	—	25	D > L	Datta and Sinha-Roy (1975)
A. virginica	E	S	—	30/15	L = D	Baskin and Baskin (unpublished results)
Ipomoea aquatica	E	S	—	25	—	Datta and Biswa (1970)
Lotus uliginosus	E	S	—	20/15	L > D	Grime *et al.* (1981)
Neptunia oleraceae	FA	S	—	Room	—	Sharma *et al.* (1984)
Nelumbo lutea	E	S	—	35/23	—	Francko (1986)
N. nucifera	E	S	—	Room	—	Ohga (1926)
Physiological dormancy						
Agostis stolonifera	E	C	270	30/20	—	Shipley and Parent (1991)
Alisma plantago-aquatica	E	C	30	20	L = D	Forsberg (1966)
A. subcordata	E	C	210	25/15	—	Leck (1996), Kaul (1978)
Alternanthera sessilis	E	W?[h]	—	30	L > D	Datta and Biswas (1968)
Amaranthus cannabinus	E	C	180	25/15	—	Leck (1996)
Ammannia coccinea	E	C	100	35/20	L	Baskin *et al.* (unpublished results)
Aster laurentianus	E	C	180	20/10	L > D	Galinato and van der Valk (1986)
Bacopa acuminata	E	C	100	35/20	L	Baskin and Baskin (unpublished results)
Bidens cernua	E	C	270	30/20	L	Hogue (1976), Shipley and Parent (1991)

(continues)

TABLE 11.8—Continued

Species	H[a]	Tr[b]	D[c]	T[d]	L:D[e]	Reference
B. frondosa	E	C	270	30/20	—	Shipley and Parent (1991)
B. laevis	E	C	112	35/20	L > D	Leck *et al.* (1994)
Butomus umbellatus	E	C	210	19/15[g]	—	Muenscher (1936a)
Callitriche longipedunculata	S	W	—	18	L > D	McLaughlin (1974)
Carex canescens	E	C	30	20/10	L	Schuetz and Milberg (1997)
C. comosa	E	C	84	35/20	L > D	Baskin *et al.* (1996b)
C. crinita	E	C	56	35/20	L	Baskin *et al.* (1996)
C. elongata	E	C	120	25	L > D	Schuetz (1995)
C. folliculata	E	C	270	30/20	—	Shipley and Parent (1991)
C. laevigata	E	C	150	20/15	—	Grime *et al.* (1981)
C. lurida	E	C	240	25/15	—	Leck (1996)
C. nigra	E	C	30	20/15	L > D	Grime *et al.* (1981)
C. paniculata	E	C	120	25	L > D	Schuetz (1995)
C. projecta	E	C	270	30/20	—	Shipley and Parent (1991)
C. pseudocyperus	E	C	120	25	L	Schuetz (1995)
C. remota	E	C	120	25	L > D	Schuetz (1995)
C. retrorsa	E	C	270	30/20	—	Shipley and Parent (1991)
C. stricta	E	C	84	30/15	L > D	Baskin *et al.* (1996b)
C. tuckermanii	E	C	270	30/20	—	Shipley and Parent (1991)
C. vulpinoidea	E	C	45	30/15	L	Baskin and Baskin (unpublished results)
Ceratophyllum demersum	S	C	150	19/15[g]	—	Muenscher (1936a)
Chenopodium rubrum	E	C	180	25/15	L > D	Galinato and van der Valk (1986)
Cladium mariscus	E	C	90	30/20	L > D	Goossens and Devillez (1974)
Coreopsis rosea	E	C	270	30/20	—	Shipley and Parent (1991)
Cyperus aristatus	E	C	84	30/15	L	Baskin and Baskin (1978)
C. diandrus	E	C	270	30/20	—	Shipley and Parent (1991)
C. erythrorhizos	E	C	90	35/20	L	Baskin *et al.* (1993a)
C. flavicomus	E	C	90	35/20	L	Baskin *et al.* (1993a)
C. odoratus	E	C	84	30/15	L	Baskin *et al.* (1989)
C. rivularis	E	C	270	30/20	—	Shipley and Parent (1991)
C. scoparia	E	—	—	25/21	—	Larson and Stearns (1990)
Dalzellia zeylanica	S	—[i]	—	26	—	Uniyal and Mohan Ram (1996)
Diplachne fusca	E	W	225	28/17	L	McIntyre *et al.* (1989a,b)
Dulichium aurundinaceum	E	C	—	20	L	Shipley *et al.* (1989)
Echinodorus rostratus	E	C?	—	Spring/summer	—	Kaul (1978, 1985)
Eclipta prostrata	E	W	30	30/15	L	Baskin *et al.* (unpublished results)
Eichhornia crassipes	FU	W	30	g.h.[j]	L > D	Barton and Hotchkiss (1951)
			52	g.h.[k]	L	Obeid and El Seed (1976)
Elatine americana	E	C	210	19/15[g]	—	Muenscher (1936a)
Eleocharis acicularis	S	C	60	15	—	Yeo (1986)
E. coloradoensis	S	W	21	22	—	Yeo and Thurston (1979)
E. obtusa	E	C	270	30/20	—	Shipley and Parent (1991)
Eragrostis hypnoides	E	C	112	35/20	L	Baskin *et al.* (unpublished results)
Eriocaulon septangulare	E	C	210	19/15[g]	—	Muenscher (1936a)
Eriophorum latifolium	E	C	42	22/12	L > D	Maas (1989)
Eupatorium perfoliatum	E	C	270	30/20	—	Shipley and Parent (1991)
Filipendula ulmaria	E	C	42	22/12	L > D	Maas (1989)
Fimbristylis autumnalis	E	C	90	35/20	L	Baskin *et al.* (1993a)
F. vahlii	E	C	90	35/20	L	Baskin *et al.* (1993a)
Geum rivale	E	W?	—	22	L > D	Grime *et al.* (1981), Graves and Taylor (1988)
Glyceria striata	E	C	150	19/15[g]	—	Muenscher (1936a)
Gratiola aurea	E	C	270	30/20	—	Shipley and Parent (1991)
G. neglecta	E	W	112	30/15	L	Baskin *et al.* (unpublished results)
G. virginica	E	W	56	30/15	L	Baskin *et al.* (unpublished results)
G. viscidula	E	C	84	30/15	L	Baskin *et al.* (1989)
Heliotropium supinum	E	W	56	25/9	—	Mall (1954)
Heteranthera dubia	E	C	150	30	L	Muenscher (1936a), Marler (1969)

(continues)

TABLE 11.8—*Continued*

Species	H[a]	Tr[b]	D[c]	T[d]	L:D[e]	Reference
H. limosa	E	C	84	35/20	L	Baskin *et al.* (unpublished results)
Hottonia inflata	FA	W	84	25/15	L	Baskin *et al.* (1996a)
H. palustris	FA	W	30	20/15	L > D	Brock *et al.* (1989)
Howellia aquatilis	E/S	W	100	20/5	D > L	Lesica (1992)
Hygrophila auriculata	S	W	90	28	L = D	Amritphale *et al.* (1989)
Impatiens glandulifera	E	C	30	20/15	—	Grime *et al.* (1981)
Juncus articulatus	E	C	150	19/15[g]	L	Muenscher (1936a), Grime *et al.* (1981)
J. canadensis	E	C	270	30/20	—	Shipley and Parent (1991)
J. effusus	E	C	270	30/20	L > D	Grime *et al.* (1981), Shipley and Parent (1991)
J. filiformis	E	C	270	30/20	L	Richards (1943), Shipley and Parent (1991)
J. marginatus	E	C	100	25/15	L	Baskin and Baskin (1988, unpublished results)
Jussiaea decurrens	E	C	100	35/20	L	Baskin *et al.* (unpublished results)
J. suffruticosa	E	W?[h]	—	40	L	Wulff and Medina (1969), Wulff *et al.* (1972)
Leersia oryzoides	E	C	270	30/20	—	Shipley and Parent (1991)
Lemna gibba	FA	W	Summer	25	—	Witztum (1977)
Leptochloa panicoides	E	W	63	35/20	L	Baskin *et al.* (1993b)
Leucospora multifida	E	C	63	30/15	L	Baskin *et al.* (1994)
Littorella uniflora	S	W?[i]	28	20/8	L	Arts *et al.* (1990)
Lobelia dortmanna	E	C	30–60	25	L	Farmer and Spence (1987)
		C	180	18–20	—	Szmeja (1987)
Lophotocarpus calycinus	E	C?	—	Spring/summer	—	Kaul (1978, 1985)
Ludwigia alternifolia	E	C	84	35/20	L > D	Baskin *et al.* (unpublished results)
L. decurrens	E	C	56	35/20	L	Baskin *et al.* (unpublished results)
Lycopus americanus	E	C	270	30/20	—	Shipley and Parent (1991)
L. europaeus	E	W?[b]	—	25/10	L	Thompson (1969)
		C	240	20/15	L > D	Grime *et al.* (1981)
Lysimachia vulgaris	E	C	42	22/12	L > D	Maas (1989)
Lythrum salicaria	E	C	270	30/20	L	Grime *et al.* (1981)
			20		L > D	Shamsi and Whitehead (1974)
Mentha aquatica	E	C	30	20/15	L > D	Grime *et al.* (1981)
Mimulus ringens	E	C	270	30/20	—	Shipley and Parent (1991)
Mollugo hirta	E	W[m]	56	25/9	—	Mall (1954)
Myriophyllum spicatum	S	C	120	20	L > D	Hartleb *et al.* (1993)
Najas flexilis	S	C	210	19/15[g]	—	Muenscher (1936a)
N. marina	S	C	—[n]	24	L = D	van Vierssen (1982c)
				20–25	D > L	Forsberg (1965), Agami and Waisel (1984)
Nuphar luteum	FA	C	45	25	—	Beal and Southall (1977)
Nymphaea alba	FA	C	30	20	L = D	Forsberg (1966)
N. odorata	FA	C	150	Room	—	Else and Riemer (1984)
N. tuberosa	FA	C	210	19/15[g]	—	Muenscher (1936a)
Nymphoides peltata	FA	C	70	15	L > D	Smits *et al.* (1990)
Orcuttia californica	E	W	210	23/13	L > D	Keeley (1988)
Panicum hirsutum	E	W	180	26–32	—	Orozco-Segovia and Vazquez-Yanes (1980)
P. laxum	E	W	14	24/19	L	Cole (1977)
P. longifolium	E	C	270	30/20	—	Shipley and Parent (1991)
Peltandra virginica	E	C	210	19/15[g]	—	Muenscher (1936a)
		C	28	Room	—	West and Whigham (1975–1976)
Penthorum sedoides	E	C	28	35/20	L	Baskin *et al.* (1989)
Phalaris aurundinacea	E	C	270	30/20	L > D	Vose (1962), Grime *et al.* (1981), Shipley and Parent (1991)
Phragmites australis	E	C	180	30/20	L > D	Galinato and van der Valk (1986)
P. communis	E	W	210	Room	—	Harris and Marshall (1960)

(continues)

TABLE 11.8—*Continued*

Species	H[a]	Tr[b]	D[c]	T[d]	L:D[e]	Reference
Pistia stratiotes	FU	W	42	25	L	Pieterse *et al.* (1981)
				Room	L > D	Harley (1990)
Polygonum amphibium	E	C	210	30/20	L > D	Justice (1944)
P. arifolium	E	C	150	25/15	—	Leck (1996)
P. coccineum	E	C	147	30/20	L > D	Justice (1944)
P. hydropiper	E	C	90	20/15	—	Grime *et al.* (1981)
P. hydropiperoides	E	C	135	30/20	L > D	Justice (1944)
P. punctatum	E	C	210	25/15	—	Leck (1996)
Polypleurum stylosum	S	W?[h]	—	26	—	Sehgal *et al.* (1993)
Pontederia cordata	E	C	150	19/15[g]	L = D	Muencher (1936a), Leck and Simpson (1993)
Potamogeton americanus	S	C	150	Room	—	Muenscher (1936b)
P. angustifolius	S	C	150	Room	—	Muenscher (1936b)
P. capillaceus	S	C	150	Room	—	Muenscher (1936b)
P. confervoides	S	C	150	Room	—	Muenscher (1936b)
P. epihydrus	FA	C	150	Room	—	Muenscher (1936b)
P. foliosus	S	C	210	19/15[g]	—	Muenscher (1936a)
			90	Room	—	Muenscher (1936b)
P. lucens	S	C	270	20	L	Forsberg (1966)
P. natans	S	C	180	20	—	Lohammar (1954)
P. obtusifolius	S	C	90	Room	—	Muenscher (1936b)
P. pectinatus	S	C	210	19/15[g]	—	Muenscher (1936a)
				25	—	van Wijk (1989)
P. pusillus	S	C	30	25	—	Teltscherova and Hejny (1973)
P. richardi	S	W	60	22–26	L	Spence *et al.* (1971)
P. schweinfurthi	S	W	60	22–26	L	Spence *et al.* (1971)
Ranunculus sceleratus	E	C	60	20/15	L	Grime *et al.* (1981), van der Toorn and ten Hove (1982)
Rhexia mariana	E	C	56	35/20	L	Baskin *et al.* (unpublished results)
Rhynchospora capillacea	E	C	270	30/20	—	Shipley and Parent (1991)
Rorippa islandica	E	W	15	30/20	L	Matsuo *et al.* (1984)
Rotala ramosior	E	C	56	35/20	L	Baskin *et al.* (unpublished results)
Rumex palustris	E	C	Winter	Spring	L	Voesenek and Blom (1992), Voesenek *et al.* (1992)
R. sanguineus	E	W	35	20/15	L	Grime *et al.* (1981)
R. verticillatus	E	C	270	30/20	—	Shipley and Parent (1991)
Sagittaria brevirostra	E	C?	—	Spring/summer	—	Kaul (1985)
S. calycina	E	C?	—	Spring/summer	—	Kaul (1985)
S. graminea	E	C	5	30/15	—	Chabreck *et al.* (1983)
S. lancifolia	E	W?[h]	—	25	L	Collon and Velasquez (1989)
S. latifolia	E	C	150	21/16	L	Delesalle and Blum (1994), Leck and Simpson (1993)
Schoenus nigricans	E	C[o]	Winter	20/10	—	Ernst and van der Ham, 1988
Scirpus acutus	E	C	84	25/10	L > D	Isely (1944), Thullen and Eberts (1995)
S. americanus	E	C	180	30–32	L	Isely (1944)
S. articulatus	E	W	28[m]	Room	—	Datta and Roy (1970)
S. atrovirens	E	C	180	30–32	L > D	Isely (1944)
S. cyperinus	E	C	180	30–32	L > D	Isely (1944)
S. eriophorum	E	C	180	30–32	L > D	Isely (1944)
S. fluviatalis	E	C	180	30–32	L	Isely (1944)
S. lacustris	E	C	80	30/5	L > D	Clevering (1995)
S. lineatus	E	C	56	25/15	L	Baskin *et al.* (1989)
S. maritimus	E	C	80	30/5	L > D	Clevering (1995)
S. paludosus	E	C	180	30–32	L	Isely (1944)
		W	7	—	—	O'Neill (1972)
S. pedicellatus	E	C	300	30–32	L > D	Isely (1944)
S. purshianus	E	C	84	35/20	L	Baskin *et al.* (unpublished results)
S. robustus	E	W?[h]	—	32	—	George and Young (1977)

(continues)

TABLE 11.8—*Continued*

Species	H[a]	Tr[b]	D[c]	T[d]	L:D[e]	Reference
		C	180	30–32	L	Isely (1944)
S. tabernaemontani	E	C	80	30/5	L > D	Clevering (1995)
S. validus	E	C	180	30–32	L > D	Isely (1944)
Scholochloa festucacea	E	C	21–28	21	D > L	Smith (1972)
Spartina pectinata	E	C	—	20	L	Shipley *et al.* (1989)
Stratiotes aloides	FU	C	90	19	—	Smolders *et al.* (1995)
Trapa natans	FA	C	210	19/15[g]	—	Muenscher (1936a)
		C	120	18	—	Cozza *et al.* (1994)
Trapella sinensis	FA	C	10	22–23	L > D	Kawahara and Takada (1961)
Triadenum fraseri	E	C	270	30/20	—	Shipley and Parent (1991)
Trichophorum alpinum	E	C	42	22/12	L = D	Maas (1989)
Tuctoria greenei	E	W	210	23/13	L > D	Keeley (1988)
Veronica anagallis-aquatica	E	C	150	19/15[g]	—	Muenscher (1936a)
Viola palustris	E	C	90	20/15	L > D	Grime *et al.* (1981)
Xyris difformis	E	C	270	30/20	—	Shipley and Parent (1991)
Zannichellia palustris[o]	S	C	60	20,24	L > D	van Vierssen (1982a)
Z. peltata[o]	S	W	Summer	16	L > D	van Vierssen (1982b)
Zizania aquatica	E	C	182	15	—	Simpson (1966)
Z. palustris	E	C	180	17–23	—	Atkins *et al.* (1987)
Z. texana	E	C	—	21	D > L	Power and Fonteyn (1995)

[a] Habitat: E, emergent; FA, floating attached; FU, floating unattached; S, submerged.

[b] Dormancy-breaking treatment: C, cold stratification; W, warm stratification (or dry storage at about room temperatures); S, scarification.

[c] Duration (days) of dormancy-breaking treatment.

[d] Optimum germination temperature (°C).

[e] Light or dark requirements for germination. See note for Table 11.5.

[f] Data not available.

[g] Seeds were germinated at room temperatures; about 19 and 15°C during the day and night, respectively.

[h] Fresh seeds were not tested for germination. Seeds were stored at room temperatures (or 20°C) for 1 to several months, during which time dormancy loss could have occurred.

[i] Dry seeds were stored at 15°C for an unspecified period of time; thus, some afterripening may have occurred. Further studies, however, may show that seeds are nondormant.

[j] Greenhouse.

[k] In a wire "greenhouse" in Sudan.

[l] Drying promoted loss of dormancy.

[m] Fresh seeds were not tested. Presumably, they came out of dormancy during flooding.

[n] Several months.

[o] Seeds came out of dormancy during winter while attached to plants in the field in The Netherlands; cold stratification may have broken dormancy.

[p] Also may grow in saline water.

ture regimes received by seeds during periods when the soil is not flooded. Dormancy break during the dry season means seeds could germinate at the beginning of the wet season, possibly before other aquatic plants became competitive with the seedlings.

Seeds of about 80% of the aquatics with PD require cold stratification for dormancy loss, whereas those of others require warm stratification and/or dry storage at room temperatures (Table 11.8). In a few species, both warm and cold stratification treatments are required to break dormancy. Some dormancy loss occurred in buried seeds of *Leptochloa panicoides* during cold stratification in winter, but it was not complete until seeds were exposed to warm stratification in late-spring or early summer (Baskin *et al.*, 1993b).

Indehiscent druplets of *Potamogeton schweinfurthii*

and *P. richardi* appear to have nondeep PD. If the fruit wall is scarified, freshly matured fruits will germinate in light. In the natural habitat in Uganda, however, fruits with their green, fleshy pericarp tissues float on the water's surface for about 50 days before they sink to the bottom. If fruits are in light after they sink, they begin to germinate within 10 days (Spence *et al.*, 1971). Although no studies have been conducted on dormancy-breaking requirements of fruits of these species, it seems logical that PD could be broken while they float on the surface of warm tropical waters.

Ideally, we would like to correlate the growth habit of aquatics with dormancy-breaking requirements, but this is not easy to do, considering our present state of knowledge. About 75% of the species listed under physiological dormancy in Table 11.8 are emergents,

TABLE 11.9   Aquatic Plant Species Whose Seeds Have Permeable Seed Coats, but
Scarification Increased Germination Percentages

| Species | Germination (%) | | Reference |
	Nonscarified	Scarified	
Alisma plantago-aquatica	0	98	Crocker (1907)
Bidens cernua	0	100	Hogue (1976)
Eichhornia crassipes	0	98	Crocker (1907)
Heteranthera limosa	0	84	Marler (1969)
Leersia oryzoides	0	32	Rosa and Corbineau (1986)
Menyanthes trifoliata	0	80	Hewett (1964)
Najas marina	11	62	van Vierssen (1982c)
Peltandra virginica	5	98[a]	Riemer and MacMillan (1972)
Phalaris arundinacea	5	73	Vose (1962)
Polygonum amphibium	0	85	Crocker (1907), Justice (1944)
Potamogeton natans	0	51[b]	Crocker (1907)
P. pectinatus	0	53[b]	Crocker (1907)
P. pusillus	24	100	Teltscherova and Hejny (1973)
P. schweinfurthii	3	85	Spence et al. (1971)
Sagittaria variabilis	0	92	Crocker (1907)
Zizania aquatica	0	55	Cardwell et al. (1978)

[a] Seed coat was removed.
[b] Some seeds had no embryos.

suggesting that species with other habits have not been studied as intensely, due perhaps to difficulties of seed collection. Also, it appears that many more species have been studied from temperate regions than from subtropical and tropical regions. Aside from these shortcomings, seeds of some emergent, submerged, and floating-attached species require cold stratification, whereas others require warm stratification (Table 11.8). Thus, cold or warm stratification for dormancy break is not restricted to seeds of aquatics with a particular growth habit. Two floating-unattached species (*Eichhornia crassipes* and *Pistia stratiotes*) require warm stratification for loss of seed dormancy, but no floating-unattached species whose seeds require cold stratification for germination were found in our survey of the literature. Do seeds of any floating-unattached species (e.g., *Hydrocharis* or any member of the Lemnaceae) require cold stratification for loss of dormancy?

Some aquatic habitats are flooded during the unfavorable season for plant growth, which would be winter in the temperate regions, and they are not flooded when conditions are suitable for plant growth, i.e., summer. The occurrence of dormancy break while seeds are submerged means they can germinate immediately after the water recedes. If dormancy loss did not occur until after the water receded, much of the favorable (nonflooded) period for plant growth could be over before seeds became nondormant and germinated. Another adaptation of seeds to seasonally dewatered habitats (e.g., mudflats) is that high temperatures of summer do

not cause flooded seeds to reenter dormancy (Baskin et al., 1993a). Thus, regardless of when waters recede during the growing season, seeds are nondormant and can germinate.

2. Oxygen

Oxygen levels may determine whether seeds come out of dormancy. Seeds of some species such as *Rumex dentatus* (Datta and Roy, 1970) and *Polygonum plebejum* (Mall, 1954) do not come out of dormancy when flooded, whereas those of other species, including *Carex comosa*, *C. stricta* (Baskin et al., 1996b), *Cyperus erythrorhizos*, *C. flavicomus* (Baskin et al., 1993a), *Diplachne fusca* (McIntyre et al., 1989b), *Fimbristylis autumnalis*, *F. vahlii* (Baskin et al., 1993a), *Heliotropium supinum* (Mall, 1954), *Leptochloa panicoides* (Baskin et al., 1993b), *Mollugo hirta* (Mall, 1954), *Scirpus articulatus* (Datta and Roy, 1970), and *Zizania aquatica* (Simpson, 1966), come out of dormancy while they are under water. Seeds of *Leucospora multifida* require cold stratification for dormancy break, but do not come out of dormancy during winter if they are flooded (Baskin et al., 1994). However, if dormancy loss begins prior to flooding, it will continue during flooding in winter. Exposing seeds of *Alisma plantago-aquatica* to alternate flooding and nonflooding conditions promotes dormancy loss (Stockey and Hunt, 1992).

Come et al. (1991) concluded that a lack of oxygen

sometimes can break dormancy, especially in seeds requiring cold stratification. However, their conclusion was based on data from seeds of nonaquatic species. Although seeds of various aquatics come out of dormancy while they are flooded, it is not known if (1) a lack of oxygen would substitute for cold stratification or (2) dormancy break would occur at low temperatures over a range of oxygen concentrations.

Another aspect of the effect of oxygen levels on seed dormancy is its role in the induction of dormancy. Nondormant seeds of *Cyperus erythrorhizos*, *C. flavicomus*, *Fimbristylis autumnalis*, *F. vahlii* (Baskin *et al.*, 1993a), and *Leptochloa panicoides* (Baskin *et al.*, 1993b) did not reenter dormancy when they were flooded, but those of *Lobelia dortmanna* (Farmer and Spence, 1987) and *Echinochloa crus-galli* (Honek and Martinkova, 1992) did. Secondary dormancy in *L. dortmanna* and *E. crus-galli* seeds was broken by cold stratification. Seeds of *Hygrophila auriculata* exposed to nitrogen gas did not lose their light requirement for germination nor did they enter secondary dormancy (Amritphale *et al.*, 1995).

3. Drying

Drying for 2–4 weeks at room temperature promoted dormancy loss in seeds of *Littorella uniflora* (Arts *et al.*, 1990). However, this treatment induced some seeds of *Diplachne fusca* (McIntyre *et al.*, 1989b) and *Eichhornia crassipes* (Barton and Hotchkiss, 1951) into dormancy. Drying for only 3 hr significantly reduced the germination of *Nymphaea odorata* seeds (Else and Riemer, 1984). Seeds of *Potamogeton pectinatus* entered secondary dormancy when they were dried, and cold stratification was required before they would germinate (Sharp, 1940). Drying *Hydrilla verticillata* seeds for up to 1 year did not induce dormancy, and seeds germinated when placed in water (Lal and Gopal, 1993).

Seeds of *Griffithella hookeriana* tolerated drying only at low (10–15°C) temperatures (Vidyashankari and Mohan Ram, 1987), and those of *Zizania aquatica* dried for 90 days died (Simpson, 1966). Although Duvel (1906) found that seeds of *Zizania* died unless they were stored in water at low temperatures (conditions suitable for cold stratification), Kovach and Bradford (1992) showed that it is possible to dehydrate many seeds to a moisture content of about 8% and later rehydrate them without a loss of viability. About 80% of the seeds of *Z. palustris* survived when they were dehydrated at 12% RH at 30°C and then rehydrated at 15, 20, or 25°C. However, only 20% or less of the seeds survived when dehydrated at 12% RH at 5 or 10°C and then rehydrated at 5, 10, 15, 20, 25, or 30°C. With an increase in RH during dehydration, the temperature range over which seeds could be dehydrated and then rehydrated without loss

of viability increased. At 75% RH, 65–95% of the seeds survived dehydration at temperatures ranging from 5 to 30°C followed by rehydration over the same range of temperatures (Kovach and Bradford, 1992). When dried to 0.6 to 0.8 g water/g dry wt, excised embryos of *Z. palustris* survived a 2-hr exposure to −50°C, and those of *Z. texana* survived −18°C (embryos of *Z. texana* were not exposed to −50°C). Embryos of these two species died if their water content decreased below a certain critical value, which increased with decreasing temperatures (Vertucci *et al.*, 1995).

C. Germination Requirements

1. Temperature

The optimum germination temperature for nondormant seeds of aquatics ranges from 15 to 41°C, with a mean of about 24°C (Table 11.8). Seeds of only a few species germinate well at temperatures of 15°C or less. The lowest temperature at which seeds of *Geum rivale* germinated was 11 or 12°C (Graves and Taylor, 1988), and those of *Fimbristylis vahlii*, *F. autumnalis* (Baskin *et al.*, 1993a), *Cyperus odoratus* (Baskin *et al.*, 1989), and *Leucospora multifida* (Baskin *et al.*, 1994) germinated at daily alternating temperature regimes of 20/10, but not at 15/6°C. Further, seeds of *Leptochloa panicoides* germinated at 25/15 but not at 20/10°C (Baskin *et al.*, 1993b). Thus, temperatures, especially those in temperate regions, may not be optimal for the germination of some species until early summer. Cold stratification increased the temperature range for the germination of *Zizania palustris* seeds from minimum and maximum temperatures of 17 and 23°C, respectively, to 5 and 34°C, respectively (Atkins *et al.*, 1987).

Seeds of many aquatics germinate to high percentages at constant temperatures (Table 11.8), and those of others may germinate to higher percentages at alternating than at constant temperatures (e.g. Atkins *et al.*, 1987). Seeds of *Fimbristylis littoralis* (Pons and Schroder, 1986), *Lycopus europaeus* (Thompson, 1969), *Rumex palustris* (Voesenek *et al.*, 1992), *Scirpus juncoides* (Pons and Schroder, 1986), and *Typha latifolia* (Morinaga, 1926a; Lombardi *et al.*, 1997) have an almost absolute requirement for alternating temperatures for germination. The alternating temperature requirement for germination was lost when seeds of *T. latifolia* were mechanically scarified (Morinaga, 1926a) and when seeds of *Rorippa islandica* came out of dormancy during exposure to high temperatures (Matsuo *et al.*, 1984). The minimum daily temperature fluctuation required for the germination of *L. europaeus* seeds ranged from 6.5 to 15°C, depending on how high the daily maximum and minimum temperatures were and how many times the temperature cycle was repeated (Thompson, 1970).

Sensitivity to daily temperature fluctuations can function as a depth-detecting mechanism for seeds that germinate under water (Pons and Schroder, 1986). When water was continuously stirred, depth (down to 35 cm) played no role in the germination of *Fimbristylis littoralis* seeds in light; however, germination did not exceed 50% until the daily temperature fluctuation was 11.9°C. In water that was not stirred, seeds germinated to 100% at the surface, where the amplitude of daily temperature fluctuations was 16°C, but they germinated to 50% at 5 cm, where the fluctuation was 10°C (Pons and Schroder, 1986).

Buried seeds of some aquatic species in temperate regions undergo seasonal changes in their temperature requirements for germination (Fig. 11.8). Dormancy loss occurs during cold stratification in winter, and seeds are nondormant by spring, at which time they germinate to high percentages over a range of temperatures in light (and sometimes in darkness, depending on the species). Seeds are conditionally dormant (germinate at high but not at low temperatures) or dormant by autumn, but they are nondormant by the following spring. Seasonal cycles in temperature requirements for germination have been observed in seeds of *Carex canescens, C. cespitosa, C. elongata, C. paniculata, C. pseudocyperus, C. remota* (Schuetz, 1997), *Cyperus inflexus* (Baskin and Baskin, 1978), *C. flavicomus, Fimbristylis vahlii, F. autumnalis* (Baskin *et al.*, 1993a), *Gratiola viscidula* (Baskin *et al.*, 1989), *Leptochloa panicoides* (Baskin *et al.*, 1993b), *Leucospora multifida* (Baskin *et al.*, 1994), and *Scirpus lineatus* (Baskin *et al.*, 1989).

Buried seeds of some aquatics do not undergo seasonal changes in temperature requirements for germination. Seeds may be dormant at maturity, but after this

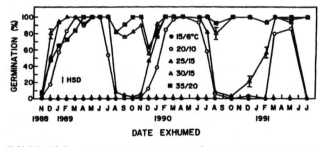

FIGURE 11.8 Germination percentages (mean ± SE, if ≥5%) of seeds of the summer annual *Leucospora multifida* incubated for 14 days at a 14-hr daily photoperiod following 0–32 months of burial under nonflooded conditions in a nonheated greenhouse in Lexington, Kentucky. HSD, Tukey's honestly significant difference test at 5% level of significance. Seeds have an annual nondormancy/conditional dormancy cycle. Seeds do not germinate in darkness. Modified from Baskin *et al.* (1994), with permission.

dormancy is broken they can germinate to high percentages over a range of temperatures at all times of the year. This has been observed in seeds of *Cyperus odoratus* (Baskin *et al.*, 1989), *Hottonia inflata* (Fig. 11.9), and *Penthorum sedoides* (Baskin *et al.*, 1989) and to a certain extent in *C. erythrorhizos* (Baskin *et al.*, 1993a). Studies need to be done in subtropical and tropical regions to determine whether buried seeds of aquatic species undergo seasonal changes in temperature requirements for germination.

2. Light

Light:dark requirements for germination have been determined for 134 species (Table 11.8). Seeds of 67 species (50%) require light for germination, 52 (39%) germinate to higher percentages in light than in darkness, 9 (7%) germinate equally well in light and darkness, and 6 (4%) germinate to higher percentages in darkness than in light. Thus, seeds of most aquatics are more likely to germinate if incubated in light than in darkness. Scarification of *Typha latifolia* seeds stimulated germination in darkness (Sifton, 1959) and increased the germination rate, but not percentage, of *Pistia stratiotes* seeds in both light and darkness (Harley, 1990).

One consequence of dormancy-breaking treatments is that the ability of seeds to germinate in darkness may increase. After 90 days of warm stratification, especially at 40°C, seeds of *Hygrophila auriculata* germinated to 95% in darkness (Amritphale *et al.*, 1989). Seeds of *Carex stricta* cold stratified during winter under both flooded and nonflooded conditions germinated to 70–100% in darkness over a range of thermoperiods (Fig. 11.10), but those of *C. comosa* cold stratified under flooded conditions could germinate only in darkness (Fig. 11.11). Seeds of *Bidens laevis* cold stratified in light and darkness germinated to 87 and 0%, respectively, at 35/20°C. However, seeds cold stratified in light and tested in darkness at 35/20°C germinated to 27%, whereas those cold stratified in darkness and tested in light germinated to 93% (Leck *et al.*, 1994). Thus, if seeds are cold stratified in light during winter but become covered by sediment or litter in spring, some of them still may be capable of germinating.

Seasonal changes in light:dark requirements for germination have been found in *Carex cespitosa, C. elongata, C. paniculata, C. remota* (Schuetz, 1997), and *C. stricta* (Fig. 11.10). Seeds exhumed after burial under flooded and under nonflooded conditions germinated to high percentages in light at 25/15, 30/15, and 35/20°C throughout the year, but they germinated to high percentages in darkness at these temperatures only in late winter and spring. Seeds of *C. comosa* exhumed

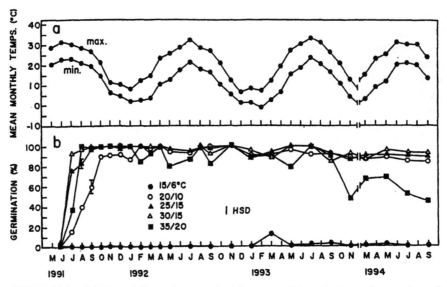

FIGURE 11.9 (a) Mean daily maximum and minimum monthly temperatures in a nonheated greenhouse in Lexington, Kentucky. (b) Germination percentages (mean ± SE, if ≥5%) of seeds of the aquatic winter annual *Hottonia inflata* incubated for 15 days at a 14-hr daily photoperiod following 0–39 months of burial under flooded conditions in a nonheated green-house. HSD, Tukey's honestly significant difference test at 5% level of significance. From Baskin *et al.* (1996a), with permission.

after burial under nonflooded conditions germinated to high percentages in light at 30/15 and 35/20°C throughout the year, but no seeds germinated in darkness (Fig. 11.11). Seeds of *C. comosa* exhumed after burial under flooded conditions germinated to relatively high percentages in light and darkness at 30/15 and 35/20°C throughout the year. Peaks of germination for flooded and nonflooded seeds in light and in darkness at 25/15 and 20/10°C occurred in spring. No seeds of *C. stricta* or *C. comosa* germinated while they were buried under either flooded or nonflooded conditions. Thus, some factor associated with the burial environment other than darkness per se prevented germination. Volatile germination inhibitors resulting from anaerobic respiration (Holm, 1972) and/or improper aeration (Bibbey, 1948) are possible reasons why buried seeds did not germinate. However, seeds of *Scirpus lacustris, S. tabernaemontani,* and *S. maritimus* buried to a depth of 0.5 cm in soil and then flooded by 10 cm of water germinated to 43, 50, and 92%, respectively, and those buried to a depth of 2.0 cm in soil and then flooded with 10 cm of water germinated to 22, 39, and 75%, respectively (Clevering, 1995).

Not surprisingly, the light response of seeds of aquatic plants is controlled by the phytochrome system, with red promoting and far-red light inhibiting germination. A reversibility between red and far-red irradiation has been documented in seeds of *Heteranthera limosa* (Marler, 1969), *Hygrophila auriculata* (Amritphale *et*

al., 1989, 1995), *Jussiaea suffruticosa* (Wulff and Medina, 1969), *Lobelia dortmanna* (Farmer and Spence, 1987), *Potamogeton richardi* (Spence *et al.*, 1971), *Rumex palustris* (Voesenek *et al.*, 1992), and *Typha latifolia* (Bonnewell *et al.*, 1983). Blue light also inhibits the germination of seeds of *T. latifolia* (Sifton, 1959; Gopal and Sharma, 1983) and *Myriophyllum spicatum* (Coble and Vance, 1987), and this inhibition is overcome in *T. latifolia* seeds by yellow and, to some extent, red light (Gopal and Sharma, 1983).

Exposure of light-requiring seeds of aquatic plants to continuous light sometimes inhibits germination. Seeds of *Eichhornia crassipes* incubated in continuous light at 30°C did not germinate (Obeid and El Seed, 1976), and seeds of *Lemna perpusilla* germinated to only 2% in continuous light at 26°C (Posner and Hillman, 1962). However, seeds of *L. perpusilla* given 1, 2, 4, or 10 days of darkness prior to being placed in continuous light germinated to 12, 89, 95, and 96%, respectively (Posner and Hillman, 1962).

Illuminance ("light intensity") and duration of exposure can influence germination percentages. The germination of *Typha latifolia* seeds incubated in continuous light increased with increases in illuminance from 5 to 1000 lux, at which point they germinated to 100%. However, three 9-hr exposures to 1500 lux also resulted in 100% germination (Gopal and Sharma, 1983). In continuous light of 2000 lux, seeds of *Trapella sinensis* germinated to 100% after 5 days, whereas 11 days were re-

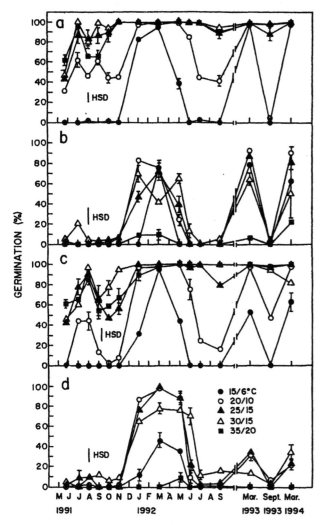

FIGURE 11.10 Germination percentages (mean ± SE, if ≥5%) of seeds of the perennial *Carex stricta* incubated for 15 days at a 14-hr daily photoperiod following 0–33 months of burial in a nonheated greenhouse in Lexington, Kentucky. (a) Nonflooded in greenhouse and tested in light. (b) Nonflooded in greenhouse and tested in darkness. (c) Flooded in greenhouse and tested in light. (d) Flooded in greenhouse and tested in darkness. HSD, Tukey's honestly significant difference test at 5% level of significance. From Baskin *et al.* (1996b), with permission.

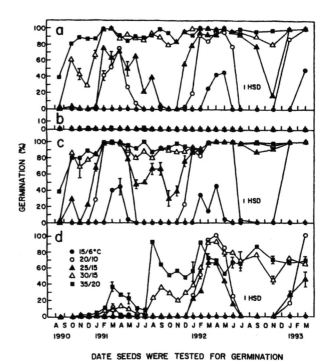

FIGURE 11.11 Germination percentages (mean ± SE, if ≥5%) of seeds of the perennial *Carex comosa* incubated for 15 days at a 14-hr daily photoperiod following 0–30.5 months of burial in a nonheated greenhouse in Lexington, Kentucky. (a) Nonflooded in greenhouse and tested in light. (b) Nonflooded in greenhouse and tested in darkness. (c) Flooded in greenhouse and tested in light. (d) Flooded in greenhouse and tested in darkness. HSD, Tukey's honestly significant difference test at 5% level of significance. From Baskin *et al.* (1996b), with permission.

quired for 100% germination at 15 lux (Kawahara and Takada, 1961). At 15,500 lux, germination percentages of seeds of *Jussiaea suffruticosa* increased as temperatures increased from 20 to 40°C. However, in continuous light of 750–7400 lux, germination percentages were relatively low at 20 and 30°C and high at 25, 35, and 40°C (Wulff *et al.,* 1972).

Seeds of *Limosella aquatica* exposed to bright light for 20, 9, or 2.5 hr each day germinated to 72–96% within 5 days, whereas those exposed to dim daylight for 15 hr each day germinated to only about 2% after

8 days (Salisbury, 1967). Salisbury (1967) also found that seeds of this species exposed to daylight on cloudy days germinated to only about 6%. He speculated that low light levels would prevent most flooded seeds from germinating and that optimum conditions for germination occurred when water receded, leaving seeds on wet mud fully exposed to sunlight. In contrast, cold-stratified seeds of *Lobelia dortmanna* germinated to 100% at a very low light level (4.0 μmol photons m⁻² sec⁻¹, 400–700 nm) at 25°C, although no germination occurred in darkness (Farmer and Spence, 1987). Very low irradiance (ca. 0.5 W m⁻² was optimal) also promoted the germination of *Panicum laxum* seeds, which require flooding for maximum germination (Cole, 1977).

Irradiance/illuminance decreases with an increase in water depth (Fig. 11.12), and the maximum depth of light penetration varies with the wavelength. In the portion of the solar spectrum that is photomorphogenic, far-red light penetrates to the shallowest depths and blue light to the deepest. Light-requiring seeds can germinate under water, but percentages decrease as the depth increases. At a depth of 25 cm, *Pistia stratiotes*

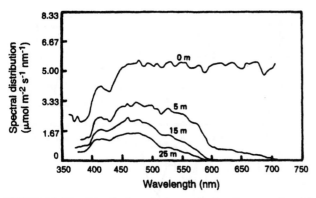

FIGURE 11.12 Distribution of photons in the photomorphogenic portion of the solar spectrum in marine water (Gulf of Mexico) at depths of 0, 5, 15, and 25 m. Modified from Kirk (1983), with permission.

seeds germinated to 78% at an irradiance of 4000 erg cm^{-2} sec^{-1} (4 W m^{-2}), but at 900 erg cm^{-2} sec^{-1} (0.9 W m^{-2}), only 9% of them germinated. In relatively deep water, germination of this species was inhibited, even in light. Thus, at 50 cm and 2000 erg cm^{-2} sec^{-1} (2 W m^{-2}), the germination of *P. stratiotes* seeds was 4%, and at 100 cm and 900 erg cm^{-2} sec^{-1} (0.9 W m^{-2}), no seeds germinated (Pieterse *et al.*, 1981). Seeds of *Myriophyllum spicatum* exposed to darkness and to filtered light of wavelengths of 445, 520, 700, and 725 nm and to white light (energy level for all seeds in filtered or white light was 4.35 μmol m^{-2} sec^{-1}) germinated to 5, 31, 76, 83, 97, and 83%, respectively (Coble and Vance, 1987). Thus, darkness and blue light (445 nm) greatly inhibited germination. At a depth of 50 cm, seeds of *M. spicatum* germinated to only 8.3% in white light at an energy level of 4.5 μmol m^{-2} sec^{-1}, but germination increased to 42% when the energy level was increased to 9.0 μmol m^{-2} sec^{-1} (Coble and Vance, 1987). These data suggested that the inhibition of germination was due to a reduction in red light reaching the seeds. Because blue light inhibits germination and water deeper than about 0.5 m filters out much of the red light, leaving the blue (Coble and Vance, 1987), germination at depths greater than 0.5 m may be unlikely for seeds of many species.

In view of the light requirement for the germination of seeds of most aquatics, it seems logical that germination would decrease if seeds become covered by sediment or litter. The germination of *Myriophyllum spicatum* seeds covered with 2 cm or more of sediment was reduced from 46% to 15% or less (Hartleb *et al.*, 1993), and the germination of *Typha* × *glauca* seeds covered by 0.4 cm of sediment was reduced from 80% to about 10% (Wang *et al.*, 1994). A sediment depth of 1 cm significantly reduced the germination of a number of aquatics, including *Carex* spp., *Echinochloa crus-galli*, *Eleocharis acicularis*, *Leersia oryzoides*, *Polygonum la-*

pathifolium, *Typha* spp. (Jurik *et al.*, 1994), *Aster laurentianus*, *Phragmites australis*, and *T. glauca* (Galinato and van der Valk, 1986). The presence of dead plant litter on the soil surface in Australian rice fields at the time of flooding reduced seed germination of the weed *Diplachne fusca* by 98% (McIntyre *et al.*, 1989b). In Manitoba, Canada, the presence of litter at the time of water drawdown in a freshwater marsh complex decreased the number of species and the number of individuals of a species that became established from seeds in the soil seed bank. Litter may have decreased germination by reducing light at the soil surface or by decreasing the amplitude of daily temperature fluctuations (van der Valk, 1986).

Light interacts with temperature to control the germination of some species. At low irradiance, seeds of *Typha latifolia* germinated to higher percentages at alternating temperature regimes than at constant temperature regimes (Sifton, 1959). The light requirement for the germination of *Cladium mariscus* seeds decreased with an increase in the difference between day and night temperatures (Goossens and Devillez, 1974). Regardless of the level of irradiance, seeds of *Lythrum salicaria* given a 1-min light exposure at 31°C germinated to 30–40%, whereas they germinated to only 1–5% at 25°C (Rao, 1925). An increase in the dark imbibition temperature from 25 to 35°C increased the responsiveness of *Jussiaea suffruticosa* seeds to 6-, 8-, 24-, 48-, and 72-hr light exposures (Wulff and Medina, 1969). Seeds of *Najas marina* incubated at low redox potentials germinated equally well (80%) in light and darkness at 24°C. However, at 20°C germination was higher in darkness than in light, and at 12 and 16°C germination occurred in darkness but not in light (van Vierssen, 1982c).

3. Oxygen

Seeds of most species, including aquatics, require oxygen for germination (Morinaga, 1926c; Come *et al.*, 1991). However, a few studies have been done to determine if seeds of aquatics can germinate in the complete absence of oxygen. For germination tests in the absence of oxygen, seeds are placed in flasks or other sealed containers that are flushed with nitrogen or hydrogen gas to replace the oxygen. One of the most famous species whose seeds germinate in the complete absence of oxygen is rice, *Oryza sativa* (Takahashi, 1905). Not too surprising, seeds of barnyard grass, *Echinochloa crus-galli* var. *oryzicola* and var. *crus-galli*, which are serious weeds in rice fields, also germinate in the absence of oxygen (Kennedy *et al.*, 1980). Seeds of rice (Tsuji, 1972), barnyard grass (Rumpho and Kennedy, 1981), and *E. turnerana* (Conover and Geiger, 1984) germinated in an oxygen-free environment produce colorless coleoptiles [coleoptile emerges before the coleorhiza (radicle)], and the development of seminal roots and leaves is inhibited greatly. Deoxygenated water promoted coleoptile extension in rice seedlings, but had no effect on elongation of the coleoptile (and prevented mesocotyl extension) in *Echinochloa* (Pearce and Jackson, 1991). After coleoptiles are exposed to oxygen, they turn green and begin to photosynthesize (Bozarth and Kennedy, 1985; VanderZee and Kennedy, 1982). Seeds of *Scirpus juncoides* also germinate in 0% oxygen, and the colorless seedlings exhibit no root growth (Pons and Schroder, 1986). *Peltandra virginica* seeds germinate in almost a complete absence of oxygen (Edwards, 1933), and fruits of *Nelumbo nucifera* germinate in an atmosphere of 100% nitrogen (Ohga, 1926). However, fruits of *N. nucifera* contain about 0.2 cc of gas with an oxygen concentration of 18.33–18.88%, which probably is enough oxygen to promote germination (Ohga, 1926).

The concentration of dissolved oxygen in water primarily depends on temperature and depth (Fig. 11.13), but other factors such as turbulence, activity of organisms, and presence of pollutants can be important (Sculthorpe, 1967). Thus, the oxygen concentration varies with the position within a body of water, but at the same position it changes over time. For example, in temperate regions, the oxygen concentration in lakes and rivers is maximal during the cool season and minimal during the warm season. However, regardless of the season, seeds germinating in natural bodies of water probably are exposed to some oxygen, although the concentration could be quite low.

Seeds of many aquatics apparently can survive long periods of flooding, but they do not germinate until the water recedes. Seeds of these species may form persistent seed banks in/on the flooded soil, and most of them require light for germination (Table 11.10). A light requirement would be important in preventing the germination of seeds in muddy water; however, if the water is clear and shallow, some light would reach seeds on the surface of sediments. Little is known about the actual effects of flooding on germination of these seeds. Is germination underwater prevented by low light levels, decreased oxygen, or an interaction of these two factors?

Dormancy loss in seeds may not necessarily be optimal under flooded conditions. In *Heteranthera limosa* and *Gratiola neglecta,* for example, dormancy loss was more complete in seeds cold stratified under nonflooded conditions than in those cold stratified under flooded conditions; however, nondormant seeds germinated to higher percentages in cups of water than on moist sand in petri dishes (Baskin *et al.,* unpublished results). Lack of a flooding requirement for loss of dormancy, but an ability to germinate under flooded conditions, is an

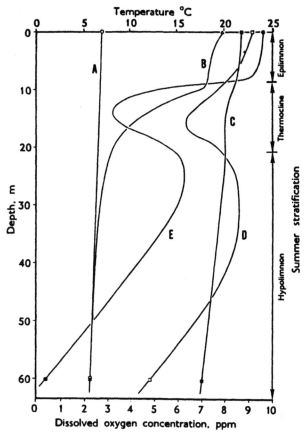

FIGURE 11.13 Temperatures and dissolved oxygen concentrations at various depths in temperate lakes. (A) Winter temperatures. (B) Summer temperatures, showing thermal stratification. (C,D, and E) Oxygen concentrations that might occur during the development of thermal stratification from winter (C) to summer (D and E) in a typical nutrient-rich lake or during increasing productivity of a given lake over a number of years. From Sculthorpe (1967), with permission.

adaptation of species growing in habitats flooded during the favorable season for germination and growth. Such habitats include seasonally wet lowlands in tropical/subtropical areas that fill with water in the monsoon season (Mall, 1954; Datta and Roy, 1970), seasonal (vernal) pools in regions with a Mediterranean climate that have water in the winter (Keeley, 1988), and pools in temperate regions that become filled with water by late winter or early spring (Baskin *et al.,* unpublished results). Loss of dormancy during the season when bodies of water are dry allows seeds to germinate immediately after the habitat becomes flooded. If dormancy loss did not occur until after seeds were flooded, much of the favorable period for plant growth (i.e., the period of flooding) could be over before seeds came out of dormancy and germinated.

Flooding decreases seed germination in some species (Pons, 1982; Moore and Keddy, 1988; Gerritsen and Greening, 1989; Finlayson *et al.,* 1990; Coops and van

TABLE 11.10   Aquatic Species Whose Seeds Have Been Found in Persistent Seed Banks and for Which Data on Light:Dark (L:D) Requirements for Germination are Available

Species	L:D[a]	Reference
Ammannia coccinea	L	Baskin *et al.* (unpublished results)
Bidens laevis	L > D	Leck and Simpson (1987)
Cladium mariscus	L > D	Wilson *et al.* (1993)
Cyperus aristatus	L	Baskin *et al.* (unpublished results)
C. flavicomus	L	Baskin *et al.* (1993a)
Eragrostis hypnoides	L	Baskin *et al.* (unpublished results)
Fimbristylis autumnalis	L	Keddy and Reznicek (1982), Baskin *et al.* (1993a)
F. vahlii	L	Baskin *et al.* (1993a)
Heteranthera limosa	L	Baskin *et al.* (unpublished results)
Juncus effusus	L > D	Leck and Simpson (1987)
Leptochloa panicoides	L	Baskin *et al.* (1993b)
Leucospora multifida	L	Baskin *et al.* (1994)
Limosella aquatica	L	Milberg and Stridh 1994
Lythrum salicaria	L	Welling and Becker (1990)
Phalaris arundinacea	L > D	Leck and Simpson (1994)
Ranunculus sceleratus	L	Leck and Simpson (1987), Poiani and Johnson (1988)
Rotala ramosior	L	Baskin *et al.* (unpublished results)
Rumex palustris	L	Voesenek and Blom (1992)
Sagittaria latifolia	L	Leck and Simpson (1994)
Scirpus acutus	L > D	Poiani and Johnson (1988)
S. validus	L > D	Isely (1994)
Scolochloa festucacea	D > L	Poiani and Johnson (1988)
Typha latifolia	L	Leck and Simpson (1987)
Zannichellia palustris	L > D	Poiani and Johnson (1988)

[a] See note for Table 11.5.

der Velde, 1995; Ponzio *et al.*, 1995), but it promotes germination in seeds of others (Table 11.11). Seeds of *Zizania texana* germinated to 85, 79, 8, and 3% in water with 0.1, 1.0, 4.0, and 5.0 ppm dissolved oxygen, respectively (Power and Fonteyn, 1995), whereas maximum germination of *Scirpus juncoides* occurred at about 5% oxygen (= 50,000 ppm) (Pons and Schroder, 1986). It is hard to explain why scarified seeds of *Typha latifolia* lose their low oxygen requirement for germination (Morinaga, 1926a,b). Cutting a hole in the seed coat should increase (not decrease) the flow of oxygen to the embryo!

Seeds of some species, including *Heteranthera limosa* (Marler, 1969), *Hottonia palustris* (Brock *et al.*, 1989), *Hydrilla verticillata* (Lal and Gopal, 1993), *Hymenocallis occidentalis* (Flint and Moreland, 1943), *Potamogeton schweinfurthii*, *P. richardi* (Spence *et al.*, 1971), and *Typha latifolia* (Sifton, 1959; Bonnewell *et al.*, 1983), germinate to high percentages only under the low oxy-

gen levels associated with flooding; they also require exposure to light. Flooded seeds of *Nuphar lutea* germinated equally well in light and darkness (Smits *et al.*, 1990), whereas those of *Pistia stratiotes* (Pieterse *et al.*, 1981; Harley, 1990) and *Polygonum amphibium* (Justice, 1944) germinated to higher percentages in light than in darkness. As previously mentioned, flooded seeds of *Najas marina* incubated at 12 or 16°C germinated only in darkness (van Vierssen, 1982c). A light requirement for the germination of flooded seeds would ensure that they germinated only in relatively shallow and/or clear water, where light is available for photosynthesis and seedling growth.

Under anaerobic conditions, nondormant (cold-stratified) seeds of *Nuphar luteum* and *Nymphaea alba* produced up to 6–7 mmol of ethanol [gdw]$^{-1}$, whereas those of *Nymphoides peltata* produced only 1.5 mmol of ethanol [gdw]$^{-1}$ (Smits *et al.*, 1995). The germination of *N. luteum* and *N. alba* seeds is promoted by ethanol, but that of *N. peltata* seeds is not. Further, seeds of the former two species germinated readily under anaerobic conditions, but those of the latter species did not. Seeds of all three species germinated under aerobic conditions, but those of *N. peltata* germinated faster than those of the other two species. Smits *et al.* (1995) suggested that seeds of *N. luteum* and *N. alba* germinated in soft, organic sediments in darkness under relatively deep water and anaerobic conditions, whereas those of *N. peltata* germinated in shallow water where light and oxygen were available.

4. pH

Only a limited amount of information is available, but it appears that seeds of aquatics germinate to high percentages over a wide range of pH values. The maximum germination rate of mechanically scarified seeds of *Zizania aquatica* was at pH 7.0; however, seeds germinated to 97–100% at pH values ranging from 6.0 to 8.7 (Simpson, 1966). Seeds of *T. latifolia* germinated to 70–84% in light at pH values of 4.0, 7.0, and 12.0 (Rivard and Woodard, 1989).

After soil seed bank samples from acidified moorland pools in the Netherlands were placed in water that subsequently was limed to increase pH, seeds of the soft water species *Elatine hexandra*, *Eleocharis acicularis*, *Littorella uniflora*, and *Lobelia dortmanna* germinated (Bellemakers *et al.*, 1996).

5. Soil Texture

A gradient of particle sizes ranging from 0.13 to 16 mm with the water table 1 or 4 cm below the soil surface was used to study the germination of cold-stratified

TABLE 11.11   Aquatic Plant Species Whose Seeds Germinate to Higher Percentages under Flooded (F) than under Nonflooded (NF) Conditions

| Species | Germination (%) | | Reference |
	F	NF	
Alisma subcordata	62	0	Moore and Keddy (1988)
Callitriche palustris	484[a]	392[a]	Milberg and Stridh (1994)
Diplachne fusca	100[b]	10[b]	McIntyre *et al.* (1989a)
Eichhornia crassipes	100	9	Obeid and El Seed (1976)
Eleocharis robbinsii	6[a]	1[a]	Gerritsen and Greening (1989)
Heteranthera limosa	96[c]	0	Marler (1969)
Limnocharis flava	48	0	Pons (1982)
Lobelia dortmanna	28	10	Moore and Keddy (1988)
Maidenia rubra	635[a]	60[a]	Finlayson *et al.* (1990)
Monochoria vaginalis	83	67	Pons (1982)
Nymphaea odorata	210[a]	10[a]	Gerritsen and Greening (1989)
Orcuttia californica	48[c]	30	Keeley (1988)
Panicum laxum	95	48	Cole (1977)
Podostemum ceratophyllum	100	0?	Philbrick (1984)
Pontederia cordata	64	0	Leck (1996)
Rorippa nasturtium	15	0	Morinaga (1926c)
Sagittaria lancifolia	100	68	Collon and Velasquez (1989)
S. latifolia	95	0	Leck (1996)
Scirpus juncoides	88	25	Pons and Schroder (1986)
Scolochloa festucacea	100	16	Smith (1972)
Typha angustifolia	88	8	Coops and van der Velde (1995)
T. latifolia	89	61	Sifton (1959)
	33	0	Moore and Keddy (1988)
Tuctoria greenei	70[c]	3	Keeley (1988)
Xyris smalliana	6[a]	1[a]	Gerritsen and Greening (1989)
Zizania texana	85	50	Power and Fonteyn (1995)

[a] Seedling density (m²) in soil samples.
[b] Calculated relative values based on seedling densities in field plots.
[c] Low oxygen levels were obtained using nitrogen gas.

seeds of various aquatics (Keddy and Constabel, 1986). When the water table was at a depth of 4 cm, the germination of seeds of *Acorus calamus, Alisma plantago-aquatica, Bidens cernua, B. vulgata, Cyperus aristatus, Lythrum salicaria, Sagittaria latifolia, Scirpus americanus,* and *Typha angustifolia* decreased with an increase in particle size. However, particle size had no effect on the germination of *Polygonum punctatum* seeds. When the water table was at a depth of 1 cm, only seeds of *C. aristatus, L. salicaria,* and *S. latifolia* showed a significant decrease in germination with an increase in particle size.

6. Other Factors

As the number of *Nymphaea odorata* seeds placed in water-filled glass vials was increased from 20 to 100, germination percentages increased from 0 to 52%. These data suggest that the seeds themselves produced a stimulatory compound, possibly ethylene (Else and Riemer, 1984). These authors found that ethephon promoted the germination of uncrowded seeds, and maximum germination (18%) occurred at the highest concentration used (100 ppm). In contrast, increasing the number of *Nuphar advena* seeds per water-filled vial resulted in a decrease in germination percentages (Riemer, 1985). Was the oxygen supply depleted?

Less than 1.0% of the seeds of *Typha latifolia* germinated when exposed to aqueous leaf extracts of the species, suggesting that autotoxic compounds produced by mature plants could prevent seed germination and seedling establishment in natural populations (McNaughton, 1968). In contrast, Grace (1983) was unable to demonstrate any allelopathic effects when seeds of *T. latifolia* were exposed to aqueous leaf extracts or pieces of leaves of the species. In another study, green leaf extracts of *T. latifolia* reduced germination in darkness from 50 to 15%; however, the addition of ashes from burned leaves to seeds in distilled water or to seeds in green leaf extracts increased germination in darkness to 73 and 40%, respectively (Rivard and Woodard, 1989). Aqueous extracts of leaves, roots, and rhizomes of *T. angustata* and various other aquatic plant species did not inhibit the germination of *T. angustata* seeds. Further, seedlings of *T. angustata* and *T. elephantina* were observed in natural populations in India (Sharma and Gopal, 1978). Chemical leachates from litter of *Justicia americana* reduced the germination of *T. latifolia* seeds from 67 to 22%, but had no effect on the germination of *Polygonum lapathifolium* or *P. psyllium* seeds (Carter *et al.*, 1985). However, no seedling growth of *P. lapathifolium* was reduced (Carter and Grace, 1986).

The effects of various water pollutants on the germination ecology of aquatic species has received little attention. Walsh *et al.* (1991a) found no effects of effluents from a coke plant, paper mill, or sewage treatment plant on the germination of *Echinochloa crus-galli* or *Sesbania macrocarpa* seeds. In another study, however, effluents from a sewage treatment plant and a metal plating works inhibited the germination of *E. crus-gallii* seeds in darkness, but not in light (Walsh *et al.*, 1991b). Solutions of lead nitrate, copper sulfate, and cadmium chloride did not inhibit the germination of *Juncus acutus* seeds, but seedling growth was strongly reduced, varying with the concentration of the heavy metals (Stefani *et al.*, 1991). The germination of *Sagittaria latifolia* seeds decreased with an increase in salinity from 0 to 0.8% (0.14 *M*) NaCl, and inhibition was greater at 27/16°C than at 21/16°C (Delesalle and Blum, 1994).

VII. HALOPHYTES

A halophyte is a plant that not only grows vegetatively when rooted in salty soil, but it is able to flower and produce seeds in such a habitat. Chapman (1942) used the word halophyte to describe all plants growing in an environment with more than 0.5% NaCl. Euhalophytes grow optimally only when the NaCl level is greater than 0.5%; there are relatively few euhalophytes. Miohalophytes are found in habitats with greater than 0.5% NaCl, but their optimum growth occurs when NaCl is less than 0.5%. Apparently, miohalophytes grow in high salt areas because this habitat prevents the growth of salt-sensitive plants (glycophytes) that outcompete halophytes in nonsaline habitats (Chapman, 1942; Bakker *et al.*, 1985).

Only a few mosses, liverworts, ferns, and gymnosperms are salt tolerant, but this characteristic is found in 499 genera in 129 families of angiosperms (Aronson, unpublished data cited in Flowers *et al.*, 1986). The great diversity of halophytes is attributed to (1) habitats variation with respect to kinds and concentration of salts, degree and duration of flooding, and/or amount of annual precipitation and geographical location (Waisel, 1972) and (2) species variation in life forms and mecha-

nisms of adapting to saline conditions (Zahran, 1982). Halophytes can be annuals, herbaceous perennials, subshrubs, shrubs, or trees (Sen and Rajpurohit, 1982; Zahran, 1982).

Waisel (1972) divided halophytes into two categories: (1) those requiring salt for survival and/or maximum growth and (2) those resisting salt. As roots of plants absorb salts from the soil solution and water is transpired, salts accumulate in the plant. In nonhalophytes, an increase in salt content often leads to death of the plant, but halophytes have evolved various mechanisms to deal with the problem. Salt-enduring halophytes can tolerate a high level of salts in their cells. In salt-excluding halophytes, salts are secreted from the shoot, accumulated in special hairs or retransported from the shoot back to the roots. Salt-evading halophytes either do not absorb salts from the soil solution or the salts are not transported to the leaves (Waisel, 1972).

Halophytes grow in coastal (Chapman, 1960, 1977) and inland (Chapman, 1976; Ungar, 1967b; Zahran, 1982) salt marshes, in salt deserts (Chapman, 1960), submerged in seawater (Waisel, 1972), on muds of tropical estuaries and deltas (Walsh, 1974), and on coastal beaches (Macdonald and Barbour, 1974; Ignaciuk and Lee, 1980), cliffs (Okusanya, 1979; Malloch *et al.*, 1985), and dunes (Schat, 1983; Doing, 1985).

This section discusses the germination ecology of halophytes growing (1) in salt marshes and salt deserts, (2) on coastal beaches, bases of cliffs by oceans, and the seaward side of foredunes, (3) submerged in bodies of salt water, and (4) in mangrove communities, which replace salt marshes in most tropical coastal areas (Walsh, 1974). Because (1) the roots of plants growing on high dunes can reach fresh water (Chapman, 1960) and (2) soil salinity levels on dunes are low (Boyce, 1954), the germination ecology of these species is covered under psammophytes (see Section VIII).

A. Salt Marshes and Salt Deserts

Salinity (as a percentage of NaCl) in coastal salt marshes ranges from 0.8 to 2.4% (Beeftink, 1977). In inland regions with low rainfall and high evaporation, soluble salts accumulate in the upper horizons of the soil, and maximum yearly salinity levels can reach 8% (Waisel, 1972). Halophytes in salt deserts grow in either saline-nonsodic (also called saline, white alkali, or solonchak) or saline-sodic (saline-alkali) soils (Chapman, 1960). Saline-nonsodic and saline-sodic soils have a high soluble salt content. The most important cation in these salts is sodium, calcium, or magnesium, and the most important anion is chloride, sulfate, or bicarbonate (Black, 1968). In saline-nonsodic soils, sodium occupies less than 15% of the cation exchange capacity (CEC),

TABLE 11.12 Families and Number of Genera with Halophytes (1) Growing in Salt Marshes
and Salt Deserts or (2) on Coastal Sea Beaches, Bases of Cliffs, and Foredunes[a] and Type(s) of
Seed Dormancy Known to Occur in Each Family[b]

Family	No. of genera (species)		Type(s) of dormancy known in family
	Marshes, deserts	Beaches, cliffs, and dunes	
Acanthaceae	1 (4)	1 (1)	Physiological
Aizoaceae	8 (21)	11 (17)	Physiological
Amaranthaceae	3 (4)	5 (8)	Physiological
Amaryllidaceae	0	2 (3)	Morphophysiological
Apiaceae	2 (3)	13 (15)	Morphophysiological, morphological
Apocynaceae	1 (1)	0	Physiological
Araliaceae	0	1 (1)	Morphophysiological
Asclepiadaceae	1 (2)	1 (1)	Physiological
Asteraceae	24 (34)	37 (62)	Physiological
Batidaceae	1 (2)	0	?
Boraginaceae	1 (1)	7 (10)	Physiological
Brassicaceae	3 (4)	10 (15)	Physiological
Campanulaceae	2 (2)	1 (1)	Physiological
Caryophyllaceae	2 (6)	9 (14)	Physiological
Chenopodiaceae	42 (162)	14 (40)	Physiological
Combretaceae	0	1 (1)	?
Convolvulaceae	3 (6)	4 (7)	Physical
Crassulaceae	1 (2)	1 (2)	Physiological
Cyperaceae	10 (29)	10 (25)	Physiological
Ehretiaceae	0	1 (1)	Physiological
Euphorbiaceae	2 (2)	6 (17)	Physiological
Fabaceae	5 (5)	21 (28)	Physical
Frankeniaceae	1 (7)	1 (3)	?
Gentianaceae	1 (1)	0	Physiological
Globulariaceae	0	1 (1)	?
Goodeniaceae	2 (2)	1 (3)	Physiological
Hydrophyllaceae	0	1 (1)	Physiological, morphophysiological
Iridaceae	1 (1)	1 (1)	Morphophysiological
Juncaceae	1 (10)	1 (2)	Physiological
Juncaginaceae	1 (4)	0	Physiological
Lamiaceae	0	1 (1)	Physiological
Liliaceae	0	3 (3)	Morphophysiological
Lythraceae	0	2 (2)	Physiological
Malvaceae	6 (7)	2 (2)	Physical

(*continues*)

whereas in saline-sodic soils it occupies more than 15% of the CEC (Brady, 1974).

The germination ecology of halophytes in salt marshes and in salt deserts is discussed together because (1) genera and even species are shared between the two habitats and (2) physiognomy of the plants in both habitats is similar (Chapman, 1960). Although seed germination has been studied in a number of halophytes, many species have not been investigated. Seed dormancy in those families with halophytes for which no studies have been done can be inferred from what is known about it in other members of the family (Table 11.12). A list of species occurring in salt marshes and salt deserts throughout the world was compiled from numerous sources (Table 11.12). This list includes only species described as halophytes and/or those growing in the low portions of coastal marshes or in saline areas of deserts.

1. Types of Seed Dormancy

A few halophytes of salt marshes and salt deserts have nondormant seeds and therefore do not require any special pretreatments for a high percentages of them to germinate when placed on a moist substrate (Table 11.13). No halophytes of salt marshes and salt deserts have been reported to have seeds with morphological (MD) or morphophysiological dormancy (MPD), and

TABLE 11.12—*Continued*

Family	No. of genera (species)		Type(s) of dormancy known in family
	Marshes, deserts	Beaches, cliffs, and dunes	
Nolanaceae	0	1 (1)	?
Nyctaginaceae	2 (2)	3 (5)	Physiological
Onagraceae	0	1 (3)	Physiological
Papaveraceae	0	1 (1)	Morphophysiological
Parnassiaceae	0	1 (1)	Physiological
Poaceae	29 (66)	32 (51)	Physiological
Plantaginaceae	1 (4)	1 (4)	Physiological
Plumbaginaceae	4 (28)	3 (17)	Physiological
Polemoniaceae	1 (1)	0	Physiological
Polygalaceae	0	1 (1)	Physiological
Polygonaceae	2 (4)	2 (7)	Physiological
Portulacaceae	2 (2)	1 (2)	Physiological
Primulaceae	2 (3)	2 (3)	Physiological
Restionaceae	1 (1)	0	Physiological
Rosaceae	1 (1)	1 (1)	Physiological
Rubiaceae	1 (1)	4 (6)	Physiological
Scorphulariaceae	2 (2)	5 (5)	Physiological
Solanaceae	2 (4)	3 (5)	Physiological
Surianaceae	0	1 (1)	?
Tamaricaceae	2 (16)	0	Nondormant
Typhaceae	2 (2)	0	Nondormant, physiological
Tiliaceae	1 (1)	0	Physical–physiological
Urticaceae	0	1 (1)	Physiological
Valerianaceae	0	1 (1)	Physiological
Verbenaceae	1 (1)	1 (1)	Physiological
Zygophyllaceae	1 (4)	2 (4)	Physiological
Total	180 (465)	238 (409)	

[a]Compiled from: Shmueli, 1948; Chapman, 1960, 1977; Yamamoto, 1964; Lesko and Walker, 1969; Barbour and Davis, 1970; Flowers, 1972; Wiebe and Walter, 1972; Goodman, 1973; Reimold and Queen, 1974; Ramati et al, 1976; Albert and Popp, 1977; Danin, 1978; Okusanya, 1979; Groves, 1981; Pate and McComb, 1981; Rush and Epstein, 1981; Sen and Rajpurohit, 1982, West, 1983; Glenn and O'Leary, 1984; Doing, 1985; Woodell, 1985; Schat and Scholten, 1985; Evenari et al., 1986; Glenn, 1987; Dawson, 1988; Ellenberg, 1988; Ungar, 1991b; Coupland, 1992; Mariko et al., 1992; van der Maarel, 1993; Zomlefer, 1994.
[b]See tables in Chapters 9 and 10.

with the exception of a few species in the Apiaceae, it is unlikely that many will be found (see Table 11.12). Physical dormancy is reported in seeds of only a few halophytes belonging to the Convolvulaceae, Fabaceae, and Malvaceae (Table 11.13). It is expected that other halophytic members of these families, as well as the Tiliaceae, also would have physical dormancy. Most salt marsh and salt desert halophytes belong to families known to have physiological dormancy (PD) (Table 11.12); thus, it is not surprising that seeds of most species investigated thus far have this type.

2. Dormancy-Breaking Requirements

If seeds of halophytes in the Apiaceae have MPD and germinate in spring, cold stratification probably is a part of the dormancy-breaking treatment. However,

if seeds have MPD and germinate in autumn, they would require only warm stratification.

Environmental conditions required to break dormancy in halophyte seeds with physical dormancy have not been investigated. About 90% of the seeds of *Kosteletzkya virginica* germinated following 4 years of dry storage in darkness at 5°C (Poljakoff-Mayber et al., 1992). It is not known if dryness and/or the low temperatures caused the seed coats to become permeable.

It is difficult to make general statements concerning requirements to break PD in seeds of some halophytes because investigators did not test freshly matured seeds prior to storing them for 1 to many months either at 3–6°C or at room temperatures (Table 11.13). Thus, the amount of dormancy loss that occurred during periods of storage cannot be determined. The rate of dormancy loss has been determined for seeds of *Hordeum maritimum* tested at 10, 20, and 30°C after 15, 30, 45, 60, 90,

TABLE 11.13 Seed Dormancy in Halophytes of Salt Marshes and Salt Deserts

Species	Tr[a]	Days[b]	Temperature[c]	L:D[d]	Reference
Nondormant					
Atriplex polycarpa			9–15, 15	D > L	Sankary and Barbour (1972), Cornelius and Hylton (1969)
Baccharis halimifolia			18, 25/18	L	Panetta (1979)
Cotula coronopifolia			5, 30/15	—[e]	van der Toorn and ten Hove, (1982)
Halopeplis perfoliata			32/10	—	Mahmoud *et al.* (1983)
Juncus kraussii			14–30	—	Zedler *et al.* (1990)
Limonium axillare			35/11	—	Mahmoud *et al.* (1983)
Tamarix aphylla			18–28	—	Waisel (1960)
T. pentandra			19–41	—	Wilgus and Hamilton (1962)
Triglochin bulbosa			30/20	L > D	Naidoo and Naicker (1992)
T. stricta			30/20	L > D	Naidoo and Naicker (1992)
Physical dormancy					
Cressa cretica			10–20	—	Khan (1991)
Kosteletzkya virginica			28–30	L = D	Poljakoff-Mayber *et al.* (1992, 1994)
Melilotus indica			25/15	—	Maranon *et al.* (1989)
M. messanensis			25/15	—	Maranon *et al.* (1989)
M. segetalis			25/15	—	Maranon *et al.* (1989)
Prosopis farcta			15–40	—	Dafni and Negbi (1978)
Physiological dormancy					
Althenia filiformis	W	135	20	—	Onnis and Pelosini (1976)
Artemisia fukudo	—[h,i]	—	25/20	—	Mariko *et al.* (1992)
A. maritima	C[l]	30	25	—	Bakker and de Vries (1992)
A. triangularis	—[h,l]	—	30/20	—	Khan and Ungar (1984a)
Atriplex canescens	—[f]	—	13–24	L = D	Springfield (1970)
A. dimorphostegia	—[g]	—	28	D > L	Koller (1957)
A. griffithii	—[h,i]	—	25/10	—	Khan and Rizvi (1994)
A. gmelinii	—[h,i]	—	25/20	—	Mariko *et al.* (1992)
A. inflata	—[g]	—	5–25, 15	L = D	Beadle (1952), Uchiyama (1987)
A. nummularia	—[g]	—	5–25	L = D	Beadle (1952)
A. patula	C[l]	30	27/16	—	Shumway and Bertness (1992)
A. prostrata	C[l]	30	25/15	—	Bakker and de Vries (1992)
A. semibaccata	—[g]	—	5–20	L = D	Beadle (1952)
A. spongiosa	—[g]	—	10–25	L = D	Beadle (1952)
A. vesicaria	—[g]	—	5–25	L = D	Beadle (1952)
Aster tripolium	C[l]	30	15/10, 27/16	—	Mariko *et al.* (1992), Shumway and Bertness (1992)
Carex lyngbyei	C	100	17/8	—	Hutchinson and Smythe (1986)
Ceratoides lanata	—[i,j]	—	30/18	—	Workman and West (1967)
Cochlearia danica	—[h]	—	10	—	Bakker and De Vries (1992)
Diplachne fusca	W	840	31/11	—	Morgan and Myers (1989)
Distichlis spicata	C[l]	30	27/4, 27/16	—	Amen *et al.* (1970), Shumway and Bertness (1992)
Erianthus ravennae	—[k]	—	30/20	—	Lombardi *et al.* (1991)
Glaux maritima	—[h,i]	—	25/20	—	Mariko *et al.* (1992)
Haloxylon salicornicum	—[i,k]	—	30	L = D	Kaul and Shankar (1988)
Hordeum jubatum	—[h,l]	—	25/15	—	Ungar (1974), Badger and Ungar (1989)
H. maritimum	W	60	10	—	Onnis and Lombardi (1994)
Iva annua	—[h,i]	—	15	L = D	Ungar and Hogan (1970)
I. frutescens	C[l]	30	27/16	—	Shumway and Bertness (1992)
Juncus gerardii	C[l]	30	27/16	—	Shumway and Bertness (1992)
Kochia americana	—[i,k]	—	30/19	—	Clark and West (1969)
K. littorea	—[h,i]	—	25/20	—	Mariko *et al.* (1992)
Lepilaena cylindrocarpa	C	Winter	20	—	Vollebergh and Congdon (1986)
Limonium nashii	C[l]	30	27/16	—	Shumway and Bertness (1992)
L. vulgare	C[l]	30	30, 35/25	—	Bakker *et al.* (1985)
Phragmites communis	—[h,i]	—	35/20	—	Mariko *et al.* (1992)
Plantago coronopus	C[l]	30	20, 25/15	—	Bakker *et al.* (1985)
P. maritima	C[l]	30	25	—	Bakker and de Vries (1992)

(continues)

TABLE 11.13—*Continued*

Species	Tr[a]	Days[b]	Temperature[c]	L:D[d]	Reference
Puccinellia ciliata	—[m]	—	17–24	—	Myers and Couper (1989)
P. festucaeformis	W	90	10	—	Onnis and Miceli (1975)
P. nuttalliana	—[h,i]	—	20/5	L = D	Macke and Ungar (1971)
Ruppia megacarpa	W	Summer	Fall	—	Brock (1982)
R. polycarpa	C	Winter	20	—	Vollebergh and Congdon (1986)
R. tuberosa	W	Summer	Fall	—	Brock (1982)
Salicornia bigelovii	—[h,i]	—	5, 16	—	Rivers and Weber (1971)
S. brachystachya	—[h,i]	—	25/15	—	Huiskes *et al.* (1985)
S. dolichostachya	—[h,i]	—	25/10	—	Huiskes *et al.* (1985)
S. emerici	C[i]	40	25/12	—	Grouzis (1973)
S. europaea	C[i]	7–30	25, 15/5, 35/25	—	Ungar (1977), Philipupillai and Ungar (1984), Bakker *et al.* (1985)
S. fruticosa	C[i]	40	25/12	—	Grouzis (1973)
S. pacifica	—[h,i]	—	15/5	—	Khan and Weber (1986)
S. patula	C[i,l]	25	23	L	Berger (1985)
Sarcobatus vermiculatus	—[k]	—	10	—	Romo and Eddleman (1985)
Scirpus americanus	—[h,l]	—	35/20	—	Palmisano (1971)
S. olneyi	—[h,i]	—	35/20	—	Palmisano (1971)
S. robustus	C	120	30/20	L	Dietert and Shontz (1978)
Sesuvium portulacastrum	—[h,l]	—	35/20	—	Palmisano (1971)
Solidago sempervirens	C[i]	30	27/16	—	Shumway and Bertness (1992)
Spartina anglica	C[i]	60	20	—	Marks and Truscott (1985)
S. alterniflora	C[i,n]	40	26, 35/18	—	Mooring *et al.* (1971), Plyer and Carrick (1993)
S. foliosa	C	63	35/20	—	Seneca (1974)
S. patens	C[i]	30	27/16	—	Shumway and Bertness (1992)
Spergularia marina	C[h,i]	21	15/5	L, L > D	Okusanya and Ungar (1983), Ungar (1984, 1991a)
S. maritima	C[i]	30	25, 30/20	—	Bakker *et al.* (1985)
S. media	C	30	15	L = D	Ungar and Binet (1975)
S. salina	C[i]	30	15/5	—	Bakker *et al.* (1985)
Sporobolus arabicus	W	—[o]	35	L = D	Sheikh and Mahmood (1986)
Suaeda depressa	C	21	25/3	—	Ungar and Capilupo (1969)
S. fruticosa	—[p]	—	25	—	Sheikh and Mahmood (1986)
S. linearis	C[i]	30	27/16	—	Shumway and Bertness (1992)
S. maritima	C[i]	30	25/15	—	Bakker *et al.* (1985)
Triglochin maritima	C[e,g]	30	20–30	L > D	Bakker *et al.* (1985), Davy and Bishop (1991), Binet (1959)
Zygophyllum dumosum	—[i,k]	—	10–25	—	Agami (1986)
Z. qatarense	W	Summer[q]	22/10	—	Ismail and El-Ghazaly (1990)
Zannichellia palustris	C	30	20	L > D	Lombardi *et al.* (1996)
Z. pedunculata	C	60	16–20	L > D	van Vierssen (1982a)

[a]Dormancy-breaking treatment: W, warm stratification (or dry storage at room temperatures); C, cold stratification.

[b]Length of the dormancy-breaking treatment.

[c]Optimum germination temperature (°C) or test temperature that resulted in a high germination percentage.

[d]Light or dark requirements for germination. See note for Table 11.5.

[e]Data not available.

[f]Seeds afterripened during 10 months of dry storage, presumably at room temperatures.

[g]Seeds germinated when fruit bracts were removed.

[h]Seeds stored dry at 3–6°C for 1 or more months.

[i]Freshly matured seeds were not tested for germination. Presumably, seeds were dormant, but some could have been nondormant.

[j]Seeds were stored dry at room temperatures for 60 days.

[k]Seeds stored dry at 20°C or room temperatures for 1 or more months.

[l]Cold stratification increased germination at high salinity.

[m]Age of seeds and storage conditions not given.

[n]Dormancy was broken by surgically altering the scutellum.

[o]Seeds were scarified to overcome the need for an afterripening period.

[p]Acid scarification increased percentages and temperature range of germination.

[q]Seeds buried in soil in field in Qatar.

105, and 120 days of dry storage at 18–21°C; they were nondormant after 60 days (Onnis and Lombardi, 1994).

A further complication in our understanding the dormancy-breaking requirements of seeds is that some investigators have not tested them after a long period of dry storage. Further, at the end of the storage period, seeds were given a cold stratification treatment of 20–60 days (Table 11.13), and they germinated to high percentages. Were the fresh seeds dormant? Would short periods of cold stratification have broken dormancy in freshly matured seeds? Were seeds dormant at the end of the period of dry storage?

In some studies (e.g., Seneca, 1974; Ungar and Binet, 1975) where fresh seeds have been tested and then given cold stratification, only 30–60 days of cold stratification were required to break dormancy. These results suggest that seeds have nondeep PD. Further, the high germination percentages of seeds stored dry and then given relatively short periods of cold stratification indicate that dormancy is nondeep and easily broken. Another indication of nondeep PD is that seed treatments such as scarification (Malcolm, 1964; Joshi and Iyengar, 1977; Ungar, 1991b) or exposure to GA (Boucard and Ungar, 1973; Ungar, 1984; Khan and Ungar, 1985) resulted in germination. These data indicate that the embryo is physiologically dormant and thus has a low growth potential. However, after PD is broken in nature during cold or warm stratification, the embryo has enough growth potential to push through the covering layers. If seeds had intermediate or deep PD, warm stratification would not break dormancy, and dormant seeds would require exposure to long periods of cold stratification to come out of dormancy (see Chapter 3).

Cold stratification breaks dormancy in most halophyte seeds with PD that have been studied thus far, or at least it does not cause them to enter secondary dormancy (Table 11.13). Usually, seeds are cold stratified in fresh water; therefore, we do not know the maximum salinity at which seeds can respond to cold stratification. Much research is needed on the effects of salinity on dormancy loss during cold stratification before we truly understand the germination ecology of halophytes. In regions with a winter-wet, summer-dry climate, however, seeds of halophytes come out of dormancy during summer and germinate in autumn (e.g., Kingsbury et al., 1976). Seeds of autumn-germinating halophytes would be expected to require exposure to high summer temperatures to come out of dormancy. In fact, low winter temperatures might cause nondormant seeds of these species to enter secondary dormancy. Because seeds that come out of dormancy during warm stratification can do so during dry storage (Chapter 3), high salinity in summer may have little effect on dormancy loss.

3. Germination Requirements

Optimum temperatures for the germination of nondormant (either initially nondormant or after dormancy is broken) seeds of salt marsh and salt desert halophytes (Table 11.13) range from 5°C in *Salicornia bigelovii* (Rivers and Weber, 1971) to 35/25°C in *Limonium vulgare* and *Salicornia europaea* (Bakker et al., 1985). The mean optimum germination temperature is about 21°C. Alternating temperature regimes stimulate the germination of *Suaeda depressa* (Williams and Ungar, 1972) and *Spartina alterniflora* (Mooring et al., 1971) seeds.

Very little is known about changes in the temperature requirements for the germination of halophyte seeds as they come out of dormancy. Many halophytes belong to families, including the Amaranthaceae, Apiaceae, Asclepiadaceae, Asteraceae, Chenopodiaceae, Cyperaceae, Euphorbiaceae, Juncaceae, Poaceae, Polygonaceae, Portulacaceae, Rosaceae, and Scrophulariaceae, whose seeds come out of PD during cold stratification, and as this happens the minimum temperature for germination decreases (Baskin and Baskin, 1988; Baskin et al., 1993). Thus, it is likely that seeds of many halophytes would exhibit a decrease in the minimum temperature for germination as dormancy is broken. Also, the optimum temperature for germination could be lowered. The optimum temperature for the germination of fresh seeds of *Distichlis spicata* was 40/10°C (Cluff et al., 1983), but after 4 weeks of cold stratification it was 27/4°C (Amen et al., 1970) or 27/16°C (Shumway and Bertness, 1992). In seeds of halophytes that germinate in autumn, it is expected that dormancy is broken during exposure to high temperatures and that the maximum temperature for germination increases. However, no data are available for these species.

Light:dark requirements for germination have been determined for 23 species (Table 11.13). Seeds of 4 species require light for germination, 4 germinate to higher percentages in light than in darkness, 13 germinate equally well in light and darkness and 2 germinate to higher percentages in darkness than in light. Sankary and Barbour (1972) suggested that being covered by soil (and thus in darkness) would prevent desiccation of *Atriplex polycarpa* seeds and enhance germination. Whereas seeds of *Spergularia media* cold stratified for 10 days germinated to higher percentages in light than in darkness, those receiving 30 days of cold stratification germinated equally well in light and darkness (Ungar and Binet, 1975). A high far-red:red ratio inhibited the germination of *Baccharis halimifolia* seeds at a constant temperature regime but not at an alternating temperature regime (Panetta, 1979).

Salinity of the soil water solution is a major factor in the germination of halophytes, and there is much

TABLE 11.14   NaCl Concentration at which Germination of Seeds of Salt Marsh
and Salt Desert Species Was Reduced from 75–100% to about 10%

Species	NaCl concentration (M)	Reference
Aster tripolium	0.60	Woodell (1985)
	0.43	Bakker et al. (1985)
	0.34	Chapman (1960)
Atriplex canescens	0.21	Springfield (1970)
A. griffithii	0.35	Khan and Rizvi (1994)
A. halimus	0.34	Zid and Boukhris (1977)
A. patula	0.34	Ungar (1996)
A. prostrata	0.26	Patridge and Wilson (1987)
A. polycarpa	0.26	Chatterton and McKell (1969)
A. triangularis	0.51	Khan and Ungar (1984b)
Ceratoides lanata	0.17	Clark and West (1971)
	0.34	Workman and West (1967)
Cochlearia danica	0.43	Bakker et al. (1985)
Cotula coronopifolia	0.34	Partridge and Wilson (1987)
Cressa cretica	0.85	Khan (1991)
Distichlis spicata	0.09	Cluff et al. (1983)
Glaux maritima	0.09	Rozema, 1975
Halopeplis amplexicaulis	0.50	Tremblin and Binet (1982)
Hordeum jubatum	0.17, 0.31	Ungar (1974), Badger and Ungar (1989)
Iva annua	0.13, 0.17	Ungar (1967b), Ungar and Hogan (1970)
Juncus maritimus	0.17	Chapman (1960)
Kochia americana	1.02	Clark and West (1969)
Limonium bellidifolium	0.60	Woodell (1985)
L. vulgare	0.60	Woodell (1985)
Melaleuca ericifolia	0.17	Ladiges et al. (1981)
Melilotus indica	0.05	Maranon et al. (1989)
M. messanensis	>0.20	Maranon et al. (1989)
M. segetalis	0.20	Maranon et al. (1989)
Mesembryanthemum australe	0.34	MacKay and Chapman (1954)
Myrica cerifera	0.17	Young et al. (1994)
Phragmites communis	0.34	Chapman (1960)
Plagianthus divaricatus	0.09	Partridge and Wilson (1987)
Plantago coronopus	0.26	Partridge and Wilson (1987)
Polypogon monspeliensis	0.34	Partridge and Wilson (1987)
Prosopis farcta	0.60	Dafni and Negbi (1978)
	0.34	Bazzaz (1973)
Puccinellia distans	0.45	Harivandi et al. (1982)
P. festucaeformis	0.75	Onnis and Miceli (1975)

(continues)

variation among species in the ability of seeds to germinate under saline conditions. The NaCl concentration that reduces germination to about 10% (Table 11.14) ranges from 0.09 (0.5%) to 1.7 M (10%) with a mean of 0.36 ± 0.03 M (2.12 ± 0.18%). In general, seeds of halophytes germinate to higher percentages in distilled water than in NaCl solutions, and even slight increases in the salt concentration can decrease germination (Chapman, 1942; Binet, 1965; Clark and West, 1971; Ungar, 1974; Boucaud and Ungar, 1976; Zid and Boukhris, 1977; Tremblin and Binet, 1982; Partridge and Wilson, 1987; Khan and Rizvi, 1994; Khan and Ungar, 1996). However, in a few species, including Atriplex undulata

(Mahmood and Malik, 1986), Desmostachya bipinnata (Mahmood et al., 1996), Suaeda depressa (Ungar, 1962), S. fruticosa (Jhamb and Sen, 1984), Salsola baryosma (Mohammed and Sen, 1990), Salicornia bigelovii (Rivers and Weber, 1971), S. brachiata (Joshi and Iyengar, 1982), Ruppia tuberosa (Brock, 1982), Carex lyngbyei (Hutchinson and Smythe, 1986), and Limonium axillare (Mahmoud et al., 1983), seeds germinate to higher percentages in low NaCl concentrations (0.25-0.5%) than in distilled water. The NaCl enhancement of germination can vary with temperature and light:dark conditions (Binet, 1965).

Seed age, size/morphology, and season of maturation

TABLE 11.14—*Continued*

Species	NaCl concentration (*M*)	Reference
P. lemmoni	0.45	Harivandi *et al.* (1982)
Rumex crispus	0.43	Bakker *et al.* (1985)
Salicornia bigelovii	0.05	Rivers and Weber (1971)
	0.06	Troyo-Dieguez and Solis-Camara (1992)
S. brachystachya	0.24[a]	Huiskes *et al.* (1985)
S. dolichostachya	0.24[a]	Huiskes *et al.* (1985)
S. emerici	0.26	Grouzis (1973)
S. europaea	0.85	Ungar (1962, 1967a), Philipupillai and Ungar (1984)
S. herbacea	1.70	Chapman (1960)
S. pacifica	0.68	Khan and Weber (1986)
S. patula	0.34	Berger (1985)
Salsola kali	0.60	Woodell (1985)
Sarcobatus vermiculatus	0.06	Romo and Eddleman (1985)
	0.09–0.15	Romo and Haferkamp (1987a)
Schoenus nitens	0.17	Partridge and Wilson (1987)
Scirpus americanus	0.09	Palmisano (1971)
S. olneyi	0.09	Palmisano (1971)
S. robustus	0.20,[a] 0.17	Palmisano (1971), Dietert and Shontz (1978)
Selliera radicans	0.09	Partridge and Wilson (1987)
Sesuvium portulacastrum	0.20	Palmisano (1971)
	>0.60	Martinez *et al.* (1992)
Spartina alterniflora	0.68	Mooring *et al.* (1971)
Sporobolus virginicus	0.26	Breen *et al.* (1977)
	>0.60	Martinez *et al.* (1992)
Spergularia marina	0.17	Ungar (1962, 1991)
	0.34	Partridge and Wilson (1987)
S. salina	0.09	Bakker *et al.* (1985)
Suaeda depressa	0.85	Ungar (1962)
S. japonica	0.90	Yokoishi and Tanimoto (1994)
S. linearis	0.17	Ungar (1962)
S. nudiflora	0.26	Joshi and Iyengar (1977)
Tamarix pentandra	0.85[a]	Ungar (1967)
Triglochin striatum	0.09	Partridge and Wilson (1987)
	0.17	Bakker *et al.* (1985)
Zannichellia pedunculata	0.13	van Vierssen (1982a)
Zygophyllum dumosa	1.51	Agami (1986)
Z. qatarense	0.07	Ismail (1990)

[a]Highest concentration of NaCl tested; seeds potentially could germinate to 10% at an even higher concentration than this.

can affect salt tolerance. Whereas seeds of *Tamarix aphylla,* which are nondormant at maturity, germinated to about 45% at 1% (0.17 *M*) NaCl at the time of dispersal, they failed to germinate at this salt concentration after 10 days of storage at room temperature. Seeds germinated to 75% in distilled water after 10 days of storage (Waisel, 1960). In dimorphic seeds of *Atriplex triangularis* (Khan and Ungar, 1984a,b) and *Salicornia europaea* (Philipupillai and Ungar, 1984), large seeds are more tolerant of increased NaCl levels than small ones. The large (central in the inflorescence) seeds of *S. patula* are more tolerant of salinity than the small (lateral in the inflorescence) ones (Berger, 1985). Seeds of *Spergularia marina* that matured in late summer and autumn (August–October) germinated to higher percentages at increased salinity [0.35 *M* (2.09%), 0.40 *M*

(2.30%)] than those that matured in early and midsummer (June–July) or late autumn (November) (Okusanya and Ungar, 1983).

With loss of dormancy, the ability of seeds to germinate at increased levels of salinity increases. Seeds of *Salicornia europaea* cold-stratified for 4 weeks germinated to 82 and 43% at 3% (0.52 *M*) and 5% (0.85 *M*) NaCl, respectively, whereas nonstratified seeds germinated to 4 and 0%, respectively (Philipupillai and Ungar, 1984). Cold stratification also greatly increased germination percentages of *Spergularia marina* seeds in 1.0% (0.17 *M*) and 1.5% (2.6 *M*) NaCl (Ungar, 1984) and lateral seeds of *S. patula* in 170 (0.9%) and 340 (1.9%) m*M* NaCl (Berger, 1985). Seeds of *Puccinellia festucaeformis* afterripened for 20 months germinated to higher percentages at 0.06 *M* (0.3%) to 0.5 *M* (2.8%) NaCl than those afterripened for 8 months (Onnis and Miceli, 1975).

GA, which can substitute for cold stratification in seeds with nondeep PD (Chapter 3), increases the ability of seeds to germinate at increased levels of salinity (Ungar and Binet, 1975; Boucaud and Ungar, 1976; Ungar, 1977, 1984; Khan and Weber, 1986; Khan and Rizvi, 1994; Ozturk *et al.*, 1993, 1994). Kinetin also promotes the germination of seeds exposed to salt stress (Bozcuk, 1981; Khan and Ungar, 1985; Ozturk *et al.*, 1994).

The ability of seeds to germinate at increased levels of salinity is partly dependent on the test temperature (Ungar, 1978). In a number of species, including *Althenia filiformis* (Onnis and Mazzanti, 1971), *Arthrocnemum halocnemoides* (Malcolm, 1964), *Atriplex griffithii* (Khan and Rizvi, 1994), *A. nummularia* (Uchiyama, 1987), *Beta vulgaris* (Francois and Goodin, 1972), *Halopeplis perfoliata*, (Mahoud *et al.*, 1983), *Haloxylon recurvum* (Khan and Ungar, 1996), *Hordeum jubatum* (Badger and Ungar, 1989), *Iva annua* (Ungar and Hogan, 1970), *Limonium axillare* (Mahmoud *et al.*, 1983), *Puccinellia festucaeformis* (Onnis *et al.*, 1981), *Sarcobatus vermiculatus* (Romo and Eddleman, 1985), *Salicornia brachystachya* (Huiskes *et al.*, 1985), *Salicornia pacifica* var. *utahensis* (Khan and Weber, 1986), *Spergularia marina* (Ungar, 1984), and *Zannichellia pedunculata* (van Vierssen, 1982a), germination percentages of seeds incubated at increased salinity levels decreased with an increase in temperatures. Seeds of *Salicornia europaea* germinated to higher percentages at 15/5°C than at 25/5 or 25/15°C (Philipupillai and Ungar, 1984), but they germinated to higher percentages at 32°C than at 13 or 21°C (Ungar, 1967a). Both lateral and central seeds of *Salicornia patula* germinated to higher percentages at 23 and 28°C than at 13, 18, 33, or 25/12°C with increased salinity (Berger, 1985). Seeds of *Salicornia bigelovii* incubated on filter paper moistened with a 8.08% (1.39 *M*) sea salt in distilled water solution germinated to 63 and 0% at 15.5 and 26.6°C, respectively (Rivers and Weber, 1971). Thus, if (1) cold stratification during winter breaks dormancy and increases seed tolerance to salinity and (2) low temperatures promote germination of seeds exposed to saline conditions, it is expected that seeds of many halophytes would germinate in early spring while temperatures are quite low.

Precipitation plays an important role in the germination of halophytes because it decreases salinity of the soil solution by diluting and/or leaching of salts. In some subtropical regions, rains, followed by decreases in salinity, occur in summer (Rajpurohit and Sen, 1979) and seeds of halophytes germinate. Monsoon rains occur in Pakistan in July and August, temperatures drop 10–15°C, and seeds of *Atriplex griffithii* germinate (Khan and Rizvi, 1994). Rainfall causes a decrease in

salinity during the cool season in most saline habitats in temperate regions, and seeds germinate in late autumn, winter, or early spring (Beadle, 1952; Ward, 1967; Williams, 1979; Dietert and Shontz, 1978; Tremblin and Binet, 1982; Bulow-Olsen, 1983; Brock, 1982; Flowers *et al.*, 1986; Hutchinson and Smythe, 1986; SaadEddin and Doddema, 1986; Vollebergh and Congdon, 1986; Romo and Haferkamp, 1987a; Wertis and Ungar, 1986; Cluff and Roundy, 1988; Ismail, 1990; Ismail and El-Ghazaly, 1990; Telenius, 1993). In coastal salt marshes, where seeds are subjected to daily tidal influxes of seawater, there also can be decreases in salinity during the cool season (Phleger and Bradshaw, 1966; Chapman, 1976).

Not only is the soil of salt marshes and salt deserts too saline at certain times of the year to permit germination, but fruits (Hocking, 1982), bracts (Twitchell, 1955; Beadle, 1952; Koller, 1957), bracteoles around the seed (Uchiyama, 1987), and seed coats (Khan *et al.*, 1985) may contain concentrations of NaCl that are high enough to inhibit germination. NaCl is removed by leaching during rains, and seeds germinate while soil salinity is relatively low. Bracteole leachates of *Atriplex nummularia* inhibit germination, but the compound(s) and/or ion(s) have not been identified (Campbell and Matthewson, 1992).

Whereas nondormant seeds of glycophytes may die when exposed to high salinity (Partridge and Wilson, 1987), those of halophytes do not (SaadEddin and Doddema, 1986; Mahmoud *et al.*, 1983). Nondormant seeds of many halophytes held at high salinities will germinate to high percentages when they subsequently are transferred to fresh water (MacKay and Chapman, 1954; Ungar, 1962, 1991b; Boorman, 1968; Ungar and Capilupo, 1969; Macke and Ungar, 1971; Williams and Ungar, 1972; Ungar and Binet, 1975; Breen *et al.*, 1977; Jhamb and Sen, 1984; Partridge and Wilson, 1987; Uchiyama, 1987; Joshi and Misra, 1990; Zedler *et al.*, 1990; Davy and Bishop, 1991; Naidoo and Naicker, 1992; Garcia-Tiburcio and Troyo-Dieguez, 1993; Khan and Ungar, 1996, 1997; Keiffer and Ungar, 1995, 1997).

However, NaCl pretreatments do not always enhance germination (Partridge and Wilson, 1987; Khan and Ungar, 1997). Seeds of various halophytes were held at 3.0% (0.52 *M*), 5.0% (0.85 *M*) or 10% (1.73 *M*) NaCl solutions for 30, 60, 90, 365, and 730 days and then placed in distilled water (Keiffer and Ungar, 1997). All *Hordeum jubatum* seeds at 10% NaCl for >365 days failed to germinate, and 5 and 10% NaCl for 730 days reduced the germination of *Spergularia marina* seeds.

In a number of halophytes, experiments using various kinds of salts, including NaCl, Na_2SO_4, $NaHCO_3$, and ethylene glycol, polyethylene glycol, and mannitol to obtain a range of osmotic potentials have shown that there is only an osmotic inhibition of germination. That is, the salts were not toxic, and they prevented germination due to osmotic stress (Macke and Ungar, 1971; Ungar and Hogan, 1970; Romo and Haferkamp, 1987b; Ungar and Capilupo, 1969; Myers and Couper, 1989; Myers and Morgan, 1989). Increased osmotic stress could inhibit germination by (1) preventing seeds from becoming fully imbibed (Uhvits, 1946; Boorman, 1968), (2) decreasing the mobilization of food reserves (Filho et al., 1983), (3) delaying activation and/or synthesis of ribonuclease activity (Filho et al., 1983), and/or (4) depressing glutamate dehydrogenase synthesis (Boucaud and Billard, 1978).

However, some salts inhibit germination more than others at the same level of osmotic stress (Ungar, 1978), i.e., there is an ionic toxicity effect. $MgCl_2$ and KCl reduced the germination of Sporobolus airoides seeds more than NaCl and $CaCl_2$ (Hyder and Yasmin, 1972), and $CaCl_2$ and KCl reduced the germination of Salsola baryosma seeds more than NaCl (Mohammed and Sen, 1990). NaCl reduced the germination of Sesuvium sesuvioides seeds more than Na_2SO_4, whereas $MgSO_4$ reduced it more than Na_2SO_4 (Mohammed and Sen, 1990). Sulfate ions reduced the germination of Diplachne fusca (Myers and Morgan, 1989) and Sarcobatus vermiculatus (Romo and Eddleman, 1985) seeds more than chloride ions, and chloride salts inhibited the germination of Ceratoides lanata seeds more than did sulfate salts (Clark and West, 1971). The germination of Haloxylon salicornicum seeds was higher in NaCl and $CaCl_2$ than in $MgSO_4$ or $NaHCO_3$, and $NaHCO_3$ was more inhibitory than $MgSO_4$ (Kaul and Shankar, 1988). At the same level of osmotic stress, NaCl, KCl, $CaCl_2$, $MgCl_2$, Na_2SO_4, KNO_3, and $NaNO_3$ reduced the germination of Puccinellia festucaeformis seeds incubated at 30/20°C from 62–85% to 2–18%, whereas $MgSO_4$ reduced germination only from 76 to 70% (Onnis et al., 1981). Seeds of Suaeda japonica germinated to lower percentages in NaCl and KCl than in sodium gluconate or potassium gluconate, indicating a sensitivity to chloride ions (Yokoishi and Tanimoto, 1994).

Salts often are present in mixtures in natural habitats (e.g., Flowers, 1934); consequently, there could be osmotic and/or ionic toxicity effects of salts on seed germination (Mohammed and Sen, 1990; Jhamb and Sen, 1984). Further, the presence of one salt may reduce the inhibitory effects (whether they are osmotic or toxic) of another salt. In the glycophyte Securigera securidaca, $MgSO_4$ promotes the germination of seeds treated with NaCl or Na_2SO_4, and $MgCl_2$ promotes the germination of seeds treated with NaCl or $CaCl_2$ (Al-Jibury et al., 1986; Al-Jibury and Clor, 1986). Thus, the presence of Mg^{2+} in natural salt mixtures might allow seeds of halophytes to germinate at higher NaCl concentrations in

the field than they do in "pure" NaCl solutions in the laboratory.

Low levels of heavy metals such as mercury, cadmium (Mrozek, 1980), and lead (Mrozek and Funicelli, 1982) stimulated the germination of seeds of the halophyte Spartina alterniflora. However, germination rates and percentages, as well as long-term seed viability, declined with an increase in metal concentration, especially if salinity also increased.

Seeds of halophytes are sensitive to the amount of soil water present, even in the absence of salts. A decrease in soil moisture from 50 to 25% of water holding capacity (WHC) reduced the germination of various species: Desmostachya bipinnata (61% germination at 50% WHC was reduced to 28% at 25% WHC), Kochia indica (88 to 1%), Polypogon monspeliensis (25 to 1%), Sporobolus arabicus (18 to 6%) and Suaeda fruticosa (40 to 6%) (Mahmood et al., 1996).

Seeds also may be exposed to high salinity during flooding, and flooding per se could influence germination. Seeds of Sporobolus virginicus failed to germinate when submerged in 0%, 1.0% (0.17 M), 2.0% (0.34 M), and 3.0% (0.52 M) NaCl (Breen et al., 1977). However, flooding in 0% NaCl had no inhibitory effects on the germination of Juncus bufonius, J. maritimus, or J. alpino-articulatus, and it stimulated the germination of J. gerardii (Rozema, 1975). Increases in salinity decreased germination, but seeds of the four Juncus species were more tolerant of high salinity in flooded than in nonflooded conditions (Rozema, 1975). In the absence of O_2, seeds of Limonium vulgare germinated to 18 and 0% in freshwater and seawater, respectively, and at 20% O_2 they germinated to 75 and 20%, respectively (Boorman, 1968). Coleoptile, but not root, emergence occurred in Spartina alterniflora seeds at 0% O_2, whereas neither coleoptile nor root emergence occurred in Phragmites australis seeds at 0% O_2 (Wijte and Gallagher, 1996). Coleoptile and root emerged from seeds of both species at 2.5% O_2. In 10% CO_2, the germination of S. alterniflora and P. australis seeds was inhibited at 60 g (1.03 M) and 40 g (0.68 M) NaCl/liter, respectively.

An important question concerning the germination ecology of halophytes is what happens to seeds that fail to germinate the first, second, third, or later germination season after dispersal? Do these seeds exhibit annual dormancy/nondormancy or conditional dormancy/nondormancy cycles, or do they remain nondormant? Seeds of some halophytes form persistent seed banks (Ungar, 1987a), but very little is known about their dormancy states while they are in the seed bank. Detailed studies on the changes (or lack thereof) in temperature and light:dark requirements for the germination of buried seeds exposed to seasonal temperature cycles are needed badly.

Seeds of *Baccharis halimifolia* stored in the field on the soil surface or buried 5 cm below it and exhumed and tested at 25/17.5°C after 3, 6, 9, and 12 months germinated to 96–99% (Panetta, 1979), indicating that seeds either remained nondormant or only entered conditional dormancy. Soil samples collected from a population site of *Atriplex triangularis* in June, which was after the spring germination season and before seed dispersal, contained 923 seeds m^{-2}, which did not germinate until after they were cold stratified (Wertis and Ungar, 1986). These data indicate that seeds failing to germinate in spring had entered secondary dormancy. Thus, there is a good possibility that seeds of *A. triangularis* have an annual nondormancy/dormancy cycle.

Seeds of *Atriplex dimorphostegia* stored dry (at room temperatures?) and tested for germination in darkness at 20 and 26°C at 30-day intervals for 510 days germinated to the lowest percentages in July (summer) and to the highest percentages between November and April (Koller, 1957), indicating that seeds have an endogenous dormancy cycle. An annual rhythm also occurs in seeds of *Mesembryanthemum nodiflorum* stored dry at room temperature, with germination percentage highest in December (winter) and April and lowest in June and September (Gutterman, 1980–1981).

The germination ecology of some species seems to vary among populations. Field observations along the shore of Hudson Bay in Canada showed that while most seeds of *Salicornia europaea* germinate in June, some germination occurs throughout the summer (Jefferies *et al.*, 1983). In England, however, seeds of this species germinate only in spring and early summer, apparently exhausting the seed bank each year (Jefferies *et al.*, 1981). In Ohio, seeds of *S. europaea* germinate from February through June (Ungar *et al.*, 1979), and the species forms a small persistent seed bank (Ungar, 1987b).

B. Beaches, Cliffs, and Foredunes

As the distance from the sea increases, salinity decreases (Randall, 1970), e.g., salinity of salt marsh, cliff, and fore-dune soils in Japan was 0.20 (1.15%), 0.09 (0.52%), and 0.04 *M* (0.23%), respectively (Mariko *et al.*, 1992). To avoid the possible inclusion of glycophytes, this discussion covers only species growing on beaches, at the bases of cliffs, and on the lowermost (0.5–1.0 m) portion of the seaward side of foredunes. In these habitats, plants must be tolerant of high soil salinity, as well as salt spray, and thus they are presumed to be halophytes.

1. Types of Seed Dormancy

To aid in the literature search for data on the germination ecology of beach, cliff, and foredune species, a list of plant species growing in these habitats was prepared. The family and genus to which each species belongs are listed in Table 11.12, along with the type(s) of seed dormancy known to occur in the family. Thus, the type of seed dormancy in species that have not been studied can be inferred from that in other members of the family for which it is known. Surprisingly, germination data are available for only about 7% of the species growing on beaches, cliffs, and foredunes; thus, much research remains to be done on this group of plants.

Seeds of a few species germinate to high percentages at the time of maturity and thus are considered to be nondormant (Table 11.15). Seeds of various species, including *Inula crithmoides, Spergularia rupicola, Matthiola triscuspidata, Allium staticiforme,* and *Brassica tourneforti,* listed under PD in Table 11.15 were not tested until after long periods of dry storage; consequently, there is a possibility that some of them were dormant at maturity.

Seeds of *Pancratium maritimum* (Amaryllidaceae) also germinated to near 100% without any pretreatments (Keren and Evenari, 1974). Therefore, we assume that seeds of *P. maritimum* have MD because members of this family have underdeveloped linear embryos (Martin, 1946).

The six species listed under morphophysiological dormancy in Table 11.15 belong to the Apiaceae, except *Glaucium flavum,* which is a member of the Papaveraceae. It is expected that other halophytic members of the Apiaceae and halophytic members of the Araliaceae, Hydrophyllaceae, Iridaceae, and Liliaceae have either MD or MPD. Data are too limited to determine the type of MPD for any of the five species. In *G. flavum,* however, 80% or more of the freshly matured seeds germinated in darkness at 5, 10, and 15°C but to less than 40% at 20, 25, or 30°C (Thanos *et al.,* 1989). Germination percentages and maximum temperatures for germination increased after 20 days of cold stratification, and seeds germinated to 50% or more at 15–35°C (not tested at 5 or 10°C). These data indicate that some freshly matured seeds had only MD, whereas others had MPD, with the physiological part of the MPD being nondeep. Thus, the seed collection of *G. flavum* was a mixture of seeds with MD and MPD, as it was in the glycophyte *Conium maculatum* (Baskin and Baskin, 1990).

Germination studies show that seeds of the halophytes *Calystegia soldanella* (Convolvulaceae), *Lathyrus japonicus, L. maritimus* (Fabaceae), and *Lavatera arborea* (Malvaceae) have impermeable seed coats and thus physical dormancy (Table 11.15). However, numer-

TABLE 11.15 Seed Dormancy In Halophytes of Beaches, Sea Cliffs, and Foredunes

Species	Tr[a]	Days[b]	Temperature[c]	L:D[d]	Reference
Nondormant					
Centaurium littorale			18/8, 25/15	L	Schat (1983)
Messerschmidia argentea			Room?[e]	—[f]	Lesko and Walker (1969)
Rumex crispus			25/15	L = D	Walmsley and Davy (1997)
Salsola kali			30/10	—	Ignacuik and Lee (1980)
Scaevola taccada			Room[e]	—	Lesko and Walker (1969)
Morphological					
Pancratium maritimum			20	D > L	Keren and Evenari (1974)
Morphophysiological					
Angelica japonica	C	14	15/10	—	Mariko *et al.* (1992)
Crithmum maritimum	W[g]	120	15/5, 20	L	Okusanya (1977, 1979), Marchioni-Ortu and Bocchieri (1984)
Daucus carota	C[g]	—[h]	15, 25/15	L > D	Okusanya (1979)
Eryngium maritimum	C	122	20/10	L > D	Walmsley and Davy (1997)
Glaucium flavum	C	20	15	L > D	Thanos *et al.* (1989)
Ligusticum scoticum	C[g]	180	15	D > L	Okusanya (1979)
Physical					
Calystegia soldanella	S		25/10, 25	—	Yamamoto (1964), Mariko *et al.* (1992)
Lathyrus japonicus	S		25/15	L = D	Walmsley and Davy (1997)
L. maritimus	S		25	—	Yamamoto (1964)
Lavatera arborea	S		Room?	L > D	Okusanya (1979)
Physiological					
Allium staticiforme	W[g]	—[i]	20/13	D > L	Thanos *et al.* (1991)
Amaranthus pumilus	C	84	30/15	L	Baskin and Baskin (unpublished results)
Atriplex glabriuscula	C	—[e]	30/10	—	Ignacuik and Lee (1980)
A. laciniata	C	—[e]	30/20	—	Ignacuik and Lee (1980)
A. leucophylla	—[g]	—[j]	25	—	de Jong and Barbour (1979)
Brassica tournefortii	W[g]	—[i]	20/13	D > L	Thanos *et al.* (1991), Delipetrou *et al.* (1993)
Cakile edentula	C[g]	120?[k]	20/10	D > L	Maun and Payne (1989), Adair *et al.* (1990)
C. maritima	—[g]	—[i]	26	D > L	Barbour (1970), Thanos *et al.* (1991)
Honckenya peploides	C	122	15/5	L > D	Walmsley and Davy (1997)
Inula crithmoides	C[g]	—[h]	15	L	Okusanya (1979)
Matthiola tricuspidata	W[g]	—[i]	5–20	D	Thanos *et al.* (1994)
Parnassia palustris	C	75	15/5, 25/5	L	Schat (1983)
Samolus valerandi	C	75	15/5, 25/5	L	Schat (1983)
Spergularia rupicola	—[g]	—[h]	10/25	L	Okusanya (1979)
Spinifex hirsutus	W	—[i]	25/20	D > L	Harty and McDonald (1972)
S. sericeus	—[g]	—[m]	35/15	D > L	Maze and Whalley (1992)

[a]Dormancy-breaking treatment: W, warm stratification; C, cold stratification.
[b]Length of the dormancy-breaking treatment.
[c]Optimum germination temperatures (°C) or test temperature that resulted in a high germination percentage.
[d]Light or dark requirements for germination. See note for Table 11.5.
[e]Seeds stored dry at 5°C for an unspecified period of time.
[f]Data not available.
[g]Fresh seeds were not tested; presumably, they were dormant.
[h]Seeds stored dry at 18 and 5°C for an unspecified period of time.
[i]Seeds were stored dry at 20–21°C at room temperatures for 1 to several months.
[j]Seeds stored dry at room temperatures for about 18 months.
[k]Seeds were cold stratified outdoors in London, Ontario, Canada.
[l]Seeds stored dry, presumably at room temperatures for 12 or more months.
[m]Seeds stored dry at room temperatures for 6 months.

ous additional genera and species in these families (Table 11.12) probably have physical dormancy.

Although the majority of species growing on beaches, at bases of cliffs, and on foredunes belong to families known to have PD (Table 11.12), germination studies have been done on seeds of only a few of them (Table 11.15). Loss of dormancy during dry storage (Table 11.15) and GA stimulation of germination of seeds of some species, including *Honckenya peploides* (Tirmizi, 1988) and *Cakile maritima* (Ignaciuk and Lee, 1980), indicate that nondeep PD may be common among halophytes. However, GA had no effect on germination of *Atriplex glabriuscula* or *A. laciniata* seeds (Ignaciuk and Lee, 1980).

2. Dormancy-Breaking Requirements

Seeds with MD do not germinate until the embryo has grown to a critical length, which varies with the species (Chapter 5). Darkness and temperatures of 15–25°C are optimum conditions for the germination of *Pancratium maritimum* seeds, with high percentages of germination occurring within 1–2 weeks after the start of imbibition (Keren and Evenari, 1974). Although embryo growth and germination take place under these conditions, the critical embryo length for germination or the effects of salinity on embryo growth are not known.

No detailed studies have been done on the requirements for dormancy break and growth of embryos in seeds of halophytes with MPD. Cold stratification promoted germination in seeds of *Ligusticum scoticum* (Okusanya, 1979), *Glaucium flavum* (Thanos et al., 1989), and *Angelica japonica* (Mariko et al., 1992), whereas dry storage at room temperatures promoted germination in those of *Crithmum maritimum* (Marchioni-Ortu and Bocchieri, 1984). Seeds of *Eryngium maritimum* germinated to a maximum of only 40% in light at 20/10°C after 16–19 weeks of cold stratification (Walmsley and Davy, 1997). The effects of warm plus cold stratification need to be determined on seed dormancy break in this species.

The natural environmental conditions required to break dormancy in halophyte seeds with physical dormancy have not been determined. In view of the high amplitude of daily temperature fluctuations on beaches and dunes in spring and/or summer (e.g., Maun, 1981), effects of temperature fluctuations on breaking dormancy need to be investigated.

Although seeds of halophytes with PD may come out of dormancy in dry storage, caution should be used in extrapolating data to the field situation, especially if seeds germinate in spring or summer in the habitat. There is a chance that seeds given cold stratification, either in the laboratory or during winter in the field in temperate regions, would have a wider temperature range for germination (Thanos et al., 1989) and also be more tolerant of soil salinity (Marchioni-Ortu and Bacchieri, 1984) than those allowed to afterripen in dry storage.

Some beach, cliff, and foredune species in temperate regions may germinate in autumn, in which case seeds come out of dormancy during summer. Thus, keeping seeds such as those of *Matthiola triscuspidata,* an annual Brassicaceae that grows along the shores of the Mediterranean Sea, in dry storage at 20°C (Thanos et al., 1994)

somewhat simulates what happens in nature. Species growing on tropical or subtropical shores also may require exposure to high temperatures for dormancy loss to occur, e.g., *Spinifex hirsutus* (Harty and McDonald, 1972). If seeds germinate in spring, they may come out of dormancy during cold stratification in winter. Thus, information on germination phenology would be very helpful in planning germination experiments of seeds of beach, cliff, and foredune species.

3. Germination Requirements

The mean optimum germination temperature for seeds of species that are nondormant at maturity (Table 11.15) is 19.5±1.2°C (mean±SE), 21.9 ± 1.3°C for scarified seeds of species with physical dormancy, 15.3 ± 1.3°C for those with MPD, and 17.7 ± 1.3°C for those with PD. These temperatures indicate that seeds with MPD could germinate earlier in spring in temperate regions than those initially lacking dormancy or those with physical or physiological dormancy that become nondormant. Cold stratification increased the temperature range for the germination of *Glaucium flavum* (Thanos et al., 1989) and *Samolus valerandi* (Schat, 1983) seeds. Seeds of *Honckenya peploides* (Tirmizi, 1988), *Atriplex glabriuscula, A. laciniata, Salsola kali* (Ignaciuk and Lee, 1980), *Crithmum maritimum* (Okusanya, 1977), and *Spinifex hirsutus* (Harty and McDonald, 1972) germinated to higher percentages at alternating temperature than at constant temperatures.

Light:dark requirements for germination have been determined for 23 species (Table 11.15). Seeds of 7 species require light for germination, 5 germinate to higher percentages in light than in darkness, 2 germinate equally well in light and darkness, 8 germinate to higher percentages in darkness than in light, and 1 requires darkness. Seeds of *Glaucium flavum* imbibed in darkness at 25°C entered secondary dormancy, which was broken by a red light pulse or by cold stratification (Thanos et al., 1989). Further, nonstratified seeds of *G. flavum* germinated to 80% or more in darkness at 5, 10, and 15°C and in light at 20°C. Cold-stratified seeds of *G. flavum* germinated equally well in light and darkness at 25/15 and 15/5°C, but they germinated to a higher percentage in darkness than in light at 20/10°C (Walmsley and Davy, 1997). Burial at a soil depth of 16 cm significantly reduced the germination of *Atriplex laciniata* and *Salsola kali* seeds compared to those at 1, 4, and 8 cm; however, maximum germination at 1 cm did not exceed 80% in either species (Lee and Ignaciuk, 1985). Seeds were not sown on the soil surface, so it is unknown how much darkness vs depth per se contributed to the inhibition of germination. Seedlings of *Spinifex sericeus* emerged when seeds were buried at a depth of 12.5 cm (Maze and Whalley, 1992).

Photoinhibition of germination prevented seeds of *Allium staticiforme, Brassica tournefortii, Cakile maritima,* and *Otanthus maritimus* from germinating on the soil/sand surface (Thanos *et al.,* 1991), whereas those covered by soil/sand (and thus in darkness) could germinate (Thanos *et al.,* 1994; Keren and Evenari, 1974). If seeds germinate under sand, roots of the seedlings would be protected from desiccation, thus chances for successful establishment are increased. However, a light requirement for germination means that seeds would have to be on (or very near) the soil surface before they could germinate; thus, there is a chance that seedlings on sandy beaches or foredunes would be killed by drought before they became established. It is expected that light-requiring- seeds germinate at a time when habitat temperatures and moisture stress are relatively low. The mean optimum germination temperature for light-requiring and light-favored seeds of the species in Table 11.15 is 16.3 ± 1.0°C and that of dark-requiring and dark-favored seeds is 18.8 ± 1.5°C. More data on temperature and light:dark requirements for germination are needed to better understand the ecology of species growing in beach and foredune habitats.

The level of salinity required to reduce the seed germination of beach, cliff, and foredune species to about 10% (Table 11.16) ranges from 0.06 (0.35%) to 0.6 (3.53%) *M* with a mean (±SE) of 0.33 ± 0.03 (1.94 ± 0.18%) *M*; the mean for salt marsh and salt desert species is 0.36 ± 0.3 *M*. These values give only a little support to Woodell's (1985) conclusion that species growing in the most saline habitats germinate at higher salinities than those from less saline habitats.

The ability of seeds to germinate under saline conditions is partly a function of temperature. Seeds of *Crithmum maritimum* germinated to higher percentages at 20/10 than at 10 or 20°C at 0.12 (0.69%) *M* NaCl (Bocchieri and Marchioni-Ortu, 1984). Seeds of *Atriplex glabriuscula* and *A. laciniata* exposed to 0.24 (1.38%) *M* NaCl germinated at alternating but not at constant temperatures, and they germinated to higher percentages at 30/10 than at 30/20°C (Ignaciuk and Lee, 1980). At 40% seawater (= 0.24 *M* NaCl), seeds of *Daucus carota* ssp. *gummifer* and *Spergularia rupicola* germinated to higher percentages at 15°C than at 10 or 25°C (Okusanya, 1979). Another ecological aspect of the response of seeds to salinity is that their sensitivity may vary throughout the year. For example, seeds of *Crithmum maritimum* were less sensitive to 0.06 (0.35%) *M* NaCl in winter than in spring and summer (Marchioni-Ortu and Bocchieri, 1984).

Germination percentages of seeds of several species, including *Crithmum maritimum* (Marchioni-Ortu and Bocchieri, 1984), *Cakile maritima* (Barbour, 1970), *Honckenya peploides* (Tirmizi, 1988), *Centaurium littorale, Samolus valerandi, Parnassia palustris* (Schat, 1983), *Pancratium maritimum* (Keren and Evenari, 1974), *Scaevola taccada,* and *Messerschmidia argentea* (Lesko and Walker, 1969), increased when they were transferred from saline to nonsaline conditions. Because low temperatures and decreased salinity (as would occur

TABLE 11.16 NaCl Concentration at which Germination of Seeds of Beach, Cliff, and Foredune Species Was Reduced from 75–100% to about 10%

Species	NaCl concentration (*M*)	Reference
Atriplex glabriuscula	0.24	Ignacuik and Lee (1980)
A. laciniata	0.60	Ignacuik and Lee (1980)
Cakile maritima	0.60, 0.34	Barbour (1970), Ignacuik and Lee (1980)
Centaurium littorale	0.25[a]	Schat and Scholten (1985)
Crambe maritima	0.30–0.60	Woodell (1985)
Crithmum maritimum	0.25	Marchioni-Ortu and Bocchieri (1984)
Daucus carota	0.24–0.30	Okusanya (1979)
Honckenya peploides	0.20	Tirmizi (1988)
Inula crithmoides	0.30–0.45	Okusanya (1979)
Limonium binervosum	0.30	Woodell (1985)
Messerschmidia argentea	0.12	Lesko and Walker (1969)
Pancratium maritimum	0.45[a]	Keren and Evenari (1974)
Parnassia palustris	0.06	Schat (1983)
Salsola kali	0.60	Ignacuik and Lee (1980)
Samolus valerandi	0.25	Schat and Scholten (1985)
Scaevola taccada	0.30	Lesko and Walker (1969)
Spergularia rupicola	0.30	Okusanya (1979)

[a]Highest concentration of NaCl tested; seeds potentially could germinate to 10% at an even higher concentration than this.

following winter precipitation) promote germination, it is not surprising that newly germinated seedlings of various species have been found in the field during the cool season: *Cakile maritima,* in winter (Barbour, 1970); *Glaucium flavum,* in winter (Thanos *et al.,* 1989); *Atriplex glabriuscula, A. laciniata,* and *Salsola kali,* in spring (Ignaciuk and Lee, 1980); *C. maritima,* in winter or spring (Ignaciuk and Lee, 1980); and *Elymus mollis* and *Honckenya peploides,* in spring (Houle, 1996).

C. Marine Angiosperms

Angiosperms living submerged in seawater are called euhalophytes and belong to the families Cymodoceaceae (5 genera: *Amphibolis, Cymodocea, Halodule, Syringodium, Thalassia*), Hydrocharitaceae (*Halophila, Enhalus*), Posidoniaceae (*Posidonia*), Ruppiaceae (*Ruppia*), Zannichelliaceae (*Althenia*), and Zosteraceae (*Phyllospadix, Zostera*) (Waisel, 1972). With the exception of the genera *Ruppia* and *Althenia,* which can grow in saline marshes, the 12 genera are restricted to marine habitats (Waisel, 1972). Because the six families of marine angiosperms are members of the Monocotyledoneae, plants are called seagrasses. The genera *Cymodocea, Enhalus, Halodule, Halophila, Syringodium,* and *Thalassia* are found in tropical seas; *Amphibolis, Phyllospadix,* and *Posidonia* are characteristic of temperate and cold waters; and *Zostera* occurs in cold, temperate, and tropical oceans (Luning and Asmus, 1991). *Ruppia* and *Althenia* occur in temperate and subtropical waters (Willis, 1966).

Seagrasses may grow on tidal flats, in the lower part of the littoral zone, or in tidal estuaries, but they primarily are found in the sublittoral zone (den Hartog, 1970); the mean salinity of seawater is 3.5% (0.61 *M*) (Smayda, 1983). Depending on the species, seagrasses occur to depths of 90 m, with the maximum depth being the point where·solar irradiance is about 11% of that at the surface (Duarte, 1991). In tropical and subtropical regions, temperature fluctuations between summer and winter are only 4–5°C in seagrass beds (e.g., Phillips *et al.,* 1981), whereas in the temperate zones winter temperatures may be 20°C lower than those of summer (Setchell, 1929; Phillips *et al.,* 1981).

1. Types of Seed Dormancy

Germination ecology has been studied in only a small number of marine angiosperms, and of these about half have seeds that are nondormant at maturity (Table 11.17). In the genera *Cymodocea* and *Halophila,* some species have dormant and others have nondormant seeds (Table 11.17).

Several seagrasses have seeds that are dormant at the time of maturation. Because these dormant seeds have large, fully developed embryos (Gibbs, 1902; Miki, 1933; Martin, 1946; Taylor, 1957), they would not have MD or MPD. Further, because the seed coats of dormant seeds are permeable to water (e.g., McMillan, 1988a,b; Lewis and Phillips, 1980; Buia and Mazzella, 1991), they would not have physical dormancy. This leaves PD as the explanation for the lack of germination in freshly matured seeds of seagrasses.

Harrison (1991) suggested that seeds of *Zostera marina* have both physiological and physical dormancy. Freshly matured seeds failed to germinate unless the seed coat was scarified; however, germination occurred after they were cold stratified without scarification. These data indicate that cold stratification increased the growth potential of the embryo enough to overcome the mechanical resistance of the seed coats. Thus, seeds have only PD; physical dormancy is not known to occur in the Zosteraceae (Table 3.5). In seeds of *Z. capricornii,* germination was promoted by scarification (Conacher *et al.,* 1994), indicating that the embryo had low growth potential and probably nondeep PD. Not only do seeds of *Z. marina* lack impermeable seed coats, but they also may be recalcitrant. None of the dried seeds of this species collected from a beach along the Sea of Cortez, Mexico, germinated at conditions promoting germination of those collected in the sea (McMillan, 1983a). After 25 days of drying at 20°C, <10% of the *Z. marina* seeds germinated, and none produced viable seedlings (Hootsman *et al.,* 1987). Seeds of *Z. hornemanniana* were killed by exposure to room temperatures for 60 min on a dry substrate (Tutin, 1938).

2. Dormancy-Breaking Requirements

Depending on the species, PD in seeds of seagrasses is broken during warm or cold stratification (Table 11.17). If warm stratification breaks dormancy, seeds germinate in autumn, and if cold stratification breaks it, seeds germinate in spring and/or early summer. Phenological observations reveal that seeds of *Zostera marina* germinate (1) in autumn and early winter in the Atlantic Ocean near New York (Churchill, 1983), Massachusetts (Addy, 1947), and the Sea of Cortez along western Sonora, Mexico (Phillips and Backman, 1983); (2) in winter (December to February) in the Pacific Ocean near Japan (Miki, 1933); (3) in spring in intertidal sites of the southwestern part of the Netherlands (Harrison, 1991, 1993) and in Puget Sound of the Pacific Ocean along the Washington coast (Phillips *et al.,* 1983); and (4) from autumn until early summer (September to May), with peaks occurring in autumn and spring, in the Atlantic Ocean, Chesapeake Bay, Virginia (Orth

TABLE 11.17 Seed Dormancy in Marine Angiosperms

Species	Tr[a]	Days[b]	Temperature[c]	L:D[d]	Reference
Nondormant					
Cymodocea ciliata			Room	—[e]	Isaac (1969)
C. rotundata			Spring, summer	—	McMillan *et al.* (1982)
Halophila decipiens			24	L	McMillan (1988a,b)
Posidonia oceanica			Room	—	Buia and Mazzella (1991)
Thalassia testudinum			Summer	—	Lewis and Phillips (1980)
Zannichellia obtusifolia			20	—	Grillas *et al.* (1991)
Physiological dormancy					
Cymodocea nodosa	W	Winter[f] 240	18–24, 12–25	—	Caye and Meinesz (1986), Pirc *et al.* (1986), Buia and Mazzella (1991)
Halophila engelmannii	C	Winter[g]	24–27	L	McMillan (1987, 1988c)
H. spinulosa	W	35	25, 30	—	Birch (1981)
Phyllospadix iwatensis	C	126	5–6	—	Kuo *et al.* (1990)
Ruppia maritima	C	Winter[h] 14	10–15, 23	—	Verhoeven (1979), Koch and Dawes (1991)
Syringodium filiforme	C	Winter[i]	Spring, 26	—	McMillan (1981)
Zostera marina	C	Winter[j,k]	Spring, 5–15, 30	L > D	Hootsman *et al.* (1987), Phillips *et al.* (1983), Harrison (1991, 1993)
	W	Summer[l–n]	Autumn, 10–13, 3, 9–16, 18–20		Churchill (1983), Orth and Moore (1983), Phillips and Backman (1983), McMillan (1983a)
Z. noltii	C	Winter[o]	Spring	—	Goubin and Loques (1991), Harrison (1993)

[a]Dormancy-breaking treatment: W, warm stratification; C, cold stratification.
[b]Length of dormancy-breaking treatment.
[c]Optimum germination temperatures (°C) or test temperature that resulted in a high germination percentage.
[d]Light or dark requirements for germination. See note for Table 11.5.
[e]Data not available.
[f]Bay of Golfe-Juan, Mediterranean Sea.
[g]Gulf of Mexico, Texas.
[h]The Netherlands.
[i]Key West, Florida.
[j]Intertidal zone in southwestern part of the Netherlands.
[k]Pacific Ocean, Puget Sound, Washington.
[l]Atlantic Ocean, Long Island, New York.
[m]Atlantic Ocean, Chesapeake Bay, Virginia.
[n]Sea of Cortez, Mexico.
[o]Mediterranean Sea, near Corsica.

and Moore, 1983). Thus, depending on geographical location, warm or cold stratification may be important in breaking dormancy in seeds of *Z. marina*. In the Pacific Ocean near Japan, young seedlings of *Z. marina* were collected in December–February; *Z. nana,* March–June; *Z. caespitosa,* December–February; *Z. caulescens,* February; *Phyllospadix japonicus,* December–January; and *P. iwatensis,* May–June (Miki, 1933). Thus, seeds germinated after they had been cold stratified.

Laboratory/field studies show that seeds of *Zostera marina* in The Netherlands (Harrison, 1991) and Prince Edward Island, Canada (Taylor, 1957), germinate to high percentages following cold stratification; however, data on rates of dormancy loss during cold stratification are not available. However, seeds of *Z. marina* from the Sea of Cortez held at 18–20°C (simulated water

temperatures during time of seed germination) gradually came out of dormancy and germinated at these temperatures (McMillan, 1983a). Thus, seeds from Mexico come out of dormancy during warm stratification. It would be interesting to take seeds of *Z. marina* from the Sea of Cortez and place them in seawater at actual or simulated summer temperatures (32°C in July) for various periods of time and then test them at 18–20°C. This kind of study would tell us how fast dormancy loss occurs during summer. Seeds of *Halodule wrightii* and *Syringodium filiforme* from Texas and the U.S. Virgin Islands held at 24–27°C germinated over a period of more than 3 years (McMillan, 1983b, 1991).

3. Germination Requirements

The limited amount of data show that seeds lacking dormancy at the time of maturation have a higher mean (± SE) optimum germination temperature (22.0 ± 1.6°C) than those with PD (17.3 ± 1.9 °C). Thus, seeds initially lacking dormancy could germinate in summer, but those with PD at maturity probably would not germinate until autumn or spring, after dormancy break occurs in summer and winter, respectively. Depending on the geographical location, temperatures in summer could be too high for germination (e.g., McMillan, 1983a), and those in winter could be too low (e.g., Kuo et al., 1990) for the germination of seeds after dormancy is broken.

Little is known about the light:dark requirements for germination in seagrasses. Seeds of *Halophila decipiens* (McMillan, 1988a,b) and *H. engelmannii* (McMillan, 1987, 1988c) require light for germination, and those of *Zostera marina* germinate to higher percentages in light than in darkness (Harrison, 1991).

Seeds of many seagrasses, including *Cymodocea nodosa* (Cayne and Meinesz, 1986; Cayne et al., 1992), *C. rotundata* (McMillan et al., 1982), *Halodule wrightii* (McMillan, 1981), *Ruppia maritima* (Koch and Seelinger, 1988; Koch and Dawes, 1991), *Zostera marina* (Phillips, 1971; Hootsman et al., 1987), *Z. noltii* (Loques et al., 1990), and *Z. capricorni* (Conacher et al., 1994), germinate to higher percentages in diluted seawater than in full-strength seawater. However, other studies on *Z. marina* show that seeds can germinate in undiluted seawater (Harrison, 1991; Phillips et al., 1983; Orth and Moore, 1983). The dormancy state of seeds at the time the test was done may be the reason why some *Z. marina* seeds germinated in undiluted seawater and others did not. Cold-stratified seeds of this species show an increase in ability to germinate at high salinities (Hootsman et al., 1987; Harrison, 1991). Thus, *Z. marina* seeds capable of germinating in seawater may be nondormant, whereas those requiring diluted seawater may be conditionally dormant.

If nondormant seeds of seagrasses required diluted seawater for germination, they could only germinate following influxes of fresh water. The logical time to expect a decrease in salinity of seas is when the adjacent land mass is receiving high levels of precipitation. Decreases in salinity can occur from autumn to spring in temperate regions (e.g., Goubin and Loques, 1991) and during the wet season in the tropics (e.g., John and Lawson, 1991). However, reductions in salinity may not always be the key to understanding the germination phenology of seagrasses. For example, McMillan and Soong (1989) observed a dramatic reduction in seed reserves at a population site of *Halophila decipiens* during the dry season in Panama. Was the decline in the number of seeds in the sediments due to germination? If so, germination occurred at a time when freshwater influxes from the adjacent land masses would have been low, and thus salinity should have been high.

Oxygen concentration is another environmental factor that may play a role in the germination ecology of seagrasses. Seeds of *Zostera marina* buried in sediments at low oxygen levels [−50 to −350 Eh(mV)] and flooded with seawater germinated in October (autumn), starting when water temperatures dropped to 15°C (Moore et al., 1993). Seeds in oxygenated water [400 Eh(mV)] in light germinated to a maximum of about 40% in October, with an additional 20–40% of the seeds germinating in winter. However, seeds in oxygenated water in darkness did not reach 40% germination until January, with an additional 15–20% of the seeds germinating in spring. Thus, the ability to germinate in oxygenated water on the surface of sediments depends partly on whether light reaches the seeds. Also, there is a possibility that oxygen becomes less inhibitory for germination as seeds of *Z. marina* come out of PD. After a 5-week period of warm stratification, during which dormancy loss apparently occurred, seeds of *Halophila spinulosa* germinated in oxygenated and deoxygenated seawater (Birch, 1981). Germination requirements for seeds of *Amphibolis antarctica* and *A. griffithii* are unknown; however, they are viviparous and germinate in winter while attached to the mother plants (Ducker et al., 1977). Thus, seeds of these two species germinate above the sediment in oxygenated waters.

D. Mangroves

Tropical evergreen forests also occur in intertidal areas, where the trees are partially submerged during high tides by water with a salt concentration of about 35 parts per thousand (3.5%) and a osmotic potential of −2.5 MPa (Walter, 1979). These plant communities are called "mangal," and the constituent species are called "mangroves" (Tomlinson, 1986). Walter (1979) classified mangrove communities as hydrobiomes and placed them in the vegetation zone with an equatorial diurnal climate (tropical rain forest) as well as in the zone with a summer-wet, winter-dry climate (tropical deciduous forests). Although mangal is best developed near the equator, it is found at 32°N in Bermuda and at 38°S in Australia (Walter, 1979).

Mangal is dominated by trees, and the major genera belong to the Arecaceae, Avicenniaceae, Combretaceae, Euphorbiaceae, Fabaceae, Malvaceae, Rhizophoraceae, and Sonneratiaceae (Tomlinson, 1986). The understory layer is poorly developed, but some shrubs, lianas, herbaceous species, and even epiphytes occur in mangrove forests (Soegiarto, 1984; Tomlinson, 1986). Although mangal is a type of tropical forest, mangroves are discussed under "halophytes" because the species grow in a high salt environment.

Mangrove trees exhibit varying degrees of tolerance to salinity and flooding, and thus different species grow in distinct zones with the most tolerant ones being next to the sea. However, in a reciprocal transplant study, propagules of *Avicennia, Laguncularia, Pelliciera,* and *Rhizophora* grew and survived in habitats of adults of other species as well as they did in habitats of adults of their own species (Rabinowitz, 1978a). Thus, Rabinowitz (1978a) suggested that zonation is controlled by "tidal sorting of the propagules according to size and by differential ability of propagules to establish in deep water." The relative competitive ability of a species also may be an important factor in determining the zone in which it grows (Walter, 1979).

1. Types of Seed Dormancy

Although the seed and seedling stages in the life cycle of some mangroves have been described (Tomlinson, 1986), little is known about the dormancy state of mangrove seeds between the time of maturation and germination. However, preliminary information on the type(s) of dormancy to be expected can be obtained by looking up each mangrove family in Table 9.1. Most of the families to which mangroves belong have some species with nondormant seeds and others with either physical or physiological dormancy (Table 11.18).

Members of the Arecaceae have underdeveloped embryos (Martin, 1946); thus, mangroves belonging to this family have either MD or MPD. Growth occurs in embryos of *Nypa fruticans* seeds prior to germination (Tomlinson, 1971), and seeds sown in soil in Bangladesh began germinating after only 4–6 days; germination was completed in 20–25 days (Siddiqi *et al.*, 1991). The rapid rate of germination in this study indicates that *N. fruticans* seeds have MD.

Nothing is known about the dormancy-breaking requirements in mangroves whose seeds have morphophysiological, physical, or physiological dormancy.

2. Viviparous Germination

Mangroves have been of great interest to biologists for decades because of viviparous germination in seeds of some species. Vivipary means that the seed germinates while it is still attached to the mother plant. Further, the seedling grows, emerges from the fruit, and is eventually dispersed. (However, see Chapter 10,VIII,c)

Studies on *Rhizophora mangle* in Puerto Rica show that an embryo germinates about 70 days after pollination, when the fruit is about 1.8 cm long (Sussex, 1975). Does this mean that some type of dormancy is being broken? During germination of *R. mangle* seeds, growth of the endosperm forces the micropyle open and essentially pushes the embryo to the outside of the integument (Juncosa, 1982). During subsequent seedling growth, the fused cotyledons enlarge, and the radicle breaks through the fruit wall (Fig. 11.14). According to Sussex (1975), the radicle emerges from the fruit wall when the fruit is about 2.5 cm in length and 100 days old. Cotyledons extend only a short distance from the edge of the fruit (Tomlinson, 1986), but the hypocotyl undergoes extensive elongation. *Rhizophora mangle* seedlings are about 23 cm long at the time of dispersal (Sussex, 1975); however, those of *R. mucronata* are 30–60 cm in length (Ridley, 1930). Cotyledons remain attached to the mother plant, and the seedling (dispersal unit) consists of a bud (plumule), hypocotyl, and radicle. The small plumule is protected by stipules of a pair of aborted leaves (Juncosa, 1982).

In addition to *Rhizophora*, vivipary occurs in three other genera of Rhizophoraceae, *Bruguiera, Ceriops,* and *Kandelia* (Table 11.18). Cotyledons in *Rhizophora* and *Ceriops* are fused into a tube-like structure, and *Bruguiera* seedlings have a tubular structure with three to five lobes at the tip (de Vogel, 1980). At the time of seedling dispersal, the fused cotyledons of *Rhizophora* and *Ceriops* remain attached to the fruit on the mother plant (de Vogel, 1980); however, in *Bruguiera* (Burger, 1972) and possibly in *Kandelia* (see drawing on page 359 in Tomlinson, 1986) the fruit and cotyledons are dispersed along with the seedling.

Seeds germinate in some mangroves before they are dispersed (Table 11.18), but the seedling does not emerge from the fruit prior to dispersal; this is called cryptovivipary (Tomlinson, 1986). In *Avicennia* spp., the fruit coat splits at or shortly after the time of dispersal, releasing an embryo with thick, fleshy cotyledons folded in opposite directions. In *Aegiceras* and *Aegialitis*, the radicle and hypocotyl, respectively, elongate in the fruit, and at dispersal the torpedo-shaped propagules are released. The plumule emerges from fruits of *Nypa* at about the time of dispersal, leaving a haustorium inside the seed that continues to absorb food from the endosperm. *Pelliciera* fruits consist of a single seed that has only fragments of the seed coat at maturity. Prior to dispersal, the hypocotyl extends into a space above the two large, fleshy cotyledons (Tomlinson, 1986).

The cryptoviviparous propagules (seeds) of *Avicennia marina* are recalcitrant. They have a moisture content of about 170% (dry mass basis) at the time of dispersal but lose viability when dehydrated to 110% (Berjak *et al.*, 1984). However, fully mature seeds could be stored in a dry air steam for up to 10 days before they began to die (Pammenter *et al.*, 1984), but premature seeds were intolerant of any dehydration (Farrant

TABLE 11.18   Families and Genera of Mangrove Trees and Shrubs[a]
and Type of Dormancy Known to Occur in the Family (Based on Information in Table 9.1)

Family	Genera	Tree/shrub	Type of seed dormancy known in the family[b]	Type of seed germination[c]
Anacardiaceae	Gluta	S or T	ND, PY	N, H[a,d]
Apocynaceae	Cerbera	T	ND, PD	N, H[a]
Arecaceae	Calamus	Rattan	MD, MPD	—
	Nypa	T		C[a]
	Oncosperma	T		—
	Phoenix	T		—
Avicenniaceae	Avicennia	T	ND	C[a]
Bignoniaceae	Amphitecna	T	ND	N[e]
	Dolichandrone			N, E[a]
Bombacaceae	Camptostemon	T	ND, PY	N, D[e]
Celastraceae	Cassine	T	ND	N, E[f]
Clusiaceae	Calophyllum	T	ND, PD	N, H[a]
Combretaceae	Conocarpus	T	ND, PD	N, E[a]
	Laguncularia	T		N, E[a]
	Lumnitzera	T		N, E[g]
	Terminalia	T		N, H[a]
Ebenaceae	Diospyros		ND, PD	N, E[e]
Euphorbiaceae	Excoecaria	T	ND, PD	N, E[g]
	Glochidion	S		N, E[e]
	Hippomane	T		N, E[e,h]
Fabaceae	Caesalpinia	S	ND, PY	N, H[e,i]
	Cynometra	T		N, E[a]
	Dalbergia	S		N, E[e,i]
	Derris	S		N, E[g,i]
	Inocarpus	T		N[e]
	Intsia	T		N, E[e]
	Mora	T		N, H[e,h]
	Pongamia	T		N[e]
Flacourtiaceae	Scolopia	T	ND, PD	N[e]
Goodeniaceae	Scaevola	S		N, E[i]
Lecythidaceae	Barringtonia	T	ND, PD	N, H[e]
Lythraceae	Pemphis	T	ND	N[e]
Malvaceae	Hibiscus	T	ND, PY	N, E[d]
	Thespesia	T		N, E[d]
	Pavonia	S		N, E?

(continues)

et al., 1993). Seeds of *A. marina* do not require additional water for germination after dispersal (Farrant *et al.,* 1992); therefore, they begin to elongate immediately following dispersal, regardless of the salinity of the habitat. Seeds of *A. marina* placed in soil moistened with 0, 10, 25, 50, 75, or 100% seawater "germinated" (split the pericarp) in 2, 4, 4, 5, 8, and 7 days, respectively (Downton, 1982). Why do they not "germinate" on the mother plant? Embryos of *Rhizophora mangle* maintained a water content of 80% while they were growing on the mother plant, but the water content declined to 50% just prior to dispersal (Sussex, 1975).

In a summary of the explanations for the prevalence of vivipary in mangrove species, Joshi (1933, 1934) discussed the following ideas. (1) When the earth had massive swamps, vivipary was an adaptation for seedling establishment. Vivipary persists in mangal today because these communities are relicts from the past. (2) Vivipary is an adaptation for dispersal and establishment of a species in a wet, saline habitat. (3) Vivipary allows seeds to germinate before they are exposed to salt water, which might kill them. In a more recent survey of the hypotheses to account for vivipary in mangroves, Tomlinson (1986) concluded that more research was needed on seeds and seedlings before an explanation becomes obvious.

3. Nonviviparous Mangroves

Seeds of most mangroves do not germinate until after they are dispersed, and seedlings are either epigeal or hypogeal (Table 11.18). However, little is known about

TABLE 11.18—*Continued*

Family	Genera	Tree/shrub	Type of seed dormancy known in the family[b]	Type of seed germination[c]
Melastomataceae	*Ochthocharis*	S	PD	N, E?
Meliaceae	*Xylocarpus*	T	PD, ND	N, H[g]
Myristicaceae	*Myristica*	T	MPD	N, H[i]
Myrsinaceae	*Aegiceras*	T		C[a]
	Ardisia	S		N, E[d,i]
	Myrsine	T		N, E?
Myrtaceae	*Osbornea*	T	ND, PD	N, E[a]
Pellicieraceae	*Pelliciera*	T		C[a]
Plumbaginaceae	*Aegialitis*	T		C[a]
Rhizophoraceae	*Bruguiera*	T	PD	V[a]
	Ceriops	T		V[a]
	Kandelia	T		V[a]
	Rhizophora	T		V[a]
Rubiaceae	*Scyphiphora*	T	PD, ND	N, E[a]
	Ixora	S		N, E[d,f]
	Rustia	S		N[d]
Rutaceae	*Merope*	S	ND, PD	N[d]
Sapindaceae	*Allophyllus*	S	ND, PY	N, E[d]
Sapotaceae	*Pouteria*	T	ND, PD	N, H[d,f]
Sonneratiaceae	*Sonneratia*	T		N, E[g]
Sterculiaceae	*Heritiera*	T	ND, PY	N, H[a]
Tiliaceae	*Brownlowia*	S	ND, PY	N, E[d]

[a]From Tomlinson (1986).

[b]ND, nondormant; MD, morphological dormancy; MPD, morphophysiological dormancy; PD, physiological dormancy; and PY, physical dormancy.

[c]V, vivipary; C, cryptovivipary; N, normal (i.e., germinates after dispersal); E, epigeal; H, hypogeal; D, Durian.

[d]From Burger (1972).

[e]From Ng (1978).

[f]From Duke (1965).

[g]From Bhosale and Mulik (1990).

[h]From Duke (1969).

[i]From de Vogel (1980).

the seed germination requirements of nonviviparous mangroves.

Successful seedling establishment of mangroves requires that conditions be suitable for the penetration of roots into the soil, i.e., the seed (or propagule) must be in contact with the soil and physicochemical factors have to be appropriate for root growth. However, physicochemical characteristics of the soil may be relatively unimportant in some species because the large seeds (or dispersal units) could contain enough water and food for germination and considerable root growth of the seedling. Examples of mangrove species with large seeds (or dispersal units) include *Barringtonia asiatica* (Lecythidaceae), 10–15 cm in diameter; *Calophyllum inophyllum* (Clusiaceae), 2–4 cm; *Cerbera manghas* (Apocynaceae), 3–4 cm; *Gluta velutina* (Anacardiaceae), 7 cm; *Heritiera littoralis* (Sterculiaceae), 5–6 cm; *Laguncularia racemosa* (Combretaceae), 2 cm; *Mora oleifera* (Fabaceae), 12 cm; and *Xylocarpus granatum* (Meliaceae), 6 cm (Tomlinson, 1986). Because large

seeds, such as those of *Cerbera, Heritiera,* and *Laguncularia,* can float (Tomlinson, 1986), one wonders if they can germinate while they are floating?

Not all mangroves have large seeds, e.g., Cassine viburnifolia (Celastraceae), 1 cm in diameter; *Dolichandrone spathacea* (Bignoniaceae), 1.5 cm; *Excoecaria agallocha* (Euphorbiaceae), 0.3 cm; *Osbornia octodonta* (Myrtaceae), 0.5–0.7 cm; and *Sonneratia caseolaris* (Sonneratiaceae), 0.3 cm (Tomlinson, 1986). Can seeds of the small-seeded species float and, if so, can they germinate while floating?

Although some seasonal variation in soil salinity occurs in mangal habitats in response to rainfall patterns, year-to-year fluctuations may be just as great as seasonal ones (Nazrul-Islam, 1993). Thus, it seems unlikely that the timing of germination of mangrove seeds is controlled by seasonal changes in salinity, but we do not know. Studies are needed on the effects of salinity and other environmental factors, including oxygen concentrations, on the germination of mangrove seeds. Also,

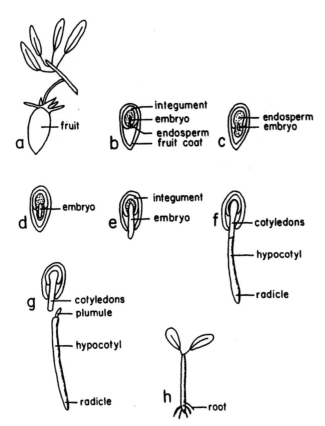

FIGURE 11.14 Germination in *Rhizophora mangle*. Based on information from Troup (1921), Sussex (1975), Juncosa (1982), and Tomlinson (1986).

it would be interesting to compare responses of large- and small-seeded species.

4. Buoyancy of Seeds and Propagules

An important characteristic of seeds and propagules of many mangroves is that they can float. Seeds have a variety of morphological adaptations that facilitate flotation, including fibrous fruit walls; thick, corky seed coats (Tomlinson, 1986); air spaces between seed and fruit wall; and short, corky "wings" (Ridley, 1930). Even the large seedlings of *Rhizophora* spp. float, and they have been seen floating in masses in the sea. Ridley (1930) reported that most *R. mangle* seedlings float horizontally, but some are vertical. Rabinowitz (1978b) found that *R. mangle* seedlings float horizontally following dispersal, after which they either float vertically, or they sink and then regain buoyancy and float horizontally. Crossland (1903) described the self-planting of vertically floating seedlings into soft mud as the tide fell. Lawrence (1949) observed that seedlings of *R. mangle* deposited in a horizonal position on a wet surface could assume an erect position due to rooting and upward bending of the hypocotyl.

Propagules of *Avicennia* were still floating after 80 days in fresh and salt water; however, they required a period of at least 5 days without flooding for establishment. Propagules of *Laguncularia, Rhizophora,* and *Pelliciera* floated for 23, 20–100, and 1 day(s), respectively, in fresh water and for 31, 20–100, and 6 days, respectively, in salt water before sinking. Roots were produced after 11, 40, and 9 days, respectively, in fresh water and after 16, 40, and 30 days, respectively, in salt water (Rabinowitz, 1978b).

VIII. PSAMMOPHYTES OF NONSALINE SOILS

Plants with an affinity for sandy habitats are called psammophytes, and they grow on beaches and dunes in coastal regions throughout the world (van der Maarel, 1993), near large bodies of fresh water (Maun, 1993), and on inland dunes in various deserts and in semiarid regions (Bowers, 1982). The subject of this discussion, however, is the germination ecology of psammophytes growing in nonsaline habitats. Thus, we have included only the psammophytes growing on coastal hinddunes, dunes near fresh water, and inland dunes that presently may not be associated with a body of water. Psammophytes growing on coastal beaches and foredunes are exposed to high soil salinity, as well as to salt spray (Salisbury, 1952; Lee, 1993); consequently, they were included in the section on halophytes (see Section VII). However, psammophytes growing on dunes behind the foredunes are subjected to very little salinity. Not only does salt spray decrease with an increase in distance from the sea, but rain water flushes salts from the dune sands (Kearney, 1904; Oosting and Billings, 1942; Etherington, 1967). Thus, plants growing on coastal hind dunes may be xerophytes; they are not halophytes (Kearney, 1904; Pignatti, 1993).

Species growing in dune slacks (low areas between ridges of dunes), on freshwater beaches, and in dune shrub and forest communities have not been included in this section. Because the water table in dune slacks can be at or above the soil surface for part or all of the year (Boorman, 1993; van der Meulen and van der Maarel, 1993), these species are covered under aquatics or halophytes, depending on salinity of the water. Species growing on low portions of beaches along the shores of freshwater lakes are covered under aquatics. Further, trees and shrubs that become established in advanced stages of dune succession are components of the regional (climax) vegetation; consequently, they are covered in Chapter 9 or 10.

Sand dunes occur throughout the world and thus in a diversity of climatic regimes. However, regardless of geographical location, dune habitats have certain characteristics in common that influence seed germination and seedling establishment of psammophytes: (1) movement of sand, (2) low soil moisture, (3) high insolation, and (4) nutrient deficiency (Purer, 1936; Moreno-Casasola, 1986; Pignatti, 1993; Maun, 1993; Stalter, 1993). The uniqueness of dune habitats is further emphasized by the presence of endemic taxa in most of the major dune areas of the world (e.g., Bowers, 1982; van der Maarel, 1993).

A. Types of Dormancy

Freshly matured seeds of a few psammophytes are reported to be nondormant (Table 11.19); however, detailed studies may show that they are conditionally dormant. Species whose seeds have conditional dormancy at maturity should be listed under physiological dormancy (PD).

No psammophytes are listed under morphological (MD) or morphophysiological (MPD) dormancy in Table 11.19. However, a survey of the dune species throughout the world (van der Maarel, 1993) shows that a number of them belong to families, including the Amaryllidaceae, Apiaceae, Arecaceae, Iridaceae, Liliaceae, and Santalaceae, whose seeds have underdeveloped embryos (Table 3.3). Thus, it seems reasonable that some psammophytes could have MD or MPD.

Physical dormancy has been studied in seeds of only a few psammophytes (Table 11.19). However, because dune species belong to various families, including the Cistaceae, Convolvulaceae, Fabaceae, Geraniaceae, Malvaceae, Sapindaceae, Sterculiaceae, and Tiliaceae, whose seeds have impermeable seed coats (Table 3.5), it is expected that further research will show that a number of other psammophytes have physical dormancy.

The majority of psammophytes studied thus far have seeds with PD (Table 11.19), and many additional species may have seeds with this type of dormancy. Because (1) PD is the only type of dormancy known to occur in the Aizoaceae, Asteraceae, Amaranthaceae, Asclepiadaceae, Boraginaceae, Brassicaceae, Caryophyllaceae, Chenopodiaceae, Cyperaceae, Euphorbiaceae, Nyctaginaceae, Poaceae, Polygonaceae, Rubiaceae, and Scrophulariaceae (Table 11.12) and (2) many dune species belong to these families (van der Maarel, 1993), we conclude that the list of psammophytes whose seeds have PD is very incomplete.

Various pieces of evidence suggest that seeds of many psammophytes have nondeep PD. (1) Seeds in over half the species listed in Table 11.19 come out of dormancy during warm stratification or dry storage at room temperatures. (2) Only relatively short periods of exposure to low temperatures are needed to promote germination in those species whose seeds require cold stratification for loss of dormancy. (3) Scarification promotes germination in seeds of a number of species, including *Agropyron psammophilum* (Zhang and Maun, 1989a), *Ammophila arenaria* (van der Putten, 1990), *Calandrinia arenaria* (Balboa, 1983), *Cakile edentula* (Gedge and Maun, 1992), *Corispermum hyssopifolium* (Gedge and Maun, 1992), *Croton punctatus* (van der Valk, 1974), *Leymus arenarius* (Greipsson and Davy, 1994), *Panicum virgatum* (Zhang and Maun, 1989b), and *Uniola paniculata* (Wagner, 1964; Westra and Loomis, 1966). (4) GA promotes the germination of *Calandrinia arenaria* seeds (Balboa, 1983).

B. Dormancy-Breaking Requirements

No studies have been done on dormancy-breaking requirements of seeds of psammophytes belonging to families known to have underdeveloped embryos. Also, little is known about how physical dormancy in seeds of psammophytes is broken in nature. Seeds of *Lupinus arboreus* germinated to 45% after 95 days of exposure to daily alternating temperature regimes of 32/4°C (Davidson and Barbour, 1977). Studies need to be done to determine if the great difference between day and night temperatures at the sand surface on dunes (Salisbury, 1952) plays a role in breaking physical dormancy.

Physiological dormancy is broken in seeds of psammophytes by relatively short periods of cold stratification or by exposure to high temperatures (Table 11.19). In most studies where PD has been broken by high temperatures, seeds have not been imbibed at simulated summer temperatures; instead, they have been stored dry at about 20–25°C. Considering the high temperatures and dryness of the surface layers of sand during summer, dry storage of seeds (even at 20–25°C) simulates natural conditions better than warm stratification per se.

Although the ecology of germination has been studied in only a small fraction of psammophytes whose seeds have PD, it is noteworthy that seeds of more species come out of dormancy in summer than in winter. Dormancy loss during exposure to high temperatures means seeds could germinate at the beginning of the cool season in autumn in temperate regions or at the beginning of the wet season in subtropical/tropical regions. Many of the psammophytes whose seeds come out of dormancy at high temperatures are winter annuals (Rozijn and van Andel, 1985; Pemadasa and Lovell, 1975), but some are perennials (Martinez *et al.*, 1992; Bell, 1993). Not only do seeds of winter annual psammophytes come out of dormancy during summer, but low

TABLE 11.19 Dormancy-Breaking and Germination Requirements
of Seeds of Psammophytes of Nonsaline Soils

Species	Tr[a]	D[b]	Temperature[c]	L:D[d]	Reference
Nondormant					
Agrostis vinealis			22/12	L = D	Pons (1989)
Amaranthus greggii			25/20	L = D	Martinez *et al.* (1992)
Corynephorus canescens			22/12	L = D	Pons (1989)
Oenothera glazioviana			15–30	L > D	Kachi (1990)
Senecio jacobaea			20	L	van der Meijden and van der Waals-Kooi (1979)
Trachypogon gouini			35	L = D	Martinez *et al.* (1992)
Physical dormancy					
Ipomoea stolonifera	S		35	—[e]	Martinez *et al.* (1992)
Lupinus arboreus	W[f]	95	32/4	—	Davidson and Barbour (1977)
L. littoralis	S		25/15	—	Kumler (1969)
Strophostyles helvola	S		15/10, 25/15	L = D	van der Valk (1974)
S. umbellata	S		25/20	—	Erickson and Young (1995)
Physiological dormancy					
Aira caryophyllea	W	56	10	·L = D	Pemadasa and Lovell (1975)
A. praecox	W	56	10	L = D	Pemadasa and Lovell (1975)
Ammophila arenaria	C	42	30/20	L > D	van der Putten (1990)
A. breviligulata	C	30	29/18	—	Seneca (1969)
Andropogon glomeratus	W	—[g]	32/20	L > D	Martinez *et al.* (1992)
Cakile edentula	C	Winter	Spring	—	Payne and Maun (1986)
Calamovilfa longifolia	C	56	25/10	—	Maun (1981)
Cerastium atrovirens	W	112	15	L = D	Pemadasa and Lovell (1975)
C. semidecandrum	W	42	5	—	Rozijn and van Andel (1985)
Coreopsis lanceolata	—[h]	300	25	L > D	Banovitz and Scheiner (1994)
Corispermum hyssopifolium	C	42[i]	35/20	—	van Asdall and Olmsted (1963)
Croton punctatus	C	30–60	30/20	D > L	van der Valk (1974)
Cynoglossum officinale	C	Winter	8.5/2.5	—	Freijsen *et al.* (1980)
Eragrostis trichodes	C	14	30/20	—	Ahring *et al.* (1963)
Erophila verna	W	112	15/5	L > D	Baskin and Baskin (1970), Pemadasa and Lovell (1975)

(continues)

temperatures during winter cause them to enter secondary dormancy. Seeds of *Spergula morisonii* buried in sand dunes in The Netherlands in June were nondormant by autumn (September), dormant by spring (March), and nondormant again by the following autumn (Pons, 1989). Dormancy loss during exposure to low temperatures means seeds could germinate at the beginning of the warm season in temperate and subarctic/arctic regions. Most of the psammophytes whose seeds come out of dormancy at low temperatures are perennials (Seneca, 1969; van der Valk, 1974; Maun, 1981; Greipsson and Davy, 1994), but some are summer annuals (van Asdall and Olmsted, 1963; Payne and Maun, 1984).

The close adaptation of psammophytes to the environmental conditions of their dune habitat is illustrated further by differences in the dormancy of seeds of the same species collected at various latitudes. Freshly matured seeds of *Ammophila breviligulata* collected from sand dunes in Michigan (45°30' N latitude) required cold stratification for germination at 29/18°C, whereas those collected in North Carolina (36°30" N latitude) did not (Seneca and Cooper, 1971). Freshly matured seeds of *Uniola paniculata* collected from various locations in Florida (about 28–30° N latitude) germinated to 53–97% at 29/18°C, whereas those collected in North Carolina germinated to only 6–26% (Seneca, 1972). Length of the cold stratification period required for high germination percentages of seeds of *U. paniculata* and *Iva imbricata* varied clinally. Seeds collected in North

TABLE 11.19—*Continued*

Species	Tr[a]	D[b]	Temperature[c]	L:D[d]	Reference
	W	42	5	—	Rozijn and van Andel (1985)
Iva imbricata	C	30–60	25/15	L > D	van der Valk (1974)
Leymus arenarius	C	14	30/20	D > L	Greipsson and Davy (1994)
Lithospermum caroliniense	C[j]	49	25/15	—	Westelaken and Maun (1985)
Mibora minima	W	112	15/5	L = D	Pemadasa and Lovell (1975)
Oenothera drummondii	W	180[k]	20–25	L > D	Bell (1993)
Palafoxia lindenii	W	—[g]	32/20	L = D	Martinez *et al.* (1992)
Panicum amarulum	C	30	29/18	—	Seneca (1969)
P. maximum	W	—[g]	35	L = D	Martinez *et al.* (1992)
P. virgatum	C	56	27/17	L > D	Zhang and Maun (1989b)
Pappophorum vaginatum	W	—[g]	35	L > D	Martinez *et al.* (1992)
Physalis viscosa	C	30–60	25/15	L > D	van der Valk (1974)
Saxifraga tridactylites	W	112	10	L = D	Pemadasa and Lovell (1975)
Solidago sempervirens	C	30–60	15/10	L > D	van der Valk (1974)
Spergula morisonii	W	90	22/12	L > D	Pons (1989)
Trachyandra divaricata	W	180[k]	20–25	D > L	Bell (1993)
Uniola paniculata	W[l]	120	35/18	—	Hester and Mendelssohn (1987)
	C	120	g.h.[m]	—	Wagner (1964)
	C	30	29/18	—	Seneca (1969)
Vulpia fasciculata	W	56	15–20	—	Watkinson (1978)
V. membranacea	W	112	23	L = D	Pemadasa and Lovell (1975)

[a]Dormancy-breaking treatment: C, cold stratification; W, warm stratification (or dry storage at about room temperatures).
[b]Duration (days) of dormancy-breaking treatment.
[c]Optimum germination temperature (°C) or test temperature that resulted in high germination percentages.
[d]Light or dark requirements for germination. See note for Table 11.5.
[e]Information not available.
[f]Alternating day and night temperatures of 32 and 4°C, respectively.
[g]Seeds stored dry at room temperatures for 7–14 months.
[h]Seeds afterripened during dry storage at room temperatures.
[i]Seeds stored dry at room temperatures for 4 months prior to cold stratification.
[j]Only 2% germination after 7 weeks of cold stratification; thus, a longer period of cold stratification may be needed (or a different test temperature).
[k]Seeds stored dry at room temperatures for 6 months.
[l]Fresh seeds were not tested; they were stored dry at room temperatures for 5 months before the study was initiated.
[m]Greenhouse.

Carolina required a long stratification period to come out of dormancy, whereas those collected in Florida required a short period (Colosi and Seneca, 1977).

C. Germination Requirements

1. Temperature

The maximum temperature at which germination is possible increases as seeds of winter annuals come out of dormancy during summer (Chapter 4; Pemadasa and Lovell, 1975). Germination is prevented in the habitat in summer, even if there are rains, because temperatures are above those required for germination. Seeds germinate in autumn because temperatures required for germination overlap with those in the habitat. In autumn, however, some seeds can germinate at relatively high temperatures, although the optimum germination temperature is quite low. Nondormant seeds of *Mibora min-*

ima germinated to about 70, 55, and 20% at 15/5, 20/10, and 25/15°C, respectively (Pemadasa and Lovell, 1975). Rains in early autumn would stimulate some seeds of *M. minima* to germinate while temperatures were about 25/15°C, which is above the optimum germination temperature for the species. The remainder of the seeds would not germinate until temperatures declined to about 20/10 or 15/5°C, which are closer to optimal for germination than is 25/15°C. The germination of psammophytes in desert dune areas is delayed until a sufficient amount of rainfall "triggers" germination, and in the Sonoran Desert of California and Sonora, Mexico, this may not happen until winter (Bowers, 1996).

Data on the temperature requirements for germination as seeds of dune summer annuals and perennials come out of dormancy are sparse. However, cold stratification appears to lower the temperature requirement for germination in seeds of *Ammophila breviligulata* (Seneca, 1969), *Corisperum hyssopifolium* (van Asdell and Olmsted, 1963), and *A. arenaria* (van der Putten, 1990). Further studies may show that some seeds of summer annual and perennial dune species can germinate in early spring, while temperatures are still below the optimum for germination of most seeds. Germination phenology studies in sand dunes (in Poland) showed that seeds of some species germinated mostly in either autumn or spring, but those of a few species, including *Cerastium semidecandrum, Plantago indica,* and *Veronica dilleni,* germinated about equally well in spring and autumn (Symonides, 1993).

Alternating (vs constant) temperatures promote the germination in seeds of *Ammophila arenaria* (van der Putten, 1990), *Leymus arenarius* (Greipsson and Davy, 1994), *Palafoxia lindenii,* and *Panicum maximum* (Martinez *et al.,* 1992).

2. Light

In the majority of psammophytes studied, seeds germinate to higher percentages in light than in darkness (Table 11.19). Seeds of *Croton punctatus* (van der Valk, 1974), *Leymus arenarius* (Greipsson and Davy, 1994), and *Trachyandra divaricata* (Bell, 1993) germinate to higher percentages in darkness than in light. Seeds of *Trachyandra divaricata* germinated to 0% at 570 nm (orange light), 0% at 640 nm (red light), about 10% at 430–490 nm (blue light), and 15% in darkness (Bell, 1993). The inhibition of germination by red light (in sunlight) would prevent *T. divaricata* seeds from germinating on the surface of the sand, where soil moisture conditions are unlikely to be favorable for successful seedling establishment (Bell, 1993; Greipsson and Davy, 1994).

The light sensitivity of *Vulpia membranacea* seeds depends on the temperature. Seeds germinated to about 95 and 85% at 5 and 23°C, respectively, in darkness, but they germinated to 90% at 23 and to 25% at 5°C at a daily 12/12 hr photoperiod (Pemadasa and Lovell, 1975). Perhaps light inhibition of germination at about 23°C has not been selected for because germination is delayed in the field by high temperatures until mid-October (Pemadasa and Lovell, 1975). By this time, temperatures probably are near 23°C and the surface layers of sand are sufficiently moist for seedling survival.

3. Moisture

Moisture and light:dark requirements for the germination of psammophyte seeds are related. The ability of seeds to germinate in darkness means they can germinate beneath the surface layer of sand, where moisture levels would be higher than they are at the surface. If seeds germinate on the surface, they may die from drought stress before the radicle grows into the damp, relatively deep layers of sand (e.g., Buckley, 1982).

Seeds of *Senecio jacobaea* require light for germination, but they also are sensitive to soil moisture. Seeds germinated to 8.5, 53.0, 85.5, 89.0, and 88.0% at soil moisture levels of 1.8, 3.6, 7.3, 14.5, and 29.1% in light at 20°C (van der Meijden and van der Waals-Kooi, 1979). Thus, germination is inhibited if seeds are in full light on the sand surface, where they are subjected to low moisture levels. These authors found that seeds on the sand surface germinated to only 18%, but those covered by 1 mm of sand germinated to 77%. The light level and quality under 1 mm of sand were not determined.

Another aspect of moisture relations of seeds is that drying at high temperatures may delay germination after moisture is no longer limiting for germination. Drying seeds of *Senecio jacobaea* at 40% RH for 2 or 6 days at 25–35°C delayed germination 14.4 days, whereas drying them at 15–25°C delayed germination only 4.1 days (van der Meijden and van der Waals-Kooi, 1979).

4. Depth

One of the consequences of sand movement on dunes is that seeds can become buried, sometimes to depths of 20–30 cm (van der Valk, 1974; Maun, 1981). Many psammophytes are tolerant of sand accretion, and a list of them has been compiled by Moreno-Casasola (1986).

Burial by sand could have a major impact on seed germination, especially on seedling emergence. In fact, depth of burial may have a greater effect on emergence than it has on germination per se (Pemadasa and Lovell, 1975; Watkinson, 1978). The maximum depth of seedling emergence has been determined for a number of psammophytes (Table 11.20). At least some seedlings of several species emerge from depths of ≥10 cm. However, 10 cm is the optimum depth for seedling emergence of *Strophostyles helvola.* Seedling emergence for permeable seeds of this legume sown at depths of 2, 6, 10, 12, 18, and 24 cm was 29, 83, 84, 72, 61, and 37%, respectively (Yanful and Maun, 1996).

The emergence of seedlings is related to the amount of food reserves available for growth in darkness. Consequently, seedlings from the heaviest seeds of a species are more likely to emerge from a given depth (Weller, 1985; Zhang and Maun, 1991) and to grow and develop faster (Zhang, 1996) than those from the lightest seeds. Interestingly, in 16 annual species on sand dunes in the United Kingdom, there was a significant negative relationship between seed weight and the relative abundance of a species in the community (Rees, 1995).

The ultimate in germination of buried seeds in sandy habitats may be the geocarpic seeds produced by *Alexgeorgea* spp. (Restionaceae) in Western Australia. Seeds range in size from about 162 to 640 mg, depending

TABLE 11.20 Depth of Emergence of Seedlings of Psammophytes
from Seeds Germinating under Sand

Species	Depth (cm)	Reference
Abronia maritima	8	Johnson (1985)
Acacia ligulata	3.8[a]	Buckley (1982)
Agropyron psammophilum	12	Zhang and Maun (1990a)
Aira caryophyllea	1	Pemadasa and Lovell (1975)
A. praecox	1	Pemadasa and Lovell (1975)
Amaranthus greggii	2[b]	Martinez *et al.* (1992)
Ammophila breviligulata	6	Maun and Lapierre (1986)
Andropogon glomeratus	2	Martinez *et al.* (1992)
Astragalus magdalenae	8	Bowers (1996)
Cakile edentula	12, 10	Maun and Lapierre (1986), Tyndall *et al.* (1986)
Calamovilfa longifolia	6, 8	Maun (1981), Maun and Riach (1981)
Calotis erinacea	2.28[a]	Buckley (1982)
Cerastium atrovirens	1	Pemadasa and Lovell (1975)
Chamaesyce wheeleri	2.44[a]	Buckley (1982)
Corispermum hyssopifolium	8	Maun and Lapierre (1986)
Crotalaria cunninghamii	3.93[a]	Buckley (1982)
Croton punctatus	11	van der Valk (1974)
Elymus canadensis	10	Maun and Lapierre (1986)
E. farctus	13	Harris and Davy (1986)
Helianthus niveus	5	Bowers (1996)
Ipomoea stolonifera	2[b]	Martinez *et al.* (1992)
Iva imbricata	4	van der Valk (1974)
Lithospermum caroliniense	3[b]	Weller (1985)
Mibora minima	1	Pemadasa and Lovell (1975)
Palafoxia arida	5	Bowers (1996)
P. lindenii	2[b]	Martinez *et al.* (1992)
Panicum maximum	2[b]	Martinez *et al.* (1992)
P. virgatum	16	Zhang and Maun (1990b)
Pappophorum vaginatum	2[b]	Martinez *et al.* (1992)
Physalis viscosa	3	van der Valk (1974)
Solidago sempervirens	1	van der Valk (1974)
Strophostyles helvola	24	Yanful and Maun (1996)
Trachypogon gouini	2[b]	Martinez *et al.* (1992)
Triplasis purpurea	6	Tyndall *et al.* (1986)
Uniola paniculata	4	Tyndall *et al.* (1986)
Vulpia fasciculata	3	Watkinson (1978)
V. membranacea	1	Pemadasa and Lovell (1975)

[a]Mean seed burial depth of successfully emerging seedlings in the field.
[b]Because germination exceeded 50% at this depth, it probably was not the maximum depth from which emergence was possible.

on the species, and they are produced 10–15 cm below the soil surface in subterranean inflorescences on female plants (Meney *et al.*, 1990). Seeds germinate *in situ*, usually in winter, after the rhizome to which they are attached dies.

Responses to the amplitude of daily temperature fluctuations could play a role in controlling the germination of buried seeds, especially if they have the ability to germinate in darkness. *Leymus arenarius* seeds germinated to higher percentages in darkness at alternating temperatures than at constant temperatures. An alternating temperature requirement for germination is viewed as a mechanism whereby germination is prevented in seeds buried too deep in sand for seedlings to emerge (Greipsson and Davy, 1994).

After seedlings from buried seeds emerge, they could be buried again, or those from nonburied seeds could become buried. Successful establishment, therefore, may depend on the ability of young seedlings to tolerate burial in sand for varying periods of time (Huiskes, 1977; Payne and Maun, 1984; Harris and Davy, 1987). The survival of buried seedlings partly depends on their ability to produce elongated stems or leaves (Sykes and Wilson, 1989; Zhang and Maun, 1990a,b; Maun, 1994).

5. Salinity

Although coastal hinddunes usually are considered to be nonsaline, they receive some salt spray, especially during storms. If high winds are not followed by rain, salinity at the surface layers of sand increases (Boyce, 1954). Also, in seasonally arid regions, salinity may build up during the dry season (Sykes and Wilson, 1989). Thus, seeds of psammophytes can be exposed to periods of increased soil salinity, which raises questions about its effect on germination. Germination was reduced to 50% in seeds of *Ammophila breviligulata* and *Uniola paniculata* by 0.09 M (0.5%) NaCL (Seneca, 1969); *Panicum amarulum,* 0.17 M (1.0%) (Seneca, 1969); *Leymus arenarius* (most populations), 0.30 M (1.8%) (Greipsson and Davy, 1994); *Palafoxia lindenii,* 0.15 M (0.9%); *Ipomoea stolonifera,* 0.3 M; *Andropogon glomeratus,* 0.6 M (3.6%); and *Trachypogon gouini,* 0.6 M (Martinez *et al.,* 1992). However, more than 50% of the seeds of *Amaranthus greggii, Pappophorum vaginatum,* and *Panicum maximum* germinated at 0.6 M NaCl (Martinez *et al.,* 1992). Scarified seeds of *Strophostyles umbellata* germinated at a salinity of 0.34 M (2.04%) NaCl, but stomatal conductance and the xylem pressure potential of plants were significantly reduced by a salinity of 0.03 M (0.18%) (Erickson and Young, 1995). Thus, it appears that seeds of some psammophytes are very sensitive to salinity, whereas those of others are quite tolerant of it.

Germination of seeds of many psammophytes in temperate regions during the cool moist season (Wagner, 1964; Kachi and Hirose, 1990; Maun, 1994) and in subtropical/tropical regions following rains (Martinez *et al.,* 1992) means that germination occurs after salts have been flushed from the surface layers of sand. Thus, seeds may not be exposed to increased salinity at the actual time of germination; however, little is known about the effects of increased salinity on the loss of dormancy in seeds of psammophytes.

6. Nitrates

The addition of nitrates to the germination substrate improved germination in seeds of *Cynoglossum officinale* (Freijsen *et al.,* 1980) and *Amaranthus greggii* (Martinez *et al.,* 1992), decreased it in seeds of *Andropogon glomeratus* (at 0.3%) (Martinez *et al,* 1992), and had no effect on those of *Leymus arenarius* (Greipsson and Davy, 1994), *Palafoxia lindenii, Pappophorum vaginatum, Panicum maximum,* or *Trachypogon gouini* (Martinez *et al.,* 1992). The role of nitrates in the germination ecology of psammophytes needs to be investigated in relation to germination phenology and annual precipitation, which could result in decreases in nitrate levels at the time of germination.

REFERENCES

Abu-Irmaileh, B. E. (1981). Response of hemp broomrape (*Orobanche ramosa*) infestation to some nitrogenous compounds. *Weed Sci.* **29,** 8–10.

Abu-Irmaileh, B. E. (1994). Nitrogen reduces branched broomrape (*Orobanche ramosa*) seed germination. *Weed Sci.* **42,** 57–60.

Abu-Shakra, S., Miah, A. A., and Saghir, A. R. (1970). Germination of seed of branched broomrape (*Orobanche ramosa* L.). *Hort. Res.* **10,** 119–124.

Adair, J. A., Higgins, T. R., and Brandon, D. L. (1990). Effects of fruit burial depth and wrack on the germination and emergence of the strandline species *Cakile edentula* (Brassicaceae). *Bull. Torrey Bot. Club* **117,** 138–142.

Addy, C. E. (1947). Germination of eelgrass seed. *J. Wildl. Manage.* **11,** 279.

Agami, M. (1986). The effects of different soil water potentials, temperature and salinity on germination of seeds of the desert shrub *Zygophyllum dumosum. Physiol. Plant.* **67,** 305–309.

Agami, M., and Waisel, Y. (1984). Germination of *Najas marina* L. *Aquat. Bot.* **19,** 37–44.

Ah-Lan, L., and Prakash, N. (1973). Life history of *Nepenthes gracilis. Malaysian J. Sci.* **2**(A), 45–53.

Ahring, R. M., Dunn, N. L., Jr., and Harlan, J. R. (1963). Effect of various treatments in breaking seed dormancy in sand lovegrass, *Eragrostis trichodes* (Nutt.) Wood. *Crop Sci.* **3,** 131–133.

Albert, R., and Popp, M. (1977). Chemical composition of halophytes from the Neusiedler Lake Region in Austria. *Oecologia* **27,** 157–170.

Al-Jibury, L. K., and Clor, M. A. (1986). Interaction between sodium, calcium and magnesium chlorides affecting germination and seedling growth of *Securigera securidaca* Linn. *Ann. Arid. Zone* **25,** 105–110

Al-Jibury, L. K., Clor, M. A., and Talabany, D. (1986). Effects of certain salts and their combinations on germination and seedling development of *Securigera securidaca* Linn. *Arab Gulf J. Sci. Res.* **4,** 5–12.

Allred, K. R., and Tingey, D. C. (1964). Germination and spring emergence of dodder as influenced by temperature. *Weeds* **12,** 45–48.

Al-Menoufi, O. A., and Zaitoun, F. M. F. (1987). Studies on *Orobanche* spp. 9. Effect of seed age and treatment with growth regulators on seed germination. *In* "Parastic Flowering Plants" (H. C. Weber and W. Forstreuter, eds.), pp. 29–36. Proc. 4th Int. Symp. Parasitic Flowering Plants, Marburg, FRG.

Al-Menoufi, O. A., Othman, M. A. S., and El-Safwani, N. A. (1987). Studies on *Orobanche* spp. 10. Chemical identity of germination stimulant of *Orobanche* seed. *In* "Parasitic Flowering Plants" (H. C. Weber and W. Forstreuter, eds.), pp. 37–52. Proc. 4th Int. Symp. Parasitic Flowering Plants, Marburg, FRG.

Amen, R. D., Carter, G. E., and Kelly, R. J. (1970). The nature of seed dormancy and germination in the salt marsh grass *Distichlis spicata. New Phytol.* **69,** 1005–1013.

Amritphale, D., Gutch, A., and Hsiao, A. I. (1995). Phytochrome-mediated germination control of *Hygrophila auriculata* seeds following dry storage augmented by temperature pulse, hormones, anaerobiosis or osmoticum imbibition. *Environ. Exp. Bot.* **35,** 187–192.

Amritphale, D., Lyengar, S., and Sharma, R. K. (1989). Effect of light and storage temperature on seed germination in *Hygrophila auriculata* (Schumach.) Haines. *J. Seed Technol.* **13,** 39–43.

Arditti, J. (1961). *Cynoches ventricosum* Batem. var. *chlorochilon* (Klotzsch) P. H. Allen *Comb. Nov.* (*Cynoches chlorochilon* Klotzsch, 1838). *Ceiba* **9,** 11–22.

Arditti, J. (1967). Factors affecting the germination of orchid seeds. *Bot. Rev.* **33,** 1–97.

Arditti, J. (1977). Clonal propagation of orchids by means of tissue culture: A manual. *In* "Orchid Biology: Reviews and Perspectives" (J. Arditti, ed.), Vol. I, pp. 203–293. Cornell Univ. Press, Ithaca, NY.

Arditti, J. (1979). Aspects of the physiology of orchids. *Adv. Bot. Res.* **7,** 421–655.

Arditti, J. (1982). Orchid seed germination and seedling culture: A manual. *In* "Orchid Biology: Reviews and Perspectives" (J. Arditti, ed.), Vol. II, pp. 245–370. Cornell Univ. Press, Ithaca, NY.

Arditti, J., Ernst, R., Yam, T. W., and Glabe, C. (1990). The contributions of orchid mycorrhizal fungi to seed germination: A speculative review. *Lindleyana* **5,** 249–255.

Arditti, J., and Harrison, C. R. (1977). Vitamin requirements and metabolism in orchids. *In* "Orchid Biology: Reviews and Perspectives" (J. Arditti, ed.), Vol. I, pp. 158–175. Cornell Univ. Press, Ithaca, NY.

Arditti, J., Michaud, J. D., and Oliva, A. P. (1981). Seed germination of North American orchids. I. Native California and related species of *Calypso, Epipactis, Goodyera, Piperia,* and *Platanthera. Bot. Gaz.* **142,** 442–453.

Arditti, J., Oliva, A. P., and Michaud, J. D. (1982). Practical germination of North American and related orchids-II. *Goodyera oblongifolia* and *G. tesselata. Am. Orchid Soc. Bull.* **51,** 394–397.

Arekal, G. D., and Shivamurthy, G. R. (1976). 'Seed' germination in *Balanophora abbreviata. Phytomorphology* **26,** 135–138.

Arts, G. H. P., and van der Heijden, R. A. J. M. (1990). Germination ecology of *Littorella uniflora* (L.) Aschers. *Aquat. Bot.* **37,** 139–151.

Ashton, F. M., and Santana, D. (1976). *Cuscuta* spp. (dodder): A literature review of its biology and control. *Univ. California Coop. Ext. Bull.* 1880.

Atkins, T. A., Thomas, A. G., and Stewart, J. M. (1987). The germination of wild rice seed in response to diurnally fluctuating temperatures and afterripening period. *Aquat. Bot.* **29,** 245–259.

Atsatt, P. R., Hearn, T. F, Nelson, R. L., and Heineman, R. T. (1978). Chemical induction and repression of haustoria in *Orthocarpus purpurascens* (Scrophulariaceae). *Ann. Bot.* **42,** 1177–1184.

Babiker, A. G. T., Butler, L. G., Ejeta, G., and Woodson, W. R. (1993a). Enhancement of ethylene biosynthesis and germination by cytokinins and 1-aminocyclopropane-1-carboxylic acid in *Striga asiatica* seeds. *Physiol. Plant.* **89,** 21–26.

Babiker, A. G. T., Ejeta, G., Butler, L. G., and Woodson, W. R. (1993b). Ethylene biosynthesis and strigol-induced germination of *Striga asiatica. Physiol. Plant.* **88,** 359–365.

Babiker, A. G. T., and Hamdoun, A. M. (1982). Factors affecting the activity of GR7 in stimulating germination of *Striga hermonthica* (Del.) Benth. *Weed Res.* **22,** 111–115.

Babiker, A. G. T., and Hamdoun, A. M. (1983). Factors affecting the activity of ethephon in stimulating seed germination of *Striga hermonthica* (Del.) Benth. *Weed Res.* **23,** 125–131.

Babiker, A. G. T., Hamdoun, A. M., and Mansi, M. G. (1987). Influence of some soil and environmental factors on response of *Striga hermonthica* (Del.) Benth. seeds to selected germination stimulants. *In* "Parasitic Flowering Plants" (H. C. Weber and W. Forstreuter, eds.), pp. 53–66. Proc. 4th Int. Symp. Parasitic Flowering Plants, Marburg, FRG.

Babiker, A. G. T., Parker, C., and Suttle, J. C. (1992). Induction of *Striga* seed germination by thidiazuron. *Weed Res.* **32,** 243–248.

Baccarini, A. (1966). Autotrophic incorporation of $C^{14}O_2$ in *Cuscuta australis* in relation to its parasitism. *Experientia* **22,** 46–47.

Badger, K. S., and Ungar, I. A. (1989). The effects of salinity and temperature on the germination of the inland halophyte *Hordeum jubatum. Can. J. Bot.* **67,** 1420–1425.

Bagonneaud-Berthome, V., Arnaud, M. C., and Fer, A. (1995). A new experimental approach to the chemical control of *Striga* using simplified models *in vitro. Weed Res.* **35,** 25–32.

Baird, W. V., and Riopel, J. L. (1986a). Life history studies of *Conopholis americana* (Orobanchaceae). *Am. Midl. Nat.* **116,** 140–151.

Baird, W. V., and Riopel, J. L. (1986b). The developmental anatomy of *Conopholis americana* (Orobanchaceae) seedlings and tubercles. *Can. J. Bot.* **64,** 710–717.

Bajaj, Y. P. S. (1967). In vitro studies on the embryos of two mistletoes, *Amyema pendula* and *Amyema miquelii. New Zeal. J. Bot.* **5,** 49–56.

Bajaj, Y. P. S. (1968). Some factors affecting growth of embryos of *Dendrophthoe falcata* in cultures. *Can. J. Bot.* **46,** 429–433.

Bajaj, Y. P. S. (1970). Growth responses of excised embryos of some mistletoes. *Z. Pflazenphysiol.* **63,** 408–415.

Bakker, D., ter Borg, S. J., and Otzen, D. (1966). Ecological research at the Plant Ecology Laboratory, State University, Groningen. *Wentia* **15,** 1–24.

Bakker, J. P., and de Vries, Y. (1992). Germination and early establishment of lower salt-marsh species in grazed and mown salt marsh. *J. Veg. Sci.* **3,** 247–252.

Bakker, J. P., Dijkstra, M., and Russchen, P. T. (1985). Dispersal, germination and early establishment of halophytes and glycophytes on a grazed and abandoned salt-marsh gradient. *New Phytol.* **101,** 291–308.

Bakshi, T. S. (1959). Ecology and morphology of *Pterospora andromedea. Bot. Gaz.* **120,** 203–216.

Balboa, O. (1983). Seed dormancy in four species in a Chilean sanddune: a laboratory study. *Acta Oecol.* **4,** 355–361.

Ballard, W. W. (1987). Sterile propagation of *Cypripedium reginae* from seeds. *Am. Orchid Soc. Bull.* **56,** 935–946.

Banovetz, S. J., and Scheiner, S. M. (1994). Secondary seed dormancy in *Coreopsis lanceolata. Am. Midl. Nat.* **131,** 75–83.

Barbour, M. G. (1970). Germination and early growth of the strand plant *Cakile maritima. Bull. Torrey Bot. Club* **97,** 13–22.

Barbour, M. G., and Davis, C. B. (1970). Salt tolerance of five California salt marsh plants. *Am. Midl. Nat.* **84,** 263–266.

Barlow, B. A. (1983). Biogeography of Loranthaceae and Viscaceae. *In* "The Biology of Mistletoes" (M. Calder and P. Bernhardt, eds.), pp. 19–46. Academic Press, Paris.

Barthlott, W. (1975). Morphologie der Samen von Orchideen im Hinblick auf taxonomische und funktionelle Aspekte. *In* "Proc. 8th World Orchid Conference" (K. Senghas, ed.), pp. 444–455. Harvard Univ. Printing Office, Cambridge, MA.

Barton, L. V., and Hotchkiss, J. E. (1951). Germination of seeds of *Eichhornia crassipes* Solms. *Contrib. Boyce Thomp. Inst.* **16,** 215–220.

Baskin, C. C., and Baskin, J. M. (1988). Germination ecophysiology of herbaceous plant species in a temperate region. *Am. J. Bot.* **75,** 286–305.

Baskin, C. C., Baskin, J. M., and Chester, E. W. (1993a). Seed germination ecophysiology of four summer annual mudflat species of Cyperaceae. *Aquat. Bot.* **45,** 41–52.

Baskin, C. C., Baskin, J. M., and Chester, E. W. (1993b). Germination ecology of *Leptochloa panicoides,* a summer annual grass of seasonally dewatered mudflats. *Acta Oecol.* **14,** 693–704.

Baskin, C. C., Baskin, J. M., and Chester, E. W. (1994). Annual dormancy cycle and influence of flooding in buried seeds of mudflat populations of the summer annual *Leucospora multifida. Ecoscience* **1,** 47–53.

Baskin, C. C., Baskin, J. M., and Chester, E. W. (1996a). Seed germination ecology of the aquatic winter annual *Hottonia inflata. Aquat. Bot.* **54,** 51–57.

Baskin, C. C., Baskin, J. M., and Leck, M. A. (1993). Afterripening pattern during cold stratification of achenes of ten perennial Asteraceae from eastern North America, and evolutionary implication. *Plant Species Biol.* **8,** 61–65.

Baskin, C. C., Chester, E. W., and Baskin, J. M. (1996b). Effect of flooding on annual dormancy cycles in buried seeds of two wetland *Carex* species. *Wetlands* **16,** 84–88.

Baskin, J. M., and Baskin, C. C. (1970). Germination ecophysiology of *Draba verna. Bull. Torrey Bot. Club* **97,** 209–216.

Baskin, J. M., and Baskin, C. C. (1978). Seasonal changes in the germination response of *Cyperus inflexus* seeds to temperature and their ecological significance. *Bot. Gaz.* **139,** 231–235.

Baskin, J. M., and Baskin, C. C. (1983). Germination ecology of *Veronica arvensis. J. Ecol.* **71**, 57–68.

Baskin, J. M., and Baskin, C. C. (1984a). Effect of temperature during burial on dormant and non-dormant seeds of *Lamium amplexicaule* L. and ecological implications. *Weed Res.* **24**, 333–339.

Baskin, J. M., and Baskin, C. C. (1984b). Role of temperature in regulating timing of germination in soil seed reserves of *Lamium purpureum* L. *Weed Res.* **24**, 341–349.

Baskin, J. M., and Baskin, C. C. (1986). Seasonal changes in the germination responses of buried witchgrass (*Panicum capillare*) seeds. *Weed Sci.* **34**, 22–24.

Baskin, J. M., and Baskin, C. C. (1986). Temperature requirements for after-ripening in seeds of nine winter annuals. *Weed Res.* **26**, 375–380.

Baskin, J. M., and Baskin, C. C. (1990). Seed germination ecology of poison hemlock, *Conium maculatum. Can. J. Bot.* **68**, 2018–2024.

Baskin, J. M., Baskin, C. C., Parr, P. D., and Cunningham, M. (1991). Seed germination ecology of the rare hemiparasite *Tomanthera auriculata* (Scrophulariaceae). *Castanea* **56**, 51–58.

Baskin, J. M., Baskin, C. C., and Spooner, D. M. (1989). Role of temperature, light and date seeds were exhumed from soil on germination of four wetland perennials. *Aquat. Bot.* **35**, 387–394.

Bazzaz, F. A. (1973). Seed germination in relation to salt concentration in three populations of *Prosopis farcta. Oecologia* **13**, 73–80.

Beadle, N. C. W. (1952). Studies in halophytes. I. The germination of the seed and establishment of the seedlings of five species of *Atriplex* in Australia. *Ecology* **33**, 49–62.

Beal, E. O., and Southall, R. M. (1977). The taxonomic significance of experimental selection by vernalization in *Nuphar* (Nymphaeaceae). *Syst. Bot.* **2**, 49–60.

Bebawi, F. F. (1987). Fertilization influence on the management of forage sorghum infested with witchweed. *In* "Parasitic Flowering Plants" (H. C. Weber and W. Forstreuter, eds.), pp. 67–78. Proc. 4th Int. Symp. Parasitic Flowering Plants, Marburg, FRG.

Bebawi, F. F., Awad, A. E., and Khalid, S. A. (1986). Germination, host preference, and phenolic content of witchweed (*Striga hermonthica*) seed populations. *Weed Sci.* **34**, 529–532.

Bebawi, F. F., and Eplee, R. E. (1986). Efficacy of ethylene as a germination stimulant of *Striga hermonthica* seed. *Weed Sci.* **34**, 694–698.

Bebawi, F. F., Eplee, R. E., Harris, C. E., and Norris, R. S. (1984). Longevity of witchweed (*Striga asiatica*) seed. *Weed Sci.* **32**, 494–497.

Beckman, K. M., and Roth, L. F. (1968). The influence of temperature on longevity and germination of seed of western dwarf mistletoe. *Phytopathology* **58**, 147–150.

Beeftink, W. G. (1977). The coastal salt marshes of western and northern Europe: An ecological and phytosociological approach. *In* "Ecosystems of the World" (V. J. Chapman, ed.), Vol. 1, pp. 109–155. Elsevier, Amsterdam.

Begon, M., Harper, J. L., and Townsend, C. R. (1996). "Ecology Individuals, Populations and Communities" 3rd Ed. Blackwell, Boston.

Bell, D. T. (1993). The effect of light quality on the germination of eight species from sandy habitats in Western Australia. *Aust. J. Bot.* **41**, 321–326.

Bellemakers, M. J. S., Maessen, M., Verheggen, G. M., and Roelofs, J. G. M. (1996). Effects of liming on shallow acidified moorland pools: A culture and seed bank experiment. *Aquat. Bot.* **54**, 37–50.

Berger, A. (1985). Seed dimorphism and germination behaviour in *Salicornia patula. Vegetatio* **61**, 137–143.

Berjak, P., Dini, M., and Pammenter, N. W. (1984). Possible mechanisms underlying the differing dehydration responses in recalcitrant and orthodox seeds: Desiccation-associated subcellular

changes in propagules of *Avicennia marina. Seed Sci. Technol.* **12**, 365–384.

Bhandari, N. N., and Vohra, S. C. A. (1983). Embryology and affinities of Viscaceae. *In* "The Biology of Mistletoes" (M. Calder and P. Bernhardt, eds.), pp. 69–86. Academic Press, Paris.

Bhatnagar, S. P., and Johri, B. M. (1983). Embryology of Loranthaceae. *In* "The Biology of Mistletoes" (M. Calder and P. Bernhardt, eds.), pp. 47–67. Academic Press, Paris.

Bhosale, L. J., and Mulik, N. G. (1990). Strategies of seed germination in mangroves. *In* "International Symposium on Environmental Influences on Seed and Germination Mechanism: Recent Advances in Research and Technology" (D. N. Sen, S. Mohammed, P. K. Kasera, and T. P. Thomas, eds.), pp. 89–90. Univ. Jodhpur, Jodhpur, India. [Abstract No. 64]

Bibbey, R. O. (1948). Physiological studies on weed seed germination. *Plant Physiol.* **23**, 467–484.

Binet, P. (1959). Dormances primaire et secondaire des semences de *Triglochin maritimum* L.: Action du froid et de la lumiere. *Bull. Soc. Linn. Normandie* **10**, 131–142.

Binet, P. (1965). Action de la temperature et de la salinite sur la germination des graines de *Glaux maritima* L. *Bull. Soc. Bot. France* **112**, 346–350.

Birch, W. R. (1981). Morphology of germinating seeds of the seagrass *Halophila spinulosa* (R.Br.) Aschers. (Hydrocharitaceae). *Aquat. Bot.* **11**, 79–90.

Bishchof, F., and Koch, W. (1973). Einige beitrage zur biologie von *Orobanche aegyptiaca* L. *In* "Parasitic Weeds Research Group" pp. 48–54. Proc. European Weed Res. Coun. Symp. Parasitic Weeds, Wageningen, The Netherlands.

Black, C. A. (1968). "Soil-Plant Relationships," 2nd Ed. Wiley, New York.

Blakeman, J. P., Mokahel, M. A., and Hadley, G. (1976). Effect of mycorrhizal infection on respiration and activity of some oxidase enzymes of orchid protocorms. *New Phytol.* **77**, 697–704.

Bocchieri, E., and Marchioni-Ortu, A. (1984). *Crithmum maritimum* L.: Comportamento alla germinazione de popolazioni sarde di "sabbia" e de "rupe." *Rendiconti Seminario Facolta Sci. Univ. Cagliari* **54**, 74–83.

Bonga, J. M. (1972). *Arceuthobium pusillum*: moisture requirements for germination and radicle growth. *Can. J. Bot.* **50**, 2143–2147.

Bonga, J. M., and Chakraborty, C. (1968). In vitro culture of a dwarf mistletoe, *Arceuthobium pusillum. Can. J. Bot.* **46**, 161–164.

Bonnewell, V., Koukkari, W. L., and Pratt, D. C. (1983). Light, oxygen, and temperature requirements for *Typha latifolia* seed germination. *Can. J. Bot.* **61**, 1330–1336.

Boorman, L. A. (1968). Some aspects of the reproductive biology of *Limonium vulgare* Mill., and *Limonium humile* Mill. *Ann. Bot.* **32**, 803–824.

Boorman, L. A. (1993). Dry coastal ecosystems of Britain: Dunes and shingle beaches. *In* "Ecosystems of the World" (E. van der Maarel, ed.), Vol. 2A pp. 197–228. Elsevier, Amsterdam.

Botha, P. J. (1948). The parasitism of *Alectra vogelii* Benth. with special reference to the germination of its seeds. *S. Afr. J. Bot.* **14**, 63–80.

Botha, P. J. (1950). The germination of the seeds of angiospermous root-parasites. Part II. The effect of time of pre-exposure, temperature of pre-exposure and concentration of the host factor on the germination of the seed of *Alectra vogelii* Benth. *S. Afr. J. Bot.* **16**, 23–29.

Boucaud, J., and Billard, J. P. (1978). Caracterisation de la glutamate deshydrogenase chez un halophyte obligatoire: le *Suaeda maritima* var. *macrocarpa. Physiol. Plant.* **44**, 31–37.

Boucaud, J., and Ungar, I. A. (1973). The role of hormones in controlling the mechanically induced dormancy of *Suaeda* spp. *Physiol. Plant.* **29**, 97–102.

Boucard, J., and Ungar, I. A. (1976). Hormonal control of germination under saline conditions of three halophytic taxa in the genus *Suaed. Physiol. Plant.* **37**, 143–148.

Bowers, J. E. (1982). The plant ecology of inland dunes in western North America. *J. Arid. Environ.* **5**, 199–220.

Bowers, J. E. (1996). Seedling emergence on Sonoran Desert dunes. *J. Arid Environ.* **33**, 63–72.

Boyce, S. G. (1954). The salt spray community. *Ecol. Monogr.* **24**, 50–67.

Bozarth, C. S., and Kennedy, R. A. (1985). Photosynthetic development of anaerobically grown rice (*Oryza sativa*) after exposure to air. *Plant Physiol.* **78**, 514–518.

Bozcuk, S. (1981). Effects of kinetin and salinity on germination of tomato, barley and cotton seeds. *Ann. Bot.* **48**, 81–84.

Brady, N. C. (1974). "The Nature and Properties of Soils," 8th Ed. MacMillan, New York.

Breen, C. M., Everson, C., and Rogers, K. (1977). Ecological studies on *Sporobolus viginicus* (L.) Kunth with particular reference to salinity and inundation. *Hydrobiologia* **54**, 135–140.

Brock, M. A. (1982). Biology of the salinity tolerant genus *Ruppia* L. in saline lakes in South Australia. I. Morphological variation within and between species and ecophysiology. *Aquat. Bot.* **13**, 219–248.

Brock, T. C. M., Mielo, H., and Oostermeijer, G. (1989). On the life cycle and germination of *Hottonia palustris* L. in a wetland forest. *Aquat. Bot.* **35**, 153–166.

Brown, R. (1965). The germination of angiospermous parasite seeds. *In* "Handbuch der Pflanzenphysiologie" (W. Ruhland, ed.), Vol. 15/2, pp. 925–932. Springer-Verlag, New York.

Brown, R., and Edwards, M. (1944). The germination of the seed of *Striga lutea*. I. Host influence and the progress of germination. *Ann. Bot.* **8**, 131–148.

Brown, R., and Edwards, M. (1946). The germination of the seed of *Striga lutea*. II. The effect of time of treatment and of concentration of the host stimulant. *Ann. Bot.* **10**, 133–142.

Brown, R., Greenwood, A. D., Johnson, A. W., and Long, A. G. (1951a). The stimulant involved in the germination of *Orobanche minor* Sm. 1. Assay techniques and bulk preparation of the stimulant. *Biochem. J.* **48**, 559–564.

Brown, R., Greenwood, A. D., Johnson, A. W., Long, A. G., and Tyler, G. J. (1951b). The stimulant involved in the germination of *Orobanche minor* Sm. 2. Chromatographic purification of crude concentrates. *Biochem. J.* **48**, 564–568.

Brown, R., Johnson, A. W., Robinson, E., and Todd, A. R. (1949). The stimulant involved in the germination of *Striga hermonthica*. *Proc. Roy. Soc. Lond. Ser. B* **136**, 1–12.

Brown, R., Johnson, A. W., Robinson, E., and Tyler, G. J. (1952). The *Striga* germination factor. 2. Chromatographic purification of crude concentrates. *Biochem. J.* **50**, 596–600.

Buckley, R. C. (1982). Seed size and seedling establishment in tropical arid dunecrest plants. *Biotropica* **14**, 314–315.

Buia, M. C., and Mazzella, L. (1991). Reproductive phenology of the Mediterranean seagrasses *Posidonia oceanica* (L.) Delile, *Cymodocea nodosa* (Ucria) Aschers., and *Zostera noltii* Hornem. *Aquat. Bot.* **40**, 343–362.

Bulow-Olsen, A. (1983). Germination response to salt in *Festuca rubra* in a population from a salt marsh. *Holarct. Ecol.* **6**, 194–198.

Burgeff, H. (1936). "Samenkeimung der Orchideen und Entwicklung ihrer Keimpflanzen." Fischer-Verlag, Jena.

Burgeff, H. (1959). Mycorrhiza of orchids. *In* "The orchids: A Scientific Survey" (C. L. Withner, ed.), pp. 361–395. The Ronald Press Co., New York.

Burger, D. (1972). "Seedlings of Some Tropical Trees and Shrubs Mainly of Southeast Asia." Centre for Agricultural Publishing and Documentation, Wageningen, The Netherlands.

Calder, D. M. (1983). Mistletoes in focus: An introduction. *In* "The Biology of Mistletoes" (M. Calder and P. Bernhardt, eds.), pp. 1–18. Academic Press, Paris.

Campbell, E. E., and Matthewson, W. J. (1992). Optimizing germination in *Atriplex nummularia* (Lind.) for commerical cultivation. *S. Afr. J. Bot.* **58**, 478–481.

Campion-Bourget, F. (1983). La germination des graines des especes francaises de *Rhinanthus* L. *Rev. Cytol. Biol. Veget. Bot.* **6**, 15–94.

Cannon, W. A. (1904). Observations on the germination of *Phoradendron villosum* and *P. californicum*. *Bull. Torrey Bot. Club* **31**, 435–443.

Cantlon, J. E., Curtis, E. J. C., and Malcolm, W. M. (1963). Studies of *Melampyrum lineare*. *Ecology* **44**, 466–474.

Cappelletti, C. (1935). Osservazioni sulla germinazione asimbiotica e simbiotica di alcune orchidee. *Nuovo Giornale Bot. Ital.* **42**, 436–457.

Cardwell, V. B., Oelke, E. A., and Elliott, W. A. (1978). Seed dormancy mechanisms in wild rice (*Zizania aquatica*). *Agron. J.* **70**, 481–484.

Carpenter, L. R., Nelson, E. E., and Stewart, J. L. (1979). Development of dwarf mistletoe infections on western hemlock in coastal Oregon. *For. Sci.* **25**, 237–243.

Carter, M. F., and Grace, J. B. (1986). Relative effects of *Justicia americana* litter on germination, seedlings and established plants of *Polygonum lapathifolium*. *Aquat. Bot.* **23**, 341–349.

Carter, M. F., Lane, F. E., and Grace, J. B. (1985). Effects of litter on aquatic macrophyte germination and growth. *Proc. Arkansas Acad. Sci.* **39**, 29–33.

Castejon-Munoz, M., Romero-Munoz, F., and Garcia-Torres, L. (1993). Effect of planting date on broomrape (*Orobanche cernua* Loefl.) infections in sunflower (*Helianthus annuus* L.). *Weed Res.* **33**, 171–176.

Caye, G., Bulard, C., Meinesz, A., and Loques, F. (1992). Dominant role of seawater osmotic pressure on germination in *Cymodocea nodosa*. *Aquat. Bot.* **42**, 187–193.

Caye, G., and Meinesz, A. (1986). Experimental study of seed germination in the seagrass *Cymodocea nodosa*. *Aquat. Bot.* **26**, 79–87.

Cechin, I., and Press, M. C. (1993). Nitrogen relations of the sorghum-*Striga hermonthica* host-parasite association: Germination, attachment and early growth. *New Phytol.* **124**, 681–687.

Cezard, R. (1973). Quelques aspects particuliers de la biologie des Orobanches. *In* "Parasitic Weeds Research Group," pp. 55–67. Proc. European Weed Res. Coun. Symp. Parasitic Weeds, Wageningen, The Netherlands.

Chabreck, R. H., Pilcher, B. K., and Ensiminger, A. B. (1983). Growth, production, and wildlife use of delta duckpotatoes in Louisiana. *Proc. Ann. Conf. Southeast. Assoc. Fish Wildl. Agencies* **37**, 56–66.

Chancellor, R. J. (1973). Germination and dormancy of *Odontites verna* (Bell.) Dum. *In* "Parasitic Weeds Research Group," pp. 260–268. Proc. European Weed Res. Coun. Symp. Parasitic Weeds, Wageningen, The Netherlands.

Chang, M., and Lynn, D. G. (1986). The haustorium and the chemistry of host recognition in parasitic angiosperms. *J. Chem. Ecol.* **12**, 561–579.

Chang, M., Netzly, D. H., Butler, L. G., and Lynn, D. G. (1986). Chemical regulation of distance: Characterization of the first natural host germination stimulant for *Striga asiatica*. *J. Am. Chem. Soc.* **108**, 7858–7860.

Chapman, V. J. (1942). The new perspective in the halophytes. *Quart. Rev. Biol.* **17**, 291–311.

Chapman, V. J. (1960). "Salt Marshes and Salt Deserts of the World." Leonard Hill [Books] Limited, London.

Chapman, V. J. (1976). "Coastal Vegetation," 2nd Ed. Pergamon Press, Oxford.

Chapman, V. J. (ed.) (1977). "Ecosystems of the World," Vol. 1. Elsevier, Amsterdam.

Chatterton, N. J., and McKell, C. M. (1969). *Atriplex polycarpa.* I. Germination and growth as affected by sodium chloride in water cultures. *Agron. J.* **61,** 448–450.

Chennaveeraiah, M. S., Nataraja, K., and Chikkannaiah, P. S. (1971). In vitro culture of the seeds of a root parasite: *Aeginetia indica* Linn. *Curr. Sci.* **40,** 668–669.

Chester, E. W. (1992). The vascular flora of Lake Barkley (lower Cumberland River) seasonally dewatered flats. *In* "Proceedings of the Contributed Paper Sessions of the Fourth Annual Symposium on the Natural History of Lower Tennessee and Cumberland River Valleys" (D. H. Snyder, ed.), pp. 67–79. Center for Field Biology, Austin Peay State University, Clarksville, TN.

Choudhuri, G. N. (1966). Seed germination and flowering in *Vallisneria spiralis. Northwest Sci.* **40,** 31–35.

Christy, E. J., and Sharitz, R. R. (1980). Characteristics of three populations of a swamp annual under different temperature regimes. *Ecology* **61,** 454–460.

Chuang, T.-L., and Heckard, L. R. (1971). Observations on root-parasitism in *Cordylanthus* (Scrophulariaceae). *Am. J. Bot.* **58,** 218–228.

Churchill, A. C. (1983). Field studies on seed germination and seedling development in *Zostera marina* L. *Aquat. Bot.* **16,** 21–29.

Ciotola, M., Watson, A. K., and Hallett, S. G. (1995). Discovery of an isolate of *Fusarium oxysporum* with potential to control *Striga hermonthica* in Africa. *Weed Res.* **35,** 303–309.

Clark, L. D., and West, N. E. (1969). Germination of *Kochia americana* in relation to salinity. *J. Range Manage.* **22,** 286–287.

Clark, L. D., and West, N. E. (1971). Further studies of *Eurotia lanata* germination in relation to salinity. *Southw. Nat.* **15,** 371–375.

Clay, K., Dement, D., and Rejmanek. M. (1985). Experimental evidence for host races in mistletoe (*Phoradendron tomentosum*). *Am. J. Bot.* **72,** 1225–1231.

Clements, M. A. (1988). Orchid mycorrhizal associations. *Lindleyana* **3,** 73–86.

Clements, M. A., Muir, H., and Cribb, P. J. (1986). A preliminary report on the symbiotic germination of European terrestrial orchids. *Kew Bull.* **41,** 437–445.

Clevering, O. A. (1995). Germination and seedling emergence of *Scirpus lacustris* L. and *Scirpus maritimus* L. with special reference to the restoration of wetlands. *Aquat. Bot.* **50,** 63–78.

Clifford, H. T., and Smith, W. K. (1969). Seed morphology and classification of Orchidaceae. *Phytomorphology* **19,** 133–139.

Cluff, G. J., Evans, R. A., and Young, J. A. (1983). Desert saltgrass seed germination and seedbed ecology. *J. Range Manage.* **36,** 419–422.

Cluff, G. J., and Roundy, B. A. (1988). Germination responses of desert saltgrass to temperature and osmotic potential. *J. Range Manage.* **41,** 150–153.

Coble, T. A., and Vance, B. D. (1987). Seed germination in *Myriophyllum spicatum* L. *J. Aquat. Plant Manage.* **25,** 8–10.

Cole, N. H. A. (1977). Effect of light, temperature, and flooding on seed germination of the neotropical *Panicum laxum* Sw. *Biotropica* **9,** 191–194.

Collon, E. G., and Velasquez, J. (1989). Dispersion, germination and growth of seedlings of *Sagittaria lancifolia* L. *Folia Geobot. Phytotax.* **24,** 37–49.

Colosi, J. C. and Seneca, E. D. (1977). The seed germination patterns

of two wide-ranging coastal foredune dominants. *Bull. Ecol. Soc. Am.* **58**(2), 25. [Abstract]

Come, D., Corbineau, F., and Soudain, P. (1991). Beneficial effects of oxygen deprivation on germination and plant development. *In* "Plant Life under Oxygen Deprivation" (M. B. Jackson, D. D. Davies, and H. Lambers, eds.), pp. 69–83. SPB Academic Publishing, The Hague.

Conacher, C. A., Poiner, I. R., Butler, J., Pun, S., and Tree, D. J. (1994). Germination, storage and viability testing of seeds of *Zostera capricorni* Aschers. from a tropical bay in Australia. *Aquat. Bot.* **49,** 47–58.

Conover, D. G., and Geiger, D. R. (1984). Germination of Australian channel millet [*Echinochloa turnerana* (Domin) J. M. Black] seeds. II. Effects of anaerobic conditions, continuous flooding, and low water potential. *Aust. J. Plant Physiol.* **11,** 395–408.

Cook, C. D. K., Gut, B. J., Rix, E. M., Schneller, J., and Seitz, M. (1974). "Water Plants of the World. A Manual for the Identification of the Genera of Freshwater Macrophytes." Junk, The Hague.

Cook, C. E., Whichard, L. P., Turner, B., Wall, M. E., and Egley, G. H. (1966). Germination of witchweed (*Striga lutea* Lour.): Isolation and properties of a potent stimulant. *Science* **154,** 1189–1190.

Cook, C. E., Whichard, L. P., Wall, M. E., Egley, G. H., Coggon, P., Luhan, P. A., and McPhail, A. T. (1972). Germination stimulants. II. The structure of strigol: A potent seed germination stimulant for witchweed (*Striga lutea* Lour.). *J. Am. Chem. Soc.* **94,** 6198–6199.

Coops, H., and van der Velde, G. (1995). Seed dispersal, germination and seedling growth of six helophyte species in relation to water-level zonation. *Freshwater Biol.* **34,** 13–20.

Cordas, D. I. (1973). Effects of branched broomrape on tomatoes in California fields. *Plant Dis. Report.* **57,** 926–927.

Corker, B. (1986). Germination and viability of seeds of the pitcher plant, *Nepenthes mirabilis* Druce. *Malayan Nat. J.* **39,** 259–264.

Cornelius, D. R., and Hylton, L. O. (1969). Influence of temperature and leachate on germination of *Atriplex polycarpa. Agron. J.* **61,** 209–211.

Corner, E. J. H. (1976). "The Seeds of Dicotyledons." 2 Vols. Cambridge Univ. Press, Cambridge, UK.

Coupland, R. T. (ed.). (1992). "Ecosystems of the World," Vol. 8A. Elsevier, Amsterdam.

Cozza, R., Galanti, G., Bitonti, M. B., and Innocenti, A. M. (1994). Effect of storage at low temperature on the germination of the waterchestnut (*Trapa natans* L.). *Phyton (Austria)* **34,** 315–320.

Crocker, W. (1907). Germination of seeds of water plants. *Bot. Gaz.* **44,** 375–380.

Crossland, C. (1903). Note on the dispersal of mangrove seedlings. *Ann. Bot.* **17,** 267–270.

Curtis, E. J. C., and Cantlon, J. E. (1963). Germination of *Melampyrum lineare*: Interrelated effects of afterripening and gibberellic acid. *Science* **140,** 406–408.

Curtis, J. T. (1936). The germination of native orchid seeds. *Am. Orchid Soc. Bull.* **5,** 42–47.

Curtis, J. T. (1939). The relation of specificity of orchid mycorrhizal fungi to the problem of symbiosis. *Am. J. Bot.* **26,** 390–399.

Curtis, J. T. (1943). Germination and seedling development in five species of *Cypripedium* L. *Am. J. Bot.* **30,** 199–206.

Curtis, J. T. (1947). Studies on the nitrogen nutrition of orchid embryos. I. Complex nitrogen sources. *Am. Orchid Soc. Bull.* **16,** 655–660.

Dafni, A., and Negbi, M. (1978). Variability in *Prosopis farcta* in Israel: Seed germination as affected by temperature and salinity. *Israel J. Bot.* **27,** 147–159.

Dale, J. E., and Egley, G. H. (1971). Stimulation of witchweed germination by run-off water and plant tissues. *Weed Sci.* **19,** 678–681.

Danin, A. (1978). Species diversity of semishrub xerohalophyte communities in the Judean Desert of Israel. *Israel J. Bot.* **27**, 66–76.

Datta, S. C., and Biswas, K. K. (1968). Influence of temperature, light and stimulators and their interaction on the germination and seedling growth of *Alternanthera sessilis*. *Osterr. Bot. Z.* **115**, 391–399.

Datta, S. C., and Biswas, K. K. (1970). Germination-regulating mechanisms in aquatic angiosperms. I. *Ipomoea aquatica* Fors[s]k. *Broteria* **39**, 175–185.

Datta, S. C., and Roy, A. K. (1970). Comparative studies on the seed germination of two plants of low-lying lands. *Broteria* **39**, 169–174.

Datta, S. C., and Sinha-Roy, S. P. (1975). Germination-regulating mechanisms in aquatic angiosperms. II. *Aeschynomene aspera*, L. *Broteria* **44**, 81–91.

Davidson, E. D., and Barbour, M. G. (1977). Germination, establishment, and demography of coastal bush lupin (*Lupinus arboreus*) at Bodega Head, California. *Ecology* **58**, 592–600.

Davy, A. J., and Bishop, G. F. (1991). Biological flora of the British Isles No. 172. *Triglochin maritima* L. (*Triglochin maritimum* L.). *J. Ecol.* **79**, 531–555.

Dawoud, D. A., and Sauerborn, J. (1994). Impact of drought stress and temperature on the parasitic weeds *Striga hermonthica* and *Alectra vogelii* in their early growth stages. *Exp. Agric.* **30**, 249–257.

Dawson, J. (1988). "Forest Vines to Snow Tussocks." Victoria Univ. Press, Wellington, New Zealand.

Dawson, T. E., and Ehleringer, J. R. (1991). Ecological correlates of seed mass variation in *Phoradendron juniperinum*, a xylem-tapping mistletoe. *Oecologia* **85**, 332–342.

De Jong, T. M., and Barbour, M. G. (1979). Contributions to the biology of *Atriplex leucophylla*, a C₄ California beach plant. *Bull. Torrey Bot. Club* **106**, 9–19.

de la Harpe, A. C., Visser, J. H., and Grobbelaar, N. (1981). Photosynthetic characteristics of some South African parasitic flowering plants. *Z. Pflanzenphysiol.* **103**, 265–275.

DeLaubenfels, D. J. (1959). Parasitic conifer found in New Caledonia. *Science* **30**, 97.

Delesalle, V. A., and Blum, S. (1994). Variation in germination and survival among families of *Sagittaria latifolia* in response to salinity and temperature. *Int. J. Plant Sci.* **155**, 187–195.

Delipetrou, P., Georghiou, K., and Thanos, C. A. (1993). On the photoinhibition of seed germination in the maritime plant *Brassica tournefortii*. *In* "Fourth International Workshop on Seeds: Basic and Applied Aspects of Seed Biology" (D. Come and F. Corbineau, eds.), Vol. 2, pp. 473–478. ASFIS, Paris.

del Rio, C. M., Hourdequin, M., Silva, A., and Medel, R. (1995). The influence of cactus size and previous infection on bird deposition of mistletoe seeds. *Aust. J. Ecol.* **20**, 571–576.

del Rio, C. M., Silva, A., Medel, R., and Hourdequin, M. (1996). Seed dispersers as disease vectors: Bird transmission of mistletoe seeds to plant hosts. *Ecology* **77**, 912–921.

den Hartog, C. (1970). "The Sea-Grasses of the World." North-Holland, Amsterdam.

De Pauw, M. A., and Remphrey, W. R. (1993). In vitro germination of three *Cypripedium* species in relation to time of seed collection, media, and cold treatment. *Can. J. Bot.* **71**, 879–885.

De Pauw, M. A., Remphrey, W. R., and Palmer, C. E. (1995). The cytokinin preference for in vitro germination and protocorm growth of *Cypripedium candidum*. *Ann. Bot.* **75**, 267–275.

De Vogel, E. F. (1980). "Seedlings of Dicotyledons." Centre for Agricultural Publishing and Documentation, Wageningen, The Netherlands.

Dick, M. W. (1981). Resources and regulators of fungal populations. *In* "The Fungal Community: Its Organization and Role in the Ecosystem" (D. T. Wicklow and G. C. Carroll, eds.), pp. 263–278. Dekker, New York.

Dietert, M. F., and Shontz, J. P. (1978). Germination ecology of a Maryland population of saltmarsh bulrush (*Scirpus robustus*). *Estuaries* **1**, 164–170.

Dinelli, G., Bonetti, A., and Tibiletti, E. (1993). Photosynthetic and accessory pigments in *Cuscuta campestris* Yuncker and some host species. *Weed Res.* **33**, 253–260.

Doing, H. (1985). Coastal fore-dune zonation and succession in various parts of the world. *Vegetatio* **61**, 65–75.

Dowding, E. S. (1929). The vegetation of Alberta. III. The sandhill areas of central Alberta with particular reference to the ecology of *Arceuthobium americanum* Nutt. *J. Ecol.* **17**, 82–105.

Downie, D. G. (1940). Notes on the germination of *Corallorhiza innata*. *Trans. Proc. Bot. Soc. Edinburgh* **33** (Part I), 36–51.

Downie, D. G. (1943). On the germination and growth of *Goodyera repens*. *Trans. Proc. Bot. Soc. Edinburgh* **33** (Part IV), 380–382.

Downie, D. G. (1949). The germination of *Goodyera repens* (L.) R. Br. in fungal extract. *Trans. Proc. Bot. Soc. Edinburgh* **35**(Part II), 120–125.

Downton, W. J. S. (1982). Growth and osmotic relations of the mangrove *Avicennia marina*, as influenced by salinity. *Aust. J. Plant Physiol.* **9**, 519–528.

Dressler, R. L. (1981). "The Orchids: Natural History and Classification." Harvard Univ. Press, Cambridge, MA.

Dressler, R. L. (1993). "Phylogeny and Classification of the Orchid Family." Dioscorides Press, Portland, OR.

Duarte. C. M. (1991). Seagrass depth limits. *Aquat. Bot.* **40**, 363–377.

Ducker, S. C., Foord, N. J., and Knox, R. B. (1977). Biology of Australian seagrasses: The genus *Amphibolis* C. Agardh (Cymodoceaceae). *Aust. J. Bot.* **25**, 67–95.

Duke, J. A. (1965). Keys for the identification of seedlings of some prominent woody species in eight forest types in Puerto Rico. *Ann. Missouri Bot. Gard.* **52**, 314–350.

Duke, J. A. (1969). On tropical seedlings. I. Seeds, seedlings, systems, and systematics. *Ann. Missouri Bot. Gard.* **56**, 125–161.

Duvel, J. W. T. (1906). The storage and germination of wild rice seed. *USDA Bur. Plant Indust. Bull. No. 90. Misc. Papers.* pp. 1–13 + 2 plates.

Edwards, T. I. (1933). The germination and growth of *Peltandra virginica* in the absence of oxygen. *Bull. Torrey Bot. Club* **60**, 573–581.

Edwards, W. G. H. (1972). *Orobanche* and other plant parasite factors. *Ann. Proc. Phytochem. Soc.* **8**, 235–248.

Egbers, W. S., Pierterse, A. H., and Verkleij, J. A. C. (1991). Germination of freshly harvested seed of *Striga hermonthica* after storage under various temperature and relative humidity conditions. *In* "Proc. 5th Int. Symp. Parasitic Weeds" (J. K. Ransom, L. J. Musselman, A. D. Worsham, and C. Parker, eds.), pp. 407–414. Nairobe, Kenya.

Egley, G. H. (1965). Studies on the witchweed (*Striga asiatica*) seed germination stimulant exuded from plant roots. *Proc. South. Weed Conf.* **18**, 660–661. [Abstract]

Egley, G. H. (1972). Influence of the seed envelope and growth regulators upon seed dormancy in witchweed (*Striga lutea* Lour.). *Ann. Bot.* **36**, 755–770.

Egley, G. H. (1990). Isolation and identification of witchweed seed germination stimulants from natural sources. *Monogr. Ser. Weed Sci. Soc. Am.* **5**, 37–45.

Egley, G. H., and Dale, J. E. (1970). Ethylene, 2-choroethylphosphonic acid, and witchweed germination. *Weed Sci.* **18**, 586–589.

Elias, P. (1987). Chlorophyll contents in leaves of a mistletoe *Loranthus europaeus* Jacq. *In* "Parasitic Flowering Plants." (H. C. Weber and W. Forstreuter, eds.), pp. 171–173. Proc. 4th Int. Symp. Parasitic Flowering Plants, Marburg, FRG.

Ellenberg, H. (1988). "Vegetation Ecology of Central Europe, 4th ed." Cambridge Univ. Press, Cambridge, UK. [English translation by G. K. Strutt]

Else, M. J., and Riemer, D. N. (1984). Factors affecting germination of seeds of fragrant waterlily (*Nymphaea odorata*). *J. Aquat. Plant Manage.* **22,** 22–25.

England, C. M., and Rice, E. L. (1957). A comparison of the soil fungi of a tall-grass prairie and of an abandoned field in central Oklahoma. *Bot. Gaz.* **118,** 186–190.

Eplee, R. E. (1975). Ethylene: A witchweed seed germination stimulant. *Weed Sci.* **23,** 433–436.

Eplee, R. E., and Westbrooks, R. G. (1990). Movement and survival of witchweed seeds in the soil. *Monogr. Ser. Weed Sci. Soc. Am.* **5,** 81–84.

Erickson, D. L., and Young, D. R. (1995). Salinity response, distribution, and possible dispersal of a barrier island strand glycophyte, *Strophostyles umbellata* (Fabaceae). *Bull. Torrey Bot. Club* **122,** 95–100.

Ernst, R. (1967). Effect of carbohydrate selection on the growth rate of freshly germinated *Phalaenopsis* and *Dendrobium* seed. *Am. Orchid Soc. Bull.* **36,** 1068–1073.

Ernst, W. H. O., and van der Ham, N. F. (1988). Population structure and rejuvenation potential of *Schoenus nigricans* in coastal wet dune slacks. *Acta Bot. Neerl.* **37,** 451–465.

Etherington, J. R. (1967). Studies on nutrient cycling and productivity in oligotrophic ecosystems. I. Soil potassium and wind-blown sea-spray in a South Wales dune grassland. *J. Ecol.* **55,** 743–752.

Evenari, M., Noy-Meir, I., and Goodall, D. W. (eds.) (1986). "Ecosystems of the World," Vol. 12B. Elsevier, Amsterdam.

Farah, A. F. (1987). Some ecological aspects of *Cistanche phelypaea* (L.) Cout. (Orobanchaceae) in Al-Hasa Oasis, Saudi Arabia. *In* "Parasitic Flowering Plants" (H. C. Weber and W. Forstreuter, eds.), pp. 187–196. Proc. 4th Int. Symp. Parasitic Flowering Plants, Marburg, FRG.

Farina, M. P. W., Thomas, P. E. L., and Channon, P. (1985). Nitrogen, phosphorus and potassium effects on the incidence of *Striga asiatica* (L.) Kuntze in maize. *Weed Res.* **25,** 443–447.

Farmer, A. M., and Spence, D. H. N. (1987). Flowering, germination and zonation of the submerged aquatic plant *Lobelia dortmanna* L. *J. Ecol.* **75,** 1065–1076.

Farrant, J. M., Berjak, P., and Pammenter, N. W. (1993). Studies on the development of the desiccation-sensitive (recalcitrant) seeds of *Avicennia marina* (Forssk.) Virh.: The acquisition of germinability and response to storage and dehydration. *Ann. Bot.* **71,** 405–410.

Farrant, J. M., Pammenter, N. W., and Berjak, P. (1992). Development of the recalcitrant (homoiohydrous) seeds of *Avicennia marina*: Anatomical, ultrastructural and biochemical events associated with development from histodifferentiation to maturation. *Ann. Bot.* **70,** 75–86.

Ferreira, D. P., and Small, J. G. C. (1974). Preliminary studies on seed germination of *Drosera aliciae* Hamet. *S. Afr. J. Bot.* **40,** 65–73.

Filho, E. G., Prisco, J. T., de Paiva Campos, F. de A., and Filho, J. E. (1983). Effects of NaCl salinity in vivo and in vitro on ribonuclease activity of *Vigna unguiculata* cotyledons during germination. *Physiol. Plant.* **59,** 183–188.

Fineran, B. A., and Hocking, P. J. (1983). Features of parasitism, morphology and haustorial anatomy in Loranthaceous root parasites. *In* "The Biology of Mistletoes" (M. Calder and P. Bernhardt, eds.), pp. 205–227. Academic Press, Paris.

Finlayson, C. M., Cowie, I. D., and Bailey, B. J. (1990). Sediment seedbanks in grassland on the Magela Creek floodplain, northern Australia. *Aquat. Bot.* **38,** 163–176.

Fischer, N. H., Weidenhamer, J. D., and Bradow, J. M. (1989). Dihy-

droparthenolide and other sesquiterpene lactones stimulate witchweed germination. *Phytochemistry* **28,** 2315–2317.

Fischer, N. H., Weidenhamer, J. D., and Bradow, J. M. (1990). Stimulation of witchweed germination by sequiterpene lactones: A structure-activity study. *Phytochemistry* **29,** 2479–2483.

Fisher, J. T. (1983). Water relations of mistletoes and their hosts. *In* "The Biology of Mistletoes" (M. Calder and P. Bernhardt, eds.), pp. 161–184. Academic Press, Paris.

Flint, L. H., and Morehead, C. F. (1943). Note on photosynthetic activity in seeds of the spider lily. *Am. J. Bot.* **30,** 315–317.

Flowers, S. (1934). Vegetation of the Great Salt Lake region. *Bot. Gaz.* **95,** 353–418.

Flowers, T. J. (1972). The effect of sodium chloride on enzyme activities from four halophyte species of Chenopodiaceae. *Phytochemistry* **11,** 1881–1886.

Flowers, T. J., Hajibagheri, M. A., and Clipson, N. J. W. (1986). Halophytes. *Quart. Rev. Biol.* **61,** 313–337.

Forsberg, C. (1965). Sterile germination of oospores of *Chara* and seeds of *Najas marina. Physiol. Plant.* **18,** 128–137.

Forsberg, C. (1966). Sterile germination requirements of seeds of some water plants. *Physiol. Plant.* **19,** 1105–1109.

Francke, H.-L. (1934). Beitrage zur Kenntnis der Mykorrhiza von *Montropa hypopitys* L. Analyse und Synthese der Symbiose. *Flora* **129,** 1–52.

Francko, D. A. (1986). Studies on *Nelumbo lutea* (Willd.). Pers. I. Techniques for axenic liquid seed culture. *Aquat. Bot.* **26,** 113–117.

Francois, L. E., and Goodin, J. R. (1972). Interaction of temperature and salinity on sugar beet germination. *Agron. J.* **64,** 272–273.

Frei, Sr. J. K., and Dodson, C. H. (1972). The chemical effect of certain bark substrates on the germination and early growth of epiphytic orchids. *Bull. Torrey Bot. Club* **99,** 301–307.

Freijsen, A. H. J., Troelstra, S. R., and van Kats, M. J. (1980). The effect of soil nitrate on the germination of *Cynoglossum officinale* L. (Boraginaceae) and its ecological significance. *Acta Oecol.* **1,** 71–79.

French, R. C., and Sherman, L. J. (1976). Factors affecting dormancy, germination, and seedling development of *Aeginetia indica* L. (Orobanchaceae). *Am. J. Bot.* **63,** 558–570.

Furman, T. E., and Trappe, J. M. (1971). Phylogeny and ecology of mycotrophic achlorophyllous angiosperms. *Quart. Rev. Biol.* **46,** 219–225.

Gaertner, E. E. (1950). Studies of seed germination, seed identification, and host relationships in dodders, *Cuscuta* spp. *N. Y. Agric. Exp. Sta. Mem.* 294.

Gaertner, E. E. (1956). Dormancy in the seed of *Cuscuta europea. Ecology* **37,** 389.

Galinato, M. I., and van der Valk, A. G. (1986). Seed germination traits of annuals and emergents recruited during drawdowns in the Delta Marsh, Manitoba, Canada. *Aquat. Bot.* **26,** 89–102.

Gams, W., and Domsch, K. H. (1969). The spatial and seasonal distribution of microscopic fungi in arable soils. *Trans. Br. Mycol. Soc.* **52,** 301–308.

Garcia-Tiburcio, H., and Troyo-Dieguez, E. (1993). Efectos del decremento instantaneo de la salinidad sobre la germinacion de *Salicornia bigelowii* Torr. en laboratorio. *Phyton (Buenos Aires)* **54,** 127–137.

Garcia-Torres, L., Lopez-Granados, F., and Castejon-Munoz, M. (1994). Pre-emergence herbicides for the control of broomrape (*Orobanche cernua* Loefl.) in sunflower (*Helianthus annuus* L.). *Weed Res.* **34,** 395–402.

Garcia-Torres, L., Mesa-Garcia, J., and Romero-Munoz, F. (1987). Agronomic problems and chemical control of broomrapes (*Orobanche* spp.) in Spain: A studies review. *In* "Parasitic Flowering

Plants" (H. C. Weber and W. Forstreuter, eds.), pp. 241–247. Proc. 4th Int. Symp. Parasitic Flowering Plants, Marburg, FRG.

Gardner, W. A. (1921). Effect of light on germination of light-sensitive seeds. *Bot. Gaz.* **71,** 249–288.

Garmon, H. (1890). The broom-rape of hemp and tobacco. *Kentucky Agric. Exp. Sta. Bull.* No. 24.

Garrard, A. (1955). The germination and longevity of seeds in an equatorial climate. *Gard. Bull. Singapore* **14,** 534–545.

Gaudet, J. J. (1977). Natural drawdown on Lake Naivasha, Kenya, and the formation of papyrus swamps. *Aquat. Bot.* **3,** 1–47.

Gedge, K. E., and Maun, M. A. (1992). Effects of simulated herbivory on growth and reproduction of two beach annuals, *Cakile edentula* and *Corispermum hyssopifolium. Can. J. Bot.* **70,** 2467–2475.

George, H. A., and Young, J. A. (1977). Germination of alkali bulrush seed. *J. Wildl. Manage.* **41,** 790–793.

Gerritsen, J., and Greening, H. S. (1989). Marsh seed banks of the Okefenokee Swamp: Effects of hydrologic regime and nutrients. *Ecology* **70,** 750–763.

Gibbs, R. E. (1902). *Phyllospadix* as a beach-builder. *Am. Nat.* **36,** 101–109.

Gill, L. S., and Hawksworth, F. G. (1961). The mistletoes: A literature review. *USDA Tech. Bull. No.* 1242.

Glenn, E. P. (1987). Relationship between cation accumulation and water content of salt-tolerant grasses and a sedge. *Plant, Cell Environ.* **10,** 205–212.

Glenn, E. P., and O'Leary, J. W. (1984). Relationship between salt accumulation and water content of dicotyldeonous halophytes. *Plant, Cell Environ.* **7,** 253–261.

Glimcher, J. (1938). The germination of *Viscum cruciatum* Sieb. *Palestine J. Bot.* **1,** 103–105.

Godschalk, S. K. B. (1983). Mistletoe dispersal by birds in South Africa. *In* "The Biology of Mistletoes" (M. Calder and P. Bernhardt, eds.), pp. 117–128. Academic Press, Paris.

Goodman, P. J. (1973). Physiological and ecotypic adaptations of plants to salt desert conditions in Utah. *J. Ecol.* **61,** 473–494.

Goossens, M., and Devillez, F. (1974). Les conditions thermiques d'incubation et la germination des akenes non dormants de *Cladium mariscus. Acad. Roy. Belg. Bull. Cl. Sci.* **60,** 350–367.

Gopal, B., and Sharma, K. P. (1983). Light regulated seed germination in *Typha angustata* Bory et Chaub. *Aquat. Bot.* **16,** 377–384.

Gore, A. J. P. (1983). Introduction. *In* "Ecosystems of the World" (A. J. P. Gore, ed.), pp. 1–34. Elsevier, Amsterdam.

Goubin, C., and Loques, F. (1991). Germinating *Zostera noltii* Hornemann found in the Etang de Diana, Corsica. *Aquat. Bot.* **42,** 75–79.

Gous, J., Visser, J. H., and Grobbelaar, N. (1980). Some aspects of the bidirectional translocation of ^{14}C-labelled metabolites between *Alectra vogelii* Benth. and *Voandzeia subterranea* (L.) Thou. *Z. Pflanzenphysiol.* **99,** 225–233.

Govier, R. N., Brown, J. G. S., and Pate, J. S. (1968). Hemiparasitic nutrition in Angiosperms. II. Root haustoria and leaf glands of *Odontites verna* (Bell.) Dum. and their relevance to the abstraction of solutes from the host. *New Phytol.* **67,** 963–972.

Govier, R. N., and Harper, J. L. (1964). Hemiparasitic weeds. *British Weed Cont. Conf.* **7** (Vol. 2), 577–582.

Govier, R. N., Nelson, M. D., and Pate, J. S. (1967). Hemiparasitic nutrition in Angiosperms. I. The transfer of organic compounds from host to *Odontites verna* (Bell.) Dum. (Scrophulariaceae). *New Phytol.* **66,** 285–297.

Grace, J. B. (1983). Autotoxic inhibition of seed germination by *Typha latifolia*: an evaluation. *Oecologia* **59,** 366–369.

Graves, J. D., Press, M. C., Smith, S., and Stewart, G. R. (1992). The carbon canopy economy of the association between cowpea and the parasitic angiosperm *Striga gesnerioides. Plant, Cell Environ.* **15,** 283–288.

Graves, J. D., and Taylor, K. (1988). A comparative study of *Geum rivale* L. and *G. urbanum* L. to determine those factors controlling their altitudinal distribution. III. The response of germination to temperature. *New Phytol.* **110,** 391–397.

Graves, J. D., Wylde, A., Press, M. C., and Stewart, G. R. (1990). Growth and carbon allocation in *Pennisetum typhoides* infected with the parasitic angiosperm *Striga hermonthica. Plant, Cell Environ.* **13,** 367–373.

Green, S. (1967). Notes on the distribution of *Nepenthes* species in Singapore. *Gard. Bull. Singapore* **22,** 53–65.

Greipsson, S., and Davy, A. J. (1994). Germination of *Leymus arenarius* and its significance for land reclamation in Iceland. *Ann. Bot.* **73,** 393–401.

Grelen, H. E., and Mann, W. F., Jr. (1973). Distribution of senna seymeria (*Seymeria cassioides*), a root parasite on southern pines. *Econ. Bot.* **27,** 339–342.

Grillas, P., Wijck, C. V., and Bonis, A. (1991). Life history traits: A possible cause for the higher frequency of occurrence of *Zannichellia pedunculata* than of *Zannichellia obtusifolia* in temporary marshes. *Aquat. Bot.* **42,** 1–13.

Grime, J. P., Mason, G., Curtis, A. V., Rodman, J., Band, S. R., Mowforth, M. A. G., Neal, A. M., and Shaw, S. (1981). A comparative study of germination characteristics in a local flora. *J. Ecol.* **69,** 1017–1059.

Groenewald, E. G., Gouws, J., and Visser, J. H. (1979). *In vitro* studies on the effect of host rhizosphere on the seed germination and haustorium formation of the root parasite *Alectra vogelii* Benth. *S. Afr. J. Sci.* **75,** 42–43.

Groom, P. (1895). On a new saprophytic Monocotyledon. *Ann. Bot.* **9,** 45–58.

Grouzis, M. (1973). Exigences ecologiques comparees d'une salicorne vivace et d'une salicorne annuelle: Germination et croissance des stades jeunes. *Oecol. Plant.* **8,** 367–375.

Groves, R. H. (ed.) (1981). "Australian Vegetation." Cambridge Univ. Press, Cambridge, UK.

Gutterman, Y. (1980–1981). Annual rhythm and position effect in the germinability of *Mesembryanthemum nodiflorum. Israel J. Bot.* **29,** 93–97.

Hadley, G. (1970a). The interaction of kinetin, auxin and other factors in the development of north temperate orchids. *New Phytol.* **69,** 549–555.

Hadley, G. (1970b). Non-specificity of symbiotic infection in orchid mycorrhiza. *New Phytol.* **69,** 1015–1023.

Hadley, G. (1982). Orchid mycorrhiza. *In* "Orchid Biology: Reviews and Perspective" (J. Arditti, ed.), Vol. II, pp. 84–117. Cornell Univ. Press, Ithaca, NY.

Hadley, G., and Harvais, G. (1968). The effect of certain growth substances on asymbiotic germination and development of *Orchis purpurella. New Phytol.* **67,** 441–445.

Hameed, K. M., Saghir, A. R., and Foy, C. L. (1973). Influence of root exudates on *Orobanche* seed germination. *Weed Res.* **13,** 114–117.

Harivandi, M. A., Butler, J. D., and Soltanpour, P. N. (1982). Effects of sea water concentrations on germination and ion accumulation in alkaligrass (*Puccinellia* spp.). *Commun. in Soil Sci. Plant Anal.* **13,** 507–517.

Harley, J. L., and Smith, S. E. (1983). "Mycorrhizal Symbiosis." Academic Press, London.

Harley, K. L. S. (1990). Production of viable seeds by water lettuce, *Pistia stratiotes* L., in Australia. *Aquat. Bot.* **36,** 277–279.

Harris, D., and Davy, A. J. (1986). Regenerative potential of *Elymus farctus* from rhizome fragments and seed. *J. Ecol.* **74,** 1057–1067.

Harris, D., and Davy, A. J. (1987). Seedling growth in *Elymus farctus* after episodes of burial with sand. *Ann. Bot.* **60,** 587–593.

Harris, S. W., and Marshall, W. H. (1960). Experimental germination of seed and establishment of seedlings of *Phragmites communis*. *Ecology* **41,** 395.

Harrison, C. R. (1977). Ultrastructural and histochemical changes during the germination of *Cattleya aurantiaca* (Orchidaceae). *Bot. Gaz.* **138,** 41–45.

Harrison, C. R., and Arditti, J. (1978). Physiological changes during the germination of *Cattleya aurantiaca* (Orchidaceae). *Bot. Gaz.* **139,** 180–189.

Harrison, P. G. (1991). Mechanisms of seed dormancy in an annual population of *Zostera marina* (eelgrass) from The Netherlands. *Can. J. Bot.* **69,** 1972–1976.

Harrison, P. G. (1993). Variations in demography of *Zostera marina* and *Z. noltii* on an intertidal gradient. *Aquat. Bot.* **45,** 63–77.

Hartleb, C. F., Madsen, J. D., and Boylen, C. W. (1993). Environmental factors affecting seed germination in *Myriophyllum spicatum* L. *Aquat. Bot.* **45,** 15–25.

Hartman, G. L., and Tanimonure, O. A. (1991). Seed populations of *Striga* species in Nigeria. *Plant Dis.* **75,** 494–496.

Harty, R. L., and McDonald, T. J. (1972). Germination behaviour in beach spinifex (*Spinifex hirsutus* Labill.). *Aust. J. Bot.* **20,** 241–251.

Harvais, G. (1972). The development and growth requirements of *Dactylorhiza purpurella* in asymbiotic cultures. *Can. J. Bot.* **50,** 1223–1229.

Harvais, G. (1973). Growth requirements and development of *Cypripedium reginae* in axenic culture. *Can. J. Bot.* **51,** 327–332.

Harvais, G. (1974). Notes on the biology of some native orchids of Thunder Bay, their endophytes and symbionts. *Can. J. Bot.* **52,** 451–460.

Harvais, G., and Hadley, G. (1967). The development of *Orchis purpurella* in asymbiotic and inoculated cultures. *New Phytol.* **66,** 217–230.

Harvais, G., and Raitsakas, A. (1975). On the physiology of a fungus symbiotic with orchids. *Can. J. Bot.* **53,** 144–155.

Hassawy, G. S. (1973). *Cuscuta* species in Iraq: Their hosts and seed germination. *In* "Parasitic Weeds Research Group," pp. 280–288. Proc. European Weed Res. Coun. Symp. Parasitic Weeds, Wageningen, The Netherlands.

Hauck, C., Muller, S., and Schildknecht, H. (1992). A germination stimulant for parasitic flowering plants from *Sorghum bicolor*, a genuine host plant. *J. Plant Physiol.* **139,** 474–478.

Hawksworth, F. G. (1961). "Dwarfmistletoe of Ponderosa Pine in the Southwest." *USDA Tech. Bull. No.* 1246.

Hawksworth, F. G. (1965). Life tables for two species of dwarfmistletoe. I. Seed dispersal, interception, and movement. *For. Sci.* **11,** 142–151.

Hawksworth, F. G. (1987). Paleobotany and evolution of the dwarf mistletoes (*Arceuthobium*). *In* "Parasitic Flowering Plants." (H. C. Weber and W. Forstreuter, eds.), pp. 309–316. Proc. 4th Int. Symp. Parasitic Flowering Plants, Marburg, FRG.

Healey, P. L., Michaud, J. D., and Arditti, J. (1980). Morphometry of orchid seeds. III. Native California and related species of *Goodyera, Piperia, Platanthera* and *Spiranthes*. *Am. J. Bot.* **67,** 508–518.

Heather, J. B., Mittal, R. S. D., and Sih, C. J. (1974). The total synthesis of dl-strigol. *J. Am. Chem. Soc.* **96,** 1976–1977.

Heinricher, E. (1915). Bietrage zur Biologie der Zwergmistel, *Arceuthobium oxycedri*, besonders zur Kenntnis des anatomischen Baues und der Mechanik ihrer explosiven Beeren. *Akad. Wissen. Wien (Abt. 1)* **124,** 181–230. + 4 plates.

Heinricher, E. (1917). Die erste Aufzucht einer Rafflesiacee, *Cytinus hypocistic* L., aus Samen. *Ber. Deuts. Bot. Gesell.* **35,** 505–512.

Hester, M. W., and Mendelssohn, I. A. (1987). Seed production and germination response of four Louisiana populations of *Uniola paniculata* (Gramineae). *Am. J. Bot.* **74,** 1093–1101.

Hewett, D. G. (1964). Biological flora of the British Isles: *Menyanthes trifoliata* L. *J. Ecol.* **52,** 723–735.

Hinds, T. E., Hawksworth, F. G., and McGinnies, W. J. (1963). Seed discharge in *Arceuthobium*: A photographic study. *Science* **140,** 1236–1238.

Hiron, R. W. P. (1973). An investigation into the processes involved in germination of *Orobanche crenata* using a new bio-assay technique. *In* "Parasitic Weeds Research Group," pp. 76–87. Proc. European Weed Res. Coun. Symp. Parasitic Weeds, Wageningen, The Netherlands.

Hocking, P. J. (1982). Salt and mineral nutrient levels in fruits of two strand species, *Cakile maritima* and *Arctotheca populifolia*, with special reference to the effect of salt on the germination of *Cakile*. *Ann. Bot.* **50,** 335–343.

Hocking, P. J., and Fineran, B. A. (1983). Aspects of the nutrition of root parasitic Loranthaceae. *In* "The Biology of Mistletoes" (M. Calder and P. Bernhardt, eds.), pp. 229–258. Academic Press, Paris.

Hogue, E. J. (1976). Seed dormancy of nodding beggarticks (*Bidens cernua* L.). *Weed Sci.* **24,** 375–378.

Holm, R. E. (1972). Volatile metabolites controlling germination in buried weed seeds. *Plant Physiol.* **50,** 293–297.

Holmes, G. W., Worley, D. J., Evans, L. R., and Oshima, N. (1968). Seed germination in dwarf mistletoes. *Colorado-Wyoming Acad. Sci.* **5,** 48. [Abstract]

Homes, J., and Vanseveren-Van Espen, N. (1973). Quelques formes de plastes induites par le milieu de culture dans des protocormes d'orchidees cultives in vitro. *Bull. Soc. Roy. Bot. Belgique* **106,** 117–121.

Honek, A., and Martinkova, Z. (1992). The induction of secondary seed dormancy by oxygen deficiency in a barnyard grass *Echinochloa crus-galli*. *Experientia* **48,** 904–906.

Hootsmans, M. J. M., Vermaat, J. E., and van Vierssen, W. (1987). Seed-bank development, germination and early seedling survival of two seagrass species from The Netherlands: *Zostera marina* L. and *Zostera noltii* Hornem. *Aquat. Bot.* **28,** 275–285.

Houle, G. (1996). Environmental filters and seedling recruitment on a coastal dune in subarctic Quebec (Canada). *Can. J. Bot.* **74,** 1507–1513.

Housley, T. L., Ejeta, G., Cherif-Ari, O., Netzly, D. H., and Butler, L. G. (1987). Progress towards an understanding of sorghum resistance to *Striga*. *In* "Parasitic Flowering Plants" (H. C. Weber and W. Forstreuter, eds.), pp. 411–419. Proc. 4th Int. Symp. Parasitic Flowering Plants, Marburg, FRG.

Hsiao, A. I., Worsham, A. D., and Moreland, D. E. (1979). Factors affecting conditioning and germination of witchweed [*Striga asiatica* (L.) Kuntze] seeds under laboratory conditions. *In* "The Second International Symposium on Parasitic Weeds" (L. J. Musselman, A. D. Worsham, and R. E. Eplee, eds.), pp. 193–201. North Carolina State Univ., Raleigh.

Hsiao, A. I., Worsham, A. D., and Moreland, D. E. (1981). Regulation of witchweed (*Striga asiatica*) conditioning and germination by dl-strigol. *Weed Sci.* **29,** 101–104.

Hsiao, A. I., Worsham, A. D., and Moreland, D. E. (1983). Leaching and degradation of dl-strigol in soil. *Weed Sci.* **31,** 763–765.

Hsiao, A. I., Worsham, A. D., and Moreland, D. E. (1987). Effects of drying and dl-strigol on seed conditioning and germination of *Striga asiatica* (L.) Kuntze. *Weed Res.* **27,** 321–328.

Hsiao, A. I., Worsham, A. D., and Moreland, D. E. (1988a). Effects of temperature and dl-strigol on seed conditioning and germination of witchweed (*Striga asiatica*). *Ann. Bot.* **61,** 65–72.

Hsiao, A. I., Worsham, A. D., and Moreland, D. E. (1988b). Effects of chemicals often regarded as germination stimulants on seed conditioning and germination of witchweed (*Striga asiatica*). *Ann. Bot.* **62**, 17–24.

Hudler, G., and French, D. W. (1976). Dispersal and survival of seed of eastern dwarf mistletoe. *Can. J. For. Res.* **6**, 335–340.

Huiskes, A. H. L. (1977). The natural establishment of *Ammophila arenaria* from seed. *Oikos* **29**, 133–136.

Huiskes, A. H. L., Stienstra, A. W., Koutstaal, B. P., Markusse, M. M., and van Soelen, J. (1985). Germination ecology of *Salicornia dolichostachya* and *Salicornia brachystachya*. *Acta Bot. Neerl.* **34**, 369–380.

Hutchinson, I., and Smythe, S. R. (1986). The effect of antecedent and ambient salinity levels on seed germination in populations of *Carex lyngbyei* Hornem. *Northwest Sci.* **60**, 36–41.

Hutchison, J. M., and Ashton, F. M. (1979). Effect of desiccation and scarification on the permeability and structure of the seed coat of *Cuscuta campestris*. *Am. J. Bot.* **66**, 40–46.

Hutchison, J. M., and Ashton, F. M. (1980). Germination of field dodder (*Cuscuta campestris*). *Weed Sci.* **28**, 330–333.

Hyder, S. Z., and Yasmin, S. (1972). Salt tolerance and cation interaction in alkali sacaton at germination. *J. Range Manage.* **25**, 390–392.

Ichihashi, S. (1989). Seed germination of *Ponerorchis graminifolia*. *Lindleyana* **4**, 161–163.

Igbinnosa, I., and Okonkwo, S. N. O. (1992). Stimulation of germination of seeds of cowpea witchweed (*Striga gesnerioides*) by sodium hypochlorite and some growth regulators. *Weed Sci.* **40**, 25–28.

Ignaciuk, R., and Lee, J. A. (1980). The germination of four annual strand-line species. *New Phytol.* **84**, 581–591.

Isaac, F. M. (1969). Floral structure and germination in *Cymodocea ciliata*. *Phytomorphology* **19**, 44–51.

Isely, D. (1944). A study of conditions that affect the germination of *Scirpus* seeds. *Cornell Univ. Agric. Exp. Sta. Mem.* 257.

Ismail, A. M. A. (1990). Germination ecophysiology in populations of *Zygophyllum qatarense* Hadidi from contrasting habitats: Effect of temperature, salinity and growth regulators with special reference to fusicoccin. *J. Arid. Environ.* **18**, 185–194.

Ismail, A. M. A., and El-Ghazaly, G. A. (1990). Phenological studies on *Zygophyllum qatarense* Hadidi from contrasting habitats. *J. Arid Environ.* **18**, 195–205.

Jackson, M. B., and Parker, C. (1991). Induction of germination by a strigol analogue requires ethylene action in *Striga hermonthica* but not in *S. forbesii*. *J. Plant Physiol.* **138**, 383–386.

Jacobsohn, R., Uriely, E., and Dagan, J. (1987). Preliminary experiments on broomrape control with VAPAM. *In* "Parasitic Flowering Plants" (H. C. Weber and W. Forstreuter, eds.), pp. 421–426. Proc. 4th Int. Symp. Parasitic Flowering Plants, Marburg, FRG.

Jain, R., and Foy, C. L. (1987). Influence of various nutrients and growth regulators on germination and parasitism of *Orobanche aegyptiaca*. *In* "Parasitic Flowering Plants" (H. C. Weber and W. Forstreuter, eds.), pp. 427–436. Proc. 4th Int. Symp. Parasitic Flowering Plants, Marburg, FRG.

Jain, R., and Foy, C. L. (1992). Nutrient effects on parasitism and germination of Egyptian broomrape (*Orobanche aegyptiaca*). *Weed Technol.* **6**, 269–275.

Jefferies, R. L., Davy, A. J., and Rudmik, T. (1981). Population biology of the salt marsh annual *Salicornia europaea* agg. *J. Ecol.* **69**, 17–31.

Jefferies, R. L., Jensen, A., and Bazely, D. (1983). The biology of the annual *Salicornia europaea* agg. at the limits of its range in Hudson Bay. *Can. J. Bot.* **61**, 762–773.

Jhamb, R. B., and Sen, D. N. (1984). Seed germination behaviour of halophytes in Indian Desert: 1. *Suaeda fruticosa* (Linn.) Forsk. *Curr. Sci.* **53**, 100–101.

John, D. M., and Lawson, G. W. (1991). Littoral ecosystems of tropical western Africa. *In* "Ecosystems of the World" (A. C. Mathieson

and P. H. Nienhuis, eds.), Vol. 24, pp. 297–322. Elsevier, Amsterdam.

Johnson, A. F. (1985). Ecologia de *Abronia maritima*, especie pionera de las dunas del oeste de Mexico. *Biotica* **10**, 19–34.

Johnson, A. W., Rosebery, G., and Parker, C. (1976). A novel approach to *Striga* and *Orobanche* control using synthetic germination stimulants. *Weed Res.* **16**, 223–227.

Johri, B. M. (1987a). Embryology of *Cuscuta* L. (Cuscutaceae). *In* "Parasitic Flowering Plants" (H. C. Weber and W. Forstreuter, eds.), pp. 445–447. Proc. 4th Int. Symp. Parasitic Flowering Plants, Marburg, FRG.

Johri, B. M. (1987b). Reproductive biology of mistletoes (Loranthaceae and Viscaceae). *In* "Parasitic Flowering Plants" (H. C. Weber and W. Forstreuter, eds.), pp. 449–456. Proc. 4th Int. Symp. Parasitic Flowering Plants, Marburg, FRG.

Johri, B. M., and Bajaj, Y. P. S. (1965). Growth responses of globular proembryos of *Dendrophthoe falcata* (L.f.) Ettings. in culture. *Phytomorphology* **15**, 292–300.

Johri, B. M., and Bhojwani, S. S. (1970). Embryo morphogenesis in the stem parasite *Scurrula pulverulenta*. *Ann. Bot.* **34**, 685–690.

Joshi, A. C. (1933). A suggested explanation of the prevalence of vivipary on the sea-shore. *J. Ecol.* **21**, 209–212.

Joshi, A. C. (1934). A supplementary note on "A suggested explanation of the prevalence of vivipary on the sea-shore." *J. Ecol.* **22**, 306–307.

Joshi, A. J., and Iyengar, E. R. R. (1977). Germination of *Suaeda nudiflora* Moq. *Geobios* **4**, 267–268.

Joshi, A. J., and Iyengar, E. R. R. (1982). Effect of salinity on the germination of *Salicornia brachiata* Roxb. *Indian J. Plant Physiol.* **25**, 65–69.

Joshi, A. J., and Misra, M. (1990). Effects of seawater on seed germination and seedling growth of halophytes. *In* "International Symposium on Environmental Influences on Seed and Germination Mechanism: Recent Advances in Research and Technology" (D. N. Sen, S. Mohammed, P. K. Kasera, and T. P. Thomas, eds.), pp. 93–94. Univ. Jodhpur, Jodhpur, India. [Abstract]

Juncosa, A. M. (1982). Developmental morphology of the embryo and seedling of *Rhizophora mangle* L. (Rhizophoraceae). *Am. J. Bot.* **69**, 1599–1611.

Juniper, B. E., Robins, R. J., and Joel, D. M. (1989). "The Carnivorous Plants." Academic Press, London.

Jurik, T. W., Wang, S.-C., and van der Valk, A. G. (1994). Effects of sediment load on seedling emergence from wetland seed banks. *Wetlands* **14**, 159–165.

Justice, O. L. (1944). Viability and dormancy in seeds of *Polygonum amphibium* L., *P. coccineum* Muhl. and *P. hydropiperoides* Michx. *Am. J. Bot.* **31**, 369–377.

Kachi, N. (1990). Germination traits and seed-bank dynamics of a biennial plant, *Oenothera glazioviana* Micheli. *Ecol. Res.* **5**, 185–194.

Kachi, N., and Hirose, T. (1990). Optimal time of seedling emergence in a dune-population of *Oenothera glazioviana*. *Ecol. Res.* **5**, 143–152.

Kadry, A. E. R., and Tewfic, H. (1956a). A contribution to the morphology and anatomy of seed germination in *Orobanche crenata*. *Bot. Notiser* **109**, 385–399.

Kadry, A. E. R., and Tewfic, H. (1956b). Seed germination in *Orobanche crenata* Forssk. *Svensk Bot. Tidskr.* **50**, 270–286.

Kasasian, L. (1973). Miscellaneous observations on the biology of *Orobanche crenata* and *O. aegyptiaca*. *In* "Parasitic Weeds Research Group," pp. 68–75. Proc. European Weed Res. Coun. Symp. Parasitic Weeds, Wageningen, The Netherlands.

Kaul, A., and Shankar, V. (1988). Ecology of seed germination of the chenopod shrub *Haloxylon salicornicum*. *Trop. Ecol.* **29**, 110–115.

Kaul, R. B. (1978). Morphology of germination and establishment of aquatic seedlings in Alismataceae and Hydrocharitaceae. *Aquat. Bot.* **5,** 139–147.

Kaul, R. B. (1982). Floral and fruit morphology of *Nepenthes lowii* and *N. villosa,* montane carnivores of Borneo. *Am. J. Bot.* **69,** 793–803.

Kaul, R. B. (1985). Reproductive phenology and biology in annual and perennial Alismataceae. *Aquat. Bot.* **22,** 153–164.

Kawahara, A., and Takada, H. (1961). The germination of *Trapella* seed. I. Some factors influencing stimulation of germination. *Indian J. Plant Physiol.* **4,** 156–168 + 1 plate.

Kearney, T. H. (1904). Are plants of sea beaches and dunes true halophytes? *Bot. Gaz.* **37,** 424–436.

Keddy, P. A., and Constabel, P. (1986). Germination of ten shoreline plants in relation to seed size, soil particle size and water level: An experimental study. *J. Ecol.* **74,** 133–141.

Keddy, P. A., and Reznicek, A. A. (1982). Seed banks and persistent relict shoreline flora. *BioScience* **32,** 132–133.

Keeley, J. E. (1988). Anaerobiosis as a stimulus to germination in two vernal pool grasses. *Am. J. Bot.* **75,** 1086–1089.

Keiffer, C. H., and Ungar, I. A. (1995). Germination responses of halophyte seeds exposed to prolonged hypersaline conditions. *In* "Biology of Salt Tolerant Plants" (M. A. Khan and I. A. Ungar, eds.), pp. 43–50. Univ. Karachi, Pakistan.

Keiffer, C. H., and Ungar, I. A. (1997). The effect of extended exposure to hypersaline conditions on the germination of five inland halopyte species. *Am. J. Bot.* **84,** 104–111.

Kennedy, R. A., Barrett, S. C. H., VanderZee, D., and Rumpho, M. E. (1980). Germination and seedling growth under anaerobic conditions in *Echinochloa crus-galli* (barnyard grass). *Plant, Cell Environ.* **3,** 243–248.

Keren, A., and Evenari, M. (1974). Some ecological aspects of distribution and germination of *Pancratium maritimum* L. *Israel J. Bot.* **23,** 202–215.

Khalaf, K. A. (1992). Evaluation of the biological activity of flax as a trap crop against *Orobanche* parasitism of *Vicia faba. Trop. Agric. (Trinidad)* **69,** 35–38.

Khan, M. A. (1991). Studies on germination of *Cressa cretica* L. seeds. *Pakistan J. Weed Sci. Res.* **4,** 89–98.

Khan, M. A., and Rizvi, Y. (1994). Effect of salinity, temperature, and growth regulators on the germination and early seedling growth of *Atriplex griffithii* var. *stocksii. Can. J. Bot.* **72,** 475–479.

Khan, M. A., and Ungar, I. A. (1984a). The effect of salinity and temperature on the germination of polymorphic seeds and growth of *Atriplex triangularis* Willd. *Am. J. Bot.* **71,** 481–489.

Khan, M. A., and Ungar, I. A. (1984b). Seed polymorphism and germination responses to salinity stress in *Atriplex triangularis* Willd. *Bot. Gaz.* **145,** 487–494.

Khan, M. A., and Ungar, I. A. (1985). The role of hormones in regulating the germination of polymorphic seeds and early seedling growth of *Atriplex triangularis* under saline conditions. *Physiol. Plant.* **63,** 109–113.

Khan, M. A., and Ungar, I. A. (1996). Influence of salinity and temperature on the germination of *Haloxylon recurvum* Bunge. ex. Boiss. *Ann. Bot.* **78,** 547–551.

Khan, M. A., and Ungar, I. A. (1997). Effect of thermoperiod on recovery of seed germination of halophytes from saline conditions. *Am. J. Bot.* **84,** 279–283.

Khan, M. A., and Weber, D. J. (1986). Factors influencing seed germination in *Salicornia pacifica* var. *utahensis. Am. J. Bot.* **73,** 1163–1167.

Khan, M. A., Weber, D. J., and Hess, W. M. (1985). Elemental distribution in seeds of the halophytes *Salicornia pacifica* var. *utahensis* and *Atriplex canescens. Am. J. Bot.* **72,** 1672–1675.

Kimber, A., Korschgen, C. E., and van der Valk, A. G. (1995). The distribution of *Vallisneria americana* seeds and seedling light requirements in the Upper Mississippi River. *Can. J. Bot.* **73,** 1966–1973.

King, B. L. (1989). Seed germination ecology of *Aureolaria virginica* (L.) Penn. (Scrophulariaceae). *Castanea* **54,** 19–28.

Kingsbury, R. W., Radlow, A., Mudie, P. J., Rutherford, J., and Radlow, R. (1976). Salt stress responses in *Lasthenia glabrata,* a winter annual composite endemic to saline soils. *Can. J. Bot.* **54,** 1377–1385.

Kirk, J. T. O. (1983). "Light and Photosynthesis in Aquatic Ecosystems." Cambridge Univ. Press, Cambridge, UK.

Klaren, C. H., and van de Dijk, S. J. (1976). Water relations of the hemiparasite *Rhinanthus serotinus* before and after attachment. *Physiol. Plant.* **38,** 121–125.

Kleifeld, Y., Goldwasser, Y., Herzlinger, G., Joel, D. M., Golan, S., and Kahana, D. (1994). The effect of flax (*Linum usitatissimum* L.) and other crops as trap and catch crops for control of Egyptian broomrape (*Orobanche aegyptiaca* Pers.). *Weed Res.* **34,** 37–44.

Knudson, L. (1922). Nonsymbiotic germination of orchid seeds. *Bot. Gaz.* **73,** 1–25.

Knudson, L. (1924). Further observations on nonsymbiotic germination of orchid seeds. *Bot. Gaz.* **77,** 212–219.

Knudson, L. (1925). Physiological study of the symbiotic germination of orchid seeds. *Bot. Gaz.* **79,** 345–379.

Knudson, L. (1950). Germination of seeds of *Vanilla. Am. J. Bot.* **37,** 241–247.

Knutson, D. M. (1984). "Seed Development, Germination Behavior and Infection Characteristics of Several Species of *Arceuthobium.*" USDA For. Serv. Gen. Tech. Rep. RM-11: 77–84.

Koch, E. W., and Dawes, C. J. (1991). Influence of salinity and temperature on the germination of *Ruppia maritima* L. from the North Atlantic and Gulf of Mexico. *Aquat. Bot.* **40,** 387–391.

Koch, E. W., and Seeliger, U. (1988). Germination ecology of two *Ruppia maritima* L. populations in southern Brazil. *Aquat. Bot.* **31,** 321–327.

Koch, L., and Schulz, D. (1975). Uber Samen und Samenkeimung der *Phalaenopsis heideperle. Die Orchidee* **26,** 27–30.

Koller, D. (1957). Germination-regulating mechanisms in some desert seeds. IV. *Atriplex dimorphostegia* Kar. et Kir. *Ecology* **38,** 1–13.

Kondo, K. (1971). Germination and developmental morphology of seeds in *Utricularia cornuta* Michx. and *Utricularia juncea* Vahl. *Rhodora* **73,** 541–547.

Kovach, D. A., and Bradford, K. J. (1992). Imbibitional damage and desiccation tolerance of wild rice (*Zizania palustris*) seeds. *J. Exp. Bot.* **43,** 747–757.

Kubat, R. (1987). Report of the first investigations of parasitium in Opiliaceae (Santalales). *In* "Parasitic Flowering Plants" (H. C. Weber and W. Forstreuter, eds.), pp. 489–492. Proc. 4th Int. Symp. Parasitic Flowering Plants, Marburg, FRG.

Kuijt, J. (1960). Morphological aspects of parasitism in the dwarf mistletoes (*Arceuthobium*). *Univ. California Publ. Bot.* **30,** 137–404 + plates 34–48.

Kuijt, J. (1963). On the ecology and parasitism of the Costa Rican tree mistletoe, *Gaiadendron punctatum* (Ruiz & Pavon) G. Don. *Can. J. Bot.* **41,** 927–938.

Kuijt, J. (1964). Critical observations on the parasitism of New World mistletoes. *Can. J. Bot.* **42,** 1243–1278.

Kuijt, J. (1965). The anatomy of haustoria and related organs of *Gaiadendron* (Loranthaceae). *Can. J. Bot.* **43,** 687–694.

Kuijt, J. (1967). On the structure and origin of the seedling of *Psittacanthus schiedeanus* (Loranthaceae). *Can. J. Bot.* **45,** 1497–1506.

Kuijt, J. (1969). "The Biology of Parasitic Flowering Plants." Univ. California Press, Berkeley.

Kuijt, J. (1970). Seedling establishment in *Psittacanthus* (Loranthaceae). *Can. J. Bot.* **48**, 705–711.

Kuijt, J. (1989). A note on the germination and establishment of *Phoradendron californicum* (Viscaceae). *Madrono* **36**, 175–179.

Kumar, L. S. S., and Solomon, S. (1940). The influence of light on the germination of species of *Striga*. *Curr. Sci.* **9**, 541.

Kumar, U., and Rangaswamy, N. S. (1977). Regulation of seed germination and polarity in seedling development in *Orobanche aegyptiaca* by growth substances. *Biol. Plant.* **19**, 353–359.

Kumler, M. L. (1969). Plant succession on the sand dunes of the Oregon coast. *Ecology* **50**, 695–704.

Kuo, J., Iizumi, H., Nilsen, B. E., and Aioi, K. (1990). Fruit anatomy, seed germination and seedling development in the Japanese seagrass *Phyllospadix* (Zosteraceae). *Aquat. Bot.* **37**, 229–245.

Kust, C. A. (1963). Dormancy and viability of witchweed seeds as affected by temperature and relative humidity during storage. *Weeds* **11**, 247–250.

Kust, C. A. (1966). A germination inhibitor in *Striga* seeds. *Weeds* **14**, 327–329.

Ladiges, P. Y., Foord, P. C., and Willis, R. J. (1981). Salinity and waterlogging tolerance of some populations of *Melaleuca ericifolia* Smith. *Aust. J. Ecol.* **6**, 203–215.

La Garde, R. V. (1929). Non-symbiotic germination of orchids. *Ann. Missouri Bot. Gard.* **16**, 499–514 + 1 plate.

Lakshmi, B., and Jayachandra. (1979). Physiological variations in *Striga asiatica*. *In* "The Second International Symposium on Parasitic Weeds" (L. J. Musselman, A. D. Worsham, and R. E. Eplee, eds.), pp. 132–143. North Carolina State Univ., Raleigh.

Lal, C., and Gopal, B. (1993). Production and germination of seeds in *Hydrilla verticillata*. *Aquat. Bot.* **45**, 257–261.

Lamont, B. (1982a). Gas content of berries of the Australian mistletoe *Amyema preissii* and the effect of maturity, viscin, temperature and carbon dioxide on germination. *J. Exp. Bot.* **33**, 790–798.

Lamont, B. (1982b). Host range and germination requirements of some South African mistletoes. *S. Afr. J. Sci.* **78**, 41–42.

Lamont, B. (1983a). Mineral nutrition of mistletoes. *In* "The Biology of Mistletoes" (M. Calder and P. Bernhardt, eds.), pp. 185–204. Academic Press, Paris.

Lamont, B. (1983b). Germination of mistletoes. *In* "The Biology of Mistletoes" (M. Calder and P. Bernhardt, eds.), pp. 129–143. Academic Press, Paris.

Lamont, B., and Perry, M. (1977). The effects of light, osmotic potential and atmospheric gases on germination of the mistletoe *Amyema preissii*. *Ann. Bot.* **41**, 203–209.

Larson, J. L., and Stearns, F. W. (1990). Factors influencing seed germination in *Carex scoparia* Schk. *Wetlands* **10**, 277–283.

Lawrence, D. B. (1949). Self-erecting habit of seedling red mangroves (*Rhizophora mangle* L.). *Am. J. Bot.* **36**, 426–427.

Lawrence, G. H. M. (1951). "Taxonomy of Vascular Plants." MacMillan, New York.

Leake, J. R. (1994). Tansley Review No. 69: The biology of mycoheterotrophic ('saprophytic') plants. *New Phytol.* **127**, 171–216.

Leck, M. A. (1996). Germination of macrophytes from a Delaware River tidal freshwater wetland. *Bull. Torrey Bot. Club* **123**, 48–67.

Leck, M. A., Baskin, C. C., and Baskin, J. M. (1994). Germination ecology of *Bidens laevis* (Asteraceae) from a tidal freshwater wetland. *Bull. Torrey Bot. Club* **121**, 230–239.

Leck, M. A., and Simpson, R. L. (1987). Seed bank of a freshwater tidal wetland: Turnover and relationship to vegetation change. *Am. J. Bot.* **74**, 360–370.

Leck, M. A., and Simpson, R. L. (1993). Seeds and seedlings of the Hamilton Marshes, a Delaware River tidal freshwater wetland. *Proc. Acad. Nat. Sci. Philadelphia* **144**, 267–281.

Leck, M. A., and Simpson, R. L. (1994). Tidal freshwater wetland zonation: Seed and seedling dynamics. *Aquat. Bot.* **47**, 61–75.

Lee, J. A. (1993). Dry coastal ecosystems of West Africa. *In* "Ecosystems of the World" (E. van der Maarel, ed.), Vol. 2B, pp. 59–69. Elsevier, Amsterdam.

Lee, J. A., and Ignaciuk, R. (1985). The physiological ecology of strandline plants. *Vegetatio* **62**, 319–326.

Lesko, G. L., and Walker, R. B. (1969). Effect of sea water on seed germination in two Pacific atoll beach species. *Ecology* **50**, 730–734.

Lesica, P. (1992). Autecology of the endangered plant *Howellia aquatilis:* Implications for management and reserve design. *Ecol. Appl.* **2**, 411–421.

Lewis, R. R., III, and Phillips, R. C. (1980). Occurrence of seeds and seedlings of *Thalassia testudinum* Banks ex Konig in the Florida Keys (U.S.A). *Aquat. Bot.* **9**, 377–380.

Liddell, R. W. (1944). Germinating native orchid seed. *Am. Orchid Soc. Bull.* **12**, 344–345.

Liddy, J. (1983). Dispersal of Australian mistletoes: The Cowiebank study. *In* "The Biology of Mistletoes" (M. Calder and P. Bernhardt, eds.), pp. 101–116. Academic Press, Paris.

Light, M. H. S. (1989). Germination in the *Cypripedium/Paphiopedilum* alliance. *Can. Orchid J.* **5**, 11–19.

Light, M. H. S. (1992). Raising masdevallias from seed. *Orchid Rev.* **100**, 264–268.

Linden, B. (1980). Aseptic germination of seeds of northern terrestrial orchids. *Ann. Bot. Fennici* **17**, 174–182.

Linden, B. (1992). Two new methods for pretreatment of seeds of northern orchids to improve germination in axenic culture. *Ann. Bot. Fennici* **29**, 305–313.

Lindquist, B. (1958). A greenhouse culture of *Disa uniflora* Berg. in Gothenburg. *Am. Orchid Soc. Bull.* **27**, 652–657.

Lindquist, B. (1965). The raising of *Disa uniflora* seedlings in Gothenburg. *Am. Orchid Soc. Bull.* **34**, 317–319.

Linke, K.-H., and Vogt, W. (1987). A method and its application for observing germination and early development of *Striga* (Scrophulariaceae) and *Orobanche* (Orobanchaceae). *In* "Parasitic Flowering Plants" (H. C. Weber and W. Forstreuter, eds.), pp. 501–509. Proc. 4th Int. Symp. Parasitic Flowering Plants, Marburg, FRG.

Lloyd, F. E. (1942). "The Carnivorous Plants." Chronica Botanica Co., Waltham, MA.

Logan, D. C., and Stewart, G. R. (1991). Role of ethylene in the germination of the hemiparasite *Striga hermonthica*. *Plant Physiol.* **97**, 1435–1438.

Logan, D. C., and Stewart, G. R. (1995). Thidiazuron stimulates germination and ethylene production in *Striga hermonthica:* Comparison with the effects of GR-24, ethylene and 1-aminocyclopropane-1-carboxylic acid. *Seed Sci. Res.* **5**, 99–108.

Lohammar, A. G. (1954). Matsmaltningens inverkan pa *Potamogeton*-fronas groning. *Fauna och Flora* **49**, 17–32.

Lolas, P. C. (1994). Herbicides for control of broomrape (*Orobanche ramosa* L.) in tobacco (*Nicotiana tabacum* L.). *Weed Res.* **34**, 205–209.

Lombardi, T., Bedini, S., and Onnis, A. (1996). The germination characteristics of a population of *Zannichellia palustris* subsp. *pedicellata*. *Aquat. Bot.* **54**, 287–296.

Lombardi, T., Bertacchi, A., Onnis, A., and Stefani, A. (1991). *Erianthus ravennae* (L.) P. Beauv. (Gramineae): Effetto della salinita su germinazione e crescita iniziale. *Atti Soc. Toscana Sci. Natl. Mem. Ser. B* **98**, 159–170.

Lombardi, T., Fochetti, T., Bertacchi, A., and Onnis, A. (1997). Germination requirements in a population of *Typha latifolia*. *Aquat. Bot.* **56**, 1–10.

Lopez-Granados, F., and Garcia-Torres, L. (1993). Seed bank and other demographic parameters of broomrape (*Orobanche crenata*

Forsk.) populations in faba bean (*Vicia faba* L.). *Weed Res.* **33**, 319–327.

Lopez-Garnados, F., and Garcia-Torres, L. (1996). Effects of environmental factors on dormancy and germination of crenate broomrape (*Orobanche crenata*). *Weed Sci.* **44**, 284–289.

Loques, F., Caye, G., and Mainesz, A. (1990). Germination in the marine phanerogam *Zostera noltii* Hornemann at Golfe Juan, French Mediterranean. *Aquat. Bot.* **38**, 249–260.

Luning, K., and Asmus, R. (1991). Physical characteristics of littoral ecosystems, with special reference to marine plants. *In* "Ecosystems of the World" (A. C. Mathieson and P. H. Nienhuis, eds.), Vol. 24, pp. 7–26. Elsevier, Amsterdam.

Lynn, D. G., Steffens, J. C., Kamut, V. S., Graden, D. W., Shabanowitz, J., and Riopel, J. L. (1981). Isolation and characterization of the first host recognition substance for parasitic angiosperms. *J. Am. Chem. Soc.* **103**, 1868–1870.

Lyshede, O. B. (1992). Studies on mature seeds of *Cuscuta pedicellata* and *C. campestris* by electron microscopy. *Ann. Bot.* **69**, 365–371.

Maas, D. (1989). Germination characteristics of some plant species from calcareous fens in southern Germany and their implications for the seed bank. *Holarct. Ecol.* **12**, 337–344.

Maas, P. J. M. (1979). Neotropical saprophytes. *In* "Tropical Botany" (K. Larsen and L. B. Holm-Nielsen, eds.), pp. 365–370. Academic Press, London.

Mabee, H. F., and Garner, J. H. B. (1974). Seasonal variations of soil fungi isolated from the rhizosphere of *Liriodendron tulipifera* L. *In* "Phenology and Seasonality Modeling" (H. Lieth, ed), pp. 185–190. Springer-Verlag, New York.

Macdonald, K. B., and Barbour, M. G. (1974). Beach and salt marsh vegetation of the North American Pacific coast. *In* "Ecology of Halophytes" (R. J. Reimold and W. H. Queen, eds.), pp. 175–233. Academic Press, New York.

MacDougal, D. T., and Dufrenoy, J. (1946). Criteria of nutritive relations of fungi and seed-plants in mycorrhizae. *Plant Physiol.* **21**, 1–10.

Machado, M. A., and Zetsche, K. (1990). A structural, functional and molecular analysis of plastids of the holoparasites *Cuscuta reflexa* and *Cuscuta europaea*. *Planta* **181**, 91–96.

MacKay, J. B., and Chapman, V. J. (1954). Some notes on *Suaeda australis* Moq. var. *nova zelandica* var. nov. and *Mesembryanthemum australe* Sol. ex Forst. f. *Trans. Roy. Soc. New Zeal.* **82**, 41–47.

Macke, A. J., and Ungar, I. A. (1971). The effects of salinity on germination and early growth of *Puccinellia nuttalliana*. *Can. J. Bot.* **49**, 515–520.

MacLeod, D. (1961). Photosynthesis in *Cuscuta*. *Experientia* **17**, 542–543.

Mahmood, K., and Malik, K. A. (1986). Studies on salt tolerance of *Atriplex undulata*. *In* "Prospectus for Biosaline Research" (R. Ahmad and A. San Pietro, eds.), pp. 149–155. Proc. U.S.-Pakistan Biosaline Research Workshop, Dept. Botany, Univ. Karachi, Karachi, Pakistan.

Mahmood, K., Malik, K. A., Lodhi, M. A. K., and Sheikh, K. H. (1996). Seed germination and salinity tolerance in plant species growing on saline wastelands. *Biol. Plant.* **38**, 309–315.

Mahmoud, A., El Sheikh, A. M., and Abdul Baset, S. (1983). Germination of two halophytes: *Halopeplis perfoliata* and *Limonium axillare* from Saudi Arabia. *J. Arid Environ.* **6**, 87–98.

Malcolm, C. V. (1964). Effect of salt, temperature and seed scarification on germination of two varieties of *Arthrocnemum halocnemoides*. *J. Roy. Soc. W. Aust.* **47**, 72–74.

Malcolm, W. M. (1966). Root parasitism of *Castilleja coccinea*. *Ecology* **47**, 179–186.

Mall, L. P. (1954). Germination of seeds of three common weeds of dry phase of low-lying lands. *Proc. Natl. Acad. Sci. Allahabad, India, Part B* **24**, 197–204.

Mallet, A. I. (1973). Studies in the chemistry of the *Orobanche crenata* germination factor present in the roots of *Vicia faba* and other host plants. *In* "Parasitic Weeds Research Group" pp. 89–98. Proc. European Weed Res. Coun. Symp. Parasitic Weeds, Wageningen, The Netherlands.

Malloch, A. J. C., Bamidele, J. F., and Scott, A. M. (1985). The phytosociology of British sea-cliff vegetation with special reference to the ecophysiology of some maritime cliff plants. *Vegetatio* **62**, 309–317.

Mandaville, J. P. (1990). "Flora of Eastern Saudi Arabia." Kegan Paul International, London.

Mandossian, A. J. (1966). Germination of seeds in *Sarracenia purpurea* (pitcher plant). *Mich. Bot.* **5**, 67–79.

Mangnus, E. M., Stommen, P. L. A., and Zwanenburg, B. (1992a). A standardized bioassay for evaluation of potential germination stimulants for seeds of parasitic weeds. *J. Plant Growth Regul.* **11**, 91–98.

Mangnus, E. M., van Vliet, L. A., Vandenput, D. A. L., and Zwanenburg, B. (1992b). Structural modifications of strigol analogues. Influence of the B and C rings on the bioactivity of the germination stimulant GR24. *J. Agric. Food Chem.* **40**, 1222–1229.

Mangnus, E. M., and Zwanenburg, B. (1992). Tentative molecular mechanism for germination stimulation of *Striga* and *Orobanche* seeds by strigol and its synthetic analogues. *J. Agric. Food Chem.* **40**, 1066–1070.

Manning, J. C., and van Staden, J. (1987). The development and mobilisation of seed reserves in some African orchids. *Aust. J. Bot.* **35**, 343–353.

Maranon, T., Garcia, L. V., and Troncoso, A. (1989). Salinity and germination of annual *Melilotus* from the Guadalquiver delta (SW Spain). *Plant Soil* **119**, 223–228.

Marchioni-Ortu, A., and Bocchieri, E. (1984). A study of the germination responses of a Sardinian population of sea fennel (*Crithmum maritimum*). *Can. J. Bot.* **62**, 1832–1835.

Mariko, S., Kachi, N., Ishikawa, S.-I., and Furukawa, A. (1992). Germination ecology of coastal plants in relation to salt environment. *Ecol. Res.* **7**, 225–233.

Marks, T. C., and Truscott, A. J. (1985). Variation in seed production and germination of *Spartina anglica* within a zoned saltmarsh. *J. Ecol.* **73**, 695–705.

Marler, J. E. (1969). "A Study of the Germination Process of Seeds of *Heteranthera limosa*." Ph.D dissertation, Lousiana State Univ., Baton Rouge.

Marshall, J. D., and Ehleringer, J. R. (1990). Are xylem-tapping mistletoes partially heterotrophic? *Oecologia* **84**, 244–248.

Martin, A. C. (1946). The comparative internal morphology of seeds. *Am. Midl. Nat.* **36**, 513–660.

Martinez, M. L., Valverde, T., and Moreno-Casasola, P. (1992). Germination response to temperature, salinity, light and depth of sowing in ten tropical dune species. *Oecologia* **92**, 343–353.

Masselink, A. K. (1980). Germination and seed population dynamics in *Melampyrum pratense* L. *Acta Bot. Neerl.* **29**, 451–468.

Masuhara, G., and Katsuya, K. (1989). Effects of mycorrhizal fungi on seed germination and early growth of three Japanese terrestrial orchids. *Sci. Hort.* **37**, 331–337.

Masuhara, G., and Katsuya, K. (1994). *In situ* and *in vitro* specificity between *Rhizoctonia* spp., and *Spiranthes sinensis* (Persoon) Ames. var. *amoena* (M. Bieberstein) Hara (Orchidaceae). *New Phytol.* **127**, 711–718.

Matsuo, K., Noguchi, K., and Nara, M. (1984). Ecological studies on *Rorippa islandica* (Oeder) Borb. 1. Dormancy and external conditions inducing seed germination. *Weed Res. (Japan)* **29**, 220–225. [In Japanese with English summary]

Maun, M. A. (1981). Seed germination and seedling establishment of *Calamovilfa longifolia* on Lake Huron sand dunes. *Can. J. Bot.* **59**, 460–469.

Maun, M. A. (1993). Dry coastal ecosystems along the Great Lakes. *In* "Ecosystem of the World" (E. van der Maarel, ed.), Vol. 2B, pp. 299–316. Elsevier, Amsterdam.

Maun, M. A. (1994). Adaptations enhancing survival and establishment of seedlings on coastal dune systems. *Vegetatio* **111**, 59–70.

Maun, M. A., and Lapierre, J. (1986). Effects of burial by sand on seed germination and seedling emergence of four dune species. *Am. J. Bot.* **73**, 450–455.

Maun, M. A., and Payne, A. M. (1989). Fruit and seed polymorphism and its relation to seedling growth in the genus *Cakile. Can. J. Bot.* **67**, 2743–2750.

Maun, M. A., and Riach, S. (1981). Morphology of caryopses, seedlings and seedling emergence of the grass *Calamovilfa longifolia* from various depths in sand. *Oecologia* **49**, 137–142.

Mauseth, J. D., Montenegro, G., and Walckowiak, A. M. (1985). Host infection and flower formation by the parasite *Tristerix aphyllus* (Loranthaceae). *Can. J. Bot.* **63**, 567–581.

Maze, K. M., and Whalley, R. D. B. (1992). Germination, seedling occurrence and seedling survival of *Spinifex sericeus* R. Br. (Poaceae). *Aust. J. Ecol.* **17**, 189–194.

McCormac, J. S., and Windus, J. L. (1993). Fire and *Cuscuta glomerata* Choisy in Ohio: A connection? *Rhodora* **95**, 158–165.

McDaniel, S. (1971). "The Genus *Sarracenia* (Sarraceniaceae)." *Bull. Tall Timbers Res. Sta. No. 9.*

McIntyre, S., Mitchell, D. S., and Ladiges, P. Y. (1989a). Seedling mortality and submergence in *Diplachne fusca:* A semi-aquatic weed of rice fields. *J. Appl. Ecol.* **26**, 537–549.

McIntyre, S., Mitchell, D. S., and Ladiges, P. Y. (1989b). Germination and seedling emergence in *Diplachne fusca*: A semi-aquatic weed of rice fields. *J. Appl. Ecol.* **26**, 551–562.

McLaughlin, E. G. (1974). Autecological studies of three species of *Callitriche* native in California. *Ecol. Monogr.* **44**, 1–16.

McLuckie, J. (1923). Studies in parasitism: A contribution to the physiology of the Loranthaceae of New South Wales. *Bot. Gaz.* **75**, 333–367 + plates 14 and 15.

McMillan, C. (1981). Seed reserves and seed germination for two seagrasses, *Halodule wrightii* and *Syringodium filiforme,* from the western Atlantic. *Aquat. Bot.* **11**, 279–296.

McMillan, C. (1983a). Seed germination for an annual form of *Zostera marina* from the Sea of Cortez, Mexico. *Aquat. Bot.* **16**, 105–110.

McMillan, C. (1983b). Seed germination in *Halodule wrightii* and *Syringodium filiforme* from Texas and the U.S. Virgin Islands. *Aquat. Bot.* **15**, 217–220.

McMillan, C. (1987). Seed germination and seedling morphology of the seagrass, *Halophila engelmannii* (Hydrocharitaceae). *Aquat. Bot.* **28**, 179–188.

McMillan, C. (1988a). Seed germination and seedling development of *Halophila decipiens* Ostenfeld (Hydrocharitaceae) from Panama. *Aquat. Bot.* **31**, 169–176.

McMillan, C. (1988b). The seed reserve of *Halophila decipiens* Ostenfeld (Hydrocharitaceae) in Panama. *Aquat. Bot.* **31**, 177–182.

McMillan, C. (1988c). The seed reserve of *Halophila engelmannii* (Hydrocharitaceae) in Redfish Bay, Texas. *Aquat. Bot.* **30**, 253–259.

McMillan, C. (1991). The longevity of seagrass seeds. *Aquat. Bot.* **40**, 195–198.

McMillan, C., Bridges, K. W., Kock, R. L., and Falanruw, M. (1982). Fruit and seedlings of *Cymodocea rotundata* in Yap, Micronesia. *Aquat Bot.* **14**, 99–105.

McMillan, C., and Soong, K. (1989). An annual cycle of flowering, fruiting and seed reserve for *Halophila decipiens* Ostenfeld (Hydrocharitaceae) in Panama. *Aquat. Bot.* **34**, 375–379.

McNaughton, S. J. (1968). Autotoxic feedback in relation to germination and seedling growth in *Typha latifolia. Ecology* **49**, 367–369.

Mead, J. W., and Bulard, C. (1975). Effects of vitamins and nitrogen sources on asymbiotic germination and development of *Orchis laxiflora* and *Ophrys sphegodes. New Phytol.* **74**, 33–40.

Mead, J. W., and Bulard, C. (1979). Vitamins and nitrogen requirements of *Orchis laxiflora* Lamk. *New Phytol.* **83**, 129–136.

Meney, K. A., Pate, J. S., and Dixon, K. W. (1990). Comparative morphology, anatomy, phenology and reproductive biology of *Alexgeorgea* spp. (Restionaceae) from south-western Western Australia. *Aust. J. Bot.* **38**, 523–541.

Meredith, T. C., and Grubb, P. J. (1993). Biological Flora of the British Isles. *Peucedanum palustre* (L.) Moench. *J. Ecol.* **81**, 813–826.

Mesa-Garcia, J., and Garcia-Torres, L. (1986). Effect of planting date on parasitism of broadbean (*Vicia faba*) by crenate broomrape (*Orobanche crenata*). *Weed Sci.* **34**, 544–550.

Mhehe, G. L. (1987). A novel chemical approach to the control of witchweed [*Striga asiatica* (L.) Kuntze] and other *Striga* spp. (Scrophulariaceae). *In* "Parasitic Flowering Plants" (H. C. Weber and W. Forstreuter, eds.), pp. 563–574. Proc. 4th Int. Symp. Parasitic Flowering Plants, Marburg, FRG.

Midgley, A. R. (1926). Effect of alternate freezing and thawing on the impermeability of alfalfa and dodder seeds. *Agron. J.* **18**, 1087–1098.

Miki, S. (1933). On the sea-grasses in Japan (I) *Zostera* and *Phyllospadix,* with special reference to morphological and ecological characters. *Bot. Mag. Tokyo* **47**, 842–862.

Milberg, P., and Stridh, B. (1994). Frobanken hos nagra ettariga amfibiska vaxter vid Vikarsjon i Halsingland. *Svensk Bot. Tidskr.* **88**, 237–240.

Miyazaki, S., and Nagamatsu, T. (1965). Studies on the promotion of the early growth *in vitro* of Orchid. I. *Agric. Bull. Saga Prefectorial Univ.* **21**, 131–149.

Mohammed, S., and Sen, D. N. (1990). Germination behaviour of some halophytes in Indian desert. *Indian J. Exp. Biol.* **28**, 545–549.

Moore, D. R. J., and Keddy, P. A. (1988). Effects of a water-depth gradient on the germination of lakeshore plants. *Can. J. Bot.* **66**, 548–552.

Moore, K. A., Orth, R. J., and Nowak, J. F. (1993). Environmental regulation of seed germination in *Zostera marina* L. (eelgrass) in Chesapeake Bay: Effects of light, oxygen and sediment burial. *Aquat. Bot.* **45**, 79–91.

Mooring, M. T., Cooper, A. W., and Seneca, E. D. (1971). Seed germination response and evidence for height ecophenes in *Spartina alterniflora* from North Carolina. *Am. J. Bot.* **58**, 48–55.

Moreno-Casasola, P. (1986). Sand movement as a factor in the distribution of plant communities in a coastal dune system. *Vegetatio* **65**, 67–76.

Morgan, M. D. (1990). Seed germination characteristics of *Iris virginica. Am. Midl. Nat.* **124**, 209–213.

Morgan, W. C., and Myers, B. A. (1989). Germination of the salt-tolerant grass *Diplachne fusca.* I. Dormancy and temperature responses. *Aust. J. Bot.* **37**, 225–237.

Morinaga, T. (1926a). Effect of alternating temperatures upon the germination of seeds. *Am. J. Bot.* **13**, 141–158.

Morinaga, T. (1926b). The favorable effect of reduced oxygen supply upon the germination of certain seeds. *Am. J. Bot.* **13**, 159–166.

Morinaga, T. (1926c). Germination of seeds under water. *Am. J. Bot.* **13**, 126–140.

Moubasher, A. H., and El-Dohlob, S. M. (1970). Seasonal fluctuations of Egyptian soil fungi. *Trans. Br. Mycol. Soc.* **54**, 45–51.

Mrozek, E., Jr. (1980). Effect of mercury and cadmium on germination of *Spartina alterniflora* Loisel. seeds at various salinities. *Environ. Exp. Bot.* **20,** 367–377.

Mrozek, E., Jr., and Funicelli, N. A. (1982). Effect of zinc and lead on germination of *Spartina alterniflora* Loisel. seeds at various salinities. *Environ. Exp. Bot.* **22,** 23–32.

Muenscher, W. C. (1936a). "Storage and Germination of Seeds of Aquatic Plants." *Cornell Univ. Agric. Exp. Sta. Bull.* 652.

Muenscher, W. C. (1936b). The germination of seeds of *Potamogeton. Ann. Bot.* **50,** 805–821.

Muir, H. J. (1989). Germination and mycorrhizal fungus compatibility in European orchids. *In* "Modern Methods in Orchid Conservation: The Role of Physiology, Ecology and Management" (H. W. Pritchard, ed.), pp. 39–56. Cambridge Univ. Press, Cambridge, UK.

Muir, J. A. (1977). Dwarf mistletoe seed dispersal and germination in southwestern Alberta. *Can. J. For. Res.* **7,** 589–594.

Muller, S., Hauck, C., and Schildknecht, H. (1992). Germination stimulants produced by *Vigna unguiculata* Walp. cv Saunders upright. *J. Plant Growth Regul.* **11,** 77–84.

Muller, S., van der Merwe, A., Schildknecht, H., and Visser, J. H. (1993). An automated system for large-scale recovery of germination stimulants and other root exudates. *Weed Sci.* **41,** 138–143.

Mumera, L. M., and Below, F. E. (1993). Role of nitrogen in resistance to *Striga* parasitism of maize. *Crop Sci.* **33,** 758–763.

Musselman, L. J. (1969). Observations on the life history of *Aureolaria grandiflora* and *Aureolaria pedicularia* (Scrophulariaceae). *Am. Midl. Nat.* **82,** 307–311.

Musselman, L. J. (1972). Root parasitism of *Macranthera flammea* and *Tomanthera auriculata* (Scrophulariaceae). *J. Elisha Mitchell Sci. Soc.* **88,** 57–60.

Musselman, L. J. (1980). The biology of *Striga, Orobanche,* and other root-parasitic weeds. *Annu. Rev. Phytopath.* **18,** 463–489.

Musselman, L. J., and Dickison, W. C. (1975). The structure and development of the haustorium in parasitic Scrophulariaceae. *Bot. J. Linn. Soc.* **70,** 183–212 + 9 plates.

Musselman, L. J., and Mann, W. F., Jr. (1977a). Cataphyll behavior in *Ximenia americana* seedlings (Olacaceae). *Beitr. Biol. Pflanzen* **53,** 121–125.

Musselman, L. J., and Mann, W. F., Jr. (1977b). Seed germination and seedlings of *Krameria lanceolata* (Krameriaceae). *Sida* **7,** 224–225.

Musselman, L. J., and Mann, W. F., Jr. (1978). "Root parasites of southern forests." *USDA For. Serv. Gen. Tech. Rep.* SO-20.

Musselman, L. J., and Mann, W. F., Jr. (1979). Notes on seed germination and parasitism of seedlings of *Buckleya distichophylla* (Santalaceae). *Castanea* **44,** 108–113.

Myers, B. A., and Couper, D. I. (1989). Effects of temperature and salinity on the germination of *Puccinellia ciliata* (Bor) cv. Menemen. *Aust. J. Agric. Res.* **40,** 561–571.

Myers, B. A., and Morgan, W. C. (1989). Germination of the salt-tolerant grass *Diplachne fusca.* II. Salinity responses. *Aust. J. Bot.* **37,** 239–251.

Nagaveni, H. C., and Srimathi, R. A. (1980). Studies on germination of the sandal seeds, *Santalum album* Linn. II. Chemical stimulant for germination. *Indian For.* **106,** 792–799.

Naidoo, G., and Naicker, K. (1992). Seed germination in the coastal halophytes *Triglochin bulbosa* and *Triglochin stricta. Aquat. Bot.* **42,** 217–229.

Nakamura, S. J. (1976). Atmospheric conditions required for the growth of *Galeola septentrionalis* seedlings. *Bot. Mag. Tokyo* **89,** 211–218.

Nakamura, S. J., Uchida, T., and Hamada, M. (1975). Atmospheric conditions controlling the seed germination of an achlorophyllous orchid, *Galeola septentrionalis. Bot. Mag. Tokyo* **88,** 103–109.

Nash, S. M., and Wilhelm, S. (1960). Stimulation of broomrape seed germination. *Phytopathology* **50,** 772–774.

Nazrul-Islam, A. K. M. (1993). Environment and vegetation of Sundarban mangrove forest. *In* "Towards the Rational Use of High Salinity Tolerant Plants" (H. Leith and A. Al Masoom, eds.), Vol. 1, pp. 81–88. Kluwer, Dordrecht.

Netzly, D. H., and Butler, L. G. (1986). Roots of sorghum exude hydrophobic droplets containing biologically active components. *Crop Sci.* **26,** 775–778.

Netzly, D. H., Riopel, J. L., Ejeta, G., and Butler, L. G. (1988). Germination stimulants of witchweed (*Striga asiatica*) from hydrophobic root exudate of sorghum (*Sorghum bicolor*). *Weed Sci.* **36,** 441–446.

Ng, F. S. P. (1978). Strategies of establishment in Malayan forest trees. *In* "Tropical Trees as Living Systems" (P. B. Tomlinson and M. H. Zimmermann, eds.), pp. 129–162. Cambridge Univ. Press, Cambridge, UK.

Nickrent, D. L., Musselman, L. J., Riopel, J. L., and Eplee, R. E. (1979). Haustorial initiation and non-host penetration in witchweed (*Striga asiatica*). *Ann. Bot.* **43,** 233–236.

Nikolaeva, M. G. (1969). Physiology of deep dormancy in seeds. Izdatel'stvo "Nauka," Leningrad. [Translation from Russian by Z. Shapiro, National Science Foundation, Washington, DC.]

Nishimura, G. (1982). Japanese orchids. *In* "Orchid Biology: Reviews and Perspectives" (J. Arditti, ed.), Vol. II, pp. 331–336. Cornell Univ. Press, Ithaca, NY.

Nishimura, G. (1991). Comparative morphology of cotyledonous orchid seedlings. *Lindleyana* **6,** 140–146.

Noggle, G. R., and Wynd, F. L. (1943). Effects of vitamins on germination and growth of orchids. *Bot. Gaz.* **104,** 455–459.

Nun, N. B., and Mayer, A. M. (1993). Preconditioning and germination of *Orobanche* seeds: Respiration and protein synthesis. *Phytochemistry* **34,** 39–45.

Nwoke, F. I. O., and Okonkwo, S. N. C. (1978). Structure and development of the primary hastorium in *Alectra vogelii* Benth. (Scrophulariaceae). *Ann. Bot.* **42,** 447–454.

Nwoke, F. I. O., and Okonkwo, S. N. O. (1980). Photocontrol of seed germination in the hemiparasite *Buchnera hispida* (Scrophulariaceae). *Physiol. Plant.* **49,** 388–392.

Obeid, M., and El Seed, M. G. (1976). Factors affecting dormancy and germination of seeds of *Eichhornia crassipes* (Mart.) Solms from the Nile. *Weed Res.* **16,** 71–80.

Oelke, E. A., and Albrecht, K. A. (1980). Influence of chemical seed treatments on germination of dormant wild rice seeds. *Crop Sci.* **20,** 595–598.

Oesau, V. A. (1973). Keimung und Wurzelwachstum von *Melampyrum arvense* L. (Scrophulariaceae). *Beitr. Biol. Pflanzen* **49,** 73–100.

Oesau, V. A. (1975). Untersuchungen zur Keimung und Entwicklung des Wurzelsystems in der Gattung *Melampyrum* L. (Scrophulariaceae). *Bietr. Biol. Pflanzen* **51,** 121–147.

Ohga, I. (1926). The germination of century-old and recently harvested Indian lotus fruits, with special reference to the effect of oxygen supply. *Am. J. Bot.* **13,** 754–759.

Okonkwo, S. N. C. (1975). *In vitro* post-germination growth and development of embryos of *Alectra* (Scrophulariaceae). *Physiol. Plant.* **34,** 378–383.

Okonkwo, S. N. C. (1987). Developmental studies on witchweeds. *In* "Parasitic Weeds in Agriculture" (L. J. Musselman, ed.), Vol. 1, pp. 64–75. CRC Press, Boca Raton, FL.

Okonkwo, S. N. C., and Nwoke, F. I. O. (1974). Seed germination in *Buchnera hispida* Buch.-Ham. ex D. Don. *Ann. Bot.* **38,** 409–417.

Okonkwo, S. N. C., and Nwoke, F. I. O. (1975a). Observations on haustorial development in *Striga gesnerioides* (Scrophulariaceae). *Ann. Bot.* **39**, 979–981.

Okonkwo, S. N. C., and Nwoke, F. I. O. (1975b). Bleach-induced germination and breakage of dormancy of seeds of *Alectra vogelii*. *Physiol. Plant.* **35**, 175–180.

Okonkwo, S. N. C., and Nwoke, F. I. O. (1978). Initiation, development and structure of the primary haustorium in *Striga gesnerioides* (Scrophulariaceae). *Ann. Bot.* **42**, 455–463.

Okonkwo, S. N. C., and Raghavan, V. (1982). Studies on the germination of seeds of the root parasites, *Alectra vogelii* and *Striga gesnerioides*. I. Anatomical changes in the embryos. *Am. J. Bot.* **69**, 1636–1645.

Okusanya, O. T. (1977). The effect of sea water and temperature on the germination behaviour of *Crithmum maritimum*. *Physiol. Plant.* **41**, 265–267.

Okusanya, O. T. (1979). An experimental investigation into the ecology of some maritime cliff species. *J. Ecol.* **67**, 293–304.

Okusanya, O. T., and Ungar, I. A. (1983). The effects of time of seed production on the germination response of *Spergularia marina*. *Physiol. Plant.* **59**, 335–342.

Oliva, A. P., and Arditti, J. (1984). Seed germination of North American orchids. II. Native California and related species of *Aplectrum*, *Cypripedium*, and *Spiranthes*. *Bot. Gaz.* **145**, 495–501.

O'Neill, E. J. (1972). Alkali bulrush seed germination and culture. *J. Wildl. Manage.* **36**, 649–652.

Onnis, A., and Lombardi, T. (1994). Seed germination in two different *Hordeum maritimum* With. and *H. murinum* L. (Gramineae) populations. *Atti. Soc. Toscana Sci. Mem., Ser. B* **101**, 121–135.

Onnis, A., and Mazzanti, M. (1971). *Althenia filiformis* petit: Azione della temperatura e dell'acqua de mare sulla germinazione. *Gior. Bot. Italiano* **105**, 131–143.

Onnis, A., and Miceli, P. (1975). *Puccinellia festucaeformis* (Host) Parl.: Dormienza e influenza della salinita sulla germinazione. *Giorn. Bot. Italiano* **109**, 27–37.

Onnis, A., and Pelosini, F. (1976). *Althenia filiformis* Petit: Ecologia e significato dell'andamento della germinazione in relazione alle variazioni di temperatura e salinita del substrato nel periodo estivo-autunnale. *Giorn. Bot. Italiano* **110**, 127–136.

Onnis, A., Pelosini, F., and Stefani, A. (1981). *Puccinellia festucaeformis* (Host) Parl.: Germinazione e crescita iniziale in funzione della salinita del substrato. *Giorn. Bot. Italiano* **115**, 103–116.

Oosting, H. J., and Billings, W. D. (1942). Factors effecting vegetation zonation on coastal dunes. *Ecology* **23**, 131–142.

Orozco-Segovia, A. D. L., and Vazquez-Yanes, C. (1980). La germinacion de *Panicum hirsutum* Swartz: Una arvense de cultivos de zonas inundables. *Bol. Soc. Bot. Mexico* **39**, 91–106.

Orth, R. J., and Moore, K. A. (1983). Seed germination and seedling growth of *Zostera marina* L. (eelgrass) in the Chesapeake Bay. *Aquat. Bot.* **15**, 117–131.

Ozturk, M., Esiyok, D., Ozdemir, F., Olcay, G., and Oner, M. (1994). Studies on the effects of growth substances on the germination and seedling growth of *Brassica oleracea* L. var. *acephala* (Karalahana). *J. Fac. Sci. Ege Univ.* **16**(B), 63–70.

Ozturk, M., Gemici, M., Yilmazer, C., and Ozdemir, F. (1993). Alleviation of salinity stress by GA₃, KIN and IAA on seed germination of *Brassica campestris* L. *Doga-Tr. J. Bot.* **17**, 47–52.

Palmisano, A. W. (1971). The effect of salinity on the germination and growth of plants important to wildlife in the Gulf Coast marshes. *Proc. Southeast. Game Fish Comm. Conf.* **25**, 215–223.

Pammenter, N. W., Farrant, J. M., and Berjak, P. (1984). Recalcitrant seeds: Short-term storage effects in *Avicennia marina* (Fors[s]k.) Vierh. may be germination-associated. *Ann. Bot.* **54**, 843–846.

Panda, M. M., and Choudhury, N. K. (1992). Effect of irradiance and

nutrients on chlorophyll and carotenoid content and Hill reaction activity in *Cuscuta reflexa*. *Photosynthetica* **26**, 586–592.

Panetta, F. D. (1979). Germination and seed survival in the woody weed, groundsel bush (*Baccharis halimifolia* L.). *Aust. J. Agric. Res.* **30**, 1067–1077.

Parker, C., and Riches, C. R. (1993). "Parasitic Weeds of the World: Biology and Control." CAB International, Wallingford, UK.

Partridge, T. R., and Wilson, J. B. (1987). Germination in relation to salinity in some plants of salt marshes in Otago, New Zealand. *New Zeal. J. Bot.* **25**, 255–261.

Pate, J. S., and McComb, A. J. (eds.) (1981). "The Biology of Australian Plants." Univ. Western Australia Press, Nedlands.

Pattee, H. E., Allred, K. R., and Wiebe, H. H. (1965). Photosynthesis in dodder. *Weeds* **13**, 193–195.

Patterson, D. T. (1987). Environmental factors affecting witchweed growth and development. *In* "Parasitic Weeds in Agriculture" (L. J. Musselman, ed.), Vol. 1, pp. 28–41. CRC Press, Boca Raton, FL.

Patterson, D. T., Musser, R. L., Flint, E. P., and Eplee, R. E. (1982). Temperature responses and potential for spread of witchweed (*Striga lutea*) in the United States. *Weed Sci.* **30**, 87–93.

Pavlista, A. D., Worsham, A. D., and Moreland, D. E. (1979a). Witchweed seed germination. I. Effects of some chemical and physical treatments. *In* "The Second International Symposium on Parasitic Weeds" (L. J. Musselman, A. D. Worsham, and R. E. Eplee, eds.), pp. 219–227. North Carolina State Univ., Raleigh.

Pavlista, A. D., Worsham, A. D., and Moreland, D. E. (1979b). Witchweed seed germination. II. Stimulatory and inhibitory effects of strigol, GR7, and the effects of organic solvents. *In* "The Second International Symposium on Parasitic Weeds" (L. J. Musselman, A. D. Worsham, and R. E. Eplee, eds.), pp. 228–237. North Carolina State Univ., Raleigh.

Payne, A. M., and Maun, M. A. (1984). Reproduction and survivorship of *Cakile edentula* var. *lacustris* along the Lake Huron shoreline. *Am. Midl. Nat.* **111**, 86–95.

Pearce, D. M. E., and Jackson, M. B. (1991). Comparison of growth responses of barnyard grass (*Echinochloa oryzoides*) and rice (*Oryza sativa*) to submergence, ethylene, carbon dioxide and oxygen shortage. *Ann. Bot.* **68**, 201–209.

Pemadasa, M. A., and Lovell, P. H. (1975). Factors controlling germination of some dune annuals. *J. Ecol.* **63**, 41–59.

Pepperman, A. B., Connick, W. J., Jr., Vail, S. L., Worsham, A. D., Pavlista, A. D., and Moreland, D. E. (1982). Evaluation of precursors and analogs of strigol as witchweed (*Striga asiatica*) seed germination stimulants. *Weed Sci.* **30**, 561–566.

Pesch, C., and Pieterse, A. H. (1982). Inhibition of germination in *Striga* by means of urea. *Experientia* **38**, 559–560.

Philbrick, C. T. (1984). Aspects of floral biology, breeding system, and seed and seedling biology in *Podostemum ceratophyllum* (Podostemaceae). *Syst. Bot.* **9**, 166–174.

Philbrick, C. T., and Novelo R., A. (1994). Seed germination of Mexican Podostemaceae. *Aquat. Bot.* **48**, 145–151.

Philipupillai, J., and Ungar, I. A. (1984). The effect of seed dimorphism on the germination and survival of *Salicornia europaea* L. populations. *Am. J. Bot.* **71**, 542–549.

Phillips, R. C. (1971). Seed germination in *Zostera marina*. *Am. J. Bot.* **58**(No. 5, Part III), 459. [Abstract]

Phillips, R. C., and Backman, T. W. (1983). Phenology and reproductive biology of eelgrass (*Zostera marina* L.) at Bahia Kino, Sea of Cortez, Mexico. *Aquat. Bot.* **17**, 85–90.

Phillips, R. C., Grant, W. S., and McRoy, C. P. (1983). Reproductive strategies of eelgrass (*Zostera marina* L.). *Aquat. Bot.* **16**, 1–20.

Phillips, R. C., McMillan, C., and Bridges, K. W. (1981). Phenology and reproductive physiology of *Thalassia testudinum* from the western tropical Atlantic. *Aquat. Bot.* **11**, 263–277.

Phleger, F. B., and Bradshaw, J. S. (1966). Sedimentary environments in a marine marsh. *Science* **154**, 1551–1553.

Piehl, M. A. (1965). The natural history and taxonomy of *Comandra* (Santalaceae). *Mem. Torrey Bot. Club* **22**, 1–80 + 16 plates.

Pieterse, A. H. (1979). The broomrapes (Orobanchaceae): A review. *Abstr. Trop. Agric.* **5**, 9–35.

Pieterse, A. H. (1981). Germination of *Orobanche crenata* Forsk. seeds in vitro. *Weed Res.* **21**, 279–287.

Pieterse, A. H. (1991). The effect of nitrogen fertilizers on the germination of seeds of *Striga hermonthica* and *Orobanche crenata*. *In* "Progress in *Orobanche* Research" (K. Wegmann and L. J. Musselman, eds.), pp. 115–124. Eberhard-Karls-Universitat, Tubingen, FRG.

Pieterse, A. H., de Lange, L., and Verhagen, L. (1981). A study on certain aspects of seed germination and growth of *Pistia stratiotes* L. *Acta Bot. Neerl.* **30**, 47–57.

Pieterse, A. H., Schoenmakers, F., Scheppers, P., Yehouenou, A., Ransom, J. K., Odhiambo, G., Egbers, S., Baltus, P. C. W., ter Borg, S. J., and Verkleij, J. A. C. (1993). A comparative study of the parasitic weed *Striga hermonthica* (Del.) Benth. regarding seed dormancy and sowing date of the host crop in Benin and Kenya. *In* "Studies on the Biology and Ecology of the Parasitic Weed *Striga* in Connection with Integrated Control Schemes" (A. H. Pieterse, ed.), pp. 49–67. CIMMYT, Kari, Kenya.

Pietropaolo, J., and Pietropaolo, P. (1986). "Carnivorous Plants of the World." Timber Press, Portland, OR.

Pignatti, S. (1993). Dry coastal ecosystems of Italy. *In* "Ecosystems of the World" (E. van der Maarel, ed.), Vol. 2A, pp. 379–390. Elsevier, Amsterdam.

Pirc, H., Buia, M. C., and Mazzella, L. (1986). Germination and seedling development of *Cymodocea nodosa* (Ucria) Ascherson under laboratory conditions and "in situ." *Aquat. Bot.* **26**, 181–188.

Plyler, D. B., and Carrick, K. M. (1993). Site-specific seed dormancy in *Spartina alterniflora* (Poaceae). *Am. J. Bot.* **80**, 752–756.

Poiani, K. A., and Johnson, W. C. (1988). Evaluation of the emergence method in estimating seed bank composition of prairie wetlands. *Aquat. Bot.* **32**, 91–97.

Poljakoff-Mayber, A., Somers, G. F., Werker, E., and Gallagher, J. L. (1992). Seeds of *Kosteletzkya virginica* (Malvaceae): Their structure, germination, and salt tolerance. I. Seed structure and germination. *Am. J. Bot.* **79**, 249–256.

Poljakoff-Mayber, A., Somers, G. F., Werker, E., and Gallagher, J. L. (1994). Seeds of *Kosteletzkya virginica* (Malavaceae): Their structure, germination, and salt tolerance. II. Germination and salt tolerance. *Am. J. Bot.* **81**, 54–59.

Pons, T. L. (1982). Factors affecting weed seed germination and seedling growth in lowland rice in Indonesia. *Weed Res.* **22**, 155–161.

Pons, T. L. (1989). Dormancy, germination and mortality of seeds in heathland and inland sand dunes. *Acta Bot. Neerl.* **38**, 327–335.

Pons, T. L., and Schroder, H. F. J. M. (1986). Significance of temperature fluctuation and oxygen concentration for germination of the rice field weeds *Fimbristylis littoralis* and *Scirpus juncoides*. *Oecologia* **68**, 315–319.

Ponzio, K. J., Miller, S. J., and Lee, M. A. (1995). Germination of sawgrass, *Cladium jamaicense* Crantz, under varying hydrological conditions. *Aquat. Bot.* **51**, 115–120.

Porter, J. N. (1942). The mycorrhiza of *Zeuxine strateumatica*. *Mycologia* **34**, 380–390.

Posner, H. B., and Hillman, W. S. (1962). Aseptic production, collection and germination of seeds of *Lemna perpusilla* 6746. *Physiol. Plant.* **15**, 700–708.

Power, P., and Fonteyn, P. J. (1995). Effects of oxygen concentration and substrate on seed germination and seedling growth of Texas wildrice (*Zizania texana*). *Southw. Nat.* **40**, 1–4.

Press, M. C., Graves, J. D., and Stewart, G. R. (1990). Physiology of the interaction of angiosperm parasites and their higher plant hosts. *Plant, Cell Environ.* **13**, 91–104.

Purer, E. A. (1936). Studies of certain coastal sand dune plants of southern California. *Ecol. Monogr.* **6**, 1–87.

Purves, S., and Hadley, G. (1976). The physiology of symbiosis in *Goodyera repens*. *New Phytol.* **77**, 689–696.

Rabinowitz, D. (1978a). Early growth of mangrove seedlings in Panama, and an hypothesis concerning the relationship of dispersal and zonation. *J. Biogeogr.* **5**, 113–133.

Rabinowitz, D. (1978b). Dispersal properties of mangrove propagules. *Biotropica* **10**, 47–57.

Rajpurohit, K. S., and Sen, D. N. (1979). Seasonal variation in chloride ion percentage of plants and soils of Pachpadra Salt Basin in Indian Desert. *Indian J. Bot.* **2**, 17–23.

Ramati, A., Liphschitz, N., and Waisel, Y. (1976). Ion localization and salt secretion in *Sporobolus arenarius* (Gou.) Duv.-Jouv. *New Phytol.* **76**, 289–294.

Ramsbottom, J. (1922–1923). Orchid mycorrhiza. *Trans. Br. Mycol. Soc.* **8**, 28–61 + plates II–VII.

Randall, R. E. (1970). Salt measurement on the coast of Barbados, West Indies. *Oikos* **21**, 65–70.

Rangan, T. S. (1965). Morphogenesis of the embryo of *Cistanche tubulosa* Wight in vitro. *Phytomorphology* **15**, 180–182.

Rangaswamy, N. S. (1967). Morphogenesis of seed germination in angiosperms. *Phytomorphology* **17**, 477–487.

Rangaswamy, N. S., and Rangan, T. S. (1966). Effects of seed germination-stimulants on the witchweed *Striga euphrasioides* (Vahl) Benth. *Nature* **210**, 440–441.

Rangaswamy, N. S., and Rao, P. S. (1963). Experimental studies on *Santalum album* L.: Establishment of tissue culture of endosperm. *Phytomorphology* **13**, 450–454.

Rao, L. (1925). Quantitative Untersuchungen uber die Wirkung des Lichtes auf die Samenkeimung von *Lythrum salicaria*. *Jahrb. Wiss. Bot.* **64**, 249–280.

Rao, P. N., and Rama Rao, P. V. (1990). Not embryos but seedlings in seeds of *Cuscuta*. *Indian J. Bot.* **13**, 159–160.

Rao, P. N., and Reddy, A. R. S. (1991). Seed formation and seed germination in China dodder *Cuscuta chinensis* Lamk. *In* "Proc. Int. Seed Symp." (D. N. Sen and S. Mohammed, eds.), pp. 9–12. Jodhpur, India.

Rao, P. N., Reddy, B. V. N., and Raghavaswamy, B. V. (1985). Seed germination of *Striga angustifolia* in root elutes of 20 sorghum cultivars. *Proc. Asian-Pacific Weed Sci. Soc. Conf.* **10**, 469–473.

Rasmussen, H. N. (1990). Cell differentiation and mycorrhizal infection in *Dactylorhiza majalis* (Rchb.f.) Hunt & Summerh. (Orchidaceae) during germination *in vitro*. *New Phytol.* **116**, 137–147.

Rasmussen, H. N. (1995). "Terrestrial Orchids from Seed to Mycotrophic Plant." Cambridge Univ. Press, Cambridge, UK.

Rasmussen, H. N., Andersen, T. F., and Johansen, B. (1990a). Light stimulation and darkness requirement for the symbiotic germination of *Dactylorhiza majalis* (Orchidaceae) *in vitro*. *Physiol. Plant.* **79**, 226–230.

Rasmussen, H. N., Andersen, T. F., and Johansen, B. (1990b). Temperature sensitivity of *in vitro* germination and seedling development of *Dactylorhiza majalis* (Orchidaceae) with and without a mycorrhizal fungus. *Plant, Cell Environ.* **13**, 171–177.

Rasmussen, H. N., and Whigham, D. F. (1993). Seed ecology of dust seeds *in situ*: A new study technique and its application in terrestrial orchids. *Am. J. Bot.* **80**, 1374–1378.

Redbo-Torstensson, P. (1994). The demographic consequences of nitrogen fertilization of a population of sundew, *Drosea rotundifolia*. *Acta Bot. Neerl.* **43**, 175–188.

Rediske, J. H., and Shea, K. R. (1961). The production and translocation of photosynthate in dwarfmistletoe and lodgepole pine. *Am. J. Bot.* **48**, 447–452.

Rees, M. (1995). Community structure in sand dune annuals: Is seed weight a key quantity? *J. Ecol.* **83**, 857–863.

Reid, D. C., and Parker, C. (1979). Germination requirements of *Striga* species. *In* "The Second International Symposium on Parasitic Weeds" (L. J. Musselman, A. D. Worsham, and R. E. Eplee, eds.), pp. 202–210. North Carolina State Univ., Raleigh.

Reid, N. (1987). Safe sites for *Amyema quandang* (Lindl.) Van Tiegh. *In* "Parasitic Flowering Plants" (H. C. Weber and W. Forstreuter, eds.), pp. 691–699. Proc. 4th Int. Symp. Parasitic Flowering Plants, Marburg, FRG.

Reid, N. (1989). Dispersal of mistletoes by honeyeaters and flowerpeckers: Components of seed dispersal quality. *Ecology* **70**, 137–145.

Reid, N. (1990). Mutualistic interdependence between mistletoes (*Amyema quandang*), and spiny-cheeked honeyeaters and mistletoebirds in an arid woodland. *Aust. J. Ecol.* **15**, 175–190.

Reid, N. (1991). Coevolution of mistletoes and frugivorous birds? *Aust. J. Ecol.* **16**, 457–469.

Reimold, R. J., and Queen, W. H. (eds.) (1974). "Ecology of Halophytes." Academic Press, New York.

Reyburn, A. N. (1978). The effects of pH on the expression of a darkness-requiring dormancy in seeds of *Cypripedium reginae* Walt. *Am. Orchid Soc. Bull.* **47**, 798–802.

Richards, P. W. (1943). Biological flora of the British Isles: *Juncus filiformis* L. *J. Ecol.* **31**, 60–65.

Richardson, K. A., Peterson, R. L., and Currah, R. S. (1992). Seed reserves and early symbiotic protocorm development of *Platanthera hyperborea* (Orchidaceae). *Can. J. Bot.* **70**, 291–300.

Richter, A., and Popp, M. (1992). The physiological importance of accumulation of cyclitols in *Viscum album* L. *New Phytol.* **121**, 431–438.

Ridley, H. N. (1930). "The Dispersal of Plants throughout the World." L. Reeve and Co., Ltd. Ashford, Kent, England.

Riemer, D. N. (1984). "Introduction to Freshwater Vegetation." Avi Publ. Co., Inc. Westport, CT.

Riemer, D. N. (1985). Seed germination in spatterdock (*Nuphar advena* Ait.). *J. Aquat. Plant Manage.* **23**, 46–47.

Riemer, D. N., and MacMillan, W. W. (1972). Seed germination in arrowarum (*Peltandra virginica* (L.) Kunth). *Proc. Northe. Weed Cont. Conf.* **26**, 183–188.

Rigby, J. F. (1959). Light as a control in the germination and development of several mistletoe species. *Proc. Linn. Soc. New South Wales* **84**, 335–337.

Riopel, J. L. (1979). Experimental studies on induction of haustoria in *Agalinis purpurea*. *In* "The Second International Symposium on Parasitic Weeds" (L. J. Musselman, A. D. Worsham, and R. E. Eplee, eds.), pp. 165–173. North Carolina State Univ., Raleigh.

Riopel, J. L., and Baird, W. V. (1987). Morphogenesis of the early development of primary haustoria in *Striga asiatica*. *In* "Parasitic Weeds in Agriculture" (L. J. Musselman, ed.), Vol. I, pp. 107–125. CRC Press, Boca Raton, FL.

Riopel, J. L., Baird, W. V., Chang, M., and Lynn, D. G. (1990). Haustorial development in *Striga asiatica*. *Monogr. Ser. Weed Sci. Soc. Am.* **5**, 27–36.

Riopel, J. L., and Musselman, L. J. (1979). Experimental initiation of haustoria in *Agalinis purpurea* (Scrophulariaceae). *Am. J. Bot.* **66**, 570–575.

Rivard, P. G., and Woodard, P. M. (1989). Light, ash, and pH effects on the germination and seedling growth of *Typha latifolia* (cattail). *Can. J. Bot.* **67**, 2783–2787.

Rivers, W. G., and Weber, D. J. (1971). The influence of salinity and temperature on seed germination in *Salicornia bigelovii*. *Physiol. Plant.* **24**, 73–75.

Romo, J. T., and Eddleman, L. E. (1985). Germination response of greasewood (*Sarcobatus vermiculatus*) to temperature, water potential and specific ions. *J. Range Manage.* **38**, 117–120.

Romo, J. T., and Haferkamp, M. R. (1987a). Effects of osmotic potential, potassium chloride, and sodium chloride on germination of greasewood (*Sarcobatus vermiculatus*). *Great Basin Nat.* **47**, 110–116.

Romo, J. T., and Haferkamp, M. R. (1987b). Forage kochia germination response to temperature, water stress, and specific ions. *Agron. J.* **79**, 27–30.

Room, P. M. (1971). Some physiological aspects of the relationship between cocoa, *Theobroma cacao*, and the mistletoe *Tapinanthus bangwensis* (Engl. and K. Krause). *Ann. Bot.* **35**, 169–174.

Room, P. M. (1973). Ecology of the mistletoe *Tapinanthus bangwensis* growing on cocoa in Ghana. *J. Ecol.* **61**, 729–742.

Rosa, M. L., and Corbineau, F. (1986). Quelques aspects de la germination des caryopses de *Leersia oryzoides* (L.) Sw. *Weed Res.* **26**, 99–104.

Rozema, J. (1975). The influence of salinity, inundation and temperature on the germination of some halophytes and nonhalophytes. *Oecol. Plant.* **10**, 341–353.

Rozijn, N. A. M. G., and van Andel, J. (1985). Analysis of the germination syndrome of dune annuals. *Flora* **177**, 175–185.

Rumpho, M. E., and Kennedy, R. A. (1981). Anaerobic metabolism in germinating seeds of *Echinochloa crus-galli* (barnyard grass). *Plant Physiol.* **68**, 165–168.

Rush, D. W., and Epstein, E. (1981). Comparative studies on the sodium, potassium, and chloride relations of a wild halophytic and a domestic salt-sensitive tomato species. *Plant Physiol.* **68**, 1308–1313.

SaadEddin, R., and Doddema, H. (1986). Anatomy of the 'extreme' halophyte *Arthrocnemum fruticosum* (L.) Moq. in relation to its physiology. *Ann. Bot.* **57**, 531–544.

Saghir, A. R. (1979). Strigol analogues and their potential for *Orobanche* control. *In* "The Second International Symposium on Parasitic Weeds" (L. J. Musselman, A. D. Worsham, and R. E. Eplee, eds.), pp. 238–244. North Carolina State Univ., Raleigh.

Sahai, A., and Shivanna, K. R. (1982). Seed germination and seedling morphogenesis in parasitic angiosperms of the families Scrophulariaceae and Orobanchaceae. *Seed Sci. Technol.* **10**, 565–583.

Sahai, A., and Shivanna, K. R. (1985). Seed germination and seedling growth in *Sopubia delphinifolia*—a hemi-root parasite: Requirements for seedling growth and the role of cotyledons. *Ann. Bot.* **55**, 785–791.

Sahni, B. (1933). Explosive fruits in *Viscum japonicum* Thunb. *J. Indian Bot. Soc.* **12**, 96–101.

Salisbury, E. J. (1952). "Downs & Dunes, Their Plant Life and Its Environment." G. Bell & Sons, Ltd., London.

Salisbury, E. J. (1967). The reproduction and germination of *Limosella aquatica*. *Ann. Bot.* **31**, 147–162.

Salisbury, E. J. (1970). The pioneer vegetation of exposed muds and its biological features. *Philos. Trans. Roy Soc. Lond., Series B. Biol. Sci.* **259**, 207–255.

Salle, G. (1983). Germination and establishment of *Viscum album* L. *In* "The Biology of Mistletoes" (M. Calder and P. Bernhardt, eds.), pp. 145–159. Academic Press, Paris.

Sankary, M. N., and Barbour, M. G. (1972). Autecology of *Atriplex polycarpa* from California. *Ecology* **53**, 1155–1162.

Sargent, S. (1995). Seed fate in a tropical mistletoe: The importance of host twig size. *Funct. Ecol.* **9,** 197–204.

Sastri, R. L. N. (1962). Studies in Lauraceae. III. Embryology of *Cassytha. Bot. Gaz.* **123,** 197–206.

Saunders, A. R. (1933). Studies in Phanerogamic Parasitism with Particular Reference to *Striga lutea* Lour. *Union of South Africa. Dept. of Agric. Sci. Bull. No.* 128.

Savina, G. L. (1974). Fertilization in Orchidaceae. *In* "Fertilization in Higher Plants" (H. F. Linskens, ed.), pp. 197–204. North-Holland, Amsterdam.

Scharpf, R. F. (1970). "Seed Viability, Germination and Radicle Growth of Dwarf Mistletoe in California." *USDA For. Serv. Res. Paper PSW-59.*

Scharpf, R. F., and Parmeter, J. R., Jr. (1962). The collection, storage, and germination of seeds of a dwarfmistletoe. *J. For.* **60,** 551–552.

Schat, H. (1983). Germination ecology of some dune slack pioneers. *Acta Bot. Neerl.* **32,** 203–212.

Schat, H., and Scholten, M. (1985). Comparative population ecology of dune slack species: The relation between population stability and germination behaviour in brackish environments. *Vegetatio* **61,** 189–195.

Schnell, D. E. (1976). "Carnivorous Plants of the United States and Canada." John F. Blair, Winston-Salem, NC.

Schuetz, W. (1995). Keimungsokologie von 5 horstbildendenden *Carex*-arten nasser standorte. *Verhandl. Gesell. Okol.* **24,** 155–160.

Schuetz, W. (1997). Primary dormancy and annual dormancy cycles in seeds of six temperate wetland sedges. *Aquat. Bot.* (in press).

Schuetz, W., and Milberg, P. (1997). Seed dormancy in *Carex canescens:* Regional differences and ecological consequences. *Oikos* **78,** 420–428.

Sculthorpe, C. D. (1967). "The Biology of Aquatic Vascular Plants." Edward Arnold, London.

Sehgal, A., Mohan Ram, H. Y., and Bhatt, J. R. (1993). In vitro germination, growth, morphogenesis and flowering of an aquatic angiosperm, *Polypleurum stylosum* (Podostemaceae). *Aquat. Bot.* **45,** 269–283.

Sen, D. N., and Rajpurohit, K. S. (1982). "Contributions to the Ecology of Halophytes." Junk, The Hague.

Seneca, E. D. (1969). Germination response to temperature and salinity of four dune grasses from the outer banks of North Carolina. *Ecology* **50,** 45–53.

Seneca, E. D. (1972). Germination and seedling response of Atlantic and Gulf Coasts populations of *Uniola paniculata. Am. J. Bot.* **59,** 290–296.

Seneca, E. D. (1974). A preliminary germination study of *Spartina foliosa,* California cordgrass. *Wasmann J. Biol.* **32,** 215–219.

Seneca, E. D., and Cooper, A. W. (1971). Germination and seedling response to temperature, daylength, and salinity by *Ammophila breviligulata* from Michigan and North Carolina. *Bot. Gaz.* **132,** 203–215.

Setchell, W. A. (1929). Morphological and phenological notes on *Zostera marina* L. *Univ. California Publ. Bot.* **14,** 389–452.

Shamsi, S. R. A., and Whitehead, F. H. (1974). Comparative ecophysiology of *Epilobium hirsutum* L. and *Lythrum salicaria* L. I. General biology, distribution and germination. *J. Ecol.* **62,** 279–290.

Sharma, K. P., and Gopal, B. (1978). Seed germination and occurrence of seedlings of *Typha* species in nature. *Aquat. Bot.* **4,** 353–358.

Sharma, K. P., Khan, T. I., and Bhardwaj, N. (1984). Temperature-regulated seed germination in *Neptunia oleracea* Lour. and its ecological significance. *Aquat. Bot.* **20,** 185–188.

Sharp, W. M. (1940). Propagation of *Potamogeton* and *Sagittaria* from seeds. *Proc. N. Amer. Wildl. Conf.* **4,** 351–358.

Sheikh, K. H., and Mahmood, K. (1986). Some studies on field distribution and seed germination of *Suaeda fruticosa* and *Sporobolus*

arabicus with reference to salinity and sodicity of the medium. *Plant Soil* **94,** 333–340.

Shipley, B., Keddy, P. A., Moore, D. R. J., and Lemky, K. (1989). Regeneration and establishment strategies of emergent macrophytes. *J. Ecol.* **77,** 1093–1110.

Shipley, B., and Parent, M. (1991). Germination responses of 64 wetland species in relation to seed size, minimum time to reproduction and seedling relative growth rate. *Funct. Ecol.* **5,** 111–118.

Shmueli, E. (1948). The water balance of some plants of the Dead Sea salines. *Palestine J. Bot.* **4,** 117–143.

Shumway, S. W., and Bertness, M. D. (1992). Salt stress limitation of seedling recruitment in a salt marsh plant community. *Oecologia* **92,** 490–497.

Shushan, S. (1959). Developmental anatomy of an orchid, *Cattleya* x *Trimos. In* "The Orchids: A Scientific Survey" (C. L. Withner, ed.), pp. 45–72. The Ronald Press Company, New York.

Siame, B. A., Weerasuriya, Y., Wood, K., Ejeta, G., and L. G. Butler, L. G. (1993). Isolation of strigol, a germination stimulant for *Striga asiatica,* from host plants. *J. Agric. Food Chem.* **41,** 1486–1491.

Siddiqi, N. A., Shahidullah, M., and Shahjalal, M. A. H. (1991). Studies on seeds, germination success and raised seedlings of *Nypa fruticans. Indian For.* **117,** 553–559.

Sifton, H. B. (1959). The germination of light-sensitive seeds of *Typha latifolia* L. *Can. J. Bot.* **37,** 719–739.

Silim, S. N., Saxena, M. C., and Erskine, W. (1991). Effect of sowing date on the growth and yield of lentil in a rainfed Mediterranean environment. *Exp. Agric.* **27,** 145–154.

Simpson, G. M. (1966). A study of germination in the seed of wild rice (*Zizania aquatica*). *Can. J. Bot.* **44,** 1–9.

Smayda, T. J. (1983). The phytoplankton of estuaries. *In* "Ecosystems of the World" (B. H. Ketchum, ed.), Vol. 26, pp. 65–102. Elsevier, Amsterdam.

Smith, A. L. (1972). Factors influencing germination of *Scolochloa festucacea* caryopses. *Can. J. Bot.* **50,** 2085–2092.

Smith, C. M. (1931). Development of *Dionaea muscipula.* II. Germination of seed and development of seedling to maturity. *Bot. Gaz.* **91,** 377–394.

Smith, M. C., and Webb, M. (1996). Estimation of the seedbank of *Striga* spp. (Scrophulariaceae) in Malian fields and implications for a model of biocontrol of *Striga hermonthica. Weed Res.* **36,** 85–92.

Smith, S. E. (1973). Asymbiotic germination of orchid seeds on carbohydrates of fungal origin. *New Phytol.* **72,** 497–499.

Smith, S. E. (1966). Physiology and ecology of orchid mycorrhizal fungi with reference to seedling nutrition. *New Phytol.* **65,** 488–499.

Smith, S. E. (1967). Carbohydrate translocation in orchid mycorrhizas. *New Phytol.* **66,** 371–378.

Smith, S. E. (1973). Asymbiotic germination of orchid seeds on carbohydrates of fungal origin. *New Phytol.* **72,** 497–499.

Smits, A. J. M., Schmitz, G. H. W., van der Velde, G., and Voesenek, L. A. C. J. (1995). Influence of ethanol and ethylene on the seed germination of three nymphaeid water plants. *Freshwater Biol.* **34,** 39–46.

Smits, A. J. M., van Avesaath, P. H., and van der Velde, G. (1990). Germination requirements and seed banks of some nymphaeid macrophytes: *Nymphaea alba* L., *Nuphar lutea* (L.) Sm. and *Nymphoides peltata* (Gmel.) O. Kuntze. *Freshwater Biol.* **24,** 315–326.

Smolders, A. J. P., den Hartog, C., and Roelofs, J. G. M. (1995). Germination and seedling development in *Stratiotes aloides* L. *Aquat. Bot.* **51,** 269–279.

Smreciu, E. A., and Currah, R. S. (1989). Symbiotic germination of seeds of terrestrial orchids of North America and Europe. *Lindleyana* **1,** 6–15.

Soegiarto, A. (1984). The mangrove ecosystem in Indonesia, its problems and management. *In* "Physiology and Management of Mangroves" (H. J. Teas, ed.), pp. 69–78. Junk, The Hague.

Solomon, S. (1953). Studies in the physiology of phanerogamic parasitism with special reference to *Striga lutea* Lour. and *S. densiflora* Benth. on *Andropogon sorghum* Hack. *Proc. Indian Acad. Sci. B* **36**, 198–214 + 2 plates.

Spence, D. H. N., Milburn, T. R., Ndawula-Senyimba, M., and Roberts, E. (1971). Fruit biology and germination of two tropical *Potamogeton* species. *New Phytol.* **70**, 197–212.

Spoerl, E. (1948). Amino acids as sources of nitrogen for orchid embryos. *Am. J. Bot.* **35**, 88–95.

Springfield, H. W. (1970). Germination characteristics of *Atriplex canescens* seeds. *Proc. Int. Grassl. Cong.* **11**, 586–589.

Srimathi, R. A., and Rao, P. S. (1969). Accelerated germination of sandal seed. *Indian For.* **95**, 158–159.

Stalter, R. (1993). Dry coastal ecosystems of the Gulf Coast of the United States of America. *In* "Ecosystems of the World" (E. van der Maarel, ed.), Vol. 2B, pp. 375–387. Elsevier, Amsterdam.

Stefani, A., Arduini, I., and Onnis, A. (1991). *Juncus acutus*: Germination and initial growth in presence of heavy metals. *Ann. Bot. Fennici* **28**, 37–43.

Steffens, J. C., Lynn, D. G., Kamat, V. S., and Riopel, J. L. (1982). Molecular specificity of haustorial induction in *Agalinis purpurea* (L.) Raf. (Scrophulariaceae). *Ann. Bot.* **50**, 1–7.

Steffens, J. C., Roark, J. L., Lynn, D. G., and Riopel, J. L. (1983). Host recognition in parasitic angiosperms: Use of correlation spectroscopy to identify long-range coupling in an haustorial inducer. *J. Am. Chem Soc.* **105**, 1669–1671.

Stewart, G. R., and Press, M. C. (1990). The physiology and biochemistry of parasitic angiosperms. *Annu. Rev. Plant Physiol. Plant Mol. Biol.* **41**, 127–151.

Stevens, R. A., and Eplee, R. E. (1979). *Striga* germination stimulants. *In* "The Second International Symposium on Parasitic Weeds" (L. J. Musselman, A. D. Worsham, and R. E. Eplee, eds.), pp. 211–218. North Carolina State Univ., Raleigh.

Stockey, A., and Hunt, R. (1992). Fluctuating water conditions identify niches for germination in *Alisma plantago-aquatica*. *Acta Oecol.* **13**, 227–229.

Stoessl, A., and Arditti, J. (1984). Orchid phytoalexins. *In* "Orchid Biology, Reviews and Perspectives" (J. Arditti, ed.), Vol. III, pp. 153–175. Cornell Univ. Press, Ithaca, NY.

Stojanovic, D., and Mijatovic, K. (1973). Distribution, biology and control of *Cuscuta* spp. in Yugoslavia. *In* "Parasitic Weeds Research Group," pp. 269–279. Proc. European Weed Res. Coun. Symp. Parasitic Weeds, Wageningen, The Netherlands.

Stoutamire, W. P. (1964). Seeds and seedlings of native orchids. *Michigan Bot.* **3**, 107–119.

Stoutamire, W. P. (1974). Terrestrial orchid seedlings. *In* "The Orchids: Scientific Studies" (C. Withner, ed.), pp. 101–127. Wiley, New York.

Stoutamire, W. P. (1983). Early growth in North American terrestrial orchid seedlings. *In* "Proceedings of Symposium II and Lectures, North American Terrestrial Orchids" (E. H. Plaxton, ed.), pp. 14–24. Michigan Orchid Society, Southfield.

Suessenguth, K. (1927). Uber die Gattung *Lennoa*. Ein Beitrag zur Kenntnis exotischer Parasiten. *Flora* **122**, 264–305.

Sunderland, N. (1960). The production of the *Striga* and *Orobanche* germination stimulants by maize roots. I. The number and variety of stimulants. *J. Exp. Bot.* **11**, 236–245.

Sussex, I. (1975). Growth and metabolism of the embryo and attached seedling of the viviparous mangrove, *Rhizophora mangle*. *Am. J. Bot.* **62**, 948–953.

Swamy, R. D., and Mohan Ram, H. Y. (1967). Cultivation of embryos

of *Drosophyllum lusitanicum* Link: An insectivorous plant. *Experientia* **23**, 675.

Swamy, R. D., and Mohan Ram, H. Y. (1969). Studies on growth and flowering in axenic cultures of insectivorous plants. I. Seed germination and establishment of cultures of *Utricularia inflexa*. *Phytomorphology* **19**, 363–371.

Sykes, M. T., and Wilson, J. B. (1989). The effect of salinity on the growth of some New Zealand sand dune species. *Acta Bot. Neerl.* **38**, 173–182.

Symonides, E. (1993). Seed characteristics in relation to early successional stages in sand dunes. *In* "Fourth International Workshop on Seeds: Basic and Applied Aspects of Seed Biology" (D. Come and F. Corbineau, eds.), pp. 449–454. Universite Pierre et Marie Curie, Paris.

Szmeja, J. (1987). The seasonal development of *Lobelia dortmanna* L. and annual balance of its population size in an oligotrophic lake. *Aquat. Bot.* **28**, 15–24.

Tainter, F. H. (1967). The histology of germinating embryos of the eastern dwarfmistletoe (*Arceuthobium pusillum* Peck). *J. Minnesota Acad. Sci.* **34**, 88–90.

Takahashi, T. (1905). Is germination possible in absence of air? *Bull. Coll. Agric. Imperial Univ.* **4**, 439–442.

Taylor, A. R. A. (1957). Studies on the development of *Zostera marina* L. II. Germination and seedling development. *Can. J. Bot.* **35**, 681–695.

Telenius, A. (1993). The demography of the short-lived perennial halophyte *Spergularia maritima* in a sea-shore meadow in southwestern Sweden. *J. Ecol.* **81**, 61–73.

Teltscherova, L., and Hejny, S. (1973). The germination of some *Potamogeton* species from south-Bohemian fishponds. *Folia Geobot. Phytotax.* **8**, 231–239.

ter Borg, S. (1987). Qualitative and quantitative aspects of the interaction between *Rhinanthus* and *Orobanche* species and their hosts. *In* "Parasitic Flowering Plants" (H. C. Weber and W. Forstreuter, eds.), pp. 109–120. Proc. 4th Int. Symp. Parasitic Flowering Plants, Marburg, FRG.

Teryokhin, E. S., and Nikiticheva, Z. I. (1982). Biology and evolution of embryo and endosperm in parasitic flowering plants. *Phytomorphology* **32**, 335–339.

Thanos, C. A., Georghiou, K., Douma, D. J. and Marangaki, C. J. (1991). Photoinhibition of seed germination in Mediterranean maritime plants. *Ann. Bot.* **68**, 469–475.

Thanos, C. A., Georghiou, K., and Skarou, F. (1989). *Glaucium flavum* seed germination—an ecophysiological approach. *Ann. Bot.* **63**, 121–130.

Thanos, C. A, Georghiou, K., and Delipetrou, P. (1994). Photoinhibition of seed germination in the maritime plant *Matthiola tricuspidata*. *Ann. Bot.* **73**, 639–644.

Thien, L. B., Ellgaard, E. G., Devall, M. S., Ellgaard, S. E., and Ramp, P. F. (1994). Population structure and reproductive biology of *Saururus cernuus* L. (Saururaceae). *Plant Spec. Biol.* **9**, 47–55.

Thompson, P. A. (1969). Germination of *Lycopus europaeus* L. in response to fluctuating temperatures and light. *J. Exp. Bot.* **20**, 1–11.

Thompson, P. A. (1970). An analysis of the effect of alternating temperatures on germination of *Lycopus europaeus* L. *J. Exp. Bot.* **21**, 808–823.

Thullen, J. S., and Eberts, D. R. (1995). Effects of temperature, stratification, scarification, and seed origin on the germination of *Scirpus acutus* Muhl. seeds for use in constructed wetlands. *Wetlands* **15**, 298–304.

Timko, M. P., Florea, C. S., and Riopel, J. L. (1989). Control of germination and early development in parasitic angiosperms. *In* "Recent Advances in the Development and Germination of

Seeds" (R. B. Taylorson, ed.), pp. 225–240. Plenum Press, New York.

Tingey, D. C., and Allred, K. R. (1961). Breaking dormancy in seeds of *Cuscuta approximata. Weeds* **9,** 429–436.

Tirmizi, S. A. S. (1988). Factors influencing germination and dormancy of *Honckenya peploides* (L.) Ehrh. Part I. Improvement of germination. *Pakistan J. Sci. Ind. Res.* **31,** 842–847.

Tocher, R. D., Gustafson, S. W., and Knutson, D. M. (1984). Water metabolism and seedling photosynthesis in dwarf mistletoes. *In* "Biology of Dwarf Mistletoes" (F. G. Hawksworth and R. F. Scharpf, Tech. Coords.), pp. 62–69. USDA For. Serv. Gen. Tech. Rep. RM-11.

Tomlinson, P. B. (1971). The shoot apex and its dichotomous branching in the *Nypa* palm. *Ann. Bot.* **35,** 865–879.

Tomlinson, P. B. (1986). "The Botany of Mangroves." Cambridge Univ. Press, Cambridge, UK.

Tremblin, G., and Binet, P. (1982). Installation d'*Halopeplis amplexicaulis* (Vahl) Ung. dans une sebkha algerienne. *Acta Oecol.* **3,** 373–379.

Troup, R. S. (1921). "The Silviculture of Indian Trees." Clarendon Press, Oxford.

Troyo-Dieguez, E., and Solis-Camara, A. M. B. (1992). Germinacion de *Salicornia bigelovii* Torr. (Chenopodiaceae) bajo diferentes concentraciones de agua marina. *Southw. Nat.* **37,** 22–27.

Tsivion, Y. (1978). Loading of assimilates and some sugars into the translocation system of *Cuscuta. Aust. J. Plant Physiol.* **5,** 851–857.

Tsuji, H. (1972). Respiratory activity in rice seedlings germinated under strictly anaerobic conditions. *Bot. Mag. Tokyo* **85,** 207–218.

Tutin, T. G. (1938). The autecology of *Zostera marina* in relation to its wasting disease. *New Phytol.* **37,** 50–71.

Twitchell, L. F. T. (1955). Germination of fourwing saltbush seed as afffected by soaking and chloride removal. *J. Range Manage.* **8,** 218–220.

Tyndall, R. W., Teramura, A. H., Mulchi, C. L., and Douglas, L. W. (1986). Seed burial effect on species presence along a mid-Atlantic beach. *Can. J. Bot.* **64,** 2168–2170.

Uchiyama, Y. (1987). Salt tolerance of *Atriplex nummularia. Tech. Bull. Trop. Agric. Res. Center Japan* **22,** 1–69.

Uhvits, R. (1946). Effect of osmotic pressure on water absorption and germination of alfalfa seeds. *Am. J. Bot.* **33,** 278–285.

Ungar, I. A. (1962). Influence of salinity on seed germination in succulent halophytes. *Ecology* **43,** 763–764.

Ungar, I. A. (1967a). Influence of salinity and temperature on seed germination. *Ohio J. Sci.* **67,** 120–123.

Ungar, I. A. (1967b). Vegetation-soil relationship on saline soils in northern Kansas. *Am. Midl. Nat.* **78,** 98–120.

Ungar, I. A. (1974). The effect of salinity and temperature on seed germination and growth of *Hordeum jabatum. Can. J. Bot.* **52,** 1357–1362.

Ungar, I. A. (1977). Salinity, temperature, and growth regulator effects on seed germination of *Salicornia europaea* L. *Aquat. Bot.* **3,** 329–335.

Ungar, I. A. (1978). Halophyte seed germination. *Bot. Rev.* **44,** 233–264.

Ungar, I. A. (1984). Alleviation of seed dormancy in *Spergularia marina. Bot. Gaz.* **145,** 33–36.

Ungar, I. A. (1987a). Population ecology of halophyte seeds. *Bot. Rev.* **53,** 301–334.

Ungar, I. A. (1987b). Population characteristics, growth, and surival of the halophyte *Salicornia europaea. Ecology* **68,** 569–575.

Ungar, I. A. (1991a). Seed germination responses and the seed bank dynamics of the halophyte *Spergularia marina* (L.) Griseb. *In* "Marvels of seeds" (D. N. Sen and S. Mohammed, eds.), pp. 81–86. Univ. Jodphur, India.

Ungar, I. A. (1991b). "Ecophysiology of Vascular Halophytes." CRC Press, Boca Raton, FL.

Ungar, I. A. (1996). Effect of salinity on seed germination, growth, and ion accumulation of *Atriplex patula* (Chenopodiaceae). *Am. J. Bot.* **83,** 604–607.

Ungar, I. A., Benner, D. K., and McGraw, D. C. (1979). The distribution and growth of *Salicornia europaea* on an inland salt pan. *Ecology* **60,** 329–336.

Ungar, I. A., and Binet, P. (1975). Factors influencing seed dormancy in *Spergularia media* (L.) C. Presl. *Aquat. Bot.* **1,** 45–55.

Ungar, I. A., and Capilupo, F. (1969). An ecological life history study of *Suaeda depressa* (Pursh) Wats. *Adv. Front. Plant Sci.* **23,** 137–155.

Ungar, I. A., and Hogan, W. C. (1970). Seed germination in *Iva annua* L. *Ecology* **51,** 150–154.

Uniyal, P. L., and Mohan Ram, H. Y. (1996). *In vitro* germination and seedling development of *Dalzellia zeylanica* (Gardner) Wight (Podostemaceae). *Aquat. Bot.* **54,** 59–71.

Uttaman, P. (1950). A study on the germination of *Striga* and on the mechanism and nature of parasitism of *Striga lutea* (Lour.) on rice. *Proc. Indian Acad. Sci.* **32**(B), 133–142.

Vallance, K. B. (1950). Studies on the germination of the seeds of *Striga hermonthica.* I. The influence of moisture-treatment, stimulant-dilution, and after-ripening on germination. *Ann. Bot.* **14,** 347–363.

Vallance, K. B. (1951). Germination of seeds of *Bartsia odontites. Nature* **167,** 732.

Vallance, K. B. (1952). The germination of the seeds of *Rhinanthus crista-galli. Ann. Bot.* **16,** 409–420.

van Asdall, W., and Olmsted, C. E. (1963). *Corispermum hyssopifolium* on the Lake Michigan dunes, its community and physiological ecology. *Bot. Gaz.* **124,** 155–172.

van der Maarel, E. (ed.) (1993). "Ecosystems of the world," Vol. 2A. Elsevier, Amsterdam.

van der Meijden, E., and van der Waals-Kooi, R. E. (1979). The population ecology of *Senecio jacobaea* in a sand dune system. I. Reproductive strategy and the biennial habit. *J. Ecol.* **67,** 131–153.

van der Merwe, C. A. (1993). Seed germination requirements of *Alectra sessiliflora. S. Afr. J. Bot.* **59,** 459–460.

van der Meulen, F., and van der Maarel, E. (1993). Dry coastal ecosystems of the central and southwestern Netherlands. *In* "Ecosystems of the World" (E. van der Maarel, ed.), pp. 271–306. Elsevier, Amsterdam.

van der Putten, W. H. (1990). Establishment of *Ammophila arenaria* (marram grass) from culms, seeds and rhizomes. *J. Appl. Ecol.* **27,** 188–199.

van der Toorn, J., and ten Hove, H. J. (1982). On the ecology of *Cotula coronopifolia* L., and *Ranunculus sceleratus* L. II. Experiments on germination, seed longevity, and seedling survival. *Acta. Oecol.* **3,** 409–418.

van der Valk, A. G. (1974). Environmental factors controlling the distribution of forbs on coastal foredunes in Cape Hatteras National Seashore. *Can. J. Bot.* **52,** 1057–1073.

van der Valk, A. G. (1986). The impact of litter and annual plants on recruitment from the seed bank of a lacustrine wetland. *Aquat. Bot.* **24,** 13–26.

VanderZee, D., and Kennedy, R. A. (1982). Plastid development in seedlings of *Echinochloa crus-galli* var. *oryzicola* under anoxic germination conditions. *Planta* **155,** 1–7.

van Hezewijk, M. J. (1994). "Germination Ecology of *Orobanche crenata:* Implications for Cultural Control Measures." Ph.D thesis, Vrije Universiteit, Amsterdam.

van Hezewijk, M. J., Linke, K.-H., Lopez-Granados, F., Al-Menoufi, O. A., Garcia-Torres, L., Saxena, M. C., Verkleij, J. A. C., and

Pieterse, A. H. (1994). Seasonal changes in germination response of buried seeds of *Orobanche crenata* Fors[s]k. *Weed Res.* **34**, 369–376.

van Hezewijk, M. J., Pieterse, A. H., Saxena, M. S., and ter Borg, S. J. (1987). Relationship between sowing date and *Orobanche* (broomrape) development on faba bean (*Vicia faba* L.) and lentil (*Lens culinaris* Medikus) in Syria. *In* "Parasitic Flowering Plants" (H. C. Weber and W. Forstreuter,. eds.), pp. 377–390. Proc. 4th Int. Symp. Parasitic Flowering Plants, Marburg, FRG.

van Hezewijk, M. J., van Beem, A. P., and Verkleij, J. A. C. (1993). Germination of *Orobanche cernata* seeds, as influenced by conditioning temperature and period. *Can. J. Bot.* **71**, 786–792.

van Hezewijk, M. J., Verkleij, J. A. C. (1996). The effect of nitrogenous compounds on *in vitro* germination of *Orobanche crenata* Fors[s]k. *Weed Res.* **36**, 395–404.

van Hezewijk, M. J., Verkleij, J. A. C., and Pieterse, A. H. (1991). Temperature dependence of germination in *Orobanche crenata*. *In* "Progress in *Orobanche* Research" (K. Wegmann and L. J. Musselman, eds.), pp. 125–133. Eberhard-Karls-Universitat, Tubingen, FRG.

van Vierssen, W. (1982a). The ecology of communities dominated by *Zannichellia* taxa in western Europe. I. Characterization and autecology of the *Zannichellia* taxa. *Aquat. Bot.* **12**, 103–155.

van Vierssen, W. (1982b). On the identity and autecology of *Zannichellia peltata* Bertol. in western Europe. *Aquat. Bot.* **13**, 367–383.

van Vierssen, W. (1982c). Some notes on the germination of seeds of *Najas marina* L. *Aquat. Bot.* **12**, 201–203.

van Waes, J. M., and Debergh, P. C. (1986). *In vitro* germination of some western European orchids. *Physiol. Plant.* **67**, 253–261.

van Wijk, R. J. (1989). Ecological studies on *Potamogeton pectinatus* L. III. Reproductive strategies and germination ecology. *Aquat. Bot.* **33**, 271–299.

Varrelman, F. A. (1937). *Cuscuta* not a complete parasite. *Science* **85**, 101.

Verhoeven, J. T. A. (1979). The ecology of *Ruppia*-dominated communities in western Europe. I. Distribution of *Ruppia* representatives in relation to their autecology. *Aquat. Bot.* **6**, 197–268.

Vertucci, C. W., Crane, J., Porter, R. A., and Oelke, E. A. (1995). Survival of *Zizania* embryos in relation to water content, temperature and maturity status. *Seed Sci. Res.* **5**, 31–40.

Vickery, J. W. (1933). Vegetative reproduction in *Drosera peltata* and *D. auriculata*. *Proc. Linn. Soc. New South Wales* **58**, 245–269.

Vidyashankari, B., and Mohan Ram, H. Y. (1987). In vitro germination and origin of thallus in *Griffithella hookeriana* (Podostemaceae). *Aquat. Bot.* **28**, 161–169.

Visser J. H. (1975). Germination stimulants of *Alectra vogelii* Benth seed. *Z. Pflanzenphysiol.* **74**, 464–469.

Visser, J. H., and Botha, P. J. (1974). Chromatographic investigation of the *Striga* seed germination stimulant. *Z. Pflanzenphysiol.* **72**, 352–358.

Visser, J. H., Herb, R., and Schildknecht, H. (1987). Recovery and preliminary chromatographic investigation of germination stimulants produced by *Vigna unguiculata* Walp. cv. Saunders upright. *J. Plant Physiol.* **129**, 375–381.

Visser, J. H., and Wentzel, L. F. (1980). Quantitative estimation of *Alectra* and *Striga* seed in soil. *Weed Res.* **20**, 77–81.

Voesenek, L. A. C. J., and Blom, C. W. P. M. (1992). Germination and emergence of *Rumex* in river flood-plains. I. Timing of germination and seedbank characteristics. *Acta Bot. Neerl.* **41**, 319–329.

Voesenek, L. A. C. J., de Graaf, M. C. C., and Blom, C. W. P. M. (1992). Germination and emergence of *Rumex* in river flood-plains. II. The role of perianth, temperature, light and hypoxia. *Acta Bot. Neerl.* **41**, 331–343.

Voigtlander, C. W., and Poppe, W. L. (1989). The Tennessee River.

In "International Large River Symposium" (C. P. Dodge, ed.). *Can. Spec. Publ. Fish. Aquat. Sci.* **106**, 372–384.

Vollebergh, P. J., and Congdon, R. A. (1986). Germination and growth of *Ruppia polycarpa* and *Lepilaena cylindrocarpa* in ephemeral saltmarsh pools, Westernport Bay, Victoria. *Aquat. Bot.* **26**, 165–179.

Vose, P. B. (1962). Delayed germination in reed canary-grass *Phalaris arundinacea* L. *Ann. Bot.* **26**, 197–206.

Wagner, R. H. (1964). The ecology of *Uniola paniculata* L. in the dune-strand habitat of North Carolina. *Ecol. Monogr.* **34**, 79–96.

Waisel, Y. (1960). Ecological studies on *Tamarix aphylla* (L.) Karst. I. Distribution and reproduction. *Phyton (Buenos Aires)* **15**, 7–17.

Waisel, Y. (1972). "Biology of Halophytes." Academic Press, New York.

Walmsley, C. A., and Davy, A. J. (1997). Germination characteristics of shingle beach species, effects of seed ageing and their implications for vegetation restoration. *J. Appl. Ecol.* **34**, 131–142.

Walsh, G. E. (1974). Mangroves: A review. *In* "Ecology of halophytes" (R. J. Reimold and W. H. Queen, eds.), pp. 51–174. Academic Press, New York.

Walsh, G. E., Weber, D. E., Nguyen, M. T., and Esry, L. K. (1991a). Response of wetland plants to effluents in water and sediment. *Environ. Exp. Bot.* **31**, 351–358.

Walsh, G. E., Weber, D. E., Simon, T. L., and Brashers, L. K. (1991b). Toxicity tests of effluents with marsh plants in water and sediment. *Environ. Toxicol. Chem.* **10**, 517–525.

Walter, H. (1979). "Vegetation of the Earth and Ecological Systems of the Geo-biosphere," 2nd Ed. Springer-Verlag, Berlin. Translated from the third, revised German edition by Joy Wieser.

Wang, S.-C., Jurik, T. W., and van der Valk, A. G. (1994). Effects of sediment load on various stages in the life and death of cattail (*Typha x glauca*). *Wetlands* **14**, 166–173.

Warcup, J. H. (1971). Specificity of mycorrhizal association in some Australian terrestrial orchids. *New Phytol.* **70**, 41–46.

Warcup, J. H. (1973). Symbiotic germination of some Australian terrestrial orchids. *New Phytol.* **72**, 387–392.

Warcup, J. H. (1985). *Rhizanthella gardneri* (Orchidaceae), its *Rhizoctonia* endophyte and close association with *Melaleuca uncinata* (Myrtaceae) in western Australia. *New Phytol.* **99**, 273–280.

Ward, J. M. (1967). Studies in ecology on a shell barrier beach. Section III. Chemical factors of the environment. *Vegetatio* **15**, 77–112.

Watkinson, A. R. (1978). The demography of a sand dune annual: *Vulpia fasciculata*. II. The dynamics of seed populations. *J. Ecol.* **66**, 35–44.

Weatherhead, M. A., Zee, S. Y., and Barretto, G. (1986). Some observations on the early stages of development of *Eulophia yushuiana*. *Mem. Hong Kong Nat. Hist. Soc.* **17**, 85–90.

Webb, D. H., Dennis, W. M., and Bates, A. L. (1988). An analysis of the plant community of mudflats of TVA mainstream reservoirs. *In* "Proceedings of the First Annual Symposium on the Natural History of Lower Cumberland and Tennessee River Valleys." (D. H. Snyder, ed.), pp. 177–198. The Center for Field Biology of Land Between The Lakes, Austin Peay State University, Clarksville, TN.

Wee, Y. C. (1978). Concerning *Nepenthes* seedlings. *Malayan Nat. J.* **32**, 105–106.

Weerasuriya, Y., Siame, B. A., Hess, D., Ejeta, F., and Butler, L. G. (1993). Influence of conditions and genotype on the amount of *Striga* germination stimulants exuded by roots of several host crops. *J. Agric. Food Chem.* **41**, 1492–1496.

Weller, S. G. (1985). Establishment of *Lithospermum caroliniense* on sand dunes: The role of nutlet mass. *Ecology* **65**, 1893–1901.

Welling, C. H., and Becker, R. L. (1990). Seed bank dynamics of

Lythrum salicaria L.: Implications for control of this species in North America. *Aquat. Bot.* **38,** 303–309.

Went, F. W. (1957). The plants of Krakatoa. *In* "Plant life." G. Piel, ed.) pp. 137–145. Simon and Schuster, New York.

Wertis, B. A., and Ungar, I. A. (1986). Seed demography and seedling survival in a population of *Atriplex triangularis* Willd. *Am. Midl. Nat.* **116,** 152–162.

West, D., and Whigham, D. F. (1975–1976). Seed germination of arrow arum (*Peltandra virginica* L.). *Bartonia No.* **44,** 44–49.

West, N. E. (ed.) (1983). "Ecosystems of the World," Vol. 5. Elsevier, Amsterdam.

Westelaken, I. L., and Maun, M. A. (1985). Reproductive capacity, germination and survivorship of *Lithospermum caroliniense* on Lake Huron sand dunes. *Oecologia* **66,** 238–245.

Westra, R. N., and Loomis, W. E. (1966). Seed dormancy in *Uniola paniculata. Am. J. Bot.* **53,** 407–411.

Whitney, P. J. (1979). Broomrape seed germination stimulants and inhibitors from host roots. *In* "The Second International Symposium on Parasitic Weeds" (L. J. Musselman, A. D. Worsham, and R. E. Eplee, eds.)., pp. 182–192. North Carolina State Univ., Raleigh.

Wicker, E. F. (1974). "Ecology of Dwarf Mistletoe Seed." *USDA For. Serv. Res. Paper* INT-154.

Widden, P. (1981). Patterns of phenology among fungal populations. *In* "The Fungal Community: Its Organization and Role in the Ecosystem" (D. L. Wicklow and G. C. Carroll, eds.), pp. 387–401. Dekker, New York.

Widden, P., and Parkinson, D. (1973). Fungi from Canadian coniferous forest soils. *Can. J. Bot.* **51,** 2275–2290.

Wiebe, H. H., and Walter, H. (1972). Mineral ion composition of halophytic species from northern Utah. *Am. Midl. Nat.* **87,** 241–245.

Wiesner, J. (1894). Pflanzenphysiologische mittheilungen aus buitenzorg. *Sitzungsb. Akad. Wiss. Wien Math. Naturwiss. Kl. Abt. 1.* **103,** 401–437.

Wiesner, J. (1897). Ueber die Ruheperiode und uber einige Keimungsbedingungen der Samen von *Viscum album. Ber. Deut. Bot. Gesell.* **15,** 503–515.

Wijte, A. H. B. M., and Gallagher, J. L. (1996). Effect of oxygen availability and salinity on early life history stages of salt marsh plants. I. Different germination strategies of *Spartina alterniflora* and *Phragmites australis* (Poaceae). *Am. J. Bot.* **83,** 1337–1342.

Wilgus, F., and Hamilton, K. C. (1962). Germination of salt cedar seed. *Weeds* **10,** 332–333.

Wilkinson, E., and Musselman, L. J. (1979). The parasitic gymnosperm, *Parasitaxus ustus. In* "The Second International Symposium on Parasitic Weeds" (L. J. Musselman, A. D. Worsham, and R. E. Eplee, eds.), p. 17. North Carolina State Univ., Raleigh. [Abstract]

Wilkinson, K. G., Doxon, K. W., and Sivasithamparam, K. (1989). Interaction of soil bacteria, mycorrhizal fungi and orchid seed in relation to germination of Australian orchids. *New Phytol.* **112,** 429–435.

Williams, C. E., and Zuck, R. K. (1986). Germination of seeds of *Epifagus virginiana* (Orobanchaceae). *Michigan Bot.* **25,** 103–106.

Williams, C. N. (1960). Growth movements of the radicle of *Striga. Nature* **188,** 1043–1044.

Williams, C. N. (1961a). Tropism and morphogenesis of *Striga* seedlings in the host rhizosphere. *Ann. Bot.* **25,** 407–415.

Williams, C. N. (1961b). Growth and morphogenesis of *Striga* seedlings. *Nature* **189,** 378–381.

Williams, D. G. (1979). The comparative ecology of two perennial chenopods. *In* "Studies of the Australian Arid Zone" (R. D. Graetz and K. M. W. Howes, eds.). Vol. IV, pp. 29–40. Div.

Land Resources Manage., Commonwealth Scientific and Industrial Research Organization, Australia.

Williams, M. D., and Ungar, I. A. (1972). The effect of environmental parameters on the germination, growth, and development of *Suaeda depressa* (Pursh) Wats. *Am. J. Bot.* **59,** 912–918.

Williamson, B., and Hadley, G. (1969). DNA content of nuclei in orchid protocorms symbiotically infected with *Rhizoctonia. Nature* **222,** 582–583.

Williamson, B., and Hadley, G. (1970). Penetration and infection of orchid protocorms by *Thanatephorus cucumeris* and other *Rhizoctonia* isolates. *Phytopathology* **60,** 1092–1096.

Willis, J. C. (1966). "A Dictionary of the Flowering Plants and Ferns." Cambridge Univ. Press, London.

Wilson, S. D., Moore, D. R. J., and Keddy, P. A. (1993). Relationships of marsh seed banks to vegetation patterns along environmental gradients. *Freshwater Biol.* **29,** 361–370.

Wirth, M., and Withner, C. L. (1959). Embryology and development in the Orchidaceae. *In* "The Orchids: A Scientific Survey" (C. L. Withner, ed.). pp. 155–188. The Ronald Press Co., New York.

Withner, C. L. (1953). Germination of "cyps." *Orchid J.* **2,** 473–477.

Withner, C. L. (1955). Ovule culture and growth of *Vanilla* seedlings. *Am. Orchid Soc. Bull.* **24,** 380–392.

Withner, C. L. (1959). Orchid physiology. *In* "The Orchids: A Scientific Survey," (C. L. Withner, ed.). pp. 315–360. The Ronald Press Co., New York.

Witztum, A. (1977). An ecological niche for *Lemna gibba* L. that depends on seed formation. *Israel J. Bot.* **26,** 36–38.

Wolswinkel, P. (1974). Complete inhibition of setting and growth of fruits of *Vicia faba* L., resulting from the draining of the phloem system by *Cuscuta* species. *Acta Bot. Neerl.* **23,** 48–60.

Woodell, S. R. J. (1985). Salinity and seed germination patterns in coastal plants. *Vegetatio* **61,** 223–229.

Workman, J. P., and West, N. E. (1967). Germination of *Eurotia lanata* in relation to temperature and salinity. *Ecology* **48,** 659–661.

Worsham, A. D. (1961). "Germination of *Striga asiatica* (L.) Kuntze (Witchweed) Seed and Studies on the Chemical Nature of the Germination Stimulant." Ph.D thesis, North Carolina State Univ., Raleigh.

Worsham, A. D. (1987). Germination of witchweed seeds. *In* "Parasitic Weeds in Agriculture Striga" (L. J. Musselman, ed.), Vol. 1, pp. 46–61. CRC Press, Boca Raton, FL.

Worsham, A. D., and Egley, G. H. (1990). Physiology of witchweed seed dormancy and germination. *In* "Witchweed Research and Control in the United States" (P. F. Sand, R. E. Eplee, and R. G. Westbrooks, eds.), pp. 11–26. Weed Science Society of America Monograph Series, Champaign, IL.

Worsham, A. D., Moreland, D. E., and Klingman, G. C. (1959). Stimulation of *Striga asiatica* (witchweed) seed germination by 6-substituted purines. *Science* **130,** 1654–1656.

Worsham, A. D., Moreland, D. E., and Klingman, G. C. (1964). Characterization of the *Striga asiatica* (Witchweed) Germination stimulant from *Zea mays* L. *J. Exp. Bot.* **15,** 556–567.

Wright, E., and Bollen, W. B. (1961). Microflora of Douglas-fir forest soil. *Ecology* **42,** 825–828.

Wulff, R., Arias, I., Ponce, M., and Munoz, V. (1972). A bimodal temperature response and effect of light intensity in the photocontrol of germination of seeds in *Jussiaea suffruticosa. Planta* **107,** 369–373.

Wulff, R., and Medina, E. (1969). Germination of seeds in *Jussiaea suffruticosa. Plant Cell Physiol.* **10,** 503–511.

Wynd, F. L. (1933a). Sources of carbohydrate for germination and growth of orchid seedlings. *Ann. Missouri. Bot. Gard.* **20,** 569–581.

Wynd, F. L. (1933b). The sensitivity of orchid seedlings to nutritional ions. *Ann. Missouri Bot. Gard.* **20,** 223–237.

Wynd, F. L. (1933c). Nutritional solutions for orchids. *Ann. Missouri Bot. Gard.* **20,** 363–372

Yaduraju, N. T., and Setty, T. K. P. (1979). Effect of time and dose of nitrogen application on *Striga asiatica* incidence in sorghum. *In* "The Second International Symposium on Parasitic Weeds" (L. J. Musselman, A. D. Worsham, and R. E. Eplee, eds.), pp. 285–289. North Carolina State Univ., Raleigh.

Yamamoto, M. (1964). Water absorption in strand plant seeds. *Bot. Mag. Tokyo* **77,** 228–235.

Yan, Z. (1993). Germination and seedling development of two mistletoes, *Amyema preissii* and *Lysiana exocarpi*: Host specificity and mistletoe-host compatibility. *Aust. J. Ecol.* **18,** 419–429.

Yanful, M., and Maun, M. A. (1996). Effects of burial of seeds and seedlings from different seed sizes on the emergence and growth of *Strophostyles helvola*. *Can. J. Bot.* **74,** 1322–1330.

Yeo, P. F. (1961). Germination, seedlings, and the formation of haustoria in *Euphrasia*. *Watsonia* **5,** 11–22.

Yeo, R. R. (1986). Dormancy in slender spikerush seed. *J. Aquat. Plant Manage.* **24,** 11–16.

Yeo, R. R. and Thurston, J. R. (1979). Survival of seed and tubers of dwarf spikerush (*Eleocharis coloradoensis*) after exposure to extreme temperatures. *Weed Sci.* **27,** 434–436.

Yokoishi, T., and Tanimoto, S. (1994). Seed germination of the halophyte *Suaeda japonica* under salt stress. *J. Plant Res.* **107,** 385–388.

Yoshikawa, F., Worsham, A. D., Moreland, D. E., and Eplee, R. E. (1978). Biochemical requirements for seed germination and shoot development of witchweed (*Striga asiatica*). *Weed Sci.* **26,** 119–122.

Young, D. R., Erickson, D. L., and Semones, S. W. (1994). Salinity and the small-scale distribution of three barrier island shrubs. *Can. J. Bot.* **72,** 1365–1372.

Zahran, M. A. (1982). Ecology of the halophytic vegetation of Egypt. *In* "Contributions to the Ecology of Halophytes" (D. N. Sen and K. S. Rajurohit, eds.), pp. 3–20. Dr. W. Junk, The Hague.

Zahran, M. K. (1982). Weed and *Orobanche* control in Egypt. *In* "Faba Bean Improvement" (G. Hawtin and C. Webb, eds.), pp. 191–198. Martinus Hijhoff Publishers, Dordrecht.

Zedler, J. B., Paling, E., and McComb, A. (1990). Differential responses to salinity help explain the replacement of native *Juncus kraussii* by *Typha orientalis* in western Australian salt marshes. *Aust. J. Ecol.* **15,** 57–72.

Zhang, J. (1996). Seed mass effects across environments in an annual dune plant. *Ann. Bot.* **77,** 555–563.

Zhang, J., and Maun, M. A. (1989a). Effect of partial removal of endosperm on seedling sizes of *Panicum virgatum* and *Agropyron psammophilum*. *Oikos* **56,** 250–255.

Zhang, J., and Maun, M. A. (1989b). Seed dormancy of *Panicum virgatum* L. on the shoreline sand dunes of Lake Erie. *Am. Midl. Nat.* **122,** 77–87.

Zhang, J., and Maun, M. A. (1990a). Effects of sand burial on seed germination, seedling emergence, survival, and growth of *Agropyron psammophilum*. *Can. J. Bot.* **68,** 304–310.

Zhang, J., and Maun, M. A. (1990b). Sand burial effects on seed germination, seedling emergence and establishment of *Panicum virgatum*. *Holoarct. Ecol.* **13,** 56–61.

Zhang, J., and Maun, M. A. (1991). Establishment and growth of *Panicum virgatum* L. seedlings on a Lake Erie sand dune. *Bull. Torr. Bot. Club* **118,** 141–153.

Zid, E., and Boukhris, M. (1977). Quelques aspects de la tolerance de l'*Atriplex halimus* L. au chlorure de sodium. Multiplication, croissance, composition minerale. *Oecol. Plant.* **12,** 351–362.

Zomlefer, W. B. (1994). "Guide to Flowering Plant Families." Univ. North Carolina Press, Chapel Hill.

Biogeographical and Evolutionary Aspects of Seed Dormancy

I. PURPOSE

Our survey of seed dormancy and germination in the major vegetation regions of the world (Chapters 9 and 10) and of plants with specialized life cycles or habitats (Chapter 11) reveals that freshly matured seeds of numerous species are nondormant, whereas those of others are dormant. Various questions come to mind. (1) From a theoretical evolutionary ecology perspective, under what conditions has seed dormancy evolved? (2) Were the first seeds dormant or nondormant, and what kind of embryos did they have? (3) Is there any correlation between type of seed dormancy and phylogenetic position of the family? (4) What are the origins and evolutionary relationships of the various types of seed dormancy? (5) Under what kinds of environmental conditions (and when) did the various kinds of dormancy evolve?

Obviously, these are difficult, if not impossible, questions to answer, but even partial answers would (1) help focus future research efforts of seed ecologists and biologists, (2) encourage modelers to consider all types of dormancy, not just nondeep physiological dormancy, in their work, (3) provide additional insight into the great evolutionary flexibility of the seed dormancy and germination stages of the life cycle, and (4) increase our appreciation of why angiosperms have been so successful.

The primary objective of this chapter is to formulate hypotheses concerning the origins and evolutionary relationships of the various types of seed dormancy. However, before meaningful hypotheses can be generated, information is needed on various subjects: (1) world biogeography of seed dormancy types, (2) theoretical reasons why seed dormancy has evolved, (3) paleoclimatic data, (4) fossil record of embryos and seeds, (5)

phylogenetic position of families whose seeds have each type of dormancy, (6) first appearance in the fossil record of families with each type of dormancy, (7) family tree of seed phylogeny, and (8) type of embryo and presumed age of family. Thus, after available information on these topics is surveyed, hypotheses will be presented.

II. WORLD BIOGEOGRAPHY OF SEED DORMANCY TYPES

A. Dormancy vs Nondormancy

Information on seed dormancy of trees, shrubs, lianas, and herbaceous species growing in the major vegetation types of the world is presented in Chapters 9 and 10. Now, drawing on these data for a total of 3580 species, it is possible to look at seed dormancy and nondormancy from a global perspective and compare their relative importance across precipitation and temperature gradients.

In tropical rain forests, which have the highest precipitation and temperatures of any vegetation type on earth, nondormancy is more important (61%) than all other types of seed dormancy combined (39%) (Fig. 12.1). Further, in semievergreen forests, the proportion of species with nondormant seeds (46%) is only slightly less than that of species with dormant seeds (54%). With the exception of tropical evergreen rain forests and semievergreen forests, however, all vegetation types on earth clearly have a much higher proportion of species with dormant seeds than with nondormant seeds.

With decreases in precipitation and temperature (increased distance from the equator) in the tropical/subtropical region, nondormancy decreases and dormancy

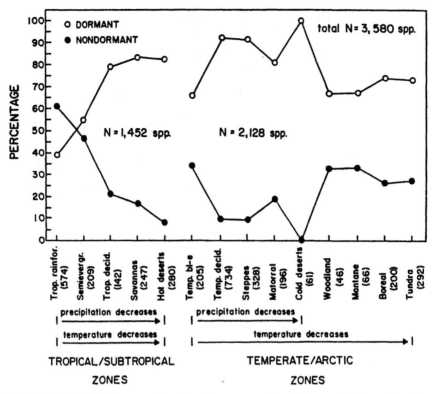

FIGURE 12.1   World biogeography of seed dormancy and nondormancy. Number of species is given in parentheses.

increases (Fig. 12.1). When temperate broad-leaved evergreen forests, deciduous forests, steppes, and cold deserts are compared in the temperate/arctic zones, nondormancy also decreases with decreases in precipitation and temperature. The increase of nondormancy in the matorral is due to a relatively high number of species with nondormant seeds that are held in serotinous cones or fruits. Thus far, no species has been found in cold deserts whose seeds are nondormant at maturity. The reason for the increase in nondormancy in woodland, montane, boreal/subalpine, and tundra vegetation is unknown. However, one possibility is that temperatures where these vegetation types occur are below those required for cold stratification during much of the winter, and thus seeds do not receive enough cold stratification to come out of dormancy. Therefore, if seeds are nondormant at the time of maturation in late summer or early autumn, low habitat temperatures would prevent germination. When temperatures increase in spring, seeds can germinate immediately without having to use up part of the short growing season for dormancy loss.

B. Geographical Distribution of Seed Dormancy Types

For all species combined (i.e., trees, shrubs, lianas, and herbs) and for all types of vegetation in both tropi-

cal/subtropical and temperate/arctic regions, physiological dormancy is the most important kind of dormancy (Fig. 12.2). Maximum importance of physiological dormancy is found in both tropical/subtropical and temperate/arctic zones in the driest types of vegetation, i.e., in hot and cold deserts, respectively.

In the tropical/subtropical region, the importance of physical dormancy peaks in tropical deciduous forests, whereas in temperate/arctic regions this occurs in the matorral. In both regions, physical dormancy reaches a maximum at relatively high temperatures and about half way down the precipitation gradient. The combination of physiological and physical dormancy types is never very important (maximum of only 4% in matorral vegetation), but it is more common in temperate/arctic types of vegetation than in tropical/ subtropical types of vegetation. In the tropical/subtropical region, the combination of physiological and physical dormancy is reported only in hot deserts, and only 1% of the species have this type. In the temperate/arctic region, the combination of dormancy types has been found in all vegetation types, except steppes, cold deserts, boreal/subalpine, and tundra vegetation.

Morphological dormancy is never very important (maximum of 4% in temperate broad-leaved evergreen forests), and it tends to be most common in both tropical/subtropical and temperate/arctic regions in vegeta-

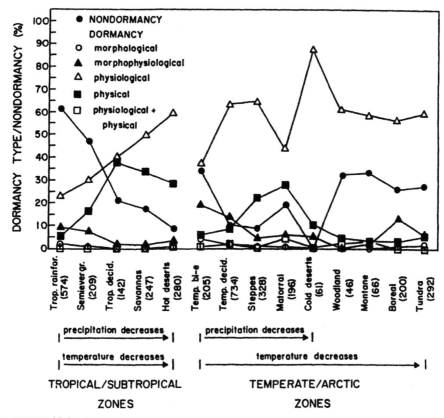

FIGURE 12.2 World biogeography of morphological, morphophysiological, physical, physical plus physiological, and physiological dormancies and of nondormancy in seeds. Number of species is given in parentheses.

tion types with the highest precipitation and temperatures, i.e., tropical rain forest and temperate broad-leaved evergreen forests, respectively. Morphophysiological dormancy also is most common in vegetation types with high precipitation and high temperatures, i.e., tropical rain forests and semievergreen forests in tropical/subtropical zones and temperate broad-leaved evergreen and deciduous forests in temperate/arctic zones.

C. Environmental Gradients, Seed Dormancy Types, and Life Forms

In the tropical/subtropical region, nondormant seeds are found in all kinds of species (i.e., trees, shrubs, lianas, and herbs) in all types of vegetation; however, with a decrease in rainfall and temperatures there is a decrease in the proportion of nondormant seeds in the various categories of life-forms (Fig. 12.3). Morphological and morphophysiological dormancies occur in trees, lianas and herbaceous species (herbs), whereas morphophysiological dormancy is known from all categories of species, except shrubs. A few trees and shrubs in evergreen rain forests have seeds with physical dormancy, but the

proportion of trees and shrubs with physical dormancy increases in tropical/subtropical vegetation types with a decrease in rainfall and temperature. Also, with a decrease in rainfall and temperature, physical dormancy is found in some lianas and herbs but not in any succulents. Physiological dormancy is found in all life-forms (except lianas in evergreen rainforests) in all types of tropical/subtropical vegetation.

In the temperate/arctic region, nondormant seeds are found in all vegetation types, except cold deserts, and in many vegetation types more woody species have nondormant seeds than herbaceous ones (Fig. 12.4). Morphological dormancy occurs in trees, shrubs, lianas, and herbs in temperate broad-leaved evergreen forests; in shrubs, vines, and herbs of deciduous forests; and in herbs of steppes, matorral, boreal, and tundra vegetation. Morphological dormancy has not been reported for any species in cold deserts, woodland, or montane vegetation. Morphophysiological dormancy is known in all types of species (trees, shrubs, lianas, herbs) in temperate broad-leaved evergreen and deciduous forests and in shrubs and herbs of steppe, matorral, and boreal vegetation, but only in herbs of cold deserts and tundra. Morphophysiological dormancy has not been reported

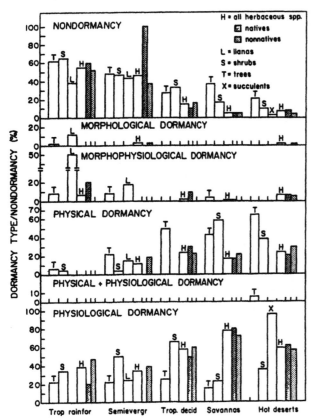

FIGURE 12.3 Relative importance of morphological, morphophysiological, physical, physical plus physiological, and physiological dormancies and of nondormancy in seeds of various kinds of plants in tropical/subtropical types of vegetation.

in woodland or montane species, but this may reflect lack of research rather than an absence of this kind of dormancy in these vegetation types.

Physical dormancy occurs in one or more life-forms in each of the nine temperate/arctic vegetation types. Physical dormancy peaks in shrubs of steppes and matorral; however, as temperatures decline, physical dormancy decreases, especially in woody species. The combination of physical and physiological dormancy types is found mostly in shrubs or trees.

Physiological dormancy is very important in all temperate/arctic vegetation types, but it is more common in herbaceous than in woody species in matorral, woodland, and tundra vegetation.

1. Trees

With a decrease in rainfall and temperature in the tropical/subtropical region, the proportion of trees in a vegetation type with nondormant seeds decreases and the proportion with physical dormancy increases; the proportion with physiological dormancy stays about the

same (Fig. 12.3). In hot deserts, no trees are reported to have seeds with physiological dormancy, but some have a combination of physical and physiological dormancy.

In the temperate/arctic vegetation types in which trees occur, the proportion of trees with nondormant seeds decreases with decreases in precipitation (Fig. 12.4) Further, with a decrease in trees with nondormant seeds there is an increase in physiological dormancy, rather than in physical dormancy, as occurs in tropical/subtropical trees.

2. Shrubs

In tropical/subtropical zones, the proportion of shrubs with nondormant seeds decreases with decreases in rainfall and temperature, and physical dormancy increases (Fig. 12.3). Unlike trees, however, the pattern of physical dormancy in shrubs does not exhibit a smooth increase. Shrubs with physical dormancy are not important in semievergreen forests and are not reported in tropical deciduous forests, but they are relatively common in savannas and hot deserts. Physiological dormancy is found in some shrubs in all tropical/subtropical vegetation types, but it reaches a maximum in tropical deciduous forests.

As precipitation decreases from temperate broad-leaved evergreen forests to cold deserts, the proportion of shrubs in a vegetation type with nondormant seeds or with physical or physiological dormancy remains about the same (Fig. 12.4). All shrubs of cold deserts have physiological dormancy. With greatly reduced temperatures (especially in tundra vegetation), nondormancy increases in seeds of shrubs and physiological dormancy decreases. It seems likely that additional montane species could have nondormant seeds.

3. Lianas

Germination studies have been done for seeds of lianas in evergreen and semievergreen tropical forests. Lianas in evergreen rain forests mostly have either nondormant seeds or seeds with morphophysiological dormancy, whereas those in semievergreen forests have nondormant seeds or morphophysiological, physical, or physiological dormancy (Fig. 12.3). Thus, with a decrease in rainfall and temperature, morphophysiological dormancy in lianas decreases and physiological and physical dormancies increase.

Lianas are important in temperate broad-leaved evergreen and in deciduous forests, and most of them have seeds with physiological dormancy (Fig. 12.4). However, some lianas in both types of forests have seeds with morphological and morphophysiological dormancies,

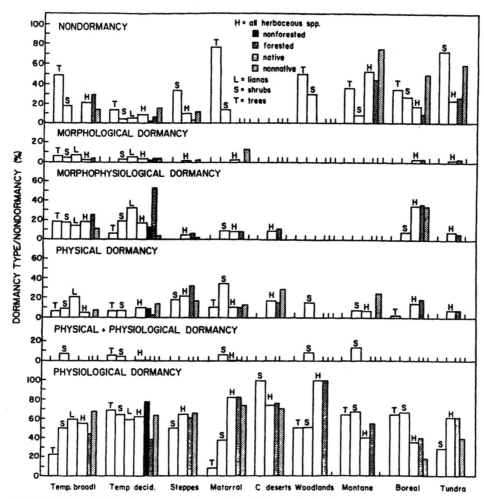

FIGURE 12.4   Relative importance of morphological, morphophysiological, physical, physical plus physiological, and physiological dormancies and of nondormancy in seeds of various kinds of plants in temperate/arctic types of vegetation.

and some in temperate broad-leaved evergreen and deciduous forests have seeds with physical dormancy and nondormancy, respectively.

4. Herbaceous Species

With decreases in precipitation and temperature in the tropical/subtropical region, the proportion of herbaceous species in a vegetation type with nondormant seeds decreases (Fig. 12.3). Correspondingly, there is some increase in herbaceous species with physical dormancy, but the greatest increase is in physiological dormancy. Thus, whereas physical dormancy in trees increases as rainfall and temperature decrease in the tropical/subtropical region, herbaceous species under the same conditions show a greater increase in physiological dormancy rather than in physical dormancy.

When herbs are subdivided into native and nonnative categories, the general pattern of distribution across types of seed dormancy is the same as for all herbaceous species combined, except in two cases (Fig. 12.3). (1) In the region where evergreen rain forests occur, none of the weeds has morphophysiological dormancy, whereas 20% of the native herbs have this type; physiological dormancy is more common in nonnative species than in native species. (2) In the region of semievergreen forests, all native species have nondormant seeds, whereas the nonnatives have either nondormant seeds or seeds with physical or physiological dormancy.

With decreases in precipitation and temperature in temperate/arctic regions, only minor changes occur in the proportion of herbs with nondormant seeds or in those with morphological, morphophysiological, or physical dormancy (Fig. 12.4). These changes are not coincident with obvious increases (or decreases) in physiological dormancy. In fact, with the exception of montane vegetation, where 53% of the herbs have nondormant seeds (and 40% have physiological dormancy), physiological dormancy accounts for the condition found in freshly matured seeds of 50% or more of the herbs in all nine temperate/arctic vegetation types. As precipitation and temperature decrease, nondormancy disappears in herbaceous species of matorral, cold desert, and woodland vegetation but not in montane, boreal, and tundra vegetation.

In the region of temperate broad-leaved evergreen forests, nondormancy and morphophysiological dormancies are more common in native herbs than in nonnative herbs, but physiological dormancy is more common in nonnative herbs than in native herbs (Fig. 12.4). In deciduous forests, 53% of the woodland herbs have morphophysiological dormancy and another 37% have physiological dormancy. However, 12% of the native herbs of open, well-lighted habitats within the region of deciduous forests have morphophysiological dormancy and 78% physiological dormancy, and 5% of the weeds have morphophysiological dormancy and 63% physiological dormancy.

In steppes, matorral, and cold deserts, more than 60% of both native and nonnative herbs have physiological dormancy (Fig. 12.4). All native herbs in woodland vegetation have physiological dormancy; no data are available for weedy species. Seeds of native herbs in montane vegetation are either nondormant (45%) or have physiological dormancy (55%), whereas those of nonnative herbs are either nondormant (75%) or have physical dormancy (25%). In boreal/subalpine and tundra vegetation, physiological dormancy is more common in native species than in nonnative species, and nondormancy is more common in nonnative species than in native species.

D. Specialized Habitats or Life Cycles

Using information from Chapter 11, we can examine the question of whether species with specialized habitats or life cycles have seed dormancy vs nondormancy? Also, if seed dormancy is present, what kind(s) is the most important?

1. Aquatic/Wetland Habitats

More species of emergent, floating attached, floating unattached, and submerged aquatics have dormant seeds than nondormant seeds, and physiological dormancy is the most common type (Table 12.1). None of the aquatics have morphological dormancy, and only a few emergents have morphophysiological dormancy. Physical dormancy is reported in only a few emergent and floating attached species.

2. Carnivorous Plants

Some species have nondormant seeds, but most have dormant seeds, with physiological dormancy being the most common (Table 12.1). A few species have morphological or morphophysiological dormancy, but physical dormancy has not been reported.

3. Halophytes

More marine angiosperms have nondormant seeds than species of beaches and cliffs or of salt deserts and salt marshes; however, all these habitats have more species with dormant seeds than with nondormant seeds (Table 12.1). Physiological dormancy is the most important kind of dormancy. Some beach and cliff species have morphophysiological and physical dormancies, and some salt marsh and salt desert species have physical dormancy. According to our estimates based on family nondormancy/seed dormancy type(s), 50% of the mangrove species have nondormant seeds, 31% have physiological dormancy, and the remainder have morphological, morphophysiological, or physical dormancy.

4. Orchids and Holoparasites

In some species, seeds have undifferentiated embryos and, thus, morphological dormancy (Table 12.1). Also, seeds of many species appear to have some physiological dormancy; consequently, these have morphophysiological dormancy. Seeds of holoparasites with chlorophyll (*Cuscuta*) have physical dormancy.

5. Hemiparasites

Seeds of host-stimulated hemiparasitic species (*Striga*, *Alectra*) have only physiological dormancy. A few (3%) rooted, green hemiparasitic species have nondormant seeds, but most (81%) have physiological dormancy; morphophysiological dormancy also is present in some (16%) species. Seeds of nonrooted hemiparasitic species (mistletoes) have underdeveloped, differentiated embryos; however, the morphologically dormant seeds of some mistletoes germinate immediately when placed under favorable conditions. Physiological dormancy also occurs in seeds of some mistletoes; therefore, they have morphophysiological dormancy.

6. Psammophytes

Fourteen percent of psammophytes have nondormant seeds, 84% of them have seeds with physiological dormancy, and 12% have physical dormancy.

TABLE 12.1 Relative Importance (%) of Nondormancy and the Various Types of Seed Dormancy in Species with Specialized Types of Habitats or Life Cycles

Type of habitat or life cycle	No. of species	Type of dormancy[a]				
		ND	MD	MPD	PD	PY
Aquatic/wetland						
Emergents	165	5	0	6	85	4
Floating attached	13	8	0	0	84	8
Floating unattached	4	25	0	0	75	0
Submerged	36	31	0	0	69	0
Carnivorous	15	13	20	13	54	0
Halophytes						
Beaches and cliffs	32	16	3	19	50	12
Mangroves	46[b]	50[b]	2[b]	4[b]	31[b]	13[b]
Marine angiosperms	14	43	0	0	57	0
Salt marshes and salt deserts	91	11	0	0	82	7
Orchids	—[c]		X[d]	X[d,e]		
Parasites						
Holoparasites	—			X[d,e]		
Holoparasites with chlorophyll	9	0	0	0	0	100
Hemiparasites						
Host-stimulated	6	0	0	0	100	0
Rooted	31	3	0	16	81	0
Nonrooted (mistletoes)	—		X[f]	X[e,f]		
Psammophytes	43	14	0	0	74	12

[a] ND, nondormancy; MD, morphological dormancy; MPD, morphophysiological dormancy; PD, physiological dormancy; PY, physical dormancy.

[b] Numbers are for types of dormancy known to occur in families (not species) to which mangroves belong.

[c] Data not available.

[d] Seeds have undifferentiated embryos.

[e] Available evidence indicates that some species also have physiological dormancy; thus, seeds would have a special type of morphophysiological dormancy.

[f] Seeds have underdeveloped, differentiated embryos.

7. Conclusion

Specialized habitats have some species with nondormant seeds, whereas others in the same habitat have dormant seeds, with physiological dormancy being the most important. In highly specialized life cycles, e.g., orchids and parasitic plants, few if any species have nondormant seeds. Morphological dormancy is always a characteristic of seeds of orchids, holoparasites without chlorophyll, and mistletoes, and it may (or may not) be combined with physiological dormancy. Morphological and/or morphophysiological dormancy is not necessarily a requirement for the parasitic life cycle per se, because holoparasites with chlorophyll and host-stimulated hemiparasites have physical and physiological dormancy, respectively. One reason for the decrease in the diversity of dormancy types among orchids and most kinds of parasitic plants is that they occur in relatively few plant families compared to species in specialized habitats.

III. THEORETICAL CONSIDERATIONS ON THE EVOLUTION OF SEED DORMANCY

Theoretical/evolutionary biologists have identified various ecological situations that may have led to the selection of seed dormancy. Seed dormancy could (1) ensure persistence (e.g., seed bank) of species in risky environments, (2) prevent seedlings from competing with the mother plant or siblings, (3) be an adaptation for survival during a season when environmental conditions are unfavorable for seedling establishment, (4) play a role in the timing of germination so that fitness (seed production) of the resulting plant is maximized, and (5) be one of several life cycle traits inherited together that maximized fitness of a species in its habitat.

A. Persistence in Risky Environments

Chapter 7,VI discussed Cohen's (1966, 1967, 1968) models for optimizing reproduction in a randomly varying environment. Ecologists know that long-term seed banks allow an annual species to persist at a population site (without immigration of seeds) where the unpredictable environment may prevent seed set in some years. What Cohen did was help us understand the size the seed bank needs to be for an annual species to survive in the unpredictable desert environment under various combinations of conditions. Since seed banks result because germination is delayed, the assumption is made that the seeds are dormant. Thus, the evolution of seed dormancy is important for the survival of species in randomly varying environments.

Since Cohen published his models, other theoretical ecologists have added ideas, such as density-dependent mortality of seedlings and seed dispersal, to them to fine-tune our understanding of the theoretical aspects of the evolution of seed dormancy.

1. Density-Dependent Mortality

Bulmer (1984) added the density-dependent mortality of seedlings to Cohen's (1966) model for the evolution of delayed germination of seeds of an annual species in a variable environment. This addition to the model showed that the variability in fitness of seedlings due to density-dependent effects of the presence of other seedlings resulted in selection pressure for delayed germination. Thus, regardless of what causes temporal fluctuations in population size (environmental stress or seedling competition), there is a selection for delayed germination.

2. Predation

Brown and Venable (1991) showed that seed predation may influence the evolution of seed dormancy. Selective effects of seed predation on dormancy depend on (1) presence or absence of a seed bank, (2) high predation of fresh vs buried seeds, (3) predation levels in favorable vs unfavorable years for seed production, and (4) how the seed yield of individual plants (and populations) influences the amount of predation from year to year. Several predation regimes can select for the production of more seeds in favorable years than in unfavorable years. High seed production in favorable years (masting) by perennial species results in predator satiation, whereas the presence of seed dormancy (i.e., seed bank) also allows an annual species to mast. Thus, an increase in predation causes an increase in seed dormancy, even if predation is density-independent.

3. Dispersal

Venable and Lawlor (1980) added seed dispersal to Cohen's model and determined the effect of dispersal on the optimum germination fraction. That is, what is the maximum number of seeds that could germinate in an unpredictable environment and still leave an adequate seed bank for population persistence in the event that no seeds were set? The optimum germination fraction increased with an increase in seed dispersal, and at low levels of dispersal it increased greatly with an increase in environmental quality of the patches into which seeds were dispersed. As predicted by the model, seeds of various species of Asteraceae and Brassicaceae that normally are dispersed away from the mother plant were nondormant, whereas those that are not dispersed were dormant (Venable and Lawlor, 1980).

In a study of *Heterosperma pinnatum* (Asteraceae), a desert annual with heteromorphic achenes, Venable *et al.* (1995) found that populations varied in the proportion of awned (central) vs. nonawned (peripheral) achenes in the heads. There was a strong genetic component to the determination of achene proportions among and within populations. Awned achenes (which are easily dispersed) come out of dormancy faster than nonawned ones, and populations with a high portion of awned achenes tend to be found in disturbed (successional) sites in areas with relatively high rainfall. Populations with a high portion of nonawned achenes tend to be located in the "fairly open vegetation" of semiarid areas. Thus, awned achenes are more likely to reach ephemeral openings in areas with the highest rainfall than nonawned ones, but seedlings from nonawned achenes are more tolerant of harsh conditions in semiarid areas than are those from awned achenes.

Some models on effects of dispersal of animals are of interest to seed ecologists. Kuno (1981) concluded that in spatiotemporally unstable habitats the reproductive rate for a whole animal population may be increased by dispersal. Further, "...individuals that disperse may be able to leave more progeny on the average than those that do not." Metz *et al.* (1983) agreed that dispersal could result in an increase in total size of the population, but they concluded that the number of progeny per individual did not increase.

Consequently, Bulmer (1984) examined the effects of dispersal of seeds on optimal germination rates (=optimal germination fraction) using two competing genotypes and varying numbers of patches with good or bad environments in the models. In an exponential model with S (number of seeds produced by surviving plant) = 10, the optimal germination rate (OGR) increased with an increase in the number of patches. According to Bulmer (1984), "... selection for delayed germination is derived almost entirely from fluctuations in the physical environment and disappears when seeds disperse contemporaneously over a large number of independent environments." With $S = 100$, OGR increased with an increase in number of patches, but OGR at any given number of patches was less than what it was when $S = 10$. For example, with 10 patches, OGR was 0.81 and 0.48 for $S = 10$ and $S = 100$, respectively.

Levin *et al.* (1984) also developed a model to help explain the evolution of dispersal and seed dormancy. Using mainly simulations, they showed that as the probability of a seed (or dispersal unit) being dispersed to a new site increases, the optimal level of dispersal increases. With an increase in dormancy, however, the optimal level of dispersal decreases. In a subsequent model, Cohen and Levin (1985) investigated optimal dispersal and dormancy strategies and the interaction between selection for dormancy and dispersal. As the size of the seed bank increases (i.e., as dormancy increases), the selective advantage of dispersal decreases. Consequently, as size of the seed bank decreases, the selective advantage of dispersal increases. If both dispersal and dormancy are subject to selection, the relative importance of each in a joint optimal strategy is determined by the effectiveness of dispersal, survival of dormant seeds, and environmental conditions. Dispersal is favored over seed dormancy if the environment is highly favorable for plant growth and seed set occurs only rarely.

Klinkhamer *et al.* (1987) modeled the effects of delayed germination and dispersal on population dynamics of annual plants under density-independent (i.e., no interactions between organisms) and density-dependent conditions. In the density-independent model, the optimal dispersal fraction for a fixed level of germination each year increases only slightly with an increase in the number of seeds in the soil that germinate (Fig. 12.5a). However, the optimal germination fraction for a fixed level of dispersal each year increases greatly with an increase in the number of seeds dispersed (Fig. 12.5b). Under density-dependent conditions, the optimal dispersal fraction increases greatly with an increase in the fraction (above 0.5) of seeds in the soil that germinates (Fig. 12.5c; compare with Fig. 12.5a). The optimal germination fraction increases with an increase in the fraction (above 0.5) of seeds that disperses (Fig. 12.5d; compare with Fig. 12.5b). If seed dormancy and dispersal are optimized simultaneously, dormancy is favored if population growth is low and fluctuations in subpopulations are correlated. However, dispersal is favored if population growth is high and fluctuations in subpopulations are not correlated. If seed dormancy and dispersal are not optimized simultaneously and if either dormancy or dispersal is held at a fixed level, then (1) the optimal dispersal fraction decreases with increasing dormancy (= decrease in germination) and (2) the optimal germination fraction increases with increasing dispersal.

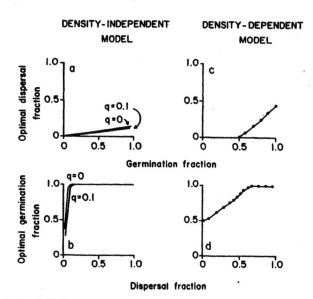

FIGURE 12.5   Models for an evolutionarily stable strategy for annual plant species. (a) Relation between optimal dispersal fraction and (fixed) germination fraction. (b) Relation between optimal germination fraction and (fixed) dispersal fraction. (c) Relation between optimal dispersal fraction and germination fraction. (d) Relation between optimal germination fraction and dispersal fraction. q is the correlation between r_i values (number of progeny per seedling) of the different patches. From Klinkhamer *et al.* (1987), with permission.

4. Bet Hedging

According to Philippi and Seger (1989), bet hedging is a trade-off between the mean and variance of fitness; therefore, "... phenotypes with reduced mean fitness may be at a selective advantage under certain conditions." In a bet-hedging strategy, the geometric mean fitness (over generations) is increased by reductions in the variance of mean fitness of each generation (Philippi and Seger, 1989).

The term "bet hedging" has been used to describe the situation when seeds produced by a desert annual have delayed germination. That is, although some seeds produced by a single plant germinate the first germination season after maturation, others do not germinate until the second or a later season. Cohen's (1966) original model and subsequent models dealing with delayed germination as an adaptation to the unpredictable desert habitat are about bet hedging. In fact, some recent attempts to add to empirical knowledge of the selection for delayed germination in unpredictable environments have been described in terms of learning more about the bet-hedging germination of desert annuals.

In a study of bet-hedging germination of six species of desert annuals in the Chihuahuan Desert in Arizona, seeds collected in the field and sown in flats of soil were given good (= soil kept well watered) and bad (no water) conditions for germination during the normal (first) winter germination season (Philippi, 1993a). All seeds were given good conditions for germination the second winter. In all six species, some seeds that did not germinate under good conditions the first year did so under good conditions the second year. The germination of *Lepidium lasiocarpum, Haplopappus gracilis, Plantago purshii,* and *Chaenactis stevioides* seeds was less the second year than the first. However, the germination of *Microseris linearifolia* and *Eriogonum abertianum* seeds was higher the second year than the first year. These results show that desert annuals produce seeds that germinate over a period of 2 (or more?) years, even if conditions are favorable for germination each year (Philippi, 1993a). Reasons for the delay in germination are unknown; however, one possibility is that some seeds fail to come out of dormancy the first summer following maturation, but do so the second (or some subsequent summer, e.g., as in *Eriastrum diffusum*) (Baskin *et al.* 1993b; see Fig. 9.5). In the field, the germination of nondormant seeds may be prevented because the soil is too dry. Philippi (1993a) showed that seeds kept dry the first year germinated under wet conditions the second year.

Assumptions of bet-hedging germination are that (1) there are no heritable differences between seeds that germinate in the first year and those that germinate in subsequent years and (2) a single plant produces some seeds that germinate the first year and others with delayed germination (Philippi, 1993b). In a study of the desert winter annual *Lepidium lasiocarpum,* Philippi (1993b) found that (1) some seeds that do not germinate under good conditions the first winter do so under the same conditions the second winter, (2) a single plant produces some seeds that germinate the first year and others that do not germinate until at least the second year, and (3) there is no variation in the first-year germination of seeds produced by plants resulting from seeds germinating the first and second year (Philippi, 1993b). Thus, the species has a bet-hedging germination strategy.

5. Evolutionary Stable Strategy Germination Strategies

In an evolutionary stable strategy (ESS), the genotype with a particular characteristic (or action) can resist invasion by rare mutants with an alternative mutant form; however, the strategy has to be common in the population (Bulmer, 1994). With a decrease in the frequency of the ESS trait in the population, a point is reached when a mutant strategy could invade.

An ESS germination fraction is another way of asking what proportion of the seeds of an annual species in (on) soil at the population site in a randomly varying environment should germinate each year. Given that the site has good and bad years with respect to favorability of conditions for survival and growth to seed set, the objective in determining the ESS germination fraction is to find the value that would give germinated and nongerminated seeds the same "average absolute fitness" (Ellner, 1985a). According to Ellner (1985a), an ESS germination fraction is "... characterized as having enough dormancy to do moderate decreases in population density but not so much that 'good' years aboveground are under exploited...."

Ellner's (1985a,b) models for population dynamics of desert annuals in randomly varying environments are for density-regulated populations, whereas Cohen (1966) assumed that populations were density independent. Ellner analyzed his models to determine the ESS germination fraction, and it is interesting to compare his results with those of Cohen. In Cohen's and Ellner's models, an increase in seed survivorship predicts a decrease in the proportion of seeds germinating, but with an increase in the frequency of favorable years Cohen's model predicts an increase in germination and Ellner's a decrease. An increase in the mean yield of seeds causes the germination fraction to increase in Cohen's model, but in Ellner's model the ESS germination fraction decreases with an increased variability in seed yield.

6. Population Dynamic Triangle

Kadmon and Shmida (1990) suggested that many of the variables in models are too difficult to be measured easily, especially under field conditions. Thus, they proposed that research on seed dormancy and dispersal be approached in terms of seedling recruitment each year. Further, they devised three components for seedling recruitment: component I, density of seedlings from locally produced seeds of the current year (i.e., if no seeds in the seed bank germinate and there is no immigration of seeds); component II, seedlings resulting from the germination of seeds in the seed bank; and component III, seedlings resulting from the germination of seeds dispersed to the population site. By comparing these three components with each other and with seedling density in patch (i) in year ($t + 1$), the relative importance of seed bank and net dispersal of seeds into the patch as the source of the seedlings growing in the patch each year can be determined.

Kadmon and Shmida (1990) developed a population dynamic triangle (Fig. 12.6). By obtaining information at a population site over time, one would have a cluster of points in the triangle and thus a picture of the spatiotemporal demographics of the population (Fig. 12.7). Various types of population demographic patterns are

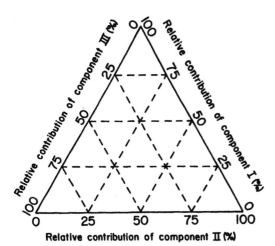

FIGURE 12.6 The population dynamic triangle. Each side represents a 0–100% contribution of a given component of recruitment. Component I, contribution from local reproduction in the preceding year; component II, contribution from seeds in the soil seed bank; and component III, the net contribution from dispersal. From Kadmon and Shmida (1990), with permission.

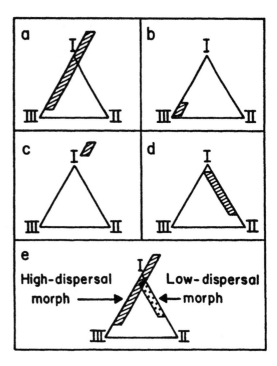

FIGURE 12.8 Some expected types of clusters in the population dynamic triangle: (a) a fugitive type, (b) a sink population type, (c) a source population type, (d) a low-risk strategy type, and (e) expected clusters for species having two dispersal morphs. From Kadmon and Shmida (1990), with permission.

to be expected (Fig. 12.8). The fugitive type of pattern (Fig. 12.8a) is for species growing in disturbed habitats. Over time, a disturbed site is first a recipient of dispersed seeds, but after a population of plants is established, it is a donor of dispersed seeds. In a sink population (Fig. 12.8b), there is a high contribution from dispersed seeds and a low contribution from locally produced ones. In a source population (Fig. 12.8c), there is no contribution from dispersed seeds and a high contribution from locally produced ones. In a low-risk strategy (Fig. 12.8d), there is a high contribution from the seed bank after bad years and a low one after good years. Also, if seed dormancy is high, dispersal is expected to be low. In the two-morph type (Fig. 12.8e), one morph (the one with a dispersal mechanism) would be a fugitive type (Fig. 12.8a), whereas the other morph (with no dispersal mechanism) would be a low-risk strategy type (Fig. 12.8d).

In Israel, the main source of seedlings for a population of *Stipa capensis* occurring in a wadi was local seed production, i.e., component I (Kadmon and Shmida, 1990). However, dispersal (component II) was the main source of seedlings for a population occurring on a slope. Seed banks (component III) were a minor source of seedlings in both population sites.

7. Models for Various Kinds of Plants in Nondesert Habitats

Much time and energy have been devoted to understanding the theoretical aspects of delayed germination in desert annuals; however, little is known about what is happening (has happened) in other kinds of habitats. For example, from a theoretical perspective why should seeds of species growing in mesic habitats have dormant seeds?

Ample evidence shows that seed dormancy and germination characteristics of many species vary, depending on the position on the mother plant where seeds develop (Table 8.5) and that preconditioning is a common phenomenon (see Tables 8.3, 8.4, and 8.6). In many species, if seeds (especially those with nondeep physiological dormancy) are exposed to appropriate

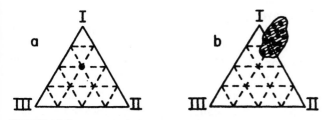

FIGURE 12.7 Hypothetical patterns in the population dynamic triangle. (a) Result of a single year. Contributions to the population from components I, II, and III are 50, 25, and 25% respectively. (b) Results from a 14-year study expressed as a cluster of points. Note that in most years the dispersal component is negative. From Kadmon and Shmida (1990), with permission.

dormancy-breaking conditions, a high percentage (sometimes 100%) of them come out of dormancy the first year and germinate if temperature, light : dark, and moisture requirements are fulfilled (e.g., Baskin and Baskin, 1981, 1990a; Baskin et al., 1994a). However, in some species, a small portion of the seed crop fails to come out of dormancy until the second year or later (e.g., Baskin and Baskin, 1990b; Baskin et al., 1989). Nevertheless, persistent seed banks of numerous species occur in mesic habitats (Chapter 7). Thus, models are needed for what size the seed banks of annuals, short- and long-lived monocarpic perennials, and polycarpic perennials should be for long-term persistence in environments where favorable conditions for germination and plant growth are seasonally predictable, but disturbance is unpredictable. Further, when there are gradients from environmental (1) unpredictability to predictability or (2) unfavorability to favorability, at what point should the control of dormancy be under genetic vs maternal control? For example, in a study on *Bromus tectorum*, Beckstead et al. (1996) concluded that, "... the more extreme yet predictable environments select for seed germination and after-ripening patterns that are genetically fixed, while populations from more favorable environments tended to show more between-year variation, suggesting more phenotypic plasticity."

Rees (1994) modeled the effects of adult longevity, timing of reproduction, and population age/stage structure on the evolution of seed dormancy in constant and variable environments. In general, he found that the ESS in constant environments is for no seed dormancy regardless of adult longevity, timing of reproduction, or the cost of reproduction on subsequent survival, which may help explain why a high proportion of the species in tropical rain forests have nondormant seeds. In a variable environment, increased adult longevity selected against seed dormancy (Rees, 1993); however, the models showed that there can be exceptions (Rees, 1994). For example, if reproduction occurs in a single dominant age class and if reproduction and/or seedling establishment is unsuccessful in some years, selection for seed dormancy will occur. Also, variation in adult survival and the probability of successful seedling establishment will select for seed dormancy. When the cost of reproduction in polycarpic plants was added to the model, plant longevity decreased, thereby selecting for more seed dormancy (Rees, 1994).

B. Escape from Competition

Seed banks (or delayed germination) are not only an adaptation for persistence in unpredictable environments, they also ensure long-term survival if the seed set is reduced or prevented by high densities of plants in the population. Thus, seed banks are one of several seed traits that function as an adaptation to escape the effects of local crowding (Venable, 1989).

1. Competition and Seed Dormancy

Leon (1985) published a model "... showing that competition without uncertainty may also favor dormancy." He explained that fitness could be increased when a seed (1) germinates and the resulting plant produces seeds or (2) does not germinate until the following year and risks dying in the soil. Therefore, the best strategy would be to balance the marginal gains coming from both possibilities.

However, Ellner (1987) showed that dormancy is not selected in Leon's model when there is competition without uncertainty. Using Leon's model and assumption of a stable equilibrium, Ellner (1987) found that the ESS would be for no seeds to be dormant when competition occurs without uncertainty. If Leon's assumption of a stable equilibrium is ignored and if competition causes fluctuations in the number of seeds produced per year, dormancy may be favored. Competition in a constant environment can select for dormancy if it is localized in the population (not population-wide) and if it causes fluctuations in population density. If competition occurs throughout the population in a variable environment, it intensifies the effects of environmental fluctuations, and dormancy is favored.

2. Sibling Competition

In a temporally constant environment with population density at a stable equilibrium, competition among siblings favors dormancy, and Ellner's (1986) model predicts an increase in dormancy with an increase in competition between siblings. The reason suggested for increased seed dormancy is that it would prevent siblings from competing with each other. Therefore, the ultimate reproductive success of the mother plant would be increased (Ellner, 1986, 1987). Seeds from upper nodes of the grasses *Sporobolus vaginiflorus* and *Triplasis purpurea* have a greater chance of dispersal than those from lower nodes, and they are less dormant than those with the reduced chance of dispersal (Cheplick, 1996).

Nilsson et al. (1994) evaluated the adaptive significance of a seed bank and a nonseed bank strategy when sibling competition is intense by looking at the predicted number of "grandchildren" produced by a mother plant in each situation. Their model predicts that the reproductive success (i.e., grandchildren) of the plant with a seed bank strategy will be up to four times greater than that of the plant with a nonseed bank strategy. With an increase in seed survival, the proportion of seeds that should remain dormant to maximize the relative reproductive success of the seed bank type of the mother plant increases. However, even if seed survival is low, the optimum proportion of seeds produced by a seed-bank mother plant that has delayed germination is high (Nilsson *et al.*, 1994).

Plants of the desert annual *Lepidium lasiocarpum* grown under high soil moisture conditions produced large numbers of seeds (up to 2700 per plant) (Philippi, 1993b). These large plants had a high proportion of seeds in which germination was delayed until the second year; thus, high moisture appears to have increased dormancy. Moisture stress during seed maturation has been shown to both increase and decrease dormancy, depending on the species (see Chapter 8,V).

3. Parent vs Offspring or Westoby vs Ellner

Westoby (1981) and Ellner (1986) agree that mother plants sometimes produce seeds that differ in their level of dormancy. Consequently, some seeds produced by a plant may germinate the first germination season following dispersal, but the germination of other seeds produced by the same plant is delayed until the second (or later) germination season. However, Westoby and Ellner have different ideas about the factor that has selected for delayed germination. Westoby (1981) views delayed germination as a bet-hedging strategy for survival in an unpredictable environment, whereas Ellner (1986) thinks it is the result of strong sibling competition.

Both Westoby and Ellner mention the fact that seeds have genetically diverse tissues: (1) The *embryo* (offspring) is diploid and has a set of chromosomes from the mother and the father. (2) The *endosperm* is usually triploid and has two sets of chromosomes from the mother and one from the father. (3) *Seed coats* are diploid, and both sets of chromosomes are from the mother. Further, both authors suggest that there is a conflict between the mother plant and the embryo as to when germination should occur. They suggest that the highest fitness for the mother is for germination to be delayed, whereas the highest fitness for the embryo is for germination to occur immediately. Their view is that the mother wins the contest by manipulating the seed coats, thereby making germination relatively easy or difficult.

As shown in Chapter 8,VI, preconditioning results in changes in seed size, color, and shape, and in *Chenopodium,* day length causes variations in seed coat thickness. However, in *C. album* (Karssen, 1970) and *Amaranthus retroflexus* (Kigel *et al.,* 1977) seed dormancy is not correlated with the thickness of the seed coat. Thus, little, or no, evidence is available to show that delayed germination in seeds is due to the differential production of seed coat characteristics by the mother plant. It would be informative to measure seed coat thickness in seeds of desert annuals that germinate in the first vs subsequent years after they are produced.

Another problem with most of the models discussed thus far in this chapter is that the assumption is made that the embryo is nondormant and that germination is controlled only by the seed coats. Because most models appear to be devoted to understanding the germination ecology of seeds with nondeep physiological dormancy, we will focus our attention on seeds with this type of dormancy. Seeds with nondeep physiological dormancy have embryos with shallow (or nondeep) dormancy, and seed coats are permeable (Chapter 3). If the embryo is excised from the seed, it grows normally. Permeable seed coats play a role in preventing the germination of seeds with nondeep physiological dormancy, but evidence showing that the seed coats change during the dormancy-breaking period is limited. However, after the appropriate dormancy-breaking treatment has been applied for a sufficient period of time, the embryo comes out of dormancy. As dormancy loss occurs, the growth potential of the embryo increases until there is enough "push power" for it to break through the seed coats. Changes in the growth potential of embryos have been demonstrated via changes in the ability of embryos to grow in solutions with increased osmotic potentials (Chapter 3), suggesting that the physiological state of the embryo may be the primary reason why a seed with permeable seed coats does or does not germinate. Thus, the mother wins for only a short period of time, and then the offspring wins; the offspring is the ultimate winner!

There is no denying that seeds produced by the same plant may germinate in different years; however, to understand why this happens, we need to know more about the loss of dormancy. For example, we do not know how much moisture (level and duration) is required for the loss of dormancy of seeds of desert annuals in the field. Are the effects of proper moisture levels for loss of dormancy cumulative? Do seeds from the same plant vary in the total number of hours (or weeks) of favorable moisture conditions required for a loss of dormancy? This might explain why some seeds germinate under the same conditions the second but not the first germination season following maturation (Philippi, 1993b). Do seeds require exposure to a certain temperature range for moisture to promote a loss of dormancy?

Models of population dynamics of desert annuals have helped us understand much about the evolution of seed dormancy and dispersal. However, to continue to make progress in this field, correct information is needed about basic seed ecology; consequently, seed ecologists have much work to do to provide modelers with the facts they need. Perhaps future models will be concerned with rates of dormancy loss as a measure of how favorable (or unfavorable) the environment is during a given year.

C. Survival in Unfavorable Seasons

Levins (1969) discussed the adaptive significance of dormancy within the context of strategic analysis. His explanation involves four basic principles. (1) A strategy is how some limited resource is allocated among several possibilities. The resource may be something tangible like a compound or it may be "... some fitness parameter which varies with the environment in such a way that it cannot be maximal simultaneously in all environments." (2) Because fitness cannot be maximized in all environments at the same time, the optimum strategy is to allocate resources so that the combination of fitness in different environments is maximized. The appropriate combination of fitness depends on the amplitude and frequency of environmental fluctuations. In a coarse-grained or heterogenous environment, conditions are such that an organism (e.g., an annual plant species) spends part of its life cycle in an alternative state (i.e., seeds), i.e., the environment is not favorable for the growth of annuals all year due to below-freezing temperatures or to a dry season. (3) If environmental extremes exceed the tolerances of the individual and if the environment is coarse-grained, the optimum norm of reaction is a mixed strategy, i.e., there are different phenotypes and they vary as a function of the environment. The norm of reaction is subject to natural selection, and total selection is greater in the common environment than in the rare environment. (4) The optimum strategy in uncertain environments is to be active during the time of year when environmental extremes do not exceed tolerances of the plant and to be dormant during the time when these tolerances are exceeded (Levins, 1969).

Schaal and Leverich (1981) view seed dormancy as "... an unavoidable life history characteristic, and [it] is adaptive in a predictable environment with an unfavorable period." Specifically, they were talking about winter annuals that usually die at the beginning of the hot, dry season. Seedlings from seeds that germinate in late spring or summer would be killed. Thus, because plants are intolerant of summer environmental conditions dormancy is an unavoidable life history characteristic. However, there is a risk that dormant seeds in the soil will die (for various reasons) during summer. Germination occurs in autumn when the probability of death as a seedling falls below the probability of a seed dying in the soil (Schaal and Leverich, 1981).

D. Optimizing the Time for Germination

A discussion of the theoretical aspects of the evolution of seed dormancy would not be complete without considering the timing of germination. It long has been realized that germination-regulating mechanisms probably have adaptive value (e.g., Koller, 1964). Thus, it is not surprising that this phase of the life cycle has received some attention from theoretical ecologists.

1. When to Germinate?

For an annual species, seed germination at the very beginning of the favorable season gives the resulting plant the maximum length of time for growth and reproduction. However, if environmental conditions early in the growing season are unstable seedlings may not survive. If early germinating seedlings are not killed, the resulting plants may be larger (and produce more seeds) than those from late-germinating seeds (Baskin and Baskin, 1972b).

Leon's (1985) models for timing of germination in fine-grained (short-term fluctuations), uncertain environments predict that the optimum germination strategy is to "... germinate synchronously on a day which maximizes the product of the reproductive success gained by germinating that day, multiplied by the chance that the favorable season begins before it." That is, full germination would occur on a certain day (synchronic germination).

In coarse-grained, uncertain environments, Leon's (1985) models predict that if $S(t)$ (S is the set of environmental states that can serve as signals; t is the onset of favorable good season) is concave, linear, or slightly convex, there is no reason to risk germination prior to the most common day of the onset of the favorable season. However, if $S(t)$ is strongly convex, there is a reason to risk early germination. Thus, diachronic germination or germination that is spread out over time may be the result.

In a model looking at the regulating effects of competition, Leon (1985) found the germination probability function that "... gives the same fitness to individuals germinating at any of a set of different days." Thus, "... time-spreading [of germination] is a strategy induced by competition"; however, a high constant rate of mortality of the seeds would favor germination at the beginning of the favorable season (Leon, 1985).

2. Effects on Fitness

Studies have been done in which newly germinated seedlings were marked at various times during the germination season in the field and plants were monitored until they flowered and set seeds. Thus, effects of the germination date on the percentage of seedling survival and number of seeds produced per surviving plant could be determined. For a high percentage of seedling survival, the best and worst times for emergence in many

TABLE 12.2   Effects of Timing of Germination on Seedling Survival and on Number of Seeds
Produced per Surviving Plant[a]

Species	For a high percentage of seedling survival		For a high number of seeds produced per plant		Reference
	Best germination time	Worst germination time	Best germination time	Worst germination time	
Ambrosia trifida	2–1	3	1	3	Abul-Fatih and Bazzaz (1979)
Alyssum alyssoides	2, 3	1	—[b]	—	Baskin and Baskin (1974)
Androsace septentrionalis	1	3	—	—	Symonides (1977)
Avena sterilis	1	3	1	3	Fernandez-Quintanilla et al. (1986)
Bromus tectorum	1, 1, 2[c]	3, 2, 1[c]	—	—	Mack and Pyke (1984)
	2, 1, 1[d]	3, 1, 3[d]	—	—	Mack and Pyke (1984)
	1, 3, 1[e]	3, 2, 3[e]	—	—	Mack and Pyke (1984)
Corynephorus canescens	2	3	—	—	Symonides (1977)
Echium plantagineum	3	1	—	—	Burdon et al. (1983)
Erucastrum gallicum	1[f]	3[f]	1[f]	3[f]	Klemow and Raynal (1983)
Helenium amarum	1	2	1	2	Baskin and Baskin (1973)
Heterotheca pinnatum	3	1	1	3	Venable et al. (1995)
Lactuca serriola	3	1	2	1	Marks and Prince (1981)
Leavenworthia exigua	2	1, 3	—	—	Baskin and Baskin (1972a)
L. stylosa	2	1	1	2	Baskin and Baskin (1972b)
Ludwigia leptocarpa	1	2	1	2	Dolan and Sharitz (1984)
Lychnis flos-cuculi	1	3	—	—	Biere (1991)
Mimosa pigra	1	3	—	—	Lonsdale and Abrecht (1989)
Papaver dubium	1, 3[g]	3, 1[g]	1	3	Arthur et al. (1973)
Pastinaca sativa	3	1	—	—	Baskin and Baskin (1979)
Phacelia purshii	2, 3	1	2	3	Baskin and Baskin (1976)
Plantago coronopus	1, 2	3, 3	—	—	Waite (1984)
Saxifraga osloensis	1	3	1	3	Nilsson (1995)
Senecio vulgaris	2	1, 3	2	1	Putwain et al. (1982)
Rumex crispus	3	1	3	1	Weaver and Cavers (1979)
R. obtusifolius	3	1	3	1	Weaver and Cavers (1979)
Thlaspi arvense	2–3	1	1	3	Klebesadel (1969)
	—	—	1	3	Milberg and Andersson (1994)
Tragopogon heterospermus	1	3	—	—	Symonides (1977)
Viola blanda	1	3	—	—	Cook (1980)
Vulpia fasciculata	2	3	1	3	Watkinson (1981)
Xanthium canadense	—	—	3[h]	1	Shitaka and Hirose (1993)

[a] 1, first seeds to germinate; 2, intermediate seedling cohort(s); 3, last seeds to germinate.
[b] No data.
[c] Moist site in eastern Washington in three different years.
[d] Mesic site in eastern Washington in three different years.
[e] Dry site in eastern Wastington in three different years.
[f] Based on number of flowering plants.
[g] Varies with the year.
[h] Ratio of final reproductive yield to vegetation biomass at end of growing season.

species are at the beginning and at the end of the germination season, respectively (Table 12.2). However, there are exceptions to this general statement, and in *Bromus tectorum* the best time for emergence (that results in high seedling survival) varies from year to year at a site and from site to site in the same year (Mack and Pyke, 1984). Time of germination was not correlated with fitness in *Crepis tectorum* (Andersson, 1992).

For a high number of seeds to be produced per surviving plant (high fitness), the best time for seedling emergence in many species is at the beginning of the germination season, and the worst time for emergence is at the end of the germination season (Table 12.2). In fact, Silvertown (1985) said that "... an early-germinating strategy is the ESS." With a population model that incorporated seasonal dormancy, i.e., some seeds germinated in autumn and others in spring, Silvertown (1988) predicted that parental fitness would decline with an increase in spring germination. Data collected by Fernandez-Quintanilla *et al.* (1986) for survivorship and fecundity of *Avena sterilis* cohorts germinated in the field in central Spain from mid-October to mid-April fit Silvertown's (1985) model.

In *Papaver dubium,* some seeds germinate in autumn and others germinate in spring; however, there is essentially no heritability in the population for autumn vs spring germination (Arthur *et al.,* 1973). If winters are mild, high percentages of the seedlings survive, and each plant produces a high number of seeds. If winters are harsh, only a low percentage of the autumn-emerging seedlings survive. However, the relatively few plants that survive bad winters produce more seeds per plant than those resulting from seeds germinating in spring.

In some species, including *Lactuca serriola, Phacelia purshii, Rumex crispus, R. obtusifolius,* and *Xanthium canadense,* more seeds are produced per plant when plants result from seeds germinating in the mid- and late phases of the germination season than are produced by plants resulting from seeds that germinate at the beginning of the germination season (Table 12.2). Thus, these species do not fit Silvertown's (1988) model. One possible reason for reduced fitness of the surviving plants from the earliest germinating cohorts in these species is that environmental conditions at the beginning of the germination season were so severe that they permanently stunted the plants.

In those species where a decline in fitness is correlated with a delay in germination, increased competition is thought to be the reason for a decrease in fitness. That is, each day that germination is delayed seeds of competitors are not only germinating, but seedlings are growing (Silvertown, 1988; Benjamin, 1990). Seedlings of *Plantago lanceolata* became established in a hayfield in the Netherlands, but those of *P. major* did not (van der Toorn and Pons, 1988). Seeds of *P. lanceolata* germinated earlier than those of *P. major.* Thus, seedlings of *P. lanceolata* had a period of growth before the grasses began to grow, but those of *P. major* did not. The general decrease in fitness when seedlings emerge late in the germination season may be an important factor in selection for secondary dormancy. Is low-temperature induction of secondary dormancy in seeds of winter annuals and high-temperature induction of dormancy in seeds of summer annuals (Chapter 4) related to a decrease in fitness that results when seeds of winter and summer annuals germinate at the end of their respective germination season? This possibility has not been explored by theoretical ecologists.

E. Seed Dormancy and the Evolution of Other Traits

Seed dormancy can be correlated with other life-history traits. For example, no dormancy is found in seeds of the winter annual form of *Senecio sylvaticus,* which is frost tolerant, although it is present in seeds of the frost-sensitive form (Ernst, 1989). Ritland (1983) modeled the joint evolution of delayed germination and delayed flowering times of annuals growing in variable (desert) environments. In this model, a fraction of the seeds germinate each year, whereas others remain dormant in the soil, surviving at a random rate during the year. Of the seeds that germinate, some flower and set seeds relatively early in the season, whereas others delay flowering until late in the season. Early flowering plants produce fewer seeds than late-flowering ones, which may be killed before they flower. Reproductive gains come from the random increase of reproduction due to (1) germination and (2) delayed flowering time. The optimum germination fraction depends only on the mean and variance of the first reproductive gain, whereas the optimal delayed flowering fraction is a ratio of means and variances of both gains. Thus, with zero covariance and joint optimality, the amount of seed dormancy is largely determined by growing conditions prior to flowering, which implies that flowering time is a sufficient character for adapting to conditions during flowering. Optimum delayed germination and flowering fractions are interdependent, and the optimal mean of each character is more sensitive to the other's than to its own variance.

If traits are genetically correlated, independent evolution may be prevented, but if they are not genetically correlated the evolution of traits can be independent (Lande, 1982; Rathcke and Lacey, 1985). Using a diallelic locus with additive effects for each character, Ritland (1983) looked at the evolution of delayed germination and delayed flowering. If both characters were genetically variable and if alleles at the second locus were at stochastic equilibrium, those adapted to more environmental variability were favored at the first locus. "Thus, genetic variation of one character, when uncorrelated with genetic variation of the second character, has the analogous effect of increasing the environmental variation the second character is subject to" (Ritland, 1983).

IV. PALEOCLIMATE AS A BACKGROUND FOR THE EVOLUTION OF SEED DORMANCY

If seed dormancy has evolved to optimize reproduction in varying environments, then we need to know something about climatic conditions during the time since seed-producing plants first appeared in the fossil record. DiMichele *et al.* (1987) have emphasized the role of extrinsic stress (e.g., extended periods of reduced precipitation) in increasing the rate of speciation. That is, during and following major periods of environmental stress new species appear, and often there also are major extinctions of species. For example, conifers appeared in the fossil record of the early part of the Middle Pennsylvanian, but did not become abundant until the beginning of a second dry interval in the Upper Pennsylvanian (DiMichele *et al.*, 1987). It seems reasonable that the evolution of new species, or an increase in abundance of those already present, might involve the development of seed dormancy. In the following description of world precipitation and temperature patterns since the Triassic and Cretaceous, respectively, it is apparent that seed-producing plants have been subjected to great climatic changes.

A. Rainfall Patterns

Atmospheric circulation is an important determinant of rainfall patterns, which results from (1) thermal contrasts between the equator and the poles, (2) thermal contrasts between land masses and oceans, and (3) rotation of the earth (Parrish and Curtis, 1982). Thus, the location of land masses ultimately has an influence on rainfall patterns. Because the position of land masses has changed over time, attention will be given to continental drift as well as rainfall patterns.

1. Paleozoic

The present day Northern Hemisphere continents were in the equatorial zone during the Paleozoic, and the Southern Hemisphere ones were largely at high latitudes during this time. Indirect information on rainfall during the Paleozoic comes from (1) geographical distribution and kinds of evaporite deposits, such as hematite, gypsum, anhydrite, phosphorites, and bauxite and (2) location and extent of coal beds. Because the first fossil seeds are from the Devonian (see Section V), we will begin with this period. In the Devonian, several land masses collided to form Laurussia, which was located north of the continent called Gondwana (Ziegler, 1981). It was relatively arid during the Early and Middle Devonian, but the Late Devonian was humid, as illustrated by the formation of coal (Frakes, 1979).

In the Carboniferous, the distance between Laurussia and Gondwana decreased (Ziegler, 1981), and the vast number and extent of coal deposits, especially in the Northern Hemisphere, indicate that high rainfall was

common (Frakes, 1979). During the Permian, all the land masses came together, and by the earliest Triassic a supercontinent called Pangaea was formed (Parrish *et al.*, 1982). During the Permian, much coal was deposited in the Southern Hemisphere, indicating a wet climate, while many evaporites were deposited, especially in North America, indicating a dry climate (Frakes, 1979). However, Asia may have had a relatively wet climate (Frakes, 1979).

2. Mesozoic

Using paleogeography and distribution of deposits of coal and various types of evaporites, Parrish *et al.* (1982) constructed a series of maps showing not only continental drift but also the regions with high and low precipitation from the Early Triassic to Middle Miocene. Their work is summarized briefly.

In the Triassic and Early Jurassic, the equatorial region and the interior of Pangaea were very dry, whereas coastal areas and the Arctic Ocean had high rainfall. By the latest Jurassic, Pangaea had separated into two land masses, which were divided by the Tethys Sea. The interior of Gondwana (southern continent) and the northern continental mass were relatively dry, whereas areas at both poles and along the coast were wet.

By Middle Cretaceous, South America and India had separated from Africa. The interior of North America, Asia, and regions in southwestern Africa and westcentral South America had low rainfall, whereas the remainder of the world's land areas was relatively wet. By latest Cretaceous, areas of highest rainfall were located in equatorial regions (north and south of the equator in South America and in Africa), in western North America, and in coastal areas of Australia.

3. Cenozoic

By the Middle Eocene, Australia had separated from Antarctica. Also, India and Africa were moving northward, and North and South America had separated. The west coast of Australia, equatorial regions, western North America, coastal Asia, and India had the highest rainfall. By the Middle Miocene, India had collided with Asia, and Australia had moved north; also, North and South America were connected. During this time, rainfall decreased in western Australia and eastern Africa.

The fossil record for the Tertiary Period indicates that decreases in rainfall were associated with major vegetation changes. By the Late Paleocene and Early Eocene, a wooded savanna had developed in South America (e.g., Patagonia), and this was a time of speciation of mammals, especially herbivores (Solbrig, 1976;

Webb, 1978). In the Great Plains of North America (e.g., North Dakota), fossil soils (paleosols) indicate that annual rainfall in the Late Eocene (38 MYBP) was >1000 mm, but by the Oligocene (32 MYBP) it had decreased to 500–900 mm (Retallack, 1992). Thus, the moist forests of the Late Eocene were replaced by dry woodlands at 33 MYBP, by wooded grasslands with gallery forests along streams at 32 MYBP, and by grasslands at 30 MYBP (Retallack, 1992). However, Leopold *et al.* (1992) noted that grasses were not important in the fossil record of the Oligocene in North America, central and eastern USSR, or China. Thus, the open ("low-biomass") Oligocene vegetation was dominated by shrubs, and grasses did not become important until the Miocene (Leopold *et al.,* 1992).

In the Miocene and Pliocene, mountain ranges were uplifted on various continents, giving rise to rainshadow effects and semideserts and deserts (Solbrig, 1976; Stewart and Rothwell, 1993).

4. Speculations on Seed Dormancy

A modern map of the annual precipitation on earth (e.g., Walter, 1973) reveals that some areas have low amounts of precipitation each year, whereas others have high amounts. When paleo and modern climatic maps are compared, it becomes apparent that some areas (e.g., northern South America, equatorial Africa, and southeast Asia) have had high annual rainfall since the Cretaceous, when angiosperms first appeared in the fossil record. Further, these high rainfall regions presently have tropical rain forests (Fig. 9.1) with a high number of species whose seeds are nondormant (Fig. 12.1).

Some regions (e.g., southcentral Africa, westcentral South America, and central Asia) have had relatively low amounts of annual rainfall since the middle Cretaceous, whereas other regions (e.g., western and central Australia, and central North America) have had variable amounts. In regions with a long history of low rainfall and in those that have had periods of high rainfall followed by low rainfall, seed dormancy is very common among extant species (Fig. 12.1; also see later). Thus, we conclude that environmental conditions have been more conducive for the development of seed dormancy in regions with low rainfall than in those with high rainfall.

B. Temperature

Paleotemperatures of the oceans are inferred from studies on (1) sedimentary rocks and the biogeographical distribution patterns of the fossilized organisms they contain and (2) the ratio of ^{18}oxygen/^{16}oxygen in the calcium carbonate deposited by marine organisms. An inverse linear relationship exists between the uptake of O^{18} (vs O^{16}) and temperature. Thus, with a decrease in temperature the O^{18}/O^{16} ratio increases, whereas with an increase in temperature the ratio decreases (Epstein *et al.,* 1951). Temperatures measured by oxygen isotopes are called isotopic temperatures (Savin, 1977); however, isotopic temperatures for rocks older than the Cretaceous may not be reliable (Frakes, 1979). Land temperatures have been inferred from information on deposits of coal and evaporites.

Wolfe (1971, 1981) has devised a way to determine paleotemperatures for land areas. He compared characteristics of fossil leaves (size, type of margin, texture, organization, shape, and presence of drip tips) and wood (growth rings) with those of extant plants growing in various types of communities all over the world. Then, after he found a community where the characteristics of the fossil leaves and wood matched those of extant plants, he determined (or looked up) the temperatures for that community. Thus, Wolfe used characteristics of fossil plant material to infer what the temperatures were when the plants were alive.

1. Paleozoic

In terms of mean global temperatures, much of the Devonian was characterized by a "gentle cooling" (Frakes, 1979). The presence of coral reefs in the Carboniferous, especially in the Mississippian, indicates that the climate was warm, but cooling and glaciation occurred in the Pennsylvanian. Thus, the climate of the Early Permian probably was colder than it is today; however, by Mid- and Late Permian, temperatures apparently had increased (Frakes, 1979).

2. Cretaceous

The stages of the Cretaceous Period (from old to young) are Berriasian, Valanginian, Hauterivian, Barremian, Aptian, Albian, Cenomanian, Turonian, Coniacian, Santonian, Campanian, and Maastrichtian (Taylor, 1981). According to Savin (1977), the oldest rocks for which isotopic paleotemperatures are available are from the Late Cretaceous Period. Marine depositional temperatures were relatively high (up to 17°C) during the Albian (ca. 105 MYBP) and then declined in Late Albian or in the Cenomanian. Another temperature increase (to about 14°C) occurred in the Turonian and Coniacian stages (ca. 90–80 MYBP), which was followed by a decrease (down to roughly 9°C) during Santonian through the Maastrichtian stages (ca 80–65 MYBP), which is the end of the Cretaceous (Savin,

1977). Thus, during the Late Cretaceous there was a net cooling of the earth. From the warm Albian to the end of the Cretaceous there was a net decrease in temperature of "... as much as 8° or 10°C in both northern Europe and the tropical Pacific Ocean" (Savin, 1977). It should be noted that this period of cooling came before the mass extinctions of organisms that occurred at the Cretaceous–Tertiary boundary (Savin, 1977).

A model simulating atmospheric conditions in the Cretaceous, using realistic geography of land masses for the period, predicted temperatures in tropical regions that were 1–2°C above those occurring in the tropics today (Barron and Washington, 1982). Although the equator-to-pole temperature gradient was less in the Cretaceous than it is today, the model did not predict that atmospheric circulation was sluggish. Further, the model predicted that there was no poleward displacement of the subtropical high (Barron and Washington, 1982).

Upchurch and Wolfe (1987) are unsure about using information from fossil leaves and wood to determine land temperatures during the Aptian and Early Cenomanian stages of the Cretaceous (when angiosperms were evolving rapidly) because there are no close modern examples of the type of vegetation that existed during these stages. Data from fossil leaves and wood can be used to infer that temperatures increased in the Santonian, decreased a few degrees in the Campanian and Maastrichtian, but returned to Santonian levels by Late Maastrichtian (Wolfe and Upchurch, 1987). Wolfe and Upchruch (1987) concluded that "... most Late Cretaceous plants evolved in a climate characterized by absence of freezing and low to moderate amounts of precipitation."

Collinson (1990) also used data from plant fossils and concluded that land temperatures at high latitudes in the northern hemisphere during the Late Cretaceous, Paleocene, and Eocene may have been as much as 30°C higher than they are today, whereas at low latitudes in the northern hemisphere they were 5–10°C higher than today. Paleoclimatic and paleovegetation data from the southern hemisphere indicate that tropical and paratropical vegetation reached 55–65° latitude in both the northern and southern hemispheres, suggesting that maximum temperatures in the southern hemisphere during the Late Cretaceous to Eocene were comparable to those in the northern hemisphere (Collinson, 1990).

3. Tertiary

Epochs during the Tertiary Period (from old to young) are Paleocene, Eocene, Oligocene, Miocene, and Pliocene. According to Savin (1977), isotopic paleotemperatures (i.e., temperatures of the oceans) mostly are for the Pacific Basin. Throughout the Tertiary, isotopic temperatures at high and intermediate latitudes show a general decline; however, the curve is not smooth. There are periods with rapid temperature declines and others with increases in temperature (Savin, 1977; Frakes, 1979). Data from fossil marine fauna also indicate that a major cooling occurred in the Late Eocene or Early Oligocene and during the Miocene (Savin, 1977).

Studies on Tertiary fossil leaves (Wolfe, 1971, 1978, 1992, 1994a) indicate that the mean annual temperature was relatively high in the Early Eocene (ca. 27°C at paleolatitude 45°N in western North America). Fluctuations in the mean annual temperature of 5–8°C occurred throughout the Eocene, but the mean annual range of temperatures was low. At the Eocene–Oligocene boundary (33 MYBP), the mean annual temperature decreased 7–8°C, but the mean annual range of temperatures increased 12–13°C, mainly due to a decrease in the mean temperature of the coldest month (Wolfe, 1994a).

Climatic changes at the Eocene–Oligocene boundary had profound effects on plants, as well as on animals (Berggren and Prothero, 1992). Many species of plants became extinct, and great changes occurred in the vegetation. In North America (Wolfe, 1978, 1992) and western and central Europe (Collinson, 1992), broad-leaved evergreen subtropical forests of the Eocene were replaced by dense coniferous forests or by microthermal broad-leaved deciduous forests (Wolfe, 1985). In the southern hemisphere, the Late Eocene–Early Oligocene boundary does not seem to be well represented in the fossil record of Africa, South America, or Australia (Berggren and Prothero, 1992). However, the pollen record in southeastern Australia shows a decrease in warm temperature genera during the low-temperature period of the Early Oligocene; Oligocene deposits occur only in the southeastern part of the country (Kemp, 1978).

After the sharp decline in temperatures at the Eocene–Oligocene boundary, temperatures gradually increased in the Late Oligocene and Early Middle Miocene and then declined in the Early Middle Miocene. Cooling in the Early Middle Miocene was greater at high latitudes than at low latitudes; thus, the pole-to-equator temperature gradient increased in the oceans (Savin, 1977). Both summer and winter temperatures decreased in Beringia at about 12–13 MYBP, but summer temperatures decreased more than those of winter (Wolfe, 1994b). A warm period in the Late Miocene was followed by a cool period (Savin, 1977). Kemp (1978) notes that the Late Miocene cooling period was correlated with a decrease in precipitation in Australia. Ice sheets first formed on Antarctica during the Late

Eocene–Early Oligocene (Bartek, 1992), and they covered the continent by Middle (White, 1990) or Late (Frakes, 1979) Miocene.

4. Quaternary

Epochs during the Quaternary Period are Pleistocene and Recent. The Pleistocene was characterized by cyclic periods of glaciation, and on most continents there were four major periods of glaciation. In addition to continental ice sheets, glaciers formed on many mountains, and in some cases they flowed down to join glaciers at low elevations. The maximum extend of the last glaciation was about 18,000 years ago, but ice at low elevations and low latitudes now has melted. Rapid warming occurred during the various interglacial periods and following the last glaciation, with temperatures over the continents increasing by as much as 10–15°C (Frakes, 1979).

There have been three periods during the Recent Epoch in which glaciers expanded somewhat and then retreated (Frakes, 1979).

5. Speculations on Seed Dormancy

The history of temperatures on earth since the Devonian has been characterized by repeated increases and decreases. When these temperature shifts caused (or were associated with) environmental stress such as low rainfall (discussed earlier), seed dormancy may have developed in some species. For example, it seems likely that seed dormancy would have evolved in species growing in regions where climatic changes resulted in a dry season each year, i.e., seed dormancy could play an important role in timing the life cycle so that seedling growth occurred during the wet season of the year.

However, if climatic changes resulted in a season with temperatures below freezing, seeds dormancy could prevent germination until the beginning of the warm season, i.e., if seedlings and/or adult plants were susceptible to freezing, seed dormancy broken by low temperatures would ensure that germination occurred at the beginning of the frost-free season.

According to Wolfe (1985), deciduous forests appeared in eastern North America, during the Oligocene. The origin of deciduous forests was facilitated by (1) the general decrease in temperatures during the Early Oligocene, (2) an increase in the differences between summer and winter temperatures, and (3), very importantly, the beginning of intense Arctic fronts during winter that lowered temperatures well below freezing; these still occur today (Wolfe, 1985). Did the conditions that led to the development of broad-leaved deciduous forests also favor the evolution of seed dormancy?

V. FOSSIL HISTORY OF SEEDS AS IT RELATES TO DORMANCY

A. Fossil Seeds

1. Gymnosperm

The oldest seeds in the fossil record belong to a group of primitive gymnosperms commonly called seed ferns. To understand the structure of these ancient seeds, it is informative to look at an ovule of a modern cycad (Fig. 12.9a). In gymnosperms, the ovule is not protected by a fruit wall (pericarp); thus, the term "naked seeds." However, ovules of several Paleozoic and Mesozoic seed fern groups were enclosed in cup-like structures called cupules. Cupules of the various groups had separate origins and widely differing structures. Depending on the species, cupules within each group had varying shapes, sizes, and degrees of fusion of their parts, and the number of ovules per cupule ranged from one to many (Fig. 12.10). Cupuoles or ovules also were borne in a variety of positions on fern-like leaves. Some were terminal on dissected frond tips, while others occurred laterally on the abaxial surface of pinnules. Some were

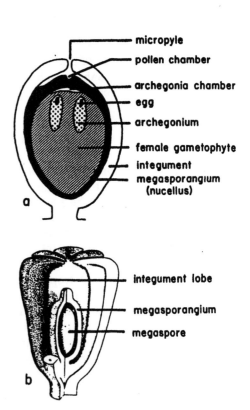

FIGURE 12.9 (a) Stylized drawing of a longitudinal section of an ovule of a modern-day cycad. (b) Reconstruction of a primitive seed-like structure from the early Carboniferous of France that may represent an evolutionary stage between pteridophytic and gymnospermous reproduction. From Galtier and Rowe (1989), with permission.

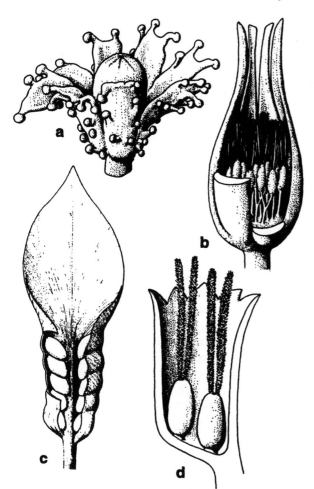

FIGURE 12.10 Paleozoic seed bearing organs. (a) *Lagenostoma* seed contained in *Calymmatotheca* cupule. (b) *Calathospermum* cupule containing numerous stalked ovules. (c) *Archaeocycas* megasporophyll showing the enrolled nature of lamina. (d) *Gnetopsis* cupule containing two ovules with extended integumentary processes. From Taylor and Millary (1979), with permission.

solitary and either sessile or on the ends of short stalks, while others were grouped in highly branched specialized regions (Dilcher, 1979; Stewart and Rothwell, 1993).

The megasporangium (or nucellus) in gymnosperms is relatively large and contains the megaspore, which divides and gives rise to the female gametophyte (Fig. 12.9a). The female gametophyte has one (or more) archegonium(ia), each containing an egg. Integuments are fused around the megasporangium in modern gymnosperms; however, a space called the pollen chamber is located between the integuments and the distal end of the megasporangium. In a very primitive gymnosperm from the early Carboniferous of France, the integuments are not fused and there is no pollen chamber (Fig. 12.9b). In gymnosperms more advanced than the fossil

from France, integuments ranged from not being fused at all to being entirely fused (Figs. 12.11a–12.11d); however, a pollen chamber was present (Fig. 12.11e). A funnel-like structure called a lagenostome at the top of the pollen chamber played a role in pollen reception (Figs. 12.11f and 12.11g). The seed fern *Oclloa* from the Carboniferous in Peru had a lagenostome, but unlike other seed ferns with a lagenostome, its ovules did not occur in a cupule (Erwin *et al.*, 1994).

For years, the oldest seeds were thought to belong to *Archaeosperma arnoldii;* these seeds were collected in McKean County, Pennsylvania, from the Oswayo Formation (Famennian Stage, 363–367 MYBP), Late Devonian (Pettitt and Beck, 1968). Ovules of this species occurred in cupules, and each cupule was quite

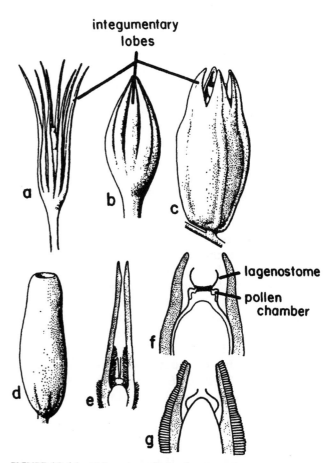

FIGURE 12.11 Paleozoic seeds. (a–d) A progressive fusion of integumentary lobes to form a micropyle: (a) *Genomosperma kidstoni,* (b) *G. latens,* (c) *Eurystoma angulare,* and (d) *Stamnostoma huttonense.* (e) Distal end of *Salpingostoma dasu* ovule showing an extended end of nucellus and inwardly facing integumentary hairs. (f) Distal end of *Eurystoma angulare* ovule showing the pollen-receiving structure. (g) Distal end of *Physostoma elegans* ovule showing the pollen-receiving structure. From Taylor and Millay (1979), with permission.

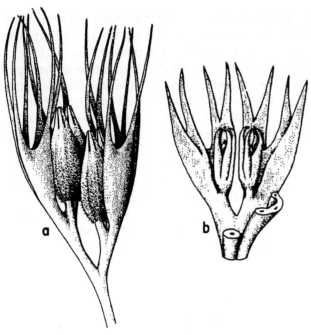

FIGURE 12.12   (a) Diagrammatic representation of *Archaeosperma arnoldii* cupule complex showing cupules arranged in pairs and each cupule containing two seeds. From Pettitt and Beck (1968), with permission. (b) Diagrammatic representation of *Elkinsia polymorpha* cupule containing preovules and showing a forked-branching pattern. From Stewart and Rothwell (1993), with permission.

open and contained two ovules (Fig. 12.12a). Seeds were 2.4 mm long × 1.4–1.7 mm at their widest point, and the integument was divided into lobes at the distal end of the megasporangium.

In 1981, Gillespie and colleagues published an account of cupulate ovules collected in Randolph County, West Virginia, from the Hampshire Formation (Famennian stage Fa2c), Late Devonian. According to Gillespie *et al.* (1981), *Archaeosperma arnoldii* was from Famennian Stage Fa2d. Thus, the new fossils from West Virginia, which subsequently were named *Elkinsia polymorpha* (Rothwell *et al.*, 1989), were older than those of *A. arnoldii.* In *E. polymorpha,* the integumentary lobes were not fused the distal two-thirds of their length (Fig. 12.12b). Seeds of *E. polymorpha* were 5.6 mm long × 2 mm at their widest point (Gillespie *et al.,* 1981).

Gillespie *et al.* (1981) commented that the cupulate seeds of *Hydrasperma tenuis* (Matten *et al.,* 1980) and *Spermolithus devonicus* (Chaloner *et al.,* 1977) from Ireland might be older than those of *Elkinsia polymorpha.* However, Rothwell and Scheckler (1988) subsequently spoke of the *H. tenuis* and *S. devonicus* fossils as belonging to the Devonian Tournaisian or Late Strunian, which would make them younger than *E. polymorpha.*

Although seeds of the earliest gymnosperms (i.e., pre-Carboniferous) are quite small, mostly 3–7 mm long × 1–2 mm wide (Rothwell and Scheckler, 1988), some of the medullosan seed ferns in the Carboniferous had large seeds. Seeds of *Pachytesta incrassata* were 10–11 cm long × 5–6 cm wide, and those of *P. gigantea* were 8–9 cm long × 5–6 cm wide (Hoskins and Cross, 1946). However, not all medullosan seed ferns had large seeds; those of *P. berryvillensis* were only 6.1–6.8 mm long × 5.0–5.2 mm wide (Taylor and Eggert, 1969).

2. Angiosperms

Double fertilization evolved prior to the divergence of angiosperms and Gnetales from a common ancestor, as evidenced by its presence in these two groups of plants. This phenomenon occurs in the extant members of the Gnetales *Ephedra* (Friedman, 1990, 1992, 1994) and *Gnetum* (Carmichael and Friedman, 1996); information is not available for *Welwitschia.* The product of the second fertilization in *Ephedra* and *Gnetum* is a diploid embryo; thus, two diploid embryos are formed in the female gametophyte after the two sperm nuclei are released from the pollen tube. In *Ephedra,* one sperm nucleus unites with the egg and the other with the ventral canal nucleus (Friedman, 1990, 1992). An egg is not formed in *Gnetum;* thus, the sperm nuclei unite with two free haploid nuclei at the micropylar end of the female gametophyte to form two embryos (Carmichael and Friedman, 1996). Only one of the embryos survives to seed maturity in *Ephedra* and in *Gnetum* (Friedman and Carmichael, 1996). The female gametophyte supplies nourishment to the developing embryo in both *Ephedra* and *Gnetum,* but development of the female gametophyte is not completed in *Gnetum* until after fertilization has occurred (Carmichael and Friedman, 1996).

The product of the second fertilization in angiosperms is a nonembryo polyploid tissue called endosperm, which supplies food to the developing embryo. The development of endosperm and reduction of the female gametophyte to form the embryo sac occurred in angiosperms after their divergence from gymnosperms (Friedman, 1992). Thus, in development of the angiosperm seed, there has been "... a tendency to shift the dependency of the early embryo directly to the parent sporophyte and away from an intermediate gametophyte generation or some modified form of that generation" (Steeves, 1983).

Numerous attempts have been made to find fossils of angiosperms that date from the Jurassic and even the Triassic. Although many intriguing plant fossils have been found in rocks of these periods, none has been judged to be unquestionably the remains of an angiosperm (Hughes, 1976; Tiffney, 1984). However, some Triassic and Jurassic fossils have characteristics of both gymnosperms (mostly) and angiosperms (Stewart and Rothwell, 1993). Thus, true angiosperms are not found in the fossil record until the Lower Cretaceous (Hickey and Doyle 1977; Doyle, 1978; Tiffney, 1984).

Fossil fruits and seeds from the Lower Cretaceous said to be angiosperms include those of *Onoana, Nyssidium,* and *Kenella* (Hughes, 1976). However, in a summary of Cretaceous seed and fruit fossils with presumed angiosperm affinities, Tiffney (1984) concluded that there was insufficient evidence for members of these three genera to be called angiosperms. Further, there also seemed to be questions about whether Aptian (107.5–112 MYBP; a stage of the Cretaceous) fossils of *Carpolithus, Onoana,* or *Prototrapa* and Albian (105 MYBP; a stage of the Cretaceous) fossils of *Araliaecarpum, Caricopsis,* or *Carpolithus* were angiosperms (Tiffney, 1984). The first fossils with "clear angiospermous affinities" are *Caspiocarpus paniculiger* (dehiscent follicle and seed) and *Ranunculaecarpus quinquiecarpellatus* (dehiscent follicle) from Albian Stage deposits in Kazakhstan and the Kolyma River, USSR, respectively (Tiffney, 1984). A number of genera, including *Carpites, Laurus, Platanus,* and "*Salix,*" with angiosperm affinities have been collected in Cenomanian (97.5 MYBP; a stage of the Cretaceous) deposits (Tiffney, 1984).

Fossil angiosperm seeds from the Albian and Cenomanian stages are small: *Ranunculaecarpus quinquiecarpellatus,* 1.5 mm long × 0.6 mm wide and *Carpites liriophylli,* 1.4 mm long × 0.6 mm wide (Tiffney, 1984). Further, these small seeds were produced in follicles or capsules (Tiffney, 1984, 1986). A major radiation of angiosperms occurred in the Late Cretaceous and Early Tertiary, which is when many of the modern families and genera first appeared in the fossil record (Tiffney, 1981, 1986). Many, but not all, of these new genera had large (up to 50,000–100,000 mm³) seeds (Tiffney, 1986).

Haig and Westoby (1991) note that for both fossil and extant species, the smallest gymnosperm seeds are larger than the smallest angiosperm seeds. These authors suggest that the time of fertilization in angiosperms is more efficient with regard to the allocation of resources than it is in gymnosperms. In gymnosperms, a female gametophyte that contains food reserves for the embryo is produced prior to fertilization, whereas in angiosperms the food-supplying tissue (endosperm) is not produced until after fertilization. Thus, if fertilization does not occur, fewer resources are lost via ovule abortion in angiosperms than in gymnosperms. Seeds of angiosperms can be smaller than those of gymnosperms because the costs of pollination are reduced substantially in angiosperms (Haig and Westoby, 1991). How-

ever, there is a limit to how small seeds can be, which may be determined partly by "accessory costs" (i.e., the costs of pollen capture and ovules that abort). With a decrease in seed size, accessory costs increase; consequently, allocation of food reserves to the developing embryo decreases. Minimum seed size is the point at which any further decrease in resources allocated to the embryo would reduce chances of seedling survival (Haig and Westoby, 1991).

B. Fossil Embryos

1. Gymnosperms

Embryos in Paleozoic gymnosperm seeds are conspicuous by their absence (Arnold, 1938; Taylor, 1981), but there are exceptions. Stidd and Cosentino (1976) reported cellular remains of an undifferentiated embryo in one archegonium of *Nucellangium* (a Codaitales gymnosperm) from the Middle Pennsylvanian. However, it is hard to see anything that looks very much like an embryo in their photographs of the archegonium. A linear embryo was found in a conifer seed from the Lower Permian, but although it contained some tracheids it did not have cotyledons (Miller and Brown, 1973). Gymnospermous seeds of Permian age from the Beardmore Glacier region of Antarctica had two archegonia in each megagametophyte, and each archegonium had an early embryo in the same stage of development; a first example of polyembryony from the late Paleozoic (Smoot and Taylor, 1986)! Early embryos are longer than wide (Smoot and Taylor, 1986; Taylor and Taylor, 1987), and thus they appear to be linear.

Long (1974–1975) observed tracheids and presumably two cotyledons, hypocotyl, root tips, and root cap in a small linear embryo in a Lagenostomales seed fern from the lower Carboniferous. However, the oldest embryo that clearly has cotyledons is in a Walchian conifer seed from deposits of either the uppermost Pennsylvanian or lowermost Permian (Mapes *et al.,* 1989). Stockey (1975) found embryos complete with a shoot apex, cotyledons, and root meristem in seeds of *Araucaria mirabilis* (probably from Middle to Late Jurassic) from Patagonia (Argentina). Other embryos have been found in fossil seeds of Araucariaceae: (1) *Pararaucaria patagonica,* Jurassic, linear, fully developed embryo (Stockey, 1977); (2) *Araucaria sphaerocarpa,* Jurassic, well-developed embryo with two relatively long cotyledons (Stockey, 1980a); (3) *A. brownii,* Jurassic, well-developed embryo with four (?) cotyledons (Stockey, 1980b); and (4) *A. nihongia,* Upper Cretaceous, embryos with two, long straight cotyledons (Stockey *et al.,* 1992).

Fossil seedlings of Araucariaceae from the Cerro Cuadrado Petrified Forest (Jurassic) in Patagonia have been described as "fig-like" or "corm-like" structures (Stockey and Taylor, 1978). Anatomical studies of fossil and extant members of this family indicate that germination was probably hypogeal and the hypocotyl increased greatly in diameter after germination, similar to extant members of the section *Bunya* of the family (Stockey and Taylor, 1978). The extant bunya pine (*Araucaria bidwillii*, a member of the Australian rain forest community) has cryptogeal germination, i.e., the cotyledons are fused into a tube that elongates, pushing the root and stem meristems into the soil. Then, the hypocotyl expands greatly in diameter to form a tuber-like structure (Burrows *et al.*, 1992), comparable to fossil seedlings from Argentina. Eventually, the cotyledon tube separates from the swollen hypocotyl, and the stem meristem gives rise to a shoot (Burrows *et al.*, 1992).

Large, linear embryos with two cotyledons were found in seeds of Mesozoic cycadeoideas (Bennettitae) by Wieland (1916). Fossil seeds of *Diploporous torreyoides* (an extinct member of the Taxaceae) from the Middle Eocene of Oregon have a large linear embryo that is 77% the length of the seed (Manchester, 1994).

Cones of the genus *Cycadeoidea* from the Cretaceous of the Black Hills, South Dakota, had ovules in the pregametophytic stage of development (Crepet and Delevoryas, 1972). Middle Eocene ovulate cones from British Columbia, Canada, had seeds, but neither embryo nor megagametophyte tissues was present (Stockey, 1984).

2. Angiosperms

Much effort has been devoted to finding and studying fossils of angiosperm flowers (e.g., Dilcher, 1979; Basinger and Christophel, 1985; Erwin and Stockey, 1990; Manchester, 1992; Nixon and Crepet, 1993; Crane and Herendeen, 1996). Also, angiosperm seeds have received considerable attention, but this work frequently has been done from a floristic perspective, which means that lists have been prepared of the different kinds of fossil seeds/fruits/reproductive structures in various deposits, e.g., Paleocene Sentinel Butte Shale from South Dakota (Crane *et al.*, 1990); Middle Eocene Princeton Chert from southern British Columbia, Canada (Cevallos-Ferriz *et al.*, 1991, 1993; Cevallos-Ferriz and Stockey, 1988a,b, 1990; Pigg *et al.*, 1993); Lower Oligocene clays on the Isle of Wight, England (Collinson, 1983); Early Oligocene Brandon Lignite in Vermont (Tiffney, 1977, 1993); and Middle Miocene in Poland (Lesiak, 1994). Much less attention has been given to the kind of embryo found in angiosperm seeds than to the external characteristics of the seeds (e.g., Collinson,

1980; Tiffney, 1980; Millan, 1994; Delevoryas and Mickle, 1995).

Lack of information on the internal structure of seeds may mean that only the external anatomy was examined or that embryos were not preserved. Sections through three Rosaceae (*Prunus*) seeds did not reveal an embryo (Cevallos-Ferriz and Stockey, 1991). Presence of an embryo is inferred in seeds of *Joffrea speirsii* (Cercidiphyllaceae) because fossil seeds with an emerging radicle have been found (Crane and Stockey, 1985). A large number of fossil *Cercidiphyllum*-like seedlings have been found in the Paleocene in Alberta, Canada; germination was epigeal (Stockey and Crane, 1983).

In the Paleocene Sentinel Butte Shale in South Dakota, Crane *et al.* (1990) found *Psidium* (Myrtaceae) seeds with large, curved embryos. Many extant members of this family have seeds with large, folded embryos and no endosperm (Martin, 1946).

Some information has been obtained on the internal structure of seeds from fossils collected in the Middle Eocene Princeton Chert in British Columbia. Fossil seeds of *Keratosperma allenbyensis* (Araceae) have a coiled linear embryo (Cevallos-Ferriz and Stockey, 1988a). In extant members of the Araceae, some genera have seeds with a linear embryo and an endosperm, but others have a large, fully developed linear embryo and no endosperm (Martin, 1946). Fossil seeds of extinct *Decodon allenbyensis* (Lythraceae) have a fully developed embryo, but endosperm cells were not identified (Cevallos-Ferriz and Stockey, 1988b). Seeds of extant *Decodon verticillatus* have an investing embryo and lack endosperm (Martin, 1946). A section through a fossil seed of *Ampelocissus similkameenensis* (Vitaceae) revealed no embryo or endosperm (Cevallos-Ferriz and Stockey, 1990). However, the fossil seed has the same shape as that found in seeds of the extant *Ampelopsis cordata,* which have a large, fleshy endosperm and a relatively small spatulate embryo (Martin, 1946). Seeds of *Allenbya collinsonae* (Nymphaeaceae) had cells judged to be perisperm, but an embryo was not found; however, the size of the cavity inside the seed suggested that the embryo was small (Cevallos-Ferriz and Stockey, 1989). Extant members of the Nymphaeaceae have small, broad embryos (Martin, 1946). Seeds of *Princetonia allenbyensis* (belonging to an undetermined family in the Magnoliopsida) also had a tissue judged to be perisperm and a small embryo cavity, but no embryo (Stockey and Pigg, 1991). Endosperm and embryos were not found in fossil fruits of *Prunus* sp. (Rosaceae), but a heart-shaped structure, which may have been a proembryo, was found in fruits of *Paleorosa similkameenensis,* another member of the rose family (Cevallos-Ferriz *et al.*, 1993). Remnants of a nucellus, but no other tissue, have been found in fruits of *Paleomyrtinaea,* a member of the Myrtaceae (Pigg *et al.,* 1993).

A fossil seed of *Ensete oregonense* (Musaceae) from the Middle Eocene in Oregon has a capitate embryo and much endosperm (Manchester and Kress, 1993); it is very similar to Martin's (1946) drawing for this family.

C. Speculations on Dormancy

1. Gymnosperms

Many of the fossil gymnosperm "seeds" are really just ovules (Arnold, 1938). Eames (1955) discussed the problem of ovules vs seeds in *Ginkgo biloba,* and he reviewed various studies, including his own, showing that fertilization of the egg may occur either before or after the dispersal of ovules from the trees. Time of fertilization even varied for ovules on the same tree; however, after fertilization occurred embryo development was continuous until it was completed. Thus, embryo growth in *G. biloba* can occur while the ovule is on the tree, or it may be delayed until after the ovule has been dispersed (Eames, 1955; Holt and Rothwell, 1997).

The general absence of embryo-containing ovules in Paleozoic fossils may mean that fertilization took place after ovules were dispersed, i.e., pollen remained in the pollen chamber for a time following pollination, thus delaying fertilization. The presence of well-developed female gametophytes with archegonia in some Paleozoic fossils (Long, 1944; Rothwell, 1971) may indicate that ovules were buried shortly after dispersal, before embryo growth was initiated. The presence of immature embryos in some fossil seeds (Miller and Brown, 1973; Long, 1974–1975) may indicate that fertilization and early stages of embryo development occurred in some species prior to dispersal and burial.

Ovules were attached to the tips of scales in ovulate cones of *Cordaianthus duquesnensis* (a Cordaitean gymnosperm) from the Upper Pennsylvanian (Rothwell, 1982). Further, distal ovules were more immature (smaller) than proximal ones, and evidence showed that ovules already may have been dispersed from the proximal end of the cone. Rothwell (1982) suggested that there was no seed dormancy and that ovules were dispersed as soon as they reached full size. Further, because embryo growth probably was continuous (once it started), a delay in dispersal would have meant that germination occurred while ovules were in the cone. Thus, the only way an ovule could have been an effective disseminule was for dispersal to occur immediately after it reached full size (Rothwell, 1982). Rothwell (1982) did not mention embryos being present in seeds from the ovulate cone, and thus it is assumed that embryo

development occurred after ovule dispersal in this species.

The occurrence of well-developed embryos in seeds in Paleozoic fossil conifer cones has been interpreted by Mapes *et al.* (1989) to mean that there was a "... significant delay between fertilization and seed germination in the earliest conifers." Further, these authors suggest that this period of quiescence "... may be a first step in the evolution of seed dormancy" (Mapes *et al.,* 1989). Well-developed embryos also have been found in fossil cones of *Pararaucaria patagonica* (Stockey, 1977) and *Araucaria mirabilis* (Stockey, 1978), suggesting that seeds were dormant when they were fossilized.

Based on available fossil evidence, it appears that early gymnosperms lacked seed dormancy per se and that any delay in germination was due to the timing of fertilization. In relatively more advanced gymnosperms, e.g., conifers, the presence of a fully developed embryo in seeds still in a cone (Stockey, 1977, 1978; Mapes *et al.,* 1989) means that fertilization and embryo growth occurred prior to dispersal. Thus, a delay in germination of these seeds following dispersal must be attributed to seed dormancy and/or unfavorable conditions for germination.

2. Angiosperms

Based on our present knowledge of embryos in fossil angiosperm seeds, no conclusion is possible concerning seed dormancy in early angiosperms.

Perhaps fossil angiosperm seeds have not received as much attention as fossil flowers because it is not widely recognized how valuable this kind of information would be to researchers trying to figure out the origins of the various types of seed dormancy. For example, there are many things about seeds and embryos of fossil angiosperms that would be of great importance in understanding evolution of the various kinds of seed dormancy. (1) Do fossil representatives of families that have morphophysiological or morphological dormancy today (see Table 3.3) have underdeveloped embryos? (2) In families that have some extant members with underdeveloped embryos and others with fully developed embryos, such as the Apiaceae, Araceae, Berberidaceae, and Hydrophyllaceae, what kind(s) of embryo was found in the fossils? (3) In families whose extant members have dwarf or micro seeds, such as the Ericaceae, Saxifragaceae, and Scrophulariaceae, does the fossil record show any changes in seed size and/or the relative size of the embryo and endosperm? (4) In families, whose present-day members have seeds with large, bent embryos, such as the Anacardiaceae, Brassicaceae, and Fabaceae, does the fossil record indicate an increase in seed or embryo size and a decrease in amount of endosperm? (5) In families whose extant members have seeds with impermeable seed coats, such as the Fabaceae and Malvaceae, do fossil seeds have the same seed coat anatomy as those of extant species?

VI. SEED DORMANCY AND PHYLOGENETIC POSITION OF FAMILY

A. Angiosperms

Information on seed dormancy at the family level (from Chapters 9, 10, and 11) was plotted on the phylogenetic tree for flowering plants published by Takhtajan (1980). Takhtajan (1980) lists 93 orders of flowering plants, and some data on germination are available for all of them, except the Balanopales, Bazbeyales, Calycerales, Cercidiphyllales, Cyclanthales, Didymelales, Eriocaulales, Eucommiales, Eupeleales, Hydratellales, Leitneriales, Pandanales, Restionales, and Trochodendrales. However, seeds of Trochodendraceae have tiny embryos and much endosperm (Grushvitzky, 1967; Heywood, 1993); consequently, they may have morphological or morphophysiological dormancy.

1. Nondormancy

Families with nondormant seeds are concentrated in orders in the upper portion of the phylogenetic tree (Fig. 12.13). Nondormancy is absent in the subclass Ranunculidae and rare in the Magnoliidae and Arecidae. In the Arecidae and Alismatidae, species belonging to families (and orders) with nondormant seeds are aquatics. Basal orders of the Liliidae have families with nondormant seeds, but those of the Dilleniidae and Hamamelididae do not. No seed germination data are available for families belonging to the basal orders of the Hamamelididae. In the Rosidae and Asteridae, nondormancy is found in basal orders as well as in intermediate and apical ones.

Many orders have both tropical/subtropical and temperate/arctic families with nondormant seeds, e.g., Myrtales, Gentianales, Caryophyllales, and Scrophulariales. However, there are several orders, especially in the upper portion of the Rosidae, in which all families with nondormant seeds are found only in tropical/subtropical types of vegetation. In some orders, such as the Dipsacales and Saxifragales, families with nondormant seeds are found only in temperate/arctic types of vegetation.

2. Morphological Dormancy

Tropical/subtropical and temperate/arctic families with morphological dormancy belong to orders at the base of the subclasses Liliidae, Magnoliidae, Ranunculidae, and Rosidae (Fig. 12.14). Some species with specialized habitats and/or life cycles (=specialized) also belong to these same subclasses, but they are at a higher position than species occurring in the tropical/subtropical and/or temperate/arctic types of vegetation.

3. Morphophysiological Dormancy

Families with morphophysiological dormancy occurring only in the tropical/subtropical region belong to the subclass Magnoliidae, or they occur in the order Dilleniales at the base of the Dilleniidae (Fig. 12.15). Families with morphophysiological dormancy occurring in both tropical/subtropical and temperate/arctic types of vegetation belong to orders at the base of the Arecidae, Liliidae, Magnoliidae, and Rosidae and near the middle of the Asteridae. Further, families with morphophysiological dormancy (or with morphophysiological and physiological dormancies) occurring only in temperate/arctic vegetation types belong to six subclasses (Asteridae, Dilleniidae, Liliidae, Magnoliidae, Ranunculidae, and Rosidae), In the Asteridae, morphophysiological dormancy is found in the Polemoniales, near the top of the subclass. Specialized species are found in the Ranunculidae and Liliidae and in the order Santalales in the Rosidae.

4. Physiological Dormancy

The position of physiological dormancy on the phylogenetic tree (Fig. 12.16) is very similar to that of nondormancy (Fig. 12.13), i.e., it is concentrated in the upper portion. Further, most orders that have families with nondormant seeds also have families with physiologically dormant seeds; however, there are a few exceptions. For example, the Primulales and Sapindales have physiological dormancy but not nondormancy, and the Casuarinales and Zingiberales have nondormancy but not physiological dormancy.

Families with physiological dormancy reported only in tropical/subtropical types of vegetation belong to orders in the Asteridae, Caryophyllidae, Dilleniidae, and Rosidae, but not in the Hamamelididae, Liliidae, Magnoliidae, or Ranunculidae. However, families occurring in both tropical/subtropical and temperate/arctic or only in temperate/arctic types of vegetation belong to orders in all eight of these subclasses. Many aquatics whose seeds have physiological dormancy belong to orders in the Alismatidae, but aquatics with physiological dormancy also occur in the Asteridae, Caryophyllidae, Dilleniidae, Liliidae, and Rosidae. None of the orders in the Arecidae are reported to have seeds with physiological dormancy.

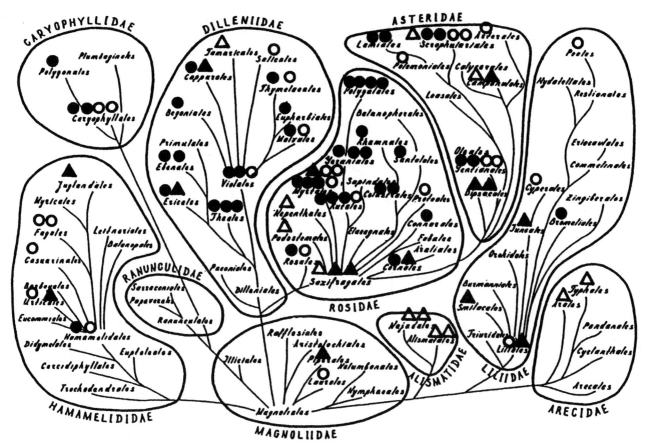

FIGURE 12.13 Nondormancy in seeds and phylogenetic position of family. Each symbol represents one family: ●, in tropical/subtropical zones; ○, in tropical/subtropical and temperate/arctic zones; ▲, in temperate/arctic zones; and △, specialized habitat and/or life cycle. Phylogenetic scheme used with permission from Takhtajan (1980).

5. Physical Dormancy

With the exception of the order Nelumbonales in the Magnoliidae, all orders containing seeds or fruits with physical dormancy are in a high position on the phylogenetic tree and are found only in the Asteridae, Dilleniidae, or Rosidae (Fig. 12.17). Members of the Cistaceae (Violales) and Geraniaceae (Geraniales) in temperate/arctic types of vegetation and of Bombacaceae (Malvales) in tropical/subtropical vegetation have physical dormancy. Except for the Cistaceae, Geraniaceae, and Bombacaceae, other families with physical dormancy belonging to orders in the Asteridae, Dilleniidae, or Rosidae have some members in tropical/subtropical vegetation with nondormant seeds and others with physical dormancy, but in temperate/arctic vegetation members of these families have only physical dormancy.

B. Gymnosperms

Information on seed dormancy at the family level (from Chapters 9 and 10) was plotted on a cladogram showing relationships of modern conifer families and fossil groups (Fig. 12.18). Morphophysiological dormancy is found in the Taxaceae clade, next to the hypothetical ancestor. Morphophysiological and physiological dormancies and nondormancy are present in the Pinaceae–Araucariaceae–Cephalotaxaceae–Podocarpaceae clade, whereas physiological dormancy and nondormancy are present in the Cupressaceae–Taxodiaceae clade. It should be noted that relationships among conifer families are not well understood and that cladistic analyses are still being done (e.g., Nixon *et al.*, 1994).

VII. SEED DORMANCY AND PRESUMED AGE OF PLANT FAMILIES

Is it possible to infer anything about the evolution of seed dormancy types from information on the time of first appearance of various plant families in the fossil record? The first step in attempting to answer this question was to combine the list of the types of seed dor-

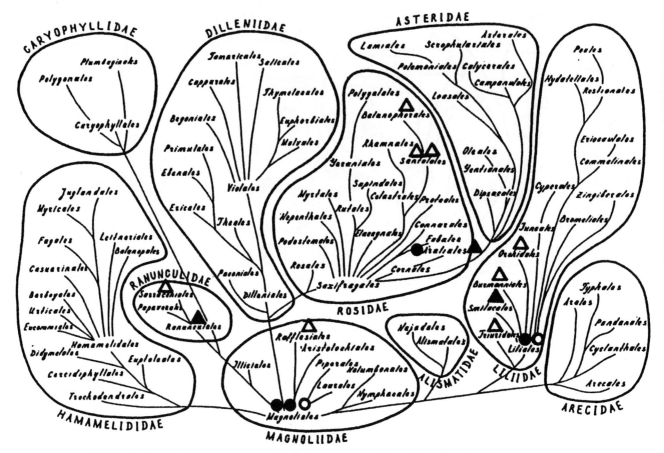

FIGURE 12.14 Morphological dormancy in seeds and phylogenetic position of family. Each symbol represents one family: ●, in tropical/subtropical zones; ○, in tropical/subtropical and temperate/arctic zones; ▲, in temperate/arctic zones; and △, specialized habitat and/or life cycle. Phylogenetic scheme used with permission from Takhtajan (1980).

mancy (or nondormancy) in plant families with information compiled by Muller (1981) and Taylor (1990) on time of first appearance of angiosperm plant families in the fossil record; a list of 114 families was obtained. Similarly, information on the first appearance of extant gymnosperm families (Thomas and Spicer, 1987; Stewart and Rothwell, 1993) for which germination data are available resulted in a list of 10 families. The only family of gymnosperms not included in Chapter 9 or 10 is the Ginkgoaceae. Some seeds of *Ginkgo biloba* have morphophysiological dormancy, whereas others have morphological dormancy (West *et al.,* 1970).

A. Time of Family Origin

1. Angiosperms

Plotting the time of first appearance in the fossil record of the 114 angiosperm families on Takhtajan's (1980) phylogenetic tree for flowering plants showed a general pattern of decrease in age from the bottom to the top of the tree as a whole and also within several of the subclasses. However, families within some orders showed much variation in time of first appearance in the fossil record, and in some cases there is as much difference within an order as there is within a subclass. For example, families in the Caryophyllales first appeared in the Upper Cretaceous, Eocene, Oligocene, or Miocene, whereas in the Saxifragales they first appeared in the Eocene, Oligocene, or Pliocene.

Although position on the phylogenetic tree is somewhat helpful in predicting the type of dormancy (see earlier), the time of first appearance of a family in the fossil record does not predict the kind of dormancy found in extant members of the family. For example, families whose seeds have morphophysiological dormancy today first appeared in the Cretaceous, Paleocene, Eocene, Oligocene, Miocene, or Pliocene, and those with physical dormancy today first appeared in the Cretaceous, Paleocene, Eocene, Oligocene, or Miocene.

2. Gymnosperms

Time of first appearance of extant families of gymnosperms (Thomas and Spicer, 1987; Stewart and Roth-

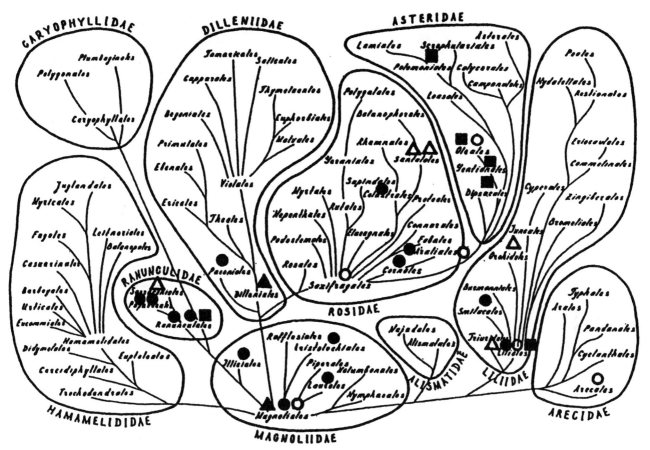

FIGURE 12.15 Morphophysiological dormancy in seeds and phylogenetic position of family. Each symbol represents one family: ●, in temperate/arctic zones; ○ in tropical/subtropical and temperate/arctic zones; ▲, in tropical/subtropical zones; △, specialized habitat or life cycle, and ■, family in temperate/arctic zones with both morphophysiological dormancy and physiological dormancy. Phylogenetic scheme used with permission from Takhtajan (1980).

well, 1993) and type(s) of dormancy are (1) Araucariaceae, Jurassic, nondormancy and physiological dormancy; (2) Cephalotaxaceae, Jurassic, morphophysiological dormancy; (3) Cycadaceae, Permian, morphophysiological dormancy; (4) Ephedraceae, Triassic?, physiological dormancy; (5) Ginkgoaceae, Permian, morphological and morphophysiological dormancies; (7) Pinaceae, Jurassic, nondormancy and physiological dormancy; (8) Podocarpaceae, Triassic, morphological and morphophysiological dormancy; (8) Taxaceae, Jurassic, morphophysiological dormancy; (9) Taxodiaceae, Jurassic, nondormancy and physiological dormancy; and (11) Welwitschiaceae, Triassic?, physiological dormancy. Thus, the type of dormancy is not correlated with time of first appearance of a family in the fossil record. For example, morphophysiological dormancy is found in families first appearing in the Permian, Triassic, and Jurassic, and physiological dormancy is found in those first appearing in the Triassic and Jurassic.

B. Time of Dormancy Origin

1. Angiosperms

In both tropical and temperate regions, families that appear first in the fossil record of the Cretaceous have all the known kinds of seed dormancy, as well as nondormancy (Table 12.3). Does this mean that all types of seed dormancy were present at, or shortly after, the time of origin of angiosperms? No extant families in tropical/subtropical zones whose seeds have morphological or physical dormancy first appear in the fossil record after the Eocene, and none in the temperate/arctic zones whose seeds are nondormant or have morphological or physical dormancy first appear in the fossil record after the Miocene.

There are two possibilities concerning the time of origin of the various kinds of seed dormancy and of nondormancy. (1) When families first evolved, they had the type of seed dormancy (or nondormancy) they have today. Thus, the various types of dormancy are very old

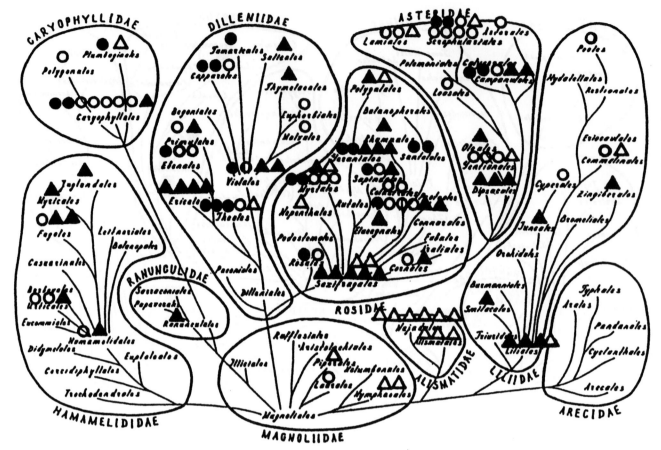

FIGURE 12.16 Physiological dormancy in seeds and phylogenetic position of family. Each symbol represents one family: ●, in tropical/subtropical zones; ○, in tropical/subtropical and temperate/arctic zones; ▲, in temperate/arctic zones; and △, specialized habitat or life cycle. Phylogenetic scheme used with permission from Takhtajan (1980).

and would have been a part of the "inheritance" from the ancestral family. (2) Regardless of when a family appeared, seed dormancy has developed in response to environmental conditions. Thus, one set of conditions has selected for nondormancy, another for physiological dormancy, another for morphological dormancy, and so on. For example, under the proper conditions for development of physiological dormancy, this type of dormancy would have developed no matter when a family originated.

Although it seems logical that some changes might occur in seed dormancy and germination characteristics after the evolution of a family, two pieces of evidence suggest that the types of seed dormancy are very old. (1) Even when both tropical/subtropical and temperate/arctic members of a family are considered, a family rarely has more than one type of seed dormancy. However, nondormancy and one type of dormancy (usually physiological dormancy) are found within many families. If dormancy in a family was determined only by

environmental factors after the family evolved, families should have several types of dormancy, as members of a given plant family may grow in a diversity of habitats in many climatic zones. However, this usually is not the case. (2) In some orders (see Figs. 12.13–12.17), all the families have either nondormant seeds or the same type of seed dormancy (or nondormancy and one type of dormancy). Thus, the type of dormancy may be one of the characteristics that members of the order have in common, in which case it must have been present for a long period of time. For example, in the order Malvales members of four extant families, the Bombacaceae, Tiliaceae, Sterculiaceae, and Malvaceae have physical dormancy, and these families first appear in the fossil record of the Cretaceous, Cretaceous, Paleocene, and Eocene, respectively.

Families whose extant members have specialized habitats or life cycles also appear in the fossil record at different times: Nymphaeaceae, physiological dormancy, Cretaceous; Nelumbonaceae, physical dor-

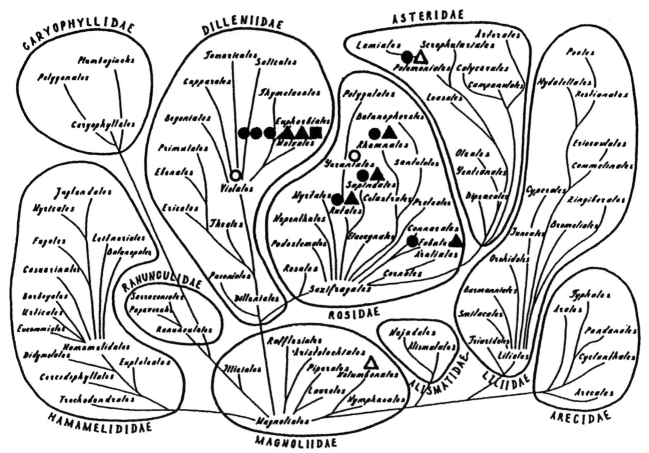

FIGURE 12.17 Physical dormancy in seeds and phylogenetic position of family. Each symbol represents one family: ●, nondormancy and/or physical dormancy in tropical/subtropical zones and physical dormancy in temperate/arctic zones; ○, physical dormancy in temperate/arctic zones; ▲, physical plus physiological dormancy in temperate/arctic zones; ■, nondormancy and/or physical dormancy in tropical/subtropical zone; and △, specialized habitat or life cycle. Phylogenetic scheme used with permission from Takhtajan (1980).

mancy, Oligocene; Droseraceae, physiological dormancy, Eocene; Lentibulariaceae, Nepenthaceae, Potamogetonaceae and Trapaceae, physiological dormancy, Miocene; Santalaceae, morphological and morphophysiological dormancies, Paleocene; Balanophoraceae, morphological dormancy, Miocene; and Orobanchaceae, morphophysiological dormancy, Pliocene.

2. Gymnosperms

If the type of dormancy found in extant families of gymnosperms also was present at the time they first appeared in the fossil record, we can push back the time of origin of physiological, morphological, and morphophysiological dormancies and of nondormancy from the Upper to at least the Lower Mesozoic. There is no evidence that seeds of either fossil or extant gymnosperms had (have) physical dormancy. Thus, if morphological, morphophysiological, and physiological dor-

mancies were present in the Lower Mesozoic, and physical dormancy did not evolve until the Upper Mesozoic, physical dormancy would be the youngest type.

VIII. SEED DORMANCY AND FAMILY TREE OF SEED PHYLOGENY

From studies on seeds from 1287 genera of plants, Martin (1946) constructed a family tree of seed phylogeny (Fig. 12.19). Underdeveloped embryos of the rudimentary type are at the base of the main trunk of the tree. In Martin's examples of families with rudimentary embryos, seeds of the Aquifoliaceae, Araliaceae, and Papaveraceae have morphophysiological dormancy, and those in the Magnoliaceae and Ranunculaceae have morphological or morphophysiological dormancy.

Seeds with linear embryos are more advanced than those with rudimentary embryos, and Martin (1946)

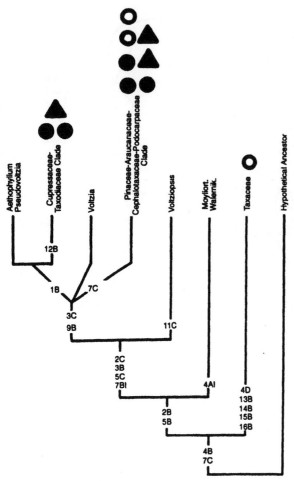

FIGURE 12.18 Cladogram showing general relationships of the modern conifer families and fossil groups [from Miller (1988), with permission] and types of seed dormancy. Each symbol represents one family. ●, physiological dormancy; ○, morphophysiological dormancy; and ▲, nondormant.

listed Amaryllidaceae, Liliaceae, Pontederiaceae, Solanaceae, Vacciniaceae, and Umbelliferae (Apiaceae) as examples of families with linear embryos. Linear embryos in the Amaryllidaceae, Apiaceae, and Liliaceae are underdeveloped, and seeds have morphological or morphophysiological dormancy, depending on the species. Linear embryos in seeds of the Pontederiaceae are as long as the seed, those of the Vacciniaceae are almost the full length of the seed, and those of the Solanaceae are coiled inside the seed. Thus, linear embryos in seeds of Pontederiaceae, Solanaceae, and Vacciniaceae are fully developed and no growth is required prior to germination. Seeds in these three families have physiological dormancy.

Martin (1946) described embryos of gymnosperms as being linear, except for those of *Ephedra* (Gnetaceae), which are spatulate. *Welwitschia*, which also belongs to the Gnetaceae, has a well-developed linear embryo (Bower, 1881; Pearson, 1910); seeds apparently have physiological dormancy. Members of the Cycadaceae, Ginkgoaceae, Taxaceae (Martin, 1946), and Podocarpaceae (Sinnott, 1913) have underdeveloped linear embryos, and they have morphological or morphophysiological dormancy. Seeds in the Araucariaceae (see Fig. 4 in Burrows *et al.*, 1992) and Pinaceae have fully developed linear embryos (Martin, 1946), and they have physiological dormancy. Embryos in some Pinaceae, e.g., *Thuja, Libocedrus, Chamaecyparis, Cupressus,* and *Juniperus,* are quite large and have a spatulate-like appearance (Martin, 1946).

According to Martin (1946), angiosperm seeds with spatulate embryos originated from two lines. One line is from linear seeds, and the resulting spatulate em-

TABLE 12.3   Number of Extant Angiosperm Families with Nondormant Seeds and Various Types of Seed Dormancy and Relative Importance (%) of Various Time Periods when Families with Nondormancy or a Particular Type of Dormancy First Appear in the Fossil Record

Dormancy type/nondormancy	No. of families	Time period						
		Cret.	Paleo.	Eocene	Oligo.	Mio.	Plio.	Pleisto.
Tropical/subtropical zones								
Nondormant	56	32	16	34	11	5	2	0
Morphological	6	50	17	33	0	0	0	0
Morphophysiological	8	38	0	38	0	12	12	0
Physiological	47	26	13	28	17	13	3	0
Physical	8	50	25	25	0	0	0	0
Temperate/arctic zones								
Nondormant	35	46	14	11	20	9	0	0
Morphological	4	25	0	25	0	50	0	0
Morphophysiological	18	44	6	27	6	11	6	0
Physiological	66	35	12	21	13	17	2	0
Physical	10	30	10	40	10	10	0	0

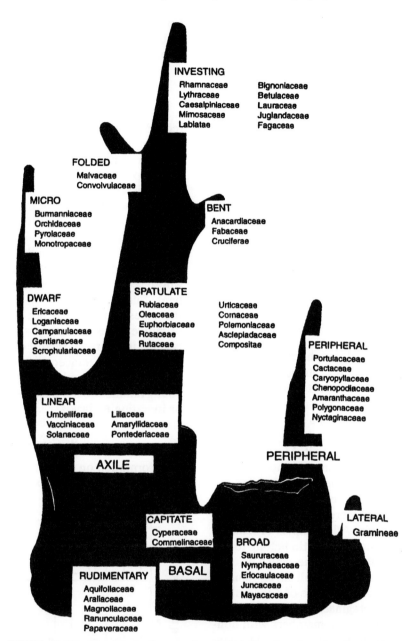

FIGURE 12.19 Martin's family tree of seed phylogeny based on data from 1287 genera of plants. From Martin (1946), with permission.

bryos have thick cotyledons. The second line is from rudimentary seeds, and the resulting spatulate embryos have thin cotyledons. Of the families listed by Martin (1946) as having seeds with spatulate embryos, the Cornaceae, Polemoniaceae, and Urticaceae have physiological dormancy, the Ascelpiadaceae, Asteraceae, Euphorbiaceae, Rosaceae, Rubiaceae, and Rutaceae have physiological dormancy and nondormancy, and the Oleaceae have physiological and morphophysiological dormancies.

Micro seeds are at the tip of a branch on the phylogenetic tree. Thus, micro seeds are more advanced than dwarf ones, which have fully developed embryos. According to Martin (1946), "... the Micro type represents a more advanced state of the compression and suppression evident in the Dwarf type and the relative classification of the two is based on this premise." Micro seeds occur in the Burmanniaceae, Orchidaceae, Monotropaceae, and Pyrolaceae, and they have undifferentiated embryos and thus morphological dormancy. Some micro seeds also have physiological dormancy, giving the combination of morphological and physiological dormancies in some species.

The central portion of the tree (Fig. 12.19) has large seeds with large embryos. Some families (e.g., Anacardiaceae, Convolvulaceae, Fabaceae, Malvaceae, Rhamnaceae) whose seeds have bent, folded, or investing embryos have physical dormancy, whereas others with these types of embryos have physiological dormancy and nondormancy.

Seeds with broad embryos (Fig. 12.19) were considered by Martin (1946) to represent either a stranded lateral line of evolution that preceded the origin of monocots or an intermediate group between the rudimentary type of seeds and various dicots and monocots with definitely starchy endosperm. Martin seemed to favor the latter hypothesis. Members of the Saururaceae, Nymphaeaceae, and Juncaceae have broad embryos and physiological dormancy, but no information is available for other families with broad embryos such as the Eriocaulaceae or Mayacaceae.

Martin (1946) also described peripheral, lateral, and capitate embryos as "phylogenetic orphans." Seeds with capitate embryos have some similarity to those of monocots with either linear or broad embryos. Seeds with peripheral embryos are on the end of a side branch near the base of the family tree of seed phylogeny. Although the peripheral group is a coherent unit, Martin considered it to be a "... blind alley leading no where beyond itself." Further, Martin thought the parental series giving rise to seeds with peripheral embryos was not found in extant species; thus, there were missing links in the evolution of peripheral seeds. All examples of families with peripheral embryos listed by Martin belong to the Caryophyllales, except the Polygonaceae, which belongs to the Polygonales. Further, seeds of all seven families have physiological dormancy, and those of Amaranthaceae, Caryophyllaceae, Nyctaginaceae, Polygonaceae, and Portulacaceae also have some members with nondormant seeds. The Poaceae is the only family mentioned by Martin as having lateral embryos; seeds of grasses have physiological dormancy or nondormancy. Families with capitate embryos include the Commelinaceae, Cyperaceae, Dioscoreaceae, and Musaceae (Martin, 1946), and physiological dormancy and nondormancy are found among members of these families.

IX. TYPE OF EMBRYO AND PRESUMED AGE OF PLANT FAMILIES

Does the time when a plant family first appeared in the fossil record correlate with the type of embryo found in extant members of the family? That is, do the oldest

families have rudimentary or linear (underdeveloped) embryos, whereas relatively young ones have folded or investing embryos?

A. Angiosperms

Information on time of first appearance in the fossil record (Muller, 1981; Taylor, 1990) and type of embryo (Martin, 1946; Corner, 1976) was obtained for 105 families. For families known to have each type of embryo, the proportion of those families that first appeared in the fossil record of each time period was calculated (Fig. 12.20). The only types of embryos not included in Fig. 12.20 are broad, Nymphaeaceae, Cretaceous; lateral, Poaceae, Paleocene; and capitate, Cyperaceae, Eocene.

Some of the plant families that first appeared in the fossil record in the Cretaceous have extant members whose seeds have rudimentary, linear, dwarf, spatulate, bent, folded, investing, or peripheral embryos. Thus, the oldest families do not necessarily have the most primitive types of embryos. In plant families first appearing in the fossil record of various time periods that followed the Cretaceous, a range of embryo types is found in extant members of the families. Thus, relatively young families do not necessarily have seeds with folded or investing embryos.

B. Gymnosperms

Information on embryos, type of dormancy, and first appearance of various extant gymnosperm families in

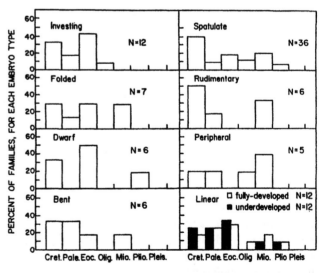

TIME OF FIRST APPEARANCE IN FOSSIL RECORD

FIGURE 12.20 Type of embryo and relative importance (%) of various time periods when families first appear in the fossil record. Families with broad, capitate, and lateral embryos are not included: broad, Nymphaceae, Cretaceous; capitate, Eocene; and lateral, Poaceae, Paleocene.

the fossil record already has been summarized. However, it should be emphasized here that extant gymnosperms with underdeveloped linear embryos belong to families (Cycadaceae, Ginkgoaceae, Podocarpaceae, Taxaceae) that first appeared in the fossil record of the Permian, Triassic, or Jurassic, whereas those with spatulate (Ephedraceae) or spatulate-like (Pinaceae) embryos belong to families that first appeared in the Triassic or Jurassic. Thus, some seeds with underdeveloped linear embryos belong to families that first appeared in the Permian, but there is an overlap in the times (Triassic and Jurassic) when some families with underdeveloped linear embryos and those with spatulate embryos first appeared.

X. ORIGINS AND RELATIONSHIPS OF SEED DORMANCY TYPES

A. Seed/Embryo Types

Solving the problem of the origins of seed dormancy types in angiosperms would be facilitated greatly by a knowledge of the type of embryo found in seeds of the first angiosperms and their ancestors. However, this information is not available. Cladistic studies indicate that angiosperms are the sister group of the Bennettitales and Gnetales (Doyle and Donoghue, 1987). These authors agree with some paleobotanists of the early 1900s that angiosperms and the Bennettitales were derived from a common ancestor. Further, the Gnetales appear to have been derived from the Bennettitales.

There is a greater diversity in the types of embryos in seeds of extant angiosperms than in gymnosperms. Fourteen different types of embryos are found among angiosperms (Fig. 12.21), whereas only three (underdeveloped linear, fully developed linear, and spatulate) are known for gymnosperms. Thus, some types of seeds/embryos must have been derived after angiosperms evolved.

An underdeveloped, linear embryo is found in seeds of many extant as well as in those of some fossil gymnosperms, and it occurs in angiosperm families in the basal orders of the Magnoliidae and Liliidae. Thus, at least some members of the ancestral stock that gave rise to the angiosperms may have had an underdeveloped linear embryo (Fig. 12.21). Significantly, rudimentary embryos are not known in extant gymnosperms, and they have not been reported in seeds of fossil gymnosperms. Martin (1946) did not attempt to include gymnosperms in the family tree of seed phylogeny because of the, "... absence of Basal forms in present-day Gymnosperms."

The apparent absence of rudimentary embryos

among gymnosperms raises a question concerning the relative position of seeds with rudimentary vs linear embryos in an evolutionary scheme. Did some members of the ancestral gymnosperm stock that gave rise to angiosperms have rudimentary embryos, which are not present in extant gymnosperms? The discovery of fossil gymnosperm seeds with rudimentary embryos would help us better understand the origins of angiosperms in general and of rudimentary embryos in angiosperms in particular.

Another possible explanation for the presence of rudimentary embryos in angiosperms, but not in gymnosperms, is that the underdeveloped, linear embryo gave rise to the rudimentary embryo after the evolution of angiosperms. In some angiosperms families such as the Magnoliaceae, Papaveraceae, and Ranunculaceae, many members have seeds with rudimentary embryos, but a few have seeds with linear embryos. In the Apiaceae and Berberidaceae, many members have seeds with linear embryos, but some have seeds with rudimentary embryos (Martin, 1946). Thus, there appears to be at least a close relationship between rudimentary and linear embryos in various families. According to Martin (1946), the rudimentary embryo is below the linear one on the family tree of seed phylogeny (Fig. 12.19). Accordingly, in the subclasses Magnoliidae and Ranunculidae families with rudimentary embryos (or with mostly rudimentary embryos) are more basal than those with linear embryos; however, in the Rosidae and Liliidae families with linear embryos are more basal than those with rudimentary embryos. Thus, there is a small hint that some rudimentary embryos could have been derived from linear ones, but more work is needed on this problem.

The presence of large, fully developed, linear embryos and the absence of endosperm in seeds of the Lauraceae (order Laurales, subclass Magnoliidae) raises the possibility that some angiosperm ancestors may have had a fully developed linear embryo. However, there are three reasons to think that the angiosperm ancestral stock lacked spatulate embryos. (1) Families in the Magnoliidae and Ranunculidae and those in the basal orders of the Arecidae, Liliidae, Rosidae, Dilleniidae, and Hamamelididae do not have spatulate embryos; they have underdeveloped (rudimentary or linear) embryos. (2) Spatulate embryos occupy a relatively high position on the family tree of seed phylogeny (Fig. 12.19), which may indicate that spatulate embryos evolved after angiosperms started to diversify. (3) In gymnosperms, a true spatulate embryo is found only in the Gnetaceae (Martin, 1946). Fossil seeds of the Bennettitales have a fully developed linear embryo (Wieland, 1916); thus, the spatulate embryo may have developed in gymnosperms after the origin of angiosperms.

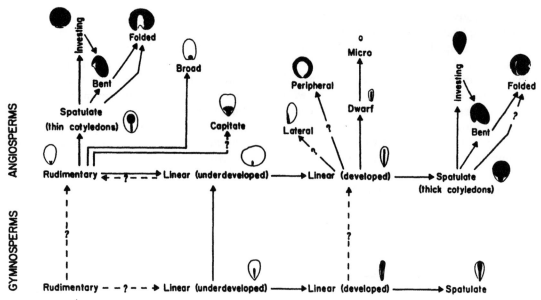

FIGURE 12.21 Suggested origins and evolutionary relationships of the various kinds of gymnosperm and angiosperm seeds. Pictures of embryos are from Martin (1946), with permission.

Although Martin (1946) listed the Cyperaceae under the capitate type of embryo (Fig. 12.21), he noted that some genera in this family, including *Mariscus, Hemicarpha, Lipocarpha,* and *Eriophorum,* have embryos showing a relationship to broad or rudimentary ones. Further, embryos in seeds of two genera in the Cyperaceae, *Dulichium* and *Kyllinga,* tend to be somewhat linear. In the Zingiberales, the basal family is the Musaceae (Kress, 1990), and its seeds have a capitate embryo (see photographs in Manchester and Kress, 1993; Kress, 1995). Other families in this order, including the Strelitziaceae, Heliconiaceae, Costaceae, Zingiberaceae, Cannaceae, and Marantaceae, have seeds with fully developed linear embryos (see Manchester and Kress, 1993). Thus, capitate embryos seem to be related to rudimentary and broad, and perhaps linear, embryos, but the details of these relationships have not been determined.

Seeds with linear embryos have given rise to dwarf and micro seeds and to those with large expanded cotyledons (spatulate, folded, bent, and investing) (Martin, 1946). The Hydrophyllaceae illustrates nicely the pivotal position of the linear embryo. Seeds of *Hydrophyllum canadense* have underdeveloped, linear embryos, those of *Phacelia strictiflora* have a linear embryo that is larger than the one in *H. canadense* seeds (may be fully developed), those of *P. californica* have a spatulate embryo, and those of *Eriodictylon angustifolium* are fully developed and linear, but the seed is dwarf (Martin, 1946).

Dwarf seeds occur in a number of families, including the Campanulaceae, Cephalotaceae, Droseraceae, Ericaceae, Gentianaceae, Hydrophyllaceae, Loganiaceae,

Orobanchaceae, Penthoraceae, Polemoniaceae, Primulaceae, Rubiaceae, Scrophulariaceae, Saxifragaceae, and Solanaceae (Martin, 1946). In members of these families with dwarf seeds, embryos are fully developed and linear. Other members of all these families, except Cephalotaceae, Orobanchaceae, and Penthoraceae, have nondwarf seeds with spatulate and/or linear embryos (Martin, 1946). In nine families, dwarf seeds have only linear (developed) embryos, whereas other members of the same family have spatulate ones. Why? Perhaps after dwarf seeds evolved in these families the environmental conditions that selected for dwarf seeds changed and thus families were subjected to conditions that eventually led to the development of spatulate embryos.

Seeds with underdeveloped, linear embryos seem to be the logical ancestor for those with fully developed (elongated), linear embryos, such as those found in seeds of the Capparidaceae, Casuarinaceae, Combretaceae, Myrsinaceae, Platanaceae, Resedaceae, Rhizophoraceae, Solanaceae, and Typhaceae (Martin, 1946). In various families, including the Apiaceae, Boraginaceae, Euphorbiaceae, Hydrophyllaceae, Oleaceae, Rubiaceae, and Rutaceae, some species have seeds with either underdeveloped or fully developed, linear embryos, whereas others have seeds with spatulate embryos (Martin, 1946). The implication of the presence of linear and spatulate embryos in the same family is that spatulate embryos were derived from linear ones.

Martin (1946) thought the spatulate embryo in some seeds was derived from the linear type, but in others it was developed directly from the rudimentary type. Development of a spatulate embryo from a rudimentary embryo may explain the occurrence of some small spatulate embryos in the Araliaceae (*Hedera*) and Apiaceae (*Coriandrum, Foeniculum, Petroselinum,* and *Pimpinella*) (Martin, 1946). There is a tendency for the cotyledons to be thick in spatulate embryos derived from a linear embryo and to be thin in those derived from a rudimentary embryo. The spatulate embryo from the rudimentary line has given rise to folded, bent, and investing embryos. Thin cotyledons are found in (1) Caesalpiniaceae, Sterculiaceae, and Theaceae with spatulate, investing, and folded embryos; (2) Rhamnaceae with investing and bent embryos; and (3) Aceraceae with folded and bent embryos (Martin, 1946).

According to Martin's (1946) data, many families with thick cotyledons, including the Betulaceae, Curcurbitaceae, Elaeagnaceae, Lythraceae, Meliaceae, Thymelaeaceae, Verbenaceae, and Zygophyllaceae, have spatulate and investing embryos. Therefore, the investing embryo may have been derived directly from the spatulate type. A few families with thick cotyledons, including the Cistaceae and Loasaceae, have spatulate and bent embryos, but none has spatulate or folded embryos. Thus, bent embryos also may have been derived directly from the spatulate type with thick cotyledons, but this does not seem to be the case for folded embryos. Martin (1946) presented a series of drawings for various genera of the Fabaceae, nicely showing a progression from an investing to a bent embryo. Thus, bent also may have been derived from investing embryos in some families.

The combination of spatulate, investing, and bent embryos in families such as the Lamiaceae and Malaceae (with thick cotyledons) suggests that spatulate may have given rise to both investing and bent embryos. Another possibility is that the investing embryo in these two families was derived from the spatulate embryo and then gave rise to the bent embryo. The presence of folded, investing, and bent embryos in the Ulmaceae (with thick cotyledons) may represent a progression from investing to bent to folded embryos. However, investing and bent embryos may have been developed from the spatulate embryo and, in turn, folded from bent embryos.

B. Seed Dormancy Types

1. In Angiosperm Ancestors

A discussion of the relationships between the various types of seed dormancy would be enhanced by information about the type(s) found in the ancestral gymnosperm stock that gave rise to the angiosperms, but this information is not available. All we can do is consider fossil seeds in terms of what is known about morphology and types of dormancy in extant species. For example, if a fossil seed has an underdeveloped embryo, we can surmise that the seed had morphological dormancy. However, we do not know if a fossil seed with an underdeveloped embryo also had physiological dormancy and thus morphophysiological dormancy.

Seeds of the ancestral gymnospermous stock of angiosperms may have had morphological, morphophysiological, or physiological dormancy, or they may have been nondormant (Fig. 12.22). That is, the kind of dormancy (or lack thereof) in seeds of the ancestral stock would have depended on the size (degree of development) of the embryo and whether physiological dormancy was present. However, in view of the prevalence of underdeveloped embryos in families in the basal orders of the various subclasses of angiosperms, morphological and/or morphophysiological dormancy may have been the most important (and perhaps the only) type found in the ancestral stock.

If seeds of the ancestors had morphological or morphophysiological dormancy, an increase in embryo size would have resulted in seeds with fully developed linear embryos that were nondormant or physiologically dormant, respectively. Another way in which seeds with fully developed linear embryos that were physiologically dormant could have arisen was via the development of physiological dormancy after enlargement of the embryo.

2. In Angiosperms

If seeds of the first angiosperms had morphological dormancy, an induction of physiological dormancy would have resulted in morphophysiological dormancy (Fig. 12.22). It appears that there are two ways in which seeds with fully developed, physiologically dormant embryos may have originated from morphologically or morphophysiologically dormant seeds. (1) Seeds with morphological dormancy underwent embryo development, resulting in a loss of the morphological dormancy; thus, seeds would have been nondormant. Then, physiological dormancy was induced in the nondormant seeds. (2) Seeds with morphophysiological dormancy underwent embryo development, resulting in a loss of morphological dormancy; thus, seeds would have had only physiological dormancy.

In angiosperm families with rudimentary or underdeveloped, linear embryos, some species in each family have morphological dormancy, whereas others have morphophysiological dormancy. Morphological dormancy and morphophysiological dormancy each occur in 11 families (Chapters 9–11); thus, there is a close

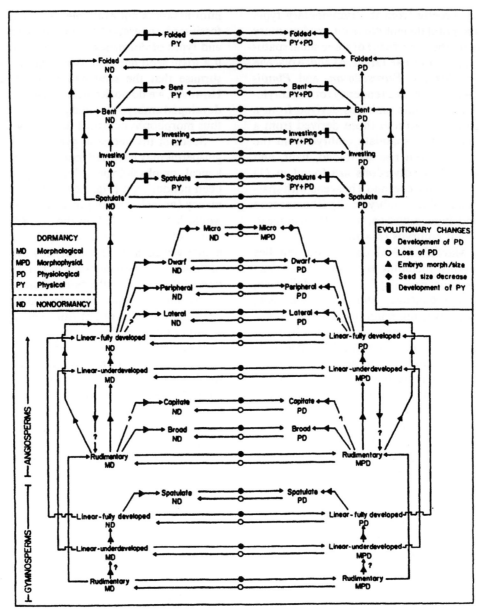

FIGURE 12.22 Suggested origins and evolutionary relationships of the various types of seed dormancy.

relationship between morphological and morphophysiological dormancies. Some members of each family with advanced embryos (i.e., a type other than rudimentary or underdeveloped linear) have nondormant seeds, whereas others have physiological dormancy. In fact, nondormancy and physiological dormancy each occur in at least 73 families of angiosperms (Chapters 9–11). Thus, there is a close relationship between nondormancy and physiological dormancy.

The occurrence of nondormancy and physiological dormancy in families with advanced types of embryos raises the question of whether dormant or nondormant seeds gave rise to seeds with the various kinds of fully developed embryos. There are two ways in which physiologically dormant seeds with an advanced embryo type could have been derived (Fig. 12.22). (1) Physiologically dormant seeds underwent embryo changes (linear gave rise to spatulate, etc.), resulting in physiologically dormant seeds with a new embryo type. (2) Embryo changes occurred in nondormant seeds (i.e., a new embryo type developed), and then physiological dormancy developed.

The problem with hypothesizing about the origin of physiologically dormant seeds with a "new" type of embryo is how do we account for the presence of nondormant seeds with the same kind of new embryo in the same family? After a new type of physiologically dormant seed developed in a family, did seeds in some members of the family lose their physiological dormancy? Perhaps some seeds in various families could have lost their physiological dormancy if there was immigration into a nonrisky environment or if the environment became risk-free. At present, however, no body of theoretical work is available to explain how this would (could) happen. The new kind of embryo (or all the kinds of embryos) may have evolved before environmental conditions selected for physiological dormancy; however, physiological dormancy was not induced in all members of a family with the new embryo type, i.e., some remained nondormant.

In some families (Balsaminaceae, Oxalidaceae, Passifloraceae, Polygalaceae, Thymelaeaceae, and Turneraceae), the tropical species have nondormant seeds, but the temperate ones have physiological dormancy; all these families have spatulate embryos. Further, in other families, including the Acanthaceae (with bent and spatulate embryos), Bignoniaceae (investing), Celastraceae (spatulate), Clusiaceae (dwarf), Ebenaceae (spatulate), Euphorbiaceae (spatulate), Hamamelidaceae (spatulate), Portulacaceae (peripheral), Rubiaceae (spatulate), Rutaceae (spatulate), and Sapotaceae (spatulate), tropical members have nondormant seeds or physiological dormancy, whereas temperate members have physiological dormancy (Chapters 9 and 10). If temperate climates with great seasonal temperature variations are more recent than tropical ones (see Wolfe, 1985), the implication of families having nondormant seeds in tropical species and physiological dormancy in temperate ones is that physiological dormancy may have evolved after the seed/embryo type developed. However, in many families, including the Amaranthaceae (peripheral embryo), Apocynaceae (spatulate), Asclepidaceae (spatulate), Asteraceae (spatulate), Boraginaceae (spatulate), Caryophyllaceae (peripheral), Cyperaceae (capitate), Elaeocarpaceae (fully developed, linear), Fagaceae (investing, folding), Lythraceae (investing, spatulate), Meliaceae (spatulate), Moraceae (bent), Myrtaceae (folded, bent), Poaceae (lateral), Rosaceae (spatulate), Scrophulariaceae (spatulate), and Solanaceae (developed, linear), species with nondormant and physiologically dormant seeds are found in both tropical and temperate regions. Thus, in these families there is no hint as to whether physiologically dormant or nondormant seeds gave rise to new types of seeds/embryos. Perhaps both physiologically dormant and nondormant

seeds have given rise to new seed/embryo types, depending on the family and environmental conditions.

Physical dormancy is found in seeds with spatulate, investing, bent, and folded embryos; however, not all seeds with these types of embryos have physical dormancy: they may be nondormant or have physiological dormancy. In tropical/subtropical regions, some members of the Anacardiaceae, Bombacaceae, Convolvulaceae, Fabaceae, Malvaceae, Rhamnaceae, Sapindaceae, Sterculiaceae, and Tiliaceae have nondormant seeds, whereas others have seeds with physical dormancy (Chapter 9). Thus, there is a close tie between nondormancy and physical dormancy in these families.

The reason some members of a family have permeable seeds and others have impermeable seeds is that the palisade layer of lignified cells is present in some species but absent in the seed or fruit coats of others (Corner, 1976). Thus, physical dormancy has evolved as a result of the development of impermeable seed or fruit coats. In families with both nondormancy and physical dormancy, it seems reasonable that species with physical dormancy did not give rise to those with nondormant seeds because that would mean loss of a complicated structure (impermeable seed coat) after it had evolved in a family. In fact, Corner (1976) says, "What does involve progressive evolution is the conversion of thin-walled and unspecialized cells into thick-walled palisade cells." Thus, ancestors of extant species with physical dormancy may have had nondormant seeds.

In some families whose seeds have physical dormancy, e.g., Cistaceae, Geraniaceae, and Nelumbonaceae, no members of the family are known to have nondormant seeds (Chapters 10 and 11). Further, these families have a temperate or temperate/subtropical geographical distribution pattern (Heywood, 1993). Are there members of these families, especially in the Cistaceae and Geraniaceae still to be discovered whose seeds are nondormant or have only physiological dormancy? Did the evolution of physical dormancy differ in temperate and tropical/subtropical regions?

C. Development of Seed Dormancy Types and Nondormancy

The purpose of this section is to formulate hypotheses on the origin of each type of seed dormancy and of nondormancy. To do this, material already presented in this chapter is reviewed from an ecological perspective. Formulation of these hypotheses has been approached with great caution, and the ideas presented by Gould and Lewontin (1979) have been kept in mind. That is, the fact that a trait seems to be adaptive does not prove that it is a product of natural selection; it may or may not be. Also, the whole organism (plant) has to be considered, not just a single characteristic (i.e., type of seed dormancy). It is hoped that this attempt at trying to figure out what led to the development of the various types of dormancy and of nondormancy will help focus future research efforts in seed germination ecology.

1. Morphological Dormancy

This type of dormancy is found in seeds with underdeveloped embryos that must differentiate and/or grow before germination is possible. The fact that morphological dormancy is lower on the angiosperm phylogenetic tree than morphophysiological dormancy or nondormancy (Figs. 12.13–12.17) indicates that morphological dormancy is quite old. Further, the origin of morphological dormancy appears to have its roots in the ancient gymnosperms; thus, it may be the oldest type of dormancy known.

The presence of elaborate pollen chambers in fossil gymnosperm ovules suggests that a delay in fertilization may have been a mechanism for indirectly regulating the timing of germination. In extant gymnosperms such as *Cephalotaxus, Torreya,* and *Pinus,* the period of time between pollination and fertilization is about 1 year (Singh and Johri, 1972). Rothwell (1982) concluded that embryo growth in the Paleozoic gymnosperm *Cordaianthus duquesnensis* was continuous after fertilization finally occurred; consequently, seeds may have germinated as soon as the embryo reached full size.

Another way in which germination is delayed in extant gymnosperms is that a long period of time is required for development of the embryo after fertilization has occurred. In *Gnetum gnemon,* ". . . the primordium of the embryo has been formed at the time which the seed falls, but it only develops further at a later period" (Goebel, 1905). Bower (1882) was unable to find an embryo in seeds of this species obtained from Java; however, seeds sown in soil at Kew in London germinated after 8 months. Seeds of *Cycas rumphii* are dispersed 6 months after fertilization, at which time cotyledons are starting to differentiate (de Silva and Tambiah, 1952). After dispersal, the embryo requires an additional 4 months to reach full maturity.

Embryos in some extant gymnosperms are differentiated (but underdeveloped) at the time of dispersal (Martin, 1946), and germination does not occur until they grow to a critical size (e.g., Kuo-Huang *et al.,* 1996). Embryos in freshly matured seeds grow, and there is no delay between embryo growth and germination per se (Chamberlain, 1919; Dyer, 1965); however, embryo growth and consequently germination are delayed when temperature and soil moisture conditions are unfavorable. Thus, morphological underdevelopment of the embryo plays a role in controlling the timing of germination in some extant gymnosperms. The presence of small, differentiated, linear embryos in a few Paleozoic fossil gymnosperm seeds (Miller and Brown, 1973; Long, 1974–1975) hints that postdispersal embryo growth, or morphological dormancy, may have played a role in the germination ecology of ancient gymnosperms.

What kind of conditions could have contributed to a change from seeds in which growth of the newly formed embryo was continuous until germination occurred to ones in which embryo growth could be arrested (without death of the seed), resulting in a delay of germination? Climatic data for the Carboniferous and Permian, which correspond to the ages of fossil gymnosperm seeds with small linear embryos, indicate that much of the earth's land mass received high rainfall, but some areas were arid (Frakes, 1979). As discussed in Section IV, much of the Carboniferous was warm, but a cooling period occurred in the Late Pennsylvanian and Early Permian; this was followed by a warm period in the Mid- and Late Permian (Frakes, 1979).

In an environment with high rainfall and high temperatures, with little or no seasonal variation in either factor (e.g., the humid coal-producing regions during the Carboniferous and Permian), it seems reasonable that more seedlings of a species might become established if germination occurred immediately after maturation/dispersal than if it was delayed. A delay in germination might mean (1) an increase in seed predation or (2) that seedlings did not appear until after those of competing species had become established. However, if (1) all seeds in a given cohort germinated immediately following dispersal or (2) all seedlings subsequently were killed, there would be no ungerminated seeds available until the next crop of seeds matured. That is, lack of dormancy would mean that germination of seeds in a given cohort could not be spread out over a period of time.

In a uniform environment, there are two ways in which germination could be spread out over time. (1) Plants produce seeds over a period of several months. This may explain the fossil ovulate cones of *Cordaianthus duquesnensis* in which some mature ovules had been dispersed, while many immature ones had not (Rothwell, 1982). (2) Embryos stop growing before they reach full size, thus seeds would have morphological dormancy at the time of maturation/dispersal. Morphological dormancy may account for the relatively small embryos found in fossil gymnosperm seeds by Miller and Brown (1973) and by Long (1974–1975).

Let us assume that morphological dormancy evolved in some gymnosperms during the Paleozoic. One consequence of this development would have been a major change in how the timing of germination was controlled. Whereas germination previously was controlled indirectly by the timing of fertilization, morphological dormancy permitted it to be controlled by seed–environment interactions. In addition to the delay in germination due to the period of time actually required for embryo development and/or growth, it is possible that morphological dormancy may have delayed germination for extended periods of time in some gymnosperms. From extant angiosperm species whose seeds have morphological dormancy, we know that germination can be delayed for up to several months, depending on the species, if temperature, light:dark, and/or soil moisture conditions are unfavorable for embryo growth, i.e., growth is delayed until environmental conditions become nonlimiting (Chapter 3).

It seems reasonable that morphological dormancy may have increased the possibility of gymnosperms being able to expand into regions with a dry season. However, establishment in a seasonally arid habitat would have required development of (1) specific environmental conditions for embryo growth and germination and (2) tolerance of seeds to desiccation, i.e., not recalcitrant. Presumably, the environmental conditions (e.g., period of increased soil moisture) promoting embryo growth and germination in seeds with morphological dormancy were the same as those that were optimal for seedling establishment and survival.

We will never know if our hypothesis concerning the development of morphological dormancy in seeds of ancient gymnosperms is correct; however, this mental exercise reveals two things. (1) Morphological dormancy is a way of delaying germination, at least for short periods of time, in warm, humid environments with little seasonal variation. (2) If seeds with morphological dormancy also have specific environmental requirements for embryo growth and germination, a rather simple (as compared to seeds with physiological dormancy) way of controlling the timing of germination results.

Because seeds with underdeveloped rudimentary or linear embryos are lower on the family tree of seed phylogeny than micro seeds (Fig. 12.21), the assumption is made that micro seeds are a secondarily derived form of morphological dormancy (see Martin, 1946). In Chapter 11,III, the idea was presented that micro seeds may have evolved in parasitic plants in response to the scattered distribution patterns of suitable host plants (Teryokhin and Nikiticheva, 1982). This idea also applies to orchids that become established where appropriate fungi occur. With an increase in the number of seeds, there is an increase in the chance that a few of them will reach a suitable host or microsite. A reduction in seed and embryo size allows the same quantity of resources to be placed in an increased number of seeds (Haig and Westoby, 1991). However, in micro seeds the embryo has become so small that it is not differentiated into specific parts; thus, we see an extreme form of morphological dormancy. The time when micro seeds evolved is unknown; they are not known in gymnosperms. Finding fossil seeds of members of the Burmanniaceae, Monotropaceae, Orchidaceae, or Pyrolaceae will be quite a challenge due to their small size, but such discoveries would help elucidate the origins of micro seeds.

2. Physiological Dormancy

This type of dormancy is a characteristic of the embryo, but unlike morphologically dormant seeds, which potentially can germinate at any time if environmental factors become nonlimiting for embryo growth, those with physiological dormancy cannot germinate until they have received a dormancy-breaking treatment. Thus, seeds with morphological dormancy could be stimulated to germinate by relatively brief periods of favorable environmental conditions, long before habitats conditions are suitable for seedling establishment. However, seeds with physiological dormancy not only have to come out of dormancy, but the nondormant seeds have to be subjected to appropriate germination-promoting conditions. Consequently, the timing of germination potentially can be more precisely controlled in seeds with physiological dormancy than in those with morphological dormancy.

The mechanism for control of timing of germination in seeds with physiological dormancy means that generally seeds germinate at the beginning of the season that is most favorable for subsequent seedling establishment, growth, and eventual maturation of the plant. As previously discussed (Chapter 4), physiological dormancy is broken during summer (high temperatures) or winter (low temperatures), depending on the species, and seeds germinate in autumn or spring, respectively. Seedlings of autumn-germinating species may be intolerant of the high temperatures and drought conditions of summer, whereas those of spring-germinating species are intolerant of the low (below freezing) temperatures of winter. However, it is the high temperatures of summer that overcome physiological dormancy in autumn-germinating seeds and cold stratification during winter that breaks physiological dormancy in spring-germinating seeds. Because loss of physiological dormancy occurs during the season that is unfavorable for seedling establishment, seeds are nondormant at the beginning of the favorable season. Consequently, none of the time during the favorable season is required for loss of physiological dormancy; as a result, seedlings have the entire period for growth.

There is much that we do not know about physiological dormancy. (1) When did it evolve? Physiological dormancy occurs in extant gymnosperms, but was it present in seeds of ancient gymnosperms? Was physiological dormancy present in seeds of any of the gymnosperms when angiosperms evolved? (2) Physiological dormancy is found in seeds with every known type of embryo. However, did physiological dormancy develop after each type of seed evolved, or did it develop in a common ancestor prior to the development of various seed types? Cold stratification was required for 100% germination of seeds of *Dioscorea* (capitate embryos) from eastern Asia, eastern North America, southwestern Asia (Caucasus Mountains), and southeastern Europe (Black Sea Region) (Terui and Okagami, 1993). These authors suggested that a cold stratification requirement for germination (i.e., physiological dormancy) was present before disjunction of the genus began to occur in the Late Miocene. (3) What are the biochemical differences between physiologically dormant seeds and those that have come out of this state? From a biochemical perspective, is physiological dormancy the same in all types of seeds? Just because the same environmental conditions (e.g., cold stratification) break physiological dormancy in many kinds of seeds does not necessarily mean that physiological dormancy is the same in all of them; it probably is not. (4) What conditions lead to the development of deep, intermediate, and the five types of nondeep physiological dormancy?

In seeds with deep physiological dormancy, a relatively long period of cold stratification is required for loss of dormancy, and its length is not decreased by a period of dry storage (or warm stratification) at high temperatures. In seeds with intermediate physiological dormancy, dry storage at room temperatures may decrease the cold stratification requirement for dormancy loss. Some seeds with nondeep physiological dormancy come out of dormancy during dry storage at high temperatures, whereas those of others require cold stratification. Further, gibberellic acid (GA) may promote germination in seeds with nondeep or intermediate physiological dormancy, but it has no effect on those with deep physiological dormancy (Chapter 3). Thus, it seems possible that the biochemistry of deep physiological dormancy could be different from that of intermediate dormancy and nondeep physiological dormancy. However, this has not been investigated.

By combining what is known about the temperature requirements for dormancy break and the germination of physiologically dormant seeds of species from different vegetation zones and habitats, a model showing the relationships and proposed origins of deep, intermediate, and the five kinds of nondeep physiological dormancy was developed (Fig. 12.23).

At the base of the model is a species (or groups of species) that lived in a warm, wet habitat and whose seeds were nondormant, thus germinating over a range of temperatures. With changes in the environment (decrease in temperature or rainfall), which resulted in a habitat with a favorable and an unfavorable season for seedling establishment, there was selection pressure for germination to occur at the very beginning of the favorable season. In habitats with a dry season, the favorable period for seedling growth would have begun when the rains started, and in those with a cold season, it would have begun when temperatures rose above freezing.

In a climate with annual warm, wet and warm, dry seasons (e.g., savannas or semievergreen tropical forests), the favorable season for seedling establishment is the wet season. Thus, seeds come out of dormancy during exposure to high temperatures and germinate at high temperatures when the wet season begins (Chapter 9,IV,VI). Much additional research needs to be done on the temperature requirements for the germination of seeds from tropical areas with a dry season. However, available information suggests that there are no changes in the temperature requirements for germination as seeds come out of dormancy; thus, this is type 4 of nondeep physiological dormancy (i.e., as seeds come out of dormancy they germinate only at high temperatures).

We hypothesize that type 4, probably through many steps, gave rise to type 2 nondeep physiological dormancy (decrease in minimum temperature for germination as dormancy is broken) during the development of climates with low winter temperatures that alternate with high summer temperatures. One step in the progression of changes from type 4 to type 2 possibly is illustrated by what is found today in the temperature requirements for the germination of autumn-maturing achenes of Asteraceae in southcentral Texas (Baskin *et al.*, unpublished results). Freshly matured achenes of *Brickellia dentata, Verbesina virginica,* and *Viguiera dentata* germinated to 60% or more at 25/15, 30/15, and 35/20°C (or only at 25/15°C), and cold stratification at 5°C decreased the minimum temperature for germination (down to 15/6°C). Thus, seeds exhibited a type 2 pattern as they came out of dormancy during cold stratification. However, nonstratified seeds incubated on moist sand at 20/10, 25/15, and 30/15°C for 10 weeks showed increases in germination at each temperature equal to those of seeds stratified at 5°C for 8 weeks and then tested at each temperature.

In the development of type 2, there has been a lowering of the minimum temperature for germination. Thus, nondormant seeds can germinate in early spring in temperate regions while temperatures are still quite

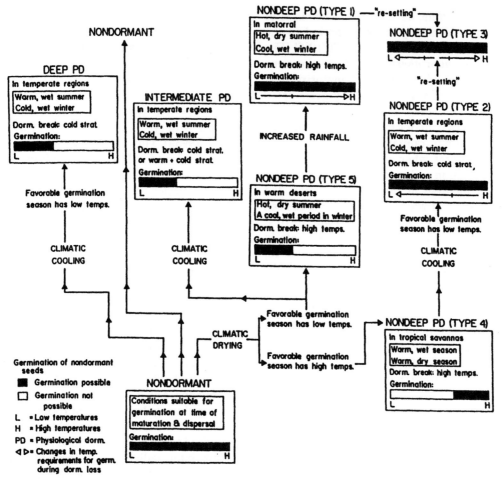

FIGURE 12.23 Suggested origins and evolutionary relationships of the various types of physiological dormancy.

low. Many of the species with type 2 have seeds that can germinate at high (30/15, 35/20°C) but not at low (15/6, 20/10°C) temperatures when they are freshly matured in autumn. However, seeds do not germinate in autumn in temperate regions because temperatures in the habitat are below those required for germination (Chapter 4). The assumption (mostly not tested) is that seedlings of type 2 species are intolerant of freezing.

The ability of achenes of some Texas Asteraceae species to germinate at high temperatures at maturity in autumn can be used as a starting point to formulate an explanation for the origin of type 2 species. We can imagine what would be necessary for the Texas Asteraceae to persist (perhaps as summer annuals) in a climate with a cold (below freezing many times) winter. (1) Seeds should not germinate in autumn because low temperatures of winter probably would kill the seedlings. The high temperature requirement for the germination of freshly matured seeds would prevent germination in autumn in regions more temperate than Texas. (2) Seeds

should germinate in early spring, otherwise seedlings would not have the entire warm season in which to grow.

We hypothesize that the selective pressure for increased fitness resulting from seeds that germinate in spring has led to a lowering of the temperature requirement for the germination of nondormant seeds. Further, in the evolution of type 2, the ability of seeds to germinate at high temperatures has not been lost (Baskin *et al.*, 1993a). Thus, seeds first germinate at high and then at decreasing temperatures as the loss of dormancy progresses. That is, the ability to germinate at low temperatures has been added to the ability to germinate at high temperatures.

In warm deserts with hot, dry summers and a brief cool, moist period in winter, the favorable season for seedling establishment is in winter. Thus, seeds come out of dormancy during exposure to high temperatures and germinate only at low temperatures, when (and if) the soil is moist; this is type 5 nondeep physiological dormancy. Seeds of *Eriastrum diffusum,* from the Chihuahuan Desert in Arizona, have type 5, and even after a 4-month summer dormancy-breaking period in the desert, they germinated only at 15/6 and 20/10°C (Fig. 9.5). Seeds of *Eriogonum abertianum*, from the same desert, are nondormant when they reach maturity, near the end of the dry season; however, they germinated only at 15/6 and 20/10°C (Fig. 9.6).

Across a gradient of climates from deserts to temperate forests, an increase in rainfall is associated with an increase in temperatures when soil moisture becomes sufficient for germination and seedling survival of winter annuals. With an increase in rainfall in a region, the maximum temperature at which nondormant seeds can germinate increases, e.g., in deserts in Arizona, 20/10°C (Baskin *et al.*, 1993b); in seasonally arid habitats in southcentral Texas, 30/15 and 35/20°C (Baskin *et al.*, 1992); and in temperate eastern North America, 35/20°C (Baskin and Baskin, 1974).

Observations on differences in maximum germination temperatures for winter annuals in different regions suggest that type 1 may have been derived from type 5, and this happened as amounts of rainfall increased, resulting in an increase in the length of the favorable season. That is, selection pressure due to increased fitness resulting from seedlings germinating in early autumn led to an increase in the maximum temperature for the germination of nondormant seeds. Further, in the evolution of type 1 the ability of seeds to germinate at low temperatures has been retained. Thus, seeds first germinate at low and then at increasing temperatures as loss of dormancy progresses.

It should be emphasized that seeds with nondeep physiological dormancy do not absolutely require a type 1 pattern of temperature responses to germinate in autumn in the temperate zone, nor do they require a type 2 pattern to germinate in spring. In temperate eastern North America, six species whose seeds come out of dormancy in summer exhibit type 2 and germinate in autumn, and eight species whose seeds come out of dormancy in winter exhibit type 1 and germinate in spring (Baskin *et al.*, 1994b). These seemingly inconsistent species may represent changes in the type of life cycles after the formation of type 1 or type 2, but a retention of previous responses to temperature as dormancy is being broken. For example, *Lamium amplexicaule* is a type 1 species, but some of its seeds may germinate in spring. The species behaves mostly as a winter annual, with seeds coming out of dormancy in summer and germinating in autumn. However, in some years with high summer/autumn rainfall, plants resulting from seeds that germinate in late summer produce seeds before winter. These seeds exhibit some dormancy loss during cold stratification and gain the ability to germinate at low temperatures (a type 1 response).

Type 3 nondeep physiological dormancy (an increase in the maximum and a decrease in the minimum temperatures for germination as dormancy is broken) may represent a "resetting" of type 1 or 2 in response to climatic changes, after the formation of types 1 and 2 in a species. Seeds of some species with type 3 behave ecologically like they had type 1 (Baskin *et al.*, 1994b), whereas others with type 3 behave like they had type 2 (Baskin *et al.*, 1993a). Thus, type 3 is related, at least ecologically, to types 1 and 2. It would be informative to have data on the geographical variations in the germination ecology of species whose seeds have type 3.

Seeds with deep and intermediate physiological dormancy are similar in that they both have a low temperature (e.g., 5, 15/6°C) requirement for germination, after they become nondormant. However, seeds with intermediate dormancy are more like those with nondeep than they are those with deep physiological dormancy for two reasons: (1) GA can promote germination and (2) high temperatures can play a role in dormancy break. Further, with regard to low temperature requirements for germination, seeds with intermediate physiological dormancy are more like those with type 5 than they are those with the other types of nondeep physiological dormancy. Seeds with types 1 and 5 come out of dormancy during dry storage at room temperatures, which decreases the length of the cold stratification treatment required to break intermediate physiological dormancy. Thus, intermediate physiological dormancy may have been derived from the line that led to types 5 and 1, possibly after a low-temperature requirement for germination had developed. However, in response to decreases in winter temperatures, the favorable season for seedling survival may have been shifted from autumn to early spring. If such a shift occurred in the timing of the favorable season for germination, no change in the temperature requirement for germination would have resulted because seeds already could germinate at low temperatures.

Deep physiological dormancy also may have developed in response to decreases in winter temperatures, which resulted in a climate with warm summers and cold winters. Thus, the favorable growing season for seedlings began in early spring. We suggest that deep physiological dormancy originated from species with nondormant seeds and that the selective advantage of early spring germination led to the development of a low temperature requirement for germination.

With regard to the effects of GA on the germination of seeds, it appears that this plant growth regulator can substitute for the loss of dormancy that occurs while seeds are stored dry at high temperatures (i.e., afterripening). GA promotes germination in type 1 seeds that require high temperatures for dormancy loss (Chen and

Park, 1973; Evans and Fratianne, 1977; Hilhorst and Karssen, 1988). In seeds with intermediate physiological dormancy, GA substitutes for the period of storage at room temperatures that can decrease the length of the cold stratification period. For example, seeds of *Melampyrum lineare* require warm dry storage followed by cold stratification before they will germinate; however, GA substitutes for the dry storage but not the cold stratification requirement for dormancy loss (Curtis and Cantlon, 1963). GA substitutes for the cold stratification requirement for dormancy loss in seeds of some species (Bewley and Black, 1994), and further studies may reveal that these species have seeds with type 2 nondeep physiological dormancy (e.g, Baskin and Baskin, 1970).

The model in Fig. 12.23 is an attempt to show how different types of physiological dormancy could have developed in a family or how different families could have a different type(s) of physiological dormancy. For example, deep (Nikolaeva, 1969), intermediate (Crocker and Barton, 1931), and types 1 and 2 nondeep physiological dormancy (Baskin and Baskin, 1985a, 1990c; Roberts and Neilson, 1982) and nondormancy (Bonde, 1965) are known in the Rosaceae. Nondormancy and types 1, 2, and 3 nondeep physiological dormancy are reported in the Asteraceae (e.g., Baskin and Baskin, 1988; Baskin *et al.*, 1994b). Is the absence of some types of physiological dormancy in the Rosaceae and Asteraceae due to the evolutionary history of these families or does it reflect insufficient research?

The time frame for the evolution of the various kinds of physiological dormancy is unknown. However, because (1) all seeds known to have deep physiological dormancy have either spatulate or folded embryos, (2) spatulate and folded are advanced embryo types (Chapter 3; Martin, 1946), and (3) seeds with intermediate and/or nondeep physiological dormancy can have every known type of embryo, we suggest that deep physiological dormancy is a relatively recent development. Further, it is not known if the climatic cooling that supposedly led to the development of deep physiological dormancy, intermediate physiological dormancy, and type 2 nondeep physiological dormancy should be represented as a single period or as three different ones. Climatic cooling is shown at the same level in Fig. 12.23, and a good candidate for the time of the cooling period is the Eocene–Oligocene boundary. In fact, some (sometimes all) extant members of *Crataegus, Sorbus, Prunus* (Rosaceae), *Acer* (Aceraceae), and *Fraxinus* (Oleaceae) have deep physiological dormancy (Chapter 3), and they first appear in the fossil record of North America in the Oligocene (Taylor, 1990). However, there also was a cooling period during the middle of the Cretaceous, end of the Cretaceous, and Middle/Late Eocene. As information on the fossil history of various

plant families increases, correlations between the time of appearance of various genera and the type of dormancy in extant species should help elucidate the time of development of the dormancy type.

3. Morphophysiological Dormancy

Deep, intermediate, and nondeep physiological dormancies have developed in seeds with various kinds of underdeveloped embryos in apparently the same way (and for the same reasons) they developed in seeds with fully developed embryos (see Fig. 12.23). Although germination can be delayed in seeds with morphological dormancy, the timing of germination depends on the beginning of favorable environmental conditions for embryo growth. However, successful seedling establishment occurs only if the period for embryo growth and germination is followed by conditions suitable for seedling establishment and growth. In climates with warm summers alternating with cold winters, autumns usually have the soil moisture and temperature conditions required for the germination of seeds with morphological dormancy (e.g., Baskin and Baskin, 1986). However, if seeds germinate in autumn, seedlings will be killed unless they are tolerant of the below-freezing temperatures of winter. If seeds with morphological dormancy also have physiological dormancy, germination may be prevented in autumn. Further, if seeds germinated at the end of the wet season in a region with a wet and dry season each year, seedlings might be killed during the dry season. Thus, it seems reasonable that if a dry or a cold (temperatures below freezing) season developed in a region, the induction of physiological dormancy in seeds with morphological dormancy would prevent them from germinating until the beginning of the favorable wet or warm season, respectively.

In temperate zone species whose seeds have one of the complex types of morphophysiological dormancy, embryo growth and loss of physiological dormancy occur during cold stratification in winter (Table 5.1). However, in species whose seeds have one of the deep simple types of morphophysiological dormancy or intermediate simple morphophysiological dormancy, embryos grow in autumn (at 15–20°C), but physiological dormancy is not broken until seeds receive cold stratification during winter. In seeds with deep simple epicotyl morphophysiological dormancy, radicle emergence occurs in autumn, but shoot emergence is delayed until spring (Chapter 5). The delay of shoot emergence until spring may be related to the intolerance of shoots to temperatures below freezing, but this has not been tested in species with deep simple epicotyl (or any other type of) morphophysiological dormancy. One exception is found in seeds with nondeep simple morphophysiological dormancy, which come out of physiological dormancy during summer. Consequently, embryo growth and germination occur in autumn, and the frost-tolerant plants behave as winter annuals (e.g., Baskin and Baskin, 1990d).

We conclude that the development of physiological dormancy in seeds with morphological dormancy in both temperate and tropical zones can be viewed as a block to embryo growth. Further, the block (physiological dormancy) becomes effective before, during, or after embryo growth, depending on the species. Regardless of when the block occurs, however, physiological dormancy effectively prevents germination of seeds with morphological dormancy until the beginning of the favorable season for seedling establishment and growth.

The time of origin of morphophysiological dormancy is unknown, but it can be inferred that it has been present since the Tertiary. Many families and genera in northern temperate regions have an Arcto–Tertiary distribution pattern (Li, 1972; Wood, 1972; Tiffney, 1985a). That is, members of a genus occur in eastern and western North America, Asia, and Europe (or in some combination of these four areas). Disjunct distribution patterns may represent elements of the Arcto–Tertiary geoflora that have become separated due to climatic changes (Chaney, 1947; Wen and Stuessy, 1993) or the result of long-distance migration and divergence of species (Tiffney, 1985a,b).

Molecular phylogenetic techniques (e.g., chloroplast DNA restriction site data) make it possible to estimate divergence times of species. Eastern Asian (EA) and eastern North America (ENA) species have various divergence times: *Campsis grandiflora* (EA) and *C. radicans* (ENA), 24.4 million years ago (MYBP), Oligocene–Miocene boundary (Wen and Jensen, 1995); and *Symplocarpus renifolius* (EA) and *S. foetidus* (ENA), 6.1 MYBP, late Miocene (Wen *et al.*, 1996). Internal transcribed spacer (ITS) sequences of nuclear ribosomal DNA show that the divergence of *Panax trifolius* (ENA) from its Asian relatives occurred much earlier than the divergence between *P. quinquefolium* (ENA) and its Asian relatives (Wen and Zimmer, 1996). The estimated divergence time of *Caulophyllum robustum* (EA) and *C. thalictroides* (ENA) based on allozyme data is 6.0–6.5 MYBP, late Miocene (Lee *et al.*, 1996).

A number of genera with an Arcto–Tertiary distribution pattern, including *Cryptotaenia*, *Osmorhiza*, and *Sanicula* (Apiaceae), *Arisaema* (Araceae), *Aralia*, *Panax* (Araliaceae), *Asarum* (Aristolochiaceae), *Podophyllum*, *Jeffersonia*, *Caulophyllum* (Berberidaceae), *Dicentra* (Fumariaceae), *Clintonia*, *Smilacina*, *Trillium*, *Erythronium*, *Veratrum* (Liliaceae), *Stylophorum* (Papaveraceae), and *Hydrastis*, *Actaea*, *Cimicifuga*, *Isopyrum* (Ranunculaceae) (Li, 1992; Wood, 1972; Tiffney

1985a,b) have seeds with morphophysiological dormancy (Chapter 5). Thus, this type of seed dormancy may have been present in these genera since the Tertiary.

Seeds of *Jeffersonia dubia* from Asia (Grushvitzky, 1967) and those of *J. diphylla* from eastern North America (Baskin and Baskin, 1989) have deep simple morphophysiological dormancy. Also, seeds of *Panax ginseng* from Asia (Nikolaeva, 1977) and those of *P. quinquefolium* (Stoltz and Snyder, 1985) from eastern North America have deep simple morphophysiological dormancy. Occurrence of the same type of morphophysiological dormancy in Asian and North American disjuncts that are members of genera with an Arcto–Tertiary distribution pattern suggests that this type of dormancy is at least as old as the Tertiary.

However, seeds of *Osmorhiza occidentalis*, *O. chilensis*, and *Erythronium grandiflorum* from western North America have deep complex morphophysiological dormancy (Baskin *et al.*, 1995), whereas those of *O. longistylis* (Baskin and Baskin, 1984b), *O. claytonii* (Baskin and Baskin, 1991), and *E. albidum* (Baskin and Baskin, 1985b) from eastern North America have nondeep complex morphophysiological dormancy. Thus, disjunct species in genera with an Arcto–Tertiary distribution pattern can have different types of morphophysiological dormancy; however, it is unknown when these differences developed.

4. Physical Dormancy

Germination phenology studies of species whose seeds have physical dormancy indicate that for the most part they have a definite germination season (Chapter 6), which presumably is during the favorable time for seedling establishment. Because (1) there is a relatively high proportion of species with physical seed dormancy in tropical deciduous forests, savannas, hot deserts, steppes, and matorral (Fig. 12.2) and (2) these vegetation zones have a dry season and a wet season, it seems reasonable that the favorable season for seedling establishment is the wet season. In temperate/arctic regions, the favorable time for seedling establishment for most species with physically dormant seeds is at the beginning of the warm season, but in winter annuals (e.g. Fabaceae, Geraniaceae) with physically dormant seeds germination occurs in autumn. Regardless of when the favorable season for seedling establishment begins, the presence of impermeable seed or fruit coats prevents germination in response to precipitation during the unfavorable season.

What could have led to the development of physical dormancy? If physiological dormancy prevents germination during an unfavorable season for seedling establishment, why do seeds of some species have physical rather then physiological dormancy? Physical dormancy sometimes is viewed as an adaptation that extends the period of seed longevity under warm, humid conditions (Christiansen *et al.*, 1960; Potts, *et al.*, 1978; Puckridge and French, 1983). One point to keep in mind about seeds with physiological vs physical dormancy is that dormancy loss in seeds with physiological dormancy generally takes a much longer period of time than it does in those with physical dormancy. Thus, impermeable seed (fruit) coats function as protective barriers that prevent imbibition of water until a relatively short time prior to germination. Would seeds with physical dormancy die if subjected to a long period of high temperatures and high relative humidities after they became permeable? Data for seeds of crop species such as *Glycine max* (Potts *et al.*, 1978) and *Gossypium hirsutum* (Christiansen *et al.*, 1960) indicate that permeable seeds die quickly if subjected to an extended period of wet, warm weather.

Is death due to seed metabolic activities per se or to attack from fungi? Seeds of the legume *Dipteryx alata* from Brazil have permeable seed coats. Freshly matured (15-day-old) seeds germinated to 55% at 33°C after 7 days, whereas 2-month-old seeds stored dry at room temperatures germinated to 96% in 4 days (Melhem, 1975). These data indicate that seeds had some physiological dormancy at maturity that was lost during afterripening. However, after 1, 2, 3, and 4 years of dry storage, 10, 35, 80, and 100% of the seeds were dead (Melhem, 1975).

In seven families (Anacardiaceae, Bombacaceae, Cucurbitaceae, Fabaceae, Malvaceae, Sapindaceae, and Sterculiaceae) known to have species with physical dormancy (Table 3.5), some species also have recalcitrant seeds (von Teichman and van Wyk, 1994; see this reference for a list of 45 dicotyledonous families with at least some species with recalcitrant seeds). Are there any species with recalcitrant seeds in other families with physical dormancy, including the Cannaceae, Cistaceae, Convolvulaceae, Geraniaceae, Musaceae, Nelumbonaceae, Rhamnaceae, and Tiliaceae? Why did physical dormancy evolve in so many families with recalcitrant seeds? von Teichman and van Wyk (1991a, 1994) consider recalcitrance to be an ancestral character state in dicotyledonous species. Further, recalcitrance may have been present in Paleozoic pteridosperms, and it is found in seeds of some extant gymnosperms, e.g., cycads (von Teichman and van Wyk, 1994).

Further insight into the relationship between recalcitrance and physical dormancy is found in the Anacardiaceae. Anatomical studies of seeds (fruits) in this family led to the conclusions that (1) the most primitive state is found in the tribe Anacardieae, (2) the tribe Spondiadeae is more advanced than the Anacardieae, and (3) the tribe Rhoeae is more advanced than the Spondiadeae (von Teichman and van Wyk, 1991b). Members of the Anacardieae have large, recalcitrant, pachychalazal (=massive or extensive chalaza) seeds with no endosperm, starch stored in the cotyledons, an undifferentiated seed coat, and an endocarp with less than four distinct layers, some of which may be sclereids (von Teichman *et al.*, 1988; Wannan and Quinn, 1990, 1991). Members of the Spondiadeae have large, partially pachychalazal orthodox seeds with no endosperm, proteins, and lipids stored in cotyledons, an undifferentiated seed coat, and an endocarp composed of lignified, irregularly oriented sclereids (von Teichman, 1988a,b, 1990; von Teichman and van Wyk, 1988, 1996; Wannan and Quinn, 1990, 1991). Members of the Rhoeae have small, partially pachychalazal orthodox seeds with no endosperm, proteins, and lipids stored in the cotyledons, a differentiated seed coat, and, in most members, a four-layered endocarp with palisade-shaped sclereids (von Teichman, 1993; Wannan and Quinn, 1990, 1991). The genus *Rhus*, which belongs to the Rhoeae, has a four-layered endocarp with palisade-shaped sclereids (von Teichman and Robbertse, 1986; von Teichman and van Wyk, 1991b; von Teichman, 1989). Further, members of this genus have physical dormancy (Table 10.16). Thus, as fruit size has been reduced in the Anacardiaceae, there has been a loss of recalcitrance and the endocarp has become impermeable to water. However, it should be noted that a few members of the Rhoeae have only parenchyma cells in the endocarp (Wannan and Quinn, 1990). The endocarp of *Heeria argentea* (Rhoeae) has two layers of sclereid cells and the large pachychalazal fruits are recalcitrant with large cotyledons that store starch (von Teichman and van Wyk, 1996). Thus, the most advanced tribe of the Anacardiaceae has at least one member with large recalcitrant fruits. Is this a case of secondary derivation of recalcitrance, a genus being in the incorrect tribe, or retention of an "old" character? Corner (1992) argues that pachychalaza is an advanced rather than a primitive trait of dicotyledons and that it evolved in parallel in several families. Thus, he might suggest that recalcitrance in *H. argentea* is an advanced trait.

The time and environmental conditions during the origin of physical dormancy are unknown. The Tiliaceae, Sapindaceae, Fabaceae, and Bombacaceae first appear in the fossil record in the Cretaceous; Sterculiaceae in the Paleocene; and Anacardiaceae, Malvaceae, and Nelumbonaceae in the Eocene (Muller, 1981; Taylor, 1990). By Middle Eocene, the Fabaceae was quite diverse (Herendeen, 1992). Axelrod (1992) suggested that climatic drying (e.g., 48–47 MYBP) and/or cooling (e.g., 15 MYBP) was important in promoting species diversification in this family. However, the appearance of one of these families in the fossil record of a particular period of time does not mean that this is when physical dormancy originated. The first members of these families may have had nondormant seeds. As discussed earlier in Section IV, there were major periods of decreased precipitation in the Eocene and Oligocene, and perhaps these are good candidates for the time when physical dormancy could have originated. Because physical dormancy is associated with a specific seed or fruit coat anatomy, fossil evidence may become available someday that will allow us to determine when this type of dormancy first appeared in various families. If we knew when physical dormancy originated, then it would be possible to correlate the time of origin with paleoclimatic conditions. Sections through fossil fruits of *Rhus rooseae* (Anacardiaceae) from the Middle Eocene show what appears to be a dense layer of macrosclerids in the endocarp (Manchester, 1994); this is the layer responsible for the impermeability of fruits of extant *Rhus* species. Thus, impermeable fruit coats appear to have been present in *Rhus* by the Middle Eocene.

5. Nondormancy

Nondormancy is found in seeds of two or more kinds of plants (trees, shrubs, lianas, herbs) in all of the 14 major vegetation types on earth, except cold deserts (Figs. 12.3 and 12.4). Thus, nondormancy persists under a wide array of environmental conditions. In the evolution of angiosperm seeds, there has been a general decrease in the amount of endosperm per seed with a corresponding increase in the size of the embryo, especially cotyledons (Martin, 1946). Thus, nondormancy may have been derived from seeds with morphological dormancy as a consequence of increased embryo size.

Regardless of the origin of nondormancy, the big question is how does it persist? In warm, wet tropical regions, lack of seed dormancy permits rapid germination as soon as seeds are dispersed. Because there are no annual cycles of favorable and unfavorable seasons for seedling establishment in warm, wet tropical regions, nondormancy of freshly matured seeds comes as no surprise. However, nondormancy also is found in seeds of species growing in regions with an annual dry season and in those with an annual cold season. What controls the timing of germination in species with nondormant seeds so that it occurs during the favorable season? There are at least two ways in which the timing of germination is controlled in seeds with nondormancy:

1. Seeds mature at the end of the unfavorable season or at the beginning of the favorable season. Further, seeds are dispersed as soon as they mature, and germination and seedling establishment occur immediately. For example, in semievergreen forests in Panama, many species with nondormancy disperse their seeds during the early part of the wet season and germinate within 2 weeks (Garwood, 1982). One example from the temperate regions is found in some winter annual Asteraceae whose achenes are matured and dispersed in late summer or early autumn, followed by germination within 1–2 weeks (e.g., Baskin and Baskin, 1983).

2. The temperature range for the germination of seeds with nondormancy may be such that seeds cannot germinate during early winter (e.g., Baskin and Baskin, 1973) or midsummer (Baskin *et al.,* unpublished results), depending on the species. Thus, germination is prevented until habitat temperatures increase or decrease enough to be within the range of those required for germination. Although seeds of these species may be subjected to high summer or low winter temperatures, their temperature requirements for germination do not change (Baskin and Baskin, 1973; Baskin *et al.* unpublished results).

6. Physical plus Physiological Dormancy

We conclude that physiological dormancy was induced in seeds with physical dormancy. One reason why physiological dormancy may have been added to physically dormant seeds is that the climate changed (or species migrated into regions with a different climatic regime) and seeds with physical dormancy became nondormant at a time that was unfavorable for seedling establishment. Let us suppose that a region had warm summers and mild winters. Physical dormancy was broken during summer and seedlings could become established in autumn; seedlings survived during mild winters. However, climatic changes in the region resulted in warm summers and cold winters. Physical dormancy was broken during summer and germination occurred in autumn; seedlings were killed by below-freezing temperatures in winter. Induction of physiological dormancy that was broken during winter would prevent permeable seeds from germinating in autumn.

In a few species with physical and physiological dormancies, both types are broken during summer and seeds germinate in autumn; these are usually winter annuals. Because physiological dormancy is lost during dry storage at high temperatures (afterripening), the embryo is nondormant long before physical dormancy is broken and the favorable (autumn) season for seedling establishment begins (Chapter 6). Thus, induction of physiological dormancy that is broken during summer in seeds with physical dormancy was (is) a way to prevent germination before seed coats became impermeable. That is, if permeable seeds were dispersed in spring while soil moisture was still adequate for germination, seeds might germinate and the seedlings subsequently would be killed by drought conditions of summer. Only a few winter annuals whose seeds also have physical dormancy are known to have physiological dormancy; more research is needed on them.

D. Future Studies

Certain aspects of the origins of the various types of seed dormancy and nondormancy probably will always remain a mystery. One of the problems is that there is no way to know the ecophysiological responses of fossil seeds; thus, we have no way to test some of our assumptions regarding their germination ecology. Another problem is that there are "missing links" in the fossil record, e.g., no gymnosperm seeds with rudimentary embryos have been found (maybe they never existed), and fossils that could help clarify the origins and relationships of seeds with lateral, broad, peripheral, and capitate embryos are not available. It seems reasonable to hope, however, that the fossil record might someday provide information on the time of origin of micro seeds, impermeable seed/fruit coats, and the various types of embryos.

Much remains to be learned about the germination ecology of extant species. At present, there are many plant families for which no information is available (not even for a single species) on the type of embryo, kind of dormancy (or lack thereof), or dormancy-breaking and germination requirements of the seeds. Little is known about tropical/subtropical seeds with nondeep physiological dormancy. Is the geographical distribution of the various types of nondeep physiological dormancy determined solely by climatic conditions or has the past history of movement of continents played a role?

Studies also are needed on the effects of GA on the germination of seeds with different types of nondeep physiological dormancy, keeping in mind the temperature requirements for dormancy loss and germination. Is it possible to induce secondary dormancy in seeds with deep or intermediate physiological dormancy? Can we find ways to make buried weed seeds germinate *in situ* or to store recalcitrant seeds under dry conditions? What kind of morphophysiological dormancy (if any) occurs in seeds of Arcto–Tertiary genera besides *Panax* and *Jeffersonia*?

An impressive amount of information on seed germination ecology has accumulated since the days when Theophrastus worked with seeds, and he, no doubt, would be pleased. However, numerous unanswered questions quickly come to mind when the present state of our knowledge of seed germination ecology, biogeog-

raphy, and evolution is reviewed. Thus, much exciting research remains to be done.

REFERENCES

Abul-Fatih, H. A., and Bazzaz, F. A. (1979). The biology of *Ambrosia trifida* L. II. Germination, emergence, growth and survival. *New Phytol.* **83**, 817–827.

Andersson, S. (1992). Phenotypic selection in a population of *Crepis tectorum* ssp. *pumila* (Asteraceae). *Can. J. Bot.* **70**, 89–95.

Arnold, C. A. (1938). Paleozoic seeds. *Bot. Rev.* **4**, 205–235.

Arthur, A. E., Gale, J. S., and Lawrence, M. J. (1973). Variation in wild populations of *Papaver dubium*. VII. Germination time. *Heredity* **30**, 189–197.

Axelrod, D. I. (1992). Climatic pulses, a major factor in legume evolution. *In* "Advances in Legume Systematics: Part 4. The Fossil Record" (P. S. Herendeen and D. L. Dilcher, eds.), pp. 259–279. The Royal Botanic Gardens, Kew.

Barron, E. J., and Washington, W. M. (1982). Cretaceous climate: A comparison of atmospheric simulations with the geologic record. *Palaeogeogr. Palaeoclim. Palaeoecol.* **40**, 103–133.

Bartek, L. R., Sloan, L. C., Anderson, J. B., and Ross, M. I. (1992). Evidence from the Antarctic continental margin of Late Paleogene ice sheets: A manifestation of plate reorganization and synchronous changes in atmospheric circulation over the emerging southern ocean? *In* "Eocene-Oligocene Climatic and Biotic Evolution" (D. R. Prothero and W. A. Berggren, eds.), pp. 131–159. Princeton Univ. Press, Princeton, NJ.

Basinger, J. F., and Christophel, D. C. (1985). Fossil flowers and leaves of the Ebenaceae from the Eocene of southern Australia. *Can. J. Bot.* **63**, 1825–1843.

Baskin, C. C., and Baskin, J. M. (1988). Germination ecophysiology of herbaceous plant species in a temperate region. *Am. J. Bot.* **75**, 286–305.

Baskin, C. C., Baskin, J. M., and Chester, E. W. (1994a). Annual dormancy cycle and influence of flooding in buried seeds of mudflat populations of the summer annual *Leucospora multifida*. *Ecoscience* **1**, 47–53.

Baskin, C. C., Baskin, J. M., and Leck, M. A. (1993a). Afterripening pattern during cold stratification of achenes of ten perennial Asteraceae from eastern North America, and evolutionary implication. *Plant Species Biol.* **8**, 61–65.

Baskin, C. C., Baskin, J. M., and Van Auken, O. W. (1992). Germination response patterns to temperature during afterripening of achenes of four Texas winter annual Asteraceae. *Can. J. Bot.* **70**, 2354–2358.

Baskin, C. C., Baskin, J. M., and Van Auken, O. W. (1994b). Germination response patterns during dormancy loss in achenes of six perennial Asteraceae from Texas, USA. *Plant Species Biol.* **9**, 113–117.

Baskin, C. C., Chesson, P. L., and Baskin, J. M. (1993b). Annual seed dormancy cycles in two desert winter annuals. *J. Ecol.* **81**, 551–556.

Baskin, C. C., Meyer, S. E., and Baskin, J. M. (1995). Two types of morphophysiological dormancy in seeds of two genera (*Osmorhiza* and *Erythronium*) with an Arcto-Tertiary distribution pattern. *Am. J. Bot.* **82**, 293–298.

Baskin, J. M., and Baskin, C. C. (1970). Replacement of chilling requirement in seeds of *Ruellia humilis* by gibberellic acid. *Planta* **94**, 250–252.

Baskin, J. M., and Baskin, C. C. (1972a). The ecological life cycle of the cedar glade endemic *Leavenworthia exigua* var. *exigua*. *Can. J. Bot.* **50**, 1711–1723.

Baskin, J. M., and Baskin, C. C. (1972b). Influence of germination date on survival and seed production in a natural population of *Leavenworthia stylosa*. *Am. Midl. Nat.* **88**, 318–323.

Baskin, J. M., and Baskin, C. C. (1973). Ecological life cycle of *Helenium amarum* in central Tennessee. *Bull. Torrey Bot. Club* **100**, 117–124.

Baskin, J. M., and Baskin, C. C. (1974). Germination and survival in a population of the winter annual *Alyssum alyssoides*. *Can. J. Bot.* **52**, 2439–2445.

Baskin, J. M., and Baskin, C. C. (1976). Some aspects of the autecology and population biology of *Phacelia purshii*. *Am. Midl. Nat.* **96**, 431–442.

Baskin, J. M., and Baskin, C. C. (1979). Studies on the autecology and population biology of the weedy monocarpic perennial, *Pastinaca sativa*. *J. Ecol.* **67**, 601–610.

Baskin, J. M., and Baskin, C. C. (1981). Seasonal changes in the germination responses of buried *Lamium amplexicaule* seeds. *Weed Res.* **21**, 299–306.

Baskin, J. M., and Baskin, C. C. (1983). Ecophysiology of seed germination and flowering in common broomweed, *Amphiachyris dracunculoides* (DC.) Nutt. *J. Range Manage.* **36**, 619–622.

Baskin, J. M., and Baskin, C. C. (1984a). Effect of temperature during burial on dormant and non-dormant seeds of *Lamium amplexicaule* L. and ecological implications. *Weed Res.* **24**, 333–339.

Baskin, J. M., and Baskin, C. C. (1984b). Germination ecophysiology of the woodland herb *Osmorhiza longistylis* (Umbelliferae). *Am. J. Bot.* **71**, 687–692.

Baskin, J. M., and Baskin, C. C. (1985a). Role of dispersal date and changes in physiological responses in controlling timing of germination in achenes of *Geum canadense*. *Can. J. Bot.* **63**, 1654–1658.

Baskin, J. M., and Baskin, C. C. (1985b). Seed germination ecophysiology of the woodland spring geophyte *Erythronium albidum*. *Bot. Gaz.* **146**, 130–136.

Baskin, J. M., and Baskin, C. C. (1986). Germination ecophysiology of the mesic deciduous forest herb *Isopyrum biternatum*. *Bot. Gaz.* **147**, 152–155.

Baskin, J. M., and Baskin, C. C. (1989). Seed germination ecophysiology of *Jeffersonia diphylla*, a perennial herb of mesic deciduous forests. *Am. J. Bot.* **76**, 1073–1080.

Baskin, J. M., and Baskin, C. C. (1990a). The role of light and alternating temperatures on germination of *Polygonum aviculare* seeds exhumed on various dates. *Weed Res.* **30**, 397–402.

Baskin, J. M., and Baskin, C. C. (1990b). Seed germination biology of the narrowly endemic species *Lesquerella stonensis* (Brassicaceae). *Plant Species Biol.* **5**, 205–213.

Baskin, J. M., and Baskin, C. C. (1990c). Role of temperature and light in the germination ecology of buried seeds of *Potentilla recta*. *Ann. Appl. Biol.* **117**, 611–616.

Baskin, J. M., and Baskin, C. C. (1990d). Germination ecophysiology of seeds of the winter annual *Chaerophyllum tainturieri*: A new type of morphophysiological dormancy. *J. Ecol.* **78**, 993–1004.

Baskin, J. M., and Baskin, C. C. (1991). Nondeep complex morphophysiological dormancy in seeds of *Osmorhiza claytonii* (Apiaceae). *Am. J. Bot.* **78**, 588–593.

Baskin, J. M., Baskin, C. C., and Spooner, D. M. (1989). Role of temperature, light and date seeds were exhumed from soil on germination of four wetland perennials. *Aquat. Bot.* **34**, 387–394.

Beckstead, J., Meyer, S. E., and Allen, P. S. (1996). *Bromus tectorum* germination: Between-population and between-year variation. *Can. J. Bot.* **74**, 875–882.

Benjamin, L. R. (1990). Variation in time of seedling emergence within populations: A feature that determines individual growth and development. *Adv. Agron.* **44**, 1–25.

Berggren, W. A., and Prothero, D. R. (1992). Eocene-Oligocene climatic and biotic evolution: An overview. *In* "Ecocene-Oligocene

Climatic and Biotic Evolution" (D. R. Prothero and W. A. Berggren, eds.), pp. 1–28. Princeton Univ. Press, Princeton, NJ.

Bewley, J. D., and Black, M. (1994). "Seeds: Physiology of development and germination." 2nd ed. Plenum Press, New York.

Biere, A. (1991). Parental effects in *Lychnis flos-cuculi*. II. Selection on time of emergence and seedling performance in the field. *J. Evol. Biol.* **3**, 467–486.

Bold, H. C. (1957). "Morphology of Plants," 3rd Ed. Harper & Row, New York.

Bonde, E. K. (1965). Further studies on the germination of seeds of Colorado alpine plants. *Univ. Colorado Stud. No. 18.*

Bower, F. O. (1881). On the germination and histology of the seedling of *Welwitschia mirabilis*. *Quart. J. Microscop. Sci.* **21**, 15–30 + plates III and IV.

Bower, F. O. (1882). The germination and embryogeny of *Gnetum gnemon*. *Quart. J. Microscop. Sci.* **22**, 278–298 + plate XXV.

Brown, J. S., and Venable, D. L. (1991). Life history evolution of seedbank annuals in response to seed predation. *Evol. Ecol.* **5**, 12–29.

Bulmer, M. G. (1984). Delayed germination of seeds: Cohen's model revisited. *Theor. Pop. Biol.* **26**, 367–377.

Bulmer, M. (1994). "Theoretical Evolutionary Ecology." Sinauer, Sunderland, MA.

Burdon, J. J., Marshall, D. R., and Brown, A. H. D. (1983). Demographic and genetic changes in populations of *Echium plantagineum*. *J. Ecol.* **71**, 667–679.

Burrows, G. E., Boag, T. S., and Stockey, R. A. (1992). A morphological investigation of the unusual cryptogeal germination strategy of bunya pine (*Araucaria bidwillii*): An Australian rain forest conifer. *Int. J. Plant Sci.* **153**, 503–512.

Carmichael, J. S., and Friedman, W. E. (1996). Double fertilization in *Gnetum gnemon* (Gnetaceae): Its bearing on the evolution of sexual reproduction within the Gnetales and the anthophyte clade. *Am. J. Bot.* **83**, 767–780.

Cevallos-Ferriz, S. R. S., Erwin, D. M., and Stockey, R. A. (1993). Further observations on *Paleorosa similkameenensis* (Rosaceae) from the Middle Eocene Princeton chert of British Columbia, Canada. *Rev. Palaeobot. Palynol.* **78**, 277–291.

Cevallos-Ferriz, S. R. S., and Stockey, R. A. (1988a). Permineralized fruits and seeds from the Princeton chert (Middle Eocene) of British Columbia: Araceae. *Am. J. Bot.* **75**, 1099–1113.

Cevallos-Ferriz, S. R. S., and Stockey, R. A. (1988b). Permineralized fruits and seeds from the Princeton chert (Middle Eocene) of British Columbia: Lythraceae. *Can. J. Bot.* **66**, 303–312.

Cevallos-Ferriz, S. R. S., and Stockey, R. A. (1989). Permineralized fruits and seeds from the Princeton chert (Middle Eocene) of British Columbia: Nymphaeaceae. *Bot. Gaz.* **150**, 207–217.

Cevallos-Ferriz, S. R. S., and Stockey, R. A. (1990). Permineralized fruits and seeds from the Princeton chert (Middle Eocene) of British Columbia: Vitaceae. *Can. J. Bot.* **68**, 288–295.

Cevallos-Ferriz, S. R. S., and Stockey, R. A. (1991). Fruits and seeds from the Princeton chert (Middle Eocene) of British Columbia: Rosaceae (Prunoideae). *Bot. Gaz.* **152**, 369–379.

Cevallos-Ferriz, S. R. S., Stockey, R. A., and Pigg, K. B. (1991). The Princeton chert: Evidence for *in situ* aquatic plants. *Rev. Palaeobot. Palynol.* **70**, 173–185.

Chaloner, W. G., Hill, A. J., and Lacey, W. S. (1977). First Devonian platyspermic seed and its implications in gymnosperm evolution. *Nature* **265**, 233–235.

Chamberlain, C. J. (1919). "The Living Cycads." Univ. Chicago Press, IL.

Chaney, R. W. (1947). Tertiary centers and migration routes. *Ecol. Monogr.* **17**, 139–148.

Chen, S. S. C., and Park, W.-M. (1973). Early actions of gibberellic acid on the embryo and on the endosperm of *Avena fatua* seeds. *Plant Physiol.* **52**, 174–176.

Cheplick, G. P. (1996). Do seed germination patterns in cleistogamous annual grasses reduce the risk of sibling competition? *J. Ecol.* **84**, 247–255.

Christiansen, M. N., Moore, R. P., and Rhyne, C. L. (1960). Cotton seed quality preservation by a hard seed coat characteristic which restricts internal water uptake. *Agron. J.* **52**, 81–84.

Cohen, D. (1966). Optimizing reproduction in a randomly varying environment. *J. Theoret. Biol.* **12**, 119–129.

Cohen, D. (1967). Optimizing reproduction in a randomly varying environment when a correlation may exist between the conditions at the time a choice has to be made and the subsequent outcome. *J. Theoret. Biol.* **16**, 1–14.

Cohen, D. (1968). A general model of optimal reproduction in a randomly varying environment. *J. Ecol.* **56**, 219–228.

Cohen, D., and Levin, S. A. (1985). The interaction between dispersal and dormancy strategies in varying and heterogeneous environments. *In* "Lecture Notes in Biomathematics: Mathematical Topics in Population Biology, Morphogenesis and Neurosciences" (E. Teramoto and M. Yamaguti, eds.), pp. 110–122. Springer-Verlag, Berlin.

Collinson, M. E. (1980). Recent and Tertiary seeds of the Nymphaeaceae *sensu lato* with a revision of *Brasenia ovula* (Bong.) Reid and Chandler. *Ann. Bot.* **46**, 603–632.

Collinson, M. E. (1983). Palaeofloristic assemblages and palaeoecology of the Lower Oligocene Bembridge Marls, Hamstead Ledge, Isle of Wight. *Bot. J. Linn. Soc.* **86**, 177–225.

Collinson, M. E. (1990). Plant evolution and ecology during the Early Cainozoic diversification. *Adv. Bot. Res.* **17**, 1–98.

Collinson, M. E. (1992). Vegetational and floristic changes around the Eocene/Oligocene boundary in western and central Europe. *In* "Eocene-Oligocene Climatic and Biotic Evolution" (D. R. Prothero and W. A. Berggren, eds.), pp. 437–450. Princeton Univ. Press, NJ.

Cook, R. E. (1980). Germination and size-dependent mortality in *Viola blanda*. *Oecologia* **47**, 115–117.

Corner, E. J. H. (1976). "The Seeds of Dicotyledons." Cambridge Univ. Press, Cambridge, UK.

Corner, E. J. H. (1992). The pachychalaza in dicotyledons: Primitive or advanced? *Bot. J. Linn. Soc.* **108**, 15–19.

Crane, P. R., and Herendeen, P. S. (1996). Cretaceous floras containing angiosperm flowers and fruits from eastern North America. *Rev. Palaeobot. Palynol.* **90**, 319–337.

Crane, P. R., Manchester, S. R., and Dilcher, D. L. (1990). "A Preliminary Survey of Fossil Leaves and Well-Preserved Reproductive Structures from the Sentinel Butte Formation (Paleocene) near Almont, North Dakota." *Fieldiana: Geology, New Ser.* No. 20.

Crane, P. R., and Stockey, R. A. (1985). Growth and reproductive biology of *Joffrea speirsii* gen. et sp. nov., a *Cercidiphyllum*-like plant from the Late Paleocene of Alberta, Canada. *Can. J. Bot.* **63**, 340–364.

Crepet, W. L., and Delevoryas, T. (1972). Investigations of North American cycadeoids: Early ovule ontogeny. *Am. J. Bot.* **59**, 209–215.

Crocker, W., and Barton, L. V. (1931). After-ripening, germination, and storage of certain Rosaceous seeds. *Contrib. Boyce Thomp. Inst.* **3**, 385–404.

Curtis, E. J. C., and Cantlon, J. E. (1963). Germination of *Melampyrum lineare*: Interrelated effects of afterripening and gibberellic acid. *Science* **140**, 406–408.

Delevoryas, T., and Mickle, J. E. (1995). Upper Cretaceous Magnoliaceous fruit from British Columbia. *Am. J. Bot.* **82**, 763–768.

de Silva, B. L. T., and Tambiah, M. S. (1952). A contribution to the

life history of *Cycas rumphii* Miq. *Ceylon J. Sci. Sect. A Bot.* **12**, 223–249 + 5 plates.

Dilcher, D. L. (1979). Early angiosperm reproduction: An introductory report. *Rev. Palaeobot. Palynol.* **27**, 291–328.

DiMichele, W. A., Phillips, T. L., and Olmstead, R. G. (1987). Opportunistic evolution: Abiotic environmental stress and the fossil record of plants. *Rev. Paleobot. Palynol.* **50**, 151–178.

Dolan, R. W., and Sharitz, R. R. (1984). Population dynamics of *Ludwigia leptocarpa* (Onagraceae) and some factors affecting size hierachies in a natural population. *J. Ecol.* **72**, 1031–1041.

Doyle, J. A. (1978). Origin of angiosperms. *Annu. Rev. Ecol. Syst.* **9**, 365–392.

Doyle, J. A., and Donoghue, M. J. (1987). The origin of angiosperms: A cladistic approach. *In* "The Origins of Angiosperms and Their Biological Consequences" (E. M. Friis, W. G. Chaloner, and P. R. Crane, eds.), pp. 17–49. Cambridge Univ. Press, Cambridge, UK.

Dyer, R. A. (1965). The cycads of southern Africa. *Bothalia* **8**, 405–515.

Eames, A. J. (1955). The seed and ginkgo. *J. Arnold Arb.* **36**, 165–170.

Ellner, S. (1985a). ESS germination strategies in randomly varying environments. I. Logistic-type models. *Theor. Pop. Biol.* **28**, 50–79.

Ellner, S. (1985b). ESS germination strategies in randomly varying environments. II. Reciprocal yield-law models. *Theor. Pop. Biol.* **28**, 80–116.

Ellner, S. (1986). Germination dimorphisms and parent-offspring conflict in seed germination. *J. Theor. Biol.* **123**, 173–185.

Ellner, S. (1987). Competition and dormancy: A reanalysis and review. *Am. Nat.* **130**, 798–803.

Epstein, S., Buchsbaum, R., Lowenstam, H., and Urey, H. C. (1951). Carbonate-water isotopic temperature scale. *Bull. Geol. Soc. Am.* **62**, 417–425.

Ernst, W. H. O. (1989). Selection of winter and summer annual life forms in populations of *Senecio sylvatics* L. *Flora* **182**, 221–231.

Erwin, D. M., Pfefferkorn, H. W., and Alleman, V. (1994). Early seed plants in the Southern Hemisphere. I. Associated ovulate and microsporangiate organs from the Carboniferous of Peru. *Rev. Palaeobot. Palynol.* **80**, 19–38.

Erwin, D. M., and Stockey, R. A. (1990). Sapindaceous flowers from the Middle Eocene Princeton chert (Allenby Formation) of British Columbia, Canada. *Can. J. Bot.* **68**, 2025–2034.

Evans, R. C., and Fratianne, D. G. (1977). Interactions of applied hormones in the germination of *Lepidium virginicum* seeds. *Ohio J. Sci.* **77**, 236–239.

Fernandez-Quintanilla, C., Navarrete, L., Andujar, J. L. G., Fernandez, A., and Sanchez, M. J. (1986). Seedling recruitment and age-specific survivorship and reproduction in populations of *Avena sterilis* L. ssp. *ludoviciana* (Durieu) Nyman. *J. Appl. Ecol.* **23**, 945–955.

Frakes, L. A. (1979). "Climates throughout Geologic Time." Elsevier, Amsterdam.

Friedman, W. E. (1990). Double fertilization in *Ephedra*, a nonflowering seed plant: Its bearing on the origin of angiosperms. *Science* **247**, 951–954.

Friedman, W. E. (1992). Evidence of a pre-angiosperm origin of endosperm: Implications for the evolution of flowering plants. *Science* **255**, 336–339.

Friedman, W. E. (1994). The evolution of embryogeny in seed plants and the developmental origin and early history of endosperm. *Am. J. Bot.* **81**, 1468–1486.

Friedman, W. E., and Carmichael, J. S. (1996). Double fertilization in Gnetales: Implications for understanding reproduction diversification among seed plants. *Int. J. Plant Sci.* **157**(Suppl.), 77–94.

Galtier, J., and Rowe, N. P. (1989). A primitive seed-like structure and its implications for early gymnosperm evolution. *Nature* **340**, 225–227.

Garwood, N. C. (1982). Seasonal rhythm of seed germination in a semideciduous tropical forest. *In* "The Ecology of a Tropical Forest: Seasonal Rhythms and Long-Term Changes" (E. G. Leigh, Jr., A. S. Rand, and D. M. Windsor, eds.), pp. 173–185. Smithsonian Institution Press, Washington, DC.

Gillespie, W. H., Rothwell, G. W., and Scheckler, S. E. (1981). The earliest seeds. *Nature* **293**, 462–464.

Goebel, K. (1905). "Organography of Plants, Especially of the Archegoniatae and Spermophyta." Clarendon Press, Oxford.

Gould, S. J., and Lewontin, R. C. (1979). The spandrels of San Marco and the Panglossian paradigm: A critique of the adaptationist programme. *Proc. Roy. Soc. Lond. B* **205**, 581–598.

Grushvitzky, I. V. (1967). After-ripening of seeds of primitive tribes of angiosperms, conditions and peculiarities. *In* "Physiologie, Okologie und Biochemie der Keimung" (H. Borriss, ed.), Vol. 1, pp. 320–335. Ernst-Moritz-Arndt Universitat, Greifswald.

Haig, D., and Westoby, M. (1991). Seed size, pollination costs and angiosperm success. *Evol. Ecol.* **5**, 231–247.

Herendeen, P. S. (1992). The fossil history of the Leguminosae from the Eocene of southeastern North America. *In* "Advances in Legume Systematics: Part 4. The Fossil Record" (P. S. Herendeen and D. L. Dilcher, eds.), pp. 85–160. The Royal Botanic Gardens, Kew.

Heywood, V. H. (ed.) (1993). "Flowering Plants of the World." Oxford Univ. Press, New York.

Hickey, L. J., and Doyle, J. A. (1977). Early Cretaceous fossil evidence for angiosperm evolution. *Bot. Rev.* **43**, 3–104.

Hilhorst, H. W. M., and Karssen, C. M. (1988). Dual effect of light on the gibberellin- and nitrate-stimulated seed germination of *Sisymbrium officinale* and *Arabidopsis thaliana. Plant Physiol.* **86**, 591–597.

Holt, B. F., and Rothwell, G. W. (1997). Is *Ginkgo biloba* (Ginkgoaceae) really an oviparous plant? *Am. J. Bot.* **84**, 870–872.

Hoskins, J. H., and Cross, A. T. (1946). Studies in the Trigonocarpales. Part II. Taxonomic problems and a revision of the genus *Pachytesta. Am. Midl. Nat.* **36**, 331–361.

Hughes, N. F. (1976). "Palaeobiology of Angiosperm Origins: Problems of Mesozoic Seed-Plant Evolution." Cambridge Univ. Press, Cambridge, UK.

Kadmon, R., and Shmida, A. (1990). Spatiotemporal demographic processes in plant populations: An approach and a case study. *Am. Nat.* **135**, 382–397.

Karssen, C. M. (1970). The light promoted germination of the seeds of *Chenopodium album* L. III. Effect of the photoperiod during growth and development of the plants on the dormancy of the produced seeds. *Acta Bot. Neerl.* **19**, 81–94.

Kemp, E. M. (1978). Tertiary climatic evolution and vegetation history in the southeast Indian Ocean region. *Paleogeogr. Palaeoclim. Palaeoecol.* **24**, 169–208.

Kigel, J., Ofir, M., and Koller, D. (1977). Control of the germination responses of *Amaranthus retroflexus* L. seeds by their parental photothermal environment. *J. Exp. Bot.* **28**, 1125–1136.

Klebesadel, L. J. (1969). Life cycles of field pennycress in the Subarctic as influenced by time of seed germination. *Weed Sci.* **17**, 563–566.

Klemow, K. M., and Raynal, D. J. (1983). Population biology of an annual plant in a temporally variable habitat. *J. Ecol.* **71**, 691–703.

Klinkhamer, P. G. L., de Jong, T. J., Metz, J. A. J., and Val, J. (1987). Life history tactics of annual organisms: The joint effects of dispersal and delayed germination. *Theor. Pop. Biol.* **32**, 127–156.

Koller, D. (1964). The survival value of germination-regulating mechanisms in the field. *Herb. Abst.* **34**, 1–7.

Kress, W. J. (1990). The phylogeny and classification of the Zingiberales. *Ann. Missouri Bot. Gard.* **77**, 698–721.

Kress, W. J. (1995). Phylogeny of the Zingiberanae: Morphology and molecules. *In* "Monocotyledons; Systematics and Evolution"

(P. J. Rudall, P. J. Cribb, D. F. Cutler, and C. J. Humphries, eds.), pp. 443–460. Royal Botanic Garden, Kew.

Kuno, E. (1981). Dispersal and the persistence of populations in unstable habitats: A theoretical note. *Oecologia* **49**, 123–126.

Kuo-Huang, L.-L., Chien, C.-T., and Lin, T.-P. (1996). Ultrastructural study on *Taxus mairei* seed during the germination promotion by combination of warm and cold stratification. *Am. J. Bot.* **83** (Suppl. to No. 6), 45. [Abstract]

Lande, R. (1982). A quantitative genetic theory of life history evolution. *Ecology* **63**, 607–615.

Lee, N. S., Sang, T., Crawford, D. J., Yeau, S. H., and Kim, S.-C. (1996). Molecular divergence between disjunct taxa in eastern Asia and eastern North America. *Am. J. Bot.* **83**, 1373–1378.

Leon, J. A. (1985). Germination strategies. *In* "Evolution" (P. J. Greenwood, P. H. Harvey, and M. Slatkin, eds.), pp. 129–142. Cambridge Univ. Press, Cambridge, UK.

Leopold, E. B., Liu, G., and Clay-Poole, S. (1992). Low-biomass vegetation in the Oligocene. *In* "Ecocene-Oligocene Climatic and Biotic Evolution" (D. Prothero and W. A. Berggren, eds.), pp. 399–420. Princeton Univ. Press, Princeton, NJ.

Lesiak, M. A. (1994). Plant macrofossils from the Middle Miocene of Lipnica Mala (Orawa-Nowy Targ Basin, Poland). *Acta Palaeobot.* **34**, 27–81.

Levin, S. A., Cohen, D., and Hastings, A. (1984). Dispersal strategies in patchy environments. *Theor. Pop. Biol.* **26**, 165–191.

Levins, R. (1969). Dormancy as an adaptive strategy. "Symp. Soc. Exp. Biol. No. XXIII. Dormancy and Survival." Academic Press, New York.

Li, H.-L. (1972). Eastern Asia-eastern North America species-pairs in wide-ranging genera. *In* "Floristics and Paleofloristics of Asia and Eastern North America" (A. Graham, ed.), pp. 65–78. Elsevier, Amsterdam.

Long, A. G. (1944). On the prothallus of *Lagenostoma ovoides* Will. *Ann. Bot.* **8**, 105–117 + 2 plates.

Long, A. G. (1974–1975). Further observations on some Lower Carboniferous seeds and cupules. *Trans. Roy. Soc. Edinburgh* **69**, 267–293 + 6 plates.

Lonsdale, W. M., and Abrecht, D. G. (1989). Seedling mortality in *Mimosa pigra*, an invasive tropical shrub. *J. Ecol.* **77**, 371–385.

Mack, R. N., and Pyke, D. A. (1984). The demography of *Bromus tectorum:* The role of microclimate, grazing and disease. *J. Ecol.* **72**, 731–748.

Manchester, S. R. (1992). Flowers, fruits, and pollen of *Florissantia*, an extinct Malvalean genus from the Eocene and Oligocene of western North America. *Am. J. Bot.* **79**, 996–1008.

Manchester, S. R. (1994). Fruits and seeds of the Middle Eocene Nut Beds Flora, Clarno Formation, Oregon. *Palaeontograph. Am.* **58**, 1–205.

Manchester, S. R., and Kress, W. J. (1993). Fossil bananas (Musaceae): *Ensete oregonense* sp. nov. from the Eocene of western North America and its phytogeographic significance. *Am. J. Bot.* **80**, 1264–1272.

Mapes, G., Rothwell, G. W., and Haworth, M. T. (1989). Evolution of seed dormancy. *Nature* **337**, 645–646.

Marks, M., and Prince, S. (1981). Influence of germination date on survival and fecundity in wild lettuce *Lactuca serriola. Oikos* **36**, 326–330.

Martin, A. C. (1946). The comparative internal morphology of seeds. *Am. Midl. Nat.* **36**, 513–660.

Matten, L. C., Lacey, W. S., May, B. I., and Lucas, R. C. (1980). A megafossil flora from the uppermost Devonian near Allyheigue, Co. Kerry, Ireland. *Rev. Paleobot. Palynol.* **29**, 241–251.

Melhem, T. S. (1975). Fisiologia da germinacao das sementes de *Dipteryx alata* Vog. (Leguminosae-Lotoideae). *Hoehnea* **5**, 59–90.

Metz, J. A. J., de Jong, T. J., and Klinkhamer, P. G. L. (1983). What are the advantages of dispersing; a paper by Kuno explained and extended. *Oecologia* **57**, 166–169.

Milberg, P., and Andersson, L. (1994). Effect of emergence date on seed production and seed germinability in *Thlaspi arvense*. *Swed. J. Agric. Res.* **24**, 143–146.

Millan, J. H. (1994). The identification and classification scheme for seeds from Gondwanaland: Update and critical analysis. *Ana. Acad. Brasil. Cien.* **66**, 475–489.

Miller, C. N., Jr. (1988). The origin of modern conifer families. *In* "Origin and Evolution of Gymnosperms" (C. B. Beck, ed.), pp. 448–486. Columbia Univ. Press, New York.

Miller, C. N., and Brown, J. T. (1973). Paleozoic seeds with embryos. *Science* **179**, 184–185.

Muller, J. (1981). Fossil pollen records of extant angiosperms. *Bot. Rev.* **47**, 1–142.

Nikolaeva, M. G. (1969)." Physiology of Deep Dormancy in Seeds." Izdatel'stvo Nauka. Leningrad. [Translated from Russian by Z. Shapiro, National Science Foundation, Washington, DC.]

Nikolaeva, M. G. (1977). Factors controlling the seed dormancy pattern. *In* "The Physiological and Biochemistry of Seed Dormancy and Germination" (A. A. Khan, ed.), pp. 51–74. North-Holland, Amsterdam.

Nilsson, P., Fagerstrom, T., Tuomi, J., and Astrom, M. (1994). Does seed dormancy benefit the mother plant by reducing sib competition? *Evol. Ecol.* **8**, 422–430.

Nilsson, T. (1995). Density dependent processes and the importance of periodic germination in the winter annual plant *Saxifraga osloensis*. *Ecography* **18**, 131–137.

Nixon, K. C., and Crepet, W. L. (1993). Late Cretaceous fossil flowers of Ericalean affinity. *Am. J. Bot.* **80**, 616–623.

Nixon, K. C., Crepet, W. L., Stevenson, D., and Friis, E. M. (1994). A reevaluation of seed plant phylogeny. *Ann. Missouri Bot. Gard.* **81**, 484–533.

Parrish, J. T., and Curtis, R. L. (1982). Atmospheric circulation, up-welling, and organic-rich rocks in the Mesozoic and Cenozoic eras. *Palaeogeogr. Palaeoclim. Palaeoecol.* **40**, 31–66.

Parrish, J. T., Ziegler, A. M., and Scotese, C. R. (1982). Rainfall patterns and the distribution of coals and evaporites in the Mesozoic and Cenozoic. *Palaeogeogr. Palaeoclim. Palaeoecol.* **40**, 67–101.

Pearson, H. H. W. (1910). On the embryo of *Welwitschia*. *Ann. Bot.* (Old Series) **24**, 759–766 + 5 plates.

Pettitt, J. M., and Beck, C. B. (1968). *Archaeosperma arnoldii*: Cupulate seed from the Upper Devonian of North America. *Contrib. Mus. Paleontol. Univ. Michigan* **22**, 139–154.

Philippi, T., and Seger, J. (1989). Hedging one's evolutionary bets, revisited. *Trends Ecol. Evol.* **4**, 41–44.

Philippi, T. (1993a). Bet-hedging germination of desert annuals: Beyond the first year. *Am. Nat.* **142**, 474–487.

Philippi, T. (1993b). Bet-hedging germination of desert annuals: Variation among populations and maternal effects in *Lepidium lasiocarpum*. *Am. Nat.* **142**, 488–507.

Pigg, K. B., Stockey, R. A., and Maxwell, S. L. (1993). *Paleomyrtinaea*, a new genus of permineralized myrtaceous fruits and seeds from the Eocene of British Columbia and Paleocene of North Dakota. *Can. J. Bot.* **71**, 1–9.

Potts, H. C., Duangpatra, J., Hairston, W. G., and Delouche, J. C. (1978). Some influences of hardseededness on soybean seed quality. *Crop Sci.* **18**, 221–224.

Puckridge, D. W., and French, R. J. (1983). The annual legume pasture in cereal-ley farming systems of southern Australia: A review. *Agric. Ecosyst. Environ.* **9**, 229–267.

Putwain, P. D., Scott, K. R., and Holliday, R. J. (1982). The nature of resistance to triazine herbicides: Case histories of phenology and population studies. *In* "Herbicide Resistance in Plants" (H. M. LeBaron and J. Gressel, eds.), pp. 99–115. Wiley, New York.

Rathcke, B., and Lacey, E. P. (1985). Phenological patterns of terrestrial plants. *Annu. Rev. Ecol. Syst.* **16**, 179–214.

Rees, M. (1993). Trade-offs among dispersal strategies in British plants. *Nature* **366**, 150–152.

Rees, M. (1994). Delayed germination of seeds: A look at the effects of adult longevity, the timing of reproduction, and population age/stage structure. *Am. Nat.* **144**, 43–64.

Retallack, G. J. (1992). Paleosols and changes in climate and vegetation across the Eocene/Oligocene boundary. *In* "Eocene-Oligocene Climatic and Biotic Evolution" (D. R. Prothero and W. A. Berggren, ed.), pp. 382–398. Princeton Univ. Press, Princeton, NJ.

Retallack, G., and Dilcher, D. L. (1981). A coastal hypothesis for the dispersal and rise to dominance of flowering plants. *In* "Paleobotany, Paleoecology, and Evolution" (K. J. Niklas, eds.), pp. 27–77. Praeger, New York.

Ritland, K. (1983). The joint evolution of seed dormancy and flowering time in annual plants living in variable environments. *Theor. Pop. Biol.* **24**, 213–243.

Roberts, H. A., and Neilson, J. E. (1982). Seasonal changes in the temperature requirements for germination of buried seeds of *Aphanes arvensis* L. *New Phytol.* **92**, 159–166.

Rothwell, G. W. (1971). Ontogeny of the Paleozoic ovule, *Callospermarion pusillum*. *Am. J. Bot.* **58**, 706–715.

Rothwell, G. W. (1982). *Cordaianthus duquesnensis* sp. nov., anatomically preserved ovulate cones from the Upper Pennsylvanian of Ohio. *Am. J. Bot.* **69**, 239–247.

Rothwell, G. W., and Scheckler, S. E. (1988). Biology of ancestral gymnosperms. *In* "Origin and Evolution of Gymnosperms" (C. B. Beck, ed.), pp. 83–134. Columbia Univ. Press, New York.

Rothwell, G. W., Scheckler, S. E., and Gillespie, W. H. (1989). *Elkinsia* gen. nov., a Late Devonian gymnosperm with cupulate ovules. *Bot. Gaz.* **150**, 170–189.

Savin, S. M. (1977). The history of the earth's surface temperature during the past 100 million years. *Annu. Rev. Earth Planet. Sci.* **5**, 319–355.

Schaal, B. A., and Leverich, W. J. (1981). The demographic consequences of two-stage life cycles: Survivorship and the time of reproduction. *Am. Nat.* **118**, 135–138.

Shitaka, Y., and Hirose, T. (1993). Timing of seed germination and the reproductive effort in *Xanthium canadense*. *Oecologia* **95**, 334–339.

Silvertown, J. (1985). When plants play the field. *In* " Evolution" (P. J. Greenwood, P. H. Harvey, and M. Slatkin, eds.), pp. 143–153. Cambridge Univ. Press, Cambridge, UK.

Silvertown, J. (1988). The demographic and evolutionary consequences of seed dormancy. *In* "Plant Population Ecology" (A. J. Davy, M. J. Hutchings, and A. R. Watkinson, eds.), pp. 205–219. Blackwell, Oxford.

Singh, H., and Johri, B. M. (1972). Development of gymnosperm seeds. *In* "Seed Biology" (T. T. Kozlowski, ed.), Vol. I, pp. 21–75. Academic Press, New York.

Sinnott, E. W. (1913). The morphology of the reproductive structures in the Podocarpineae. *Ann. Bot.* **27**, 39–82 + plates 5–9.

Smoot, E. L., and Taylor, T. N. (1986). Evidence of simple polyembryony in Permian seeds from Antarctica. *Am. J. Bot.* **73**, 1079–1081.

Solbrig, O. T. (1976). The origin and floristic affinities of the South American temperate desert and semidesert regions. *In* "Evolution of Desert Biota" (D. W. Goodall, ed.), pp. 7–49. Univ. Texas Press, Austin.

Steeves, T. A. (1983). The evolution and biological significance of seeds. *Can. J. Bot.* **61**, 3550–3560.

Stewart, W. N., and Rothwell, G. W. (1993). "Paleobotany and the Evolution of Plants." Cambridge Univ. Press, Cambridge, UK.

Stidd, B. M., and Cosentino, K. (1976). *Nucellangium:* Gametophytic structure and relationship to Cordaites. *Bot. Gaz.* **137**, 242–249.

Stockey, R. A. (1975). Seeds and embryos of *Araucaria mirabilis*. *Am J. Bot.* **62**, 856–868.

Stockey, R. A. (1977). Reproductive biology of the Cerro Cuadrado (Jurassic) fossil conifers: *Pararaucaria patagonica*. *Am. J. Bot.* **64**, 733–744.

Stockey, R. A. (1978). Reproductive biology of Cerro Cuadrado fossil conifers: Ontogeny and reproductive strategies in *Araucaria mirabilis* (Spegazzini) Windhausen. *Palaeontographica* **166**, 1–15 + 6 plates.

Stockey, R. A. (1980a). Anatomy and morphology of *Araucaria sphaerocarpa* Carruthers from the Jurassic inferior oolite of Bruton, Somerset. *Bot. Gaz.* **141**, 116–124.

Stockey, R. A. (1980b). Jurassic araucarian cone from southern England. *Palaeontology* **23**, 657–666 + plates 83–86.

Stockey, R. A. (1984). Middle Eocene *Pinus* remains from British Columbia. *Bot. Gaz.* **145**, 262–274.

Stockey, R. A., and Crane, P. R. (1983). *In situ Cercidiphyllum*-like seedlings from the Paleocene of Alberta, Canada. *Am. J. Bot.* **70**, 1564–1568.

Stockey, R. A., Nishida, H., and Nishida, M. (1992). Upper Cretaceous araucarian cones from Hokkaido: *Araucaria nihongii* sp. nov. *Rev. Palaeobot. Palynol.* **72**, 27–40.

Stockey, R. A., and Pigg, K. B. (1991). Flowers and fruits of *Princetonia allenbyensis* (Magnoliopsida; family indet.) from the Middle Eocene Princeton chert of British Columbia. *Rev. Palaeobot. Palynol.* **70**, 163–172.

Stockey, R. A., and Taylor, T. N. (1978). On the structure and evolutionary relationships of the Cerro Cuadrado fossil conifer seedlings. *Bot. J. Linn. Soc.* **76**, 161–176.

Stoltz, L. P., and Snyder, J. C. (1985). Embryo growth and germination of American ginseng seed in response to stratification temperatures. *Hortscience* **20**, 261–262.

Symonides, E. (1977). Mortality of seedlings in natural psammophyte populations. *Ekol. Polska* **25**, 635–651.

Takhtajan, A. L. (1980). Outline of the classification of flowering plants (Magnoliophyta). *Bot. Rev.* **46**, 225–359.

Taylor, D. W. (1990). Paleobiogeographic relationships of angiosperms from the Cretaceous and Early Tertiary of the North American area. *Bot. Rev.* **56**, 279–417.

Taylor, T. N. (1981). "Paleobotany. An Introduction to Fossil Plant Biology." McGraw-Hill, New York.

Taylor, T. N., and Eggert, D. A. (1969). On the structure and relationships of a new Pennsylvanian species of the seed *Pachytesta*. *Palaeontology* **12**, 382–387.

Taylor, T. N., and Millay, M. A. (1979). Pollination biology and reproduction in early seed plants. *Rev. Palaeobot. Palynol.* **27**, 329–355.

Taylor, T. N., and Taylor, E. L. (1987). Structurally preserved fossil plants from Antarctica III. Permian seeds. *Am. J. Bot.* **74**, 904–913.

Terui, K., and Okagami, N. (1993). Temperature effects on seed germination of east Asian and Tertiary relict species of *Dioscorea* (Dioscoreaceae). *Am. J. Bot.* **80**, 493–499.

Teryokhin, E. S., and Nikiticheva, Z. I. (1982). Biology and evolution of embryo and endosperm in parasitic flowering plants. *Phytomorphology* **32**, 335–339.

Thomas, B. A., and Spicer, R. A. (1987). "The Evolution and Palaeobiology of Land Plants." Croom Helm, London.

Tiffney, B. H. (1977). Fruits and seeds of the Brandon lignite: Magnoliaceae. *Bot. J. Linn. Soc.* **75**, 299–323.

Tiffney, B. H. (1980). Fruits and seeds of the Brandon lignite, V. Rutaceae. *J. Arnold Arb.* **61**, 1–40.

Tiffney, B. H. (1981). Diversity and major events in the evolution of land plants. *In* "Paleobotany, Paleoecology, and Evolution" (K. J. Niklas, ed.), pp. 193–230. Praeger, New York.

Tiffney, B. H. (1984). Seed size, dispersal syndromes, and the rise of the angiosperms: Evidence and hypothesis. *Ann. Missouri Bot. Gard.* **71**, 551–576.

Tiffney, B. H. (1985a). Perspectives on the origin of the floristic similarity between eastern Asia and eastern North America. *J. Arnold Arb.* **66**, 73–94.

Tiffney, B. H. (1985b). The Eocene North Atlantic land bridge: Its importance in Tertiary and modern phytogeography of the northern hemisphere. *J. Arnold Arb.* **66**, 243–273.

Tiffney, B. H. (1986). Evolution of seed dispersal syndromes according to the fossil record. *In* "Seed Dispersal" (D. R. Murray, ed.), pp. 274–305. Academic Press, Sydney.

Tiffney, B. H. (1993). Fruits and seeds of the Tertiary Brandon lignite. VII. *Sargentodoxa* (Sargentodoxaceae). *Am. J. Bot.* **80**, 517–523.

Upchurch, G. R., Jr., and Wolfe, J. A. (1987). Mid-Cretaceous to early Tertiary vegetation and climate: Evidence from fossil leaves and woods. *In* "The Origins of Angiosperms and Their Biological Consequences" (E. M. Friis, W. G. Chaloner, and P. R. Crane, eds.), pp. 75–105. Cambridge Univ. Press, New York.

van der Toorn, J., and Pons, T. L. (1988). Establishment of *Plantago lanceolata* L., and *Plantago major* L. among grass. II. Shade tolerance of seedlings and selection on time of germination. *Oecologia* **76**, 341–347.

Venable, D. L. (1989). Modeling the Evolutionary Ecology of Seed Banks. *In* "Ecology of Soil Seed Banks" (M. A. Leck, V. T. Parker, and R. L. Simpson, eds.), pp. 67–87. Academic Press, San Diego.

Venable, D. L., Dyreson, E., and Morales, E. (1995). Population dynamic consequences and evolution of seed traits of *Heterosperma pinnatum* (Asteraceae). *Am. J. Bot.* **82**, 410–420.

Venable, D. L., and Lawlor, L. (1980). Delayed germination and dispersal in desert annuals: Escape in space and time. *Oecologia* **46**, 272–282.

von Teichman, I. (1988a). Development and structure of the seed-coat of *Lannea discolor* (Sonder) Engl. (Anacardiaceae). *Bot. J. Linn. Soc.* **96**, 105–117.

von Teichman, I. (1988b). Notes on the ontogeny and structure of the seed-coat of *Sclerocarya birrea* (Richard) Hochst. subsp. *caffra* (Sonder) Kokwaro (Anacardiaceae). *Bot. J. Linn. Soc.* **98**, 153–158.

von Teichman, I. (1989). Reinterpretation of the pericarp of *Rhus lancea* (Anacardiaceae). *S. Afr. J. Bot.* **55**, 383–384.

von Teichman, I. (1990). Pericarp and seed coat structure in *Tapirira guianensis* (Spondiadeae: Anacardiaceae). *S. Afr. J. Bot.* **56**, 435–439.

von Teichman, I. (1993). Development and structure of the seed of *Ozoroa paniculosa* (Anacardiaceae) and taxonomic notes. *Bot. J. Linn. Soc.* **111**, 463–470.

von Teichman, I., and Robbertse, P. J. (1986). Development and structure of the pericarp and seed of *Rhus lancea* L. fil. (Anacardiaceae), with taxonomic notes. *Bot. J. Linn. Soc.* **93**, 291–306.

von Teichman, I., Robbertse, P. J., and Schoonraad, E. (1988). The structure of the seed of *Mangifera indica* L., and notes on seed characters of the tribe Mangifereae (Anacardiaceae). *S. Afr. J. Bot.* **54**, 472–476.

von Teichman, I., and van Wyk, A. E. (1988). The ontogeny and structure of the pericarp and seed-coat of *Harpephyllum caffrum* Bernh. ex Krauss (Anacardiaceae). *Bot. J. Linn. Soc.* **98**, 159–176.

von Teichman, I., and van Wyk, A. E. (1991a). Trends in the evolution of dicotyledonous seeds based on character associations, with special reference to pachychalazy and recalcitrance. *Bot. J. Linn. Soc.* **105**, 211–237.

von Teichman, I., and van Wyk, A. E. (1991b). Taxonomic position of *Rhus problematodes* (Anacardiaceae): Evidence from fruit and seed structure. *S. Afr. J. Bot.* **57**, 29–33.

von Teichman, I., and van Wyk, A. E. (1994). Structural aspects and trends in the evolution of recalcitrant seeds in dicotyledons. *Seed Sci. Res.* **4**, 225–239.

von Teichman, I., and van Wyk, A. E. (1996). Taxonomic significance of pericarp and seed structure in *Heeria argentea* (Thunb.) Meisn. (Anacardiaceae), including reference to pachychalazy and recalcitrance. *Bot. J. Linn. Soc.* **122**, 335–352.

Waite, S. (1984). Changes in the demography of *Plantago coronopus* at two coastal sites. *J. Ecol.* **72**, 809–826.

Walter, H. (1973). "Vegetation of the Earth in Relation to Climate and the Eco-physiological Conditions." Translated from the third, revised German edition by Joy Wieser. Springer-Verlag, New York.

Wannan, B. S., and Quinn, C. J. (1990). Pericarp structure and generic affinities in the Anacardiaceae. *Bot. J. Linn. Soc.* **102**, 225–252.

Wannan, B. S., and Quinn, C. J. (1991). Floral structure and evolution in the Anacardiaceae. *Bot. J. Linn. Soc.* **107**, 349–385.

Watkinson, A. R. (1981). The population ecology of winter annuals. *In* "The Biological Aspects of Rare Plants Conservation" (H. Synge, ed.), pp. 253–264. Chichester, UK.

Weaver, S. E., and Cavers, P. B. (1979). The effects of date of emergence and emergence order on seedling survival rates in *Rumex crispus* and *R. obtusifolius. Can. J. Bot.* **57**, 730–738.

Webb, S. D. (1978). A history of savanna vertebrates in the New World. Part II. South America and the great interchange. *Annu. Rev. Ecol. Syst.* **9**, 393–426.

Wen, J., and Jansen, R. K. (1995). Morphological and molecular comparisons of *Campsis grandiflora* and *C. radicans* (Bignoniaceae), an eastern Asian and eastern North American vicariad species pair. *Plant Syst. Evol.* **196**, 173–183.

Wen, J., Jansen, R. K., and Kilgore, K. (1996). Evolution of the eastern Asian and eastern North American disjunct genus *Symplocarpus* (Araceae): Insights from chloroplast DNA restriction site data. *Biochem. Syst. Ecol.* **24**, 735–747.

Wen, J., and Stuessy, T. F. (1993). The phylogeny and biogeography of *Nyssa* (Cornaceae). *Syst. Bot.* **18**, 68–79.

Wen, J., and Zimmer, E. A. (1996). Phylogeny and biogeography of *Panax* L. (the ginseng genus), Araliaceae: Inferences from ITS sequences of nuclear ribosomal DNA. *Mol. Phylogenet. Evol.* **6**, 167–177.

West, W. C., Frattarelli, F. S., and Russin, K. J. (1970). Effect of stratification and gibberellin on seed germination of *Ginkgo biloba. Bull. Torrey Bot. Club* **97**, 380–384.

Westoby, M. (1981). How diversified seed germination behavior is selected. *Am. Nat.* **118**, 882–885.

White, M. E. (1990). "The Flowering of Gondwana." Princeton Univ. Press, Princeton, NJ.

Wieland, G. R. (1916). American fossil cycads. Vol. II. Taxonomy *Carnegie Inst. Wash. Publ. No.* **34.**

Wolfe, J. A. (1971). Tertiary climatic fluctuations and methods of analysis of Tertiary floras. *Palaeogeogr. Palaeoclim. Palaeoecol.* **9**, 27–57.

Wolfe, J. A. (1978). A paleobotanical interpretation of Tertiary climates in the Northern Hemisphere. *Am. Scient.* **66**, 694–703.

Wolfe, J. A. (1981). Paleoclimatic significance of the Oligocene and Neogene floras of the northwestern United States. *In* "Paleobotany, Paleoecology, and Evolution" (K. J. Niklas, ed.), pp. 79–101. Praeger, New York.

Wolfe, J. A. (1985). Distribution of major vegetation types during the Tertiary. *In* "The Carbon Cycle and Atmospheric CO_2: Natural Variations Archean to Present" (E. T. Sundquist and W. S. Broecker, eds.), pp. 357–375. Geophysical Monograph 32, American Geophysical Union, Washington, DC.

Wolfe, J. A. (1992). Climatic, floristic, and vegetational changes near the Eocene/Oligocene boundary in North America. *In* "Eocene-Oligocene Climatic and Biotic Evolution" (D. R. Prothero and W. A. Berggren, eds.), pp. 421–436. Princeton Univ. Press, Princeton, NJ.

Wolfe, J. A. (1994a). Tertiary climatic changes at middle latitudes of western North America. *Palaeogeogr. Palaeoclim. Palaeoecol.* **108**, 195–205.

Wolfe, J. A. (1994b). An analysis of Neogene climates in Beringia. *Palaeogeogr. Palaeoclim. Palaeoecol.* **108**, 207–216.

Wolfe, J. A., and Upchurch, G. R., Jr. (1987). North American nonmarine climates and vegetation during the Late Cretaceous. *Palaeogeogr. Palaeoclim. Palaeoecol.* **61**, 33–77.

Wood, C. E., Jr. (1972). Morphology and phytogeography: The classical approach to the study of disjunctions. *Ann. Missouri Bot. Gard.* **59**, 107–124.

Ziegler, A. M. (1981). Paleozoic paleogeography. *In* "Paleoreconstruction of the Continents: Geodynamics Series" (M. W. McElhinny and D. A. Valencio, eds.), Vol. 2, pp. 31–37. America Geophysical Union, Washington, DC.

Subject Index

Taxonomic Index

Printed and bound by CPI Group (UK) Ltd, Croydon, CR0 4YY

08/05/2025

01864912-0001